YAMAHA Outboards
1997-03 REPAIR MANUAL
ALL 2-STROKE ENGINES

Stark County District Library
Main Library
715 Market Ave N
Canton, OH 44702
330.452.0665
www.starklibrary.org

MAR 2008

D1560704

Managing Partners	Dean F. Morgantini, S.A.E.
	Barry L. Beck
Executive Editor	Kevin M. G. Maher, A.S.E.
Manager-Marine/Recreation	James R. Marotta, A.S.E.
Production Managers	Melinda Possinger
	Ronald Webb

Manufactured in USA
© 2004 Seloc Publications
104 Willowbrook Lane
West Chester, PA 19382
ISBN 0-89330-065-9
3456789012 5432109876

www.selocmarine.com
1-866-SELOC55

CONTENTS

1 GENERAL INFORMATION, SAFETY AND TOOLS

HOW TO USE THIS MANUAL	1-2
BOATING SAFETY	1-4
BOATING EQUIPMENT (NOT REQUIRED BUT RECOMMENDED)	1-10
SAFETY IN SERVICE	1-13
TROUBLESHOOTING	1-13
SHOP EQUIPMENT	1-17
TOOLS	1-19
FASTENERS, MEASUREMENT, AND CONVERSIONS	1-27
SPECIFICATIONS	1-28

2 MAINTENANCE AND TUNE-UP

GENERAL INFORMATION	2-2
LUBRICATION	2-5
ENGINE MAINTENANCE	2-10
BOAT MAINTENANCE	2-27
TUNE-UP	2-30
TIMING AND SYNCHRONIZATION	2-44
CLEARING A SUBMERGED MOTOR	2-87
SPECIFICATIONS	2-88

3 FUEL SYSTEM

FUEL AND COMBUSTION BASICS	3-2
FUEL TANK AND LINES	3-6
CARBURETED FUEL SYSTEM	3-11
CARBURETOR SERVICE	3-17
ELECTRONIC FUEL INJECTION SYSTEMS	3-66
FUEL PUMP (LOW PRESSURE) SERVICE	3-99
SPECIFICATIONS	3-105

4 IGNITION AND ELECTRICAL

UNDERSTANDING AND TROUBLESHOOTING ELECTRICAL SYSTEMS	4-2
IGNITION SYSTEMS	4-8
CHARGING CIRCUIT	4-38
CRANKING CIRCUIT	4-43
ELECTRICAL SWITCH/SOLENOID SERVICE	4-53
SPECIFICATIONS	4-57
WIRING DIAGRAMS	4-65

5 LUBRICATION AND COOLING

PRECISION BLEND OIL INJECTION SYSTEMS	5-2
OIL INJECTION WARNING SYSTEM	5-20
COOLING SYSTEM	5-27
SPECIFICATIONS	5-49

CONTENTS

6 POWERHEAD

POWERHEAD MECHANICAL	6-2
POWERHEAD REFINISHING	6-67
POWERHEAD BREAK-IN	6-79
SPECIFICATIONS	6-80

7 LOWER UNIT

LOWER UNIT	7-2
GEARCASE OVERHAUL	7-10
JET DRIVE	7-78

8 TRIM AND TILT

TRIM & TILT SYSTEMS	8-2
HYDRO TILT LOCK SYSTEM	8-2
SINGLE TILT RAM POWER TRIM/TILT SYSTEMS	8-3
LARGE MOTOR POWER TRIM/TILT SYSTEM	8-13

9 REMOTE CONTROL

REMOTE CONTROLS	9-2

10 HAND REWIND STARTER

HAND REWIND STARTER	10-2

MASTER INDEX

MASTER INDEX	10-22

MARINE TECHNICIAN TRAINING

INDUSTRY SUPPORTED PROGRAMS
OUTBOARD, STERNDRIVE & PERSONAL WATERCRAFT

- Dyno Testing • Boat & Trailer Rigging • Electrical & Fuel System Diagnostics
- Powerhead, Lower Unit & Drive Rebuilds • Powertrim & Tilt Rebuilds
- Instrument & Accessories Installation

For information regarding housing, financial aid and employment opportunities in the marine industry, contact us today:

CALL TOLL FREE
1-800-528-7995

TRAIN IN SUNNY FLORIDA!

An Accredited Institution

Name _____
Address _____
City _____ State ____ Zip ____
Phone _____

MARINE MECHANICS INSTITUTE
A Division of CTI
9751 Delegates Drive • Orlando, Florida 32837
2844 W. Deer Valley Rd. • Phoenix, AZ 85027

MEMBER NMMA

FINANCIAL ASSISTANCE AVAILABLE FOR THOSE WHO QUALIFY!

SAFETY NOTICE

Proper service and repair procedures are vital to the safe, reliable operation of all marine engines, as well as the personal safety of those performing repairs. This manual outlines procedures for servicing and repairing engines and drive systems using safe, effective methods. The procedures contain many NOTES, CAUTIONS and WARNINGS which should be followed, along with standard procedures, to minimize the possibility of personal injury or improper service which could damage the vehicle or compromise its safety.

It is important to note that repair procedures and techniques, tools and parts for servicing these engines, as well as the skill and experience of the individual performing the work, vary widely. It is not possible to anticipate all of the conceivable ways or conditions under which the engine may be serviced, or to provide cautions as to all possible hazards that may result. Standard and accepted safety precautions and equipment should be used during cutting, grinding, chiseling, prying, or any other process that can cause material removal or projectiles.

Some procedures require the use of tools specially designed for a specific task. Before substituting another tool or procedure, you must be completely satisfied that neither your personal safety, nor the performance of the vessel, will be endangered. All procedures covered in this manual requiring the use of special tools will be noted at the beginning of the procedure by means of an **OEM symbol**

Additionally, any procedure requiring the use of an electronic tester or scan tool will be noted at the beginning of the procedure by means of a **DVOM symbol**

Although information in this manual is based on industry sources and is complete as possible at the time of publication, the possibility exists that some manufacturers made later changes which could not be included here. While striving for total accuracy, Seloc Publishing cannot assume responsibility for any errors, changes or omissions that may occur in the compilation of this data. We must therefore warn you to follow instructions carefully, using common sense. If you are uncertain of a procedure, seek help by inquiring with someone in your area who is familiar with these motors before proceeding.

PART NUMBERS

Part numbers listed in this reference are not recommendations by Seloc Publishing for any particular product brand name, simply iterations of the manufacturer's suggestions. They are also references that can be used with interchange manuals and aftermarket supplier catalogs to locate each brand supplier's discrete part number.

SPECIAL TOOLS

Special tools are recommended by the manufacturers to perform a specific job. Use has been kept to a minimum, but, where absolutely necessary, they are referred to in the text by the part number of the manufacturer if at all possible; and also noted at the beginning of each procedure with one of the following symbols: **OEM** or **DVOM.**

The **OEM** symbol usually denotes the need for a unique tool purposely designed to accomplish a specific task, it will also be used, less frequently, to notify the reader of the need for a tool that is not commonly found in the average tool box.

The **DVOM** symbol is used to denote the need for an electronic test tool like an ohmmeter, multi-meter or, on cetain later engines, a scan tool.

These tools can be purchased, under the appropriate part number, from your local dealer or regional distributor, or an equivalent tool can be purchased locally from a tool supplier or parts outlet. Before substituting any tool for the one recommended, read the SAFETY NOTICE at the top of this page.

Providing the correct mix of service and repair procedures is an endless battle for any publisher of "How-To" information. Users range from first time do-it yourselfers to professionally trained marine technicians, and information important to one is frequently irrelevant to the other. The editors at Seloc Publishing strive to provide accurate and articulate information on all facets of marine engine repair, from the simplest procedure to the most complex. In doing this, we understand that certain procedures may be outside the capabilities of the average DIYer. Conversely we are aware that many procedures are unnecessary for a trained technician.

SKILL LEVELS

In order to provide all of our users, particularly the DIYers, with a feeling for the scope of a given procedure or task before tackling it we have included a rating system denoting the suggested skill level needed when performing a particular procedure. One of the following icons will be included at the beginning of most procedures:

EASY. These procedures are aimed primarily at the DIYer and can be classified, for the most part, as basic maintenance procedures; battery, fluids, filters, plugs, etc. Although certainly valuable to any experience level, they will generally be of little importance to a technician.

MODERATE. These procedures are suited for a DIYer with experience and a working knowledge of mechanical procedures. Even an advanced DIYer or professional technician will occasionally refer to these procedures. They will generally consist of component repair and service procedures, adjustments and minor rebuilds.

DIFFICULT. These procedures are aimed at the advanced DIYer and professional technician. They will deal with diagnostics, rebuilds and internal engine/drive components and will frequently require special tools.

SKILLED. These procedures are aimed at highly skilled technicians and should not be attempted without previous experience. They will usually consist of machine work, internal engine work and gear case rebuilds.

Please remember one thing when considering the above ratings—they are a guide for judging the complexity of a given procedure and are subjective in nature. Only you will know what your experience level is, and only you will know when a procedure may be outside the realm of your capability. First time DIYer, or life-long marine technician, we all approach repair and service differently so an easy procedure for one person may be a difficult procedure for another, regardless of experience level. All skill level ratings are meant to be used as a guide only! Use them to help make a judgement before undertaking a particular procedure, but by all means read through the procedure first and make your own decision—after all, our mission at Seloc is to make boat maintenance and repair easier for everyone whether you are changing the oil or rebuilding an engine. Enjoy boating!

ALL RIGHTS RESERVED

No part of this publication may be reproduced, transmitted or stored in any form or by any means, electronic or mechanical, including photocopy, recording, or by information storage or retrieval system, without prior written permission from the publisher.

ACKNOWLEDGMENTS

Seloc Publishing expresses appreciation to the following companies who supported the production of this book:
- Yamaha Motor Corporation—Cypress, CA
- Marine Mechanics Institute—Orlando, FL
- Belks Marine—Holmes, PA

Thanks to John Hartung and Judy Belk of Belk's Marine for for there assistance, guidance, patience and access to some of the motors photographed for this manual.

Seloc Publishing would like to express thanks to the fine companies who participate in the production of all our books:
- Hand tools supplied by Craftsman are used during all phases of our vehicle teardown and photography.
- Many of the fine specialty tools used in our procedures were provided courtesy of Lisle Corporation.
- Much of our shop's electronic testing equipment was supplied by Universal Enterprises Inc. (UEI).

ANCHORS	1-10
AVOIDING THE MOST COMMON MISTAKES	1-3
AVOIDING TROUBLE	1-2
BAILING DEVICES	1-10
BASIC OPERATING PRINCIPLES	1-13
2-STROKE MOTORS	1-14
4-STROKE MOTORS	1-16
COMBUSTION	1-16
BOATING EQUIPMENT (NOT REQUIRED BUT RECOMMENDED)	**1-10**
ANCHORS	1-10
BAILING DEVICES	1-10
COMPASS	1-11
FIRST AID KIT	1-10
OAR/PADDLE (SECOND MEANS OF PROPULSION)	1-10
TOOLS AND SPARE PARTS	1-12
VHF-FM RADIO	1-10
BOATING SAFETY	**1-4**
COURTESY MARINE EXAMINATIONS	1-10
REGULATIONS FOR YOUR BOAT	1-4
REQUIRED SAFETY EQUIPMENT	1-6
BOLTS, NUTS AND OTHER THREADED RETAINERS	1-27
CAN YOU DO IT?	1-2
CHEMICALS	1-17
CLEANERS	1-18
LUBRICANTS & PENETRANTS	1-18
SEALANTS	1-18
COMPASS	1-11
COMPASS PRECAUTIONS	1-12
INSTALLATION	1-11
SELECTION	1-11
COURTESY MARINE EXAMINATIONS	1-10
DIRECTIONS AND LOCATIONS	1-2
ELECTRONIC TOOLS	1-23
BATTERY CHARGERS	1-24
BATTERY TESTERS	1-23
GAUGES	1-24
MULTI-METERS (DVOMS)	1-24
FASTENERS, MEASUREMENTS AND CONVERSIONS	**1-27**
BOLTS, NUTS AND OTHER THREADED RETAINERS	1-27
STANDARD AND METRIC MEASUREMENTS	1-27
TORQUE	1-27
FIRST AID KIT	1-10
HAND TOOLS	1-19
HAMMERS	1-22
PLIERS	1-21
SCREWDRIVERS	1-22
SOCKET SETS	1-9
WRENCHES	1-21
HOW TO USE THIS MANUAL	**1-2**
AVOIDING THE MOST COMMON MISTAKES	1-3
AVOIDING TROUBLE	1-2
CAN YOU DO IT?	1-2
DIRECTIONS AND LOCATIONS	1-2
MAINTENANCE OR REPAIR?	1-2
PROFESSIONAL HELP	1-3
PURCHASING PARTS	1-3
WHERE TO BEGIN	1-2
MAINTENANCE OR REPAIR?	1-2
MEASURING TOOLS	1-25
DEPTH GAUGES	1-26
DIAL INDICATORS	1-25
MICROMETERS & CALIPERS	1-25
TELESCOPING GAUGES	1-26
OAR/PADDLE (SECOND MEANS OF PROPULSION)	1-10
OTHER COMMON TOOLS	1-22
PROFESSIONAL HELP	1-3
PURCHASING PARTS	1-3
REGULATIONS FOR YOUR BOAT	1-4
CAPACITY INFORMATION	1-4
CERTIFICATE OF COMPLIANCE	1-5
DOCUMENTING OF VESSELS	1-4
HULL IDENTIFICATION NUMBER	1-4
LENGTH OF BOATS	1-4
NUMBERING OF VESSELS	1-4
REGISTRATION OF BOATS	1-4
SALES AND TRANSFERS	1-4
VENTILATION	1-5
VENTILATION SYSTEMS	1-5
REQUIRED SAFETY EQUIPMENT	1-6
FIRE EXTINGUISHERS	1-6
PERSONAL FLOTATION DEVICES	1-7
SOUND PRODUCING DEVICES	1-9
TYPES OF FIRES	1-6
VISUAL DISTRESS SIGNALS	1-9
WARNING SYSTEM	1-7
SAFETY IN SERVICE	**1-13**
DO'S	1-13
DON'TS	1-13
SAFETY TOOLS	1-17
EYE AND EAR PROTECTION	1-17
WORK CLOTHES	1-17
WORK GLOVES	1-17
SHOP EQUIPMENT	**1-17**
SAFETY TOOLS	1-17
CHEMICALS	1-17
SPECIAL TOOLS	1-23
SPECIFICATIONS	**1-28**
CONVERSION FACTORS	1-28
METRIC BOLTS - TYPICAL TORQUE VALUES	1-28
U.S. STANDARD BOLTS - TYPICAL TORQUE VALUES	1-28
STANDARD AND METRIC MEASUREMENTS	1-27
TOOLS	**1-19**
ELECTRONIC TOOLS	1-23
HAND TOOLS	1-19
MEASURING TOOLS	1-25
OTHER COMMON TOOLS	1-22
SPECIAL TOOLS	1-23
TOOLS AND SPARE PARTS	1-12
TORQUE	1-27
TROUBLESHOOTING	**1-13**
BASIC OPERATING PRINCIPLES	1-13
VHF-FM RADIO	1-10
WHERE TO BEGIN	1-2

1

GENERAL INFORMATION, SAFETY AND TOOLS

HOW TO USE THIS MANUAL	1-2
BOATING SAFETY	1-4
BOATING EQUIPMENT (NOT REQUIRED BUT RECOMMENDED)	1-10
SAFETY IN SERVICE	1-13
TROUBLESHOOTING	1-13
SHOP EQUIPMENT	1-17
TOOLS	1-19
FASTENERS, MEASUREMENTS AND CONVERSIONS	1-27
SPECIFICATIONS	1-28

1-2 GENERAL INFORMATION, SAFETY AND TOOLS

HOW TO USE THIS MANUAL

This manual is designed to be a handy reference guide to maintaining and repairing your Yamaha Outboard. We strongly believe that regardless of how many or how few year's experience you may have, there is something new waiting here for you.

This manual covers the topics that a factory service manual (designed for factory trained mechanics) and a manufacturer owner's manual (designed more by lawyers than boat owners these days) covers. It will take you through the basics of maintaining and repairing your outboard, step-by-step, to help you understand what the factory trained mechanics already know by heart. By using the information in this manual, any boat owner should be able to make better informed decisions about what they need to do to maintain and enjoy their outboard.

Even if you never plan on touching a wrench (and if so, we hope that we can change your mind), this manual will still help you understand what a mechanic needs to do in order to maintain your engine.

Can You Do It?

If you are not the type who is prone to taking a wrench to something, NEVER FEAR. The procedures provided here cover topics at a level virtually anyone will be able to handle. And just the fact that you purchased this manual shows your interest in better understanding your outboard.

You may even find that maintaining your outboard yourself is preferable in most cases. From a monetary standpoint, it could also be beneficial. The money spent on hauling your boat to a marina and paying a tech to service the engine could buy you fuel for a whole weekend of boating. And, if you are really that unsure of your own mechanical abilities, at the very least you should fully understand what a marine mechanic does to your boat. You may decide that anything other than maintenance and adjustments should be performed by a mechanic (and that's your call), but if so you should know that every time you board your boat, you are placing faith in the mechanic's work and trusting him or her with your well-being, and maybe your life.

It should also be noted that in most areas a factory-trained mechanic will command a hefty hourly rate for off site service. If the tech comes to you this hourly rate is often charged from the time they leave their shop to the time that they return home. When service is performed at a boat yard, the clock usually starts when they go out to get the boat and bring it into the shop and doesn't end until it is tested and put back in the yard. The cost savings in doing the job yourself might be readily apparent at this point.

Of course, if even you're already a seasoned Do-It-Yourselfer or a Professional Technician, you'll find the procedures, specifications, special tips as well as the schematics and illustrations helpful when tackling a new job on a motor.

■ **To help you decide if a task is within your skill level, procedures will often be rated using a wrench symbol in the text. When present, the number of wrenches designates how difficult we feel the procedure to be on a 1-4 scale. For more details on the wrench icon rating system, please refer to the information under Skill Levels at the beginning of this manual.**

Where to Begin

Before spending any money on parts, and before removing any nuts or bolts, read through the entire procedure or topic. This will give you the overall view of what tools and supplies will be required to perform the procedure or what questions need to be answered before purchasing parts. So read ahead and plan ahead. Each operation should be approached logically and all procedures thoroughly understood before attempting any work.

Avoiding Trouble

Some procedures in this manual may require you to "label and disconnect . . ." a group of lines, hoses or wires. Don't be lulled into thinking you can remember where everything goes — you won't. If you reconnect or install a part incorrectly, the motor may operate poorly, if at all. If you hook up electrical wiring incorrectly, you may instantly learn a very expensive lesson.

A piece of masking tape, for example, placed on a hose and another on its fitting will allow you to assign your own label such as the letter "A", or a short name. As long as you remember your own code, you can reconnect the lines by matching letters or names. Do remember that tape will dissolve when saturated in some fluids (especially cleaning solvents). If a component is to be washed or cleaned, use another method of identification. A permanent felt-tipped marker can be very handy for marking metal parts; but remember that some solvents will remove permanent marker. A scribe can be used to carefully etch a small mark in some metal parts, but be sure NOT to do that on a gasket-making surface.

SAFETY is the most important thing to remember when performing maintenance or repairs. Be sure to read the information on safety in this manual.

Maintenance or Repair?

Proper maintenance is the key to long and trouble-free engine life, and the work can yield its own rewards. A properly maintained engine performs better than one that is neglected. As a conscientious boat owner, set aside a Saturday morning, at least once a month, to perform a thorough check of items that could cause problems. Keep your own personal log to jot down which services you performed, how much the parts cost you, the date, and the amount of hours on the engine at the time. Keep all receipts for parts purchased, so that they may be referred to in case of related problems or to determine operating expenses. As a do-it-yourselfer, these receipts are the only proof you have that the required maintenance was performed. In the event of a warranty problem (on new motors), these receipts can be invaluable.

It's necessary to mention the difference between maintenance and repair. Maintenance includes routine inspections, adjustments, and replacement of parts that show signs of normal wear. Maintenance compensates for wear or deterioration. Repair implies that something has broken or is not working. A need for repair is often caused by lack of maintenance.

For example: draining and refilling the gearcase oil is maintenance recommended by all manufacturers at specific intervals. Failure to do this can allow internal corrosion or damage and impair the operation of the motor, requiring expensive repairs. While no maintenance program can prevent items from breaking or wearing out, a general rule can be stated: MAINTENANCE IS CHEAPER THAN REPAIR.

Directions and Locations

◆ See Figure 1

Two basic rules should be mentioned here. First, whenever the Port side of the engine (or boat) is referred to, it is meant to specify the left side of the engine when you are sitting at the helm. Conversely, the Starboard means your right side. The Bow is the front of the boat and the Stern or Aft is the rear.

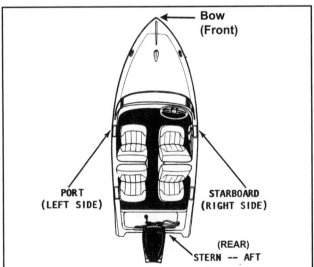

Fig. 1 Common terminology used for reference designation on boats of all size. These terms are used through out the text

GENERAL INFORMATION, SAFETY AND TOOLS 1-3

Most screws and bolts are removed by turning counterclockwise, and tightened by turning clockwise. An easy way to remember this is: righty-tighty; lefty-loosey. Corny, but effective. And if you are really dense (and we have all been so at one time or another), buy a ratchet that is marked ON and OFF (like Snap-on® ratchets), or mark your own. This can be especially helpful when you are bent over backwards, upside down or otherwise turned around when working on a boat-mounted component.

Professional Help

Occasionally, there are some things when working on an outboard that are beyond the capabilities or tools of the average Do-It-Yourselfer (DIYer). This shouldn't include most of the topics of this manual, but you will have to be the judge. Some engines require special tools or a selection of special parts, even for some basic maintenance tasks.

Talk to other boaters who use the same model of engine and speak with a trusted marina to find if there is a particular system or component on your engine that is difficult to maintain.

You will have to decide for yourself where basic maintenance ends and where professional service should begin. Take your time and do your research first (starting with the information contained within) and then make your own decision. If you really don't feel comfortable with attempting a procedure, DON'T DO IT. If you've gotten into something that may be over your head, don't panic. Tuck your tail between your legs and call a marine mechanic. Marinas and independent shops will be able to finish a job for you. Your ego may be damaged, but your boat will be properly restored to its full running order. So, as long as you approach jobs slowly and carefully, you really have nothing to lose and everything to gain by doing it yourself.

On the other hand, even the most complicated repair is within the ability of a person who takes their time and follows the steps of a procedure. A rock climber doesn't run up the side of a cliff, he/she takes it one step at a time and in the end, what looked difficult or impossible was conquerable. Worry about one step at a time.

Purchasing Parts

◆ See Figures 2 and 3

When purchasing parts there are two things to consider. The first is quality and the second is to be sure to get the correct part for your engine. To get quality parts, always deal directly with a reputable retailer. To get the proper parts always refer to the model number from the information tag on your engine prior to calling the parts counter. An incorrect part can adversely affect your engine performance and fuel economy, and will cost you more money and aggravation in the end.

Fig. 2 By far the most important asset in purchasing parts is a knowledgeable and enthusiastic parts person

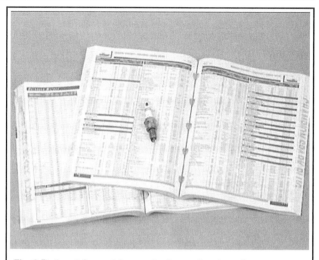

Fig. 3 Parts catalogs, giving application and part number information, are provided by manufacturers for most replacement parts

Just remember a tow back to shore will cost plenty. That charge is per hour from the time the towboat leaves their home port, to the time they return to their home port. Get the picture...$$$?

So whom should you call for parts? Well, there are many sources for the parts you will need. Where you shop for parts will be determined by what kind of parts you need, how much you want to pay, and the types of stores in your neighborhood.

Your marina can supply you with many of the common parts you require. Using a marina as your parts supplier may be handy because of location (just walk right down the dock) or because the marina specializes in your particular brand of engine. In addition, it is always a good idea to get to know the marina staff (especially the marine mechanic).

The marine parts jobber, who is usually listed in the yellow pages or whose name can be obtained from the marina, is another excellent source for parts. In addition to supplying local marinas, they also do a sizeable business in over-the-counter parts sales for the do-it-yourselfer.

Almost every boating community has one or more convenient marine chain stores. These stores often offer the best retail prices and the convenience of one-stop shopping for all your needs. Since they cater to the do-it-yourselfer, these stores are almost always open weeknights, Saturdays, and Sundays, when the jobbers are usually closed.

The lowest prices for parts are most often found in discount stores or the auto department of mass merchandisers. Parts sold here are name and private brand parts bought in huge quantities, so they can offer a competitive price. Private brand parts are made by major manufacturers and sold to large chains under a store label. And, of course, more and more large automotive parts retailers are stocking basic marine supplies.

Avoiding the Most Common Mistakes

There are 3 common mistakes in mechanical work:

1. Following the incorrect order of assembly, disassembly or adjustment. When taking something apart or putting it together, performing steps in the wrong order usually just costs you extra time; however, it CAN break something. Read the entire procedure before beginning disassembly. Perform everything in the order in which the instructions say you should, even if you can't immediately see a reason for it. When you're taking apart something that is very intricate, you might want to draw a picture of how it looks when assembled at one point in order to make sure you get everything back in its proper position. When making adjustments, perform them in the proper order; often, one adjustment affects another, and you cannot expect satisfactory results unless each adjustment is made only when it cannot be changed by subsequent adjustments.

1-4 GENERAL INFORMATION, SAFETY AND TOOLS

■ **Digital cameras are handy. If you've got access to one, take pictures of intricate assemblies during the disassembly process and refer to them during assembly for tips on part orientation.**

2. Overtorquing (or undertorquing). While it is more common for overtorquing to cause damage, undertorquing may allow a fastener to vibrate loose causing serious damage. Especially when dealing with plastic and aluminum parts, pay attention to torque specifications and utilize a torque wrench in assembly. If a torque figure is not available, remember that if you are using the right tool to perform the job, you will probably not have to strain yourself to get a fastener tight enough. The pitch of most threads is so slight that the tension you put on the wrench will be multiplied many times in actual force on what you are tightening.

3. Cross-threading. This occurs when a part such as a bolt is screwed into a nut or casting at the wrong angle and forced. Crossthreading is more likely to occur if access is difficult. It helps to clean and lubricate fasteners, then to start threading with the part to be installed positioned straight inward. Always start a fastener, etc. with your fingers. If you encounter resistance, unscrew the part and start over again at a different angle until it can be inserted and turned several times without much effort. Keep in mind that some parts may have tapered threads, so that gentle turning will automatically bring the part you're threading to the proper angle, but only if you don't force it or resist a change in angle. Don't put a wrench on the part until it has been tightened a couple of turns by hand. If you suddenly encounter resistance, and the part has not seated fully, don't force it. Pull it back out to make sure it's clean and threading properly.

BOATING SAFETY

In 1971 Congress ordered the U.S. Coast Guard to improve recreational boating safety. In response, the Coast Guard drew up a set of regulations. Aside from these federal regulations, there are state and local laws you must follow. These sometimes exceed the Coast Guard requirements. This section discusses only the federal laws. State and local laws are available from your local Coast Guard. As with other laws, "Ignorance of the boating laws is no excuse." The rules fall into two groups: regulations for your boat and required safety equipment on your boat.

Regulations For Your Boat

Most boats on waters within Federal jurisdiction must be registered or documented. These waters are those that provide a means of transportation between two or more states or to the sea. They also include the territorial waters of the United States.

DOCUMENTING OF VESSELS

A vessel of five or more net tons may be documented as a yacht. In this process, papers are issued by the U.S. Coast Guard as they are for large ships. Documentation is a form of national registration. The boat must be used solely for pleasure. Its owner must be a citizen of the U.S., a partnership of U.S. citizens, or a corporation controlled by U.S. citizens. The captain and other officers must also be U.S. citizens. The crew need not be.

If you document your yacht, you have the legal authority to fly the yacht ensign. You also may record bills of sale, mortgages, and other papers of title with federal authorities. Doing so gives legal notice that such instruments exist. Documentation also permits preferred status for mortgages. This gives you additional security, and it aids in financing and transfer of title. You must carry the original documentation papers aboard your vessel. Copies will not suffice.

REGISTRATION OF BOATS

If your boat is not documented, registration in the state of its principal use is probably required. If you use it mainly on an ocean, a gulf, or other similar water, register it in the state where you moor it.

If you use your boat solely for racing, it may be exempt from the requirement in your state. Some states may also exclude dinghies, while others require registration of documented vessels and non-power driven boats.

All states, except Alaska, register boats. In Alaska, the U.S. Coast Guard issues the registration numbers. If you move your vessel to a new state of principal use, a valid registration certificate is good for 60 days. You must have the registration certificate (certificate of number) aboard your vessel when it is in use. A copy will not suffice. You may be cited if you do not have the original on board.

NUMBERING OF VESSELS

A registration number is on your registration certificate. You must paint or permanently attach this number to both sides of the forward half of your boat. Do not display any other number there.

The registration number must be clearly visible. It must not be placed on the obscured underside of a flared bow. If you can't place the number on the bow, place it on the forward half of the hull. If that doesn't work, put it on the superstructure. Put the number for an inflatable boat on a bracket or fixture. Then, firmly attach it to the forward half of the boat. The letters and numbers must be plain block characters and must read from left to right. Use a space or a hyphen to separate the prefix and suffix letters from the numerals. The color of the characters must contrast with that of the background, and they must be at least three inches high.

In some states your registration is good for only one year. In others, it is good for as long as three years. Renew your registration before it expires. At that time you will receive a new decal or decals. Place them as required by state law. You should remove old decals before putting on the new ones. Some states require that you show only the current decal or decals. If your vessel is moored, it must have a current decal even if it is not in use.

If your vessel is lost, destroyed, abandoned, stolen, or transferred, you must inform the issuing authority. If you lose your certificate of number or your address changes, notify the issuing authority as soon as possible.

SALES AND TRANSFERS

Your registration number is not transferable to another boat. The number stays with the boat unless its state of principal use is changed.

HULL IDENTIFICATION NUMBER

A Hull Identification Number (HIN) is like the Vehicle Identification Number (VIN) on your car. Boats built between November 1, 1972 and July 31, 1984 have old format HINs. Since August 1, 1984 a new format has been used.

Your boat's HIN must appear in two places. If it has a transom, the primary number is on its starboard side within two inches of its top. If it does not have a transom or if it was not practical to use the transom, the number is on the starboard side. In this case, it must be within one foot of the stern and within two inches of the top of the hull side. On pontoon boats, it is on the aft crossbeam within one foot of the starboard hull attachment. Your boat also has a duplicate number in an unexposed location. This is on the boat's interior or under a fitting or item of hardware.

LENGTH OF BOATS

For some purposes, boats are classed by length. Required equipment, for example, differs with boat size. Manufacturers may measure a boat's length in several ways. Officially, though, your boat is measured along a straight line from its bow to its stern. This line is parallel to its keel.

The length does not include bowsprits, boomkins, or pulpits. Nor does it include rudders, brackets, outboard motors, outdrives, diving platforms, or other attachments.

CAPACITY INFORMATION

◆ See Figure 4

Manufacturers must put capacity plates on most recreational boats less than 20 feet long. Sailboats, canoes, kayaks, and inflatable boats are usually exempt. Outboard boats must display the maximum permitted horsepower of their engines. The plates must also show the allowable maximum weights of the people on board. And they must show the allowable maximum combined weights of people, engine(s), and gear. Inboards and stern drives need not show the weight of their engines on their capacity plates. The capacity plate must appear where it is clearly visible to the operator when underway. This information serves to remind you of the capacity of your boat under normal circumstances. You should ask yourself, "Is my boat loaded above its recommended capacity" and, "Is my boat overloaded for the present sea and wind conditions?" If you are stopped by a legal authority, you may be cited if you are overloaded.

GENERAL INFORMATION, SAFETY AND TOOLS

CERTIFICATE OF COMPLIANCE

◆ See Figure 4

Manufacturers are required to put compliance plates on motorboats greater than 20 feet in length. The plates must say, "This boat," or "This equipment complies with the U. S. Coast Guard Safety Standards in effect on the date of certification." Letters and numbers can be no less than one-eighth of an inch high. At the manufacturer's option, the capacity and compliance plates may be combined.

VENTILATION

A cup of gasoline spilled in the bilge has the potential explosive power of 15 sticks of dynamite. This statement, commonly quoted over 20 years ago, may be an exaggeration; however, it illustrates a fact. Gasoline fumes in the bilge of a boat are highly explosive and a serious danger. They are heavier than air and will stay in the bilge until they are vented out.

Because of this danger, Coast Guard regulations require ventilation on many powerboats. There are several ways to supply fresh air to engine and gasoline tank compartments and to remove dangerous vapors. Whatever the choice, it must meet Coast Guard standards.

■ The following is not intended to be a complete discussion of the regulations. It is limited to the majority of recreational vessels. Contact your local Coast Guard office for further information.

General Precautions

Ventilation systems will not remove raw gasoline that leaks from tanks or fuel lines. If you smell gasoline fumes, you need immediate repairs. The best device for sensing gasoline fumes is your nose. Use it! If you smell gasoline in a bilge, engine compartment, or elsewhere, don't start your engine. The smaller the compartment, the less gasoline it takes to make an explosive mixture.

Ventilation for Open Boats

In open boats, gasoline vapors are dispersed by the air that moves through them. So they are exempt from ventilation requirements.

To be "open," a boat must meet certain conditions. Engine and fuel tank compartments and long narrow compartments that join them must be open to the atmosphere." This means they must have at least 15 square inches of open area for each cubic foot of net compartment volume. The open area must be in direct contact with the atmosphere. There must also be no long, unventilated spaces open to engine and fuel tank compartments into which flames could extend.

Ventilation for All Other Boats

Powered and natural ventilation are required in an enclosed compartment with a permanently installed gasoline engine that has a cranking motor. A compartment is exempt if its engine is open to the atmosphere. Diesel powered boats are also exempt.

VENTILATION SYSTEMS

There are two types of ventilation systems. One is "natural ventilation." In it, air circulates through closed spaces due to the boat's motion. The other type is "powered ventilation." In it, air is circulated by a motor-driven fan or fans.

Natural Ventilation System Requirements

A natural ventilation system has an air supply from outside the boat. The air supply may also be from a ventilated compartment or a compartment open to the atmosphere. Intake openings are required. In addition, intake ducts may be required to direct the air to appropriate compartments.

The system must also have an exhaust duct that starts in the lower third of the compartment. The exhaust opening must be into another ventilated compartment or into the atmosphere. Each supply opening and supply duct, if there is one, must be above the usual level of water in the bilge. Exhaust openings and ducts must also be above the bilge water. Openings and ducts must be at least three square inches in area or two inches in diameter. Openings should be placed so exhaust gasses do not enter the fresh air intake. Exhaust fumes must not enter cabins or other enclosed, non-ventilated spaces. The carbon monoxide gas in them is deadly.

Intake and exhaust openings must be covered by cowls or similar devices. These registers keep out rain water and water from breaking seas. Most often, intake registers face forward and exhaust openings aft. This aids the flow of air when the boat is moving or at anchor since most boats face into the wind when properly anchored.

Power Ventilation System Requirements

◆ See Figure 5

Powered ventilation systems must meet the standards of a natural system, but in addition, they must also have one or more exhaust blowers. The blower duct can serve as the exhaust duct for natural ventilation if fan blades do not obstruct the air flow when not powered. Openings in engine compartment, for carburetion are in addition to ventilation system requirements.

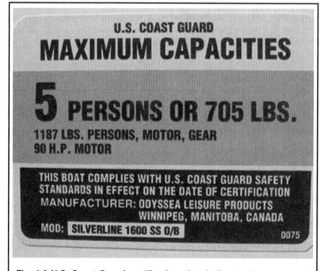

Fig. 4 A U.S. Coast Guard certification plate indicates the amount of occupants and gear appropriate for safe operation of the vessel (and allowable engine size for outboard boats)

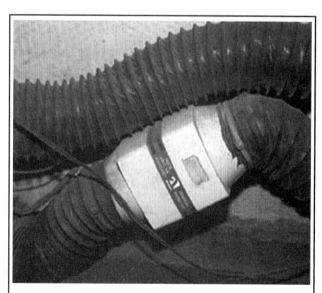

Fig. 5 Typical blower and duct system to vent fumes from the engine compartment

1-6 GENERAL INFORMATION, SAFETY AND TOOLS

Required Safety Equipment

Coast Guard regulations require that your boat have certain equipment aboard. These requirements are minimums. Exceed them whenever you can.

TYPES OF FIRES

There are four common classes of fires:
- Class A—fires are of ordinary combustible materials such as paper or wood.
- Class B—fires involve gasoline, oil and grease.
- Class C—fires are electrical.
- Class D—fires involve ferrous metals

One of the greatest risks to boaters is fire. This is why it is so important to carry the correct number and type of extinguishers onboard.

The best fire extinguisher for most boats is a Class B extinguisher. Never use water on Class B or Class C fires, as water spreads these types of fires. Additionally, you should never use water on a Class C fire as it may cause you to be electrocuted.

FIRE EXTINGUISHERS

◆ See Figure 6

If your boat meets one or more of the following conditions, you must have at least one fire extinguisher aboard. The conditions are:
- Inboard or stern drive engines
- Closed compartments under seats where portable fuel tanks can be stored
- Double bottoms not sealed together or not completely filled with flotation materials
- Closed living spaces
- Closed stowage compartments in which combustible or flammable materials are stored
- Permanently installed fuel tanks
- Boat is 26 feet or more in length.

Contents of Extinguishers

Fire extinguishers use a variety of materials. Those used on boats usually contain dry chemicals, Halon, or Carbon Dioxide (CO_2). Dry chemical extinguishers contain chemical powders such as Sodium Bicarbonate—baking soda.

Carbon dioxide is a colorless and odorless gas when released from an extinguisher. It is not poisonous but caution must be used in entering compartments filled with it. It will not support life and keeps oxygen from reaching your lungs. A fire-killing concentration of Carbon Dioxide can be lethal. If you are in a compartment with a high concentration of CO_2, you will have no difficulty breathing. But the air does not contain enough oxygen to support life. Unconsciousness or death can result.

HALON EXTINGUISHERS

Some fire extinguishers and "built-in" or "fixed" automatic fire extinguishing systems contain a gas called Halon. Like carbon dioxide it is colorless and odorless and will not support life. Some Halons may be toxic if inhaled.

To be accepted by the Coast Guard, a fixed Halon system must have an indicator light at the vessel's helm. A green light shows the system is ready. Red means it is being discharged or has been discharged. Warning horns are available to let you know the system has been activated. If your fixed Halon system discharges, ventilate the space thoroughly before you enter it. There are no residues from Halon but it will not support life.

Although Halon has excellent fire fighting properties; it is thought to deplete the earth's ozone layer and has not been manufactured since January 1, 1994. Halon extinguishers can be refilled from existing stocks of the gas until they are used up, but high federal excise taxes are being charged for the service. If you discontinue using your Halon extinguisher, take it to a recovery station rather than releasing the gas into the atmosphere. Compounds such as FE 241, designed to replace Halon, are now available.

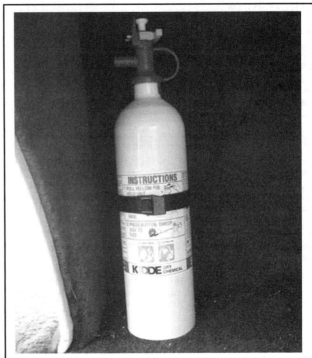

Fig. 6 An approved fire extinguisher should be mounted close to the operator for emergency use

Fire Extinguisher Approval

Fire extinguishers must be Coast Guard approved. Look for the approval number on the nameplate. Approved extinguishers have the following on their labels: "Marine Type USCG Approved, Size..., Type..., 162.208/," etc. In addition, to be acceptable by the Coast Guard, an extinguisher must be in serviceable condition and mounted in its bracket. An extinguisher not properly mounted in its bracket will not be considered serviceable during a Coast Guard inspection.

Care and Treatment

Make certain your extinguishers are in their stowage brackets and are not damaged. Replace cracked or broken hoses. Nozzles should be free of obstructions. Sometimes, wasps and other insects nest inside nozzles and make them inoperable. Check your extinguishers frequently. If they have pressure gauges, is the pressure within acceptable limits? Do the locking pins and sealing wires show they have not been used since recharging?

Don't try an extinguisher to test it. Its valves will not reseat properly and the remaining gas will leak out. When this happens, the extinguisher is useless.

Weigh and tag carbon dioxide and Halon extinguishers twice a year. If their weight loss exceeds 10 percent of the weight of the charge, recharge them. Check to see that they have not been used. They should have been inspected by a qualified person within the past six months, and they should have tags showing all inspection and service dates. The problem is that they can be partially discharged while appearing to be fully charged.

Some Halon extinguishers have pressure gauges the same as dry chemical extinguishers. Don't rely too heavily on the gauge. The extinguisher can be partially discharged and still show a good gauge reading. Weighing a Halon extinguisher is the only accurate way to assess its contents.

If your dry chemical extinguisher has a pressure indicator, check it frequently. Check the nozzle to see if there is powder in it. If there is, recharge it. Occasionally invert your dry chemical extinguisher and hit the base with the palm of your hand. The chemical in these extinguishers packs and cakes due to the boat's vibration and pounding. There is a difference of opinion about whether hitting the base helps, but it can't hurt. It is known that caking of the chemical powder is a major cause of failure of dry chemical extinguishers. Carry spares in excess of the minimum requirement. If you have guests aboard, make certain they know where the extinguishers are and how to use them.

GENERAL INFORMATION, SAFETY AND TOOLS 1-7

Using a Fire Extinguisher

A fire extinguisher usually has a device to keep it from being discharged accidentally. This is a metal or plastic pin or loop. If you need to use your extinguisher, take it from its bracket. Remove the pin or the loop and point the nozzle at the base of the flames. Now, squeeze the handle, and discharge the extinguisher's contents while sweeping from side to side. Recharge a used extinguisher as soon as possible.

If you are using a Halon or carbon dioxide extinguisher, keep your hands away from the discharge. The rapidly expanding gas will freeze them. If your fire extinguisher has a horn, hold it by its handle.

Legal Requirements for Extinguishers

You must carry fire extinguishers as defined by Coast Guard regulations. They must be firmly mounted in their brackets and immediately accessible.

A motorboat less than 26 feet long must have at least one approved hand-portable, Type B-1 extinguisher. If the boat has an approved fixed fire extinguishing system, you are not required to have the Type B-1 extinguisher. Also, if your boat is less than 26 feet long, is propelled by an outboard motor, or motors, and does not have any of the first six conditions described at the beginning of this section, it is not required to have an extinguisher. Even so, it's a good idea to have one, especially if a nearby boat catches fire, or if a fire occurs at a fuel dock.

A motorboat 26 feet to less than 40 feet long, must have at least two Type B-1 approved hand-portable extinguishers. It can, instead, have at least one Coast Guard approved Type B-2. If you have an approved fire extinguishing system, only one Type B-1 is required.

A motorboat 40 to 65 feet long must have at least three Type B-1 approved portable extinguishers. It may have, instead, at least one Type B-1 plus a Type B-2. If there is an approved fixed fire extinguishing system, two Type B-1 or one Type B-2 is required.

WARNING SYSTEM

Various devices are available to alert you to danger. These include fire, smoke, gasoline fumes, and carbon monoxide detectors. If your boat has a galley, it should have a smoke detector. Where possible, use wired detectors. Household batteries often corrode rapidly on a boat.

There are many ways in which carbon monoxide (a by-product of the combustion that occurs in an engine) can enter your boat. You can't see, smell, or taste carbon monoxide gas, but it is lethal. As little as one part in 10,000 parts of air can bring on a headache. The symptoms of carbon monoxide poisoning—headaches, dizziness, and nausea—are like seasickness. By the time you realize what is happening to you, it may be too late to take action. If you have enclosed living spaces on your boat, protect yourself with a detector.

PERSONAL FLOTATION DEVICES

Personal Flotation Devices (PFDs) are commonly called life preservers or life jackets. You can get them in a variety of types and sizes. They vary with their intended uses. To be acceptable, PFDs must be Coast Guard approved.

Type I PFDs

A Type I life jacket is also called an offshore life jacket. Type I life jackets will turn most unconscious people from facedown to a vertical or slightly backward position. The adult size gives a minimum of 22 pounds of buoyancy. The child size has at least 11 pounds. Type I jackets provide more protection to their wearers than any other type of life jacket. Type I life jackets are bulkier and less comfortable than other types. Furthermore, there are only two sizes, one for children and one for adults.

Type I life jackets will keep their wearers afloat for extended periods in rough water. They are recommended for offshore cruising where a delayed rescue is probable.

Type II PFDs

A Type II life jacket is also called a near-shore buoyant vest. It is an approved, wearable device. Type II life jackets will turn some unconscious people from facedown to vertical or slightly backward positions. The adult size gives at least 15.5 pounds of buoyancy. The medium child size has a minimum of 11 pounds. And the small child and infant sizes give seven pounds. A Type II life jacket is more comfortable than a Type I but it does not have as much buoyancy. It is not recommended for long hours in rough water. Because of this, Type IIs are recommended for inshore and inland cruising on calm water. Use them only where there is a good chance of fast rescue.

Type III PFDs

◆ See Figure 7

Type III life jackets or marine buoyant devices are also known as flotation aids. Like Type IIs, they are designed for calm inland or close offshore water where there is a good chance of fast rescue. Their minimum buoyancy is 15.5 pounds. They will **not** turn their wearers face up.

Type III devices are usually worn where freedom of movement is necessary. Thus, they are used for water skiing, small boat sailing, and fishing among other activities. They are available as vests and flotation coats. Flotation coats are useful in cold weather. Type IIIs come in many sizes from small child through large adult.

Life jackets come in a variety of colors and patterns—red, blue, green, camouflage, and cartoon characters. From purely a safety standpoint, the best color is bright orange. It is easier to see in the water, especially if the water is rough.

Type IV PFDs

◆ See Figures 8 and 9

Type IV ring life buoys, buoyant cushions and horseshoe buoys are Coast Guard approved devices called throwables. They are made to be thrown to people in the water, and should not be worn. Type IV cushions are often used as seat cushions. But, keep in mind that cushions are hard to hold onto in the water, thus, they do not afford as much protection as wearable life jackets.

The straps on buoyant cushions are for you to hold onto either in the water or when throwing them, they are **NOT** for your arms. A cushion should never be worn on your back, as it will turn you face down in the water.

Type IV throwables are not designed as personal flotation devices for unconscious people, non-swimmers, or children. Use them only in emergencies. They should not be used for, long periods in rough water.

Ring life buoys come in 18, 20, 24, and 30 in. diameter sizes. They usually have grab lines, but you will need to attach about 60 feet of polypropylene line to the grab rope to aid in retrieving someone in the water. If you throw a ring, be careful not to hit the person. Ring buoys can knock people unconscious

Type V PFDs

Type V PFDs are of two kinds, special use devices and hybrids. Special use devices include boardsailing vests, deck suits, work vests, and others. They are approved only for the special uses or conditions indicated on their labels. Each is designed and intended for the particular application shown on its label. They do not meet legal requirements for general use aboard recreational boats.

Hybrid life jackets are inflatable devices with some built-in buoyancy provided by plastic foam or kapok. They can be inflated orally or by cylinders of compressed gas to give additional buoyancy. In some hybrids the gas is released manually. In others it is released automatically when the life jacket is immersed in water.

The inherent buoyancy of a hybrid may be insufficient to float a person unless it is inflated. The only way to find this out is for the user to try it in the water. Because of its limited buoyancy when deflated, a hybrid is recommended for use by a non-swimmer only if it is worn with enough inflation to float the wearer.

If they are to count against the legal requirement for the number of life jackets you must carry, hybrids manufactured before February 8, 1995 must be worn whenever a boat is underway and the wearer must not go below decks or in an enclosed space. To find out if your Type V hybrid must be worn to satisfy the legal requirement, read its label. If its use is restricted it will say, "REQUIRED TO BE WORN" in capital letters.

Hybrids cost more than other life jackets, but this factor must be weighed against the fact that they are more comfortable than Types I, II or III life jackets. Because of their greater comfort, their owners are more likely to wear them than are the owners of Type I, II or III life jackets.

1-8 GENERAL INFORMATION, SAFETY AND TOOLS

Fig. 7 Type III PFDs are recommended for inshore/inland use on calm water (where there is a good chance of fast rescue)

Fig. 8 Type IV buoyant cushions are thrown to people in the water. If you can squeeze air out of the cushion, it should be replaced

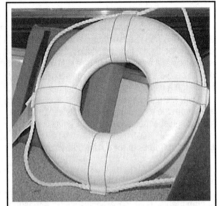

Fig. 9 Type IV throwables, such as this ring life buoy, are not designed for unconscious people, non-swimmers, or children

The Coast Guard has determined that improved, less costly hybrids can save lives since they will be bought and used more frequently. For these reasons, a new federal regulation was adopted effective February 8, 1995. The regulation increases both the deflated and inflated buoyancys of hybrids, makes them available in a greater variety of sizes and types, and reduces their costs by reducing production costs.

Even though it may not be required, the wearing of a hybrid or a life jacket is encouraged whenever a vessel is underway. Like life jackets, hybrids are now available in three types. To meet legal requirements, a Type I hybrid can be substituted for a Type I life jacket. Similarly Type II and III hybrids can be substituted for Type II and Type III life jackets. A Type I hybrid, when inflated, will turn most unconscious people from facedown to vertical or slightly backward positions just like a Type I life jacket. Type II and III hybrids function like Type II and III life jackets. If you purchase a new hybrid, it should have an owner's manual attached that describes its life jacket type and its deflated and inflated buoyancys. It warns you that it may have to be inflated to float you. The manual also tells you how to don the life jacket and how to inflate it. It also tells you how to change its inflation mechanism, recommended testing exercises, and inspection or maintenance procedures. The manual also tells you why you need a life jacket and why you should wear it. A new hybrid must be packaged with at least three gas cartridges. One of these may already be loaded into the inflation mechanism. Likewise, if it has an automatic inflation mechanism, it must be packaged with at least three of these water sensitive elements. One of these elements may be installed.

Legal Requirements

A Coast Guard approved life jacket must show the manufacturer's name and approval number. Most are marked as Type I, II, III, IV or V. All of the newer hybrids are marked for type.

You are required to carry at least one wearable life jacket or hybrid for each person on board your recreational vessel. If your vessel is 16 feet or more in length and is not a canoe or a kayak, you must also have at least one Type IV on board. These requirements apply to all recreational vessels that are propelled or controlled by machinery, sails, oars, paddles, poles, or another vessel. Sailboards are not required to carry life jackets.

You can substitute an older Type V hybrid for any required Type I, II or III life jacket provided:
 1. Its approval label shows it is approved for the activity the vessel is engaged in
 2. It's approved as a substitute for a life jacket of the type required on the vessel
 3. It's used as required on the labels
 and
 4. It's used in accordance with any requirements in its owner's manual (if the approval label makes reference to such a manual.)

A water skier being towed is considered to be on board the vessel when judging compliance with legal requirements.

You are required to keep your Type I, II or III life jackets or equivalent hybrids readily accessible, which means you must be able to reach out and get them when needed. All life jackets must be in good, serviceable condition.

General Considerations

The proper use of a life jacket requires the wearer to know how it will perform. You can gain this knowledge only through experience. Each person on your boat should be assigned a life jacket. Next, it should be fitted to the person who will wear it. Only then can you be sure that it will be ready for use in an emergency. This advice is good even if the water is calm, and you intend to boat near shore.

Boats can sink fast. There may be no time to look around for a life jacket. Fitting one on you in the water is almost impossible. Most drownings occur in inland waters within a few feet of safety. Most victims had life jackets, but they weren't wearing them.

Keeping life jackets in the plastic covers they came wrapped in, and in a cabin, assure that they will stay clean and unfaded. But this is no way to keep them when you are on the water. When you need a life jacket it must be readily accessible and adjusted to fit you. You can't spend time hunting for it or learning how to fit it.

There is no substitute for the experience of entering the water while wearing a life jacket. Children, especially, need practice. If possible, give your guests this experience. Tell them they should keep their arms to their sides when jumping in to keep the life jacket from riding up. Let them jump in and see how the life jacket responds. Is it adjusted so it does not ride up? Is it the proper size? Are all straps snug? Are children's life jackets the right sizes for them? Are they adjusted properly? If a child's life jacket fits correctly, you can lift the child by the jacket's shoulder straps and the child's chin and ears will not slip through. Non-swimmers, children, handicapped persons, elderly persons and even pets should always wear life jackets when they are aboard. Many states require that everyone aboard wear them in hazardous waters.

Inspect your lifesaving equipment from time to time. Leave any questionable or unsatisfactory equipment on shore. An emergency is no time for you to conduct an inspection.

Indelibly mark your life jackets with your vessel's name, number, and calling port. This can be important in a search and rescue effort. It could help concentrate effort where it will do the most good.

Care of Life Jackets

Given reasonable care, life jackets last many years. Thoroughly dry them before putting them away. Stow them in dry, well-ventilated places. Avoid the bottoms of lockers and deck storage boxes where moisture may collect. Air and dry them frequently.

Life jackets should not be tossed about or used as fenders or cushions. Many contain kapok or fibrous glass material enclosed in plastic bags. The bags can rupture and are then unserviceable. Squeeze your life jacket gently. Does air leak out? If so, water can leak in and it will no longer be safe to use. Cut it up so no one will use it, and throw it away. The covers of some life jackets are made of nylon or polyester. These materials are plastics. Like many plastics, they break down after extended exposure to the ultraviolet light in sunlight. This process may be more rapid when the materials are dyed with bright dyes such as "neon" shades.

Ripped and badly faded fabrics are clues that the covering of your life jacket is deteriorating. A simple test is to pinch the fabric between your thumbs and forefingers. Now try to tear the fabric. If it can be torn, it should definitely be destroyed and discarded. Compare the colors in protected places to those exposed to the sun. If the colors have faded, the materials have been weakened. A life jacket covered in fabric should ordinarily last several boating seasons with normal use. A life jacket used every day in direct sunlight should probably be replaced more often.

SOUND PRODUCING DEVICES

All boats are required to carry some means of making an efficient sound signal. Devices for making the whistle or horn noises required by the Navigation Rules must be capable of a four-second blast. The blast should be audible for at least one-half mile. Athletic whistles are not acceptable on boats 12 meters or longer. Use caution with athletic whistles. When wet, some of them come apart and loose their "pea." When this happens, they are useless.

If your vessel is 12 meters long and less than 20 meters, you must have a power whistle (or power horn) and a bell on board. The bell must be in operating condition and have a minimum diameter of at least 200mm (7.9 in.) at its mouth.

VISUAL DISTRESS SIGNALS

◆ See Figure 10

Visual Distress Signals (VDS) attract attention to your vessel if you need help. They also help to guide searchers in search and rescue situations. Be sure you have the right types, and learn how to use them properly.

It is illegal to fire flares improperly. In addition, they cost the Coast Guard and its Auxiliary many wasted hours in fruitless searches. If you signal a distress with flares and then someone helps you, please let the Coast Guard or the appropriate Search And Rescue (SAR) Agency know so the distress report will be canceled.

Recreational boats less than 16 feet long must carry visual distress signals on coastal waters at night. Coastal waters are:
• The ocean (territorial sea)
• The Great Lakes
• Bays or sounds that empty into oceans
• Rivers over two miles across at their mouths upstream to where they narrow to two miles.

Recreational boats 16 feet or longer must carry VDS at all times on coastal waters. The same requirement applies to boats carrying six or fewer passengers for hire. Open sailboats less than 26 feet long without engines are exempt in the daytime as are manually propelled boats. Also exempt are boats in organized races, regattas, parades, etc. Boats owned in the United States and operating on the high seas must be equipped with VDS.

A wide variety of signaling devices meet Coast Guard regulations. For pyrotechnic devices, a minimum of three must be carried. Any combination can be carried as long as it adds up to at least three signals for day use and at least three signals for night use. Three day/night signals meet both requirements. If possible, carry more than the legal requirement.

■ **The American flag flying upside down is a commonly recognized distress signal. It is not recognized in the Coast Guard regulations, though. In an emergency, your efforts would probably be better used in more effective signaling methods.**

Types of VDS

VDS are divided into two groups; daytime and nighttime use. Each of these groups is subdivided into pyrotechnic and non-pyrotechnic devices.

Daytime Non-Pyrotechnic Signals

A bright orange flag with a black square over a black circle is the simplest VDS. It is usable, of course, only in daylight. It has the advantage of being a continuous signal. A mirror can be used to good advantage on sunny days. It can attract the attention of other boaters and of aircraft from great distances. Mirrors are available with holes in their centers to aid in "aiming." In the absence of a mirror, any shiny object can be used. When another boat is in sight, an effective VDS is to extend your arms from your sides and move them up and down. Do it slowly. If you do it too fast the other people may think you are just being friendly. This simple gesture is seldom misunderstood, and requires no equipment.

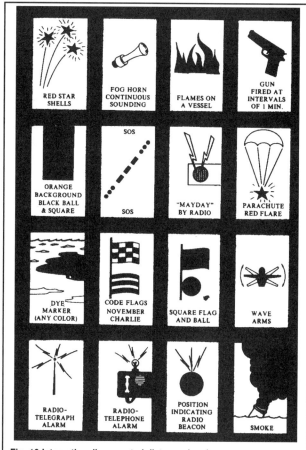

Fig. 10 Internationally accepted distress signals

Daytime Pyrotechnic Devices

Orange smoke is a useful daytime signal. Hand-held or floating smoke flares are very effective in attracting attention from aircraft. Smoke flares don't last long, and are not very effective in high wind or poor visibility. As with other pyrotechnic devices, use them only when you know there is a possibility that someone will see the display.

To be usable, smoke flares must be kept dry. Keep them in airtight containers and store them in dry places. If the "striker" is damp, dry it out before trying to ignite the device. Some pyrotechnic devices require a forceful "strike" to ignite them.

All hand-held pyrotechnic devices may produce hot ashes or slag when burning. Hold them over the side of your boat in such a way that they do not burn your hand or drip into your boat.

Nighttime Non-Pyrotechnic Signals

An electric distress light is available. This light automatically flashes the international morse code SOS distress signal (••• — •••). Flashed four to six times a minute, it is an unmistakable distress signal. It must show that it is approved by the Coast Guard. Be sure the batteries are fresh. Dated batteries give assurance that they are current.

Under the Inland Navigation Rules, a high intensity white light flashing 50-70 times per minute is a distress signal. Therefore, use strobe lights on inland waters only for distress signals.

Nighttime Pyrotechnic Devices

◆ See Figure 11

Aerial and hand-held flares can be used at night or in the daytime. Obviously, they are more effective at night.

Currently, the serviceable life of a pyrotechnic device is rated at 42 months from its date of manufacture. Pyrotechnic devices are expensive. Look at their dates before you buy them. Buy them with as much time remaining as possible.

1-10 GENERAL INFORMATION, SAFETY AND TOOLS

Fig. 11 Moisture-protected flares should be carried onboard any vessel for use as a distress signal

Like smoke flares, aerial and hand-held flares may fail to work if they have been damaged or abused. They will not function if they are or have been wet. Store them in dry, airtight containers in dry places. But store them where they are readily accessible.

Aerial VDSs, depending on their type and the conditions they are used in, may not go very high. Again, use them only when there is a good chance they will be seen.

A serious disadvantage of aerial flares is that they burn for only a short time; most burn for less than 10 seconds. Most parachute flares burn for less than 45 seconds. If you use a VDS in an emergency, do so carefully. Hold hand-held flares over the side of the boat when in use. Never use a road hazard flare on a boat; it can easily start a fire. Marine type flares are specifically designed to lessen risk, but they still must be used carefully.

Aerial flares should be given the same respect as firearms since they are firearms! Never point them at another person. Don't allow children to play with them or around them. When you fire one, face away from the wind. Aim it downwind and upward at an angle of about 60 degrees to the horizon. If there is a strong wind, aim it somewhat more vertically. Never fire it straight up. Before you discharge a flare pistol, check for overhead obstructions that might be damaged by the flare. An obstruction might deflect the flare to where it will cause injury or damage.

Disposal of VDS

Keep outdated flares when you get new ones. They do not meet legal requirements, but you might need them sometime, and they may work. It is illegal to fire a VDS on federal navigable waters unless an emergency exists. Many states have similar laws.

Emergency Position Indicating Radio Beacon (EPIRB)

There is no requirement for recreational boats to have EPIRBs. Some commercial and fishing vessels, though, must have them if they operate beyond the three-mile limit. Vessels carrying six or fewer passengers for hire must have EPIRBs under some circumstances when operating beyond the three-mile limit. If you boat in a remote area or offshore, you should have an EPIRB. An EPIRB is a small (about 6 to 20 in. high), battery-powered, radio transmitting buoy-like device. It is a radio transmitter and requires a license or an endorsement on your radio station license by the Federal Communications Commission (FCC). EPIRBs are either automatically activated by being immersed in water or manually by a switch.

Courtesy Marine Examinations

One of the roles of the Coast Guard Auxiliary is to promote recreational boating safety. This is why they conduct thousands of Courtesy Marine Examinations each year. The auxiliarists who do these examinations are well-trained and knowledgeable in the field.

These examinations are free and done only at the consent of boat owners. To pass the examination, a vessel must satisfy federal equipment requirements and certain additional requirements of the coast guard auxiliary. If your vessel does not pass the Courtesy Marine Examination, no report of the failure is made. Instead, you will be told what you need to correct the deficiencies. The examiner will return at your convenience to redo the examination.

If your vessel qualifies, you will be awarded a safety decal. The decal does not carry any special privileges, it simply attests to your interest in safe boating.

BOATING EQUIPMENT (NOT REQUIRED BUT RECOMMENDED)

Although not required by law, there are other pieces of equipment that are good to have onboard.

Oar/Paddle (Second Means of Propulsion)

All boats less than 16 feet long should carry a second means of propulsion. A paddle or oar can come in handy at times. For most small boats, a spare trolling or outboard motor is an excellent idea. If you carry a spare motor, it should have its own fuel tank and starting power. If you use an electric trolling motor, it should have its own battery.

Bailing Devices

All boats should carry at least one effective manual bailing device in addition to any installed electric bilge pump. This can be a bucket, can, scoop, hand-operated pump, etc. If your battery "goes dead" it will not operate your electric pump.

First Aid Kit

◆ See Figure 12

All boats should carry a first aid kit. It should contain adhesive bandages, gauze, adhesive tape, antiseptic, aspirin, etc. Check your first aid kit from time to time. Replace anything that is outdated. It is to your advantage to know how to use your first aid kit. Another good idea would be to take a Red Cross first aid course.

Anchors

◆ See Figure 13

All boats should have anchors. Choose one of suitable size for your boat. Better still, have two anchors of different sizes. Use the smaller one in calm water or when anchoring for a short time to fish or eat. Use the larger one when the water is rougher or for overnight anchoring.

Carry enough anchor line, of suitable size, for your boat and the waters in which you will operate. If your engine fails you, the first thing you usually should do is lower your anchor. This is good advice in shallow water where you may be driven aground by the wind or water. It is also good advice in windy weather or rough water, as the anchor, when properly affixed, will usually hold your bow into the waves.

VHF-FM Radio

Your best means of summoning help in an emergency or in case of a breakdown is a VHF-FM radio. You can use it to get advice or assistance from the Coast Guard. In the event of a serious illness or injury aboard your boat, the Coast Guard can have emergency medical equipment meet you ashore.

■ Although the VHF radio is the best way to get help, in this day and age, cell phones are a good backup source, especially for boaters on inland waters. You probably already know where you get a signal when boating, keep the phone charged, handy and off (so it doesn't bother you when boating right?). Keep phone numbers for a local dockmaster, coast guard, tow service or maritime police unit handy on board or stored in your phone directory.

GENERAL INFORMATION, SAFETY AND TOOLS 1-11

Fig. 12 Always carry an adequately stocked first aid kit on board for the safety of the crew and guests

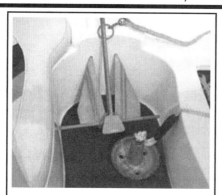

Fig. 13 Choose an anchor of sufficient weight to secure the boat without dragging

Fig. 14 Don't hesitate to spend a few extra dollars for a reliable compass

Compass

SELECTION

◆ See Figure 14

The safety of the boat and her crew may depend on her compass. In many areas, weather conditions can change so rapidly that, within minutes, a skipper may find himself socked in by a fog bank, rain squall or just poor visibility. Under these conditions, he may have no other means of keeping to his desired course except with the compass. When crossing an open body of water, his compass may be the only means of making an accurate landfall.

During thick weather when you can neither see nor hear the expected aids to navigation, attempting to run out the time on a given course can disrupt the pleasure of the cruise. The skipper gains little comfort in a chain of soundings that does not match those given on the chart for the expected area. Any stranding, even for a short time, can be an unnerving experience.

A pilot will not knowingly accept a cheap parachute. By the same token, a good boater should not accept a bargain in lifejackets, fire extinguishers, or compass. Take the time and spend the few extra dollars to purchase a compass to fit your expected needs. Regardless of what the salesman may tell you, postpone buying until you have had the chance to check more than one make and model.

Lift each compass, tilt and turn it, simulating expected motions of the boat. The compass card should have a smooth and stable reaction.

The card of a good quality compass will come to rest without oscillations about the lubber's line. Reasonable movement in your hand, comparable to the rolling and pitching of the boat, should not materially affect the reading.

INSTALLATION

◆ See Figure 15

Proper installation of the compass does not happen by accident. Make a critical check of the proposed location to be sure compass placement will permit the helmsman to use it with comfort and accuracy. First, the compass should be placed directly in front of the helmsman, and in such a position that it can be viewed without body stress as he sits or stands in a posture of relaxed alertness. The compass should be in the helmsman's zone of comfort. If the compass is too far away, he may have to bend forward to watch it; too close and he must rear backward for relief.

Second, give some thought to comfort in heavy weather and poor visibility conditions during the day and night. In some cases, the compass position may be partially determined by the location of the wheel, shift lever and throttle handle.

Third, inspect the compass site to be sure the instrument will be at least two feet from any engine indicators, bilge vapor detectors, magnetic instruments, or any steel or iron objects. If the compass cannot be placed at least two feet (six feet would be better but on a small craft, let's get real two feet is usually pushing it) from one of these influences, then either the compass or the other object must be moved, if first order accuracy is to be expected.

Once the compass location appears to be satisfactory, give the compass a test before installation. Hidden influences may be concealed under the cabin top, forward of the cabin aft bulkhead, within the cockpit ceiling, or in a wood-covered stanchion.

Move the compass around in the area of the proposed location. Keep an eye on the card. A magnetic influence is the only thing that will make the card turn. You can quickly find any such influence with the compass. If the influence cannot be moved away or replaced by one of non-magnetic material, test to determine whether it is merely magnetic, a small piece of iron or steel, or some magnetized steel. Bring the north pole of the compass near the object, then shift and bring the south pole near it. Both the north and south poles will be attracted if the compass is demagnetized. If the object attracts one pole and repels the other, then the compass is magnetized. If your compass needs to be demagnetized, take it to a shop equipped to do the job PROPERLY.

After you have moved the compass around in the proposed mounting area, hold it down or tape it in position. Test everything you feel might affect the compass and cause a deviation from a true reading. Rotate the wheel from hard over-to-hard over. Switch on and off all the lights, radios, radio direction finder, radio telephone, depth finder and, if installed, the shipboard intercom. Sound the electric whistle, turn on the windshield wipers, start the engine (with water circulating through the engine), work the throttle, and move the gear shift lever. If the boat has an auxiliary generator, start it.

If the card moves during any one of these tests, the compass should be relocated. Naturally, if something like the windshield wipers causes a slight deviation, it may be necessary for you to make a different deviation table to use only when certain pieces of equipment are operating. Bear in mind, following a course that is off only a degree or two for several hours can make considerable difference at the end, putting you on a reef, rock or shoal.

Check to be sure the intended compass site is solid. Vibration will increase pivot wear.

Fig. 15 The compass is a delicate instrument which should be mounted securely in a position where it can be easily observed by the helmsman

1-12 GENERAL INFORMATION, SAFETY AND TOOLS

Now, you are ready to mount the compass. To prevent an error on all courses, the line through the lubber line and the compass card pivot must be exactly parallel to the keel of the boat. You can establish the fore-and-aft line of the boat with a stout cord or string. Use care to transfer this line to the compass site. If necessary, shim the base of the compass until the stile-type lubber line (the one affixed to the case and not gimbaled) is vertical when the boat is on an even keel. Drill the holes and mount the compass.

COMPASS PRECAUTIONS

◆ See Figures 16, 17 and 18

Many times an owner will install an expensive stereo system in the cabin of his boat. It is not uncommon for the speakers to be mounted on the aft bulkhead up against the overhead (ceiling). In almost every case, this position places one of the speakers in very close proximity to the compass, mounted above the ceiling.

You probably already know that a magnet is used in the operation of the speaker. Therefore, it is very likely that the speaker, mounted almost under the compass in the cabin will have a very pronounced effect on the compass accuracy.

Consider the following test and the accompanying photographs as proof:

First, the compass was read as 190 degrees while the boat was secure in her slip.

Next, a full can of soda in an aluminum can was placed on one side and the compass read as 204 degrees, a good 14 degrees off.

Next, the full can was moved to the opposite side of the compass and again a reading was observed, this time as 189 degrees, 11 degrees off from the original reading.

Finally, the contents of the can were consumed, the can placed on both sides of the compass with NO effect on the compass reading.

Two very important conclusions can be drawn from these tests.

• Something must have been in the contents of the can to affect the compass so drastically.

• Keep even innocent things clear of the compass to avoid any possible error in the boat's heading.

■ **Remember, a boat moving through the water at 10 knots on a compass error of just 5 degrees will be almost 1.5 miles off course in only ONE hour. At night, or in thick weather, this could very possibly put the boat on a reef, rock or shoal with disastrous results.**

Tools and Spare Parts

◆ See Figures 19 and 20

Carry a few tools and some spare parts, and learn how to make minor repairs. Many search and rescue cases are caused by minor breakdowns that boat operators could have repaired. Carry spare parts such as propellers, fuses or basic ignition components (like spark plugs, wires or even ignition coils) and the tools necessary to install them.

Fig. 16 This compass is giving an accurate reading, right?

Fig. 17 ... well think again, as seemingly innocent objects may cause serious problems ...

Fig. 18 ... a compass reading off by just a few degrees could lead to disaster

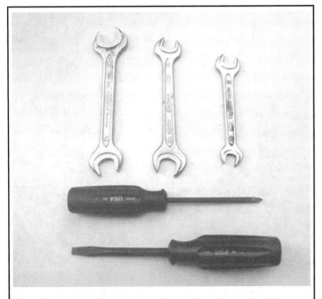

Fig. 19 A few wrenches, a screwdriver and maybe a pair of pliers can be very helpful to make emergency repairs

Fig. 20 A flashlight with a fresh set of batteries is handy when repairs are needed at night. It can also double as a signaling device

GENERAL INFORMATION, SAFETY AND TOOLS 1-13

SAFETY IN SERVICE

It is virtually impossible to anticipate all of the hazards involved with maintenance and service, but care and common sense will prevent most accidents.

The rules of safety for mechanics range from "don't smoke around gasoline," to "use the proper tool(s) for the job." The trick to avoiding injuries is to develop safe work habits and to take every possible precaution. Whenever you are working on your boat, pay attention to what you are doing. The more you pay attention to details and what is going on around you, the less likely you will be to hurt yourself or damage your boat.

Do's

- Do keep a fire extinguisher and first aid kit handy.
- Do wear safety glasses or goggles when cutting, drilling, grinding or prying, even if you have 20-20 vision. If you wear glasses for the sake of vision, wear safety goggles over your regular glasses.
- Do shield your eyes whenever you work around the battery. Batteries contain sulfuric acid. In case of contact with the eyes or skin, flush the area with water or a mixture of water and baking soda; then seek immediate medical attention.
- Do use adequate ventilation when working with any chemicals or hazardous materials.
- Do disconnect the negative battery cable when working on the electrical system. The secondary ignition system contains EXTREMELY HIGH VOLTAGE. In some cases it can even exceed 50,000 volts. Furthermore, an accidental attempt to start the engine could cause the propeller or other components to rotate suddenly causing a potentially dangerous situation.
- Do follow manufacturer's directions whenever working with potentially hazardous materials. Most chemicals and fluids are poisonous if taken internally.
- Do properly maintain your tools. Loose hammerheads, mushroomed punches and chisels, frayed or poorly grounded electrical cords, excessively worn screwdrivers, spread wrenches (open end), cracked sockets, or slipping ratchets can cause accidents.
- Likewise, keep your tools clean; a greasy wrench can slip off a bolt head, ruining the bolt and often harming your knuckles in the process.
- Do use the proper size and type of tool for the job at hand. Do select a wrench or socket that fits the nut or bolt. The wrench or socket should sit straight, not cocked.
- Do, when possible, pull on a wrench handle rather than push on it, and adjust your stance to prevent a fall.
- Do be sure that adjustable wrenches are tightly closed on the nut or bolt and pulled so that the force is on the side of the fixed jaw. Better yet, avoid the use of an adjustable if you have a fixed wrench that will fit.
- Do strike squarely with a hammer; avoid glancing blows.
- Do use common sense whenever you work on your boat or motor. If a situation arises that doesn't seem right, sit back and have a second look. It may save an embarrassing moment or potential damage to your beloved boat.

Don'ts

- Don't run the engine in an enclosed area or anywhere else without proper ventilation—EVER! Carbon monoxide is poisonous; it takes a long time to leave the human body and you can build up a deadly supply of it in your system by simply breathing in a little every day. You may not realize you are slowly poisoning yourself.
- Don't work around moving parts while wearing loose clothing. Short sleeves are much safer than long, loose sleeves. Hard-toed shoes with neoprene soles protect your toes and give a better grip on slippery surfaces. Jewelry, watches, large belt buckles, or body adornment of any kind is not safe working around any craft or vehicle. Long hair should be tied back under a hat.
- Don't use pockets for toolboxes. A fall or bump can drive a screwdriver deep into your body. Even a rag hanging from your back pocket can wrap around a spinning shaft.
- Don't smoke when working around gasoline, cleaning solvent or other flammable material.
- Don't smoke when working around the battery. When the battery is being charged, it gives off explosive hydrogen gas. Actually, you shouldn't smoke anyway, it's bad for you. Instead, save the cigarette money and put it into your boat!
- Don't use gasoline to wash your hands; there are excellent soaps available. Gasoline contains dangerous additives that can enter the body through a cut or through your pores. Gasoline also removes all the natural oils from the skin so that bone dry hands will suck up oil and grease.
- Don't use screwdrivers for anything other than driving screws! A screwdriver used as a prying tool can snap when you least expect it, causing injuries. At the very least, you'll ruin a good screwdriver.

TROUBLESHOOTING

Troubleshooting can be defined as a methodical process during which one discovers what is causing a problem with engine operation. Although it is often a feared process to the uninitiated, there is no reason to believe that you cannot figure out what is wrong with a motor, as long as you follow a few basic rules.

To begin with, troubleshooting must be systematic. Haphazardly testing one component, then another, **might** uncover the problem, but it will more likely waste a lot of time. True troubleshooting starts by defining the problem and performing systematic tests to eliminate the largest and most likely causes first.

Start all troubleshooting by eliminating the most basic possible causes. Begin with a visual inspection of the boat and motor. If the engine won't crank, make sure that the kill switch or safety lanyard is in the proper position. Make sure there is fuel in the tank and the fuel system is primed before condemning the carburetor or fuel injection system. On electric start motors, make sure there are no blown fuses, the battery is fully charged, and the cable connections (at both ends) are clean and tight before suspecting a bad starter, solenoid or switch.

The majority of problems that occur suddenly can be fixed by simply identifying the one small item that brought them on. A loose wire, a clogged passage or a broken component can cause a lot of trouble and are often the cause of a sudden performance problem.

The next most basic step in troubleshooting is to test systems before components. For example, if the engine doesn't crank on an electric start motor, determine if the battery is in good condition (fully charged and properly connected) before testing the starting system. If the engine cranks, but doesn't start, you know already know the starting system and battery (if it cranks fast enough) are in good condition, now it is time to look at the ignition or fuel systems. Once you've isolated the problem to a particular system, follow the troubleshooting/testing procedures in the section for that system to test either subsystems (if applicable, for example: the starter circuit) or components (starter solenoid).

Basic Operating Principles

◆ See Figures 21 and 22

Before attempting to troubleshoot a problem with your motor, it is important that you understand how it operates. Once normal engine or system operation is understood, it will be easier to determine what might be causing the trouble or irregular operation in the first place. System descriptions are found throughout this manual, but the basic mechanical operating principles for both 2-stroke engines (like most of the outboards covered here) and 4-stroke engines (like some outboards and like your car) are given here. A basic understanding of both types of engines is useful not only in understanding and troubleshooting your outboard, but also for dealing with other motors in your life.

All motors covered by this manual (and probably MOST of the motors you own) operate according to the Otto cycle principle of engine operation. This means that all motors follow the stages of intake, compression, power and exhaust. But, the difference between a 2- and 4-stroke motor is in how many times the piston moves up and down within the cylinder to accomplish this. On 2-stroke motors (as the name suggests) the four cycles take place in 2 movements (one up and one down) of the piston. Again, as the name suggests, the cycles take place in 4 movements of the piston for 4-stroke motors.

1-14 GENERAL INFORMATION, SAFETY AND TOOLS

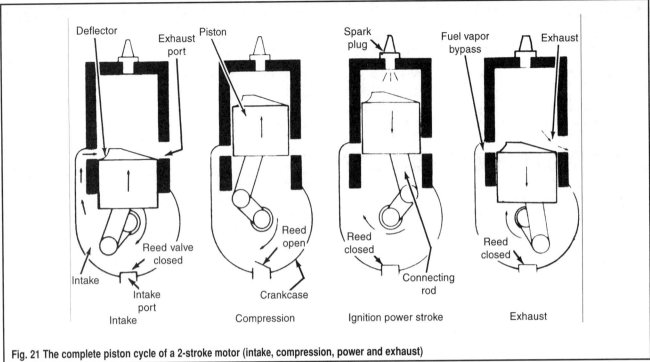

Fig. 21 The complete piston cycle of a 2-stroke motor (intake, compression, power and exhaust)

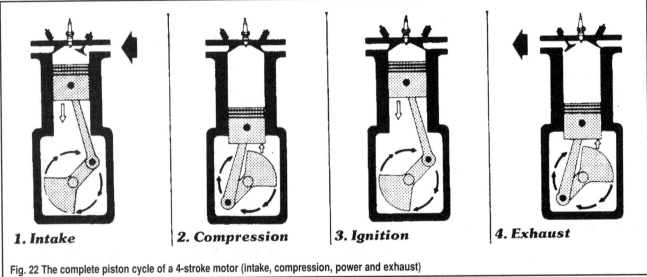

Fig. 22 The complete piston cycle of a 4-stroke motor (intake, compression, power and exhaust)

2-STROKE MOTORS

The 2-stroke engine differs in several ways from a conventional four-stroke (automobile or marine) engine.
1. The intake/exhaust method by which the fuel-air mixture is delivered to the combustion chamber.
2. The complete lubrication system.
3. The frequency of the power stroke.

Let's discuss these differences briefly (and compare 2-stroke engine operation with 4-stroke engine operation.)

Intake/Exhaust

◆ See Figures 23, 24, 25 and 26

Two-stroke engines utilize an arrangement of port openings to admit fuel to the combustion chamber and to purge the exhaust gases after burning has been completed. The ports are located in a precise pattern in order for them to be open and closed off at an exact moment by the piston as it moves up and down in the cylinder. The exhaust port is located slightly higher than the fuel intake port. This arrangement opens the exhaust port first as the piston starts downward and therefore, the exhaust phase begins a fraction of a second before the intake phase.

Actually, the intake and exhaust ports are spaced so closely together that both open almost simultaneously. For this reason, some 2-stroke engines utilize deflector-type pistons. This design of the piston top serves two purposes very effectively.

First, it creates turbulence when the incoming charge of fuel enters the combustion chamber. This turbulence results in a more complete burning of the fuel than if the piston top were flat. The second effect of the deflector-type piston crown is to force the exhaust gases from the cylinder more rapidly. Although this configuration is used in many older outboards, it is generally found only on some of the smaller Yamaha motors. The majority of Yamaha motors are of the Loop (L) charged.

Loop charged motors, or as they are commonly called "loopers", differ in how the air/fuel charge is introduced to the combustion chamber. Instead of the charge flowing across the top of the piston from one side of the cylinder to the other (CV) the use a looping action on top of the piston as the charge

GENERAL INFORMATION, SAFETY AND TOOLS 1-15

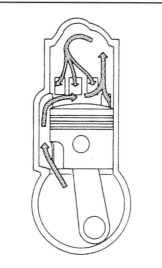

Fig. 23 The intake and exhaust cycles of a two-stroke engine—Cross flow (CV) design shown

Fig. 24 Cross-sectional view of a typical loop-charged cylinder, showing charge flow while piston is moving downward

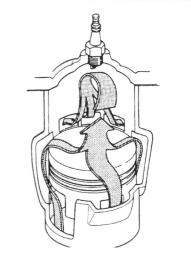

Fig. 25 Cutaway view of a typical loop-charged cylinder, depicting exhaust leaving the cylinder as the charge enters through 3 ports in the piston

is forced through irregular shaped openings cut in the piston's skirt. In a LV motor, the charge is forced out from the crankcase by the downward motion of the piston, through the irregular shaped openings and transferred upward by long, deep grooves in the cylinder wall. The charge completes its looping action by entering the combustion chamber, just above the piston, where the upward motion of the piston traps it in the chamber and compresses it for optimum ignition power.

Unlike the knife-edged deflector top pistons used in CV motors, the piston domes on Loop motors are relatively flat.

These systems of intake and exhaust are in marked contrast to individual intake and exhaust valve arrangement employed on four-stroke engines (and the mechanical methods of opening and closing these valves).

■ It should be noted here that there are some 2-stroke engines that utilize a mechanical valve train, though it is very different from the valve train employed by most 4-stroke motors. Rotary 2-stroke engines use a circular valve or rotating disc that contains a port opening around part of one edge of the disc. As the engine (and disc) turns, the opening aligns with the intake port at and for a predetermined amount of time, closing off the port again as the opening passes by and the solid portion of the disc covers the port.

Lubrication

A 2-stroke engine is lubricated by mixing oil with the fuel. Therefore, various parts are lubricated as the fuel mixture passes through the crankcase and the cylinder. In contrast, four-stroke engines have a crankcase containing oil. This oil is pumped through a circulating system and returned to the crankcase to begin the routing again.

Power Stroke

The combustion cycle of a 2-stroke engine has four distinct phases.
1. Intake
2. Compression
3. Power
4. Exhaust

The four phases of the cycle are accomplished with each up and down stroke of the piston, and the power stroke occurs with each complete revolution of the crankshaft. Compare this system with a four-stroke engine. A separate stroke of the piston is required to accomplish each phase of the cycle and the power stroke occurs only every other revolution of the crankshaft. Stated another way, two revolutions of the four-stroke engine crankshaft are required to complete one full cycle, the four phases.

Physical Laws

◆ See Figure 27

The 2-stroke engine is able to function because of two very simple physical laws.

One: Gases will flow from an area of high pressure to an area of lower pressure. A tire blowout is an example of this principle. The high-pressure air escapes rapidly if the tube is punctured.

Two: If a gas is compressed into a smaller area, the pressure increases, and if a gas expands into a larger area, the pressure is decreased.

If these two laws are kept in mind, the operation of the 2-stroke engine will be easier understood.

Actual Operation

◆ See Figure 21

■ The engine described here is of a carbureted type. EFI and HPDI motors operate similarly for intake of the air charge and for exhaust of the unburned gasses. Obviously though, the very nature of fuel injection changes the actual delivery of the fuel/oil charge.

Beginning with the piston approaching top dead center on the compression stroke: the intake and exhaust ports are physically closed (blocked) by the piston. During this stroke, the reed valve is open (because as the piston moves upward, the crankcase volume increases, which reduces the crankcase pressure to less than the outside atmosphere (creates a vacuum under the piston). The spark plug fires; the compressed fuel-air mixture is ignited; and the power stroke begins.

As the piston moves downward on the power stroke, the combustion chamber is filled with burning gases. As the exhaust port is uncovered, the

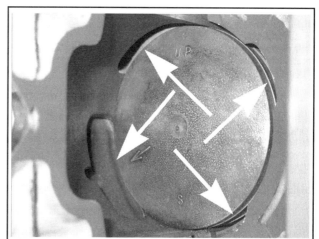

Fig. 26 The combustion chamber of a typical looper, notice the piston is far enough down the cylinder bore to reveal intake and exhaust ports

1-16 GENERAL INFORMATION, SAFETY AND TOOLS

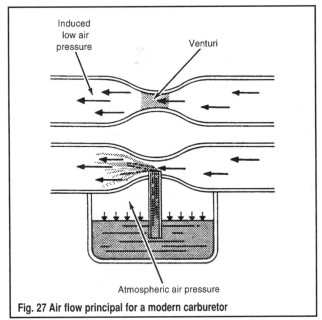

Fig. 27 Air flow principal for a modern carburetor

gases, which are under great pressure, escape rapidly through the exhaust ports. The piston continues its downward movement. Pressure within the crankcase (again, under the piston) increases, closing the reed valves against their seats. The crankcase then becomes a sealed chamber so the air-fuel mixture becomes compressed (pressurized) and ready for delivery to the combustion chamber. As the piston continues to move downward, the intake port is uncovered. The fresh fuel mixture rushes through the intake port into the combustion chamber striking the top of the piston where it is deflected along the cylinder wall. The reed valve remains closed until the piston moves upward again.

When the piston begins to move upward on the compression stroke, the reed valve opens because the crankcase volume has been increased, reducing crankcase pressure to less than the outside atmosphere. The intake and exhaust ports are closed and the fresh fuel charge is compressed inside the combustion chamber.

Pressure in the crankcase (beneath the piston) decreases as the piston moves upward and a fresh charge of air flows through the carburetor picking up fuel. As the piston approaches top dead center, the spark plug ignites the air-fuel mixture, the power stroke begins and one complete Otto cycle has been completed.

4-STROKE MOTORS

◆ See Figure 22

The 4-stroke motor may be easier to understand for some people either because of its prevalence in automobile and street motorcycle motors today or perhaps because each of the four strokes corresponds to one distinct phase of the Otto cycle. Essentially, a 4-stroke engine completes one Otto cycle of intake, compression, ignition/power and exhaust using two full revolutions of the crankshaft and four distinct movements of the piston (down, up, down and up).

Intake

The intake stroke begins with the piston near the top of its travel. As crankshaft rotation begins to pull the piston downward, the exhaust valve closes and the intake opens. As volume of the combustion chamber increases, a vacuum is created that draws in the air/fuel mixture from the intake manifold.

Compression

Once the piston reaches the bottom of its travel, crankshaft rotation will begin to force it upward. At this point the intake valve closes. As the piston rises in the bore, the volume of the sealed combustion chamber (both intake and exhaust valves are closed) decreases and the air/fuel mixture is compressed. This raises the temperature and pressure of the mixture and increases the amount of force generated by the expanding gases during the Ignition/Power stroke.

Ignition/Power

As the piston approaches top dead center (the highest point of travel in the bore), the spark plug will fire, igniting the air/fuel mixture. The resulting combustion of the air/fuel mixture forces the piston downward, rotating the crankshaft (causing other pistons to move in other phases/strokes of the Otto cycle on multi-cylinder motors).

Exhaust

As the piston approaches the bottom of the Ignition/Power stroke, the exhaust valve opens. When the piston begins its upward path of travel once again, any remaining unburned gasses are forced out through the exhaust valve. This completes one Otto cycle, which begins again as the piston passes top dead center, the intake valve opens and the Intake stroke starts.

COMBUSTION

Whether we are talking about a 2- or 4-stroke engine, all Otto cycle, internal combustion engines require three basic conditions to operate properly,
1. Compression
2. Ignition (Spark)
3. Fuel

A lack of any one of these conditions will prevent the engine from operating. A problem with any one of these will manifest itself in hard-starting or poor performance.

Compression

An engine that has insufficient compression will not draw an adequate supply of air/fuel mixture into the combustion chamber and, subsequently, will not make sufficient power on the power stroke. A lack of compression in just one cylinder of a multi-cylinder motor will cause the motor to stumble or run irregularly.

But, keep in mind that a sudden change in compression is unlikely in 2-stroke motors (unless something major breaks inside the crankcase, but that would usually be accompanied by other symptoms such as a loud noise when it occurred or noises during operation). On 4-stroke motors, a sudden change in compression is also unlikely, but could occur if the timing belt or chain was to suddenly break. Remember that the timing belt/chain is used to synchronize the valve train with the crankshaft. If the valve train suddenly ceases to turn, some intake and some exhaust valves will remain open, relieving compression in that cylinder.

Ignition (Spark)

Traditionally, the ignition system is the weakest link in the chain of conditions necessary for engine operation. Spark plugs may become worn or fouled, wires will deteriorate allowing arcing or misfiring, and poor connections can place an undue load on coils leading to weak spark or even a failed coil. The most common question asked by a technician under a no-start condition is: "do I have spark and fuel" (as they've already determined that they have compression).

A quick visual inspection of the spark plug(s) will answer the question as to whether or not the plug(s) is/are worn or fouled. While the engine is shut **OFF** a physical check of the connections could show a loose primary or secondary ignition circuit wire. An obviously physically damaged wire may also be an indication of system problems and certainly encourages one to inspect the related system more closely.

If nothing is turned up by the visual inspection, perform the Spark Test provided in the Ignition System section to determine if the problem is a lack of or a weak spark. If the problem is not compression or spark, it's time to look at the fuel system.

Fuel

If compression and spark is present (and within spec), but the engine won't start or won't run properly, the only remaining condition to fulfill is fuel. As usual, start with the basics. Is the fuel tank full? Is the fuel stale? If the engine has not been run in some time (a matter of months, not weeks) there is a good chance that the fuel is stale and should be properly disposed of and replaced.

- Depending on how stale or contaminated (with moisture) the fuel is, it may be burned in an automobile or in yard equipment, though it would be wise to mix it well with a much larger supply of fresh gasoline to prevent moving your driveability problems to that motor. But it is better to get the lawn tractor stuck on stale gasoline than it would be to have your boat motor quit in the middle of the bay or lake.

For hard starting motors, is the choke or primer system operating properly. Remember that the choke/prime should only be used for **cold** starts. A true cold start is really only the first start of the day, but it may be applicable to subsequent starts on cooler days, if the engine sat for more than a few hours and completely cooled off since the last use. Applying the primer to the motor for a hot start may flood the engine, preventing it from starting properly. One method to clear a flood is to crank the motor while the engine is at wide-open throttle (allowing the maximum amount of air into the motor to compensate for the excess fuel). But, keep in mind that the throttle should be returned to idle immediately upon engine start-up to prevent damage from over-revving.

Fuel delivery and pressure should be checked before delving into the carburetor(s) or fuel injection system. Make sure there are no clogs in the fuel line or vacuum leaks that would starve the motor of fuel.

Make sure that all other possible problems have been eliminated before touching the carburetor. It is rare that a carburetor will suddenly require an adjustment in order for the motor to run properly. It is much more likely that an improperly stored motor (one stored with untreated fuel in the carburetor) would suffer from one or more clogged carburetor passages sometime after shortly returning to service. Fuel will evaporate over time, leaving behind gummy deposits. If untreated fuel is left in the carburetor for some time (again typically months more than weeks), the varnish left behind by evaporating fuel will likely clog the small passages of the carburetor and cause problems with engine performance. If you suspect this, remove and disassemble the carburetor following procedures under Fuel System.

SHOP EQUIPMENT

Safety Tools

WORK GLOVES

◆ See Figure 28

Unless you think scars on your hands are cool, enjoy pain and like wearing bandages, get a good pair of work gloves. Canvas or leather gloves are the best. And yes, we realize that there are some jobs involving small parts that can't be done while wearing work gloves. These jobs are not the ones usually associated with hand injuries.

A good pair of rubber gloves (such as those usually associated with dish washing) or vinyl gloves is also a great idea. There are some liquids such as solvents and penetrants that don't belong on your skin. Avoid burns and rashes. Wear these gloves.

And lastly, an option. If you're tired of being greasy and dirty all the time, go to the drug store and buy a box of disposable latex gloves like medical professionals wear. You can handle greasy parts, perform small tasks, wash parts, etc. all without getting dirty! These gloves take a surprising amount of abuse without tearing and aren't expensive. Note however, that some people are allergic to the latex or the powder used inside some gloves, so pay attention to what you buy.

EYE AND EAR PROTECTION

◆ See Figures 29 and 30

Don't begin any job without a good pair of work goggles or impact resistant glasses! When doing any kind of work, it's all too easy to avoid eye injury through this simple precaution. And don't just buy eye protection and leave it on the shelf. Wear it all the time! Things have a habit of breaking, chipping, splashing, spraying, splintering and flying around. And, for some reason, your eye is always in the way!

If you wear vision-correcting glasses as a matter of routine, get a pair made with polycarbonate lenses. These lenses are impact resistant and are available at any optometrist.

Often overlooked is hearing protection. Engines and power tools are noisy! Loud noises damage your ears. It's as simple as that! The simplest and cheapest form of ear protection is a pair of noise-reducing ear plugs. Cheap insurance for your ears! And, they may even come with their own, cute little carrying case.

More substantial, more protection and more money is a good pair of noise reducing earmuffs. They protect from all but the loudest sounds. Hopefully those are sounds that you'll never encounter since they're usually associated with disasters.

WORK CLOTHES

Everyone has "work clothes." Usually these consist of old jeans and a shirt that has seen better days. That's fine. In addition, a denim work apron is a nice accessory. It's rugged, can hold some spare bolts, and you don't feel bad wiping your hands or tools on it. That's what it's for.

When working in cold weather, a one-piece, thermal work outfit is invaluable. Most are rated to below freezing temperatures and are ruggedly constructed. Just look at what local marine mechanics are wearing and that should give you a clue as to what type of clothing is good.

Chemicals

There is a whole range of chemicals that you'll find handy for maintenance and repair work. The most common types are: lubricants, penetrants and sealers. Keep these handy. There are also many chemicals that are used for detailing or cleaning.

When a particular chemical is not being used, keep it capped, upright and in a safe place. These substances may be flammable, may be irritants or might even be caustic and should always be stored properly, used properly and handled with care. Always read and follow all label directions and be sure to wear hand and eye protection!

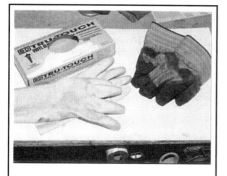

Fig. 28 Three different types of work gloves. The box contains latex gloves

Fig. 29 Don't begin major repairs without a pair of goggles for your eyes and earmuffs to protect your hearing

Fig. 30 Things have a habit of, splashing, spraying, splintering and flying around during repairs

1-18 GENERAL INFORMATION, SAFETY AND TOOLS

LUBRICANTS & PENETRANTS

◆ See Figure 31

Anti-seize is used to coat certain fasteners prior to installation. This can be especially helpful when two dissimilar metals are in contact (to help prevent corrosion that might lock the fastener in place). This is a good practice on a lot of different fasteners, BUT, NOT on any fastener that might vibrate loose causing a problem. If anti-seize is used on a fastener, it should be checked periodically for proper tightness.

Lithium grease, chassis lube, silicone grease or a synthetic brake caliper grease can all be used pretty much interchangeably. All can be used for coating rust-prone fasteners and for facilitating the assembly of parts that are a tight fit. Silicone and synthetic greases are the most versatile.

■ **Silicone dielectric grease is a non-conductor that is often used to coat the terminals of wiring connectors before fastening them. It may sound odd to coat metal portions of a terminal with something that won't conduct electricity, but here is it how it works. When the connector is fastened the metal-to-metal contact between the terminals will displace the grease (allowing the circuit to be completed). The grease that is displaced will then coat the non-contacted surface and the cavity around the terminals, SEALING them from atmospheric moisture that could cause corrosion.**

Silicone spray is a good lubricant for hard-to-reach places and parts that shouldn't be gooped up with grease.

Penetrating oil may turn out to be one of your best friends when taking something apart that has corroded fasteners. Not only can they make a job easier, they can really help to avoid broken and stripped fasteners. The most familiar penetrating oils are Liquid Wrench® and WD-40®. A newer penetrant, PB Blaster® works very well (and has become a mainstay in our shops). These products have hundreds of uses. For your purposes, they are vital!

Before disassembling any part, check the fasteners. If any appear rusted, soak them thoroughly with the penetrant and let them stand while you do something else (for particularly rusted or frozen parts you may need to soak them a few days in advance). This simple act can save you hours of tedious work trying to extract a broken bolt or stud.

SEALANTS

◆ See Figures 32 and 33

Sealants are an indispensable part for certain tasks, especially if you are trying to avoid leaks. The purpose of sealants is to establish a leak-proof bond between or around assembled parts. Most sealers are used in conjunction with gaskets, but some are used instead of conventional gasket material.

The most common sealers are the non-hardening types such as Permatex® No.2 or its equivalents. These sealers are applied to the mating surfaces of each part to be joined, then a gasket is put in place and the parts are assembled.

■ **A sometimes overlooked use for sealants like RTV is on the threads of vibration prone fasteners.**

One very helpful type of non-hardening sealer is the "high tack" type. This type is a very sticky material that holds the gasket in place while the parts are being assembled. This stuff is really a good idea when you don't have enough hands or fingers to keep everything where it should be.

The stand-alone sealers are the Room Temperature Vulcanizing (RTV) silicone gasket makers. On some engines, this material is used instead of a gasket. In those instances, a gasket may not be available or, because of the shape of the mating surfaces, a gasket shouldn't be used. This stuff, when used in conjunction with a conventional gasket, produces the surest bonds.

RTV does have its limitations though. When using this material, you will have a time limit. It starts to set-up within 15 minutes or so, so you have to assemble the parts without delay. In addition, when squeezing the material out of the tube, don't drop any glops into the engine. The stuff will form and set and travel around a cooling passage, possibly blocking it. Also, most types are not fuel-proof. Check the tube for all cautions.

CLEANERS

◆ See Figures 34 and 35

There are two basic types of cleaners on the market today: parts cleaners and hand cleaners. The parts cleaners are for the parts; the hand cleaners are for you. They are **NOT** interchangeable.

There are many good, non-flammable, biodegradable parts cleaners on the market. These cleaning agents are safe for you, the parts and the environment. Therefore, there is no reason to use flammable, caustic or toxic substances to clean your parts or tools.

As far as hand cleaners go; the waterless types are the best. They have always been efficient at cleaning, but they used to all leave a pretty smelly odor. Recently though, most of them have eliminated the odor and added stuff that actually smells good. Make sure that you pick one that contains lanolin or some other moisture-replenishing additive. Cleaners not only remove grease and oil but also skin oil.

■ **Most women already know to use a hand lotion when you're all cleaned up. It's okay. Real men DO use hand lotion too! Believe it or not, using hand lotion BEFORE your hands are dirty will actually make them easier to clean when you're finished with a dirty job. Lotion seals your hands, and keeps dirt and grease from sticking to your skin.**

Fig. 31 Keep a supply of anti-seize, penetrating oil, lithium grease, electronic cleaner and silicone spray

Fig. 32 Sealants are essential for preventing leaks

Fig. 33 On some engines, RTV is used instead of gasket material to seal components

GENERAL INFORMATION, SAFETY AND TOOLS 1-19

Fig. 34 Citrus hand cleaners not only work well, but they smell pretty good too. Choose one with pumice for added cleaning power

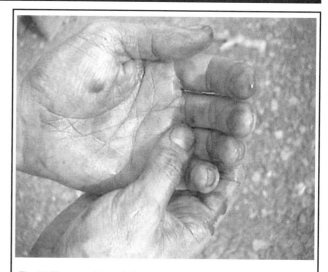

Fig. 35 The use of hand lotion seals your hands and keeps dirt and grease from sticking to your skin

TOOLS

◆ See Figure 36

Tools; this subject could fill a completely separate manual. The first thing you will need to ask yourself, is just how involved do you plan to get. If you are serious about maintenance and repair you will want to gather a quality set of tools to make the job easier, and more enjoyable. BESIDES, TOOLS ARE FUN!!!

Almost every do-it-yourselfer loves to accumulate tools. Though most find a way to perform jobs with only a few common tools, they tend to buy more over time, as money allows. So gathering the tools necessary for maintenance or repair does not have to be an expensive, overnight proposition.

When buying tools, the saying "You get what you pay for ..." is absolutely true! Don't go cheap! Any hand tool that you buy should be drop forged and/or chrome vanadium. These two qualities tell you that the tool is strong enough for the job. With any tool, go with a name that you've heard of before, or, that is recommended buy your local professional retailer. Let's go over a list of tools that you'll need.

Most of the world uses the metric system. However, some American-built engines and aftermarket accessories use standard fasteners. So, accumulate your tools accordingly. Any good DIYer should have a decent set of both U.S. and metric measure tools.

■ Don't be confused by terminology. Most advertising refers to "SAE and metric", or "standard and metric." Both are misnomers. The Society of Automotive Engineers (SAE) did not invent the English system of measurement; the English did. The SAE likes metrics just fine. Both English (U.S.) and metric measurements are SAE approved. Also, the current "standard" measurement IS metric. So, if it's not metric, it's U.S. measurement.

Hand Tools

SOCKET SETS

◆ See Figures 37 thru 43

Socket sets are the most basic hand tools necessary for repair and maintenance work. For our purposes, socket sets come in three drive sizes: 1/4 inch, 3/8 inch and 1/2 inch. Drive size refers to the size of the drive lug on the ratchet, breaker bar or speed handle.

A 3/8 inch set is probably the most versatile set in any mechanic's toolbox. It allows you to get into tight places that the larger drive ratchets can't and gives you a range of larger sockets that are still strong enough for heavy-duty work. The socket set that you'll need should range in sizes from

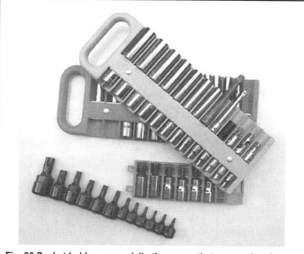

Fig. 36 Socket holders, especially the magnetic type, are handy items to keep tools in order

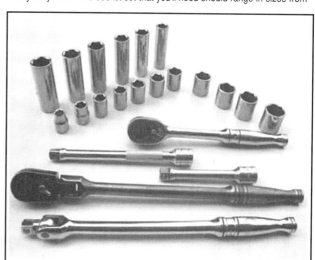

Fig. 37 A 3/8 in. socket set is probably the most versatile tool in any mechanic's tool box

1-20 GENERAL INFORMATION, SAFETY AND TOOLS

1/4 inch through 1 inch for standard fasteners, and a 6mm through 19mm for metric fasteners.

You'll need a good 1/2 inch set since this size drive lug assures that you won't break a ratchet or socket on large or heavy fasteners. Also, torque wrenches with a torque scale high enough for larger fasteners are usually 1/2 inch drive.

Plus, 1/4 inch drive sets can be very handy in tight places. Though they usually duplicate functions of the 3/8 in. set, 1/4 in. drive sets are easier to use for smaller bolts and nuts.

As for the sockets themselves, they come in shallow (standard) and deep lengths as well as 6 or 12 point. The 6 and 12 points designation refers to how many sides are in the socket itself. Each has advantages. The 6 point socket is stronger and less prone to slipping which would strip a bolt head or nut. 12 point sockets are more common, usually less expensive and can operate better in tight places where the ratchet handle can't swing far.

Standard length sockets are good for just about all jobs, however, some stud-head bolts, hard-to-reach bolts, nuts on long studs, etc., require the deep sockets.

Most marine manufacturers use recessed hex-head fasteners to retain many of the engine parts. These fasteners require a socket with a hex shaped driver or a large sturdy hex key. To help prevent torn knuckles, we would recommend that you stick to the sockets on any tight fastener and leave the hex keys for lighter applications. Hex driver sockets are available individually or in sets just like conventional sockets.

More and more, manufacturers are using Torx® head fasteners, which were once known as tamper resistant fasteners (because many people did not have tools with the necessary odd driver shape). Since Torx® fasteners have become commonplace in many DIYer tool boxes, manufacturers designed newer tamper resistant fasteners that are essentially Torx® head bolts that contain a small protrusion in the center (requiring the driver to contain a small hole to slide over the protrusion. Tamper resistant fasteners are often used where the manufacturer would prefer only knowledgeable mechanics or advanced Do-It-Yourselfers (DIYers) work.

Torque Wrenches

◆ See Figure 44

In most applications, a torque wrench can be used to ensure proper installation of a fastener. Torque wrenches come in various designs and most stores will carry a variety to suit your needs. A torque wrench should be used any time you have a specific torque value for a fastener. Keep in mind that because there is no worldwide standardization of fasteners, so charts or figure found in each repair section refer to the manufacturer's fasteners. Any general guideline charts that you might come across based on fastener size (they are sometimes included in a repair manual or with torque wrench packaging) should be used with caution. Just keep in mind that if you are using the right tool for the job, you should not have to strain to tighten a fastener.

BEAM TYPE

◆ See Figures 45 and 46

The beam type torque wrench is one of the most popular styles in use. If used properly, it can be the most accurate also. It consists of a pointer attached to the head that runs the length of the flexible beam (shaft) to a scale located near the handle. As the wrench is pulled, the beam bends and the pointer indicates the torque using the scale.

CLICK (BREAKAWAY) TYPE

◆ See Figures 47 and 48

Another popular torque wrench design is the click type. The clicking mechanism makes achieving the proper torque easy and most use a ratcheting head for ease of bolt installation. To use the click type wrench you pre-adjust it to a torque setting. Once the torque is reached, the wrench has a reflex signaling feature that causes a momentary breakaway of the torque wrench body, sending an impulse to the operator's hand. But be careful, as continuing the turn the wrench after the momentary release will increase torque on the fastener beyond the specified setting.

Fig. 38 A swivel (U-joint) adapter (left), a wobble-head adapter (center) and a 1/2 in.-to-3/8 in. adapter (right)

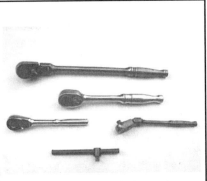

Fig. 39 Ratchets come in all sizes and configurations from rigid to swivel-headed

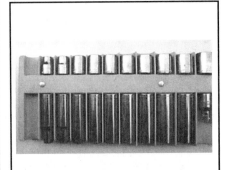

Fig. 40 Shallow sockets (top) are good for most jobs. But, some bolts require deep sockets (bottom)

Fig. 41 Hex-head fasteners require a socket with a hex shaped driver

Fig. 42 Torx® drivers . . .

Fig. 43 . . . and tamper resistant drivers are required to remove special fasteners

GENERAL INFORMATION, SAFETY AND TOOLS

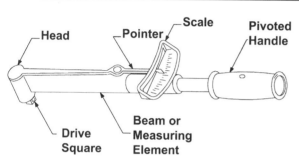

Fig. 44 Three types of torque wrenches. Top to bottom: a 3/8 in. drive beam type that reads in inch lbs., a 1/2 in. drive clicker type and a 1/2 in. drive beam type

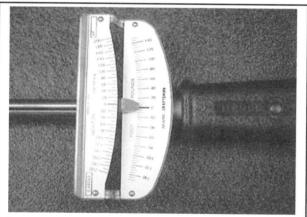

Fig. 45 Parts of a beam type torque wrench

Fig. 46 A beam type torque wrench consists of a pointer attached to the head that runs the length of the flexible beam (shaft) to a scale located near the handle

Breaker Bars

◆ See Figure 49

Breaker bars are long handles with a drive lug. Their main purpose is to provide extra turning force when breaking loose tight bolts or nuts. They come in all drive sizes and lengths. Always take extra precautions and use the proper technique when using a breaker bar (pull on the bar, don't push, to prevent skinned knuckles).

WRENCHES

◆ See Figures 50 thru 54

Basically, there are 3 kinds of fixed wrenches: open end, box end, and combination.

Open-end wrenches have 2-jawed openings at each end of the wrench. These wrenches are able to fit onto just about any nut or bolt. They are extremely versatile but have one major drawback. They can slip on a worn or rounded bolt head or nut, causing bleeding knuckles and a useless fastener.

■ Line wrenches are a special type of open-end wrench designed to fit onto more of the fastener than standard open-end wrenches, thus reducing the chance of rounding the corners of the fastener.

Box-end wrenches have a 360° circular jaw at each end of the wrench. They come in both 6 and 12 point versions just like sockets and each type has some of the same advantages and disadvantages as sockets.

Combination wrenches have the best of both. They have a 2-jawed open end and a box end. These wrenches are probably the most versatile.

As for sizes, you'll probably need a range similar to that of the sockets, about 1/4 in. through 1 in. for standard fasteners, or 6mm through 19mm for metric fasteners. As for numbers, you'll need 2 of each size, since, in many instances, one wrench holds the nut while the other turns the bolt. On most fasteners, the nut and bolt are the same size so having two wrenches of the same size comes in handy.

■ Although you will typically just need the sizes we specified, there are some exceptions. Occasionally you will find a nut that is larger. For these, you will need to buy ONE expensive wrench or a very large adjustable. Or you can always just convince the spouse that we are talking about SAFETY here and buy a whole (read expensive) large wrench set.

One extremely valuable type of wrench is the adjustable wrench. An adjustable wrench has a fixed upper jaw and a moveable lower jaw. The lower jaw is moved by turning a threaded drum. The advantage of an adjustable wrench is its ability to be adjusted to just about any size fastener.

The main drawback of an adjustable wrench is the lower jaw's tendency to move slightly under heavy pressure. This can cause the wrench to slip if it is not facing the right way. Pulling on an adjustable wrench in the proper direction will cause the jaws to lock in place. Adjustable wrenches come in a large range of sizes, measured by the wrench length.

PLIERS

◆ See Figure 55

Pliers are simply mechanical fingers. They are, more than anything, an extension of your hand. At least 3 pairs of pliers are an absolute necessity—standard, needle nose and slip joint.

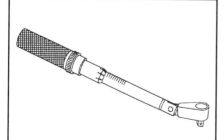

Fig. 47 A click type or breakaway torque wrench—note this one has a pivoting head

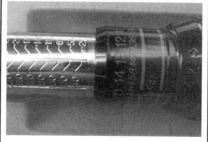

Fig. 48 Setting the torque on a click type wrench involves turning the handle until the specification appears on the dial

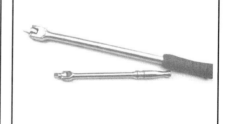

Fig. 49 Breaker bars are great for loosening large or stuck fasteners

1-22 GENERAL INFORMATION, SAFETY AND TOOLS

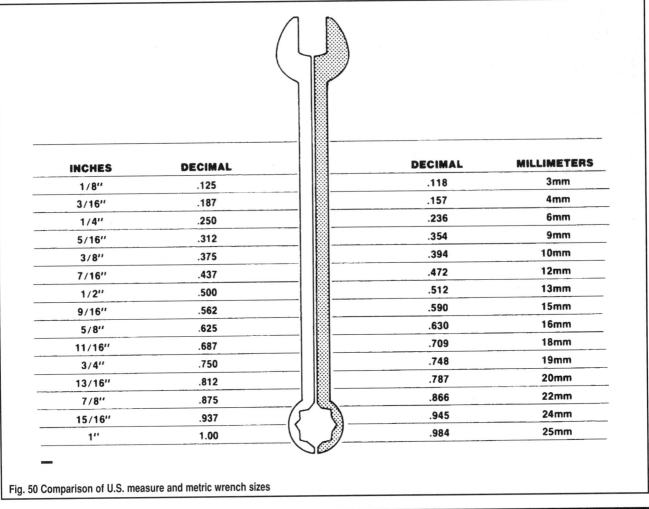

INCHES	DECIMAL	DECIMAL	MILLIMETERS
1/8"	.125	.118	3mm
3/16"	.187	.157	4mm
1/4"	.250	.236	6mm
5/16"	.312	.354	9mm
3/8"	.375	.394	10mm
7/16"	.437	.472	12mm
1/2"	.500	.512	13mm
9/16"	.562	.590	15mm
5/8"	.625	.630	16mm
11/16"	.687	.709	18mm
3/4"	.750	.748	19mm
13/16"	.812	.787	20mm
7/8"	.875	.866	22mm
15/16"	.937	.945	24mm
1"	1.00	.984	25mm

Fig. 50 Comparison of U.S. measure and metric wrench sizes

In addition to standard pliers there are the slip-joint, multi-position pliers such as ChannelLock® pliers and locking pliers, such as Vise Grips®.

Slip joint pliers are extremely valuable in grasping oddly sized parts and fasteners. Just make sure that you don't use them instead of a wrench too often since they can easily round off a bolt head or nut.

Locking pliers are usually used for gripping bolts or studs that can't be removed conventionally. You can get locking pliers in square jawed, needle-nosed and pipe-jawed. Locking pliers can rank right up behind duct tape as the handy-man's best friend.

SCREWDRIVERS

You can't have too many screwdrivers. They come in 2 basic flavors, either standard or Phillips. Standard blades come in various sizes and thickness for all types of slotted fasteners. Phillips screwdrivers come in sizes with number designations from 1 on up, with the lower number designating the smaller size. Screwdrivers can be purchased separately or in sets.

HAMMERS

◆ See Figure 56

You need a hammer for just about any kind of work. You need a ball-peen hammer for most metal work when using drivers and other like tools. A plastic hammer comes in handy for hitting things safely. A soft-faced dead-blow hammer is used for hitting things safely and hard. Hammers are also VERY useful with non air-powered impact drivers.

Other Common Tools

There are a lot of other tools that every DIYer will eventually need (though not all for basic maintenance). They include:
- Funnels
- Chisels
- Punches
- Files
- Hacksaw
- Portable Bench Vise
- Tap and Die Set
- Flashlight
- Magnetic Bolt Retriever
- Gasket scraper
- Putty Knife
- Screw/Bolt Extractors
- Prybars

Hacksaws have just one use—cutting things off. You may wonder why you'd need one for something as simple as maintenance or repair, but you never know. Among other things, guide studs to ease parts installation can be made from old bolts with their heads cut off.

A tap and die set might be something you've never needed, but you will eventually. It's a good rule, when everything is apart, to clean-up all threads, on bolts, screws or threaded holes. Also, you'll likely run across a situation in which you will encounter stripped threads. The tap and die set will handle that for you.

Gasket scrapers are just what you'd think, tools made for scraping old gasket material off of parts. You don't absolutely need one. Old gasket material can be removed with a putty knife or single edge razor blade.

GENERAL INFORMATION, SAFETY AND TOOLS 1-23

Fig. 51 Always use a backup wrench to prevent rounding flare nut fittings

Fig. 52 Note how the flare wrench jaws are extended to grip the fitting tighter and prevent rounding

Fig. 53 Several types and sizes of adjustable wrenches

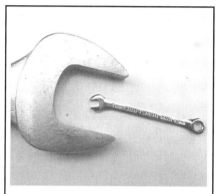
Fig. 54 You may find a nut that requires a particularly large or small wrench (that is usually available at your local tool store)

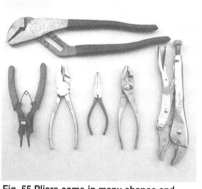
Fig. 55 Pliers come in many shapes and sizes. You should have an assortment on hand

Fig. 56 Three types of hammers. Top to bottom: ball peen, rubber dead-blow, and plastic

However, putty knives may not be sharp enough for some really stubborn gaskets and razor blades have a knack of breaking just when you don't want them to, inevitably slicing the nearest body part! As the old saying goes, "always use the proper tool for the job". If you're going to use a razor to scrape a gasket, be sure to always use a blade holder.

Putty knives really do have a use in a repair shop. Just because you remove all the bolts from a component sealed with a gasket doesn't mean it's going to come off. Most of the time, the gasket and sealer will hold it tightly. Lightly inserting a putty knife at various points between the two parts will break the seal without damage to the parts.

A small — 8-10 in. (20-25cm) long — prybar is extremely useful for removing stuck parts.

■ **Never use a screwdriver as a prybar! Screwdrivers are not meant for prying. Screwdrivers, used for prying, can break, sending the broken shaft flying!**

Screw/bolt extractors are used for removing broken bolts or studs that have broken off flush with the surface of the part.

Special Tools

◆ See Figure 57

Almost every marine engine around today requires at least one special tool to perform a certain task. In most cases, these tools are specially designed to overcome some unique problem or to fit on some oddly sized component.

When manufacturers go through the trouble of making a special tool, it is usually necessary to use it to ensure that the job will be done right. A special tool might be designed to make a job easier, or it might be used to keep you from damaging or breaking a part.

Don't worry, MOST maintenance procedures can either be performed without any special tools OR, because the tools must be used for such basic things, they are commonly available for a reasonable price. It is usually just the low production, highly specialized tools (like a super thin 7-point star-shaped socket capable of 150 ft. lbs. (203 Nm) of torque that is used only on the crankshaft nut of the limited production what-dya-callit engine) that tend to be outrageously expensive and hard to find. Hopefully, you will probably never need such a tool.

Special tools can be as inexpensive and simple as an adjustable strap wrench or as complicated as an ignition tester. A few common specialty tools are listed here, but check with your dealer or with other boaters for help in determining if there are any special tools for YOUR particular engine. There is an added advantage in seeking advice from others, chances are they may have already found the special tool you will need, and know how to get it cheaper (or even let you borrow it).

Electronic Tools

BATTERY TESTERS

The best way to test a non-sealed battery is using a hydrometer to check the specific gravity of the acid. Luckily, these are usually inexpensive and are available at most parts stores. Just be careful because the larger testers are usually designed for larger batteries and may require more acid than you will be able to draw from the battery cell. Smaller testers (usually a short, squeeze bulb type) will require less acid and should work on most batteries.

Electronic testers are available and are often necessary to tell if a sealed battery is usable. Luckily, many parts stores have them on hand and are willing to test your battery for you.

1-24 GENERAL INFORMATION, SAFETY AND TOOLS

BATTERY CHARGERS

◆ See Figure 58

If you are a weekend boater and take your boat out every week, then you will most likely want to buy a battery charger to keep your battery fresh. There are many types available, from low amperage trickle chargers to electronically controlled battery maintenance tools that monitor the battery voltage to prevent over or undercharging. This last type is especially useful if you store your boat for any length of time (such as during the severe winter months found in many Northern climates).

Even if you use your boat on a regular basis, you will eventually need a battery charger. The charger should be used anytime the boat is going to be in storage for more than a few weeks or so. Never leave the dock or loading ramp without a battery that is fully charged.

Also, some smaller batteries are shipped dry and in a partial charged state. Before placing a new battery of this type into service it must be filled and properly charged. Failure to properly charge a battery (which was shipped dry) before it is put into service will prevent it from ever reaching a fully charged state.

MULTIMETERS (DVOMS)

◆ See Figure 59

Multimeters or Digital Volt Ohmmeter (DVOMs) are an extremely useful tool for troubleshooting electrical problems. They can be purchased in either analog or digital form and have a price range to suit any budget. A multimeter is a voltmeter, ammeter and ohmmeter (along with other features) combined into one instrument. It is often used when testing solid state circuits because of its high input impedance (usually 10 megaohms or more). A brief description of the multimeter main test functions follows:

- Voltmeter—the voltmeter is used to measure voltage at any point in a circuit or to measure the voltage drop across any part of a circuit. Voltmeters usually have various scales and a selector switch to allow the reading of different voltage ranges. The voltmeter has a positive and a negative lead. To avoid the possibility of damage to the meter, whenever possible, connect the negative lead to the negative (-) side of the circuit (to ground or nearest the ground side of the circuit) and connect the positive lead to the positive (+) side of the circuit (to the power source or the nearest power source). Luckily, most quality DVOMs can adjust their own polarity internally and will indicate (without damage) if the leads are reversed. Note that the negative voltmeter lead will always be black and that the positive voltmeter will always be some color other than black (usually red).
- Ohmmeter—the ohmmeter is designed to read resistance (measured in ohms) in a circuit or component. Most ohmmeters will have a selector switch which permits the measurement of different ranges of resistance (usually the selector switch allows the multiplication of the meter reading by 10, 100, 1,000 and 10,000). Some ohmmeters are "auto-ranging" which means the meter itself will determine which scale to use. Since the meters are powered by an internal battery, the ohmmeter can be used like a self-powered test light. When the ohmmeter is connected, current from the ohmmeter flows through the circuit or component being tested. Since the ohmmeter's internal resistance and voltage are known values, the amount of current flow through the meter depends on the resistance of the circuit or component being tested. The ohmmeter can also be used to perform a continuity test for suspected open circuits. In using the meter for making continuity checks, do not be concerned with the actual resistance readings. Zero resistance, or any ohm reading, indicates continuity in the circuit. Infinite resistance indicates an opening in the circuit. A high resistance reading where there should be little or none indicates a problem in the circuit. Checks for short circuits are made in the same manner as checks for open circuits, except that the circuit must be isolated from both power and normal ground. Infinite resistance indicates no continuity, while zero resistance indicates a dead short.

※※ WARNING

Never use an ohmmeter to check the resistance of a component or wire while there is voltage applied to the circuit.

- Ammeter—an ammeter measures the amount of current flowing through a circuit in units called amperes or amps. At normal operating voltage, most circuits have a characteristic amount of amperes, called "current draw" which can be measured using an ammeter. By referring to a specified current draw rating, then measuring the amperes and comparing the two values; one can determine what is happening within the circuit to aid in diagnosis. An open circuit, for example, will not allow any current to flow, so the ammeter reading will be zero. A damaged component or circuit will have an increased current draw, so the reading will be high. The ammeter is always connected in series with the circuit being tested. All of the current that normally flows through the circuit must also flow through the ammeter; if there is any other path for the current to follow, the ammeter reading will not be accurate. The ammeter itself has very little resistance to current flow and, therefore, will not affect the circuit, but, it will measure current draw only when the circuit is closed and electricity is flowing. Excessive current draw can blow fuses and drain the battery, while a reduced current draw can cause motors to run slowly, lights to dim and other components to not operate properly.

GAUGES

Compression Gauge

◆ See Figure 60

An important element in checking the overall condition of your engine is to check compression. This becomes increasingly more important on outboards with high hours. Compression gauges are available as screw-in types and hold-in types. The screw-in type is slower to use, but eliminates the possibility of a faulty reading due to pressure escaping by the seal. A compression reading will uncover many problems that can cause rough running. Normally, these are not the sort of problems that can be cured by a tune-up.

Vacuum/Pressure Gauge/Pump

◆ See Figures 61, 62 and 63

Vacuum gauges are handy for discovering air leaks, late ignition or valve timing, and a number of other problems. A hand-held vacuum/pressure pump

Fig. 57 Almost every marine engine around today requires at least one special tool to perform a certain task

Fig. 58 The Battery Tender® is more than just a battery charger, when left connected, it keeps your battery fully charged

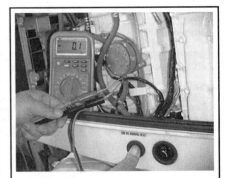

Fig. 59 Multimeters, such as this one from UEI, are an extremely useful tool for troubleshooting electrical problems

GENERAL INFORMATION, SAFETY AND TOOLS

Fig. 60 Cylinder compression test results are extremely valuable indicators of internal engine condition

Fig. 61 Vacuum gauges are useful for troubleshooting including testing some fuel pumps

Fig. 62 You can also use the gauge on a hand-operated vacuum pump for tests

can be purchased at many automotive or marine parts stores and can be used for multiple purposes. The gauge can be used to measure vacuum or pressure in a line, while the pump can be used to manually apply vacuum or pressure to a solenoid or fitting. The hand-held pump can also be used to power-bleed trailer and tow vehicle brakes.

Measuring Tools

Eventually, you are going to have to measure something. To do this, you will need at least a few precision tools.

MICROMETERS & CALIPERS

Micrometers and calipers are devices used to make extremely precise measurements. The simple truth is that you really won't have the need for many of these items just for routine maintenance. But, measuring tools, such as an outside caliper can be handy during repairs. And, if you decide to tackle a major overhaul, a micrometer will absolutely be necessary.

Should you decide on becoming more involved in boat engine mechanics, such as repair or rebuilding, then these tools will become very important. The success of any rebuild is dependent, to a great extent on the ability to check the size and fit of components as specified by the manufacturer. These measurements are often made in thousandths and ten-thousandths of an inch.

Micrometers

◆ See Figures 64 and 65

A micrometer is an instrument made up of a precisely machined spindle that is rotated in a fixed nut, opening and closing the distance between the end of the spindle and a fixed anvil. When measuring using a micrometer, don't overtighten the tool on the part as either the component or tool may be damaged, and either way, an incorrect reading will result. Most micrometers are equipped with some form of thumbwheel on the spindle that is designed to freewheel over a certain light touch (automatically adjusting the spindle and preventing it from overtightening).

Outside micrometers can be used to check the thickness of parts such shims or the outside diameter of components like the crankshaft journals. They are also used during many rebuild and repair procedures to measure the diameter of components such as the pistons. The most common type of micrometer reads in 1/1000 of an inch. Micrometers that use a vernier scale can estimate to 1/10 of an inch.

Inside micrometers are used to measure the distance between two parallel surfaces. For example, in powerhead rebuilding work, the "inside mike" measures cylinder bore wear and taper. Inside mikes are graduated the same way as outside mikes and are read the same way as well.

Remember that an inside mike must be absolutely perpendicular to the work being measured. When you measure with an inside mike, rock the mike gently from side to side and tip it back and forth slightly so that you span the widest part of the bore. Just to be on the safe side, take several readings. It takes a certain amount of experience to work any mike with confidence.

Metric micrometers are read in the same way as inch micrometers, except that the measurements are in millimeters. Each line on the main scale equals 1mm. Each fifth line is stamped 5, 10, 15 and so on. Each line on the thimble scale equals 0.01 mm. It will take a little practice, but if you can read an inch mike, you can read a metric mike.

Calipers

◆ See Figures 66, 67 and 68

Inside and outside calipers are useful devices to have if you need to measure something quickly and absolute precise measurement is not necessary. Simply take the reading and then hold the calipers on an accurate steel rule. Calipers, like micrometers, will often contain a thumbwheel to help ensure accurate measurement.

DIAL INDICATORS

◆ See Figure 69

A dial indicator is a gauge that utilizes a dial face and a needle to register measurements. There is a movable contact arm on the dial indicator. When

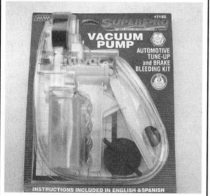

Fig. 63 Hand-held vacuum/pressure pumps are available at most parts stores

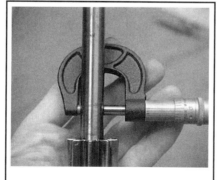

Fig. 64 Outside micrometers measure thickness, like shims or a shaft diameter

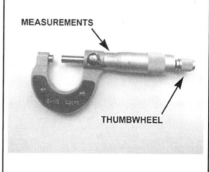

Fig. 65 Be careful not to over-tighten the micrometers always use the thumbwheel

1-26 GENERAL INFORMATION, SAFETY AND TOOLS

the arm moves, the needle rotates on the dial. Dial indicators are calibrated to show readings in thousandths of an inch and typically, are used to measure end-play and runout on various shafts and other components.

Dial indicators are quite easy to use, although they are relatively expensive. A variety of mounting devices are available so that the indicator can be used in a number of situations. Make certain that the contact arm is always parallel to the movement of the work being measured.

TELESCOPING GAUGES

◆ See Figure 70

A telescope gauge is really only used during rebuilding procedures (NOT during basic maintenance or routine repairs) to measure the inside of bores.

It can take the place of an inside mike for some of these jobs. Simply insert the gauge in the hole to be measured and lock the plungers after they have contacted the walls. Remove the tool and measure across the plungers with an outside micrometer.

DEPTH GAUGES

◆ See Figure 71

A depth gauge can be inserted into a bore or other small hole to determine exactly how deep it is. One common use for a depth gauge is measuring the distance the piston sits below the deck of the block at top dead center. Some outside calipers contain a built-in depth gauge so you can save money and buy just one tool.

Fig. 66 Calipers are the fast and easy way to make precise measurements

Fig. 67 Calipers can also be used to measure depth . . .

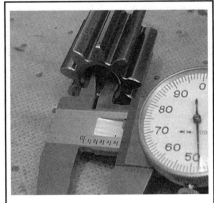

Fig. 68 . . . and inside diameter measurements, to 0.001 in. accuracy

Fig. 69 This dial indicator is measuring the end-play of a crankshaft during a powerhead rebuild

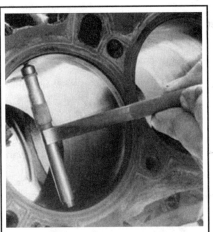

Fig. 70 Telescoping gauges are used during powerhead rebuilding procedures to measure the inside diameter of bores

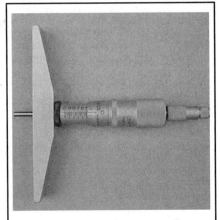

Fig. 71 Depth gauges are used to measure the depth of bore or other small holes

GENERAL INFORMATION, SAFETY AND TOOLS 1-27

FASTENERS, MEASUREMENTS AND CONVERSIONS

Bolts, Nuts and Other Threaded Retainers

◆ See Figures 72 and 73

Although there are a great variety of fasteners found in the modern boat engine, the most commonly used retainer is the threaded fastener (nuts, bolts, screws, studs, etc). Most threaded retainers may be reused, provided that they are not damaged in use or during the repair.

■ Some retainers (such as stretch bolts or torque prevailing nuts) are designed to deform when tightened or in use and should not be reused.

Whenever possible, we will note any special retainers which should be replaced during a procedure. But you should always inspect the condition of a retainer when it is removed and you should replace any that show signs of damage. Check all threads for rust or corrosion that can increase the torque necessary to achieve the desired clamp load for which that fastener was originally selected. Additionally, be sure that the driver surface itself (on the fastener) is not compromised from rounding or other damage. In some cases a driver surface may become only partially rounded, allowing the driver to catch in only one direction. In many of these occurrences, a fastener may be installed and tightened, but the driver would not be able to grip and loosen the fastener again. (This could lead to frustration down the line should that component ever need to be disassembled again).

If you must replace a fastener, whether due to design or damage, you must always be sure to use the proper replacement. In all cases, a retainer of the same design, material and strength should be used. Markings on the heads of most bolts will help determine the proper strength of the fastener. The same material, thread and pitch must be selected to assure proper installation and safe operation of the motor afterwards.

Thread gauges are available to help measure a bolt or stud's thread. Most part or hardware stores keep gauges available to help you select the proper size. In a pinch, you can use another nut or bolt for a thread gauge. If the bolt you are replacing is not too badly damaged, you can select a match by finding another bolt that will thread in its place. If you find a nut that will thread properly onto the damaged bolt, then use that nut as a gauge to help select the replacement bolt. If however, the bolt you are replacing is so badly damaged (broken or drilled out) that its threads cannot be used as a gauge, you might start by looking for another bolt (from the same assembly or a similar location) which will thread into the damaged bolt's mounting. If so, the other bolt can be used to select a nut; the nut can then be used to select the replacement bolt.

In all cases, be absolutely sure you have selected the proper replacement. Don't be shy, you can always ask the store clerk for help.

✴✴ WARNING

Be aware that when you find a bolt with damaged threads, you may also find the nut or tapped bore into which it was threaded has also been damaged. If this is the case, you may have to drill and tap the hole, replace the nut or otherwise repair the threads. Never try to force a replacement bolt to fit into the damaged threads.

Torque

Torque is defined as the measurement of resistance to turning or rotating. It tends to twist a body about an axis of rotation. A common example of this would be tightening a threaded retainer such as a nut, bolt or screw. Measuring torque is one of the most common ways to help assure that a threaded retainer has been properly fastened.

When tightening a threaded fastener, torque is applied in three distinct areas, the head, the bearing surface and the clamp load. About 50 percent of the measured torque is used in overcoming bearing friction. This is the friction between the bearing surface of the bolt head, screw head or nut face and the base material or washer (the surface on which the fastener is rotating). Approximately 40 percent of the applied torque is used in overcoming thread friction. This leaves only about 10 percent of the applied torque to develop a useful clamp load (the force that holds a joint together). This means that friction can account for as much as 90 percent of the applied torque on a fastener.

Standard and Metric Measurements

Specifications are often used to help you determine the condition of various components, or to assist you in their installation. Some of the most common measurements include length (in. or cm/mm), torque (ft. lbs., inch lbs. or Nm) and pressure (psi, in. Hg, kPa or mm Hg).

In some cases, that value may not be conveniently measured with what is available in your toolbox. Luckily, many of the measuring devices that are available today will have two scales so U.S. or Metric measurements may easily be taken. If any of the various measuring tools that are available to you do not contain the same scale as listed in your specifications, use the conversion factors that are provided in the Specifications section to determine the proper value.

The conversion factor chart is used by taking the given specification and multiplying it by the necessary conversion factor. For instance, looking at the first line, if you have a measurement in inches such as "free-play should be 2 in." but your ruler reads only in millimeters, multiply 2 in. by the conversion factor of 25.4 to get the metric equivalent of 50.8mm. Likewise, if a specification was given only in a Metric measurement, for example in Newton Meters (Nm), then look at the center column first. If the measurement is 100 Nm, multiply it by the conversion factor of 0.738 to get 73.8 ft. lbs.

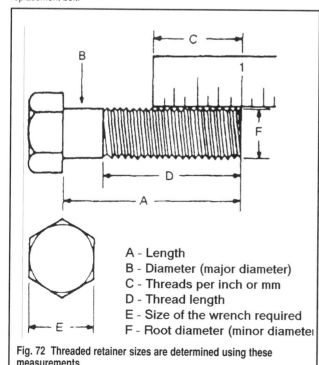

A - Length
B - Diameter (major diameter)
C - Threads per inch or mm
D - Thread length
E - Size of the wrench required
F - Root diameter (minor diameter)

Fig. 72 Threaded retainer sizes are determined using these measurements

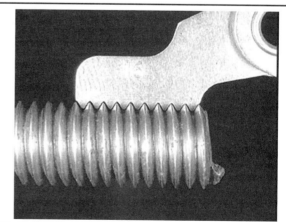

Fig. 73 Thread gauges measure the threads-per-inch and the pitch of a bolt or stud's threads

1-28 GENERAL INFORMATION, SAFETY AND TOOLS

SPECIFICATIONS

Metric Bolts

Relative Strength Marking	4.6, 4.8				8.8			
	Frequent Maximum Torque			Infrequent Maximum Torque				
Bolt Markings / Usage								
Bolt Size — Thread Size x Pitch (mm)	Ft-Lb	Kgm	Nm	Ft-Lb	Kgm	Nm		
6 x 1.0	2–3	.2–.4	3–4	3–6	.4–.8	5–8		
8 x 1.25	6–8	.8–1	8–12	9–14	1.2–1.9	13–19		
10 x 1.25	12–17	1.5–2.3	16–23	20–29	2.7–4.0	27–39		
12 x 1.25	21–32	2.9–4.4	29–43	35–53	4.8–7.3	47–72		
14 x 1.5	35–52	4.8–7.1	48–70	57–85	7.8–11.7	77–110		
16 x 1.5	51–77	7.0–10.6	67–100	90–120	12.4–16.5	130–160		
18 x 1.5	74–110	10.2–15.1	100–150	130–170	17.9–23.4	180–230		
20 x 1.5	110–140	15.1–19.3	150–190	190–240	26.2–46.9	260–320		
22 x 1.5	150–190	22.0–26.2	200–260	250–320	34.5–44.1	340–430		
24 x 1.5	190–240	26.2–46.9	260–320	310–410	42.7–56.5	420–550		

SAE Bolts

SAE Grade Number	1 or 2			5			6 or 7		
Bolt Markings — Manufacturers' marks may vary—number of lines always two less than the grade number.	Frequent Maximum Torque			Frequent Maximum Torque			Infrequent Maximum Torque		
Usage — Bolt Size (Inches)—(Thread)	Ft-Lb	kgm	Nm	Ft-Lb	kgm	Nm	Ft-Lb	kgm	Nm
1/4—20	5	0.7	6.8	8	1.1	10.8	10	1.4	13.5
—28	6	0.8	8.1	10	1.4	13.6			
5/16—18	11	1.5	14.9	17	2.3	23.0	19	2.6	25.8
—24	13	1.8	17.6	19	2.6	25.7			
3/8—16	18	2.5	24.4	31	4.3	42.0	34	4.7	46.0
—24	20	2.75	27.1	35	4.8	47.5			
7/16—14	28	3.8	37.0	49	6.8	66.4	55	7.6	74.5
—20	30	4.2	40.7	55	7.6	74.5			
1/2—13	39	5.4	52.8	75	10.4	101.7	85	11.75	115.2
—20	41	5.7	55.6	85	11.7	115.2			
9/16—12	51	7.0	69.2	110	15.2	149.1	120	16.6	162.7
—18	55	7.6	74.5	120	16.6	162.7			
5/8—11	83	11.5	112.5	150	20.7	203.3	167	23.0	226.5
—18	95	13.1	128.8	170	23.5	230.5			
3/4—10	105	14.5	142.3	270	37.3	366.0	280	38.7	379.6
—16	115	15.9	155.9	295	40.8	400.0			
7/8—9	160	22.1	216.9	395	54.6	535.5	440	60.9	596.5
—14	175	24.2	237.2	435	60.1	589.7			
1—8	236	32.5	318.6	590	81.6	799.9	660	91.3	894.8
—14	250	34.6	338.9	660	91.3	849.8			

CONVERSION FACTORS

LENGTH-DISTANCE

Inches (in.)	x 25.4	= Millimeters (mm)
Feet (ft.)	x .305	= Meters (m)
Miles	x 1.609	= Kilometers (km)

Millimeters x .0394 = Inches
Meters x 3.281 = Feet
Kilometers x .0621 = Miles

VOLUME

Cubic Inches (in3)	x 16.387	= Cubic Centimeters
IMP Pints (IMP pt.)	x .568	= Liters (L)
IMP Quarts (IMP qt.)	x 1.137	= Liters (L)
IMP Gallons (IMP gal.)	x 4.546	= Liters (L)
IMP Quarts (IMP qt.)	x 1.201	= US Quarts (US qt.)
IMP Gallons (IMP gal.)	x 1.201	= US Gallons (US gal.)
Fl. Ounces	x 29.573	= Milliliters
US Pints (US pt.)	x .473	= Liters (L)
US Quarts (US qt.)	x .946	= Liters (L)
US Gallons (US gal.)	x 3.785	= Liters (L)

Cubic Centimeters x .061 = in3
Liters x 1.76 = IMP pt.
Liters x .88 = IMP qt.
Liters x .22 = IMP gal.
Liters x .833 = IMP qt.
Liters x .833 = IMP gal.
Milliliters x .034 = Ounces
Liters x 2.113 = Pints
Liters x 1.057 = Quarts
Liters x .264 = Gallons

MASS-WEIGHT

Ounces (oz.)	x 28.35	= Grams (g)
Pounds (lb.)	x .454	= Kilograms (kg)

Grams x .035 = Ounces
Kilograms x 2.205 = Pounds

PRESSURE

Pounds Per Sq. In. (psi)	x 6.895	= Kilopascals (kPa)
Inches of Mercury (Hg)	x .4912	= psi
Inches of Mercury (Hg)	x 3.377	= Kilopascals (kPa)
Inches of Water (H2O)	x .07355	= Inches of Mercury
Inches of Water (H2O)	x .03613	= psi
Inches of Water (H2O)	x .248	= Kilopascals (kPa)

Kilopascals x .145 = psi
psi x 2.036 = Hg
Kilopascals x .2961 = Hg
Inches of Mercury x 13.783 = H_2O
psi x 27.684 = H_2O
Kilopascals x 4.026 = H_2O

TORQUE

Pounds-Force Inches (in-lb)	x .113	= Newton Meters (N·m)
Pounds-Force Feet (ft-lb)	x 1.356	= Newton Meters (N·m)

Newton Meters x 8.85 = in-lb
Newton Meters x .738 = ft-lb

VELOCITY

Miles Per Hour (MPH)	x 1.609	= Kilometers Per Hour (KPH)

Kilometers Per Hour x .621 = MPH

POWER

Horsepower (Hp)	x .745	= Kilowatts

Kilowatts x 1.34 = Horsepower

FUEL CONSUMPTION*

Miles Per Gallon IMP (MPG)	x .354	= Kilometers Per Liter (Km/L)
Kilometers Per Liter (Km/L)	x 2.352	= IMP MPG
Miles Per Gallon US (MPG)	x .425	= Kilometers Per Liter (Km/L)
Kilometers Per Liter (Km/L)	x 2.352	= US MPG

*It is common to covert from miles per gallon (mpg) to liters/100 kilometers (l/100 km), where mpg (IMP) x l/100 km = 282 and mpg (US) x l/100 km = 235.

TEMPERATURE

Degree Fahrenheit (°F) = (°C x 1.8) + 32
Degree Celsius (°C) = (°F − 32) x .56

2

MAINTENANCE & TUNE-UP

GENERAL INFORMATION 2-2
LUBRICATION 2-5
ENGINE MAINTENANCE 2-10
BOAT MAINTENANCE 2-27
TUNE-UP 2-30
TIMING AND SYNCHRONIZATION 2-44
CLEARING A SUBMERGED MOTOR 2-87
SPECIFICATIONS.................. 2-88

Index

ANODES (ZINCS) 2-25
  INSPECTION 2-26
  SERVICING 2-26
BATTERIES......................... 2-27
  MAINTENANCE 2-27
  STORAGE 2-29
  TESTING 2-28
BEFORE/AFTER EACH USE 2-4
  VISUALLY INSPECTING THE BOAT
    AND MOTOR 2-4
BOAT MAINTENANCE **2-27**
  BATTERIES 2-27
  FIBERGLASS HULL 2-29
  INTERIOR 2-30
CLEARING A SUBMERGED MOTOR..... **2-87**
COMPRESSION TESTING 2-32
  COMPRESSION CHECK 2-32
  LOW COMPRESSION................ 2-33
COOLING SYSTEM 2-11
  FLUSHING JET DRIVES............ 2-13
  FLUSHING THE COOLING SYSTEM ... 2-11
ELECTRICAL SYSTEM CHECKS 2-38
  CHECKING THE BATTERY........... 2-38
  CHECKING THE INTERNAL
    WIRING HARNESS 2-39
  CHECKING THE STARTER MOTOR..... 2-39
ELECTRONIC IGNITION SYSTEMS 2-38
ENGINE COVERS.................... 2-10
  REMOVAL & INSTALLATION 2-10
ENGINE IDENTIFICATION 2-2
  ENGINE MODEL & SERIAL NUMBERS ... 2-3
ENGINE MAINTENANCE **2-10**
  ANODES (ZINCS)................. 2-25
  COOLING SYSTEM 2-11
  ENGINE COVERS 2-10
  ENGINE OIL..................... 2-13
  FUEL FILTER 2-15
  JET DRIVE IMPELLER 2-24
  PROPELLER..................... 2-18
ENGINE OIL 2-13
  CHECKING....................... 2-15
  FILLING 2-14
  OIL RECOMMENDATIONS 2-13
FIBERGLASS HULL 2-29
FUEL FILTER 2-15
FUEL SYSTEM CHECKS 2-39
GEARCASE (LOWER UNIT) OIL......... 2-7
  CHECKING LEVEL & CONDITION 2-7
  DRAINING AND FILLING 2-8
  OIL RECOMMENDATIONS 2-7
GENERAL INFORMATION **2-2**
  BEFORE/AFTER EACH USE 2-4
  ENGINE IDENTIFICATION 2-2
  MAINTENANCE COVERAGE IN
    THIS MANUAL 2-2
  MAINTENANCE EQUALS SAFETY..... 2-2
  OUTBOARDS ON SAIL BOATS 2-2
IDLE SPEED ADJUSTMENT............. 2-43
INTERIOR......................... 2-30
INTRODUCTION TO TUNE-UPS......... 2-30
JET DRIVE BEARING 2-9
  BEARING LUBRICATION 2-9
  GREASE REPLACEMENT 2-9
  RECOMMENDED LUBRICANT 2-9
JET DRIVE IMPELLER 2-24
  CLEARANCE 2-24
  INSPECTION 2-24
LUBRICANTS 2-6
LUBRICATING THE MOTOR............ 2-6
LUBRICATION **2-5**
  GEARCASE (LOWER UNIT) OIL...... 2-7
  JET DRIVE BEARING.............. 2-9
  LUBRICANTS 2-6
  LUBRICATING THE MOTOR......... 2-6
  LUBRICATION INSIDE THE BOAT ... 2-7
  POWER TRIM/TILT RESERVOIR 2-9
LUBRICATION INSIDE THE BOAT 2-7
MAINTENANCE COVERAGE IN
  THIS MANUAL 2-2
MAINTENANCE EQUALS SAFETY 2-2
OUTBOARDS ON SAIL BOATS........... 2-2
POWER TRIM/TILT RESERVOIR 2-9
  FLUID LEVEL/CONDITION.......... 2-9
  RECOMMENDED LUBRICANT 2-9
PREPPING THE MOTOR 2-44
  TACHOMETER CONNECTIONS......... 2-45
PROPELLER........................ 2-18
  GENERAL INFORMATION 2-18
  INSPECTION 2-21
  REMOVAL & INSTALLATION 2-21
RE-COMMISSIONING................. 2-86
SPARK PLUG WIRES................. 2-38
  REMOVAL & INSTALLATION 2-38
  TESTING 2-38
SPARK PLUGS...................... 2-33
  HEAT RANGE 2-34
  INSPECTION & GAPPING 2-37
  REMOVAL & INSTALLATION 2-34
  READING SPARK PLUGS 2-36
SPECIFICATIONS................. **2-88**
  CAPACITIES 2-96
  ENGINE 2-91
  LUBRICATION 2-95
  MAINTENANCE INTERVALS 2-90
  TUNE-UP 2-88
STORAGE **2-84**
  RE-COMMISSIONING............... 2-86
  WINTERIZATION 2-84
SYNCHRONIZATION 2-44
TIMING........................... 2-44
TIMING AND SYNCHRONIZATION..... **2-44**
  GENERAL INFORMATION 2-44
  PREPPING THE MOTOR 2-44
  SYNCHRONIZATION 2-44
  TIMING......................... 2-44
  2 HP MODEL 2-45
  3 HP MODEL 2-45
  4/5 HP (83 AND 103CC) MODELS .. 2-46
  6/8 HP MODELS.................. 2-47
  9.9/15 HP MODELS............... 2-49
  20/25 HP (395CC) MODELS 2-51
  20/25 HP (430CC) MODELS 2-53
  25/30 HP (496CC 2-
    CYLINDER) MODELS............. 2-54
  40 HP (2-CYLINDER) MODEL 2-54
  48 HP (2-CYLINDER) MODELS 2-57
  25/30 HP (496CC 3-CYLINDER) MODELS . 2-61
  28J-50 HP (698CC) MODELS 2-66
  50-70 HP (849CC) AND 65J-90 HP
    (1140CC) MODELS.............. 2-68
  E60 (849CC) AND E75, E75A, 85A, E60J
    (1140CC) MOTORS 2-70
  V4 AND V6 CARBURETED MOTORS.... 2-74
  V6 EFI (OX66) AND HPDI MOTORS.. 2-79
TUNE-UP **2-30**
  COMPRESSION TESTING 2-32
  ELECTRICAL SYSTEM CHECKS 2-38
  ELECTRONIC IGNITION SYSTEMS ... 2-38
  FUEL SYSTEM CHECKS 2-39
  IDLE SPEED ADJUSTMENT.......... 2-43
  INTRODUCTION TO TUNE-UPS....... 2-30
  SPARK PLUG WIRES 2-38
  SPARK PLUGS 2-33
  TUNE-UP SEQUENCE 2-31
TUNE-UP SEQUENCE 2-31
WINTERIZATION.................... 2-84
  PREPPING FOR STORAGE 2-84

MAINTENANCE & TUNE-UP

GENERAL INFORMATION (WHAT EVERYONE SHOULD KNOW ABOUT MAINTENANCE)

At Seloc, we estimate that 75% of engine repair work can be directly or indirectly attributed to lack of proper care for the engine. This is especially true of care during the off-season period. There is no way on this green earth for a mechanical engine, particularly an outboard motor, to be left sitting idle for an extended period of time, say for six months, and then be ready for instant satisfactory service.

Imagine, if you will, leaving your car or truck for six months, and then expecting to turn the key, having it roar to life, and being able to drive off in the same manner as a daily occurrence.

Therefore it is critical for an outboard engine to either be run (at least once a month), preferably, in the water and properly maintained between uses or for it to be specifically prepared for storage and serviced again immediately before the start of the season.

Only through a regular maintenance program can the owner expect to receive long life and satisfactory performance at minimum cost.

Many times, if an outboard is not performing properly, the owner will "nurse" it through the season with good intentions of working on the unit once it is no longer being used. As with many New Year's resolutions, the good intentions are not completed and the outboard may lie for many months before the work is begun or the unit is taken to the marine shop for repair.

Imagine, if you will, the cause of the problem being a blown head gasket. And let us assume water has found its way into a cylinder. This water, allowed to remain over a long period of time, will do considerably more damage than it would have if the unit had been disassembled and the repair work performed immediately. Therefore, if an outboard is not functioning properly, do not stow it away with promises to get at it when you get time, because the work and expense will only get worse the longer corrective action is postponed. In the example of the blown head gasket, a relatively simple and inexpensive repair job could very well develop into major overhaul and rebuild work.

Maintenance Equals Safety

OK, perhaps no one thing that we do as boaters will protect us from risks involved with enjoying the wind and the water on a powerboat. But, each time we perform maintenance on our boat or motor, we increase the likelihood that we will find a potential hazard before it becomes a problem. Each time we inspect our boat and motor, we decrease the possibility that it could leave us stranded on the water.

In this way, performing boat and engine service is one of the most important ways that we, as boaters, can help protect ourselves, our boats, and the friends and family that we bring aboard.

Outboards On Sail Boats

Owners of sailboats pride themselves in their ability to use the wind to clear a harbor or for movement from Port A to Port B, or maybe just for a day sail on a lake. For some, the outboard is carried only as a last resort - in case the wind fails completely, or in an emergency situation or for ease of docking.

Therefore, in some cases, the outboard is stowed below, usually in a very poorly ventilated area, and subjected to moisture and stale air - in short, an excellent environment for "sweating" and corrosion.

If the owner could just take the time at least once every month, to pull out the outboard, clean it up, and give it a short run, not only would he/she have "peace of mind" knowing it will start in an emergency, but also maintenance costs will be drastically reduced.

Maintenance Coverage In This Manual

At Seloc, we strongly feel that every boat owner should pay close attention to this section. We also know that it is one of the most frequently used portions of our manuals. The material in this section is divided into sections to help simplify the process of maintenance. Be sure to read and thoroughly understand the various tasks that are necessary to keep your outboard in tip-top shape.

Topics covered in this section include:
- General Information (What Everyone Should Know About Maintenance) - an introduction to the benefits and need for proper maintenance. A guide to tasks that should be performed before and after each use.
- Lubrication Service - after the basic inspections that you should perform each time the motor is used, the most frequent form of periodic maintenance you will conduct will be the Lubrication Service. This section takes you through each of the various steps you must take to keep corrosion from slowly destroying your motor before your very eyes.
- Engine Maintenance - the various procedures that must be performed on a regular basis in order to keep the motor and all of its various systems operating properly.
- Boat Maintenance - the various procedures that must be performed on a regular basis in order to keep the boat hull and its accessories looking and working like new.
- Tune-Up - also known as the pre-season tune-up, but don't let the name fool you. A complete tune-up is the best way to determine the condition of your outboard while also preparing it for hours and hours of hopefully trouble-free enjoyment. And if you use your boat enough during a single season, a second or even third tune-up could be required.
- Winter Storage and Spring Commissioning Checklists - use these sections to guide you through the various parts of boat and motor maintenance that protect your valued boat through periods of storage and return it to operating condition when it is time to use it again.
- Specification Charts - located at the end of the section are quick-reference, easy to read charts that provide you with critical information such as General Engine Specifications, Maintenance Intervals, Lubrication Service (intervals and lubricant types) and Capacities.

Engine Identification

◆ See Figures 1 thru 5

From 1997 to 2003 Yamaha produced an extremely large number of models with regards to horsepower ratings, as well a large number of trim and option variances on each of those models. In this service guide, we've included all of the models, including the 1-3 cylinder inline models, as well as V4 and V6 motors. We chose to do this because of the many similarities these motors have to each other. But, enough differences exist that many procedures will apply only to a sub-set of these motors. When this occurs, we'll either refer to the differences within a procedure or, if the differences are more significant, we'll break the motors out and give separate procedures. In order to prevent confusion, we try to sort and name the models in a way that is most easily understood.

Fig. 1 A model ID tag, and often a date of manufacture tag, is found on the port . . .

MAINTENANCE & TUNE-UP 2-3

In many cases, it is simply not enough to refer to a motor as a 25 hp model, since in these years Yamaha produced as many as 4 different motors with that rating (the 395cc 2-cylinder, the 430cc 2-cylinder, the 496cc 2-cylinder AND the 496cc 3-cylinder). This makes proper engine identification important for everything from ordering parts to even just using the procedures in this manual.

Throughout this manual we will make reference to motors the easiest way possible. In some cases procedures may apply to all motors, in other cases, they may apply to all 1-cylinder or all 2-cylinder motors (or all 3-cylinder motors, or perhaps all V6 motors, as applicable). When it is necessary to distinguish between different types of motors with the same number of cylinders, we'll differentiate using the Hp rating or, since different motors may have the same rating, we'll use the Hp rating plus the size (and in the case of the 496cc 25hp motor, the number of cylinders).

In most cases, mechanical procedures will be similar or the same across different Hp ratings of the same engine family (of the same size). So it won't be uncommon to see a title or a procedure refer to 9.9/15 hp (246cc) motors or 105J-225 hp (2596cc) motors. In both cases, we would be referring to all the motors of a particular family, including all B (Inshore), C, P (Pro), S (Saltwater) or V (V-Max) motors or other special models.

Starting in 1997 Yamaha began using fuel injection systems on some of their larger outboards. By 2003 there are three possible systems (2 of which) we cover in this repair guide. Two of them are known as Electronic Fuel Injection (EFI) systems. The EFI (OX66) system was the first marine fuel injection system introduced by Yamaha. It is a multi-port manifold injected system which uses an automotive style oxygen sensor to continually adjust the air/fuel mixture (unique in the marine industry). Yamaha 4-stroke engines (covered elsewhere) also utilize an EFI system which is a multi-port manifold injected system, however it does not use oxygen sensor feedback. The third style of injection (used only on the largest of motors) is the High Pressure Direct Injection (HPDI) which, as the name suggests, uses extreme high pressure air charges to inject fuel directly into the combustion chamber. Throughout this repair guide we may refer to both of the systems as a group, calling them simply Fuel Injected Motors, or we may specify EFI, meaning all but HPDI, EFI (OX66) meaning the EFI or HPDI.

To help with proper engine identification, all of the engines covered by this manual are listed in the General Engine and General Engine System Specifications charts at the end of this section. In these charts, the engines are listed with their respective engine families, by horsepower rating, number of cylinders, engine type (inline or V), years of production and displacement (cubic inches and cubic centimeters or CCs).

But, whether you are trying to tell which version of a particular horsepower rated motor you have in order to follow the correct procedure or are trying to order replacement parts, the absolute best method is to start by referring to the engine serial number tag. For all models covered here this ID tag (in the accompanying figure) is located on the side of the engine clamp or swivel/tilt brackets (port or starboard side depending upon the year and model). Most models are also equipped with a date of manufacture tag (located on the opposite side of the clamp or swivel/tilt bracket). Lastly, most models are also equipped with an Emissions Control Information label as well.

ENGINE MODEL & SERIAL NUMBERS

◆ See Figures 1 thru 5

The engine model numbers are the manufacturer's key to engine changes. These alpha-numeric codes identify the year of manufacture, the horsepower rating, gearcase shaft length and various model/option differences (such as Saltwater, Pro-Series or V-max and starting/trim tilt options such as manual start/manual tilt or electric start power trim/tilt). If any correspondence or parts are required, the engine model number must be used for proper identification.

Remember that the model number establishes the model year for which the engine was produced, which is often not the year in which the motor was first installed on a boat. Also, keep in mind that a date of manufacture may be the year prior to the designated model year.

The engine model number tag also contains information used by the manufacturer internally as an engine family designation and a serial number (a unique sequential identifier given ONLY to that one motor).

When present, the emissions control information label states that the motor is in compliance with EPA emissions regulations for the model year of that engine. And, more importantly, it gives tune-up specifications that are vital to proper engine performance (that minimize harmful emissions). The specifications on this label may reflect changes that are made during production runs and are often not later reflected in a company's service

Fig. 2 . . .and/or starboard side of most engine clamp or swivel/tilt brackets

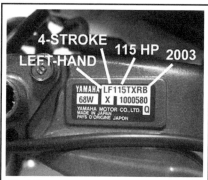

Fig. 3 The model ID tag provides critical information to identify and service the engine (4-stroke shown, but 2-stroke tags are mounted in the same way)

Fig. 4 Keep in mind that the date of manufacture is often the year BEFORE the model year

Fig. 5 When present, the emission control information label supercedes specifications listed elsewhere

2-4 MAINTENANCE & TUNE-UP

literature. For this reason, specifications on the label always supercede those of a print or electronic manual. Typical specifications that are found on this label may include:
- Spark plug type and gap.
- Fuel recommendations.
- Idle speed settings
- Possibly engine ignition timing (such as wide-open throttle and/or idle timing) specifications

Deciphering The Model Code

◆ See Figure 6

Engines built since the 1991 model year contain a model code similar to that of earlier year production models, however there are more variations and therefore more possible codes. Most 1991 and later models will contain a 6-9 digit code. The code may or may not begin with a one or two digit alpha model description. This tells you what series (Inshore, C, Pro, Saltwater, etc) to which the engine belongs. The next one, two or three digits will be numbers, representing the horsepower rating. The digit following the horsepower rating will be a single digit alpha code (E, M, T or P) identifying the starting and trim/tilt system on the motor. Following the starting and trim/tilt system identifier will be a single alpha identifier (S, L, X, U or J) representing gearcase shaft length (or type in the case of J for Jet Drive). Next, a single-digit, alpha identifier is used for the year. Finally, in some cases, a single check digit is used by the manufacturer to designate the 3-cylinder version of the 25 hp motor.

Refer to the accompanying illustration to interpret the various alpha digits found throughout the model code.

Before/After Each Use

As stated earlier, the best means of extending engine life and helping to protect yourself while on the water is to pay close attention to boat/engine maintenance. This starts with an inspection of systems and components before and after each time you use your boat.

A list of checks, inspections or required maintenance can be found in the Maintenance Intervals Chart at the end of this section. Some of these inspections or tasks are performed before the boat is launched, some only after it is retrieved and the rest, both times.

VISUALLY INSPECTING THE BOAT AND MOTOR

◆ See Figures 7 and 8

Both before each launch and immediately after each retrieval, visually inspect the boat and motor as follows:

1. **Check the fuel and oil levels** according to the procedures in this manual. Do NOT launch a boat without properly topped off fuel and oil tanks. It is not worth the risk of getting stranded or of damage to the motor. Likewise, upon retrieval, check the oil and fuel levels while it is still fresh in your mind. This is a good way to track fuel consumption (one indication of engine performance). Compare the fuel consumption to the oil consumption (a dramatic change in proportional use may be an early sign of trouble).

2. **Check for signs of fuel or oil leakage.** Probably as important as making sure enough fuel and oil is onboard, is the need to make sure that no dangerous conditions might arise due to leaks. Thoroughly check all hoses, fittings and tanks for signs of leakage. Oil leaks may cause the boat to become stranded, or worse, could destroy the motor if undetected for a significant amount of time. Fuel leaks can cause a fire hazard, or worse, an explosive condition. This check is not only about properly maintaining your boat and motor, but about helping to protect your life.

✱✱ CAUTION

On fuel injected motors (ESPECIALLY HPDI) fuel is pumped at high pressure through various lines under the motor cowl. The smallest leak will allow for fuel to spray in a fine, atomized and highly combustible stream from the damaged hose/fitting. It is critical that you remove the cowling and turn the key to the ON position (to energize the fuel pump and begin building system pressure) for a quick check before starting the motor (to ensure that no leaks are present). Even so, leaks may not show until the motor is operating so it is a good idea to either leave the cover off until the motor is running or to remove it again later in the day to double-check that you are leak free. Of course, if you DO remove the cover with the engine running take GREAT care to prevent contact with any moving parts.

3. **Inspect the boat hull and engine cases** for signs of corrosion or damage. Don't launch a damaged boat or motor. And don't surprise yourself dockside or at the launch ramp by discovering damage that went unnoticed last time the boat was retrieved. Repair any hull or case damage now.

YAMAHA MODEL IDENTIFICATION DECODER

Model Description	Prop Shaft HP	Trim/Tilt Starting Method	Shaft Length	Method of Control	Model Year	Code Variant ①
(1 or 2 digits / Alpha)	(1-3 digits / Numerals)	(1 digit - Alpha)	(1 digit - Alpha)	(1 digit - Alpha)	(1 digit - Alpha)	(1 digit / Numeral)
B = Inshore Series						
C = C Series	A number between	1986-90 ②	S = 15" (38cm)	1991-03 only	N = 1984	2 = 3 cyl (1997)
D = Twin Prop	2 = 2 hp	E = Electric Start		H = Tiller	K = 1985	
DX = Twin Prop / EFI	and		L = 20" (51cm)		J = 1986	3 = 3 cyl
E = Enduro Series	250 = 250 hp	T = Power Trim or		R = Remote	H = 1987	
F = 4-Stroke		Power Trim/Tilt	X = 25" (64cm)		G = 1988	1984-90
FT = 4-Stroke / High Thrust					F = 1989	JD = Jet Drive
L = Left-Hand Rotation		M = Tiller	U = 30" (76cm)		D = 1990	
LF = Left-Hand / 4-Stroke					P = 1991	
LX = Left-Hand / EFI		1991-03 ②	1991-03		Q = 1992	
LZ = Left-Hand / HPDI		E = Manual Tilt	J = Jet Drive		R = 1993	
P = Pro Series Model		w/ Electric Start			S = 1994	
PX = Pro Series / EFI					T = 1995	
S = Saltwater Series		M = Manual Tilt			U = 1996	
SX = Saltwater Series / EFI		w/ Manual Start			V = 1997	
T = 4-Stroke / High Thrust					W = 1998	
V = V-Max Series		T = Power Trim/Tilt			X = 1999	
VX = V-Max / EFI		w/ Electric Start			Y = 2000	
VZ = V-Max / HPDI					Z = 2001	
Z = HPDI		P = Power Tilt			A = 2002	
		w/ Electric Start			B = 2003	

① A special model variation code is used on 25 hp motors to distinguish between 2 cyl models (no code) and 3 cyl models (code present)
② For models through 1990 this portion of the code may be 1, 2 or 3 Alpha designators, all 1991 or later use only 1 digit at this part of the code

Fig. 6 Yamaha Model codes

MAINTENANCE & TUNE-UP

Fig. 7 Rope or fishing line entangled behind the propeller can cut through the seal, allowing water in or lubricant out

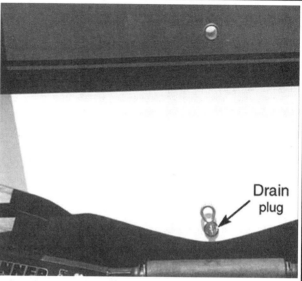

Fig. 8 Always make sure the transom plug is installed and tightened securely before a launch

4. **Check the battery** connections to make sure they are clean and tight. A loose or corroded connection will cause charging problems (damaging the system or preventing charging). There's only one thing worse than a dead battery dockside/launch ramp and that's a dead battery in the middle of a bay, river or worse, the ocean. Whenever possible, make a quick visual check of battery electrolyte levels (keeping an eye on the level will give some warning of overcharging problems). This is especially true if the engine is operated at high speeds for extended periods of time.

5. **Check the propeller (impeller on jet drives) and gearcase.** Make sure the propeller shows no signs of damage. A broken or bent propeller may allow the engine to over-rev and it will certainly waste fuel. The gearcase should be checked before and after each use for signs of leakage. Check the gearcase oil for signs of contamination if any leakage is noted. Also, visually check behind the propeller for signs of entangled rope or fishing lines that could cut through the lower gearcase propeller shaft seal. This is a common cause of gearcase lubricant leakage, and eventually, water contamination that can lead to gearcase failure. Even if no gearcase leakage is noted when the boat is first retrieved, check again next time before launching. A nicked seal might not seep fluid right away when still swollen from heat immediately after use, but might begin seeping over the next day, week or month as it sat, cooled and dried out.

6. **Check all accessible fasteners for tightness.** Make sure all easily accessible fasteners appear to be tight. This is especially true for the propeller nut, any anode retaining bolts, all steering or throttle linkage fasteners and the engine clamps or mounting bolts. Don't risk loosing control or becoming stranded due to loose fasteners. Perform these checks before heading out, and immediately after you return (so you'll know if anything needs to be serviced before you want to launch again.)

7. **Check operation of all controls including the throttle/shifter, steering and emergency stop/start switch and/or safety lanyard.** Before launching, make sure that all linkage and steering components operate properly and move smoothly through their range of motion. All electrical switches (such as power trim/tilt) and especially the emergency stop system(s) must be in proper working order. While underway, watch for signs that a system is not working or has become damaged. With the steering, shifter or throttle, keep a watchful eye out for a change in resistance or the start of jerky/notchy movement.

8. **Check the water pump intake grate and water indicator.** The water pump intake grate should be clean and undamaged before setting out. Remember that a damaged grate could allow debris into the system that could destroy the impeller or clog cooling passages. Once underway, make sure the cooling indicator stream is visible at all times. Make periodic checks, including one final check before the motor is shut down each time. If a cooling indicator stream is not present at any point, troubleshoot the problem before further engine operation.

9. **If used in salt, brackish or polluted waters thoroughly rinse the engine (and hull), then flush the cooling system** according to the procedure in this section.

■ Keep in mind that the cooling system can use attention, even if used in fresh waters. Sand, silt or other deposits can help clog passages, chemicals or pollutants can speed corrosion. It's a good idea to flush your motor after every use, regardless of where you use it.

10. **Visually inspect all anodes** after each use for signs of wear, damage or to make sure they just plain didn't fall off (especially if you weren't careful about checking all the accessible fasteners the last time you launched).

11. **For Pete's sake, make sure the plug is in!** We shouldn't have to say it, but unfortunately we do. If you've been boating for any length of time, you've seen or heard of someone whose backed a trailer down a launch ramp, forgetting to check the transom drain plug before submerging (literally) the boat. Always make sure the transom plug is installed and tight before a launch.

LUBRICATION

An outboard motor's greatest enemy is corrosion. Face it, oil and water just don't mix and, as anyone who has visited a junkyard knows, metal and water aren't the greatest of friends either. To expose an engine to a harsh marine environment of water and wind is to expect that these elements will take their toll over time. But, there is a way to fight back and help prevent the natural process of corrosion that will destroy your beloved boat motor.

Various marine grade lubricants are available that serve two important functions in preserving your motor. Lubricants reduce friction on metal-to-metal contact surfaces and, they also displace air and moisture, therefore slowing or preventing corrosion damage. Periodic lubrication services are your best method of preserving an outboard motor. Marine lubricants are designed for the harsh environment to which outboards are exposed and are designed to stay in place, even when submerged in water (and we've got some bathing suits with marine grease stains to attest to this).

Lubrication takes place through various forms. For all engines, internal moving parts are lubricated by engine oil, in the case of these motors, through oil contained in the fuel/oil mixture. Also, on all motors the gearcase is filled with gear oil that lubricates the driveshaft, propshaft, gears and other internal gearcase components. The gear oil for all motors should be periodically checked and replaced following the appropriate Engine Maintenance procedures. Perform these services based on time or engine use, as outlined in the Maintenance Intervals chart at the end of this section.

For motors equipped with power trim/tilt, the fluid level and condition in the reservoir should be checked periodically to ensure proper operation. Proper fluid level not only ensures that the system will function properly, but also helps lubricate and protect the internal system components from corrosion.

2-6 MAINTENANCE & TUNE-UP

Most other forms of lubrication occur through the application of grease (Yamaha all-purpose Marine grease, either applied by hand (an old toothbrush can be helpful in preventing a mess) or using a grease gun to pump the lubricant into grease fittings (also known as zerk fittings). When using a grease gun, do not pump excessive amounts of grease into the fitting. Unless otherwise directed, pump until either the rubber seal (if used) begins to expand or until the grease just begins to seep from the joints of the component being lubricated (if no seal is used).

To ensure your motor is getting the protection it needs, perform a visual inspection of the various lubrication points at least once a week during regular seasonal operation (this assumes that the motor is being used at least once a week). Follow the recommendations given in the Lubrication Chart at the end of this section and perform the various lubricating services at least every 60 days when the boat is operated in fresh water or every 30 days when the boat is operated in salt, brackish or polluted waters. We said **at least** meaning you should perform these services more often, if a need is discovered by your weekly inspections.

■ Jet drive models require one form of lubrication EVERY time that they are used. The jet drive bearing should be greased, following the procedure given in this section, after every day of boating. But don't worry, it only takes a minute once you've done it before.

Lubricants

◆ See Figures 9 and 10

✳✳ WARNING

Lubricants recommended in the lubrication procedures are NOT interchangeable as each is designed to perform under different conditions.

YAMAHA OR YAMALUBE ALL PURPOSE MARINE GREASE

All-purpose marine grease is a general outboard lubricant/grease, chemically formulated to resist salt water. This lubricant is recommended for application to bearings, bushings, and oil seals.

YAMALUBE 2-STROKE LUBRICANT

Yamalube lubricant is a two-stroke engine oil. It is a petroleum based, clean burning lubricant. Yamalube reduces carbon deposits and ensures maximum protection against engine wear. No oil additives are recommended by the manufacturer. Yamalube contains ashless detergent to minimize piston rings from sticking.

YAMAHA OR YAMALUBE GEARCASE LUBRICANT

Yamaha or Yamalube Gearcase Lubricant contains high viscosity additives to protect the lower unit gears at high speed operation. The lubricant will extend gear life, reduce gear noise, minimize friction, and has a cooling affect on the lower unit moving parts.

YAMAHA OR YAMALUBE POWER TRIM AND TILT FLUID

Yamaha or Yamalube power trim and tilt fluid is a highly refined hydraulic fluid. This product has a high detergent content and additives to keep seals pliable. A high grade automatic transmission fluid, Dexron® II, may also be used if Yamalube is not available.

Lubricating the Motor

◆ See Figures 11 thru 16

The first thing you should do upon purchasing a new or "new to you" motor is to remove the engine top cover and look for signs of grease. Note all components that have been freshly greased (or if the motor has been neglected that shows signs of wear or dirt/contamination that has collected on the remnants of old applied grease). If the motor shows signs of dirt, corrosion or wear, clean those components thoroughly and apply a fresh coat of grease.

Fig. 9 Yamaha recommended lubricants and additives will not only keep the unit within the limits of the warranty but will also contribute to dependable performance and reduced maintenance costs

Fig. 10 Yamaha products, available from your local dealer, will do much to keep the outboard unit looking sharp and running right

Thereafter, follow the recommendations in the accompanying lubrication chart (as they apply to the motor on which you are working) and grease all necessary surfaces regularly to keep them clean and well lubricated. As a general rule of thumb any point where two metallic mechanical parts connect and push, pull, turn, slide, pivot on each other should be greased. For most motors this will include shift and throttle cables and/or linkage, steering and swiveling points and items such as the cowl clamp bolts (on smaller motors) and top cover or cowling clamp levers.

■ For more information on greasing and lubrication points, check your owner's manual. Most Yamaha owner's manuals will provide one or more illustrations to help you properly identify all necessary greasing points.

Points such as the swivel bracket and/or the tilt tube will normally be equipped with grease (zerk) fittings. For these, use a grease gun to carefully pump small amounts of grease into the fittings, displacing some of the older grease and lubricating the internal surfaces of the swivel and tilt tubes. Some engine cowl levers require a dab of lubricant be applied manually over sliding surfaces, but many are also equipped with grease fittings for lubrication using a grease gun. A few of the larger Yamaha outboards actually use a cable release system for the cowling, the cable ends and latches should all be greased periodically to prevent binding and wear.

Items without a grease fitting, such as the steering ram, cable ends, shifter and carburetor linkage all must normally be greased by hand using a small dab of lubricant. Be sure not to over apply grease as it is just going to get over everything and exposed grease will tend to attract and hold dirt or other particles of general crud. For this reason it is always a good idea to wipe away the old grease before applying fresh lubricant to these surfaces.

MAINTENANCE & TUNE-UP 2-7

Fig. 11 Various lubrication points on the powerhead should be maintained regularly to ensure a long service life including all rotating. . .

Fig. 12 . . . or sliding linkage points and cable ends

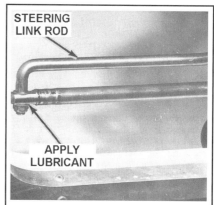

Fig. 13 Apply Yamaha All Purpose Grease to the steering link rod joint

Fig. 14 The swivel bracket usually contains one or two grease fittings

Fig. 15 The propeller should be removed periodically to clean and re-grease the prop shaft splines

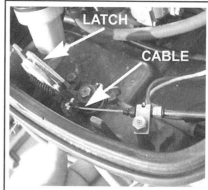

Fig. 16 Any cowling clamp latches and/or cables should also be lubricated with Yamaha All Purpose Grease

Lubrication Inside the Boat

The following points inside the boat will also usually benefit from lubrication with Yamalube All Purpose lubricant:
- Remote control cable ends next to the hand nut. DO NOT over-lubricate the cable.
- Steering arm pivot socket.
- Exposed shaft of the cable passing through the cable guide tube.
- Steering link rod to steering cable.

Gearcase (Lower Unit) Oil

◆ See Figures 17 and 18

Regular maintenance and inspection of the lower unit is critical for proper operation and reliability. A lower unit can quickly fail if it becomes heavily contaminated with water or excessively low on oil. The most common cause of a lower unit failure is water contamination.

Water in the lower unit is usually caused by fishing line or other foreign material, becoming entangled around the propeller shaft and damaging the seal. If the line is not removed, it will eventually cut the propeller shaft seal and allow water to enter the lower unit. Fishing line has also been known to cut a groove in the propeller shaft if left neglected over time. This area should be checked frequently.

OIL RECOMMENDATIONS

Use only Yamaha Gear Case Lube or an equivalent SAE-90W gearcase oil.

■ **Remember, it is this lower unit lubricant that prevents corrosion and lubricates the internal parts of the drive gears. Lack of lubrication due to water contamination or the improper type of oil can cause catastrophic lower unit failure.**

CHECKING GEARCASE OIL LEVEL & CONDITION

◆ See Figure 19

Visually inspect the gearcase before and after each use for signs of leakage. At least monthly, or as needed, remove the gearcase level plug in order to check the lubricant level and condition as follows:

1. Position the engine in the upright position with the motor shut off for at least 1 hour. Whenever possible, checking the level overnight cold will give a true indication of the level without having to account for heat expansion.
2. Disconnect the negative battery cable or remove the propeller for safety.

✳✳ CAUTION

Always observe extreme care when working anywhere near the propeller. Take steps to ensure that no accidental attempt to start the engine occurs while work is being performed or remove the propeller completely to be safe.

3. Position a small drain pan under the gearcase, then unthread the drain/filler plug at the bottom of the housing and allow a small sample (a teaspoon or less) to drain from the gearcase. Quickly install the drain/filler plug and tighten securely.
4. Examine the gear oil as follows:
   a. Visually check the oil for obvious signs of water. A small amount of moisture may be present from condensation, especially if a motor has been stored for some time, but a milky appearance indicates that either the fluid has not been changed in ages or the gearcase allowing some water to intrude. If significant water contamination is present, the first suspect is the propeller shaft seal.

2-8 MAINTENANCE & TUNE-UP

b. Dip an otherwise clean finger into the oil, then rub a small amount of the fluid between your finger and your thumb to check for the presence of debris. The lubricant should feel smooth. A **very** small amount of metallic shavings may be present, but should not really be felt. Large amounts of grit or metallic particles indicate the need to overhaul the gearcase looking for damaged/worn gears, shafts, bearings or thrust surfaces.

■ If a large amount of lubricant escapes when the level/vent plug is removed in the next step, either the gearcase was seriously overfilled on the last service, the crankcase is still too hot from running the motor in gear (and the fluid is expanded) or a large amount of water has entered the gearcase. If the later is true, some water should escape before the oil and/or the oil will be a milky white in appearance (showing the moisture contamination).

5. Next, remove the level/vent plug from the top of the gearcase and ensure the lubricant level is up to the bottom of the level/vent plug opening. A very small amount of fluid may be added through the level plug, but larger amounts of fluid should be added through the drain/filler plug opening to make certain that the case is properly filled. If necessary, add gear oil until fluid flows from the level/vent opening. If much more than 1 oz. (29 ml) is required to fill the gearcase, check the case carefully for leaks. Install the drain/filler plugs and/or the level/vent plug and then tighten both securely.

■ One trick that makes adding gearcase oil less messy is to install the level/vent plug BEFORE removing the pump from the drain/filler opening and threading the drain/filler plug back into position.

6. Once fluid is pumped into the gearcase, let the unit sit in a shaded area for at least 1 hour for the fluid to settle. Recheck the fluid level and, if necessary, add more lubricant.

7. Install the propeller and/or connect the negative battery cable, as applicable.

DRAINING AND FILLING

◆ See Figures 20, 21 and 22

1. Place a suitable container under the lower unit. It is usually a good idea to place the outboard in the tilted position so the drain plug is at the lowest position on the gearcase, this will help ensure the oil drains fully.

2. Loosen the oil level/vent plug on the lower unit. This step is important! If the oil level/vent plug cannot be loosened or removed, you cannot refill the gearcase with fluid and purge it of air.

■ Never remove the vent or filler plugs when the lower unit is hot. Expanded lubricant will be released through the hole.

3. Remove the drain/filler plug from the lower end of the gear housing followed by the oil level/vent plug.

4. Allow the lubricant to completely drain from the lower unit.

5. If applicable, check the magnet end of the drain screw for metal particles. Some amount of metal is considered normal wear is to be

Fig. 17 This lower unit was destroyed because the bearing carrier froze due to lack of lubrication

Fig. 18 Fishing line entangled behind the prop can actually cut through the seal

Fig. 19 Yamaha often labels the vent/level plug on their gearcases

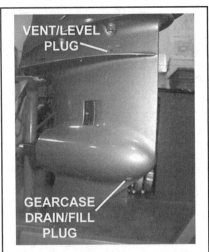

Fig. 20 The gearcase oil vent/level plug is on top, while the drain/fill plug is on the bottom

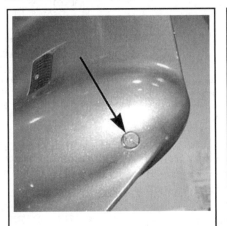

Fig. 21 Better view of a typical Yamaha gearcase drain/fill plug (designed for a large flat-head screwdriver)

Fig. 22 When draining the lower unit, ensure it is fully tilted and a drain pan of adequate capacity is place to catch the lubricant

MAINTENANCE & TUNE-UP 2-9

expected but if there are signs of metal chips or excessive metal particles, the gearcase needs to be disassembled and inspected.

6. Inspect the lubricant for the presence of a milky white substance, water or metallic particles. If any of these conditions are present, the lower unit should be serviced immediately.
7. Place the outboard in the proper position for filling the lower unit (straight up and down). The lower unit should not list to either port or starboard and should be completely vertical.
8. Insert the lubricant tube into the oil drain hole at the bottom of the lower unit and inject lubricant until the excess begins to come out the oil level hole.

■ **The lubricant must be filled from the bottom to prevent air from being trapped in the lower unit. Air displaces lubricant and can cause a lack of lubrication or a false lubricant level in the lower unit.**

9. Oil should be squeezed in using a tube or with the larger quantities, by using a pump kit to fill the gearcase through the drain plug.

■ **One trick that makes adding gearcase oil less messy is to install the level/vent plug BEFORE removing the pump from the drain/filler opening and threading the drain/filler plug back into position.**

10. Using new gaskets/washers (if equipped) install the oil level/vent plug first, then install the oil fill plug.
11. Wipe the excess oil from the lower unit and inspect the unit for leaks.
12. Place the used lubricant in a suitable container for transportation to an authorized recycling facility.

Jet Drive Bearing

◆ See Figure 23

Jet drive models covered by this manual require special attention to ensure that the driveshaft bearing remains properly lubricated.

Yamaha recommends that you lubricate the jet drive bearing using a grease gun after EACH days use. However, at an absolute minimum, use the grease fitting every 10 hours (in fresh water) or 5 hours (in salt water). Also, after every 50 hours of fresh water operation or every 25 hours of salt/brackish/polluted water operation, the drive bearing grease must be replaced. Follow the appropriate procedure:

RECOMMENDED LUBRICANT

Use Yamaha All-Purpose Marine grease or an equivalent water-resistant NLGI No. 1 lubricant.

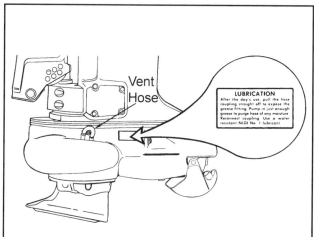

Fig. 23 Jet drive models require lubrication of the bearing after each day of use, sometimes they are equipped with a label on the housing to remind the owner

DAILY BEARING LUBRICATION

◆ See Figures 24 and 25

A grease fitting is located under a vent hose on the lower port side of the jet drive. Disconnect the hose from the fitting, then use a grease gun to apply enough grease to the fitting to **just** fill the vent hose. Basically, grease is pumped into the fitting until the old grease just starts to come out from the passages through the hose coupling - then reconnect the hose to the fitting.

■ **Do not attempt to just grasp the vent hose and pull, as it is a tight fit and when it does come off, you'll probably go flying if you didn't prepare for it. The easier method of removing the vent hose from the fitting is to deflect the hose to one side and snap it free from the fitting.**

GREASE REPLACEMENT

◆ See Figure 24, 25 and 26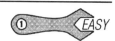

A grease fitting is located under a vent hose on the lower port side of the jet drive. This grease fitting is utilized at the end of each day's use to add fresh grease to the jet drive bearing. But, every 50 or 25 hours and/or 30 or 15 days (depending if use is in fresh or salt/brackish/polluted waters), the grease should be completely replaced. This is very similar to the daily greasing, except that a lot more grease it used. Disconnect the hose from the fitting (by deflecting it to the side until it snaps free from the fitting), then use a grease gun to apply enough grease to the fitting until grease exiting the assembly fills the vent hose. Then, continue to pump grease into the fitting to force out all of the old grease (you can tell this has been accomplished when fresh grease starts to come out of the vent instead of old grease, which will be slightly darker due to minor contamination from normal use). When nothing but fresh grease comes out of the vent the fresh grease has completely displaced the old grease and you are finished. Be sure to securely connect the vent hose to the fitting.

Each time this is performed, inspect the grease for signs of moisture contamination or discoloration. A gradual increase in moisture content over a few services is a sign of seal wear that is beginning to allow some seepage. Very dark or dirty grease may indicate a worn seal (inspect and/or replace the seal, as necessary to prevent severe engine damage should the seal fail completely).

■ **Keep in mind that some discoloration of the grease is expected when a new seal is broken-in. The discoloration should go away gradually after one or two additional grease replacement services.**

Whenever the jet drive bearing grease is replaced, take a few minutes to apply some of that same water-resistant marine grease to the pivot points of the jet linkage.

Power Trim/Tilt Reservoir

RECOMMENDED LUBRICANT

Yamaha or Yamalube power trim and tilt fluid is a highly refined hydraulic fluid. This product has a high detergent content and additives to keep seals pliable. A high grade automatic transmission fluid, Dexron® II may also be used if the Yamalube fluid is not available.

CHECKING FLUID LEVEL/CONDITION

◆ See Figures 27, 28 and 29

The fluid in the power trim/tilt reservoir should be checked periodically to ensure it is full and is not contaminated. To check the fluid, tilt the motor upward to the full tilt position, then manually engage the tilt support for safety and to prevent damage. Loosen and remove the filler plug using a suitable socket or wrench and make a visual inspection of the fluid. It should seem clear and not milky. The level is proper if, with the motor at full tilt, the level is even with the bottom of the filler plug hole.

2-10 MAINTENANCE & TUNE-UP

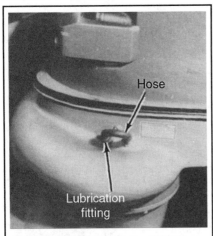

Fig. 24 The jet drive lubrication fitting is found under the vent hose

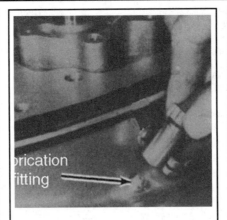

Fig. 25 Attach a grease gun to the fitting for lubrication

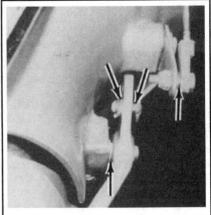

Fig. 26 Also, coat the pivot points of the jet linkage with grease periodically

Fig. 27 On most medium sized motors, the PTT systems use only a single trim/tilt rod

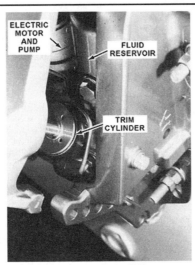

Fig. 28 More common, is the PTT system which uses one tilt and two trim rods. . .

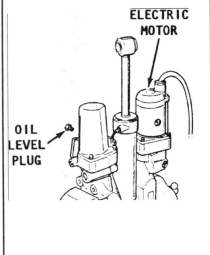

Fig. 29 . . . the fill plug is on the front of the reservoir

ENGINE MAINTENANCE

Engine Covers (Top Cover and Cowling)

REMOVAL & INSTALLATION

◆ See Figures 30 thru 35

Removal of the top cover is necessary for the most basic of maintenance and inspection procedures. The cover should come off before and after each use in order to perform these basic safety checks. Unlike some outboard manufacturers that use large and intricate multi-piece lower covers/cowlings, most Yamaha outboards simple use a low profile 1-piece cowling that does not require removal except during powerhead R&R. However there are a few exceptions on larger motors that are equipped with fore or aft or port and starboard lower covers which cover part of the intermediate housing or steering bracket. Removal of those lower covers is normally pretty straight forward in that most fasteners are exposed and obvious.

Although top cover removal will vary slightly from motor to motor, most have one, two or three latches located at the front and rear of the cowling. The latches must be released in order to free the cover from the cowling. On smaller motors, if only one lever is used the other end of the cover is usually secured by a tab that is held in place by friction once the lever is secured. On larger motors (mostly V6 models) one lever may be connected by cable to a release mechanism at both ends of the top cover. On most larger motors there are two or three separate latches.

Probably the most common form of cover retention on Yamaha outboards is the use of 2 separate levers, one at the front and one at the rear of the outboard. On most inline motors these levers are horizontal when latched and rotated 1/4 turn downward (or upward in some cases) to release. On most V-configuration motors these latches are lifted up and outward to release or pushed down and inward to lock.

As mentioned in the lubrication section all cover latch mechanisms should be wiped clean and re-greased periodically to ensure proper operation and to prevent un-necessary wear or damage.

Whenever the top cover is removed, check the cover seal (a rubber insulator normally used between the cover and cowling) for wear or damage. Take the opportunity to perform a quick visual inspection for leaking hoses, chaffing wires etc. Get to know where things are placed under the cover so you'll recognize instantly if something is amiss. When installing the cover, care must be taken to ensure wiring and hoses are in their original positions to prevent damage. Once the cover is installed, make sure the latches grab it securely to prevent it from coming loose and flying off while underway.

■ On a few of the smaller Yamaha outboards the cover is equipped with one or more drain holes. If equipped, check them periodically to ensure they are not blocked by debris.

MAINTENANCE & TUNE-UP

Fig. 30 Most covers are secured by 1 or 2 latches...

Fig. 31 ...some are rotated 1/4 turn upward...

Fig. 32 ...or downward to release (while others are pulled straight upward)

Fig. 33 Some V6 motors use a single latch, which when pulled outward...

Fig. 34 ...pulls on a cable that...

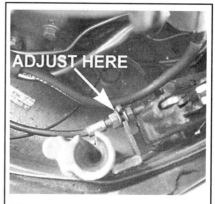
Fig. 35 ...actuates the latch on the other end of the cowling

Cooling System

FLUSHING THE COOLING SYSTEM

◆ See Figures 36 thru 40

The most important service that you can perform on your motor's cooling system is to flush it periodically using fresh, clean water. This should be done immediately following any use in salt, brackish or polluted waters in order to prevent mineral deposits or corrosion from clogging cooling passages. Even if you do not always boat in salt or polluted waters, get used to the flushing procedure and perform it often to ensure no silt or debris clogs your cooling system over time.

■ Flush the cooling system after any use in which the motor was operated through suspended/churned-up silt, debris or sand.

Although the flushing procedure should take place right away (dockside or on the trailer), be sure to protect the motor from damage due to possible thermal shock. If the engine has just been run under high load or at continued high speeds, allow time for it to cool to the point where the powerhead can be touched. Do not pump very cold water through a very hot engine, or you are just asking for trouble. If you trailer your boat short distances, the flushing procedure can probably wait until you arrive home or wherever the boat is stored, but ideally it should occur within an hour of use in salt water. Remember that the corrosion process begins as soon as the motor is removed from the water and exposed to air.

The flushing procedure is not used only for cooling system maintenance, but it is also a tool with which a technician can provide a source of cooling water to protect the engine (and water pump impeller) from damage anytime the motor needs to be run out of the water. **Never** start or run the engine out of the water, even for a few seconds, for any reason. Water pump impeller damage can occur instantly and damage to the engine from overheating can follow shortly thereafter. If the engine must be run out of the water for tuning or testing, always connect an appropriate flushing device **before** the engine is started and leave it turned on until **after** the engine is shut off.

✴✴ WARNING

ANYTIME the engine is run, the first thing you should do is check the cooling stream or water indicator. All Yamahas are equipped with some form of a cooling stream indicator towards the aft portion of the lower engine cover. Anytime the engine is operating, a steady stream of water should come from the indicator, showing that the pump is supplying water to the engine for cooling. If the stream is ever absent, stop the motor and determine the cause before restarting.

As we stated earlier, flushing the cooling system consists of supplying fresh, clean water to the system in order to clean deposits from the internal passages. If the engine is running, the water does not normally have to be pressurized, as it is delivered through the normal water intake passages and the water pump (the system can self flush if supplied with clean water). Smaller, portable engines can be flushed by mounted them in a test tank (a sturdy, metallic 30 gallon drum or garbage pail filled with clean water).

These days many of the Yamaha models are equipped with a built-in flushing adapter. Normally flushing adapters are found on most V configuration models located almost midway back on the port side of the cowling, threaded upward into the underside of the cowling itself. However, keep in mind that, when present, the flushing adapter is designed for cleaning the cooling system and should not be used when running the engine on a hose, as it may not allow sufficient water to reach the impeller (and could therefore allow water pump damage).

When present the Yamaha flush adapters consist of a short length of hose which runs from a fitting on the powerhead to a female hose connection which is threaded onto a fitting on the outboard cowling. The adapter is used by unthreaded the female connection and attaching the male end of a garden hose to it. This allows you to supply fresh water from the garden

2-12 MAINTENANCE & TUNE-UP

hose directly into the powerhead cooling passages, bypassing the water pump. It is critical that you thread the hose back onto the cowling fitting before returning the motor to service. If the adapter hose is left unblocked, a certain amount of water from the water pump may bypass the powerhead during engine operation which could lead to overheating.

For almost all Yamaha motors, if you are going to run it on a hose at any point, you'll

need a generic (ear-muff type) flush adapters. The generic adapters fit over the engine water intakes on the gearcase (and resemble a pair of strange earmuffs with a hose fitting on one side). This will work on all Yamahas whose water intake is on the sides of the gearcase. It may not work on the smallest Yamahas whose water intake is in the anti-ventilation plate, however those models are small enough to run in a sturdy metal trash can.

■ Jet drive models are equipped with a flushing port mounted under a flat head screw directly above the jet drive bearing grease fitting. For more information, please refer to FLUSHING JET DRIVES, later in this section.

■ When running the engine on a flushing adapter using a garden hose, make sure the hose delivers about 20-40 psi (140-300 kPa) of pressure.

Some of the smaller, portable motors covered by this manual utilize a water intake that is directly above the propeller. On these models the propeller must usually be removed before a clamp style flush adapter can be connected to the motor (unless the adapter is very thin and mounted so close to the anti-ventilation plate that it will not be hit by the propeller).

✴✴ CAUTION

For safety, the propeller should be removed ANYTIME the motor is run on the trailer or on an engine stand. We realize that this is not always practical when flushing the engine on the trailer, but cannot emphasize enough how much caution must be exercised to prevent injury to you or someone else. Either take the time to remove the propeller or take the time to make sure no-one or nothing comes close enough to it to become injured. Serious personal injury or death could result from contact with the spinning propeller.

1. Check the engine top case and, if necessary remove it to check the powerhead and ensure it is cooled enough to flush without causing thermal shock.
2. Prepare the engine for flushing, depending on the method you are using, as follows:
   a. If using a test tank, make sure the tank is made of sturdy material, then securely mount the motor to the tank. If necessary, position a wooden plank between the tank and engine clamp bracket for thickness. Fill the tank so the water level is at least 4 in. (10cm) above the anti-ventilation plate (above the water inlet).

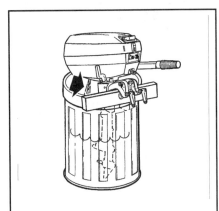

Fig. 36 All models may be flushed in a test tank. Smaller ones, in a garbage pail

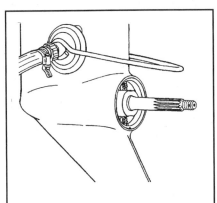

Fig. 37 The easiest way to flush most models is using a clamp-type adapter

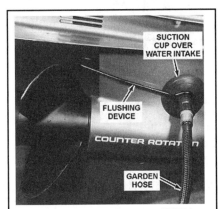

Fig. 38 A source of cooling water MUST be used anything the engine is run out of the water

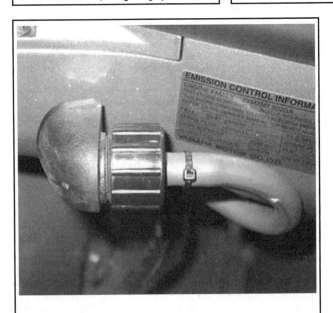

Fig. 39 Most V motors are equipped with some form of a flush adapter. . .

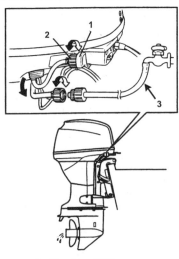

1 - Cowling Fitting
2 - Flushing Adapter
3 - Garden Hose

Fig. 40 . . . usually found on the underside of the cowling

MAINTENANCE & TUNE-UP

b. If using a flushing adapter of either the generic clamp-type or specific port-type for your model attach the water hose to the flush test adapter and connect the adapter to the motor following the instructions that came with the adapter. If the motor is to be run (for flushing or testing), be sure to use the ear-muff style adapter, position the outboard vertically and remove the propeller, for safety. Also, be sure to position the water hose so it will not contact with moving parts (tie the hose out of the way with mechanic's wire or wire ties, as necessary).

■ When using a clamp-type adapter, position the suction cup(s) over water intake grate(s) in such a way that they form tight seals. A little pressure seepage should not be a problem, but look to the water stream indicator once the motor is running to be sure that sufficient water is reaching the powerhead.

3. Unless using a test tank, turn the water on, making sure that pressure does not exceed 45 psi (300 kPa).

4. If using a test tank or if the motor must be run for testing/tuning procedures, start the engine and run in neutral until the motor reaches operating temperature. The motor will continue to run at fast idle until warmed.

✱✱ WARNING

As soon as the engine starts, check the cooling system indicator stream. It must be present and strong as long as the motor is operated. If not, stop the motor and rectify the problem before proceeding. Common problems could include insufficient water pressure or incorrect flush adapter installation.

5. Flush the motor for at least 5-10 minutes or until the water exiting the engine is clear. When flushing while running the motor, check the engine temperature (using a gauge or carefully by touch) and stop the engine immediately if steam or overheating starts to occur. Make sure that carbureted motors slow to low idle for the last few minutes of the flushing procedure.

6. Stop the engine (if running), **then** shut the water off.

■ If the motor is to be stored for any length of time more than a week or two, consider fogging the motor before it is shut off. Flushing and fogging a motor is one of the most important things you can do to ensure long life of the pistons, rods, crankshaft and bearings.

7. Remove the adapter from the engine or the engine from the test tank, as applicable.

8. If flushing did not occur with the motor running (so the motor would already by vertical), be sure to place it in the full vertical position allowing the cooling system to drain. This is especially important if the engine is going to be placed into storage and could be exposed to freezing temperatures. Water left in the motor could freeze and crack the powerhead or gearcase.

FLUSHING JET DRIVES

◆ See Figure 41

Regular flushing of the jet drive will prolong the life of the powerhead, by clearing the cooling system of possible obstructions.

Jet drives are equipped with a plug on the port side just above the lubrication hose.

1. Remove the plug and gasket, install the flush adapter, connect the garden hose and turn on the water supply.

2. Start and operate the powerhead at a fast idle for about 15 minutes. Disconnect the flushing adapter and replace the plug.

■ The procedure just described will only flush the powerhead cooling system, not the jet drive and impeller. To flush the jet drive unit, direct a stream of high pressure water through the intake grille.

Engine Oil

OIL RECOMMENDATIONS

◆ See Figure 42

Use only an NMMA (National Marine Manufacturers Association) certified TC-W3 or equivalent 2-stroke lubricant. Of course, Yamaha recommends Yamalube 2-stroke engine oil lubricant, since it is specially formulated to match the needs of Yamaha motors. Yamalube is a petroleum based, clean burning lubricant that reduces carbon deposits and ensures maximum protection against engine wear. Yamalube contains ashless detergent to minimize piston rings from sticking.

■ Yamaha does not recommend the use of any oil additives.

In all cases, use a high quality TC-W3 oils designed to ensure optimal engine performance and to minimize combustion chamber deposits, to avoid detonation and prolong spark plug life. Use only 2-stroke type outboard oil. Never use automotive motor oil.

■ Remember, it is this oil, mixed with the gasoline that lubricates the internal parts of the 2-stroke engine. Lack of lubrication due to the wrong mix or improper type of oil can cause catastrophic powerhead failure.

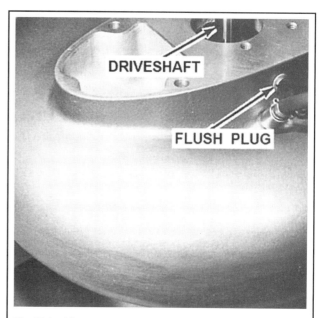

Fig. 41 Jet drives are equipped with a flushing plug on the side of the unit

Fig. 42 This scuffed piston is an example of the damage caused by improper 2-stroke oil or mixture

MAINTENANCE & TUNE-UP

FILLING

There are two methods of adding 2-stroke oil to an outboard. The first is the pre-mix method used on most low horsepower and on some commercial outboards. The second is the Precision Blend system that automatically injects the correct quantity of oil into the engine based on throttle position and operating conditions. In both cases, the fuel ratio should be considered. This is even true on automatic oiling systems for special operation, such as during break-in.

Fuel:Oil Ratio

The proper fuel:oil ratio will depend upon the engine and operating conditions. Most Yamahas above 15 hp are equipped (or may be equipped) with the Precision Blend oil injection system. The system is usually designed to maintain a 50:1 ratio without adding anything to the fuel tank. When the injection system is not used most Yamahas require either a fuel:oil ratio of 100:1 for smaller motors or 50:1 for larger motors.

■ **Check the Capacities Specifications chart at the end of this section for more details on your motor. But realize that these are the specs listed in Yamaha service literature. Because your engine may differ slightly from service manual specification, refer to your owner's manual or a reputable dealer to be certain that you use the proper mixture for your motor.**

Regardless of the normal operational fuel:oil ratio, it should be doubled during the first 10 hours of break-in for a new or rebuilt powerhead. That means motors which normally run 100:1 ratio should run 50:1 ratio for 10 hours. Subsequently, motors that normally run a 50:1 ratio should run a 25:1 ratio during this break-in period.

■ **Motors equipped with the precision blend oiling system should run a 50:1 ratio of pre-mix in the boat or portable fuel tank during break-in. This, added to the normal output of the oiling system will produce the proper 25:1 or 50:1 break-in ratio required to ensure proper seating of the pistons and rings during break-in. Pre-mix should be discontinued on Precision Blend motors once the 10 hour period has expired (as long as proper oil system operation is verified by observing a change in oil level before and after the 10 hour period).**

Pre-Mix

◆ See Figure 43

Mixing the engine lubricant with gasoline before pouring it into the tank is by far the simplest method of lubrication for 2-stroke outboards. However, this method is the messiest and causes the most amount of harm to our environment.

The most important part of filling a pre-mix system is to determine the proper fuel:oil ratio. Yamaha motors normally use either a 100:1 or a 50:1 ratio (that is 100 or 50 parts of fuel respectively to 1 part of oil). Consult the information in this section on Fuel:Oil Ratio and your owner's manual to determine what the appropriate ratio should be for your engine.

The procedure itself is uncomplicated, but you've got a couple options depending on how the fuel tank is set-up for your boat. To fill an empty portable tank, add the appropriate amount of oil to the tank, then add gasoline and close the cap. Rock the tank from side-to-side to gently agitate the mixture, thereby allowing for a thorough mixture of gasoline and oil. When just topping off built-in or larger portable tanks, it is best to use a separate 3 or 6 gallon (11.4 or 22.7 L) mixing tank in the same manner as the portable tank noted earlier. In this way a more exact measurement of fuel can occur in 3 or 6 gallon increments (rather than just directly adding fuel to the tank and realizing that you've just added 2.67 gallons of gas and need to ad, uh, a little less than 8 oz of oil for a 50:1 ratio, but exactly how many ounces would that be?) Use of a mixture tank will prevent the need for such mathematical equations. Of course, the use of a mixing tank may be inconvenient or impossible under certain circumstances, so the next best method for topping off is to take a good guess (but be a little conservative to prevent an excessively rich oil ratio). Either add the oil and gasoline at the same time, or add the oil first, then add the gasoline to ensure proper mixing. For measurement purposes, it would obviously be more exact to add the

Fig. 43 For portable tanks, either add the oil and gasoline at the same time, or add the oil first, then add the gasoline to ensure proper mixing

gasoline first, then add a suitable amount of oil to match it. The problem with adding gasoline first is that unless the tank could be thoroughly agitated afterward (and that would be **really** difficult on built-in tanks), the oil might not mix properly with the gasoline. Don't take that unnecessary risk.

To determine the proper amount of oil to add to achieve the desired fuel:oil ratio, refer to the Fuel:Oil Ratio chart at the end of this section.

Oil Injection

◆ See Figures 44, 45 and 46

The Yamaha Precision Blend oil injection system utilizes a mechanically driven oil pump mounted to the powerhead that is connected to the throttle by way of a linkage arm. The system is powered by the crankshaft, which drives a gear in the pump, creating oil pressure. As the throttle lever is advanced to increase engine speed, the linkage arm also moves, opening a valve that allows more oil to flow into the oil pump.

Most mechanical-injection systems incorporate some form of a low-oil warning alarm that is also connected to an engine-overheating sensor. Some versions of the system might include a built-in speed limiter. This sub-system is designed to reduce engine speed automatically when oil problems occur. This important feature goes a long way toward preventing severe engine damage in the event of an oil injection problem.

The procedure for filling these systems is simple. Most Yamaha motors utilize a powerhead mounted oil reservoir. Many V6 motors are also equipped with a remote, boat mounted, oil tank (a larger tank designed to hold more oil than a powerhead mounted unit). In either case, the main reservoir tank contains a filler cap that is removed in order to add oil to the tank. Be sure to check the oil level EVERY time the motor is operated. Whenever oil is added, place a piece of tape on the tank to mark the level and watch how fast it drops in relation to engine usage (hours and fuel consumption). Watch for changes in usage patterns that could indicate under or over oiling. Especially with a system that suddenly begins to deliver less oil, you could save yourself significant engine damage by discovering a problem that could have starved the motor for lubrication.

Should the oil hose become disconnected or suffer a break/leak, the oil prime might be lost. If so, the system should be primed **before** priming the fuel system and starting the engine. More details on servicing the oiling system are found in the Lubrication section of this manual.

■ **It is highly advisable to carry a few spare bottles of 2-stroke oil with you onboard. Even in the event of an oil system failure, oil can be added to a fuel tank (in the proper ratio) in order to limp the boat and motor safely home.**

MAINTENANCE & TUNE-UP 2-15

CHECKING FOR WATER OR CONTAMINANTS

◆ See Figure 47

To protect the powerhead from potential damage should contaminants enter the oil system (instead of oil), you should ALWAYS perform a quick check of the oil tank before every outing. Powerhead mounted oil tanks are normally equipped with a water and contaminant trap (either a short length of dead-end hose on the bottom of the tank or a trap/drain hose that is run upward attaching to the filler neck).

If water or contaminants are found, they must be removed in order to protect the motor. If large amounts are present, the tank should be drained and thoroughly cleaned. Also, if large amounts of water or contaminants are present, you'd do well to discover the source. What gremlin is sneaking onto your boat at night, opening the cap and putting them there?

Fuel Filter

◆ See Figure 48

A fuel filter is designed to keep particles of dirt and debris from entering the carburetor(s) or fuel injectors clogging the tiny internal passages. A small speck of dirt or sand can drastically affect the ability of the fuel system to deliver the proper amount of air and fuel to the engine. If a filter becomes clogged, it will quickly impede the flow of gasoline. This could cause lean fuel mixtures, hesitation and stumbling and idle problems in carburetors. Although a clogged fuel passage in a fuel injected engine could also cause lean symptoms and idle problems, dirt can also prevent a fuel injector from closing properly. A fuel injector that is stuck partially open by debris will cause the engine to run rich due to the unregulated fuel constantly spraying from the pressurized injector.

Regular cleaning or replacement of the fuel filter (depending on the type or types used) will decrease the risk of blocking the flow of fuel to the engine, which could leave you stranded on the water. It will also decrease the risk of damage to the small passages of a carburetor or fuel injector that could require more extensive and expensive replacement. Keep in mind that fuel filters are usually inexpensive and replacement is a simple task. Service your fuel filter on a regular basis to avoid fuel delivery problems.

We tried to find a pattern to tell us what type of filter will be installed on a powerhead, based on the year or model of the engine. However, Yamaha does not seem to have a whole lot of rhyme or reason when it comes to choosing a fuel filter for their outboards (or if they did, we couldn't figure it out). Loosely, we'll say that most smaller Yamaha outboards use disposable, inline filters (or fuel tank strainers on the smallest of outboards), while larger motors tend to use a serviceable filter element that can be removed from a housing/cup assembly for cleaning or replacement. However there are exceptions to this and some larger motors may be equipped with a non-serviceable inline filter. The HPDI motors are normally equipped with both a serviceable filter (located inline before the mechanical fuel-pump) and a non-serviceable filter (located inline before the vapor separator assembly).

Also, keep in mind that the type of fuel filter used on your boat/engine will vary not only with the year and model, but also with the accessories and rigging. Because of the number of possible variations it is impossible to accurately give instructions based on model. Instead, we will provide instructions for the different types of filters the manufacturer used on various families of motors or systems with which they are equipped. To determine what filter(s) are utilized by your boat and motor rigging, trace the fuel line

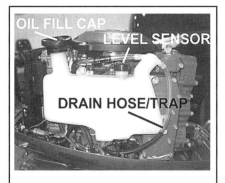

Fig. 44 Most Yamaha motors use a powerhead mounted oil reservoir...

Fig. 45 ... but many V6 motors utilize a boat mounted remote oil tank

Fig. 46 On some models there is an access cover in the engine top case

Fig. 47 Check the oil reservoir trap for water or contaminants before EVERY outing

Fig. 48 Boats with integrated fuel tanks will usually be rigged with an automotive style spin-on fuel filter/water separator

MAINTENANCE & TUNE-UP

from the tank to the fuel pump and then from the pump to the carburetor(s) or EFI/HPDI fuel vapor separator tank. Most Yamaha outboards utilize a serviceable filter element that is placed inside a cap and housing found inline just before the fuel pump. However, some models may instead (or ALSO in the case of HPDI models) be equipped with a non-serviceable inline filter in a disposable housing which is replaced by simply removing the clamps, disconnecting the hoses and installing a new filter. When installing a new disposable inline filter, make sure the arrow on the filter points in the direction of fuel flow.

Some motors have a fuel filter mounted in the fuel tank itself. On small motors with integral fuel tanks (such as the 2-5 hp motors) the only fuel filter element is normally mounted on the fuel petcock or the fuel outlet fitting which must be removed from the tank in order to service it. For larger motors, in addition to the fuel filter mounted on the engine, a filter is usually found inside or near the fuel tank. Because of the large variety of differences in both portable and fixed fuel tanks, it is impossible to give a detailed procedure for removal and installation. However, keep in mind that most in-tank filters are simply a screen on the pickup line inside the fuel tank. Filters of this type rarely require service or attention, but if the tank is removed for cleaning the filter will usually only need to be cleaned and returned to service (assuming they are not torn or otherwise damaged).

■ **Most EFI/HPDI motors are equipped with additional fuel filter elements either on the high-pressure electric fuel pump or the pressure regulator. However these elements are not maintenance items and should only need attention during repair or replacement of these components.**

Depending upon the boat rigging a fuel filter/water separator may be found inline between a boat mounted fuel tank and the motor. Boat mounted fuel filter/water separators are normally of the spin-on filter type (resembling automotive oil filters) but will vary greatly with boat rigging. Replacement of filters/separators is the same as a typical automotive oil filter replacement, just make sure to have a small drain basin handy to catch escaping fuel and make sure to coat the rubber gasket with a small dab of engine oil during installation.

FUEL FILTER SERVICE

✴✴ CAUTION
Observe all applicable safety precautions when working around fuel. Whenever servicing the fuel system, always work in a well-ventilated area. Do not allow fuel spray or vapors to come in contact with a spark or open flame. Do not smoke while working around gasoline. Keep a dry chemical fire extinguisher near the work area. Always keep fuel in a container specifically designed for fuel storage; also, always properly seal fuel containers to avoid the possibility of fire or explosion.

Integral Fuel Tank Models

A small serviceable filter element is usually found on the fuel petcock or fuel tank outlet fitting on integral tank models. This should include the 2-5 hp motors. To inspect, clean and/or replace this filter element, proceed as follows:

1. Drain the fuel in the tank into a suitable container.
2. It is usually necessary or just plain easier to work on with the fuel tank removed from the powerhead. Remove the mounting bolts and reposition the tank as necessary to disconnect a fuel line. In some cases it is easier to leave the fuel line connected to the tank and instead disconnect it from the other (carburetor) end.
3. Loosen the fuel petcock or fuel outlet fitting clamp screw or nut.
4. Remove the petcock from the clamp and tank.
5. Clean the filter assembly in solvent and blow it dry with compressed air. If excessively dirty or contaminated with water, replace the filter.
6. Install the petcock on the clamp and tank then connect the hose.
7. Tighten the fuel petcock clamp screw or nut.
8. Reposition the fuel tank and secure using the mounting bolts.
9. Check the fuel filter installation for leakage.

Disposable Inline Filters

As noted earlier, some Yamaha outboards are be equipped with a disposable inline filter. Generally speaking a few of the motors (including the 6/8 hp and some of the fuel injected motors, specifically most HPDI models) are equipped with a disposable, inline filter. For all carbureted models, the filter is found inline just before the mechanical fuel pump. For HPDI model the non-serviceable filter is normally found inline just before the vapor separator tank.

This type of filter is a sealed canister type (usually plastic) and cannot be cleaned, so service is normally limited to replacement. Because of the relative ease and relatively low expense of a filter (when compared with the time and hassle of a carburetor overhaul) we encourage you to replace the filter at least annually. On HPDI models this filter is a second line of defense, as fuel must pass through a serviceable filter prior to reaching the vapor separator filter, therefore annual replacement may not be necessary, if you've kept up with maintaining the primary filter.

✴✴ CAUTION
Before servicing the high or medium pressure fuel circuits of the HPDI system, please refer to the information on Releasing Fuel System Pressure in the High Pressure Direct Injection System section. Failure to release system pressure before disconnecting fuel lines or fittings could result in extremely dangerous pressurized fuel spray which would create a significant danger of fire or explosion, as well as danger of injury from the spray itself.

When replacing the filter, release the hose clamps (on carbureted models they are usually equipped with spring-type clamps that are released by squeezing the tabs using a pair of pliers) and slide them back on the hose, past the raised portion of the filter inlet/outlet nipples. On HPDI models the clamps are usually crimped into position and must be carefully cut to release them (and obviously, replaced with new crimp style clamps once the filter is installed again).

Once a clamp is released, position a small drain pan or a shop towel under the filter and carefully pull the hose from the nipple. Allow any fuel remaining in the filter and fuel line to drain into the drain pan or catch fuel with the shop towel. Repeat on the other side, noting which fuel line connects to which portion of the filter (for assembly purposes). Inline filters are usually marked with an arrow indicating fuel flow. The arrow should point towards the fuel line that runs to the motor (not the fuel tank).

Before installation of the new filter, make sure the hoses are in good condition and not brittle, cracking and otherwise in need of replacement. During installation, be sure to fully seat the hoses, then place the clamps over the raised portions of the nipples to secure them. Spring clamps will weaken over time, so replace them if they've lost their tension. If wire ties or adjustable clamps were used, be careful not to overtighten the clamp. If the clamp cuts into the hose, it's too tight; loosen the clamp or cut the wire tie (as applicable) and start again.

✴✴ CAUTION
Before returning the outboard to service, use the primer bulb to pressurize the system and check the filter/fittings for leaks. On HPDI motors, refer to the section on High Pressure Direct Injection to determine how to properly pressurize the system and check for leaks.

Serviceable Canister-Type Inline Filters

◆ See Figure 50

It's probably safe to say that most Yamaha motors (except for the smallest of the motors) are equipped with a serviceable canister-type inline filter. These can be identified by their design and shape, which varies from the typical inline filter. A typical, disposable inline filter will have a simple round canister to which the fuel lines attach at either end. The serviceable inline filters used by Yamaha usually have 2 fuel inlets at the top of the filter housing and the serviceable element is located in a bowl or cap that is threaded up from underneath the housing. When servicing these filters it may be possible to access and unthread the bowl without disconnecting any hoses, but if so, you'll have to hold the housing steady to prevent stressing and damaging the fuel lines. However, on some motors access to the bowl is impossible without first disconnecting one or more fuel hoses and repositioning the assembly.

MAINTENANCE & TUNE-UP

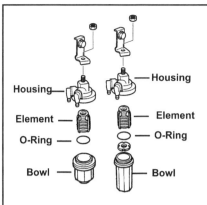

Fig. 50 Exploded view of a common serviceable Yamaha canister-type fuel filter

Fig. 51 Typical Yamaha serviceable canister-type filter assembly

Fig. 52 Access the element by unthreading the bowl from the housing

Serviceable filter elements should be cleaned carefully using solvent and inspected for clogging, tears or damage. If problems are found, normally just the element itself will require replacement. You really should always replace the O-ring, its cheap insurance, but you CAN reuse it in a pinch, as long as it is not torn, cut or deformed. While you're servicing the filter always inspect and replace any damaged hose or clamp as you would with any other fuel filter.

Inline Engines

◆ See Figures 50, 51 and 52

■ It is usually possible to service the filter element without removing the canister from the motor, however if access is tight, follow the steps for removing the hose(s) from the assembly so it can be repositioned.

Although most filter assemblies are pretty straight forward, containing a filter element and a single O-ring seal, some models may differ containing additional components such as a water indicator float, a spring, or additional seals. Keep track of the components as they are removed to ensure you will know how to properly assemble them.

1. If necessary for better access, remove the assembly from the powerhead as follows:

   a. On models with metal clips, slide each hose retaining clip off the filter assembly cover nipples with a pair of pliers. Disconnect the hoses from the cover and plug the hoses to prevent fuel leakage.

   b. On models with plastic clips, unsnap the plastic clips holding the hoses to the filter assembly cover. Disconnect the hoses from the cover and plug the hoses with golf tees to prevent fuel leakage.

   c. Remove the nut securing the filter cover to its mounting bracket. Remove the cover and canister assembly.

2. Unscrew the canister from the filter assembly cover. Remove the filter element from the canister.

3. Drain the canister and wipe the inside dry with a clean lint-free cloth or paper towel.

4. Inspect the cover O-ring. If necessary, remove and discard the old O-ring or gasket, as required.

To Install:

5. Clean the filter screen with solvent to remove any particles. If the filter screen is clogged or damaged, replace it.

6. Install the undamaged O-ring, or a new O-ring seal/gasket to the filter canister, as required.

7. Install the filter and the filter canister. Tighten the canister securely.

8. If the filter housing assembly was repositioned for access, install it as follows:

   a. Connect the inlet and outlet hoses to the canister cover.

   b. Install the nut securing the filter cover to its mounting bracket and tighten securely.

   c. On models with metal clips, slide each hose retaining clip onto the filter assembly cover nipples with a pair of pliers.

   d. On models with plastic clips, snap the plastic clips holding the hoses to the filter assembly cover.

9. Check the fuel filter installation for leakage by priming the fuel system with the fuel line primer bulb.

V4 and V6 Engines

◆ See Figures 53 and 54

Although most filter assemblies are pretty straight forward, containing a filter element and a single O-ring seal, some models may differ containing additional components such as a water indicator float, a spring, or additional seals. Keep track of the components as they are removed to ensure you will know how to properly assemble them.

The filter assembly found on V4 and V6 motors is very similar to the serviceable filter used on many inline engines, however the major difference

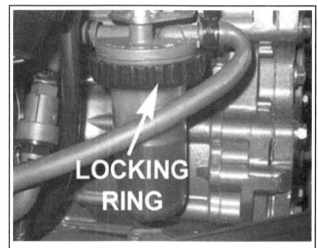

Fig. 53 Typical canister-type filter used on the V4 and V6 motors

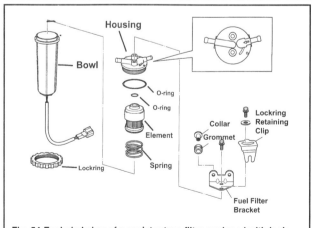

Fig. 54 Exploded view of a canister-type filter equipped with lockring and lock-ring retaining clip

2-18 MAINTENANCE & TUNE-UP

comes in the use of a lock-ring to secure the filter bowl. In addition some models (including most or all EFI/HPDI models) utilize a lock-ring retaining clip which bolts to the top of the housing to prevent the lock-ring from possibly loosening in service.

1. On EFI/HPDI motors, or models equipped with a lock-ring retaining clip, loosen the bolt and remove the retaining clip so the lock-ring can be loosened.
2. If equipped, unplug the bowl sensor harness connector.
3. Unscrew the ring nut securing the canister to the filter assembly cover.
4. Unscrew the canister from the filter assembly cover.
5. Remove the filter element and spring from the canister (along with any other components, such as a water indicator float).
6. Drain the canister and wipe the inside dry with a clean lint-free cloth or paper towel.
7. Inspect the cover O-ring. If necessary, remove and discard the old O-ring or gasket, as required.

To Install:

8. Clean the filter screen with solvent to remove any particles. If the filter screen is clogged or damaged, replace it.
9. Install the undamaged O-ring, or a new O-ring seal/gasket to the filter canister, as required.
10. Install the spring and filter into the filter canister (along with any other components as noted during removal).
11. Install the canister assembly onto the filter assembly cover and screw on the ring nut. Tighten the ring nut securely.
12. If equipped, connect the bowl wiring harness.
13. On EFI/HPDI motors or models so equipped, install the lock-ring retaining clip and secure using the fastener.
14. Check the fuel filter installation for leakage by priming the fuel system with the fuel line primer bulb.

Propeller

◆ See Figures 55, 56 and 57

As you know, the propeller is actually what moves the boat through the water. The propeller operates in water in much the same manner as a wood screw or auger passing through wood. The propeller "bites" into the water as it rotates. Water passes between the blades and out to the rear in the shape of a cone. This "biting" through the water is what propels the boat.

All Yamaha outboards are equipped, from the factory, with a through-the-propeller exhaust, meaning that exhaust gas is routed out through the propeller.

GENERAL PROPELLER INFORMATION

Diameter and Pitch

◆ See Figures 58 and 59

Only two dimensions of the propeller are of real interest to the boat owner: diameter and pitch. These two dimensions are stamped on the propeller hub and always appear in the same order, the diameter first and then the pitch. Propellers furnished with the outboard by Yamaha have a letter designation following the pitch size. This letter indicates the propeller type. For instance, the numbers and letter 9-7/8 x 10-1/2 - F stamped on the back of one blade indicates the propeller diameter to be 9-7/8 in., with a pitch of 10-1/2 in., and it is a Type F.

The diameter is the measured distance from the tip of one blade to the tip of the other.

The pitch of a propeller is the angle at which the blades are attached to the hub. This figure is expressed in inches of water travel for each revolution of the propeller. In our example of a 9-7/8 in. x 10-1/2 in., the propeller should travel 10-1/2 inches through the water each time it revolves. If the propeller action was perfect and there was no slippage, then the pitch multiplied by the propeller rpm would be the boat speed.

Most outboard manufacturers equip their units with a standard propeller, having a diameter and pitch they consider to be best suited to the engine and boat. Such a propeller allows the engine to run as near to the rated rpm and horsepower (at full throttle) as possible for the boat design.

The blade area of the propeller determines its load-carrying capacity. A two-blade propeller is used for high-speed running under very light loads.

A four-blade propeller is installed in boats intended to operate at low speeds under very heavy loads such as tugs, barges or large houseboats. The three-blade propeller is the happy medium covering the wide range between high performance units and load carrying workhorses.

Propeller Selection

There is no one propeller that will do the proper job in all cases. The list of sizes and weights of boats is almost endless. This fact, coupled with the many boat-engine combinations, makes the propeller selection for a specific purpose a difficult task. Actually, in many cases the propeller may be changed after a few test runs. Proper selection is aided through the use of charts set up for various engines and boats. These charts should be studied and understood when buying a propeller. However, bear in mind that the charts are based on average boats with average loads; therefore, it may be necessary to make a change in size or pitch, in order to obtain the desired results for the hull design or load condition.

Propellers are available with a wide range of pitch. Remember, a low pitch design takes a smaller bite of water than a high pitch propeller. This means the low pitch propeller will travel less distance through the water per

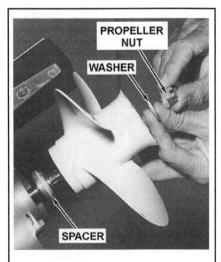

Fig. 55 Most Yamaha propellers are secured using a washer, castle nut and cotter pin

Fig. 56 An installed view of a nut retained propeller

Fig. 57 A severely damaged propeller (it's dead Jim)

MAINTENANCE & TUNE-UP 2-19

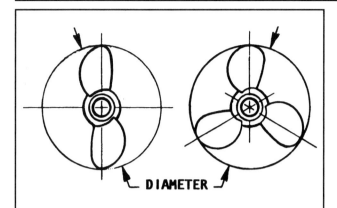

Fig. 58 Diameter and pitch are the two basic dimensions of a propeller. Diameter is measured across the circumference of a circle scribed by the propeller blades.

revolution. However, the low pitch will require less horsepower and will allow the engine to run faster.

All engine manufacturers design their units to operate with full throttle at, or in the upper end of the specified operating rpm. If the powerhead is operated at the rated rpm, several positive advantages will be gained.
- Spark plug life will be increased.
- Better fuel economy will be realized.
- Steering effort is often reduced.
- The boat and power unit will provide best performance.

Therefore, take time to make the proper propeller selection for the rated rpm of the engine at full throttle with what might be considered an average load. The boat will then be correctly balanced between engine and propeller throughout the entire speed range.

A reliable tachometer must be used to measure powerhead speed at full throttle, to ensure that the engine achieves full horsepower and operates efficiently and safely. To test for the correct propeller, make a test run in a body of smooth water with the lower unit in forward gear at full throttle. If the reading is above the manufacturer's recommended operating range, try propellers of greater pitch, until one is found allowing the powerhead to operate continually within the recommended full throttle range.

If the engine is unable to deliver top performance and the powerhead is properly tuned, then the propeller may not be to blame. Operating conditions have a marked effect on performance. For instance, an engine will lose rpm when run in very cold water. It will also lose rpm when run in salt water, as compared with fresh water. A hot, low-barometer day will also cause the engine to lose power.

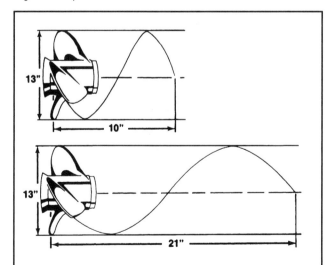

Fig. 59 This diagram illustrates the pitch dimension of a propeller. The pitch is the theoretical distance a propeller would travel through water if there were no friction

Cavitation

◆ See Figure 60

Cavitation is the forming of voids in the water just ahead of the propeller blades. Marine propulsion designers are constantly fighting the battle against the formation of these voids, due to excessive blade tip speed and engine wear. The voids may be filled with air or water vapor, or they may actually be a partial vacuum. Cavitation may be caused by installing a piece of equipment too close to the lower unit, such as the knot indicator pickup, depth sounder or bait tank pickup.

Vibration

The propeller should be checked regularly to ensure that all blades are in good condition. If any of the blades become bent or nicked, this condition will set up vibrations in the drive unit and motor. If the vibration becomes very serious, it will cause a loss of power, efficiency, and boat performance. If the vibration is allowed to continue over a period of time, it can have a damaging effect on many of the operating parts.

Vibration in boats can never be completely eliminated, but it can be reduced by keeping all parts in good working condition and through proper maintenance and lubrication. Vibration can also be reduced in some cases by increasing the number of blades. For this reason, many racers use two-blade propellers, while luxury cruisers have four- and five-blade propellers installed.

Shock Absorbers

◆ See Figure 61

The shock absorber in the propeller plays a very important role in protecting the shafting, gears and engine against the shock of a blow, should the propeller strike an underwater object. The shock absorber allows the propeller to stop rotating at the instant of impact, while the power train continues turning.

How much impact the propeller is able to withstand, before causing the shock absorber to slip, is calculated to be more than the force needed to propel the boat, but less than the amount that could damage any part of the power train. Under normal propulsion loads of moving the boat through the water, the hub will not slip. However, it will slip if the propeller strikes an object with a force that would be great enough to stop any part of the power train.

If the power train was to absorb an impact great enough to stop rotation, even for an instant, something would have to give, resulting in severe damage. If a propeller is subjected to repeated striking of underwater objects, it will eventually slip on its clutch hub under normal loads. If the propeller should start to slip, a new shock absorber/cushion hub will have to be installed by a propeller repair shop.

Propeller Rake

◆ See Figure 62

If a propeller blade is examined on a cut extending directly through the center of the hub, and if the blade is set vertical to the propeller hub, the propeller is said to have a zero degree (0°.) rake. As the blade slants back, the rake increases. Standard propellers have a rake angle from 0° to 15°.

A higher rake angle generally improves propeller performance in a cavitating or ventilating situation. On lighter, faster boats, a higher rake often will increase performance by holding the bow of the boat higher.

Progressive Pitch

◆ See Figure 63

Progressive pitch is a blade design innovation that improves performance when forward and rotational speed is high and/or the propeller breaks the surface of the water.

Progressive pitch starts low at the leading edge and progressively increases to the trailing edge. The average pitch over the entire blade is the number assigned to that propeller. In the illustration of the progressive pitch, the average pitch assigned to the propeller would be 21.

2-20 MAINTENANCE & TUNE-UP

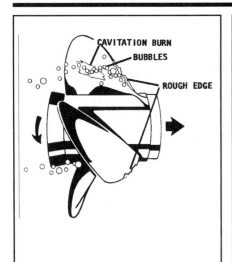

Fig. 60 Cavitation (air bubbles) can damage a prop

Fig. 61 A damaged rubber hub could cause the propeller to slip

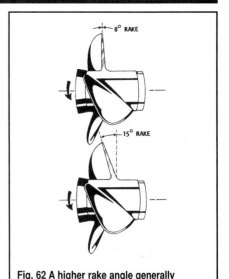

Fig. 62 A higher rake angle generally improves propeller performance

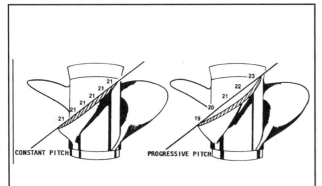

Fig. 63 Comparison of a constant and progressive pitch propeller. Notice how the pitch of the progressive propeller (right) changes to give the blade more thrust

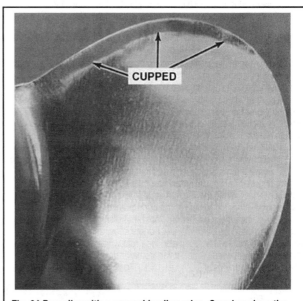

Fig. 64 Propeller with a cupped leading edge. Cupping gives the propeller a better hold in the water

Cupping

◆ See Figure 64

If the propeller is cast with an edge curl inward on the trailing edge, the blade is said to have a cup. In most cases, cupped blades improve performance. The cup helps the blades to HOLD and not break loose, when operating in a cavitating or ventilating situation.

A cup has the effect of adding to the propeller pitch. Cupping usually will reduce full-throttle engine speed about 150 to 300 rpm below that of the engine equipped with the same pitch propeller without a cup to the blade. A propeller repair shop is able to increase or decrease the cup on the blades. This change, as explained, will alter powerhead rpm to meet specific operating demands. Cups are rapidly becoming standard on propellers.

In order for a cup to be the most effective, the cup should be completely concave (hollowed) and finished with a sharp corner. If the cup has any convex rounding, the effectiveness of the cup will be reduced.

Rotation

◆ See Figure 65

Propellers are manufactured as right-hand (RH) rotation or left-hand (LH) rotation. The standard propeller for outboard units is RH rotation.

A right-hand propeller can easily be identified by observing it. Observe how the blade of the right-hand propeller slants from the lower left to upper right. The left-hand propeller slants in the opposite direction, from lower right to upper left.

When the RH propeller is observed rotating from astern the boat, it will be rotating clockwise when the outboard unit is in forward gear. The left-hand propeller will rotate counterclockwise.

High Performance Propellers

◆ See Figures 66 and 67

The term high performance is usually associated with, or has the connotation of, something used only for racing. The Yamaha high performance propeller does not fit this category and is not considered an aftermarket item.

The Yamaha high performance propeller is made of stainless steel with sophisticated designed blades, and carries an embossed P for positive identification.

The accompanying illustration of a high performance propeller clearly shows the unique design of the long blades and other features. This propeller is the weed less type, having extra sharp blades.

Installation of a high performance propeller requires raising the transom height, trim out after planing, and installation of a special design trim tab. Installation of a water pressure gauge is highly recommended, because raising the transom height will affect the amount of water entering the lower unit through the intake holes. An inadequate amount of water taken in will certainly cause powerhead cooling problems.

The high performance propeller has standard attaching hardware with the exception of the inner spacer, which is a special three-pronged design.

MAINTENANCE & TUNE-UP 2-21

INSPECTION

◆ See Figures 68, 69 and 70

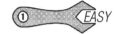

The propeller should be inspected before and after each use to be sure the blades are in good condition. If any of the blades become bent or nicked, this condition will set up vibrations in the motor. Remove and inspect the propeller. Use a file to trim nicks and burrs. Take care not to remove any more material than is absolutely necessary.

✷✷ CAUTION

Never run the engine with serious propeller damage, as it can allow for excessive engine speed and/or vibration that can damage the motor. Also, a damaged propeller will cause a reduction in boat performance and handling.

Also, check the rubber and splines inside the propeller hub for damage. If there is damage to either of these, take the propeller to your local marine dealer or a "prop shop". They can evaluate the damaged propeller and determine if it can be saved by re-hubbing.

Additionally, the propeller should be removed AT LEAST every 100 hours of operation or at the end of each season, whichever comes first for cleaning, greasing and inspection. Whenever the propeller is removed, apply a fresh coating of Yamaha All-Purpose Marine Grease or an equivalent water-resistant grease to the propeller shaft and the inner diameter of the propeller hub. This is necessary to prevent possible propeller seizure onto the shaft that could lead to costly or troublesome repairs. Also, whenever the propeller is removed, any material entangled behind the propeller should be removed before any damage to the shaft and seals can occur. This may seem like a waste of time at first, but the small amount of time involved in removing the propeller is returned many times by reduced maintenance and repair, including the replacement of expensive parts.

■ Propeller shaft greasing and debris inspection should occur more often depending upon motor usage. Frequent use in salt, brackish or polluted waters would make it advisable to perform greasing more often. Similarly, frequent use in areas with heavy marine vegetation, debris or potential fishing line would necessitate more frequent removal of the propeller to ensure the gearcase seals are not in danger of becoming cut.

REMOVAL & INSTALLATION

The propeller is secured to the gearcase propshaft either by a drive pin on only the smallest of the smallest portable motors, or by a castellated hex nut on all other Yamahas. It is pretty easy to tell the difference, just take a look at the propeller, if you see a nut, that's how it is secured. If however, you're working on a 2 hp motor, you should see a cotter pin going through the nose cone of the propeller itself, telling you that it is a shear pin prop.

For models secured by a hex nut, the propeller is driven by a splined connection to the shaft and the rubber drive hub found inside the propeller. The rubber hub provides a cushioning that allows softer shifts, but more importantly, it provides some measure of protection for the gearcase components in the event of an impact. On 2 hp motors, where the propeller is retained by a drive pin, impact protection is provided by the drive pin itself. The pin is designed to break or shear when a specific amount of force is applied because the propeller hits something. In both cases (rubber hubs or shear pins) the amount of force necessary to break the hub or shear the pin is supposed to be just less than the amount of force necessary to cause gearcase component damage. In this way, the hope is that the propeller and hub or shear pin will be sacrificed in the event of a collision, but the more expensive gearcase components will survive unharmed. Although these systems do supply a measure of protection, this, unfortunately, is not always the case and gearcase component damage will still occur with the right impact or with a sufficient amount of force.

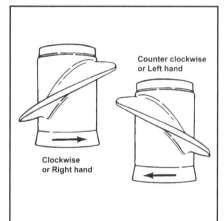

Fig. 65 Note the blade angle is reversed on right and left-hand propellers

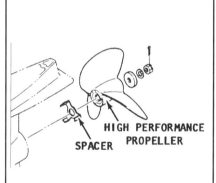

Fig. 66 Exploded view of a high performance prop mounting (note 3-prong spacer)

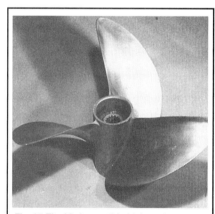

Fig. 67 The blades on this high performance prop are narrow and have more pitch than a standard prop

Fig. 68 This propeller is long overdue for repair or replacement

Fig. 69 Although minor damage can be dressed with a file. . .

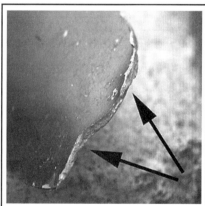

Fig. 70 . . . a propeller specialist should repair large nicks or damage

2-22 MAINTENANCE & TUNE-UP

✴✴ WARNING

Do not use excessive force when removing the propeller from the hub as excessive force can result in damage to the propeller, shaft and, even other gearcase components. If the propeller cannot be removed by normal means, consider having a reputable marine shop remove it. The use of heat or impacts to free the propeller will likely lead to damage.

■ Clean and lubricate the propeller and shaft splines using a high-quality, water-resistant, marine grease every time the propeller is removed from the shaft. This will help keep the hub from seizing to the shaft due to corrosion (which would require special tools to remove without damage to the shaft or gearcase.)

Many outboards are equipped with aftermarket propellers. Because of this, the attaching hardware may differ slightly from what is shown. Contact a reputable propeller shop or marine dealership for parts and information on other brands of propellers.

Shear Pin Props (2 Hp Motors)

◆ See Figures 71 and 72

■ On these models there is normally a small holder on the tiller handle that contains one or more extra drive pins and cotter pins. ALWAYS replace these pins when used on the water, that way you won't be stranded next time.

The propeller on the smallest of the portable Yamaha outboards (2 hp) is secured to the propshaft using a drive pin. Be sure to always keep a spare drive pin handy when you are onboard the boat. Remember that a sheared drive pin will leave you stranded on the water. A damaged shear pin can also contribute to motor damage, exposing it to over-revving while trying to produce thrust. The pin itself is usually locked in position by the propeller/cone that is in turn fastened by a cotter pin. ALWAYS replace the cotter pin once it has been removed. Remember that should the cotter pin fail, you could be diving to recover your propeller.

1. Disconnect the spark plug lead from the plug for safety.

✴✴ CAUTION

Don't ever take the risk of working around the propeller if the engine could accidentally be started. Always take precautions such as disconnecting the spark plug lead.

2. Cut the ends off the cotter pin using a pair of wire cutters (as that is easier than trying to straighten them in most cases) or straighten the ends using a pair of pliers (whichever you prefer). Next, free the pin by grabbing the head with a pair of needlenose pliers. Either tap on the pliers gently with a hammer to help free the pin from the propeller cap or carefully use the pliers as a lever by carefully prying back against the propeller cone. Discard the cotter pin once it is removed.

3. Grasp and gently pull the propeller off the drive pin and the propeller shaft.

4. Grasp and remove the drive pin using the needle-nose pliers.

■ If the drive pin is difficult to remove, use a small punch or a new drive pin as a driver and gently tap the pin free from the shaft.

To Install:

5. Clean the propeller hub and shaft splines, then apply a fresh coating of a water-resistant, marine grease.

6. Insert the drive pin into the propeller shaft.

7. Align the propeller, then carefully slide it over the shaft.

8. Install a new cotter pin and then spread the pin ends in order to form tension and secure them.

9. Reconnect the spark plug lead.

Castellated Nut Props

◆ See Figures 73 thru 81

On almost all Yamaha outboards the propeller is held in place over the shaft splines by a large castellated nut. The nut is so named because, when viewed from the side, it appears similar to the upper walls or tower of a castle.

For safety, the nut is locked in place by a cotter pin that keeps it from loosening while the motor is running. The pin passes through a hole in the propeller shaft, as well as through the notches in the sides of the castellated nut. Install a new cotter pin anytime the propeller is removed and, perhaps more importantly, make sure the cotter pin is of the correct size and is made of materials designed for marine use.

Whenever working around the propeller, check for the presence of black rubber material in the drive hub (don't confuse bits of black soot/carbon deposits from the exhaust as rubber hub material) and spline grease. Presence of this material normally indicates that the hub has turned inside the propeller bore (have the propeller checked by a propeller repair shop). Keep in mind that a spun hub will not allow proper torque transfer from the motor to the propeller and will allow the engine to over-rev in attempting to produce thrust. If the propeller has spun on the hub it has been weakened and is more likely to fail completely in use.

1. For safety, disconnect the negative cable (if so equipped) and/or disconnect the spark plug leads from the plugs (ground the leads to prevent possible ignition damage should the motor be cranked at some point before the leads are reconnected to the spark plugs).

✴✴ CAUTION

Don't ever take the risk of working around the propeller if the engine could accidentally be started. Always take precautions such as disconnecting the spark plug leads and, if equipped, the negative battery cable.

2. Cut the ends off the cotter pin using wire cutters (as that is usually easier than trying to straighten them in most cases) or straighten the ends of the pin using a pair of pliers, whichever you prefer. Next, free the pin by grabbing the head with a pair of needle-nose pliers. Either tap on the pliers gently with a hammer to help free the pin from the nut or carefully use the pliers as a lever by prying back against the castellated nut. Discard the cotter pin once it is removed.

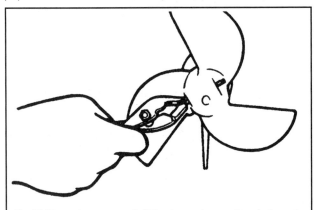

Fig. 71 Shear pin props are held in place using a cotter pin through the nose cone

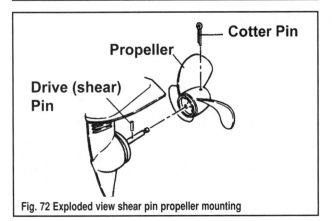

Fig. 72 Exploded view shear pin propeller mounting

MAINTENANCE & TUNE-UP

3. Place a block of wood between the propeller and the anti-ventilation housing to lock the propeller and shaft from turning, then loosen and remove the castellated nut. Note the orientation and then remove the washer and/or splined spacer from the propeller shaft.

4. Slide the propeller from the shaft. If the prop is stuck, use a block of wood to prevent damage and carefully drive the propeller from the shaft.

■ If the propeller is completely seized on the shaft, have a reputable marine or propeller shop free it. Don't risk damage to the propeller or gearcase by applying excessive force.

5. Note the direction in which the thrust washer is facing (since some motors or aftermarket props may use a thrust washer equipped with a fishing line trap that must face the proper direction if it is to protect the gearcase seal). Remove the thrust washer from the propshaft (if the washer appears stuck, tap lightly to free it from the propeller shaft).

6. On Vmax models with Twin Rotating Props (TRP), bend any teeth from the toothed washer away from the inner (forward) propeller retaining nut). Next, use a box-end wrench or open-end socket and a breaker bar (along with the same block of wood to hold the prop) to loosen the forward propeller retaining nut. Remove the nut, toothed washer, forward propeller and forward thrust washer from the shaft (also noting the orientation, the bevel normally faces forward).

7. Clean the thrust washer(s), propeller and shaft splines of any old grease. Small amounts of corrosion can be removed carefully using steel wool or fine grit sandpaper.

8. Inspect the shaft for signs of damage including twisted splines or excessively worn surfaces. Rotate the shaft while looking for any deflection.

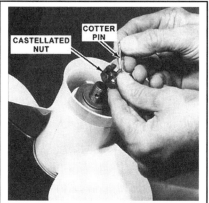
Fig. 73 Most Yamaha props are secured by a nut and cotter pin

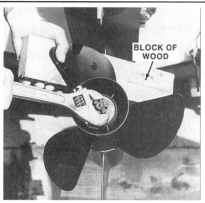
Fig. 74 To remove the nut use a block of wood to keep the prop from turning

Fig. 75 A stuck propeller can be freed with heat...

Fig. 76 ... but you'll likely destroy the rubber hub

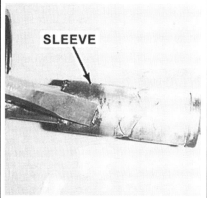
Fig. 77 If so, use a chisel to careful remove the frozen sleeve

Fig. 78 For installation position the inner spacer first...

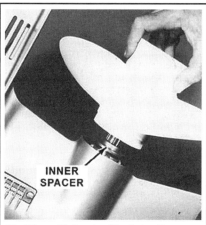
Fig. 79 .. with the shoulder facing the prop

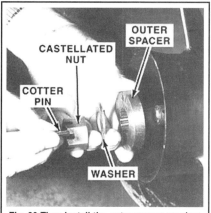
Fig. 80 Then install the outer spacer, washer (if used), nut and cotter pin

Fig. 81 Assembled and ready to go!

2-24 MAINTENANCE & TUNE-UP

Replace the propeller shaft if these conditions are found. Inspect the thrust washer for signs of excessive wear or cracks and replace, if found.

To Install:

9. Apply a fresh coating of Yamaha All-Purpose Marine Grease or an equivalent water-resistant grease to all surfaces of the propeller shaft and to the splines inside the propeller hub.

■ Some people prefer to use Anti-Seize on the hub splines, which is acceptable in most applications, but if used you should double-check after the first 10-20 hours of service to make sure the anti-seize is holding up to operating conditions.

10. On Vmax models with TRP, install the forward thrust washer (with the bevel facing forward as noted during removal), followed by the forward propeller, toothed washer and propeller nut. Tighten the nut to 47 ft. lbs. (65 Nm), then bend the teeth of the washer in position to secure the nut.

11. Position the thrust washer over the propshaft in the direction noted during removal. (Generally speaking, the flat shoulder should face rearward toward the propeller while the bevel faces forward toward the gearcase).

12. Carefully slide the propeller onto the propshaft, rotating the propeller to align the splines. Push the propeller forward until it seats against the thrust washer.

13. Install the splined and/or plain spacer onto the propeller shaft, as equipped.

14. Place a block of wood between the propeller and housing to hold the prop from turning, then thread the castellated nut onto the shaft with the cotter pin grooves facing outward.

15. Tighten the castellated nut to specification using a suitable torque wrench. Install a new cotter pin through the grooves in the nut that align with the hole in the propshaft. If the cotter pin hole and the grooves do not align, tighten (or loosen) the nut very slightly, just enough to align them. Once the cotter pin is inserted, spread the ends sufficiently to lock the pin in place. Propeller nut torque specifications are as follows:
- For 3 and 4/5 hp motors: there is no specification, snug the nut making sure the cotter pin holes are exposed
- 6-15 hp motors: 12.5 ft. lbs. (17 Nm)
- 20 hp and larger inline motors: 22-25 ft. lbs. (30-35 Nm)
- V4 and V6 motors: 40 ft. lbs. (55 Nm)

16. Connect the spark plug leads and/or the negative battery cable, as applicable.

Jet Drive Impeller

◆ See Figure 82

A jet drive motor uses an impeller enclosed in a jet drive housing instead of the propeller used by traditional gearcases. Outboard jet drives are designed to permit boating in areas prohibited to a boat equipped with a conventional propeller outboard drive system. The housing of the jet drive barely extends below the hull of the boat allowing passage in ankle deep water, white water rapids, and over sand bars or in shoal water which would foul a propeller drive.

The outboard jet drive provides reliable propulsion with a minimum of moving parts. It operates, simply stated, as water is drawn into the unit through an intake grille by an impeller. The impeller is driven by the driveshaft off the powerhead's crankshaft. Thrust is produced by the water that is expelled under pressure through an outlet nozzle that is directed away from the stern of the boat.

As the speed of the boat increases and reaches planing speed, only the very bottom of the jet drive where the intake grille is mounted facing downward remains in contact with the water.

The jet drive is provided with a reverse-gate arrangement and linkage to permit the boat to be operated in reverse. When the gate is moved downward over the exhaust nozzle, the pressure stream is deflected (reversed) by the gate and the boat moves sternward.

Conventional controls are used for powerhead speed, movement of the boat, shifting and power trim and tilt.

INSPECTION

◆ See Figure 83

The jet impeller is a precisely machined and dynamically balanced aluminum spiral. Close observation will reveal drilled recesses at exact

Fig. 82 Instead of a traditional gearcase, jet drives use an impeller (it's sort of an enclosed propeller) mounted in a jet drive housing that just barely extends below the boat's hull

locations used to achieve this delicate balancing. Excessive vibration of the jet drive may be attributed to an out-of-balance condition caused by the jet impeller being struck excessively by rocks, gravel or from damage caused by cavitation "burn".

The term cavitation "burn" is a common expression used throughout the world among people working with pumps, impeller blades, and forceful water movement. These "burns" occur on the jet impeller blades from cavitation air bubbles exploding with considerable force against the impeller blades. The edges of the blades may develop small dime-size areas resembling a porous sponge, as the aluminum is actually "eaten" by the condition just described.

Excessive rounding of the jet impeller edges will reduce efficiency and performance. Therefore, the impeller and intake grate (that protects it from debris) should be inspected at regular intervals.

Before and after each use, make a quick visual inspection of the intake grate and impeller, looking for obvious signs of damage. Always clear any debris such as plastic bags, vegetation or other items that sometimes become entangled in the water intake grate before starting the motor. If the intake grate is damaged, do not operate the motor, or you will risk destroying the impeller if rocks or other debris are drawn upward by the jet drive. If possible, replace a damaged grate before the next launch. This makes inspection after use all that much more important. Imagine the disappointment if you only learn of a damaged grate while inspecting the motor immediately prior to the next launch.

An obviously damaged impeller should be removed and either repaired or replaced depending on the extent of the damage. If rounding is detected, the impeller can be placed on a work bench and the edges restored to as sharp a condition as possible, using a file. Draw the file in only one direction. A back-and-forth motion will not produce a smooth edge. Take care not to nick the smooth surface of the jet impeller. Excessive nicking or pitting will create water turbulence and slow the flow of water through the pump. For more details on impeller replacement or service, please refer to the information on Jet Drives in the Gearcase section of this manual.

CHECKING IMPELLER CLEARANCE

◆ See Figure 84 and 85

Proper operation of the jet drive depends upon the ability to create maximum thrust. In order for this to occur the clearance between the outer edge of the jet drive impeller and the water intake housing cone wall should

MAINTENANCE & TUNE-UP 2-25

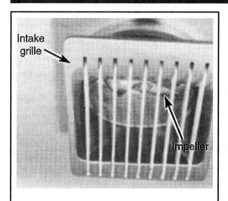

Fig. 83 Visually inspect the intake grate and impeller with each use

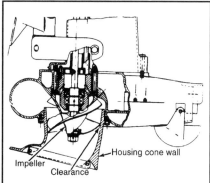

Fig. 84 Jet drive impeller clearance is the gap between the edges of the impeller and its housing

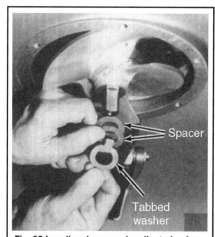

Fig. 85 Impeller clearance is adjusted using shims below and above the impeller

be maintained at approximately 1/32 in. (0.79mm). This distance can be checked visually by shining a flashlight up through the intake grille and estimating the distance between the impeller and the casing cone, as indicated in the accompanying illustrations. But, it is not humanly possible to accurately measure this clearance by eye. Close observation between outings is fine to maintain a general idea of impeller condition, but, at least annually, the clearance must be measured using a set of long feeler gauges.

✱✱ CAUTION

Whenever working around the impeller, ALWAYS disconnect the negative battery cable and/or disconnect the spark plug leads to make sure the engine cannot be accidentally started during service. Failure to heed this caution could result in serious personal injury or death in the event that the engine is started.

When checking clearance, a feeler gauge larger than the clearance specification should not fit between the tips of the impeller and the housing. A gauge within specification should fit, but with a slight drag. A smaller gauge should fit without any interference whatsoever. Check using the feeler gauge at various points around the housing, while slowly rotating the impeller by hand.

After continued normal use, the clearance will eventually increase. In anticipation of this the manufacturer mounts the impeller deep in a tapered housing, and positions spacers beneath the impeller to hold it in position. The spacers are used to position the impeller along the driveshaft with the desired clearance between the jet impeller and the housing wall. When clearance has increased, spacers are removed from underneath the impeller and repositioned behind it, dropping the impeller slightly in the housing and thereby decreasing the clearance again.

If adjustment is necessary, refer to the Jet Drive procedures under Gearcase in this manual for impeller removal, shimming and installation procedures. Follow the appropriate parts of the Removal & Disassembly, as well as Assembly procedures for impeller service.

Anodes (Zincs)

◆ See Figures 86, 87 and 88

The idea behind anodes (also known as sacrificial anodes) is simple: When dissimilar metals are dunked in water and a small electrical current is leaked between or amongst them, the less-noble metal (galvanically speaking) is sacrificed (corrodes).

The zinc alloy of which most anodes are made is designed to be less noble than the aluminum alloy of which your outboard is constructed. If there's any electrolysis, and there almost always is, the inexpensive zinc anodes are consumed in lieu of the expensive outboard motor.

■ **We say the zinc allow of which MOST anodes are made because the anodes recommended for fresh water applications may also be made of a different material. Some manufacturers recommend that you use magnesium anodes for motors used solely in freshwater applications. Magnesium offers a better protection against corrosion for aluminum motors, however the ability to offer this protection generally makes it too active a material for use in salt water applications. Because zinc is the more common material many people (including us) will use the word zinc interchangeably with the word anode, don't let this confuse you as we are referring equally to all anodes, whether they are made of zinc or magnesium.**

These zincs require a little attention in order to make sure they are capable of performing their function. Anodes must be solidly attached to a clean mounting site. Also, they must not be covered with any kind of paint, wax or marine growth.

Fig. 86 Extensive corrosion of an anode suggests a problem or a complete disregard for maintenance

Fig. 87 Although most Yamaha's use a trim tab anode, others are normally found on the gearcase and powerhead

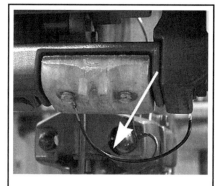

Fig. 88 Lead wires are used to protect bracketed components

MAINTENANCE & TUNE-UP

INSPECTION

◆ See Figures 86, 87 and 88

Visually inspect the anodes, especially gearcase mounted ones, before and after each use. You'll want to know right away if it has become loose or fallen off in service. Periodically inspect them closely to make sure they haven't eroded too much. At a certain point in the erosion process, the mounting holes start to enlarge, which is when the zinc might fall off. Obviously, once this happens your engine no longer has any protection. Generally, a zinc anode is considered worn if it has shrunken to 2/3 or less than the original size. To help judge this, buy a spare and keep it handy (in the boat or tow vehicle for comparison).

If you use your outboard in salt water or brackish water, and your zincs never seem to wear, inspect them carefully. Paint, wax or marine growth on zincs will insulate them and prevent them from performing their function properly. They must be left bare and must be installed onto bare metal of the motor. If the zincs are installed properly and not painted or waxed, inspect around them for sings of corrosion. If corrosion is found, strip it off immediately and repaint with a rust inhibiting paint. If in doubt, replace the zincs.

On the other hand, if your zinc seems to erode in no time at all, this may be a symptom of the zincs themselves. Each manufacturer uses a specific blend of metals in their zincs. If you are using zincs with the wrong blend of metals, they may erode more quickly or leave you with diminished protection.

At least annually or whenever an anode has been removed or replaced, check the mounting for proper electrical contact using a multi-meter. Set the multi-meter to check resistance (ohms), then connect one meter lead to the anode and the other to a good, unpainted or un-corroded ground on the motor. Resistance should be very low or zero. If resistance is high or infinite, the anode is insulated and cannot perform its function properly.

SERVICING

◆ See Figures 89 thru 97

Depending on your boat, motor and rigging, you may have anywhere from one to four (or even more) anodes. Regardless of the number, there are some fundamental rules to follow that will give your boat and motor's sacrificial anodes the ability to do the best job protecting your boat's underwater hardware that they can.

■ **On most models the trim tab serves as the gearcase anode. If the anode must be replaced, make an alignment mark between the anode and gearcase before removal, then transfer the mark to the replacement anode to preserve trim tab adjustment. If the boat pulls to one side after replacement and did not previously, refer to the Trim Tab Adjustment procedure in the Gearcase section to correct this condition.**

All motors covered by this manual are equipped with at least one gearcase anode, normally mounted in, on, or near the anti-ventilation plate. Many of the motors covered by this manual also have a powerhead mounted anode and/or an engine clamp bracket anode. Location of the powerhead zincs will vary slightly from motor-to-motor including mounting bosses specifically cast in the motor. Most multi-cylinder motors are equipped with one or more anodes on the engine mount clamp bracket.

■ **Many Yamaha motors are equipped with anodes in the water jackets surrounding the cylinders. These anodes can only be accessed by removing the cylinder head. Obviously this is not a common maintenance practice, but could easily be justified every 5 years or 500 hours of operation or so. It will give you the opportunity to check the pistons and combustion chambers for wear, scoring or carbon deposits as well.**

Fig. 89 Some trim tab anodes are bolted from below...

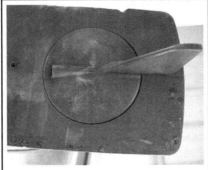

Fig. 90 ... while others are fastened from the top

Fig. 91 Remove the rubber cover...

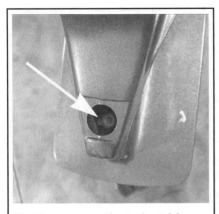

Fig. 92 ... to access the anode retaining bolt

Fig. 93 Many motors have one...

Fig. 94 ... or more anodes on the transom bracket

MAINTENANCE & TUNE-UP 2-27

Some people replace zincs annually. This may or may not be necessary, depending on the type of waters in which you boat and depending on whether or not the boat is hauled with each use or left in for the season. Either way, it is a good idea to remove zincs at least annually in order to make sure the mounting surfaces are still clean and free of corrosion.

The first thing to remember is that zincs are electrical components and like all electrical components, they require good clean connections. So after you've undone the mounting hardware you want to get the zinc mounting sites clean and shiny.

Get a piece of coarse emery cloth or some 80-grit sandpaper. Thoroughly rough up the areas where the zincs attach (there's often a bit of corrosion residue in these spots). Make sure to remove every trace of corrosion as it could insulate the zinc from the motor.

Zincs are attached with stainless steel machine screws that thread into the mounting for the zincs. Over the course of a season, this mounting hardware is inclined to loosen. Mount the zincs and tighten the mounting hardware securely. Tap the zincs with a hammer hitting the mounting screws squarely. This process tightens the zincs and allows the mounting hardware to become a bit loose in the process. Now, do the final tightening. This will insure your zincs stay put for the entire season.

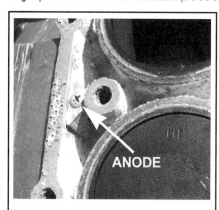

Fig. 95 Many Yamahas use anodes in cooling passages under the cylinder head...

Fig. 96 ... they are only accessible when the cylinder head/chamber cover is removed

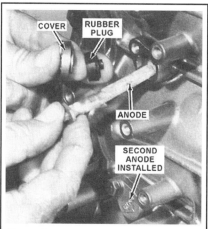

Fig. 97 Other Yamahas have anodes mounted under covers in small powerhead bores

BOAT MAINTENANCE

Batteries

◆ See Figures 98 and 99

Batteries require periodic servicing, so a definite maintenance program will help ensure extended life. A failure to maintain the battery in good order can prevent it from properly charging or properly performing its job even when fully charged. Low levels of electrolyte in the cells, loose or dirty cable connections at the battery terminals or possibly an excessively dirty battery top can all contribute to an improperly functioning battery. So battery maintenance, first and foremost, involves keeping the battery full of electrolyte, properly charged and keeping the casing/connections clean of corrosion or debris.

If a battery charges and tests satisfactorily but still fails to perform properly in service, one of three problems could be the cause.

1. An accessory left on overnight or for a long period of time can discharge a battery.

2. Using more electrical power than the stator assembly or lighting coil can replace would slowly drain the battery during motor operation, resulting in an undercharged condition.

3. A defect in the charging system. A faulty stator assembly or lighting coil, defective regulator or rectifier or high resistance somewhere in the system could cause the battery to become undercharged.

■ For more information on marine batteries, please refer to BATTERY in the Ignition and Electrical Systems section.

MAINTENANCE

◆ See Figures 99 thru 102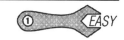

Electrolyte Level

The most common and important procedure in battery maintenance is checking the electrolyte level. On most batteries, this is accomplished by removing the cell caps and visually observing the level in the cells. The bottom of each cell normally is equipped with a split vent which will cause the surface of the electrolyte to appear distorted when it makes contact. When the distortion first appears at the bottom of the split vent, the electrolyte level is correct. Smaller marine batteries are sometimes equipped with translucent cases that are printed or embossed with high and low level markings on the side. On some of these, shining a flashlight through the battery case will help make it easier to determine the electrolyte level.

During hot weather and periods of heavy use, the electrolyte level should be checked more often than during normal operation. Add distilled water to bring the level of electrolyte in each cell to the proper level. Take care not to overfill, because adding an excessive amount of water will cause loss of electrolyte and any loss will result in poor performance, short battery life and will contribute quickly to corrosion.

Fig. 98 Explosive hydrogen gas is released from the batteries in a discharged state. This one exploded when something ignited the gas. Explosions can be caused by a spark from the battery terminals or jumper cables

2-28 MAINTENANCE & TUNE-UP

■ Never add electrolyte from another battery. Use only distilled water. Even tap water may contain minerals or additives that will promote corrosion on the battery plates, so distilled water is always the best solution.

Although less common in marine applications than other uses today, sealed maintenance-free batteries also require electrolyte level checks, through the window built into the tops of the cases. The problem for marine applications is the tendency for deep cycle use to cause electrolyte evaporation and electrolyte cannot be replenished in a sealed battery. Although, more and more companies are producing a maintenance-free batteries for marine applications and their success should be noted.

The second most important procedure in battery maintenance is periodically cleaning the battery terminals and case.

Cleaning

Dirt and corrosion should be cleaned from the battery as soon as it is discovered. Any accumulation of acid film or dirt will permit a small amount of current to flow between the terminals. Such a current flow will drain the battery over a period of time.

Clean the exterior of the battery with a solution of diluted ammonia or a paste made from baking soda and water. This is a base solution that will neutralize any acid that may be present. Flush the cleaning solution off with plenty of clean water.

※※ WARNING

Take care to prevent any of the neutralizing solution from entering the cells as it will quickly neutralize the electrolyte (ruining the battery).

Poor contact at the terminals will add resistance to the charging circuit. This resistance may cause the voltage regulator to register a fully charged battery and thus cut down on the stator assembly or lighting coil output adding to the low battery charge problem.

At least once a season, the battery terminals and cable clamps should be cleaned. Loosen the clamps and remove the cables, negative cable first. On batteries with top mounted posts, if the terminals appear stuck, use a puller specially made for this purpose to ensure the battery casing is not damaged. NEVER pry a terminal off a battery post. Battery terminal pullers are inexpensive and available in most parts stores.

Clean the cable clamps and the battery terminal with a wire brush until all corrosion, grease, etc., is removed and the metal is shiny. It is especially important to clean the inside of the clamp thoroughly (a wire brush or brush part of a battery post cleaning tool is useful here), since a small deposit of foreign material or oxidation there will prevent a sound electrical connection and inhibit either starting or charging. It is also a good idea to apply some dielectric grease to the terminal, as this will aid in the prevention of corrosion.

After the clamps and terminals are clean, reinstall the cables, negative cable last, do not hammer the clamps onto battery posts. Tighten the clamps securely but do not distort them. To help slow or prevent corrosion, give the clamps and terminals a thin external coating of grease after installation.

Check the cables at the same time that the terminals are cleaned. If the insulation is cracked or broken or if its end is frayed, that cable should be replaced with a new one of the same length and gauge.

TESTING

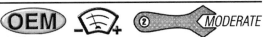

◆ See Figure 103

A quick check of the battery is to place a voltmeter across the terminals. Although this is by no means a clear indication, it gives you a starting point when trying to troubleshoot an electrical problem that could be battery related. Most marine batteries will be of the 12 volt DC variety. They are constructed of 6 cells, each of which is capable of producing slightly more than two volts, wired in series so that total voltage is 12 and a fraction. A fully charged battery will normally show more than 12 and slightly less than 13 volts across its terminals. But keep in mind that just because a battery reads 12.6 or 12.7 volts does NOT mean it is fully charged. It is possible for it to have only a surface charge with very little amperage behind it to maintain that voltage rating for long under load. A discharged battery will read some value less than 12 volts, but can normally be brought back to 12 volts through recharging. Of course a battery with one or more shorted or un-

Fig. 99 Ignoring a battery (and corrosion) to this extent is asking for it to fail

Fig. 100 Place a battery terminal tool over posts, then rotate back and forth . . .

Fig. 101 . . . until the internal brushes expose a fresh, clean surface on the post

Fig. 102 Clean the insides of cable ring terminals using the tool's wire brush

MAINTENANCE & TUNE-UP

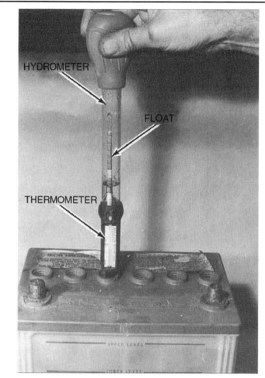

Fig. 103 A hydrometer is the best method for checking battery condition

chargeable cells will also read less than 12, but it cannot be brought back to 12+ volts after charging. For this reason, the best method to check battery condition on most marine batteries is through a specific gravity check or a load test.

A hydrometer is a device that measures the density of a liquid when compared to water (specific gravity). Hydrometers are used to test batteries by measuring the percentage of sulfuric acid in the battery electrolyte in terms of specific gravity. When the condition of the battery drops from fully charged to discharged, the acid is converted to water as electrons leave the solution and enter the plates, causing the specific gravity of the electrolyte to drop.

It may not be common knowledge but hydrometer floats are calibrated for use at 80°F (27°C). If the hydrometer is used at any other temperature, hotter or colder, a correction factor must be applied.

■ Remember, a liquid will expand if it is heated and will contract if cooled. Such expansion and contraction will cause a definite change in the specific gravity of the liquid, in this case the electrolyte.

A quality hydrometer will have a thermometer/temperature correction table in the lower portion, as illustrated in the accompanying illustration. By measuring the air temperature around the battery and from the table, a correction factor may be applied to the specific gravity reading of the hydrometer float. In this manner, an accurate determination may be made as to the condition of the battery.

When using a hydrometer, pay careful attention to the following points:
1. Never attempt to take a reading immediately after adding water to the battery. Allow at least 1/4 hour of charging at a high rate to thoroughly mix the electrolyte with the new water. This time will also allow for the necessary gases to be created.
2. Always be sure the hydrometer is clean inside and out as a precaution against contaminating the electrolyte.
3. If a thermometer is an integral part of the hydrometer, draw liquid into it several times to ensure the correct temperature before taking a reading.
4. Be sure to hold the hydrometer vertically and suck up liquid only until the float is free and floating.
5. Always hold the hydrometer at eye level and take the reading at the surface of the liquid with the float free and floating.
6. Disregard the slight curvature appearing where the liquid rises against the float stem. This phenomenon is due to surface tension.

7. Do not drop any of the battery fluid on the boat or on your clothing, because it is extremely caustic. Use water and baking soda to neutralize any battery liquid that does accidentally drop.
8. After drawing electrolyte from the battery cell until the float is barely free, note the level of the liquid inside the hydrometer. If the level is within the charged (usually green) band range for all cells, the condition of the battery is satisfactory. If the level is within the discharged (usually white) band for all cells, the battery is in fair condition.
9. If the level is within the green or white band for all cells except one, which registers in the red, the cell is shorted internally. No amount of charging will bring the battery back to satisfactory condition.
10. If the level in all cells is about the same, even if it falls in the red band, the battery may be recharged and returned to service. If the level fails to rise above the red band after charging, the only solution is to replace the battery.

■ An alternate way of testing a battery is to perform a load test using a special Carbon-Pile Load Tester. These days most automotive and many marine parts stores contain a tester and will perform the check for free (hoping that your battery will fail and they can sell you another). Essentially a load test involves placing a specified load (current drain/draw) on a fully-charged battery and checking to see how it performs/recovers. This is the only way to test the condition of a sealed maintenance-free battery.

STORAGE

If the boat is to be laid up (placed into storage) for the winter or anytime it is not going to be used for more than a few weeks, special attention must be given to the battery. This is necessary to prevent complete discharge and/or possible damage to the terminals and wiring. Before putting the boat in storage, disconnect and remove the batteries. Clean them thoroughly of any dirt or corrosion and then charge them to full specific gravity readings. After they are fully charged, store them in a clean cool dry place where they will not be damaged or knocked over, preferably on a couple blocks of wood. Storing the battery up off the deck, will permit air to circulate freely around and under the battery and will help to prevent condensation.

Never store the battery with anything on top of it or cover the battery in such a manner as to prevent air from circulating around the filler caps. All batteries, both new and old, will discharge during periods of storage, more so if they are hot than if they remain cool. Therefore, the electrolyte level and the specific gravity should be checked at regular intervals. A drop in the specific gravity reading is cause to charge them back to a full reading.

In cold climates, care should be exercised in selecting the battery storage area. A fully-charged battery will freeze at about 60°F below zero. The electrolyte of a discharged battery, almost dead, will begin forming ice at about 19°F above zero.

■ For more information on batteries and the engine electrical systems, please refer to the Ignition and Electrical section of this manual.

Fiberglass Hull

INSPECTION AND CARE

◆ See Figures 104, 105 and 106

Fiberglass reinforced plastic hulls are tough, durable and highly resistant to impact. However, like any other material they can be damaged. One of the advantages of this type of construction is the relative ease with which it may be repaired.

A fiberglass hull has almost no internal stresses. Therefore, when the hull is broken or stove-in, it retains its true form. It will not dent to take an out-of-shape set. When the hull sustains a severe blow, the impact will be either absorbed by deflection of the laminated panel or the blow will result in a definite, localized break. In addition to hull damage, bulkheads, stringers and other stiffening structures attached to the hull may also be affected and therefore, should be checked. Repairs are usually confined to the general area of the rupture.

2-30 MAINTENANCE & TUNE-UP

Fig. 104 The best way to care for a fiberglass hull is to wash it thoroughly

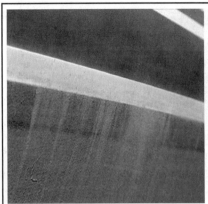

Fig. 105 If marine growth is a problem, apply a coating of anti-foul bottom paint

Fig. 106 Fiberglass, vinyl and rubber care products, like those from Meguiar's protect your boat

■ The best way to care for a fiberglass hull is to wash it thoroughly, immediately after hauling the boat while the hull is still wet. The next best way to care for your hull is to give it a waxing a couple of times per season. Your local marina or boat supply store should be able to help you find some high quality boat soaps and waxes.

A foul bottom can seriously affect boat performance. This is one reason why racers, large and small, both powerboat and sail, are constantly giving attention to the condition of the hull below the waterline.

In areas where marine growth is prevalent, a coating of vinyl, anti-fouling bottom paint should be applied if the boat is going to be left in the water for extended periods of time such as all or a large part of the season. If growth has developed on the bottom, it can be removed with a diluted solution of muriatic acid applied with a brush or swab and then rinsed with clear water. Always use rubber gloves when working with Muriatic acid and take extra care to keep it away from your face and hands. The fumes are toxic. Therefore, work in a well-ventilated area; or if outside, keep your face on the windward side of the work.

■ If marine growth is not too severe you may avoid the unpleasantness of working with muriatic acid by trying a power washer instead. Most marine vegetation can be removed by pressurized water and a little bit of scrubbing using a rough sponge (don't use anything that will scratch or damage the surface).

Barnacles have a nasty habit of making their home on the bottom of boats that have not been treated with anti-fouling paint. Actually they will not harm the fiberglass hull but can develop into a major nuisance.

If barnacles or other crustaceans have attached themselves to the hull, extra work will be required to bring the bottom back to a satisfactory condition. First, if practical, put the boat into a body of fresh water and allow it to remain for a few days. A large percentage of the growth can be removed in this manner. If this remedy is not possible, wash the bottom thoroughly with a high-pressure fresh water source and use a scraper. Small particles of hard shell may still hold fast. These can be removed with sandpaper.

Interior

INSPECTION AND CARE

No one wants to walk around in bare feet on a boat whose deck or carpet is covered in fish guts right? It's not just a safety hazard, it's kind of nasty. Taking time to wash down and clean your boat's interior is just as important to the long term value of your boat as it is to your enjoyment. So take time, after every outing to make sure your baby is clean on the inside too.

Always try to find gentle cleaners for your vinyl and plastic seats. Harsh chemicals and abrasives will do more harm then good. Take care with guests aboard, as more than one brand of sun-tan lotion has been know to cause stains. Some people get carried away, forbidding things like cheesy coated chips/snacks or mustards on board. Don't let keeping your boat clean so much of an obsession that you forget to enjoy it, just keep a bottle of cleaner handy for quick spill clean-ups. And keeping it handy will ensure you'll be more likely to wipe things down after a fun-filled outing.

Be sure to always test a cleaner on a hidden or unexposed area of your carpet or vinyl before soaking things down with it. If it does not harm or damage the color of your finish, you're good to go.

When we trailer our boats, we sometimes find it more convenient to hit a spray-it-yourself car wash on the way home. This gives us a chance to spray down the boat hull, trailer and tow vehicle before we get home and turn our attention to engine flushing and wiping down/cleaning the interior.

If you're lucky enough to have snap out marine carpet, remove it and give it a good wash down once in a while. This allows you to spray down the deck as well. Hang the carpet to dry and reinstall once it is ready. If you've got permanently installed marine carpet, you can spray it down too; just make sure you can give it a chance to dry before putting the cover back on.

■ For permanently installed marine carpet, try renting a rug steam-cleaner at least once a season and give it a good deep cleaning. We like to do it at the beginning and the end of each season!

TUNE-UP

Introduction to Tune-Ups

A proper tune-up is the key to long and trouble-free outboard life and the work can yield its own rewards. Studies have shown that a properly tuned and maintained outboard can achieve better fuel economy than an out-of-tune engine. As a conscientious boater, set aside a Saturday morning, say once a month, to check or replace items which could cause major problems later. Keep your own personal log to jot down which services you performed, how much the parts cost you, the date and the number of hours on the engine at the time. Keep all receipts for such items as oil and filters, so that they may be referred to in case of related problems or to determine operating expenses. These receipts are the only proof you have that the required maintenance was performed. In the event of a warranty problem on newer engines, these receipts will be invaluable.

The efficiency, reliability, fuel economy and enjoyment available from boating are all directly dependent on having your outboard tuned properly. The importance of performing service work in the proper sequence cannot be over emphasized. Before making any adjustments, check the specifications. Never rely on memory when making critical adjustments.

Before tuning any outboard, insure it has satisfactory compression. An outboard with worn or broken piston rings, burned pistons or scored cylinder walls, will not perform properly no matter how much time and expense is spent on the tune-up. Poor compression must be corrected or the tune-up will not give the desired results.

The extent of the engine tune-up is usually dependent on the time lapse since the last service. In this section, a logical sequence of tune-up steps will be presented in general terms. If additional information or detailed service work is required, refer to the section of this manual containing the appropriate instructions.

MAINTENANCE & TUNE-UP 2-31

Tune-Up Sequence

A tune-up can be defined as pre-determined series of procedures (adjustments, tests and worn component replacements) that are performed to bring the engine operating parameters back to original condition. The series of steps are important, as the later procedures (especially adjustments) are dependant upon the earlier procedures. In other words, a procedure is performed only when subsequent steps would not change the result of that procedure (this is mostly for adjustments or settings that would be incorrect after changing another part or setting). For instance, fouled or excessively worn spark plugs may affect engine idle. If adjustments were made to the idle speed or mixture of a carbureted engine **before** these plugs were cleaned or replaced, the idle speed or mixture might be wrong after replacing the plugs. The possibilities of such an effect become much greater when dealing with multiple adjustments such as timing, idle speed and/or idle mixture. Therefore, be sure to follow each of the steps given here. Since many of the steps listed here are full procedures in themselves, refer to the procedures of the same name in this section for details.

■ Computer controlled ignition and fuel components on more and more modern outboards have lessoned the amount of steps necessary for a pre-season tune-up, but not completely eliminated the need for replacing worn components. EFI and HPDI motors may not allow for many (or any) timing or mixture adjustments, however they still have mechanical and electrical components that wear making compression checks, spark plug/wire replacement, and component inspection steps all that much more important.

A complete pre-season tune-up should be performed at the beginning of each season or when the motor is removed from storage. Operating conditions, amount of use and the frequency of maintenance required by your motor may make one or more additional tune-ups necessary during the season. Perform additional tune-ups as use dictates.

1. Before starting, inspect the motor thoroughly for signs of obvious leaks, damage and loose or missing components. Make repairs, as necessary.

✱✱ CAUTION

We can't emphasize enough how important is this first step, ESPECIALLY on HPDI motors. The extreme high-pressure fuel system of these motors makes fuel system/line integrity a very important safety issue.

2. Check all accessible bolts and fasteners and tighten any that are loose.
3. For many decades, a standard part of the outboard tune-up procedure was to re-torque the cylinder head/cover bolts. Typically manufacturers advised you to follow the applicable steps of the cylinder the cylinder head/cover removal and installation procedures. Each of the bolts would be loosened slightly using the reverse of the tightening sequence, then re-torqued using one or more passes of the tightening sequence, as directed. Refer to the procedures under Powerhead for details. However, Yamaha is somewhat obscure about the need to do this on their modern outboards. Yamaha specifically mentions that you should NOT do this on some 4-stroke motors. But they do specifically mention that you SHOULD do this to the 3, 4/5 hp and 25/35 hp (3-cylinder), 50/60/70 hp (849cc) and 65J-90 hp (1140cc) motors. Also, they do not specifically include or exclude the cylinder head bolts from the maintenance requirement of checking/tightening ALL BOLTS AND NUTS listed in most of the service charts for all of their other motors. So we'll leave it to your judgment whether or not you want to include the cylinder head/cover bolts on other Yamaha motors, but it is probably not a bad idea to include it annually anyway (if in doubt, check with your local marine dealer).
4. Perform a compression check to make sure the motor is mechanically ready for a tune-up. An engine with low compression on one or more cylinder should be overhauled, not tuned. A tune-up will not be successful without sufficient engine compression. Refer to the Compression Testing in this section.

■ If this tune-up is occurring immediately after removing an engine from storage, be sure to start and run the motor using the old plugs first (if possible) while burning off the fogging oil. Then install the new spark plugs once the compression test is completed!

5. Since the spark plugs must be removed for the compression check, take the opportunity to inspect them thoroughly for signs of oil fouling, carbon fouling, damage due to detonation, etc. Clean and re-gap the plugs or, better yet, install new plugs as no amount of cleaning will precisely match the performance and life of new plugs. Refer to Spark Plugs, in this section. Also, this is a good time to check the spark plug wires as well. Please refer to Spark Plug Wires, in this section.

■ We don't care how little you use your motor, there is usually no excuse for not installing new plugs at the beginning of each season. If only because when storing a motor the fogging oil will go a long way to fouling even a decent set of spark plugs. Remember, the secondary ignition circuit is the most likely performance problem that occurs on an outboard.

6. Visually inspect all ignition system components for signs of obvious defects. Look for signs of burnt, cracked or broken insulation. Replace wires or components with obvious defects. If spark plug condition suggests weak or no spark on one or more cylinders, perform ignition system testing to eliminate possible worn or defective components. Refer to the Ignition System Inspection procedures in this section and the Ignition and Electrical System section.
7. Visually inspect all engine wiring and, if equipped, the battery and starter motor. A quick starter motor draw test can tell you a lot about the condition of your electrical starting system.
8. Remove and clean (on serviceable filters) or replace the inline filter and/or fuel pump filter, as equipped. Refer to the Fuel Filter procedures in this section. Perform a thorough inspection of the fuel system, hoses and components. Replace any cracked or deteriorating hoses. If carburetor adjustment or overhaul is necessary, perform these procedures before proceeding.
9. Pressurize the fuel system according to the procedures found in the Fuel System section, then check carefully for leaks. Again, this is important on all motors, but even more so on EFI motors and critical on HPDI engines!
10. Perform engine Timing and Synchronization adjustments as described in this section.

■ Although many of the motors covered here allow for certain ignition timing and, if applicable, carburetor adjustment procedures, none of them require the level of tuning attention that was once the norm. Many of the motors are equipped with electronic ignition systems that limit or eliminate timing adjustments. Carburetors used on many of these Yamahas are U.S. EPA regulated and contain few mixture adjustments. The air/fuel mixture is completely computer controlled on fuel injected motors and allows for no adjustment.

11. Except for jet drive models, remove the propeller in order to thoroughly check for leaks at the shaft seal. Inspect the propeller or impeller condition, look for nicks, cracks or other signs of damage and repair or replace, as necessary. If available, install a test wheel to run the motor in a test tank after completion of the tune-up. If no test wheel is available, lubricate the shaft/splines, then install the propeller or rotor. Refer to the procedures for Propeller or Jet Drive Impeller in this section, as applicable.
12. Change the gearcase oil as directed under the Gearcase Oil procedures in this section. If you are conducting a pre-season tune-up and the oil was changed immediately prior to storage this is not necessary. But, be sure to check the oil level and condition. Drain the oil anyway if significant contamination is present.

■ Anytime large amounts of water or debris is present in the gearcase oil, be sure to troubleshoot and repair the problem before returning the gearcase to service. The presence of water may indicate problems with the seals, while debris could a sign that overhaul is required.

13. Perform a test run of the engine to verify proper operation of the starting, fuel, oil and cooling systems. Although this can be performed using a flush/test adapter or even on the boat itself (if operating with a normal load/passengers), the preferred method is the use of a test tank. Keep in mind that proper operation without load and at low speed (on a flush/test adapter) doesn't really tell you how well the motor will run under load. If possible, run the engine, in a test tank using the appropriate test wheel. Monitor the cooling system indicator stream to ensure the water pump is working properly. Once the engine is fully warmed, slowly advance the engine to wide-open throttle, then note and record the maximum engine speed. Refer to the Tune-Up Specifications chart to compare engine speeds

2-32 MAINTENANCE & TUNE-UP

with the test propeller minimum rpm specifications. If engine speeds are below specifications, yet engine compression was sufficient at the beginning of this procedure, recheck the fuel and ignition system adjustments (as well as the condition of the prop if you're not using a test wheel).

Compression Testing

The quickest way to gauge the condition of an internal combustion engine is through a compression check. In order for an internal combustion engine to work properly, it must be able to generate sufficient compression in the combustion chamber to take advantage of the explosive force generated by the expanding gases after ignition.

If the combustion chambers or reed valves are worn or damaged in some fashion as to allow pressure to escape, the engine cannot develop sufficient horsepower. Under these circumstances, combustion will not occur properly meaning that the air/fuel mixture cannot be set to maximize power and minimize emissions. Obviously, it is useless to try to tune an engine with extremely low or erratic compression readings, since a simple tune-up will not cure the problem. An engine with poor compression on one or more cylinders should be overhauled.

The pressure created in the combustion chamber may be measured with a gauge that remains at the highest reading it measures during the action of a one-way valve. This gauge is inserted into the spark plug hole and held or threaded in position while the motor is cranked. A compression test will uncover many mechanical problems that can cause rough running or poor performance.

If the powerhead shows any indication of overheating, such as discolored or scorched paint, inspect the cylinders visually through the spark plug hole or transfer ports for possible scoring. It is possible for a cylinder with satisfactory compression to be scored slightly. Also, check the water pump, since a faulty water pump can cause an overheating condition.

TUNE-UP COMPRESSION CHECK

◆ See Figures 107, 108 and 109

■ A compression check requires a compression gauge and a spark plug port adapter that matches the plug threads of your motor.

When analyzing the results of a compression check, generally the actual amount of pressure measured during a compression check is not quite as important as the variation from cylinder-to-cylinder on the same motor. For multi-cylinder powerheads, Yamaha states that variations of up to 30 psi (207 kPa) may be considered normal. However, on single cylinder powerheads, a drop of about 15 psi (103 kPa) from the normal compression pressure you established when it was new is cause for concern (you did do a compression test on it when it was new, didn't you?).

■ On V-models the variations should appear between cylinders in a vertical bank, i.e. the top versus the middle or the top and middle versus the bottom cylinders. However, less of a variation should exist between the same cylinder in the port and starboard banks (i.e. the top on both banks, the middle on both banks or the bottom on both banks).

Ok, for the point of arguments sake let's say you bought the engine used or never checked compression the first season or so, assuming it wasn't something you needed to worry about. You're not alone, many of us have done that. But now that you're reading this it is your chance to take the data and note it for future.

For many years Yamaha did not publish much in the way of specific compression specifications for the amount of compression each of their engines should generate, a general rule of thumb that can be applied is that generally internal combustion engines should generate at least 100 psi (690 kPa). However, a quick check of the specifications that Yamaha HAS published to date shows a range of about 78 psi (556 kPa) to 153 psi (1079 kPa) on 2-stroke motors.

■ To see if Yamaha has published a specification for your particular engine, please check the Engine Specifications charts in the Powerhead section. If a specification is available it will be listed under Compression.

If a specification is not available for your motor, put the most weight on a comparison of the readings from the other cylinders on the same motor (or readings when the motor was new).

When taking readings during the compression check, repeat the procedure a few times for each cylinder, recording the highest reading for that cylinder. Then, for all multi-cylinder motors, the compression reading on the lowest cylinder should within 30 psi (207 kPa) of the highest reading (and still within specification if one is given in the Engine Specifications chart for that motor). If not, consider 15-30% to be the absolute limit. If the reading in the lowest cylinder is less than 70 % of the reading in the highest cylinder, it's time to find and remedy the cause.

■ If the powerhead has been in storage for an extended period, the piston rings may have relaxed. This will often lead to initially low and misleading readings. Always run an engine to normal operating temperature to ensure that the readings are accurate.

■ If you've never removed the spark plugs from this cylinder head before, break each one loose and retighten them before starting the motor in order to make sure they will not seize in the head once it is warmed. Better yet, remove each one and coat the threads very lightly with some fresh anti-seize compound.

Yamaha has modified the cylinder head design on some V4 and V6 models in an attempt to keep the temperature of each head the same. This action has changed the shape of the combustion chamber, and therefore the volume and compression pressure of each cylinder.

As a general rule, the pressure between pairs of cylinders which share the same crankshaft throw, should be approximately the same. Cylinder No. 1 should be the same as cylinder No. 2; cylinder No. 3 should be the same as cylinder No. 4; and so on.

Typically, on a V6 powerhead, cylinder No. 1 and No. 2 will have the highest compression pressure. Cylinder No. 5 and No. 6 will have the lowest compression pressure.

Prepare the engine for a compression test as follows:

1. Run the engine until it reaches operating temperature. The engine is at operating temperature a few minutes after the powerhead becomes warm to the touch and the stream of water exiting the cooling indicator becomes warm. If the test is performed on a cold engine, the readings will be considerably lower than normal, even if the engine is in perfect mechanical condition.

2. Label and disconnect the spark plug wires. Always grasp the molded cap and pull it loose with a twisting motion to prevent damage to the connection.

3. Clean all dirt and foreign material from around the spark plugs, and then remove all the plugs. Keep them in order by cylinder for later evaluation.

Fig. 107 Removing the high tension lead. Always use a twist and pull motion on the boot to prevent damage to the wire

MAINTENANCE & TUNE-UP 2-33

■ On most Yamaha motors you can disable the ignition system by leaving the safety lanyard disconnected but still use the starter motor to turn the motor. This is handy for things like compression tests or distributing fogging oil. To be certain use a spark plug gap tester on one lead and crank the motor using the keyswitch. If no spark is present, you're good to go, if not, you'll have to ground the spark plug leads to the cylinder head.

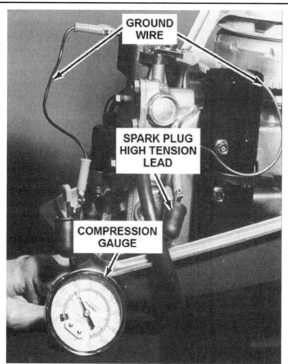

Fig. 108 The ignition must be disabled or all spark plugs must be grounded while making compression tests

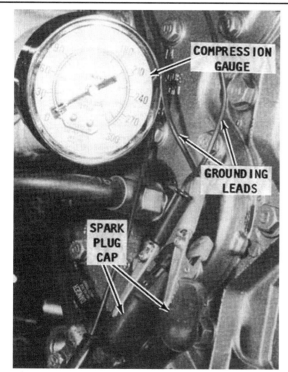

Fig. 109 The compression tester is threaded into an open spark plug port

4. Ground the spark plug leads to the engine to render the ignition system inoperative while performing the compression check.

✱✱ CAUTION

Grounding the spark plug leads not only protects the ignition system from potential damage that may be caused by the excessive load placed on operating the system with the wires disconnected, but more importantly, protects you from the dangers of arcing. The ignition system operates at extremely high voltage and could cause serious shocks. Also, keep in mind that you're cranking an engine with open spark plug ports which could allow any remaining fuel vapors to escape become ignited by arcing current.

5. Insert a compression gauge into the No. 1, top, spark plug opening.
6. Move the throttle to the wide open position in order to make sure the throttle plates are not restricting air flow. If necessary you may have to spin the propeller shaft slowly by hand while advancing the throttle in order to get the shifter into gear.
7. Crank the engine with the starter through at least 4-5 complete strokes with the throttle at the wide-open position, to obtain the highest possible reading. Record the reading.

■ On electric start motors, it is very important to use a freshly charged cranking battery as a weakened battery will cause a slower than normal cranking speed, reducing the compression reading.

8. Repeat the test and record the compression for each cylinder.
9. A variation between cylinders is far more important than the actual readings. A variation of more than 30 psi (207 kPa), between cylinders indicates the lower compression cylinder is defective. Not all engines will exhibit the same compression readings. In fact, two identical engines may not have the same compression. Generally, the rule of thumb is that the lowest cylinder should be within 15-30% of the highest (difference between the two readings).
10. If compression is low in one or more cylinders, the problem may be worn, broken, or sticking piston rings, scored pistons or worn cylinders.

LOW COMPRESSION

Compression readings that are generally low indicate worn, broken, or sticking piston rings, scored pistons or worn cylinders, and usually indicate an engine that has a lot of hours on it. Low compression in two adjacent cylinders (with normal compression in the other cylinders) indicates a blown head gasket between the low-reading cylinders. Other problems are possible (broken ring, hole burned in a piston), but a blown head gasket is most likely.

■ Use of an engine cleaner, available at any automotive parts house, will help to free stuck rings and to dissolve accumulated carbon. Follow the directions on the container.

To test a cylinder or motor further, add a few drops of engine oil to the cylinder and recheck.
• If compression is higher with oil added to the cylinder, suspect a worn or damaged piston.
• If compression is the same as without oil, suspect one or more defective rings or a defective piston, but you can also suspect oil seals or reed valves.
• If compression is well above specification, suspect the carbon deposits are on the cylinder head and/or piston crown. Significant carbon deposits will lead to pre-ignition and other performance problems and should be cleaned (either using an additive or by removing the cylinder head for manual cleaning).

Spark Plugs

◆ See Figure 110

The spark plug performs four main functions:
• First and foremost, it provides spark for the combustion process to occur.
• It also removes heat from the combustion chamber.
• Its removal provides access to the combustion chamber (for inspection or testing) through a hole in the cylinder head.

2-34 MAINTENANCE & TUNE-UP

- It acts as a dielectric insulator for the ignition system.

It is important to remember that spark plugs do not create heat, they help remove it. Anything that prevents a spark plug from removing the proper amount of heat can lead to pre-ignition, detonation, premature spark plug failure and even internal engine damage, especially in 2-stroke engines.

In the simplest of terms, the spark plug acts as the thermometer of the engine. Much like a doctor examining a patient, this "thermometer" can be used to effectively diagnose the amount of heat present in each combustion chamber.

Spark plugs are valuable tuning tools, when interpreted correctly. They will show symptoms of other problems and can reveal a great deal about the engine's overall condition. Evaluating the appearance of the spark plug's firing tip, gives visual cues to determine the engine's overall operating condition, get a feel for air/fuel ratios and even diagnose driveability problems.

As spark plugs grow older, they lose their sharp edges as material from the center and ground electrodes slowly erodes away. As the gap between these two points grows, the voltage required to bridge this gap increases proportionally. The ignition system must work harder to compensate for this higher voltage requirement and hence there is a greater rate of misfires or incomplete combustion cycles. Each misfire means lost horsepower, reduced fuel economy and higher emissions. Replacing worn out spark plugs with new ones (with sharp new edges) effectively restores the ignition system's efficiency and reduces the percentage of misfires, restoring power, economy and reducing emissions.

■ Although spark plugs can typically be cleaned and re-gapped if they are not excessively worn, no amount of cleaning or re-gapping will return most spark plugs to original condition and it is usually best to just go ahead and replace them.

How long spark plugs last will depend on a variety of factors, including engine compression, fuel used, gap, center/ground electrode material and the conditions in which the outboard is operated.

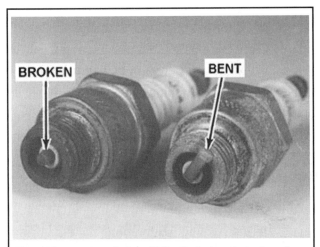

Fig. 110 Damaged spark plugs. Notice the broken electrode on the left plug. The electrode must be found and retrieved prior to returning the powerhead to service

SPARK PLUG HEAT RANGE

◆ See Figures 111, 112 and 113

Spark plug heat range is the ability of the plug to dissipate heat from the combustion chamber. The longer the insulator (or the farther it extends into the engine), the hotter the plug will operate; the shorter the insulator (the closer the electrode is to the engine's cooling passages) the cooler it will operate.

Selecting a spark plug with the proper heat range will ensure that the tip maintains a temperature high enough to prevent fouling, yet cool enough to prevent pre-ignition. A plug that absorbs little heat and remains too cool will quickly accumulate deposits of oil and carbon since it won't be able to burn them off. This leads to plug fouling and consequently to misfiring. A plug that

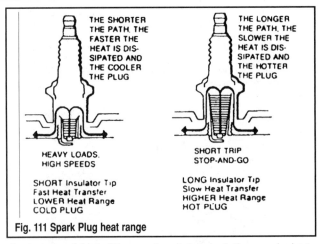

Fig. 111 Spark Plug heat range

absorbs too much heat will have no deposits but, due to the excessive heat, the electrodes will burn away quickly and might also lead to pre-ignition or other ignition problems.

Pre-ignition takes place when plug tips get so hot that they glow sufficiently to ignite the air/fuel mixture before the actual spark occurs. This early ignition will usually cause a pinging during heavy loads and if not corrected, will result in severe engine damage. While there are many other things that can cause pre-ignition, selecting the proper heat range spark plug will ensure that the spark plug itself is not a hot-spot source.

■ The manufacturer recommended spark plugs are listed in the Tune-Up Specifications chart. Recommendations may also be present on one or more labels affixed to your motor. When the label disagrees with the chart, we'd normally defer to the label as it may reflect a change that was made mid-production and not reflected in the Yamaha service literature.

REMOVAL & INSTALLATION

◆ See Figures 114 thru 119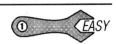

■ New technologies in spark plug and ignition system design have greatly extended spark plug life over the years. But, spark plug life will still vary greatly with engine tuning, condition and usage. In general, 2-stroke motors are a little tougher on plugs, especially if great care is not taken to maintain proper oil/fuel mixtures on pre-mix motors.

Typically spark plugs will require replacement once a season. The electrode on a new spark plug has a sharp edge but with use, this edge becomes rounded by wear, causing the plug gap to increase. As the gap increases, the plug's voltage requirement also increases. It requires a greater voltage to jump the wider gap and about two to three times as much voltage to fire a plug at high speeds than at idle.

■ Fouled plugs can cause hard-starting, engine mis-firing or other problems. You don't want that happening on the water. Take time, at least once a month to remove and inspect the spark plugs. Early signs of other tuning or mechanical problems may be found on the plugs that could save you from becoming stranded or even allow you to address a problem before it ruins the motor.

Tools needed for spark plug replacement include: a ratchet, short extension, spark plug socket (there are two types; either 13/16 inch or 5/8 inch, depending upon the type of plug), a combination spark plug gauge and gapping tool and a can of anti-seize type compound.

1. When removing spark plugs from multi-cylinder motors, work on one at a time. Don't start by removing the plug wires all at once, because unless you number them, they may become mixed up. Take a minute before you begin and number the wires with tape.
2. For safety, disconnect the negative battery cable or turn the battery switch **OFF**.
3. If the engine has been run recently, allow the engine to thoroughly cool (unless performing a compression check and then you should have already broken them loose once when cold and retightened them before

MAINTENANCE & TUNE-UP 2-35

Fig. 112 Many Yamahas have a label which lists spark plug type and gap

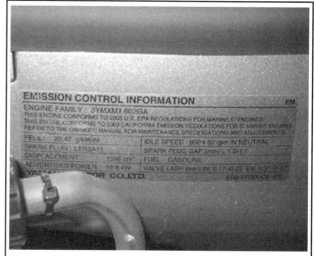

Fig. 113 Though these days, at least in the US, most have an Emission Control label which contains that information

warming the motor, so they should have less of a tendency to stick). Attempting to remove plugs from a hot cylinder head could cause the plugs to seize and damage the threads in the cylinder head, especially on aluminum heads!

■ To ensure an accurate reading during a compression check, the spark plugs must be removed from a hot engine. But, DO NOT force a plug if it feels like it is seized. Instead, wait until the engine has cooled, remove the plug and coat the threads lightly with anti-seize then reinstall and tighten the plug, then back off the tightened position a little less than 1/4 turn. With the plug(s) installed in this manner, re-warm the engine and conduct the compression check.

4. Carefully twist the spark plug wire boot to loosen it, then pull the boot using a twisting motion to remove it from the plug. Be sure to pull on the boot and not on the wire, otherwise the connector located inside the boot may become separated from the high-tension wire.

■ If removal is difficult (or on motors where the spark plug boot is hard to grip because of access) a spark plug wire removal tool is recommended as it will make removal easier and help prevent damage to the boot and wire assembly. Most tools have a wire loom that fits under the plug boot so the force of pulling upward is transmitted directly to the bottom of the boot.

Fig. 114 The spark plugs are threaded into the center of the cylinder cover

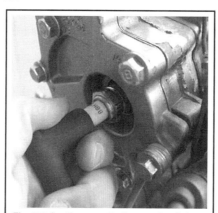

Fig. 115 Gently grasp the boot and pull the wire from the plug

Fig. 116 Then unthread the plug using a ratchet and socket

Fig. 117 ALWAYS thread plugs by hand to prevent cross-threading...

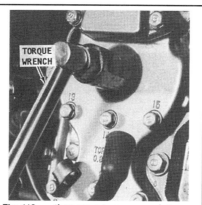

Fig. 118 ... then use a torque wrench to tighten the plug to spec

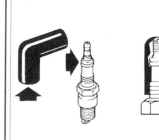

Fig. 119 To prevent corrosion, apply a small amount of grease to the plug and boot during installation

MAINTENANCE & TUNE-UP

5. Using compressed air (and safety glasses), blow debris from the spark plug area to assure that no harmful contaminants are allowed to enter the combustion chamber when the spark plug is removed. If compressed air is not available, use a rag or a brush to clean the area. Compressed air is available from both an air compressor or from compressed air in cans available at photography stores. In a pinch, blow up a balloon and use the escaping air to blow debris from the spark plug port(s).

■ **Remove the spark plugs when the engine is cold, if possible, to prevent damage to the threads. If plug removal is difficult, apply a few drops of penetrating oil to the area around the base of the plug and allow it a few minutes to work.**

6. Using a spark plug socket that is equipped with a rubber insert to properly hold the plug, turn the spark plug counterclockwise to loosen and remove the spark plug from the bore.

※※ WARNING

Avoid the use of a flexible extension on the socket. Use of a flexible extension may allow a shear force to be applied to the plug. A shear force could break the plug off in the cylinder head, leading to costly and/or frustrating repairs. In addition, be sure to support the ratchet with your other hand - this will also help prevent the socket from damaging the plug.

7. Evaluate each cylinder's performance by comparing the spark condition. Check each spark plug to be sure they are from the same plug manufacturer and have the same heat range rating. Inspect the threads in the spark plug opening of the block and clean the threads before installing the plug.
8. When purchasing new spark plugs, always ask the dealer if there has been a spark plug change for the engine being serviced. Sometimes manufacturers will update the type of spark plug used in an engine to offer better efficiency or performance.
9. Always use a new gasket (if applicable). The gasket must be fully compressed on clean seats to complete the heat transfer process and to provide a gas tight seal in the cylinder.
10. Inspect the spark plug boot for tears or damage. If a damaged boot is found, the spark plug boot and possibly the entire wire will need replacement.
11. Check the spark plug gap prior to installing the plug. Most spark plugs do not come gapped to the proper specification.
12. Apply a thin coating of anti-seize on the thread of the plug. This is extremely important on aluminum head engines to prevent corrosion and heat from seizing the plug in the threads (which could lead to a damaged cylinder head upon removal).
13. Carefully thread the plug into the bore by hand. If resistance is felt before the plug completely bottoms, back the plug out and begin threading again.

※※ WARNING

Do not use the spark plug socket to thread the plugs. Always carefully thread the plug by hand or using an old plug wire/boot to prevent the possibility of cross-threading and damaging the cylinder head bore. An old plug wire/boot can be used to thread the plug if you turn the wire by hand. Should the plug begin to cross-thread the wire will twist before the cylinder head would be damaged. This trick is useful when accessories or a deep cylinder head design prevents you from easily keeping fingers on the plug while it is threaded by hand.

14. Carefully tighten the spark plug to specification using a torque wrench, as follows:
   - 25/30 hp (496cc) 3-cylinder engines: 14 ft. lbs. (20 Nm)
   - All other engines: 18 ft. lbs. (25 Nm)

■ **Whenever possible, spark plugs should be tightened to the factory torque specification. If a torque wrench is not available, and the plug you are installing is equipped with a crush washer, tighten the plug until the washer seats, then tighten it an additional 1/4 turn to crush the washer.**

15. Apply a small amount of a silicone dielectric grease or Yamaha All-Purpose Marine grease to the ribbed, ceramic portion of the spark plug lead and inside the spark plug boot to prevent sticking, then install the boot to the spark plug and push until it clicks into place. The click may be felt or heard. Gently pull back on the boot to assure proper contact.

16. If applicable, connect the negative battery cable or turn the battery switch **ON**.
17. Test run the outboard (using a test tank or flush fitting) and insure proper operation.

READING SPARK PLUGS

◆ See Figures 120 thru 125

Reading spark plugs can be a valuable tuning aid. By examining the insulator firing nose color, you can determine much about the engine's overall operating condition.

In general, a light tan/gray color tells you that the spark plug is at the optimum temperature and that the engine is in good operating condition.

Dark coloring, such as heavy black wet or dry deposits usually indicate a fouling problem. Heavy, dry deposits can indicate an overly rich condition, too cold a heat range spark plug, possible vacuum leak, low compression, overly retarded timing or too large a plug gap.

If the deposits are wet, it can be an indication of a breached head gasket or an extremely rich condition, depending on what liquid is present at the firing tip.

Also look for signs of detonation, such as silver specs, black specs or melting or breakage at the firing tip.

Compare your plugs to the illustrations shown to identify the most common plug conditions.

Fouled Spark Plugs

A spark plug is "fouled" when the insulator nose at the firing tip becomes coated with a foreign substance, such as fuel, oil or carbon. This coating makes it easier for the voltage to follow along the insulator nose and leach back down into the metal shell, grounding out, rather than bridging the gap normally.

Fuel, oil and carbon fouling can all be caused by different things but in any case, once a spark plug is fouled, it will not provide voltage to the firing tip and that cylinder will not fire properly. In many cases, the spark plug cannot be cleaned sufficiently to restore normal operation. It is therefore recommended that fouled plugs be replaced.

Signs of fouling or excessive heat must be traced quickly to prevent further deterioration of performance and to prevent possible engine damage.

Overheated Spark Plugs

When a spark plug tip shows signs of melting or is broken, it usually means that excessive heat and/or detonation was present in that particular combustion chamber or that the spark plug was suffering from thermal shock.

Since spark plugs do not create heat by themselves, one must use this visual clue to track down the root cause of the problem. In any case, damaged firing tips most often indicate that cylinder pressures or temperatures were too high. Left unresolved, this condition usually results in more serious engine damage.

Detonation refers to a type of abnormal combustion that is usually preceded by pre-ignition. It is most often caused by a hot spot formed in the combustion chamber.

As air and fuel is drawn into the combustion chamber during the intake stroke, this hot spot will "pre-ignite" the air fuel mixture without any spark from the spark plugs.

Detonation

Detonation exerts a great deal of downward force on the pistons as they are being forced upward by the mechanical action of the connecting rods. When this occurs, the resulting concussion, shock waves and heat can be severe. Spark plug tips can be broken or melted and other internal engine components such as the pistons or connecting rods themselves can be damaged.

Left unresolved, engine damage is almost certain to occur, with the spark plug usually suffering the first signs of damage.

■ **When signs of detonation or pre-ignition are observed, they are symptom of another problem. You must determine and correct the situation that caused the hot spot to form in the first place.**

MAINTENANCE & TUNE-UP 2-37

Fig. 120 A normally worn spark plug should have light tan or gray deposits on the firing tip (electrode)

Fig. 121 A carbon-fouled plug, identified by soft, sooty black deposits, may indicate an improperly tuned powerhead

Fig. 122 This spark plug has been left in the powerhead too long, as evidenced by the extreme gap. Plugs with such an extreme gap can cause misfiring and stumbling accompanied by a noticeable lack of power

Fig.123 An oil-fouled spark plug indicates a powerhead with worn piston rings or a malfunctioning oil injection system that allows excessive oil to enter the combustion chamber

Fig. 124 A physically damaged spark plug may be evidence of severe detonation in that cylinder. Watch the cylinder carefully between services, as a continued detonation will not only damage the plug but will most likely damage the powerhead

Fig. 125 A bridged or almost bridged spark plug, identified by the build-up between the electrodes caused by excessive carbon or oil build up on the plug

INSPECTION & GAPPING

◆ See Figures 126 and 127

A particular spark plug might fit hundreds of powerheads and although the factory will typically set the gap to a pre-selected setting, this gap may not be the right one for your particular powerhead.

Insufficient spark plug gap can cause pre-ignition, detonation, even engine damage. Too much gap can result in a higher rate of misfires, noticeable loss of power, plug fouling and poor economy.

■ Refer to the Tune-Up Specifications chart for spark plug gaps.

Check spark plug gap before installation. The ground electrode (the L-shaped one connected to the body of the plug) must be parallel to the center electrode and the specified size wire gauge must pass between the electrodes with a slight drag.

Do not use a flat feeler gauge when measuring the gap on a used plug, because the reading may be inaccurate. A round-wire type gapping tool is the best way to check the gap. The correct gauge should pass through the electrode gap with a slight drag. If you're in doubt, try a wire that is one size

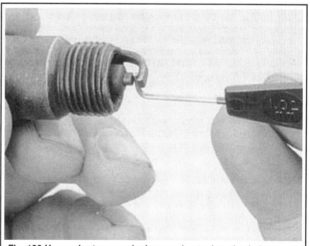
Fig. 126 Use a wire-type spark plug gapping tool to check the distance between center and ground electrodes

2-38 MAINTENANCE & TUNE-UP

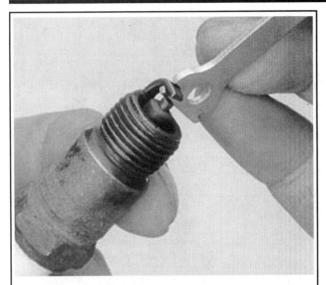

Fig. 127 Most plug gapping tools have an adjusting fitting used to bend the ground electrode

smaller and one larger. The smaller gauge should go through easily, while the larger one shouldn't go through at all.

Wire gapping tools usually have a bending tool attached. **USE IT!** This tool greatly reduces the chance of breaking off the electrode and is much more accurate. Never attempt to bend or move the center electrode. Also, be careful not to bend the side electrode too far or too often as it may weaken and break off within the engine, requiring removal of the cylinder head to retrieve it.

Spark Plug Wires

TESTING

Each time you remove the engine cover, visually inspect the spark plug wires for burns, cuts or breaks in the insulation. Check the boots on the coil and at the spark plug end. Replace any wire that is damaged.

Once a year, usually when you change your spark plugs, check the resistance of the spark plug wires with an ohmmeter. Wires with excessive resistance will cause misfiring and may make the engine difficult to start. In addition worn wires will allow arcing and misfiring in humid conditions.

Remove the spark plug wire from the engine. Test the wires by connecting one lead of the ohmmeter to the coil end of the wire and the other lead to the spark plug end of the wire. Resistance should measure approximately 7000 ohms per foot of wire.

■ **Keep in mind that just because a spark plug wire passes a resistance test doesn't mean that it is in good shape. Cracked or deteriorated insulation will allow the circuit to misfire under load, especially when wet. Always visually check wires to cuts, cracks or breaks in the insulation. If found, run the engine in a test tank or on a flush device either at night (looking for a bluish glow from the wires that would indicate arcing) or while spraying water on them while listening for an engine stumble.**

Yamaha only publishes specifications for resistance of the spark plug caps used on the 80J-140 hp V4 motors. These models contain resistor caps on the ends of the spark plug leads, the caps should have 4000-6000 ohms resistance.

■ **Yamaha technical literature mentions that some other models (like the 2 hp) use spark plug resistor caps, however we could not locate any testing specifications other than those listed above.**

Regardless of resistance tests and visual checks, it is never a bad idea to replace spark plug leads at least every couple of years, and to keep the old ones around for spares. Think of spark plug wires as a relatively low cost item that whose replacement can also be considered maintenance.

REMOVAL & INSTALLATION

When installing a new set of spark plug wires, replace the wires one at a time so there will be no confusion. Coat the inside of the boots with a dielectric grease or Yamaha All-Purpose Marine grease to prevent sticking and corrosion. Install the boot firmly over the spark plug until it clicks into place. The click may be felt or heard. Gently pull back on the boot to assure proper contact. Repeat the process for each wire.

■ **It is vitally important to route the new spark plug wire the same as the original and install it in a similar manner on the powerhead. Improper routing of spark plug wires may cause powerhead performance problems.**

Electronic (CDI/CDI Micro/TCI/TCI Micro) Ignition Systems

INSPECTION

◆ See Figure 128

Modern electronic ignition systems have become one of the most reliable components on an outboard. There is very little maintenance involved in the operation of these ignition systems and even less to repair if they fail. Most systems are sealed and there is no option other than to replace failed components.

1. Just as a tune-up is pointless on an engine with no compression, installing new spark plugs will not do much for an engine with a damaged ignition system. At each tune-up, visually inspect all ignition system components for signs of obvious defects. Look for signs of burnt, cracked or broken insulation. Replace wires or components with obvious defects. If spark plug condition suggests weak or no spark on one or more cylinders, perform ignition system testing to eliminate possible worn or defective components.

If trouble is suspected, it is very important to narrow down the problem to the ignition system and replace the correct components rather than just replace parts hoping to solve the problem. Electronic components can be very expensive and are usually not returnable.

Refer to the "Ignition and Electrical Systems" section for more information on troubleshooting and repairing ignition systems.

These days, all Yamaha outboards are equipped with some form of a Capacitor Discharge Ignition (CDI) System or Transistor Controlled Ignition (TCI) which are very similar systems. The only possible adjustment on most CDI/TCI systems would be timing (Idle, Carb Pickup and/or WOT) and that varies by engine/model. For more details, please refer to the Timing and Synchronization procedure for your motor.

Various engines are equipped with the Yamaha Microcomputer Ignition System (YMIS) or the TCI Micro system. YMIS is essentially a standard CDI type ignition with computer controls. The TCI Micro system is similar to the CDI Micro/YMIS system. In both cases, there are normally no adjustable components in this system, but for more details on a particular engine, please refer to Timing and Synchronization in this section.

Electrical System Checks

CHECKING THE BATTERY

Difficulty in starting accounts for almost half of the service required on boats each year. Some years ago, a survey by Champion Spark Plug Company indicated that roughly one third of all boat owners experienced a "won't start" condition in a given year. When an engine won't start, most people blame the battery when, in fact, it may be that the battery has run down in a futile attempt to start an engine with other problems.

Maintaining your battery in peak condition may be though of as either tune-up or maintenance material. Most wise boaters will consider it to be both. A complete check up of the electrical system in your boat at the beginning of the boating season is a wise move. Continued regular maintenance of the battery will ensure trouble free starting on the water.

MAINTENANCE & TUNE-UP 2-39

Details on battery service procedures are included under Batteries in Boat Maintenance. The following is a list of basic electrical system service procedures that should be performed as part of any tune-up.
- Check the battery for solid cable connections
- Check the battery and cables for signs of corrosion damage
- Check the battery case for damage or electrolyte leakage
- Check the electrolyte level in each cell
- Check to be sure the battery is fastened securely in position
- Check the battery's state of charge and charge as necessary
- Check battery voltage while cranking the starter. Voltage should remain above 9.5 volts
- Clean the battery, terminals and cables
- Coat the battery terminals with dielectric grease or terminal protector

CHECKING THE STARTER MOTOR

◆ See Figure 129

The starter motor system generally includes the battery, starter motor, solenoid, ignition switch and in most cases, a relay.

The frequency of starts governs how often the motor should be removed and reconditioned. The manufacturer recommends removal and overhaul every 1000 hours.

When checking the starter motor circuit during a tune-up, ensure the battery has the proper rating and is fully charged. Many starter motors are needlessly overhauled, when the battery is actually the culprit.

Connect one lead of a voltmeter to the positive terminal of the starter motor. Connect the other meter lead to a good ground on the engine. Check the battery voltage under load by turning the ignition switch to the **START** position and observing the voltmeter reading. If the reading is 9.5 volts or greater, and the starter motor fails to operate, repair or replace the starter motor.

CHECKING THE INTERNAL WIRING HARNESS

◆ See Figure 130

Corrosion is probably a boater's worst enemy. It is especially harmful to wiring harnesses and connectors. Small amounts of corrosion can cause havoc in an electrical system and make it appear as if major problems are present.

The following are a list of checks that should be performed as part of any tune-up.
- Perform a through visual check of all wiring harnesses and connectors on the vessel
- Check for frayed or chafed insulation, loose or corroded connections between wires and terminals
- Unplug all suspect connectors and check terminal pins to be sure they are not bent or broken, then lubricate and protect all terminal pins with dielectric grease to provide a water tight seal
- Check any suspect harness for continuity between the harness connection and terminal end. Repair any wire that shows no continuity (infinite resistance)

Fuel System Checks

FUEL INSPECTION

 EASY

Most often, tune-ups are performed at the beginning of the boating season. If the fuel system in your boat was properly winterized, there should be no problem with starting the outboard for the first time in the spring. If problems exist, perform the following checks.

1. If the condition of the fuel is in doubt, drain, clean, and fill the tank with fresh fuel.
2. Visually check all fuel lines for kinks, leaks, deterioration or other damage.

> ※※ **CAUTION**
>
> We cannot over-emphasize how important it is to visually check the fuel lines and fittings for any sign of leakage. This is even more important on the high-pressure fuel circuits of the EFI and especially the HPDI motors. It is a good idea to prime the all fuel systems using the primer bulb and, on EFI/HPDI systems, check the lines/fittings of the high-pressure fuel circuit after the motor has been operated (meaning the system has been under normal operating pressures).

3. Disconnect the fuel lines and blow them out with compressed air to dislodge any contamination or other foreign material.
4. Check the line between the fuel pump and the carburetor (or the vapor separator tank on fuel injected motors) while the powerhead is operating and the line between the fuel tank and the pump when the powerhead is not operating. A leak between the tank and the pump many times will not appear when the powerhead is operating, because the suction created by the pump drawing fuel will not allow the fuel to leak. Once the powerhead is shut down and the suction no longer exists, fuel may begin to leak.

> ※※ **CAUTION**
>
> DO NOT do anything around the fuel system without first reviewing the warnings and safety precautions in the Fuel System section. Remember that fuel is highly combustible and all potential sources of ignition from cigarettes and open flame to sparks must be kept far away from the work area.

LOW-PRESSURE FUEL PUMP INSPECTION

◆ See Figure 131    MODERATE

ALL Yamaha motors use a low-pressure fuel pump to draw fuel from a boat mounted or portable fuel tank and feed either the carburetor float bowls or fuel vapor separator tank of fuel injected motors. The low-pressure pumps

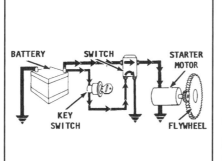

Fig. 129 Functional diagram of a typical cranking circuit

Fig. 128 Typical Yamaha ignition control unit, used on models equipped with the CDI Micro (YMIS) system

Fig. 130 Any time electrical gremlins are present, always check the harness connectors for pins which are bent, broken or corroded

MAINTENANCE & TUNE-UP

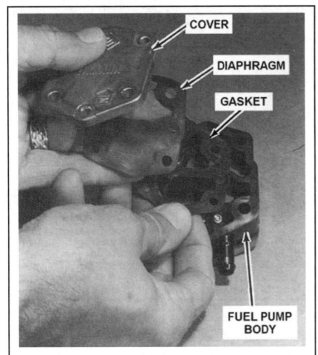

Fig. 131 Arrangement of the vacuum operated fuel pump parts. A tiny hole in the diaphragm can affect fuel pump performance

operate in virtually the same manner for all Yamaha motors, using a displacement-diaphragm design to alternately create a vacuum in the line from the fuel tank and generate pressure in the line going to the float bowl or vapor separator.

On carbureted motors, if the powerhead operates as if the load on the boat is being constantly increased and decreased, even though an attempt is being made to hold a constant powerhead speed, the problem can most likely be attributed to the fuel pump.

Many times, a defective fuel pump diaphragm is mistakenly diagnosed as a problem in the ignition system. The most common problem is a tiny pin-hole in the diaphragm or a bent check valve inside the fuel pump. Such a small hole will permit gas to enter the crankcase and wet foul the spark plug at idle-speed. During high-speed operation, gas quantity is limited, the plug is not fouled and will therefore fire in a satisfactory manner.

If the fuel pump fails to perform properly, an insufficient fuel supply will be delivered to the carburetor or the vapor separator tank. This lack of fuel will cause the engine to run lean, lose rpm or possibly even contribute to piston scoring.

Pressure Check

◆ See Figures 132 and 133

■ If an integral fuel pump carburetor is installed, the fuel pressure cannot be checked.

1. Mount the outboard unit in a test tank, or on the boat in a body of water.

✱✱ CAUTION

Never operate the engine at high speed with a flush device attached. The engine, operating at high speed with such a device attached, would runaway from lack of a load on the propeller, causing extensive damage.

2. Install the fuel pressure gauge in the fuel line between the fuel pump and the carburetor or the vapor separator tank.
3. Start the engine and check the fuel pressure.

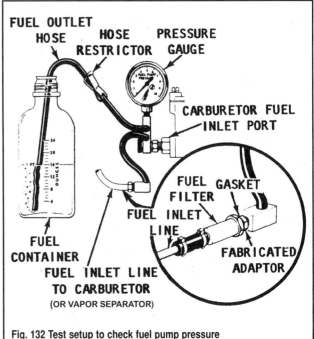

Fig. 132 Test setup to check fuel pump pressure

Fig. 133 Lack of adequate fuel, possibly caused by a defective fuel pump, caused the hole burned into the top of this piston

■ Remember, the powerhead will not start without the emergency tether in place behind the kill switch knob.

✱✱ CAUTION

Water must circulate through the lower unit to the engine any time the engine is run to prevent damage to the water pump in the lower unit. Just five seconds without water will damage the water pump.

4. Operate the powerhead at full throttle and check the pressure reading. The gauge should indicate at least 2 psi (14 kPa).

HIGH-PRESSURE FUEL PUMP INSPECTION

In addition to the all important visual inspection of the fuel system lines and fittings on outboards with fuel injection a quick check of the high-

MAINTENANCE & TUNE-UP 2-41

pressure fuel circuits will verify the ability of the system to operate properly. This check should be performed with each tune-up or at least annually at the beginning of the season.

Pressure Check

EFI engines utilize a 2-stage fuel system, the low-pressure circuit (fuel tank-to-pump and pump-to-vapor separator), as well as a high-pressure circuit (vapor separator-to-fuel injectors).

HPDI engines actually use a 3 stage fuel system, low-pressure (fuel tank-to-pump and pump-to-vapor separator), a first stage high-pressure (or medium pressure circuit, vapor separator-to-mechanical high-pressure pump) and second stage high-pressure (extreme high-pressure circuit, high-pressure pump-to-fuel injectors).

The low-pressure circuit of all fuel-injected motors works in the same manner as the low-pressure circuit used on carbureted models.

The high-pressure circuit of all fuel injected Yamaha motors works in a similar fashion using a submerged electric fuel pump located within the vapor separator tank to achieve fuel pressure somewhere in the 30-40 psi (207-278 kPa) range depending upon the year and model. Most models are equipped with a fuel test port on the top of the separator tank, but there are exceptions where the port is remote mounted.

The extreme high-pressure circuit of HPDI engines takes this already high-pressure fuel and achieves operating pressures up to 1000 psi (6895 kPa) on some direct injection motors.

EFI Motors

◆ See Figure 134

  1. Locate the fuel pressure check valve/test fitting on the motor. In most cases it will be a Schrader type valve fitting located on the top of the vapor separator tank (under a protective cap).
  2. Remove the protective cap from the pressure test fitting, then cover the fitting with a clean, dry shop rag (to protect against possible fuel spray) and connect a fuel pressure gauge to the test fitting/check valve.
  3. Provide a source of cooling water (using a test tank or flush fitting) and start the engine, allowing it to run at idle for about a minute. Observe and record the fuel pressure indicated on the gauge.
  4. The fuel pressure should be 35.6 psi (250 kPa) for OX66 motors.
  5. Shut the engine off and, making sure a shop rag is in place to catch any escaping fuel spray, carefully disconnect the gauge.
  6. If fuel pressure is below specification, refer to the Fuel System section for further diagnosis.

HPDI Motors

◆ See Figures 134 and 135

  1. Remove the protective cap form the fuel pressure check valve/test fitting (a Schrader type valve fitting located on the top of the vapor separator tank).
  2. Cover the fitting with a clean, dry shop rag (to protect against possible fuel spray) and connect a fuel pressure gauge to the test fitting/check valve.
  3. For 225/250 hp V6 engines, turn the ignition switch **ON** (without starting the motor) and observe the pressure. Within 5 seconds the electric pump should run generating a line pressure of about 50.8 psi (350 kPa), but then pressure should lower slightly to about 43.5 psi (300 kPa).
  4. Provide a source of cooling water (using a test tank or flush fitting) and start the engine, allowing it to run at idle for about a minute (on 225/250 hp motors, run the engine for about 5 minutes). Observe and record the fuel pressure indicated on the gauge.
  5. Shut the engine off and, making sure a shop rag is in place to catch any escaping fuel spray, carefully disconnect the gauge.
  6. The first stage high-pressure (medium pressure for these HPDI motors) fuel circuit should generate about 39.8-51.2 psi (280-360 kPa) while the engine is running on all except the 225/250 hp motors, on which pressure must be at least 50.8 psi (350 kPa).
  7. If fuel pressure is below specification, refer to the Fuel System section for further diagnosis.
  8. To check the extreme-high pressure circuit (2nd stage high-pressure fuel circuit) on 150-200 hp motors, proceed as follows:
    a. Remove the flywheel cover for access.
    b. Locate and disconnect the fuel pressure sensor wiring harness (it's mounted on the mechanical fuel pump assembly).
    c. Install the Yamaha 3-pin test harness (#YB-06769) or fabricate a test harness using 3 jumper leads to reconnect the pins in the disconnected harness. The key to fabricating a test harness is to make sure you can safely connect the 3 sets of pins without them shorting to each other or ground AND making sure that you can safely probe the sensor output voltage using a DVOM.
    d. Using the source of cooling water, start and run the engine at idle speed.
    e. Measure the fuel pressure sensor output voltage across the Pink and Black wires, it should be 2.8-3.2 volts. If voltage is as specified the sensor circuit and the extreme high-pressure fuel circuit should both be operating within the required specification (however, keep in mind that a malfunctioning sensor could, in theory, hide a malfunctioning fuel circuit).
  9. Once the test is completed, reconnect the wiring harness and install the flywheel cover.

Checking/Changing HPDI High-Pressure Pump Gear Oil

◆ See Figures 136, 137 and 138

The extreme-high pressure fuel circuit of the HPDI system is feed by a belt-driven, mechanical fuel pump. The pump contains a reservoir of gear oil for internal lubrication. Although the pump requires VERY little maintenance, you should check the gear oil level, at least annually, with the pre-season tune-up.

Checking pump gear oil is a very straightforward procedure. For access you'll have to remove the flywheel cover. Once exposed, follow the drive belt

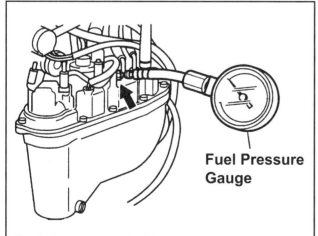

Fig. 134 On most motors the high-pressure fuel circuit is checked using a test fitting on top of the vapor separator tank

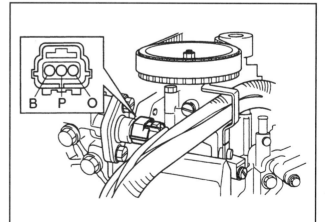

Fig. 135 On 150-200 hp HPDI motors you can check the extreme high-pressure fuel circuit by monitoring fuel pressure sensor output voltage

2-42 MAINTENANCE & TUNE-UP

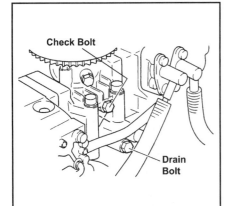

Fig. 136 The HPDI mechanical fuel pump uses a built-in oil reservoir with drain (1) and level check (2) plugs - 225/250 hp model shown (others similar)

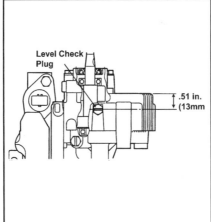

Fig. 137 On 150-200 hp models the pump gear oil level should be about 0.51 in. (13mm) below the top of the pump body

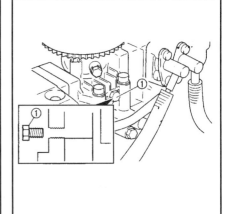

Fig. 138 This drawing shows the oil level, just even with the bottom of the check plug bore on 225/250 hp motors

from the crankshaft pulley found just above the flywheel to the fuel pump's driven gear. Next, look down the side of the pump body, there will be two slotted-head (and/or hex-head) screws on the pump body. The upper one is for fluid level, the lower one is for draining

Carefully loosen and remove the upper plug. If properly filled, oil will start to run out of the plug. If not, stick a thin tool (such as a screwdriver or Allen key) into the hole to check the level. If necessary, add a small amount of SAE 90W gearcase lubricant to top off the system and install the fluid level check bolt.

■ On 150-200 hp motors the proper fluid level is measured as approximately 0.51 in. (13mm) below the surface of the mechanical fuel pump body and the cover. This measurement places the level almost dead center on the level checking screw.

About every 5 years or thousand hours of operation you should change the pump gear oil. This is done by first removing the upper (fluid level/check) plug (because you never remove a drain plug until the level plug has at least been loosened to ensure you'll be able to refill the reservoir). Then, place a small drain pan or funnel/drain tube leading to a drain pan under the lower (drain) plug. Remove the drain plug from the pump body and allow the gear oil to completely drain.

The recommended method of refilling the fuel pump is to squeeze or pump fresh gear oil in through the drain plug until gear oil starts to run out of the level/check plug. Install the check plug, squeeze in a tiny bit more (to make up for what will run out while you're installing the drain plug) and finally, install the drain plug securely to the pump body.

After checking or changing the pump gear oil, wipe away all traces of oil using a clean shop rag and operate the powerhead, then check the surfaces around the plugs again to make sure there is no leakage.

Checking the HPDI Drive Belt

◆ See Figure 139

A drive belt is used to operate mechanical fuel pump which powers the HPDI extreme high-pressure fuel circuit. The belt, driven off a crankshaft mounted pulley (found just above the flywheel) runs around the pump pulley and is secured using a tensioner pulley.

The toothed belt is similar in design and use to that of a timing belt and should therefore be subject to the same level of checking and scrutiny. At least once a season (preferably at the pre-season tune-up) remove the flywheel cover and visually inspect the belt for signs of excessive wear or damage.

Regardless of visual inspection, be sure to replace the belt after 1000 hours or 5 years of service. Keep in mind that the belt is necessary to operate the fuel system, so failure will keep the motor from running, potentially stranding the craft.

For more details on belt service, please refer to the Fuel System section.

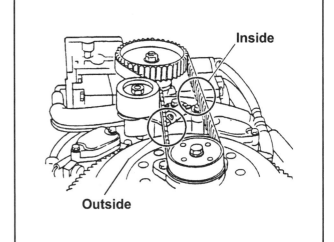

Fig. 139 At least once a year, visually check the inside and outside of the HPDI fuel pump drive belt for wear or damage

CHECKING/CLEANING THE CARBURETORS OR THROTTLE BODIES

Visually inspect the carburetors or throttle bodies at each tune-up. Make sure there are no signs of gunk in the throttle bores. The throttle valves should move smoothly without sticking or binding. Linkage points should be lubricated as necessary.

Periodic carburetor adjustments are usually not necessary, as long as the outboard is running correctly. As a matter of fact, in recent service literature Yamaha finally came out and stated that "carburetor adjustments are not necessary on a properly operating motor" so they've finally admitted what we knew all along, "if it ain't broke, don't fix it." When it does come time for carburetor adjustments the mixture screw (when equipped) does not usually require adjustment unless the motor has been moved to/from altitude.

The timing and synchronization adjustments are covered elsewhere in this section and should at least be checked annually, even if not actual adjustment is made to the linkages.

■ Many carburetor adjustments require that the outboard unit is running in a test tank or with the boat in a body of water (and a helper to navigate the craft). For maximum performance, the idle rpm and other carburetor adjustments should be made under actual operating (load) conditions.

MAINTENANCE & TUNE-UP 2-43

Idle Speed Adjustment (Minor Adjustments)

■ The engine idle speed is normally computer controlled on EFI/HPDI motors, no adjustments are necessary or possible. For more details, please refer to Timing and Synchronization in this section or the information on fuel injection found in the Fuel System section.

Although the full timing and synchronization adjustments should be checked at each tune-up, idle speed can be adjusted independently on carbureted models using the carburetor throttle stop screw which physically determines the positioning of the throttle plate. When the screw is turned one direction (usually clockwise) the throttle plate is held open more, increasing idle speed. When the screw is turned in the opposite direction (normally counterclockwise) the plate is allowed to close further, lowering idle speed.

■ Don't rely on the boat mounted tachometer when setting idle speed, use a high quality shop tachometer to ensure accurate readings. For more details on tachometers refer to the information under Timing and Synchronization.

IDLE SPEED ADJUSTMENT

◆ See Figures 140 thru 143

■ Remember, the powerhead should not start without the emergency tether in place behind the kill switch knob.

✳✳ CAUTION

Water must circulate through the lower unit to the engine any time the engine is run to prevent damage to the water pump in the lower unit. Just five seconds without water will damage the water pump. Never operate the engine at high speed with a flush device attached. The engine, operating at high speed with such a device attached, would runaway from lack of a load on the propeller, causing extensive damage.

■ The 2 hp model has only one carburetor adjustment screw - the idle speed screw on the starboard side of the carburetor. This screw controls the amount of air entering the powerhead instead of fuel.

The idle speed is regulated by the throttle stop screw. The screw sets the position of the throttle plate inside the carburetor throat.

■ For additional information on your engine/carburetor, please refer to the Timing and Synchronization section and the Fuel System section.

1. Remove the cowling and attach a tachometer.
2. Start the engine and allow it to warm to operating temperature.
3. Note the idle speed on the tachometer. If the idle speed is not within specification, rotate the idle speed screw until the idle speed falls within specification. The idle speed specification is noted in the Tune-Up Specifications chart.
4. Rotating the idle speed screw clockwise increases powerhead speed, and rotating the screw counterclockwise decreases powerhead speed.

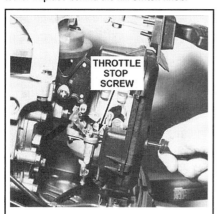

Fig. 140 Location of the throttle stop screw typically found on 2-cylinder powerheads

Fig. 141 Location of the throttle stop screw - 25/30 hp 3-cylinder powerhead

Fig. 142 Location of the throttle stop screw - all other 3-cylinder powerheads except the 25/30 hp

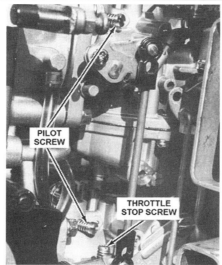

Fig. 143 The idle speed is regulated by the throttle stop screw which sets the position of the throttle plate inside the carburetor throat - typical V4 and V6 powerheads

MAINTENANCE & TUNE-UP

TIMING AND SYNCHRONIZATION

General Information

Any time the fuel system or the ignition system on a powerhead is serviced to replace a faulty part or any adjustments are made for any reason, powerhead timing and synchronization must be carefully checked and verified.

Depending on the engine (predominantly on carbureted motors without microcomputer controls), adjustment of the timing and synchronization can be extremely important to obtain maximum efficiency. The powerhead cannot perform properly and produce its designed horsepower output if the fuel and ignition systems have not been precisely adjusted. We say, depending on the engine because some of the models covered by this manual are equipped with a single carburetor or a computer controlled system (including the EFI/HPDI motors) which requires few, if any periodic adjustments once installed and properly set-up.

As a matter of fact, because of the EPA regulated carburetors used on most of the later model Yamahas, very few adjustments are possible on most carburetors. Periodic mixture adjustments should not be necessary. However, any carburetor will require initial set-up and adjustment after disassembly or rebuilding. Also, any motor equipped with multiple carburetors will require synchronization with each other after the carburetors have been removed or separated. The multiple throttle valves used on EFI/HPDI motors normally require some form of synchronization as well if they are removed or the linkage is disconnected for some other reason.

Although some of the motors covered by this manual utilize fully electronically controlled ignition and timing systems, most of the motors allow for SOME form of timing adjustment. Care should be taken to ensure settings are correct during each tune-up.

■ Before making any adjustments to the ignition timing or synchronizing the ignition to the fuel system, both systems should be verified to be in good working order.

Timing

All outboard powerheads have some type of synchronization between the fuel and ignition systems. Many of the Yamahas covered here, except those equipped with a micro-computer controlled for the TCI/CDI systems or the EFI/HPDI systems, are equipped with a mechanical advance type Capacitor Discharge Ignition (CDI) system and use a series of link rods between the carburetor and the ignition base plate assembly. At the time the throttle is opened, the ignition base plate assembly is rotated by means of the link rod, thus advancing the timing.

On models equipped with a micro-computer, the control module decides when to advance or retard the timing, based on input from various sensors (usually a crankshaft position sensor). Therefore, there is no link rod between the magneto control lever and the stator assembly.

Many models have timing marks on the flywheel and CDI base. A timing light is normally used to check the ignition timing dynamically - with the powerhead operating. An alternate method is to check the static timing - with the powerhead not operating. This second method requires the use of a dial indicator gauge.

Various models have unique methods of checking ignition timing. These differences are explained in detail later in this section.

Synchronization

In simple terms, synchronization is timing the carburetion or throttle valves to the ignition (and to each other). As the throttle is advanced to increase powerhead rpm, the carburetor/throttle valve and the ignition systems are both advanced equally and at the same rate.

Any time the fuel system or the ignition system on a powerhead is serviced to replace a faulty part or any adjustments are made for any reason, powerhead timing and synchronization must be carefully checked and verified. For this reason the timing and synchronizing procedures have been separated from all others and presented alone in this section.

Before making adjustments with the timing or synchronizing, the ignition system should be thoroughly checked and the fuel system verified to be in good working order.

Prepping the Motor for Timing and Synchronization

Timing and synchronizing the ignition and fuel systems on an outboard motor are critical adjustments. The following equipment is essential and is called out repeatedly in this section. This equipment must be used as described, unless otherwise instructed by the equipment manufacturer. Naturally, the equipment is removed following completion of the adjustments.

For many of the adjustments, the manufacturer recommends the use of a test propeller instead of the normal propeller in order to put a specific load on the engine and propeller shaft. The use of the test propeller prevents the engine from excessive rpm while applying a load of pre-set value.

- **Dial Indicator** - Top dead center (TDC) of the No. 1 (top) piston must be precisely known before the timing adjustment can be made on many models. TDC can only be determined through installation of a dial indicator into the No. 1 spark plug opening.
- **Timing Light** - During many procedures in this section, the timing mark on the flywheel must be aligned with a stationary timing mark on the engine while the powerhead is being cranked or is running. Only through use of a timing light connected to the No. 1 spark plug lead, can the timing mark on the flywheel be observed while the engine is operating.
- **Tachometer** - A tachometer connected to the powerhead must be used to accurately determine engine speed during idle and high-speed adjustment. Engine speed readings range from about 0-6,000 rpm in increments of 100 rpm. Choose a tachometer with solid state electronic circuits which eliminates the need for relays or batteries and contribute to their accuracy. For maximum performance, the idle rpm should be adjusted under actual operating conditions. Under such conditions it might be necessary to attach a tachometer closer to the powerhead than the one installed on the control panel.

■ An auxiliary tachometer can be connected by attaching it to the tachometer leads in the control panel. These leads are usually Black and Green. Connect the Black lead to the ground terminal of the auxiliary tachometer and the Green lead to the input or hot terminal of the auxiliary tachometer.

- **Flywheel Rotation** - The instructions may call for rotating the flywheel until certain marks are aligned with the timing pointer. When the flywheel must be rotated, always move the flywheel in the indicated direction (the normal direction of rotation, normally clockwise, but if in doubt you can always double-check by bumping the motor gently using the starter). If the flywheel should be rotated in the opposite direction, the water pump impeller vanes would be twisted. Should the powerhead be started with the pump tangs bent back in the wrong direction, the tangs may not have time to bend in the correct direction before they are damaged. The least amount of damage to the water pump will affect cooling of the powerhead
- **Test Tank** - Since the engine must be operated at various times and engine speeds during some procedures, a test tank or moving the boat into a body of water, is necessary. If installing the engine in a test tank, outfit the engine with an appropriate test propeller

✱✱ CAUTION

Water must circulate through the lower unit to the powerhead anytime the powerhead is operating to prevent damage to the water pump in the lower unit. Just a few seconds without water will damage the water pump impeller.

■ Remember the powerhead should not start without the emergency tether in place behind the kill switch knob.

✱✱ CAUTION

Never operate the powerhead above a fast idle with a flush attachment connected to the lower unit. Operating the powerhead at a high rpm with no load on the propeller shaft could cause the powerhead to runaway causing extensive damage to the unit.

MAINTENANCE & TUNE-UP 2-45

■ Many adjustment procedures involve checking for proper operation BEFORE touching the current settings. In these cases, remember that no adjustment is necessary if the motor passes the checking portion of the procedure.

TACHOMETER CONNECTIONS

◆ See Figures 144 and 145

A tachometer is installed as standard equipment in the dash of many boats. However these tachometers are normally not easy to read when working on the motor and they are normally not labeled as accurately as a shop tachometer. If adjustments need to be made with the outboard running it is usually necessary to attach a tachometer closer to the powerhead than the one installed on the control panel.

Many of the outboards covered in this manual use a CDI system firing a twin lead ignition coil twice for each crankshaft revolution. If an induction tachometer is installed to measure powerhead speed, the tachometer will probably indicate double the actual crankshaft rotation.

1. On manual start models except the 2 hp model, connect the two tachometer leads to the two green leads from the stator. Either tachometer lead may be connected to either green lead.
2. On electric start models, open the remote control box. Locate the Black and Green leads or on models equipped with a tachometer, disconnect the Black and Green leads from the tachometer. Connect the Black lead to the ground terminal of the auxiliary tachometer and the Green lead to the input or hot terminal of the auxiliary tachometer.
3. On the 2 hp model, connect the positive tachometer lead to the primary negative terminal of the coil (usually a small black lead), and the negative tachometer lead to a suitable ground.
4. If a tachometer is purchased from Yamaha or another manufacturer, it should be calibrated for the model matching the particular model powerhead. Remove the rubber plug from the back of the meter. Observe the ring with 4P, 6P, and 12P embossed around the ring. These numbers indicate the number of possible poles used on the flywheel magnetos. Determine the number of poles on the flywheel (refer to the Tune-Up Specifications chart in this section) and using a slotted screwdriver, move the arrow until it points toward the desired pole setting.

2 Hp Model

IGNITION TIMING

The ignition system on these models provides automatic ignition advance. Ignition timing is not adjustable.

IDLE SPEED

 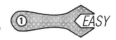

◆ See Figure 146

1. Mount the engine in a test tank or move the boat to a body of water.
2. If necessary, remove cowling and connect a tachometer to the powerhead.

■ The idle speed adjustment screw extends through the front of the engine cowling, however it may be necessary to remove the sides of the cowling in order to attach the tachometer.

3. Start the engine and allow it to reach operating temperature.
4. Check idle speed. The powerhead should idle at the specified rpm in the Tune-up Specifications chart.
5. If adjustment is necessary, rotate the idle adjustment screw until the powerhead idles at the required rpm. Generally, turning the screw **IN** will raise the idle speed, while turning the screw **OUT** will lower the idle speed.
6. Stop the engine and remove the tachometer.

3 Hp Model

IDLE SPEED

◆ See Figures 147 and 148

1. Mount the engine in a test tank or move the boat to a body of water.
2. Remove the cowling and connect a tachometer to the powerhead.
3. Turn the pilot screw in until it lightly seats and then back out the specified number of turns, as indicated in the Carburetor Set-Up Specifications chart, found in the Fuel System section.
4. Start the engine and allow it to warm to operating temperature.
5. Check engine speed at idle. The powerhead should idle at the rpm specified in the Tune-up Specifications chart.
6. Place the engine in gear and check engine trolling speed in the same manner.
7. If adjustment is necessary, rotate the idle speed adjustment screw (not the pilot screw) until the powerhead idles at the required rpm. Normally, turning the screw **IN** will raise the idle speed, while turning the screw **OUT** will lower the idle speed.
8. Check and adjust the Ignition Timing, as detailed in this section.

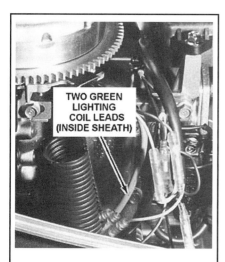

Fig. 144 Two green lighting coil leads are located inside a sheath on the powerhead. One or more of these leads may be used to connect a tachometer to the powerhead

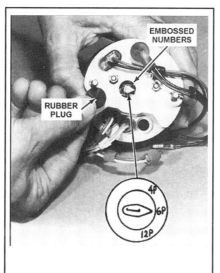

Fig. 145 Setting a tachometer for the correct calibration

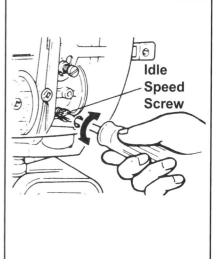

Fig. 146 Idle speed adjustment on 2 hp models

2-46 MAINTENANCE & TUNE-UP

IGNITION TIMING

◆ See Figure 149

The ignition system on these models provides automatic ignition advance. Ignition timing is not adjustable, however it should still be checked periodically to ensure the system is operating properly.

1. Check the idle speed and adjust as necessary.
2. In addition to the tachometer already installed for idle speed adjustment, connect a timing light to the spark plug lead.

✳✳ WARNING

Remember, the motor MUST be operated in a test tank or with the boat launched in a body of water allowing it to run under load when running above idle. Failure to heed this warning will likely result in damage to the motor from over-speeding which will occur when the motor is raised toward WOT without a proper load on the propshaft.

3. Aim the timing light at the timing windows (located on the side of the flywheel housing). If the timing mark can be seen through the left window at idle and the right window at full throttle, the timing is correct.
4. Timing cannot be adjusted. If timing is incorrect, a fault has occurred in the CDI system.
5. Stop the engine and remove the tachometer and timing light.

CHECKING/ADJUSTING THE THROTTLE CABLE

With the throttle grip in the fully closed position, the throttle lever on the carburetor should also be in the fully closed position. If necessary, loosen the inner cable screw and adjust the inner cable to the proper length in order to allow both positions to just fully close; then tighten the cable screw. Turn the throttle grip making sure the throttle lever moves to the fully opened position and then returns to the fully closed position.

4/5 Hp (83 and 103cc) Models

IDLE SPEED

◆ See Figures 150 and 151

1. Mount the engine in a test tank or move the boat to a body of water.
2. Remove the cowling and connect a tachometer to the powerhead.
3. Turn the pilot screw (the vertical screw on top of the carburetor body) in until it lightly seats and then back out the specified number of turns, as indicated in the Carburetor Set-Up Specifications chart, found in the Fuel System section.
4. Start the engine and allow it to warm to operating temperature.

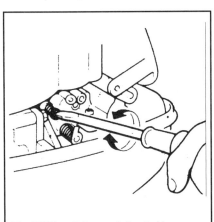

Fig. 147 The pilot screw is located in a horizontal position on the carburetor

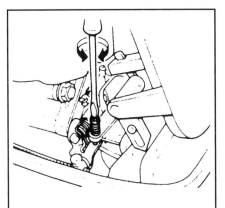

Fig. 148 The idle speed screw is located in a vertical position on the carburetor

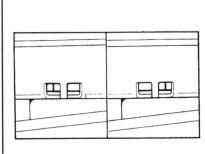

Fig. 149 The timing is correct if the timing mark can be seen through the left window at idle and the right window at full throttle

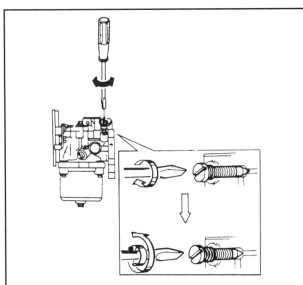

Fig. 150 The pilot screw is located in a vertical position on top of the carburetor

Fig. 151 The idle speed screw is located in a horizontal position contacting the throttle lever

MAINTENANCE & TUNE-UP 2-47

5. Check engine speed at idle. The powerhead should idle at the rpm specified in the Tune-up Specifications chart.
6. Place the engine in gear and check engine trolling speed in the same manner.
7. If adjustment is necessary, rotate the idle adjustment screw (the horizontal screw that contacts the throttle lever, not the pilot screw) until the powerhead idles at the required rpm. Turn the screw **inward** to increase idle speed or **outward** to decrease idle speed.

IGNITION TIMING

◆ See Figure 152

The ignition system on these models provides automatic ignition advance. Ignition timing is not adjustable. However, ignition timing should be checked periodically to ensure proper powerhead operation.
1. Check the idle speed and adjust as necessary.
2. In addition to the tachometer already installed for idle speed adjustment, connect a timing light to the spark plug lead.
3. Start and run the engine, making sure it is fully warmed.
4. Place the engine in gear.

✶✶ WARNING

Remember, the motor MUST be operated in a test tank or with the boat launched in a body of water allowing it to run under load when running above idle. Failure to heed this warning will likely result in damage to the motor from over-speeding which will occur when the motor is raised toward WOT without a proper load on the propshaft.

5. Aim the timing light at the timing windows. If the timing mark can be seen through the left window at with the powerhead operating between 1150 and 1700 rpm and the right window at 4500 rpm or more, the timing is correct.
6. Timing cannot be adjusted. If timing is incorrect, a fault has occurred in the CDI system.

THROTTLE LINKAGE ADJUSTMENT

◆ See Figure 153

1. The engine idle speed must be properly adjusted before attempting of adjust the throttle linkage. If not already done, follow the steps under Idle Speed, in this section, before proceeding.
2. Remove the air intake silencer cover so you can clearly see the carburetor throttle lever.
3. With the powerhead not operating, rotate the throttle grip back and forth between the idle and Wide-Open Throttle (WOT) position a couple of times.
4. Slowly rotate the grip to WOT, watching when the throttle lever on the carburetor contacts the WOT stopper. You want the lever to come in contact with the stopper just as the twist grip reaches with WOT position.
5. If it does not operate as desired, adjust the cable by first loosening the throttle cable lock screw (on the throttle lever) and then carefully pulling the cable out of the lever.
6. Turn the throttle grip to the **SLOW** (idle) position, then insert the inner wire into the hole in the throttle lever and lock it with the screw.
7. Next, pull out the outer wire and hook it onto the wire hook on the carburetor.
8. Check adjustment by again rotating the grip back and forth from idle to WOT a few times. Again, watch the throttle lever making sure it contacts the stopper just as the both the throttle valve and hand grip takes the WOT position.
9. Once the adjustment is satisfactory, install the air intake silencer cover and the cowling.

6/8 Hp Models

IDLE SPEED

1. Mount the engine in a test tank or move the boat to a body of water.
2. Remove the cowling and connect a tachometer to the No. 1 cylinder spark plug lead.
3. Turn the pilot screw (the horizontal screw threaded into the side of the carburetor body) until it lightly seats and then back out the specified number of turns, as indicated in the Carburetor Set-Up Specifications chart, found in the Fuel System section.
4. Start the engine and allow it to warm to operating temperature.
5. Check engine speed at idle. The powerhead should idle at the rpm specified in the Tune-up Specifications chart.
6. Place the engine in gear and check engine trolling speed in the same manner.
7. If adjustment is necessary, rotate the idle adjustment screw (the vertical screw that contacts the carburetor throttle lever, not the pilot screw) until the powerhead idles at the required rpm. Turn the screw **inward** to increase idle speed or **outward** to decrease idle speed.
8. Check and, if necessary, adjust the Ignition Timing, as detailed in this section.

Fig. 152 View through the window showing the timing mark on the flywheel

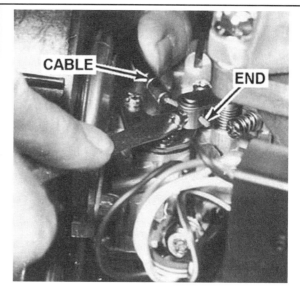

Fig. 153 Loosening the screw on the barrel retaining end of the throttle cable

2-48 MAINTENANCE & TUNE-UP

IGNITION TIMING

◆ See Figures 154, 155 and 156

The manufacturer provides a Static Timing Check/Adjustment procedure which can be performed without the use of a tachometer or timing light, but which requires the use of a Top Dead Center gauge or dial gauge which can be used to determine the top of piston travel. Alternately, we've also provided a Dynamic Timing Check/Adjustment procedure which we've found is also effective (and is preferred by some techs).

Static Timing Check/Adjustment

1. Remove the engine cowling for access.
2. Disconnect the link rod between the CDI magneto base and the magneto control lever at the ball joint.
3. Slowly rotate the flywheel clockwise by hand to align the timing plate with the 35 degree BTDC mark on the flywheel indicator.
4. Align the marks on the magneto base and the flywheel by turning the magneto base. Make sure the magneto base contacts the vertical pointer (stopper plate).
5. If adjustment is necessary, remove both spark plugs and install a dial gauge or TDC indicator in to the hole of the No. 1 (top) cylinder.
6. Slowly turn the flywheel clockwise by hand until the No. 1 piston comes up to TDC (the highest point of travel as indicated by the dial gauge).
7. Set the timing plate to TDC.
8. Next, turn the flywheel to align the timing plate with the 35 degree BTDC mark on the flywheel indicator.
9. Loosen the vertical pointer (stopper plate) set bolt, then align the marks on the magneto base and flywheel by turning the magneto base.
10. Adjust the vertical pointer (stopper plate) so it is in contact with the magneto base stopper (press the stopper plate against the full-advanced side of the magneto base).
11. Tighten the vertical pointer (stopper plate) bolt and reconnect the link rod.

Dynamic Check/Adjustment

1. Check the idle speed and adjust as necessary.
2. In addition to the tachometer already installed for idle speed adjustment, connect a timing light to the spark plug lead.
3. Disconnect the link rod between the CDI magneto base and the magneto control lever at the ball joint.

■ This link rod will remain disconnected until all adjustments have been completed.

4. Start the engine and allow it to reach operating temperature. Rotate the CDI magneto base counterclockwise until the base stopper on the left contacts the vertical timing pointer. This is the full-advanced position.

※※ WARNING

Remember, the motor MUST be operated in a test tank or with the boat launched in a body of water allowing it to run under load when running above idle. Failure to heed this warning will likely result in damage to the motor from over-speeding which will occur when the motor is raised toward WOT without a proper load on the propshaft.

5. Aim the timing light at the vertical timing pointer. The pointer should align between the 34 and 36° BTDC embossed marks on the flywheel edge.
6. If the pointer does not align, as described, loosen the bolt on the vertical pointer bracket and move the pointer and the stopper on the magneto base plate together, until the vertical pointer is properly aligned. Hold the magneto base plate and the pointer together with the pointer on the mark and tighten the bolt.
7. Shut down the powerhead.

THROTTLE LINKAGE ADJUSTMENT

◆ See Figure 157

To check the throttle linkage adjustment set the shift lever in the full-forward position and fully open the throttle grip. Check that the magneto base stopper is fully in contact with the vertical pointer (stopper plate) and the throttle valve is fully opened. If adjustment is necessary, proceed as follows:

1. Disconnect the link rod between the CDI magneto base and the magneto control lever at the ball joint.
2. Set the shifter lever to the forward position and manually bring the stopper on the full-advanced side of the magneto base to contact the magneto base stopper.
3. Fully open the throttle grip, then loosen the locknut and turn the cable adjuster on the throttle "pull" cable (the outer of the 2 cables, which as the name suggests is used to pull the throttle open) until the carburetor throttle valve is fully opened. Then tighten the locknut for the throttle "pull" cable.
4. Now, loosen the locknut on the "push" throttle cable (the inner of the 2 cables, which is used to close the throttle valve) until there is about 0.12 in. (3mm) free-play on the throttle grip. Then tighten the locknut for the "push" cable.
5. Finally, adjust the magneto link rod so that the control lever comes into contact with the magneto base. Make sure the link rod is centered over the ball joint on the control lever, then carefully reconnect the link.
6. Verify that the throttle valve is in the fully opened position.

Fig. 154 Make sure the link rod remains disconnected until all adjustments have been completed

Fig. 155 A timing light is needed for a dynamic check

Fig. 156 The full-advanced position is achieved when the magneto base is rotated counterclockwise until the stopper contacts the vertical timing pointer

MAINTENANCE & TUNE-UP 2-49

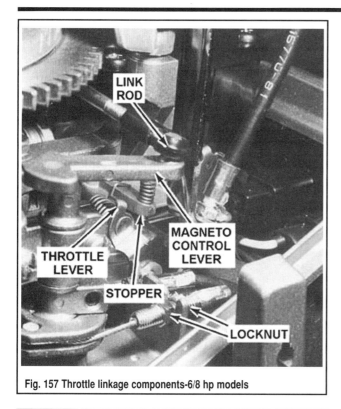

Fig. 157 Throttle linkage components-6/8 hp models

9.9/15 Hp Models

IGNITION TIMING

◆ See Figures 158 thru 161

The complete timing check and adjustment procedure takes place with the engine NOT running. The good news is that means you won't need a tachometer or timing light, and nor will you need to mount it in a test tank or launch the boat. The bad news is that you WILL need a dial gauge. Ok, maybe that's not a bad thing cause you'll be able to do other things with it, like check rotor run-out on the tow vehicle (or on the crankshaft if you ever rebuild the powerhead).

1. Remove the cowling for access and locate the timing marks on the flywheel and the pointer on the starter housing.
2. Slowly turn the flywheel in the normal direction of rotation (clockwise) until the timing mark for WOT operation (refer to the Tune-Up Specifications chart in this section) aligns with the pointer on the starter cover.
3. Turn the magneto control lever so it contacts the WOT stopper, then check the link joint timing indicator to make sure it aligns with the flywheel mark. If it does, timing is properly set and no adjustment is necessary, you can stop here. If not, continue the procedure to properly adjust/set the ignition timing.
4. Loosen the locknut on the link joint, then disconnect the link joint from the magneto control lever.
5. Remove the spark plug from the No. 1 (top) cylinder and install a dial gauge to determine TDC. Slowly turn the flywheel clockwise by hand until the piston reaches TDC. The dial indicator will increase in value until TDC is reached, then as the piston reverses direction it will begin to decrease again. With the piston at TDC zero the dial gauge.
6. Rotate the flywheel counterclockwise (yes, this is one of the few times we'll ever tell you to do that, but do it SLOWLY) until the dial gauge indicate the piston is 0.166 in. (4.22mm) Before Top Dead Center (BTDC).
7. At this point, turn the magneto control lever so that it contacts the WOT stopper. Now adjust the link joint length until the link joint timing indicator aligns with the flywheel mark and tighten the locknut.
8. Now check the idle timing adjustment by turning the flywheel CLOCKWISE (again, slowly) until the idle timing mark (not the WOT mark) aligns with the cover pointer. In this position, rotate the magneto lever to that

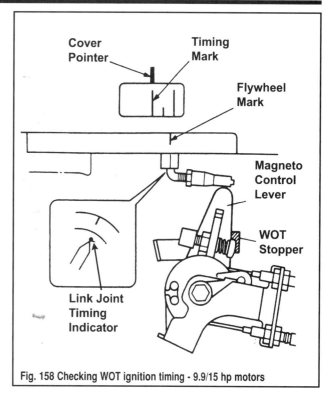

Fig. 158 Checking WOT ignition timing - 9.9/15 hp motors

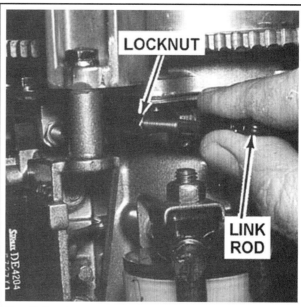

Fig. 159 To adjust the timing, start by loosening the locknut and disconnecting the link joint. . .

the idle timing screw contacts the idle timing stopper. Finally, at this point the link joint timing indicator should align with the marking on the flywheel (see the accompanying illustration). If this setting is correct, no further checking/adjustment is necessary. However, if the marks do not align as indicated continue the procedure in order to properly set the idle timing.

9. Turn the flywheel clockwise until the dial gauge indicates that the piston is 0.005 in. (0.12mm) After Top Dead Center (ATDC).
10. Turn the magneto control lever so that the idle timing screw contacts the idle timing stopper. Now adjust the screw so that the link joint timing indicator aligns properly with the flywheel mark.
11. Remove the dial gauge and install the spark plug.
12. Check the throttle linkage adjustment.

MAINTENANCE & TUNE-UP

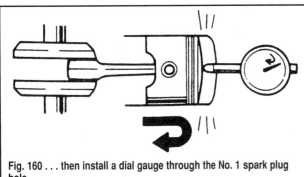

Fig. 160 . . . then install a dial gauge through the No. 1 spark plug hole

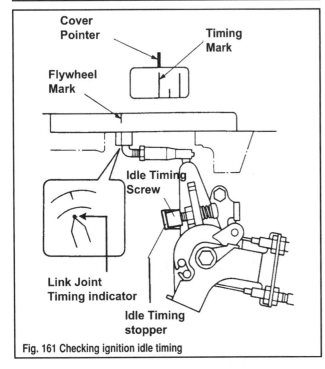

Fig. 161 Checking ignition idle timing

THROTTLE LINKAGE ADJUSTMENT

◆ See Figures 161 thru 164

■ Refer to the numbers in the accompanying illustration for component identification.

1. Check and adjust the ignition timing, as necessary.
2. Check the throttle linkage adjustment by fully closing the throttle grip and checking to make sure the idle timing screw contacts the idle timing stopper. If so, stop right here, since the adjustment is fine. If not, continue with the procedure to properly adjust the throttle linkage.

3. Locate the throttle cables where they connect to the magneto control lever bracket. Loosen the locknuts for both the pull and push cables.
4. Turn the magneto control lever until the idle timing screw contacts the idle timing stopper, then rotate the adjuster for the lower cable (1) on the bracket until there is 0.04 in. (1mm) of free-play between the pulley stoppers and the free acceleration lever. Then tighten the locknut for the lower cable (1).
5. Next, turn the adjuster for the upper cable (2) until there is 0.04 in. (1mm) of free-play on the throttle cable at the pulley, then tighten the locknut.
6. Recheck the throttle operation checking that the screw contacts the stopper and that it operates smoothly without binding.

CHECKING/ADJUSTING THE STARTER LOCKOUT

◆ See Figure 165

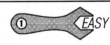

This model contains a starter lockout safety feature to prevent the motor from being started while in gear. The system should be checked at each tune-up to ensure that it is adjusted and functioning properly. Checking is a simple matter of placing the motor in gear and gently attempting to start the motor. If the starter will not rotate, the system is functioning and no further attention is required. However, if the motor rotates in gear, then adjust the cable as follows:

1. Begin with the engine not running and shifter positioned in **Neutral**.
2. Loosen the lockout cable locknut (at the cable mounting bracket).
3. Turn the adjuster nut (located at the cable mounting bracket, on the opposite side of the locknut) until the end of the stopper aligns with the marking on the starter case.
4. Once positioned properly, retighten the locknut to hold the cable in this position.
5. Verify that the lockout is now working properly.

IDLE SPEED

◆ See Figure 166

1. Mount the engine in a test tank or move the boat to a body of water.
2. Remove the cowling and connect a tachometer to the powerhead.
3. Loosen the acceleration rod lock screw.
4. Turn the pilot screw (the threaded horizontally into the top side of the carburetor body) inward until it JUST lightly seats and then back out the specified number of turns, as indicated in the Carburetor Set-Up Specifications chart, found in the Fuel System section.
5. Start the engine and allow it to warm to operating temperature.
6. Check engine speed at idle. The powerhead should idle at the rpm specified in the Tune-up Specifications chart.
7. Place the engine in gear and check engine trolling speed in the same manner.
8. If adjustment is necessary, rotate the idle adjustment screw (the spring loaded screw threaded vertically downward into contact with the throttle lever, not the pilot screw) until the powerhead idles at the required rpm. Turning the screw inward will increase idle speed, while turning the screw outward will decrease speed.
9. Pull the acceleration rod through the throttle lever bore until the idle timing screw contacts the idle timing stopper, then tighten the acceleration rod lock screw.

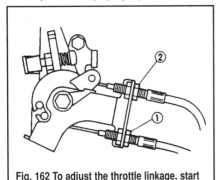

Fig. 162 To adjust the throttle linkage, start by loosening the cable locknuts...

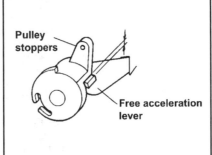

Fig. 163 ...turn the lower cable to adjust free-play at the pulley stoppers...

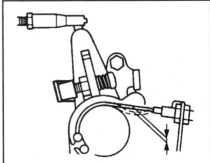

Fig. 164 ...then turn the upper cable to adjust free-play on the cable itself

MAINTENANCE & TUNE-UP 2-51

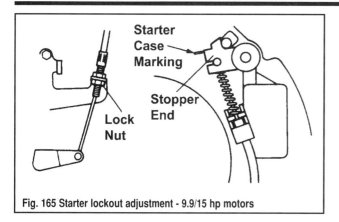

Fig. 165 Starter lockout adjustment - 9.9/15 hp motors

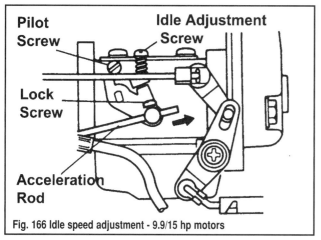

Fig. 166 Idle speed adjustment - 9.9/15 hp motors

20/25 Hp (395cc) Models

IGNITION TIMING

◆ See Figures 167, 168 and 169

The complete timing check and adjustment procedure takes place with the engine NOT running. The good news is that means you won't need a tachometer or timing light, and nor will you need to mount it in a test tank or launch the boat. The bad news is that you WILL need a dial gauge. Ok, maybe that's not a bad thing cause you'll be able to do other things with it, like check rotor run-out on the tow vehicle (or on the crankshaft if you ever rebuild the powerhead).

■ Refer to the letters in the accompanying illustration for component identification.

1. Remove the cowling for access and locate the timing marks on the flywheel.
2. To check the WOT ignition timing (fully-advanced setting) slowly turn the flywheel by hand in the normal direction of rotation (clockwise) and align the timing pointer with the specified timing mark on top of the flywheel. Please refer to the Tune-Up Specifications chart in this section for timing specs.
3. With the timing pointer aligned with the appropriate specification mark on the flywheel, turn the magneto control lever so that it contacts the full advanced stopper. At this point check the timing indicator (a) under the flywheel and make sure it aligns with the marking (b) on the flywheel (for clarification, please refer to the accompanying illustration). If adjustment is necessary, continue with the procedure.
4. Loosen the locknut on the link at the top of the magneto control lever and disconnect the link joint from the lever.
5. Remove the spark plug from the No. 1 cylinder, then attach a dial gauge to the spark plug hole in order to measure piston height.
6. Slowly turn the flywheel by hand until the piston reaches TDC, then zero the dial gauge. Now turn the flywheel very slowly **Counterclockwise** (YES, this is one of the few times we'll ever tell you to do that) until the dial gauge shows the piston is 0.13 in. (3.34mm) BTDC.
7. With the piston at this position, manually turn the magneto control lever to the WOT position and adjust the link joint length so that the timing indicator aligns with the mark under the flywheel. Connect the link joint and tighten the locknut to secure the adjustment.
8. Next, check the idle timing by turning the flywheel clockwise until the timing pointer aligns with the appropriate idle timing mark on top of the flywheel. In this position, turn the magneto control lever the opposite direction, so it contacts the idle stopper and check if the timing indicator (a) aligns with the mark under the flywheel. If not, continue with the procedure in order to set the idle timing.
9. If idle adjustment is necessary, simply turn the idle stopper screw until the magneto lever moves sufficiently to align the timing indicator (a) with the mark (b) below the flywheel.

CHECKING/ADJUSTING THE CARBURETOR LINK ROD

◆ See Figure 170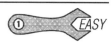

The 2 carburetors used on this model are connected by a link that ensures both throttle valves will open and close at the same time. When performing adjustments, check the link to make sure that both throttle valves open and close together. If not, adjust the link rod as follows:

1. Loosen the idle speed adjustment screw enough to fully close the throttle valves (count the number of turns you loosen it in order to preserve current idle speed adjustment).
2. Loosen the link rod lock-screw and reposition the link so that the throttle valves are fully closed, then tighten the lock-screw.
3. Open and close the throttle a few times making sure the throttle valves move together (from idle to WOT positions).
4. Turn the idle speed adjustment screw inward the same number of turns you counted while backing it out.
5. Continue the timing and synchronization checks/adjustments, making sure to check/adjust the idle speed.

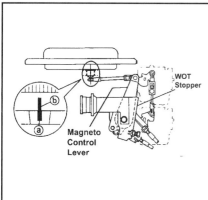

Fig. 167 Checking/adjusting WOT ignition Timing - 20/25 hp (395cc) motors

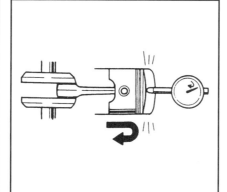

Fig. 168 Use a dial gauge to find TDC and then to measure piston depth

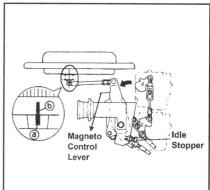

Fig. 169 Checking/adjusting idle ignition Timing - 20/25 hp (395cc) motors

2-52 MAINTENANCE & TUNE-UP

CHECKING/ADJUSTING THE CARBURETOR PICKUP TIMING

◆ See Figures 171 and 172

■ Refer to the letters in the accompanying illustration for component identification.

1. Check/adjust the ignition timing and carburetor link.
2. Check for proper carburetor pickup timing by slowly turning the flywheel clockwise until the appropriate timing mark on top of the flywheel aligns with the timing pointer. Please refer to the Tune-Up Specifications chart in this section for timing specs.
3. With the timing pointer aligned with the appropriate specification mark on the flywheel, turn the magneto control lever so that it just contacts the throttle roller. At this point check the timing indicator (a) under the flywheel and make sure it aligns with the marking on the flywheel (for clarification, please refer to the accompanying illustration). If adjustment is necessary, continue with the procedure.
4. Loosen the throttle roller adjusting screw and (with the flywheel still in the appropriate pickup timing position), then turn the magneto control lever so the mark on the underside of the flywheel aligns with the indicator mark. Now set the roller so it contacts the magneto lever and tighten the roller adjusting screw.

CHECKING/ADJUSTING THE THROTTLE CABLE

◆ See Figure 173

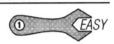

Set the shifter into the forward position; then operate the throttle, checking for smooth, correct operation. Visually inspect the throttle control cable for signs of damage or excessive wear and replace, if necessary. With the throttle in the WOT position, check the indicator mark to see if it aligns with the throttle roller. If not, adjust the throttle cable, as follows:

1. With the shift lever still in the forward position, move the magneto control lever to the WOT stop.
2. Locate the 2 throttle cables in the bracket at the base of the linkage assembly, the top cable is the PULL or accelerator cable, the bottom is the PUSH or decelerator cable. Loosen the locknuts on both cables.
3. Turn both cable adjusters until the throttle roller is aligned with the indicator mark, then loosen the cable adjuster on the PUSH (bottom) cable until there is 0.12 in. (3mm) of free-play at the throttle grip. Then tighten both locknuts.
4. Verify that throttle operation (specifically the movement of the magneto control lever) is smooth. Repair or readjust as necessary.

CHECKING/ADJUSTING THE DIAPHRAGM

◆ See Figure 174

Fully close the throttle to the idle position and check that the diaphragm plunger is fully retracted and the throttle control lever is closed all the way. If not, or if the diaphragm operation is rough, it should be adjusted. Adjustment is a relatively simple matter. Loosen the 2 bolts on the side of the bracket, then place the magneto control lever in the full retard (idle) position and retighten the bolts.

CHECKING/ADJUSTING THE NEUTRAL OPENING LIMIT

◆ See Figure 175

These motors use a type of mechanical limiter to prevent engine overspeed in **Neutral** by limiting the throttle position. The checking and adjustment procedures are pretty much the same, involving operating the motor in Neutral (in a test tank, using a flushing device or on a body of water) and making sure the engine will not go beyond about 3500-4100 rpm. To check and/or adjust the neutral opening limit, proceed as follows:

■ Refer to the numbers in the accompanying illustration for component identification.

1. Mount the engine in a test tank or launch the boat on a body of water.
2. Remove the cowling and attach a tachometer.
3. Set the shift lever to **Neutral** and start the engine.
4. Allow the engine to operate and come up to normal operating temperature.
5. Set the throttle so the engine speed is 3500-3800 rpm. It should not go any further in **Neutral**.

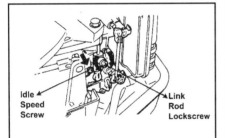

Fig. 170 Idle speed and carburetor link rod adjustment screws - 20/25 hp (395cc) motors

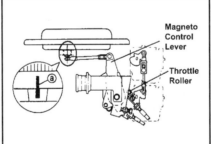

Fig. 171 Checking/adjusting the carburetor pickup timing - 20/25 hp (395cc) motors

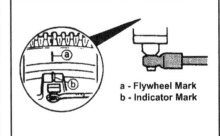

Fig. 172 Make sure the mark under the flywheel mark aligns with the indicator mark

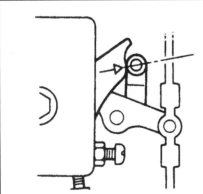

Fig. 173 Checking the throttle cable adjustment - 20/25 hp (395cc) motors

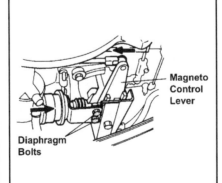

Fig. 174 Adjusting the diaphragm - 20/25 hp (395cc) motors

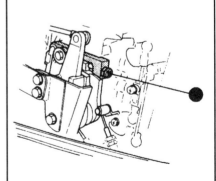

Fig.175 Adjusting the neutral opening limit - 20/25 hp (395cc) motors

MAINTENANCE & TUNE-UP 2-53

6. If adjustment is necessary, turn the neutral speed control screw (on the linkage just inboard of the magneto control lever, for clarification please refer to the accompanying illustration). Turning the screw INWARD will DECREASE speed, while turning the screw OUTWARD will INCREASE speed.

IDLE SPEED

◆ See Figure 170

1. Check and/or adjust the Neutral Opening Limit. Leave the engine mounted in the test tank or keep the boat launched.
2. Turn the pilot screw (mounted diagonally downward into the top side of the carburetor body) in until it lightly seats and then back out the specified number of turns, as indicated in the Carburetor Set-Up Specifications chart, found in the Fuel System section.
3. Start the engine and allow it to warm to operating temperature.
4. Check engine speed at idle. The powerhead should idle at the rpm specified in the Tune-up Specifications chart.
5. Place the engine in gear and check engine trolling speed in the same manner.
6. If adjustment is necessary, rotate the idle adjustment screw, (NOT the pilot screw) until the powerhead idles at the required rpm. The idle adjustment screw is mounted vertically, just behind the carburetor link and contacts the throttle valve. Turning the screw INWARD will increase idle speed, while turning the screw OUTWARD will decrease idle speed.

CHECKING/ADJUSTING THE STARTER LOCKOUT

◆ See Figure 176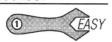

This model contains a starter lockout safety feature to prevent the motor from being started while in gear. The system should be checked at each tune-up to ensure that it is adjusted and functioning properly. Checking is a simple matter of placing the motor in gear and gently attempting to start the motor. If the starter will not rotate, the system is functioning and no further attention is required. However, if the motor rotates in gear, then adjust the cable as follows:
1. Begin with the engine not running and shifter positioned in **Neutral**.
2. Loosen the locknut at the top of the cable (it's the small nut), near the starter housing.
3. Next, use the large adjuster nut (located just below the adjuster nut) to change the length of the cable until the starter stop plunger line is centered in the sight hole.
4. Once positioned properly, retighten the locknut to hold the cable in this position.
5. Verify that the lockout is now working properly.

CHECKING/ADJUSTING THE OIL PUMP LINK ROD

◆ See Figure 177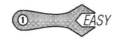

The oil pump link is properly adjusted when there is 0.02 in. (0.5mm) of clearance, measured between the oil pump lever and the WOT stopper when the carburetor throttle valve is at WOT. To check this dimension open the throttle valve and measure the distance using a feeler gauge. If adjustment is necessary, proceed as follows:
1. Loosen the locknut at the top of the oil pump link, then disconnect the link joint from the throttle control lever.
2. Open the carburetor throttle valve to the WOT position and position the oil pump lever 0.02 in. (0.5mm) from the WOT stopper. Gently push the lever against a feeler gauge to find the correct distance, then hold the lever and gauge in that position (an assistant is really handy here).
3. Adjust the link joint until the hole aligns with the oil pump set pin, then reconnect the link joint.
4. Verify that the throttle valve opens fully and that the proper gap now exists between the pump lever and WOT stopper, then tighten the locknut.

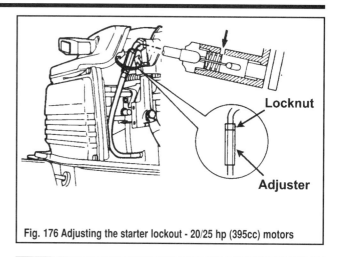

Fig. 176 Adjusting the starter lockout - 20/25 hp (395cc) motors

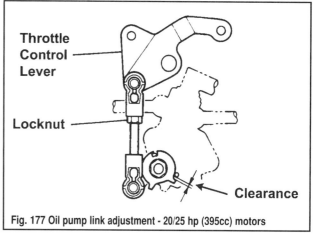

Fig. 177 Oil pump link adjustment - 20/25 hp (395cc) motors

20/25 hp (430cc) Models

■ This includes most of the C25 models.

IGNITION TIMING

1. Pry the link rod free from the ball joint on the magneto control lever.
2. Remove the park plug and install a dial indicator in the No. 1 cylinder spark plug hole.
3. Rotate the flywheel clockwise until the piston is at top dead center (TDC).
4. Check the timing pointer with the flywheel timing scale. If alignment is incorrect, loosen the timing pointer nut and move the pointer as required to align the pointer with the 0° mark on the flywheel.
5. Rotate the flywheel clockwise until the timing plate is aligned to the full advance timing as specified in the Tune-up Specifications chart.
6. Set the magneto base to the full-advanced position and set the stopper.
7. Loosen the locknut on the link rod and adjust the length of the rod until it can be snapped back into place on the ball joint of the magneto control lever, without any movement at the other end of the rod.

CARBURETOR LINKAGE ADJUSTMENT

1. Loosen the carburetor control link set screw.
2. Set the magneto control lever in the full-advanced position.
3. Adjust the guide collar so that the roller contacts the high point of the accelerator cam.
4. Hold down the control ring and tighten the set screw securely.

2-54 MAINTENANCE & TUNE-UP

THROTTLE LINKAGE ADJUSTMENT

1. Place the lower unit in **FORWARD** gear and loosen the joint link set screw.
2. Adjust the accelerator control link rod so that the center to center length is 2.72 in. (69mm).

IDLE SPEED

1. Mount the engine in a test tank or move the boat to a body of water.
2. Remove the cowling and connect a tachometer to the powerhead.
3. Turn the pilot screw in until it lightly seats and then back out the specified number of turns, as indicated in the Carburetor Set-Up Specifications chart, found in the Fuel System section.
4. Start the engine and allow it to warm to operating temperature.
5. Check engine speed at idle. The powerhead should idle at the rpm specified in the Tune-up Specifications chart.
6. Place the engine in gear and check engine trolling speed in the same manner.
7. If adjustment is necessary, rotate the idle adjustment screw (not the pilot screw) until the powerhead idles at the required rpm.

25/30 Hp (496cc 2-Cylinder) Models

IGNITION TIMING

1. Mount the engine in a test tank or move the boat to a body of water.
2. Remove the cowling and connect a tachometer and a timing light to the powerhead.
3. Start the engine and allow it to warm to operating temperature. Place the lower unit in Neutral.
4. Increase engine speed to 4500-5500 and aim the timing light at the indicator on the starter case. It should align with the specified timing full throttle timing figure in the Tune-up Specifications chart.
5. If the timing mark is not aligned properly, stop the engine, shift the lower unit into **FORWARD** and rotate the flywheel until the timing marks align properly.
6. Rotate the magneto base and align the timing mark with the ignition mark on the rotor.
7. If the magneto base stopper is not in contact with the full open stopper on the cylinder body, loosen the holding bolts and adjust until proper contact is made.
8. Ensure the full open (T) mark on the accelerator cam aligns with the center of the cam roller.
9. If alignment is not as specified, loosen the accelerator cam bolts and align the full open (T) mark with the center of the carburetor throttle roller. Tighten the holding bolts.
10. Remove the starter.
11. Loosen the rod adjusting screw and adjust the rod so the throttle is fully open and the open stopper is pushed up against the stopper. Tighten the screw.
12. Install the starter.

IDLE SPEED

1. Mount the engine in a test tank or move the boat to a body of water.
2. Remove the cowling and connect a tachometer to the powerhead.
3. Turn the pilot screw in until it lightly seats and then back out the specified number of turns, as indicated in the Carburetor Set-Up Specifications chart, found in the Fuel System section.
4. Start the engine and allow it to warm to operating temperature.
5. Check engine speed at idle. The powerhead should idle at the rpm specified in the Tune-up Specifications chart.
6. Place the engine in gear and check engine trolling speed in the same manner.
7. If adjustment is necessary, rotate the idle adjustment screw (not the pilot screw) until the powerhead idles at the required rpm.

THROTTLE CABLE

1. With the lower unit in **FORWARD** gear, twist the throttle grip to the wide-open throttle position.
2. The magneto base stopper should be in contact with the full open stopper.
3. If adjustment is necessary, loosen the adjust locknut on the pull side of the throttle cable. Turn the adjusting bolt until all slack is taken up and tighten the locknut.
4. Loosen the adjust locknut on the push side of the throttle cable. Turn the adjusting bolt until 0.12 in (3mm) slack is present with the throttle grip in the **Slow** position. Tighten the locknut.
5. Ensure the throttle shaft full open stopper contacts the full open stopper on the carburetor with the throttle in the wide-open throttle position.
6. If the throttle stopper is not a specified, loosen the locking screw and adjust it to specification.
7. Now adjust the throttle control link.

THROTTLE CONTROL LINK

■ The throttle cable must be adjusted prior to adjusting the throttle control link.

1. Twist the throttle grip to the wide-open throttle position.
2. With the lower unit in **FORWARD** gear, ensure the magneto base stopper is in contact with the stopper on the cylinder and the throttle control lever in contact with the stopper on the bottom cowling.
3. If these condition do not exist, proceed as follows.
4. Disconnect the link rod between the CDI magneto base and the magneto control lever at the ball joint.
5. Rotate the magneto base plate until the stopper is in contact with the stopper on the cylinder.
6. Twist the throttle grip to the wide-open throttle position.
7. Adjust the connector on the end of the link rod so the center to center length is 1.81 in. (46mm).
8. Connect the link rod between the CDI magneto base and the magneto control lever at the ball joint without moving the magneto base.

40 Hp (2-Cylinder) Model

■ This includes most C40 models.

DYNAMIC TIMING

◆ See Figures 178, 179 and 180

1. Mount the engine in a test tank or on a boat in a body of water.
2. Obtain a timing light and clip the pickup lead to the No. 1 spark plug lead.
3. Connect a tachometer to the powerhead per the instructions with the instrument.
4. Start the engine and allow it to warm to operating temperature.
5. Push the magneto control lever downward until the lower screw tip barely makes contact with the stopper. This action fully advances the timing. Allow the powerhead to operate at approximately 4,500 rpm.
6. Aim the timing light at the timing pointer. The pointer should align halfway between the 21-23° BTDC marks embossed on the flywheel. If the marks align, the full-advanced timing is correctly set.
7. Shut down the powerhead. Pry off the link from the ball joint at the magneto control lever ball joint. Restart the powerhead. Pull the magneto control lever all the way up. Aim the timing light at the timing pointer and use the free end of the link rod to rotate the magneto base until the timing pointer aligns properly.

MAINTENANCE & TUNE-UP 2-55

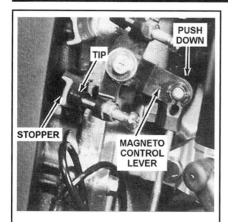

Fig. 178 Push the magneto control lever downward until the lower screw tip barely makes contact with the stopper

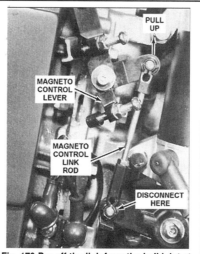

Fig. 179 Pry off the link from the ball joint at the magneto control lever ball joint

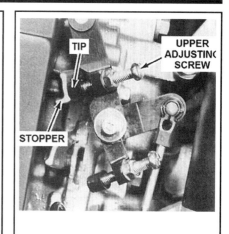

Fig. 180 Check to be sure the upper adjusting screw tip barely makes contact with the stopper

8. Shut down the powerhead. Adjust the length of the link rod to snap back onto the ball joint of the magneto control lever without moving the magneto base or the magneto control lever. Snap the link rod back onto the ball joint of the magneto control lever.

9. Start the powerhead and allow it to idle. Check to be sure the upper adjusting screw tip barely makes contact with the stopper.

10. Aim the timing light at the timing pointer. The pointer should align half way between the 1-3° ATDC marks embossed on the flywheel. If the marks align, the fully retarded timing is correctly set and the timing procedures are completed. If the marks do not align, then proceed as follows: With the powerhead still running, continue to aim the timing light at the pointer and at the same time adjust the upper adjusting screw until the pointer aligns properly. Shut down the powerhead.

STATIC TIMING

Full Advance Adjustment

◆ See Figures 181 thru 184

1. Remove all three spark plugs from the powerhead. Install a dial indicator into the No. 1 cylinder opening.

2. Rotate the flywheel clockwise until the dial indicator indicates the piston is at TDC (top dead center). Check the timing pointer to be sure it aligns with the TDC mark embossed on the flywheel. If the mark is misaligned, loosen the set screw on the timing plate and align the pointer with the flywheel mark. Tighten the screw to hold the adjustment.

3. Rotate the flywheel clockwise until the timing pointer aligns with 21-23° on the flywheel.

4. Rotate the lower adjustment screw until the tip contacts the stopper. Tighten the locknut to hold this new adjusted position.

Full Retard Adjustment

◆ See Figures 185 and 186

1. Rotate the flywheel clockwise until the timing pointer aligns with the 1-3° ATDC mark embossed on the flywheel.

2. Rotate the upper adjustment screw until the tip contacts the stopper. Tighten the locknut to hold this new adjusted position.

CARBURETOR LINK

◆ See Figures 187 and 188

1. Pull off the accelerator lever rod. This rod connects the three throttle levers together and is a set length. Loosen but do not remove the throttle valve screws on the top and center carburetors, by rotating the screws clockwise. Yes, they are rotated clockwise, because they have left hand threads. This fact is emphasized by the arrow and the word OFF embossed on each lever.

2. Loosen the idle speed adjustment screw. Snap on the accelerator rod over all three ball joints. Push down on the cam to close all throttle valves and then tighten the throttle valve screws on the top and center carburetor. This is accomplished by rotating the screws counterclockwise.

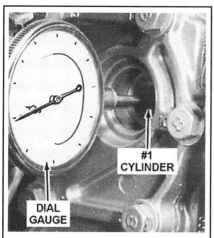

Fig. 181 Install a dial indicator into the No. 1 cylinder opening

Fig. 182 Rotate the flywheel clockwise until the dial indicator indicates the piston is at TDC (top dead center)

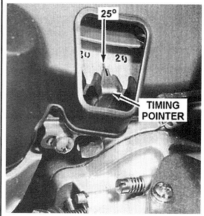

Fig. 183 Rotate the flywheel clockwise until the timing pointer aligns with 21-23° on the flywheel

MAINTENANCE & TUNE-UP

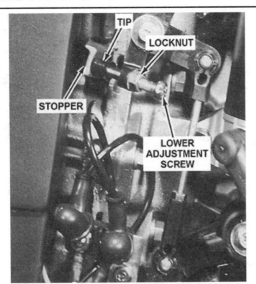

Fig. 184 Rotate the lower adjustment screw until the tip contacts the stopper

Fig. 185 There is no mark between TDC and 5° on the flywheel. Adjust the timing so the pointer falls just to the left of the 5° mark

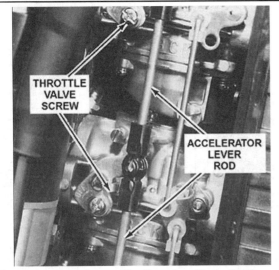

Fig. 187 Accelerator lever rod location adjacent to the throttle screws

IDLE SPEED

◆ See Figure 189

1. Mount the engine in a test tank or move the boat to a body of water.
2. Remove the cowling and connect a tachometer to the powerhead.
3. Turn the pilot screw in until it lightly seats and then back out the specified number of turns, as indicated in the Carburetor Set-Up Specifications chart, found in the Fuel System section.
4. Start the engine and allow it to warm to operating temperature.
5. Check engine speed at idle. The powerhead should idle at the rpm specified in the Tune-up Specifications chart.
6. Place the engine in gear and check engine trolling speed in the same manner.
7. If adjustment is necessary, rotate the idle adjustment screw (not the pilot screw) until the powerhead idles at the required rpm.

THROTTLE LINK

◆ See Figure 190

1. Disconnect the magneto control link.
2. Align the full-closed mark on the pulley with the mark on the bracket.
3. Rotate the magneto base clockwise until the full-closed side of the magneto base stopper No. 1 contacts the adjust bolt for the magneto base stopper No. 2.

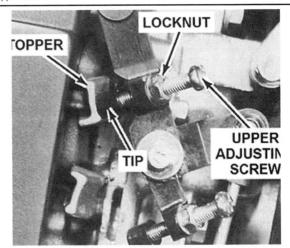

Fig. 186 Rotate the upper adjustment screw until the tip contacts the stopper and tighten the locknut

Fig. 188 Push the cam to close all throttle valves; then tighten the valve screws by rotating them counterclockwise

MAINTENANCE & TUNE-UP 2-57

Fig. 189 Pilot and idle speed screw locations on the side of the powerhead

4. Adjust the plastic snap-on connector on the end of the link rod until it can be reconnected to the control lever ball stud without changing the position of the linkage or magneto base.
5. Align the wide-open throttle mark on the pulley with the mark on the throttle bracket.
6. Adjust the throttle link joint so the wide-open throttle mark on the throttle cam aligns with the center of the carburetor throttle roller.

48 Hp (2-Cylinder) Models

IDLE SPEED

◆ See Figures 191 and 192

1. Mount the engine in a test tank or move the boat to a body of water.
2. Remove the cowling and connect a tachometer to the powerhead.
3. Start the engine and allow it to warm to operating temperature.
4. Check engine speed at idle. The powerhead should idle at the rpm specified in the Tune-up Specifications chart.
5. Place the engine in gear and check engine trolling speed in the same manner.
6. If adjustment is necessary, shut the motor down, then adjust both of the pilot screws (they are mounted horizontally into the side of the

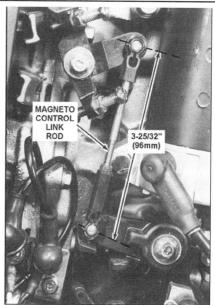

Fig. 190 Magneto link control rod should be adjusted so that it can be reconnected to the control lever ball stud without changing the position of the linkage or magneto base

carburetors). Turn the pilot screw in until it lightly seats and then back out the specified number of turns, as indicated in the Carburetor Set-Up Specifications chart, found in the Fuel System section.
7. Restart and re-warm the powerhead, then finalize the idle speed setting by rotating the idle adjustment screw (not the pilot screws) until the powerhead idles at the required rpm. The idle speed adjustment screw is located on the lower carburetor, just behind the carburetor linkage, it contacts the throttle lever.

IGNITION TIMING

◆ See Figures 193 thru 196

1. If the timing pointer has been moved even SLIGHTLY during maintenance or engine repair, then you'll have to start this procedure by adjusting the timing plate position as follows:
   a. Remove the spark plug from the No. 1 (top) cylinder and install a dial gauge into the spark plug hole in order to measure piston depth.

■ If it is difficult to turn the flywheel, go ahead and remove the spark plug from the No. 2 (lower) cylinder as this will alleviate engine compression.

   b. Turn the flywheel CLOCKWISE slowly, by hand, until the piston reaches Top Dead Center (TDC) as indicated on the dial gauge.

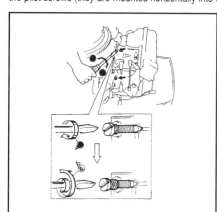

Fig. 191 Pilot screw adjustment - 48 hp 2-cylinder motors

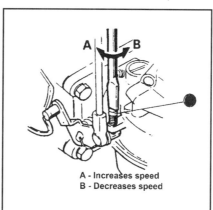

Fig. 192 Idle speed screw adjustment - 48 hp 2-cylinder motors

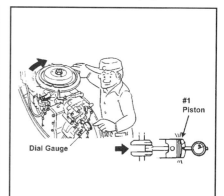

Fig. 193 Use a dial gauge to ensure the No. 1 cylinder is at TDC...

MAINTENANCE & TUNE-UP

c. With the No. 1 cylinder piston at TDC check to make sure the timing pointer is aligned with the TDC mark on the flywheel. If not, loosen the set screw and adjust the pointer position, then secure it with the set screw. Remove the dial gauge and reinstall the spark plug(s).

2. To check and adjust the idle timing, proceed as follows:

■ Refer to the numbers in the accompanying illustration for component identification.

a. Turn the flywheel CLOCKWISE slowly, by hand, until the timing pointer is facing the correct idle timing specification as listed in the Tune-Up Specifications chart in this section. At this point disconnect the magneto control rod (1) from the magneto base (2) and move the base clockwise until it contacts the idle stopper (3). If the mark on the flywheel (4) aligns with the mark on the magneto base (5), the idle timing is correct.

b. However, if the marks are not aligned as noted, loosen the locknut on the idle stopper screw and turn the screw until the marks are in alignment. Turning the screw clockwise will advance timing, while turning the screw counterclockwise will retard timing.

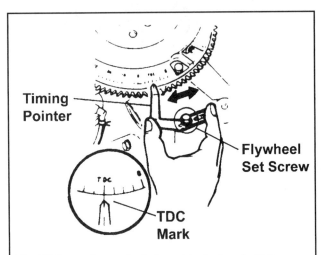

Fig. 194 Then make sure the timing pointer is aligned with the flywheel TDC mark

c. Once the adjustment is correct, tighten the locknut to hold the idle stopper screw in place.

d. You can leave the magneto control rod (1) disconnected from the magneto base (2) as this will be necessary to check/adjust the WOT timing.

3. To check and adjust the WOT timing, proceed as follows:

■ Refer to the numbers in the accompanying illustration for component identification.

a. Turn the flywheel CLOCKWISE slowly, by hand, until the timing pointer is facing the correct WOT timing specification as listed in the Tune-Up Specifications chart in this section. Make sure the magneto control rod (1) is still disconnected from the magneto base (2), or disconnect them now. Then move the base counterclockwise until it contacts the WOT stopper (3). If the mark on the flywheel (4) aligns with the mark on the magneto base (5), the WOT timing is correct.

b. However, if the marks are not aligned as noted, loosen the locknut on the WOT stopper screw and turn the screw until the marks are in alignment. Turning the screw clockwise will retard timing, while turning the screw counterclockwise will advance timing.

c. Once the adjustment is correct, tighten the locknut to hold the WOT stopper screw in place.

d. You can leave the magneto control rod (1) disconnected from the magneto base (2) as this will be necessary when starting the Carburetor & Throttle Linkage Adjustments if the magneto base length must be changed.

CARBURETOR & THROTTLE LINKAGE ADJUSTMENTS

◆ See Figures 196, 197 and 198

 MODERATE

■ The throttle link cannot be properly set unless the ignition idle and WOT timing has already been checked and adjusted.

1. Measure the length of the magneto control rod between the centers of the rod ends. The rod should be 1.95-1.98 in. (49.6-50.4mm) in length measured this way. If not it must be adjusted by turning one or both of the threaded ends, as follows:

a. If not done already, disconnect the magneto control rod from the magneto base (this was necessary for idle and WOT timing checks/adjustment).

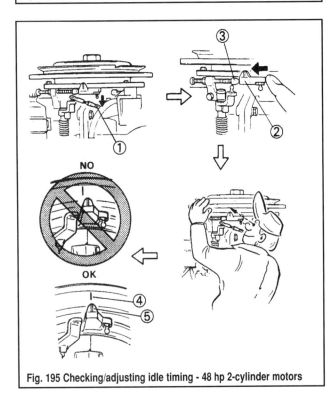

Fig. 195 Checking/adjusting idle timing - 48 hp 2-cylinder motors

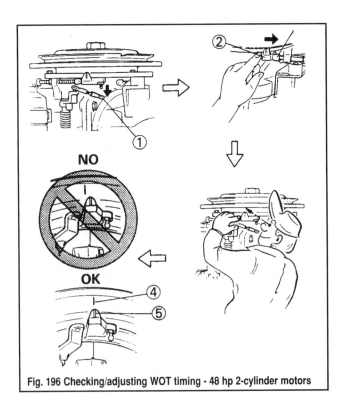

Fig. 196 Checking/adjusting WOT timing - 48 hp 2-cylinder motors

MAINTENANCE & TUNE-UP 2-59

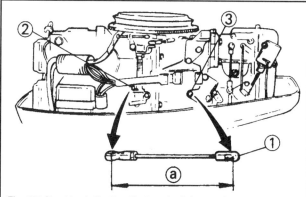

Fig. 197 Checking/adjusting the length of the accelerator link rod - 48 hp 2-cylinder motors

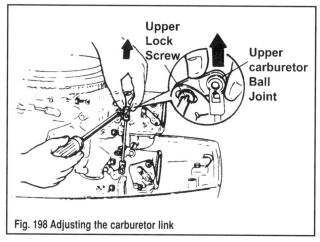

Fig. 198 Adjusting the carburetor link

b. Turn the end of the magneto control rod until the length of the rod measured between the centers of the control rod ends is 1.95-1.98 in. (49.6-50.4mm). Then reconnect it to the base of the control lever.

2. Measure the length of the accelerator link rod between the centers of the rod ends. The rod should be 5.45 in. (138.5mm) in length measured this way. If not it must be adjusted by turning the threaded end, as follows:

■ **Refer to the numbers in the accompanying illustration for component identification.**

a. Disconnect the accelerator link rod (1) from the magneto control lever (2) and accelerator cam (3).

b. Turn the threaded end of the accelerator link until the length is 5.45 in. (138.5mm) as specified between the centers of the ends, then reconnect both ends of the link to the magneto control lever and accelerator cam ball joints.

3. A carburetor link is used to make sure the throttle valves of the 2 carburetors open and close at precisely the same instant. Operate the throttle and observe the 2 throttle valves inside the carburetor throats. If adjustment is necessary, proceed as follows:

a. Loosen the carburetor idle adjust screw an even number of turns (counting them as you go to preserve idle speed adjustment). Loosen the idle adjust screw until the throttle valve closes fully.

b. Loosen the upper carburetor ball joint lock screw (it is normally a left-hand thread screw, so it would LOOSEN by turning CLOCKWISE).

c. Pull up on the upper carburetor ball joint to remove play between the upper and lower carburetors then tighten the upper lock screw (again, if it is a left-hand thread screw, it would TIGHTEN by turning COUNTERCLOCKWISE).

d. Move the accelerator lever up and down several times to make sure the upper and lower carburetors open and close simultaneously.

4. Turn the idle speed adjustment screw back inward the exact same number of turns as loosened before adjusting the link. But, to be certain, connect a tachometer and adjust the Idle Speed as detailed in this section. However if you adjusted the idle speed at the beginning of this Tune-Up, there is no need to readjust the pilot screws.

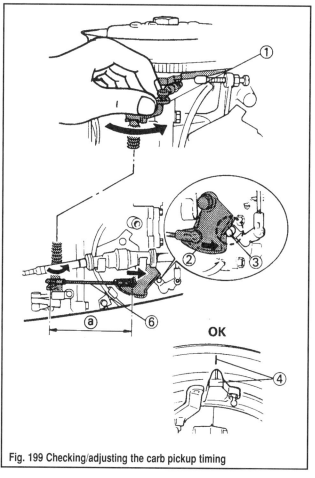

Fig. 199 Checking/adjusting the carb pickup timing

5. Check and adjust the cam roller pickup timing as follows:

■ **Refer to the numbers in the accompanying illustration for component identification.**

a. Turn the flywheel CLOCKWISE slowly, by hand, until the timing pointer is facing the correct carb pickup timing specification as listed in the Tune-Up Specifications chart in this section.

b. Turn the magneto control lever (1) counterclockwise and bring the accelerator cam (2) into light contact with the cam roller (3). At this point the throttle valve will JUST begin to open.

c. Check the mark on the flywheel to make sure it is aligned with the mark on the magneto base (4).

d. If the mark is not aligned, adjust the length (a) of the accelerator link rod (6) so that it is aligned.

CHECKING THE NEUTRAL OPENING LIMIT

◆ See Figure 200

These motors use a type of mechanical limiter to prevent engine over-speed in **Neutral** by limiting the throttle position. There are 2 ways to check operation dynamically or statically. To check the system dynamically, operate the motor in Neutral (in a test tank, using a flushing device or on a body of water) and making sure the engine will not go beyond about 2800-3800 rpm. To check the system statically, visually observe the movement of the shift slider and the magneto control lever (as shown in the accompanying illustration) as the slider is moved in and out of gear.

■ **Refer to the numbers in the accompanying illustration for component identification.**

1. To check the system statically, proceed as follows:

a. Move the shift slider (1) in turn to the **Forward**, **Neutral** and **Reverse** positions.

2-60 MAINTENANCE & TUNE-UP

b. By opening the throttle (2) check that the magneto control lever (3) is in contact with the stopper (4) in the appropriate conditions (for clarification please refer to the accompanying illustration).

2. To check the system dynamically, proceed as follows:
   a. Mount the engine in a test tank or launch the boat on a body of water.
   b. Remove the cowling and attach a tachometer.
   c. Set the shift lever to **Neutral** and start the engine.
   d. Allow the engine to operate and come up to normal operating temperature.
   e. Set the throttle so the engine speed is 2800-3800 rpm. It should not go any further in **Neutral**.

3. If the system does not work properly, inspect the neutral opening limit components and repair/replace, as necessary.

CHECKING/ADJUSTING THE THROTTLE/SHIFT CABLES

Operate the throttle and shift cables, checking for smooth, correct operation. Visually inspect the cables for signs of damage or excessive wear and replace, if necessary. If the cables show excessive slack and/or do not operate correct, adjust the throttle and/or shift cables.

■ The throttle and shift cable adjustments can be performed independently of each other. If only one cable is out of adjustment, you need only adjust THAT cable.

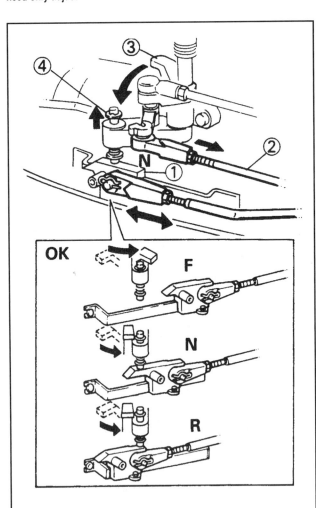

Fig. 200 Checking length neutral opening limit - 48 hp 2-cylinder motors

Shift Cable Adjustment

◆ See Figures 201, 202 and 203

■ Refer to the numbers in the accompanying illustration for component identification.

1. To adjust the shift cable, disconnect the clip (1) which secures the cable end (2) to the pin (3) on the shift slider (4).
2. Move the shift slider (4) to the **Neutral** position (the middle of the slider's travel on its shaft).
3. Now, place the shifter handle in the **Neutral** position.
4. Turn the cable end so that it is threaded onto the inner cable at least 0.31 in. (8mm) deeper than the beginning of the threads and position it as necessary to connect to the shift slider. Tighten the locknut to hold the cable end in this position.

✻✻ CAUTION

Failure to thread the cable end at least 0.31 in. (8mm) onto the cable may result in the end and cable separating in use which would cause an inability to shift (including a possible inability to obtain Neutral when underway, which could cause navigation hazards).

5. Connect the cable end to the pin on the shift slider and secure using the clip.

Throttle Cable Adjustment

◆ See Figures 204, 205 and 206

■ Refer to the numbers in the accompanying illustration for component identification.

1. To adjust the throttle cable, start by disconnecting the free accel link stopper, then remove the cable end (2) from the pin (3) on the lower magneto control lever (4).
2. Move the middle magneto control lever (5) to the fully closed position and place the throttle grip in the fully closed (idle) position.
3. Turn the cable end so that it is threaded onto the inner cable at least 0.31 in. (8mm) deeper than the beginning of the threads and position it as necessary to connect to the pin (3) on the lower magneto control lever (4). Tighten the locknut to hold the cable end in this position.

✻✻ CAUTION

Failure to thread the cable end at least 0.31 in. (8mm) onto the cable may result in the end and cable separating in use which would cause an inability to operate the throttle (including a possible inability to reduce throttle from a WOT or advanced state when underway, which could cause navigation hazards).

4. Connect the cable end to the pin on the lower magneto control lever, then connect the free accel link stopper.
5. After the cable is adjusted, turn the throttle grip from idle to WOT and back again 2-3 times. Make sure that when it is in the idle or SLOW position the middle magneto control lever (5) moves to the fully closed position. If it does not close fully, adjust the length of the cable again and make sure that the lever can close fully.

CHECKING/ADJUSTING THE STARTER LOCKOUT

◆ See Figure 207

This model contains a starter lockout safety feature to prevent the motor from being started while in gear. The system should be checked at each tune-up to ensure that it is adjusted and functioning properly. Checking is a simple matter of placing the motor in gear and gently attempting to start the motor. If the starter will not rotate, the system is functioning and no further attention is required. However, if the motor rotates in gear, then adjust the cable as follows:

MAINTENANCE & TUNE-UP 2-61

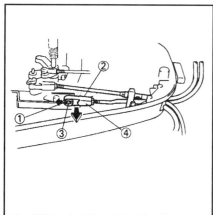

Fig. 201 To adjust the shift cable, disconnect the end (2) from the slider (4)...

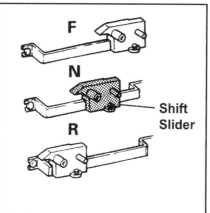

Fig. 202 ... then position the slider in Neutral...

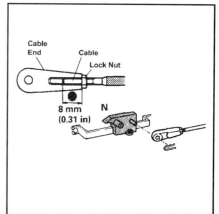

Fig. 203 ...then adjust the cable end and secure it to the slider

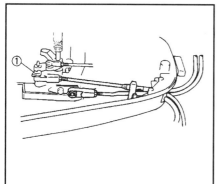

Fig. 204 Disconnect the free accel link stopper...

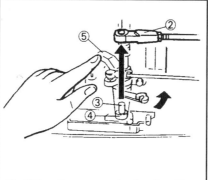

Fig. 205 ... then remove the throttle cable end...

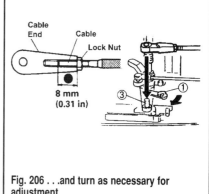

Fig. 206 ...and turn as necessary for adjustment

■ Refer to the numbers in the accompanying illustration for component identification.

1. Set the shifter handle in the **Neutral** position.
2. Loosen the bolt securing the 1st cable stay (2).
3. Align the right side edge of the 1st cable stay (2) with the same right side edge of the cable bracket (8), then tighten the bolt securing the stay.
4. Loosen the bolt securing the 2nd cable stay (6).
5. Move the 2nd cable stay (6) as necessary to align the mark (a) on the stopper (5) with the mark (b) on the guide cam (3), then tighten the bolt securing the stay.
6. Recheck operation of the system. If it still does not work correctly, try readjusting the 1st cable stay (2). Otherwise, repeat the entire adjustment procedure.

25/30 Hp (496cc 3-Cylinder) Models

FUEL ENRICHMENT VALVE ADJUSTMENT

◆ See Figure 208

At least annually, the fuel enrichment valve should be adjusted to ensure proper operation.

■ Refer to the numbers in the accompanying illustration for component identification.

1. Close the choke valve fully, then adjust the length of the pulley hook (1) until the marks on the plunger (2) and fuel enrichment valve end (3) are aligned.
2. Next, place the pulling wire hook (1) over the pin (4) on the choke lever and secure it using an O-ring (5).

CARBURETOR LINKAGE

◆ See Figures 209 thru 212

1. Loosen the throttle stop screw by turning clockwise until clearance develops between the screw and the throttle valve stop.
2. Close the throttle valves.
3. Loosen the throttle stop screws for the No. 1 and No. 2 carburetors.
4. Make sure all carburetor throttle valves are closed, then lightly depress the No. 2 carburetor throttle roller and tighten the No. 1 and No. 2 throttle stop screws.
5. Adjust the Pickup Timing followed by the Idle Speed.

IGNITION TIMING

◆ See Figures 213 thru 217

1. Remove the No. 1 (top) spark plug and install a dial indicator gauge into the cylinder on top of the piston crown. Although it is not absolutely necessary, removing the other 2 spark plugs will relieve engine compression, making it easier to rotate the engine in the upcoming steps.
2. Rotate the flywheel clockwise until the dial indicator registers TDC (top dead center) of the piston's travel.
3. Observe the timing plate on the port side of the powerhead. Check to verify the timing pointer is aligned with the TDC mark embossed on the flywheel. If the pointer is not aligned, loosen the timing plate set bolt beneath the pointer and move the pointer to the correct location. Tighten the bolt to hold the pointer aligned with the TDC mark.
4. Remove the dial indicator gauge and install the spark plug. Connect the high tension lead.

MAINTENANCE & TUNE-UP

5. Rotate the flywheel clockwise until the timing pointer aligns in the center of the WOT timing specification range (the 25° BTDC mark) embossed on the flywheel. At this point the dial gauge should indicate that the piston is about 0.14 in. (3.55mm) BTDC.

6. Move the magneto lever away from the link to align the single line timing mark on the base with the TDC mark on the flywheel. If the marks do not align, disconnect the link from the magneto base, loosen the locknut and adjust the length of the link as necessary to reinstall the link to the magneto base with the marks aligned. Retighten the locknut after adjustment.

7. Again, slowly turn the flywheel clockwise, this time until the timing pointer indicates the center of the idle timing range (the 5° ATDC mark) embossed on the flywheel.

8. Rotate the magneto lever toward the link, to align the single line timing mark on the base with the TDC mark on the flywheel. Check to verify the magneto base idle stopper makes contact with the crankcase stopper. If they do not make contact or if they do make contact but the TDC mark is not aligned with the single line mark on the base, loosen the base idle stopper set bolt, align the marks, bring the idle stopper into contact and then tighten the bolt.

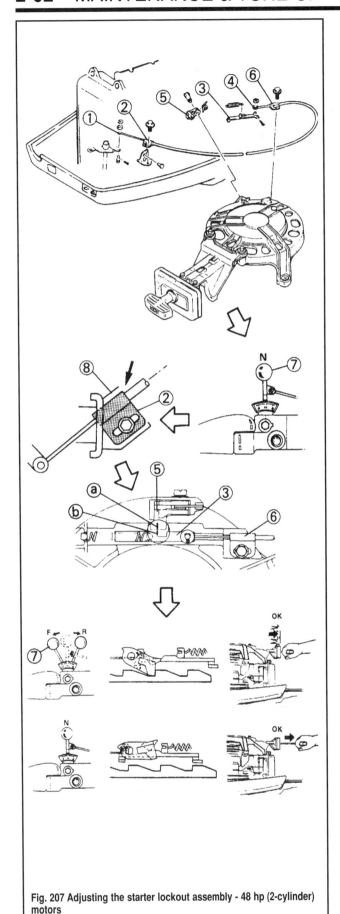

Fig. 207 Adjusting the starter lockout assembly - 48 hp (2-cylinder) motors

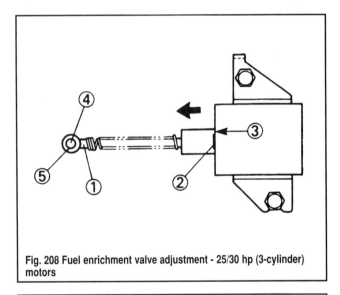

Fig. 208 Fuel enrichment valve adjustment - 25/30 hp (3-cylinder) motors

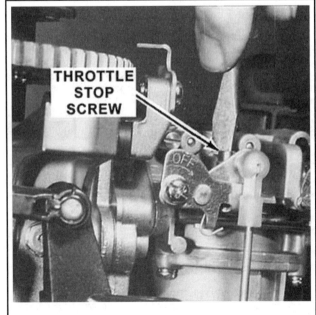

Fig. 209 Throttle stop screw location on the carburetor

MAINTENANCE & TUNE-UP

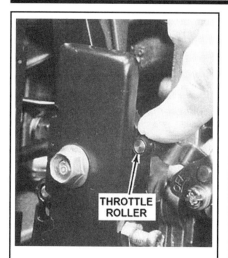

Fig. 210 Lightly press the carburetor throttle roller (closing the throttle valve)

Fig. 211 The throttle lever screws are left-hand thread. Rotate them clockwise to loosen and counterclockwise to tighten

Fig. 212 Push down on the throttle roller again to keep the throttle valves closed while the screws are tightened

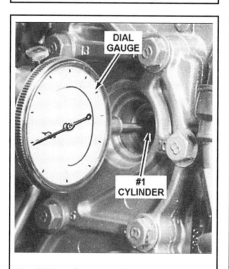

Fig. 213 Install a dial indicator gauge into the cylinder on top of the piston crown

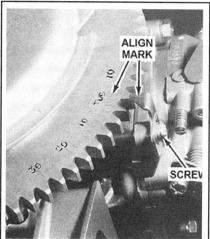

Fig. 214 Verify the timing pointer is aligned with the TDC mark on the flywheel

Fig. 215 Rotate the magneto lever this direction to check WOT timing (and adjust using the magneto link)

Fig. 216 Rotate the magneto lever this direction to check idle timing...

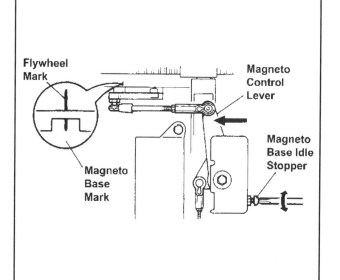

Fig. 217 ...in this position, adjust the idle timing using the idle stopper

2-64 MAINTENANCE & TUNE-UP

IDLE SPEED

◆ See Figures 218 and 219

1. Mount the engine in a test tank or move the boat to a body of water.
2. Remove the cowling and connect a tachometer to the powerhead.
3. Turn the pilot screw for each carburetor (threaded horizontally into the side of each carburetor) in until it lightly seats and then back out the specified number of turns, as indicated in the Carburetor Set-Up Specifications chart, found in the Fuel System section.
4. Start the engine and allow it to warm to operating temperature.
5. Check engine speed at idle. The powerhead should idle at the rpm specified in the Tune-up Specifications chart.
6. Place the engine in gear and check engine trolling speed in the same manner.
7. If adjustment is necessary, rotate the idle adjustment screw (NOT the pilot screw) until the powerhead idles at the required rpm. The idle adjustment screw is threaded vertically down into the throttle valve linkage on the top carburetor.

Fig. 218 Turn the pilot screw in until it lightly seats and then back out the specified number of turns

Fig. 219 Rotate the idle adjustment screw (not the pilot screw) until the powerhead idles at the required rpm

PICKUP TIMING

◆ See Figure 220 and 221

1. Check and adjust the Ignition Timing and the Idle Speed before attempting to check/adjust the pickup timing.
2. Keep the engine mounted in the test tank or the boat on the body of water used for Idle Speed adjustment.
3. In addition to the tachometer used for Idle Speed adjustment, connect a timing light to the motor.
4. Start the engine and allow it to re-warm to operating temperature. Place the lower unit in Neutral.
5. Aim the timing light at the timing pointer. Move the magneto control lever VERY slightly toward the WOT position (move the magneto lever slightly away from the link) while an assistant aims the timing light at the flywheel. Find the position that achieves the proper Carb Pickup timing specification (as listed in the Tune-Up Specifications chart in this section). Hold the control lever in that position once it is determined.
6. With your other hand, carefully loosen the throttle lever screw for the middle carburetor, then bring the throttle roller just LIGHTLY into contact with the throttle cam and retighten the throttle lever screw.
7. Shut the powerhead down and remove the tune-up equipment (tachometer and timing light).

OIL PUMP LINK ADJUSTMENT

◆ See Figure 222

1. Manually open the throttle valve to the WOT position.
2. Turn the oil pump lever toward the wide open throttle position until it is against the wide-open stopper.
3. Loosen the locknut and adjust the plastic snap-on connector on the end of the link rod until
it can be reconnected to the ball stud without changing the throttle or oil pump lever position.
4. Reconnect the link rod to the ball stud and tighten the locknut.
5. Manually operate the throttle valve and make sure the link adjustment does not prevent it from reaching the WOT position.

CHECKING/ADJUSTING THE STARTER LOCKOUT

◆ See Figure 223

This model contains a starter lockout safety feature to prevent the motor from being started while in gear. The system should be checked at each tune-up to ensure that it is adjusted and functioning properly. Checking is a simple matter of placing the motor in gear and gently attempting to start the motor. If the starter will not rotate, the system is functioning and no further attention is required. However, if the motor rotates in gear, then adjust the cable as follows:

1. Begin with the engine not running and shifter positioned in **Neutral**.
2. Locate the adjustment bracket (the point where the cable passes vertically through a powerhead bracket). Loosen the locknut and turn the adjuster nut to change the length of the cable until the starter stop plunger line is centered in the sight hole.
3. Once positioned properly, retighten the locknut to hold the cable in this position.
4. Verify that the lockout is now working properly.

TILLER MODEL THROTTLE CABLE ADJUSTMENT

◆ See Figures 224 and 225

1. Shift the lower unit into **FORWARD** gear.
2. Rotate the throttle grip to the Wide-Open Throttle (WOT) position. At this point, the center of the throttle roller should align with the WOT mark on the throttle cam.
3. If the marks do not align, loosen the adjusting bolt locknut on the pull side of the throttle cable. Turn the adjusting bolt until all slack is removed, then tighten the locknut.
4. Loosen the adjusting bolt locknut on the push side of the throttle cable. Turn the adjusting bolt until the cable has approximately 0.12 in. (3 mm) slack, then tighten the locknut.

MAINTENANCE & TUNE-UP

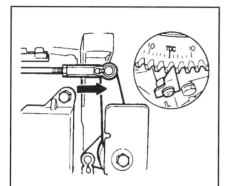

Fig. 220 Manually move the magneto lever toward WOT just enough to obtain Carb Pickup Timing...

Fig. 221 ...then adjust the throttle valve screw for the middle carburetor

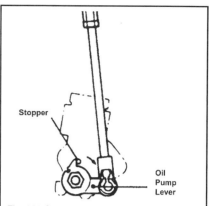

Fig. 222 Oil pump link adjustment - 25/30 hp (3-cylinder) motors

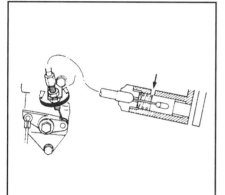

Fig. 223 When adjusting the starter lockout, make sure the end of the plunger is visible and centered in the hole

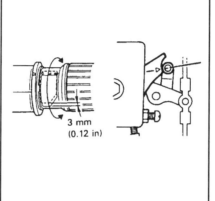

Fig. 224 Check throttle cable adjustment by rotating to WOT and checking the throttle cam mark

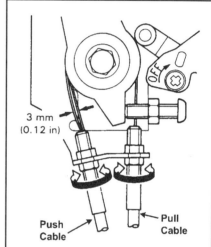

Fig. 225 Align the marks and remove slack using the pull cable, then add 0.12 in. (3mm) freeplay with the push cable

CHECKING/ADJUSTING THE NEUTRAL OPENING LIMIT

◆ See Figures 226 and 227

These motors use a type of mechanical limiter to prevent engine overspeed in **Neutral** by limiting the throttle position. In theory, you could shortcut the checking procedure and simply operate the motor (in a test tank or with the boat launched) and seeing the top rpm the motor will achieve in **Neutral**. But, otherwise, the checking and adjustment procedures involve first setting the linkage up properly and THEN making sure the engine will not go beyond about 3700-3800 rpm. To check and/or adjust the neutral opening limit, proceed as follows:

■ Refer to the numbers in the accompanying illustration for component identification.

1. Mount the engine in a test tank or launch the boat on a body of water.
2. Remove the cowling and attach a tachometer.
3. Set the shift lever to **Forward**, then set the throttle to the WOT position until the throttle hits the stops.
4. Rotate the magneto control lever until the throttle control lever (1) contacts the stopper (2).
5. Adjust the length of the link rod connecting the magneto control lever.
6. Set the shift lever to **Neutral** and start the engine.
7. Allow the engine to operate and come up to normal operating temperature.
8. Open the throttle and, while keeping the magneto control lever in contact with the adjusting screw, turn the screw (in or out, as necessary) until the engine speed is 3700-3800 rpm.

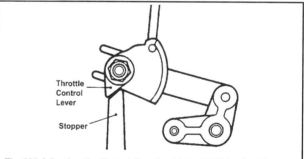

Fig. 226 Adjusting the Neutral Opening Limit - 25/30 hp (3-cylinder) motors

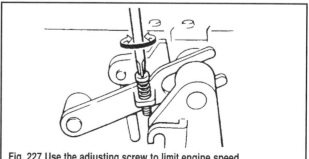

Fig. 227 Use the adjusting screw to limit engine speed

2-66 MAINTENANCE & TUNE-UP

9. Lower engine speed down to idle, then with the shifter still in **Neutral** rotate the throttle back up to speed and recheck operation. Readjust, as necessary.

28J-50 Hp (698cc) Models

IGNITION TIMING

These models are equipped with an electronic ignition advance mechanism in place of a mechanical ignition advance system. Adjustment of the throttle linkage sets the timing.

THROTTLE LINKAGE CHECKING/ADJUSTMENT

◆ See Figures 228 and 229

1. Start by checking the linkage as follows:
   a. Turn the magneto control lever so that the adjust screw contacts the full-retard stopper.
   b. Check the CDI indicator to see if it aligns with the Idle timing specification as stated in the Tune-up Specifications chart.
   c. Turn the magneto control lever so that the adjust screw contacts the full-advance adjusting screw.
   d. Check the CDI indicator to see if it aligns with the WOT timing specification as stated in the Tune-up Specifications chart.
2. To adjust full-retard timing, loosen the locknut and adjust the length of the full-retard screw to 20 mm (0.79 in.) Tighten the locknut.
3. Loosen the control rod locknut and disconnect the control rod from the CDI unit.
4. Turn the magneto control lever so its adjusting screw contacts the full-retard stopper.
5. Adjust the control rod length so the CDI unit indicator aligns with the Idle timing specification as stated in the Tune-up Specifications chart.

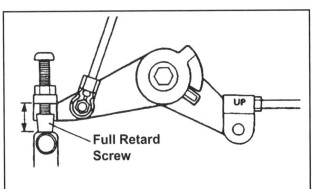

Fig. 228 Moving the full-retard or full-advance stoppers...

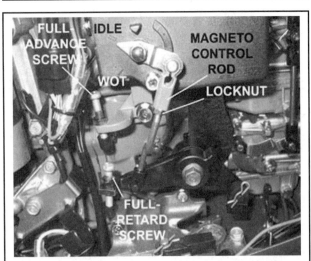

Fig. 229 ...and the positioning of the magneto control rod sets Idle and WOT timing

6. Connect the magneto control rod to the CDI unit.
7. To adjust the full-advance timing, turn the magneto control lever so that it contacts the full-advance adjusting screw
8. Adjust the full-advance adjusting screw so that the CDI unit indicator aligns with the full-advance timing specification as stated in the Tune-up Specifications chart.
9. Adjust the throttle cable.

THROTTLE CABLE

◆ See Figure 230

1. Adjust the Throttle Linkage, as detailed in this section.
2. Shift the lower unit into **FORWARD** gear.
3. Rotate the throttle grip to the wide-open throttle position. At this point, the throttle valve lever should contact the wide-open throttle stopper.
4. If it does not make contact, loosen the locknut and remove the clip.
5. Disconnect the cable joint from the magneto control lever
6. Rotate the throttle grip to the fully closed position.
7. Turn the magneto control lever so that it's adjusting screw contacts the full-retard stopper.

■ The cable joint should be screwed into the fitting by at least 0.31 in. (8 mm).

8. Adjust the position of the cable joint until its hole aligns with the set pin. Position the cable joint with the UP mark facing up and install it onto the set pin.
9. Install the clip and tighten the locknut.

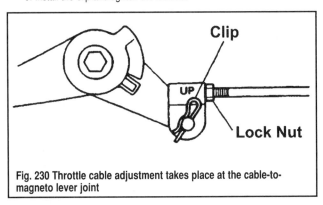

Fig. 230 Throttle cable adjustment takes place at the cable-to-magneto lever joint

IDLE SPEED

◆ See Figures 231, 232 and 233

1. Adjust the Throttle Linkage, as detailed in this section.
2. Mount the engine in a test tank or move the boat to a body of water.
3. Remove the cowling and connect a tachometer to the powerhead.
4. Start the engine and allow it to warm to operating temperature.
5. Check engine speed at idle. The powerhead should idle at the rpm specified in the Tune-up Specifications chart.
6. For all except Jet models, place the engine in gear and check engine trolling speed in the same manner.
7. If adjustment is necessary, stop the powerhead, then loosen the idle adjustment screw (located on the middle carburetor) in order to fully close the throttle valve.
8. Loosen the throttle lever screws that secure the upper and middle carburetors by turning them CLOCKWISE (they are left-hand thread).
9. Turn the pilot screws (threaded horizontally into the top side of each carburetor) in until they lightly seat and then back out the specified number of turns, as indicated in the Carburetor Set-Up Specifications chart, found in the Fuel System section.
10. Start and re-warm the engine, then the idle adjustment screw (not the pilot screw) until the powerhead idles at the required rpm. Turning the screw INWARD will INCREASE idle speed, while turning the screw OUTWARD will DECREASE idle speed.

MAINTENANCE & TUNE-UP 2-67

Fig. 231 To adjust idle speed, first loosen the idle adjustment screw

Fig. 232 ...then loosen the throttle lever securing screws for the top and middle carbs...

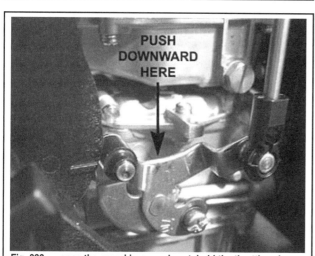

Fig. 233 ...once the speed is properly set, hold the throttle valve closed on the lower carb while you tighten the throttle lever securing screws on the top and middle carbs

11. Push lightly downward on the throttle lever of the lower carburetor and hold it in this position (holding it fully-closed) and tighten the throttle lever securing screws of the upper and middle carburetors by turning them COUNTERCLOCKWISE (again, they are left-hand thread and tighten to the LEFT).

OIL PUMP LINK

◆ See Figure 234    MODERATE

1. Open the carburetor throttle valve to the WOT position. At this point, the oil pump lever
should be approximately 0.04 in. (1.0 mm) off the WOT side stopper.
2. If adjustment is necessary, fully open the carburetor throttle valve.
3. Loosen the locknut on the link joint, then disconnect the link rod from the oil pump lever ball stud. Set the oil pump lever 0.04 in. (1 mm) off the full open side stopper (using a feeler gauge), then adjust the plastic snap-on connector on the end of the link rod until its hole aligns with the oil pump set pin.
4. Connect the link joint to the ball stud and check that the throttle valve still opens fully (and is not limited by the pump link). If the adjustment is good, tighten the locknut.
5. Rotate the throttle grip to the wide-open throttle position and recheck the clearance.

SHIFT CABLE ADJUSTMENT

◆ See Figure 235    EASY

Visually check the shift linkage movement in response to the shift handle. If the shifter does not engage properly or smoothly, adjust, as follows:
1. Set the shift control lever to the **Neutral** position.
2. Remove the retaining clip and disconnect the shift cable joint from the powerhead linkage.
3. Align the center of the linkage set pin with the mark in the middle of the shift bracket.
4. Loosen the locknut, then adjust the position of the shift cable joint (by turning the joint inward or outward on the cable threads) until its mounting hole aligns with the linkage set pin at the center of the bracket. Tighten the locknut to hold it in position on the cable.

✱✱ CAUTION

Make sure that after adjustment the cable joint is threaded over at LEAST 0.31 in. (8mm) of cable threads, otherwise there is a risk that the joint could separate in service causing a severe navigation/control hazard.

5. Position the cable joint over the control lever set pin and secure using the retaining clip.

CHECKING/ADJUSTING THE STARTER LOCKOUT

◆ See Figure 236    EASY

Manual start models contain a starter lockout safety feature to prevent the motor from being started while in gear. The system should be checked at each tune-up to ensure that it is adjusted and functioning properly. Checking is a simple matter of placing the motor in gear and gently attempting to start the motor. If the starter will not rotate, the system is functioning and no further attention is required. However, if the motor rotates in gear, then adjust the cable as follows:
1. Begin with the engine not running and shifter positioned in **Neutral**.
2. Locate the adjustment bracket (the point where the cable passes horizontally through a powerhead bracket). Loosen the locknut (on the outside, or manual starter side of the cable bracket) and turn the adjuster nut (on the inside of the bracket) to change the length of the cable until the starter stop plunger line is centered in the sight hole in the starter case.
3. Once positioned properly, retighten the locknut to hold the cable in this position.
4. Verify that the lockout is now working properly.

2-68 MAINTENANCE & TUNE-UP

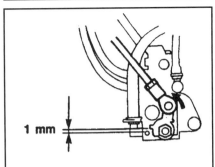

Fig. 234 Checking/adjusting the oil pump link - 28J-50 hp (698cc) motors

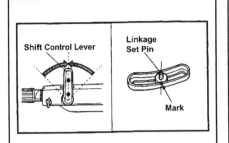

Fig. 235 Shift cable adjustment - 28J-50 hp (698cc) motors

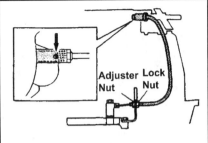

Fig. 236 Checking/adjusting the starter lockout - 28J-50 hp (698cc) motors

50-70 Hp (849cc) and 65J-90 Hp (1140cc) Models

■ The head of 50-70 hp (849cc) and 65J-90 hp (1140cc) models is slightly misleading. The timing and synchronization procedures listed under this head apply to MOST of the models produced and marketed on both these powerhead configurations, however there are some notable exceptions. These procedures DO NOT cover any versions of the E60 (a type of 60 hp, 849cc) motor marketed in the U.S. and worldwide (though they do cover the 60, P60 and C60). Also, these procedures do not cover the E75 (75 hp, 1140cc) motor marketed in the U.S. or all versions of the 75 "A" models (75 hp, 1140cc w/ A in the model suffix), 85 "A" models (85 hp, 1140cc w/ A in the model suffix) and the E65J (65 hp version of the 60J) all marketed worldwide, except the U.S. These exceptions, listed as the E60 (849cc) and E75, 75A Models, 85A Models, E60J (1140cc) motors are covered in a separate section, following these adjustments.

TIMING PLATE POSITION

◆ See Figures 237 and 238

■ This procedure MUST be performed prior to adjusting the ignition timing. And, since the Idle Speed procedure (which is also performed before checking the timing) involves running the motor (and making components hot to the touch), we thought it would be a good idea to do this FIRST.

1. Remove the spark plugs, and install a dial indicator in the No. 1 cylinder spark plug hole.

■ Although it is not absolutely necessary to remove ALL spark plugs, doing so will relieve engine compression and make turning the flywheel to an exact position much easier.

2. Slowly rotate the flywheel clockwise and stop when the piston reaches top dead center (TDC). This will be obvious using the dial gauge because the TDC is the highest point of piston travel.
3. Check the timing plate alignment with the flywheel timing scale.
4. If the end of the timing plate is not aligned with the TDC mark on the CDI magneto rotor, loosen the timing plate set screw, align the timing plate end with the TDC mark, then tighten the screw.

CARBURETOR LINKAGE

◆ See Figures 239 and 240

1. Loosen the carburetor idle adjustment screw, turning it outward until the throttle valves are closed.

■ The throttle lever securing screws have left-hand threads (as normally indicated by an arrow and embossed OFF pointing to the left). Turn the screws clockwise to loosen and counterclockwise to tighten.

2. For 50-70 hp (849cc) models, loosen the throttle lever securing screws on the upper and middle carburetors.

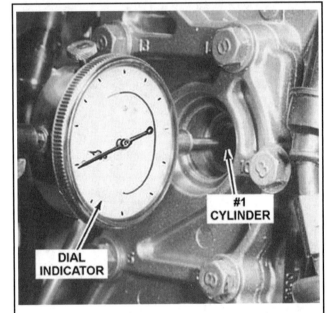

Fig. 237 Install a dial indicator in the No. 1 cylinder and slowly rotate the flywheel clockwise and stop when the piston reaches top dead center (TDC)

Fig. 238 If necessary, loosen the timing pointer set-screw, reposition the pointer and tighten the screw

MAINTENANCE & TUNE-UP 2-69

3. For 65J-90 hp (1140cc) models, loosen the throttle lever securing screws on the upper and lower carburetors.

4. While lightly pushing downward on the throttle lever of the carburetor whose throttle valve screw was NOT loosened, tighten the throttle lever securing screws of the other two carburetors. This means that on 50-70 hp (849cc) models you'll push down on the throttle valve for the lower carb, while on 65J-90 hp (1140cc) models you'll push on the middle carb lever.

5. Move the accelerator lever up and down several times to make sure all carburetors open and close simultaneously.

IDLE SPEED

◆ See Figures 240 and 241   MODERATE

1. Mount the engine in a test tank or move the boat to a body of water.
2. Remove the cowling and connect a tachometer to the powerhead.
3. Turn each of the carburetor pilot screws (threaded horizontally into the side of the carburetor cover) inward until it JUST lightly seats and then back out the specified number of turns, as indicated in the Carburetor Set-Up Specifications chart, found in the Fuel System section.
4. Start the engine and allow it to warm to operating temperature.
5. Check engine speed at idle. The powerhead should idle at the rpm specified in the Tune-up Specifications chart.
6. For all except Jet models, place the engine in gear and check engine trolling speed in the same manner.
7. If adjustment is necessary, rotate the idle adjustment screw (not the pilot screw) until the powerhead idles at the required rpm. The idle adjustment screw is found on the middle carburetor for these models and is threaded vertically downward, contacting the throttle valve linkage. Turning the screw INWARD will INCREASE idle speed, while backing it OUTWARD will DECREASE idle speed.

THROTTLE LINKAGE

◆ See Figures 242 thru 246   MODERATE

The ignition timing on these motors is advanced electronically by the CDI ignition module in response to an internal switch which is moved by throttle linkage (the throttle control link and throttle cam). The movement of this linkage is controlled by adjustable Idle and WOT stopper screws. The throttle control link and cam lengths must be checked and set before the ignition timing itself can be adjusted.

1. Check the length of the throttle sensor control link rod by measuring the length from the center of each link end (the center of the ends that fit over the link ball joints). The link should be:
- 50-70 hp: 4.70-4.74 in. (119.5-120.5mm)
- 65J-90 hp: 3.66-3.70 in. (93-94mm)

2. Check the length of the throttle cam link rod by measuring the length from the center of each link end (the center of the ends that fit over the link ball joints). The link should be:
- 50-70 hp: 3.74 in. (95mm)
- 65J-90 hp: 4.74 in. (120.5mm)

3. If the length of either link is off, loosen the locknut for that link, then disconnect the link rod from the lever ball studs and adjust one or both of the plastic snap-on connectors on the end of the link to obtain the correct length. Install the link joint(s) and tighten the locknut.

4. Adjust the length of the idle stopper screw so that when the screw contacts the stopper, the timing indicator aligns with the idle mark on the CDI unit.

■ For 65J-90 hp motors, the nut in the magneto control lever should be 0.08 in. (2 mm) inset, into the magneto control lever.

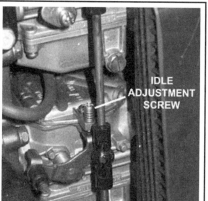
Fig. 239 Throttle valve securing screw - 50-70 hp (849cc) and 65J-90 hp (1140cc) motors

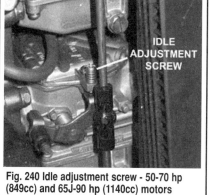
Fig. 240 Idle adjustment screw - 50-70 hp (849cc) and 65J-90 hp (1140cc) motors

Fig. 241 Pilot screw - 50-70 hp (849cc) and 65J-90 hp (1140cc) motors

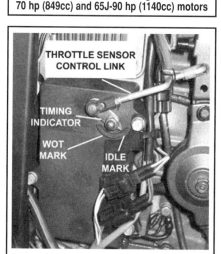
Fig. 242 Ignition timing is controlled through linkage length...

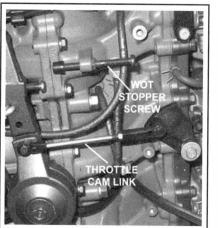
Fig. 243 ... and through positioning the WOT...

Fig. 244 ...and idle stopper screws - 65J-90 hp (1140cc) motors shown in these photos

2-70 MAINTENANCE & TUNE-UP

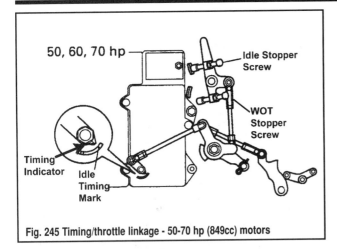

Fig. 245 Timing/throttle linkage - 50-70 hp (849cc) motors

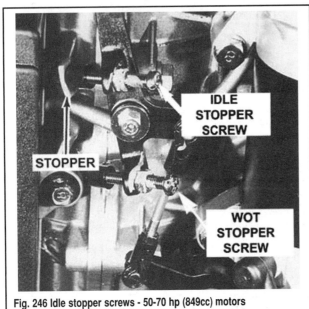

Fig. 246 Idle stopper screws - 50-70 hp (849cc) motors

5. Adjust the length of the WOT stopper screw so that when the screw contacts the stopper, the timing indicator aligns with the WOT mark on the CDI unit.

IGNITION TIMING

◆ See Figures 242 thru 246

The ignition timing on these motors is advanced electronically by the CDI ignition module in response to an internal switch which is moved by throttle linkage (the throttle control link and throttle cam). The movement of this linkage is controlled by adjustable Idle and WOT stopper screws. The throttle control link and cam lengths must be checked and set before the ignition timing itself can be adjusted.

■ Both the Timing Plate Position and the Throttle Linkage must be checked and, if necessary, adjusted before checking and adjusting the ignition timing.

1. Leave the engine mounted in the test tank or keep the boat launched.
2. In addition to the tachometer which was installed for the Idle Speed adjustment procedure, connect a timing light to the No. 1 cylinder.
3. Start the engine and allow it to warm up for approximately 5 minutes. Let the engine idle at specification.
4. Place the ignition in the idle (full-retard position) by manually pushing the linkage toward the CDI unit until the idle screw contacts the stopper.

5. Point the timing light at the timing pointer. The timing pointer should align with the Idle timing specification as stated in the Tune-up Specifications chart.
6. If the timing pointer does not align, loosen the locknut on the Idle Stopper Screw and adjust the screw position until the timing is within specification. Tighten the locknut
7. Manually move the magneto control lever to the wide-open throttle position (WOT or full-advanced ignition, by moving the linkage away from the CDI unit).
8. Increase engine to the maximum speed specified in the Tune-up Specifications chart.
9. Point the timing light at the timing pointer. The timing pointer should align with the WOT timing specification as stated in the Tune-up Specifications chart.
10. If the timing pointer does not align, loosen the locknut on the WOT Stopper Screw and adjust the timing as necessary.
11. Shut the powerhead down and remove the test/diagnostic equipment. No other adjustments will be performed with the engine running.

PICKUP TIMING

◆ See Figure 247 and 248

1. Place the ignition in the idle (full-retard position), by manually pushing the linkage toward the CDI unit until the idle screw contacts the stopper.
2. Bring the throttle cam to JUST lightly contact the throttle lever roller. The throttle valve should not open.
3. Loosen the locknut on the accelerator link joint, then disconnect the link rod from the ball stud. Adjust the plastic snap-on connector on the end of the accelerator link rod until its hole aligns with the set pin, then connect the magneto control lever to the accelerator cam.

OIL PUMP LINK

◆ See Figure 249

This adjustment obviously only applies to models equipped with the Precision Blend automatic oiling system. This normally means that models like the C75, C80 and C85 models will not require this adjustment, however the system may have been retrofitted to an engine not originally equipped with it.

1. Manually open the carburetor throttle valve to the WOT position.
2. At this point, the oil pump lever should be 0.04 in. (1 mm) off the WOT side stopper.
3. If adjustment is necessary, fully open the carburetor throttle valve. Set the oil pump lever 0.04 in. (1 mm) off the wide-open throttle side stopper and hold it there gently against a feeler gauge.
4. Loosen the locknut on the link joint, then disconnect the link rod from the oil pump lever ball stud. Adjust the plastic snap-on connector on the end of the link rod until its hole aligns with the oil pump set pin.
5. Connect the link joint and check that the throttle valve opens fully.
6. Reconnect the link rod to the ball stud and tighten the locknut.
7. Manually open the throttle to the WOT position and recheck the clearance.

E60 (849cc) and E75, 75A Models, 85A Models, E60J (1140cc) Motors

■ Although the motors listed under this heading bear a great many mechanical similarities to the other 849cc and 1140cc motors listed earlier in this section, they contain enough differences in the carburetion and ignition set-ups that require separate timing and synchronization procedures. Therefore procedures under this heading apply to all versions of the E60 (60 hp, 849cc) motor marketed in the U.S. and worldwide. Also, these procedures cover the E75 (75 hp, 1140cc) motor marketed in the U.S. or all versions of the 75 "A" models (75 hp, 1140cc w/ A in the model suffix), 85 "A" models (85 hp, 1140cc w/ A in the model suffix) and the E65J (65 hp version of the 60J) all marketed worldwide, except the U.S.

MAINTENANCE & TUNE-UP 2-71

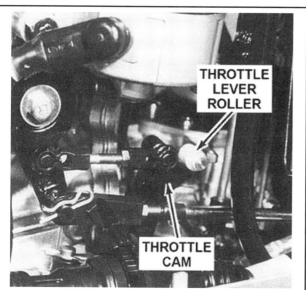

Fig. 247 Bring the throttle cam to contact the throttle lever roller lightly

Fig. 248 The accelerator link rod is located just below the starter motor on this model - 50-70 hp (849cc) shown in these photos

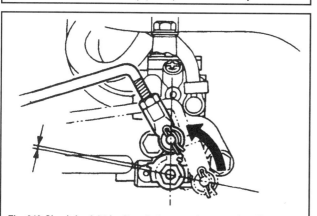

Fig. 249 Check for 0.04 in. (1mm) clearance between the oil pump lever and the WOT side stopper - 50-70 hp (849cc) and 65J-90 hp (1140cc) motors

IGNITION TIMING

◆ See Figure 250, 251 and 252

 MODERATE

Ignition timing on these models is checked statically, by visually observing the linkage and timing marks while the engine is NOT running. The WOT and Idle timing settings are first checked, and adjusted, only if out of specification.

1. Slowly turn the flywheel in the normal direction of rotation (clockwise) until the timing mark for WOT operation aligns with the pointer as follows:
- E60: 19° BTDC
- 75AM, 75AE (European), 75AET (European): 20° BTDC
- 75AEM, 75AE (Exc. European), 75AET (Exc. European): 22° BTDC
- 85A: 24° BTDC
- E60J, E75B: 22° BTDC

2. Rotate the magneto control lever clockwise until it contacts the WOT stopper.

3. Visually check the mark on the underside of the flywheel to make sure it aligns with the mark on the magneto base assembly. If it does, timing is properly set and no adjustment is necessary, you can stop here. If not, continue the procedure to properly adjust/set the ignition WOT timing.

4. Remove the spark plugs (all to relieve engine compression), then mount a dial gauge so that it can be used to measure the depth of the No. 1 piston. Slowly turn the flywheel by hand CLOCKWISE until the piston comes up to TDC; then zero the gauge.

5. Set the piston to the appropriate position (depth) before (below) TDC for WOT timing adjustment by slowly turning the flywheel CLOCKWISE until the dial gauge reads as follows:
- E60: 0.1 in. (2.42mm)
- 75AM, 75AE (European), 75AET (European): 0.11 in. (2.83mm)
- 75AEM, 75AE (Exc. European), 75AET (Exc. European): 0.13 in. (3.41mm)
- 85A: 0.16 in. (4.05mm)
- E60J, E75B: 0.13 in. (3.41mm)

6. At this point, double-check the timing pointer, it should align with the WOT timing spec, as noted earlier in this procedure. If not, loosen the set-screw, reposition the pointer and secure it again.

7. For the E60, adjust the WOT timing by turning the WOT stopper screw until the mark on the underside of the flywheel aligns with the mark on the magneto base.

8. For all except the E60, adjust the WOT timing as follows:
   a. There is a link joint at the top of the magneto control lever, disconnect it at one end before proceeding.

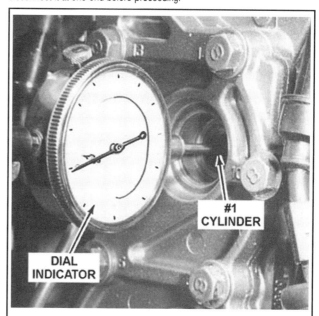

Fig. 250 Mount a dial gauge so that you can measure piston depth for the No. 1 cylinder - E60 (849cc) and E75, 75A Models, 85A Models, E60J (1140cc) motors

2-72 MAINTENANCE & TUNE-UP

b. Rotate the magneto base clockwise until the mark on the underside of the flywheel aligns with the mark on the magneto base. Now adjust the end length of the WOT stopper to specification, as follows:

- 75A (European): 1.34-1.38 in. (34-35mm)
- 75A (Exc. European): 1.26-1.30 in. (32-33mm)
- 85A: 1.18-1.22 in. (30-31mm)
- E60J, E75B: 1.12-1.16 in. (28.5-29.5mm)

c. Finally, adjust the length of the joint link (on top of the magneto control lever) so that the control lever contacts the WOT timing stopper.

9. Check the Idle Timing by slowly turn the flywheel in the normal direction of rotation (clockwise) until the timing mark for Idle operation aligns with the pointer as follows:
- E60, E60J, E75B: 2° ATDC
- 75A, 85A: 2° BTDC

10. Rotate the magneto control lever counterclockwise until it contacts the Idle stopper.

11. Visually check the mark on the underside of the flywheel to make sure it aligns with the mark on the magneto base assembly. If it does, timing is properly set and no adjustment is necessary, you can stop here. If not, continue the procedure to properly adjust/set the ignition Idle timing.

12. Adjust the Idle timing by turning the Idle stopper screw until the mark on the underside of the flywheel aligns with the mark on the magneto base.

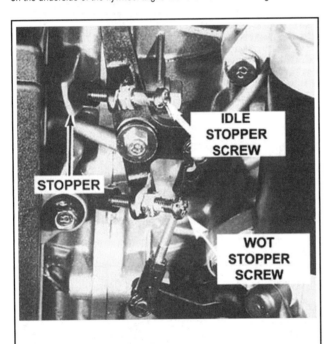

Fig. 251 Timing adjustments are made using the WOT and Idle Stopper Screws - E60 (849cc) motors

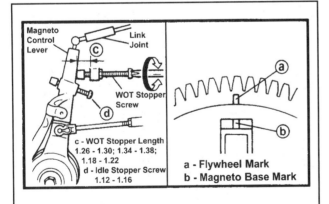

Fig. 252 Setting WOT timing - E60 (849cc) and E75, 75A Models, 85A Models, E60J (1140cc) motors

THROTTLE LINK

◆ See Figure 253

E60 models utilize a throttle link to connect upper and lower magneto control levers. The link must be checked/adjusted after Ignition Timing changes are made and before adjusting the Idle Speed.

1. To check the Throttle Link, rotate the lower magneto control lever clockwise until it contacts the surface of the crankcase, then check to make sure the upper magneto control lever is contacting the Idle stopper. If so, no adjustment is necessary. However, if not, continue with this procedure in order to adjust the length of the throttle link.
2. Disconnect the link joint from one of the magneto control levers.
3. Position the lower magneto control lever in contact with the crankcase and the upper lever in contact with the idle stopper. In this position, loosen the locknut and turn the throttle link end to adjust the link length so it will fit back over the magneto control lever ball studs.
4. Install the link over the lever ball studs.

IDLE SPEED

◆ See Figures 254 and 255

1. Mount the engine in a test tank or move the boat to a body of water.
2. Remove the cowling and connect a tachometer to the powerhead.
3. Start the engine and allow it to warm to operating temperature.
4. Check engine speed at idle. The powerhead should idle at the rpm specified in the Tune-up Specifications chart.
5. For all except Jet models, place the engine in gear and check engine trolling speed in the same manner.
6. If adjustment is necessary, shut down the powerhead and start by setting the throttle valves.
7. Loosen the carburetor idle adjustment screw (on the middle carburetor), turning it outward until the throttle valves are closed.

■ The throttle lever securing screws have left-hand threads (as normally indicated by an arrow and embossed OFF pointing to the left). Turn the screws clockwise to loosen and counterclockwise to tighten.

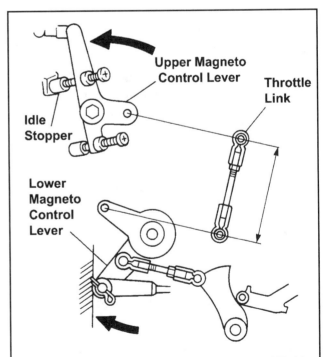

Fig. 253 Throttle link adjustment - E60 (849cc) and E75, 75A Models, 85A Models, E60J (1140cc) motors

MAINTENANCE & TUNE-UP 2-73

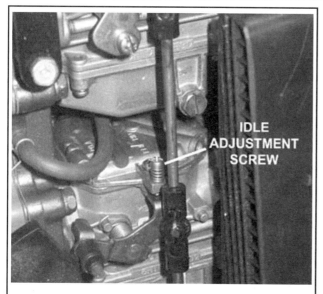

Fig. 254 Idle adjustment screw - E60 (849cc) and E75, 75A Models, 85A Models, E60J (1140cc) motors

Fig. 255 Pilot screw - E60 (849cc) and E75, 75A Models, 85A Models, E60J (1140cc) motors

8. Loosen the throttle lever securing screws on the upper and middle carburetors.

9. While lightly pushing downward on the throttle lever of the carburetor whose throttle valve screw was NOT loosened tighten the throttle lever securing screws of the other two carburetors. This means that you'll push down on the throttle valve for the lower carb, while tightening the screws on the upper and middle carbs.

10. Move the accelerator lever up and down several times to make sure all carburetors open and close simultaneously.

11. Turn each of the carburetor pilot screws (threaded horizontally into the side of the carburetor cover) inward until it JUST lightly seats and then back out the specified number of turns, as indicated in the Carburetor Set-Up Specifications chart, found in the Fuel System section.

12. Start the engine and allow it to re-warm to operating temperature.

13. Rotate the idle adjustment screw (not the pilot screw) until the powerhead idles at the required rpm. The idle adjustment screw is found on the middle carburetor for these models and is threaded vertically downward, contacting the throttle valve linkage. Turning the screw INWARD will INCREASE idle speed, while backing it OUTWARD will DECREASE idle speed.

PICKUP TIMING

◆ See Figures 256 and 257

1. Manually place the ignition in the Idle timing (full-retard) position by rotating the magneto control lever counterclockwise against the idle stopper.

2. Bring the accelerator cam to JUST contact the throttle lever roller lightly. The throttle valve should not open.

3. Loosen the locknut on the link joint, then disconnect the link rod from the ball stud on the accelerator cam. Adjust the plastic snap-on connector on the end of the link rod until the length is proper to reconnect the link, then tighten the locknut and install the link rod.

THROTTLE CABLE ADJUSTMENT

The throttle cable should operate smoothly without sticking or binding. To check the cable adjustment, operate the throttle, feeling for smooth operation. Set it to the WOT position, then check the carburetor throttle valves to ensure they have opened fully. If so, no further attention is required. However, if the carburetor throttle valves do not open fully, adjust the throttle cables. The manufacturer provides a procedure for adjusting the E60 as follows:

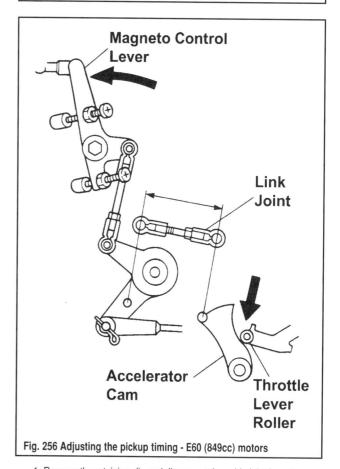

Fig. 256 Adjusting the pickup timing - E60 (849cc) motors

1. Remove the retaining clip and disconnect the cable joint from the magneto control lever.

2. Fully close the throttle.

3. Manually turn the magneto control lever counterclockwise to the fully closed or idle timing position. In this position the control lever will contact the idle (full retard) stopper.

4. Loosen the locknut and adjust the length of the throttle cable joint until it will fit properly back on the magneto control lever ball joint. Make sure that the embossed **UP** on the link joint is positioned on top after adjustment.

MAINTENANCE & TUNE-UP

✳✳ CAUTION

The end of the cable joint MUST be threaded at least 0.31 in. (8mm) onto the cable otherwise it could loosen in service causing a loss of throttle control (at potentially very serious navigation hazard).

5. Tighten the locknut, then install the link joint and secure using the retaining clip.
6. Reconfirm proper adjustment and throttle operation.

SHIFT CABLE ADJUSTMENT

◆ See Figure 258

Visually check the shift linkage movement in response to the shift handle. If the shifter does not engage properly or smoothly, adjust, as follows:
1. Set the shift control lever to the **Neutral** position.
2. Remove the retaining clip and disconnect the shift cable joint from the powerhead linkage.
3. Align the center of the linkage set pin with the mark in the middle of the shift bracket.
4. Loosen the locknut, then adjust the position of the shift cable joint (by turning the joint inward or outward on the cable threads) until its mounting hole aligns with the linkage set pin at the center of the bracket. Tighten the locknut to hold it in position on the cable.

✳✳ CAUTION

Make sure that after adjustment the cable joint is threaded over at LEAST 0.31 in. (8mm) of cable threads, otherwise there is a risk that the joint could separate in service causing a severe navigation/control hazard.

5. Position the cable joint over the control lever set pin and secure using the retaining clip.

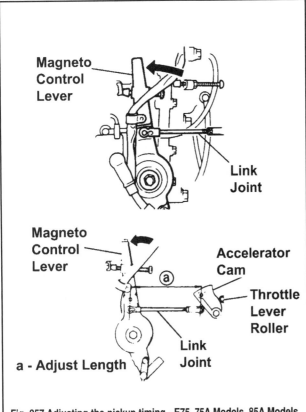

Fig. 257 Adjusting the pickup timing - E75, 75A Models, 85A Models, E60J (1140cc) motors

CHECKING/ADJUSTING THE STARTER LOCKOUT

◆ See Figure 259

Manual start models of this motor contain a starter lockout to prevent the motor from being started while in gear. The system should be checked at each tune-up to ensure that it is adjusted and functioning properly. Checking is a simple matter of placing the motor in gear and gently attempting to start the motor. If the starter will not rotate, the system is functioning and no further attention is required. However, if the motor rotates in gear, then adjust the cable as follows:
1. Begin with the engine not running and shifter positioned in **Neutral**.
2. Loosen the screw securing the cable adjusting plate to the starter housing.
3. Reposition the adjusting plate until the mark on the stopper (a) aligns with the mark on the cam guide (b), then retighten the adjusting plate screw.
4. Verify that the lockout is now working properly.

V4 and V6 Carbureted Motors

IGNITION TIMING

◆ See Figures 260 thru 264

On these Yamaha outboards Ignition Timing is checked by positioning the No. 1 piston in various positions and checking the corresponding timing marks and magneto linkage. WOT timing is then adjusted, if necessary, by changing the length of the magneto control link (the link that connects the magneto lever to the timer base/pulser coil).
1. If equipped, remove the flywheel cover for access.
2. Slowly rotate the flywheel by hand (CLOCKWISE) in the normal direction of rotation until the timing pointer on the powerhead is aligned with WOT timing specification for the motor. Please refer to the Tune-Up Specifications chart in this section for details on timing specs.

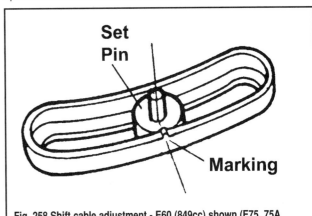

Fig. 258 Shift cable adjustment - E60 (849cc) shown (E75, 75A Models, 85A Models, E60J motors, similar)

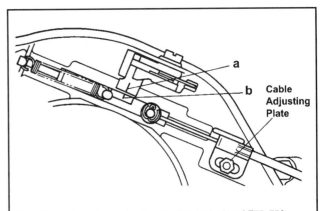

Fig. 259 Starter lockout adjustment - E60 (849cc) and E75, 75A Models, 85A Models, E60J (1140cc) motors

MAINTENANCE & TUNE-UP 2-75

3. With the timing pointer aligned with the proper WOT timing specification, manually rotate the magneto control lever CLOCKWISE until it contacts the WOT stopper. Now follow the linkage on top of the magneto control lever under the flywheel (where it contacts the timing base/pulser coil assembly). The timing mark on the underside of the flywheel should align with the mark on the timing base/pulser coil. If so, skip the WOT timing adjustment step and go on to check the Idle Timing.

4. If WOT timing must be adjusted, proceed as follows:

    a. Make sure the lower unit in **Neutral**.

    b. Remove all 4 or 6 spark plugs to ease compression and allow the motor to be rotated freely.

    c. On most models (including all 1999 and later models) it is a good idea to remove the air intake silencer for better access.

    d. Install a dial indicator in the No. 1 cylinder spark plug hole so that it can measure piston movement.

    e. Slowly rotate the flywheel clockwise until it reaches TDC (the absolute top of piston travel), then zero the dial gauge.

    f. Continue to turn the flywheel almost another complete rotation until you reach the specified distance BEFORE Top Dead Center (BTDC). The distance varies by model as follows:

    - 80J, 100, 130 and 140 hp motors - 0.12 in. (3.05mm) BTDC
    - 115A - 0.13 in. (3.33mm) BTDC
    - All other 115 Models - 0.15 in. (3.91mm) BTDC
    - 150A, L150A, C150TR and 175A - 0.09 in. (2.28mm) BTDC
    - 150F, L150F, D150H, D150TR, 175D, S175D and S175TR - 0.12 in. (3.05mm) BTDC
    - 105J, S150F, LS150F, S150TR, L150TR, 175F, P175TR, 200G and P200TR - 0.13 in. (3.33mm) BTDC
    - 200A and L200A - 0.08 in. (2.05mm) BTDC
    - 150G, P150TR, 200F, L200F and 200TR - 0.10 in. (2.53mm) BTDC
    - S200F, LS200F, S200TR, L200TR and 225 - 0.11 in. (2.78mm) BTDC

    g. Check the timing pointer on the powerhead to see with what specification on the flywheel it is currently aligned. Compare this to the WOT timing spec in the Tune-Up Specifications chart. If necessary, loosen the set-

Fig. 260 To check WOT timing, turn the flywheel to align the spec with the pointer.

Fig. 261 . . .then manually rotate the magneto control lever clockwise, against the WOT stopper. . .

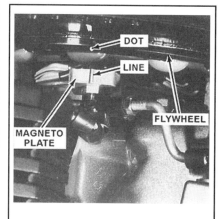

Fig. 262 . . . and check the timing marks under the flywheel for alignment

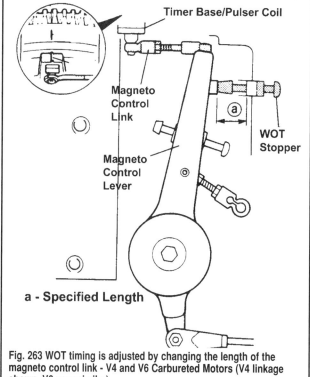

Fig. 263 WOT timing is adjusted by changing the length of the magneto control link - V4 and V6 Carbureted Motors (V4 linkage shown, V6 very similar)

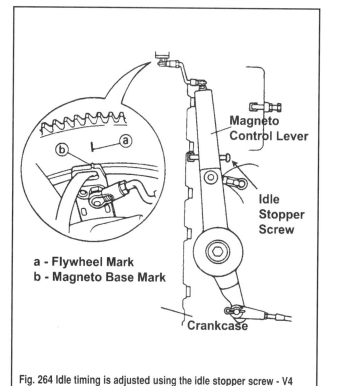

Fig. 264 Idle timing is adjusted using the idle stopper screw - V4 and V6 Carbureted Motors (V6 linkage shown, V4 similar)

2-76 MAINTENANCE & TUNE-UP

screw for the timing pointer and slide it one way or the other slightly to align it with the proper WOT timing spec.

   h. Now adjust the length of the WOT stopper screw to achieve the specified length between the end of the WOT stopper and the point where the threads enter the stopper mounting bracket (on the stopper side of the bracket). The length varies with model as follows:

- 80J, 100, 130 and 140 hp motors - 1.14 in. (29mm)
- E115A - 0.96 in. (24.5mm)
- C115, C115TR and 115B - 0.87 in. (22mm)
- All other 115 Models - 1.02 in. (26mm)
- 150A, L150A, C150TR, 175A, 175D, S175D and S175TR - 0.93 in. (23.5mm)
- 150F, L150F, 105J, S150F, LS150F, S150TR and L150TR - 0.85 in. (21.5mm)
- D150H and D150TR - 1.62 in. (41.2mm)
- 150G and P150TR - 1.71 in. (43.5mm)
- 175F, P175TR, 200G and P200TR - 1.57 in. (40mm)
- 200A and L200A - 0.98 in. (25mm)
- 200F, L200F and 200TR - 0.94 in. (24mm)
- S200F, LS200F, S200TR, L200TR and 225 - 1.67 in. (42.5mm)

   i. Disconnect the magneto control link from the top of the control lever, then turn the control lever clockwise until it contacts the WOT stopper.

   j. WOT timing is adjusted by loosening the locknut and adjusting the length of the magneto control link so that it can reconnect to the control lever when the timing mark on the underside of the flywheel aligns with the timing base/pulser coil marking.

   k. Once the adjustment is correct, tighten the locknut and reconnect the link to the top of the magneto control lever.

  5. Next, it is time to check the Idle Timing by turning the flywheel clockwise again until the proper Idle Timing specification on the flywheel aligns with the timing pointer.

  6. Manually rotate the magneto control lever counterclockwise until the Idle Stopper (threaded through the control lever itself, a little below where the lever contacts the WOT stopper) contacts the crankcase.

  7. Just like with the WOT timing check, in this position the timing mark on the magneto timer base/pulse coil should align with mark on the underside of the flywheel. If so, you're done with the timing procedure. However, if not, you'll have to adjust the Idle Timing, by turning the Idle Stop screw inward or outward until the marks align properly.

  8. If the air intake silencer assembly was removed, keep it off and proceed to the Synchronizing the Carburetors procedure.

SYNCHRONIZING THE CARBURETORS

◆ See Figures 265, 266 and 267

■ Check and/or adjust the ignition timing prior to performing this procedure.

  1. If not done already for Ignition Timing adjustment, remove the air silencer for access and so that you can observe the throttle valves.

  2. Manually rotate the lower (V4) or middle (V6) throttle lever and observe the opening and closing of all the throttle valves. They must open and close at the same instants in order for the motor to idle properly. If there are idle problems, or you observe one or more throttle valves out of synchronization with the rest, continue with this procedure in order to adjust them.

  3. Loosen the idle adjust screw on the lower (V4) or middle (V6) carburetor until it no longer contacts the throttle arm stopper.

■ You can make Idle Speed adjustment easier on yourself if you count the number of turns that you loosen the idle adjust screw.

  4. Next, loosen the throttle valve screw(s) of the upper (V4) or upper and lower (V6) carburetors by turning each CLOCKWISE and making sure the throttle valves are fully closed.

  5. Push VERY lightly inward on the throttle valve for the lower (V4) or middle (V6) carburetor to make sure it is fully closed and hold it while retighten the throttle valve screw or screws, as applicable. Be sure to tighten the throttle valve screws COUNTERCLOCKWISE, since they have left-hand threads.

  6. Slowly tighten the idle adjust screw until it contacts the throttle arm stopper. From this position, tighten it another 1 to 1-1/8 turns further as a starting position (unless you counted the number of turns at the beginning of this procedure, then tighten it the same number of turns you loosened it earlier).

  7. Move the throttle link up and down several times to make sure the carburetors open and close simultaneously.

  8. Adjust the engine Idle Speed.

IDLE SPEED

◆ See Figures 266, 268 and 269

  1. Be sure to check/set the Ignition Timing and Synchronize the Carburetors, before proceeding.

  2. Mount the engine in a test tank or move the boat to a body of water.

  3. Remove the cowling and connect a tachometer to the powerhead.

  4. Start the engine and allow it to warm to operating temperature.

  5. Check engine speed at idle. The powerhead should idle at the rpm specified in the Tune-up Specifications chart.

  6. For all except Jet models, place the engine in gear and check engine trolling speed in the same manner.

  7. If adjustment is necessary, shut down the powerhead.

  8. Locate the carburetor pilot screws threaded horizontally into a boss at the top of the carburetor. Since each carburetor body contains 2 complete carburetor circuits, there are 2 pilot screws, one on either side, facing diagonally outward from their mounting bosses. Turn each of the carburetor pilot screws inward until it JUST lightly seats and then back out the specified number of turns, as indicated in the Carburetor Set-Up Specifications chart, found in the Fuel System section.

  9. Loosen the throttle roller adjusting screw.

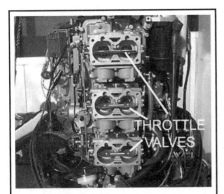

Fig. 265 Carburetor synchronization means all carb throttle valves open and close at the same time

Fig. 266 Idle adjust screw

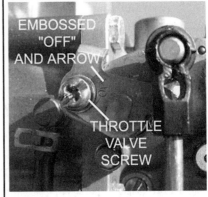

Fig. 267 Throttle valve screws are left-hand thread

MAINTENANCE & TUNE-UP

Fig. 268 Carburetor pilot screws - V4 and V6 Carbureted Motors

10. Start the engine and allow it to re-warm to operating temperature. Make sure the magneto control lever is in the idle position (with the idle stopper contacting the crankcase, see Ignition Timing for clarification).

11. Rotate the idle adjust screw (not the pilot screw) until the powerhead idles at the required rpm. The idle adjustment screw is found on the lower (V4) or middle (V6) carburetor for these models and is threaded vertically downward, contacting the throttle valve linkage. Turning the screw INWARD will INCREASE idle speed, while backing it OUTWARD will DECREASE idle speed.

12. Retighten the throttle roller adjusting screw and check the Carburetor Pickup Timing.

CARBURETOR PICKUP TIMING ADJUSTMENT

◆ See Figures 269, 270 and 271

■ Check/adjust the Ignition Timing, Carburetor Synchronization and Idle Speed prior to starting this procedure.

1. Start by checking the current adjustment by manually rotating the magneto control linkage counterclockwise so the idle stopper contacts the crankcase, then check that the mark on the throttle cam aligns with the center of the throttle roller. If so, no adjustment is necessary. If the roller and mark are not aligned, continue with the procedure.

2. Check and, if necessary, adjust the length of the throttle cam control link. The proper length varies by model, as follows:
- 80J-140 140 hp motors - 2.09 in. (53mm)
- 105J, all 150 hp motors (except 150G, D150H, P150TR or D150 TR), all 175 hp motors (except 175F or P175TR) and all 200 hp motors (except 200G, S200F, LS200F, P200TR, S200TR or L200TR) - 1.67 in. (42.5mm)
- 150G, D150H, P150TR, D150 TR, 175F, P175TR, 200G, S200F, LS200F, P200TR, S200TR, L200TR and 225 hp motors - 2.07 in. (52.5mm)

3. If necessary, disconnect the throttle cam control link, loosen the locknut and adjust the link to specification, then retighten the locknut. Reconnect the link.

4. With the magneto control lever idle stopper still contacting the crankcase, loosen the throttle roller adjusting screw and reposition the linkage so the roller aligns with the mark on the throttle cam. Hold the roller in this position and retighten the adjusting screw.

■ If alignment cannot be achieved, loosen the locknut and adjust the length of the throttle cam control link. Tighten the locknut.

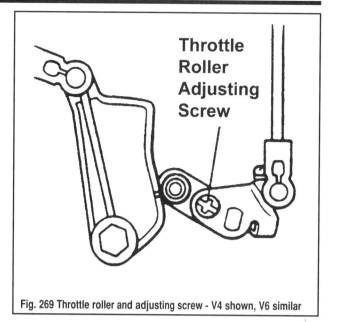

Fig. 269 Throttle roller and adjusting screw - V4 shown, V6 similar

Fig. 270 Carburetor pickup timing is correct if the alignment mark on the throttle cam aligns with the center of the throttle roller - V4 and V6 Carbureted Motors (typical V6 shown, others similar)

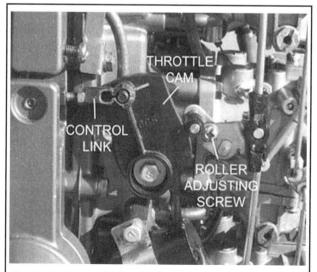

Fig. 271 Throttle cam-to-throttle roller alignment is changed using the throttle roller adjusting screw (once the throttle control link is set to a specific length)

MAINTENANCE & TUNE-UP

ADJUSTING THE REMOTE CONTROL SHIFT CABLE

◆ See Figure 272

Visually check the shift linkage movement in response to the shift handle. If the shifter does not engage properly or smoothly, adjust, as follows:

1. Remove the retaining clip and disconnect the shift cable joint from the powerhead linkage.
2. Loosen the shift cable joint locknut.
3. Set the remote control lever to the **Neutral** position.
4. Move the powerhead linkage set pin to align it with the **Neutral** mark on the lower cowling.
5. Adjust the position of the shift cable joint (by turning the joint inward or outward on the cable threads) until its mounting hole aligns with the linkage set pin (while it is aligned with the **Neutral** mark on the cowling). Tighten the locknut to hold it in position on the cable.

☆☆ CAUTION

Make sure that after adjustment the cable joint is threaded over at LEAST 0.31 in. (8mm) of cable threads, otherwise there is a risk that the joint could separate in service causing a severe navigation/control hazard.

6. Position the cable joint over the control lever set pin and secure using the retaining clip.
7. Operate the control lever to confirm proper operation and smooth adjustment.

ADJUSTING THE REMOTE CONTROL THROTTLE CABLE

■ Be sure to follow the Ignition Timing and Synchronizing the Carburetors procedures before attempting to adjust the throttle cable.

1. Visually check the throttle cable adjustment by moving the remote throttle to the WOT position. Then check the magneto control lever to make sure it is touching the WOT stopper (for clarification, refer to the Ignition Timing procedure). If so, no further adjustment is necessary. However, if not, continue with the procedure to reset the throttle cable.
2. Loosen the locknut on the throttle cable joint, then remove the retaining clip and pull it from the set pin.
3. Move the remote throttle lever to the **Neutral/Idle** position.
4. Manually move the magneto control lever counterclockwise so the idle stopper contacts the crankcase.
5. Turn the cable joint inward or outward on the cable threads until it aligns with the set pin while both the remote throttle and the powerhead throttle lever are in the idle position. Then secure using the locknut.

☆☆ CAUTION

Make sure that after adjustment the cable joint is threaded over at LEAST 0.31 in. (8mm) of cable threads, otherwise there is a risk that the joint could separate in service causing a severe navigation/control hazard.

6. Install the cable joint back over the set pin and secure using the retaining clip.
7. Open and close the throttle a couple of times using the remote while visually checking to be sure the throttle valves on the carburetor open and close smoothly and to the full extent of their travel. If necessary, repeat the adjustment procedure.

CHECKING/ADJUSTING THE STARTER LOCKOUT (MANUAL START MODELS)

◆ See Figure 273

Manual start models of this motor contain a starter lockout safety feature to prevent the motor from being started while in gear. The system should be checked at each tune-up to ensure that it is adjusted and functioning properly. Checking is a simple matter of placing the motor in gear and gently attempting to start the motor. If the starter will not rotate, the system is functioning and no further attention is required. However, if the motor rotates in gear, then adjust the cable as follows:

1. Begin with the engine not running and shifter positioned in **Neutral**.
2. Loosen the screw securing the cable adjusting plate to the starter housing.
3. Reposition the adjusting plate until the mark on the cable (a) aligns with the mark on the cam flywheel cover (b), then retighten the adjusting plate screw.
4. Verify that the lockout is now working properly.

OIL PUMP LINK

◆ See Figure 274

Obviously this adjustment is only for models equipped with the Yamaha Precision Blend automatic oil injection system.

1. Check the oil pump link adjustment by placing the throttle in the idle position and making sure that the oil pump lever is touching the stopper when the carburetor throttle valves are closed. If so, no further attention is necessary. However, if the pump lever is not in the correct position, continue with the procedure to adjust the link.
2. Remove the retaining clip and washer, then disconnect the control link from the oil pump.
3. Make sure the throttle valves are fully closed. If necessary, loosen the idle adjust screw (counting the number of turns so you can reposition it after the adjustment).
4. Loosen the oil control link locknut.
5. Hold the oil injection pump lever in the full-closed position, then adjust the control link length until the joint fits over the lever and tighten the locknut.
6. Reinstall the link onto the oil pump using the washer and retaining clip.
7. If you loosened the idle adjust screw, retighten it the same number of turns you counted earlier. Check and, if necessary, re-adjust the Idle Speed, as detailed earlier in this section.

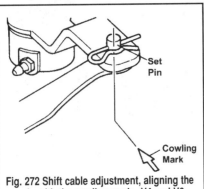

Fig. 272 Shift cable adjustment, aligning the set pin with the cowling mark - V4 and V6 Motors

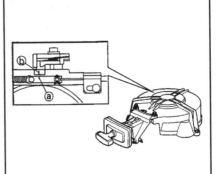

Fig. 273 Starter lockout adjustment - Manual Start V4 Motors

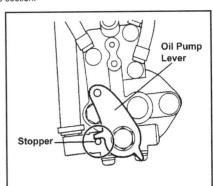

Fig. 274 Oil pump link adjustment - V4 and V6 Motors

MAINTENANCE & TUNE-UP

V6 EFI (OX66) and HPDI Motors

SYNCHRONIZING THE THROTTLE VALVES

◆ See Figures 275, 276 and 277

If the throttle is operating properly there is NO reason to perform this adjustment. Un-necessary tampering usually just leads to performance problems. Improper throttle valve adjustment will normally lead to an unstable idle. If necessary, remove the air intake silencer assembly and visually inspect the throttle valves to make sure that they open and close at the same instants. If you can see signs that the throttle valves open unevenly, adjust the valve synchronization, as follows:

■ The throttle valve procedure for HPDI and 250 hp EFI Vmax models is a little more involved, not more difficult, but has a few more steps. Also, following the throttle valve adjustment on these models, unlike the other EFI motors, you should skip to Throttle Position Sensor Adjustment, and come back to Idle Speed afterwards.

1. If not done already, remove the air intake silencer for access.
2. Disconnect the throttle lever rod and, for the HPDI and 250 hp EFI Vmax, the oil pump rod.
3. Loosen the idle adjust screw until it is no longer touching the stopper.

■ For all EFI models, except the 250 hp Vmax, loosen the idle adjust screw only using full turns and count the number of turns outward to help resetting the idle screw after the throttle valves have been synchronized.

4. Check and make sure that all the throttle valves are now fully closed once the idle screw is off the stopper. If so, no further adjustment is necessary.

■ On HPDI and 250 hp EFI motors it may be necessary to loosen the No. 4 throttle valve screw before the throttle valves will be completely closed. Of course, at that point, you might as well finish the adjustment procedure.

5. For EFI models, loosen the throttle valve adjusting screws for all cylinders EXCEPT No. 4. On the HPDI and 250 hp EFI Vmax, loosen ALL throttle valve adjusting screws, including the No. 4 valve screw.

■ Because the throttle valve screws have left-hand thread, they are loosened by turning CLOCKWISE.

6. Using gentle finger pressure, hold the No. 1 throttle valve in the fully closed position, then tighten the No. 1 throttle valve adjusting screw by turning it COUNTERCLOCKWISE. Repeat this step for cylinders No. 2, 3, 5, and 6.
7. For all EFI models except the 250 hp Vmax, turn the idle adjust screw back inward the same number of turns you counted when loosening it. On these models, you've finished the procedure.
8. For the HPDI and 250 hp EFI Vmax motors finish the procedure as follows:
   a. Reconnect the oil pump link rod.
   b. Turn the idle adjust screw back inward until the No. 4 throttle valve JUST starts to open, then tighten the screw an additional 1/2-1 turns.

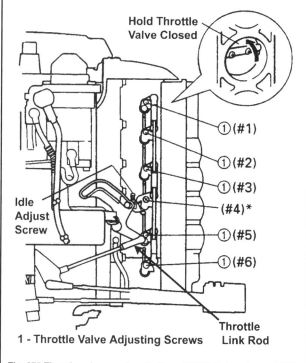

Fig. 275 Throttle valve synchronization - V6 EFI Motors (except 250 Hp Vmax)

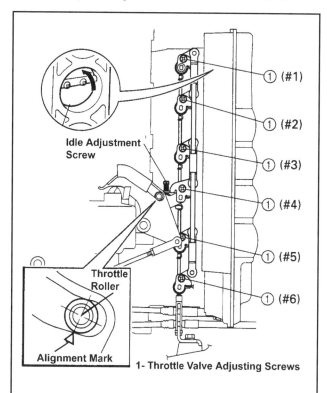

Fig. 276 Throttle valve synchronization - HPDI and 250 Hp EFI Vmax Motors

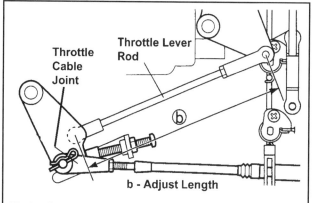

Fig. 277 On all HPDI and the 250 Hp EFI Vmax, you'll also have to adjust the throttle lever rod length

2-80 MAINTENANCE & TUNE-UP

c. Align the center of the throttle roller with the mark on the throttle cam, then tighten the No. 4 throttle adjusting screw by turning it COUNTERCLOCKWISE.

d. Open and close the throttle valve 2-3 times and double-check that the center of the roller is still aligned with the mark.

e. For 3.3L HPDI models, connect a 3-pin test harness to the Throttle Position Sensor (TPS) wiring harness. Then turn the ignition switch **ON** and check the TPS output voltage across the Pink and Orange wires of the harness using a Digital Volt Ohmmeter (DVOM). Turn the idle adjust screw in or out as necessary until the TPS output is 0.58-0.62 volts. In theory, you can skip the TPS adjustment procedure for this model since you've just checked/adjusted it now.

f. Disconnect the throttle cable joint and adjust the lever rod to 6.2 in. (157mm) for EFI motors, 6.4 in. (163mm) for 2.5L HPDI motors or to 6.0 in. (151.5mm) for 3.3L HPDI motors. Once the length is set, reconnect the cable joint.

g. Install the air intake silencer and proceed to Throttle Position Sensor Adjustment before setting the Idle Speed.

THROTTLE POSITION SENSOR ADJUSTMENT

◆ See Figure 278

With each tune-up, or at least annually, you should check the Throttle Position Sensor (TPS) output voltage with the throttle valves fully closed. If the reading is out of specification, the physical positioning of the sensor can be adjusted in order to bring the reading into spec.

However, the sensor does not generate a voltage, so much as the variable resistor within the sensor will change a reference signal from the Engine Control Module (ECM). Therefore, the circuit must be intact in order to test the sensor signal. You've got 2 options. One option is to attempt to back-probe the connector (insert the probes through the rear of the connector while it is still attached to the sensor). However this method risks damaging the connector or the wiring insulation (and could lead to problems with the circuit later). The better method is to disconnect the wiring harness and use 3 jumper wires (one for each terminal) to reconnect the harness. The jumpers must not contact each other (or you risk damage to the ECM from a short), however, some point on the jumper must be exposed so that you can probe the completed circuit using a DVOM. Yamaha makes 3-pin test harness for this application (#90890-06757).

1. Make sure the throttle valves are properly adjusted (check the Throttle Valve Synchronization) before proceeding.
2. If equipped, remove the flywheel cover for additional access.
3. If it is not still removed from checking the Throttle Valve Synchronization, remove the air intake assembly for access.
4. Disconnect the throttle link from the throttle valve for the No. 1 (top) throttle body.
5. Disconnect the wiring from the TPS and connect a test harness or jumper wires to complete the circuit.

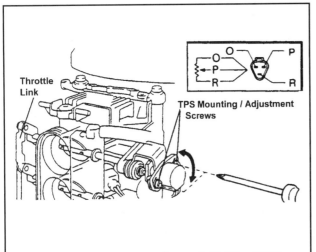

Fig. 278 Throttle position sensor adjustment - V6 EFI and HPDI motors

■ The TPS sensor wiring is routed down the side of the powerhead, below the sensor itself (which is mounted adjacent to and connected to the top throttle body). Follow the harness from where it exits the sensor, down to the wiring harness connector, noting the routing for installation purposes. After testing, make sure the sensor wiring is safely tucked in the same position, to prevent possible damage from moving components.

6. Connect a Digital Volt Ohmmeter (DVOM) set to read DC volts to the Pink and Orange wires of the jumper/test harness.
7. Turn the ignition switch **ON**, then measure the output voltage with the throttle valves fully closed. The meter should show 0.48-0.52 volts.
8. If the sensor voltage is out of range, loosen the 2 mounting screws and reposition the sensor (turn it slightly one direction or the other) to obtain the correct voltage readings. Once the sensor is properly repositioned, tighten the screws again to hold it in position.
9. Reconnect the throttle link to the top throttle body.
10. Operate the throttle valve 2-3 times and double-check that the closed throttle voltage remains in spec.
11. For 3.3L HPDI motors, make sure the TPS voltage is 0.58-0.62 with the throttle link connected to the top throttle body. If necessary, use the idle adjust screw to obtain the correct reading.
12. Check/adjust the Idle Speed, as detailed earlier in this section.

IDLE SPEED

◆ See Figures 279 and 280

■ HOLD IT. On HPDI and 250 hp EFI Vmax models, the Throttle Position Sensor must be checked and/or adjusted before the idle speed. Refer to Throttle Position Sensor Adjustment before proceeding.

1. Be sure to check/set the Throttle Valve Synchronization and TPS Sensor Adjustment, before proceeding.
2. Mount the engine in a test tank or move the boat to a body of water.
3. Remove the cowling and connect a tachometer to the powerhead.
4. Start the engine and allow it to warm to operating temperature.
5. Check engine speed at idle. The powerhead should idle at the rpm specified in the Tune-up Specifications chart.
6. If applicable, place the engine in gear and check engine trolling speed in the same manner.

■ Yamaha does not provide a trolling engine speed spec for all EFI or HPDI models.

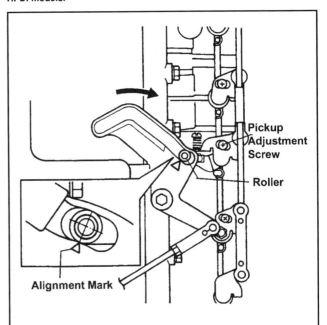

Fig. 279 Before adjusting idle speed be sure to center the throttle roller. . .

MAINTENANCE & TUNE-UP 2-81

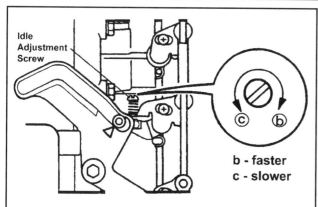

Fig. 280 . . .then turn the idle adjust screw as necessary to obtain the desired idle

7. Check the alignment of the mark on the throttle cam and the throttle roller. If the roller is not centered on the mark, loosen the throttle valve adjust screw (it is normally left-hand thread, meaning it would be loosened by turning CLOCKWISE) for the No. 4 cylinder. This screw is also known as the pickup adjustment screw. Reposition and align the mark on the throttle cam with the center of the throttle roller.

■ On the HPDI and 250 hp EFI Vmax models you really should have already accomplished this while Synchronizing the Throttle Valves, so just double-check the setting, and only readjust, if necessary. However, except of 3.3L HPDI models you should still loosen the screw slightly in order to properly adjust the idle speed.

8. Turn the idle adjust screw as necessary to obtain the correct idle speed. Turning the screw INWARD will INCREASE idle speed, while turning the screw OUTWARD will DECREASE idle speed.
9. Except for 3.3L HPDI motors, once the idle speed is set correctly, press down gently on the throttle control lever cam roller and, while holding it in this position, tighten the pickup adjustment screw.

ADJUSTING THE REMOTE CONTROL SHIFT CABLE

EFI Motors and Some 2.6L HPDI Motors

◆ See Figure 272

This is the most common shift control cable configuration for EFI and HPDI models. It is found on all EFI motors, as well as most 2.6L HPDI models including the Z150Q, VZ150, Z175H, VZ175 and some versions of the Z200TR, Z200NETO, LZ200TR and LZ200NETO models. If the linkage on your motors differs, follow the procedures for 2.6L HPDI Motors, Except Z150Q, VZ150, Z175H and VZ175 Models, found later in this section.

Visually check the shift linkage movement in response to the shift handle. If the shifter does not engage properly or smoothly, adjust, as follows:
1. Remove the retaining clip and disconnect the shift cable joint from the powerhead linkage.
2. Loosen the shift cable joint locknut.
3. Set the remote control lever to the **Neutral** position.
4. Move the powerhead linkage set pin to align it with the **Neutral** mark on the lower cowling.
5. Adjust the position of the shift cable joint (by turning the joint inward or outward on the cable threads) until its mounting hole aligns with the linkage set pin (while it is aligned with the **Neutral** mark on the cowling). Tighten the locknut to hold it in position on the cable.

✸✸ CAUTION

Make sure that after adjustment the cable joint is threaded over at LEAST 0.31 in. (8mm) of cable threads, otherwise there is a risk that the joint could separate in service causing a severe navigation/control hazard.

6. Position the cable joint over the control lever set pin and secure using the retaining clip.
7. Operate the control lever to confirm proper operation and smooth adjustment.

2.6L HPDI Motors, Except Z150Q, VZ150, Z175H and VZ175 Models

◆ See Figures 281, 282 and 283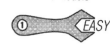

This procedure is for the alternate shift linkage used by certain 2.6L engines, except most Z150Q, VZ150, Z175H, VZ175 and some versions of the Z200TR, Z200NETO, LZ200TR and LZ200NETO models. If the linkage on your motors differs, follow the procedures for EFI Motors and Some 2.6L HPDI Motors, found earlier in this section.

Visually check the shift linkage movement in response to the shift handle. If the shifter does not engage properly or smoothly, adjust, as follows:
1. Remove the retaining clip and disconnect the shift cable joint from the powerhead linkage.
2. Loosen the shift cable joint locknut.
3. Move the remote control lever through **Reverse**, **Neutral** and **Reverse** several times, then set the remote control lever to the **Neutral** position.
4. Manually push the cable joint all the way inward and mark the position where the cable enters the cable casing.
5. Now manually pull the cable joint all the way outward and mark the position where the cable exists the casing.
6. At this point you'll have two lines on the cable, not all that far apart. These marks represent the amount of available cable free-play. Now make a third mark, centered right between the two, which will represent the center of the free-play.
7. Gently push the cable back into the housing until the center free-play mark is right at the opening of the housing (meaning that the cable is centered in the middle of its free-play).
8. Manually position the shift rod set pin to the center of the slider bracket. There is a detent in the slider inline with a retaining screw and the set pin in this position.
9. Adjust the position of the shift cable joint (by turning the joint inward or outward on the cable threads) until its mounting hole aligns with the linkage set pin (while it is aligned with the center of the slider). Tighten the locknut to hold it in position on the cable.

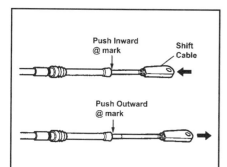

Fig. 281 Make marks on the cable pushed inward and pulled outward. . .

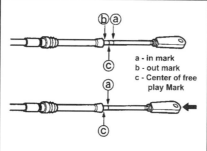

Fig. 282 . . . then make a mark centered between them, and push the cable inward to that mark

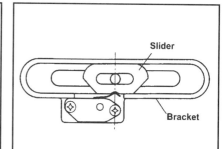

Fig. 283 The slider bracket should be centered on the detent

2-82 MAINTENANCE & TUNE-UP

✱✱ CAUTION

Make sure that after adjustment the cable joint is threaded over at LEAST 0.31 in. (8mm) of cable threads, otherwise there is a risk that the joint could separate in service causing a severe navigation/control hazard.

10. Position the cable joint over the control lever set pin and secure using the retaining clip.
11. Operate the control lever to confirm proper operation and smooth adjustment.

3.3L HPDI Motors

◆ See Figures 284 and 285

Visually check the shift linkage movement in response to the shift handle. If the shifter does not engage properly or smoothly, adjust, as follows:
1. Set the remote control lever to the **Neutral** position.
2. Remove the retaining clip and disconnect the shift cable joint from the powerhead linkage.
3. Loosen the shift cable joint locknut.
4. Manually move the powerhead linkage (shift lever) until the lever pin aligns with the line of the shift position switch plate (this is the **Neutral** position).
5. Now align the center of the set pin for the cable with the **Neutral** alignment mark on the lower cowling.
6. Adjust the position of the shift cable joint (by turning the joint inward or outward on the cable threads) until its mounting hole aligns with the linkage set pin (while it is aligned with the **Neutral** mark on the cowling). Tighten the locknut to hold it in position on the cable.

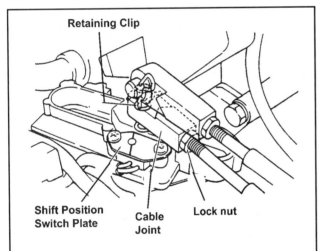

Fig. 284 Adjust the shift lever until the lever pin aligns with the line of the shift position switch plate - 3.3L HPDI Motors

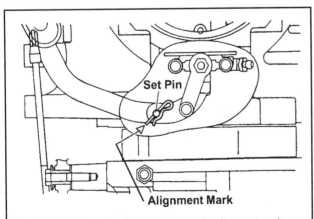

Fig. 285 Align the center of the set pin with the alignment mark on the lower cowling - 3.3L HPDI Motors

✱✱ CAUTION

Make sure that after adjustment the cable joint is threaded over at LEAST 0.31 in. (8mm) of cable threads, otherwise there is a risk that the joint could separate in service causing a severe navigation/control hazard.

7. Position the cable joint over the control lever set pin and secure using the retaining clip.
8. Operate the control lever to confirm proper operation and smooth adjustment.

ADJUSTING THE REMOTE CONTROL THROTTLE CABLE

◆ See Figures 286 and 287

■ Be sure to follow the Synchronizing the Throttle Valve and Idle Speed adjustment procedures before attempting to adjust the throttle cable.

1. Check the throttle cable adjustment by moving the remote throttle to the **Neutral/Idle** position, then visually check the throttle valves in the throttle bodies to make sure they are fully closed. Slowly advance the throttle to WOT and check that the throttle valves open completely. If so, no further adjustment is necessary. However, if not, continue with the procedure to reset the throttle cable.
2. For HPDI and 250 hp EFI Vmax motors, loosen the locknut and throttle linkage stopper screw.
3. Loosen the locknut on the throttle cable joint, then remove the retaining clip and pull it from the set pin.
4. Move the remote throttle lever to the **Neutral/Idle** position.
5. Align the center of the throttle control lever cam roller with the alignment mark.
6. For HPDI and 250 hp EFI Vmax motors, tighten throttle linkage stopper screw until it just contacts the throttle lever, then tighten the locknut.
7. Turn the cable joint inward or outward on the cable threads until it aligns with the set pin, then secure using the locknut.

✱✱ CAUTION

Make sure that after adjustment the cable joint is threaded over at LEAST 0.31 in. (8mm) of cable threads, otherwise there is a risk that the joint could separate in service causing a severe navigation/control hazard.

8. Install the cable joint back over the set pin and secure using the retaining clip.
9. Open and close the throttle a couple of times using the remote while visually checking to be sure the throttle valves open and close smoothly and to the full extent of their travel. If necessary, repeat the adjustment procedure.

ADJUSTING THE CRANKSHAFT POSITION SENSOR

◆ See Figure 288

All EFI OX66 and HPDI motors utilize a Crankshaft Position Sensor (CPS) to provide information regarding engine positioning to the Engine Control Module (ECM). The sensor is used for fuel and ignition mapping purposes. Although all these motors use the sensor, only the HPDI models and the 250 hp EFI Vmax mention checking or adjusting the sensor position. This means that the either the gap is fixed on other EFI models and not adjustable, or this is just an oversight in the Yamaha service literature. If you have an EFI model, other than the 250 hp Vmax, you may want to locate the sensor and check for yourself if it is adjustable. If so, follow this procedure to check the gap.

On all models with adjustable sensors, at least annually, double-check the Crankshaft Position Sensor (CPS) gap. Remember that the proper gap must exist between the sensor and the flywheel in order for the HPDI or EFI system to work properly.

Check the sensor adjustment by trying to insert a 0.02-0.06 in. (0.5-1.5mm) feeler gauge between the sensor and the flywheel. When measuring

MAINTENANCE & TUNE-UP 2-83

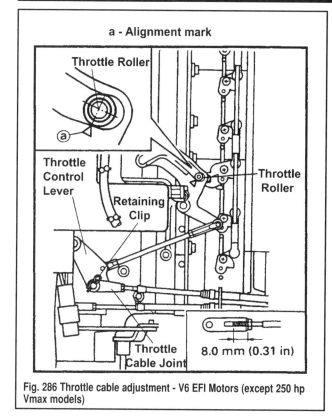

Fig. 286 Throttle cable adjustment - V6 EFI Motors (except 250 hp Vmax models)

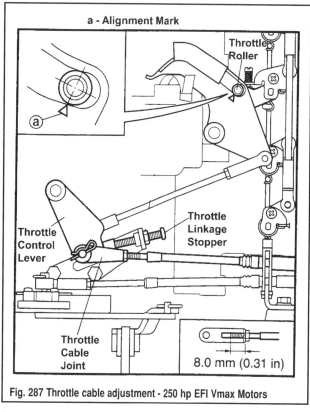

Fig. 287 Throttle cable adjustment - 250 hp EFI Vmax Motors

a gap with a feeler gauge, remember that the gap is equal to the size gauge that will fit through the gap with a slight drag. The next larger size gauge should not fit, while the next smaller gauge should fit without touching or dragging. If adjustment is necessary, loosen the 2 CPS mounting screws and gently slide the sensor against a 0.02-0.06 in. (0.5-1.5mm) feeler gauge inserted between the flywheel and sensor. Tighten the mounting screws and double-check the gap using the gauge set.

OIL PUMP LINK

◆ See Figure 289

 MODERATE

1. Check the oil pump link adjustment by placing the throttle in the idle position and checking the positioning of the oil pump lever and stopper when the throttle valves are fully closed. On most models (except the 2.6L HPDI motors) there should be 0.04 in. (1mm) of clearance between the oil pump lever and the stopper, however on 2.6L HPDI motors the lever should be contacting the stopper. If so, no further attention is necessary. However, if the pump lever is not in the correct position, continue with the procedure to adjust the link.

2. Remove the retaining clip and washer, then disconnect the control link from the oil pump.

3. Make sure the throttle valves are fully closed.

4. Loosen the oil control link locknut.

5. For all except 2.6L HPDI motors, hold the oil injection pump lever gently against a 0.04 in. (1mm) feeler gauge inserted between the lever and the stopper, then adjust the control link length until the joint fits over the lever and tighten the locknut.

6. For 2.6L HPDI motors, hold the oil injection pump lever gently against the stopper, then adjust the control link length until the joint fits over the lever and tighten the locknut.

7. Reinstall the link onto the oil pump using the washer and retaining clip.

8. Double-check the gap (or lack thereof on 2.6L HPDI motors) between the lever and stopper using the feeler gauge. Readjust, if necessary.

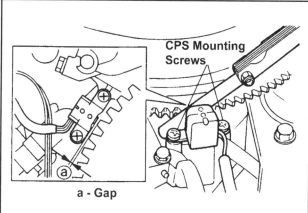

Fig. 288 Crankshaft position sensor gap adjustment - HPDI and 250 hp EFI Vmax Motors

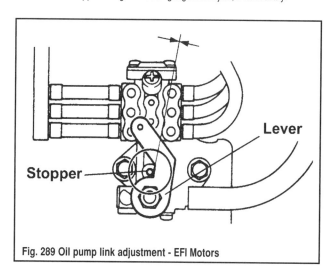

Fig. 289 Oil pump link adjustment - EFI Motors

MAINTENANCE & TUNE-UP

IGNITION TIMING

Ignition timing on these powerheads is adjusted (advanced/retarded) automatically by the Engine Control Module to match rpm operating conditions. Because there are no adjustments for timing, any problems or irregularities should be diagnosed and repaired as the result of defective mechanical or electrical components. No attempt should be made to perform adjustments in order to achieve the proper timing specifications, let the electronics do their job, but give them the tools that they need.

However, most motors are equipped with a timing pointer and timing marks on the flywheel. If desired, you can use a timing light to confirm that the ECM is properly controlling timing. To do this, make sure the motor is either mounted in a test tank or the boat is launched as the motor must operate under load. Always check the ignition timing with the motor fully warmed to normal operating temperature. For details on timing specifications, please refer to the Tune-Up Specifications chart in this section. If the timing is out of specification, double-check your procedures and test conditions, then suspect a fault with the ignition/fuel injection/engine management system.

STORAGE (WHAT TO DO BEFORE AND AFTER)

Winterization

◆ See Figure 290

Taking extra time to store the boat and motor properly at the end of each season or before any extended period of storage will greatly increase the chances of satisfactory service at the next season. Remember, that next to hard use on the water, the time spent in storage can be the greatest enemy of an outboard motor. Ideally, outboards should be used regularly. If weather in your area allows it, don't store the motor, enjoy it. Use it, at least on a monthly basis. It's best to enjoy and service the boat's steering and shifting mechanism several times each month. If a small amount of time is spent in such maintenance, the reward will be satisfactory performance, increased longevity and greatly reduced maintenance expenses.

But, in many cases, weather or other factors will interfere with time for enjoying a boat and motor. If you must place them in storage, take time to properly winterize the boat and outboard. This will be your best shot at making time stand still for them.

For many years there was a widespread belief simply shutting off the fuel at the tank and then running the powerhead until it stops constituted prepping the motor for storage. Right? Well, WRONG!

First, it is not possible to remove all fuel in the carburetor or fuel injection system by operating the powerhead until it stops. Considerable fuel will remain trapped in the carburetor float chamber (or vapor separator tank of fuel injected motors) and other passages, especially in the lines leading to carburetors or injectors. The only guaranteed method of removing all fuel from a carbureted motor is to take the physically drain the carburetors from the float bowls. And, though you should drain the fuel from the vapor separator tank on most fuel injected motors, you still will not be able to remove all of it from the sealed high-pressure lines.

Depending upon the length of storage you can also use fuel stabilizer as opposed to draining the fuel system, but if the motor is going to be stored for more than a couple of months at a time, draining the system is really the better option.

■ Up here in the northeastern U.S., we always start adding fuel stabilizer to the fuel tank with every fuel fill up starting sometime in September. That helps to make sure that we'll be at least partially protected if the weather takes a sudden turn and we haven't had a chance to complete winterization yet.

Proper storage involves adequate protection of the unit from physical damage, rust, corrosion and dirt. The following steps provide an adequate maintenance program for storing the unit at the end of a season.

■ If your outboard requires one or more repairs, PERFORM THEM NOW or during the off-season. Don't wait until the sun is shining, the weather is great and you want to be back on the water. That's not the time to realize you have to pull of the gearcase and replace the seals. Don't put a motor that requires a repair into storage unless you plan on making the repair during the off-season. It's too easy to let it get away from you and it will cost you in down time next season.

PREPPING FOR STORAGE

Where to Store Your Boat and Motor

Ok, a well lit, locked, heated garage and work area is the best place to store you precious boat and motor, right? Well, we're probably not the only ones who wish we had access to a place like that, but if you're like most of us, we place our boat and motor wherever we can.

Of course, no matter what storage limitations are placed by where you live or how much space you have available, there are ways to maximize the storage site.

If possible, select an area that is dry. Covered is great, even if it is under a carport or sturdy portable structure designed for off-season storage. Many people utilize canvas and metal frame structures for such purposes. If you've got room in a garage or shed, that's even better. If you've got a heated garage, God bless you, when can we come over? If you do have a garage or shed that's not heated, an insulated area will help minimize the more extreme temperature variations and an attached garage is usually better than a detached for this reason. Just take extra care to make sure you've properly inspected the fuel system before leaving your boat in an attached garage for any amount of time.

If a storage area contains large windows, mask them to keep sunlight off the boat and motor otherwise, use a high-quality, canvas cover over the boat, motor and if possible, the trailer too. A breathable cover is best to avoid the possible build-up of mold or mildew, but a heavy duty, non-breathable cover will work too. If using a non-breathable cover, place wooden blocks or length's of 2 x 4 under various reinforced spots in the cover to hold it up off the boat's surface. This should provide enough room for air to circulate under the cover, allowing for moisture to evaporate and escape.

Whenever possible, avoid storing your boat in industrial buildings or parks areas where corrosive emissions may be present. The same goes for storing

Fig. 290 Add fuel stabilizer to the system anytime it will be stored without complete draining

MAINTENANCE & TUNE-UP 2-85

your boat too close to large bodies of saltwater. Hey, on the other hand, if you live in the Florida Keys, we're jealous again, just enjoy it and service the boat often to prevent corrosion from causing damage.

Finally, when picking a place to store your motor, consider the risk or damage from fire, vandalism or even theft. Check with your insurance agent regarding coverage while the boat and motor is stored.

Storage Checklist (Preparing the Boat and Motor)

◆ See Figures 291 thru 294

The amount of time spent and number of steps followed in the storage procedure will vary with factors such as the length of planed storage time, the conditions under which boat and motor are to be stored and your personal decisions regarding storage.

But, even considering the variables, plans can change, so be careful if you decide to perform only the minimal amount of preparation. A boat and motor that has been thoroughly prepared for storage can remain so with minimum adverse affects for as short or long a time as is reasonably necessary. The same cannot be said for a boat or motor on which important winterization steps were skipped.

■ **If possible always store your Yamaha vertically on the boat or on a suitable engine stand. However, if you are to store a motor lying on its side, NO motor should be placed on its side until after ALL water has drained, otherwise water may enter a cylinder through an exhaust port causing corrosion (or worse, may become trapped in a passage and freeze causing cracks in the powerhead or gearcase!). Check your owner's manual for more details if you need to store a motor on its side. However if you do store it this way, be sure to return the motor vertically a few days before intended service and check the combustion chambers for oil before cranking the motor.**

1. Thoroughly wash the boat motor and hull. Be sure to remove all traces of dirt, debris or marine life. Check the water stream fitting, water inlet(s) and, on jet models, the impeller grate for debris. If equipped, inspect the speedometer opening at the leading edge of the gearcase or any other gearcase drains for debris (clean debris with low-pressure compressed air or a piece of thin wire).

■ **As the season winds down and you are approaching your last outing, start treating the fuel system then to make sure it thoroughly mixes with all the fuel in the tank. By the last outing of the season you should already have a protected fuel system.**

2. Stabilize the engine's fuel supply using a high quality fuel stabilizer (of course the manufacturer recommends using Yamaha Fuel Conditioner and Stabilizer) and take this opportunity to thoroughly flush the engine cooling system at the same time as follows:

a. Add an appropriate amount of fuel stabilizer to the fuel tank (for Yamaha stabilizer it is normally one ounce for each gallon of untreated fuel) and top off to minimize the formation of moisture through condensation in the fuel tank.

b. Attach a flushing attachment as a cooling water/flushing source. For details, please refer to the information on Flushing the Cooling System, in this section.

c. Start and run the engine at fast idle approximately 10-15 minutes. This will ensure the entire fuel supply system contains the appropriate storage mixtures.

✳✳ WARNING

Yamaha warns not to use a fogging oil that contains Silicon, Phosphorus or Lead on EFI or HPDI motors that are equipped with an oxygen sensor. Also, although you could use a Silicon based spray on external components for these motors, be sure not to spray it near the air intake or the oxygen sensor to prevent possible damage. If you are uncertain whether or not your motor contains an oxygen sensor, please refer to the Fuel System section or the Wiring Diagrams in the Ignition and Electrical System section.

d. Just prior to stopping the motor fog the engine using Yamaha Stor-Rite Engine Fogging Oil (or equivalent fogging spray). Spray the oil alternately into each of the carburetor or throttle body (EFI or HPDI) throats (you'll likely have to remove an air intake silencer/flame arrestor for access). When properly fogged the motor will smoke excessively, will stumble and will almost stall.

e. Stop the engine and remove the flushing source, keeping the outboard perfectly vertical. Allow the cooling system to drain completely, especially if the outboard might be exposed to freezing temperatures during storage.

✳✳ WARNING

NEVER keep the outboard tilted when storing in below-freezing temperatures as water could remain trapped in the cooling system. Any water left in cooling passages might freeze and could cause severe engine damage by cracking the powerhead or gearcase.

3. Drain and refill the engine gearcase while the oil is still warm (for details, refer to the Gearcase Oil procedures in this section). Take the opportunity to inspect for problems now, as storage time should allow you the opportunity to replace damaged or defective seals. More importantly, remove the old, contaminated gear oil now and place the motor into storage with fresh oil to help prevent internal corrosion.

4. Finish fogging the motor manually through the spark plug ports as follows:

a. Tag and disconnect the spark plug leads, then remove the spark plugs as described under Spark Plugs.

b. Spray a generous amount of fogging oil into the spark plug ports. Yamaha recommends a 5-10 second long spray of their Yamaha Stor-Rite Engine Fogging Oil for each cylinder.

Fig. 291 Be sure to fog the motor through the spark plug ports. . .

Fig. 292 . . .and drain the carb float bowls before storage

Fig. 293 Multiple carburetors will mean multiple float bowls to drain

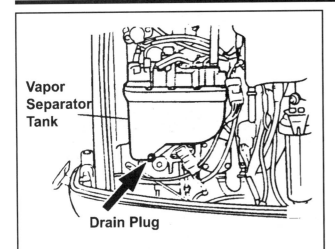

Fig. 294 HPDI and some EFI motors have a drain plug on the vapor separator tank, just like carburetor float bowls, they should be drained

■ On most Yamaha motors you can disable the ignition system by leaving the safety lanyard disconnected but still use the starter motor to turn the motor. This is handy for things like compression tests or distributing fogging oil. To be certain use a spark plug gap tester on one lead and crank the motor using the keyswitch. If no spark is present, you're good to go.

   c. Turn the flywheel slowly by hand (clockwise, in the normal direction of rotation) to distribute the fogging oil evenly across the cylinder walls. On electric start models, the starter can be used to crank the motor over in a few short bursts, but make sure the spark plugs leads remain disconnected and grounded to the powerhead (away from the spark plug ports) to prevent accidental combustion. If necessary, re-spray into each cylinder when that cylinder's piston reaches the bottom of its travel. Reinstall and tighten the spark plugs, but leave the leads disconnected (and GROUNDED) to prevent further attempts at starting until the motor is ready for re-commissioning.

■ On motors equipped with a rope start handle, the rope can be used to turn the motor slowly and carefully using the rope starter. For other models, turn the flywheel by hand or using a suitable tool, but be sure to ALWAYS turn the engine in the normal direction of rotation (normally clockwise on these motors). Also, keep in mind that the flywheel on most Yamaha outboards (even electric start models) is notched to accept an emergency starter rope. You can always use a knotted rope inserted into the notch and wound around the flywheel to help turn it.

   5. On carbureted motors, if the motor is to be stored for any length of time more than one off-season you really MUST drain the carburetor float bowls. Honestly, it is a pretty easy task and we'd recommend doing that for all motors, even if they are only going to be stored for a few months. To drain the float bowls locate the drain screw on the bottom of each bowl, place a small container under the bowl and remove the screw. Repeat for the remaining float bowls on multiple carburetor motors.

   6. On EFI or HPDI motors, if the vapor separator tank contains a drain screw, it is a good idea to completely drain the tank of fuel. This will help protect the fuel lines and components in the tank from possible damage by deteriorating fuel.

   7. For models equipped with portable fuel tanks, disconnect and relocate them to a safe, well-ventilated, storage area, away from the motor. Drain any fuel lines that remain attached to the tank. It's a tough call whether or not to drain a portable tank. Plastic tanks, drain em' and burn the fuel in something else. Metal tanks, well, draining them will expose them to moisture and possible corrosion, while topping them off will help prevent this, so it probably makes more sense to top them off with treated fuel.

   8. Remove the battery or batteries from the boat and store in a cool dry place. If possible, place the battery on a smart charger or Battery Tender®, otherwise, trickle charge the battery once a month to maintain proper charge.

✱✱ WARNING

Remember that the electrolyte in a discharged battery has a much lower freezing point and is more likely to freeze (cracking/destroying the battery case) when stored for long periods in areas exposed to freezing temperatures. Although keeping the battery charged offers one level or protection against freezing; the other is to store the battery in a heated or protected storage area.

   9. For models equipped with a boat mounted fuel filter or filter/water canister, clean or replace the boat mounted fuel filter at this time. If the fuel system was treated, the engine mounted fuel filters should be left intact, so the sealed system remains filled with treated fuel during the storage period.

   10. For motors with external oil tanks, if possible, leave the oil supply line connected to the motor. This is the best way to seal moisture out of the system. If the line must be disconnected for any reason (such as to remove the motor or oil tank from the boat), seal the line by sliding a snug fitting cap over the end.

   11. Perform a complete lubrication service following the procedures in this section.

   12. Except for Jet Drive models, remove the propeller and check thoroughly for damage. Clean the propeller shaft and apply a protective coating of grease.

   13. On Jet models, thoroughly inspect the impeller and check the impeller clearance. Refer to the procedures in this section.

   14. Check the motor for loose, broken or missing fasteners. Tighten fasteners and, again, use the storage time to make any necessary repairs.

   15. Inspect and repair all electrical wiring and connections at this time. Make sure nothing was damaged during the season's use. Repair any loose connectors or any wires with broken, cracked or otherwise damaged insulation.

   16. Clean all components under the engine cover and apply a corrosion preventative spray.

   17. Too many people forget the boat and trailer, don't be one of them.
   a. Coat the boat and outside painted surfaces of the motor with a fresh coating of wax, then cover it with a breathable cover
   b. If possible place the trailer on stands or blocks so the wheels are supported off the ground.
   c. Check the air pressure in the trailer tires. If it hasn't been done in a while, remove the wheels to clean and repack the wheel bearings.

   18. Sleep well, since you know that your baby will be ready for you come next season.

Re-Commissioning

REMOVAL FROM STORAGE

   The amount of service required when re-commissioning the boat and motor after storage depends on the length of non-use, the thoroughness of the storage procedures and the storage conditions.

   At minimum, a thorough spring or pre-season tune-up and a full lubrication service is essential to getting the most out of your engine. If the engine has been properly winterized, it is usually no problem to get it in top running condition again in the springtime. If the engine has just been put in the garage and forgotten for the winter, then it is doubly important to perform a complete tune-up before putting the engine back into service. If you have ever been stranded on the water because your engine has died and you had to suffer the embarrassment of having to be towed back to the marina you know how it can be a miserable experience. Now is the time to prevent that from occurring.

■ Although you should normally replace your spark plugs at the beginning of each season, we like to start and run the engine (for the spring compression check) using the old spark plugs. Why? Well, on that first start-up you're going to be burning off a lot of fogging oil, why expose the new plugs to it? Besides you have to remove the plugs in order to perform the tune-up compression check anyway so you can install the new ones at that time.

   Take the opportunity to perform any annual maintenance procedures that were not conducted immediately prior to placing the motor into storage. If the motor was stored for more than one off-season, pay special attention to inspection procedures, especially those regarding hoses and fittings. Check the engine gear oil for excessive moisture contamination. The same goes for oil tanks on 2-strokes, so equipped. If necessary, change the gearcase or engine oil to be certain no bad or contaminated fluids are used.

■ Although not absolutely necessary, it is a good idea to ensure optimum cooling system operation by replacing the water pump impeller at this time. Impeller replacement was once an annual ritual, we now hear that most impellers seem to do well for 2-3 seasons, and some more. You'll have to make your own risk assessment here, but keep an eye on your cooling indicator stream get to know how strong the spray looks (and how warm it feels) to help you make your decision.

Other items that require attention include:
1. Install the battery (or batteries) if so equipped.
2. Inspect all wiring and electrical connections. Rodents have a knack for feasting on wiring harness insulation over the winter. If any signs of rodent life are found, check the wiring carefully for damage, do not start the motor until damaged wiring has been fixed or replaced.
3. On motors with a remote oil tank, if the line was disconnected, remove the cover and reconnect the line, then prime the system to ensure proper operation once the motor is started.

■ The Precision Blend oiling system is normally equipped with a manual priming switch/button. On other models it will be necessary to run the engine on a separate pre-mix tank while the Precision Blend oiling line is disconnected from the engine to bleed air from the line and prime the system. Please refer to your owner's manual or to the Lubrication and Cooling section for more details.

4. If not done when placing the motor into storage clean and/or replace the fuel filters at this time.

■ Portable fuel tanks should be emptied and cleaned using solvent. Take the opportunity to thoroughly inspect the condition of the tank. For more details on fuel tanks, please refer to the Fuel System section.

5. If the fuel tank was emptied, or if it must be emptied because the fuel is stale fill the tank with fresh fuel. Keep in mind that even fuel that was treated with stabilizer will eventually become stale, especially if the tank is stored for more than one off-season. Pump the primer bulb and check for fuel leakage or flooding at the carburetor.
6. Attach a flush device or place the outboard in a test tank and start the engine. Run the engine at idle speed and warm it to normal operating temperature. Check for proper operation of the cooling, electrical and warning systems.

※※ **CAUTION**

Before putting the boat in the water, take time to verify the drain plug is installed. Countless number of spring boating excursions have had a very sad beginning because the boat was eased into the water only to have the boat begin to fill with it.

CLEARING A SUBMERGED MOTOR

Unfortunately, because an outboard is mounted on the exposed transom of a boat, and many of the outboards covered here are portable units that are mounted and removed on a regular basis, an outboard can fall overboard. Ok, it's relatively rare, but it happens often enough to warrant some coverage here. The best way to deal with such a situation is to prevent it, by keeping a watchful eye on the engine mounting hardware (bolts and/or clamps). But, should it occur, here's how to salvage, service and enjoy the motor again.

In order to prevent severe damage, be sure to recover an engine that is dropped overboard or otherwise completely submerged as soon as possible. It is really best to recover it immediately. But, keep in mind that once a submerged motor is recovered exposure to the atmosphere will allow corrosion to begin etching highly polished bearing surfaces of the crankshaft, connecting rods and bearings. For this reason, not only do you have to recover it right away, but you should service it right away too. Make sure the motor is serviced within about 3 hours of initial submersion.

OK, maybe now you're saying "3 hours, it will take me that long to get it to a shop or to my own garage." Well, if the engine cannot be serviced immediately (or sufficiently serviced so it can be started), re-submerge it in a tank of fresh water to minimize exposure to the atmosphere and slow the corrosion process. Even if you do this, do not delay any more than absolutely necessary, service the engine as soon as possible. This is especially important if the engine was submerged in salt, brackish or polluted water as even submersion in fresh water will not preserve the engine indefinitely. Service the engine, at the **MOST** within a few days of protective submersion.

After the engine is recovered, vigorously wash all debris from the engine using pressurized freshwater.

■ If the engine was submerged while still running, there is a good chance of internal damage (such as a bent connecting rod). Under these circumstances, don't start the motor, follow the beginning of this procedure to try turning it over slowly by hand, feeling for mechanical problems. If necessary, refer to Powerhead Overhaul for complete disassembly and repair instructions.

※※ **WARNING**

NEVER try to start a recovered motor until at least the first few steps (the ones dealing with draining the motor and checking to see it if is hydro-locked or damaged) are performed. Keep in mind that attempting to start a hydro-locked motor could cause major damage to the powerhead, including bending or breaking a connecting rod.

If the motor was submerged for any length of time it should be thoroughly disassembled and cleaned. Of course, this depends on whether water intruded into the motor or not. Check the gearcase oil (on all motors) for signs of contamination.

The extent of cleaning and disassembly that must take place depends also on the type of water in which the engine was submerged. Engines totally submerged, for even a short length of time, in salt, brackish or polluted water will require more thorough servicing than ones submerged in fresh water for the same length of time. But, as the total length of submerged time or time before service increases, even engines submerged in fresh water will require more attention. Complete powerhead disassembly and inspection is required when sand, silt or other gritty material is found inside the engine cover.

Many engine components suffer the corrosive effects of submersion in salt, brackish or polluted water. The symptoms may not occur for some time after the event. Salt crystals will form in areas of the engine and promote significant corrosion.

Electrical components should be dried and cleaned or replaced, as necessary. If the motor was submerged in salt water, the wire harness and connections are usually affected in a shorter amount of time. Since it is difficult (or nearly impossible) to remove the salt crystals from the wiring connectors, it is best to replace the wire harness and clean all electrical component connections. The starter motor, relays and switches on the engine usually fail if not thoroughly cleaned or replaced.

To ensure a thorough cleaning and inspection:
1. Remove the engine cover and wash all material from the engine using pressurized freshwater. If sand, silt or gritty material is present inside the engine cover, completely disassemble and inspect the powerhead.
2. Tag (except on single cylinder motors) and disconnect the spark plugs leads. Be sure to grasp the spark plug cap and not the wire, then twist the cap while pulling upward to free it from the plug. Remove the spark plugs. For more details, refer to the Spark Plug procedure in this section.
3. Disconnect the fuel supply line from the engine, then drain and clean all fuel lines. Depending on the circumstances surrounding the submersion, inspect the fuel tank for contamination and drain, if necessary.

※※ **WARNING**

When attempting to turn the flywheel for the first time after the submersion, be sure to turn it SLOWLY, feeling for sticking or binding that could indicate internal damage from hydro-lock. This is a concern, especially if the engine was cranked before the spark plug(s) were removed to drain water or if the engine was submerged while still running.

2-88 MAINTENANCE & TUNE-UP

4. Support the engine horizontally with the spark plug port(s) facing downward, allowing water, if present, to drain. Force any remaining the water out by slowly rotating the flywheel by hand about 20 times or until there are no signs of water. If there signs of water present, spray some fogging oil into the spark plug ports before turning the flywheel. This will help dislodge moisture and lubricate the cylinder walls.

5. On carbureted models, drain the carburetor(s). The best method to thoroughly drain/clean the carburetor is to remove and disassemble it, but the float bowl drain screws are pretty accessible on Yamahas and this is not absolutely necessary. It's not a bad idea to spray some fogging oil into the carburetors to make sure moisture is displaced. The oil will burn once the motor is started anyway. For more details on carburetor service refer to the Carburetor procedures under Fuel System.

6. Support the engine in the normal upright position. Check the engine gearcase oil for contamination. Refer to the procedures for Gearcase Oil in this section. The gearcase is sealed and, if the seals are in good condition, should have survived the submersion without contamination. But, if contamination is found, look for possible leaks in the seals, then drain the gearcase and make the necessary repairs before refilling it. For more details, refer to the section on Gearcases.

7. On motors with an injection oil tank mounted TO the motor, drain the oil and dispose of it properly. Don't risk your engine with potentially contaminated oil. Clean the tank using solvent and drain, then refill the tank with fresh, clean 2-stroke oil.

8. Remove all external electrical components for disassembly and cleaning. Spray all connectors with electrical contact cleaner, then apply a small amount of dielectric grease prior to reconnection to help prevent corrosion. For electric start models, remove, disassemble and clean the starter components. For details on the electrical system components, refer to the Ignition and Electrical section.

9. Reassemble the motor and mount the engine or place it in a test tank. Start and run the engine for 1/2 hour. If the engine won't start, remove the spark plugs again and check for signs of moisture on the tips. If necessary, use compressed air to clean moisture from the electrodes or replace the plugs.

10. Stop the engine, then recheck the gearcase oil and injection oil tank, if equipped.

11. Perform all other lubrication services.

12. Try not to let it get away from you (or anyone else) again!

SPECIFICATIONS

TUNE-UP SPECIFICATIONS CHART - 2-Stroke Motors

Model (Hp)	No. of Cyl	Engine Type	Year	Displace cu. in. (cc)	Spark Plug NGK	Spark Plug Champion	Gap Inch(mm)	Ignition Timing Degrees Idle	Ignition Timing Degrees Carb Pickup	Ignition Timing Degrees Full Advance	Idle Speed RPM Neutral	Idle Speed RPM Trolling	Tach Pole Setting	Full Throttle RPM	Test Propeller Kent Moore Part Number	Test Propeller New Engine RPM
2	1	IL 2-stroke	1997-02	2.6 (43)	B5HS ①	L90C	0.024-0.028 (0.6-0.7)	16-20	-	-	1100-1200	-	4	4000-5000	YB-1601	4400-4700
2	1	IL 2-stroke	1997-02	3.0 (50)	B7HS	L90C	0.024-0.028 (0.6-0.7)	16-20	-	-	1100-1200	-	4	4000-5000	YB-1601	4400-4700
3	1	IL 2-stroke	1997-02	4.3 (70)	B6HS-10	L86C	0.035-0.039 (0.9-1.0)	4-8	-	18-24	1150-1250	1000-1100	-	4500-5500	YB-1630	4500-5500
4	1	CL 2-stroke	1997-99	5.0 (83)	B7HS ②	L82C	0.024-0.028 (0.6-0.7)	3-9	-	25-31	1100-1200	950-1050	4	4500-5500	-	-
4	1	CL 2-stroke	1997-02	6.3 (103)	B7HS ②	L82C	0.024-0.028 (0.6-0.7)	3-9	-	25-31	1100-1200	950-1050	4	4500-5500	-	-
5	1	CL 2-stroke	1997-02	6.3 (103)	B7HS ②	L82C	0.024-0.028 (0.6-0.7)	3-9	-	25-31	1100-1200	950-1050	4	4500-5500	-	-
6	2	CL 2-stroke	1997-00	10 (165)	B7HS-10	L82C	0.035-0.039 (0.9-1.0)	3-5	-	34-36	850-950	750-850	4	4000-5000	YB-1625	4500-4700
8	2	CL 2-stroke	1997-03	10 (165)	B7HS-10	L82C	0.035-0.039 (0.9-1.0)	3-5	-	34-36	850-950	750-850	4	4500-5500	YB-1625	5300-5500
9.9	2	IL 2-stroke	1997-03	15 (246)	B7HS-10	L82C	0.035-0.039 (0.9-1.0)	4-6	-	29-31	700-800	600-700	4	4500-5500	YB-1619	5000-5200
15	2	IL 2-stroke	1997-03	15 (246)	B7HS-10	L82C	0.035-0.039 (0.9-1.0)	4-6	9-11	29-31	850-950	670-770	4	4500-5500	YB-1619	5200-5400
20	2	IL 2-stroke	1997	24 (395)	B7HS-10 ③	L82C	0.035-0.039 (0.9-1.0)	6-8 ATDC	0-1 ATDC	24-26	700-800 ④	600-700	4	5000-6000	YB-1621	5500-5700
25	2	IL 2-stroke	1997-03	24 (395)	B7HS-10 ③	L82C	0.035-0.039 (0.9-1.0)	6-8 ATDC	0-1 ATDC	24-26	700-800 ④	550-650	4	5000-6000	YB-1621	4200-4400
20	2	IL 2-stroke	1997-98	26 (430)	B7HS	L82C	0.024-0.28 (0.6-0.7)	1-3 ATDC	-	20-22	950-1050	800-900	4	4500-5500	YB-1621	4200-4400
25	2	IL 2-stroke	1997-98	26 (430)	B7HS	L82C	0.024-0.28 (0.6-0.7)	1-3 ATDC	-	23-25	1150-1250	800-900	4	4500-5500	YB-1621	4200-4400
25	2	IL 2-stroke	1997	30 (496)	B7HS-10 ⑤	L82C	0.035-0.039 (0.9-1.0)	4-8 ATDC	-	20-24	1100-1200	900-1000	6	4500-5500	YB-1621	4200-4400
30	2	IL 2-stroke	1997	30 (496)	B8HS-10 ⑥	L78C	0.035-0.039 (0.9-1.0)	4-8 ATDC	-	23-27	1100-1200	900-1000	6	4500-5500	YB-1629	5250-5450
40	2	IL 2-stroke	1997	36 (592)	B8HS ⑦	L78C	0.024-0.28 (0.6-0.8)	1-3	5	21-23	1100-1200	-	4	4500-5500	YB-1611	4800-5000
48	2	IL 2-stroke	1997-00	46 (760)	B7HS ②	L82C	0.024-0.28 (0.6-0.8)	3-5	3-5	19-21	1200-1300	850-950	4	4500-5500	YB-1611	⑧
25	3	IL 2-stroke	1997-02	30 (496)	B7HS-10 ③	L82C	0.035-0.039 (0.9-1.0)	4-6 ATDC	2-4 ATDC	24-26	700-800 ⑨	600-700 ⑨	6	4500-5500	YB-1621	4200-4400
30	3	IL 2-stroke	1997-02	30 (496)	B7HS-10 ③	L82C	0.035-0.039 (0.9-1.0)	4-6 ATDC	1-3 ATDC	24-26	700-800 ⑨	600-700 ⑨	6	4500-5500	YB-1621	5250-5450
28J	3	IL 2-stroke	1997-01	43 (698)	B7HS-10	L82C	0.035-0.039 (0.9-1.0)	6-8 ATDC	6-8 ATDC	24-28	750-850	-	6	4500-5500	-	-
35J	3	IL 2-stroke	1997-01	43 (698)	B8HS-10	L78C	0.035-0.039 (0.9-1.0)	6-8 ATDC	6-8 ATDC	24-28	750-850	-	6	4500-5500	-	-
40	3	IL 2-stroke	1997-03	43 (698)	B7HS-10 ③	L82C	0.035-0.039 (0.9-1.0)	6-8 ATDC	6-8 ATDC	24-28	750-850	550-650	6	4500-5500	YB-1611	4900-5100
50	3	IL 2-stroke	1997-03	43 (698)	B8HS-10 ⑩	L78C	0.035-0.039 (0.9-1.0)	6-8 ATDC	6-8 ATDC	24-28	750-850	550-650	6	4500-5500	YB-1611	5250-5450
50	3	IL 2-stroke	1997-03	52 (849)	BR8HS-10	L78C	0.035-0.039 (0.9-1.0)	6-8 ATDC	6-8 ATDC	21-23	750-850	550-650	6	4500-5500	YB-1620	4950-5150
60	3	IL 2-stroke	1997-03	52 (849)	B8HS-10 ⑩	L78C	0.035-0.039 (0.9-1.0)	⑪	⑪	⑫	⑬	⑬	6	4500-5500	YB-1620	4950-5150
70	3	IL 2-stroke	1997-03	52 (849)	B8HS-10 ⑩	L78C	0.035-0.039 (0.9-1.0)	6-8 ATDC	6-8 ATDC	19-21	750-850	550-650	6	5000-6000	YB-1620	5250-5450
65J	3	IL 2-stroke	1997-01	70 (1140)	B8HS-10 ⑩	L78C	0.035-0.039 (0.9-1.0)	1-3 BTDC	2 BTDC	21-23	750-850	-	12	4500-5500	-	-
75	3	IL 2-stroke	1997-00	70 (1140)	B8HS-10 ⑩	L78C	0.035-0.039 (0.9-1.0)	⑭	⑭	⑮	750-850	550-650	12	4500-5500	YB-1620	⑮
80	3	IL 2-stroke	1997-00	70 (1140)	B8HS-10 ⑩	L78C	0.035-0.039 (0.9-1.0)	7-9 ATDC	7-9 ATDC	19-21	750-850	550-650	12	4500-5500	YB-1620	4800-5500
85	3	IL 2-stroke	1997-00	70 (1140)	B8HS-10 ⑩	L78C	0.035-0.039 (0.9-1.0)	1-3 BTDC	2 BTDC	23-25	750-850	550-650	12	4500-5500	YB-1620	5100-5300
90	3	IL 2-stroke	1997-03	70 (1140)	B8HS-10 ⑩	L78C	0.035-0.039 (0.9-1.0)	7-9 ATDC	7-9 ATDC	21-23	750-850	550-650	12	4500-5500	YB-1620	5100-5300
80J	4	90 LV 2-st	1997-01	106 (1730)	BR8HS-10	QL78C	0.035-0.039 (0.9-1.0)	4-6 ATDC	3-5 ATDC	24-26	600-700	-	12	4500-5500	-	-

MAINTENANCE & TUNE-UP 2-89

TUNE-UP SPECIFICATIONS CHART - 2-Stroke Motors

Model (Hp)	No. of Cyl	Engine Type	Year	Displace cu. in. (cc)	Spark Plug NGK	Spark Plug Champion	Gap Inch(mm)	Ignition Timing Degrees Idle	Ignition Timing Degrees Carb Pickup	Ignition Timing Degrees Full Advance	Idle Speed RPM Neutral	Idle Speed RPM Trolling	Tach Pole Setting	Full Throttle RPM	Test Propeller Kent Moore Part Number	Test Propeller New Engine RPM
100	4	90 LV 2-st	1997-02	106 (1730)	BR8HS-10	QL78C	0.035-0.039 (0.9-1.0)	4-6 ATDC	3-5 ATDC	21-23	700-800	600-700	12	4500-5500	YB-1624	4900-5100
115	4	90 LV 2-st	1997-03	106 (1730)	BR8HS-10	QL78C	0.035-0.039 (0.9-1.0)	4-6 ATDC	3-5 ATDC	⑰	700-800	600-700	12	4500-5500	YB-1624	4900-5100
130	4	90 LV 2-st	1997-03	106 (1730)	BR9HS-10	QL77CJ4	0.035-0.039 (0.9-1.0)	4-6 ATDC	3-5 ATDC	21-23	700-800	600-700	12	5000-6000	YB-1624	5200-5400
140	4	90 LV 2-st	1997-02	106 (1730)	BR9HS-10	QL77CJ4	0.035-0.039 (0.9-1.0)	4-6 ATDC	3-5 ATDC	21-23	700-800	600-700	12	5000-6000	YB-1624	5200-5400
105J	6	90 LV 2-st	1997-00	158 (2596)	B8HS-10	L78C	0.035-0.039 (0.9-1.0)	5-9 ATDC	5-9 ATDC	20-24	675-725	-	12	4500-5500	-	-
150 Carb	6	90 LV 2-st	1997-03	158 (2596)	BR7HS-10 ⑱	RL82C ⑲	0.035-0.039 (0.9-1.0)	5-9 ATDC	6-8 ATDC	20-24 ⑲	675-725	545-595	12	4500-5500	YB-1626	4600-4800
150 EFI	6	90 LV 2-st	1999-03	158 (2596)	BR7HS-10	RL82C	0.035-0.039 (0.9-1.0)	1-5 ATDC	-	19-23	700-760	-	12	4500-5500	YB-1626	4600-4800
175	6	90 LV 2-st	1997-00	158 (2596)	B8HS-10	QL78C	0.035-0.039 (0.9-1.0)	5-9 ATDC	5-9 ATDC	20-24 ⑳	675-725	550-600	12	4500-5500	YB-1626	4900-5100
200 Carb	6	90 LV 2-st	1997-99	158 (2596)	B8HS-10	QL78C	0.035-0.039 (0.9-1.0)	5-9 ATDC	5-9 ATDC	19-21 ㉑	675-725	550-600	12	4500-5500 ㉒	YB-1626	5300-5500
200 EFI	6	90 LV 2-st	1999-03	158 (2596)	BR8HS-10	QL78C	0.035-0.039 (0.9-1.0)	3-7 ATDC	-	21-25	700-760	-	12	4500-5500	YB-1626	5300-5500
225	6	90 LV 2-st	1997	158 (2596)	BR8HS-10	QL82C	0.035-0.039 (0.9-1.0)	5-7 ATDC	5-7 ATDC	20-22	675-725	575-625	12	5000-6000	YB-1626	5450-5650
150 HPDI	6	76 LV 2-st	2000-03	158 (2596)	BKR6ES-11 ㉓	-	0.035-0.039 (0.9-1.0)	㉔	-	15-19	670-730	-	12	4500-5500	-	-
175 HPDI	6	76 LV 2-st	2001-03	158 (2596)	BKR7ES-11 ㉕	-	0.035-0.039 (0.9-1.0)	0-6 BTDC	-	15-19	670-730	-	12	4500-5500	YB-1624	4900-5100
200 HPDI	6	76 LV 2-st	2000-03	158 (2596)	BKR7ES-11 ㉖	-	0.035-0.039 (0.9-1.0)	1-7 BTDC	-	15-19	670-730	-	12	4500-5500	-	-
200 EFI ㉚	6	76 LV 2-st	1998	191 (3130)	BR9HS-10	-	0.035-0.039 (0.9-1.0)	1-5 ATDC	-	15-19	700-760	595-655	12	4500-5500	-	-
	6	76 LV 2-st	1999-03	191 (3130)	BR8HS-10	QL78C	0.035-0.039 (0.9-1.0)	1-5 ATDC	-	18-22	700-760	-	12	4500-5500	YB-1626	5300-5500
225 EFI ㉚	6	76 LV 2-st	1997	191 (3130)	BR9HS-10	-	0.035-0.039 (0.9-1.0)	8 ATDC	-	18	700-760	595-655	12	4500-5500	-	-
	6	76 LV 2-st	1998-03	191 (3130)	BR9HS-10	-	0.035-0.039 (0.9-1.0)	1-5 ATDC	-	㉗	700-760	595-655	12	4500-5500	-	-
250 EFI ㉚	6	76 LV 2-st	1997	191 (3130)	BR9HS-10	-	0.035-0.039 (0.9-1.0)	8 ATDC	-	18	700-760	595-655	12	4500-5500	-	-
	6	76 LV 2-st	1998-03	191 (3130)	BR9HS-10	-	0.035-0.039 (0.9-1.0)	㉘	-	㉙	700-760	595-655	12	4500-5500	-	-
225 HPDI	6	76 LV 2-st	2003	204 (3342)	BKR7EK-U1	-	0.020-0.028 (0.5-0.7)	9 ATDC	-	19 (Cyl #1)	670-730	-	12	4500-5500	-	-
250 HPDI	6	76 LV 2-st	2003	204 (3342)	BKR7EK-U1	-	0.020-0.028 (0.5-0.7)	9 ATDC	-	19 (Cyl #1)	670-730	-	12	4500-5500	-	-

NOTE: It is recommended for optimum performance and engine life to adjust the wide open throttle engine speed to the top end of the recommended range

① Specification is for USA Models, the manufacturer recommends using NGK BR5HS for Canada and Europe
② Specification is for USA Models, the manufacturer recommends using noise suppressor type plug NGK BR7HS for Canada and Europe
③ Specificed plug is for all except Europe and Canda where the manufacturer recommends the resistor type plug BR7HS-10 for this application
④ Specification is for all except 1996 Canada models which should be 750-810 rpm
⑤ Specification is for all except Canada and Europe for which the manufacturer recommends using resistor type plug BR7HS-10
⑥ Specification is for all except Canada, Europe, and South Africa for which the manufacturer recommends using resistor type plug BR8HS-10
⑦ Specification is for all except Canada and Europefor which the manufacturer recommends using resistor type plug BR8HS
⑧ Specification is 5300-5500 rpm through 1998 or 5000-5200 rpm for 1999 and later
⑨ Specifications are for all except PTT models which are 1000-1100 in Neutral or 750-850 while Trolling
⑩ The resistor type plug BR8HS-10 is also available for this application and is often specified for use outside the U.S. in Europe and Canada
⑪ Specification is 6-8 ATDC for most models, however the 1999 E60 which should be set to 1-3 ATDC
⑫ Specification is 21-23 BTDC for all motors, except for E60 models for which the WOT timing should be set to 18-20 BTDC
⑬ Specification is 750-850 idle / 550-650 trolling on most motors, except E60 models for which the trolling speed should be 850-900 rpm. Also, note that the 1999 E90 idle should be set to 1000 rpm
⑭ Specification is 7-9 ATDC idle and cam pickup for most models including C75 and P75, however specs for the E75 are 1-3 BTDC idle and cam pickup for E75 models
⑮ Specification is 19-21 BTDC for most motors, including all C75 & P75, but specs for E and A models should be set to 21-23 BTDC except for European spec motors which should also be 19-21 BTDC
⑯ Specification is 4800-5500 for most motors, including all C75 and P75, but E and A models should be 4750-4950
⑰ Specification 24-26 BTDC for most motors, except the E115A for which full advance timing should be set to 22-24 BTDC
⑱ Specification is for most carbureted 150 hp motors, the D150 which uses the NGK plug BR7HS-10 or Champion plug QL82C and the C150 which uses Champion B8HS-10 or NGK L82C
⑲ Specification is for most carbureted 150 hp motors, except the D150/P150 for which are 21-25 BTDC, the V150 which is 19-21 BTDC, or the C150 which is 18-20 BTDC
⑳ Specification is for the S175, WOT timing for the P175 should be set to 21-25 BTDC
㉑ Specification is for most 200 hp motors, except the S200/L200 which are 19-23 BTDC or the P200 which is 21-25 BTDC
㉒ Specification is for most 200 hp motors, except the P200 which is 5500-6000 rpm
㉓ Specification is listed in factory service manual, but factory tune-up guide lists different plugs based on engine serial numbers as follows: Z150TL 00~800452, LZ150 00~800554, VZ150 00~800711, Z150TX 00~851535 use BKR6ES-10 while Z150TL 800453~, LZ150 800555~, VZ150 800712~, Z150TX 851535 use BKR6EK-U1
㉔ Specification varies by model, L motors are 0-6 BTDC, X motors are 1-7 BTDC
㉕ Specification is listed in factory service manual, but factory tune-up guide lists different plugs based on engine serial numbers as follows: Z175 800101~800231, and VZ175 800105~800201 use BKR7ES-10 while Z175 800232~, and VZ175 800202~ use BKR7EK-U1
㉖ Specification is listed in factory service manual, but factory tune-up guide lists different plugs based on engine serial numbers as follows: Z200TL 00~150689, Z200TX 00~102420 and LZ200 00~100874 use BKR7ES-10 while Z200TL 105690~, Z200TX 102421~ and LZ200 100875~ use BKR7EK-U1
㉗ Specification is 15.4-19.4 BTDC for all models except V and VX motors for which the WOT timing is set to 16-20 BTDC
㉘ Specification is 1-5 ATDC for all models except 02 and later VX models for which the idle timing is set to 4 ATDC for Cyl #1 and 10 ATDC for Cyl #2
㉙ Specification is 13.8-17.8 BTDC for all models except 02 and later VX models for which the WOT timing is set to 20 BTDC for Cyl #1 and
㉚ Although specifications are provided for checking operating parameters, in most cases timing is computer controlled on these models and is not adjustable

MAINTENANCE & TUNE-UP

Maintenance Intervals Chart

Component	Each Use or As Needed	Initial 10-Hour Check (Break-In)	Initial 50-Hour Check (3 months)	Every 6mths/100hrs ①	Off Season 12mths/200hrs ①
2-Stroke engine oil (except pre-mix models)	I (top off tank as, as needed)				
Anode(s)	I	I	I	I	I
Battery condition and connections (if equipped)*	I (condition/connections), T			I	I
Battery charge / fluid level (if equipped)*	I (at least monthly)			I	I
Boat hull*	I			I	
Bolts and nuts (all accessible fasteners)*	I	I, T		I, T	I
Carburetor (clean / inspect - if equipped)		C, I	C, I	C, I	C, I
Carburetor (adjust - if equipped) ②		A		A (inline motors)	A (V motors)
Case finish and condition (inspect & wash/wax)	C, salt/brackish/polluted water	I	I	I, C	I, C
Choke solenoid (if equipped)		A			A
Cylinder head or cover bolts (as applicable) ③		Check, T		Check, T	
Electricical wiring and connectors*		I	I	I	I
Emergency stop switch, clip &/or lanyard*	I	I		I	I
Engine mounting bolts		I		I	I
Engine crankcase oil (4-st)	I	R		R	R
Engine crankcase oil filter (4-st, 8/9.9 hp & up) ④		C or R (as applicable)		C/R (as applicable)	C/R (as applicable)
Flush cooling system	salt/brackish/polluted water		P	P	P
Fuel enrichment valve (if equipped)				I, A	
Fuel filter (clean or replace, as applicable)		P (exc 2-8 hp 2-st)	P (exc 2-8 hp 2-st)	P	P
Fuel (enrichment line) filter (3.1L models)					R
Fuel hose and system components*	I	I	I (EFI & HPDI)	I	I
Fuel pump oil (mechanical pump on HPDI) ⑤					I (R 1000 hrs/5 yrs)
Fuel tank (integral or portable)		I			C (portable), I
Fuel water separator (if equipped)	I (200/225 4-st)			I	I
Gear oil	I (for signs of leakage)	R	I	R	R
HPDI drive belt					I (R 1000 hrs/5 yrs)
Idle speed	I (look/listen for changes)			A	A
Ignition timing (200/225 4-st Ins. only at 1st 10hr)		I		I	I
Impeller clearance/intake grate (jet models)	I (visually inspect impeller)	I		I	I
Jet drive bearing lubrication ⑥	L, fill vent hose after each day	L		L	L
Lubrication points	I	⑦		⑦	⑦
Neutral opening control device (if equipped)		A			A
Oil pump link (if applicable)		I, A	I, A	I, A	
Oil pump check (200/225 4-st)					I (400 hrs/2 yrs)
Oil system (include tank water drain) *	I, C	I, C	I, C	I, C	I, C
Oil system functional check		I		I	
Power trim and tilt (if equipped)	I (check fluid level monthly)	I		I	I
Propeller	I	I		I	I
Propeller shaft and cotter pin/nut (or shear pin)	I	I		I, L / T	I, L / T
Remote control*		I		I, A	I, A
Spark plugs	R (as needed)	I	I	I	I
Start in gear lockout system (if applicable)	I	Check, A			Check, A
Steering cable*	L (as needed)	L		L	L
TCI air gap (2.5 & 4 hp 4-st)		I, A		I, A	
Thermostat (if equipped)		I			I
Throttle (tiller models)		Check, A			Check, A
Throttle position sensor (EFI & HPDI)					I, A
Throttle valve switch (EFI & HPDI)					I, A
Tune-up	A (as needed)			I (annually)	Pre-season tune-up
Timing belt (4-st, 6/8-225 hp)				I	I (R 1000 hrs/5 yrs)
Timing chain (4-st, 200/225 hp)				I	(R 1000 hrs/5 yrs)
Timing chain tensioner (4-st, 200/225 hp)				I	I, R (1000 hrs/5 yrs)
Valve clearance (4-st motors up to 60 hp)		I		I	I
Valve clearance (4-st motors 75 hp & up)					I, A (400 hrs/2 yrs) ⑧
Water pressure valve (3.1L)					I
Water pump impeller					varies ⑨
Water pump intake grate and indicator	I				

A-Adjust I-Inspect and Clean, Adjust, Lubricate or Replace, as necessary R-Replace P - Perform
C-Clean L-Lubricate T-Tighten

* Denotes possible safety item (although, all maintenance inspections/service can be considered safety related when it means not being stranded on the water should a component fail.)

① Many items are listed for both every 100 hours and off season. Since many boaters use their crafts less than 100 hours a year, these items should at least be performed annually. If you find yourself right around 100 hours per season, try to time the service so it occurs immediately prior to placing the motor in storage, as some items must be reperformed if the engine is used again (even once) before storage.
② Recommendations vary regarding the need to adjust the carburetor (as opposed to only checking how it is operating). The most recent advice we've received from Yamaha is not to tamper with a carb that is running correctly
③ DO NOT retighten the cylinder head or crankcase bolts on the 200/225 4-stroke motors
④ For models that use replaceable filter elements the replacement interval may be extended up to 200 hours, but only during a single season
⑤ Originally with the 2000 model year Yamaha recommended replacing oil every 200 hours, but by 2001 publications were changed to recommend checking every 200 hours and replacing every 1000 hours or 5 years
⑥ Lubricate the jet drive bearing at the end of EVERY day's use and every 10 hours. Also, every 50 hours replace the grease using the same lubrication fitting, but adding enough to purge the old grease
⑦ Varies with use, generally every 30 days when used in salt, brackish or polluted water and every 60 days when used in fresh water (refer to Lubrication Chart for more details)
⑧ Except when operated on leaded gasoline, then inspection/adjustment interval should be reduced to 300 hours / 2 years
⑨ Water pump impeller inspection recommendations vary from no spec (most common), to inspect every 100 hours (most 4-st) or replace every 500 hours/30 months (200/225 hp 4-st)

MAINTENANCE & TUNE-UP 2-91

General Engine Specifications - 2-Stroke Motors

Model (Hp)	No. of Cyl	Engine Type	Year	Displacement cu. in. (cc)	Bore and Stroke in. (mm)	Compression Ratio	Appx Weight lb. (kg) *
2	1	IL 2-stroke	1997-02	2.6 (43)	1.54 x 1.42 (39 x 36)	7.4:1	22 (10)
2	1	IL 2-stroke	1997-02	3.0 (50)	1.65 x 1.42 (42 x 36)	8.3:1	21.6 (9.8)
3	1	IL 2-stroke	1997-02	4.3 (70)	1.81 x 1.65 (46 x 42)	6.9:1	36.3-38.5 (16.5-17.5)
4	1	CL 2-stroke	1992-99	5.0 (83)	1.97 x 1.65 (50 x 42)	7.0:1	45-47 (20-22)
4	1	CL 2-stroke	1997-02	6.3 (103)	2.13 x 1.77 (54 x 45)	6.5:1	46-47 (21-22)
5	1	CL 2-stroke	1997-02	6.3 (103)	2.13 x 1.77 (54 x 45)	6.5:1	45-46 (20-21)
6	2	CL 2-stroke	1997-00	10 (165)	1.97 x 1.65 (50 x 42)	7.0:1	60-65 (27-30)
8	2	CL 2-stroke	1997-03	10 (165)	1.97 x 1.65 (50 x 42)	7.0:1	60-65 (27-30)
9.9	2	IL 2-stroke	1997-03	15 (246)	2.20 x 1.97 (56 x 50)	6.8:1	80-86 (36-39)
15	2	IL 2-stroke	1997-03	15 (246)	2.20 x 1.97 (56 x 50)	6.8:1	80-86 (36-39)
20	2	IL 2-stroke	1997	24 (395)	2.64 x 2.20 (67 x 56)	7.2:1	106-108 (48-49)
25	2	IL 2-stroke	1997-03	24 (395)	2.64 x 2.20 (67 x 56)	7.2:1	111-115 (51-52)
20	2	IL 2-stroke	1997-98	26 (430)	2.64 x 2.40 (67 x 61)	6.6:1	109 (50)
25	2	IL 2-stroke	1997-98	26 (430)	2.64 x 2.40 (67 x 61)	7.0:1	106-112 (48-51)
25	2	IL 2-stroke	1997	30 (496)	2.83 x 2.40 (72 x 61)	6.2:1	115-122 (52-56)
30	2	IL 2-stroke	1997	30 (496)	2.83 x 2.40 (72 x 61)	7.0:1	115-126 (52-57)
40	2	IL 2-stroke	1997	36 (592)	2.95 x 2.64 (75 x 67)	6.5:1	133-165 (61-75)
48	2	IL 2-stroke	1997-00	46 (760)	3.20 x 2.80 (82 x 72)	6.5:1	183-187 (83-85)
25	3	IL 2-stroke	1997-02	30 (496)	2.34 x 2.34 (59.5 x 59.5)	6.8:1	130-146 (59-66)
30	3	IL 2-stroke	1997-02	30 (496)	2.34 x 2.34 (59.5 x 59.5)	6.8:1	130-148 (59-67)
28J	3	IL 2-stroke	1997-01	43 (698)	2.64 x 2.60 (67 x 66)	6.0:1	158 (72)
35J	3	IL 2-stroke	1997-01	43 (698)	2.64 x 2.60 (67 x 66)	6.0:1	170 (77)
40	3	IL 2-stroke	1997-03	43 (698)	2.64 x 2.60 (67 x 66)	6.0:1	162-208 (74-92)
50	3	IL 2-stroke	1997-03	43 (698)	2.64 x 2.60 (67 x 66)	6.0:1	158-202 (72-92)
50	3	IL 2-stroke	1997-03	52 (849)	2.83 x 2.74 (72.0 x 69.5)	6.33:1	225 (102)
60	3	IL 2-stroke	1997-03	52 (849)	2.83 x 2.74 (72.0 x 69.5)	6.1:1	207-247 (94-112)
70	3	IL 2-stroke	1997-03	52 (849)	2.83 x 2.74 (72.0 x 69.5)	6.1:1 ②	207-247 (94-112)
65J	3	IL 2-stroke	1997-01	70 (1140)	3.23 x 2.83 (82 x 72)	4.5:1	247-267 (112-121)
75	3	IL 2-stroke	1997-00	70 (1140)	3.23 x 2.83 (82 x 72)	③	245-277 (111-126)
80	3	IL 2-stroke	1997-00	70 (1140)	3.23 x 2.83 (82 x 72)	5.9:1	266-272 (121-124)
85	3	IL 2-stroke	1997-00	70 (1140)	3.23 x 2.83 (82 x 72)	5.1:1	245-273 (111-124)
90	3	IL 2-stroke	1997-03	70 (1140)	3.23 x 2.83 (82 x 72)	5.86:1	246-277 (112-126)
80J	4	90 LV 2-stroke	1997-98	106 (1730)	3.54 x 2.68 (90 x 68)	6.7:1	340 (154)
	4	90 LV 2-stroke	1999-01	106 (1730)	3.54 x 2.68 (90 x 68)	6.5:1	340 (154)
100	4	90 LV 2-stroke	1997-98	106 (1730)	3.54 x 2.68 (90 x 68)	6.5:1	368 (167)
	4	90 LV 2-stroke	1999-02	106 (1730)	3.54 x 2.68 (90 x 68)	6.8:1	368 (167)
115	4	90 LV 2-stroke	1997-03	106 (1730)	3.54 x 2.68 (90 x 68)	6.5:1 ④	322-377 (146-171)
130	4	90 LV 2-stroke	1997-03	106 (1730)	3.54 x 2.68 (90 x 68)	6.8:1	364-373 (165-169)
140	4	90 LV 2-stroke	1997-98	106 (1730)	3.54 x 2.68 (90 x 68)	6.5:1	340-353 (154-160)
	4	90 LV 2-stroke	1999-02	106 (1730)	3.54 x 2.68 (90 x 68)	6.8:1	340-353 (154-160)
105J	6	90 LV 2-stroke	1997-00	158 (2596)	3.54 x 2.68 (90 x 68)	6.2:1	398 (180)
150	6	90 LV 2-stroke ①	1997-03	158 (2596)	3.54 x 2.68 (90 x 68)	⑤	392-467 (178-212)
175	6	90 LV 2-stroke ①	1997-00	158 (2596)	3.54 x 2.68 (90 x 68)	⑥	392-437 (178-198)
200	6	90 LV 2-stroke ①	1997-03	158 (2596)	3.54 x 2.68 (90 x 68)	⑦	392-467 (178-212)
225	6	90 LV 2-stroke	1997	158 (2596)	3.54 x 2.68 (90 x 68)	6.4:1	397-437 (180-198)
150 HPDI	6	76 LV 2-stroke ①	2000-03	158 (2596)	3.54 x 2.68 (90 x 68)	6.6:1	472-489 (214-222)
175 HPDI	6	76 LV 2-stroke ①	2001-03	158 (2596)	3.54 x 2.68 (90 x 68)	6.5:1	474-489 (215-222)
200 HPDI	6	76 LV 2-stroke ①	2000-03	158 (2596)	3.54 x 2.68 (90 x 68)	6.4:1	472-489 (214-222)

General Engine Specifications - 2-Stroke Motors

Model (Hp)	No. of Cyl	Engine Type	Year	Displacement cu. in. (cc)	Bore and Stroke in. (mm)	Compression Ratio	Appx Weight lb. (kg) *
200 EFI	6	76 LV 2-stroke ①	1998-03	191 (3130)	3.54 x 3.23 (90 x 82)	⑧	519 (236)
225 EFI	6	76 LV 2-stroke ①	1997-03	191 (3130)	3.54 x 3.23 (90 x 82)	⑨	519-539 (236-244)
250 EFI	6	76 LV 2-stroke ①	1997-03	191 (3130)	3.54 x 3.23 (90 x 82)	⑩	519-539 (236-244)
225 HPDI	6	76 LV 2-stroke ①	2003	204 (3342)	3.54 x 3.23 (90 x 82)	6.2:1	524 (238)
250 HPDI	6	76 LV 2-stroke ①	2003	204 (3342)	3.54 x 3.23 (90 x 82)	6.2:1 ⑪	556-567 (252-257)

CL - inline cross-flow charge   IL -Inline loop-charge   # LV - degree "V" loop-charge

* NOTE: Measurements are approximate, minus fluids and propeller for most models. Generally speaking (especially on larger models), pre-mix carbureted versions are lightest and EFI or HPDI are heaviest, but there are some exceptions. See owner's manual

① Induction system varies with model, all models are loop charged, but some motors are carbureted, some are EFI and some are HPDI

② Specification is 6.2:1 for models through 1998 or 6.1:1 for 1999 and later models

③ Specification varies by model, 5.9:1 for most models including C, P or T models, but 4.5:1 for A models or 5.0:1 for European spec A models

④ Specification is for most models, but 1999 and later A models (E115AMH, E115AWH, E115AE, E115AET) are 5.7:1

⑤ Specification varies greatly with year and model from 5.8:1 to 6.8:1, most later models are 6.2:1 if carbureted or 6.3:1 for EFI. Refer to your owner's manual for more details

⑥ Specification varies with year and model. Through 1997 it is 5.9:1 for the SW or 5.6:1 for the SW, for 1998 and later motors it is 6.0:1

⑦ Specification varies with year and model. All carbureted motors are 6.0:1, but EFI motors are either 6.2:1 through 2000 or 6.4:1 for 2001 and later motors

⑧ Specification varies by cylinder, 5.4:1 for Cyl #'s 1-4 and 5.2:1 for Cyl #'s 5 and 6

⑨ Specification varies with year and model. It is 5.9:1 through 1998 and varies by cylinder between 5.8:1 and 6.2:1 for later models. Note on later models Cyl #'s 1-4 have higher compression ratio than Cyl #'s 5 & 6

⑩ Specification varies with year and model. It is 5.8:1 through 1998 and varies by cylinder between 5.7:1 and 6.2:1 for later models. Note on later models Cyl #'s 1-4 have higher compression ratio than Cyl #'s 5 & 6

⑪ Specification listed is for all except V-Max models, which is 6.0:1

MAINTENANCE & TUNE-UP

General Engine System Specifications - 2-Stroke Motors

Model (Hp)	No. of Cyl	Engine Type	Year	Displace cu. in. (cc)	Oil Injection System ①	Ignition System	Starting System	Fuel System	Battery ② CCA (AH)	Reserve
2	1	IL 2-stroke	1997-02	2.6 (43)	Pre-Mix 100:1	CDI	Manual	Carb	n/a	n/a
2	1	IL 2-stroke	1997-02	3.0 (50)	Pre-Mix 100:1	CDI	Manual	Carb	n/a	n/a
3	1	IL 2-stroke	1997-02	4.3 (70)	Pre-Mix 100:1	CDI	Manual	Carb	n/a	n/a
4	1	CL 2-stroke	1997-99	5.0 (83)	Pre-Mix 100:1	CDI	Manual	Carb	n/a	n/a
4	1	CL 2-stroke	1997-02	6.3 (103)	Pre-Mix 100:1	CDI	Manual	Carb	n/a	n/a
5	1	CL 2-stroke	1997-02	6.3 (103)	Pre-Mix 100:1	CDI	Manual	Carb	n/a	n/a
6	2	CL 2-stroke	1997-00	10 (165)	Pre-Mix 100:1	CDI	Man or Man & Elec	Carb	210 (40)	59 minutes
8	2	CL 2-stroke	1997-03	10 (165)	Pre-Mix 100:1	CDI	Man or Man & Elec	Carb	210 (40)	59 minutes
9.9	2	IL 2-stroke	1997-03	15 (246)	Pre-Mix 100:1	CDI	Man or Man & Elec	Carb	210 (40)	59 minutes
15	2	IL 2-stroke	1997-03	15 (246)	Pre-Mix 100:1	CDI	Man or Man & Elec	Carb	210 (40)	59 minutes
20	2	IL 2-stroke	1997	24 (395)	PB or 100:1 PM	CDI	Man or Man & Elec	Carb	210 (40)	59 minutes
25	2	IL 2-stroke	1997-03	24 (395)	PB or 100:1 PM	CDI	Man or Man & Elec	Carb	210 (40)	59 minutes
20	2	IL 2-stroke	1997-98	26 (430)	Pre-Mix 100:1	CDI	Man or Man & Elec	Carb	245 (40)	59 minutes
25	2	IL 2-stroke	1997-98	26 (430)	Pre-Mix 100:1	CDI	Man or Man & Elec	Carb	245 (40)	59 minutes
25	2	IL 2-stroke	1997	30 (496)	Pre-Mix 100:1	CDI	Manual	Carb	210 (40)	59 minutes
30	2	IL 2-stroke	1997	30 (496)	Pre-Mix 100:1	CDI	Man or Man & Elec	Carb	210 (40)	59 minutes
40	2	IL 2-stroke	1997	36 (592)	Pre-Mix 100:1	CDI	Man or Man & Elec	Carb	385 (70)	124 minutes
48	2	IL 2-stroke	1997-00	46 (760)	Pre-Mix 50:1	CDI	Manual	Carb	380 (70)	124 minutes
25	3	IL 2-stroke	1997-02	30 (496)	PB or 100:1	CDI	Manual and/or Elec	Carb	210 (40)	59 minutes
30	3	IL 2-stroke	1997-02	30 (496)	Precision Blend	CDI	Manual and/or Elec	Carb	210 (40)	59 minutes
28J	3	IL 2-stroke	1997-01	43 (698)	Precision Blend	CDI	Manual and/or Elec	Carb	210 (40)	59 minutes
35J	3	IL 2-stroke	1997-01	43 (698)	Precision Blend	CDI	Electric	Carb	210 (40)	59 minutes
40	3	IL 2-stroke	1997-03	43 (698)	PB or 50:1 PM	③	Manual and/or Elec	Carb	380 (70)	124 minutes
50	3	IL 2-stroke	1997-03	43 (698)	PB or 50:1 PM	③	Electric	Carb	380 (70)	124 minutes
50	3	IL 2-stroke	1997-03	52 (849)	PB or 50:1 PM	CDI Micro	Electric	Carb	380 (70)	124 minutes
60	3	IL 2-stroke	1997-03	52 (849)	PB or 50:1 PM	CDI Micro	Electric	Carb	380 (70)	124 minutes
70	3	IL 2-stroke	1997-03	52 (849)	PB or 50:1 PM	CDI Micro	Electric	Carb	380 (70)	124 minutes
65J	3	IL 2-stroke	1997-01	70 (1140)	PB or 50:1 PM	CDI Micro	Manual and/or Elec	Carb	380 (70)	124 minutes
75	3	IL 2-stroke	1997-00	70 (1140)	PB or 50:1 PM	④	Manual and/or Elec	Carb	380 (70)	124 minutes
80	3	IL 2-stroke	1997-00	70 (1140)	Precision Blend	CDI Micro	Electric	Carb	380 (70)	124 minutes
85	3	IL 2-stroke	1997-00	70 (1140)	PB or 50:1 PM	CDI Micro	Manual and/or Elec	Carb	380 (70)	124 minutes
90	3	IL 2-stroke	1997-03	70 (1140)	PB or 50:1 PM	CDI Micro	Electric	Carb	380 (70)	124 minutes
80J	4	90 LV 2-stroke	1997-01	106 (1730)	Precision Blend	CDI	Electric	Carb	380 (70)	124 minutes
100	4	90 LV 2-stroke	1997-02	106 (1730)	Precision Blend	CDI	Electric	Carb	380 (70)	124 Reserve
115	4	90 LV 2-stroke	1997-03	106 (1730)	PB or 50:1 PM	CDI	Electric	Carb	380 (70)	124 minutes
130	4	90 LV 2-stroke	1997-03	106 (1730)	Precision Blend	CDI	Electric	Carb	380 (70)	124 minutes
140	4	90 LV 2-stroke	1997-02	106 (1730)	PB or 50:1 PM	CDI	Electric	Carb	380 (70)	124 minutes
105J	6	90 LV 2-stroke	1997-00	158 (2596)	Precision Blend	CDI Micro	Electric	Carb	380 (70)	124 minutes
150	6	90 LV 2-stroke	1997-03	158 (2596)	PB or 50:1 PM	⑤	Electric	Carb or EFI	380 (70)	124 minutes
175	6	90 LV 2-stroke	1997-00	158 (2596)	PB or 50:1 PM	CDI Micro	Electric	Carb	380 (70)	124 minutes
200	6	90 LV 2-stroke	1997-03	158 (2596)	PB or 50:1 PM	CDI Micro	Electric	Carb or EFI	380 (70)	124 minutes
225	6	90 LV 2-stroke	1997	158 (2596)	PB or 50:1 PM	CDI Micro	Electric	Carb	360 (100)	124 minutes
150 HPDI	6	76 LV 2-stroke	2000-03	158 (2596)	Precision Blend	TCI Micro	Electric	HPDI	360 (100)	124 minutes

General Engine System Specifications - 2-Stroke Motors

Model (Hp)	No. of Cyl	Engine Type	Year	Displace cu. in. (cc)	Oil Injection System ①	Ignition System	Starting System	Fuel System	Battery ② CCA (AH)	Reserve
175 HPDI	6	76 LV 2-stroke	2001-03	158 (2596)	Precision Blend	TCI Micro	Electric	HPDI	360 (100)	124 minutes
200 HPDI	6	76 LV 2-stroke	2000-03	158 (2596)	Precision Blend	TCI Micro	Electric	HPDI	360 (100)	124 minutes
200 EFI	6	76 LV 2-stroke	1998-03	191 (3130)	Precision Blend	CDI Micro	Electric	EFI	512 (100)	182 minutes
225 EFI	6	76 LV 2-stroke	1997-03	191 (3130)	Precision Blend	CDI Micro	Electric	EFI	512 (100)	182 minutes
250 EFI	6	76 LV 2-stroke	1997-03	191 (3130)	Precision Blend	CDI Micro	Electric	EFI	512 (100)	182 minutes
225 HPDI	6	76 LV 2-stroke	2003	204 (3342)	Precision Blend	TCI Micro	Electric	HPDI	512 (100)	182 minutes
250 HPDI	6	76 LV 2-stroke	2003	204 (3342)	Precision Blend	TCI Micro	Electric	HPDI	512 (100)	182 minutes

AH - Amp Hours
CDI - Capacitor Discharge Ignition
CCA - Cold Cranking Amps
CDI Micro - CDI w/ Micro Computer Control

① Most common oiling system installed. Note that oiling systems could be installed or removed during boat rigging. Generally speaking (but not always) "C" models are usually pre-mix motors.
② Minimum recommended cca (mca) ratings for electric start models when motor/cables are new. Replacements must meet or exceed specification
③ CDI used on models through 1997 and again on 2001 and later models, but some literature lists CDI Micro for 1998-2000 models
④ Most models CDI including all C75 through 1996 and all E75 models, but all P75 and 1997 or later C75 utilize CDI Micro
⑤ CDI used on all models except the C150

TWO-STROKE MOTOR FUEL:OIL RATIO CHART

Desired Fuel:Oil Ratio	Amount of oil needed when mixed with:				
	3 G (11.4 L) of Gas	6 G (22.7 L) of Gas	18 G (68.1 L) of Gas	30 G (114 L) of Gas	45 G (171 L) of Gas
100:1 (1% oil)	4 fl. oz. (118 mL)	8 fl. oz. (236 mL)	24 fl. oz. (708 mL)	40 fl. oz. (1180 mL)	60 fl. oz. (1770 mL)
50:1 (2% oil)	8 fl. oz. (236 mL)	16 fl. oz. (473 mL)	48 fl. oz. (1419 mL)	80 fl. oz. (2360 mL)	120 fl. oz. (3.54 L)
25:1 (4% oil)	16 fl. oz. (473 mL)	32 fl. oz. (946 mL)	96 fl. oz. (2838 mL)	160 fl. oz. (4.73 L)	240 fl. oz. (7.1 L)

NOTE: Fuel:Oil ratios listed here are for calcuation purposes. Refer to the fuel:oil recommendations for your engine before mixing. Remember that a pre-mix system designed to produce a 50:1 ratio will produce a 25:1 ratio if a 50:1 ratio is already in the fuel tank feeding the motor.

Lubrication Chart

Component	Recommended Lubricant	Minimum Frequency ①	
		Fresh Water	**Salt, Polluted or Brackish Water**
Choke lever (2-5 hp only)	Yamaha all-purpose marine grease	every 100 hours or 60 days	every 50 hours or 30 days
Clamp bolts	Yamaha all-purpose marine grease	every 100 hours or 60 days	every 50 hours or 30 days
Cowling clamp lever journal	Yamaha all-purpose marine grease	every 100 hours or 60 days	every 50 hours or 30 days
Engine oil 2-Stroke	Yamalube - 2-stroke engine oil or equivalent TC-W3 certified outboard oil	Check every outing	Check every outing
Jet drive bearing lubrication	Yamaha all-purpose marine grease	every 10 hours	every 5 hours
Jet drive bearing grease replacement	Yamaha all-purpose marine grease	every 50 hours	every 25 hours
Lower unit	Yamaha GEAR CASE LUBE (hypoid gear oil - SAE 90)	every 100 hours or 60 days	every 50 hours or 30 days
Manual recoil starter (2-5 hp motors)	Yamaha all-purpose marine grease	every 100 hours or 60 days	every 50 hours or 30 days
Manual recoil starter (6 hp or larger motors)	Yamaha all-purpose marine grease	Upon assembly only	Upon assembly only
Propeller shaft	Yamaha all-purpose marine grease	every 100 hours or 60 days	every 50 hours or 30 days
Shift lever journal / housing	Yamaha all-purpose marine grease	every 100 hours or 60 days	every 50 hours or 30 days
Shift mechanism	Yamaha all-purpose marine grease	every 100 hours or 60 days	every 50 hours or 30 days
Steering pivot shaft	Yamaha all-purpose marine grease	every 100 hours or 60 days	every 50 hours or 30 days
Swivel bracket	Yamaha all-purpose marine grease	every 100 hours or 60 days	every 50 hours or 30 days
Throttle control lever	Yamaha all-purpose marine grease	every 100 hours or 60 days	every 50 hours or 30 days
Throttle grip housing	Yamaha all-purpose marine grease	every 100 hours or 60 days	every 50 hours or 30 days
Throttle linkage	Yamaha all-purpose marine grease	every 100 hours or 60 days	every 50 hours or 30 days
Throttle link journal	Yamaha all-purpose marine grease	every 100 hours or 60 days	every 50 hours or 30 days
Tilt mechanism	Yamaha all-purpose marine grease	every 100 hours or 60 days	every 50 hours or 30 days

NOTE: not all components used by all models. Refer to the section on Lubrication and/or your owner's manual for more details

① Lubrication points should be checked weekly or with each use, whichever is LESS frequent. Based upon individual motor/use needs frequency of actual lubrication should occur at recommended intervals or more often during season. Perform all lubrication procedures immediately prior to extended motor storage

Capacities - 2-Stroke Motors

Model (Hp)	No. of Cyl	Engine Type	Year	Displace cu. in. (cc)	Gear Oil Oz	Gear Oil mL/cc ①	Injection Oil Ratio ②
2	1	IL 2-stroke	1997-02	2.6 (43)	1.5	45	Pre-Mix 100:1
2	1	IL 2-stroke	1997-02	3.0 (50)	1.5	45	Pre-Mix 100:1
3	1	IL 2-stroke	1997-02	4.3 (70)	2.5	75	Pre-Mix 100:1
4	1	CL 2-stroke	1997-99	5.0 (83)	3.6	105	Pre-Mix 100:1
4	1	CL 2-stroke	1997-02	6.3 (103)	3.6 ③	105 ③	Pre-Mix 100:1
5	1	CL 2-stroke	1997-02	6.3 (103)	3.6 ③	105 ③	Pre-Mix 100:1
6	2	CL 2-stroke	1997-00	10 (165)	5.4	160	Pre-Mix 100:1
8	2	CL 2-stroke	1997-03	10 (165)	5.4	160	Pre-Mix 100:1
9.9	2	IL 2-stroke	1997-03	15 (246)	8.45	250	Pre-Mix 100:1
15	2	IL 2-stroke	1997-03	15 (246)	8.45	250	Pre-Mix 100:1
20	2	IL 2-stroke	1997	24 (395)	12.5	370	PB or 100:1 PM
25	2	IL 2-stroke	1997-03	24 (395)	12.5	370	PB or 100:1 PM
20	2	IL 2-stroke	1997-98	26 (430)	6.1	180	Pre-Mix 100:1
25	2	IL 2-stroke	1997-98	26 (430)	6.1	180	Pre-Mix 100:1
25	2	IL 2-stroke	1997	30 (496)	10.8	320	Pre-Mix 100:1
30	2	IL 2-stroke	1997	30 (496)	10.8	320	Pre-Mix 100:1
40	2	IL 2-stroke	1997	36 (592)	10.7	315	Pre-Mix 100:1
48	2	IL 2-stroke	1997-00	46 (760)	16.9	500	Pre-Mix 50:1
25	3	IL 2-stroke	1997-02	30 (496)	6.8	200	PB or 100:1 PM
30	3	IL 2-stroke	1997-02	30 (496)	6.8	200	PB oil injected
28J	3	IL 2-stroke	1997-01	43 (698)	n/a	n/a	Precision Blend
35J	3	IL 2-stroke	1997-01	43 (698)	n/a	n/a	Precision Blend
40	3	IL 2-stroke	1997-03	43 (698)	14.5	430	PB or 50:1 PM
50	3	IL 2-stroke	1997-03	43 (698)	14.5	430	PB or 50:1 PM
50	3	IL 2-stroke	1997-03	52 (849)	16.9	500	PB or 50:1 PM
60	3	IL 2-stroke	1997-03	52 (849)	20.6	610	PB or 50:1 PM
70	3	IL 2-stroke	1997-03	52 (849)	20.6	610	PB or 50:1 PM
65J	3	IL 2-stroke	1997-01	70 (1140)	n/a	n/a	PB or 50:1 PM
75	3	IL 2-stroke	1997-00	70 (1140)	20.6	610	PB or 50:1 PM
80	3	IL 2-stroke	1997-00	70 (1140)	20.6	610	Precision Blend
80	3	IL 2-stroke	1997-00	70 (1140)	20.6	610	PB or 50:1 PM
90	3	IL 2-stroke	1997-03	70 (1140)	20.6	610	PB or 50:1 PM
80J	4	90 LV 2-stroke	1997-01	106 (1730)	n/a	n/a	Precision Blend
100	4	90 LV 2-stroke	1997-02	106 (1730)	25.7	760	Precision Blend
115	4	90 LV 2-stroke	1997-03	106 (1730)	25.7	760	PB or 50:1 PM
130	4	90 LV 2-stroke	1997-03	106 (1730)	25.7	760	Precision Blend
140	4	90 LV 2-stroke	1997-02	106 (1730)	25.7	760	PB or 50:1 PM
105J	6	90 LV 2-stroke	1997-00	158 (2596)	n/a	n/a	Precision Blend
150	6	90 LV 2-stroke	1997-03	158 (2596)	33.1 ④	980 ④	PB or 50:1 PM
175	6	90 LV 2-stroke	1997-00	158 (2596)	33.1	980	PB or 50:1 PM
200	6	90 LV 2-stroke	1997-03	158 (2596)	33.1 ⑤	980 ⑤	PB or 50:1 PM
225	6	90 LV 2-stroke	1997	158 (2596)	33.1	980	PB or 50:1 PM

Capacities - 2-Stroke Motors

Model (Hp)	No. of Cyl	Engine Type	Year	Displace cu. in. (cc)	Gear Oil Oz	Gear Oil mL/cc ①	Injection Oil Ratio ②
150 HPDI	6	76 LV 2-stroke	2000-03	158 (2596)	33.1 ⑥	980 ⑥	Precision Blend
175 HPDI	6	76 LV 2-stroke	2001-03	158 (2596)	33.1	980	Precision Blend
200 HPDI	6	76 LV 2-stroke	2000-03	158 (2596)	33.1 ⑥	980 ⑥	Precision Blend
200 EFI	6	76 LV 2-stroke	*1998-03*	191 (3130)	33.1 ⑤	980 ⑤	Precision Blend
225 EFI	6	76 LV 2-stroke	*1997-03*	191 (3130)	38.9 ⑦	1150 ⑦	Precision Blend
250 EFI	6	76 LV 2-stroke	*1997-03*	191 (3130)	38.9 ⑦	1150 ⑦	Precision Blend
225 HPDI	6	76 LV 2-stroke	*2003*	204 (3342)	38.9 ⑦	1150 ⑦	Precision Blend
250 HPDI	6	76 LV 2-stroke	*2003*	204 (3342)	38.9 ⑦	1150 ⑦	Precision Blend

NOTE - all capacities are approximate. Always add lubricants gradually, checking the level often

n/a - not applicable

① Yamaha provides this specification as CCs which is equal to milliliters (the measurement found on most bottles of gear lube)

② Injection oil ratio based on normal operating conditions, some severe or high performance applications may need higher ratio, refer to information on Fuel Recommendations under Maintenance

③ Specification is for models through 1998, the 1999 and later models use a little less gear oil, about 3.4 oz (100mL)

④ Specification is for the majority of motors, but there are a number of exceptions. The D150 is 30.4 oz (900mL) and the L150 is 29.4 oz (870mL). The DX150 is 31.5 oz (900mL) and LX150 is 30.5 oz (870mL). The PX150 and SX150 are 34.5 oz (980cc). All counter-rotating units other than listed above are 29.4 oz (870mL).

⑤ Specification is for the majority of motors, but there are a number of exceptions. The LX200 and SX200 is 30.5 oz (870mL). The V200 is 38.9 oz (1150mL) for 1998 or 40.5 oz (1150mL) for 1999 and later. Counter-rotating units are 30.4 oz (900cc) through 1998 (except for the counter-rotating V200 which is 33.8 oz (1000cc). The 1999 and later counter-rotating units are 30.5 oz (870mL), except for LX or SX which are 34.5 oz (980mL).

⑥ Specification is for all except counter-rotating units (LZ150PETO/LZ150TR, LZ200PETO/LZ200TR) for which the spec is 29.4 oz (870mL)

⑦ Specification is for all except L (counter-rotating) models for which the specification is 33.8 oz (1000mL)

SPARK PLUG DIAGNOSIS

Tracking Arc
High voltage arcs between a fouling deposit on the insulator tip and spark plug shell. This ignites the fuel/air mixture at some point along the insulator tip, retarding the ignition timing which causes a power and fuel loss.

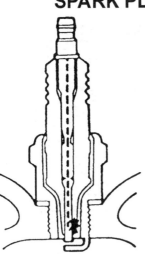

Wide Gap
Spark plug electrodes are worn so that the high voltage charge cannot arc across the electrodes. Improper gapping of electrodes on new or "cleaned" spark plugs could cause a similar condition. Fuel remains unburned and a power loss results.

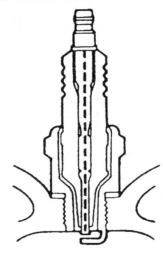

Flashover
A damaged spark plug boot, along with dirt and moisture, could permit the high voltage charge to short over the insulator to the spark plug shell or the engine. A buttress insulator design helps prevent high voltage flashover.

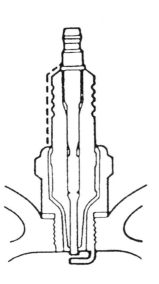

Fouled Spark Plug
Deposits that have formed on the insulator tip may become conductive and provide a "shunt" path to the shell. This prevents the high voltage from arcing between the electrodes. A power and fuel loss is the result.

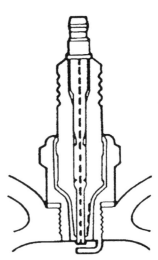

Bridged Electrodes
Fouling deposits between the electrodes "ground out" the high voltage needed to fire the spark plug. The arc between the electrodes does not occur and the fuel air mixture is not ignited. This causes a power loss and exhausting of raw fuel.

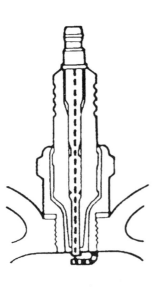

Cracked Insulator
A crack in the spark plug insulator could cause the high voltage charge to "ground out." Here, the spark does not jump the electrode gap and the fuel air mixture is not ignited. This causes a power loss and raw fuel is exhausted.

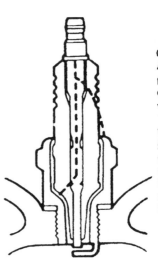

CARBURETED FUEL SYSTEM	3-11
DESCRIPTION AND OPERATION	3-12
TROUBLESHOOTING THE CARBURETED FUEL SYSTEM	3-14
CARBURETOR IDENTIFICATION	3-17
CARBURETOR SERVICE	3-17
CARBURETOR IDENTIFICATION	3-17
2 HP MODELS	3-17
3 HP MODELS	3-22
4/5 HP (83CC AND 103CC) MODELS	3-24
6/8 HP AND 9.9/15 HP MODELS	3-28
20-48 HP (2-CYLINDER) MODELS	3-35
3-CYLINDER POWERHEADS	3-40
V4 & V6 POWERHEADS	3-49
CRANKSHAFT POSITION SENSOR (CPS)/PULSER COIL	3-82
ELECTRONIC FUEL INJECTION SYSTEMS	3-66
CONTROLLED COMBUSTION SYSTEM	3-58
CRANKSHAFT POSITION SENSOR (CPS)/PULSER COIL	3-82
ENGINE CONTROL MODULE (ECM/CDI)	3-79
FUEL INJECTION AND ELECTRONIC COMPONENT LOCATIONS	3-89
FUEL INJECTION BASICS	3-55
FUEL PRESSURE REGULATOR	3-79
FUEL RAIL AND INJECTORS (EFI)	3-74
HIGH-PRESSURE (MECHANICAL) FUEL PUMP, FUEL RAILS AND INJECTORS (HPDI)	3-72
KNOCK SENSOR (3.1L EFI OX66 MOTORS)	3-86
HPDI DRIVE BELT AND SPROCKETS	3-68
HPDI INJECTOR DRIVER	3-70
NEUTRAL OR SHIFT POSITION SWITCHES	3-89
OXYGEN SENSOR (EFI OX66 AND SOME HPDI)	3-87
PRESSURE SENSORS (APS AND FPS)	3-85
SELF DIAGNOSTIC SYSTEM	3-60
SENSOR AND CIRCUIT RESISTANCE/OUTPUT TESTS	3-61
TEMPERATURE SENSORS (AIR AND WATER/ENGINE)	3-83
THROTTLE BODY AND INTAKE ASSEMBLY	3-62
THROTTLE POSITION SENSOR (TPS)	3-82
TROUBLESHOOTING ELECTRONIC FUEL INJECTION	3-59
YAMAHA EFI OX66 INJECTION	3-56
YAMAHA HIGH PRESSURE DIRECT INJECTION (HPDI)	3-57
VAPOR SEPARATOR TANK AND HIGH-PRESSURE FUEL PUMP	3-64
WATER DETECTION SENSOR (HPDI)	3-88
ENGINE CONTROL MODULE (ECM/CDI)	3-79
REMOVAL & INSTALLATION	3-80
FUEL	3-2
FUEL AND COMBUSTION BASICS	3-2
COMBUSTION	3-6
FUEL SYSTEM SERVICE CAUTIONS	3-2
FUEL	3-2
FUEL SYSTEM PRESSURIZATION	3-4
FUEL SYSTEM SERVICE CAUTIONS	3-2
FUEL INJECTION AND ELECTRONIC COMPONENT LOCATIONS	3-89
FUEL LINES AND FITTINGS	3-8
SERVICE	3-10
TESTING	3-8
FUEL PRESSURE REGULATOR	3-79
REMOVAL & INSTALLATION	3-79
TESTING	3-79
FUEL PUMP (LOW-PRESSURE) SERVICE	3-99
FUEL PUMPS	3-99
FUEL PUMPS	3-99
DESCRIPTION & OPERATION	3-99
GENERAL INFORMATION	3-100
OVERHUAL	3-103
REMOVAL & INSTALLATION	3-102
TESTING	3-100
FUEL RAIL AND INJECTORS (EFI)	3-74
REMOVAL & INSTALLATION	3-78
TESTING	3-74
FUEL SYSTEM PRESSURIZATION	3-4
FUEL SYSTEM SERVICE CAUTIONS	3-2
FUEL TANK	3-6
FUEL TANK AND LINES	3-6
FUEL LINES AND FITTINGS	3-8
FUEL TANK	3-6
HIGH-PRESSURE (MECHANICAL) FUEL PUMP, FUEL RAILS AND INJECTORS (HPDI)	3-72
PUMP & INJECTOR OVERHAUL	3-74
REMOVAL & INSTALLATION	3-72
TESTING INJECTORS	3-72
HPDI DRIVE BELT AND SPROCKETS	3-68
REMOVAL & INSTALLATION	3-68
HPDI INJECTOR DRIVER	3-70
REMOVAL & INSTALLATION	3-71
TESTING	3-70
KNOCK SENSOR (3.1L EFI OX66 MOTORS)	3-86
REMOVAL & INSTALLATION	3-87
TESTING	3-86
NEUTRAL OR SHIFT POSITION SWITCHES	3-89
OXYGEN SENSOR (EFI OX66 AND SOME HPDI)	3-87
REMOVAL & INSTALLATION	3-88
TESTING	3-87
PRESSURE SENSORS (APS AND FPS)	3-85
REMOVAL & INSTALLATION	3-86
TESTING	3-86
REED VALVES	3-104
REED VALVE BLOCK	3-104
SELF DIAGNOSTIC SYSTEM	3-60
READING	3-60
TROUBLE CODES	3-62
SENSOR AND CIRCUIT RESISTANCE/OUTPUT	3-61
SPECIFICATIONS	3-105
CARBURETOR SET-UP SPECIFICATIONS	3-105
TEMPERATURE SENSORS (AIR AND WATER/ENGINE)	3-83
REMOVAL & INSTALLATION	3-85
TESTING	3-83
THROTTLE BODY AND INTAKE ASSEMBLY	3-62
REMOVAL & INSTALLATION	3-62
THROTTLE POSITION SENSOR (TPS)	3-82
REMOVAL & INSTALLATION	3-82
TESTING	3-82
TROUBLESHOOTING ELECTRONIC FUEL INJECTION	3-59
TROUBLESHOOTING THE CARBURETED FUEL SYSTEM	3-14
VAPOR SEPARATOR TANK AND HIGH-PRESSURE FUEL PUMP	3-64
OVERHAUL	3-67
REMOVAL & INSTALLATION	3-66
TESTING	3-65
YAMAHA EFI OX66 INJECTION	3-56
YAMAHA HIGH PRESSURE DIRECT INJECTION (HPDI)	3-57
WATER DETECTION SENSOR (HPDI)	3-88
2 HP MODELS (CARBURETOR)	3-17
ASSEMBLY & INSTALLATION	3-20
CLEANING & INSPECTION	3-19
REMOVAL & DISASSEMBLY	3-18
3 HP MODELS (CARBURETOR)	3-22
CLEANING & INSPECTION	3-24
OVERHAUL	3-24
REMOVAL & INSTALLATION	3-23
4/5 HP (83CC AND 103CC) (CARBURETOR) MODELS	3-24
ASSEMBLY & INSTALLATION	3-27
CLEANING & INSPECTION	3-26
DISASSEMBLY	3-25
REMOVAL	3-24
6/8 HP AND 9.9/15 HP MODELS (CARBURETOR)	3-28
ASSEMBLY	3-32
CLEANING & INSPECTION	3-30
DISASSEMBLY	3-30
FLOAT ADJUSTMENT	3-33
INSTALLATION	3-34
REMOVAL	3-29
20-48 HP (2-CYLINDER) MODELS (CARBURETOR)	3-35
CLEANING & INSPECTION	3-40
DISASSEMBLY & ASSEMBLY	3-37
REMOVAL	3-35
3-CYLINDER POWERHEADS (CARBURETOR)	3-40
CLEANING & INSPECTION	3-48
OVERHAUL	3-44
REMOVAL & INSTALLATION	3-41
V4 & V6 POWERHEADS (CARBURETOR)	3-49
CLEANING & INSPECTION	3-55
OVERHAUL	3-52
REMOVAL & INSTALLATION	3-49

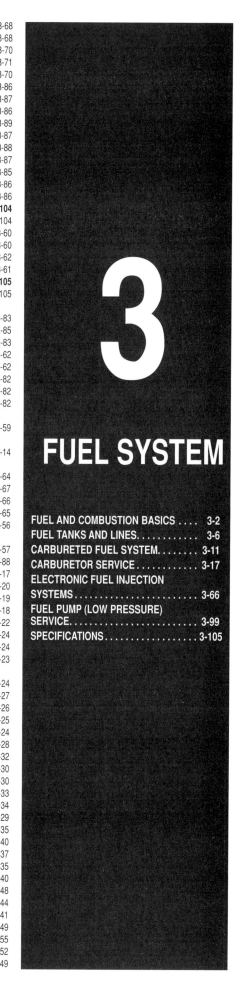

3

FUEL SYSTEM

FUEL AND COMBUSTION BASICS	3-2
FUEL TANKS AND LINES	3-6
CARBURETED FUEL SYSTEM	3-11
CARBURETOR SERVICE	3-17
ELECTRONIC FUEL INJECTION SYSTEMS	3-66
FUEL PUMP (LOW PRESSURE) SERVICE	3-99
SPECIFICATIONS	3-105

3-2 FUEL SYSTEM

FUEL AND COMBUSTION BASICS

> ✲✲ **CAUTION**
>
> If equipped, disconnect the negative battery cable ANYTIME work is performed on the engine, especially when working on the fuel system. This will help prevent the possibility of sparks during service (from accidentally grounding a hot lead or powered component). Sparks could ignite vapors or exposed fuel. Disconnecting the cable on electric start motors will also help prevent the possibility fuel spillage if an attempt is made to crank the engine while the fuel system is open.

> ✲✲ **CAUTION**
>
> Fuel leaking from a loose, damaged, or incorrectly installed hose or fitting may cause a fire or an explosion. ALWAYS pressurize the fuel system and run the motor while inspecting for leaks after servicing any component of the fuel system.

To tune, troubleshoot or repair the fuel system of an outboard motor you must first understand the carburetion or fuel injection, and ignition principles of engine operation.

If you have any doubts concerning your understanding of engine operation, it would be best to study

The Basic Operating Principles of an engine as detailed under Troubleshooting in Section 1, before tackling any work on the fuel system.

The fuel systems used on engines covered by this guide range from single carburetors to multiple carburetors or various forms of electronic fuel injection systems. Carbureted motors utilize various means of enriching fuel mixture for cold starts, including a manual choke, manual primer or electric primer solenoid. Similarly, the 2-stroke motors covered by this manual may require pre-mixing of the fuel and oil or may be equipped with the Precision Blend automatic oiling system. Refer to the General Engine System Specifications chart in the Maintenance section for more details as to what systems were commonly used on what motors, but keep in mind that systems such as the Precision Blend may have been available as accessories for some motors or could have been retrofitted after the factory. For details on the Precision Blend oiling system, please refer to the Lubrication and Cooling section.

Fuel System Service Cautions

There is no way around it. Working with gasoline can provide for many different safety hazards and requires that extra caution is used during all steps of service. To protect yourself and others, you must take all necessary precautions against igniting the fuel or vapors (which will cause a fire at best or an explosion at worst).

> ✲✲ **CAUTION**
>
> Take extreme care when working with the fuel system. NEVER smoke (it's bad for you anyhow, but smoking during fuel system service could kill you much faster!) or allow flames or sparks in the work area. Flames or sparks can ignite fuel, especially vapors, resulting in a fire at best or an explosion at worst.

For starters, disconnect the negative battery cable EVERY time a fuel system hose or fitting is going to be disconnected. It takes only one moment of forgetfulness for someone to crank the motor, possibly causing a dangerous spray of fuel from the opening. This is especially true on EFI systems that contain an electric fuel pump, where simply turning the key on (without cranking) will energize the pump and result in high-pressure fuel spray in certain fuel system lines/fittings.

Gasoline contains harmful additives and is quickly absorbed by exposed skin. As an additional precaution, always wear gloves and some form of eye protection (regular glasses help, but only safety glasses can really protect your eyes).

■ Throughout service, pay attention to ensure that all components, hoses and fittings are installed them in the correct location and orientation to prevent the possibility of leakage. Matchmark components before they are removed as necessary.

Because of the dangerous conditions that result from working with gasoline and fuel vapors
always take extra care and be sure to follow these guidelines for safety:
• Keep a Coast Guard-approved fire extinguisher handy when working.
• Allow the engine to cool completely before opening a fuel fitting. Don't allow gasoline to drip on a hot engine.
• The first thing you must do after removing the engine cover is to check for the presence of gasoline fumes. If strong fumes are present, look for leaking or damage hoses, fittings or other fuel system components and repair.
• Do not repair the motor or any fuel system component near any sources of ignition; including sparks, open flames, or anyone smoking.
• Clean up spilled gasoline right away using clean rags. Keep all fuel soaked rags in a metal container until they can be properly disposed of or cleaned. NEVER leave solvent, gasoline or oil soaked rags in the hull.
• Don't use electric powered tools in the hull or near the boat during fuel system service or after service, until the system is pressurized and checked for leaks.
• Fuel leaking from a loose, damaged or incorrectly installed hose or fitting may cause a fire or an explosion. ALWAYS pressurize the fuel system and run the motor while inspecting for leaks after servicing any component of the fuel system.

Fuel

◆ See Figure 1

Fuel recommendations have become more complex as the chemistry of modern gasoline changes. The major driving force behind the many of the changes in gasoline chemistry was the search for additives to replace lead as an octane booster and lubricant. These additives are governed by the types of emissions they produce in the combustion process. Also, the replacement additives do not always provide the same level of combustion stability, making a fuel's octane rating less meaningful.

In the 1960's and 1970's, leaded fuel was common. The lead served two functions. First, it served as an octane booster (combustion stabilizer) and second, in 4-stroke engines, it served as a valve seat lubricant. For 2-stroke engines, the primary benefit of lead was to serve as a combustion stabilizer. Lead served very well for this purpose, even in high heat applications.

For decades now, all lead has been removed from the refining process. This means that the benefit of lead as an octane booster has been eliminated. Several substitute octane boosters have been introduced in the place of lead. While many are adequate in automobile engines, most do not perform nearly as well as lead did, even though the octane rating of the fuel is the same.

Fig. 1 Damaged piston, possibly caused by; using too-low an octane fuel or by insufficient oil in the oil/fuel mixture

FUEL SYSTEM 3-3

OCTANE RATING

◆ See Figure 1

A fuel's octane rating is a measurement of how stable the fuel is when heat is introduced. Octane rating is a major consideration when deciding whether a fuel is suitable for a particular application. For example, in an engine, we want the fuel to ignite when the spark plug fires and not before, even under high pressure and temperatures. Once the fuel is ignited, it must burn slowly and smoothly, even though heat and pressure are building up while the burn occurs. The unburned fuel should be ignited by the traveling flame front, not by some other source of ignition, such as carbon deposits or the heat from the expanding gasses. A fuel's octane rating is known as a measurement of the fuel's anti-knock properties (ability to burn without exploding). Essentially, the octane rating is a measure of a fuel's stability.

Usually a fuel with a higher octane rating can be subjected to a more severe combustion environment before spontaneous or abnormal combustion occurs. To understand how two gasoline samples can be different, even though they have the same octane rating, we need to know how octane rating is determined.

The American Society of Testing and Materials (ASTM) has developed a universal method of determining the octane rating of a fuel sample. The octane rating you see on the pump at a gasoline station is known as the pump octane number. Look at the small print on the pump. The rating has a formula. The rating is determined by the R+M/2 method. This number is the average of the research octane reading and the motor octane rating.

• The Research Octane Rating is a measure of a fuel's anti-knock properties under a light load or part throttle conditions. During this test, combustion heat is easily dissipated.

• The Motor Octane Rating is a measure of a fuel's anti-knock properties under a heavy load or full throttle conditions, when heat buildup is at maximum.

In general, 2-stroke engines tend to respond more to the motor octane rating than the research octane rating, because a 2-stroke engine has a power stroke (with heat buildup) every revolution. Therefore, in a 2-stroke outboard motor, the motor octane rating of the fuel is one of the best indications of how it will perform.

VAPOR PRESSURE

Fuel vapor pressure is a measure of how easily a fuel sample evaporates. Many additives used in gasoline contain aromatics. Aromatics are light hydrocarbons distilled off the top of a crude oil sample. They are effective at increasing the research octane of a fuel sample but can cause vapor lock (bubbles in the fuel line) on a very hot day. If you have an inconsistent running engine and you suspect vapor lock, use a piece of clear fuel line to look for bubbles, indicating that the fuel is vaporizing.

One negative side effect of aromatics is that they create additional combustion products such as carbon and varnish. If your engine requires high octane fuel to prevent detonation, de-carbon the engine more frequently with an internal engine cleaner to prevent ring sticking due to excessive varnish buildup.

ALCOHOL-BLENDED FUELS

When the Environmental Protection Agency mandated a phase-out of the leaded fuels in January of 1986, fuel suppliers needed an additive to improve the octane rating of their fuels. Although there are multiple methods currently employed, the addition of alcohol to gasoline seems to be favored because of its favorable results and low cost. Two types of alcohol are used in fuel today as octane boosters, methanol (wood alcohol) or ethanol (grain alcohol).

When used as a fuel additive, alcohol tends to raise the research octane of the fuel, so these additives will have limited benefit in an outboard motor. There are, however, some special considerations due to the effects of alcohol in fuel.

• Since alcohol contains oxygen, it replaces gasoline without oxygen content and tends to cause the air/fuel mixture to become leaner.

• On older outboards, the leaching affect of alcohol will, in time, cause fuel lines and plastic components to become brittle to the point of cracking. Unless replaced, these cracked lines could leak fuel, increasing the potential for hazardous situations. However, later model outboards (such as the models covered here) utilize fuel lines which are much more resistant to alcohol in the fuel mixture.

■ Modern outboard fuel lines and plastic fuel system components have been specially formulated to resist alcohol leaching effects.

• When alcohol blended fuels become contaminated with water, the water combines with the alcohol then settles to the bottom of the tank. This leaves the gasoline (and the oil for 2-stroke models using premix) on a top layer.

• All fuels have chemical compounds added to reduce the tendency towards phase separation. If phase separation occurs, however, there is a possibility of a lean oil/fuel mixture with the potential for engine damage. With oil-injected outboards (Precision Blend models), phase separation will be less of a problem because the oil is injected separately rather than being premixed.

THE BOTTOM LINE WITH FUELS

If we could buy fuel of the correct octane rating, free of alcohol and aromatics, this would be our first choice.

Yamaha continues to recommend unleaded fuel. This is almost a redundant recommendation due to the near universal unavailability of any other type fuel.

According to the fuel recommendations that come with your outboard, there is no engine in the product line that requires more than 89 octane. Most Yamaha engines (including EFI and high performance Vmax models) need only 86 octane or less. An 89 octane rating generally means middle grade unleaded. Premium unleaded is more stable under severe conditions, but also produces more combustion products. Therefore, when using premium unleaded, more frequent de-carboning is necessary.

Regardless of the fuel octane rating you choose, try to stay with a name brand fuel. You never know for sure what kinds of additives or how much is in off brand fuel.

HIGH ALTITUDE OPERATION

At elevated altitudes there is less oxygen in the atmosphere than at sea level. Less oxygen means lower combustion efficiency and less power output. Power output is reduced three percent for every thousand feet above sea level. At ten thousand feet, power is reduced 30 percent from that available at sea level.

Re-jetting carburetors for high altitude does not restore this lost power. It simply corrects the air-fuel ratio for the reduced air density, and makes the most of the remaining available power. If you re-jet an engine, you are locked into the higher elevation. You cannot operate at sea level until you re-jet for sea level. Understand that going below the elevation jetted for your motor will damage the engine with overly lean fuel:air (or oil/fuel:air) ratios. As a general rule, jet for the lowest elevation anticipated. Spark plug insulator tip color is the best guide for high altitude jetting.

If you are in an area of known poor fuel quality, you may want to use fuel additives. Today's additives are mostly alcohol and aromatics, and their effectiveness may be limited. It is difficult to find additives without ethanol, methanol, or aromatics. If you use octane boosters frequent de-carbonings may be necessary. If possible, the best policy is to use name brand pump fuel with no additional additives except Yamaha fuel conditioner and Ring-Free(tm).

CHECKING FOR STALE/CONTAMINATED FUEL

◆ See Figures 2 and 3

Outboard motors often sit weeks at a time making them the perfect candidate for fuel problems. Gasoline has a short life, as combustibles begin evaporating almost immediately. Even when stored properly, fuel starts to deteriorate within a few months, leaving behind a stale fuel mixture that can cause hard-starting, poor engine performance and even lead to possible engine damage.

Further more, as gasoline evaporates it leaves behind gum deposits that can clog filters, lines and small passages. Although the sealed high-pressure fuel system of an EFI or HPDI motor is less susceptible to fuel evaporation, the low-pressure fuel systems of all engines can suffer the affects. However carburetors, due to their tiny passages and naturally vented designs are the most susceptible components on non-EFI motors.

As mentioned under Alcohol-Blended fuels, modern fuels contain alcohol, which is hydroscopic (meaning it absorbs water). And, over time, fuel stored

3-4 FUEL SYSTEM

in a partially filled tank or a tank that is vented to the atmosphere will absorb water. The water/alcohol settles to the bottom of the tank, promoting rust (in metal tanks) and leaving a non-combustible mixture at the bottom of a tank that could leave a boater stranded.

One of the first steps to fuel system troubleshooting is to make sure the fuel source is not at fault for engine performance problems. Check the fuel if the engine will not start and there is no ignition problem.

Stale or contaminated fuels will often exhibit an unusual or even unpleasant unusual odor.

■ **The best method of disposing stale fuel is through a local waste pickup service, automotive repair facility or marine dealership. But, this can be a hassle. If fuel is not too stale or too badly contaminated, it may be mixed with greater amounts of fresh fuel and used to power lawn/yard equipment or (as long as we're not talking about disposing pre-mix here) even an automobile (if greatly diluted so as to prevent misfiring, unstable idle or damage to the automotive engine). But we feel that it is much less of a risk to have a lawn mower stop running because of the fuel problem than it is to have your boat motor quit or refuse to start.**

Yamaha carburetors are equipped with a float bowl drain screw that can be used to drain fuel from the carburetor for storage or for inspection. The vapor separator tank on EFI and HPDI models is usually equipped with a drain screw, not unlike the float bowl screw on carburetor motors. The drain screws on Yamahas are usually pretty easy to access, but on a few motors it may be easier to drain a fuel sample from the hoses leading to or from the low-pressure fuel filter or fuel pump. Removal and installation instructions for the fuel filters are provided in the Maintenance Section, while fuel pump procedures are found in this section. To check for stale or contaminated fuel:

1. Disconnect the negative battery cable for safety. Secure it or place tape over the end so that it cannot accidentally contact the terminal and complete the circuit.

✱✱ CAUTION

Throughout this procedure, clean up any spilled fuel to prevent a fire hazard.

Fig. 2 Yamaha float bowls are equipped with drain screws which make draining the float bowls or taking a fuel sample a snap!

2. For carbureted motors, remove the float bowl drain screw (and orifice plug, if equipped), then allow a small amount of fuel to drain into a glass container.

3. For EFI motors, remove the float bowl drain screw from the bottom of the vapor separator tank or disconnect the fuel supply hose from the pump or low-pressure fuel filter (as desired).

4. If there is no fuel present in the float bowl, vapor tank or fuel line (as applicable) squeeze the fuel primer bulb to force fuel through the lines and to the opening. If necessary, you can always disconnect the fuel line between the low-pressure filter and pump or, on models with serviceable filter elements simply unscrew the housing to get a sample.

■ **If a sample cannot be obtained from the float bowls/separator tank, fuel filter or pump supply hose, there is a problem with the fuel tank-to-motor fuel circuit. Check the tank, primer bulb, fuel hose, fuel pump, fitting or carburetor float inlet needle (as necessary).**

5. Check the appearance and odor of the fuel. An unusual smell, signs of visible debris or a cloudy appearance (or even the obvious presence of water) points to a fuel that should be replaced.

6. If contaminated fuel is found, drain the fuel system and dispose of the fuel in a responsible manner, then clean the entire fuel system.

■ **If debris is found in the fuel system, clean and/or replace all fuel filters.**

7. When finished, reconnect the negative battery cable, then properly pressurize the fuel system and check for leaks.

Fuel System Pressurization

When it comes to safety and outboards, the condition of the fuel system is of the utmost importance. The system must be checked for signs of damage or leakage with every use and checked, especially carefully when portions of the system have been opened for service.

The best method to check the fuel system is to visually inspect the lines, hoses and fittings once the system has been properly pressurized.

All engines, carbureted or fuel injected utilize a low-pressure fuel circuit which delivers fuel from a portable or boat mounted tank to the powerhead via fuel hoses and the action of a diaphragm-displacement fuel pump.

EFI engines utilize a 2-stage fuel system, the low-pressure circuit (fuel tank-to-pump and pump-to-vapor separator), as well as a high-pressure circuit (vapor separator-to-fuel injectors).

HPDI engines actually use a 3 stage fuel system, low-pressure (fuel tank-to-pump and pump-to-vapor separator), a first stage high-pressure (or medium pressure circuit, vapor separator-to-mechanical high-pressure pump) and second stage high-pressure (extreme high-pressure circuit, high-pressure pump-to-fuel injectors).

The high-pressure circuit of all fuel injected Yamaha motors works in a

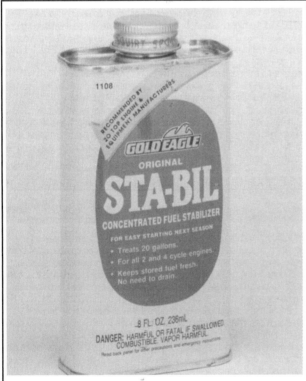

Fig. 3 Commercial additives, such as Sta-bil, may be used to help prevent "souring"

FUEL SYSTEM 3-5

similar fashion using a submerged electric fuel pump located within the vapor separator tank to achieve fuel pressure somewhere in the 30-40 psi (207-278 kPa) range depending upon the year and model. Most models are equipped with a fuel test port on the top of the separator tank, but there are exceptions where the port is mounted to the top of the fuel rail assembly.

The extreme high-pressure circuit of HPDI engines takes this already high-pressure fuel and achieves operating pressures up to 1000 psi (6895 kPa) on some direct injection 2-stroke motors.

RELIEVING FUEL SYSTEM PRESSURE (EFI & HPDI MOTORS)

◆ See Figure 4

 MODERATE

As its name implies, the high-pressure circuit of EFI and HPDI motors contains fuel under pressure that, if given the chance, will spray from a damaged/loose hose or fitting. When servicing components of the high-pressure system, the fuel pressure must first be relieved in a safe and controlled manner to help avoid the potential explosive and dangerous conditions that would result from simply opening a fitting and allowing fuel to spray uncontrolled into the work area.

Relieving fuel system pressure from the high-pressure (first stage on HPDI motors) fuel circuit is a relatively straight forward procedure on Yamaha motors, however Yamaha service literature varies slightly on a couple of models when it comes to recommendations.

All fuel injected motors are equipped with a high-pressure test port (a threaded Schrader type valve) to which you can connect a fuel pressure gauge. This port is usually found under a rubber cap on the top of the vapor separator tank, though there are a couple of exceptions for which the test port is mounted at the top of the fuel rail assembly.

The BEST method of relieving fuel system pressure is to cover this fitting with a rag and then attach a fuel pressure gauge which is equipped with a pressure bleed valve/hose. Direct the bleed hose into a suitable container, then slowly open the bleed valve, draining (spraying) fuel from the hose into the fuel storage container. Once the pressure has dropped, remove the drain screw and drain the remaining fuel from the vapor separator tank. At this point you will be able to disconnect any of the high-pressure (first stage ONLY for HPDI) fuel lines/fittings without worrying about spray, however fuel will still be present in lines and so you'll want to keep a rag handy.

Now in a few service publications Yamaha shows fuel pressure relief WITHOUT the use of a fuel gauge, by placing a rag over the fitting, then depressing the Schrader valve fitting using the tip of a screwdriver. We can't honestly recommend this method, because it is both more dangerous and messy than using a pressure gauge. However, if precautions are taken against fuel spray, it could in theory be a viable alternative to the use of a pressure gauge. But we emphasize that you'll need protection (rubber gloves, safety glasses), a thick shop rag, and a well-ventilated area, completely free of potential sources of ignition (sparks or flame).

HPDI motors also use an extreme high-pressure (second stage high-pressure) fuel circuit that is powered by a belt-driven, mechanical fuel pump. In theory, bleeding pressure from the first stage circuit shouldn't affect the high-pressure circuit, however Yamaha service literature fails to mention a specific method that WOULD reduce pressure in the second stage circuit. Therefore, we recommend taking a few additional steps to reduce system pressure if you plan on servicing the extreme high-pressure fuel circuit of HPDI motors. After relieving pressure from the first-stage unit, disconnect the fuel supply from the engine AND disable the ignition system, then crank the motor to discharge fuel from the HPDI injectors, bleeding pressure in the second stage circuit. Disabling the ignition system can either be done with the safety lanyard (leave it off) or by disconnecting and grounding the individual spark plug wires to keep them from firing.

✱✱ WARNING

Even after following these steps, some residual pressure (and definitely, some residual fuel) will be present in the fuel system. Always cover lines and fittings with a shop rag when disconnecting them to protect yourself from fuel spray. In the case of HPDI motors, DO NOT work on the extreme pressure fuel circuit without safety goggles.

PRESSURIZING THE FUEL SYSTEM (CHECKING FOR LEAKS)

Carbureted Motors

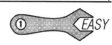 EASY

Carbureted engines are equipped with a low-pressure fuel system, making pressure release before service a relative non-issue. However, if the primer bulb is firm when you go to disconnect a fuel line, there will be some fuel leakage once the hose is disconnected, so always have a rag or handy during service. But, even a low-pressure fuel system should be checked following repairs to make sure that no leaks are present. Only by checking a fuel system under normal operating pressures can you be sure of the system's integrity.

✱✱ CAUTION

Fuel leaking from a loose, damaged or incorrectly installed hose or fitting may cause a fire or an explosion. ALWAYS pressurize the fuel system and run the motor while inspecting for leaks after servicing any component of the fuel system.

Most carbureted engines (except some integral tank models with gravity feed) utilize a fuel primer bulb mounted inline between the fuel tank and engine. On models so equipped, the bulb can be used to pressurize that portion of the fuel system. Squeeze the bulb until it and the fuel lines feel firm with gasoline. At this point check all fittings between the tank and motor for signs of leakage and correct, as necessary.

Once fuel reaches the engine it is the job of the fuel pump(s) to distribute it to the carburetor(s). No matter what system you are inspecting, start and run the motor with the engine top case removed, then check each of the system hoses, fittings and gasket-sealed components to be sure there is no leakage after service.

Fuel Injected Motors

◆ See Figure 4

The low-pressure circuit of an EFI or HPDI motor is pressurized and checked using the primer bulb, in the same manner as the low-pressure circuit of a carbureted motor. For more details, refer to the section on Carbureted Motors.

However, the vapor separator tank contains an electric high-pressure pump which is used on the high-pressure (first stage for HPDI motors) circuit to build system pressure. Normally this pump is energized for a few seconds when the ignition switch is turned on (even before the engine is started). If so, you can usually hear the pump run for a few seconds when the switch is turned on. For these models, turn the ignition switch on, wait a few seconds, and then turn the switch off again. Repeat a few times until system pressure starts to build (you can monitor this using a pressure gauge attached to the fuel pressure test port).

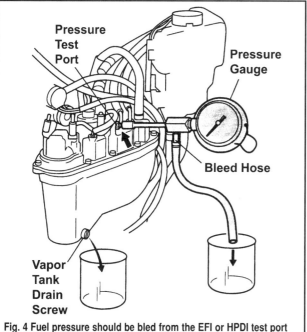

Fig. 4 Fuel pressure should be bled from the EFI or HPDI test port

3-6 FUEL SYSTEM

■ The BEST way to make sure no leaks occur during service is to visually inspect the motor while it is running at idle (that way all fuel circuits are pressurized and operating as normal).

The extreme high-pressure circuit (second stage) used by HPDI motors is pressurized by a mechanical fuel pump. The only real way to check the integrity of this circuit is to crank or start/run the motor which visually inspecting lines and fittings.

✷✷ CAUTION

As usual, use extreme caution when working around an operating engine. Keep hands, fingers, hair, loose clothing, etc as far away as possible from moving components. Severe injury or death could occur from getting caught in rotating components.

Combustion

A two cycle engine has a power stroke every revolution of the crankshaft. A four cycle engine has a power stroke every other revolution of the crankshaft. Therefore, the two cycle engine has twice as many power strokes for any given RPM. If the displacement of the two types of engines is identical, then the two cycle engine has to dissipate twice as much heat as the four cycle engine. In such a high heat environment, the fuel must be very stable to avoid detonation. If any parameters affecting combustion change suddenly (the engine runs lean for example), uncontrolled heat buildup occurs very rapidly in a two cycle engine.

ABNORMAL COMBUSTION

There are two types of abnormal combustion:
- Pre-ignition - Occurs when the air-fuel mixture is ignited by some other incandescent source other than the correctly timed spark from the spark plug.
- Detonation - Occurs when excessive heat and or pressure ignites the air/fuel mixture rather than the spark plug. The burn becomes explosive.

FACTORS AFFECTING COMBUSTION

The combustion process is affected by several interrelated factors. This means that when one factor is changed, the other factors also must be changed to maintain the same controlled burn and level of combustion stability.

Compression

Determines the level of heat buildup in the cylinder when the air-fuel mixture is compressed. As compression increases, so does the potential for heat buildup. Over time, the build-up of carbon deposits can increase compression (by decreasing the available size of the combustion chamber). Similarly, machining the cylinder head (4-stroke) or cylinder head/combustion chamber cover (to correct for warpage) can increase compression. Increases of compression often bring with them the need to increase fuel octane to prevent abnormal combustion.

Ignition Timing

Determines when the ignited gasses will start to expand in relation to the motion of the piston. If the gasses begin to expand too soon, such as they would during pre-ignition or in an overly advanced ignition timing, the motion of the piston opposes the expansion of the gasses, resulting in extremely high combustion chamber pressures and heat.

As ignition timing is retarded, the burn occurs later in relation to piston position. This means that the piston has less distance to travel under power to the bottom of the cylinder, resulting in less usable power.

Fuel Mixture

Determines how efficient the burn will be. A rich mixture burns cooler and slower than a lean one. However, if the mixture is too lean, it can't become explosive. The slower the burn, the cooler the combustion chamber, because pressure buildup is gradual.

Fuel Quality (Octane Rating)

Determines how much heat is necessary to ignite the mixture. Once the burn is in progress, heat is on the rise. The unburned poor quality fuel is ignited all at once by the rising heat instead of burning gradually as a flame front of the burn passing by. This action results in detonation (pinging).

Other Factors

In general, anything that can cause abnormal heat buildup can be enough to push an engine over the edge to abnormal combustion, if any of the four basic factors previously discussed are already near the danger point, for example, excessive carbon buildup raises the compression and retains heat as glowing embers.

FUEL TANK AND LINES

✷✷ CAUTION

If equipped, disconnect the negative battery cable ANYTIME work is performed on the engine, especially when working on the fuel system. This will help prevent the possibility of sparks during service (from accidentally grounding a hot lead or powered component). Sparks could ignite vapors or exposed fuel. Disconnecting the cable on electric start motors will also help prevent the possibility fuel spillage if an attempt is made to crank the engine while the fuel system is open.

✷✷ CAUTION

Fuel leaking from a loose, damaged or incorrectly installed hose or fitting may cause a fire or an explosion. ALWAYS pressurize the fuel system and run the motor while inspecting for leaks after servicing any component of the fuel system.

If a problem is suspected in the fuel supply, tank and/or lines, by far the easiest test to eliminate these components as possible culprits is to substitute a known good fuel supply. This is known as running a motor on a test tank (as opposed to running a motor IN a test tank, which is an entirely different concept). If possible, borrow a portable tank, fill it with fresh gasoline (or gas and oil for pre-mix 2-strokes) and connect it to the motor.

■ When using a test fuel tank, make sure the inside diameter of the fuel hose and fuel fittings is at least equal to the size of the hose used on the usual supply line (otherwise you could cause fuel starvation problems or other symptoms which will make troubleshooting nearly impossible.

Fuel Tank

◆ See Figures 5 and 6

There are 3 different types of fuel tanks that might be used along with these Yamaha motors. The very smallest motors covered by this guide may be equipped with an integral fuel tank mounted to the powerhead. But, most motors, even some of the smaller ones, may be rigged using either a portable fuel tank or a boat mounted tank. In both cases, a tank that is not mounted to the engine itself is commonly called a remote tank.

■ Although many Yamaha dealers rig boats using Yamaha fuel tanks, there are many other tank manufacturers and tank designs may vary greatly. Your outboard might be equipped with a tank from the engine manufacturer, the boat manufacturer or even another tank manufacturer. Although components used, as well as the techniques for cleaning and repairing tanks are similar for almost all fuel tanks be sure to use caution and common sense. Since design vary greatly we

FUEL SYSTEM 3-7

cannot provide step-by-step service procedures for the fuel tanks you might encounter. Instead, we've provided information about tanks and how you can decide when service is necessary. Refer to a reputable marine repair shop or marine dealership when parts are needed for aftermarket fuel tanks.

Whether or not your boat is equipped with a boat mounted, built-in tank depends mostly on the boat builder and partially on the initial engine installer. Boat mounted tanks can be hard to access (sometimes even a little hard to find if parts of the deck must be removed). When dealing with boat mounted tanks, look for access panels (as most manufacturers are smart or kind enough to install them for tough to reach tanks). At the very least, all manufacturers must provide access to fuel line fittings and, usually, the fuel level sender assembly.

No matter what type of tank is used, all must be equipped with a vent (either a manual vent or an automatic one-way check valve) which allows air in (but should prevent vapors from escaping). An inoperable vent (one that is blocked in some fashion) would allow the formation of a vacuum that could prevent the fuel pump from drawing fuel from the tank. A blocked vent could cause fuel starvation problems. Whenever filling the tank, check to make sure air does not rush into the tank each time the cap is loosened (which could be an early warning sign of a blocked vent).

If fuel delivery problems are encountered, first try running the motor with the fuel tank cap removed to ensure that no vacuum lock will occur in the tank or lines due to vent problems. If the motor runs properly with the cap removed but stalls, hesitates or misses with the cap installed, you know the problem is with the tank vent system.

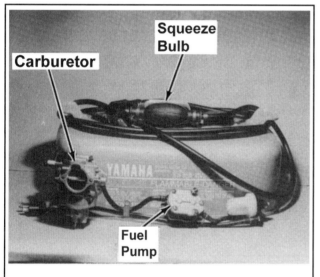

Fig. 5 Typical Yamaha portable fuel tank (and related fuel system components)

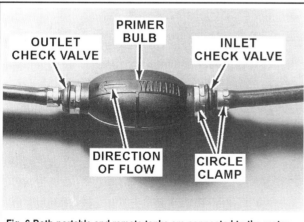

Fig. 6 Both portable and remote tanks are connected to the motor using a fuel line equipped with a primer bulb

SERVICE

Integral Fuel Tanks

◆ See Figure 7

MODERATE

On small powerheads equipped with an integral fuel tank above the powerhead, the filter in the tank may become plugged preventing proper fuel flow. To check fuel flow, disconnect the fuel line at the carburetor and fuel should flow from the line if the shut-off valve is open. If fuel is not present, the usual cause is a plugged filter in the fuel tank. It can be very difficult to determine these filter are plugged simply by a visual inspection. The general appearance that the filter is satisfactory may be a false indication. Therefore, if the filter is suspected, the best remedy is replacement. The cost is very modest and this one area is thus eliminated as a problem source.

Would you believe, many times a lack of fuel at the carburetor is caused because the vent on the fuel tank was not opened? As mentioned under Fuel Tank in this section, it is important to make sure the vent is open at all times during engine operation to prevent formation of a vacuum in the tank that could prevent fuel from reaching the fuel pump or carburetor.

To remove the tank, proceed as follows:

1. Remove the upper and lower engine covers (as applicable). For more information, please refer to Engine Covers (Top Cover and Cowling) in the Engine Maintenance section.
2. If equipped, make sure the fuel tank valve is shut off.

■ For models not equipped with a fuel tank mounted shut off valve, have a shop rag, large funnel and gas can handy to drain the fuel into when the hose is removed from the carburetor.

3. Trace the fuel line from the tank to the carburetor, then disconnect it from the tank or the carburetor fitting, whichever is easier and provides more access. For models without a shut off valve, immediately direct the hose into the funnel and gas can to prevent or minimize fuel spillage.
4. Remove the screws fastening the tank to the powerhead, then make sure no other wires, components or brackets will interfere with tank removal.
5. Carefully lift the fuel tank from the powerhead.
6. Check the tank, filler cap and gasket for wear or damage. Replace any components as necessary to correct any leakage (if found.)
7. On models with a tank-mounted filter element, remove the filter and clean or replace as necessary. For more information refer to Fuel Filters in the Engine Maintenance section.
8. If cleaning is necessary, flush the tank with a small amount of solvent or gasoline, then drain and dispose of the flammable liquid properly.
9. If equipped, check the tank cushions for excessive deterioration, wear or damage and replace, as necessary.
10. When installing the tank, be sure all fuel lines are connected properly and that the retaining screws are securely bolted in place.
11. Refill the tank and pressure test the system by opening the fuel valve, then starting and running the engine.

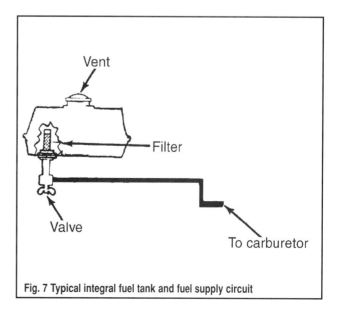

Fig. 7 Typical integral fuel tank and fuel supply circuit

3-8 FUEL SYSTEM

Portable Fuel Tanks

Modern fuel tanks are vented to prevent vapor-lock of the fuel supply system, but are normally vented by a one-way valve to prevent pollution through the evaporation of vapors. A squeeze bulb is used to prime the system until the powerhead is operating. Once the engine starts, the fuel pump, mounted on the powerhead pull fuel from the tank and feeds the carburetor(s) or fuel vapor separator tank (EFI and HPDI). The pickup unit in the tank is usually sold as a complete unit, but without the gauge and float.

To disassemble and inspect or replace tank components, proceed as follows:

1. For safety, remove the filler cap and drain the tank into a suitable container.
2. Disconnect the fuel supply line from the tank fitting.
3. Fuel pick-up units are usually 1 of 2 possible types:
    a. Threaded fittings are mounted to the tops of tanks using a Teflon sealant or tape to seal the threads. They may contain right or left-hand threads, depending upon the tank manufacturer so care should be used when removing or installing the fitting. On many plastic tanks the pick-up itself cannot be removed and the tank must be replaced if damaged.
    b. Bolted fittings are used, predominantly on metallic portable tanks. To replace the pickup unit, first remove the screws (normally 4) securing the unit in the tank. Next, lift the pickup unit up out of the tank. On models with an integrated fuel gauge on the pickup, there are normally screws securing the gauge to the bottom of the pickup unit.

■ **If the pickup unit is not being replaced, clean and check the screen for damage. It is possible to bend a new piece of screen material around the pickup and solder it in place without purchasing a complete new unit.**

4. If equipped with a level gauge assembly, check for smooth, and non-binding movement of the float arm and replace if binding is found. Check the float itself for physical damage or saturation and replace, if found.
5. Check the fuel tank for dirt or moisture contamination. If any is found use a small amount of gasoline or solvent to clean the tank. Pour the solvent in and slosh it around to loosen and wash away deposits, then pour out the solvent and recheck. Allow the tank to air dry, or help it along with the use of an air hose from a compressor.

✱✱ CAUTION

Use extreme care when working with solvents or fuel. Remember that both are even more dangerous when their vapors are concentrated in a small area. No source of ignition from flames to sparks can be allowed in the workplace for even an instant.

To Install:

6. For bolted fuel fittings:
    a. Attach the fuel gauge to the new pickup unit and secure it in place with the screws.
    b. Clean the old gasket material from fuel tank and, if being used, the old pickup unit. Position a new gasket/seal, then work the float arm down through the fuel tank opening, and at the same time the fuel pickup tube into the tank. It will probably be necessary to exert a little force on the float arm in order to feed it all into the hole. The fuel pickup arm should spring into place once it is through the hole.
    c. Secure the pickup and float unit in place with the attaching screws.
7. For threaded fuel fittings, carefully clean the threads on the fitting and the tank, removing any traces of old sealant or tape. Apply a light coating of Teflon plumbers tape or sealant to the threads, then carefully thread the fitting into position until lightly seated.
8. If removed, connect the fuel line to the tank, then pressurize the fuel system and check for leaks.

Boat Mounted Fuel Tanks

The other type of remote fuel tank sometimes used on these models (usually only on the larger Yamaha models) is a boat mounted built-in tank. Depending on the boat manufacturer, built-in tanks may vary greatly in actual shape/design and access. All should be of a one-way vented valve type to prevent a vacuum lock, but capped to prevent evaporation design.

Most boat manufacturers are kind enough to incorporate some means of access to the tank should fuel lines, fuel pickup or floats require servicing. But, the means of access will vary greatly from boat-to-boat. Some might contain simple access panels, while others might require the removal of one or more minor or even major components for access. If you encounter difficulty, seek the advice of a local dealer for that boat builder. The dealer or his/her techs should be able to set you in the right direction.

✱✱ CAUTION

Observe all fuel system cautions, especially when working in recessed portions of a hull. Fuel vapors tend to gather in enclosed areas causing an even more dangerous possibility of explosion.

Fuel Lines and Fittings

◆ See Figure 8

In order for an engine to run properly it must receive an uninterrupted and unrestricted flow of fuel. This cannot occur if improper fuel lines are used or if any of the lines/fittings are damaged. Too small a fuel line could cause hesitation or missing at higher engine rpm. Worn or damaged lines or fittings could cause similar problems (also including stalling, poor/rough idle) as air might be drawn into the system instead of fuel. Similarly, a clogged fuel line, fuel filter or dirty fuel pickup or vacuum lock (from a clogged tank vent as mentioned under Fuel Tank) could cause these symptoms by starving the motor for fuel.

If fuel delivery problems are suspected, check the tank first to make sure it is properly vented, then turn your attention to the fuel lines. First check the lines and valves for obvious signs of leakage, then check for collapsed hoses that could cause restrictions.

■ **If there is a restriction between the primer bulb and the fuel tank, vacuum from the fuel pump may cause the primer bulb to collapse. Watch for this sign when troubleshooting fuel delivery problems.**

✱✱ CAUTION

Only use the proper fuel lines containing suitable Coast Guard ratings on a boat. Failure to do so may cause an extremely dangerous condition should fuel lines fail during adverse operating conditions.

TESTING

A lack of adequate fuel supply will cause the powerhead to run lean or loose rpm. If a fuel tank other than the one supplied or recommended by Yamaha is being used you should make sure the fuel cap has an adequate air vent. Also, verify the size of the fuel line from the tank to be sure it is of sufficient size to accommodate the powerhead demands.

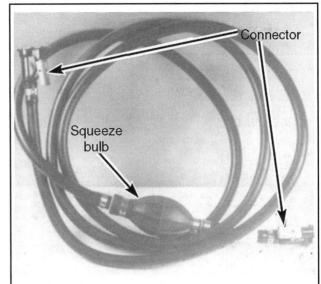

Fig. 8 Problems with the fuel lines or fittings can manifest themselves in engine performance problems, especially at WOT

FUEL SYSTEM 3-9

An adequate size line would be anywhere from 5/16-3/8 in. (7.94-9.52mm) Inner Diameter (ID). Check the fuel strainer on the end of the pickup in the fuel tank to be sure it is not tool small or clogged. Check the fuel pickup tube to make sure it is large enough and is not kinked or clogged. Be sure to check the inline filter to make sure sufficient quantities of fuel can pass through it.

If necessary, perform a Fuel Line Quick Check to see if the problem is fuel line or delivery system related.

Fuel Line Quick Check

◆ See Figure 8

Stalling, hesitation, rough idle, misses at high rpm are all possible results of problems with the fuel lines. A quick visual check of the lines for leaks, kinked or collapsed lengths or other obvious damage may uncover the problem. If no obvious cause is found, the problem may be due to a restriction in the line or a problem with the fuel pump.

If a fuel delivery problem due to a restriction or lack of proper fuel flow is suspected, operate the engine while attempting to duplicate the miss or hesitation. While the condition is present, squeeze the primer bulb rapidly to manually pump fuel from the tank to (and through) the fuel pump to the carburetors (or the vapor separator tank on fuel injected motors). If the engine then runs properly while under these conditions, suspect a problem with a clogged restricted fuel line, a clogged fuel filter or a problem with the fuel pump.

Checking Fuel Flow at Motor

◆ See Figures 8 thru 12

To perform a more thorough check of the fuel lines and isolate or eliminate the possibility of a restriction, proceed as follows:

1. For safety, disconnect the spark plug leads, then ground each of them to the powerhead to prevent sparks and to protect the ignition system.
2. Disconnect the fuel line from the engine (quick-connector on many models). Place a suitable container over the end of the fuel line to catch the fuel discharged. If equipped with a quick-connector, insert a small screwdriver into the end of the line to hold the valve open.
3. Squeeze the primer bulb and observe if there is satisfactory fuel flow from the line. If there is no fuel discharged from the line, the check valve in the squeeze bulb may be defective, or there may be a break or obstruction in the fuel line.
4. If there is a good fuel flow, reconnect the tank-to-motor fuel supply line and disconnect the fuel line from the carburetor(s) or fuel injection vapor separator tank (as applicable), directing that line into a suitable container. Crank the powerhead. If the fuel pump is operating properly, a healthy stream of fuel should pulse out of the line. If sufficient fuel does not pulse from the line, compare flow at either side of the inline fuel filter (if equipped) or check the fuel pump.
5. Continue cranking the powerhead and catching the fuel for about 15 pulses to determine if the amount of fuel decreases with each pulse or maintains a constant amount. A decrease in the discharge indicates a restriction in the line. If the fuel line is plugged, the fuel stream may stop. If there is fuel in the fuel tank but no fuel flows out the fuel line while the powerhead is being cranked, the problem may be in one of several areas:

6. Plugged fuel line from the fuel pump to the carburetor(s) or vapor separator tank (EFI and HPDI).
7. Defective O-ring in fuel line connector into the fuel tank.
8. Defective O-ring in fuel line connector into the engine.
9. Defective fuel pump.
10. The line from the fuel tank to the fuel pump may be plugged; the line may be leaking air; or the squeeze bulb may be defective.
11. Defective fuel tank.
12. If the engine does not start even though there is adequate fuel flow from the fuel line, the float inlet needle valve and the seat may be gummed together and prevent adequate fuel flow into the float bowl or fuel injection vapor separator tank.

Checking the Primer Bulb

◆ See Figures 8 and 13

The way most outboards are rigged, fuel will evaporate from the system during periods of non-use. Also, anytime quick-connect fittings on portable tanks are removed, there is a chance that small amounts of fuel will escape and some air will make it into the fuel lines. For this reason, outboards are normally rigged with some method of priming the fuel system through a hand-operated pump (primer bulb).

When squeezed, the bulb forces fuel from inside the bulb, through the one-way check valve toward the motor filling the carburetor float bowl(s) or fuel injection vapor separator tank with the fuel necessary to start the motor. When the bulb is released, the one-way check valve on the opposite end (tank side of the bulb) opens under vacuum to draw fuel from the tank and refill the bulb.

When using the bulb, squeeze it gently as repetitive or forceful pumping may flood the carburetor(s) or potentially overfill the vapor separator tank on fuel injected motors. The bulb is operating normally if a few squeezes will cause it to become firm, meaning the float bowl/tank is full, and the float valve is closed. If the bulb collapses and does not regain its shape, the bulb must be replaced.

For the bulb to operate properly, both check valves must operate properly and the fuel lines from the check valves back to the tank or forward to the motor must be in good condition (properly sealed). To check the bulb and check valves use hand operated vacuum/pressure pump (available from most marine or automotive parts stores):

1. Remove the fuel hose from the tank and the motor, then remove the clamps for the quick-connect valves at the ends of the hose.

■ Many quick-connect valves are secured to the fuel supply hose using disposable plastic ties that must be cut and discarded for removal. If equipped, spring-type or threaded metal clamps may be reused, but be sure they are in good condition first. Do not over-tighten threaded clamps and crack the valve or cut the hose.

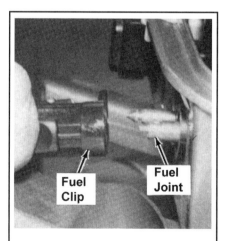

Fig. 9 Disconnect the fuel supply from the engine quick-connect to check fuel flow

Fig. 10 The fuel line can also be disconnected from the fuel pump. In either case, squeeze the primer bulb and watch for unrestricted fuel flow

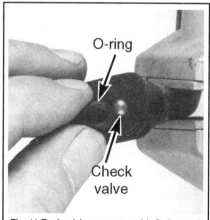

Fig. 11 Fuel quick-connector with O-ring and check valve visible (you'll have to depress the check valve in order to see fuel flow with the connector removed from the engine

3-10 FUEL SYSTEM

■ For proper orientation during testing or installation, the primer bulb is marked with an arrow that faces the engine side check valve.

4. Securely connect the pressure pump to the hose on the tank side of the primer bulb. Using the pump, slowly apply pressure while listening for air escaping from the end of the hose that connects to the motor. If air escapes, both one-way check valves on the tank side and motor side of the prime bulb are opening.

5. If air escapes prior to the motor end of the hose, hold the bulb, check valve and hose connections under water (in a small bucket or tank). Apply additional air pressure using the pump and watch for escaping bubbles to determine what component or fitting is at fault. Repair the fitting or replace the defective hose/bulb component.

6. If no air escapes, attempt to draw a vacuum from the tank side of the primer bulb. The pump should draw and hold a vacuum without collapsing the primer bulb, indicating that the tank side check valve remained closed.

7. Securely connect the pressure pump to the hose on the motor side of the primer bulb. Using the pump, slowly apply pressure while listening for air escaping from the end of the hose that connects to the tank. This time, the check valve on the tank side of the primer bulb should remain closed, preventing air from escaping or from pressurizing the bulb. If the bulb pressurizes, the motor side check valve is allowing pressure back into the bulb, but the tank side valve is operating properly.

8. Replace the bulb and/or check valves if they operate improperly.

SERVICE

◆ See Figures 13 thru 15

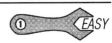

■ When replacing fuel lines, make sure the inside diameter of the fuel hose and fitting is at least the same diameter as the hose it is replacing. Also, be certain to use only marine fuel line the meets or exceeds United States Coast Guard (USCG) A1 or B1 guidelines.

Whenever work is performed on the fuel system, check all hoses for wear or damage. Replace hoses that are soft and spongy or ones that are hard and brittle. Fuel hoses should be smooth and free of surface cracks, and they should definitely not have split ends (there's a bad hair joke in there, but we won't sink that low). Do not cut the split ends of a hose and attempt to reuse it, whatever caused the split (most likely time and deterioration) will cause the new end to follow soon. Fuel hoses are safety items, don't scrimp on them, instead, replace them when necessary. If one hose is too old, check the rest, as they are likely also in need of replacement.

When replacing fuel lines only use replacement hoses or other marine fuel supply lines that meet United States Coast Guard (USCG) requirements A1 or B1 for marine applications. All lines must be of the same inner diameter as the original to prevent leakage and maintain the proper seal that is necessary for fuel system operation.

■ Using a smaller fuel hose than originally specified/equipped could cause fuel starvation problems leading to misfiring, hesitation, rough idling and possibly even engine damage.

The USCG ratings for fuel supply lines have to do with whether or not the lines have been testing regarding length of time it might take for them to succumb to flame (burn through) in an emergency situation. A line is "A" rated if it passes specific requirements regarding burn-through times, while "B" rated lines are not tested in this fashion. The A1 and B1 lines (normally recommended on Johnson/Evinrude applications) are capable of containing liquid fuel at all times. The A2 and B2 rated lines are designed to contain fuel vapor, but not liquid.

✳✳ CAUTION

To help prevent the possibility of significant personal injury or death, do not substitute "B" rated lines when "A" rated lines are required. Similarly, DO NOT use "A2" or "B2" lines when "A1" or "A2" lines are specified.

Various styles of fuel line clamps may be found on these motors. Many applications will simply secure lines with plastic wire ties or special plastic locking clamps. Although some of the plastic locking clamps may be released and reconnected, it is usually a good idea to replace them. Obviously wire ties are cut for removal, which requires that they be replaced.

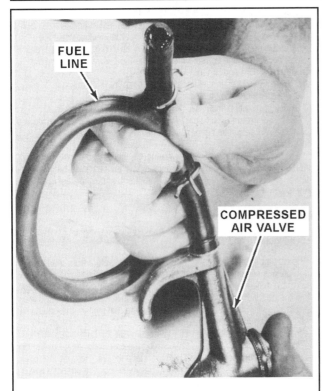

Fig. 12 Many times, restrictions such as foreign matter may be cleared from fuel lines using compressed air. But be careful to make sure the hose is pointed away from yourself or others

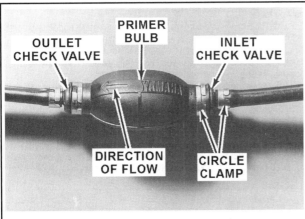

Fig. 13 The primer bulb contains an arrow that indicates the direction of fuel flow (points toward the motor)

2. Carefully remove the fuel line connector (fitting or quick-connect valve, as applicable) from the motor side of the fuel line, then place the end of the line into the filler opening of the fuel tank. Gently pump the primer bulb to empty the hose into the fuel tank.

■ Be careful when removing the quick-connect valve from the fuel line as fuel will likely still be present in the hose and will escape (drain or splash) if the valve is jerked from the line. Also, make sure the primer bulb is empty of fuel before proceeding.

3. Next, remove the fuel line connector (fitting or quick-connect valve) from the tank side of the fuel line, draining any residual fuel into the tank. Boat mounted fuel tanks often have a water separating fuel filter installed inline between the fuel tank and the motor, on these systems disconnect the tank/primer bulb line from the separator.

FUEL SYSTEM 3-11

Some applications use metal spring-type clamps, that contain tabs which are squeezed allowing the clamp to slid up the hose and over the end of the fitting so the hose can be pulled from the fitting. Threaded metal clamps are nice since they are very secure and can be reused, but do not overtighten threaded clamps as they will start to cut into the hose and they can even damage some fittings underneath the hose. Metal clamps should be replaced anytime they've lost tension (spring type clamps), are corroded, bent or otherwise damaged.

■ **The best way to ensure proper fuel fitting connection is to use the same size and style clamp that was originally installed (unless of course the "original" clamp never worked correctly, but in those cases, someone probably replaced it with the wrong type before you ever saw it).**

To avoid leaks, replace all displaced or disturbed gaskets, O-rings or seals whenever a fuel system component is removed.

On many installations, the fuel line is provided with quick-disconnect fittings at the tank and at the powerhead. If there is reason to believe the problem is at the quick-disconnects, the hose ends can be replaced as an assembly, or new O-rings may be installed. A supply of new O-rings should be carried on board for use in isolated areas where a marine store is not available (like dockside, or worse, should you need one while on the water). For a small additional expense, the entire fuel line can be replaced and eliminate this entire area as a problem source for many future seasons. (If the fuel line is replaced, keep the old one around as a spare, just in case).

If an O-ring must be replaced, use two small punches, picks or similar tools, one to push down the check valve of the connector and the other to work the O-ring out of the hole. Apply just a drop of oil into the hole of the connector. Apply a thin coating of oil to the surface of the O-ring. Pinch the O-ring together and work it into the hole while simultaneously using a punch to depress the check valve inside the connector.

The primer squeeze bulb can be replaced in a short time. A squeeze bulb assembly kit, complete with the check valves installed, may be obtained from your dealer or marine parts retailer. The replacement kit will usually also include two tie straps (or some other type of clamp) to secure the bulb properly in the line.

An arrow is clearly visible on the squeeze bulb to indicate the direction of fuel flow. The squeeze bulb must be installed correctly in the line because the check valves in each end of the bulb will allow fuel to flow in only one direction. Therefore, if the squeeze bulb should be installed backwards, in a moment of haste to get the job done, fuel will not reach the carburetor(s).

To replace the bulb, first undo the clamps on the hose at each end of the bulb. Next, pull the hose out of the check valves at each end of the bulb. New clamps are included with a new squeeze bulb.

If the fuel line has been exposed to considerable sunlight, it may have become hardened, causing difficulty in working it over the check valve. To remedy this situation, simply immerse the ends of the hose in boiling water for a few minutes to soften the rubber. The hose will then slip onto the check valve without further problems. After the lines on both sides have been installed, snap the clamps in place to secure the line. Check a second time to be sure the arrow is pointing in the fuel flow direction, towards the powerhead.

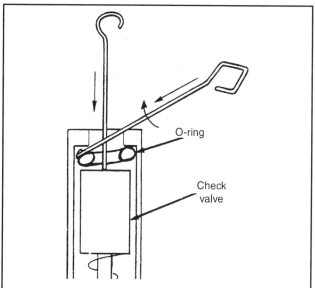

Fig. 14 Use two picks, punches or other small tool to replace quick-connect O-rings. One to push the valve and the other work the O-ring free

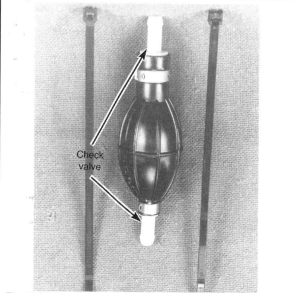

Fig. 15 A squeeze bulb kit usually includes the bulb, 2 check valves, and two tie straps (or other replacement clamps)

CARBURETED FUEL SYSTEM

✱✱ CAUTION

If equipped, disconnect the negative battery cable ANYTIME work is performed on the engine, especially when working on the fuel system. This will help prevent the possibility of sparks during service (from accidentally grounding a hot lead or powered component). Sparks could ignite vapors or exposed fuel. Disconnecting the cable on electric start motors will also help prevent the possibility fuel spillage if an attempt is made to crank the engine while the fuel system is open.

✱✱ CAUTION

Fuel leaking from a loose, damaged or incorrectly installed hose or fitting may cause a fire or an explosion. ALWAYS pressurize the fuel system and run the motor while inspecting for leaks after servicing any component of the fuel system.

Carbureted motors covered by this repair guide are equipped with anything from one single barrel carburetor to 3 two-barrel carbs or even one carburetor per cylinder. Although on initial inspection some of the larger carbureted motors may look somewhat complicated, they're actually very basic, especially when compared with other modern fuel systems (such as an automotive or marine fuel injection system).

The entire system essentially consists of a fuel tank, a fuel supply line, and a mechanical fuel pump assembly mounted to the powerhead all designed to feed the carburetor with the fuel necessary to power the motor. Cold starting is enhanced by the use of a choke plate or an enrichment circuit.

For information on fuels, tanks and lines please refer to the sections on Fuel System Basics and Fuel Tanks and Lines.

The most important fuel system maintenance that a boat owner can perform is to stabilize fuel supplies before allowing the system to sit idle for any length of time more than a few weeks (or better yet, to completely drain the carburetors. The next most important item is to provide the system with

3-12 FUEL SYSTEM

fresh gasoline if the system has stood idle for any length of time, especially if it was without fuel system stabilizer during that time.

If a sudden increase in gas consumption is noticed, or if the engine does not perform properly, a carburetor overhaul, including cleaning or replacement of the fuel pump may be required.

Description and Operation

◆ See Figures 16 and 17

GENERAL INFORMATION

The Role of a Carburetor

◆ See Figures 16 and 17

The carburetor is merely a metering device for mixing fuel and air in the proper proportions for efficient engine operation. At idle speed, an outboard engine requires a mixture of about 8 parts air to 1 part fuel. At high speed or under heavy duty service, the mixture may change to as much as 12 parts air to 1 part fuel.

Carburetors are wonderful devices that succeed in creating precise air/fuel mixture ratios based on tiny passages, needle jets or orifices and the variable vacuum that occurs as engine rpm and operating conditions vary.

Because of the tiny passages and small moving parts in a carburetor (and the need for them to work precisely to achieve exact air/fuel mixture ratios) it is important that the fuel system integrity is maintained. Introduction of water (that might lead to corrosion), debris (that could clog passages) or even the presence of unstabilized fuel that could evaporate over time can cause big problems for a carburetor. Keep in mind that when fuel evaporates it leaves behind a gummy deposit that can clog those tiny passages, preventing the carburetor (and therefore preventing the engine) from operating properly.

Float Systems

◆ See Figure 16

Ever lift the tank lid off the back of your toilet. Pretty simple stuff once you realize what's going on in there. A supply line keeps the tank full until a valve opens allowing all or some of the liquid in the tank to be drawn out through a passage. The dropping level in the tank causes a float to change position, and, as it lowers in the tank it opens a valve allowing more pressurized liquid back into the tank to raise levels again. OK, we were talking about a toilet right, well yes and no, we're also talking about the float bowl on a carburetor. The carburetor uses a more precise level, uses vacuum to draw out fuel from the bowl through a metered passage (instead of gravity) and, most importantly, store gasoline instead of water, but otherwise, they basically work in the same way.

A small chamber in the carburetor serves as a fuel reservoir. A float valve admits fuel into the reservoir to replace the fuel consumed by the engine. If the carburetor has more than one reservoir, the fuel level in each reservoir (chamber) is controlled by identical float systems.

Fuel level in each chamber is extremely critical and must be maintained accurately. Accuracy is obtained through proper adjustment of the float. This adjustment will provide a balanced metering of fuel to each cylinder at all speeds. Improper levels will lead to engine operating problems. Too high a level can promote rich running and spark plug fouling, while excessively low float bowl fuel levels can cause lean conditions, possibly leading to engine damage.

Following the fuel through its course, from the fuel tank to the combustion chamber of the cylinder, will provide an appreciation of exactly what is taking place. In order to start the engine, the fuel must be moved from the tank to the carburetor by a squeeze bulb installed in the fuel line. This action is necessary because the fuel pump does not have sufficient pressure to draw fuel from the tank during cranking before the engine starts.

The fuel for some small horsepower units is gravity fed from a tank mounted at the rear of the powerhead. Even with the gravity feed method, a small fuel pump may be an integral part of the carburetor for some manufacturers.

After the engine starts, the fuel passes through the pump to the carburetor. All systems have some type of filter installed somewhere in the line between the tank and the carburetor. Over the years many carburetors have utilized a filter as an integral part of the carburetor.

At the carburetor, the fuel passes through the inlet passage to the needle and seat, and then into the float chamber (reservoir). A float in the chamber rides up and down on the surface of the fuel. After fuel enters the chamber and the level rises to a predetermined point, a tang on the float closes the inlet needle and the flow entering the chamber is cut off. When fuel leaves the chamber as the engine operates, the fuel level drops and the float tang allows the inlet needle to move off its seat and fuel once again enters the chamber. In this manner, a constant reservoir of fuel is maintained in the chamber to satisfy the demands of the engine at all speeds.

A fuel chamber vent hole is located near the top of the carburetor body to permit atmospheric pressure to act against the fuel in each chamber. This pressure assures an adequate fuel supply to the various operating systems of the powerhead.

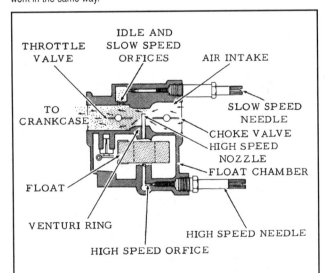

Fig. 16 Fuel flow through the venturi, showing principle and related parts controlling intake and outflow (carburetor with manual choke circuit shown)

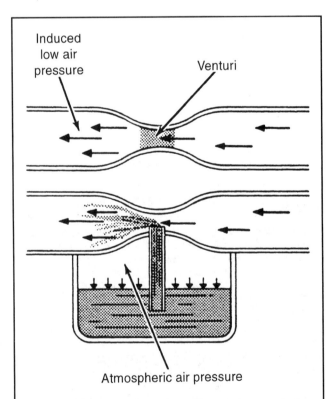

Fig. 17 Air flow principle of a modern carburetor, demonstrates how the low pressure induced behind the venturi draws fuel through the high speed nozzle

Air/Fuel Mixture

◆ See Figure 17

A suction effect is created each time the piston moves upward in the cylinder of a 2-stroke motor. This suction draws air through the throat of the carburetor. A restriction in the throat, called a venturi, controls air velocity and has the effect of reducing air pressure at this point.

The difference in air pressures at the throat and in the fuel chamber causes the fuel to be pushed out of metering jets extending down into the fuel chamber. When the fuel leaves the jets, it mixes with the air passing through the venturi. This fuel/air mixture should then be in the proper proportion for burning in the cylinders for maximum engine performance.

In order to obtain the proper air/fuel mixture for all engine speeds, high- and low-speed orifices or needle valves are installed. On most modern powerheads the high-speed needle valve has been replaced with a fixed high-speed orifice (to more discourage tampering and to help maintain proper emissions under load). There is no adjustment with the orifice type. The needle valves are used to compensate for changing atmospheric conditions. The low-speed needles, on the other hand, are still provided so that air/fuel mixture can be precisely adjusted for idle conditions other than what occurs at atmospheric sea-level. Although the low-speed needle should not normally require periodic adjustment, it can be adjusted to compensate for high-altitude (river/lake) operation or to adjust for component wear within the fuel system.

A throttle valve controls the flow of air/fuel mixture drawn into the combustion chambers. A cold powerhead requires a richer fuel mixture to start and during the brief period it is warming to normal operating temperature. A choke valve is often placed ahead of the metering jets and venturi. As this valve begins to close, the volume of air intake is reduced, thus enriching the mixture entering the cylinders.

When this choke valve is fully closed, a very rich fuel mixture is drawn into the cylinders.

■ **Some carburetors do not use a choke valve, but instead use an enriching circuit which mechanically provides more fuel to the carburetor or intake during cold start-up.**

The throat of the carburetor is usually referred to as the barrel. Carburetors with single, double, or four barrels have individual metering jets, needle valves, throttle and, if applicable, choke plates for each barrel. Single and two barrel carburetors are fed by a single float and chamber.

CARBURETOR CIRCUITS

The following section illustrates the circuit functions and locations of a typical marine carburetor.

Starting Circuit

◆ See Figure 18

The choke plate is closed, creating a partial vacuum in the venturi. As the piston rises, negative pressure in the crankcase draws the rich air-fuel mixture from the float bowl into the venturi and on into the engine.

Low Speed Circuit

◆ See Figure 19

Zero-one-eighth throttle, when the pressure in the crankcase is lowered, the air-fuel mixture is discharged into the venturi through the pilot outlet because the throttle plate is closed. No other outlets are exposed to low venturi pressure. The fuel is metered by the pilot jet. The air is metered by the pilot air jet. The combined air-fuel mixture is regulated by the pilot air screw.

Mid-Range Circuit

◆ See Figure 20

One-eighth-three-eighths throttle, as the throttle plate continues to open, the air-fuel mixture is discharged into the venturi through the bypass holes. As the throttle plate uncovers more bypass holes, increased fuel flow results because of the low pressure in the venturi. Depending on the model, there could be two, three or four bypass holes.

High Speed Circuit

◆ See Figure 21

Three-eighths-wide-open throttle, as the throttle plate moves toward wide open, we have maximum air flow and very low pressure. The fuel is metered through the main jet, and is drawn into the main discharge nozzle. Air is metered by the main air jet and enters the discharge nozzle, where it combines with fuel. The mixture atomizes, enters the venturi, and is drawn into the engine.

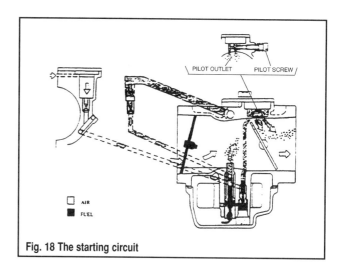

Fig. 18 The starting circuit

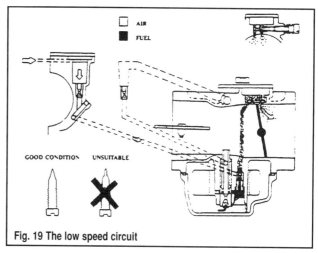

Fig. 19 The low speed circuit

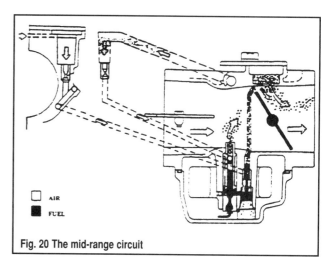

Fig. 20 The mid-range circuit

3-14 FUEL SYSTEM

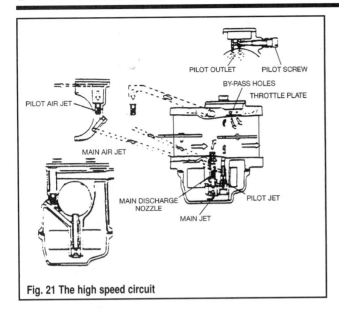

Fig. 21 The high speed circuit

BASIC FUNCTIONS

◆ See Figure 22

The carburetor systems on inline and V-engines require careful cleaning and adjustment if problems occur. These carburetors may seem a little complicated but they are not too complex to understand. All carburetors operate on the same principles.

Traditional carburetor theory often involves a number of laws and principles. To troubleshoot carburetors learn the basic principles, watch how the carburetor comes apart, trace the circuits, see what they do and make sure they are clean. These are the basic steps for troubleshooting and successful repair.

The diagram illustrates several carburetor basics. If you blow through the horizontal straw an atomized mixture (air and fuel droplets) comes out. When you blow through the straw a pressure drop is created in the straw column inserted in the liquid. In a carburetor this is mostly air and a little fuel. The actual ratio of air to fuel differs with engine conditions but is usually from 15 parts air to one part fuel at optimum cruise to as little as 7 parts air to one part fuel at full choke.

If the top of the container is covered and sealed around the straw what will happen? No flow. This is typical of a clogged carburetor bowl vent.

If the base of the straw is clogged or restricted what will happen? No flow or low flow. This represents a clogged main jet.

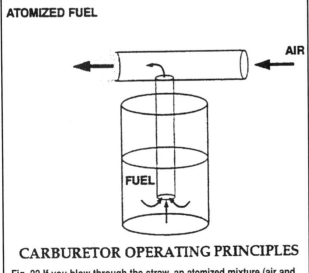

Fig. 22 If you blow through the straw, an atomized mixture (air and fuel droplets) comes out

If the liquid in the glass is lowered and you blow through the straw with the same force what will happen? Not as much fuel will flow. A lean condition occurs.

If the fuel level is raised and you blow again at the same velocity what happens? The result is a richer mixture.

Yamaha carburetors control air flow semi-independently of RPM. This is done with a throttle plate. The throttle plate works in conjunction with other systems or circuits to deliver correct mixtures within certain RPM bands. The idle circuit pilot outlet controls from 0-1/8 throttle. The series of small holes in the carburetor throat called transition holes control the 1/8-3/8 throttle range. At wide open throttle the main jet handles most of the fuel metering chores, but the low and mid-range circuits continue to supply part of the fuel.

Enrichment is necessary to start a cold engine. Fuel and air mix does not want to vaporize in a cold engine. In order to get a little fuel to vaporize, a lot of fuel is dumped into the engine. On some inline engines a choke plate is used for cold starts. This plate restricts air entering the engine and increases the fuel to air ratio.

The Prime Start enrichment system used on many late-model engines is controlled by a heated wax pellet. This pellet is heated by current from the stator. Temperature is monitored and enrichment is automatic. Other inline engines use choke plates, or electric solenoid enrichment systems.

DUAL-THROAT CARBURETORS

Dual-throat carburetor systems are used on V4 and V6 engines which require careful cleaning and adjustment if problems occur. Like the units used on their inline kin, these carburetors are not difficult to understand and operate on the same basic principles. For best results, trace and analyze one circuit at a time.

Beginning in 1996, all Saltwater series 90 degree V engines have an additional jet in the carburetor. This pull-over or enrichment jet improves mid-range response while maintaining fuel economy.

Cold start enrichment is straight forward on most models. The many V4 and V6 engines utilize a choke plate for cold starts. This plate restricts air entering the engine and increases the fuel/air ratio.

The enrichment system on the 90-degree 225 hp engines is controlled by a microprocessor. Temperature and throttle position are monitored and enrichment is automatic. A pair of injectors with different diameters is used to provide enrichment.

Troubleshooting the Carbureted Fuel System

◆ See Figure 23

Troubleshooting fuel systems requires the same techniques used in other areas. A thorough, systematic approach to troubleshooting will pay big rewards. Build your troubleshooting checklist, with the most likely offenders at the top. Use your experience to adjust your list for local conditions. At one time or another nearly everyone has been tempted to jump into the carburetor based on nothing more than a vague hunch. Pause a moment and review the facts when this urge occurs.

In order to accurately troubleshoot a carburetor or fuel system problem, you must first verify that the problem is fuel related. Many symptoms can have several different possible causes. Be sure to eliminate mechanical and electrical systems as the potential fault. Carburetion is a common cause of most engine problems, but there are many other possibilities.

One of the toughest tasks with a fuel system is the actual troubleshooting. Several tools are at your disposal for making this process very simple. A timing light works well for observing carburetor spray patterns. Look for the proper amount of fuel and for proper atomization in the two fuel outlet areas (main nozzle and bypass holes). The strobe effect of the lights helps you see in detail the fuel being drawn through the throat of the carburetor. On multiple carburetor engines, always attach the timing light to the cylinder you are observing so the strobe doesn't change the appearance of the patterns. If you need to compare two cylinders, change the timing light hookup each time you observe a different cylinder.

Pressure testing fuel pump output can determine whether sufficient fuel is being supplied for the fuel spray and if the fuel pump diaphragms are functioning correctly. A pressure gauge placed between the fuel pumps and the carburetors will test the entire fuel delivery system. Normally a fuel system problem will show up at high speed where the fuel demand is the greatest. A common symptom of a fuel pump output problem is surging at wide open throttle, while still operating normally at slower speeds. To check the fuel pump output, install the pressure gauge and accelerate the engine to

FUEL SYSTEM 3-15

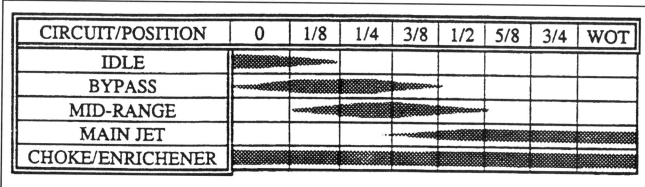

Fig. 23 Illustration of throttle position and carburetor operating circuits

wide open throttle. Observe the pressure gauge needle. It should always swing up to some value above 2 psi (14 kPa), usually something in the 5-6 psi (34-41 kPa) area and remain steady. This reading would indicate a system that is functioning properly.

If the needle gradually swings down toward zero, fuel demand is greater than the fuel system can supply. This reading isolates the problem to the fuel delivery system (fuel tank or line). To confirm this, an auxiliary tank should be installed and the engine retested. Be aware that a bad anti-siphon valve on a built-in tank can create enough restriction to cause a lean condition and serious engine damage.

If the needle movement becomes erratic, suspect a ruptured diaphragm in the fuel pump.

A quick way to check for a ruptured fuel pump diaphragm is, while the engine is at idle speed, to squeeze the primer bulb and hold steady firm pressure on it. If the diaphragm is ruptured, this will cause a rough running condition because of the extra fuel passing through the diaphragm into the crankcase. After performing this test you should check the spark plugs. If the spark plugs are OK, but the fuel pumps are still suspected, you should remove the fuel pumps and completely disassemble them. Rebuild or replace the pumps as needed.

To check the boat's fuel system for a restriction, install a vacuum gauge in the line before the fuel pump. Run the engine under load at wide open throttle to get a reading. Vacuum should read no more than 4.5 in. Hg (15.2 kPa) for engines up to and including 200 hp, and should not exceed 6.0 in. Hg (20.2 kPa) for engines greater than 200 hp.

To check for air entering the fuel system, install a clear fuel hose between the fuel screen and fuel pump. If air is in the line, check all fittings back to the boat's fuel tank.

Spark plug tip appearance is a good indication of combustion efficiency. The tip should be a light tan. A white insulator or small beads on the insulator indicate too much heat. A dark or oil fouled insulator indicates incomplete combustion. To properly read spark plug tip appearance, run the engine at the rpm you are testing for about 15 second and then immediately turn the engine OFF without changing the throttle position.

Reading spark plug tip appearance is also the proper way to test jet verifications in high altitude.

The accompanying illustration explains the relationship between throttle position and carburetion circuits.

LOGICAL TROUBLESHOOTING

The following paragraphs provide an orderly sequence of tests to pinpoint problems in the fuel system.

1. Gather as much information as you can.
2. Duplicate the condition. Take the boat out and verify the complaint.
3. If the problem cannot be duplicated, you cannot fix it. This could be a product operation problem.
4. Once the problem has been duplicated, you can begin troubleshooting. Give the entire unit a careful visual inspection. You can tell a lot about the engine from the care and condition of the entire rig. What's the condition of the propeller and the lower unit? Remove the hood and look for any visible signs of failure. Are there any signs of head gasket leakage? Is the engine paint discolored from high temperature or are there any holes or cracks in the engine block? Perform a compression (or better yet, if you have the equipment, a leak down test). While cranking the engine during the compression test, listen for any abnormal sounds. If the engine passes these simple tests we can assume that the mechanical condition of the engine is good. All other engine mechanical inspection would be too time consuming at this point.
5. Your next step is to isolate the fuel system into two sub-systems. Separate the fuel delivery components from the carburetors. To do this, substitute a known good fuel supply for the regular boat's fuel supply. Use a 6 gallon portable tank and fuel line. Connect the portable fuel supply directly to the engine fuel pump, bypassing the boat fuel delivery system. Now test the engine. If the problem is no longer present, you know where to look. If the problem is still present, further troubleshooting is required.
6. When testing the engine, observe the throttle position when the problem occurs. This will help you pinpoint the circuit that is malfunctioning. Carburetor troubleshooting and repair is very demanding. You must pay close attention to the location, position and sometimes the numbering on each part removed. The ability to identify a circuit by the operating RPM it affects is important. Often your best troubleshooting tool is a can of cleaner. This can be used to trace those mystery circuits and find that last speck of dirt. Be careful and wear safety glasses when using this method.

■ **The last step of fuel system troubleshooting is to adjust or rebuild and then adjust the carburetor. We say it is the last step, because it is the most involved repair procedures on the fuel system and should only be performed after all other possible causes of fuel system trouble have been eliminated.**

COMMON PROBLEMS

Fuel Delivery

Many times fuel system troubles are caused by a plugged fuel filter, a defective fuel pump, or by a leak in the line from the fuel tank to the fuel pump. Aged fuel left in the carburetor and the formation of varnish could cause the needle to stick in its seat and prevent fuel flow into the bowl. A defective choke may also cause problems. Would you believe, a majority of starting troubles, which are traced to the fuel system, are the result of an empty fuel tank or aged fuel.

If fuel delivery problems are suspected, refer to the testing procedures in Fuel Tank and Lines to make sure the tank vent is working properly and that there are not leaks or restrictions that would prevent fuel from getting to the pump and/or carburetor(s).

A blocked low-pressure fuel filter causes hard starting, stalling, misfire or poor performance. Typically the engine malfunction worsens with increased engine speed. This filter prevents contaminants from reaching the low-pressure fuel pump. Refer to the Fuel Filter in the section on Maintenance and Tune-Up for more details on checking, cleaning or replacing fuel filters.

Sour Fuel

◆ See Figure 24

Under average conditions (temperate climates), fuel will begin to break down in about four months. A gummy substance forms in the bottom of the fuel tank and in other areas. The filter screen between the tank and the carburetor and small passages in the carburetor will become clogged. The gasoline will begin to give off an odor similar to rotten eggs. Such a condition can cause the owner much frustration, time in cleaning components, and the expense of replacement or overhaul parts for the carburetor.

3-16 FUEL SYSTEM

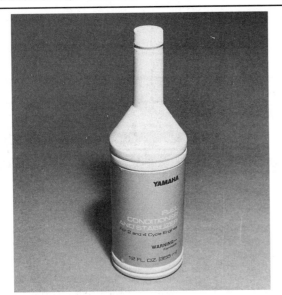

Fig. 24 The use of an approved fuel additive, such as this Yamaha Fuel Conditioner and Stabilizer, will prevent fuel from souring for up to twelve months

Even with the high price of fuel, removing gasoline that has been standing unused over a long period of time is still the easiest and least expensive preventative maintenance possible. In most cases, this old gas can be used without harmful effects in an automobile using regular gasoline.

The gasoline preservative additive Yamaha Fuel Conditioner and Stabilizer for 2 cycle engines, will keep the fuel fresh for up to twelve months. If this particular product is not available in your area, other similar additives are produced under various trade names.

Refer to the information on Fuel System Basics in this section, specifically the procedure under Fuel entitled Checking For Stale/Contaminated Fuel will provide information on how to determine if stale fuel is present in the system. If draining the system of contaminated fuel and refilling it with fresh fuel does not make a difference in the problem, look for restrictions or other problems with the fuel delivery system. If stale fuel was left in the tank/system for a long period of time and evaporation occurred, there is a good chance that the carburetor is gummed (tiny passages are clogged by deposits left behind when the fuel evaporated). If no fuel delivery problems are found, the carburetor(s) should be removed for disassembly and cleaning.

■ Although there are some commercially available fuel system cleaning products that are either added to the fuel mixture or sprayed into the carburetor throttle bores, the truth is that although they can provide some measure of improvement, there is no substitute for a thorough disassembly and cleaning. The more fuel which was allowed to evaporate, the more gum or varnish may have been left behind and the more likely that only a disassembly will be able to restore proper performance.

Choke Problems

◆ See Figure 25

When the engine is hot, the fuel system can cause starting problems. After a hot engine is shut down, the temperature inside the fuel bowl may rise to 200 degrees F (94 degrees C) and cause the fuel to actually boil. All carburetors are vented to allow this pressure to escape to the atmosphere. However, some of the fuel may percolate over the high-speed nozzle.

If the choke should stick in the open position, the engine will be hard to start. If the choke should stick in the closed position, the engine will flood, making it very difficult to start.

In order for this raw fuel to vaporize enough to burn, considerable air must be added to lean out the mixture. Therefore, the best remedy for a flooded motor is to remove the spark plugs, ground the leads, crank the powerhead through about ten revolutions, clean the plugs, reinstall the plugs, and start the engine.

■ Another common solution to a flooded motor is to open the throttle all the way while cranking (using the additional air to help clear/burn the excess fuel in the combustion chambers), but the problem with marine engines is that this puts the gearcase in FORWARD which will prevent the starter from working on most models.

If the needle valve and seat assembly is leaking, an excessive amount of fuel may enter the reed housing on 2-stroke motors in the following manner. After the powerhead is shut down, the pressure left in the fuel line will force fuel past the leaking needle valve. This extra fuel will raise the level in the fuel bowl and cause fuel to overflow into the reed housing.

A continuous overflow of fuel into the reed housing may be due to a sticking inlet needle or to a defective float, which would cause an extra high level of fuel in the bowl and overflow into the reed housing.

Rough Engine Idle

■ As with all troubleshooting procedures, start with the easiest items to check/fix and work towards the more complicated ones.

If an engine does not idle smoothly, the most reasonable approach to the problem is to perform a tune-up to eliminate such areas as:
- Faulty spark plugs
- Timing and synchronization out of adjustment

Other problems that can prevent an engine from running smoothly include:
- An air leak in the intake manifold
- Uneven compression between the cylinders
- Sticky or broken reeds

Of course any problem in the carburetor affecting the air/fuel mixture will also prevent the engine from operating smoothly at idle speed. These problems usually include:
- Too high a fuel level in the bowl
- A heavy float
- Leaking needle valve and seat
- Defective automatic choke
- Improper adjustments for idle mixture or idle speed

"Sour" fuel (fuel left in a tank without a preservative additive) will cause an engine to run rough and idle with great difficulty.

Excessive Fuel Consumption

◆ See Figures 26 and 27

Excessive fuel consumption can result from one of four conditions, or some combination of the four.
  1. Inefficient engine operation.
  2. Damaged condition of the hull, outdrive or propeller, including excessive marine growth.
  3. Poor boating habits of the operator.
  4. Leaking or out of tune carburetor.

If the fuel consumption suddenly increases over what could be considered normal, then the cause can probably be attributed to the engine or boat and not the operator (unless he/she just drastically changed the manner in which the boat is operated).

Marine growth on the hull can have a very marked effect on boat performance. This is why sail boats always try to have a haul-out as close to race time as possible. While you are checking the bottom take note of the propeller condition. A bent blade or other damage will definitely cause poor boat performance.

If the hull and propeller are in good shape, then check the fuel system for possible leaks. Check the line between the fuel pump and the carburetor while the engine is running and the line between the fuel tank and the pump when the engine is not running. A leak between the tank and the pump many times will not appear when the engine is operating, because the suction created by the pump drawing fuel will not allow the fuel to leak. Once the engine is turned off and the suction no longer exists, fuel may begin to leak.

If a minor tune-up has been performed and the spark plugs and engine timing/synchronization are properly adjusted, then the problem most likely is in the carburetor, indicating an overhaul is in order. Check for leaks at the needle valve and seat. Use extra care when making any adjustments affecting the fuel consumption, such as the float level or automatic choke.

FUEL SYSTEM 3-17

Fig. 25 Fouled spark plug, possibly caused by over-choking or a malfunctioning enrichment circuit

Fig. 26 Marine growth on the lower unit will create "drag" and seriously hamper boat performance

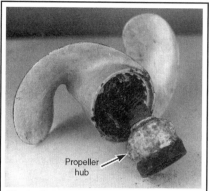

Fig. 27 A corroded hub on a small engine propeller. Hub and propeller damage will also cause poor performance

Engine Surge

If the engine operates as if the load on the boat is being constantly increased and decreased, even though an attempt is being made to hold a constant engine speed, the problem can most likely be attributed to the fuel pump (or a restriction between the tank and powerhead). Refer to Fuel Tank and Lines in this section for information on checking the lines for restrictions and checking fuel flow. Also, refer to Fuel Pump Service for more information on fuel pump testing, operation and repair.

COMBUSTION RELATED PISTON FAILURES

When an engine has a piston failure due to abnormal combustion, fixing the mechanical portion of the engine is the easiest part. The hard part is determining what caused the problem, in order to prevent a repeat failure. Think back to the four basic areas that affect combustion to find the cause of the failure.

Since you probably removed the cylinder head. Inspect the failed piston, look for excessive deposit buildup that could raise compression, or retain heat in the combustion chamber. Statically check the wide open throttle timing. Be sure that the timing is not over advanced. It is a good idea to seal these adjustments with paint to detect tampering.

Look for a fuel restriction that could cause the engine to run lean (which can destroy pre-mix 2-stroke motors due to a lack of oiling). Don't forget to check the fuel pump, fuel tank and lines, especially if a built in tank is used. Be sure to check the anti-siphon valve on built in tanks.

If everything else looks good, the final possibility is poor quality fuel.

CARBURETOR SERVICE

Carburetor Identification

◆ See Figures 28 and 29

A carburetor identification number is normally stamped somewhere on the housing (usually on a flange). Be sure to take down this number and have it handy when purchasing parts or overhaul kits. In some cases adjustment specifications will vary dependant upon this number, we've cited those instances in the Carburetor Set-Up Specifications chart at the end of this section.

2 Hp Models

◆ See Figure 30

This carburetor is a single-barrel, float feed type with a manual choke. Fuel to the carburetor is gravity fed from a fuel tank mounted at the rear of the powerhead.

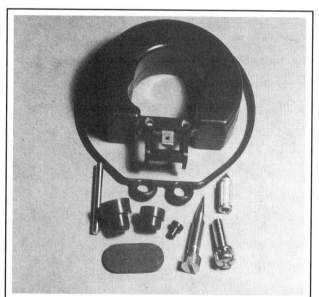

Fig. 29 Carburetor repair kits, available at your local service dealer, contain the necessary components to perform a carburetor overhaul. In most cases, an illustration showing the parts contained in the package is included in the cleaning and inspection portion for each carburetor.

Fig. 28 The carburetor identification number is normally embossed on a carburetor flange - typical carb for 1- and 2-cylinder powerheads shown

3-18 FUEL SYSTEM

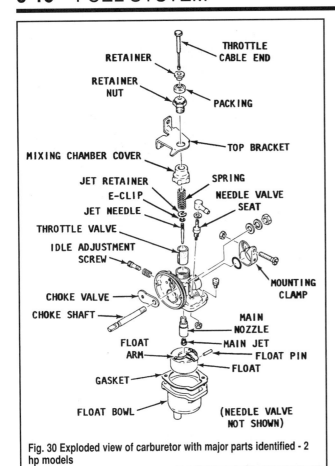

Fig. 30 Exploded view of carburetor with major parts identified - 2 hp models

REMOVAL & DISASSEMBLY

♦ See Figure 30 and 31 thru 48

MODERATE

Good shop practice dictates purchasing a carburetor repair kit and using new parts any time the carburetor is disassembled. Make an attempt to keep the work area organized and to cover parts after they have been cleaned. This practice will prevent foreign matter from entering passageways or adhering to critical parts.

1. Remove the screws on one half of the spark plug cover. Remove four more screws securing one-half of the cowling. Separate the cowling half from which the screws were removed. Remove the four screws securing the other half of the cowling and remove it from the engine. The spark plug cover will remain attached to one of the cowling halves.

■ Observe the different length screws used to secure the cowling halves to the powerhead. Remember their location as an aid during assembling.

2. Remove the two Philips screws securing the intake silencer and the screw securing the throttle control knob to the throttle lever. Remove the knob.

3. Hold the choke knob with one hand and remove the nut on the back side of the intake silencer with a wrench and the other hand. Slide the lockwasher free of the choke shaft.

4. Pull the choke knob with the choke shaft attached out through the hole in the intake silencer. The choke valve and slotted washer will come out with the shaft. Place these small items in order on the workbench as an aid during installation.

5. Carefully separate the White lead (engine stop button lead), from the quick disconnect fitting anchored to the powerhead. Disconnect the black lead, ground, at the horseshoe bracket. Thread the small screw back into the bracket as a precaution against the screw being lost. Lift the intake silencer free of the engine with the wire leads attached to the stop button.

6. Loosen the bolt on the clamp securing the carburetor to the powerhead.

7. Close the fuel valve. Be prepared to catch a small amount of fuel with a cloth when the fuel line is disconnected. Disconnect the fuel line at the carburetor. Remove the carburetor from the reed valve housing. The clamp will come off with the carburetor.

8. Pry off the circlip, and then remove the pivot pin attaching the throttle control lever to the carburetor. Remove the screw, spring, washer, and nut from the carburetor top bracket.

9. Loosen, but do not remove the top retainer brass nut on the carburetor. Hold the carburetor securely with one hand and unscrew the mixing chamber cover with the other, using a pair of pliers.

10. Lift out the throttle valve assembly.

11. Compress the spring in the throttle valve assembly to allow the throttle cable end to clear the recess in the base of the throttle valve and to slide down the slot.

12. Disassemble the throttle valve consisting of the throttle valve, spring, jet needle (with an E-clip on the second groove), jet retainer, and throttle cable end.

13. Remove and discard the two screws securing the float bowl to the carburetor body. Remove the float bowl. Lift out the float. A new pair of float bowl screws are provided in the carburetor rebuild kit.

14. Push the float pin free using a fine pointed awl.

15. Lift out the float arm and needle valve. Slide the needle valve free of the arm.

16. Unscrew the main jet from the main nozzle. Remove and discard the float bowl gasket.

Fig. 31 Step 1

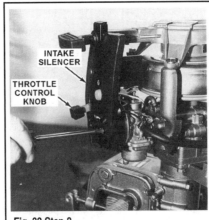

Fig. 32 Step 2

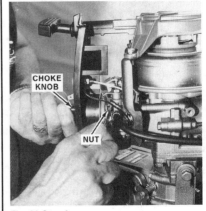

Fig. 33 Step 3

FUEL SYSTEM 3-19

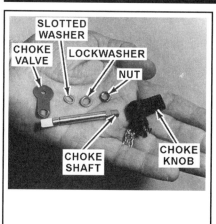
Fig. 34 Step 4

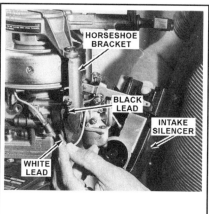
Fig. 35 Step 5

Fig. 36 Step 6

Fig. 37 Step 7

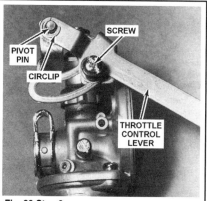
Fig. 38 Step 8

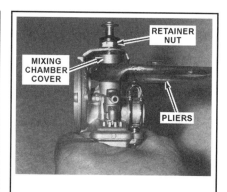
Fig. 39 Step 9

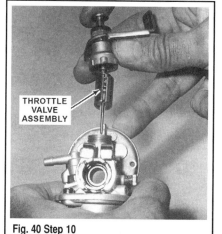
Fig. 40 Step 10

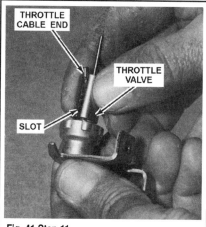
Fig. 41 Step 11

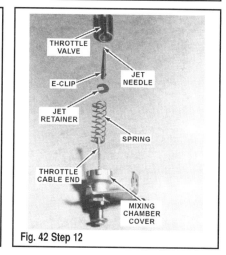
Fig. 42 Step 12

17. Count and record the number of turns required to lightly seat the idle adjustment screw. The number of turns will give a rough adjustment during installation. Back out the idle speed screw and discard the screw, but save the spring. a new screw is provided in the carburetor rebuild kit to ensure a damaged screw is not used again. this screw is most important for maximum performance.

18. Pry out and discard the carburetor sealing ring.

■ It is not necessary to remove the E-clip from the jet needle, unless replacement is required or if the powerhead is to be operated at a significantly different elevation.

CLEANING & INSPECTION

◆ See Figures 30 and 49

✱✱ CAUTION

Never dip rubber parts, plastic parts, diaphragms, or pump plungers in carburetor cleaner. These parts should be cleaned only in solvent, and then blown dry with compressed air.

Place all metal parts in a screen-type tray and dip them in carburetor cleaner until they appear completely clean, then blow them dry with compressed air.

3-20 FUEL SYSTEM

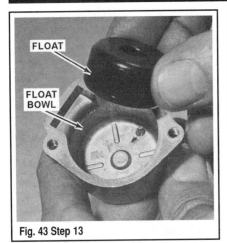
Fig. 43 Step 13

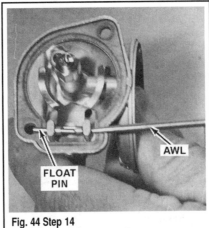
Fig. 44 Step 14

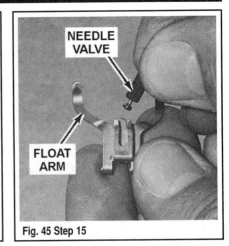
Fig. 45 Step 15

Fig. 46 Step 16

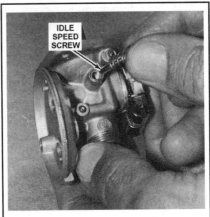
Fig. 47 Step 17

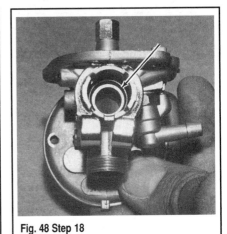
Fig. 48 Step 18

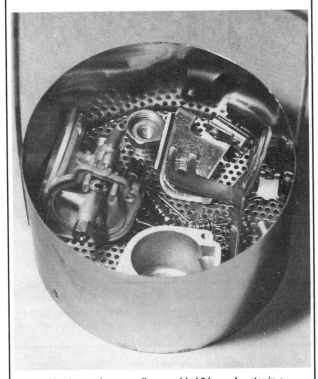
Fig. 49 Metal parts from our disassembled 2 hp carburetor in a basket ready to be immersed in carburetor cleaner

Blow out all passages in the castings with compressed air. Check all parts and passages to be sure they are not clogged or contain any deposits. Never use a piece of wire or any type of pointed instrument to clean drilled passages or calibrated holes in a carburetor.

Move the throttle shaft back and forth to check for wear. If the shaft appears to be too loose, replace the complete throttle body because individual replacement parts are not available.

Inspect the main body, airhorn, and venturi cluster gasket surfaces for cracks and burrs which might cause a leak. Check the float for deterioration. Check to be sure the float spring has not been stretched. If any part of the float is damaged, the unit must be replaced. Check the float arm needle contacting surface and replace the float if this surface has a groove worn in it.

Inspect the tapered section of the idle adjusting needles and replace any that have developed a groove.

As previously mentioned, most of the parts which should be replaced during a carburetor overhaul are included in overhaul kits available from your local marine dealer. One of these kits will contain a matched fuel inlet needle and seat. This combination should be replaced each time the carburetor is disassembled as a precaution against leakage.

ASSEMBLY & INSTALLATION

◆ See Figures 30 and 50 thru 67

1. Install a new carburetor O-ring into the carburetor body.
2. Apply an all-purpose lubricant to a new idle speed screw. Install the idle speed screw and spring.
3. Install the main jet into the main nozzle and tighten it just snug with a screwdriver.
4. Slide a new needle valve into the groove of the float arm.
5. Lower the float arm into position with the needle valve sliding into the needle valve seat. Now, push the float pin through the holes in the carburetor

FUEL SYSTEM 3-21

body and hinge using a small awl or similar tool.

6. Hold the carburetor body in a perfect upright position. Check the float hinge adjustment. Measurement should be of the total height between the float bowl gasket surface and the bottom of the float (well, it's the normally the bottom, but since the carburetor is inverted it is on TOP right now).

Carefully, bend the hinge tab, if necessary, to achieve the required measurement.

■ **For specifications, please refer to the Carburetor Set-Up Specifications chart in this section.**

7. Position a new float bowl gasket in place on the carburetor body. Install the float into the float bowl. Place the float bowl in position on the carburetor body, and then secure it with the two Phillips head screws.

8. If the E-clip on the jet needle is lowered, the carburetor will cause the powerhead to operate rich. Raising the E-clip will cause the powerhead to operate lean. At higher altitude, raise E-clip to compensate for rarefied air. Begin to assemble the throttle valve components by inserting the E-clip end of the jet needle into the throttle valve (the end with the recess for the throttle cable end). Next, place the needle retainer into the throttle valve over the E-clip and align the retainer slot with the slot in the throttle valve.

9. Thread the spring over the end of the throttle cable and insert the cable into the retainer end of the throttle valve. Compress the spring and at the same time, guide the cable end through the slot until the end locks into place in the recess.

10. Position the assembled throttle valve in such a manner to permit the slot to slide over the alignment pin while the throttle valve is lowered into the carburetor. This alignment pin permits the throttle valve to only be installed

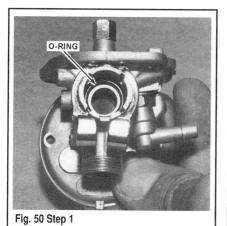
Fig. 50 Step 1

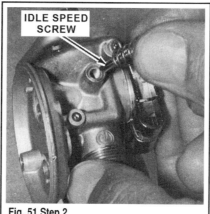
Fig. 51 Step 2

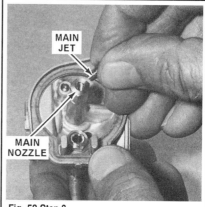
Fig. 52 Step 3

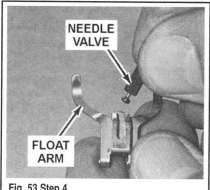
Fig. 53 Step 4

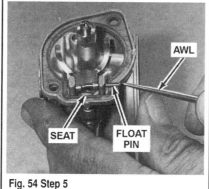
Fig. 54 Step 5

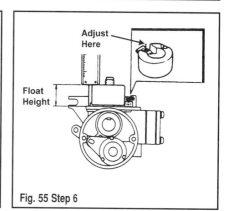
Fig. 55 Step 6

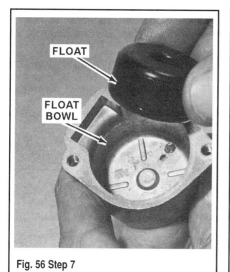
Fig. 56 Step 7

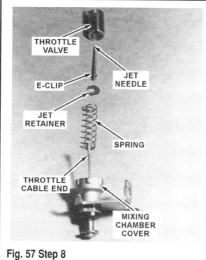
Fig. 57 Step 8

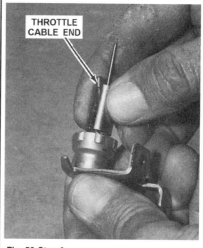
Fig. 58 Step 9

3-22 FUEL SYSTEM

one way in the correct position. Carefully tighten the mixing chamber cover with a pair of pliers.

11. Slide the pivot pin through the top carburetor bracket and then through the upper hole in the throttle control lever. Secure the pin with the circlip. Slide the spring, then the washer over the screw and align the slot in the control lever with the threaded hole in the top carburetor bracket. Install the screw until the lever moves freely with just enough friction to provide the operator with a feel of the throttle opening.

12. Position the carburetor clamp over the carburetor mounting collar, with the locking bolt and nut in place in the clamp. Slide the carburetor into place on the reed valve housing.

13. Secure the carburetor in place by tightening the bolt and nut securely.

14. Hold the intake silencer up to the carburetor. Connect the white lead (engine stop button lead) to the quick disconnect fitting anchored at the top of the powerhead. Attach the black ground lead to the screw on the horseshoe powerhead bracket.

15. Push the choke knob onto the slotted end of the choke shaft. Place the threaded end through the intake silencer and slide the choke valve, followed by the slotted washer, onto the protruding threaded end past the threads onto the slotted part of the shaft. Position the intake silencer up against the carburetor.

16. Hold the choke knob. Slide the lock washer onto the end of the choke shaft, and then thread on the nut.

■ Starting this nut is not an easy task. Holding the nut with an alligator clip may prove helpful. Tighten the nut just snugly. Connect the fuel hose to the carburetor

17. Install the throttle control knob onto the throttle lever and secure it with the Phillips head screw. Secure the intake silencer to the carburetor with the two Phillips head screws.

Slowly tighten the idle speed screw until it barely seats, then back it out the same number of turns recorded during disassembly. If the number of turns was not recorded, back the screw out 1-3/4 turns as a rough adjustment.

18. Install the two halves of the cowling around the powerhead.
19. Secure the cowling with the attaching screws. Eight screws hold the cowling halves in place plus one more for the spark plug cover.

■ As noted during disassembly, the screws are different lengths. Ensure the proper size is used to the correct location.

Mount the outboard unit in a test tank, or the boat in a body of water, or connect a flush attachment and hose to the lower unit. Start the engine and check the completed work. Allow the powerhead to warm to normal operating temperature. Adjust the idle speed to specification.

3 Hp Models

The carburetor used on 3 hp motors contains an integral diaphragm-displacement fuel pump assembly. Although both are removed or installed from the powerhead as an assembly, either unit can be overhauled without disassembly of the other (although it usually makes sense to just overhaul both at the same time). Moreover, there MAY be sufficient clearance to disassemble the fuel pump with the carburetor still installed on the motor, so pay close attention when removing components for access, if only the fuel pump requires service.

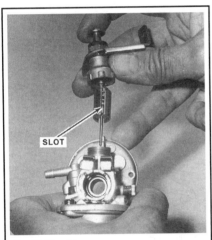

Fig. 59 Step 10

Fig. 60 Step 11

Fig. 61 Step 12

Fig. 62 Step 13

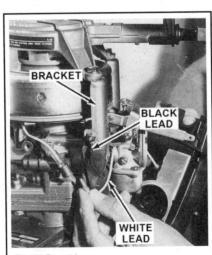

Fig. 63 Step 14

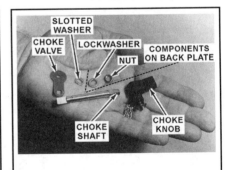

Fig. 64 Step 15

FUEL SYSTEM 3-23

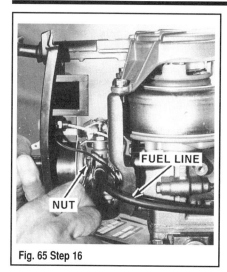

Fig. 65 Step 16

Fig. 66 Step 17

Fig. 67 Step 18

■ This carburetor is very similar (though NOT identical) to the one used on the 6/8 hp and 9.9/15 hp Yamahas. For more illustrations and photos, please refer to the section on that carburetor.

REMOVAL & INSTALLATION

◆ See Figure 71

1. Remove the flywheel for access.
2. Disconnect the choke link-rods and throttle cable from the carburetor.
3. Disconnect the fuel hose from the carburetor assembly.
4. Support the carburetor assembly and remove the 2 bolts from the face of the air intake silencer which secure the silencer and carburetor to the powerhead.
5. Carefully remove the air intake silencer and position aside, then pull the carburetor outward from the powerhead.
6. Remove and discard the old carburetor mounting gasket.

7. Clean and inspect and/or overhaul the carburetor, as applicable.

To Install:

8. Position a new gasket by aligning the bolt holes; then install the carburetor and air intake silencer assembly. It is sometimes easier to connect the carb and silencer, pushing the mounting bolts through so they protrude from the powerhead side of the carb so you can put the gasket in position over the bolts ends. In this case, you then position the entire assembly and thread the bolts into the intake manifold. Either way, make sure the gasket is properly aligned, then tighten the 2 retaining bolts securely.
9. Reconnect the fuel hose to the carburetor, making sure to fit the hose clip over the nipple to prevent possible leaks.
10. Connect the throttle cable and choke link-rods to the carburetor.
11. Refer to the Timing and Synchronization adjustments in the Maintenance and Tune-Up section to make sure the Idle Speed and Throttle Cable are both properly adjusted.

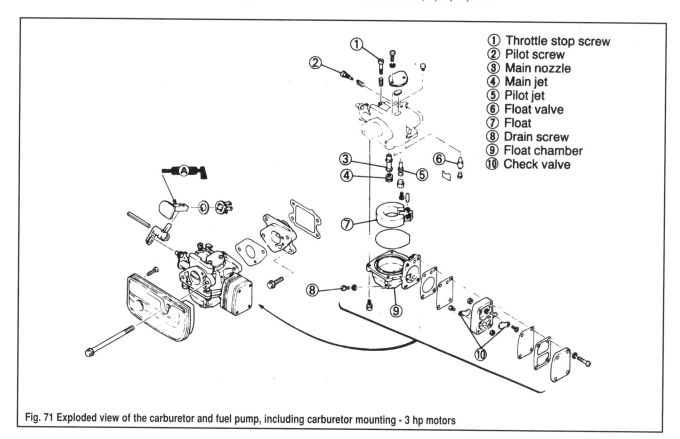

Fig. 71 Exploded view of the carburetor and fuel pump, including carburetor mounting - 3 hp motors

3-24 FUEL SYSTEM

OVERHAUL

◆ See Figure 71

Good shop practice dictates purchasing a carburetor repair kit and using new parts any time the carburetor is disassembled. Make an attempt to keep the work area organized and to cover parts after they have been cleaned. This practice will prevent foreign matter from entering passageways or adhering to critical parts.

1. Remove the carburetor from the powerhead, as detailed earlier in this section.
2. If not done already, remove the drain screw from the bottom of the carburetor float bowl and carefully drain the bowl of fuel.
3. If the fuel pump requires service, disassemble it as follows:
   a. Loosen and remove the 4 pump cover retaining bolts. Remove the cover and fuel pump body assembly from the carburetor body.
   b. The fuel pump is a sandwich consisting of 2 component sets, whose relative positions are the same on either side of the pump body. Remove the components in order as follows (paying close attention to their orientation), cover, outer gasket, outer diaphragm, pump body, inner diaphragm and inner gasket.
   c. There are 2 check valves installed on the pump body (one on either side), if necessary for replacement loosen the retaining screw (and nut which is on the opposite of the pump body from the screw), then remove the valve from the body. Again, pay close attention to valve orientation for installation purposes.
   d. Clean and inspect all components, at a minimum the gaskets and diaphragms should be replaced. The check valves should be replaced if they've lost their strengths and/or show signs of damage.
4. Loosen and remove the 2 top cover screws; then remove the top cover plate from the carburetor body. Remove and discard the old packing from under the cover plate.
5. Remove the pilot screw from the side of the carburetor body.
6. Invert the carburetor body (with the float bowl facing upward), then loosen the float bowl retaining screws. Remove the float bowl from the carburetor body, then remove and discard the old bowl O-ring.
7. Remove the screw, then grasp the end of the float hinge pin and carefully pull it free of the carburetor body and float hinge.
8. Lift gently upward to remove the float and needle valve from the carburetor body.
9. Remove the bushing, then slowly unthread the pilot jet
10. Carefully unthread the main jet, then remove main nozzle from the carburetor body.
11. Clean and inspect all components as detailed in this section. Replace any damaged, worn or defective components. Discard all O-rings or gaskets.

To Assemble:

12. Install the main nozzle, then gently screw the main jet into position until tight.
13. Install the pilot jet until seated and then install the bushing.
14. Connect the needle valve to the float, then lower the float and valve into position. Insert the hinge pin through the float and carburetor body, then install and tighten the screw.
15. With the carburetor still inverted, so the float is sitting gently on the needle valve (which is resting on the seat) measure the float height from the carburetor body-to-float bowl mating surface up to the top (actually the bottom, but it is on top now) of the float. Refer to the Carburetor Set-Up Specifications chart in this section for proper float drop/height specs. If necessary, gently bend the float hinge to achieve the proper measurement.

✳✳ WARNING

When measuring the float height DO NOT place any pressure downward on the needle valve or you could damage it (and/or you may make an incorrect adjustment).

16. Double-check that the float and valve move smoothly without sticking or binding.
17. Install the float bowl to the carburetor body using a new O-ring, then gently secure using the retaining screws. If not done already, install the drain screw to the bowl.
18. Invert the carb, then install the carburetor top cover plate to the top of the body using new packing. Tighten the screws securely.
19. Install the pilot screw until is JUST gently seats, then back it out about 1 1/4 turns as an initial low speed setting.

20. If the fuel pump was disassembled, install the components as follows:
   a. If removed, install the check valves to the pump body and secure using the bolt and nut for each. Be sure to position the valves as noted during removal.
   b. Install the NEW outer diaphragm and NEW outer gasket and pump cover to the pump body, making sure everything is aligned as noted during removal. Insert two of the retaining screws through the pump cover and body to hold everything in alignment.
   c. Install a NEW inner diaphragm and NEW inner gasket to the back of the pump body (over the 2 retaining screws), then carefully lift the assembly and install it to the carburetor by finger-tightening the 2 screws. Install the other 2 screws, then securely tighten all of the pump screws.
21. Install the carburetor and adjust for proper operation.

CLEANING & INSPECTION

◆ See Figure 71

✳✳ CAUTION

Never dip rubber or plastic parts in carburetor cleaner. These parts should be cleaned only in solvent, and then blown dry with compressed air.

Place all metal parts in a screen-type tray and dip them in carburetor cleaner until they appear completely clean, then blow them dry with compressed air.

Blow out all passages in the castings with low-pressure compressed air. Check all parts and passages to be sure they are not clogged or contain any deposits. Never use a piece of wire or any type of pointed instrument to clean drilled passages or calibrated holes in a carburetor.

Move the throttle shaft back and forth to check for wear. If the shaft appears to be too loose, replace the complete carburetor body because individual replacement parts are not available.

Inspect the main body, airhorn, and venturi cluster gasket surfaces for cracks and burrs which might cause a leak.

Check the float for deterioration. If any part of the float is damaged, the unit must be replaced.

Inspect the tapered section of the float needle replace if it has developed a groove or is no longer evenly tapered.

As previously mentioned, most of the parts which should be replaced during a carburetor overhaul are included in overhaul kits available from your local marine dealer. Replace all components included in the kit for durability.

4/5 Hp (83cc and 103cc) Models

◆ See Figure 75

The carburetor used on 4/5 hp (83cc and 103cc) motors contains an integral fuel pump. Although both are removed or installed from the powerhead as an assembly, either unit can be overhauled without disassembly of the other (although it usually makes sense to just overhaul both at the same time). Moreover, there MAY be sufficient clearance to disassemble the fuel pump with the carburetor still installed on the motor, so pay close attention when removing components for access, if only the fuel pump requires service.

REMOVAL

◆ See Figures 76 thru 80

1. If equipped, a 4 hp model, turn off the fuel supply at the base of the fuel tank by turning the fuel knob to the **OFF** position. If servicing a 5 hp model, disconnect the fuel joint from the fuel tank. Protect the disconnected ends from contamination.
2. Squeeze the wire type hose clamp on the fuel line, and then pull the hose free of the fuel inlet fitting.

If servicing a 4 hp model, plug the fuel line with a screw and slip the hose clamp in place. Tape the line to the tank to prevent loss of fuel. If servicing a 5 hp model, allow the excess fuel in the line to drain into a cloth, to prevent spilling fuel into the lower cowling.

FUEL SYSTEM 3-25

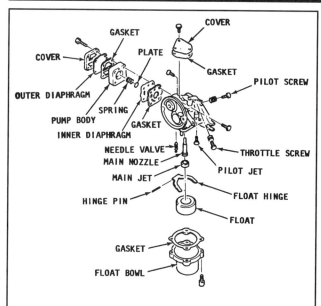

Fig. 75 Exploded view of carburetor with major parts identified - 4/5 hp (83cc and 103cc) models

3. On both the 4 hp and 5 hp models, loosen the throttle wire retaining screw and pull the wire free of the brass barrel.

4. Remove the two screws securing the silencer to the carburetor body and move the silencer to one side out of the way. Pry the choke rod from the carburetor linkage using a small blade screwdriver.

5. Remove the starboard side carburetor mounting nut. Loosen but do not remove the port side mounting nut. Grasp the carburetor and pull it forward to clear the starboard stud, and then slide it to the left for the slot in the mounting flange to clear the port stud. Remove and discard the carburetor mounting gasket.

DISASSEMBLY

DIFFICULT

◆ See Figures 75 and 81 thru 89

1. Remove the four screws securing the float bowl to the carburetor. Remove the float bowl. Lift off and discard the float bowl gasket.

2. Push out the hinge pin using an awl. Lift out the float. The needle valve will come out with the float.

3. Slide the needle valve free of the slot in the float.

4. Remove the main jet from the turret of the float bowl.

5. Unscrew and remove the main nozzle from deep inside the turret.

6. Remove the three screws securing the top cover. Remove the top cover and the gasket.

7. Back out the pilot screw and pilot jet. The number of turns need not

Fig. 76 Step 1

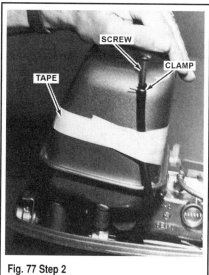

Fig. 77 Step 2

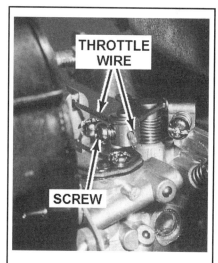

Fig. 78 Step 3

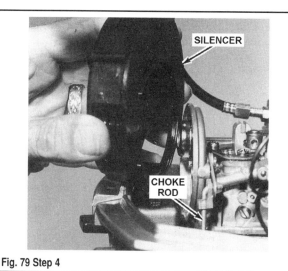

Fig. 79 Step 4

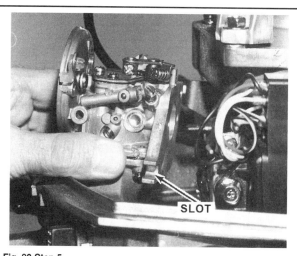

Fig. 80 Step 5

3-26 FUEL SYSTEM

be recorded because during installation a definite number of turns from the lightly seated position will be listed. Slide the pilot screw spring free.

8. Remove the four screws securing the fuel pump to the carburetor body. If the fuel pump gaskets and diaphragms are to be used again (and if you've come this far they should usually be replaced), carefully disassemble the fuel pump. Use care during disassembly and note the order of the gaskets and diaphragms. The membranes are fragile and may be easily punctured or stretched. Disassembly involves removing first, the cover, then the outer diaphragm, and finally the cover gasket.

9. Remove the fuel pump body. Take care not to lose the spring and plate as they are held under tension by the pump body. Remove the inner diaphragm, and then the pump body gasket.

CLEANING & INSPECTION

◆ See Figures 75 and 90 thru 93

✳✳ CAUTION

Never dip rubber parts, plastic parts, diaphragms, or pump plungers in carburetor cleaner. These parts should be cleaned only in solvent, and then blown dry with compressed air.

Fig. 81 Step 1

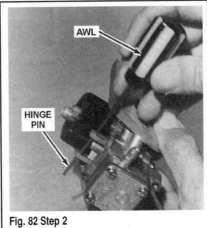

Fig. 82 Step 2

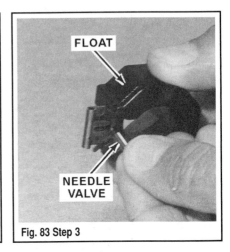

Fig. 83 Step 3

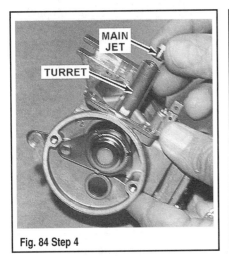

Fig. 84 Step 4

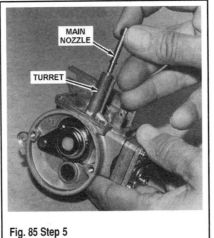

Fig. 85 Step 5

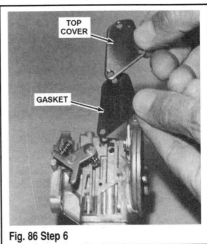

Fig. 86 Step 6

Fig. 87 Step 7

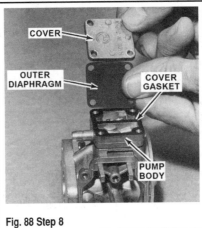

Fig. 88 Step 8

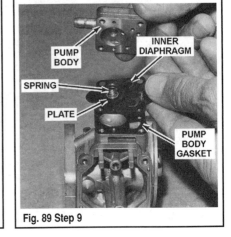

Fig. 89 Step 9

FUEL SYSTEM

Place all the metal parts in a screen-type tray and dip them in carburetor cleaner until they appear completely clean, free of gum and varnish which accumulates from stale fuel. Blow the parts dry with compressed air.

Blow out all passageways in the castings with compressed air. Check all of the parts and passages to be sure they are not clogged or contain any deposits. Never use a piece of wire or any type of pointed instrument to clean drilled passages or calibrated holes in a carburetor.

Make a thorough inspection of the fuel pump diaphragms for the tiniest pin hole. If one is discovered, the hole will only get bigger. Therefore, the diaphragms must be replaced in order to obtain full performance from the powerhead.

Carefully inspect the casting for cracks, stripped threads, or plugs for any sign of leakage. Inspect the float hinge in the hinge pin area for wear and the float for any sign of leakage.

Examine the inlet needle for wear and if there is any evidence of wear, the inlet needle must be replaced.

Always replace any and all worn parts.

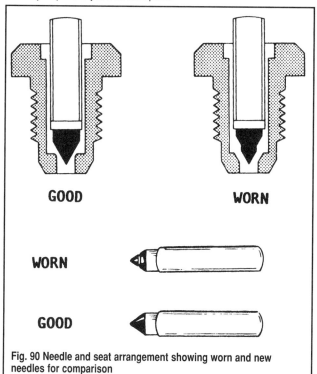

Fig. 90 Needle and seat arrangement showing worn and new needles for comparison

ASSEMBLY & INSTALLATION

◆ See Figures 75 and 94 thru 107 — DIFFICULT

1. Position first, the pump body gasket, then the inner diaphragm, next the plate, then the spring onto the carburetor body in the order given. The plate and spring must index with the round hole on the underside of the pump body. When the pump body is placed over the spring, ensure all the parts are aligned and the spring seats in an upright position between the plate and the pump body. Hold it all together for the next step.

2. Place the cover gasket, the outer diaphragm and the cover over the pump body. Visually check to be sure the holes are aligned before threading the screws through the many layers of fuel pump parts. If the screws are forced through, one of the fragile parts may be damaged. Secure the parts together with the four attaching screws.

3. Install the pilot jet snugly. Install the pilot screw and spring. Slowly rotate the pilot screw into the carburetor body until it barely seats. From this position, back it out the appropriate number of turns (refer to the Carburetor Set-Up Specifications chart in this section, but the spec is about 1-3/4 turns for 83cc motors or about 1-1/2 turns for 103cc motors). Fine idle adjustment will be made later using the throttle stop screw.

4. Install the gasket and top carburetor cover.

5. Thread the main nozzle into the float bowl turret and tighten it just snug.

6. Install the main jet over the main nozzle and tighten it snugly.

7. Slide the needle valve into the slot of the float.

8. Lower the float and needle valve assembly over the turret, engaging the needle valve into the needle seat. Position the float hinge between the mounting posts and install the hinge pin securing the float in place.

9. Hold the carburetor inverted, as shown. Measure the distance between the carburetor body to the top of the float. The float height should be about 7/8 in. or a little more than 22mm (check the Carburetor Set-Up Specifications chart in this section for details). This dimension, with the carburetor inverted, places the lower surface of the float parallel to the carburetor body. If necessary, carefully bend the tab on the float to obtain the correct measurement. Install the float bowl to the carburetor body and secure it with the four screws.

10. Install the carburetor mounting gasket. Slide the port side slotted end of the carburetor mounting flange onto the stud first. The nut and washer should not have been removed during disassembly. Move the starboard side of the carburetor mounting flange onto the left stud. Install the left nut and washer, and then tighten both nuts securely.

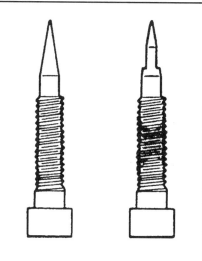

Fig. 91 Comparison of a new (left) and a worn (right) pilot screw. Note the ridge which has developed on the worn screw

Fig. 92 A carburetor service kit is available at modest cost from the local marine dealer. The kit will contain all necessary parts to perform carburetor overhaul work

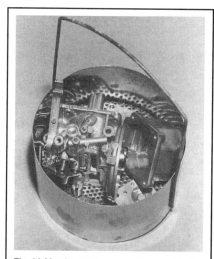

Fig. 93 Metal parts from our disassembled 4/5 hp (83cc and 103cc) motor carburetor in a basket ready to be immersed in carburetor cleaner

3-28 FUEL SYSTEM

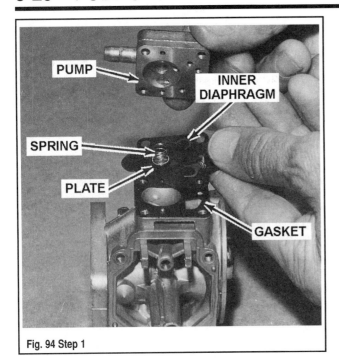

Fig. 94 Step 1

11. Snap the choke link into place. Connect the fuel line to the fuel pump inlet fitting. Install the silencer to the carburetor body and secure it in place with the two screws.

12. Thread the throttle wire through the holder first, and then through the brass barrel. Tighten the screw to retain the wire. The wire should project out from the barrel approximately 0.12-0.16 in. (3-4mm) however, you'll adjust this when your finished, so it is not critical.

13. If servicing a 4 hp model, open the fuel valve at the base of the tank by turning the fuel knob to the **ON** position. If servicing a 5 hp model, connect the fuel line from the tank to the carburetor at the fuel joint.

14. Refer to the Timing and Synchronization adjustments in the Maintenance and Tune-Up section to make sure the Idle Speed and Throttle Cable are both properly adjusted.

6/8 Hp and 9.9/15 Hp Models

◆ See Figure 119

The carburetor used on 6/8 hp and 9.9/15 hp motors contains an integral fuel pump. Although both are removed or installed from the powerhead as an assembly, either unit can be overhauled without disassembly of the other (although it usually makes sense to just overhaul both at the same time). Moreover, there MAY be sufficient clearance to disassemble the fuel pump with the carburetor still installed on the motor, so pay close attention when removing components for access, if only the fuel pump requires service.

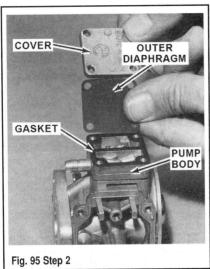

Fig. 95 Step 2

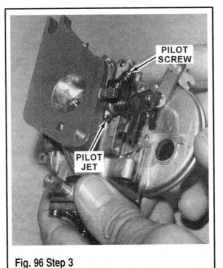

Fig. 96 Step 3

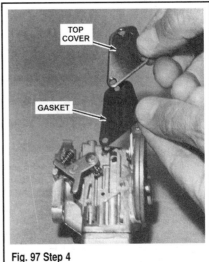

Fig. 97 Step 4

Fig. 98 Step 5

Fig. 99 Step 6

Fig. 100 Step 7

FUEL SYSTEM 3-29

REMOVAL

◆ See Figures 108 thru 116

1. Disconnect the fuel hose at the fuel joint. Protect the ends from contamination. Disconnect the fuel line at the fuel pump inlet fitting.

2. On 6/8 hp model only, pry the small choke link from the plastic fitting on the carburetor using a small slotted screwdriver.

3. Remove the three Phillips screws retaining the front panel to the lower cowling. The panel must be removed to gain access to the carburetor.

4. Pull the panel forward and set it to one side out of the way. It is not necessary to disconnect the kill switch harness or the fuel connection.

5. On 6/8 hp model only, pry the throttle link from the starboard side of the carburetor with a small slotted screwdriver. On the 9.9/15 hp units a throttle roller moving on the throttle cam is used instead of the throttle link. The roller arrangement cannot be disconnected from the carburetor.

6. It should be noted that the 9.9/15 hp carburetor has a different silencer and choke setup than the other units.

7. It should also be noted that the 9.9/15 hp carburetor has a unique linkage arrangement.

Fig. 101 Step 8

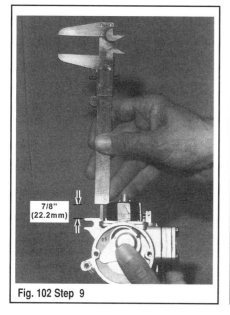

Fig. 102 Step 9

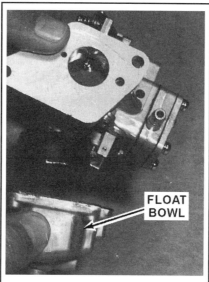

Fig. 103 Step 10

Fig. 104 Step 11

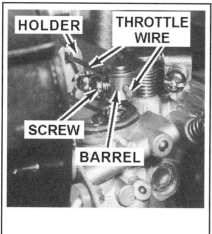

Fig. 105 Step 12

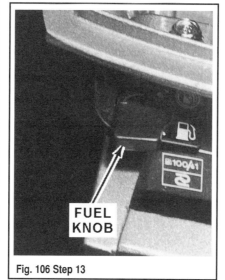

Fig. 106 Step 13

Fig. 107 Step 14

3-30 FUEL SYSTEM

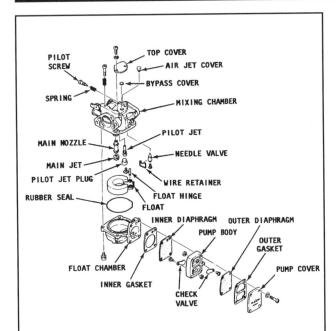

Fig. 108 Exploded view of the carburetor assembly used on the 6/8 hp motor (note that the and 9.9/15 hp carburetor is ALMOST identical, however top cover and float shapes will vary slightly)

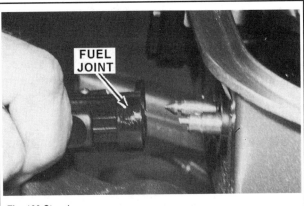

Fig. 109 Step 1

8. Reach through the opening of the front panel and remove the two long bolts which secure both the silencer and the carburetor to the powerhead. Remove the silencer. On the 9.9/15 hp model the choke rod will slide out of the choke lever and the choke roller will slide out of the choke shaft cradle. Remove the carburetor. Remove and discard the carburetor mounting gasket.

DISASSEMBLY

◆ See Figures 108 and 117 thru 123

1. Remove the pilot screw and spring. The number of turns out from a lightly seated position will be given during assembling.

Remove the two screws securing the top cover and lift off the cover. Gently pry the two rubber plugs out with an awl. Remove the oval air jet cover and the round bypass cover.

2. Remove the four screws securing the fuel pump to the carburetor body. Disassemble the pump cover, the outer gasket, the outer diaphragm, the pump body, the inner diaphragm, and finally the inner gasket in that specific order.

3. Remove the screws from both sides of the fuel pump body. Remove the check valves.

4. Remove the four screws securing the float bowl cover in place. Lift off the float bowl. Remove and discard the rubber sealing ring.

5. Remove the small Phillips screw securing the float hinge to the mounting posts. Lift out the float, the hinge pin and the needle valve. Slide the hinge pin free of the float.

6. Slide the wire attaching the needle valve to the float free of the tab.

7. Use the proper size slotted screwdriver and remove the main jet, then unscrew the main nozzle from beneath the main jet. Remove the plug, and then unscrew the pilot jet located beneath the plug.

CLEANING & INSPECTION

◆ See Figures 90, 91, 108, 124 and 125

Inspect the check valves in the fuel pump for varnish build up as well as any deformity.

Never dip rubber parts, plastic parts, diaphragms, or pump plungers in carburetor cleaner. These parts should be cleaned only in solvent, and then blown dry with compressed air..

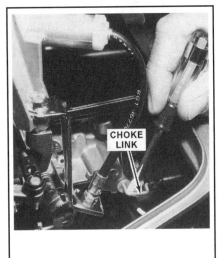

Fig. 110 Step 2

Fig. 111 Step 3

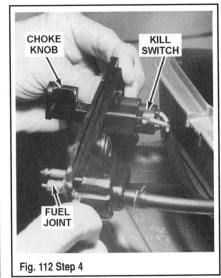

Fig. 112 Step 4

FUEL SYSTEM 3-31

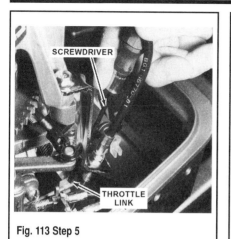
Fig. 113 Step 5

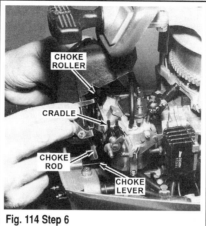
Fig. 114 Step 6

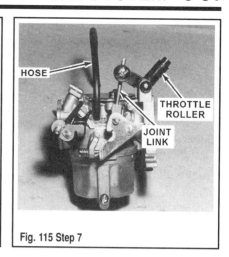
Fig. 115 Step 7

Fig. 116 Step 8

Place all metal parts in a screen-type tray and dip them in carburetor cleaner until they appear completely clean, then blow them dry with compressed air.

Blow out all passages in the castings with compressed air. Check all parts and passages to be sure they are not clogged or contain any deposits. Never use a piece of wire or any type of pointed instrument to clean drilled passages or calibrated holes in a carburetor.

Move the throttle shaft back and forth to check for wear. If the shaft appears to be too loose, replace the complete throttle body because individual replacement parts are not available.

Inspect the main body, airhorn, and venturi cluster gasket surfaces for cracks and burrs which might cause a leak. Check the float for deterioration. Check to be sure the float spring has not been stretched. If any part of the float is damaged, the unit must be replaced. Check the float arm needle contacting surface and replace the float if this surface has a groove worn in it.

Inspect the tapered section of the idle adjusting needles and replace any that have developed a groove.

As previously mentioned, most of the parts which should be replaced during a carburetor overhaul are included in overhaul kits available from your local marine dealer. One of these kits will contain a matched fuel inlet needle and seat. This combination should be replaced each time the carburetor is disassembled as a precaution against leakage.

Fig. 117 Step 1

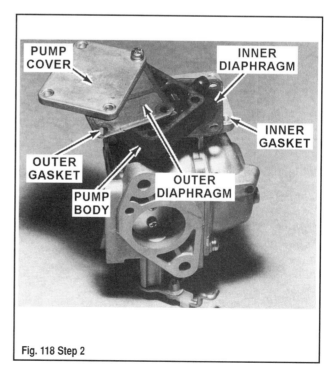
Fig. 118 Step 2

3-32 FUEL SYSTEM

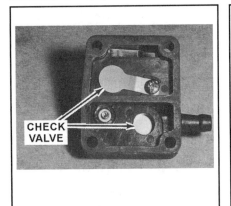

Fig. 119 Step 3

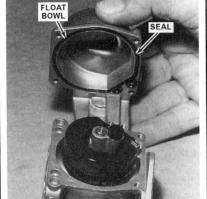

Fig. 120 Step 4

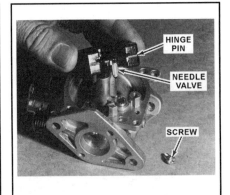

Fig. 121 Step 5

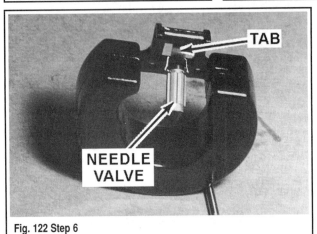

Fig. 122 Step 6

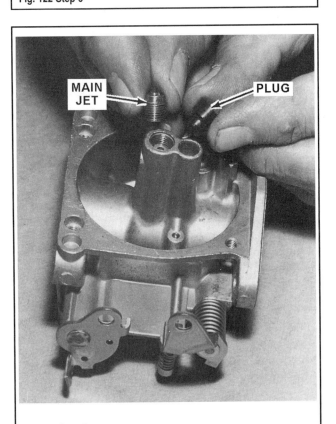

Fig. 123 Step 7

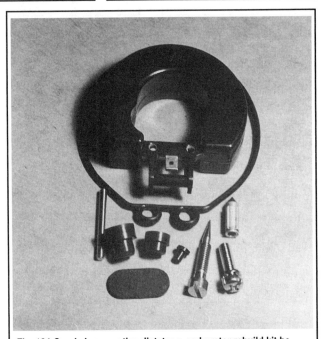

Fig. 124 Good shop practice dictates a carburetor rebuild kit be purchased and new parts, especially gaskets and O-rings be installed any time the carburetor is disassembled. This photo includes parts in a repair kit for the 6/8 hp and 9.9/15 hp carburetor (note shape of float may differ on some models)

ASSEMBLY

◆ See Figures 108 and 126 thru 128

1. Install the main nozzle into the center hole and tighten it snugly. Install the main jet on top of the nozzle and tighten it snugly also. Install the pilot jet into the center hole and then the plug. Tighten the jet and the plug securely.

2. For the 6/8 hp units, slide the wire attached to the needle valve onto the float tab. For the 9.9/15 hp units, insert the needle valve into the needle seat.

3. For the 6/8 hp units, slide the hinge pin through the float hinge. Lower the float and needle assembly down into the float chamber and guide the needle valve into the needle seat. Check to be sure the hinge pin indexes into the mounting posts. Secure the pin in place with the small Phillips screw.

For the 9.9/15 hp units, place the float hinge arm between the mounting posts and secure the arm in place with the hinge pin.

FUEL SYSTEM 3-33

Fig. 125 Remove all rubber and plastic parts before immersing metal parts of the 6/8 hp and 9.9/15 hp carburetor in cleaning solution

FLOAT ADJUSTMENT

◆ See Figures 108 and 129 thru 134  MODERATE

1. For the 6/8 hp units, invert the carburetor and allow the float to rest on the needle valve. Measure the distance between the top of the float and the mixing chamber housing. This distance should be 0.47-0.63 in. (12-16mm). This dimension, with the carburetor inverted, places the lower surface of the float parallel to the carburetor body. If the dimension is not within the limits listed, the needle valve must be replaced.

For the 9.9/15 hp units, invert the carburetor and install the float bowl gasket. Measure the distance from top surface of the gasket to the top of the hinge, as shown in the accompanying illustration. The measurement varies with year and model, for details please refer to the Carburetor Set-Up Specifications chart. If adjustment is necessary, carefully bend the float arm, as required, to obtain a satisfactory measurement. Install the float.

2. Insert the rubber sealing ring into the groove in the float bowl. Install the float bowl and secure it in place with the four Phillips screws.

3. Place the check valves, one at a time, in position on both sides of the fuel pump body. Secure each valve with the attaching screw.

4. Assemble the fuel pump components onto the carburetor body in the following order, the inner gasket, the inner diaphragm, the pump body, the outer diaphragm, the outer gasket, and finally the pump cover. Check to be sure all the parts are properly aligned with the mounting holes. Secure it all in place with the four attaching screws.

5. Install the oval air jet cover and the round bypass cover in their proper recesses. Place the top cover over them, no gasket is used. Install and tighten the two attaching screws.

6. Slide the spring over the pilot screw, and then install the screw. Tighten the screw until it barely seats, then carefully back the screw out the number of turns listed in the Carburetor Set-Up Specifications chart in this section.

7. Prior to installation the hose and joint link are connected on the 9.9/15 hp carburetors.

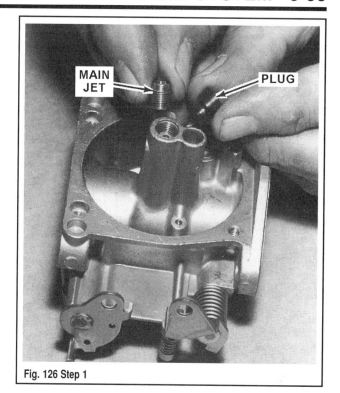

Fig. 126 Step 1

Fig. 127 Step 2

Fig. 128 Step 3

3-34 FUEL SYSTEM

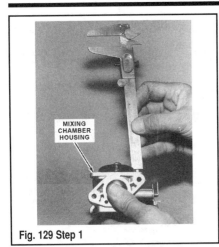

Fig. 129 Step 1

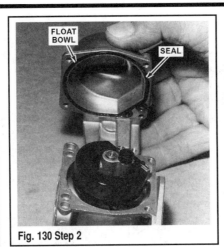

Fig. 130 Step 2

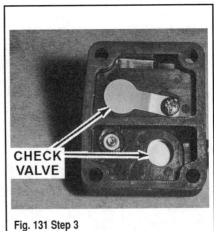

Fig. 131 Step 3

Fig. 132 Step 4

Fig. 133 Step 5

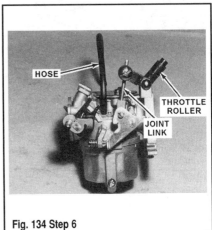

Fig. 134 Step 6

INSTALLATION

◆ See Figures 135 thru 140

 MODERATE

1. Place a new carburetor mounting gasket onto the two pins on the intake manifold.
   For the 6/8 hp units, align the carburetor and the silencer with the two mounting holes. Start the bolts, and then tighten them, through the opening in the front panel, to a torque value of 5.8 ft. lbs. (8 Nm).

For the 9.9/15 hp units, move the carburetor into place. Guide the choke rod through the slot in the choke lever and at the same time guide the throttle roller into the throttle shaft cradle. Start the mounting nuts, and then tighten them, through the opening in the front panel, to a torque value of 5.8 ft. lbs. (8 Nm).

2. For the 6/8 hp units, snap the throttle link into the plastic fitting on the starboard side of the carburetor.

3. Install the front panel and secure it in place with the three Phillips screws.

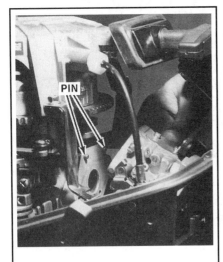

Fig. 135 Step 1

Fig. 136 Step 2

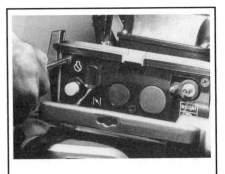

Fig. 137 Step 3

FUEL SYSTEM 3-35

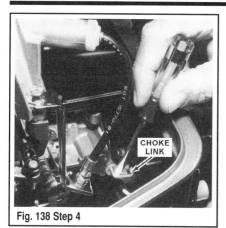
Fig. 138 Step 4

Fig. 139 Step 5

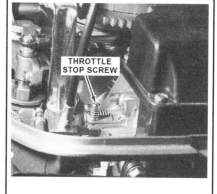
Fig. 140 Step 6

4. Snap the small choke link into the plastic fitting on the starboard side of the carburetor. Check the action of the choke knob to be sure there is no evidence of binding.
5. Connect the fuel line to the fuel pump. Connect the fuel line from the tank with the fuel joint.
6. Mount the outboard unit in a test tank, on the boat in a body of water, or connect a flush attachment and hose to the lower unit. Connect a tachometer to the powerhead.

Start the engine and check the completed work. Remember, the powerhead will not start without the emergency tether in place behind the kill switch knob. Allow the powerhead to warm to normal operating temperature. Adjust the throttle stop screw until the powerhead idles at specification.

20-48 Hp (2-Cylinder) Models

The procedures in this section are for 20-48 hp (2-cylinder) models, most of which are equipped with a separate powerhead mounted diaphragm-displacement fuel pump. However, some smaller models are may be equipped with a carburetor integrated fuel pump. On those models with a carburetor integrated fuel pump, please refer to the carburetor procedure for 6/8 hp and 9.9/15 hp Models when it comes to fuel pump components.

The carburetors use on the different 20-48 hp (2-cylinder) motor vary somewhat from motor-to-motor, so much so that an exact depiction of each possible carburetor is not possible here. However, the same basic components are contained within each and service methods are nearly identical. Refer to the exploded views as you work to find the carburetor which most closely matches your model. Just take your time and arrange or tag/label all components (matching them with the exploded views) as they are removed to insure proper assembly or installation.

REMOVAL

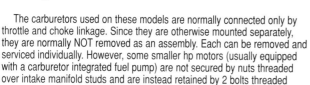

◆ See Figures 141 thru 144

The carburetors used on these models are normally connected only by throttle and choke linkage. Since they are otherwise mounted separately, they are normally NOT removed as an assembly. Each can be removed and serviced individually. However, some smaller hp motors (usually equipped with a carburetor integrated fuel pump) are not secured by nuts threaded over intake manifold studs and are instead retained by 2 bolts threaded through the air intake silencer housing. On these models it is usually easier to completely disconnect the linkage/hoses from BOTH carburetor and remove them both at the same time, using the bolts and the air intake silencer housing to hold them together.

1. For some models equipped with a hand-rewind starter (including the 48C or 48MH) it is much easier to access components if the manual starter assembly is unbolted and removed from the powerhead. If applicable, determine if this will be beneficial and then remove the hand-rewind starter assembly. For details, please refer to the Hand-Rewind Starter section.
2. On models equipped with an electric choke solenoid, disconnect the solenoid lead at the quick disconnect fitting directly beneath the carburetor. Disconnect the ground lead at the screw on the crankcase. Then disconnect and remove the choke solenoid link from the carburetor.

3. On models equipped with a manual choke, snap the choke link from the plastic fitting at the carburetor.
4. Remove the retaining screws (usually 2-4 screws) securing the air intake silencer cover to the housing on the carburetor. Remove the cover, along with the seal and/or packing (as equipped), taking care not to damage any components which you intend to reuse.
5. For models secured by nuts to intake manifold studs, remove the intake silencer housing bolts and separate the housing from the carburetors. Remove the gasket or seal(s).

■ For nut and stud mounted carburetors, although it is not absolutely necessary to remove the silencer housing at this point, it will usually provide better access to the carburetor linkage and/or hoses. Obviously, you don't want to do this yet on carbs that are secured by the same housing bolts, as the carbs would come loose from the powerhead before you've disconnected all the lines/linkage.

6. Disconnect the throttle linkage. Carefully pull the link free of one or more of the carburetor ball studs.
7. Disconnect the choke link.

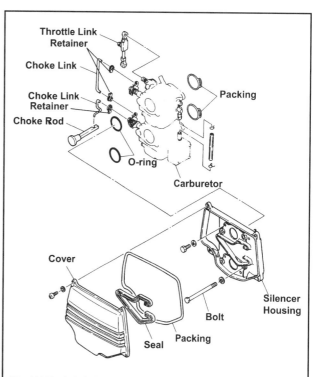
Fig. 141 Exploded view of a typical 2-cylinder carburetor mounting - 20 hp and larger models equipped with a carburetor integrated fuel pump

3-36 FUEL SYSTEM

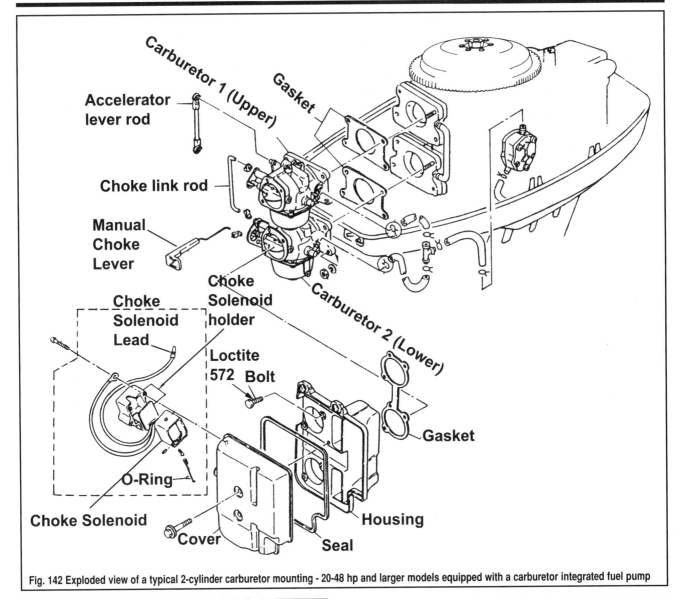

Fig. 142 Exploded view of a typical 2-cylinder carburetor mounting - 20-48 hp and larger models equipped with a carburetor integrated fuel pump

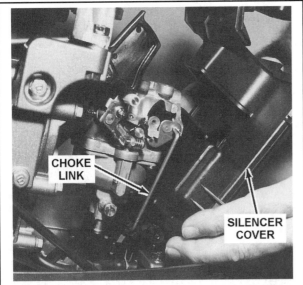

Fig. 143 On all models, you'll have to remove the air intake silencer for access to the carburetor, linkage, hoses and fasteners...

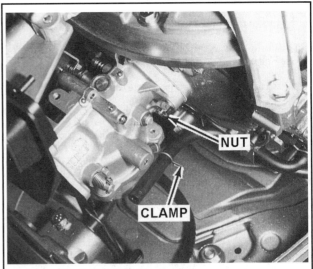

Fig. 144 ...most carburetors are mounted over intake manifold studs on the powerhead, for these models loosen and remove the retaining nuts to pull them free from the studs

FUEL SYSTEM 3-37

8. Disconnect the fuel line(s) at the carburetors. Most models utilize spring-type clamps on the fuel hoses. To remove them, squeeze the fuel hose clamp ears and move it back on the fuel line. Work the fuel line free of the inlet fitting.

9. For carbs retained by intake manifold studs and nuts, support the carburetor assembly, then remove the nuts and carefully pull the carburetor from the studs. Repeat for the other carburetor (if desired/necessary). Remove and discard the carburetor-to-intake manifold gaskets/packing.

10. For carbs retained by the intake silencer housing bolts, support the assembly and carefully loosen all housing bolts (usually 2 per carburetor). We said LOOSEN and not remove on purpose. Back each bolt out until it is JUST free of the powerhead, but don't remove it, as it is keeping the carburetor in position with relation to the air intake silencer housing. After each bolt is loosened, place some tape over the bolt heads to hold them in the air intake housing, then as the last bolt is loosened, gently tilt the housing downward away from the motor (with the carbs resting against the back of the housing). In this way, remove the housing and carburetors as an assembly, then lift each carb off the housing and the bolts. Remove and discard the old carb-to-intake seal or packing. Locate and inspect the carb-to-air intake housing O-rings.

To Install:

■ O-rings can be reused, as long as they are not cut, worn or otherwise deformed. However, gaskets do not normally fair as well and must be replaced if they are removed once they have been in service.

11. For models which are secured to the intake manifold using the air intake silencer housing bolts, insert the bolts through the housing and tape in place. Then position the housing with the carburetor side facing upward so you can install the carburetors over the bolts. Make sure the O-rings are in position, then carefully place each carburetor over the bolts. Install new carburetor-to-intake manifold packing or seals, then lift the assembly to the powerhead and carefully start the bolts (making sure the seals/packing are not dislodged). Tighten the mounting bolts securely.

12. For models secured using intake manifold studs and nuts, install each carburetor over the mounting studs using a new gasket. Install the nuts and tighten securely.

13. Reconnect the fuel line(s). Make sure all fuel lines are installed securely with the clamp over the raised portion of the nipple.

14. Reconnect the throttle and choke links.

15. For nut and stud mounted carburetors, install the air intake silencer housing using a new gasket or using the appropriate seal(s).

16. Install the air intake silencer cover using the seal and/or packing. Tighten the bolts securely, but do NOT damage the housing.

17. Reconnect the manual or electric choke linkage. For models equipped with an electric choke solenoid reposition the solenoid (if removed) and secure using the retainer(s), then reconnect the wiring.

18. If equipped and removed, install the hand-rewind starter assembly.

19. Refer to the Timing and Synchronization adjustments in the Maintenance and Tune-Up section to make sure the Idle Speed and throttle and/or choke links are all properly adjusted, as applicable.

DISASSEMBLY & ASSEMBLY

◆ See Figures 145 thru 153

Good shop practice dictates purchasing a carburetor repair kit and using new parts any time the carburetor is disassembled. Make an attempt to keep the work area organized and to cover parts after they have been cleaned. This practice will prevent foreign matter from entering passageways or adhering to critical parts.

On those models with a carburetor integrated fuel pump, the entire carburetor does not normally need to be disassembled to service the fuel pump components. For details, please refer to the fuel pump steps of the carburetor Overhaul procedure for 6/8 hp and 9.9/15 hp Models.

1. Remove the carburetor from the powerhead and drain all fuel from it using the float bowl drain screw.

2. If applicable and necessary, loosen and remove the screws securing the top cover to the carburetor, then remove the top cover and discard the gasket/packing.

3. Loosen and remove the pilot screw and spring from the bore in the top of the carburetor body (it's usually a horizontal or diagonal bore into the top of the carburetor, immediately adjacent to the intake manifold flange).

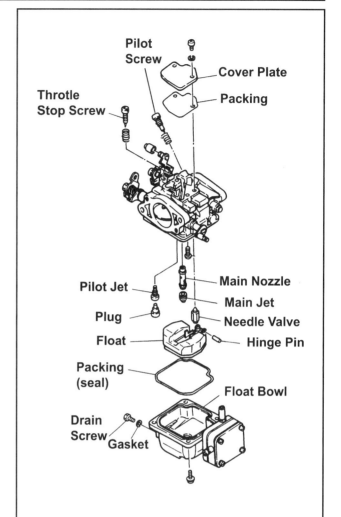

Fig. 145 Exploded view of a typical fuel pump integrated carburetor used on some 20 hp and larger 2-cylinder models

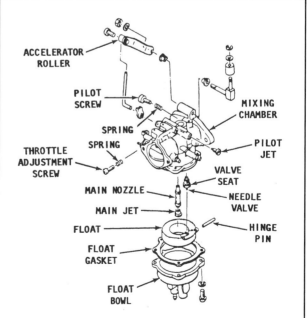

Fig. 146 Exploded view of a typical non-fuel pump integrated carburetor used on 20-48 hp 2-cylinder models

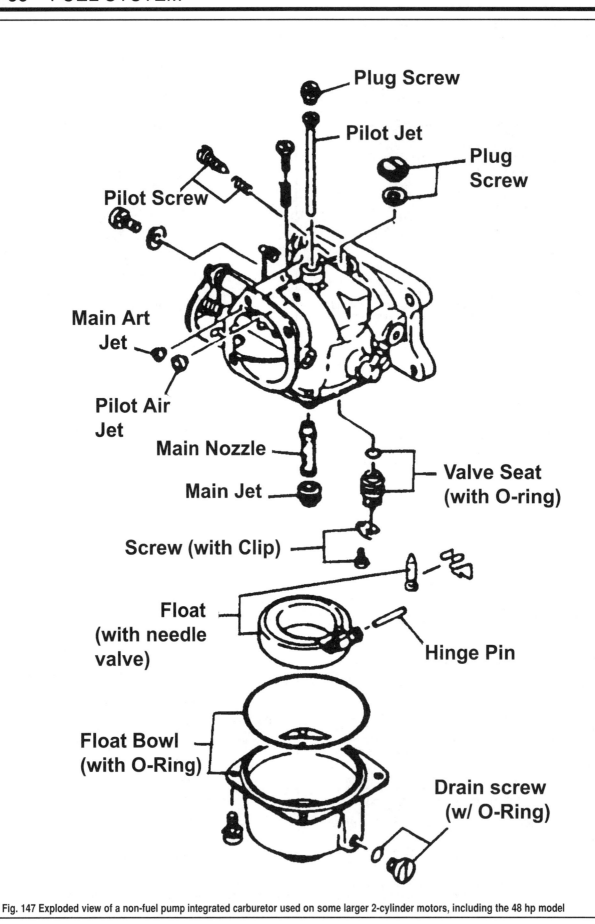

Fig. 147 Exploded view of a non-fuel pump integrated carburetor used on some larger 2-cylinder motors, including the 48 hp model

FUEL SYSTEM 3-39

4. If necessary, loosen and remove the idle adjust (throttle stop) screw and spring.

5. Invert the carburetor body (with the float bowl facing upward), then loosen the float bowl retaining screws (usually 4). Remove the float bowl from the carburetor body, then remove and discard the old bowl gasket/packing (seal).

6. Push the float pin out to one side to separate the float from the carb. Lift gently upward to remove the float and needle valve from the carburetor body.

■ On some models there may be a molded arrow in the cover. This arrow is usually meant to show a preferred direction of movement for the pin (unfortunately sometimes it is the direction of installation and other times it is the direction for removal, so if equipped push gently in the direction of the arrow, and if the pin sticks, try the other direction, don't force it).

7. If equipped with a removable needle valve seat, loosen and remove the seat retaining screw and clip, then remove the valve seat and O-ring. Anytime you disassemble the carburetor it is a good idea to replace the needle valve and seat (if replaceable) to ensure proper float valve operation.

8. Remove the carburetor fuel metering components (main jet and nozzle, plug and pilot jet and, if equipped, the pilot air jet). Refer to the exploded views to help identify each component. Be sure to keep them separated or tagged for installation or replacement purposes.

9. Clean and inspect all components as detailed in this section. Replace any damaged, worn or defective components. Discard all O-rings or gaskets.

To Assemble:

10. Install the fuel metering components (pilot air jet, the pilot jet and pilot jet plug, and then the main nozzle and main jet) to the carburetor body in the positions noted during removal.

11. If removed, install the needle valve seat and O-ring then secure using the clip and retaining screw.

12. Connect the needle valve to the float, then lower the float and valve into position. Insert the hinge pin through the float and carburetor body using a pair of needle-nose pliers and gently tap into position.

13. With the carburetor still inverted and perfectly level, so the float is sitting gently on the needle valve (which is resting on the seat) measure the float height from the carburetor body-to-float bowl mating surface up to the top (actually the bottom, but it is on top now) of the float. Refer to the Carburetor Set-Up Specifications chart in this section for the allowable range. If necessary, gently bend the float hinge to achieve the proper measurement.

✱✱ WARNING

When measuring the float height DO NOT place any pressure downward on the needle valve or you could damage it (and/or you may make an incorrect adjustment).

Fig. 148 The float bowl is normally secured by 4 screws and a gasket. . .

Fig. 149 . . . it must be removed for access to the float assembly and most metering valves

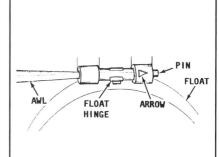

Fig. 150 The float is removed by freeing the hinge pin

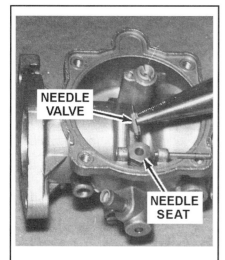

Fig. 151 The float needle valve and valve seat controls fuel flow from the pump

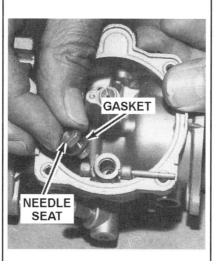

Fig. 152 On some models the needle valve seat is easily replaced

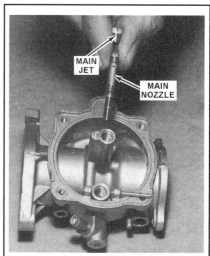

Fig. 153 Metering components (such as jets and nozzles) are installed into the carburetor body

3-40 FUEL SYSTEM

14. Double-check that the float and valve move smoothly without sticking or binding.
15. Install the float bowl to the carburetor body using a NEW gasket/packing (seal), then secure using the retaining screws. If not already done, install the drain plug and O-ring or gasket (as applicable).
16. If removed, install the idle adjust (throttle stop) screw and spring.
17. Install the pilot screw and spring. Slowly rotate the pilot screw into the carburetor body until it barely seats. From this position, back it out the appropriate number of turns (refer to the Carburetor Set-Up Specifications chart in this section).
18. If applicable (and if removed), install the carburetor top cover using a new gasket/packing, then secure using the retaining screws.
19. Install the carburetor and adjust for proper operation.

CLEANING & INSPECTION

 MODERATE

◆ See Figures 90, 91, 145, 147, 154 and 155

✱✱ CAUTION

Never dip rubber parts, plastic parts, diaphragms, or pump plungers in carburetor cleaner. These parts should be cleaned only in solvent, and then blown dry with compressed air.

Place all metal parts in a screen-type tray and dip them in carburetor cleaner until they appear completely clean, then blow them dry with compressed air.

Blow out all passages in the castings with compressed air. Check all parts and passages to be sure they are not clogged or contain any deposits. Never use a piece of wire or any type of pointed instrument to clean drilled passages or calibrated holes in a carburetor.

Move the throttle shaft back and forth to check for wear. If the shaft appears to be too loose, replace the complete throttle body because individual replacement parts are not available.

Inspect the main body, airhorn, and venturi cluster gasket surfaces for cracks and burrs which might cause a leak.

Check the float for deterioration. If any part of the float is damaged, the unit must be replaced. Check the float arm needle contacting surface and replace the float if this surface has a groove worn in it.

Inspect the tapered section of the float needle or pilot screw and replace any that have developed a groove.

■ As previously mentioned, most of the parts which should be replaced during a carburetor overhaul are included in overhaul kits available from your local marine dealer. Some of these kits will contain a matched fuel inlet needle and seat. When possible, this combination should be replaced each time the carburetor is disassembled as a precaution against leakage.

Check the main jet, pilot jet, check valve and main nozzle for signs of dirt or contamination. If they cannot be cleaned, they should be replaced. Again, NEVER clean these components using a wire or any pointer instrument, you'll change the calibration.

Always replace any and all worn parts.

3-Cylinder Powerheads

◆ See Figures 156 and 157

Three carburetors are used on the 3-cylinder outboard powerheads. Complete, detailed and illustrated procedures for these carburetors follow. Removal procedures may vary slightly due to differences in linkage. The carburetor assembly used on 25/30 hp models contains an integral fuel pump on the middle carburetor. On these smaller models cold start enrichment is achieved using choke plates and linkage. The larger 28J-90 hp motors (698cc, 849cc and 1140cc models) utilize a remote mounted fuel pump assembly. On these larger motors, cold start enrichment may take the form of either choke plates/linkage or an enrichment valve. These differences will be noted whenever they make a substantial change in a procedure.

When equipped with an electrothermal valve for cold start enrichment, it is mounted to the middle carburetor assembly and is used to provide additional fuel for cold start enrichment. When cold, the valve contains a piston which is retracted into the valve unblocking an additional fuel circuit in the carburetor (from the float bowl to the throttle bore). However, once the motor

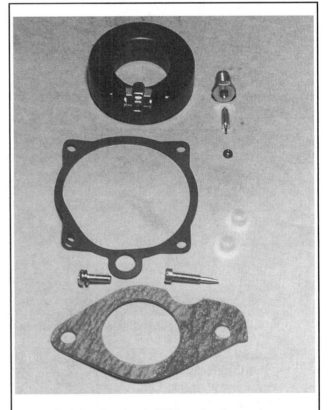

Fig. 154 The mixing chamber of the carburetor used on most larger 2-cylinder motor must not be immersed in cleaning solution. The chamber contains plastic bushings around the throttle shaft. These bushings are pressed into place and are not normally removed

Fig. 155 Typical carburetor rebuild kit used on these motors

FUEL SYSTEM 3-41

is started the power from the stator/charge coil is applied to a heater element in the valve in order to gradually build heat in the valve. This process is timed to take about the same amount of time it should take a cold motor to warm to normal operating temperature. The temperature change causes the valve piston to extend, gradually cutting the enrichment circuit as the engine warms.

REMOVAL & INSTALLATION

25/30 Hp (496cc) Models

◆ See Figures 158 thru 164

MODERATE

The middle carburetor on 25/30 hp motors is equipped with an integrated fuel pump assembly. The pump may be serviced, whether or not the carburetor is removed from the powerhead, by removing the bolts and carefully removing the cover, body and gasket/diaphragm assembly. The carburetors themselves are bolted to the powerhead using the air intake silencer housing/flame arrester assembly and 2 bolts per carburetor. As such, it is easier and safer to remove all 3 carburetors as an assembly, by keeping the bolts inserted through the flame arrester and the carburetor bodies, even after each bolt is unthreaded from the powerhead itself.

1. Loosen and remove the four screws (usually 4) securing the air silencer cover to the air silencer housing/flame arrestor. Lift the silencer cover and seal free of the flame arrestor.

2. Use a small screwdriver to carefully disconnect the powerhead choke and throttle linkage free of the carburetor linkage. You can leave the throttle and choke links connected from carburetor-to-carburetor and disconnect them only when/if you need to separate the carbs.

3. Tag and disconnect the fuel hoses from the carburetors. Most contain spring type clamps, loosen the clamps and slide them back up the hose past the raised portion of the nipple, then carefully separate the hose from the nipple.

4. Loosen, but DO NOT remove, all six bolts securing the flame arrestor/carburetor assembly to the intake manifold/reed block housing. As each bolt is freed from the powerhead, tape it into position on the flame arrester. Once the last bolt is removed from the powerhead the carburetor assembly should come free, carefully tilt the flame arrester downward, keeping the carbs on top and in position on the retaining bolts.

5. Place the assembly on a suitable work bench. If it is necessary to separate the carbs, gently ease the flame arrester from the three carburetors. Remove and discard the one piece gasket, which stretches from the top carburetor to the bottom carburetor. Remove the packing/seals from the powerhead side of each carburetor.

6. If necessary for further service, disconnect the throttle and choke linkage. Place each piece of linkage on the work bench in order as it is removed, as an aid during installation. Scribe a mark on the float bowl of each carburetor, 1, 2, 3 to ensure each is installed back in its original location.

7. Clean and inspect and/or overhaul the carburetor, as applicable.

To Install:

8. Identify each carburetor by the mark scribed on the float bowl during removal. Check to be sure the small pieces of linkage were installed onto the correct carburetor. The throttle roller belongs on the carburetor with the fuel pump attached. Place the three carburetors in line on the work bench. Install the throttle and choke linkage. Place a new one piece gasket on the carburetor throats and align the flame arrestor to the carburetor assemblies.

9. Install a new seal/packing to the powerhead side of each carburetor. Align the carburetors to the flame arrester, then insert each of the mounting bolts through the flame arrester (you can use tape to hold them from falling back out) and carburetors, then carefully lift and position the carburetor/flame arrester assembly to the powerhead. Tighten all six bolts alternately and evenly until snug.

10. Reconnect the fuel hoses, as noted during removal.

11. Reconnect the powerhead choke/throttle linkage.

12. Install the air intake silencer cover and seal, then tighten the bolts securely.

13. Refer to the Timing and Synchronization adjustments in the Maintenance and Tune-Up section to make sure the Throttle Link and Idle Speed are properly adjusted.

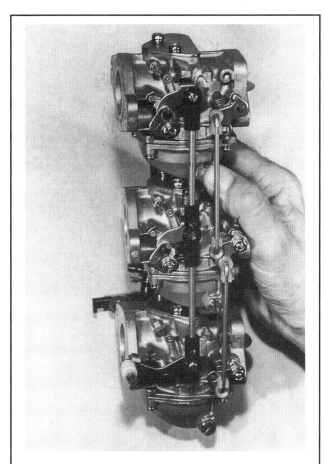

Fig. 156 The three carburetors installed on the powerhead are identical (except for the fuel pump on the middle carb for 25/30 hp models)

Fig. 157 Single carburetor disconnected from the series of carburetors and ready for service

3-42 FUEL SYSTEM

Fig. 158 On 25/30 hp (496cc) models the fuel pump is part of the middle carburetor.

Fig. 159 ... and is serviced by removing the cover/body for access to the diaphragms/gaskets...

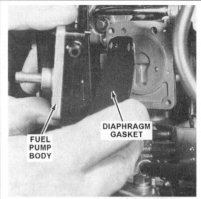

Fig. 160 ... it may be serviced with the carburetors still installed on the powerhead

Fig. 161 The air intake silencer cover is secured using screws - 25/30 hp (496cc) models

Fig. 162 Remove the cover for access to the carb mounting bolts

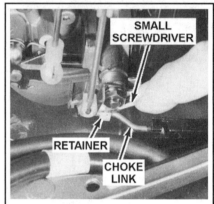

Fig. 163 Carefully disconnect the powerhead linkage and fuel lines

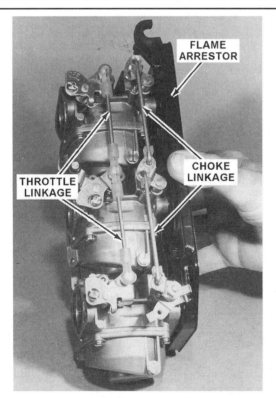

Fig. 164 Remove the carburetors and flame arrester as an assembly

28J-90 Hp (698cc, 849cc & 1140cc) Models

◆ See Figures 165 thru 174

On the 28J-90 hp motors (698cc, 849cc and 1140cc models), the 3 carburetors are normally secured to the powerhead by a common carburetor mounting plate. Like most of the other multi-carburetor Yamahas, bolts are inserted through a common mounting plate (usually a flame arrester on smaller models) and through the carburetors themselves to the powerhead. This makes removal a relatively simple matter that is usually performed by taking all 3 carburetors off at the same time, then separating them for service, if necessary.

The one variation between the carburetors is that on some models the middle carburetor may contain an electrothermal enriching valve for cold start operation. The balance of the motors use a choke plate/choke linkage for cold start enrichment.

1. Loosen and remove the screws (usually 4) securing the air silencer to the carburetor front plate. On Carburetors mounted on some units (usually the 28J-50 hp motors) have two short upper screws and two long lower screws. However, most models have equal length screws.

2. With the bolts removed, carefully pull the air intake silencer from the carburetors. Keep track of the seals, there should be one for each carburetor.

3. Carefully disconnect the powerhead choke and/or throttle linkage from the carburetor assembly. There is no need to disconnect the links that connect the individual carburetors, as they are being removed as an assembly.

4. Squeeze the wire clamp with a pair of needle-nose pliers and gently pull off the fuel intake hose from the T fitting on the fuel manifold. If the open end of the hose shows signs of weeping fuel, plug the end of the hose with a suitable screw. Again, there is no need to disconnect the fuel manifold from the individual carburetors since they are being removed as an assembly.

FUEL SYSTEM 3-43

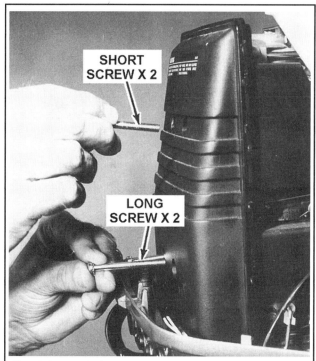

Fig. 165 Loosen the bolts and remove the air silencer from the carbs

5. If equipped with the PrecisionBlend® oil injection system, use a small screwdriver to carefully pry the oil injection link rod free of the plastic retainer on the bottom carburetor.

6. Loosen each of the carburetor assembly/mounting plate bolts, but do NOT remove them from the assembly (as they are the only things keeping the carburetors together once the bolts are free of the powerhead. If necessary, tape the bolts to the mounting plate to keep them from backing out. Once the last bolt is loosened and free of the powerhead, carefully tilt the mounting plate downward and separate the entire carburetor assembly from the powerhead.

7. Remove and discard the one piece gasket used for all three carburetors.

8. If the carburetors must be separated for replacement or service, squeeze the ends of the wire clamps at the inlet fitting of each of the three carburetors and gently pull the fuel hoses from each fitting.

■ The T-fitting in the fuel line is located between the middle and bottom carburetors.

9. Disconnect the throttle and choke linkage. Place each piece of linkage on the work bench in order as it is removed, as an aid during installation. Scribe a mark on the float bowl of each carburetor, 1, 2, 3 to ensure each is installed in its original location.

10. Clean and inspect and/or overhaul the carburetor, as applicable.

To Install:

11. Identify each carburetor by the mark scribed on the fuel bowl during removal. Check to be sure the small pieces of linkage were installed on the correct carburetor. The throttle roller MUST be installed on the lower carburetor. Place all three carburetors in line on the work bench. Install the choke and throttle linkage on the port side of the carburetors.

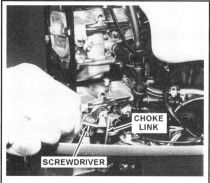

Fig.166 Disconnect the linkage...

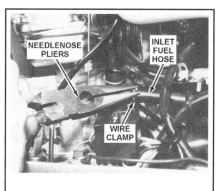

Fig. 167 ... and the fuel inlet hose

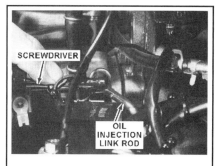

Fig. 168 If equipped, disconnect the oil pump link

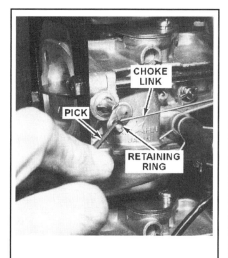

Fig. 169 If equipped, disconnect the choke link...

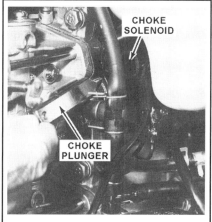

Fig. 170 ... some models utilize a choke solenoid

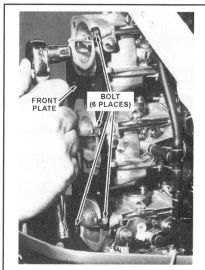

Fig. 171 Loosen the 6 carburetor/front plate mounting bolts and remove the assembly...

3-44 FUEL SYSTEM

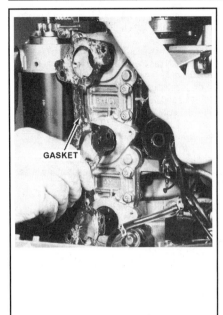

Fig.172 . . . followed by the gasket

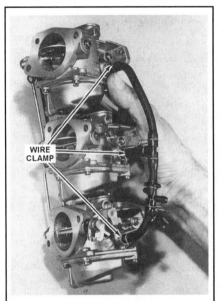

Fig. 173 To separate the carburetors, disconnect the fuel manifold. . .

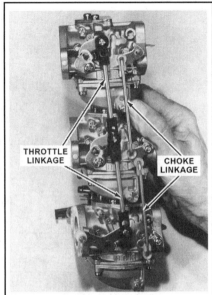

Fig. 174 . . . along with the choke and/or throttle linkages

12. Turn the carburetors over on the workbench to rest on their port sides. Push each fuel hose onto its respective fuel inlet fitting on the starboard side. The T fitting is located between the middle and bottom carburetor. Secure each hose with the wire clamp.

13. Place a new gasket in position over the reed block housing. The manufacturer recommends no sealant at this location.

14. Align the mounting plate with the carburetors, then insert of the bolts through the mounting plate (you can use tape to hold them from falling back out) and carbs. Carefully lift and position the carburetor/mounting plate assembly to the powerhead. Tighten all six bolts alternately and evenly until snug.

15. Reconnect the choke, oil pump and/or throttle linkage, as equipped.

16. Connect the fuel line from the fuel filter to the T fitting between the middle and bottom carburetors.

17. Install the air intake silencer and seal, then tighten the bolts securely. Some 28J-50 hp motors have two short upper screws and two long lower screws, but most 3-cylinder models have screws of equal length.

18. Refer to the Timing and Synchronization adjustments in the Maintenance and Tune-Up section to make sure the Carburetor Link, Oil Pump Link and Idle Speed are all properly adjusted, as applicable.

OVERHAUL

 DIFFICULT

◆ See Figures 175 thru 191

The following procedures pick up the work after the carburetors have been removed from the powerhead. The procedures for each of the three carburetors is identical, even for the 25/30 hp units with the fuel pump attached to the middle carburetor or larger units with an electrothermal valve and enrichening pump on the middle carb. However, additional steps are also provided which only apply to carbs with those components.

Good shop practice dictates purchasing a carburetor repair kit and using new parts any time the carburetor is disassembled. Make an attempt to keep the work area organized and to cover parts after they have been cleaned. This practice will prevent foreign matter from entering passageways or adhering to critical parts.

1. Remove the carburetor from the powerhead and drain all fuel from it using the float bowl drain screw.

2. Invert the carburetor body (with the float bowl facing upward), then loosen the float bowl retaining screws (usually 4). Remove the float bowl from the carburetor body (also known as a mixing chamber), then remove and discard the old bowl gasket/seal.

3. With the float bowl removed you can access the float and needle valve assembly. Some models are equipped with a screw to lock the hinge pin in place, if so, loosen, but do not remove, the screw retaining the hinge pin in its groove.

4. Grasp the float and lift gently upward to remove the float and needle valve from the carburetor body. The needle valve, attached to the tang on the float will also slide out of the needle seat.

5. Pull the hinge pin from the float. Unhook the wire clip and needle valve from the tang on the float.

6. Remove the carburetor fuel metering components (plug and pilot jet, then main jet and nozzle). Refer to the exploded views to help identify each component. Be sure to keep them separated or tagged for installation or replacement purposes. For most carburetors these components are removed as follows:

    a. Pry the small plastic plug from the center turret of the mixing chamber.
    b. Use a suitable size screwdriver and remove the pilot jet located under the plug.
    c. Remove the main jet from the center turret of the mixing chamber.
    d. Invert the mixing chamber and shake it, keeping a hand over the center turret. The main nozzle should fall free from the turret. If the nozzle refuses to fall out, gently reach in with a pick or similar instrument to raise the nozzle.

7. If equipped with a removable needle valve seat, loosen and remove the seat and O-ring. Anytime you disassemble the carburetor it is a good idea to replace the needle valve and seat (if replaceable) to ensure proper float valve operation.

■ On most carburetors you'll need a thin walled socket to remove the needle valve seat.

8. Remove the screws securing the top cover to the top of the carburetor body/mixing chamber. Lift off the cover. On some models the cover is sealed using a gasket or packing, if applicable, remove and discard the old seal.

9. Some models are equipped with a number (as many as 3) flat rubber plugs, two large and one small. Refer to the accompanying exploding views to see what plugs are used on your model. Lift off the large rubber plug closest to the throttle plate. Lift off the remaining large rubber plug and the metal plate beneath it. The first rubber plug does not and should not have a metal plate under it. Lift off the small rubber plug and plate.

FUEL SYSTEM **3-45**

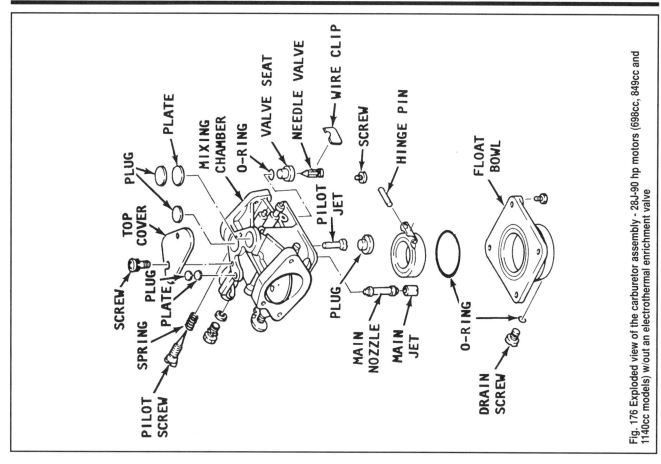

Fig. 176 Exploded view of the carburetor assembly - 28J-90 hp motors (698cc, 849cc and 1140cc models) w/out an electrothermal enrichment valve

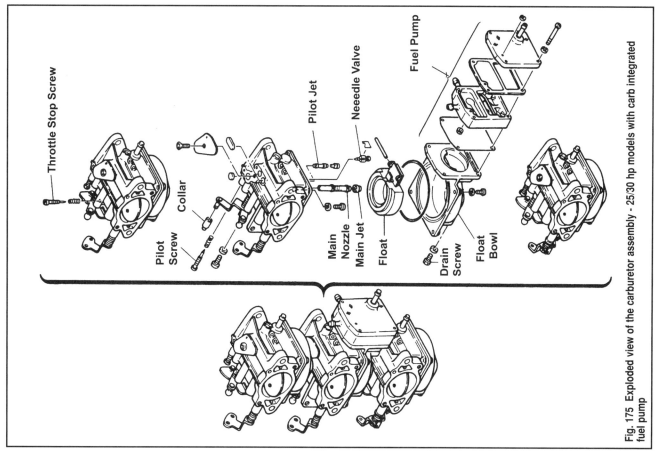

Fig. 175 Exploded view of the carburetor assembly - 25/30 hp models with carb integrated fuel pump

3-46 FUEL SYSTEM

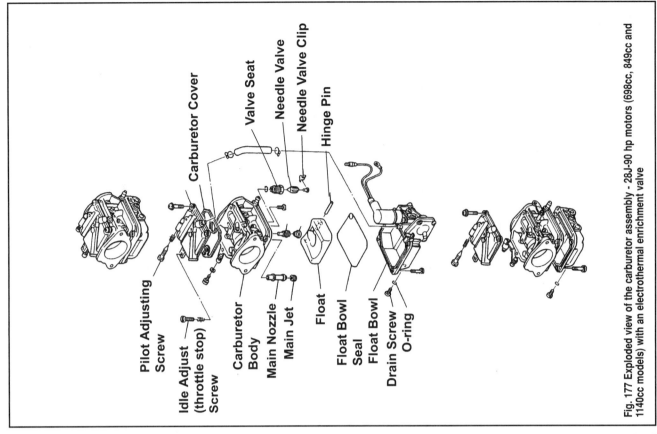

Fig. 177 Exploded view of the carburetor assembly - 28J-90 hp motors (698cc, 849cc and 1140cc models) with an electrothermal enrichment valve

10. Remove the pilot screw and spring from the carburetor. It is not necessary to count the number of turns in to a lightly seated position as a guide for installation, as the number of turns is specified in the Carburetor Set-Up Specifications chart in this section.

11. If necessary, loosen and remove the idle adjust (throttle stop) screw and spring.

12. For the middle carb on 25/30 hp motors with an integral fuel pump, if necessary remove the pump cover screws and disassemble the pump for inspection or component replacement. Remove the cover, body and diaphragm gaskets.

13. For the middle carb on 28J and larger motors with an electrothermal enrichment circuit, proceed as follows:

   a. Remove the retaining screw(s), then carefully lift the electrothermal valve from the float bowl.

   b. Loosen the enrichment pump cover screw, then remove the pump components (cover, diaphragm with gaskets and check valve assembly. If necessary, loosen the screws and remove the valves from the valve body (noting positioning for installation purposes).

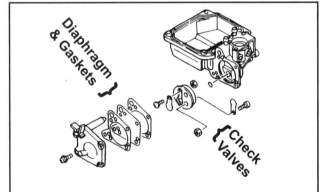

Fig. 178 Exploded view of the enrichment pump assembly - 28J-90 hp motors (698cc, 849cc and 1140cc models) with an electrothermal enrichment valve

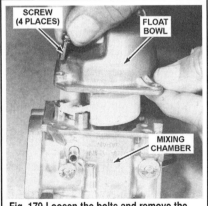

Fig. 179 Loosen the bolts and remove the float bowl...

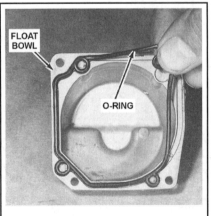

Fig. 180 ... then remove the old gasket/seal

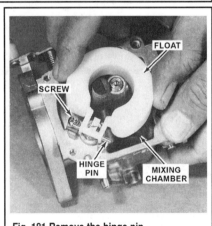

Fig. 181 Remove the hinge pin...

FUEL SYSTEM 3-47

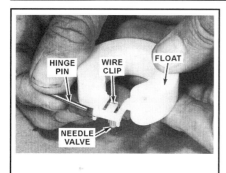

Fig. 182 . . . to free the float and needle valve assembly

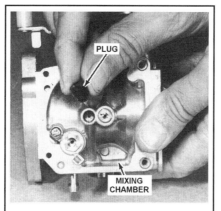

Fig. 183 Remove each of the fuel metering components. . .

Fig. 184 . . .this pilot jet was mounted under a plug

Fig. 185 This main jet was installed above...

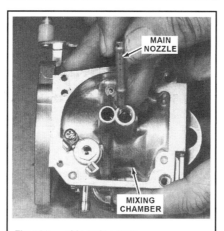

Fig. 186 . . .this main nozzle

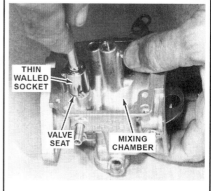

Fig. 187 On most carbs, the needle valve seat is replaceable

14. Clean and inspect all components as detailed in this section. Replace any damaged, worn or defective components. Discard all O-rings or gaskets.

To Assemble:

15. For the middle carb on 28J and larger motors with an electrothermal enrichment circuit, proceed as follows:

   a. Assemble the enrichment pump using, at a minimum, new gaskets. Install the check valve assembly, diaphragm with gaskets and the pump cover, then tighten the bolts securely.

   b. The electrothermal valve is normally sealed using an O-ring, make sure it is in good condition or, better yet, replace it to ensure there are no leaks. Install the valve and secure using the retaining screws.

16. For the middle carb on 25/30 hp motors with an integral fuel pump, install the pump body, cover and diaphragm/gaskets. Secure using the cover screws. Be careful not to damage the diaphragm/gaskets during assembly.

17. If removed, install the idle adjust (throttle stop) screw and spring.

18. Install the pilot screw and spring. Slowly rotate the pilot screw into the carburetor body until it barely seats. From this position, back it out the appropriate number of turns (refer to the Carburetor Set-Up Specifications chart in this section).

19. Some models are equipped with a number (as many as 3) flat rubber plugs in the top of the carburetor body. Proceed as applicable:

   a. Place the small metal plug in position over the smallest hole on top of the carburetor. Place the smallest rubber plug over the plate.

   b. Place the large metal plug in position over the large hole on top of the carburetor, nearest the fuel fitting, and then place one of the large rubber plugs over the plate. Place the remaining large rubber plug over the remaining hole. This plug is installed without a metal plate beneath it.

20. If applicable, install the top cover (using a new seal, if applicable) and secure using the retaining screws.

21. If removed, install a new O-ring over the shaft of the needle valve seat. Install and tighten the seat snugly.

22. Install the fuel metering components (main nozzle and main jet, then the pilot jet and pilot jet plug) to the carburetor body in the positions noted during removal. For most carburetors these components are installed in the following manner:

   a. Insert the main nozzle into the aft hole on the center turret. Position the series of holes in the nozzle to face port and starboard when installed.

   b. Install the main jet over the main nozzle. Tighten the jet until it seats snugly.

   c. Install the pilot jet into the forward hole on the center turret. Tighten the jet until it seats snugly.

   d. Install the plug over the pilot jet. Push the plug in securely. A loose plug could wedge itself between the float and the float bowl.

23. Check to be sure the wire clip is securely in position around the needle valve. Slide the clip over the tang on the float, and check to see if the needle valve can be moved freely.

24. Slide the hinge pin through the hole in the float.

25. Lower the float assembly over the center turret, guiding the needle valve into the needle seat and positioning the end of the hinge pin to the carb body (under its retaining screw, when equipped). If applicable, tighten the screw securely.

26. With the carburetor still inverted and perfectly level, so the float is sitting gently on the needle valve (which is resting on the seat) measure the float height from the carburetor body-to-float bowl mating surface up to the top (actually the bottom, but it is on top now) of the float. Refer to the Carburetor Set-Up Specifications chart in this section for the allowable range. If necessary, gently bend the float hinge to achieve the proper measurement.

✱✱ WARNING

When measuring the float height DO NOT place any pressure downward on the needle valve or you could damage it (and/or you may make an incorrect adjustment).

3-48 FUEL SYSTEM

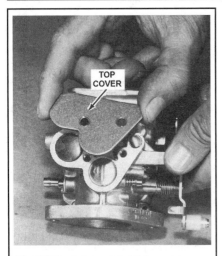

Fig. 188 On some models the top cover must be removed...

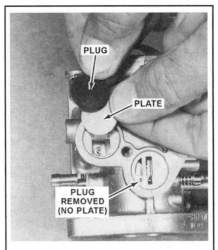

Fig. 189 ...to access various carb body passage plugs and plates...

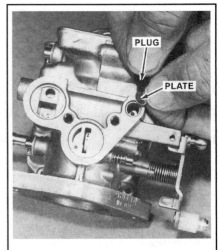

Fig. 190 ...which may be inspected/replaced as necessary

CLEANING & INSPECTION

◆ See Figures 90, 91, 175 thru 178 and 192 thru 196

> ✱✱ **CAUTION**
>
> Never dip rubber or plastic parts, in carburetor cleaner. These parts should be cleaned only in solvent, and then blown dry with compressed air.

Place all metal parts in a screen type tray and dip them in carburetor cleaner until they appear completely clean, then blow them dry with compressed air.

Blow out all passages in the castings with low-pressure compressed air. Check all parts and passages to be sure they are not clogged or contain any deposits. Never use a piece of wire or any type of pointed instrument to clean drilled passages or calibrated holes in a carburetor.

Move the throttle and, if applicable, choke shafts back and forth to check for wear. If the shaft appears to be too loose, replace the complete mixing chamber because individual replacement parts are not available.

Inspect the mixing chamber, and fuel bowl gasket surfaces for cracks and burrs which might cause a leak.

Check the float for deterioration. If any part of the float is damaged, the unit must be replaced.

Inspect the tapered section of the float needle and pilot screw for wear or damage. Replace either if they have developed a groove or are no longer evenly tapered.

Check the main jet, pilot jet and main nozzle for signs of dirt or contamination. If they cannot be cleaned, they should be replaced. Again, NEVER clean these components using a wire or any pointer instrument, you'll change the calibration.

If equipped with an integrated fuel pump or enrichment pump, check the diaphragm for signs of deterioration or damage and replace. Obviously, there can be no pin holes in the diaphragm if you expect it to work properly. Of course, like all diaphragms, they weaken over time so if there is any doubt, just replace it to be save. Make sure the spring has not deformed or lost its strength.

As previously mentioned, most of the parts which should be replaced during a carburetor overhaul are included in an overhaul kit available from your local marine dealer. One of these kits will contain a matched fuel inlet needle and seat. This combination should be replaced each time the carburetor is disassembled as a precaution against leakage.

If equipped with an electrothermal enrichment valve, check it as follows:

Fig. 191 Remove the pilot screw for inspection

27. Double-check that the float and valve move smoothly without sticking or binding.

28. Install the float bowl to the carburetor body using a NEW gasket/seal, then secure using the retaining screws. If not already done, install the drain plug and O-ring or gasket (as applicable).

■ On some models the float bowl and gasket are irregularly shaped. For these models, be sure to match the one cutaway corner with the other.

29. Install the carburetor and adjust for proper operation.

FUEL SYSTEM 3-49

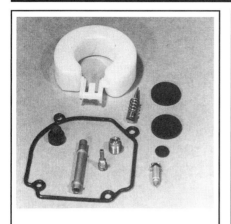

Fig. 192 Some of the parts included in the repair kit for most 3-cylinder powerhead carburetors

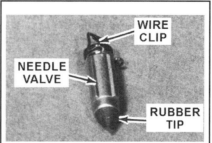

Fig. 193 A wire clip secures the needle valve to the float. If this wire should break or slip free of the float, the fuel supply will be cut off

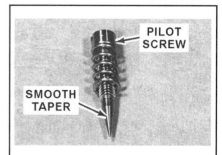

Fig. 194 Inspect the taper on the end of the pilot screw for ridges or signs of roughness. Good shop practice dictates a new pilot screw be installed each time the carburetor is overhauled

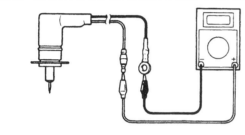

Fig. 195 If equipped with an electrothermal valve check resistance using a DVOM...

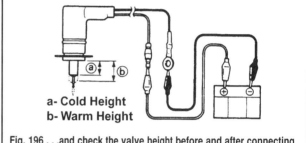

Fig. 196 ...and check the valve height before and after connecting a 12 volt battery

• Using an ohmmeter check the valve element resistance. Connect the ohmmeter across the two valve leads, it should read about 2.3-3.5 ohms resistance at an ambient temperature of about 68°F (20°C).

• Check the distance that the piston is protruding from the valve. Apply a 12 volt power source to the 2 valve leads and leave it connected for several minutes, then recheck the valve piston height. If the valve is working properly the piston height should have changed (extended) as the internal valve element gradually heated.

V4 & V6 Powerheads

Yamaha V4 powerheads use two double-barreled carburetors and V6 powerheads use three double-barreled carburetors. Complete, detailed, illustrated, procedures to remove, service, and install the carburetors follow. Removal procedures may vary slightly due to differences in linkage.

REMOVAL & INSTALLATION

◆ See Figures 197 thru 208

■ On V4 models, the enrichment valve is normally mounted to the back side of the air intake silencer, so before pulling the silencer forward for access to the breather hose(s), it is best to remove the O-ring and disconnect the enrichment solenoid pull wire from the linkage.

1. Your first step is to gain access to the carburetors which are mostly obscured by the air intake silencer. Loosen and remove the silencer retaining bolts (usually 8 for V4 models and 12 for V4 models) then pull the silencer carefully away from the carbs just enough to disconnect the breather hoses (intake manifold vent hose and, if equipped, oil tank air vent hose).

■ Pre-mix V4 models utilize a 2-piece air intake silencer cover and housing assembly. For these motors, remove the 8 bolts and cover (along with gasket), then remove the 8 bolts securing the housing and pull the housing forward for access to the intake manifold vent hose.

2. On V4 models, if not done already, remove the O-ring and disconnect the pull wire from the carb linkage. Disconnect the solenoid valve wiring. If necessary, unbolt and remove the valve from the air intake silencer.

3. With the hoses (and enrichment valve wiring for V4 motors) disconnected, remove the air intake silencer completely from the powerhead. Keep track of the silencer gasket (normally rubber).

4. Carefully disconnect the choke and/or throttle valve linkages on the port side of the carburetor assembly. Disconnect the throttle link rod which attaches to the carburetor throttle valves.

5. Carefully snip the 2 (V4) or 3 (V6) plastic wire ties, then gently pull off the fuel supply hoses from the carburetors.

6. On the starboard side of the bottom carburetor, carefully pry the oil injection link from the linkage.

7. On V6 models, if not done already, pry off the little O-ring which serves as a retainer for the choke solenoid pull wire. Take care not to lose this small part. Slip off the end of the pull wire from the carburetor linkage. The choke solenoid need not be removed on these models.

8. Identify each carburetor by inscribing or painting a 1, 2 and 3 (if applicable) on the mixing chamber cover to ensure each carburetor will be installed back into the same position from which it was removed.

9. Remove the mounting nuts, four on each carburetor, and then remove the carburetors as an assembly. Each carburetor has a separate gasket which may either come away with the carburetor, or remain on the intake manifold. Remove and discard these gaskets.

Place the carburetor assembly on the workbench and remove each piece of linkage one at a time. Arrange the linkage on the workbench as it was installed on the carburetors, as an assist during assembling.

To Install:

10. Identify each carburetor by the mark scribed on the mixing chamber during removal. Check to be sure the small pieces of linkage were installed onto the correct carburetor. On V4 models, the throttle roller is used on the lower carburetor. On V6 models, the throttle roller is used on the center carburetor.

The choke and oil injection link retainers are located on the bottom carburetor. Place the carburetors in line on the work bench. Install the throttle and choke linkage. Place a new gasket onto the studs of the intake manifold. The manufacturer recommends NO sealant at this location.

3-50 FUEL SYSTEM

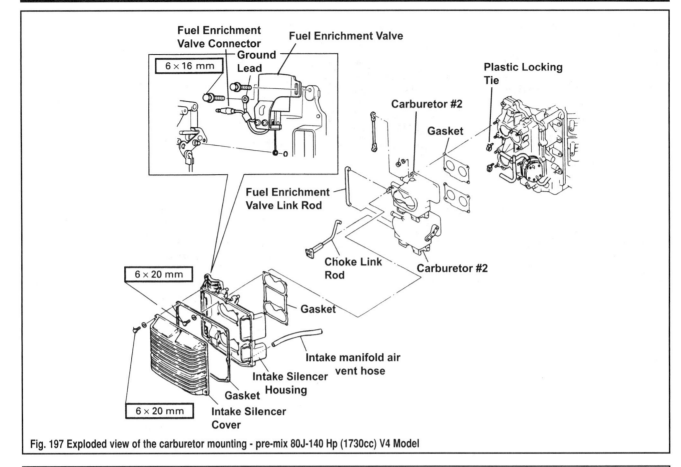

Fig. 197 Exploded view of the carburetor mounting - pre-mix 80J-140 Hp (1730cc) V4 Model

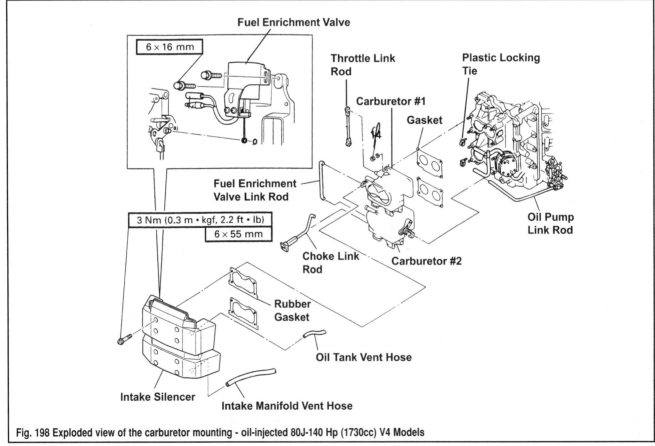

Fig. 198 Exploded view of the carburetor mounting - oil-injected 80J-140 Hp (1730cc) V4 Models

FUEL SYSTEM 3-51

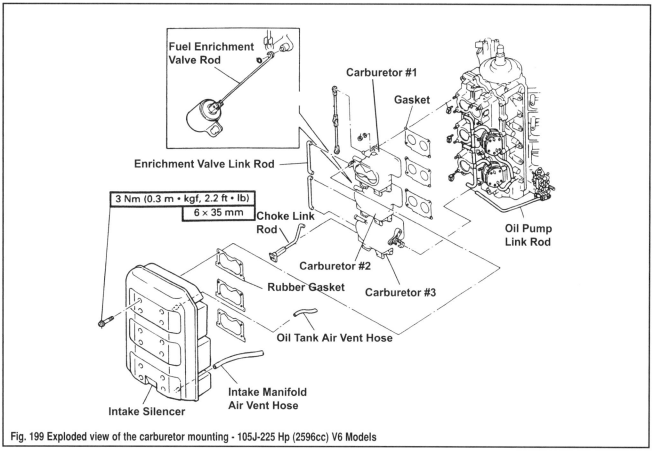

Fig. 199 Exploded view of the carburetor mounting - 105J-225 Hp (2596cc) V6 Models

11. Install each carburetor, then tighten the four mounting nuts for each carburetor to a torque value of 5.8 ft. lbs. (8 Nm).

12. On V6 models, hook the choke solenoid pull wire loop into the linkage between the two choke rods. Slide the little O-ring over the linkage ball to retain the wire loop. The choke plunger adjustment is performed later.

13. On the starboard side of the bottom carburetor, snap the oil injection link rod into the linkage. If the length of this rod was accidentally changed adjust the length to specifications.

14. Slide the hoses over the inlet fittings and secure them using new plastic wire ties.

15. On the port side of the bottom carburetor, snap the choke link onto the carburetor choke arm. Check the action of the choke linkage on the side of the carburetors by moving the choke lever in and out.

16. On V4 models, if the enrichment valve was removed from the back of the air intake silencer, install it now and secure using the retaining bolts.

17. Bring the inner silencer up to the installed carburetors. Reconnect the breather hose(s). On V4 motors, reconnect the enrichment valve wiring. Make sure the gasket(s) are in position, then seat the air intake silencer on the carburetors. Reconnect the enrichment solenoid valve pull wire and secure using the O-ring.

18. Install and tighten the 8 or 12 bolts securing the air intake silencer to the carburetors.

19. On Pre-Mix V4 models, install the air intake silencer cover and gasket, then secure using the 8 retaining bolts.

20. Refer to the Timing and Synchronization adjustments in the Maintenance and Tune-Up section to make sure the carburetors and control linkage are properly adjusted.

Fig. 200 Loosen the air intake silencer retaining bolts. . .

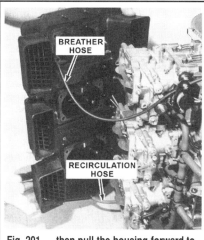

Fig. 201 . . .then pull the housing forward to disconnect the breather hose(s)

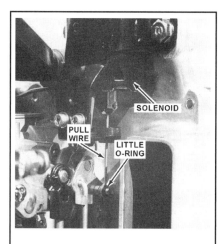

Fig. 202 Disconnect the choke and/or throttle linkage. . .

3-52 FUEL SYSTEM

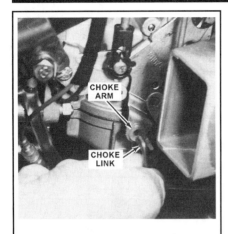

Fig. 203 Disconnect the choke and/or throttle linkage...

Fig. 204 ...and tag/disconnect the fuel lines

Fig. 205 For oil injected models, disconnect the pump link

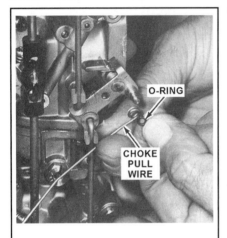

Fig. 206 V6 engines also use an enrichment (choke) pull wire

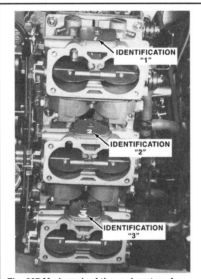

Fig. 207 Mark each of the carburetors for identification...

Fig. 208 ...then remove the nuts and remove the carburetors

OVERHAUL

 DIFFICULT

◆ See Figures 209 thru 219

The following procedures pick up the work after the carburetors have been removed from the powerhead, as outlined in the previous steps. The procedures for each of the two or three carburetors is identical.

Good shop practice dictates purchasing a carburetor repair kit and using new parts any time the carburetor is disassembled. Make an attempt to keep the work area organized and to cover parts after they have been cleaned. This practice will prevent foreign matter from entering passageways or adhering to critical parts.

1. Remove the carburetor unit from the powerhead, as detailed earlier in this section.
2. If not already done, remove the drain plugs and gaskets from the bottom sides of the float bowl and drain the fuel from the carburetor.
3. Use a small screwdriver to remove the pilot air jets.
4. Remove the Phillips screw securing the jet access cover to the carburetor. Remove the access cover.
5. Remove the 2 bypass screw plugs from the top of the carburetor mixing chamber. Remove and discard the two gaskets.
6. Remove both pilot screws and springs. It is not necessary to count the number of turns in to a lightly seated position, as a guide for installation. The number of turns in is listed in the Carburetor Set-Up Specifications chart in this section.
7. Invert the carburetor assembly, then remove the 4 securing screws and lift off the float bowl. Do not attempt to remove the gasket at this time.
8. Remove the two main jets, pilot jet plugs and pilot jets from the underneath and side of the float bowl. Remove and discard the gaskets and O-rings.
9. Slide the hinge pin outward to release the float.
10. Gently lift up the float. The needle valve, attached to the tang on the float, will also slide out of the needle seat. Unhook the wire loop and needle valve from the tang on the float. Remove the other float in a similar manner.
11. Remove and discard the float bowl gasket from the mixing chamber.
12. On some models the needle seat is replaceable. For these, obtain the correct size thin walled socket and remove the needle seat.
13. Clean and inspect all components as detailed in this section. Replace any damaged, worn or defective components. Discard all O-rings or gaskets.

To Assemble:

14. If applicable, install and tighten the needle seat snugly, using a thin walled socket.
15. Check to be sure the wire loop is securely in position around the needle valve. Slide the loop over the tang on the float, and then check to ensure the needle valve can be moved freely. Lower the float assembly into the mixing chamber guiding the needle valve into the needle seat. Install the other float in the same manner.
16. Slide the hinge pin through the mounting posts and float hinge, until the pin ends are flush with the mounting posts.
17. Hold the mixing chamber in the inverted position, (as it has been held during the past few steps). Measure the distance between the top of the float

FUEL SYSTEM 3-53

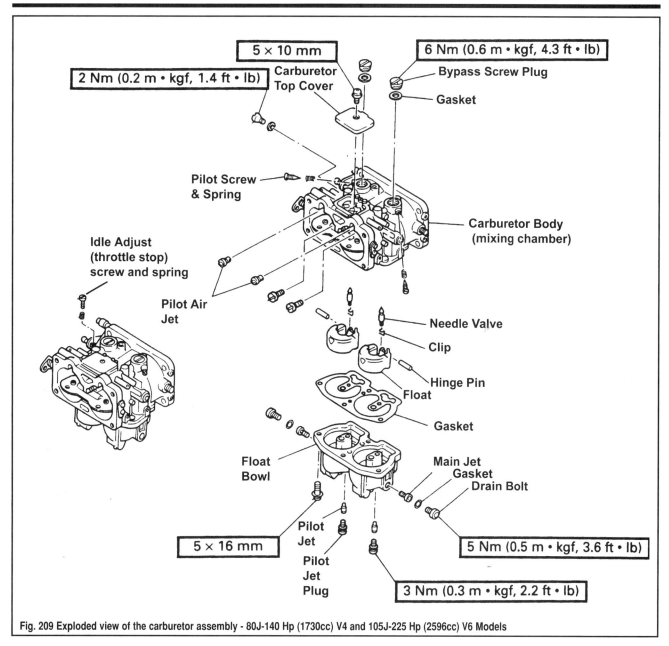

Fig. 209 Exploded view of the carburetor assembly - 80J-140 Hp (1730cc) V4 and 105J-225 Hp (2596cc) V6 Models

Fig. 210 Remove the pilot air jets from the front of the carb

Fig. 211 Remove the jet access cover and inspect the passages

Fig. 212 Remove the bypass screw plugs and gaskets/o-rings...

3-54 FUEL SYSTEM

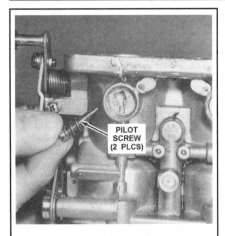
Fig. 213 ...then remove the pilot screws and springs

Fig. 214 Loosen the bolts and remove the float bowl

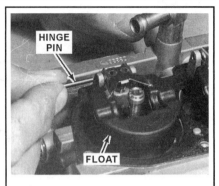
Fig. 215 Remove the hinge pin to free the float...

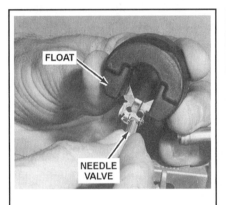
Fig. 216 ...then remove the float and needle valve

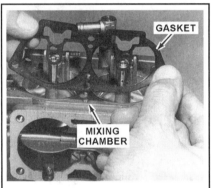
Fig. 217 Remove and discard the old float bowl gasket

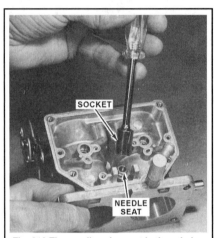
Fig. 218 The needle valve seat is threaded on some models

and the gasket. This distance should be in the float drop setting range of 0.61-0.65 in. (15.5-16.5mm) as indicated in the Carburetor Set-Up Specifications chart, found in this section.

✳✳ WARNING

When measuring the float height DO NOT place any pressure downward on the needle valve or you could damage it (and/or you may make an incorrect adjustment).

18. If the distance is not as specified, remove the float and needle valve. Gently bend the tab on the float using a small screwdriver to correct the float level measurement. Repeat the float level measurement for the other float.
19. Double-check that the float and valve move smoothly without sticking or binding.
20. Install the main jets, float bowl drain screws with gaskets, pilots jets and pilot jet plugs with new gaskets. Install the jets into the float bowl and tighten them securely. Install new gaskets around the drain screws or plugs. Install into the float bowl and tighten them securely.
21. Place a new float bowl gasket in position over the mixing chamber.
22. Lower the float bowl over the floats. Take care not to disturb the float level adjustment. Install and tighten the attaching hardware.
23. Slide new springs over the pilot screws. Install the pilot screws into the carburetor. Tighten each screw until it JUST barely seats. From this position, back out the screw the number of turns specified in the float in the Carburetor Set-Up Specifications chart, found in this section.

■ Take notice, as each model and sometimes year of manufacture could have a different pilot screw setting. Furthermore, on certain models, the port screw has a different setting from the starboard screw.

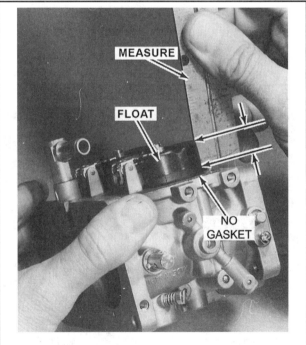
Fig. 219 Measuring float height

FUEL SYSTEM 3-55

24. Install new O-rings/gaskets around the two bypass screw plugs, and then install them into the mixing chamber.
25. Install the pilot air jets and install the jet access cover (secure using the Philips screw).
26. Install the carburetors and adjust for proper operation.

CLEANING & INSPECTION

◆ See Figures 90, 91, 209 and 220

✲✲ CAUTION

Never dip rubber or plastic parts in carburetor cleaner. These parts should be cleaned only in solvent, and then blown dry with compressed air. Actually, rubber and plastic parts should, whenever possible, be replaced.

Place all metal parts in a screen type tray and dip them in carburetor cleaner until they appear completely clean, then blow them dry with compressed air.

Blow out all passages in the castings with low-pressure compressed air. Check all parts and passages (especially the jets and bleed valves) to be sure they are not clogged or contain any deposits. Never use a piece of wire or any type of pointed instrument to clean drilled passages or calibrated holes in a carburetor.

Move the throttle and choke shafts back and forth to check for wear. If the shaft appears to be too loose, replace the complete mixing chamber because individual replacement parts are not available.

Inspect the mixing chamber, and fuel bowl gasket surfaces for cracks and burrs which might cause a leak.

Check the floats for deterioration. Check to be sure the needle valve loop has not been stretched. If any part of the float is damaged, the float must be replaced. Check the needle valve tip contacting surface and replace the needle valve if this surface has a groove worn in it.

Inspect the tapered section of the pilot screws and replace a screw if it has developed a groove.

Most of the parts which should be replaced during a carburetor overhaul are included in an overhaul kit available from your local marine dealer. Usually, one of these kits will contain a matched fuel inlet needle and seat. It is usually a good idea to replace this combination each time the carburetor is disassembled as a precaution against leakage.

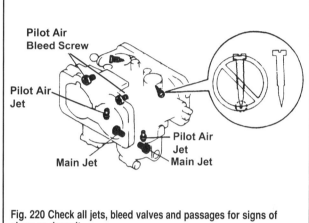

Fig. 220 Check all jets, bleed valves and passages for signs of clogs or deposit

ELECTRONIC FUEL INJECTION SYSTEMS

Beginning with the 1997 model year Saltwater Series V76 some Yamaha engines are equipped with one of 3 types of electronic fuel injection. Although the original system was introduced in 1997 it continued even after the introduction of the 2 additional systems in 2000. The largest of the Yamahas may be equipped with either this manifold-injected EFI (OX66) system or a High Pressure Direct Injection (HPDI) system.

In all cases, these systems improves low speed smoothness, reduce maintenance, improve durability, and increase fuel efficiency while lowering harmful engine emissions. In all cases an engine control computer known as the Engine or Electronic Control Module (ECM) is capable of adjusting fuel delivery and ignition timing over a wide range to match engine operation. Unlike carburetor-equipped engines, fuel-injected engines can automatically adjust for changes in altitude, air temperature, and barometric pressure. This precise control over engine operation results in crisp throttle response and maximum efficiency over a broad range of conditions.

■ **For some reason the ECM is called the CDI unit on EFI OX66 motors, however, since this unit does more than just control ignition, we think that is a little misleading and will refer to it as the ECM (the more commonly used term for fuel and ignition control units/models) throughout this repair guide. However, knowing what Yamaha calls it might be handy when dealing with a parts counterperson.**

Fuel Injection Basics

◆ See Figures 221, 222 and 223

Fuel injection is not a new invention. Even as early as the 1950s, various automobile manufacturers experimented with mechanical-type injection systems. There was even a vacuum tube equipped control unit offered for one system! This might have been the first "electronic fuel injection system." Early problems with fuel injection revolved around the control components. The electronics were not very smart or reliable. These systems have steadily improved since. Today's fuel injection technology, responding to the need for better economy and emission control, has become amazingly reliable and efficient. Computerized engine management, the brain of fuel injection, continues to get more reliable and more precise.

Components needed for a basic computer-controlled system are as follows:

• A computer-controlled engine manager, which is the Electronic Control Module (ECM), with a set of internal maps to follow (you know, if this, then do that type of information for the fuel and ignition systems).
• A set of input devices to inform the ECM of engine performance parameters.
• A set of output devices. Each device is controlled by the ECM. These devices modify fuel delivery and timing. Changes to fuel injection and timing are based on input information matched to the map programs.

This list gets a little more complicated when you start to look at specific components. Some fuel injection systems may have twenty or more input devices. On many systems, output control can extend beyond fuel and timing. The Yamaha Fuel Injection System provides more than just the basic functions, but is still straight forward in its layout. There are somewhere between eight and twelve input devices and generally between six and ten output controls (depending upon the system, year and model). The accompanying diagrams show the typical input and output devices of the various Yamaha fuel injection systems.

There are several fuel injection delivery technologies in wide use around the world and Yamaha has chosen to basically 2 forms of them for these motors. A brief discussion of the various types (even those not used by Yamaha) would be helpful in understanding the Yamaha systems.

Throttle body injection is relatively inexpensive and was used widely in early automotive systems. This is usually a low pressure system running at 15 PSI or less. Often an engine with a single carburetor was selected for throttle body injection. The carburetor was recast to hold a single injector and the original manifold was retained. Throttle body injection is not as precise or efficient as port injection.

Multi-port fuel injection is defined as an engine that uses one or more electrically activated solenoid injectors (or physically actuated injectors in the case of General Motors Central Multi-Port Injection used in some 1990's era vehicles) for each cylinder. Multi-port injection generally operates at higher pressures than throttle body systems. The Yamaha EFI systems operate with a system pressure of somewhere in the 30-40 psi (207-278 kPa) range depending upon the year and model. The Yamaha EFI system uses a type of port injection where the fuel/oil mixture is injected into the air intake track behind the throttle bodies and before the reed valves.

Another type of Multi-port fuel injection used predominantly on 2-stroke marine engines (but also found on automotive and marine diesels) is Direct

3-56 FUEL SYSTEM

Injection, which is called so because fuel is introduced from the injector DIRECTLY into the combustion chamber. The Yamaha HPDI motors are an example of this type of fuel injection system. Whereas for many years most diesel engines achieved this using fuel line pressure which overcomes spring pressure in the injector, more and more modern systems are using electronically actuated injectors. This type of system requires a higher pressure fuel charge to operate properly, and Yamaha HPDI systems operate on fuel pressures up to 1000 psi (6895 kPa) on some direct injection motors.

No matter whether the fuel is injected to the manifolds, behind valves, or directly into the combustion chamber, the real advantage of a fuel injection system is the precise control afforded the ECM by the use of high response electronic components. This gives electronic fuel injectors are great advantage over mechanical injectors when it comes to emissions and operating efficiency. Port injectors can be triggered two ways. One system uses simultaneous injection. All injectors are triggered at once. The fuel "hangs around" until the pressure drop in the cylinder pulls the fuel into the combustion chamber. This type of system looses some of the advantage that is possible with an electronic fuel injector. The second type is more precise and follows the firing order or SEQUENCE of the engine. Each cylinder gets a squirt of fuel precisely when needed. If you haven't already guessed, this second type of fuel injection is known as Sequential Multi-port Fuel Injection.

Yamaha EFI OX66 Injection

◆ See Figures 221 and 222

The Yamaha EFI OX66 system is a tuned port, sequential method of fuel injection. The injectors are located between the throttle air valves and the reed valves. The ECM controls fuel and ignition timing in two stages. In stage one, the ECM (remember, Yamaha calls it the CDI unit on these models) controls ignition timing and fuel injection volume according to information gathered about engine speed from the Crankshaft Position Sensor (CPS) and throttle opening from the Throttle Position Sensor (TPS).

In stage two, the ECM fine tunes the fuel mixture according to information from several other sensors, especially the Oxygen sensor. This second stage is sometimes referred to as closed-loop mode and is very similar to the method of fuel injection management found in nearly all modern automobiles. Closed-loop mode means that the oxygen sensor information is used to control enrichment. In an open loop condition, one or more sensors (the oxygen sensor is usually the critical input) is missing or out of range. Open-loop operation controls the engine by a fixed map or pre-programmed information in the ECM. In closed-loop operation, the ECM uses Oxygen sensor feedback to continuously monitor how good a job the ECM is doing in

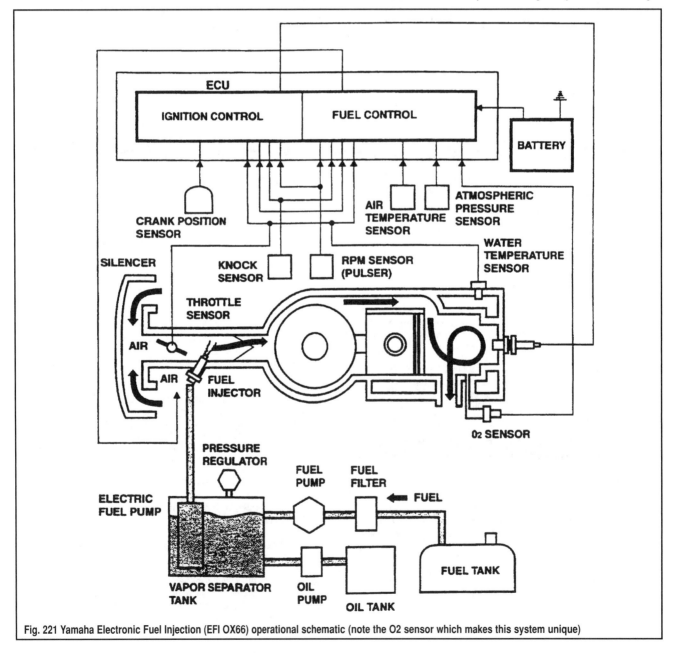

Fig. 221 Yamaha Electronic Fuel Injection (EFI OX66) operational schematic (note the O2 sensor which makes this system unique)

FUEL SYSTEM

managing combustion, then uses the information to better manage air/fuel mixtures and ignition timing. The amount of Oxygen found in exhaust gases therefore becomes the piece of the puzzle that closes the "loop" of fuel management, the results of engine management are used to make decisions on future engine management which produces results which are used for feedback and you have a circular pattern.

An additional feature of the Yamaha system is return-to-port capability or limp mode. If there is a major sensor failure that prevents the ECM from processing, the engine is run on a minimum performance map. The only input that will shut the engine down completely is battery voltage. If the battery is disconnected or the battery voltage falls below 9 volts, the fuel pump quits pumping so the engine stops.

Major components of this system include the ECM and Sensors along with the low- and high-pressure fuel delivery circuits. The electronic control sensors include:

- Throttle Position Sensor (TPS)
- Intake Air Temperature (IAT) Sensor
- Atmospheric Pressure Sensor (APS)
- Knock Sensor (3.1L engine only)
- Oxygen Sensor
- Water Temperature Sensor (WTS)
- Crankshaft Position Sensor (CPS)
- Pulser Coil

The low-pressure fuel circuit is almost identical to the fuel delivery circuit found on carbureted motors. A fuel line with primer bulb is used to connect the boat mounted or portable fuel tank to one or more powerhead mounted diaphragm-displacement fuel pump(s). The low-pressure circuit is used to deliver fuel from the fuel tank to a powerhead mounted vapor separator tank which contains a submerged high-pressure electric fuel pump. The vapor tank also receives 2-stroke engine oil from and oil pump. The high-pressure pump then draws fuel and oil from the vapor tank and forces it through high-pressure fuel lines to the fuel injectors for each cylinder.

Yamaha High Pressure Direct Injection (HPDI)

◆ See Figure 223

The Yamaha HPDI system is, as the name suggests, a direct to the combustion chamber injection system which operates under extreme high-pressure. Because the fuel/oil mixture is sprayed directly into the combustion chambers timing is critical, meaning there is no real way for a direct injection system to be anything BUT sequential. Fuel mapping decisions are based upon a combination of factors including

engine speed from the Crankshaft Position Sensor (CPS), throttle opening from the Throttle Position Sensor (TPS) and intake air from the Intake Air Temperature (IAT) sensor and Atmospheric Pressure Sensor (APS). In this way the ECM keeps track of just how much air is entering the engine (between the TPS and the temperature/pressure of the air) and can adjust air/fuel mixture ratios to suit operating conditions. A lack of Oxygen sensor makes this form of injection open-loop only (meaning the ECM receives no feedback on the combustion process).

Like most fuel injection systems, the Yamaha HPDI system incorporates a return-to-port capability or limp mode. If there is a major sensor failure that prevents the ECM from processing one or more signals, the engine is run on a minimum performance map. In most instances the ECM will substitute a fixed value for the missing sensor signal. The only input that will shut the

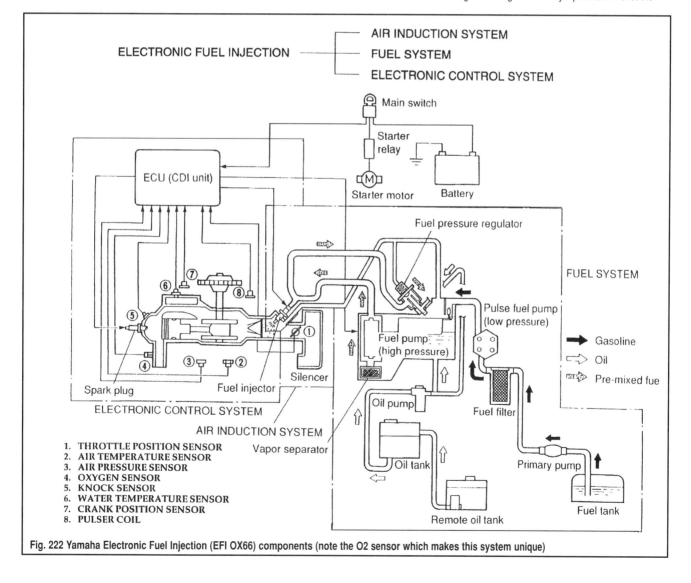

Fig. 222 Yamaha Electronic Fuel Injection (EFI OX66) components (note the O2 sensor which makes this system unique)

3-58 FUEL SYSTEM

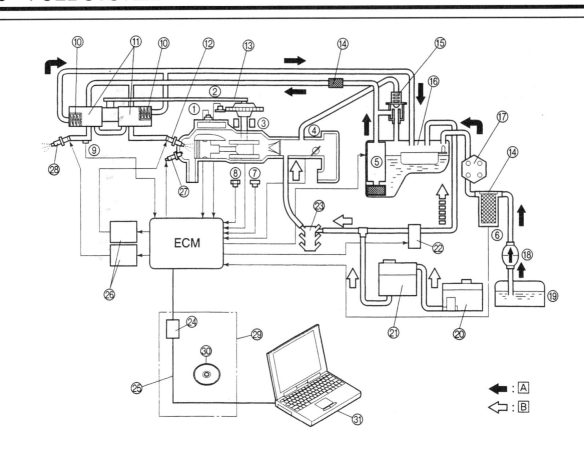

Fig. 223 Yamaha High-Pressure Direction Injection (HPDI) components and operational schematic (note the O2 sensor which makes this system unique)

① Engine temperature sensor
② Crank position sensor
③ Pulser coil
④ Throttle position sensor
⑤ Electric fuel pump
⑥ Water detection switch
⑦ Atmospheric pressure sensor
⑧ Intake air temperature sensor
⑨ Fuel pressure sensor
⑩ High-pressure fuel regulator
⑪ High-pressure fuel pump
⑫ Injectors #1, #3, #4
⑬ Drive belt
⑭ Fuel filter
⑮ Fuel pressure regulator
⑯ Vapor separator
⑰ Fuel pumps
⑱ Primer pump
⑲ Fuel tank
⑳ Remote oil tank
㉑ Oil tank
㉒ Electric oil pump
㉓ Oil pump
㉔ Adapter
㉕ Communication cable
㉖ Injector drivers
㉗ Spark plug
㉘ Injectors #2, #5, #6
㉙ Service tool (optional)
㉚ Software (CD-ROM)
㉛ Personal computer

A Gasoline
B Oil

engine down completely is battery voltage. If the battery is disconnected or the battery voltage falls below a certain point, the fuel pump quits pumping so the engine stops.

Major components of this system include the ECM and Sensors along with the low-pressure, 1st and 2nd stage high-pressure (high- and extreme high-pressure) fuel delivery circuits. The electronic control sensors include:

- Throttle Position Sensor (TPS)
- Intake Air Temperature (IAT) Sensor
- Atmospheric Pressure Sensor (APS)
- Water Temperature Sensor (WTS)
- Crankshaft Position Sensor (CPS)
- Pulser Coil
- Water Detection Sensor
- Fuel Pressure Sensor (FPS)
- Oxygen Sensor (some 150-200 hp models)

The low-pressure fuel circuit is almost identical to the fuel delivery circuit found on carbureted motors. A fuel line with primer bulb is used to connect the boat mounted or portable fuel tank to one or more powerhead mounted diaphragm-displacement fuel pump(s). The low-pressure circuit is used to deliver fuel from the fuel tank to a powerhead mounted vapor separator tank which contains a submerged high-pressure electric fuel pump. The vapor tank also receives 2-stroke engine oil from and oil pump. The high-pressure pump then draws fuel and oil from the vapor tank and forces it through high-pressure fuel lines to the single (V4) or twin (V6) powerhead mounted, belt driven mechanical extreme high-pressure fuel pump(s). The mechanical high-pressure fuel pump(s) then deliver(s) this fuel/oil mixture to the direct injectors for each cylinder.

Controlled Combustion System

Various versions of a Controlled Combustion System (CCS) are employed by fuel injected motors to reduce emissions, smooth idle and help improve shifting or low speed stability for trolling. The CCS mapping in the ECM is designed to cut one or more cylinders under certain circumstances. This reduces noise, increases slow speed stability and improves fuel economy.

FUEL SYSTEM 3-59

It is an inventive solution to the problems associated with running a large displacement engine at neutral or trolling speed. A large displacement engine is not working hard enough at low speeds to maintain peak combustion efficiency. It is also difficult to meter small amounts of fuel and air precisely through large diameter throttle valves and intakes. These problems can result in excessive fuel consumption and roughness when trolling for extended periods. The CCS effectively shrinks engine size by limiting fuel delivery to 4 or 5 cylinders, depending on the fuel system and engine RPM.

Some EFI OX66 motors will run on 5 cylinders between 500 and 850 rpm, and between 850 and 2000 rpm only 4 cylinders (if the motor is left in neutral).

To improve low-speed smoothness at reduced throttle settings, cylinder #5 does not fire between idle and 850 rpm. Cylinders #5 and #2 do not fire between 850 rpm and 2000 rpm, again depending upon throttle opening.

If the shift cut switch is activated during CCS operation, the following cylinders do not fire:
- From idle to 2000 rpm - cylinder number 5 does not fire
- From 850 rpm to 2000 rpm - cylinder number 2 does not fire and one additional cylinder does not fire
- Above 2000 rpm - cylinders number 2 and 5 do not fire when the shift cut switch is activated.

The CCS operates selectively as previously described and switches to full 6-cylinder operation under the following conditions:
- During starting
- During warm-up when the water temperature is below 113°F (45°C)
- When the oxygen sensor is not functioning
- Under acceleration
- When decelerating
- When the knock sensor is activated
- When the warning system is activated (RPM reduction, horn and red indicator activated due to overheat, low oil, or dual engine problem)
- When there is a sensor failure.

For HPDI motors partial cylinder CCS operation will commence whenever the gear shift is in Neutral. With engine speeds from idle to about 1750 rpm the ECM will shut off fuel injection and ignition to cylinders size #2 and #3. If the throttle valves are opened further while still in Neutral the ECM will shut down cylinders #5 and #6 as well.

■ **Keep the CCS system in mind when attempting to diagnose potential fuel and ignition problems, as the system may make it appear that you have one or more dead cylinders when it may not necessarily be true under normal operating conditions. Always verify a suspected dead cylinder by running the motor in gear (under load) to eliminate the CCS as a potential cause.**

Troubleshooting Electronic Fuel Injection

On carbureted outboards fuel is metered through needles and valves that react to changes in engine vacuum as the amount of air drawn into the motor increases or decreases. The amount of air drawn into carbureted motors is controlled through throttle plates that effectively increase or decrease the size of the carburetor throat (as they are rotated open or closed).

In contrast, fuel injected engines use a computer control module to regulate the amount of fuel introduced to the motor. The module or Electronic Control Module (ECM) monitors input from various engine sensors in order to receive precise data on items like engine position (where each piston is on its 2-stroke cycle), engine speed, engine and/or air temperatures, ambient air or manifold pressure and throttle position. Analyzing the data from these sensors tells the engine exactly how much air is drawn into the motor at any given moment and allows the ECM to determine how much fuel is required.

The ECM will energize (open) the fuel injectors for the precise length of time required to spray the amount of fuel needed for engine operating conditions. In actuality, the injectors are not just activated and held-open as much as they are pulsed, opened and closed rapidly for the correct total amount of time necessary to spray the desired amount of fuel. This electronically controlled, precisely metered fuel spray or "fuel injection" is the heart of a modern fuel injection system and the main difference between a fuel injected and carburetor motor.

Troubleshooting a fuel injected motor contains similarities to carbureted motors. Mechanically, the powerhead of a fuel injected motor operates in the same way as that of its carbureted counterpart. There still must be good engine compression for either engine to operate properly. Wear or physical damage will have virtually the same affect upon either motor. Furthermore, the low-pressure fuel system that supplies fuel to the reservoir in the vapor separator tank operates in the same manner as the fuel circuit that supplies gasoline to the carburetor float bowl.

The major difference in troubleshooting engine performance on EFI or HPDI motors is the presence of the ECM and electronic engine controls. The complex interrelation of the sensors used to monitor engine operation and the ECM used to control both the fuel injection and ignition systems makes logical troubleshooting all that much more important.

Before beginning troubleshooting on an EFI/HPDI motor, make sure the basics are all true. Make sure the engine mechanically has good compression (refer to the Compression Check procedure that is a part of a regular Tune-Up). Make sure the fuel is not stale. Check for leaks or restrictions in the Lines and Fittings of the low pressure fuel circuit, as directed in this section under Fuel Tank and Lines. EFI/HPDI systems cannot operate properly unless the circuits are complete and a sufficient voltage is available from the battery and charging systems. A quick-check of the battery state or charge and alternator output with the engine running will help determine if these conditions are adversely affecting EFI/HPDI operation.

■ **Loose/corroded connections or problems with the wiring harness cause a large percentage of the problems with EFI/HPDI systems. Before getting too far into engine diagnostics, check each connector to make sure they are clean and tight. Visually inspect the wiring harness for visible breaks in the insulation, burn spots or other obvious damage.**

In order to help find electronic problems with the EFI/HPDI system, the ECM contains a self-diagnostic system that constantly monitors and compares each of the signals from the various sensors. Should a value received by the ECM from one or more sensors fall outside certain pre-determined ranges the ECM will determine there is a problem with that sensor's circuit. Basically the ECM compares signals received from different sensors to each other and to real world possible values and makes a decision if it thinks one must be lying. For instance, if the ECM receives a crankshaft position sensor signal that shows the engine is rotating at a high rpm, but also receives a signal from the TPS that says the throttle is fully closed, it knows one is wrong. Similarly, if it receives a ridiculous signal, say the intake air temperature suddenly provides a signal above 338°F/170°C, the ECM will know there is something wrong with that signal. Depending on the severity of the fault or faults, the engine will continue to run, substituting fixed values for the sensors that are considered out of range. Under these circumstances, engine performance and economy may become drastically reduced.

■ **Check for the presence of diagnostic codes as described in this section BEFORE disconnecting the battery. If codes are not present, yet problems persist, use the symptom charts to help determine what further components or systems to check.**

When a fault is present the ECM will display a diagnostic code when the proper diagnostic flash adapter is attached to the motor. The adapter contains a flash indicator (essentially a small light) which will illuminate in specific patterns to communicate diagnostic codes to the technician. The codes can then be used to help determine what components and circuits should be checked for trouble. Remember that a fault code doesn't automatically mean that a component (such as a sensor) is bad, it means that the signal received from the sensor circuit is missing or out of range. This can be caused by loose or corroded connections, problems with the wiring harness, problems with mechanical components (that are actually causing this condition to be true), or a faulty sensor.

Once components or circuits that require testing have been identified, use the testing procedures found in this section (or other sections, as applicable) and the wiring diagrams to test components and circuits until the fault has been determined.

Keep in mind that although a haphazard approach might find the cause of problems, only a systematic approach will prevent wasted time and the possibility of unnecessary component replacement. In some cases, installing an electronic component into a faulty circuit that damaged or destroyed the previous component, will instantly destroy the replacement. For various reasons, including this possibility, most parts suppliers do not accept returns on electrical components.

3-60 FUEL SYSTEM

Self Diagnostic System

READING

◆ See Figures 224 thru 227

■ Certain electrical equipment such as stereos and communication radios can interfere with the electronic fuel injection system. To be certain there is no interference, shut these devices off when troubleshooting. If a check engine light illuminates immediately after installing or re-rigging an existing accessory, reroute the accessory wiring to prevent interference.

When the electronic engine control system detects a problem with one of its circuits, the ECM will display diagnostic trouble codes when the appropriate diagnostic flash adapter is connected to the wiring harness. Flash adapter and connection locations vary slightly by model. As a result of most faults, the ECM will ignore the circuit signal and enter a fail-safe mode designed to keeps the boat and motor from becoming stranded. During fail-safe operation the ECM will provide a fixed substitute value for the faulty circuit. During fail-safe operation the engine will run, but usually with reduced performance (power and economy).

Once a malfunction ceases (has been fixed or the circuit signal returns to something within the anticipated normal range), engine operation will return to normal and the diagnostic flash indicator will cease to display the trouble code for that circuit.

Proceed as follows to display stored codes:
1. Start with the engine and ignition switch turned **OFF**
2. Using the accompanying illustrations and information, locate the necessary wiring harness connector and install the diagnostic flash indicator:
 • For EFI OX66 motors, except the 2002 and later 250 hp Vmax, connect the diagnostic flash indicator (#YB-06765) to the wiring harness connector below the oil tank and above the oil pump/distribution manifold, as detailed in the accompanying illustration.
 • For 2002 later 250 hp Vmax motors with the EFI OX66 system, connect the diagnostic flash indicator (#YB-06444) to the wiring harness connector immediately adjacent to the middle low-pressure fuel pump on the powerhead, as detailed in the accompanying illustration.
 • For HPDI motors (except the 2003 or later 250 hp model), connect the diagnostic flash indicator (#YB-06765) to the wiring harness connector below the oil tank and above the oil pump/distribution manifold, as detailed in the accompanying illustration.
 • For 2003 or later 250 hp HPDI motors connect the diagnostic flash indicator (#YB-06444) to the wiring harness connector which is draped over the oil tank and the connector for the blue wires that extend out from behind the oil tank, as detailed in the accompanying illustration.

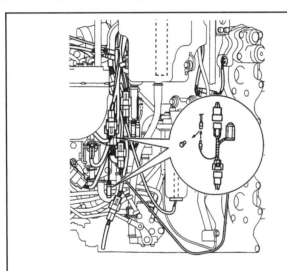

Fig. 224 Connecting the diagnostic flash indicator - EFI OX66 motors (except the 2002 and later 250 hp Vmax) and HPDI motors (except the 2003 or later 250 hp model)

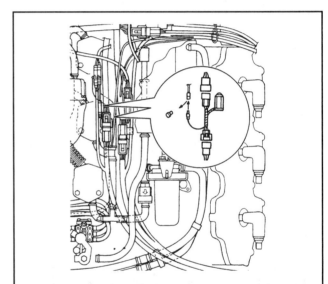

Fig. 225 Connecting the diagnostic flash indicator - 2002 or later 250 hp Vmax with EFI OX66 fuel injection

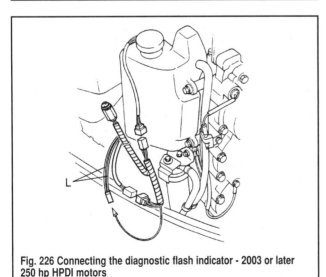

Fig. 226 Connecting the diagnostic flash indicator - 2003 or later 250 hp HPDI motors

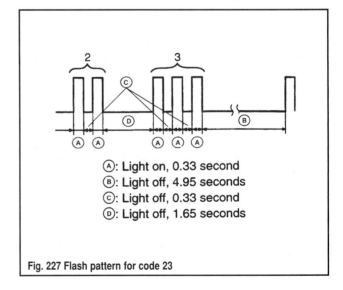

Ⓐ: Light on, 0.33 second
Ⓑ: Light off, 4.95 seconds
Ⓒ: Light off, 0.33 second
Ⓓ: Light off, 1.65 seconds

Fig. 227 Flash pattern for code 23

3. Connect a suitable source of cooling water to the motor, then start the engine and allow it idle, observe the flash pattern on the diagnostic indicator to determine if there are any trouble codes.

4. Interpret codes by counting the flashes. Only one code will be displayed at a time. The first digit of the 2-digit code will display first, with one, two, three or possibly even four flashes (the light will remain on and off for the same amount of time, about 0.33 seconds, when flashing a code digit). Count the number of times on and you've got the first digit.

■ **Some engines use a single digit code 1 to indicate normal operation. When present the single 0.3 second flash will be separated from the next single flash by a long 5 second pause.**

5. There will be a longer pause between digits of the 2-digit code (the light will remain off for about 1.65 seconds), then the short on/off flashes will indicate the second digit of the 2-digit code. This second flash may consist of any number from one to nine. A single code will continue to flash until the problem is corrected, then the next higher code (if present) will flash until corrected. This pattern will repeat until all faults are corrected.

■ **If more than one is present, codes will flash in the order of priority (lowest number to highest number) as listed in the accompanying code chart. Only a single code will be displayed until it is corrected, then the next higher code will flash.**

6. Compare the codes with the accompanying code table to determine the defective circuit. Remember that a code does NOT necessarily mean a given component is at fault, it means that the ECM sees signal that is out of the normal, predetermined operating range. The problem may also lie with the wiring, another system or another component that would make a circuit read out of specification.

■ **Make sure all connections for that circuit and related components of the electrical system are clean and tight before troubleshooting the circuit. Bad wiring or connections can cause out of range signals and set trouble codes.**

7. Test the sensor or system components as described in this section. When troubleshooting, always start with the easiest checks/fixes and work toward the more complicated.

8. Once all of the problems have been corrected, connect a source of cooling water to the flushing system, then start and run the engine to verify proper operation. If the problems are gone, the diagnostic indicator either will remain off or will flash a code 1 (depending upon the model).

9. Shut the ignition switch **off** and remove the diagnostic connector.

TROUBLE CODES

◆ See Figure 228

Sensor and Circuit Resistance/Output Tests

TESTING ELECTRONIC COMPONENTS

Before conducting resistance or voltage checks on a component/circuit, make sure all electrical connections in the system are clean and tight. Visually check the wiring for obvious breaks, defects or other problems.

Keep in mind that although a haphazard approach might find the cause of problems, only a systematic approach will prevent wasted time and the possibility of unnecessary component replacement. In some cases, installing an electronic component into a faulty circuit that damaged or destroyed the previous component, will instantly destroy the replacement. For various reasons, including this possibility, most parts suppliers do not accept returns on electrical components.

Make sure you perform each test procedure and especially, make all test connections as described. If a component tests out of range (faulty), but the reading is close to the service limit, bring the component to your marine dealer so they can verify your result before purchasing the replacement. In some cases, your dealer may be willing to check a sensor against the reading on a new one to verify whether or not your sensor is actually faulty.

Although resistance checks are relatively easy to conduct, their results can be difficult to interpret. Remember that resistance in any electrical component or circuit will vary with temperature. Also, ohmmeters will vary with quality, meaning that readings can vary on the same component between different meters. The specifications provided in this service are based upon the use of a high-quality digital multi-meter applied to a component that is currently at about 68°F (20°).

■ **More information on electrical test equipment is can be found in the General Information section, while additional information on basic electrical theory and electrical troubleshooting can be found in the Ignition and Electrical System section.**

As a general rule, the resistance of a circuit or component will increase (rise) as temperature increases. Although this is true for most resistor and windings, some manufacturers use negative temperature coefficient sensors for certain functions. A negative temperature coefficient sensor reacts in an opposite manner, meaning that its resistance decreases (lowers) as temperature increases (rises) and vice-versa.

Most components of the EFI/HPDI systems are checked using resistance or voltages checks. This means that a voltmeter or ohmmeter is connected to the component or circuit under the proper circumstances.

For resistance checks, no voltage must be applied to the circuit (except that as provided by the meter to perform the check). Typically, the circuit or component wiring is isolated, by disconnecting the wiring from the component itself or somewhere else in the circuit, then applying the meter probes across 2 terminals that connect through the component. When conducting resistance checks, you **must** isolate the ECM from the meter in order to prevent the possibility of damage to the control module. By identifying the proper circuit you can conduct tests of the entire component circuit right from the harness connectors that attach to the ECM. Otherwise, use the wiring diagrams and the test procedures to determine other points to disconnect the wiring and test the circuit/component.

We recommend you work in one of two directions. Either test the component first, then work your way back through the component wiring toward the ECM, or disconnect the harness connectors from the ECM, testing the entire circuit first, then working your way toward the component. A bad test reading at the ECM harness connector must be verified by making sure it is not the result of a bad wiring harness or connection along the way to the component. Likewise, a good reading at a component (when there is a bad reading on the circuit as noted by the presence of a trouble code or by a reading at the ECM harness) must be verified by checking the circuit wiring for trouble.

When checking the circuit wiring, isolate the harness by disconnecting it from both the component and the ECM, then using an ohmmeter to check resistance from one connector to the next on each wire. Wiggle the wire and connectors while conducting the test to see if reading fluctuate. Although no specifications are provided for wiring resistance, all sensor/switch wiring for these motors should have very little resistance. Also, be sure to check wiring for shorts to ground (by checking resistance between one terminal and a good engine ground), and for shorts to power (by checking a terminal using a voltmeter and connecting the other probe to a good engine ground or the negative battery cable).

※※ WARNING

Be VERY careful when checking circuits for electronic engine control components. Always isolate the ECM by disconnecting the wiring from it before using an ohmmeter on the wires. Keep in mind that making the wrong connections across the terminals of an ECM can fry the unit, requiring an expensive replacement.

When conducting voltage tests, the circuit must be complete in order to receive a reading. In most cases, this means the wiring must be left connected and the meter probes should be carefully inserted through the back of the connector (called back-probing) in order to get a reading without opening and disrupting circuit operation. However, back-probing should NOT be done on waterproof connectors as you will compromise the seal and jumper wires should instead by used to complete the circuit. The best method for manufacturing a jumper wire is to obtain two identical size/type connectors as the portion of the harness to which you will be inserting the jumper wires. Wire these jumper connectors together along with a second wire connected to EACH terminal. Each second wire from a terminal should route off to a female bullet connector which can be safely probed without the fear of accidental shorting by connecting 2 or more wires together.

EFI/HPDI TROUBLE CODES

Trouble Code	Fault
1	Normal
13	Incorrect pulser coil input signal
14	No or incorrect Crankshaft Position Sensor (CPS) signal (EFI OX66 and HPDI only)
15	Incorrect engine cooling Water Temperature Sensor (WTS) signal
17	Incorrect Knock Sensor input signal (EFI OX66 only)
18	Incorrect Throttle Position Sensor (TPS) signal
19	Low battery input voltage
22	Atmospheric Pressure Sensor (APS) signal out of range (EFI OX66 and HPDI only)
23	Incorrect Intake Air Temperature (IAT) sensor signal
25	Incorrect fuel pressure sensor signal (HPDI only)
26	Incorrect or no injector operational signal (HPDI only)
27	Incorrect water in fuel signal or water in fuel signal ON (HPDI only)
28	Incorrect shift position switch signal (HPDI and 4-st EFI only)
29	Intake Air Pressure (IAP) sensor signal out of normal range (4-st EFI only)
31-44	These codes are Micro-computer processing information and not necessarily faults
31	No tachometer pulse being output (EFI OX66 only)
32	Shift cutoff control - during ignition cutoff operation (EFI OX66 only)
33	Ignition timing being slightly corrected - during a cold start
35	Fuel injection period being slightly corrected - during knock control (EFI OX66 only)
36	Ignition timing being slightly corrected - during knock control (EFI OX66 only)
37	Intake passage air leakage (4-st EFI only)
39	Incorrect oil pressure sensor signal (4-st EFI only)
41	Over-rev control - during ignition cutoff operation (EFI OX66 only)
42	Overheat control / oil empty control (EFI OX66 only)
43	Buzzer activated (EFI OX66 only)
44	Engine stop switch control activated
45	Incorrect shift cut switch signal (4-st EFI only)
46	Incorrect thermoswitch signal (4-st EFI only)

NOTE: not all codes used by all model. Also remember a code means a problem with a CIRCUIT and not necessarily a component

Fig. 228 EFI/HPDI Trouble Codes

Throttle Body and Intake Assembly

Like the throttle bore of a carburetor, the throttle body assembly controls engine speed by mechanically controlling the amount of air allowed to enter the engine. Whereas a carburetor meters both fuel and air, the throttle body assembly on multi-port fuel injected motors is only used to meter incoming air. As a matter of fact, a typical throttle body assembly shares the same basic design and function (to control engine speed through the control of incoming air) as the air throttle components of a carburetor.

All EFI/HPDI motors are equipped with a throttle body assembly which uses an individual throttle bore and throttle valve for each cylinder. The throttle body assembly is found at the front of the intake manifold, behind the air intake silencer.

On most motors the throttle valves/bores are arranged in a stack on the throttle body and intake assembly. For EFI OX66 and HPDI motors, these throttle valves are part of a single throttle body unit. In all cases, the manufacturer recommends removing the throttle body and intake as an assembly.

Details on removing the vapor separator from the assembly can be found later in this section under Vapor Separator Tank and High-Pressure Fuel Pump.

REMOVAL & INSTALLATION

EFI OX66 Motors

◆ See Figures 229 and 230

On these models, there are six individual induction tracts molded in a single bank. There is one tract for each cylinder. Each induction tract has an air throttle valve and an injector. The throttle position sensor is normally mounted on the port side of the #1 throttle valve. Throttle valve synchronization procedures are similar to inline engines. These engines will require synchronization less frequently than carburetor-equipped engines. Tight synchronization and correct throttle position sensor adjustment, however, are critical as they have been with previous Yamaha V-engines.

This procedure removes the throttle body and vapor separator tank as an intact assembly for access to other components on the powerhead. The advantage of this procedure is that it allows removal of the entire high-pressure fuel system without depressurization. However, if further disassembly of the unit is required, fuel pressure must be released. This can be done before or after the assembly is removed from the powerhead, though we give it at the beginning of the procedure because if you're going to do it, there is no good reason to delay. Why work with pressurized fuel lines if you don't have to?

FUEL SYSTEM 3-63

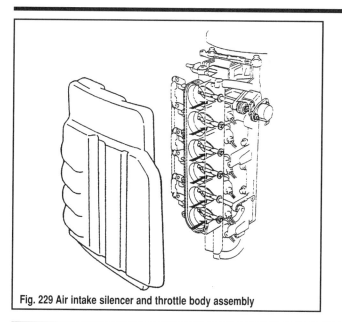

Fig. 229 Air intake silencer and throttle body assembly

1. Disconnect the negative battery cable for safety.
2. If the fuel injectors, rail, lines or vapor separator tank are going to be removed from the throttle body assembly, properly relieve the fuel system pressure, as detailed under Relieving Fuel System Pressure earlier in this section.
3. Remove the 6 screws securing the air intake silencer assembly, then pull the silencer away from the throttle body enough to access the suction and oil tank air vent hoses. Tag and disconnect the hoses from the silencer, then remove the silencer assembly from the powerhead.
4. Disconnect the throttle link rod and oil pump link rod.
5. Tag and disconnect the TPS, fuel injector harness, electric fuel pump resistor and main relay connectors.
6. Cut the wire ties securing the fuel and oil inlet hoses to the top of the vapor separator tank, then tag and disconnect those 2 hoses.
7. Using multiple passes of a spiraling, counterclockwise pattern that starts at the outer fasteners and works toward the inner fasteners (the reverse of the torque sequence), loosen the 2 nuts and 12 bolts retaining the throttle body assembly to the powerhead.
8. Remove and discard the old gasket from the mating surface.
9. If further disassembly is required refer to the Fuel Rail and Injectors and the Vapor Separator Tank and High-Pressure Fuel Pump procedures in this section.

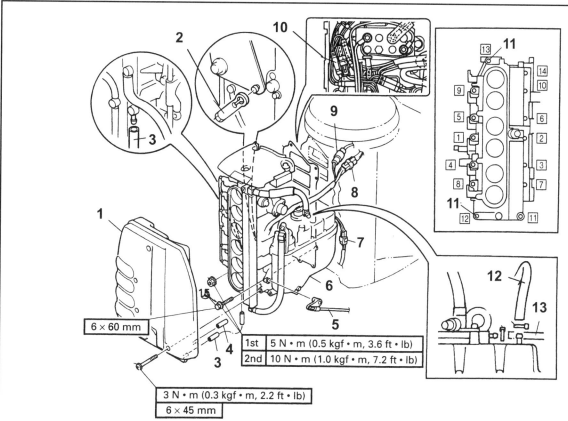

1- Intake Silencer
2- Throttle Link Rod
3- Suction Hose
4- Oil Tank air vent hose
5- Oil Pump Link rod
6- Throttle body and vapor separator assembly
7- TPS connector
8- Fuel injector connector
9- Electric fuel pump resistor connector
10- Main relay connector
11- Nut
12- Fuel inlet hose
13- Oil inlet hose

Fig. 230 Throttle body and vapor separator assembly mounting - EFI OX66 motors

FUEL SYSTEM

To Install:

10. If removed, install the Vapor Separator Tank and/or the Fuel Rail and Injectors to the Throttle Body Assembly, as detailed in those procedures.

11. Position the new throttle body gasket over the 2 mounting studs which are used for the 2 throttle body retaining nuts. Carefully align and install the throttle body assembly over the gasket and mounting studs, then thread and finger-tighten the 12 mounting bolts and 2 mounting nuts.

12. Tighten the throttle body fasteners using multiple passes of the clockwise, spiraling torque sequence that starts at the center bolts and works towards the outer fasteners. Tighten the bolts/nuts first to 3.6 ft. lbs. (5 Nm) and then to 7.2 ft. lbs. (10 Nm).

13. Reconnect the oil and fuel inlet hoses to the top of the vapor separator tank and secure using new plastic wire ties.

14. Reconnect the wiring for the main relay, electric fuel pump resistor, fuel injector harness and TPS.

15. Reconnect the oil pump link rod and throttle link rod.

16. Position the air intake silencer just in front of the throttle body assembly, then reconnect the oil tank air vent and suction hoses. Install the silencer and carefully tighten the 5 retaining bolts to 2.2 ft. lbs. (3 Nm).

17. Reconnect the negative battery cable and properly pressurize the fuel system in order to check for leaks. Start and run the motor with the top cover removed and observe the fuel lines/fittings for any signs of weepage. Double-check all lines and fittings after the first outing with the motor.

HPDI Motors

◆ See Figures 231, 232 and 233

On these models, there are 6 individual induction tracts molded into a single throttle body assembly. There is one tract for each cylinder. Each induction tract has an air throttle valve, however unlike the EFI motors, fuel is not injected to the intake assembly, as it is instead injected directly to the combustion chambers.

Yamaha recommends removing the throttle body and vapor separator tank as an assembly, however unlike the EFI models, this does not provide any advantage when it comes to system pressurization. You still need to relieve system pressure from the first stage high-pressure circuit because you need to disconnect fuel lines. Of course, if you are just removing the assembly for access to another powerhead component, the advantage comes in fewer steps. It is however possible to remove the vapor separator from the throttle body assembly without first removing the assembly from the powerhead. If that is desired, follow only the steps necessary to remove the vapor tank, leaving components like the throttle link and throttle body fasteners undisturbed.

1. Properly relieve fuel system pressure, then disconnect the negative battery cable for safety.

2. Remove the flywheel cover for access.

3. Remove the drain screw from the bottom of the vapor separator tank and drain the fuel into an approved container.

4. Loosen the 6 bolts securing the air intake silencer to the throttle body assembly, then pull the silencer away from the throttle body enough to access the suction and oil tank air vent hoses. Tag and disconnect the hoses from the silencer, then remove the silencer assembly from the powerhead.

■ **Some air intake silencers contain a rubber seal between the silencer and the throttle body assembly.**

5. Disconnect the throttle link rod and oil pump link rod.

6. Tag and disconnect the wiring for the TPS, IAT, APS, electric oil pump and electric fuel pump.

7. Cut the wire ties securing the fuel inlet, return and oil inlet pump hoses to the top of the vapor separator tank, then tag and disconnect those 4 hoses. Next locate the fuel feed hose at the top of the vapor tank (usually connected to a filter joint) and disconnect it by sliding the ring on the collar back toward the hose and pulling it free (similar to the quick-connect fittings on an air compressor hose).

8. Using multiple passes of a spiraling, counterclockwise pattern that starts at the outer fasteners and works toward the inner fasteners (the reverse of the torque sequence), loosen the 2 nuts and 11 (3.1L motors) or 12 bolts (3.3L motors) retaining the throttle body assembly to the powerhead.

9. Remove and discard the old gasket from the mating surface.

10. If further disassembly is required or to separate the vapor tank from the throttle body assembly, refer to the Vapor Separator Tank and High-Pressure Fuel Pump procedures in this section.

To Install:

11. If removed, install the Vapor Separator Tank to the Throttle Body Assembly, as detailed later in this section.

12. Position the new throttle body gasket over the 2 mounting studs which are used for the 2 throttle body retaining nuts. Carefully align and install the throttle body assembly over the gasket and mounting studs, then thread and finger-tighten the 11 (3.1L motors) or 12 (3.3L motors) mounting bolts and 2 mounting nuts.

13. Tighten the throttle body fasteners using multiple passes of the clockwise, spiraling torque sequence that starts at the center bolts and works towards the outer fasteners. Tighten the bolts/nuts first to 3.0 ft. lbs. (4 Nm) and then to 7.2 ft. lbs. (10 Nm).

14. Reconnect the oil and fuel hoses to the top of the vapor separator tank and secure using new plastic wire ties. When reconnecting the fuel feed hose the fuel joint connection should be self-explanatory, hold the ring against the collar as you slide the fitting over the joint, then release the ring and feel for a snap indicating the connecter has locked in position.

■ **Pull back gently on each of the fuel and oil connections to confirm they are secure.**

15. Reconnect the wiring for the electric oil pump, electric fuel pump, APS, IAT and TPS.

16. Reconnect the oil pump link rod and throttle link rod.

17. Position the air intake silencer just in front of the throttle body assembly, then reconnect the oil tank air vent and suction hoses. Install the silencer and carefully tighten the 5 retaining bolts to 2.2 ft. lbs. (3 Nm).

18. Reconnect the negative battery cable and properly pressurize the fuel system in order to check for leaks. Start and run the motor with the top cover removed and observe the fuel lines/fittings for any signs of weepage. Double-check all lines and fittings after the first outing with the motor.

19. Install the flywheel cover.

Vapor Separator Tank and High-Pressure Fuel Pump

The first major difference between most EFI/HPDI fuel systems and a carbureted system (as far as fuel delivery is concerned) comes at the fuel vapor separator tank. The vapor separator is mounted on the powerhead, usually on or near the throttle body assembly.

The separator tank functions as a bizarre cross between a very large float bowl and a very tiny gas tank. It receives fuel from the low-pressure pump via a float and needle valve assembly (in the same manner as a carburetor's float bowl). The level is maintained within the vapor separator tank so that it serves as a reservoir for the high-pressure electric fuel pump mounted to the separator cover.

■ **The float and needle valve are serviced in the same manner as a carburetor's float bowl. They can be accessed once the vapor separator cover/fuel pump assembly is removed.**

The high-pressure fuel system works by keeping the injectors constantly supplied with more fuel than the maximum demand requires. A pressure regulator controls this over supply by diverting excess high pressure fuel back to the separator. The HPDI fuel circuit is similar, but not the same, as the high-pressure pump in the vapor separator does not directly feed the fuel rails/injectors. Instead the high-pressure pump is used to feed the extreme high-pressure mechanical fuel pump(s) that feed(s) the fuel rails and injectors.

■ **On all models, pressurizing the fuel in the injector fuel rail retards gum formation. Often, EFI and HPDI engines will crank after a year of storage. But that is not to encourage improper storage. Remember that the fuel injectors contain tiny passages and are very sensitive to any gum formation, so don't risk it, always store the outboard properly for periods of non-use.**

To help prevent fuel problems like vapor lock or hot soak, the separator is normally vented to the air intake silencer so vapors can be drawn into the throttle body and burned when the engine is running.

The tank cover contains a replaceable high-pressure fuel pump assembly. The pump itself also contains a replaceable filter screen mounted at the fuel pickup. For 2-stroke motors the vapor tank cover also contains the fuel high-pressure regulator which can be removed for replacement or to service the filter screen.

FUEL SYSTEM 3-65

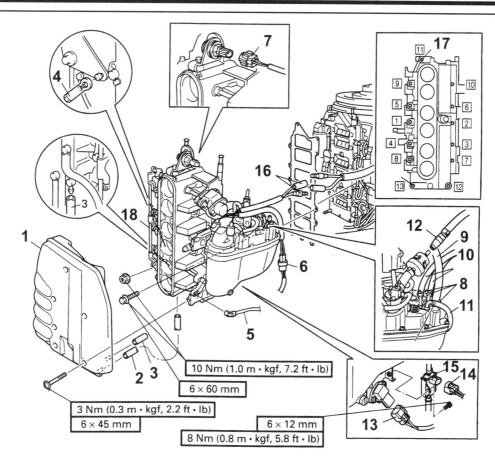

1- Intake Silencer
2- Suction hose
3- Oil Tank vent hose
4- Throttle link rod
5- Oil pump link rod
6- TPS connector
7- IAT sensor
8- Wire ties
9- Fuel inlet hose
10- Fuel return hose
11- Electric oil pump hose
12- Fuel feed hose
13- APS connector
14- Oil pump connector
15- Electric oil pump
16- Fuel pump connector
17- Nut
18- Throttle body assembly

Fig. 231 Exploded view of the throttle body and vapor separator tank assembly mounting - HPDI motors (3.1L shown, 3.3L VERY similar)

A fuel reservoir drain screw is normally found on the bottom of the tank and can be used to drain the tank of fuel.

On 2-stroke motors, the separator also functions as the mixing chamber for oil and fuel. Oil is injected into the separator tank by the engine-driven or electric oil pump (depending upon the model). The amount of oil delivered to the separator tank is dependent upon engine RPM and throttle opening. The supply side of the oil injection system is similar to previous PrecisionBlend oil injection models.

TESTING

Refer to the Maintenance and Tune-Up section for High-Pressure Fuel Pump Inspection and pressure checking procedures.

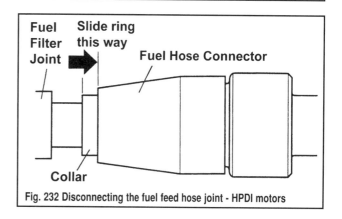

Fig. 232 Disconnecting the fuel feed hose joint - HPDI motors

3-66 FUEL SYSTEM

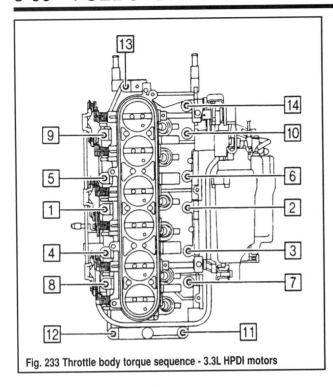

Fig. 233 Throttle body torque sequence - 3.3L HPDI motors

✱✱ CAUTION

The high pressure fuel system, as it name might imply, is capable of spraying fuel under extreme pressure. This means that the fuel will spray free under high pressure if a fitting is opened without first relieving pressure (which makes for good fuel atomization and a highly combustible condition). It will also spray fuel if the pump is actuated for any reason while a fitting is disconnected. These could lead to extremely dangerous work conditions. Do not allow ANY source of ignition (sparks, flames, etc) anywhere near the work area when servicing the fuel system.

For EFI OX66 motors the high-pressure fuel pump circuit utilizes an externally mounted resistor whose resistance itself can be checked if pump circuit problems are suspected. Locate the Brown and Blue wires for the high-pressure pump resistor and disengage the wiring from the engine harness. Connect a DVOM set to read resistance to the resistor Brown and Blue leads, the meter should show 0.53-0.57 ohms or the resistor should be replaced. Also on EFI OX66 motors if the fuel pump does not seem to operate when the switch is turned **ON** you can check the main relay by disconnecting the Brown starter relay lead (to keep the engine from starting) and using a DVOM across the Red and Red/Yellow wires of the main relay. There should be continuity when the keyswitch is turned **ON** and no continuity when the switch is turned **OFF**.

REMOVAL & INSTALLATION

◆ See Figures 234 and 335    MODERATE

On most models the vapor separator tank assembly can be removed with or without removing the throttle body and intake assembly. On most EFI models this allows you to remove the entire fuel injection unit (tank, throttle body, intake, fuel rails and injectors) without depressurizing the system.

The decision whether to remove it as an assembly or separately should come down to what is the reason for removal. If you need to service or replace the vapor separator, fuel lines or related components, then the big advantage of removing them as an assembly may be lost. However, access to the vapor separator tank and related components may be much easier with the entire unit removed from the powerhead, so it may save you some time/trouble in the long run.

So, it's up to you, but if you decide to remove the vapor tank along with the Throttle Body and Intake assembly, refer to that procedure first and then to the appropriate steps of this procedure to separate the vapor tank from the fuel injection unit.

1. Properly relieve the fuel system pressure, then disconnect the negative battery cable for safety.
2. If not done already (such as to remove the throttle body and intake assembly), tag and disconnect the wiring from the vapor tank assembly. All models have an electric fuel pump wiring harness running to the vapor tank cover. HPDI models will also have the Atmospheric Pressure Sensor (APS) harness running to the bottom of the tank.
3. Tag and disconnect the oil and fuel lines from the top of the vapor tank cover. The number of lines, locations, clamping methods all vary from model-to-model. Keep in mind that fittings with crimp style clamps are removed by cutting the crimp itself, perpendicular to the crimp. Also keep in mind that clamps should be replaced using the same style as what was originally installed (meaning you'll need a special crimping tool for crimp style clamps).

■ **On certain EFI motors it is possible to remove the vapor separator tank and fuel rail/injectors as an assembly. Either without depressurizing the system and/or without cutting the crimp style clamps usually found on the fuel rail lines. Of course, you'll still need to depressurize the system if you are going to open us the vapor separator for service. For more details about Fuel Rail and Injector service, refer to the procedures later in this section.**

4. Loosen and remove the 3 bolts securing the vapor separator tank to the throttle body assembly.
5. If the high-pressure fuel pump, pump pickup filter screen or float and needle valve assembly require service, please refer to the Overhaul procedure in this section.
6. If the fuel pressure regulator requires service, refer to Fuel Pressure Regulator, in this section.

To Install:

7. If separated for overhaul, install the cover to the vapor separator tank assembly using a new O-ring, as detailed under Overhaul
8. If removed, install the fuel pressure regulator to the vapor separator tank assembly.
9. Install the vapor tank to the throttle body assembly, intake manifold or bracket, as applicable and secure using the fasteners.
10. Reconnect the oil and fuel lines to the vapor tank cover as tagged during removal. If you run into a snag (a tag fell off or something wasn't marked) refer to the art accompanying the Throttle Body and Intake procedures to see if the hose is labeled in the artwork. Hose locations vary slightly from model-to-model. In most cases the hoses are JUST the right length to reach their fittings, so if two hoses appear kinked or bowed with extra length (or not enough length), suspect that you might have them switched. Secure the hoses using new crimp type clamps or wire ties, when applicable or using the spring-type clamps, when used.

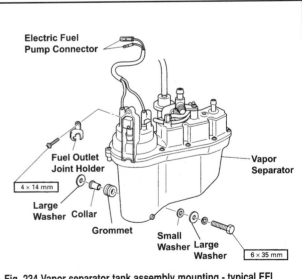

Fig. 234 Vapor separator tank assembly mounting - typical EFI model

FUEL SYSTEM 3-67

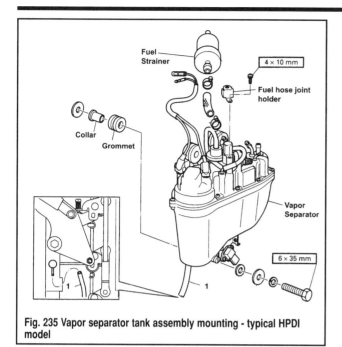

Fig. 235 Vapor separator tank assembly mounting - typical HPDI model

✲✲ CAUTION

Don't install a different style fuel hose clamp than was originally used in that location, or you're risking a leak and a severe fire/explosion hazard.

11. Unless you've removed the entire Throttle Body and Intake assembly, reconnect the harness wiring to the vapor tank components such as the high-pressure fuel pump and, on HPDI models, the APS.
12. If removed, install the Throttle Body and Intake assembly, as detailed earlier in this section.
13. Reconnect the negative battery cable and properly pressurize the fuel system in order to check for leaks. Start and run the motor with the top cover removed and observe the fuel lines/fittings for any signs of weepage. Double-check all lines and fittings after the first outing with the motor.

OVERHAUL

◆ See Figures 236 thru 239

■ The pressure regulator is mounted to the top of the vapor separator assembly. It can be removed at any time. For details, please refer to the Fuel Pressure Regulator procedures later in this section.

1. Remove the vapor separator assembly from the powerhead. For details, please refer to the procedure earlier in this section.
2. If not done already, remove the drain screw and gasket or O-ring from the bottom, side of the vapor tank, then drain the fuel into a suitable container.
3. If necessary for HPDI motors, remove the 2 screws securing the Atmospheric Pressure Sensor (APS) to the bottom, side of the vapor separator float bowl. Also, if necessary, remove the 2 bolts and the sensor/electric oil pump bracket from the bowl.

■ The APS must be removed from the vapor separator float bowl on HPDI models before the bowl is submerged in or cleaned with solvent.

4. Remove the 9 screws from the vapor separator cover flange, then carefully lift the cover separating it from the float bowl. Remove and discard the old O-ring seal from the cover-to-bowl mating surface.

■ Although the manufacturer does not specify that the bowl O-ring cannot be reused, we recommend replacing it whenever the assembly is overhauled for safety (to prevent possible leaks that might occur is a worn or deformed seal is reused).

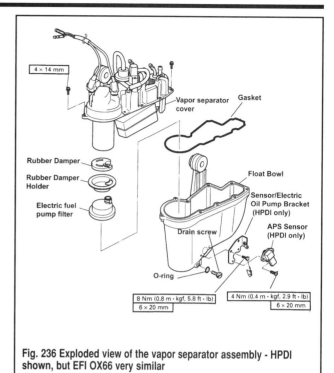

Fig. 236 Exploded view of the vapor separator assembly - HPDI shown, but EFI OX66 very similar

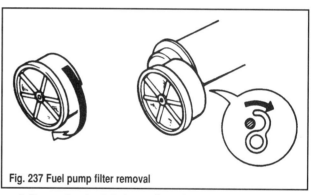

Fig. 237 Fuel pump filter removal

5. Remove the electric fuel pump filter, rubber damper holder and damper from the bottom of the electric fuel pump. The pump itself is removed by carefully twisting the filter about 1/8 turn clockwise (when looking at the bottom of the pump).
6. If necessary, remove the fuel pump and/or float assembly, as follows:
   a. For HPDI motors, remove the terminal cover.
   b. Remove the terminal caps, then loosen and remove the nuts and spring washers securing the fuel pump wiring harness terminals to the pump terminal assembly. Remove the terminals from the top of the vapor separator cover.
   c. Remove the O-rings from the electric fuel pump terminal assembly, and on EFI OX66 models, remove the flat washers.
   d. Carefully remove the electric fuel pump, guide plate, collar and O-rings from the vapor separator cover.
   e. For HPDI motors, remove the coupler holder for the terminal assembly.
   f. Remove the terminal insulators and O-rings.
   g. To remove the float assembly, remove the screw followed by the float hinge pin.
   h. Carefully lower the float and needle assembly from the vapor separator cover. If necessary, remove the clip and separate the needle valve from the float.
7. Carefully clean all metallic parts in carburetor cleaner or a suitable solvent. DO NOT immerse rubber or electrical parts in solvent. Replace any worn, damaged or questionable components. Although Yamaha does not make specific recommendations regarding O-ring replacement, we recommend that ALL O-rings be replaced to ensure proper and reliable operation.

3-68 FUEL SYSTEM

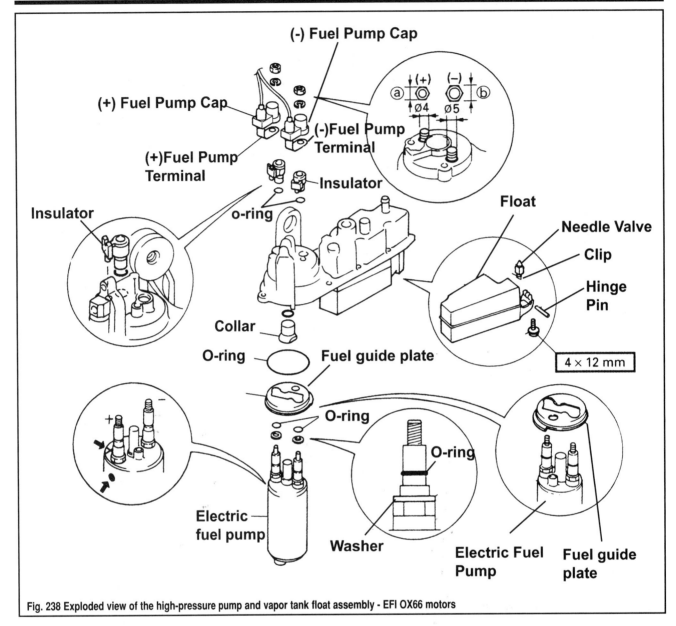

Fig. 238 Exploded view of the high-pressure pump and vapor tank float assembly - EFI OX66 motors

To Assemble:

8. If removed, install the fuel pump and/or float assembly.

   a. Secure the needle valve to the float using the clip, then carefully lower the float and valve into position in the cover. Insert the hinge pin to secure the float, then install and tighten the screw to secure the hinge pin

   b. On HPDI motors, if removed, connect the terminal assembly to the top of the fuel pump.

   c. Install the fuel pump, guide plate, collar and O-rings to the vapor separator cover.

   d. On HPDI motors, install the coupler holder for the terminal assembly.

   e. For EFI OX66 motors, install the flat washers and O-rings to the fuel pump terminals.

   f. Install the O-rings and insulators, then connect the positive and negative terminals. Secure the terminal using the spring washers and nuts, then install the terminal caps.

   g. For HPDI motors, install the terminal cover.

9. Install the rubber damper and damper holder to the bottom of the fuel pump, then install the fuel pump filter. Align the filter and then turn it counterclockwise until you feel/hear a click meaning that the filter has locked into place.

10. Install the cover to the float bowl making sure the cover O-ring seal is not dislodged or pinched during installation. Install the 9 cover screws and tighten them securely

11. On HPDI motors, if removed install the APS and electric oil pump bracket to the bottom, side of the fuel bowl. Install the APS to the bracket.

12. Install the drain screw using a new gasket or O-ring (as applicable).

13. If removed, install the Fuel Pressure Regulator to the vapor separator tank assembly.

14. Install the fuel vapor separator assembly, as detailed in this section.

HPDI Drive Belt and Sprockets

The extreme high-pressure fuel pump (2nd stage high-pressure pump) on HDI motors is a mechanical, belt driven pump run off a sprocket mounted on top of the pump itself. The drive belt should be inspected periodically and replaced, if there are any signs of wear or damage.

REMOVAL & INSTALLATION

◆ See Figures 240 thru 244

1. Disconnect the negative battery cable for safety.
2. Remove the flywheel cover from the top of the powerhead for access.

FUEL SYSTEM 3-69

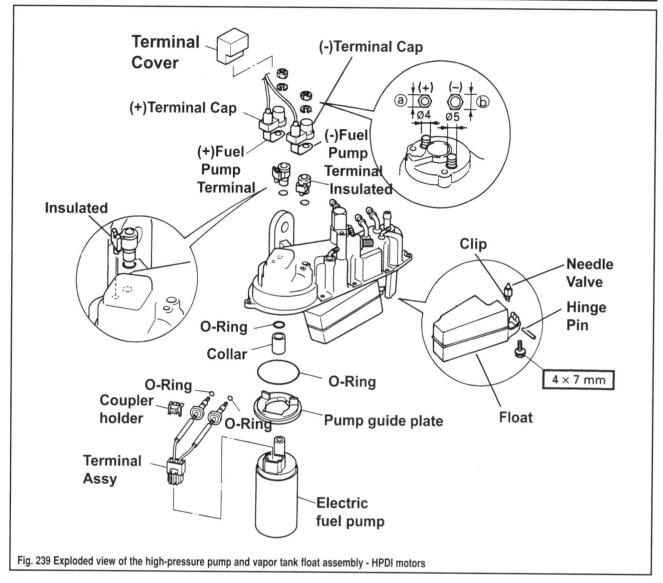

Fig. 239 Exploded view of the high-pressure pump and vapor tank float assembly - HPDI motors

3. Remove the bolt securing the drive belt tensioner to the bracket, then remove the tensioner and the spring
4. If necessary, remove the bolt securing the tensioner spring holder, then remove the holder.
5. With tension now released from the belt it should slide from the sprockets. Remove the belt, noting the direction of rotation for installation purposes (if it is going to be reused).
6. If necessary, remove the 2 bolts holding the tensioner bracket in position, then remove the bracket from the powerhead.

■ **To remove the drive sprocket, you'll need a tool to hold the flywheel. Yamaha recommends the use of their universal holder (#YU-01235) which has tabs that fit into various holes on top of the flywheel assembly. However you can fashion a tool out of metal stock to perform the same function.**

7. If the drive sprocket for the belt must be removed, hold the flywheel from turning and loosen the retaining bolt, then remove the plate (noting the orientation, as the beveled edge faces upwards). Slide the sprocket upward of the bracket, dowel pin and flywheel.

■ **To remove the driven sprocket from the top of the pump you'll need a heavy duty strap wrench to keep the sprocket from turning (such as Yamahas Sheave Holder #YS-1880-A) and a universal 2-jawed puller (such as Yamahas #YB-06540) to draw the sprocket off the pump shaft.**

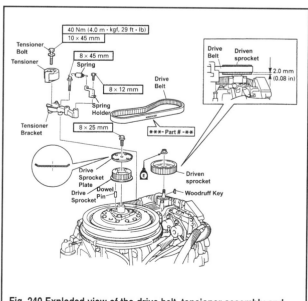

Fig. 240 Exploded view of the drive belt, tensioner assembly and sprockets - HPDI Motors

3-70 FUEL SYSTEM

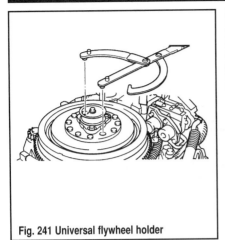

Fig. 241 Universal flywheel holder

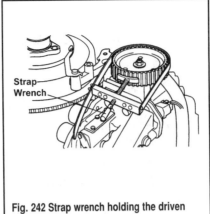

Fig. 242 Strap wrench holding the driven sprocket

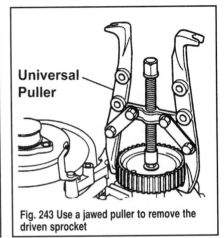

Fig. 243 Use a jawed puller to remove the driven sprocket

Fig. 244 Drive belt installation

8. If the driven sprocket must be removed from the top of the pump, loosen and remove the nut, then pull the driven sprocket off the pump shaft and remove the woodruff key.

To Install:

9. If removed, install the woodruff key to the pump driveshaft, then slide the driven sprocket over the key (aligning the keyway in the sprocket hub with the key in the shaft). Apply a light coating of clean engine oil to the threads of the retaining nut. Use a strap type wrench to hold the sprocket from turning, then install and tighten the sprocket nut securely.

10. If removed, install the drive sprocket to the bracket and dowel pin on top of the flywheel, then position the plate (with the chamfered edges facing upward). Apply a light coating of clean engine oil to the threads of the retaining bolt, then thread the retaining bolt into the flywheel bracket. Use a flywheel holder to keep the flywheel from turning and tighten the retaining bolt securely.

11. If removed, install the spring holder using the retaining bolt.

12. If removed, install the tensioner bracket and secure using the 2 retaining screws.

13. Install the drive belt tensioner and finger-tighten the retaining bolt, then loosen it about 1/4 turn.

14. Manually push the drive belt tensioner to the loosest position and carefully slide the belt over the sprockets. Start pushing the belt over the drive (flywheel) sprocket and work it counterclockwise over the driven (pump) sprocket. Once the belt is over the pump sprocket, check to make sure the bottom edge of the belt it is about 0.08 in. (2mm) above the bottom edge of the drive sprocket.

15. Install the tensioner spring.

16. Loosen the tensioner bolt and allow the spring to adjust tension on the drive belt. Check the belt tension on the run between the drive and driven sprockets, on the opposite side from the belt tensioner, there should be NO slack. If necessary, manually push the tensioner to the necessary position to remove slack on the opposite side of the belt. Once belt tension is correct, tighten the tensioner bolt to 29 ft. lbs. (40 Nm).

17. Install the flywheel cover and reconnect the negative battery cable.

HPDI Injector Driver

The HPDI system utilizes an injector driver assembly to power/actuate the fuel injectors. The driver assembly is mounted between the cylinder head banks on the powerhead.

For 150-200 hp models the ignition coils are mounted to the outside of the injector driver case, while the driver itself is mounted behind the driver case. The entire assembly can be easily removed from the powerhead for service. Obviously, individual component replacement for the ignition coils (covered under Ignition and Electrical Systems) is possible without removal of the driver case.

On 225-250 hp models the ignition coils are mounted to an ignition coil cover which is mounted over top of the injector driver/ECM mounting bracket. On these motors 2 drivers are mounted on the outside of the bracket, while the ECM and fuel pump relay are mounted to the backside of the bracket. Although the drivers can be removed and installed with the bracket in place on the powerhead, the bracket must be removed for access to the ECM or fuel pump relay.

TESTING

The injector driver is tested using a peak-reading voltmeter to read injector driver output at various speeds (cranking with the circuit open, cranking with the circuit closed, or the engine operating at/above 1500 rpm). Obviously for all but the cranking, circuit open test you'll need jumpers for the wiring harness so you can take reading while the circuit is complete.

Testing the injector driver output voltages over the following combinations of wires:

- O/R and PU/R
- O/B and PU/B
- O/Y and PU/Y
- O/G and PU/G
- O/Blue and PU/Blue
- O/W and PU/W

Injector driver output should be at least as follows, depending upon the operating condition:

- Engine cranking, open circuit: 65 volts (150-200 hp motors only, no spec for 225-250 hp motors)

FUEL SYSTEM

- Engine cranking, circuit complete: 60 volts (150-200 hp motors) or 70 volts (225-250 hp motors)
- Engine operating at or above 1500 rpm, circuit complete: 65 volts (150-200 hp motors) or 80 volts (225-250 hp motors)

REMOVAL & INSTALLATION

150-200 Hp (2596cc) Motors

◆ See Figures 245 thru 247

1. Disconnect the negative battery cable for safety.
2. Remove the flywheel cover for access.
3. Remove the 2 fuel rail covers.
4. Tag and disconnect the 6 spark plug leads from the plugs.
5. Tag and disengage the 6 ignition coil coupler connectors.
6. Remove the 2 bolts securing the lower wire harness holder, then remove the holder.
7. Remove the 2 bolts securing the upper wire harness holder, then remove the holder.
8. Remove the 3 wire harness covers from the harness, above the injector driver assembly.
9. Remove the 4 bolts securing the injector driver assembly, then reposition the assembly very slightly for better access and disconnect the injector driver coupler. This is done by pressing on the release plate while pressing up on the lock plate using both of your thumbs until it stops, then carefully pull while holding the entire driver coupler. Once the wiring is disconnected, remove the injector driver assembly from the powerhead.

⁂ WARNING

DO NOT pull on the lock plate to remove the injector driver coupler or the lock plate may break.

10. If necessary, remove the 4 bolts securing the injector driver to the back of the driver case, then remove the driver.

To Install:

11. If removed, install the injector driver to the back of the driver case using the 4 retaining bolts. Tighten the bolts securely.
12. Position the driver assembly to the powerhead and connect the injector driver coupler.
13. Secure the assembly to the powerhead using the 4 bolts.
14. Install the 3 wire harness covers to the powerhead harness above the injector driver. Install the upper and lower wire harness holders and secure each using 2 retaining bolts.

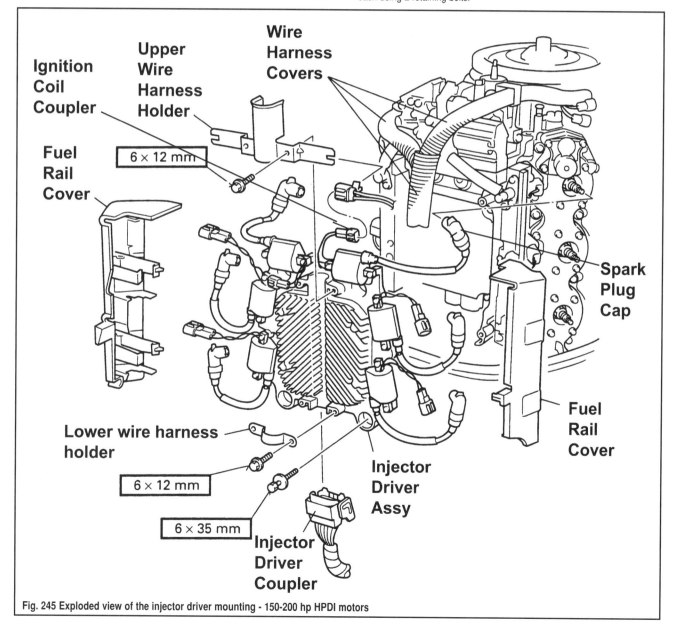

Fig. 245 Exploded view of the injector driver mounting - 150-200 hp HPDI motors

3-72 FUEL SYSTEM

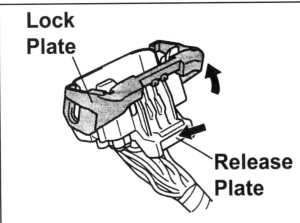

Fig. 246 Disconnecting the injector driver coupler - 150-200 hp HPDI motors

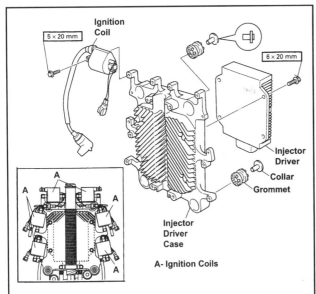

Fig. 247 Exploded view of the injector driver assembly - 150-200 hp HPDI motors

15. Engage the 6 ignition coil coupler connectors, as tagged during removal. Connect the 6 ignition coil leads to the spark plugs, again, as tagged during removal.
16. Install the 2 fuel rail covers.
17. Install the flywheel cover and reconnect the negative battery cable.

225-250 Hp (3342cc) Motors

◆ See Figure 248

1. Disconnect the negative battery cable for safety.
2. If necessary, remove the flywheel cover for access.
3. Remove the Ignition Coil Cover, as detailed in the Ignition and Electrical System section.
4. If removing the lower injector driver, loosen the 2 retaining bolts, then remove the wire harness holder from the bottom of the injector driver/ECM bracket.
5. Tag and disconnect the wiring from the lower and/or upper injector driver, as applicable.
6. Loosen and remove the 2 retaining bolts for the injector driver you are removing, then separate the driver from the mounting bracket.
7. If ECM removal is necessary, loosen and remove the 8 bolts securing the driver/ECM bracket to the powerhead and reposition the bracket slightly so you can tag and disconnect the ECM wiring. Loosen the 2 bolts and remove the ECM from the backside of the bracket.

8. If necessary, remove the bolt securing the fuel pump relay to the back of the mounting bracket and remove the relay.

To Install:

9. If necessary, install the fuel pump relay to the back of the mounting bracket and secure using the mounting bolt.
10. If removed, install the ECM driver to the back of the driver/ECM mounting bracket and secure using the 2 retaining bolts. Tighten the bolts securely. Position the driver/ECM bracket to the powerhead and reconnect the ECM wiring, as tagged during removal. Secure the assembly to the powerhead using the 4 bolts.
11. Install the upper and/or lower injector driver(s) to the mounting bracket. Secure each injector driver using the 2 mounting bolts.
12. Reconnect the driver wiring, as tagged during removal.
13. If the lower driver was removed, install the wiring harness holder and secure using the 2 retaining bolts.
14. Install the Ignition Coil Cover assembly.
15. Install the flywheel cover and reconnect the negative battery cable.

High-Pressure (Mechanical) Fuel Pump, Fuel Rails and Injectors (HPDI)

The extreme high-pressure fuel pump (2nd stage high-pressure pump) on HPDI motors is a mechanical, belt driven pump mounted to the top of the powerhead assembly and capable of generating fuel system pressures as great as 1000 psi. (6895 kPa) on some models. For most models there is a single pump feeding the dual fuel rail assembly, however 225-250 hp models use a dual mechanical pump assembly feeding the dual fuel rails. A pressure balance line is used on those motors to even out line pressure between the two rails.

Like the vapor separator and fuel rail assemblies on most EFI motors, the extreme high-pressure pump(s) can be removed as an entire assembly along with the high-pressure fuel rails and injectors. However, fuel system pressure should still be released for safety as you're working with a literal bomb with this system when it is up to full pressure. Safety goggles are more important here than almost any other time you're working on the powerhead, so WEAR THEM whether or not you're planning on opening up this system.

TESTING INJECTORS

The first and most basic test for a fuel injector is to physically touch the injector with a screwdriver while the engine is operating and feel for a light rhythmic tapping which indicates that the injector is indeed opening/closing. Unfortunately, that doesn't tell you for sure that the injector is opening and closing properly/sufficiently, but it is a start.

A better test, but still not perfect, is to check injector coil resistance across the two injector wire terminals using a high quality Digital Volt Ohmmeter (DVOM). Unfortunately, Yamaha doesn't provide a spec for this on any of the HPDI motors, except the 225-250 hp models (for them the specification is 0.9-1.1 ohms at about 68 degrees F/20 degrees C). However, on all motors, a good test would be to compare the resistance of a suspect injector with that of the other injectors on the motor (which you hopefully suspect are working properly).

■ Because of the extremely high operational pressures of this fuel system we CANNOT recommend testing/observing the fuel spray pattern of these injectors.

REMOVAL & INSTALLATION

◆ See Figures 249

1. Properly relieve the fuel system pressure, then disconnect the negative battery cable for safety.
2. Remove the flywheel cover for access.
3. Remove the HPDI Drive Belt, as detailed in this section. Remove the driven sprocket from the high-pressure fuel pump shaft.
4. For 150-200 hp motors, remove the HPDI Injector Driver assembly, as detailed in this section.

FUEL SYSTEM 3-73

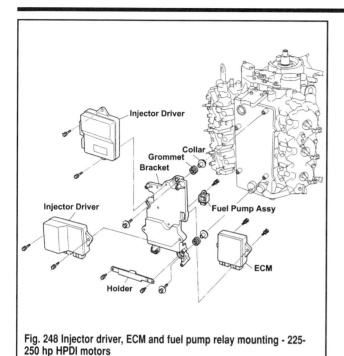

Fig. 248 Injector driver, ECM and fuel pump relay mounting - 225-250 hp HPDI motors

■ For 225-250 hp motors, Yamaha does not specify what if the ignition coil cover or injector driver must come off to access the high-pressure fuel pump components. As best as we can tell from the diagrams they DO provide you'll have to remove the Ignition Coil Cover (as detailed in the Ignition and Electrical System section), but the driver/ECM bracket assembly may remain in place.

5. Tag and disconnect the wiring harness from the fuel pressure sensor.
6. If applicable, remove the bolt and the clamp from the top side of the fuel pump.
7. Tag and disconnect the 6 wiring connectors from the fuel injectors.
8. For 150-200 hp motors, tag and disconnect the 4 ignition coil couplers, then remove the 2 bolts and 2 ignition coil coupler brackets from the fuel rails.
9. Remove the 3 (150-200 hp) or 4 (225-250 hp) bolts securing the fuel pump assembly to the top of the powerhead.
10. Remove the 6 (150-200 hp) or 8 (225-250 hp) bolts securing the fuel rails to the side of the powerhead. The top and bottom bolts on all models use collars, don't loose them.
11. Each fuel injector has a bolt and a holder keeping it in position on the powerhead; loosen the bolts of all 6 holders and remove the holders from the powerhead.
12. Carefully remove the fuel pump, fuel rails and injector assembly from the powerhead.
13. There are 2 gaskets/seals sealing each fuel injector to the powerhead. THESE SEALS MUST BE REAPLCED to ensure the injectors seal properly to the powerhead. Failure to replace these seals could lead to extremely high-pressure fuel leaks (a very dangerous situation).

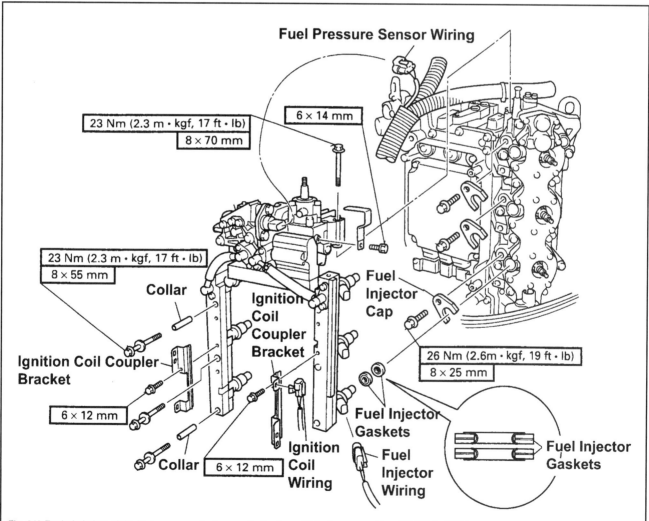

Fig. 249 Exploded view of the high-pressure fuel pump, fuel rails and injectors assembly - 150-200 hp HPDI motors shown (225-250 hp motors similar)

3-74 FUEL SYSTEM

✶✶ CAUTION

If you decide to remove one or more fuel injectors from the rails, keep in mind that there may be residual fuel pressure in the system, so be sure to wear safety goggles and work in a well-ventilated area FAR away from any/all possible sources of ignition (sparks, flame, embers, etc).

14. If necessary, remove fuel injectors from the rails by removing the rail holder clip. Each injector is sealed to the fuel rail by a gasket/seal, O-ring and gasket/seal. On some models the gaskets/seals are a different size/thickness depending upon location (usually thin at the rail side of the O-ring and thick at the injector side of the O-ring). Pay VERY close attention to each of these components as they are removed from the assembly. And, we can't say it enough, REPLACE ALL GASKETS/SEALS/O-RINGS for safety.
15. If necessary, refer to Overhaul for information on fuel pump(s) and/or fuel rail/line disassembly.

To Install:
16. If removed, install each fuel injector to the rail using NEW seals/gaskets and O-rings in the positions noted during disassembly. If you're confused or have any doubts, refer to the exploded views in the Overhaul information for the fuel pump and rails/lines. Secure the injectors to the fuel rails using the holder clips.
17. Install 2 NEW gaskets/seals on the powerhead side of each injector, then carefully position the fuel pump, rail and injector assembly to the powerhead.
18. Seat each injector in the powerhead and make sure the fuel rails and pump(s) align with their mounting points. Position the fuel injector holders and finger-tighten the retaining bolts.
19. Thread and finger-tighten the fuel rail and pump retaining bolts. Remember the upper and lower bolts for the rails have collars that must be in position first.
20. Once you're certain everything is properly aligned, tighten the fuel injector holder bolts to 19 ft. lbs. (26 Nm).
21. Tighten the fuel rail and pump retaining bolts to 17 ft. lbs. (23 Nm).
22. For 150-200 hp motors, install the 2 ignition coil coupler brackets to the fuel rails and secure using the retaining bolts, then connect the 4 ignition coil couplers, as tagged during removal.
23. Apply a light coating of injector grease to the injector wiring terminals, then connect the 6 wiring connectors to the fuel injectors, as tagged during removal.
24. If applicable, install the bolt and clamp to the top side of the fuel pump.
25. Reconnect the wiring harness to the fuel pressure sensor.

■ **If the Ignition Coil Cover assembly was removed for access on 225-250 hp motors, install it at this time.**

26. For 150-200 hp motors, install the HPDI Injector Driver assembly, as detailed in this section.
27. Install the driven sprocket to the high-pressure fuel pump shaft and then install the HPDI Drive Belt, as detailed in this section.
28. Install the flywheel cover.
29. Reconnect the negative battery cable and properly pressurize the fuel system in order to check for leaks. Start and run the motor with the top cover removed and observe the fuel lines/fittings for any signs of weepage. Double-check all lines and fittings after the first outing with the motor.

PUMP & INJECTOR OVERHAUL

◆ See Figures 250 thru 254

The high-pressure fuel pump body and fuel lines may be removed for component, seal and line replacement. If this is attempted, keep CLOSE track of the positioning of all components, comparing them to the accompanying exploded views as you proceed. This is a VERY high pressure fuel system and ALL SEALS, O-RINGS or GASKETS MUST be replaced for safety anytime they are removed. Clamps should be replaced only with the same type. Crimped clamps are removed by cutting the crimp itself, perpendicular to the crimp. But remember before removal that you'll need a proper crimping tool and replacement clamps for any that are removed.

✶✶ CAUTION

There may be residual fuel pressure in the system, so be sure to wear safety goggles and work in a well-ventilated area FAR away from any/all possible sources of ignition (sparks, flame, embers, etc).

Be sure the work surface is clean and free of dirt, debris or moisture. The fuel injectors use extremely small passages which are easily clogged by dirt or debris that is allowed to enter the system. Debris in the system can cause the injector NOT to fire (which could toast a 2-stroke cylinder) OR cause it to leak (which could ALSO toast a 2-stroke cylinder) so don't take any risks.

Fuel Rail and Injectors (EFI)

◆ See Figure 255

A fuel injector is a small, solenoid valve that is designed to open against spring pressure when power is applied to the circuit. An internal spring snaps the valve closed the instant that power is removed from the circuit.

Fuel injectors are supplied with a constant supply of high pressure fuel. Because the pressure is held constant, the amount of fuel that sprays through the injector is a function of time (the less each injector is actuated, the less fuel is delivered, the more each injector is actuated, the more fuel is delivered).

Each cylinder is equipped with an individual electronic fuel injector to deliver metered amounts of fuel, matching engine operating conditions. The exact fuel metering made possible by the fuel injection system is responsible for the EFI engine's ability to maximize both engine performance and fuel economy.

The most important fuel injection system maintenance is a combination of periodic filter changes and the use of fuel stabilizer if the motor is stored for any amount of time (more than a few weeks). This is true because the passages inside a fuel injector are very small, and are easily clogged by dirt or debris in the fuel system.

On these motors the fuel injectors are secured between the fuel rail and a throttle body/intake manifold assembly. They deliver fuel to the reed valve assemblies.

TESTING

Injector Operational Test

The fastest way to check for inoperable injectors is to listen or feel for solenoid operation. If available, use a mechanic's stethoscope, but because solid matter transmits sound and vibration, a long screwdriver can also be used to amplify and/or feel each injector.
1. Provide the motor with a source of cooling water.
2. Start and run the engine at idle.
3. Position the stethoscope or screwdriver against the body of each fuel injector.

■ **If using a screwdriver to amplify the sounds of the injector, place your ear near the handle or hold the driver lightly while feeling for the light tapping of the solenoid valve.**

4. If an injector is operating properly you will hear or feel a slight clicking from it. This tells you that the valve is opening and closing.
5. If there is a noticeably different noise or no clicking is felt at all from an individual injector, perform the Injector Resistance Test and check the wiring between the injector and ECM. If resistance is within specification, check the harness for opens or shorts. Before replacing an ECM, substitute a known good injector.

■ **For test purposes, fuel injectors can be switched from cylinder-to-cylinder to see if the problem follows the injector or remains behind. If the problem remains behind, look to the harness and signal for trouble. If the problem follows the injector, the problem IS the injector.**

FUEL SYSTEM 3-75

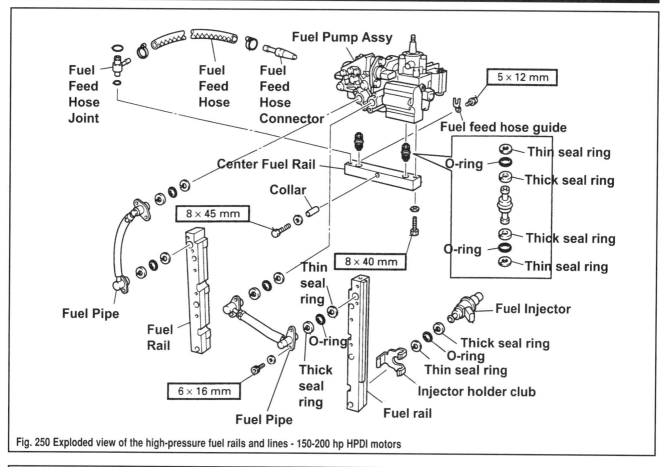
Fig. 250 Exploded view of the high-pressure fuel rails and lines - 150-200 hp HPDI motors

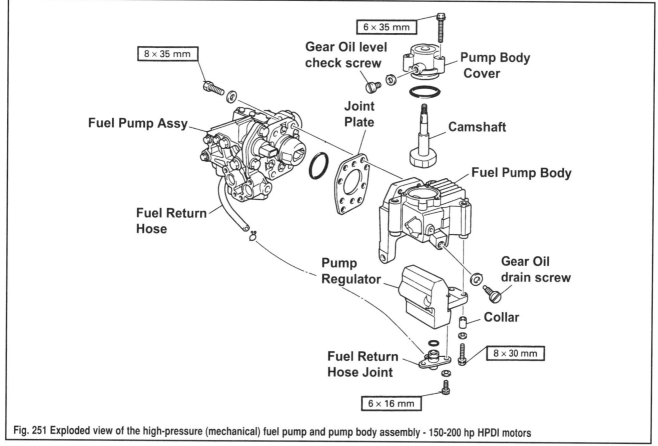
Fig. 251 Exploded view of the high-pressure (mechanical) fuel pump and pump body assembly - 150-200 hp HPDI motors

3-76 FUEL SYSTEM

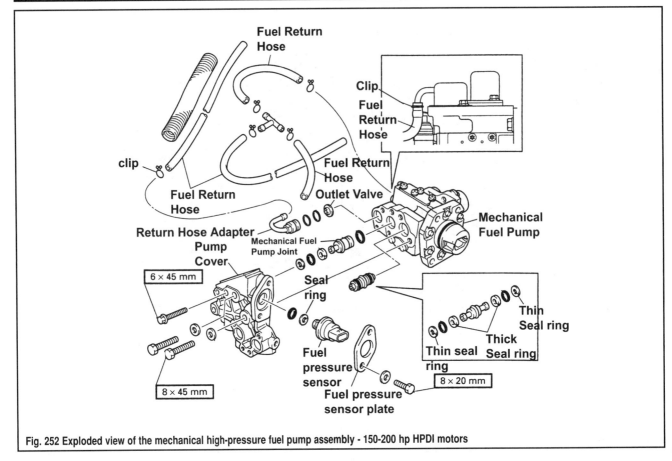

Fig. 252 Exploded view of the mechanical high-pressure fuel pump assembly - 150-200 hp HPDI motors

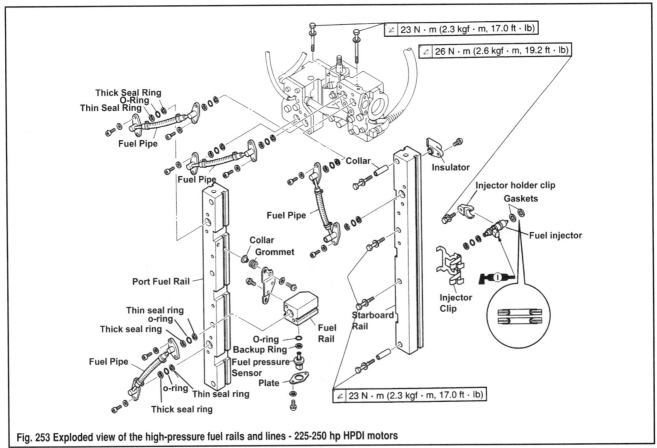

Fig. 253 Exploded view of the high-pressure fuel rails and lines - 225-250 hp HPDI motors

FUEL SYSTEM

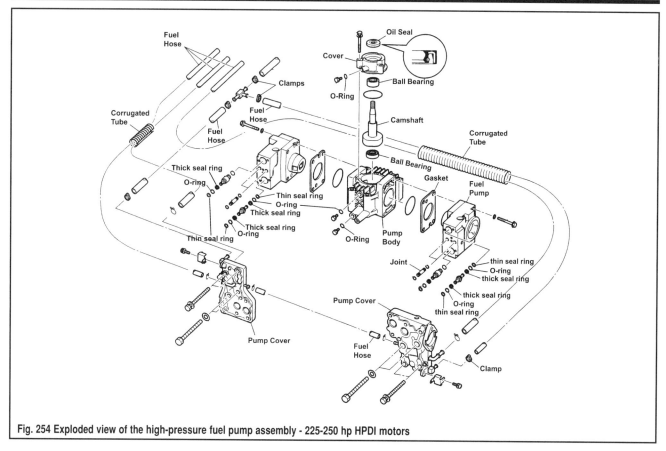

Fig. 254 Exploded view of the high-pressure fuel pump assembly - 225-250 hp HPDI motors

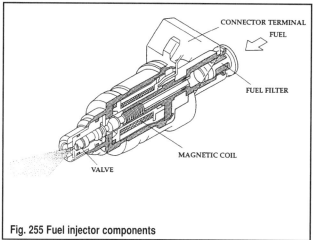

Fig. 255 Fuel injector components

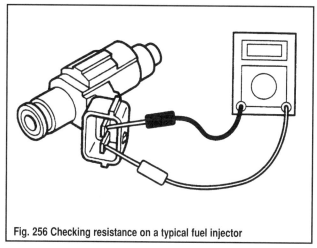

Fig. 256 Checking resistance on a typical fuel injector

Injector Resistance Test

◆ See Figure 256

Another quick-check of a fuel injector is made using an ohmmeter to measure the resistance of the winding inside the injector itself. It is important to remember that a correct reading does not mean the injector is operating. Mechanical damage or clogs within the injector could prevent it from opening or closing properly which would lead to engine performance problems.

■ **The injector does not need to be removed from the engine to check its resistance.**

1. In order to protect the test equipment, disconnect the negative battery cable.

2. Disengage the wire harness connector from the injector.

3. Set the DVOM to the resistance scale, then apply the meter probes across the 2 terminals on the top of the injector. Unfortunately Yamaha isn't consistent with the specifications they provide from motor-to-motor. The only specs we could find were for the injectors was on their V6 4-strokes motor not covered here (but for reference we'll list their specification, which should be 14-15 ohms at an ambient temperature of about 68 degrees F/20 degrees C). For other motors, take readings on all the injectors and compare them to one another to decide if one seems greatly out of line with the rest. Replace the injector if resistance seems WAY out of line.

4. If the injector is inoperable, but resistance is within specification, check the circuit and signal to help determine if it is a circuit fault or a mechanical fault within the injector itself.

5. When finished, reconnect the wiring harness to the injector.

3-78 FUEL SYSTEM

REMOVAL & INSTALLATION

◆ See Figures 257, 258 and 259  MODERATE

OK, you've got some decisions to make. For all EFI OX66 motors, Yamaha recommends removing the entire Throttle Body and Intake assembly, along with the fuel rail, injectors and vapor separator tank, THEN disassembling the unit. In some cases, this makes more sense than others (as access to fuel line fittings or certain bolts can be difficult, but NOT impossible with the assembly installed).

The truth is that removing the whole assembly will require that you replace a couple more gaskets or seals and give you the opportunity to visually inspect more components. However, it is not absolutely necessary on most models.

If you decide to remove the Throttle Body and Intake assembly, please refer to the appropriate procedure found earlier in this section for details, then follow the appropriate steps of this procedure to disassemble the fuel rail and injectors. In all cases you're going to have to depressurize the fuel system since you'll be removing injectors from the fuel rail assembly.

■ One factor which may influence your decision on how to proceed is how the fuel lines are secured to the rail(s). With 2-stroke motors there are almost always crimp-type clamps on both ends of the fuel lines, so unless there is sufficient free-play to move the fuel rail enough to free the injector without disconnecting the lines, you HAVE to either disconnect the lines OR remove/reposition the vapor separator tank.

1. Depressurize the fuel system and disconnect the negative battery cable for safety.
2. If necessary for access, remove the flywheel cover.
3. Remove the air intake silencer for access.

■ Take a good look at the fuel lines and mounting points for the fuel rail. Decide how you'd like to proceed. You can remove the entire Throttle Body and Intake for additional access. You can just remove and/or reposition the Vapor Separator Tank. OR, you can unbolt the fuel rail with the lines still attached and see if there is sufficient slack to reposition it enough to remove the fuel injectors (it has to move outward from the powerhead a little more than 1 in. (25mm), give or take.

4. There is usually at least one plastic wire tie holding wiring or fuel/vapor lines to the fuel rail assembly. Note the positioning of the wire tie(s) and components secured by the tie(s), then cut the tie and reposition the wires or lines out of the way.
5. Tag and disconnect the wiring from the fuel injector connectors. On most motors this means pulling the locking wire loom free of the connector, then gently pulling back on the connector housing.
6. Remove the fuel rail retaining bolts (usually 4).
7. If there is sufficient free-play, gently pull back on the fuel rail assembly to free the injectors from the powerhead, then reposition the rail for access to the injectors.

■ If one or more of the fuel lines are binding, holding the fuel rail in place or keeping you from repositioning it for access, you'll have to disconnect that (or those) line(s).

8. If necessary to reposition or completely remove the fuel rail disconnect one or more of the fuel lines from it. You'll usually have crimp-type clamps which are disconnected by cutting the crimp (perpendicular to the crimp itself).
9. Place a matchmark on the fuel rail immediately adjacent to each injector electrical connector to ensure the injectors are installed facing the right direction. The fuel injectors are gently pressed into the fuel rail with a rubber damper and an O-ring to seal them. Grasp each injector you wish to remove and pull it from the fuel rail.

■ Yamaha does not specify that you MUST replace the injector seal, damper or O-ring, however it is never a bad idea to ensure proper sealing and operation. If you opt NOT to replace any of these components, check them carefully for signs of wear, distortion, damage or just plain degradation from age.

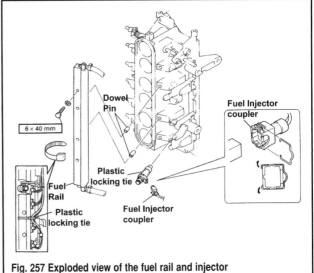

Fig. 257 Exploded view of the fuel rail and injector assembly/mounting - EFI OX66 motors

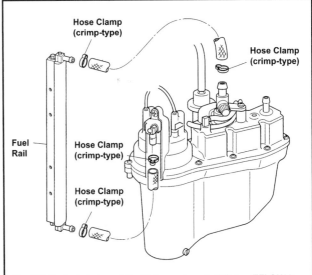

Fig. 258 Exploded view of the high-pressure fuel lines - EFI OX66 motors

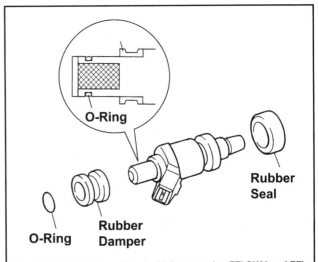

Fig. 259 Exploded view of the fuel injector seals - EFI OX66 and EFI motors

FUEL SYSTEM 3-79

To Install:

10. If the fuel injectors were removed from the fuel rail assemble each with the rubber damper and small O-ring on the fuel rail side and the thick rubber seal on the powerhead side. Gently push each injector into the fuel rail with the wiring connected facing outward toward the matchmark made before removal.

11. Position the fuel rail assembly to the powerhead and reconnect the fuel lines. Tighten fuel joint bolts securely.

12. Push gently inward toward the powerhead to seat the fuel rail, then thread and finger-tighten the retaining bolts. Make sure that all fuel lines and wires are properly routed/positioned, then tighten the fuel rail retaining bolts securely.

13. Reconnect the fuel injector wiring as tagged during removal.

14. Reposition any wires or hoses and secure using a new wire tie, as noted during removal.

15. If removed, install the air intake silencer and/or flywheel cover.

16. Reconnect the negative battery cable and properly pressurize the fuel system in order to check for leaks. Start and run the motor with the top cover removed and observe the fuel lines/fittings for any signs of weepage. Double-check all lines and fittings after the first outing with the motor.

Fuel Pressure Regulator

All EFI/HPDI fuel systems utilize a vacuum actuated pressure regulator whose job it is to control fuel pressure in the high-pressure circuit (first stage high-pressure circuit on HPDI motors). Like the electric fuel pump, the fuel pressure regulator is mounted to the vapor separator tank assembly.

The job of the electric fuel pump is to deliver fuel at a maximum pressure which meets or exceeds the demands of the motor. The job of the pressure regulator is to vary the opening on the return line to restrict flow (and thereby maintain pressure at a constant point). Essentially the pressure regulator works like a thumb placed over the end of an open garden hose, reducing the diameter of the hose and thereby raising pressure of the water stream exiting from the hose.

The design and mounting of the pressure regulator is basically the same, regardless of the mounting location. And service is a relatively simple matter of depressurizing the system, disconnecting the hoses and removing the regulator.

The extreme high-pressure circuit of the HPDI system also utilizes a pressure regulator. However the regulator for this fuel circuit is built into the mechanical high-pressure pump assembly and although from the exploded views it appears that it can be replaced, Yamaha provides no information for testing or servicing this component.

TESTING

◆ See Figure 260

In typical Yamaha fashion, they do NOT provide specific testing parameters for all EFI pressure regulators. On all motors there is a Pressure Check that can be made the high-pressure system to determine if it is operating within normal parameters. For starters, refer to the Fuel System Checks under Maintenance and Tune-Up for details on how to perform a pressure check. If fuel pressure is ABOVE the specification, check for restrictions in the fuel lines and, if none are found, the pressure regulator is likely the culprit. Similarly, if pressure is below specification you've got 2 possible culprits, the fuel pump or the pressure regulator. The quick check is to squeeze a fuel line and see if you can bring pressure up to specification by restricting fuel flow. If this works, the pressure regulator is the likely culprit, however if even after manually restricting a fuel line pressure remains below specification it is time to look at the fuel pump as the possible cause.

On 2-stroke motors Yamaha recommends that you set up the fuel Pressure Check AND connect a hand-held vacuum pump/gauge to the fuel pressure regulator (simply disconnect the vacuum line from the regulator and connect the vacuum pump in its place). With the engine running observe the fuel system operating pressure and gradually apply vacuum to the pressure regulator. Apply 7.4025 in Hg. (25 kPa) of vacuum for EFI motors or 10.3635 in. Hg. (35 kPa) of vacuum for HPDI motors. Fuel pressure should decrease as the vacuum supplied to the pressure regulator is increased. If not, the pressure regulator is faulty.

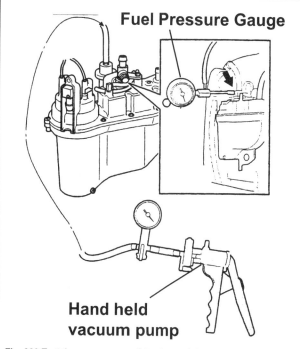

Fig. 260 Test the pressure regulator by applying vacuum and watching the fuel pressure gauge - EFI/HPDI motors

REMOVAL & INSTALLATION

◆ See Figure 261

1. Properly relieve fuel system pressure and disconnect the negative battery cable for safety.

2. On EFI OX66 motors, remove the screw and pressure check valve holder from the top of the vapor separator tank.

3. Tag and disconnect the vacuum hose from the pressure regulator.

4. Loosen the 2 retaining bolts, then carefully pull the pressure sensor both out and up from the vapor separator assembly.

5. Remove the O-ring seal from the pressure regulator, normally found under the mounting flange. Remove the fuel return joint from the bottom of the pressure regulator or the top of the vapor separator cover. Remove the 2 O-rings from the joint. Also, remove, clean and inspect the pressure regulator filter screen.

To Install:

6. If removed, install the pressure regulator filter screen to the opening in the side of the vapor separator cover, then install the fuel return joint using 2 new O-rings to the top of the cover or the bottom of the regulator.

7. Install the pressure regulator using a new O-ring on the fuel outlet under the mounting flange, secure using the 2 retaining bolts.

8. Connect the vacuum hose to the pressure regulator.

9. On EFI OX66 motors, install the pressure check valve holder and retaining screw to the top of the vapor separator tank.

10. Reconnect the negative battery cable and properly pressurize the fuel system in order to check for leaks. Start and run the motor with the top cover removed and observe the fuel lines/fittings for any signs of weepage. Double-check all lines and fittings after the first outing with the motor.

Engine Control Module (ECM/CDI)

The Engine Control Module (ECM) controls all functions of the EFI and ignition systems. Probably as a holdover from the carbureted motors, the earliest of the Yamaha fuel injection systems (EFI OX66) refer to this as the CDI Unit. However, since it also controls the fuel injection system we think this is a little misleading and will refer to it as the ECM throughout this guide.

Problems with the ECM are rare, but when they occur can cause a no-start, stumbling, misfire, hesitation, incorrect engine timing, rough idle or incorrect speed limiting through improper control of the ignition and/or fuel injection systems.

FUEL SYSTEM

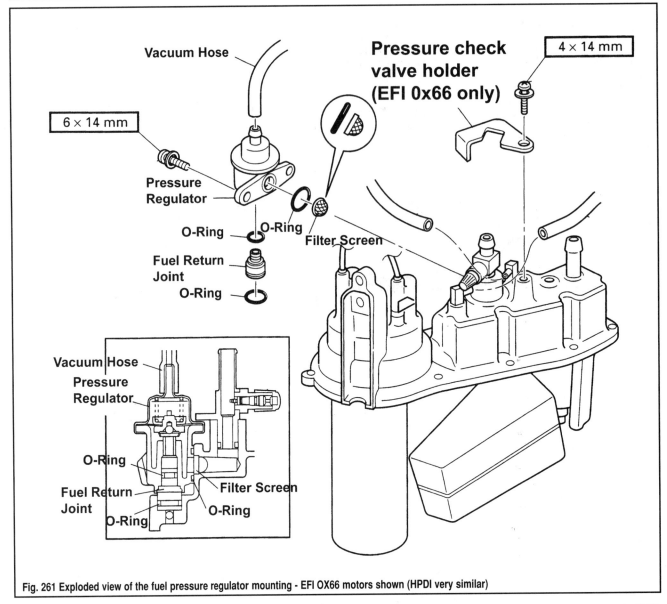

Fig. 261 Exploded view of the fuel pressure regulator mounting - EFI OX66 motors shown (HPDI very similar)

Unfortunately, solid state components like the ECM cannot be directly tested in many ways. One exception comes with checking the Capacitor Discharge Ignition (CDI) peak output voltages actuated by the ignition control circuits of the ECM. Since this test is the same for carbureted and fuel injected motors, we cover it under CDI Unit, Testing in the Ignition and Electrical System sections. However, remember that the ECM on EFI/HPDI motors is what actually performs the function of the CDI unit on carb engines.

In most cases, ECM testing involves a process of elimination, testing all other possible causes of a symptom. Condemn the ECM only if all other components that could cause a problem have been eliminated. Remember that many of the circuits used by the ECM for information or for direct control of the motor are sensitive to changes in resistance. Simple problems such as loose, dirty or corroded connectors, even pinched wires or interference caused by marine radios or other electronic accessories can cause symptoms making an otherwise good ECM seem bad.

REMOVAL & INSTALLATION

On all engines the module is mounted to the rear of the powerhead in a bracket located between the cylinder banks. However, access to the module from that point varies. On EFI OX66 motors it is a relatively easy proposition, once the ECM/CDI cover has been removed. On 200-225 hp HPDI motors is it mounted to the backside of the HPDI Injector Driver mounting bracket and is removed following that procedure in this section. But for 150-200 hp HPDI motors, you have to remove the entire High Pressure (Mechanical) Fuel Pump, Fuel Rails and Injectors for access.

Once accessed ECM removal and installation is a relatively simple matter of disconnecting the wiring and removing any fasteners which secure it to the mounting bracket.

EFI OX66 Motors

◆ See Figure 262

1. Disconnect the negative battery cable for safety.
2. For access, remove the 4 retaining bolts, then remove the ECM cover from the powerhead.
3. Remove the 2 bolts securing ground leads to the bottom of the cylinder heads (one bolt/lead on each head).
4. Remove the 2 bolts securing either the ECM coupler guide (2.6L motors) or the 2 bolts securing the ECM coupler holder and guide (3.1L motors) to the wiring harness connectors at the bottom of the ECM. Remove the guide or holder and guide.

■ Take note of guide or holder and guide orientation, especially on 3.1L motors on which some of the guides may be marked "UP" for directional orientation, but on which some are not.

FUEL SYSTEM 3-81

5. Disengage the 2 wiring connectors from the bottom of the ECM.
6. On 2.6L motors, disengage the 2 bullet connectors for the emergency switch (right near the switch at the upper port side of the ECM).
7. Disengage the wiring connectors for the fuel injector harness and the high-pressure fuel pump resistor. On 2.6L motors they are both near the bottom of the oil tank, but on 3.1L motors they are near the top of the tank.
8. Disengage the black oxygen sensor wiring at the top, starboard side of the ECM.
9. Remove the 2 bolts securing ground leads to the top of the powerhead (one bolt/lead on the powerhead just behind each cylinder head).
10. Tag and disconnect the 6 ignition coil leads from the spark plugs (yes the positioning looks obvious, but do it just be sure, especially if you decide to disassemble the unit further after removal).
11. Remove the 4 bolts securing the ECM and ignition coil bracket assembly to the powerhead.
12. If necessary, remove the ECM from the mounting bracket by removing the 4 mounting bolts and any bolts that secure ECM ground wires. Note ground wire positioning for installation purposes.

To Install:

13. If removed, install the ECM securing the ground wires as noted, then secure the ECM using the 4 mounting bolts.
14. Install the ECM bracket assembly to the powerhead and secure using the 4 mounting bolts.
15. Reconnect the 6 spark plug leads, as tagged during removal.
16. Install the 2 ground leads to the top of the powerhead, one behind each cylinder head, using the 2 retaining bolts.

■ Yamaha recommends coating the 4 ground lead bolts (the 2 on top of the powerhead and the 2 at the base of the cylinder heads) with silicone sealant on 2.6L motors, but does not mention it for 3.1L motors.

17. Reconnect the oxygen sensor, high-pressure fuel pump resistor and fuel injector harness connectors.
18. For 2.6L motors, reconnect the bullets for the emergency switch.
19. Connect the 2 main harness couplers to the bottom of the ECM, then secure using the guide or the holder and guide (as applicable). Make sure the guide (and holder) are positioned as noted during removal, then secure using 2 retaining bolts whose threads are lightly coated with Loctite®572 or an equivalent threadlock.
20. Connect the 2 ground leads to the bottom of the cylinder heads and secure using the retaining bolts.
21. Install the ECM cover to the powerhead and secure using the 4 bolts.
22. Reconnect the negative battery cable.

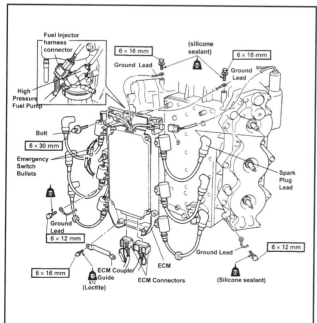

Fig. 262 Exploded view of the ECM mounting - EFI OX66 motors (2.6L shown, 3.1L very similar)

HPDI Motors

150-200 Hp (2596cc) Models

◆ See Figure 263

■ The actual removal and installation of the ECM is really only a simple procedure in and of itself, HOWEVER, it requires the much more involved and difficult procedure of removing the high-pressure pump, rails and injectors, therefore we've decided to upgrade the difficulty of the ECM procedure in this case.

1. Properly relieve the fuel system pressure and disconnect the negative battery cable for safety.
2. Remove the High-Pressure (Mechanical) Fuel Pump, Fuel Rails and Injectors as an assembly. For details, please refer to the procedure earlier in this section.
3. Remove the 2 bolts securing the ECM coupler guide over the wiring harness connectors at the bottom of the ECM. Remove the guide.
4. Disengage the 3 wiring connectors from the bottom of the ECM.
5. Either remove the 5 bolts for the ECM mounting bracket collars/grommets and remove the ECM with the bracket as an assembly, or remove the 4 bolts securing the ECM to the mounting bracket and just remove the ECM leaving the bracket behind.

■ If you decide to remove the bracket, take note of which collars mount in which positions, as not all are the same. The 2 outer collars on the top of the mounting bracket have an extended lip at the powerhead side of the collar, while the other 3 are flush.

6. Installation is essentially the reverse of the removal procedure. Make sure the retaining bolts are snug, but don't over-tighten and distort any of the fasteners or mounts.

225-250 Hp (3342cc) Models

On these models the ECM is mounted to the back of the HPDI injector driver mounting bracket. For removal and installation details, please refer to the HPDI Injector Driver procedure in this section.

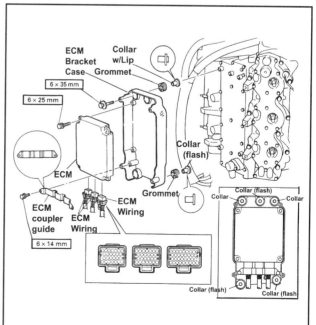

Fig. 263 Exploded view of the ECM and ECM bracket mounting - 150-200 Hp (2596cc) HPDI motors

Throttle Position Sensor (TPS)

The Throttle Position Sensor (TPS) is a rotary potentiometer, meaning it sends a variable signal to the ECM based on physical throttle (and therefore sensor) positioning.

The sensor itself receives a reference voltage (usually about 5 volts) from the ECM, and as the throttle lever is rotated, the ECM receives a return voltage signal through a separate wire. This signal will change with throttle position. As the throttle shaft opens the voltage increases, as the shaft closes voltage decreases. A third wire is used to complete the ground circuit back to the ECM.

Sensor location varies slightly from model-to-model, but on all motors, the TPS sensor is located the top or top side of the throttle body assembly, in a position where it can be directly, mechanically actuated by a throttle valve. Adjustment, which is covered under Timing and Synchronization in Maintenance and Tune-Up involves checking the output voltage (in a similar manner to the Testing covered here) while repositioning the switch.

TESTING

◆ See Figures 264

Checking the TPS output should be part of each tune-up (or should at least be performed annually). If sensor output is slightly out of spec, it can usually be unbolted and adjusted slightly to bring it within specifications, however, if it cannot be adjusted the sensor must be replaced (unless there is a problem with the ECM reference voltage).

Testing the sensor is virtually identical to the adjustment procedure, with the possible exception that sensor input voltage is checked on some models and on others the sensor output voltage is checked in both idle and WOT. Even if not specified to do so on a given model, we suggest that if you suspect problems with the TPS circuit, you should slowly open and close the throttle while watching the TPS output voltage to make sure there are no sudden spikes or drops which could lead to intermittent performance problems.

Since the sensor does not generate a voltage, but instead acts upon a reference voltage from the ECM, all testing must occur with the circuit complete. Therefore, you've got 2 options. One option is to attempt to back-probe the connector (insert the probes through the rear of the connector while it is still attached to the sensor). However this method risks damaging the connector or the wiring insulation (and could lead to problems with the circuit later). The better method is to disconnect the wiring harness and use 3 jumper wires (one for each terminal) to reconnect the harness. The jumpers must not contact each other (or you risk damage to the ECM from a short), however, some point on the jumper must be exposed so that you can probe the completed circuit using a DVOM. Yamaha makes 3-pin test harness for this application (#90890-06757).

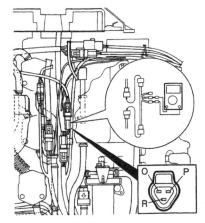

O - Orange
P - Pink
R - Red

Fig. 264 TPS circuit testing - EFI OX66 and HPDI motors

1. If necessary, remove the flywheel cover for additional access.
2. If necessary, remove the air intake assembly for access.
3. Disconnect the wiring from the TPS and connect a test harness or jumper wires to complete the circuit. Locate the harness connector as shown in the accompanying illustration.

■ The TPS sensor wiring is routed down the side of the powerhead, below the sensor itself (which is mounted adjacent to and connected to the top throttle body). Follow the harness from where it exits the sensor, down to the wiring harness connector, noting the routing for installation purposes. After testing, make sure the sensor wiring is safely tucked in the same position, to prevent possible damage from moving components.

4. Connect a Digital Volt Ohmmeter (DVOM) set to read DC volts to the appropriate terminals and turn the ignition switch **ON** without starting the motor to check for the following results depending upon motor and throttle setting:
   • For all EFI and HPDI motors (except the 3.3L HPDI), voltage output across the Orange and Pink wires of the harness should vary between 0.48 volts at idle setting (throttle closed) and 5.25 volts at WOT setting (throttle fully open).
   • For 3.3L HPDI motors, voltage output across the Orange and Pink wires of the harness should be about 0.58-0.62 at idle setting (throttle closed). Next, check the sensor input voltage the Orange and Red terminals, it should be 5 volts.
5. If the sensor voltage (not reference voltage) is out of range, refer to the appropriate adjustment procedure in the Timing and Synchronization section. If the sensor cannot be adjusted and no problems are found with the ECM, wiring or reference voltage, the sensor must be replaced.
6. If the reference voltage is out of specification, check the wiring harness thoroughly for problems, make sure the adapter/tests connections are good, then suspect the ECM only after all other options have been verified good.
7. After testing turn the ignition keyswitch **OFF**, disconnect the test harness and reconnect the TPS wiring.

REMOVAL & INSTALLATION

◆ See Figure 265, 266 and 267

Removal and installation of the TPS is relatively straightforward. On a few motors the sensor may be obscured by a cover or a bracket, so if necessary, remove the cover until you can clearly see the sensor body, wiring and retaining screws. Besides a dedicated sensor cover, depending upon positioning the air intake silencer and/or the flywheel cover may be in the way.

Once you have clear access to the sensor all you need to do is disconnect the wiring and remove the bolts/screws (there are usually 2) that secure the sensor to the bracket or throttle body. If the sensor is not being replaced, you might want to scribe a quick matchmark before removal between the sensor and mounting point. This will provide a handy reference point to start adjustment after reinstallation. Either way though, upon installation you should leave the retaining screws just slightly loosened so that you can pivot the sensor for proper adjustment. Details on TPS adjustment can be found under the appropriate Timing and Synchronization procedure in the Maintenance and Tune-Up section.

Crankshaft Position Sensor (CPS)/Pulser Coil

The ECM needs to know exactly where each piston is in its 2-stroke cycle in order to accomplish certain ignition timing and fuel injection functions. In addition, voltage must be generated by the flywheel magnets and powerhead mounted coils in order to power the ignition system.

The EFI OX66 and HPDI motors utilize both a Crankshaft Position Sensor (CPS) and Pulser Coils (with individual windings for each cylinder) to provide data to the ECM.

The ignition systems are operated in much the same manner as the carbureted Yamaha Capacitor Discharge Ignition (CDI) systems also covered in this guide. For this reason, all testing and service information on the CPS and the Pulser Coils can be found in the Ignition and Electrical System section.

FUEL SYSTEM 3-83

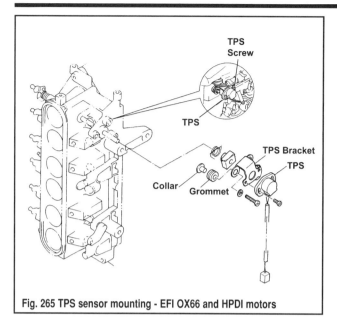

Fig. 265 TPS sensor mounting - EFI OX66 and HPDI motors

Fig. 266 Some sensors are mounted under a protective cover...

Fig. 267 ...but they're always mounted on the throttle body assembly

Temperature Sensors (Air and Water/Engine)

◆ See Figures 268, 269 and 270

■ Yamaha sometimes refers to the Water Temperature Sensor (WTS) as the Engine Temperature Sensor (ETS) or engine cooling water temperature sensor. We just wanted you to know in case you say WTS and the parts guy gives you a funny looks and says, "do you mean ETS." They are the same sensor and perform the same function, but the names vary in different Yamaha publications.

Signals from the Intake Air Temperature (IAT) and Water Temperature Sensor (WTS) sensor are used by the EFI system to manage engine operation. As their names suggest, the sensors are used to monitor air and engine/water temperatures. The IAT sensor is used by the ECM to help determine air/fuel ratios. The water temperature sensor provides essential information during cold starts and warm-up.

■ In addition to these sensors, a thermo-switch is used by the engine cooling overheat warning system. It differs from a sensor in that it is a simple ON/OFF switch. Whereas a sensor gives a varying, but constant signal, a thermo-switch either has continuity or not, and therefore returns a signal only a certain temperatures. Because the basic design of the thermo-switch is the same for all motors (carbureted and fuel injected) it is covered in the Lubrication and Cooling section of this guide.

Temperature sensors for modern fuel injection systems are normally thermistors, meaning that they are variable resistors or electrical components that change their resistance value with changes in temperature. From the test data provided by the manufacturer it appears that Yamaha sensors are usually Negative Temperature Coefficient (NTC) thermistors. Whereas the resistance of most thermistors (and most electrical circuits) increases with temperature increases (or lowers as the temperature goes down), an NTC sensor operates in an opposite manner. The resistance of an NTC thermistor goes down as temperature rises (or goes up when temperature goes down).

Sensor locations vary slightly by engine and are as follows:
• For 2.6L EFI OX66 motors, the IAT is mounted on the starboard side of the powerhead (on the top of the most forward point of the electrical junction box). The WTS is mounted to the top, starboard side of the powerhead (the wiring coupler can be found almost underneath the sensor about halfway down the electrical junction box).
• For 3.1L EFI OX66 motors, the IAT is mounted on the starboard side of the powerhead (just underneath the electrical junction box on and right above the throttle linkage). The WTS is mounted to the top, starboard side of the powerhead (the wiring coupler can be found at the base of the powerhead, directly below the sensor and just behind and below the electrical junction box).
• For 2.6L HPDI motors, the IAT is mounted at the front of the powerhead, right behind the air intake silencer assembly. The WTS is mounted to bottom, port side of the port cylinder head (don't confuse it with the thermoswitch mounted toward the top, port side of the same cylinder head).
• For 3.3L HPDI motors, the IAT is mounted on the starboard side of the powerhead (on the top of the most forward point of the electrical junction box). The WTS is mounted to the top, starboard side of the powerhead (just a little behind the CPS and a little in front of the starboard cylinder head).

TESTING

◆ See Figures 268, 269 and 270

Temperature sensors are among the easiest components of the EFI system to check for proper operation. That is because the operation of an NTC thermistor is basically straightforward. In general terms, raise the temperature of the sensor and resistance should go down. Lower the temperature of the sensor and resistance should go up. The only real concern during testing is to make clean test connections with the probe and to use accurate (high quality) testing devices including a DVOM and a relatively accurate thermometer or thermo-sensor.

A quick check of the circuit and/or sensor can be made by disconnecting the sensor wiring and checking resistance (comparing specifications to the ambient temperature of the motor and sensor at the time of the test). Keep in mind that this test can be misleading as it could mask a sensor that reads

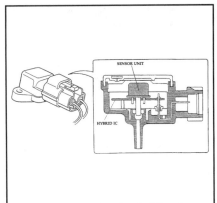

Fig. 268 Cross-section of typical Yamaha temperature sensor

Fig. 269 Typical Yamaha IAT sensor

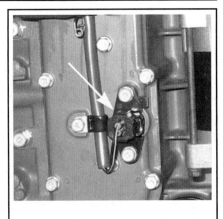

Fig. 270 Typical Yamaha WTS

incorrectly at other temperatures. Of course, a cold engine can be warmed and checked again in this manner.

More detailed testing involves removing the sensor and suspending it in a container of liquid (Yamaha recommends water), then slowly heating the liquid while watching sensor resistance changes on a DVOM. This method allows you to check for problems in the sensor as it heats across its entire operating range.

■ Testing would be easier if Yamaha decided to provide the same specs for all sensors. However, although they do provide a certain amount of resistance specs for most motors, they DO NOT supply them for ALL motors. Also, on most of the EFI OX66 motors (except the 3.1L Vmax) Yamaha provides an operating VOLTAGE specification for the IAT sensor (which must be tested differently). Lastly, though Yamaha usually provides a resistance RANGE which confirms temperature variances, THEY simply supply a range without an reference to temperatures on the IAT for the 3.1L EFI OX66 Vmax AND on the 2.6L HPDI motors.

Since the sensor specifications available from Yamaha vary, check the Test Specifications for your motor, before determining which test(s) you will follow.

Sensor Test Specifications

It appears that the calibration of the sensors used on different Yamaha motors may be the same, however, we cannot be sure of this since Yamaha does not provide a uniform set of specs for most of the EFI motors. For this reason, use the following specifications to determine which tests you can conduct with the motor on which you are working.

Intake Air Temperature (IAT) sensor testing specifications:
- **EFI OX66 Motors (except the 3.1L Vmax)**: should read about 3.4-5.3 VOLTS at an ambient engine temperature of 68°F (20°C). Because of the limited specification, only the Operational Test can be conducted on the IAT for this model. Use the Black/Yellow and Black/Yellow wires of the sensor pigtail. BUT, be sure you've got the right wiring, as the WTS uses the same color wires on this motor.
- **3.1L Vmax EFI OX66 Motors and 2.6L HPDI Motors**: should read about 1,500-4,000 ohms, unfortunately Yamaha does NOT provide a temperature for that specification. There are probably 2 possibilities, it could be a typical ambient reading OR that may be the realistic operating range of the sensor. Because of the limited specification you can use either the Quick Check or the Comprehensive Test, but be CAREFUL when interpreting the results. We really advise you trying to get the parts guy to allow you to compare by taking a reading on a sensor at the shop. Use the Black/Yellow and Black/Yellow wires of the sensor wiring. BUT, be sure you've got the right wiring, as the WTS uses the same color wires on this motor.

Water Temperature Sensor (WTS) testing specifications:
- **EFI OX66 Motors and 2.6L HPDI Motors**: should read about 128,000 ohms at 41°F (5°C), 54,000-69,000 ohms at 68°F (20°C) and 3,020-3,480 ohms at 212°F (100°C). The Quick Test and/or Comprehensive Test may be made across the Black/Yellow and Black/Yellow wires of the sensor pigtail. BUT, be sure you've got the right wiring, as the IAT uses the same color wires on this motor.

Operational Test

Yamaha only provides the voltage specifications for an operational test on the IAT for EFI OX66 motors (except the 3.1L Vmax). Were not quite sure why, but unfortunately it means you'll need a 2-wire jumper harness so you can probe both wires of the sensor circuit, while the circuit is complete with the ECM. Although you may be able to back-probe the harness connector, we don't really recommend this because it could damage the connector or the weatherproofing, leading to problems later down the road.

To test the sensor output voltage, disconnect the wiring from the sensor and install a 2-jumper wiring adapter between the connectors. Next, attach the leads from a DVOM to the jumper wires, then turn the ignition keyswitch **ON** and compare the circuit output voltage to the specifications.

If there is no reading, double-check your test connections. Also you can disconnect the harness completely and take a DC VOLTAGE reading across the wiring from the ECM to determine if a reference voltage is being supplied. If there is still no voltage reading, trace the problem back to the ECM.

Unfortunately Yamaha does not provide any further data necessary to determine if an out-of-range reading is the sensor, wiring or ECM reference signal. Of course, the wiring between the ECM and the sensor should show VERY LITTLE or NO resistance. If you decide to check that wiring ALWAYS unplug it from the ECM before attempting any resistance readings (as you could damage the ECM with improperly connecting a DVOM across its terminals).

Quick Test

When resistance specifications are available, a quick check of a temperature sensor can be made using a DVOM set to the resistance scale and applied across the sensor terminals. The DVOM can be connected to directly to the sensor, or to the sensor pigtail, as the wiring varies by model. Use a thermometer or a thermo-sensor to determine ambient engine/sensor temperature before checking resistance.

Even if the sensor tests ok cold, the sensor might read incorrectly hot (or anywhere in between). If trouble is suspected, reconnect the circuit, then start and run the engine to normal operating temperature. After the engine is fully warmed, shut the engine **OFF** and recheck the sensor hot. If the sensor checks within specification hot, it is still possible that another temperature point in between cold and fully-warmed specifications could be causing a problem, but not likely. The sensor can be removed and checked using the Comprehensive Test in this section or other causes for the symptoms can be checked. If the sensor was checked directly and looks good, but there are still problems with the circuit, be sure to check the wiring harness between the ECM and the sensor for continuity. Excessive resistance due to loose connections or damage in the wiring harness can cause the sensor signals to read out of range. Remember, never take resistance readings on the ECM harness without first disconnecting the harness from the ECM.

FUEL SYSTEM 3-85

Comprehensive Test

 MODERATE

It is important to EFI operation that the temperature sensors provides accurate signals across the entire operating range and not just when fully hot or fully cold. For this reason, when resistance specifications are available, it is best to test the sensor by watching resistance constantly as the sensor is heated from a cold temperature to the upper end of the engine's operating range. The most accurate way to do this is to suspend the sensor in a container of water, connect a DVOM and slowly heat the liquid while watching resistance on the meter.

To perform this check, you will need a high quality (accurate) DVOM, a thermometer (or thermo-sensor, some multi-meters are available with thermo-sensor adapters), a length of wire, a metal or laboratory grade glass container and a heat source (such as a hot plate or camp stove). A DVOM with alligator clip style probes will make this test a lot easier, otherwise alligator clip adapters can be used, but check before testing to make sure they do not add significant additional resistance to the circuit. This check is performed by connecting the two alligator clips together and checking for a very low or 0 resistance reading. If readings are higher than 0, record the value to subtract from the sensor resistance readings that are taken with the clips in order to compensate for the use of the alligator clips.

1. Remove the temperature switch as detailed in this section.
2. Suspend the sensor and the thermometer or thermo-sensor probe in a container of cool water or four-stroke engine oil.

■ **To ensure accurate readings make sure the temperature sensor and the thermometer are suspended in the liquid and are not touching the bottom or sides of the container (as the temperature of the container may vary somewhat from the liquid contained within and sensor or thermometer held in suspension).**

3. Set the DVOM to the resistance scale and attach the probes to the sensor or sensor pigtail terminals, as applicable.
4. Allow the temperature of the sensor and thermometer to stabilize, then note the temperature and the resistance reading. If a resistance specification is provided for low-temperatures, you may wish to add ice to the water in order to cool it down and start the test well below ambient temperatures.
5. Use the hot plate or camp stove to slowly raise the temperature. Watch the display on the DVOM closely (in case there are any sudden dips or spikes in the reading which could indicate a problem). Continue to note resistance readings as the temperature rises to 68°F (20°C) and 212°F (100°C). Of course by 212 degrees, your water should be boiling, so it is time to stop the test before you make a mess or get scalded. Up to that point, the meter should show a steady decrease in resistance that is proportional to the rate at which the liquid is heated. Extreme peaks or valleys in the sensor signal should be rechecked to see if they are results of sudden temperature increases or a possible problem with the sensor.
6. Compare the readings to the Sensor Test Specifications.

REMOVAL & INSTALLATION

 MODERATE

There are basically two ways that temperature sensors are normally mounted on Yamaha motors. Some sensors are secured to the powerhead or a mounting bracket via a retention plate and mounting screws/bolts (usually 2, one on either side of the sensor). Others contain a hex on the sensor itself and are threaded into position.

Removal and installation is therefore pretty straightforward, as long as access is not restricted. For most of the sensors mounted on top of the powerhead the flywheel cover must be removed for access and that is usually sufficient. In some rare instances, another cover or an additional component must be removed for better access. This includes some air intake silencer mounted IAT sensors to which access is difficult or impossible with the silencer installed.

Once you have clear access to the sensor all you need to do is disconnect the wiring and remove the bolts/screws (there are usually 2) that secure the sensor to the bracket or unthread the sensor, as applicable.

Pressure Sensors (APS and FPS)

◆ See Figures 271 and 272

All fuel injection systems are equipped with some form of an air pressure sensor. These motors utilize an Atmospheric Pressure Sensor (APS), while Yamaha 4-stroke motors (covered elsewhere) call it an Intake Air Pressure (IAP) sensor. Although the mounting point differs slightly from motor-to-motor and may an affect on exactly what the sensor is measuring (and for what the ECM is calibrated) the basic function performed by these sensors is the same. The sensor works like a barometer and reports barometric air pressure. This sensor allows the computer to compensate for barometric pressure changes at sea level and the normal reduction in atmospheric pressure found at high altitudes.

During normal EFI operation, a large portion of fuel mapping decisions (injector on-time strategy) is made based upon signals from the APS or IAP sensor.

In addition to this other pressure sensors may be found on some motors. The HPDI engines utilize a Fuel Pressure Sensor (FPS) to keep an eye on the extreme high-pressure circuit of the direct injection system.

Sensor locations vary slightly by engine and are as follows:
• For 2.6L EFI OX66 motors, the APS is mounted on the starboard side of the powerhead (just to the bottom, aft corner of the electrical junction box).
• For 3.1L EFI OX66 motors, the APS is mounted on the starboard side of the powerhead (just above the electrical junction box, toward the front of the powerhead)
• For HPDI motors, the APS is mounted to the bottom corner of the vapor separator tank assembly, on the port side of the powerhead.

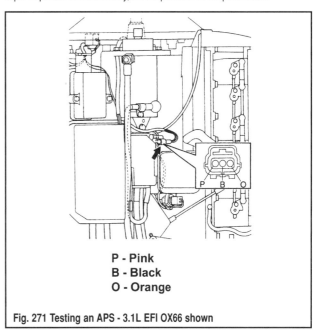

P - Pink
B - Black
O - Orange

Fig. 271 Testing an APS - 3.1L EFI OX66 shown

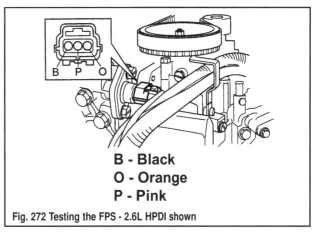

B - Black
O - Orange
P - Pink

Fig. 272 Testing the FPS - 2.6L HPDI shown

3-86 FUEL SYSTEM

- For 2.6L HPDI motors, the FPS is mounted on the extreme high-pressure (mechanical) fuel pump cover, at the rear of the powerhead. You must remove the flywheel cover for access and it is retained to the pump body by a plate mounted over the sensor body.
- For 3.3L HPDI motors, the FPS is mounted to the extreme high-pressure (mechanical) fuel line assembly, about half-way down the port fuel rail at the rear of the powerhead. You've not only got to remove the flywheel for access, but you also need to unbolt the ignition coil cover (keeping the ignition coil leads attached) and pull the cover out slightly for access.

TESTING

◆ See Figures 271 and 272

Pressure sensors operate using a reference voltage from the ECM, as a result, there is no way to test them unless the circuit is complete and operating. For the APS this means with the ignition keyswitch **ON** and the engine not running. However, the FPS must be tested with the engine running at idle (so the extreme high-pressure mechanical fuel pump will operate to produce proper system pressures.

In all cases testing requires the use of a 3-pin harness adapter or jumper wires to allow you to probe certain parts of the circuit while it is still complete and communicating with the ECM.

To test the sensor output voltage, proceed as follows:
1. Start by locating the sensor wiring. On most models this will at least involve removing the flywheel cover.
2. Disconnect the sensor wiring and install a 3-jumper wiring adapter between the connectors.

■ **Remember, the FPS must be checked with not only the keyswitch ON, but with the engine also running at idle. Be sure to provide a suitable source of cooling water when checking the FPS. The engine should NOT be running when testing the APS.**

3. Next, attach the leads from a DVOM to the jumper wires for the appropriate switch wires, then turn the ignition keyswitch **ON** and compare the circuit output voltage to the specifications as follows:
- For EFI OX66 motors, test the APS by connecting the DVOM to the Pink and Black wires, the sensor output should read 3.2-4.6 volts.
- For 2.6L HPDI motors, test the APS by connecting the DVOM to the Pink and Black wires (make sure you have the right ones, because the FPS is P-B also), the sensor output should read 3.2-4.6 volts. On these motors, test the FPS by connecting the DVOM to the Pink and Black wires for the fuel sensor, the DVOM should show 2.8-3.2 volts with the engine at idle.
- For 3.3L HPDI motors, test the APS by connecting the DVOM to the Orange and Black wires, the sensor output should be about 5 volts. On these motors, test the FPS by connecting the DVOM to the Orange and Pink wires, the DVOM should show fuel sensor output of about 3.2 volts with the engine idling at 670-730 rpm.

4. If there is no reading, double-check your test connections. If there is no reading, double-check your specifications. Also you can disconnect the harness completely and take a DC VOLTAGE reading across the wiring from the ECM to determine if a reference voltage is being supplied. If there is still no voltage reading, trace the problem back to the ECM.

■ **Unfortunately Yamaha does not provide any further data necessary to determine if an out-of-range reading is the sensor, wiring or ECM reference signal. Of course, the wiring between the ECM and the sensor should show VERY LITTLE or NO resistance. If you decide to check that wiring ALWAYS unplug it from the ECM before attempting any resistance readings (as you could damage the ECM with improperly connecting a DVOM across its terminals).**

5. If the FPS reads out of specification, the problem could be with the high-pressure fuel pump/line assembly, the sensor or the circuit.

REMOVAL & INSTALLATION

◆ See Figures 271 and 272

Removal and installation of the APS is fairly straightforward on most Yamaha motors. Locate the sensor and determine if any covers or other components need to be removed. On EFI OX66 motors, the sensors are on in the open on the side of the powerhead, though removing parts of the electrical junction box may help access. For HPDI motors, the sensors are mounted to the bottom corner of the Vapor Separator Tank assembly, and a better view of the access is available under those procedures in this section. Once accessed, the sensor is usually secured by 2 screws. Removal requires disconnecting the wiring and removing the mounting screws.

Removal and installation of the FPS sensor is a little more involved requiring depressurization of the fuel system for safety. Before proceeding properly relieve the fuel system pressure, as detailed earlier in this section. Access to the sensor will require removal of the Flywheel cover and, in the case of the 3.3L engine, removal of the Ignition Coil Cover, as well. Once accessed, the sensor is normally secured by 2 bolts and a mounting plate, disconnect the wiring and remove the mounting plate in order to remove the sensor.

Knock Sensor (3.1L EFI OX66 Motors)

◆ See Figure 273

The knock sensor works a little like a microphone. It is a piezoelectric sensor, meaning that it generates its own voltage as a result of vibration. The sensor is used to detect spark knock which occurs when the ignition timing is too far advanced for operating conditions (atmospheric pressure, temperature, cylinder compression) and the fuel being used. When the ECM receives a knock signal it will retard spark timing until the knock ceases. In this way the knock sensor helps to protect the engine from potential ignition related damage while allowing the computer to otherwise maximize ignition timing and engine performance.

The knock sensor is normally threaded into the starboard cylinder head, a little more than 1/2 way down the powerhead, and almost immediately adjacent to the middle ignition coil for that bank.

TESTING

◆ See Figure 273

Since the sensor generates its own voltage in response to vibration, testing the sensor is a relatively simple matter. Disconnect the sensor wiring, then connect a DVOM set to read AC VOLTS with one lead on the sensor wiring and the other on the sensor body. Gently tap the sensor and watch for a reading of several milli-volts on the meter. If the there is no reading, make sure your meter is capable of reading very low voltage and/or tap a little more firmly. If there is still nothing, the sensor is likely faulty and should be replaced.

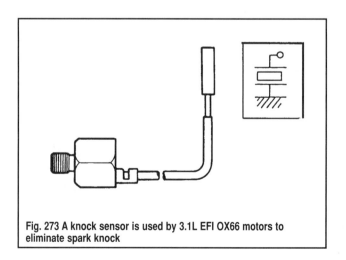

Fig. 273 A knock sensor is used by 3.1L EFI OX66 motors to eliminate spark knock

FUEL SYSTEM

REMOVAL & INSTALLATION

◆ See Figure 273

The sensor is normally threaded in place, disconnect the wiring and carefully unthread it from it's position in the starboard cylinder head.

Oxygen Sensor (EFI OX66 and some HPDI)

◆ See Figure 274

All EFI OX66 motors, as well as the 2.6L HPDI motors are equipped with an oxygen density sensor. The oxygen sensor is the fine-tuning sensor. This is the only sensor in any of the EFI systems that can monitor the combustion process. The 02 sensor is mounted in a blind pocket that is then connected to the combustion chamber. It monitors the amount of oxygen in the exhaust. This sensor reports to the computer with a voltage signal directly proportional to oxygen content. Information from this sensor is used to adjust enrichment.

The sensor will only work properly once it reaches normal operating temperature. In order to shorten the amount of time it takes for this to occur and in order to make sure it remains at operating temperature the sensor contains an internal heater element.

TESTING

◆ See Figures 275 and 276

Although the element itself can be checked with a resistance test, the sensor can only be checked by determining if it is generating a milli-volt signal. Yamaha recommends removing the sensor and holding it in a flame (under very specific conditions) to see if it is working properly. This is not an easy test to perform, as it is easy to damage the sensor instead of testing it, so use caution, especially when working around an open flame.

✹✹ CAUTION

We REALLY shouldn't have to say this, but we will. You'll be working with an open flame so make sure to keep anything that can burn AWAY from the flame. Make sure the work area is sufficiently ventilated and that there are ABSOLUTELY NO GASOLINE VAPORS present. Also, remember that batteries give off explosive hydrogen gas while discharged or while being charged, so make sure there are no automotive or marine batteries in the work area.

✹✹ WARNING

NEVER let any silicon anticorrosion solvent touch the oxygen sensor or its output will become affected.

  1. To test the sensor heater element, proceed as follows depending upon the model:
  • For EFI OX66 motors, open the electrical junction box cover (on the starboard side of the motor) and disconnect the wiring for the sensor heater (White and White wires). Connect a DVOM set to read resistance to the sensor side of the harness, the reading should be 2-100 ohms.
  • For HPDI motors, locate the sensor wiring just above the aft corner of the electrical junction box cover (and just below the aft corner of the oxygen sensor cover, on the starboard side of the motor) and disconnect the wiring for the sensor heater (Red/White and Black wires). Connect a DVOM set to read resistance to the sensor side of the harness, the reading should be 2-100 ohms.

■ **If the sensor heater element is well out of specification, the sensor must be replaced to ensure optimum performance.**

  2. To test the sensor output, remove the sensor as detailed later in this section.
  3. Carefully clean any oil from the end of the oxygen sensor using acetone or a similar solvent. Allow the sensor to thoroughly dry (do not use compressed air to dry the sensor as it may contain moisture, oil or other contaminants).

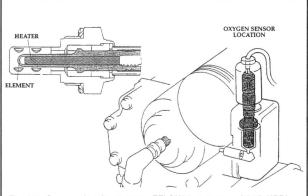

Fig. 274 Oxygen density sensor - EFI OX66 motors and 2.6L HPDI motors

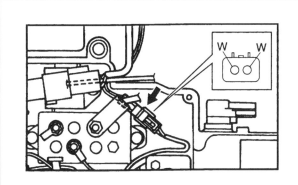

Fig. 275 Oxygen sensor heater element testing - 3.1L EFI OX66 motors shown

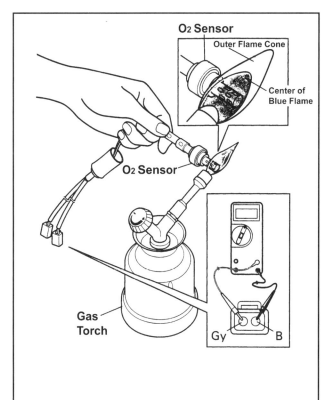

Fig. 276 Use extreme caution when testing the oxygen sensor element

3-88 FUEL SYSTEM

4. Connect a DVOM set to read milli-volts to the oxygen sensor wiring. Connect the positive lead to the Gray wire and the negative lead to the Black (EFI OX66) or Black/White (HPDI) wire.

5. Using a GAS torch (be sure NOT to use any burner with an Oxygen tank or the sensor may be damaged by the higher flame temperature, a propane plumbing torch would probably work quite well for this), carefully heat the sensor tip in the center of the blue flame for 10-15 seconds, then remove the sensor and observe the meter for a voltage change. If the instantaneous change in sensor output is 0.6 volts or greater, Yamaha considers the sensor within specification.

※※ WARNING

DO NOT heat the sensor above 1472°F (800°C) continuously for more than a minute at a time or the sensor will be damaged.

REMOVAL & INSTALLATION

◆ See Figure 277

 MODERATE

The 02 sensor is mounted in a blind pocket on the starboard side of the motor, directly above the electrical junction box.

※※ WARNING

NEVER let any silicon anticorrosion solvent touch the oxygen sensor or its output will become affected.

1. Note the positioning, then cut the wire tie securing the sensor wiring at the top of the protective rubber cap. Pull the rubber cap back up the wiring and off the sensor cover.

2. Remove the 2 bolts securing the sensor cover and then remove the cover from the powerhead.

■ The oxygen sensor is mounted in a bracket that attaches to the cylinder head with 3 bolts, a gasket and sensor joint. The purpose of the bracket is to provide a housing that allows the oxygen sensor access to the combustion gases. The sensor is threaded into a locknut on the top of the housing and can probably be separated from the bracket without removing the bracket from the powerhead. HOWEVER, since the whole purpose of the housing it to allow access to combustion gases, it is probably wiser to remove the hole bracket and inspect the housing and joint for clogging or contamination. This will help ensure proper sensor operation. There is no use installing a new sensor if the gases can't reach it, right?

3. Trace the sensor pigtail back to the harness connector and disengage the connector from the engine wiring. Note the wire routing for installation purposes.

4. Remove the 3 bolts securing the sensor mounting bracket to the powerhead and remove the bracket and sensor assembly. Remove and discard the old bracket gasket.

5. Loosen the locknut and remove the sensor from the bracket.

6. Remove the sensor joint. On some models the joint is sealed to the powerhead using a gasket and an O-ring. When equipped the gasket must be discarded, but the O-ring may be reused, if it is in good condition.

To Install:

7. If removed, install the sensor joint (with the O-ring and a NEW gasket, if applicable).

8. Install the sensor to the bracket and carefully tighten the locknut to 35 ft. lbs. (49 Nm).

9. Install the bracket to the powerhead using a new gasket; then tighten the retaining bolts to 10 ft. lbs. (14 Nm).

10. Install the sensor cover to the powerhead and tighten the 2 retaining bolts to 6.5 ft. lbs. (9 Nm).

11. Reposition the wiring as noted during removal, slide the rubber protective cover in position and secure using one or more wire ties (where applicable).

Water Detection Sensor (HPDI)

◆ See Figure 278

On HPDI motors, the motor and fuel injection system is protected by a water detection switch mounted in the low-pressure circuit filter assembly. The unit consists of a simple, ON/OFF-type switch connected to a float assembly which is weighted to remain submerged in gasoline but to float in the presence of water. If the filter cup accumulates a sufficient amount of water the float will raise to a position, closing the switch and signaling a warning to the operator.

The switch can be easily tested by removing the filter cup (refer to the Fuel Filter procedures in the Maintenance and Tune-Up section for details). Connect a DVOM to the switch wiring (it's normally either Blue/White and Black for 2.6L motors or Black and Black for 3.3L motors) and check for switch continuity. With the float in the lower position there should be NO continuity. Invert the filter cup, so the float falls to the upper position or manually raise the float in the cup and make sure the DVOM now shows continuity. If so, the switch is working properly.

Fig. 277 Oxygen sensor mounting - EFI OX66 and 2.6L HPDI motors

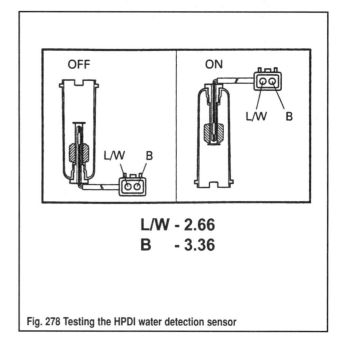

L/W - 2.66
B - 3.36

Fig. 278 Testing the HPDI water detection sensor

FUEL SYSTEM   3-89

Neutral or Shift Position Switches

◆ See Figure 279

Most fuel injected motors are equipped with one or more neutral or shift switches. Typically these switches are used to tell the ECM whether or not the motor is in gear/neutral. Different fuel and ignition mapping decisions may be made depending upon these conditions. When equipped, these switches are normally of the simple ON/OFF one switch position design. On Yamaha motors, this USUALLY means that there should be NO CONTINUITY across the switch terminals when the plunger is released (extended). Conversely, when the switch is depressed (plunger is down against the switch body) there should be continuity across the switch contacts. Using a DVOM set to read resistance you can easily check the function of these switches, it is not important that they match this description perfectly, but it is important that there is little or no resistance in one position and infinite resistance in the opposite position.

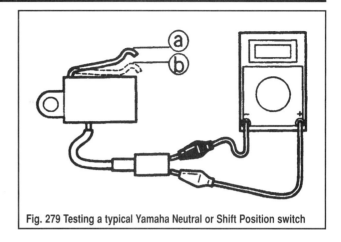

Fig. 279 Testing a typical Yamaha Neutral or Shift Position switch

Fuel Injection and Electronic Component Locations

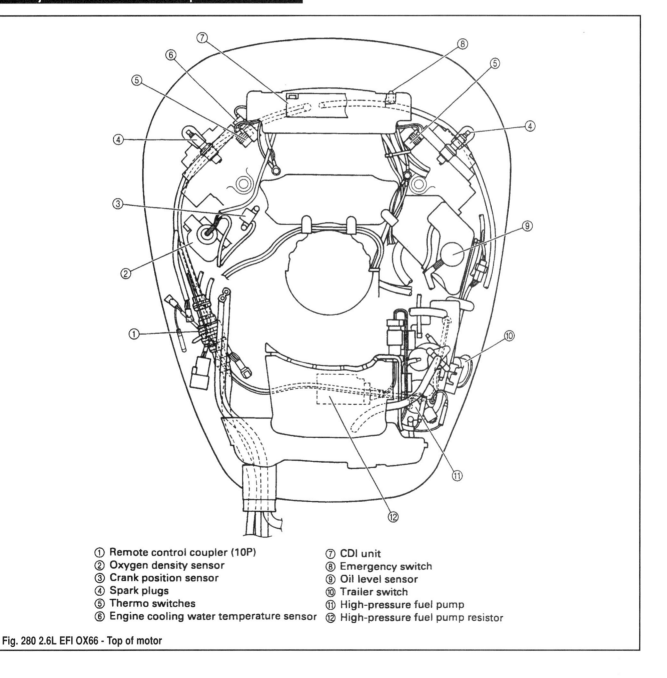

① Remote control coupler (10P)
② Oxygen density sensor
③ Crank position sensor
④ Spark plugs
⑤ Thermo switches
⑥ Engine cooling water temperature sensor
⑦ CDI unit
⑧ Emergency switch
⑨ Oil level sensor
⑩ Trailer switch
⑪ High-pressure fuel pump
⑫ High-pressure fuel pump resistor

Fig. 280 2.6L EFI OX66 - Top of motor

3-90 FUEL SYSTEM

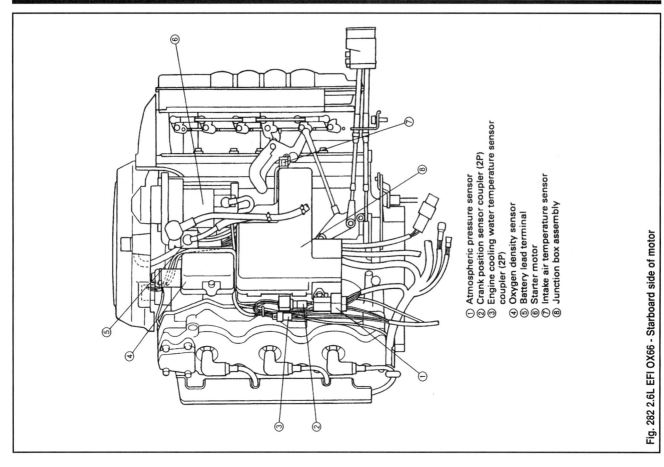

Fig. 282 2.6L EFI OX66 - Starboard side of motor

① Atmospheric pressure sensor
② Crank position sensor coupler (2P)
③ Engine cooling water temperature sensor coupler (2P)
④ Oxygen density sensor
⑤ Battery lead terminal
⑥ Starter motor
⑦ Intake air temperature sensor
⑧ Junction box assembly

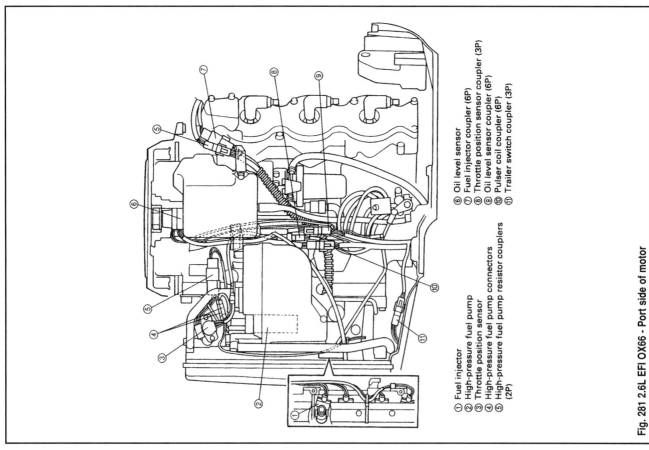

Fig. 281 2.6L EFI OX66 - Port side of motor

① Fuel injector
② High-pressure fuel pump
③ Throttle position sensor
④ High-pressure fuel pump connectors
⑤ High-pressure fuel pump resistor couplers (2P)
⑥ Oil level sensor
⑦ Fuel injector coupler (6P)
⑧ Throttle position sensor coupler (3P)
⑨ Oil level sensor coupler (6P)
⑩ Pulser coil coupler (6P)
⑪ Trailer switch coupler (3P)

FUEL SYSTEM 3-91

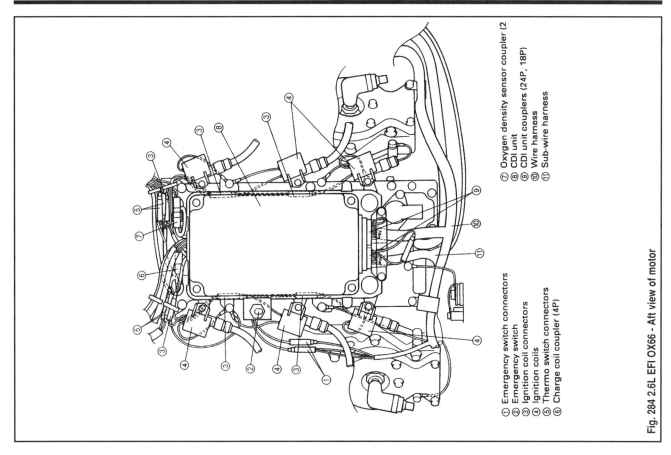

Fig. 284 2.6L EFI OX66 - Aft view of motor

① Emergency switch connectors
② Emergency switch
③ Ignition coil connectors
④ Ignition coils
⑤ Thermo switch connectors
⑥ Charge coil coupler (4P)
⑦ Oxygen density sensor coupler (2
⑧ CDI unit
⑨ CDI unit couplers (24P, 18P)
⑩ Wire harness
⑪ Sub-wire harness

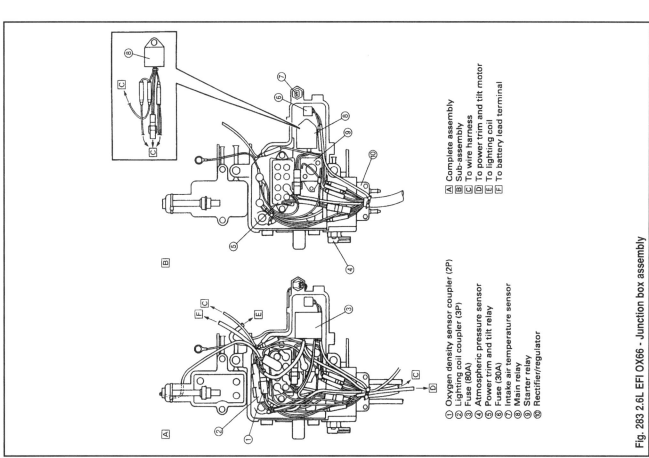

Fig. 283 2.6L EFI OX66 - Junction box assembly

① Oxygen density sensor coupler (2P)
② Lighting coil coupler (3P)
③ Fuse (80A)
④ Atmospheric pressure sensor
⑤ Power trim and tilt relay
⑥ Fuse (30A)
⑦ Intake air temperature sensor
⑧ Main relay
⑨ Starter relay
⑩ Rectifier/regulator

Ⓐ Complete assembly
Ⓑ Sub-assembly
Ⓒ To wire harness
Ⓓ To power trim and tilt motor
Ⓔ To lighting coil
Ⓕ To battery lead terminal

3-92 FUEL SYSTEM

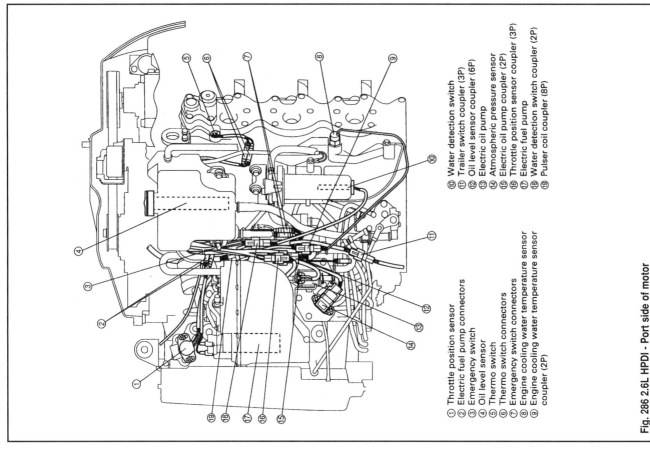

① Throttle position sensor
② Electric fuel pump connectors
③ Emergency switch
④ Oil level sensor
⑤ Thermo switch
⑥ Thermo switch connectors
⑦ Emergency switch connectors
⑧ Engine cooling water temperature sensor
⑨ Engine cooling water temperature sensor coupler (2P)
⑩ Water detection switch
⑪ Trailer switch coupler (3P)
⑫ Oil level sensor coupler (6P)
⑬ Electric oil pump
⑭ Atmospheric pressure sensor
⑮ Electric oil pump coupler (2P)
⑯ Throttle position sensor coupler (3P)
⑰ Electric fuel pump
⑱ Water detection switch coupler (2P)
⑲ Pulser coil coupler (8P)

Fig. 286 2.6L HPDI - Port side of motor

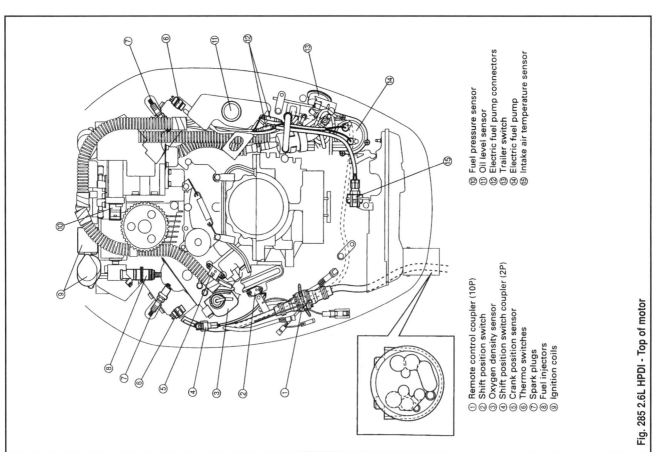

① Remote control coupler (10P)
② Shift position switch
③ Oxygen density sensor
④ Shift position switch coupler (2P)
⑤ Crank position sensor
⑥ Thermo switches
⑦ Spark plugs
⑧ Fuel injectors
⑨ Ignition coils
⑩ Fuel pressure sensor
⑪ Oil level sensor
⑫ Electric fuel pump connectors
⑬ Trailer switch
⑭ Electric fuel pump
⑮ Intake air temperature sensor

Fig. 285 2.6L HPDI - Top of motor

FUEL SYSTEM 3-93

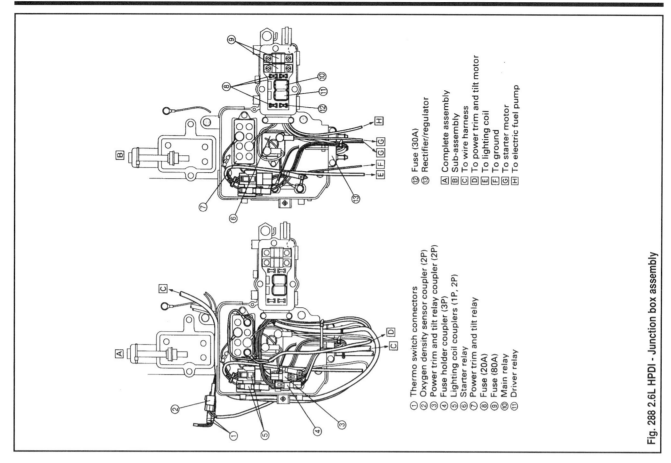

Fig. 288 2.6L HPDI - Junction box assembly

① Thermo switch connectors
② Oxygen density sensor coupler (2P)
③ Power trim and tilt relay coupler (2P)
④ Fuse holder coupler (3P)
⑤ Lighting coil couplers (1P, 2P)
⑥ Starter relay
⑦ Power trim and tilt relay
⑧ Fuse (20A)
⑨ Fuse (80A)
⑩ Main relay
⑪ Driver relay
⑫ Fuse (30A)
⑬ Rectifier/regulator

Ⓐ Complete assembly
Ⓑ Sub-assembly
Ⓒ To wire harness
Ⓓ To power trim and tilt motor
Ⓔ To lighting coil
Ⓕ To ground
Ⓖ To starter motor
Ⓗ To electric fuel pump

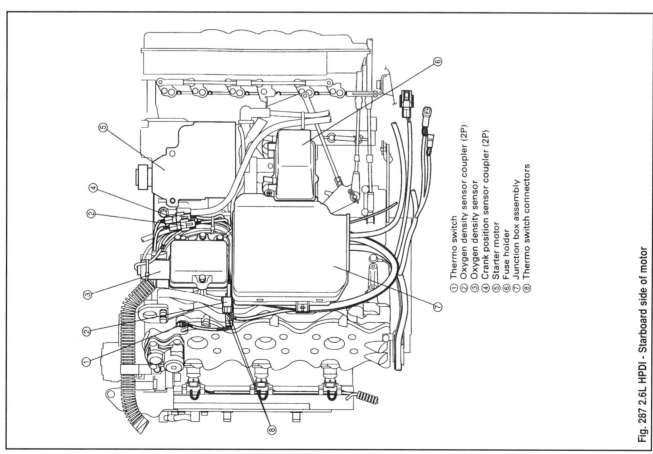

Fig. 287 2.6L HPDI - Starboard side of motor

① Thermo switch
② Oxygen density sensor coupler (2P)
③ Oxygen density sensor
④ Crank position sensor coupler (2P)
⑤ Starter motor
⑥ Fuse holder
⑦ Junction box assembly
⑧ Thermo switch connectors

3-94 FUEL SYSTEM

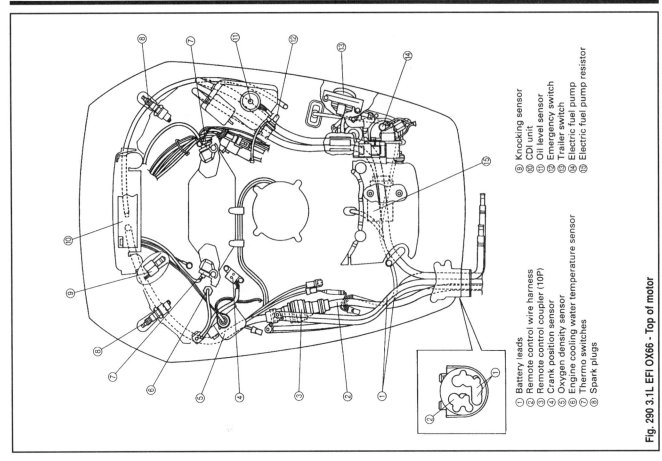

① Battery leads
② Remote control wire harness
③ Remote control coupler (10P)
④ Crank position sensor
⑤ Oxygen density sensor
⑥ Engine cooling water temperature sensor
⑦ Thermo switches
⑧ Spark plugs
⑨ Knocking sensor
⑩ CDI unit
⑪ Oil level sensor
⑫ Emergency switch
⑬ Trailer switch
⑭ Electric fuel pump
⑮ Electric fuel pump resistor

Fig. 290 3.1L EFI OX66 - Top of motor

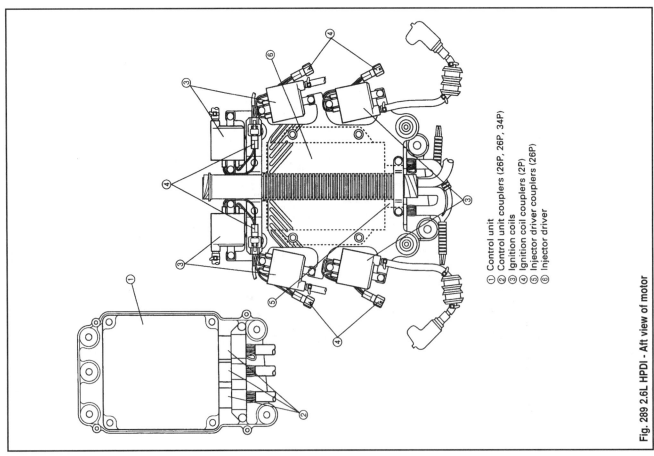

① Control unit
② Control unit couplers (26P, 26P, 34P)
③ Ignition coils
④ Ignition coil couplers (2P)
⑤ Injector driver couplers (26P)
⑥ Injector driver

Fig. 289 2.6L HPDI - Aft view of motor

FUEL SYSTEM 3-95

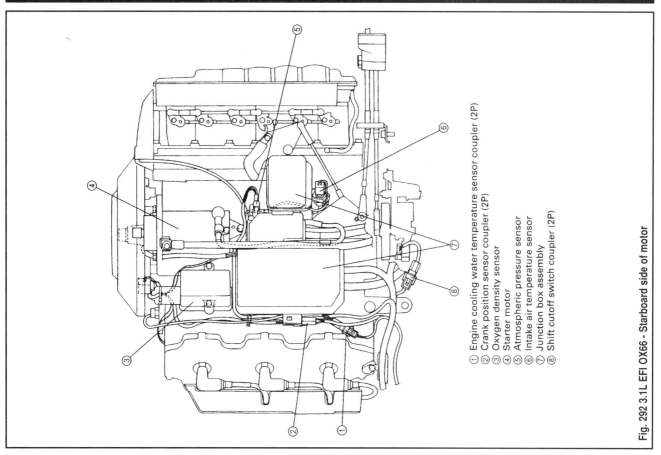

① Engine cooling water temperature sensor coupler (2P)
② Crank position sensor coupler (2P)
③ Oxygen density sensor
④ Starter motor
⑤ Atmospheric pressure sensor
⑥ Intake air temperature sensor
⑦ Junction box assembly
⑧ Shift cutoff switch coupler (2P)

Fig. 292 3.1L EFI OX66 - Starboard side of motor

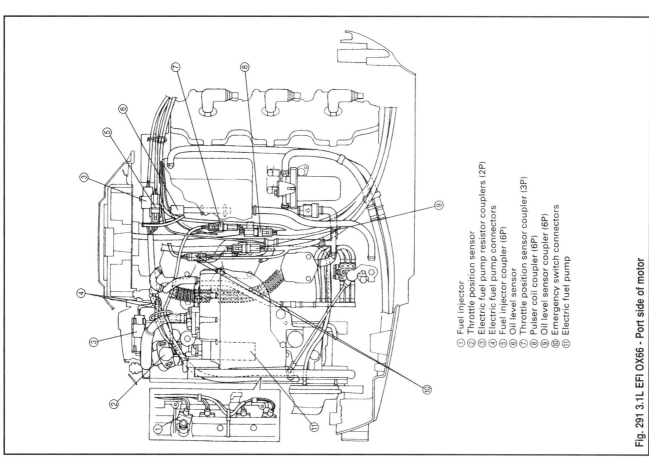

① Fuel injector
② Throttle position sensor
③ Electric fuel pump resistor couplers (2P)
④ Electric fuel pump connectors
⑤ Fuel injector coupler (6P)
⑥ Oil level sensor
⑦ Throttle position sensor coupler (3P)
⑧ Pulser coil coupler (6P)
⑨ Oil level sensor coupler (6P)
⑩ Emergency switch connectors
⑪ Electric fuel pump

Fig. 291 3.1L EFI OX66 - Port side of motor

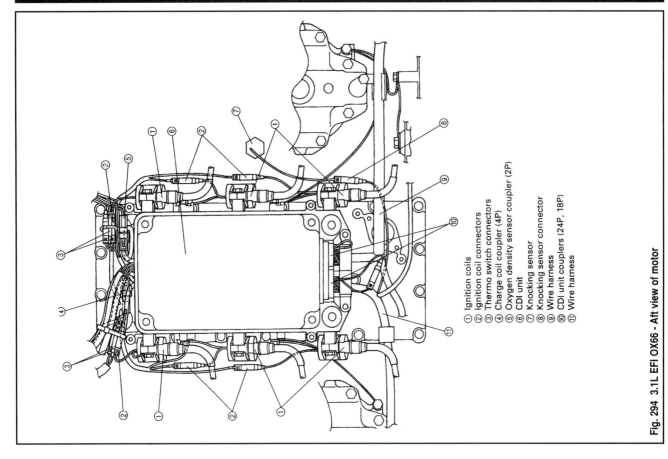

Fig. 294 3.1L EFI OX66 - Aft view of motor

① Ignition coils
② Ignition coil connectors
③ Thermo switch connectors
④ Charge coil coupler (4P)
⑤ Oxygen density sensor coupler (2P)
⑥ CDI unit
⑦ Knocking sensor
⑧ Knocking sensor connector
⑨ Wire harness
⑩ CDI unit couplers (24P, 18P)
⑪ Wire harness

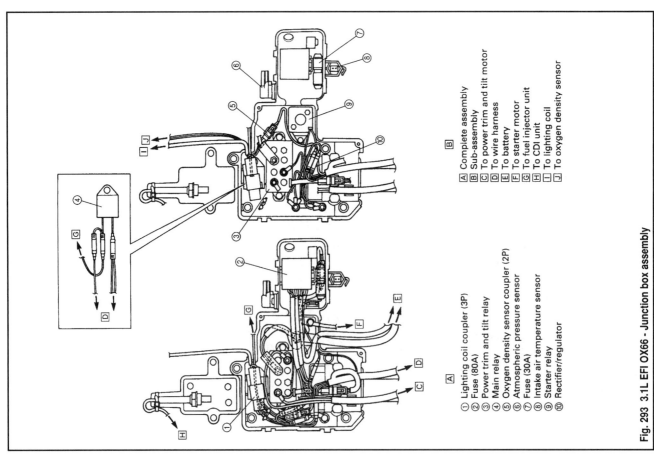

Fig. 293 3.1L EFI OX66 - Junction box assembly

Ⓐ Complete assembly
Ⓑ Sub-assembly
Ⓒ To power trim and tilt motor
Ⓓ To wire harness
Ⓔ To battery
Ⓕ To starter motor
Ⓖ To fuel injector unit
Ⓗ To CDI unit
Ⓘ To lighting coil
Ⓙ To oxygen density sensor

① Lighting coil coupler (3P)
② Fuse (80A)
③ Power trim and tilt relay
④ Main relay
⑤ Oxygen density sensor coupler (2P)
⑥ Atmospheric pressure sensor
⑦ Fuse (30A)
⑧ Intake air temperature sensor
⑨ Starter relay
⑩ Rectifier/regulator

FUEL SYSTEM 3-97

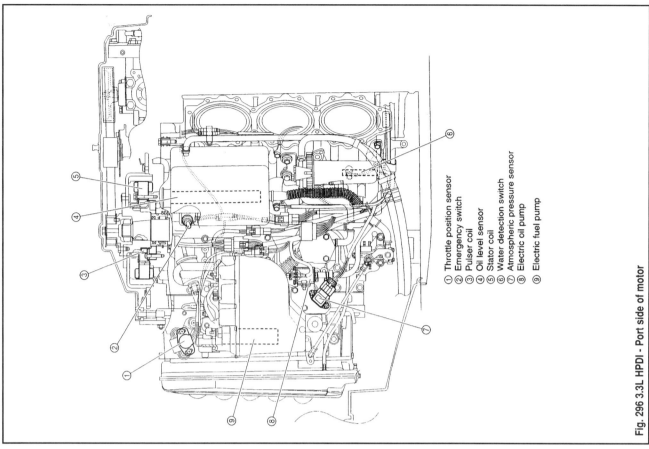

Fig. 296 3.3L HPDI - Port side of motor

① Throttle position sensor
② Emergency switch
③ Pulser coil
④ Oil level sensor
⑤ Stator coil
⑥ Water detection switch
⑦ Atmospheric pressure sensor
⑧ Electric oil pump
⑨ Electric fuel pump

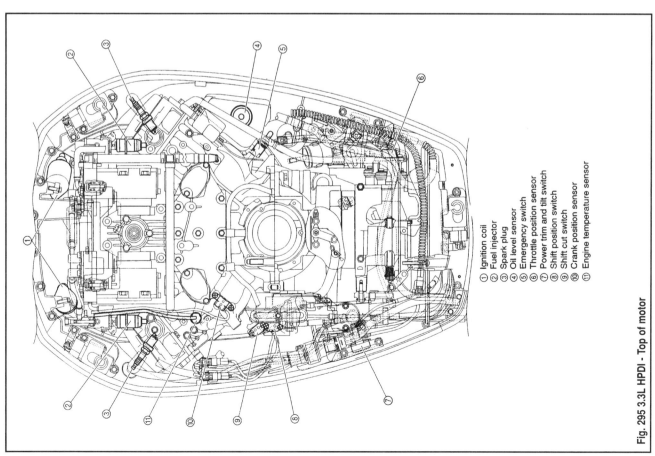

Fig. 295 3.3L HPDI - Top of motor

① Ignition coil
② Fuel injector
③ Spark plug
④ Oil level sensor
⑤ Emergency switch
⑥ Throttle position sensor
⑦ Power trim and tilt switch
⑧ Shift position switch
⑨ Shift cut switch
⑩ Crank position sensor
⑪ Engine temperature sensor

3-98 FUEL SYSTEM

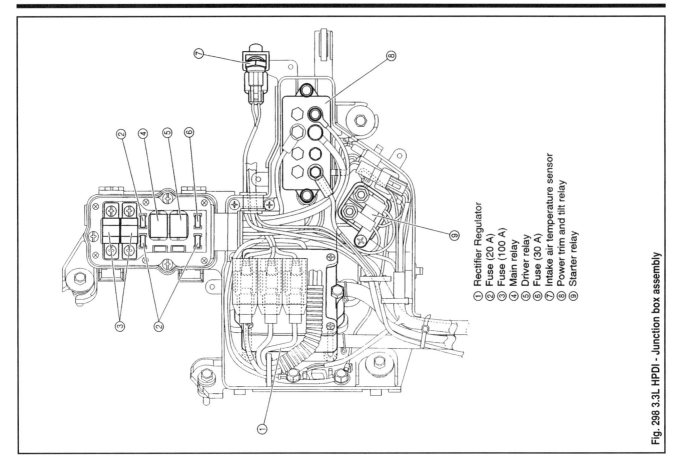

Fig. 298 3.3L HPDI - Junction box assembly

① Rectifier Regulator
② Fuse (20 A)
③ Fuse (100 A)
④ Main relay
⑤ Driver relay
⑥ Fuse (30 A)
⑦ Intake air temperature sensor
⑧ Power trim and tilt relay
⑨ Starter relay

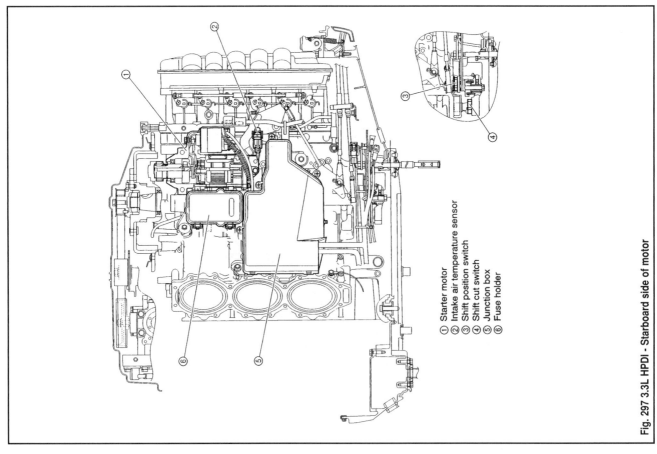

Fig. 297 3.3L HPDI - Starboard side of motor

① Starter motor
② Intake air temperature sensor
③ Shift position switch
④ Shift cut switch
⑤ Junction box
⑥ Fuse holder

FUEL SYSTEM 3-99

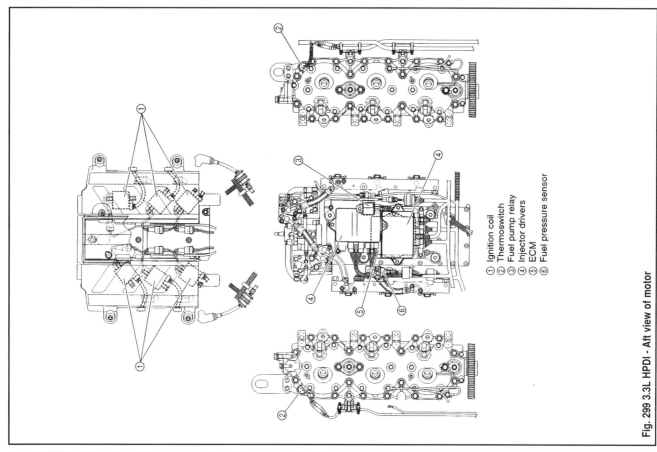

Fig. 299 3.3L HPDI - Aft view of motor
① Ignition coil
② Thermoswitch
③ Fuel pump relay
④ Injector drivers
⑤ ECM
⑥ Fuel pressure sensor

FUEL PUMP (LOW-PRESSURE) SERVICE

This section deals with the low-pressure, mechanical, diaphragm-displacement fuel pumps found on all carbureted motors and which act as a lift pump for most fuel-injected motors. The electric high-pressure pump used on all EFI/HPDI motors, and the extreme high-pressure mechanical pump used on HPDI motors are all covered in the section on Electronic Fuel Injection Systems.

Fuel Pumps

DESCRIPTION & OPERATION

■ **Most smaller hp, single cylinder Yamaha powerheads normally do not have a fuel pump. Fuel is provided to the carburetor by gravity flow from the fuel tank atop the powerhead.**

The fuel pump on most powerheads through 25 hp (and some 35 hp built on the same basic motor as the 25 hp models) utilize a fuel pump that is an integral part of the carburetor. Therefore, do not search for a separate fuel pump on these powerheads. This integral fuel pump is covered under overhaul. On 3-cylinder 25/35 hp motors, the fuel pump is normally integrated to the middle of the 3 carburetors.

Most 28J and larger models are equipped with one or more crankcase mounted fuel pumps. The following section provides detailed instructions to service the fuel pump on powerheads equipped with an external fuel pump.

Integral Pump

◆ See Figure 300

This pump consists of a series of diaphragms and check valves operated by crankcase pressure. The pump feeds fuel directly into the float bowl.

As the piston moves upward, a suction, or negative pressure, is created in the crankcase. This low pressure acts upon the inner diaphragm through the carburetor. The inner diaphragm is therefore pulled inward. When the diaphragm moves inward, the check valve opens and fuel is drawn from the bottom of the fuel pump exterior chamber into the fuel pump interior chamber.

■ **If the primer bulb is activated prior to powerhead start, the fuel pump exterior chamber will be filled with fuel.**

As the piston moves downward, a positive pressure is created in the crankcase causing the inner diaphragm to be pushed outward. The delivery check valve opens under pressure and allows fuel to flow into the upper part of the exterior chamber and on into the carburetor float bowl.

The function of the outer diaphragm is to absorb the pulsations of the fuel and allow a smooth uninterrupted fuel flow.

Fig. 300 Some 1 and 2-cylinder Yamahas utilize a pump which is integrated into the carburetor

3-100 FUEL SYSTEM

Powerhead-Mounted Diaphragm-Displacement Pump

◆ See Figures 301 and 302

The powerhead mounted fuel pump is a separate unit from the carburetor.

The pump typically consists of a spring loaded inner diaphragm, a spring loaded outer diaphragm, two valves, one for inlet (suction) and the other for outlet (discharge), and a small opening leading directly into the crankcase. The suction and compression created as the piston travels up and down in the cylinder, causes the diaphragms to flex.

As the piston moves upward, the inner diaphragm will flex inward displacing volume on its opposite side to create suction. This suction will draw fuel in through the inlet valve.

When the piston moves downward, compression is created in the crankcase. This compression causes the inner diaphragm to flex in the opposite direction. This action causes the discharge valve to lift off its seat. Fuel is then forced through the discharge valve into the carburetor.

The function of the outer diaphragm is to absorb the pulsations of the fuel and allow a smooth uninterrupted fuel flow.

Problems with the fuel pump are limited to possible leaks in the flexible neoprene suction lines, a punctured diaphragm, air leaks between sections of the pump assembly, or possibly from the valves becoming distorted or not seating properly.

Fig. 301 A powerhead mounted diaphragm-displacement fuel pump is used on most Yamaha powerheads

Fig. 302 The covers of most Yamaha diaphragm-displacement pumps are labeled to help when attaching fuel lines

GENERAL INFORMATION

◆ See Figure 303

A fuel pump is a basic mechanical device used to pull fuel from the tank and feed it to the float bowl(s) in the carburetor(s) or the fuel vapor separator assembly on EFI/HPDI motors.

For Yamaha outboards the pump is (or pumps are, on larger motors) mounted to the powerhead and utilizes crankcase positive and negative pressures to actuate a diaphragm. On the smaller motors (most up to the 25/35 hp range) the pump is usually integral with a carburetor assembly, whereas the pump on larger motors (most of the 28J and larger motors) is normally a completely separate unit bolted directly to the crankcase. Regardless of mounting position, the pump works by utilizing engine crankcase vacuum and pressure to move an internal diaphragm alternately creating vacuum and pressure in the fuel line.

For 2-stroke motors, the pumps generally contain a flexible diaphragm and two check valves (flappers or fingers) that control flow. As the piston in the cylinder goes up, crankcase pressure drops (negative pressure) and the inlet valve opens, pulling fuel from the tank. As the piston nears TDC, pressure in the pump area is neutral (atmospheric pressure). At this point both valves are closed. As the piston comes down, pressure goes up (positive pressure) and the fuel is pushed toward the carburetor bowl by the diaphragm through the now open outlet valve.

The diaphragm-displacement fuel pump is a reliable method to move fuel but can have several problems. Occasionally an engine backfire can rupture the diaphragm. In addition, the diaphragm and valves are moving parts subject to wear. The flexibility of the diaphragm material can go away over time, reducing or stopping flow. Rust or dirt can hang a valve open and reduce or stop fuel flow. And, obviously, if a diaphragm develops one or more holes it will loose the ability to generate sufficient amounts of vacuum/pressure to overcome the check valves.

TESTING

MODERATE

The problem most often seen with fuel pumps is fuel starvation, hesitation or missing due to inadequate fuel pressure/delivery. In extreme cases, this might lead to a no start condition, but that is pretty rare as the primer bulb should at least allow the operator to fill the float bowl or vapor separator tank). More likely, pump failures are not total, and the motor will start and run fine at idle, only to miss, hesitate or stall at speed when pump performance falls short of the greater demand for fuel at high rpm.

Before replacing a suspect fuel pump, be absolutely certain the problem is

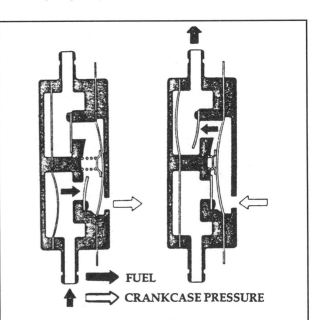

Fig. 303 A fuel pump is a basic mechanical device that pulls fuel from the tank. It utilizes crankcase positive and negative pressures to pump fuel

the pump and NOT with fuel tank, lines or filter. A plugged tank vent could create vacuum in the tank that will overpower the pump's ability to create vacuum and draw fuel through the lines. An obstructed line or fuel filter could also keep fuel from reaching the pump. Any of these conditions could partially restrict fuel flow, allowing the pump to deliver fuel, but at a lower pressure/rate. A pump delivery or pressure test under these circumstances would give a low reading that might be mistaken for a faulty pump. Before testing the fuel pump, refer to the testing procedures found under Fuel Lines and Fitting to ensure there are no problems with the tank, lines or filter.

A quick check of fuel pump operation is to gently squeeze the primer bulb with the motor running. If a seemingly rough or lean running condition (especially at speed) goes away when the bulb is squeezed, the fuel pump is suspect.

If inadequate fuel delivery is suspected and no problems are found with the tank, lines or filters, a conduct a quick-check to see how the pump affects performance. Use the primer bulb to supplement fuel pump. This is done by operating the motor under load and otherwise under normal operating conditions to recreate the problem. Once the motor begins to hesitate, stumble or stall, pump the primer bulb quickly and repeatedly while listening for motor response. Pumping the bulb by hand like this will force fuel through the lines to the vapor separator tank, regardless of the fuel pump's ability to draw and deliver fuel. If the engine performance problem goes away while pumping the bulb, and returns when you stop, there is a good chance you've isolated the low pressure fuel pump as the culprit. Depending upon the model you may be able to perform a pressure or vacuum check (Yamaha tends to recommend the later) or you'll have to disassemble the pump to physically inspect the check valves and diaphragms (your only option on carburetor integrated pumps, but it's not that difficult and can be done on ALL pumps).

✶✶ WARNING

Never run a motor without cooling water. Use a test tank, a flush/test device or launch the craft. Also, never run a motor at speed without load, so for tests running over idle speed, make sure the motor is either in a test tank with a test wheel or on a launched craft with the normal propeller installed.

Vacuum Checking the Delivery System

◆ See Figure 304

Fuel system vacuum testing is an excellent way to pinpoint air leaks, restricted fuel lines and fittings or other fuel supply related performance problems.

When a fuel starvation problem is suspected such as engine hesitation or engine stopping, perform the following fuel system test to see if you should check the fuel tank, filter(s) and lines or check the pump itself:

1. Connect the piece of clear fuel hose to a side barb of a "T" fitting.
2. Connect one end of a long piece of fuel hose to the vacuum gauge and the other end to the center barb of the "T" fitting.

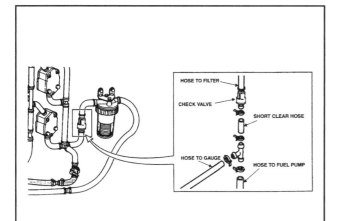

Fig. 304 Connecting a vacuum gauge inline in preparation for isolating fuel system problems

■ Use a long enough piece of fuel hose so the vacuum gauge may be read at the helm.

3. Remove the existing fuel hose from the fuel tank side of the fuel pump (that usually means between the fuel filter and the fuel pump), and connect the remaining barb of the "T" fitting to the fuel hose.
4. Connect the short piece of clear fuel hose to the fuel check valve leading from the fuel filter. If a check valve does not exist, connect the clear fuel hose directly to the fuel filter.
5. Check the vacuum gauge reading after running the engine long enough to stabilize at full power.

■ The vacuum is to not exceed 4.5 in. Hg (15.2 kPa) for up to 200 hp engines. The vacuum is to not exceed 6.0 in. Hg (20.3 kPa) for engines greater than 200 hp. On the bright side, if the vacuum DOES exceed this figures, the problem is more than likely NOT the pump!

6. An anti-siphon valve (required if the fuel system drops below the top of the fuel tank) will cause a 1.5 to 2.5 in. Hg (8.4 kPa) increase in vacuum.
7. If high vacuum is noted, move the T-fitting to the fuel filter inlet and retest.
8. Continue to the fuel filter inlet and along the remaining fuel system until a large drop/change in vacuum locates the problem.
9. A good clean water separator fuel filter will increase vacuum about 0.5 in. Hg (1.7 kPa).
10. Small internal passages inside a fuel selector valve, fuel tank pickup, or fuel line fittings may cause excessive fuel restriction and high vacuum.
11. Unstable and slowly rising vacuum readings, especially with a full tank of fuel, usually indicates a restricted vent line.

■ Bubbles in the clear fuel line section indicate an air leak, making for an inaccurate vacuum test. Check all fittings for tightened clamps and a tight fuel filter.

■ Vacuum gauges are not calibrated and some may read as much as 2 in. Hg (6.8 kPa) lower than the actual vacuum. It is recommended to perform a fuel system test while no problems exist to determine vacuum gauge accuracy.

Checking Pump Pressure

A basic low-pressure fuel pump pressure check is included in the Fuel System Checks found under the Maintenance and Tune-Up section of this guide. Unfortunately, Yamaha does not publish pump pressure specifications for their 2-stroke motors, so the information should be used to decide whether or not further diagnosis seems necessary on those motors.

You can check fuel pump delivery by measuring the amount of fuel that is expelled from a disconnected fuel pump outlet hose while the motor runs at speed for one minute. In order to safely conduct this test the motor must either be in a test tank or on a launched craft as engine speed should be maintained toward the high end of mid-range (about 3000 rpm, but check the Engine Specifications charts for exact requirements). Before starting the test you'll need to run the engine and make sure the carburetor float bowl(s) are full (since they won't continue to receive fuel once the test is started). Carefully disconnect the fuel pump outlet line from the carburetor(s) and direct it into an approved container. Start and run the engine for one minute, then shut down the powerhead and measure the amount of fuel collected using a graduated cylinder or beaker.

Fuel Pump Diaphragm and Check Valve Testing

◆ See Figures 305 and 306

For non-carburetor integrated fuel pumps you can use a simple hand-held vacuum/pressure pump and gauge to perform basic tests of the fuel pump diaphragm and check valve conditions.

Sometimes we have a hard time getting a grip on why a manufacturer chooses to do some of the things they do. When it comes to Yamaha service information, the format of what is or what is not available for a motor

3-102 FUEL SYSTEM

sometimes make no sense. In this case we bring this up because sometimes it might just be who at Yamaha wrote the given service publication or update, because they specifically give this vacuum/pressure check for all V6 and V6, but only a few of the inline motors (such as the 48 hp twin), even when there appear to be no differences in the designs of the pumps themselves.

For this reason, we feel that this test is applicable to all Yamaha, non-carburetor integrated diaphragm-displacement fuel pumps. However, if the motor on which you are working is not listed, we recommend that you DO NOT condemn it based on these test results until after you've also disassembled it (the only way that Yamaha recommends testing it in that case). However, conversely, if the pump DOES pass this test, we feel it is VERY likely that the pump is not your problem.

Attach a hand-held vacuum/pressure pump (like the Mity Vac®) to the fuel pump inlet fitting (the fitting to which the filter/tank line connects). Using the pump apply 7 psi (50 kPa) of pressure while manually restricting the outlet fitting (or fittings, there are 2 on some pumps) using your finger. If the diaphragm is in good condition it will hold the pressure for at least 10 seconds.

Next, switch to the vacuum fitting on the pump and draw 4.3 psi (30 kPa) of negative pressure (vacuum) on the fuel pump inlet fitting. This checks if the one-way check valve in the pump remains closed. It should hold vacuum for at least 10 seconds.

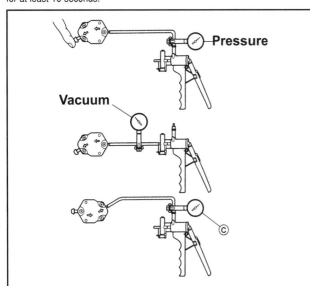

Fig. 305 Vacuum/pressure checking a typical Yamaha fuel pump

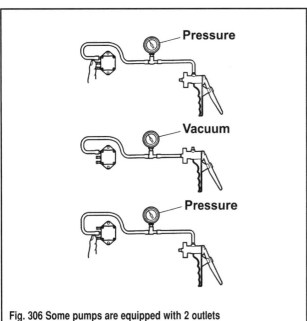

Fig. 306 Some pumps are equipped with 2 outlets

Now, move the pressure pump to the fuel pump outlet fitting. This time cover the inlet fitting (and other outlet fitting, if applicable) with your finger and apply the same amount of pressure. Again, the pressure must hold for at least 10 seconds.

■ **If pressure does not hold, verify that it is not leaking past your finger (on the opposite fitting or fittings) when applicable or from a pump/hose test connection. If leakage is occurring in the diaphragm, the fuel pump should be overhauled.**

Visually Inspecting the Pump Components

◆ See Figures 307, 308 and 309

The only way that Yamaha recommends to inspect MOST of their diaphragm-displacement fuel pumps (including all carburetor-integrated pumps) is through disassembly and visual inspection. Remove and/or disassemble the pump according to the procedures found either in this section (for non-integrated pumps) or under Carburetor Service (specifically Overhaul, for carburetor-integrated models).

Wash all metal parts thoroughly in solvent, and then blow them dry with compressed air. Use care when using compressed air on the check valves. Do not hold the nozzle too close because the check valve can be damaged from an excessive blast of air.

Inspect each part for wear and damage. Visually check the pump body/cover assembly for signs of cracks or other damage. Verify that the valve seats provide a flat contact area for the valve. Tighten all check valve connections firmly as they are replaced.

Check the diaphragms for pin holes by holding it up to the light. If pin holes are detected or if the diaphragm is not pliable, it MUST be replaced.

■ **If you've come this far and are uncertain about pump condition, replace the diaphragms and check valves, you've already got to replace any gaskets or O-rings which were removed. Once the pump is rebuilt, you can remove it from your list of potential worries for quite some time.**

REMOVAL & INSTALLATION

◆ See Figure 310

■ **Disassembly and assembly should be performed on a clean work surface. Make every effort to prevent foreign material from entering the fuel pump or adhering to the diaphragms.**

These procedures cover removal, installation and overhaul of the powerhead mounted diaphragm-displacement fuel pump found on most Yamaha motors. For models with carburetor integrated pumps, please refer to the appropriate carburetor overhaul procedure found earlier in the Carburetor Service section.

1. For safety, disconnect the negative battery cable, if equipped, and/or remove and ground the spark plug lead(s). Also for safety, on motors with portable tanks, disengage the quick-connect fitting from the motor.

■ **These actions will help prevent the possibility of sparks that could potentially ignite fuel vapors, help prevent accidental starting of the motor while you are working on it, and lastly, help prevent raw fuel from spraying through an open fitting (should someone squeeze or step on the primer bulb).**

2. Tag and disconnect the fuel lines from the pump itself. On most motors there is a spring-type clamp securing the fuel inlet and outlet lines, but some use plastic wires which must be carefully cut for removal (just make sure you don't nick and damage the fuel line itself). The inlet and outlet line fittings on the pump are normally labeled to show proper fuel flow connection.

■ **These clamps are easily disconnected by squeezing the clamp ears and slid upward along the fuel hose, until they are past the raised portion of the raised portion on the pump connection.**

FUEL SYSTEM 3-103

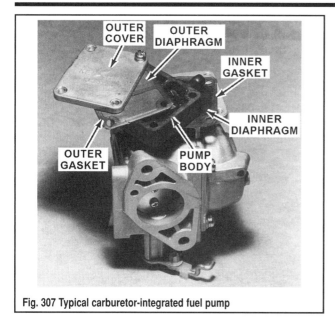

Fig. 307 Typical carburetor-integrated fuel pump

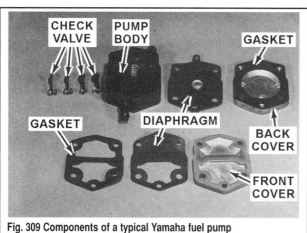

Fig. 309 Components of a typical Yamaha fuel pump

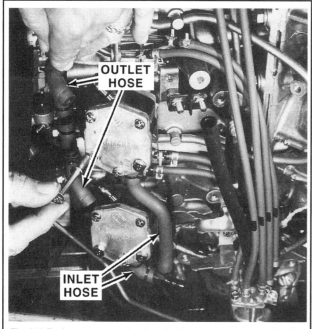

Fig. 310 Fuel pump removal or installation on a typical powerhead

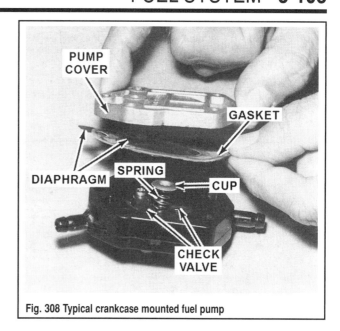

Fig. 308 Typical crankcase mounted fuel pump

3. Remove the bolts (normally 2) securing the pump itself to the crankcase. Don't confuse these bolts (normally found on either side of the fuel inlet fitting end of the pump) with the cover screws (there are usually 3 pump cover screws which are used to disassemble the pump itself).

4. Carefully remove the pump assembly from the powerhead, then remove and discard the crankcase gasket.

5. Installation is essentially the reverse of this removal. However, make sure the crankcase passage is clean and free of dirt or debris. Also, be sure to use a new gasket. Always be sure to properly pressurize the fuel system and check for leaks before attempting to start and run the motor. It is also a good idea to run the motor with the cover removed and observe the fuel lines and fittings which were disconnected, JUST TO MAKE SURE there are no leaks the first time the outboard is started after a fuel pump/system repair.

OVERHUAL

◆ See Figures 311, 312 and 313

1. Remove the pump and move it to a suitable clean work surface.

2. Remove the pump cover screws (these are usually 3 Philips head screws) that secure the pump cover and body together. Take care not to let the spring fly out or to lose the cup.

3. Separate the back cover from the pump body. If the gasket and diaphragm are to be used again, take great care in peeling them away from the surface of the cover. Remove the spring and cup. Separate the parts and keep them in order as an assist in assembling.

4. Remove the front cover from the pump body. If the diaphragm and gasket are to be used again, take great care in peeling them away from the surface of the cover. Separate the parts and keep them in order as an assist in assembling.

5. Remove the check valves and take time to note how each valve faces, because it MUST be installed in exactly the same manner, or the pump will not function.

To Assemble:

■ Proper operation of the fuel pump is essential for maximum powerhead performance. Therefore, always use new gaskets.

✳✳ WARNING

Never use any type of sealer on fuel pump gaskets.

3-104 FUEL SYSTEM

6. Place the check valves on the appropriate sides of the pump body (in the positions noted during removal) with the fold in the valve facing up. Take care not to damage the very fragile and flat surface of the valve. Secure each check valve in place with a Phillips head screw. Tighten the screw securely.

7. Place the spring, the cup (on top of the spring), the diaphragm, the gasket, and finally the inner cover on the pump body. Hold these parts together and turn the pump over.

8. Install the gasket and then the diaphragm onto the pump body. Install the front cover.

9. Check to be sure the holes for the screws are all aligned through the cover, diaphragms, and gaskets. If the diaphragms are not properly aligned, a tear would surely develop when the screws are installed.

Install the three Phillips head screws through the various parts and tighten the screws securely.

10. Properly install the pump to the powerhead and pressurize the system to make sure there are no fuel leaks.

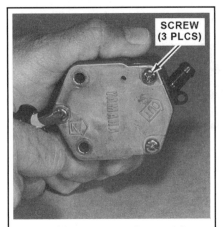

Fig. 311 Three screws usually secure the pump components

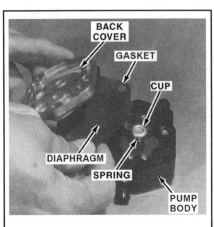

Fig. 312 Carefully pull back the pump covers to expose the diaphragms

Fig. 313 Note the check valve positioning before removal

REED VALVES

Reed Valve Block

◆ See Figure 314

On 2-stroke motors the combustible air/fuel oil mixture is drawn into the crankcase from the carburetors through reed valves. It is the job of the reed valve to open in response crankcase vacuum (allowing the air/fuel mixture to be drawn from the carburetor, through the intake manifold and into the motor) and then to close again during crankcase pressure (allowing that same mixture to be forced through the ports into the combustion chamber). The reed valves essentially take the place of the more complicated valve train used on 4-stroke motors (which is necessitated by the basic differences in engine operation).

A broken reed is usually caused by metal fatigue over a long period of time. The failure may also be due to the reed flexing too far because the reed stop was installed incorrectly or the stop has become distorted.

If the reed is broken, the loose piece must be located and removed, before the powerhead is returned to service. The piece of reed may have found its way into the crankcase, behind the bypass cover. If the broken piece cannot be located, the powerhead must be completely disassembled until it is located and removed.

An excellent check for a broken reed on an operating powerhead is to hold an ordinary business card in front of the carburetor. Under normal operating conditions, a very small amount of fine mist will be noticeable, but if fuel begins to appear rapidly on the card from the carburetor, one of the reeds is broken and causing the backflow through the carburetor onto the card. A broken reed will cause the powerhead to operate roughly and pop back through the carburetor.

✷✷ WARNING

The reeds must never be turned over in an attempt to correct a problem. Such action would cause the reed to flex in the opposite direction and break in a very short time.

The reed block is located on the engine crankcase, under the carburetors and intake manifold. Removal and installation of the block is part of Powerhead overhaul and is covered in the Powerhead section. Obviously, a complete disassembly of the powerhead itself is not necessary simply to access the reed valves, so if that is the only required service, just follow the appropriate steps of the overhaul procedure.

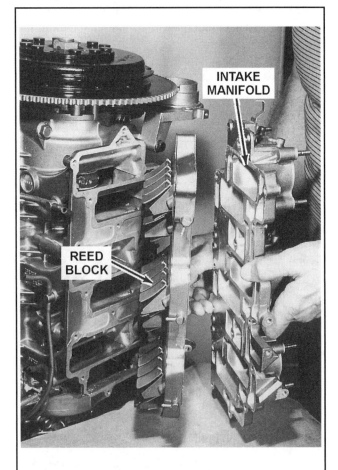

Fig. 314 The reed block on most powerheads (including the V4 or V6) may be serviced without disassembling the powerhead

FUEL SYSTEM 3-105

SPECIFICATIONS

Carburetor Set-Up Specifications - 2-Stroke Motors

Model (Hp)	No. of Cyl	Engine Type	Year	Displace cu. in. (cc)	Pilot Screw Initial Low Speed Setting	Float Height/Drop Setting In. (mm)	Valve Seat Size In. (mm)
2	1	IL 2-stroke	1997-02	2.6 (43)	-	0.66-0.70 (16.8-17.8)	0.055 (1.4)
2	1	IL 2-stroke	1997-02	3.0 (50)	-	0.66-0.70 (16.8-17.8)	0.055 (1.4)
3	1	IL 2-stroke	1997-02	4.3 (70)	1 - 1 1/2	0.47-0.63 (12-16)	0.055 (1.4)
4	1	CL 2-stroke	1997-99	5.0 (83)	1 1/2 - 2	0.85-0.89 (21.05-22.05)	0.05 (1.2)
4	1	CL 2-stroke	1997-02	6.3 (103)	1 1/4 - 1 3/4	0.85-0.89 (21.05-22.05)	0.05 (1.2)
5	1	CL 2-stroke	1997-02	6.3 (103)	1 1/4 - 1 3/4	0.85-0.89 (21.05-22.05)	0.05 (1.2)
6	2	CL 2-stroke	1997-00	10 (165)	7/8 - 1 2/8	0.47-0.63 (12-16)	0.05 (1.2)
8	2	CL 2-stroke	1997-03	10 (165)	7/8 - 1 2/8	0.47-0.63 (12-16)	0.05 (1.2)
9.9	2	IL 2-stroke	1997-03	15 (246)	1 1/4 - 1 3/4	0.49-0.61 (12.5-15.5)	0.05 (1.2)
15	2	IL 2-stroke	1997-03	15 (246)	1 1/4 - 1 3/4	0.49-0.61 (12.5-15.5)	0.05 (1.2)
20	2	IL 2-stroke	1997	24 (395)	1 3/4 - 2 1/4	0.55-0.59 (14-15)	-
25	2	IL 2-stroke	1997-03	24 (395)	1 1/4 - 2 3/4	0.55-0.59 (14-15)	-
20	2	IL 2-stroke	1997-98	26 (430)	1 1/8 - 1 5/8	0.69-0.73 (17.5-18.5)	0.055 (1.4)
25	2	IL 2-stroke	1997-98	26 (430)	1 1/4 - 1 3/4	0.69-0.73 (17.5-18.5)	0.055 (1.4)
25	2	IL 2-stroke	1997	30 (496)	1 - 1 1/2	0.571 (14.5)	0.05 (1.2)
30	2	IL 2-stroke	1997	30 (496)	1 - 1 3/4	0.571 (14.5)	0.05 (1.2)
40	2	IL 2-stroke	1997	36 (592)	1 1/2 - 2	0.65-0.89 (16.5-22.5)	0.055 (1.4)
48	2	IL 2-stroke	1997-00	46 (760)	1 1/8 - 1 5/8	0.71-0.79 (18.0-20.0)	0.06 (1.6)
25	3	IL 2-stroke	1997-02	30 (496)	1/2 - 1	0.61-0.65 (15.5-16.5)	0.04 (1.1)
30	3	IL 2-stroke	1997-02	30 (496)	①	0.57-0.61 (14.5-15.5)	0.04 (1.1)
28J	3	IL 2-stroke	1997-01	43 (698)	1 1/4 - 1 3/4	0.55-0.63 (14-16)	0.05 (1.2)
35J	3	IL 2-stroke	1997-01	43 (698)	1 3/8 - 1 7/8 ②	0.55-0.63 (14-16)	0.05 (1.2)
40	3	IL 2-stroke	1997-03	43 (698)	1 1/4 - 1 3/4	0.55-0.63 (14-16)	0.05 (1.2)
50	3	IL 2-stroke	1997-03	43 (698)	1 3/8 - 1 7/8 ②	0.55-0.63 (14-16)	0.05 (1.2)
50	3	IL 2-stroke	1997-03	52 (849)	1 1/8 - 1 5/8	0.55-0.63 (14-16)	0.05 (1.2)
60	3	IL 2-stroke	③	52 (849)	1 1/4 - 1 3/4	0.51-0.59 (13-15)	0.063 (1.6)
	3	IL 2-stroke	④	52 (849)	1 1/8 - 1 5/8	0.47-0.63 (12-16)	0.055 (1.4)
70	3	IL 2-stroke	1997-03	52 (849)	1 - 1 1/2	0.51-0.59 (13-15)	0.055 (1.4)
65J	3	IL 2-stroke	1997-01	70 (1140)	1 - 1 1/2	0.65-0.89 (19.2-19.8)	0.06 (1.6)
75	3	IL 2-stroke	⑤	70 (1140)	1 1/8 - 1 5/8	0.51-0.59 (13-15) ⑦	0.06 (1.6)
	3	IL 2-stroke	⑥	70 (1140)	1 - 1 1/2 ⑧	0.65-0.89 (16.5-22.5)	0.06 (1.6)
80	3	IL 2-stroke	1997-00	70 (1140)	1 1/8 - 1 5/8	0.51-0.59 (13-15) ⑨	0.06 (1.6)
85	3	IL 2-stroke	1997-00	70 (1140)	7/8 - 1 3/8	0.65-0.89 (16.5-22.5)	0.06 (1.6)
90	3	IL 2-stroke	1997-03	70 (1140)	1 - 1 1/2	0.51-0.59 (13-15)	0.06 (1.6)
80J	4	90 LV 2-st	1997-98	106 (1730)	3/8 - 7/8	0.61-0.65 (15.5-16.5)	0.05 (1.2)
	4	90 LV 2-st	1999-01	106 (1730)	1 - 1 1/2	0.61-0.65 (15.5-16.5)	0.05 (1.2)
100	4	90 LV 2-st	1997-02	106 (1730)	1 - 1 1/2	0.61-0.65 (15.5-16.5)	0.05 (1.2)
115	4	90 LV 2-st	1997-98	106 (1730)	3/8 - 7/8	0.61-0.65 (15.5-16.5)	0.05 (1.2)
	4	90 LV 2-st	1999-03	106 (1730)	1 - 1 1/2	0.61-0.65 (15.5-16.5)	0.05 (1.2)
130	4	90 LV 2-st	1997-03	106 (1730)	5/8 - 1 1/8	0.61-0.65 (15.5-16.5)	0.05 (1.2)
140	4	90 LV 2-st	1997-02	106 (1730)	5/8 - 1 1/8	0.61-0.65 (15.5-16.5)	0.05 (1.2)
105J	6	90 LV 2-st	1997-00	158 (2596)	3/4 - 1 1/4	0.61-0.65 (15.5-16.5)	0.05 (1.2)

Carburetor Set-Up Specifications - 2-Stroke Motors

Model (Hp)	No. of Cyl	Engine Type	Year	Displace cu. in. (cc)	Pilot Screw Initial Low Speed Setting	Float Height/Drop Setting In. (mm)	Valve Seat Size In. (mm)
150	6	90 LV 2-st	1997-03	158 (2596)	3/4 - 1 1/4 ⑩	0.61-0.65 (15.5-16.5)	0.05 (1.2)
175	6	90 LV 2-st	1997-00	158 (2596)	13/16 - 1 15/16 ⑪	0.61-0.65 (15.5-16.5)	0.05 (1.2)
200	6	90 LV 2-st	1997-03	158 (2596)	7/8 - 1 3/8 ⑫	0.61-0.65 (15.5-16.5)	0.05 (1.2)
225	6	90 LV 2-st	1997	158 (2596)	1/2 - 1 (P), 1 - 1 1/2 (S)	0.61-0.65 (15.5-16.5)	0.05 (1.2)

Initial low speed setting turn(s): back (counterclockwise) from a *lightly* seated position

① Specifications vary by carb position, top carb 1/2 -1, middle carb 1 1/2 - 2, bottom carb 3/4 - 1 1/4
② Specifications are for M model, the specification for W models is 1 1/8 - 1 5/8
③ These specifications are for most 1997-03 models, including those with the carburetor stamp mark 6H20A
④ These specifications are for 1999 C60 Models, or 60FE/60FET, C60ER/C60TR models with the carburetor stamp mark 6H210
⑤ Specifications vary more by model than by year, use these specs for C and P models (carburetor stamp marks 6H007 or 6H015)
⑥ Specifications vary more by model than by year, use these specs for A and E models
⑦ Specifications are for most C and P models (with the carb stamp mark 6H007), but for the C75TR (with the carb mark 6H015) the float should be set to 0.47-0.63 in. (12-16mm)
⑧ Specifications are for most US and World spec A and E models (with the carb stamp mark 69225), but for the 75AM or for all Europe Spec models (with the carb stamp mark 69234) the initial pilot screw setting is 7/8 - 1 3/8 turns
⑨ Specifications are for most models (with the carb stamp mark 6H007), but for the C80TR/80AET (with the carb mark 6H015) the float should be set to 0.65-0.89 in. (16.5-22.5mm)
⑩ Specifications are for most models including A and C models except as follows: for S and L models the screw setting is 1 - 1 1/2, for P or D models the settings are 13/16 - 1 5/16 port and 1 5/16 - 1 13/16 starboard
⑪ Specifications are for most models including A and D models except as follows: for S models the screw setting is 7/8 - 1 3/8, for P or V models the settings are 7/8 - 1 3/8 port and 1 3/8 - 1 7/8 starboard
⑫ Specs are for most models including A (200AET/L200AET) and 200TR, except as follows: for S models (including the S200TR/L200TR) the settings are 3/8 - 7/8 (p) & 7/8 - 1 3/8 (s). For P models, the settings are 1/2 - 1 (p) & 1 - 1 1/2 (s)

4

IGNITION AND ELECTRICAL SYSTEMS

BASIC ELECTRICAL THEORY	4-2
BATTERY	4-40
CDI/TCI SYSTEMS	4-8
CDI/TCI UNIT AND IGNITION COILS	4-30
CDI UNIT TEST CHARTS	4-18
CHARGING CIRCUIT	**4-38**
BATTERY	4-40
YAMAHA CHARGING SYSTEMS	4-38
CRANKING CIRCUIT	**4-43**
DESCRIPTION AND OPERATION	4-43
FAULTY SYMPTOMS	4-43
MAINTENANCE	4-43
STARTER MOTOR	4-46
STARTER MOTOR CIRCUIT	4-43
STARTER MOTOR RELAY/SOLENOID	4-45
CRANKSHAFT POSITION SENSOR (CPS)	4-35
ELECTRICAL COMPONENTS	4-2
ELECTRICAL SYSTEM PRECAUTIONS	4-7
ELECTRICAL SWITCH/SOLENOID SERVICE	**4-53**
KILL SWITCH	4-53
MAIN KEYSWITCH	4-53
NEUTRAL SAFETY SWITCH	4-55
START BUTTON	4-54
TETHER	4-54
WARNING BUZZER/HORN	4-56
ELECTRICAL TESTING	4-6
FLYWHEEL AND STATOR PLATE	4-23
IGNITION AND CHARGING SYSTEM TROUBLESHOOTING	4-12
IGNITION SYSTEMS	**4-8**
CDI UNIT TEST CHARTS	4-18
CDI/TCI SYSTEMS	4-8
CDI/TCI UNIT AND IGNITION COILS	4-30
CRANKSHAFT POSITION SENSOR (CPS)	4-35
FLYWHEEL AND STATOR PLATE	4-23
IGNITION AND CHARGING SYSTEM TROUBLESHOOTING	4-12
THERMO-SENSOR & THERMO-SWITCH	4-36
YAMAHA MICROCOMPUTER IGNITION SYSTEMS (YMIS)	4-33
KILL SWITCH	4-53
MAIN KEYSWITCH	4-53
NEUTRAL SAFETY SWITCH	4-55
SPECIFICATIONS	**4-57**
CHARGING SYSTEM	4-62
IGNITION SYSTEM	4-57
START BUTTON	4-54
STARTER MOTOR	4-46
STARTER MOTOR CIRCUIT	4-43
STARTER MOTOR RELAY/SOLENOID	4-45
TEST EQUIPMENT	4-4
TETHER	4-54
THERMO-SENSOR & THERMO-SWITCH	4-36
TROUBLESHOOTING ELECTRICAL SYSTEMS	4-6
UNDERSTANDING AND TROUBLESHOOTING ELECTRICAL SYSTEMS	**4-2**
BASIC ELECTRICAL THEORY	4-2
ELECTRICAL COMPONENTS	4-2
ELECTRICAL SYSTEM PRECAUTIONS	4-7
ELECTRICAL TESTING	4-6
TEST EQUIPMENT	4-4
TROUBLESHOOTING ELECTRICAL SYSTEMS	4-6
WIRE AND CONNECTOR REPAIR	4-7
WARNING BUZZER/HORN	4-56
WIRE AND CONNECTOR REPAIR	4-7
WIRING DIAGRAMS	**4-65**
INDEX	4-65
1-CYLINDER MOTORS	4-66
2-CYLINDER MOTORS	4-67
3-CYLINDER MOTORS	4-74
V4 MOTORS	4-88
V6 MOTORS	4-94
YAMAHA CHARGING SYSTEMS	4-38
YAMAHA MICROCOMPUTER IGNITION SYSTEMS (YMIS)	4-33
YMIS SELF-DIAGNOSIS	4-33
TROUBLESHOOTING	4-35

UNDERSTANDING AND TROUBLESHOOTING ELECTRICAL SYSTEMS	4-2
IGNITION SYSTEMS	4-8
CHARGING CIRCUIT	4-38
CRANKING CIRCUIT	4-43
ELECTRICAL SWITCH/SOLENOID SERVICE	4-53
SPECIFICATIONS	4-57
WIRING DIAGRAMS	4-65

4-2 IGNITION AND ELECTRICAL SYSTEMS

UNDERSTANDING AND TROUBLESHOOTING ELECTRICAL SYSTEMS

Basic Electrical Theory

◆ See Figure 1

For any 12-volt, negative ground, electrical system to operate, the electricity must travel in a complete circuit. This simply means that current (power) from the positive terminal (+) of the battery must eventually return to the negative terminal (-) of the battery. Along the way, this current will travel through wires, fuses, switches and components. If, for any reason, the flow of current through the circuit is interrupted, the component fed by that circuit would cease to function properly.

Perhaps the easiest way to visualize a circuit is to think of connecting a light bulb (with two wires attached to it) to the battery - one wire attached to the negative (-) terminal of the battery and the other wire to the positive (+) terminal. With the two wires touching the battery terminals, the circuit would be complete and the light bulb would illuminate. Electricity would follow a path from the battery to the bulb and back to the battery. It's easy to see that with wires of sufficient length, our light bulb could be mounted nearly anywhere on the boat. Further, one wire could be fitted with a switch inline so that the light could be turned on and off without having to physically remove the wire(s) from the battery.

The normal marine circuit differs from this simple example in two ways. First, instead of having a return wire from each bulb to the battery, the current travels through a single ground wire that handles all the grounds for a specific circuit. Secondly, most marine circuits contain multiple components that receive power from a single circuit. This lessens the overall amount of wire needed to power components.

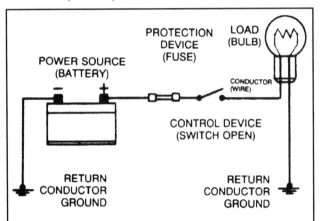

Fig. 1 This example illustrates a simple circuit. When the switch is closed, power from the positive (+) battery terminal flows through the fuse and the switch, and then to the light bulb. The electricity illuminates the bulb and the circuit is completed through the ground wire back to the negative (-) battery terminal.

HOW DOES ELECTRICITY WORK: THE WATER ANALOGY

Electricity is the flow of electrons - the sub-atomic particles that constitute the outer shell of an atom. Electrons spin in an orbit around the center core of an atom. The center core is comprised of protons (positive charge) and neutrons (neutral charge). Electrons have a negative charge and balance out the positive charge of the protons. When an outside force causes the number of electrons to unbalance the charge of the protons, the electrons will split off the atom and look for another atom to balance out. If this imbalance is kept up, electrons will continue to move and an electrical flow will exist.

Many people find electrical theory easier to understand when using an analogy with water. In a comparison with water flowing through a pipe, the electrons would be the water and the wire is the pipe.

The flow of electricity can be measured much like the flow of water through a pipe. The unit of measurement used is amperes, frequently abbreviated as amps (a). You can compare amperage to the volume of water flowing through a pipe (for water that would mean a measurement of mass usually measured in units delivered over a set amount of time such as gallons or liters per minute). When connected to a circuit, an ammeter will measure the actual amount of current flowing through the circuit. When relatively few electrons flow through a circuit, the amperage is low. When many electrons flow, the amperage is high.

Water pressure is measured in units such as pounds per square inch (psi). The electrical pressure is measured in units called volts (v). When a voltmeter is connected to a circuit, it is measuring the electrical pressure.

The actual flow of electricity depends not only on voltage and amperage, but also on the resistance of the circuit. The higher the resistance, the higher the force necessary to push the current through the circuit. The standard unit for measuring resistance is an ohm (Ω). Resistance in a circuit varies depending on the amount and type of components used in the circuit. The main factors that determine resistance are:

- Material - some materials have more resistance than others. Those with high resistance are said to be insulators. Rubber materials (or rubber-like plastics) are some of the most common insulators used, as they have a very high resistance to electricity. Very low resistance materials are said to be conductors. Copper wire is among the best conductors. Silver is actually a superior conductor to copper and is used in some relay contacts, but its high cost prohibits its use as common wiring. Most marine wiring is made of copper.
- Size - the larger the wire size being used, the less resistance the wire will have (just as a large diameter pipe will allow small amounts of water to just trickle through). This is why components that use large amounts of electricity usually have large wires supplying current to them.
- Length - for a given thickness of wire, the longer the wire, the greater the resistance. The shorter the wire, the less the resistance. When determining the proper wire for a circuit, both size and length must be considered to design a circuit that can handle the current needs of the component.
- Temperature - with many materials, the higher the temperature, the greater the resistance (positive temperature coefficient). Some materials exhibit the opposite trait of lower resistance with higher temperatures (these are said to have a negative temperature coefficient). These principles are used in many engine control sensors (especially those found on microcomputer controlled ignition and fuel injection systems).

OHM'S LAW

There is a direct relationship between current, voltage and resistance. The relationship between current, voltage and resistance can be summed up by a statement known as Ohm's law.

Voltage (E) is equal to amperage (I) times resistance (R): $E = I \times R$
Other forms of the formula are $R = E/I$ and $I = E/R$

In each of these formulas, E is the voltage in volts, I is the current in amps and R is the resistance in ohms. The basic point to remember is that if the voltage of a circuit remains the same, as the resistance of that circuit goes up, the amount of current that flows in the circuit will go down.

The amount of work that electricity can perform is expressed as power. The unit of power is the watt (w). The relationship between power, voltage and current is expressed as:

Power (W) is equal to amperage (I) times voltage (E): $W = I \times E$

This is only true for direct current (DC) circuits; the alternating current formula is a tad different, but since the electrical circuits in most vessels are DC type, we need not get into AC circuit theory.

Electrical Components

POWER SOURCE

◆ See Figure 2

Typically, power is supplied to a vessel by two devices: The battery and the stator (or battery charge coil). The stator supplies electrical current anytime the engine is running in order to recharge the battery and in order to operate electrical devices of the vessel. The battery supplies electrical power during starting or during periods when the current demand of the vessel's electrical system exceeds stator output capacity (which includes times when the motor is shut off and stator output is zero).

IGNITION AND ELECTRICAL SYSTEMS 4-3

The Battery

In most modern vessels, the battery is a lead/acid electrochemical device consisting of six 2-volt subsections (cells) connected in series, so that the unit is capable of producing approximately 12 volts of electrical pressure. Each subsection consists of a series of positive and negative plates held a short distance apart in a solution of sulfuric acid and water.

The two types of plates in each battery cell are of dissimilar metals. This sets up a chemical reaction, and it is this reaction which produces current flow from the battery when its positive and negative terminals are connected to an electrical load. Power removed from the battery in use is replaced by current from the stator and restores the battery to its original chemical state.

The Stator

Alternators and generators are devices that consist of coils of wires wound together making big electromagnets. The coil is normally referred to as a stator or battery charge coil. Either, one group of coils spins within another set (or a set of permanently charged magnets, usually attached to the flywheel, are spun around a set of coils) and the interaction of the magnetic fields generates an electrical current. This current is then drawn off the coils and fed into the vessel's electrical system.

■ Some vessels utilize a generator instead of an alternator. Although the terms are often misused and interchanged, the main difference is that an alternator supplies alternating current that is changed to direct current for use on the vessel, while a generator produces direct current. Alternators tend to be more efficient and that is why they are used on almost all modern engines.

GROUND

Two types of grounds are used in marine electric circuits. Direct ground components are grounded to the electrically conductive metal through their mounting points. All other components use some sort of ground wire that leads back to the battery. The electrical current runs through the ground wire and returns to the battery through the ground or negative (-) cable; if you look, you'll see that the battery ground cable connects between the battery and a heavy gauge ground wire.

■ A large percentage of electrical problems can be traced to bad grounds.

If you refer back to the basic explanation of a circuit, you'll see that the ground portion of the circuit is just as important as the power feed. The wires delivering power to a component can have perfectly good, clean connections, but the circuit would fail to operate if there was a damaged ground connection. Since many components ground through their mounting or through wires that are connected to an engine surface, contamination from dirt or corrosion can raise resistance in a circuit to a point where it cannot operate.

PROTECTIVE DEVICES

Problems can occur in the electrical system that will cause large surges of current to pass through the electrical system of your vessel. These problems can be the fault of the charging circuit, but more likely would be a problem with the operating electrical components that causes an excessively high load. An unusually high load can occur in a circuit from problems such as a seized electric motor (like a damaged starter) or the excessive resistance caused by a bad ground (from loose or damaged wires or connections). A short to ground that bypasses the load and allows the battery to quickly discharge through a wire can also cause current surges.

If this surge of current were to reach the load in the circuit, the surge could burn it out or severely damage it. It can also overload the wiring, causing the harness to get hot and melt the insulation. To prevent this, fuses, circuit breakers and/or fusible links are connected into the supply wires of the electrical system. These items are nothing more than a built-in weak spot in the system. When an abnormal amount of current flows through the system, these protective devices work as follows to protect the circuit:

- Fuse - when an excessive electrical current passes through a fuse, the fuse blows (the conductor melts) and opens the circuit, preventing current flow.
- Circuit Breaker - a circuit breaker is basically a self-repairing fuse. It will open the circuit in the same fashion as a fuse, but when the surge subsides, the circuit breaker can be reset and does not need replacement. Most circuit breakers on marine engine applications are self-resetting, but some that operate accessories (such as on larger vessels with a circuit breaker panel) must be reset manually (just like the circuit breaker panels in most homes).
- Fusible Link - a fusible link (fuse link or main link) is a short length of special, high temperature insulated wire that acts as a fuse. When an excessive electrical current passes through a fusible link, the thin gauge wire inside the link melts, creating an intentional open to protect the circuit. To repair the circuit, the link must be replaced. Some newer type fusible links are housed in plug-in modules, which are simply replaced like a fuse, while older type fusible links must be cut and spliced if they melt. Since this link is very early in the electrical path, it's the first place to look if nothing on the vessel works, yet the battery seems to be charged and is otherwise properly connected.

✶✶ CAUTION

Always replace fuses, circuit breakers and fusible links with identically rated components. Under no circumstances should a component of higher or lower amperage rating be substituted. A lower rated component will disable the circuit sooner than necessary (possibly during normal operation), while a higher rated component can allow dangerous amounts of current that could damage the circuit or component (or even melt insulation causing sparks or a fire).

SWITCHES & RELAYS

◆ See Figure 3

Switches are used in electrical circuits to control the passage of current. The most common use is to open and close circuits between the battery and the various electric devices in the system. Switches are rated according to the amount of amperage they can handle. If a sufficient amperage rated switch is not used in a circuit, the switch could overload and cause damage.

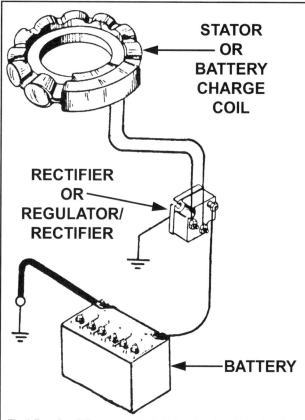

Fig. 2 Functional diagram of a typical charging circuit showing the relationship between the stator (battery charge coil), rectifier (or regulator/rectifier) and battery

4-4 IGNITION AND ELECTRICAL SYSTEMS

Some electrical components that require a large amount of current to operate use a special switch called a relay. Since these circuits carry a large amount of current, the thickness of the wire in the circuit is also greater. If this large wire were connected from the load to the control switch, the switch would have to carry the high amperage load and the space needed for wiring in the vessel would be twice as big to accommodate the increased size of the wiring harness. A relay is used to prevent these problems.

Think of relays as essentially "remote controlled switches." They allow a smaller current to throw the switch that operates higher amperages devices. Relays are composed of a coil and a set of contacts. When current is passed through the coil, a magnetic field is formed that causes the contacts to move together, closing the circuit. Most relays are normally open, preventing current from passing through the main circuit until power is applied to the coil. But, relays can take various electrical forms depending on the job for which they are intended. Some common circuits that may use relays are horns, lights, starters, electric fuel pumps and other potentially high draw circuits.

LOAD

Every electrical circuit must include a load (something to use the electricity coming from the source). Without this load, the battery would attempt to deliver its entire power supply from one pole to another. This is called a short circuit. All this electricity would take a short cut to ground and cause a great amount of damage to other components in the circuit (including the battery) by developing a tremendous amount of heat. This condition could develop sufficient heat to melt the insulation on all the surrounding wires and reduce a multiple wire cable to a lump of plastic and copper. A short can allow sparks that could ignite fuel vapors or other combustible materials in the vessel, causing an extremely hazardous condition.

WIRING & HARNESSES

The average vessel contains miles of wiring, with hundreds of individual connections. To protect the many wires from damage and to keep them from becoming a confusing tangle, they are organized into bundles, enclosed in plastic or taped together and called wiring harnesses. Different harnesses serve different parts of the vessel. Individual wires are color coded to help trace them through a harness where sections are hidden from view.

Marine wiring or circuit conductors can be either single strand wire, multi-strand wire or printed circuitry. Single strand wire has a solid metal core and is usually used inside such components as stator coil windings, motors, relays and other devices. Multi-strand wire has a core made of many small strands of wire twisted together into a single conductor. Most of the wiring in a marine electrical system is made up of multi-strand wire, either as a single conductor or grouped together in a harness. All wiring is color coded on the insulator, either as a solid color or as a colored wire with an identification stripe. A printed circuit is a thin film of copper or other conductor that is printed on an insulator backing. Occasionally, a printed circuit is sandwiched between two sheets of plastic for more protection and flexibility. A complete printed circuit, consisting of conductors, insulating material and connectors is called a printed circuit board. Printed circuitry is used in place of individual wires or harnesses in places where space is limited, such as behind 1-piece instrument clusters.

Since marine electrical systems are very sensitive to changes in resistance, the selection of properly sized wires is critical when systems are repaired. A loose or corroded connection or a replacement wire that is too small for the circuit will add extra resistance and an additional voltage drop to the circuit.

The wire gauge number is an expression of the cross-section area of the conductor. Vessels from countries that use the metric system will typically describe the wire size as its cross-sectional area in square millimeters. In this method, the larger the wire, the greater the number. Another common system for expressing wire size is the American Wire Gauge (AWG) system. As gauge number increases, area decreases and the wire becomes smaller. Using the AWG system, an 18 gauge wire is smaller than a 4 gauge wire. A wire with a higher gauge number will carry less current than a wire with a lower gauge number. Gauge wire size refers to the size of the strands of the conductor, not the size of the complete wire with insulator. It is possible, therefore, to have two wires of the same gauge with different diameters because one may have thicker insulation than the other.

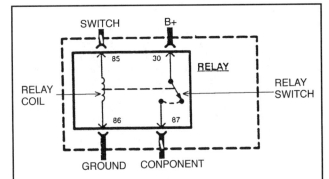

Fig. 3 Relays are composed of a coil and a switch. These two components are linked together so that when one is operated it actuates the other. The large wires in the circuit are connected from the battery to one side of the relay switch (B+) and from the opposite side of the relay switch to the load (component). Smaller wires are connected from the relay coil to the control switch for the circuit and from the opposite side of the relay coil to ground

It is essential to understand how a circuit works before trying to figure out why it doesn't. An electrical schematic shows the electrical current paths when a circuit is operating properly. Schematics break the entire electrical system down into individual circuits. In most schematics no attempt is made to represent wiring and components as they physically appear on the vessel; switches and other components are shown as simply as possible. But, this is **not** always the case on Yamaha schematics and some of the wiring diagrams provided here. So, when using a Yamaha schematic if the component in question is represented by something more than a small square or rectangle with a label, it is likely a intended to be a representation of the actual component's shape. On most schematics, the face views of harness connectors show the cavity or terminal locations in all multi-pin connectors to help locate test points.

Test Equipment

Pinpointing the exact cause of trouble in an electrical circuit is usually accomplished by the use of special test equipment, but the equipment does not always have to be expensive. The following sections describe different types of commonly used test equipment and briefly explains how to use them in diagnosis. In addition to the information covered below, be sure to read and understand the tool manufacturer's instruction manual (provided with most tools) before attempting any test procedures.

JUMPER WIRES

◆ See Figure 4

✳✳ CAUTION

Never use jumper wires made from a thinner gauge wire than the circuit being tested. If the jumper wire is of too small a gauge, it may overheat and possibly melt. Never use jumpers to bypass high resistance loads in a circuit. Bypassing resistances, in effect, creates a short circuit. This may, in turn, cause damage and fire. Jumper wires should only be used to bypass lengths of wire or to simulate switches.

Jumper wires are simple, yet extremely valuable, pieces of test equipment. They are basically test wires that are used to bypass sections of a circuit. Although jumper wires can be purchased, they are usually fabricated from lengths of standard marine wire and whatever type of connector (alligator clip, spade connector or pin connector) that is required for the particular application being tested. In cramped, hard-to-reach areas, it is advisable to have insulated boots over the jumper wire terminals in order to prevent accidental grounding. It is also advisable to include a standard marine fuse in any jumper wire. This is commonly referred to as a fused jumper. By inserting an in-line fuse holder between a set of test leads, a fused jumper wire is created for bypassing open circuits. Use a 5-amp fuse to provide protection against voltage spikes.

IGNITION AND ELECTRICAL SYSTEMS 4-5

Jumper wires are used primarily to locate open electrical circuits, on either the ground (-) side of the circuit or on the power (+) side. If an electrical component fails to operate, connect the jumper wire between the component and a good ground. If the component operates only with the jumper installed, the ground circuit is open. If the ground circuit is good, but the component does not operate, the circuit between the power feed and component may be open. By moving the jumper wire successively back from the component toward the power source, you can isolate the area of the circuit where the open is located. When the component stops functioning, or the power is cut off, the open is in the segment of wire between the jumper and the point previously tested.

You can sometimes connect the jumper wire directly from the battery to the hot terminal of the component, but first make sure the component uses a full 12 volts in operation. Some electrical components, such as sensors, are designed to operate on smaller voltages like 4 or 5 volts, and running 12 volts directly to these components can damage or destroy them.

TEST LIGHTS

◆ See Figure 5

The test light is used to check circuits and components while electrical current is flowing through them. It is used for voltage and ground tests. To use a 12-volt test light, connect the ground clip to a good ground and probe connectors the pick where you are wondering if voltage is present. The test light will illuminate when voltage is detected. This does not necessarily mean that 12 volts (or any particular amount of voltage) is present; it only means that some voltage is present. It is advisable before using the test light to touch its ground clip and probe across the battery posts or terminals to make sure the light is operating properly and to note how brightly the light glows when 12 volts is present.

※※ WARNING

Do not use a test light to probe electronic ignition, spark plug or coil wires, as the circuit is much, much higher than 12 volts. Also, never use a pick-type test light to probe wiring on electronically controlled systems unless specifically instructed to do so. Whenever possible, avoid piercing insulation with the test light pick, as you are inviting shorts or corrosion and excessive resistance. But, any wire insulation that is pierced by necessity, must be sealed with silicone and taped after testing.

Like the jumper wire, the 12-volt test light is used to isolate opens in circuits. But, whereas the jumper wire is used to bypass the open to operate the load, the 12-volt test light is used to locate the presence or lack of voltage in a circuit. If the test light illuminates, there is power up to that point in the circuit; if the test light does not illuminate, there is an open circuit (no power). Move the test light in successive steps back toward the power source until the light in the handle illuminates. The open is between the probe and the point that was previously probed.

The self-powered test light is similar in design to the 12-volt test light, but contains a 1.5 volt penlight battery in the handle. It is most often used in place of a multi-meter to check for open or short circuits when power is isolated from the circuit (thereby performing a continuity test).

The battery in a self-powered test light does not provide much current. A weak battery may not provide enough power to illuminate the test light even when a complete circuit is made (especially if there is high resistance in the circuit). Always make sure that the test battery is strong. To check the battery, briefly touch the ground clip to the probe; if the light glows brightly, the battery is strong enough for testing.

■ A self-powered test light should not be used on any electronically controlled system or component. Even the small amount of electricity transmitted by the test light is enough to damage many electronic components.

MULTI-METERS

◆ See Figure 6

Multi-meters are extremely useful for troubleshooting electrical problems. They can be purchased in both analog or digital form and have a price range to suit nearly any budget. A multi-meter is a voltmeter, ammeter and ohmmeter (along with other features) combined into one instrument. It is often used when testing solid state circuits because of its high input impedance (usually 10 mega ohms or more). A high-quality digital multi-meter or Digital Volt Ohm Meter (DVOM) helps to ensure the most accurate test results and, although not absolutely necessary for electronic components such as computer controlled ignition systems and charging systems, is highly recommended. A brief description of the main test functions of a multi-meter follows:

• Voltmeter - the voltmeter is used to measure voltage at any point in a circuit or to measure the voltage drop across any part of a circuit. Voltmeters usually have various scales and a selector switch to allow metering and display of different voltage ranges. The voltmeter has a positive and a negative lead. To avoid damage to the meter, connect the negative lead to the negative (-) side of the circuit (to ground or nearest the ground side of the circuit) and connect the positive lead to the positive (+) side of the circuit (to the power source or the nearest power source). This is mostly a concern on analog meters, as DVOMs are not normally adversely affected (as they are usually designed to take readings even with reverse polarity and display accordingly). Note that the negative voltmeter lead will always be black and that the positive voltmeter will always be some color other than black (usually red).

• Ohmmeter - the ohmmeter is designed to read resistance (measured in ohms) in a circuit or component. Most ohmmeters will have a selector switch which permits the measurement of different ranges of resistance (usually the selector switch allows the multiplication of the meter reading by 10, 100, 1,000 and 10,000). Most modern ohmmeters (especially DVOMs) are auto-ranging which means the meter itself will determine which scale to use. Since ohmmeters are powered by an internal battery, the ohmmeter can be used like a self-powered test light. When the ohmmeter is connected, current from the ohmmeter flows through the circuit or component being tested. Since the ohmmeter's internal resistance and voltage are known values, the amount of current flow through the meter depends on the resistance of the circuit or component being tested. The ohmmeter can also be used to perform a continuity test for suspected open circuits. When using the meter for continuity checks, do not be concerned with the actual resistance readings. Zero resistance, or any ohm reading, indicates continuity in the

Fig. 4 Jumper wires are simple, but valuable pieces of test equipment

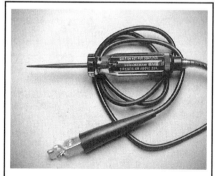

Fig. 5 A 12-volt test light is used to detect the presence of voltage in a circuit

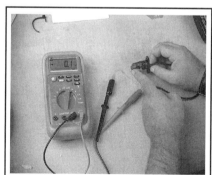

Fig. 6 Multi-meters are probably the most versatile and handy tools for diagnosing faulty electrical components or circuits

4-6 IGNITION AND ELECTRICAL SYSTEMS

circuit. Infinite resistance indicates an opening in the circuit. A high resistance reading where there should be none indicates a problem in the circuit. Checks for short circuits are made in the same manner as checks for open circuits, except that the circuit must be isolated from both power and normal ground. Infinite resistance indicates no continuity, while zero resistance indicates a dead short.

✶✶ WARNING

Never use an ohmmeter to check the resistance of a component or wire while there is voltage applied to the circuit. Voltage in the circuit can damage or destroy the meter.

• Ammeter - an ammeter measures the amount of current flowing through a circuit in units called amperes or amps. At normal operating voltage, most circuits have a characteristic amount of amperes, called current draw that can be measured using an ammeter. By referring to a specified current draw rating, and then measuring the amperes and comparing the two values, you can determine what is happening within the circuit to aid in diagnosis. An open circuit, for example, will not allow any current to flow, so the ammeter reading will be zero. A damaged component or circuit will have an increased current draw, so the reading will be high. The ammeter is always connected in series with the tested circuit. All of the current that normally flows through the circuit must also flow through the ammeter; if there is any other path for the current to follow, the ammeter reading will not be accurate. The ammeter itself has very little resistance to current flow and, therefore, will not affect the circuit, but it will measure current draw only when the circuit is closed and electricity is flowing. Excessive current draw can blow fuses and drain the battery, while a reduced current draw can cause motors to run slowly, lights to dim and other components to not operate properly.

Troubleshooting Electrical Systems

When diagnosing a specific problem, organized troubleshooting is a must. The complexity of a modern marine vessel demands that you approach any problem in a logical, organized manner. There are certain troubleshooting techniques, however, which are standard:

• **Establish when the problem occurs**. Does the problem appear only under certain conditions? Were there any noises, odors or other unusual symptoms? Isolate the problem area. To do this, make some simple tests and observations, and then eliminate the systems that are working properly. Check for obvious problems, such as broken wires and loose or dirty connections. Always check the obvious before assuming something complicated is the cause.

• **Test for problems systematically to determine the cause once the problem area is isolated**. Are all the components functioning properly? Is there power going to electrical switches and motors? Performing careful, systematic checks will often turn up most causes on the first inspection, without wasting time checking components that have little or no relationship to the problem.

• **Test all repairs after the work is done to make sure that the problem is fixed**. Some causes can be traced to more than one component, so a careful verification of repair work is important in order to pick up additional malfunctions that may cause a problem to reappear or a different problem to arise. A blown fuse, for example, is a simple problem that may require more than another fuse to repair. If you don't look for a problem that caused a fuse to blow, a shorted wire (for example) may go undetected and cause the new fuse to blow right away (if the short is still present) or during subsequent operation (as soon as the short returns if it is intermittent).

Experience shows that most problems tend to be the result of a fairly simple and obvious cause, such as loose or corroded connectors, bad grounds or damaged wire insulation that causes a short. This makes careful visual inspection of components during testing essential to quick and accurate troubleshooting.

Electrical Testing

VOLTAGE

◆ See Figure 7

This test determines the voltage available from the battery and should be the first step in any electrical troubleshooting procedure after visual inspection. Many electrical problems, especially on electronically controlled systems, can be caused by a low state of charge in the battery. Many circuits cannot function correctly if the battery voltage drops below normal operating levels.

Loose or corroded battery cable terminals can cause poor contact that will prevent proper charging and full battery current flow.

1. Set the voltmeter selector switch to the 20V position.
2. Connect the meter negative lead to the battery's negative (-) post or terminal and the positive lead to the battery's positive (+) post or terminal.
3. Turn the ignition switch **ON** to provide a small load.
4. A well charged battery should register over 12 volts. If the meter reads below 11.5 volts, the battery power may be insufficient to operate the electrical system properly. Check and charge or replace the battery as detailed under Engine Maintenance before further tests are conducted on the electrical system.

VOLTAGE DROP

◆ See Figure 8

When current flows through a load, the voltage beyond the load drops. This voltage drop is due to the resistance created by the load and also by small resistances created by corrosion at the connectors (or by damaged insulation on the wires). Since all voltage drops are cumulative, the maximum allowable voltage drop under load is critical, especially if there is more than one load in the circuit.

1. Set the voltmeter selector switch to the 20 volts position.
2. Connect the multi-meter negative lead to a good ground.
3. Operate the circuit and check the voltage prior to the first component (load).
4. There should be little or no voltage drop in the circuit prior to the first component. If a voltage drop exists, the wire or connectors in the circuit are suspect.
5. While operating the first component in the circuit, probe the ground side of the component with the positive meter lead and observe the voltage readings. A small voltage drop should be noticed. This voltage drop is caused by the resistance of the component.
6. Repeat the test for each component (load) down the circuit.
7. If an excessively large voltage drop is noticed, the preceding component, wire or connector is suspect.

RESISTANCE

◆ See Figure 9

✶✶ WARNING

Never use an ohmmeter with power applied to the circuit. The ohmmeter is designed to operate on its own power supply. The normal 12-volt electrical system voltage will damage or destroy many meters!

1. Isolate the circuit from the vessel's power source.
2. Ensure that the ignition key is **OFF** when disconnecting any components or the battery.
3. Where necessary, also isolate at least one side of the circuit to be checked, in order to avoid reading parallel resistances. Parallel circuit resistances will always give a lower reading than the actual resistance of either of the branches.
4. Connect the meter leads to both sides of the circuit (wire or component) and read the actual measured ohms on the meter scale. Make sure the selector switch is set to the proper ohm scale for the circuit being tested, to avoid misreading the ohmmeter test value.

IGNITION AND ELECTRICAL SYSTEMS 4-7

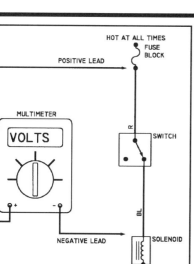

Fig. 7 A voltage check determines the amount of battery voltage available and, as such, should be the first step in any troubleshooting procedure

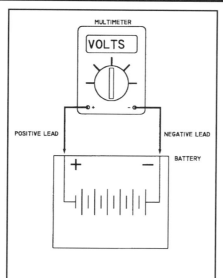

Fig. 8 Voltage drops are due to resistance in the circuit, from the load or from problems with the wiring

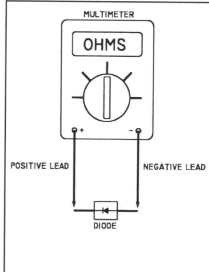

Fig. 9 Resistance tests must be conducted on portions of the circuit, isolated from battery power

■ The resistance reading of most electrical components will vary with temperature. Unless otherwise noted, specifications given are for testing under ambient conditions of 68°F (20°C). If the component is tested at higher or lower temperatures, expect the readings to vary slightly. When testing engine control sensors or coil windings with smaller resistance specifications (less than 1000 ohms) it is best to use a high quality DVOM and be especially careful of your test results. Whenever possible, double-check your results against a known good part before purchasing the replacement. If necessary, bring the old part to the marine parts dealer and have them compare the readings to prevent possibly replacing a good component.

OPEN CIRCUITS

◆ See Figure 10

This test already assumes the existence of an open in the circuit and it is used to help locate position of the open.
    5. Isolate the circuit from power and ground.
    6. Connect the self-powered test light or ohmmeter ground clip to the ground side of the circuit and probe sections of the circuit sequentially.
    7. If the light is out or there is infinite resistance, the open is between the probe and the circuit ground.
    8. If the light is on or the meter shows continuity, the open is between the probe and the end of the circuit toward the power source.

SHORT CIRCUITS

◆ See Figure 11

■ Never use a self-powered test light to perform checks for opens or shorts when power is applied to the circuit under test. The test light can be damaged by outside power.

    1. Isolate the circuit from power and ground.
    2. Connect the self-powered test light or ohmmeter ground clip to a good ground and probe any easy-to-reach point in the circuit.
    3. If the light comes on or there is continuity, there is a short somewhere in the circuit.
    4. To isolate the short, probe a test point at either end of the isolated circuit (the light should be on or the meter should indicate continuity).
    5. Leave the test light probe engaged and sequentially open connectors or switches, remove parts, etc. until the light goes out or continuity is broken.
    6. When the light goes out, the short is between the last two circuit components that were opened.

Wire And Connector Repair

Almost anyone can replace damaged wires, as long as the proper tools and parts are available. Wire and terminals are available to fit almost any need. Even the specialized weatherproof, molded and hard shell connectors used by many marine manufacturers.

Be sure the ends of all the wires are fitted with the proper terminal hardware and connectors. Wrapping a wire around a stud is not a permanent solution and will only cause trouble later. Replace wires one at a time to avoid confusion. Always route wires in the same manner of the manufacturer.

When replacing connections, make absolutely certain that the connectors are certified for marine use. Automotive wire connectors may not meet United States Coast Guard (USCG) specifications.

■ If connector repair is necessary, only attempt it if you have the proper tools. Weatherproof and hard shell connectors may require special tools to release the pins inside the connector. Attempting to repair these connectors with conventional hand tools will damage them.

Electrical System Precautions

• Wear safety glasses when working on or near the battery.
• Don't wear a watch with a metal band when servicing the battery or starter. Serious burns can result if the band completes the circuit between the positive battery terminal (or a hot wire) and ground.
• Be absolutely sure of the polarity of a booster battery before making connections. Remember that even momentary connection of a booster battery with the polarity reversed will damage charging system diodes. Connect the cables positive-to-positive, and negative (of the good battery)-to-a good ground on the engine (away from the battery to prevent the possibility of an explosion if hydrogen vapors are present from the electrolyte in the discharged battery). Connect positive cables first (starting with the discharged battery), and then make the last connection to ground on the body of the booster vessel so that arcing cannot ignite hydrogen gas that may have accumulated near the battery. • Disconnect both vessel battery cables before attempting to charge a battery.
• Never ground the alternator or generator output or battery terminal. Be cautious when using metal tools around a battery to avoid creating a short circuit between the terminals.
• When installing a battery, make sure that the positive and negative cables are not reversed.
• Always disconnect the battery (negative cable first) when charging.
• Never smoke or expose an open flame around the battery. Hydrogen gas is released from battery electrolyte during use and accumulates near the battery. Hydrogen gas is **highly** explosive.

4-8 IGNITION AND ELECTRICAL SYSTEMS

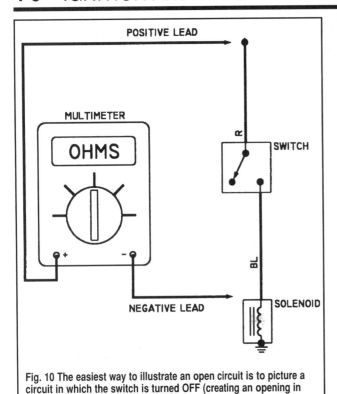

Fig. 10 The easiest way to illustrate an open circuit is to picture a circuit in which the switch is turned OFF (creating an opening in the circuit) that prevents power from reaching the load

Fig. 11 In this illustration a load (the light) is powered when it should not be (since the switch should be creating an open condition), but a short to power (battery) is powering the circuit. Shorts like this can be caused by chaffed wires with worn or broken insulation

IGNITION SYSTEMS

◆ See Figure 12

The less an outboard engine is operated, the more care it needs. Allowing an outboard engine to remain idle will do more harm than if it is used regularly. To maintain the engine in top shape and always ready for efficient operation at any time, the engine should be operated every 3 to 4 weeks throughout the year.

The carburetion and ignition principles of 2-stroke engine operation must be understood in order to perform a proper tune-up on an outboard motor. If you have any doubts concerning your understanding of engine operation, it would be best to study the operation theory before tackling any work on the ignition system.

CDI/TCI Systems

DESCRIPTION & OPERATION

◆ See Figures 13 and 14

All Yamaha motors since 1995 are equipped with some form of the Yamaha Capacitor Discharge Ignition (CDI) or Transistor Controlled Ignition (TCI) electronic ignition systems. Some models are equipped with a microcomputer controlled version of the system which means the CDI/TCI unit or ECM (as the unit is known on most fuel injected motors) utilizes additional input (sensor input such as a crank position sensor, knock sensor, temperature sensor or even oil level sensor on some 2-strokes) to help make spark timing decisions, however, the basic function of the various Yamaha ignition systems are all similar.

The first thing these ignitions have in common is that a flywheel/magneto assembly is used to generate power/signals over pulser and/or charge (or stator) coils. In all cases, the power and signals are fed to a control unit which is used to trigger the ignition coils and spark plugs.

Although the basic function of the components is the same across most systems, their actual jobs may be performed by different physical components on some motors or their functions may be combined into a single unit on some motors.

Generally speaking, charge coils are used to generate system power, however, their function may be performed by the stator coil in some cases

Fig. 12 The ignition system must be properly adjusted and synchronized for optimum powerhead performance

(either through dedicated charge coil windings in the stator coil or possibly even by the lighting coil windings when separate charge coil windings are not provided). Similarly, pulser coils are normally used to generate signals triggering the CDI/TCI unit to fire the ignition coils, however on some single cylinder models no pulser coil is needed, as the plugs are fired each time the charge coil generates power.

The CDI/TCI unit takes in power from the charge coil (or stator) and signals from the pulser coil(s) and then uses the power and signals to control function of the ignition coils. While fuel injected motors may combine the function of the CDI/TCI unit with the Electronic Control Module (ECM) which also controls the fuel injection system.

IGNITION AND ELECTRICAL SYSTEMS 4-9

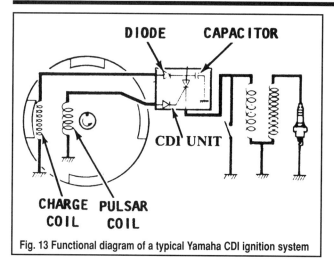

Fig. 13 Functional diagram of a typical Yamaha CDI ignition system

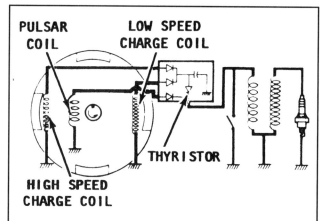

Fig. 14 Functional diagram of a Yamaha CDI ignition system with separate low- and high-speed charge coils - V4 and V6 carbureted powerheads

To understand any one Yamaha motor's ignition system, refer to the system descriptions below. If you have any doubt as to the components utilized by a given motor, please refer to the Wiring Diagrams and the Ignition System Component testing charts in this section as well.

■ Throughout this section we'll refer to the CDI unit when talking about the CDI unit, TCI unit or ECM, whichever is responsible for the ignition control module. Don't be confused if you're researching a TCI motor and we're calling it CDI, that's just to keep from having to constantly write (and for you to constantly read "CDI/TCI unit or ECM").

1-Cylinder Ignition

The CDI and TCI systems found on single-cylinder Yamaha motors are among the simplest forms of an ignition system and are composed of the following elements:
- Magneto
- Pulser coil
- Charge, or source coil
- Igniter (CDI/TCI) box
- Ignition coil
- Spark plug

Other components such as main switches, stop switches, or computer systems may be included, though, these items are not necessary for basic CDI operation.

To understand basic CDI operation, it is important to understand the basic theory of induction. Induction theory states that if we move a magnet (magnetic field) past a coil of wire (or the coil by the magnet), AC current will be generated in the coil.

The amount of current produced depends on several factors:
- How fast the magnet moves past the coil
- The size of the magnet (strength)
- How close the magnet is to the coil
- Number of turns of wire and the size of the windings

The current produced in the charge coil goes to the CDI box. On the way in, it is converted to DC current by a diode. This DC current is stored in the capacitor located inside the box. As the charge coil produces current, the capacitor stores it.

At a specific time in the magneto's revolution, the magnets go past the pulser coil (which is usually just a smaller version of the charge coil, so it has less current output). The current from the pulser also goes into the CDI box. This current signals the CDI box when to fire the capacitor (the pulser may be called a trigger coil for obvious reasons). The current from the capacitor flows out to the ignition coil and spark plug. The pulser acts much like the points in older ignitions systems.

When the pulser signal reaches the CDI box, all the electricity stored in the capacitor is released at once. This current flows through the ignition coil's primary windings.

The ignition coil is a step-up transformer. It turns the relatively low voltage entering the primary windings into high voltage at the secondary windings. This occurs due to a phenomena known as induction.

The high voltage generated in the secondary windings leaves the ignition coil and goes to the spark plug. The spark in turn ignites the air-fuel charge in the combustion chamber.

Once the complete cycle has occurred, the spinning magneto immediately starts the process over again.

Main switches, engine stop switches, and the like are usually connected on the wire in between the CDI box and the ignition coil. When the main switch or stop switch is turned to the OFF position, the switch is closed. This closed switch short-circuits the charge coil current to ground rather than sending it through the CDI box. With no charge coil current through the CDI box, there is no spark and the engine stops or, if the engine is not running, no spark is produced.

The ignition systems on Yamaha single-cylinder motors vary as follows:
- 2 hp motors: utilize a single charge coil (no pulser coil as the CDI unit fires each time a voltage is generated).
- 3-5 hp motors: utilize a low- and high-speed pulser coils in addition to the balance of the standard CDI ignition components.

2-Cylinder Ignition

Yamaha produces 2-cylinder outboards ranging from smaller 6 hp units, all the way up to larger 48 hp motors. These outboards are usually, though not always, equipped with one, dual-lead ignition coil.

In addition to the dual-lead ignition coil, the system typically uses one pulser coil (though there are a few motors that use dual pulser coils), one charge coil and a CDI box.

■ Normally, the 20/25 hp (395cc) and 25/30 hp (496cc) motors are equipped with dual pulser coils.

When equipped with only one pulser coil, both spark plugs are fired at the same time. Although both cylinders spark at the same time, only one cylinder is actually producing power. The crankshaft is a 180 degree type which means that as piston number one is at top dead center, piston number two is at bottom dead center. The piston at TDC is compressing a fuel charge that the spark then ignites. At the same moment, the piston at BDC isn't compressing a fuel charge. Actually, there are still exhaust gases going out. The spark in this cylinder has no effect on power production. This combination of engine and ignition design is called waste spark system (which is something as a misnomer, because no efficiency is actually lost by the spark in the non-compressed cylinder).

After the crankshaft has rotated another 180 degrees, the two pistons have reversed position. The spark fires again to ignite the fuel charge compressed in cylinder number two and sparks to no effect in cylinder one.

Twin coil CDI operation up to the ignition coil is exactly like the basic CDI. The difference simply comes in the utilization of the dual-lead ignition coil.

On a traditional ignition coil, the current leaves the coil, goes to the spark plug, then through the cylinder head to ground.

On a dual-lead coil, the current leaves the coil, goes through one of the spark plugs, travels through the cylinder head to the second spark plug, and returns to the coil itself. This way one coil can fire two cylinders.

4-10 IGNITION AND ELECTRICAL SYSTEMS

This type of system requires about 30% more voltage than a standard system to fire the second spark plug. This is because more energy is required for the spark to jump from the bridge of the spark plug to the center electrode.

■ **If this system has weak spark, the first sign is the second spark plug will not having a hot enough spark. This plug will foul even though the other plug is working fine.**

A quick check to tell if the plug fouling on one cylinder is due to weak ignition is to switch the ignition coil leads. If the plug fouling goes to the other cylinder, weak ignition components are the problem.

3-Cylinder Ignition

Three-cylinder models can be divided into three groups:
- Units with 3 pulser coils
- Units with 2 pulser coils (without Microcomputer control)
- Units with 1 pulser coil (with Microcomputer control)

Models equipped with three pulser coils use one pulser for each cylinder. The pulsers are spaced 120 apart. As each pulser produces its own signal, the CDI box fires the ignition coil for that cylinder.

The largest of the 3-cylinder engine families (the 65J-90 hp/1140cc motors) have two items that make them unique to other Yamaha 3-cylinder engines. One, they have 2 charge coils, one called a low-speed charge coil and the other called the high-speed charge coil. And rather than having separate large coils, these motors normally have both charge coils and the lighting coil combined into a single unit called a stator.

■ **The failure of any of these coils will result in the replacement of the whole stator rather than a single coil.**

The second unique feature to the 65J-90 hp (1140cc) motor family is that it sometimes uses 2 pulser coils rather than three. Some 3-cylinder motors use 1 pulser per cylinder, but in this case, 1 pulser actually controls 2 cylinders, however, these 2 cylinders are not firing at the same time like the earlier twin models.

In order to allow 2 cylinders to fire off 1 pulser at different times, the pulser must tell the CDI box when to ignite the appropriate cylinder. This is accomplished by taking advantage of a characteristic of the induction process.

As a magnet goes by the pulser, electricity is generated. This current flows to the igniter (CDI) box. The direction the current flows is determined by which end (north or south) of the magnet travels past the coil first. When the north end goes by the coil, current flows in one direction. When the south end of the magnet goes by, the current flows in the opposite direction. This is sometimes referred to as phase or polarity. The CDI box can differentiate the direction of current flow and direct the signal to the appropriate capacitor. The CDI box, in effect, knows which cylinder to fire by the signature of the signal.

On models with microcomputer control there is 1 pulser coil, but there is also a Crankshaft Position Sensor (CPS). In some cases the pulser fires the 1 & 3 cylinders while the computer calculates positioning of the #2 cylinder for firing purposes. In these cases a failure of the microcomputer circuitry may extinguish the #2 spark. In all cases, the CPS provides data to the CDI unit which is used to control spark timing curves.

Carbureted V4 & V6 Ignition

The V4 ignition system works like a basic CDI system with two major differences. First, 1 pulser fires 2 cylinders independently of each other. Second, there are 2 charge coils, a low-speed winding and a high-speed coil winding, both of which are incorporated into the stator assembly. The pulser coils are installed on their own sensor plate, underneath the stator.

In the most basic CDI systems, 1 pulser coil sends the signal to fire one ignition coil. In the V4 system 2 pulser coils fire for separate ignition coils. This means that one two-wire pulser coil fires two separate cylinders at different times in the firing order.

In order to fire two cylinders off 1 pulser at different times, the pulser must tell the CDI box when to fire a particular cylinder. This is accomplished by inducing distinctly different electrical signals for each pulser.

When a magnet passes a pulser, electricity is generated. This current flows to the CDI box. The direction of current flow (sometimes referred to as phase) is determined by which end of the magnet, north or south, travels past the coil first. When the north end goes by the pulser coil first, current flows in one direction. When the south end of the magnet goes by first, the current flows in the opposite direction.

The CDI box senses which direction the pulser current is flowing. The direction of flow determines which of the two cylinders controlled by that pulser will fire.

By operating this way, two pulsers can control the operation of four cylinders.

In a basic CDI system, a single charge coil is used to supply the electrical power needed for the ignition.

On larger Yamaha outboards, 2 charge coils are used to supply the electrical needs of the ignition system. On these CDI systems, the low-speed, high-speed, and lighting coils are all combined into one unit called a stator. If any of the stator components are bad, the whole assembly is replaced.

The V6 ignition is very similar to the V4 system. The V6 systems have the same low- and high-speed charge coils for better electrical output at all speeds.

The first major difference is that the V6 uses either 2 pulser coils to fire 6 cylinders (injected models and 225DET pre-mix motors) or 3 pulser coils to fire 6 cylinders (pre-mix motors, except the 225DET). These pulsers operate on the same principle as the V4 pulsers.

The second major difference is that the V6 is normally equipped with a microcomputer controlled ignition. This system utilizes additional input from a thermo-switch, an oil level sensor (oil injection models) and, a crankshaft position sensor. Additionally, this ignition contains a self-diagnosis system capable of displaying trouble-codes in the case of a microcomputer ignition circuit fault. For more information on the microcomputer control portion of this system, please refer to the Yamaha Microcomputer Ignition System later in this section.

EFI/HPDI V4 & V6 Ignition

EFI OX66 motors are equipped with 6 individual pulser coils (one per cylinder) mounted to a pulser coil plate, located below the stator. Power for the ignition comes from low- and high-speed charge coil windings which are incorporated into a stator assembly along with the lighting coil windings. Microcomputer control happens at an ECM/CDI unit and includes input from a CPS, WTS, TPS, thermo-switches, oil level sensor and knock sensor (on 3.1L engines).

HPDI motors are similar to their EFI OX66 counterparts (also using 6 individual pulser coils). However, power for the system on these units comes from the stator coil lighting coil windings and not dedicated charge coil windings. Microcomputer control takes into consideration virtually all sensors of the HPDI system including the IAT, WTS, APS, FPS, TPS, water detection switch, oil level sensor and thermo-switch (no knock sensor is used on these motors).

Charge Circuit

POWER FOR THE IGNITION

◆ See Figures 15, 16 and 17

Two basic circuits are used with the CDI/TCI system; the first of these is the charge circuit. The charge circuit typically consists of the flywheel magnet, a charge coil (or lighting coil on some models), a diode (usually contained within the CDI unit), and a capacitor (also within the CDI unit). As the flywheel magnet passes by the charge coil, a voltage is induced in the coil windings. As the flywheel continues to rotate and the coil is no longer influenced by the magnet, the magnetic field collapses. Therefore, an alternating current is produced at the charge coil. This AC current is changed to DC by a diode inside the CDI unit.

■ **A diode is a solid state unit which permits current to flow in one direction but prevents flow in the opposite direction. A diode may also be known as a rectifier.**

The current then passes to a capacitor, also located inside the CDI unit, where it is stored.

Most V4 and V6 models have two charge coils (low-speed and high-speed coils). A voltage is induced in the windings of each coil. It is not clear whether the coils actually function at different speeds or just function differently, at different speeds. The magnitude of the voltage depends upon the number of windings (turns) in each coil and the engine rpm.

The voltages generated by these two charge coils are limited by a voltage regulator and stored separately in two condensers. The condensers are charged in stages, each time a magnet passes a charge coil.

IGNITION AND ELECTRICAL SYSTEMS

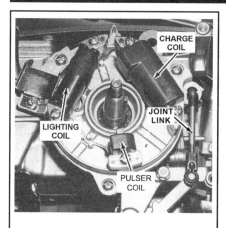

Fig. 15 Example of a Yamaha stator plate with 1 pulser, 1 charge and 1 lighting coil. The stator plate is rotated through the joint link to advance the timing

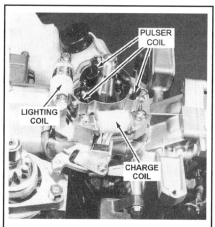

Fig. 16 Stator plate from a typical multi-cylinder motor (this one with 3 pulser coils, 1 charge and one lighting coil)

Fig. 17 The CDI unit utilizes output from the charge coil to power the ignition coils

POWER FOR ACCESSORIES

◆ See Figure 18

The battery charging system consists of the flywheel magnets, a lighting coil (the outboard equivalent of an alternator), a rectifier, voltage regulator (on some motors), and the battery.

The lighting coil, so named because it may be used to power the lights on the boat when used together with a rectifier (or regulator/rectifier), allows the powerhead to generate additional electrical current to charge the battery.

As the flywheel magnets rotate, voltage is induced in the lighting coil (either a separate coil located next to the pulser coils, or the lighting coil windings may be found in the stator coil). This alternating current passes through a series of diodes and emerges as DC current. Therefore, it may be stored in the battery. A lighting coil may be identified by its clean laminated copper windings (all other coils are normally wrapped in tape and their windings are not visible) or by the wire colors. On Yamaha motors the light coil wiring is normally Green/White and/or Green on these motors. For details on a specific motor, refer to the Charging System Testing charts and/or the Wiring Diagrams found in this section.

The rectifier converts the Alternating Current (AC) voltage into DC voltage, which may then be stored in the battery. The rectifier is a sealed unit and generally contains four diodes. If one of the diodes is defective, the entire unit must be replaced.

When equipped, the voltage regulator stabilizes the power output of the coils and extends the life of the light bulbs and battery by preventing power surges or overcharging.

Pulser Circuit

◆ See Figure 19

The second circuit used in CDI/TCI systems is the pulser circuit. The pulser circuit usually has its own flywheel magnet (or magnets), one or more pulser coils, a diode, and a thyristor. Those last 2 components (thyristor and diode) are internal components to the CDI unit. In case you were wondering, a thyristor is a solid state electronic switching device which permits voltage to flow only after it is triggered by another voltage source.

At the point in time when the ignition timing marks align, an alternating current is induced in the pulser coil, in the same manner as previously described for the charge coil. This current is then passed to a second diode located in the CDI unit where it becomes DC current and flows on to the thyristor. This voltage triggers the thyristor to permit the voltage stored in the capacitor to be discharged. The capacitor voltage passes through the thyristor and on to the primary windings of the ignition coil.

In this manner, a spark at the plug may be accurately timed by the timing marks on the flywheel relative to the magnets in the flywheel and to provide as many as 100 sparks per second for a powerhead operating at 6000 rpm.

Different ignition strategies on Yamaha motors may utilize as many as 6 pulser coils (up to one per cylinder) on V6 motors or 2 pulser coils on a single cylinder motor (on which the composite wave form is used for ignition control). Differences are spelled out as clearly as we can in earlier sections on Ignition by motor type (and in the Ignition System Component testing charts and Wiring Diagrams, in this section).

Fig. 18 The charge and lighting coil (stator) removed from a V6 powerhead for bench testing

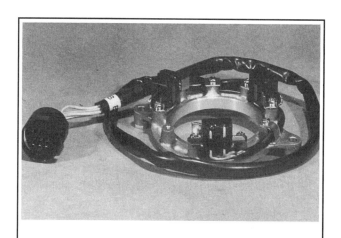

Fig. 19 A typical Yamaha 3-pulser coil assembly

4-12 IGNITION AND ELECTRICAL SYSTEMS

Ignition Coils

◆ See Figure 20

Yamaha ignition coils, usually one per cylinder or one per pair of cylinders, boost the DC voltage instantly to over 20,000 volts (and potentially as much as 50,000 volts) for spark. This completes the primary side of the ignition circuit.

Once the voltage is discharged from the ignition coil, the secondary circuit begins and only stretches from the ignition coil to the spark plugs via extremely large high tension leads. At the spark plug end, the voltage arcs in the form of a spark, across from the center electrode to the outer electrode, and then to ground via the spark plug threads. This completes the ignition circuit.

Ignition Timing

At the point in time when the ignition timing marks align, an alternating current is induced in the pulser coil, in the same manner as previously described for the charge coil. This current is then passed to a second diode located in the CDI unit where it becomes DC current and flows on to the thyristor. This voltage triggers the thyristor to permit the voltage stored in the capacitor to be discharged. The capacitor voltage passes through the thyristor and on to the primary windings of the ignition coil.

In this manner, a spark at the plug may be accurately timed by the timing marks on the flywheel relative to the magnets in the flywheel and to provide as many as 100 sparks per second for a powerhead operating at 6000 rpm.

Units equipped with an automatic advance type CDI/TCI system have no moving or sliding parts. Ignition advance is accomplished electronically - by the electric waves emitted by the pulser coils. These coils increase their wave output proportionately to engine speed increase, thus advancing the timing.

Units equipped with a mechanical advance type CDI system use a link rod between the carburetor and the ignition base plate assembly. At the time the throttle is opened, the ignition base plate assembly is rotated by means of the link rod, thus advancing the timing.

Ignition and Charging System Troubleshooting

Accurate troubleshooting of an ignition system is viewed by many as mission impossible. But this really does not have to be the case. Having a systematic approach to troubleshooting is the key. Your first priority is to understand how the system works.

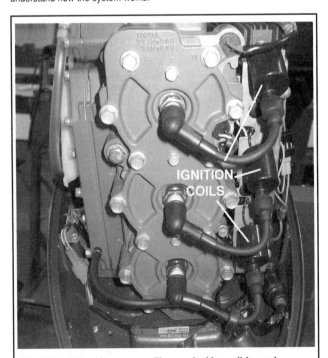

Fig. 20 Most Yamaha motors utilize one ignition coil for each cylinder

Get yourself mentally prepared for ignition troubleshooting before you touch the motor. Make notes about how the motor behaves. Is it totally dead? Misfiring? If it is dead, what were the circumstances that led to its demise? Did it run poorly before it died? If it is running, but not well, is this a permanent condition? Only when cold? Only when warm? Under load? Ask yourself a lot of questions. The more information you can gather the faster the repair can be made.

It is easy to forget to ask yourself if the root cause of the failure is truly ignition? Could it be something else? Look at the plug and check the spark. Once you have done the basics, the hands-on troubleshooting begins.

The key to successful troubleshooting is a systematic approach. Do not skip around. On new or unfamiliar equipment try writing up a check list on a 3 x 5 note card. This makes a handy reference for future troubleshooting.

Use a spark tester to observe spark quality. Does it jump the gap? Is it Blue? Orange spark usually means trouble in the ignition system.

Do not forget to check the ignition controlling components such as the main or stop switches. These can be disconnected from the system. These switches can cause failures that checks of the individual ignition components may not reveal.

Motors with microcomputer controlled ignition systems utilize other sensor inputs to help determine spark timing. A thermo-switch or oil level sensor giving an incorrect reading may prevent spark timing advance. Don't forget these and other ignition control sensors.

Always attempt to proceed with the troubleshooting in an orderly manner. The shot in the dark approach will only result in wasted time, incorrect diagnosis, replacement of unnecessary parts, and frustration.

COMPONENT TESTING

Resistance Testing Precautions

When performing an electrical resistance test on a CDI/TCI unit, rectifier unit, or any device where the resistance varies with electrical polarity or applied voltage the following points should be noted:
- There are no standard specifications for the design of either analog or digital test meters.
- When performing resistance testing, some test meter manufacturers will apply positive battery power to the Red test lead, while others apply the positive voltage to the Black lead.
- Some meters (especially digital meters) will apply a very low voltage to the test leads that may not properly activate some devices (e.g., CDI units).
- Be aware there are ohmmeters that have reverse polarity. If your test results all differ from the chart, swap + and - leads and retest.

■ **Digital meter resistance values are not reliable when testing CDI units, rectifier units, or any device containing semiconductors, transistors, and diodes.**

- Although resistance test specifications can be helpful, they are also very misleading in ignition circuits/components. A component may test within specification during a static resistance check, yet still fail during normal operating conditions. Also, remember that readings will vary both with temperature and by meter.

Spark Plugs

◆ See Figures 21, 22 and 23

1. Check the plug wires to be sure they are properly connected. Check the entire length of the wires from the plugs to the magneto under the stator plate. If the wires are to be removed from the spark plug, always use a pulling and twisting motion as a precaution against damaging the connection.

2. Attempt to remove the spark plug by hand. This is a rough test to determine if the plug is tightened properly. The attempt to loosen the plug by hand should fail. The plug should be tight and require the proper socket size tool. Remove the spark plug and evaluate its condition. Reinstall the spark plug and tighten with a torque wrench to the proper specification.

3. Use a spark tester and check for spark. If a spark tester is not available (and for Pete's sake they're cheap and can be found in almost all auto parts stores) hold the plug wire about 1/4 in. (6.4mm) from the engine (leave the plug in the port for safety). Carefully operate the starter and check for spark. A strong spark over a wide gap must be observed when testing in this manner, because under compression a strong spark is necessary in

IGNITION AND ELECTRICAL SYSTEMS   4-13

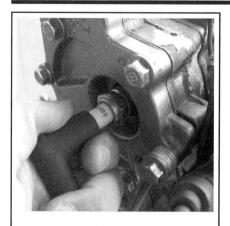
Fig. 21 Disconnect the spark plug wire...

Fig. 22 ...and check spark using a tester

Fig. 23 Remove and inspect the spark plug

order to ignite the air-fuel mixture in the cylinder. This means it is possible to think a strong spark is present, when in reality the spark will be too weak when the plug is installed. If there is no spark, or if the spark is weak, the trouble is most likely under the flywheel in the magneto.

Compression

◆ See Figure 24

Before spending too much time and money attempting to trace a problem to the ignition system, a compression check of the cylinder should be made. If the cylinder does not have adequate compression, troubleshooting and attempted service of the ignition or fuel system will fail to give the desired results of satisfactory engine performance.

For details, please refer to Compression Testing in the Maintenance and Tune-Up section.

Polarity Check

◆ See Figure 25

Coil polarity is extremely important for proper battery ignition system operation. If a coil is connected with reverse polarity, the spark plugs may demand from 30 to 40 percent more voltage to fire, or on most CDI/TCI systems, there will be no spark. Under such demanding conditions, in a very short time the coil would be unable to supply enough voltage to fire the plugs. Any one of the following three methods may be used to quickly determine coil polarity.

The polarity of the coil can be checked using an ordinary analog D.C. voltmeter set on the maximum scale. Connect the positive lead to a good ground. With the engine running, momentarily touch the negative lead to a spark plug terminal. The needle should swing upscale. If the needle swings downscale, the polarity is reversed.

If a voltmeter is not available, a pencil may be used in the following manner: Disconnect a spark plug wire and hold the metal connector at the end of the cable about 1/4 in. (6.35mm) from the spark plug terminal. Now, insert an ordinary pencil tip between the terminal and the connector. Crank the engine with the ignition switch on. If the spark feathers on the plug side and has a slight orange tinge, the polarity is correct. If the spark feathers on the cable connector side, the polarity is reversed.

The firing end of a used spark plug can give a clue to coil polarity. If the ground electrode is dished, it may mean polarity is reversed.

Pulser Coils

◆ See Figures 26 thru 30

There are basically 2 overall methods used in testing pulser coils, a static resistance check and a dynamic cranking or running output voltage check. Specifications vary by year and model so significantly that it precludes trying to list them in this procedure. In addition, the method of conducting the static check will vary with the specifications that are provided. Generally speaking cranking or running tests are more accurate and more likely to show intermittent problems. Dynamic checks are done under different circumstances, with load (circuit complete) and without load (circuit incomplete), again, depending upon the specifications provided by the manufacturer. Please refer to the Ignition System Component Testing specification charts in this section for details.

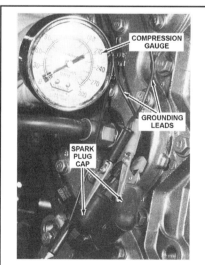
Fig. 24 Preparing to check powerhead cranking compression. Note the jumper leads used to ground the cylinders

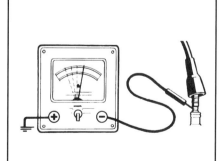
Fig. 25 Checking ignition coil polarity with an analog voltmeter

Fig. 26 Example of Low- and high-speed pulser coils mounted on 3-5 hp powerheads

4-14 IGNITION AND ELECTRICAL SYSTEMS

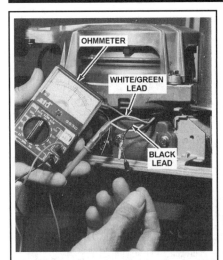
Fig. 27 Testing a high-speed pulser coil - 3-5 hp powerheads

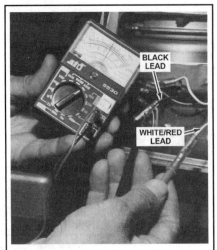
Fig. 28 Testing a low-speed pulser coil - 3-5 hp powerheads

Fig. 29 Testing a pulser coil - most 2-cylinder powerheads

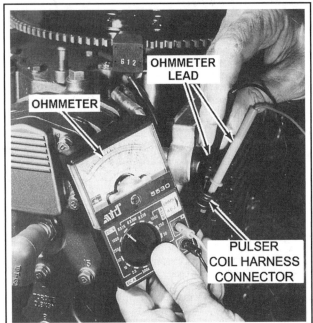
Fig. 30 Coil testing may include cranking or running voltage and/or resistance tests (this shows a typical resistance test)

■ If you're in doubt about any test connections verify them using the information in the Ignition Component Testing specifications charts AND the Wiring Diagrams, found in this section.

The basic test for a pulser coil is continuity. This measures the actual resistance from one end of the pulser coil to the other. When available, the correct specification for each coil can be found in the Ignition Testing Specifications charts.

4. Adjust the meter to read resistance (ohms).
5. Connect the tester across the pulser leads (as noted in the testing charts and/or the wiring diagrams) and note the reading.
6. Compare the coil reading to the specifications.
7. If the reading is well above or below the correct value it should be checked dynamically (if output specifications are available) and/or be replaced.

■ Remember that temperature has an affect on resistance. Most resistance specifications are given assuming a temperature of 68°F (20°C).

There are two types of pulser coils. The first type has one coil lead connected directly to ground through a Black wire or a grounded bolt hole. The other end of the coil has a color-coded wire. These type coils have one lead listed as Black and the other typically a White wire with a colored tracer.

The other type of pulser coil is not grounded at the powerhead. Instead, both leads go to the CDI box. Both coil leads are often White with a colored tracer.

When checking either type pulser coil for continuity, connect the ohmmeter to the color wires listed in the Ignition Testing Specifications chart.

A second check must be made on pulser coils that are not grounded on one end. This second check is called a short-to-ground check. Make sure that this type of coil not only has correct resistance (continuity) but also is not shorted to ground (and leaking current to ground).

8. Connect one tester lead to a coil lead.
9. Touch the other tester lead to the engine ground or the mounting point of the pulser.
10. If the pulser is good, the reading on the ohmmeter should be infinity.
11. Any ohm reading other than infinity (O.L on some digital meters) indicates a bad coil. Repair or replace the coil.

■ This second check is just as critical as the standard continuity check.

The final, and most important check, is a dynamic cranking or running output voltage check. If specifications are provided, proceed as follows:
12. Set the meter to read volts in the proper scale (as determined by the specification range in the chart).
13. Connect the meter to the pulser coil lead(s) under the appropriate conditions. For with load specifications the circuit must be complete meaning you'll have to either back-probe the wiring harness or (if this cannot be done without damaging the connectors) use jumper wires between the disconnected ends of the harness. For without load specifications you can just disconnect the harness and probe the coil sides.
14. If the coil fails to produce sufficient voltage for specifications, double-check all connections and wire colors. Once you are certain the coil is out of spec it must be replaced.

Charge Coils

 MODERATE

◆ See Figures 31, 32 and 33

The charge coil checks are essentially the same as those for the pulser coil. There are normally 2 methods used in testing charge coils, a static resistance check and a dynamic cranking or running output voltage check. Specifications vary by year and model so significantly that it precludes trying to list them in this procedure. In addition, the method of conducting the static check will vary with the specifications that are provided. Generally speaking cranking or running tests are more accurate and more likely to show intermittent problems. Dynamic checks are done under different

IGNITION AND ELECTRICAL SYSTEMS 4-15

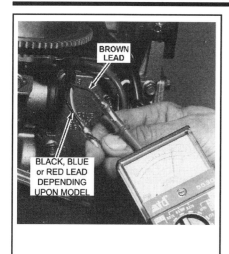

Fig. 31 Locate the charge coil leads...

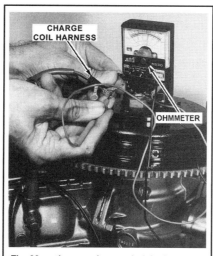

Fig. 32 ...then conduct static (ohm)...

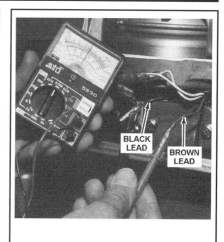

Fig. 33 ...and/or dynamic (voltage) tests

circumstances, with load (circuit complete) and without load (circuit incomplete), again, depending upon the specifications provided by the manufacturer. Please refer to the Ignition System Component Testing specification charts in this section for details.

There are two types of charge coils. One type has a charge coil lead connected to a powerhead ground. The other type charge coil has both leads go directly to the CDI box. Both charge coil types allow for a continuity test. This checks the coil's internal resistance. On charge coils that do not have a grounded lead, the short-to-ground test must also be done

■ **If you're in doubt about any test connections verify them using the information in the Ignition Component Testing specifications charts AND the Wiring Diagrams, found in this section.**

The basic test for a charge coil is continuity. This measures the actual resistance from one end of the charge coil to the other. The correct specification for each coil can be found in the Ignition System Component Testing specifications chart.
  1. Adjust the meter to read resistance (ohms).
  2. Connect the tester across the charge coil leads and note the reading.
  3. Compare the coil reading to the specifications. If the reading is well above or below the correct value it should be checked dynamically (if output specifications are available) and/or be replaced.

■ **Remember that temperature has an affect on resistance. Most resistance specifications are given assuming a temperature of 68°F (20°C).**

There are two types of charge coils. One type has one coil lead connected directly to a powerhead ground through a Black wire or a grounded bolt hole. The other end of the coil has a color-coded wire. These type coils have one lead listed as Black and the other is often a White wire with a colored tracer.

The other type of charge coil is not grounded. Instead, both leads go to the CDI box. Both coil leads are often White with a colored tracer.

When checking either type charge coil for continuity, connect the ohmmeter to the color wires listed in the specifications.

A second check must be made on charge coils that are not grounded on one end. This second check is called a short-to-ground check. Make sure that this type of coil not only has correct resistance (continuity) but also is not shorted to ground (and leaking current to ground).
  4. Connect one tester lead to a coil lead.
  5. Touch the other tester lead to the engine ground or the mounting point of the coil.
  6. If the coil is good, the reading on the ohmmeter should be infinity. Any ohm reading other than infinity (O.L on some digital meters) indicates a bad coil.
  7. Repair or replace the coil.

■ **This second check is just as critical as the standard continuity check.**

The final, and most important check, is a dynamic cranking or running output voltage check. If specifications are provided, proceed as follows:
  8. Set the meter to read volts in the proper scale (as determined by the specification range in the chart).
  9. Connect the meter to the charge coil lead(s) under the appropriate conditions. For with load specifications the circuit must be complete meaning you'll have to either back-probe the wiring harness or (if this cannot be done without damaging the connectors) use jumper wires between the disconnected ends of the harness. For without load specifications you can just disconnect the harness and probe the coil sides.
  10. If the coil fails to produce sufficient voltage for specifications, double-check all connections and wire colors. Once you are certain the coil is out of spec it must be replaced.

Lighting Coil

◆ See Figures 34 and 35

The lighting coil windings really aren't usually part of the ignition system (at least they aren't when separate charge coil windings are provided to power the ignition), however it operates in the same fashion as the charge and pulser coils, and it is replaced in the same fashion, so it makes sense to cover it here. On many newer Yamaha motors the charge coil winding are incorporated into the stator (along with the lighting coil), and on a few there are no distinct charge coil windings. On these late-model motors, the entire stator assembly must be replaced if a problem is isolated to the ignition charge circuit.

The lighting coil checks are the same as those for the pulser and charge coils. There are normally 2 methods used in testing lighting coils, a static resistance check and a dynamic cranking or running output voltage check. Specifications vary by year and model so significantly that it precludes trying to list them in this procedure. In addition, the method of conducting the static check will vary with the specifications that are provided. Generally speaking cranking or running tests are more accurate and more likely to show intermittent problems. Dynamic checks are done under different circumstances, with load (circuit complete) and without load (circuit incomplete), again, depending upon the specifications provided by the manufacturer.

■ **If you're in doubt about any test connections verify them using the information in the Charging System Testing specifications charts AND the Wiring Diagrams, found in this section.**

The basic test for a lighting coil is continuity. This measures the actual resistance from one end of the lighting coil to the other. The correct specification for each coil can be found in the Charging System Testing specifications chart.
  1. Disconnect the two wires (often, but not always, Green) between the stator and the rectifier at the rectifier. Connect the ohmmeter leads to the wires and measure the resistance.

4-16 IGNITION AND ELECTRICAL SYSTEMS

2. Compare the coil reading to the specifications. If the reading is well above or below the correct value it should be checked dynamically (if output specifications are available) and/or be replaced.

■ **Remember that temperature has an affect on resistance. Most resistance specifications are given assuming a temperature of 68°F (20°C).**

3. If the resistance is not within specification, the battery will not hold a charge and the boat accessories which depend on this coil for power may not function properly.

■ **Unless specifically directed by the test charts, never attempt to verify the charging circuit by operating the powerhead with the battery disconnected. In most cases, such action would force current (normally directed to charge the battery), back through the rectifier and damage the diodes in the rectifier.**

The final, and most important check, is a dynamic cranking or running output voltage check. If specifications are provided, proceed as follows:

4. Set the meter to read volts in the proper scale (as determined by the specification range in the chart).
5. Connect the meter to the charge coil lead(s) under the appropriate conditions. Unless otherwise directed, protect the rectifier by taking test readings with the circuit complete. You'll have to either back-probe the wiring harness or (if this cannot be done without damaging the connectors) use jumper wires between the disconnected ends of the harness.
6. If the coil fails to produce sufficient voltage for specifications, double-check all connections and wire colors. Once you are certain the coil is out of spec it must be replaced.

Ignition Coils

◆ See Figures 36, 37, 38 and 39

Although the best test for an ignition coil is on a dynamic ignition coil tester, resistance checks can also be performed to help determine condition. As with other coils, remember that a static test within specification cannot absolutely rule out a problem with the ignition coil under load in dynamic conditions.

There are two circuits in an ignition coil, the primary winding circuit and the secondary winding circuit. Whenever possible, both need to be checked.

The tester connection procedure for a continuity check will depend on how the coil is constructed. Generally, the primary circuit is the small gauge wire or wires, while the secondary circuit contains the high tension or plug lead.

When there are two primary wires running to the ignition coil, the primary circuit test is performed across both of those wire terminals (for the ignition coil). When there is only one primary wire, the circuit is checked between that wire and ground.

When an ignition coil is designed to fire only one spark plug, the secondary circuit is normally either through the spark plug lead (without the resistor cap, when equipped) and ground or ground wire terminal (depending upon the primary circuit wiring). If, however, the ignition coil contains 2 spark plug leads, the secondary circuit is checked across both leads (again, without the resistor caps, if equipped).

Some ignition coils have the primary and/or secondary circuits grounded on one end. On these type coils, only the continuity check is done. On ignition coils that are not grounded on one end (which have 2 wires going back to the CDI unit) a short-to-ground test must also be done (a check from

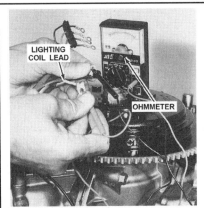

Fig. 34 Lighting coils are also tested statically (ohms)...

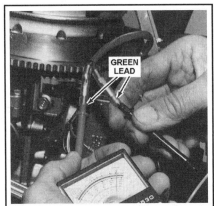

Fig. 35 ...or dynamically (voltage) - most motors use G/W and/or G leads

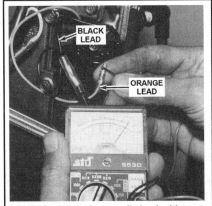

Fig. 36 Test the primary coil circuit either across the 2 small coil leads...

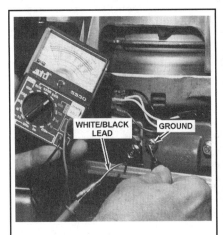

Fig. 37 ...or across the single small lead and ground (as applicable)

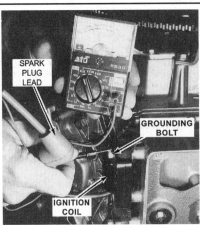

Fig. 38 Check the secondary circuit across the single spark plug lead and primary circuit ground (wire or grounding point)...

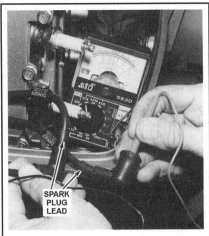

Fig. 39 ...or across the 2 spark plug leads (as applicable)

IGNITION AND ELECTRICAL SYSTEMS 4-17

the primary side of the circuit to ground in order to make sure there is no continuity). Regardless of the coil type, compare the resistance with the Ignition System Component Testing specification charts found in this section.

■ When checking the secondary side, remove the spark plug caps to make the measurement (some models, like most but not only V4 and V6 motors utilize resistor caps). When equipped with resistor caps, in some cases the cap is bad, not the coil. Bad resistor caps can be the cause of high-speed misfire. Unscrew the cap and check the resistance (usually about 5 killi-ohms, give or take a killi-ohm). Leaving the cap on during measurement could condemn an otherwise in spec ignition coil.

The other method used to test ignition coils is with a Dynamic Ignition Coil Tester. Since the output side of the ignition coil has very high voltage, a regular voltmeter can not be used. While resistance reading can be valuable, the best tool for checking dynamic coil performance is a dynamic ignition coil tester.
  1. Connect the coil to the tester according to the manufacturer's instructions.
  2. Set the spark gap according to the specifications.
  3. Operate the coil for about 5 minutes.
  4. If the spark jumps the gap with the correct spark color, the coil is probably good.

■ If you're in doubt about any test connections verify them using the information in the Ignition Component Testing specifications charts AND the Wiring Diagrams, found in this section.

CDI Unit

◆ See Figures 40, 41 and 42

There are potentially 3 methods of testing the CDI/TCI unit. There first method is available only on a handful of Yamahas. On a few motors, there are resistance specifications charts which allow a technician to bench test the internal circuitry of the CDI unit itself. However, these tests have fallen out of favor with Yamaha and are not included in most newer service publications, so any model that was released after 1997 or 1998 (or which underwent a significant ignition system change in that time frame or later) does not have a unit resistance chart.

Probably the most common and one of the most accurate methods of testing the unit is to check the output voltage. For most motors specifications are available for CDI/TCI unit output tests. This dynamic check is the preferred method of CDI unit testing, as it is more likely that faults will appear under load conditions. The test connections and specifications are provided in the Ignition System Component Testing specifications charts in this section.

■ One reason the resistance tests have likely fallen out of favor with Yamaha is their inability to show intermittent faults which only occur under load. The output tests are FAR more reliable for this reason and are the preferred method of CDI/TCI unit testing.

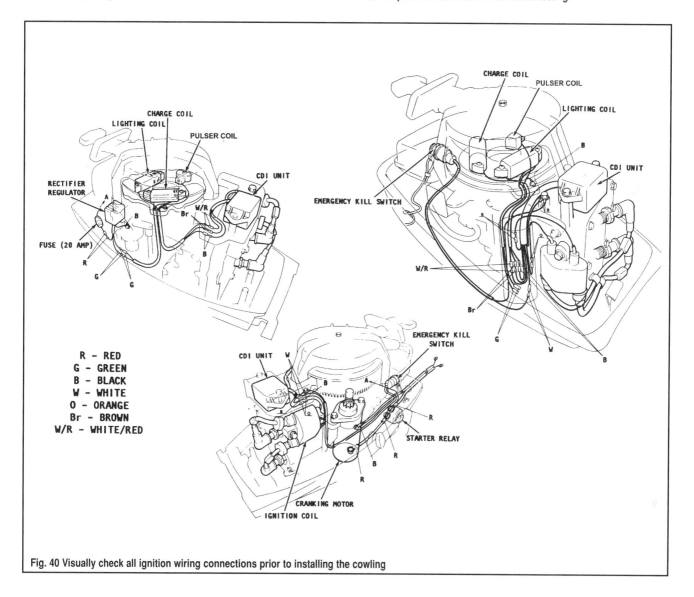

Fig. 40 Visually check all ignition wiring connections prior to installing the cowling

4-18 IGNITION AND ELECTRICAL SYSTEMS

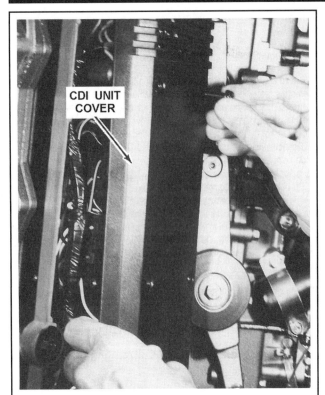

Fig. 41 On some motors, the CDI cover must be removed prior to testing the CDI unit

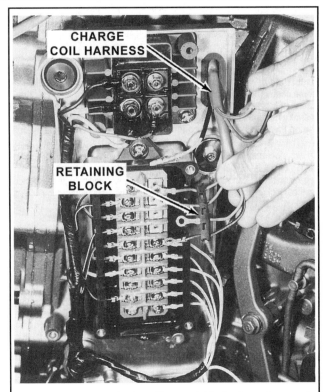

Fig. 42 Disconnect the leads as illustrated. The leads can be moved away from the unit but are still retained by a small block

The 3rd and final method of CDI/TCI unit testing is probably the most important. It is a process of elimination where the remaining ignition system components are first tested and proven good. If there are no problems found anywhere else in the system components or wiring, you are usually safe replacing the CDI/TCI unit. Just be certain that you haven't overlooked another possible component or wire when condemning the unit.

The accompanying charts outline the resistance testing procedures for those few motors on which they are available. The unit may remain installed on the powerhead, or it may be removed for testing. In either case, the testing procedures are identical.

■ No resistance testing specifications are available for ANY Yamahas introduced (or revised) after 1997 or 1998 (ish).

Select the appropriate scale on the ohmmeter. Make contact with the Red meter lead to the leads called out in the horizontal heading. Make contact with the Black meter lead to the leads called out in the vertical list of leads.

Proceed slowly and carefully in the order given. The asterisk (*) denotes the meter needle should swing toward continuity (zero ohms), and then return to stay at the specified value.

■ As usual, remember that temperature has an affect on resistance. Most resistance specifications are given assuming a temperature of 68°F (20°C). Also, keep in mind that although resistance tests may show a component to be faulty (if it tests out of specification) some problems only occur under load in dynamic conditions and a unit that tests good, may still be faulty.

CDI Unit Test Charts

◆ See Figures 43 thru 49

2 HP CDI Unit Test Chart

W: White
B: Black
Br: Brown
O: Orange

Unit: kΩ

− / +	W	B	Br	O
W		*	∞	*
B	2.2~9.5		2.2~9.5	*
Br	∞	*		*
O	7~30	2~9	7~30	

∞ No continuity
* Needle swings once and returns to home position

Fig. 43 CDI Unit Testing Chart - 2 Hp Motors

IGNITION AND ELECTRICAL SYSTEMS

3 HP CDI Unit Test Chart

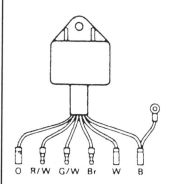

O : Orange
R/W: Red/White
G/W: Green/White
Br : Brown
W : White
B : Black

Unit: kΩ

⊖ \ ⊕	Stop W	Charge Br	Pulser G/W	Pulser R/W	Ground B	Ignition O
W		0	∞	∞	∞	∞ ✱
Br	0		∞	∞	∞	∞ ✱
G/W	23	23		∞	25	9	∞
R/W	20	20	∞		12	∞
B	4	4	∞	12		∞ ✱
O	∞	∞	∞	∞	∞	

∞ : No continuity
✱ : Needle swings once and returns to home position

Fig. 44 CDI Unit Testing Chart - 3 Hp Motors

4–5 HP CDI Unit Test Chart

Unit kΩ

B/W: Black/White
W/G: White/Green
W/R: White/Red
B : Black
Br : Brown
W : White

Tester (−) \ Tester (+)		Stop W	Charge Br	Pulser 2 (Low speed) W/G	Pulser 1 (High speed) W/R	Earth B	Ignition B/W
Stop	W		0	∞	∞	∞	∞ ✱
Charge	Br	0		∞	∞	∞	∞ ✱
Pulser 2 (Low speed)	W/G	18.4~27.6	18.4~27.6		20~30	7.2~10.8	∞
Pulser 1 (High speed)	W/R	16~24	16~24	∞		9.6~14.4	∞
Earth	B	3.2~4.8	3.2~4.8	∞	9.6~14.4		∞ ✱
Ignition	B/W	∞	∞	∞	∞	∞	

∞ : No continuity
✱ : Needle swings once and returns to home position

Fig. 45 CDI Unit Testing Chart - 4-5 Hp Motors

6-8 HP CDI UNIT TESTING
(2-Stroke)

- Digital tester can not be used for this inspection. Use analogue tester.
- C.D.I. resistance values will vary from meter to meter, especially with electronic digital meters. For some testers, polarity of leads is reversed.

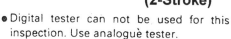

B : Black
Br : Brown
G : Green
O : Orange
W : White
W/R : White/Red

Unit: KΩ

⊖ \ ⊕	W	B	Br	W/R	O
W		∞	∞	∞	∞
B	∞		7.5~11.3	∞	•
Br	∞	63.2~94.8		∞	•
W/R	8.8~13.2	14.4~21.6	30.4~45.6		•
O	∞	∞	∞	∞	

• : Needle swings once and returns to home position.
∞ : Discontinuity

Fig. 46 CDI Unit Testing Chart - 6/8 Hp Motors

4-20 IGNITION AND ELECTRICAL SYSTEMS

20-30 HP (430cc & 496cc) CDI Unit Test Chart

W : White W/R : White/Red
B : Black W/B : White/Black
B/O : Black/Orange Br : Brown
B/W : Black/White L : Blue

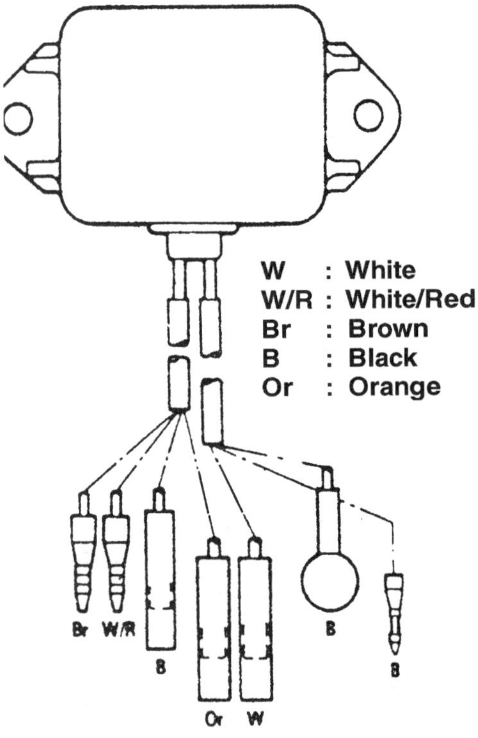

Unit: kΩ

Tester (+) → Tester (−) ↓		Stop	Ground	Ignition		Pulser		Charge	
		W	B	B/O	B/W	W/R	W/B	Br	L
Stop	W		∞	∞	∞	∞	∞	∞	∞
Ground	B	40 ~ ∞		2.0 ~ 8.0	2.0 ~ 8.0	10 ~ 30	∞	3.0 ~ 100	0
Ignition	B/O	∞	∞		∞	∞	∞	∞	∞
	B/W	∞	∞	∞		∞	∞	∞	∞
Pulser	W/R	∞	∞	∞	∞		∞	∞	∞
	W/B	∞	∞	∞	∞	∞		∞	∞
Charge	Br	80 ~ 1,000	80 ~ 300	80 ~ 1,000	80 ~ 1,000	100 ~ 1,000	∞		∞
	L	40 ~ ∞	0	2.0 ~ 8.0	2.0 ~ 8.0	10 ~ 30	∞	3.0 ~ 10.0	

∞ : No continuity

Fig. 47 CDI Unit Testing Chart - 20/25 Hp (430cc And 496cc) Motors

40 HP (2 Cyl) CDI Unit Test Chart

Unit : kΩ

W : White
W/R : White/Red
Br : Brown
B : Black
Or : Orange

Tester ⊕ → Tester ⊖ ↓		Stop White	Earth Black	Charge Brown	Pulser White/red	Ignition Orange
Stop	White		∞	∞	∞	∞
Earth	Black	9 ~ 19		2 ~ 6	∞	∞
Charge	Brown	80 ~ 160	70 ~ 150		∞	✱
Pulser	White/red	33 ~ 63	7 ~ 17	15 ~ 35		✱
Ignition	Orange	∞	∞	∞	∞	

✱ ..... Needle swings once and returns to home position.
∞ ..... No continuity

The test indicated by "✱" should be made with the condenser completely discharged, and therefore, the needle will not deflect again. If any charge remains in the condenser, the needle will not swing at all.

Fig. 48 CDI Unit Testing Chart - 40 Hp (2 Cyl) Motors

IGNITION AND ELECTRICAL SYSTEMS

48/55 HP (2-Cyl) CDI Unit Test Chart

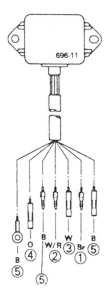

B : Black
Br : Brown
O : Orange
W : White
W/R : White/Red

Unit: kΩ

Tester (+) / Tester (−)		① Charge coil Br	② Pulser coil W/R	③ Stop W	④ Ignition coil O	⑤ Ground B
① Charge coil	Br		∞	∞	∞*	79k
② Pulser coil	W/R	38k		11k	∞*	18k
③ Stop	W	∞	∞		∞	∞
④ Ignition coil	O	∞	∞	∞		∞
⑤ Ground	B	9.4k	∞	∞	∞*	

∞*.....Needle swings once and returns to home position.
∞ ....No continuity.

Fig. 49 CDI Unit Testing Chart - 48/55 Hp Motors

Rectifier or Regulator/Rectifier

  MODERATE

◆ See Figures 50 thru 55

When an outboard is equipped with a lighting coil, unless it is an AC lighting model (which is designed to power AC accessories), it must also have a rectifier to convert the voltage to DC and/or a regulator to control the battery charging voltage. Over the years Yamaha has changed what tests it provides for the rectifier or regulator/rectifier. As they have moved away from resistance tests, for questions of accuracy they've also begun to favor output voltage checks for regulator/rectifiers. When output specifications for the regulator/rectifier are available, they are listed in the Charging System Testing specifications chart in this section. When available, this specification is the BEST way to troubleshoot the regulator/rectifier.

■ Remember that a regulator/rectifier does NOT create voltage, it only controls voltage. Whenever the Lighting Coil output should always be checked before suspecting that a regulator/rectifier is the culprit for undercharging a battery. However, since the regulator/rectifier's primary job is to prevent OVERcharging, the lighting coil is not usually the first suspect under those conditions.

When available, we've provided Resistance Test charts for a quick bench test of the rectifier or regulator/rectifier on some models. If there is no illustration covering the motor on which you are working, you'll have to use general charging system testing procedures, including whatever specifications are provided in the Charging System Testing specifications charts when trying to determine if the rectifier or regulator/rectifier is at fault.

■ The accompanying charts outline resistance testing procedures for the rectifier. The unit may remain installed on the powerhead, or it may be removed for testing. In either case, the testing procedures are identical.

1. Select the appropriate scale on an analog (needle-type) ohmmeter. Make contact with the Red meter lead to the leads called out as (+) and the Black meter lead to the leads called out as (-). Proceed slowly and carefully in the order given.

■ The manufacturer notes for some models that a digital tester will not perform properly during this test.

2. If resistance is not as specified, the rectifier is faulty and should be replaced.

■ No rectifier or regulator/rectifier resistance tests are provided for ALL fuel injected motors.

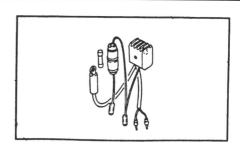

(−) \ (+)	R	G1	G2	B
R		∞	∞	∞
G1	O		∞	∞
G2	O	∞		∞
B	O	O	O	

O : Continuity
∞ : Discontinuity

Fig. 50 Rectifier troubleshooting - 6/8 hp, 48 hp 2-cylinder and 50-70 hp (849cc) 3-cylinder motors

(−) \ (+)	R	G	G/W	B
R		∞	∞	∞
G	O		∞	∞
G/W	O	∞		∞
B	O	O	O	

O : Continuity
∞ : Discontinuity

Fig. 51 Regulator/rectifier troubleshooting - 6/8 hp motors

4-22 IGNITION AND ELECTRICAL SYSTEMS

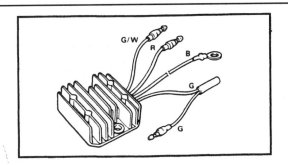

Rectifier/regulator check

Tester ⊕ / Tester ⊖	Green	Green/White	Red	Black
Green		∞	Continuity	∞
Green/White	Continuity		Continuity	Continuity
Red	∞	∞		∞
Black	Continuity	Continuity	Continuity	

∞ : Discontinuity

Fig. 52 Regulator/rectifier troubleshooting - 50-70 hp (849cc) and 65J-90 hp (1140cc), motors

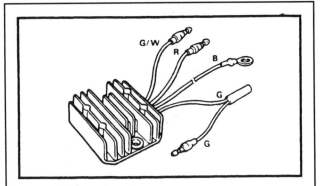

Rectifier check

⊕ \ ⊖	① G	② G/W	③ R	④ B
① G		∞	○	∞
② G/W	○		○	○
③ R	∞	∞		∞
④ B	○	○	○	

○ : Continuity
∞ : Discontinuity

Fig. 53 Regulator/rectifier troubleshooting - most pre-mix 80J-140 hp (1730cc) motors

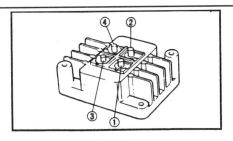

Rectifier/regulator continuity

	① Green (G1)	② Green (G2)	③ Red (R)	④ Black (B)
① Green (G1)		○	○	○
② Green (G2)	∞		○	∞
③ Red (R)	∞	∞		∞
④ Black (B)	○	○	○	

○ : Continuity
∞ : No continuity

Fig. 54 Regulator/rectifier troubleshooting - most pre-mix 105J-225 hp (2596cc) carbureted motors (except the 225DET)

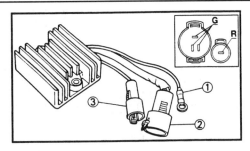

⊕ \ ⊖	① Black (B)	② Green (G)	③ Red (R)
① Black (B)		○	○
② Green (G)	∞		○
③ Red (R)	∞	∞	

○ : Continuity
∞ : Discontinuity

Fig. 55 Regulator/rectifier troubleshooting - 225DET and most oil-injected 80J-140 hp (1730cc) and 105J-225 hp (2596cc) carbureted motors

IGNITION AND ELECTRICAL SYSTEMS 4-23

Flywheel and Stator Plate

The following short lists the procedures required to pull the flywheel and remove the stator plate in order to service the ignition system components (pulser, charge, lighting coils, as applicable). Removal and installation of the stator plate itself is often necessary in order to gain access to the wiring harness retainer underneath the stator plate. Cleaning and Inspecting procedures in addition to proper assembling and installation steps are also included.

REMOVAL & INSTALLATION

1 & 2-Cylinder Powerheads

◆ See Figures 56 thru 65

Once the flywheel is removed, use the wire colors from the Wiring Diagrams or from the Ignition System Component testing charts in order to help identify the individual coils.

1. Remove the cowling from the powerhead.
2. Disconnect and ground the spark plug lead(s) for safety. On twins, be sure to tag the spark plug wires to ensure proper installation.
3. Remove the mounting hardware securing the hand rewind starter or flywheel cover to the powerhead. Remove the hand starter or flywheel cover for access to the flywheel.

■ Do not remove the handle from the hand starter rope because the rope would immediately rewind inside the starter. Such action would require considerable time and effort to correct.

4. Some motors (such as the 2 hp, 6/8 hp and certain 20/25 hp motors) are equipped with a flywheel cover and/or hand starter pulley which must be removed. If applicable, scribe a matchmark between the cover/pulley and flywheel, then carefully loosen the screws and remove the pulley/cover for access.

■ Obtain a flywheel holder or large strap wrench (but check the torque value on the flywheel nut first, refer to the installation steps, as a strap wrench isn't going to work on any but the smallest motors covered here). On these motors a flywheel holder is normally used by inserting 2 indexing pins on the tool arms through the holes in the flywheel or flywheel cover. This type of tool can usually be fabricated from a couple of pieces of heavy stock and some tempered bolts.

5. Secure the flywheel using the holder or strap wrench in order to keep it from turning (remember, you don't want to turn the flywheel counterclockwise as you could damage the vanes of the water pump impeller). Hold the flywheel from rotating and at the same time remove the flywheel nut.

■ If equipped, make sure the timing pointer will not interfere with flywheel removal (it usually will on 48 hp motors). If necessary, scribe a matchmark between it and the powerhead, then loosen and remove the bolts securing the timing pointer.

6. Obtain a threaded flywheel/steering wheel puller tool. Ensure the puller will apply force at the bolt holes in the flywheel and not from around the perimeter of the flywheel.

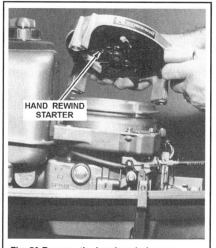

Fig. 56 Remove the hand rewind starter

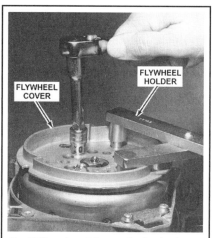

Fig. 57 If equipped, loosen the flywheel cover bolts...

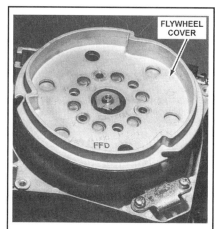

Fig. 58 ...so you can remove the cover for access to the flywheel nut

Fig. 59 Loosen and remove the flywheel nut...

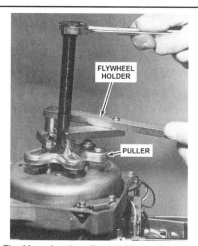

Fig. 60 ...then install a threaded puller to free the flywheel

Fig. 61 Remove the flywheel and woodruff key

4-24 IGNITION AND ELECTRICAL SYSTEMS

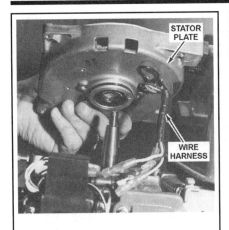

Fig. 62 On 2-strokes you usually remove the stator plate for access to the wiring

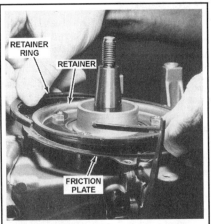

Fig. 63 Larger motors use a friction plate and retaining ring

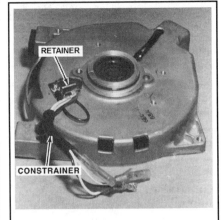

Fig. 64 If necessary remove the wiring and coils from the stator plate

✱✱ WARNING

Never attempt to use a jawed puller (which would put stress on the outside edge of the flywheel).

7. Install the puller onto the flywheel, take a strain on the puller with the proper size wrench. Now, continue to tighten the tool and at the same time, shock the crankshaft with a gentle to moderate tap using a hammer or mallet on the end of the tool. This shock will assist in breaking the flywheel loose from the crankshaft.

8. Lift the flywheel free of the crankshaft. Remove and save the Woodruff key from the recess in the crankshaft.

■ On some motors the stator coil or the separate charge/lighting coil(s) can now be unbolted and removed without further component disassembly. However, on most models, the charge/lighting and sometimes pulser coils are mounted to a stator plate that must be unbolted or repositioned for access to wiring and/or coil fasteners.

9. Tag and disconnect the stator harness at the connector fittings (most use bullet connectors).

■ Pay close attention to how the wiring harness is routed for installation purposes. It is not uncommon for someone to install the wiring incorrectly only to have the flywheel rub through the insulation, shorting the wiring and destroying the coil.

10. Remove the bolts securing the stator plate to the powerhead and lift the stator plate free of the crankshaft and powerhead. On most 6 hp and larger motors (except the 48 hp twin), the stator is bolted to a friction plate beneath a retainer. Once the stator plate is removed, this friction plate is free to rotate. Do not disturb the position of this friction plate in relation to the stator plate. Movement of the friction plate will affect powerhead timing.

■ If it is necessary to remove and replace the friction plate, take time to scribe a mark on the powerhead opposite the small timing hump, as shown in the accompanying illustration.

11. On 9.9/15 hp motors, the pulser coil assembly is mounted underneath the stator plate. If necessary, remove the pulser coil assembly, noting the positioning for installation purposes.

12. Place the stator plate on a suitable work surface. Remove the screw(s) securing the harness retainer to the plate and unwind the Black plastic harness constrainer. Now, any component mounted on top of the stator plate which tested defective may be removed.

To Install:

13. If removed, install the pulser coil assembly to the powerhead on 9.9/15 hp motors. Be sure to orient it as noted during removal.

14. On most 6 hp and larger motors (except the 48 hp twin), the stator is bolted to a friction plate beneath a retainer. For these models, place the friction plate down over the crankshaft. Rotate the plate until the hump on the outer edge is aligned with the mark scribed on the powerhead prior to

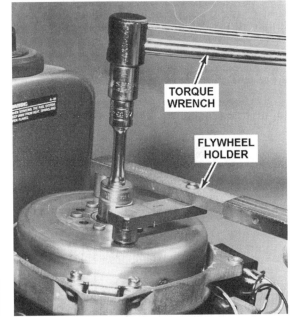

Fig. 65 Upon installation, use a torque wrench to tighten the flywheel nut

removal. Install the retainer on top of the friction plate and secure it in place with the attaching bolts. Stretch the retainer ring around the retainer with the outer edge of the retainer indexed into the ring groove.

■ On the models with the friction plate/retainer set up, when the stator plate is installed, the mounting bolts thread into the friction plate instead of into the powerhead, as on the other models. Make a final check to be sure the hump on the friction plate is still aligned with the scribed mark on the powerhead.

15. Position the stator plate in place over the crankshaft. Secure the plate with the attaching bolts and tighten securely.

16. Route the stator wire harness as noted during removal, then reconnect them as tagged (though even without tagging MOST wires connect color-to-like color, however, if the harness was ever repaired this may no longer be true).

17. Place a tiny dab of thick lubricant on the curved surface of the Woodruff key to hold it in place while the flywheel is being installed. Press the Woodruff key into place in the crankshaft recess. Wipe away any excess lubricant to prevent the flywheel from walking during powerhead operation.

18. Check the flywheel magnets to ensure they are free of any metal particles. Double check the taper in the flywheel hub and the taper on the crankshaft to verify they are clean and contain no oil.

IGNITION AND ELECTRICAL SYSTEMS 4-25

19. Slide the flywheel down over the crankshaft with the keyway in the flywheel aligned with the Woodruff key in place on the crankshaft. Rotate the flywheel carefully clockwise to be sure it does not contact any part of the stator plate or wiring.

20. Slide the washer onto the crankshaft, and then apply a thin coating of clean engine oil to the threads of the flywheel nut.

21. Thread the flywheel nut onto the crankshaft. Secure the flywheel from turning using the holding tool, then tighten the flywheel nut to the following torque value for the models listed:
- 2, 3, 4, 4/5 and 6/8 hp motors - 32 ft. lbs. (44 Nm)
- 9.9/15 hp motors - 77 ft. lbs. (105 Nm)
- 20/25 hp (395cc) motors - 72 ft. lbs. (100 Nm)
- 25/30 hp (496cc), 40 hp and 48 hp motors - 116 ft. lbs. (160 Nm)

22. If equipped, position the pulley/cover, aligning the matchmarks made earlier, then thread the retaining bolts and tighten them securely.

23. Install the hand rewind starter or flywheel cover, as applicable.

24. Reconnect the spark plug lead(s).

25. Install the cowling to the powerhead.

3-Cylinder Powerheads

◆ See Figures 66 thru 75

Some 3-cylinder models are equipped with individual charge, lighting and pulser coils, while others are equipped with a 1-piece stator assembly (which contains the charge and lighting coil windings) along with separate pulser coil(s). Once the flywheel is removed, use the wire colors from the Wiring Diagrams or from the Ignition System Component testing charts in order to help identify the individual coils.

■ **The 65J-90 hp (1140cc) motors are typically equipped with a 1-piece stator mounted above the pulser coil assembly. The remaining models are usually equipped with individual charge, lighting and pulser coils.**

The accompanying photographs illustrate the work being performed on a typical 25/30 hp powerhead. The procedural tasks are almost identical for most other models, however the pulser and charge coil setup which is illustrated varies greatly from model-to-model. The differences will be clearly identified in the text.

1. Remove the hand rewind starter or the flywheel cover, as applicable.
2. For safety either disconnect the negative battery cable or tag and disconnect the spark plugs leads (but ground them to the powerhead to prevent damage if the engine is cranked).
3. Remove the hand rewind starter or flywheel cover from the powerhead for access to the flywheel.
4. Most smaller motors (through 50 or 60 hp) are equipped with a flywheel cover or hand starter pulley which must be removed. If applicable, scribe a matchmark between the cover/pulley and flywheel, then carefully loosen the screws and remove the pulley/cover for access. A prybar can usually be used across bolt heads to keep the flywheel from turning when loosening the pulley/cover bolts.

■ **Flywheel nut torque ranges from 80-116 ft. lbs. (110-157 Nm), depending upon the model. In all cases you'll need a sturdy flywheel holding tool to loosen/tighten the nut. Yamaha recommends the use of YB-06139, but you can also fashion one out of some pieces of heavy bar stock and tempered bolts. The tool contains an adjustable, hinged bar that can be pivoted so a pin can be inserted into another hole in the flywheel without interfering with the flywheel nut. Just make sure the bolts are not too long (if the holes in the flywheel are not blind) and do not damage the coil(s) mounted to the powerhead underneath. In some cases, a large prybar may be placed across the heads of bolts inserted through the flywheel in lieu of the holding tool, but use care as it can slip off suddenly causing injury or damage.**

5. Use a tool to keep the flywheel from turning then carefully loosen then nut with a suitable socket (usually a 22mm socket). Remove the flywheel nut and washer.

6. Obtain a threaded flywheel/steering wheel puller tool. Ensure the puller will apply force at the bolt holes in the flywheel and not from around the perimeter of the flywheel.

✱✱ WARNING

Never attempt to use a jawed puller (which would put stress on the outside edge of the flywheel).

7. Install the puller onto the flywheel, take a strain on the puller with the proper size wrench. Now, continue to tighten on the tool and at the same time, shock the crankshaft with a gentle to moderate tap using a hammer or mallet on the end of the tool. This shock will assist in breaking the flywheel loose from the crankshaft.

✱✱ WARNING

A violent strike to the center bolt may cause damage to the crankshaft oil seals. Therefore, use only a gentle to moderate tap with the hammer.

8. Lift the flywheel free of the crankshaft. The flywheel may seem heavier than it actually is due to the magnetic attraction between the flywheel magnets and the laminated cores of the coils. Remove and save the Woodruff key from the recess in the crankshaft.

■ **On some models the timing pointer may interfere slightly with flywheel removal, either tilt the flywheel carefully to pull it out from under the pointer lip or scribe a matchmark to the powerhead and remove the pointer.**

9. Some models may be equipped with a timing link rod which attaches to the stator plate. When applicable (and when necessary), carefully pry the link from the ball joint under the stator plate. Take care not to alter the length of the link rod.

■ **Before proceeding, identify the coil(s) to be removed. Carefully note the wiring harness routing. Coil wires under the flywheel must be carefully positioned to prevent contact with and damage from the flywheel.**

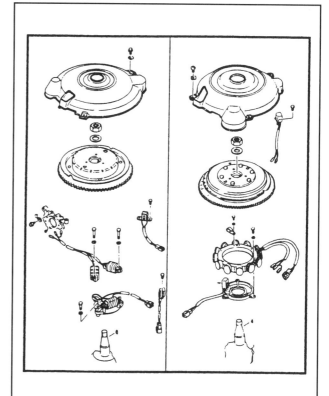

Fig. 66 Typical flywheel and coil mountings, individual coils (left) and 1-piece stator (right)

4-26 IGNITION AND ELECTRICAL SYSTEMS

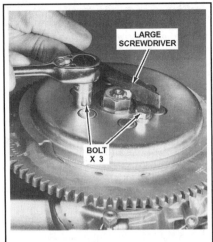

Fig. 67 If equipped loosen the fasteners...

Fig. 68 ...and remove the flywheel cover (shown) or pulley

Fig. 69 Use a holding tool and loosen the flywheel nut...

Fig. 70 ...then install a suitable threaded puller

Fig. 71 Once loosened, lift off the flywheel

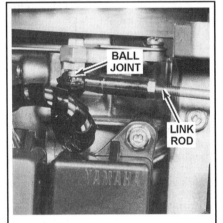

Fig. 72 If equipped, disconnect the timing link

10. Tag and disconnect the wiring for the coil(s) to be removed. If an entire stator or stator plate assembly is being removed, disconnect all of the appropriate wiring.

11. Loosen the fasteners securing the stator assembly, stator plate or individual coil (as applicable) and remove the coil from the powerhead. On 25/30 hp motors the charge coil is mounted to a stator plate along with the pulser coils, however the lighting coil is mounted separately to a bracket on top of the powerhead. On 28J-50 hp (698cc) motors all of the coils are normally mounted to a stator base assembly. Other motors either mount the coils individually to a base and/or the powerhead or use a 1-piece stator mounted to the powerhead over top of a pulser coil assembly.

12. Thoroughly clean and inspect the flywheel and stator assembly. Replace any components which are obviously defective or damaged.

To Install:

13. If equipped with a pulser coil assembly which is mounted under the stator, route the wiring (as noted during removal) and position the pulser assembly, then secure using the retaining bolts.

14. If removed, install the 1-piece stator or stator plate assembly, route the wiring (as noted during removal) and secure using the retaining bolts. The same goes for any individual coils removed from a stator base or powerhead.

15. Connect the coil or stator harness wires, as tagged during removal. This is normally an easy matter of matching color-to-color, however if part of the wiring harness has been replaced previously the colors do not always match. When equipped, multi-pin connectors are normally designed to be connected only in one manner.

16. Install the Woodruff key into the crankshaft. A tiny dab of grease will help hold the key in place while the flywheel is installed.

17. Carefully lower the flywheel over the crankshaft with the Woodruff key indexing into the slot in the flywheel. Keep in mind that the magnets may pull the flywheel sharply down into position, so watch your fingers.

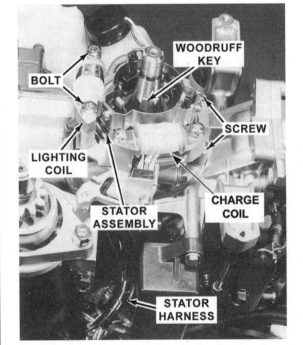

Fig. 73 Individually mounted coils on a typical 25/30 hp powerhead

IGNITION AND ELECTRICAL SYSTEMS 4-27

18. If applicable, snap the timing link rod back onto the ball joint under the stator plate. If the length of this rod has been accidentally changed, refer to the Timing and Synchronization procedures in order to adjust the length back to the specifications.

19. On models equipped with a link rod, check the action of the stator plate. The plate should move freely within the limits of travel of the magneto control lever. If any binding is felt, remove the flywheel and check the installation of the stator plate.

20. Apply a light coating of engine oil to the threads of the flywheel nut and to the surface of the flywheel washer.

21. Install the washer and flywheel nut. Keep the flywheel from rotating using a suitable holder tool and tighten the flywheel nut to the following torque value for the models listed:
- 25/30 hp and 28J-50 hp (698cc) motors - 80 ft. lbs. (110 Nm)
- 50-70 hp (849cc) and 65J-90 hp (1140cc) motors - 118 ft. lbs. (160 Nm)

22. When equipped, install the flywheel cover or pulley and secure using the mounting bolts.

23. Install the hand rewind starter or the flywheel cover, as applicable.

24. Connect the spark plug leads and/or the negative battery cable, as applicable.

V4 and V6 Powerheads

◆ See Figures 76 thru 86

All Yamaha V4 and V6 engines are equipped with a 1-piece stator coil that performs the functions of the charge and lighting coils. The stator is mounted to a stator base that contains an integral pulser coil assembly. On most models (all fuel injected engines and some carbureted engines) this stator base is mounted in a fixed position, however on some carbureted motors the base is mounted on a bushing and attached to linkage which allows for limited motion to adjust timing. On models with a fixed base, all timing functions are performed by the CDI unit or ECM.

Like most outboards, service to the flywheel and coil windings are pretty straightforward. The flywheel must be removed for access to the coil windings and care must be taken to observe and match wire routing during installation. Other than that, it is a relatively easy takes, if you have a method of holding the flywheel steady. Yamaha recommends a universal flywheel holding tool that works on most of their outboards (#YB-06139), however the 3.3L V6 motors require a slightly different tool (#YU-01235). The shape and function of both tools are very similar and it is likely just the size and dimensions that are really different. In either case the tool is basically heavy bar stock with locating pins that fit into two or three holes on the flywheel. The tool contains an adjustable, hinged bar that can be pivoted so a pin can be inserted into another hole in the flywheel without interfering with the flywheel nut. A substitute can be made from heavy bar stock and bolts which will fit into those holes (but make sure they're not deeper than the flywheel itself in case the holes are not blind, you don't want to catch on anything under the flywheel).

1. Disconnect the negative battery cable for safety.

2. Loosen the bolts securing the flywheel cover to the top of the powerhead, then carefully pull it free of the mounting. Most covers contain locating pins or grommets from which the cover must be gently pulled.

3. On HPDI motors, remove the HPDI Drive Belt and Sprockets, as detailed in the Fuel System section (actually, you only need to remove the belt and the drive sprocket from the flywheel). Once the drive sprocket is out of the way, loosen and remove the 3 bolts securing the sprocket bracket to the top of the flywheel. Remove the sprocket bracket for access to the flywheel nut.

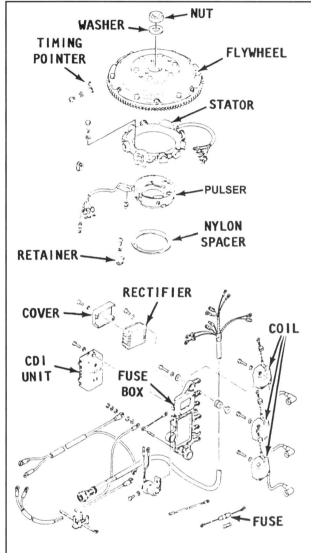

Fig. 74 Exploded view of the ignition system used on some members of the 65J-90 hp (1140cc) engine family

Fig. 75 Use a torque wrench to make sure the flywheel nut is properly tightened

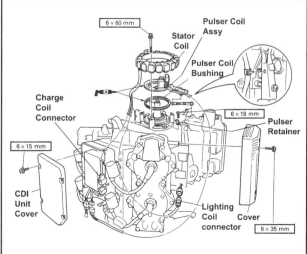

Fig. 76 Exploded view of a typical stator and pulser coil assembly—carbureted motors

4-28 IGNITION AND ELECTRICAL SYSTEMS

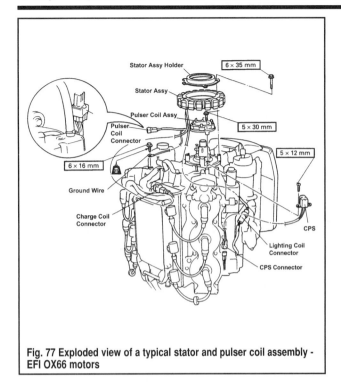

Fig. 77 Exploded view of a typical stator and pulser coil assembly - EFI OX66 motors

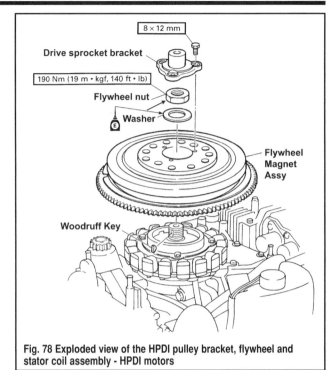

Fig. 78 Exploded view of the HPDI pulley bracket, flywheel and stator coil assembly - HPDI motors

Fig. 79 Remove the plastic flywheel cover for access

Fig. 80 Loosen the flywheel nut...

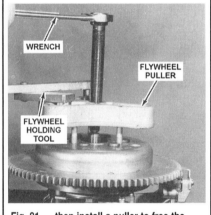

Fig. 81 ...then install a puller to free the flywheel

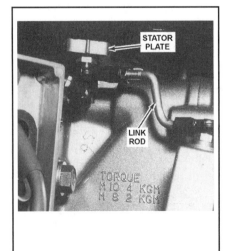

Fig. 82 If equipped, disconnect the link rod

Fig. 83 Remove any covers necessary to access wiring

Fig. 84 ...then tag and disconnect the stator/pulser wiring

IGNITION AND ELECTRICAL SYSTEMS 4-29

4. Insert a flywheel holding tool into the two holes provided in the flywheel and hold it steady while the flywheel nut is loosened with the correct size socket. Remove the flywheel nut and washer.

■ There is considerable force on the flywheel nut. It is safest to position the holder tool and the breaker bar so that you are pulling each in toward the other. Brace yourself and apply force, but avoid the use of an impact gun which could potentially damage the flywheel or crankshaft.

5. Install a threaded puller, a steering-wheel type puller may be used for this purpose, to the flywheel, making sure to keep it parallel with the flywheel. Slowly tighten the puller screw to free the flywheel from the crankshaft. If necessary, lightly tap the end of the puller screw with a mallet as this gentle shock will assist in breaking the flywheel loose from the crankshaft taper.

✳✳ WARNING

A violent strike to the center bolt may cause damage to the crankshaft oil seals. Therefore, use only a gentle to moderate tap with the hammer. Also, never attempt to use a jawed puller, which would potentially damage the flywheel by applying force to the outside edges.

6. Remove the puller and the flywheel, keeping track of the crankshaft woodruff key. Keep in mind that the strength of the magnets may make it a little more difficult to lift the flywheel, just watch your fingers to keep them from getting pinched.

7. On carbureted models which use mechanical linkage for timing advance, if the pulser coil assembly is to be removed, disconnect the link rod at the ball joint. DO NOT alter the length of the link rod, so that timing adjustments will be preserved (making Timing and Synchronization adjustment easier after the repair).

8. Trace the pulser coil assembly and/or stator wiring. As necessary, tag and disconnect the stator and/or pulser coil wiring connectors (from the CDI unit, regulator/rectifier etc). On some models (such as most carbureted and some fuel injected motors, including EFI OX66 models) you'll need to unbolt and remove the CDI unit cover for access to some of the connectors. A few models also use a regulator/rectifier cover which must be removed for access to the stator lighting coil wiring connectors.

■ Be sure to note the stator and pulser coil wire routing for installation purposes. It is critical that they are positioned in a fashion that will NOT interfere with moving components, such as the flywheel or the wiring and possibly the coils will be destroyed.

9. Matchmark the relative positions of the pulser coil assembly (stator base) and/or stator coil to the powerhead before removal. Marking their positions will help make sure wires are routed in the same positions.

10. Remove the bolts (usually 3 or 4 depending upon the model) securing the stator assembly to the powerhead. On EFI OX66 and HPDI motors the bolts are usually threaded through a stator holder which is positioned on top of the stator coil itself. Carefully life the stator coil assembly (and holder if applicable) off the powerhead.

11. If necessary, remove the stator base/pulser coil assembly retainers (usually 3 or 4 depending upon the model). On carbureted models with mechanical timing linkage there are usually 4 screws that hold pulser coil bushing retaining tabs over the edge of the pulser coil bushing. On models with fixed position pulser assemblies the 3 or 4 bolts/screws fasten the assembly directly down to the powerhead. Lift off the pulser assembly

To Install:

The following procedures pickup the work after the flywheel and stator assembly have been cleaned, inspected, serviced, and assembled.

■ When a pulser or stator coil is replaced, transfer the matchmarks made on the old coil during removal to the replacement in order to ensure proper positioning.

12. If removed, install the stator base/pulser coil assembly. For most motors it is a simple matter of positioning the assembly (With the wiring facing as noted during removal), then installing and tightening the retaining bolts. For carbureted models with mechanical timing linkage, start by placing the bushing under the coil assembly, then position it to the powerhead (with the linkage ball joint facing as noted during removal). Secure the pulser assembly using the 4 retaining tabs and screws.

13. Position the stator coil over the crankshaft, aligning the matchmarks made earlier. The 3 or 4 bolt holes are sometimes unevenly spaced which may help with alignment. Install the retainers and tighten to secure the stator coil assembly.

14. Carefully thread the wiring harnesses through the grommets or retainers, as noted during removal and connect them as tagged. If any questions occur, refer to the Wiring Diagrams in this section for clarification.

✳✳ WARNING

Remember the wiring must be positioned so there will be NO contact with moving components, otherwise the wiring and possibly the components which it connects may be damaged.

15. Install any CDI and/or regulator/rectifier covers which were removed for access to the wiring harness.

16. On carbureted models with mechanical timing linkage, reconnect the link rod at the ball joint. If the length of this rod was accidentally altered, adjust the rod back to the specified length (refer to the procedures in Timing and Synchronization for more details).

17. Install the Woodruff key into the crankshaft. A tiny dab of grease will help hold the key in place while the flywheel is installed.

18. Lower the flywheel over the crankshaft with the Woodruff key indexing into the slot in the flywheel.

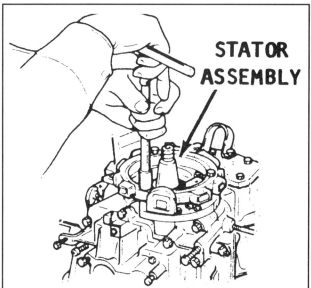

Fig. 85 Remove the pulser coil assembly and/or stator coil (shown) from the powerhead

Fig. 86 Use a torque wrench upon installation to ensure the flywheel is properly secured

4-30 IGNITION AND ELECTRICAL SYSTEMS

✲✲ CAUTION

Use care as you lower the flywheel into place not to get you fingers pinched as the magnets apply some pull to the plate.

19. For carbureted models with mechanical timing linkage, check the action of the magneto plate. It should move freely within the limits of travel of the magneto control lever. If any binding is felt, remove the flywheel and check installation of the stator plate.

20. Apply a very light coating of clean engine oil to the threads and washer mating surface of the flywheel nut, then position the washer and thread the nut onto the end of the flywheel.

21. Keep the flywheel from rotating using a flywheel holder tool and tighten the flywheel nut to specification, as follows:
- Pre-mix carbureted 2.6L motors (except the 225DET): 120 ft. lbs. (165 Nm)
- All other models: 140 ft. lbs. (190 Nm)

22. Slowly rotate the flywheel by hand feeling for binding and visually checking for interference with wiring or any other powerhead mounting components.

23. For HPDI motors, install the driven pulley bracket to the top of the flywheel and secure using the 3 retaining bolts. Then install the HPDI Drive Belt and Sprockets, as detailed in the Fuel System section.

24. Install the flywheel cover and secure using the retaining bolts.

25. Reconnect the negative battery cable.

CLEANING & INSPECTION

◆ See Figure 87

Inspect the flywheel for cracks or other damage, especially around the inside of the center hub. Check to be sure metal parts have not become attached to the magnets. Verify each magnet has good magnetism by using a screwdriver or other suitable tool.

Thoroughly clean the inside taper of the flywheel and the taper on the crankshaft to prevent the flywheel from walking on the crankshaft during operation.

Check the top seal around the crankshaft to be sure no oil has been leaking onto the stator plate. If there is any evidence the seal has been leaking, it must be replaced.

Test the stator assembly to verify it is not loose. Attempt to lift each side of the plate. There should be little or no evidence of movement.

Inspect the stator plate oil seal and the O-ring on the underside of the plate.

Some models have a retainer, a retainer ring, and a friction plate located under the stator. The retainer ring is a guard around the retainer, subject to cracking and wear. Inspect the condition of this guard and replace if it is damaged.

Fig. 87 Always check the flywheel carefully to be sure particles of metal have not stuck to the magnets

CDI/TCI Unit and Ignition Coils

The ignition coil(s) and CDI/TCI unit have to be one of the easiest components to locate on the outboard (ok, with the exception of the flywheel, propeller and spark plugs, but that's still pretty easy). To find the ignition coil for a given cylinder, simply follow the spark plug lead back from the plug directly to the coil. To locate the CDI unit, following the wiring back from the coil to the CDI. Simple, right?

Well it is, for the most part. Generally speaking, the CDI unit is normally mounted either on the side of the powerhead (inline motors) or on the rear of the powerhead, between the cylinder head banks (V4 and V6 motors) which makes sense considering the job of the CDI unit is to power/trigger the ignition coils and the job of the ignition coils is to fire the spark plugs, so keeping everything in the same general area is a good idea. However, there are some notable exceptions.

One exception is all EFI or HPDI motors, since the CDI/TCI unit on these motors is incorporated into the ECM itself. Furthermore on many fuel injected motors the ECM is mounted under one or more covers at the rear of the powerhead. We've provided detailed procedures for accessing the ECM on these models, it can be found in the Electronic Fuel Injection portion of the Fuel System section. However, other than the differences for accessing and removing the ECM, the ignition coils are pretty much serviced in the same manner as the carbureted counterparts to those motors listed here. For more details on ignition coil service, continue with the procedures in this section.

Perhaps the most overlooked portions of ignition coil service has to do with the coil mounting and wiring (which are related on some motors). The ignition coil wiring is divided into 2 parts, the power/trigger primary circuit from the CDI (the small wires) and the secondary high-tension circuit (the large wires, the spark plug leads). Some coils have a ground wire in the primary circuit, while others are grounded to the powerhead (this is where the wiring and mounting are one and the same). On these models, corrosion can cause problems with the circuit as it prevents proper circuit grounding. For all motors, make sure the coil mounting points are clean and free of corrosion and make sure that ALL wiring contains insulation that is not broken, brittle or otherwise damaged.

REMOVAL & INSTALLATION

All 1-3 Cylinder Inline Motors

◆ See Figures 88 thru 97

As stated earlier, on most 1-3 cylinder inline motors, the ignition coils and CDI unit are mounted to the side of the powerhead. But even with that said, there are exceptions. On 6/8 hp motors the CDI unit is mounted to the top of the powerhead. On 9.9/15 hp motors, the ignition coils are mounted on top of the powerhead. However, keep in mind that if you have any trouble finding the components, the rules of thumb for locating the ignition coils and CDI unit by tracing the wiring back from the spark plugs still apply.

■ **The CDI unit and coil(s) may be removed independently of each other, but in most cases, labeling the wires before proceeding is critical to ensuring proper operation once the CDI unit and/or coil(s) are reinstalled.**

1. If equipped, disconnect the negative battery cable for safety.
2. To remove the CDI unit, proceed as follows:

a. Start by locating the CDI unit itself. If necessary, follow the wiring back from the ignition coil(s) to the unit. On some models the unit is mounted under a cover, if so remove the retaining screw(s) and remove the cover for access. If the CDI unit is mounted on top of the powerhead you may need to remove the flywheel cover or hand-rewind starter for better access.

b. On most 3-cylinder motors, including most models of the 28J-50 hp (698cc), of 50-70 hp (849cc) and 65J-90 hp (1140cc) engines, the timing linkage connects directly to the CDI unit itself. Carefully disconnect the linkage at a ball joint, but do not alter the length of any link rods in order to help preserve timing adjustment.

c. Tag and disconnect the CDI unit wiring. Take close note of how each wire is routed to make sure it can be returned to the original position. On some smaller powerheads this is not a big deal as there are only a couple of wires, but as you get to the larger 3- and 4-cylinder motors there may be as many as almost a dozen wiring connections so it is a good idea to take some pictures with a digital camera (or old Polaroid).

IGNITION AND ELECTRICAL SYSTEMS 4-31

- At least one wire connection normally goes to a ground strap that is bolted to the powerhead.

  d. For multi-cylinder motors, check for white plastic sleeves on the wires leading from the CDI unit to each ignition coil. These sleeves are numbered 1, 2, 3, etc. Disconnect each wire at the quick disconnect fitting. However, if there are not sleeves, then mark each wire to identify to which coil the lead must be reconnected. If these wires are not connected properly, the powerhead cylinders will not fire in the correct sequence. The powerhead may operate, but very poorly.

- On many models the ignition coil wires from the CDI unit are all the same color, however they may not be interchangeable. This means that there may be no other way to determine which lead is for which cylinder other than the sleeves or tracing the leads. If the leads are disconnected with no sleeves, and if the leads are not identified by markings or pieces of tape, then a trial and error method must be used to find the correct firing order for the powerhead. Save yourself the potential hassle now and make sure all wires are tagged.

  3. To remove an ignition coil, proceed as follows:
  a. On multi-cylinder motors, although the coils are usually arranged in a common-sense stack (indicating which coil fires which cylinder), don't assume anything, tag the coils and wires now.
  b. Twist off the spark plug lead(s) from the plug(s).
  c. Remove the bolts securing the ignition coil to the powerhead (there are usually 2 bolts securing a coil).
  d. Repeat for the remaining coil(s).

- One of the mounting bolts of some coils will also secure a Black grounding lead with an eye connector to the powerhead. This lead may come from the coil itself or may come from a connector plug in the lower cowling.

  To Install:
  4. If removed, install each ignition coil to the powerhead using the attaching bolts (usually 2). Make sure the grounding point (lead and/or coil mounting surfaces) are clean and free of corrosion. If applicable, make sure the ground lead is positioned when installing the coil. Install and securely tighten the coil mounting bolts, then reconnect the spark plug lead(s), as tagged during removal on multi-cylinder motors.
  5. If the CDI unit was removed, proceed as follows:
  a. Position the CDI unit to the powerhead and secure using the fasteners.
  b. Carefully route the wiring as noted and tagged during removal, reconnect the ignition coil, charge coil, pulser coil and ground leads, as applicable. Make sure any ring terminal and mounting point for a ground lead is clean and free of corrosion.
  c. If applicable, connect the CDI timing linkage.
  d. If equipped, install and secure the CDI unit cover.
  e. If removed for access, install the hand-rewind starter or flywheel cover.

Fig. 88 Trace the wiring and locate the CDI unit. . .

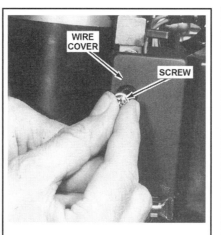

Fig. 89 . . .then, if equipped, remove the cover

Fig. 90 Tag and disconnect the wiring, including grounding bolts

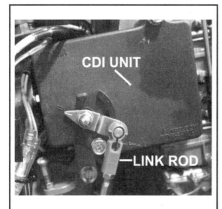

Fig. 91 If applicable, disconnect the link rod (28J-50 hp shown). . .

Fig. 92 . . .from the back of the CDI unit (65J-90 hp shown). . .

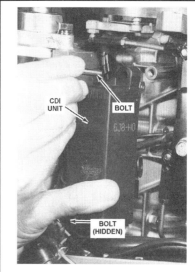

Fig. 93 . . .then remove the CDI unit

4-32 IGNITION AND ELECTRICAL SYSTEMS

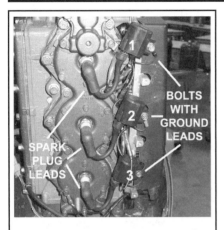

Fig. 94 Ignition coils are both easy to find and easy to remove

Fig. 95 Typical Yamaha single-lead ignition coil

Fig. 96 Typical Yamaha dual-lead ignition coils

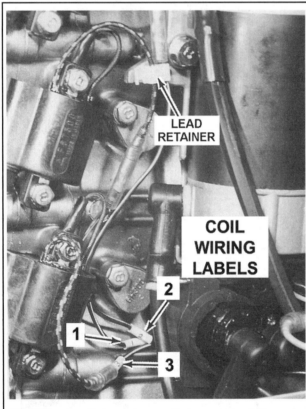

Fig. 97 Upon installation, make sure all wires are properly secured using any provided retainers and wire loops or wire ties, as applicable

V4 and V6 Powerheads

◆ See Figures 98 thru 103

On all V4 and V6 motors there is one single-lead coil per cylinder each mounted to the outside of the CDI unit bracket (most models) or ignition coil cover (HPDI motors). In all cases, the ignition coils are found in the valley between the cylinder heads.

As stated earlier, for all fuel injected motors the functions of the CDI are performed by the Electronic Control Module, so details on removal and installation of those modules are covered under Engine Control Module (ECM/CDI) and under HPDI Injector Driver in the Fuel System section. However, ignition coil service, once covers are removed for access to them, is virtually identical and you can use the appropriate portions of the procedures found here for ignition coils on EFI/HPDI motors.

■ The CDI unit and coils may be removed independently of each other, but labeling the wires before proceeding is critical to ensuring proper operation once the CDI unit and/or coils are reinstalled. However, the ignition coils on all V4 and V6 motors can easily be removed as an assembly by disconnecting their spark plug leads and unbolting the brackets on which they are mounted.

This procedure starts with CDI unit removal assuming that you wish to remove the unit as an assembly with the bracket and ignition coils. If you JUST wish to remove the CDI unit itself, ignore steps specifically related to bracket removal (such as disconnecting wiring from the bracket as opposed to the CDI unit). If however you only wish to remove one or more of the ignition coils, perform only those steps that directly apply to coil removal (coil wiring and coil mounting steps).

1. If equipped, disconnect the negative battery cable for safety.
2. Remove the CDI unit cover securing bolts and then remove the cover for access.
3. If you are removing the CDI unit bracket and ignition coils as an assembly, tag and disconnect the spark plug leads from either the plugs or the coils, whichever is desired.
4. Tag and disconnect the wiring from the CDI unit. If you are removing the entire bracket you'll have to also disconnect the wiring from the side of the bracket (there are usually one or more ground wires) and, on some models, the bullet connectors in the top of the bracket housing.

■ On some models, there is A LOT OF WIRING in and around this bracket. Take your time and label EVERYTHING. If you're just taking the whole bracket (with coils) off for access to something else, leave the CDI unit-to-coil wiring intact. As usual, pay close attention (and take pics if you can) to wire routing for installation purposes.

5. Either loosen and remove the CDI bracket mounting screws or remove the CDI unit screws, then remove the bracket/CDI unit assembly or just the CDI unit from the powerhead, as applicable.
6. To remove an ignition coil, tag and disconnect the wiring from the coil to the CDI unit (if not done already), then remove the mounting bolt(s), watching for ground leads as ring terminals may be secured by a coil mounting bolt on some models. Remove the ignition coil from the bracket.

To Install:

7. If removed, install each ignition coil to the bracket, routing the wiring as noted during removal. If there was a ground terminal connected to a mounting bolt, make sure that both the ring terminal and the mounting surface are clean and free of corrosion. Tighten the mounting bolt securely.
8. Install either the CDI unit to the bracket and/or the CDI unit/bracket assembly to the powerhead, then tighten the retaining bolts securely.
9. Reconnect the wiring and tagged during removal and route it to avoid damage. This isn't quite as critical as it is with stator wiring, but you'd be surprised at what vibration and a little chaffing can do to a wire. It is especially critical when dealing with electronic control modules as they can be damaged or destroyed if the insulation wears through and a wire shorts, so don't risk it.
10. Install the CDI or coil cover to the powerhead.
11. Reconnect the negative battery cable.

IGNITION AND ELECTRICAL SYSTEMS 4-33

Fig. 98 Loosen the CDI cover bolts...

Fig. 99 ...then remove the cover for access

Fig. 100 Tag and disconnect the wiring (there can be tons of it)

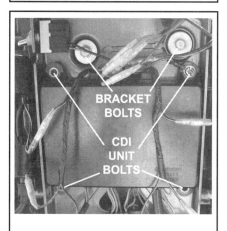
Fig. 101 Remove the CDI unit or CDI bracket bolts (as desired)

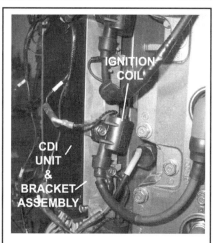
Fig. 102 On most models the ignition coils are mounted to the CDI bracket...

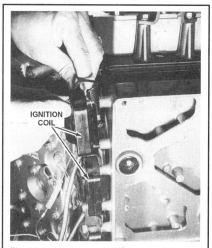
Fig. 103 ...except for HPDI motors, they can be removed without disturbing the bracket

YAMAHA MICROCOMPUTER IGNITION SYSTEMS (YMIS)

At the end of the day it is tough to put a single label on the Yamaha Microcomputer Ignition Systems. We could say that an outboard equipped with YMIS utilizes some form of electronic control over the spark timing curve of the CDI/TCI system, based on input from one or more sensors. And though that is a true statement, it also applies to some systems that Yamaha does not label as CDI Micro (such as the 80J-140 hp V4 motors, which utilize input from an oil level sensor and/or thermo switch for ignition control, but are labeled only as having a CDI system, not CDI Micro). We're a little puzzled, by this, but don't let that confuse you. If the system Wiring Diagram shows that a thermo switch or oil level sensor signal finds its way back to the CDI unit, then regardless of the system label, it may be taken into consideration when controlling spark timing and the relevant information here applies to that system as well.

Most 3-cylinder and V6 carbureted motors are equipped with a CDI Micro ignition. All V6 fuel injected motors are equipped with either a CDI Micro ignition (EFI OX66) or TCI Micro ignition (HPDI). The common denominator seems to be the utilization of a Crankshaft Position Sensor (CPS), though the carbureted motors typically also utilize an oil level switch and/or thermo switch. The EFI/HPDI motors utilize additional input from other fuel injection sensors including the WTS and, in some cases, a knock sensor.

The various forms of the YMIS allows the ignition control module to react to various inputs in order to improve engine performance.

A CPS sensor signal gives precise piston positioning information which is necessary for exact spark timing.

Thermo switches (carb) or the WTS (fuel injection) allow the engine to limit engine rpm and timing advancement both during engine warm-up and anytime there is an overheat warning signal. In both cases, this allows the ignition module to protect the powerhead or at least limit it from potential operational damage.

In a similar way to the temperature sensor/switches, the oil level sensors also allow the ignition module to protect the powerhead. In either case, when conditions occur that would trigger a warning signal (low oil level), the module can limit engine rpm and timing advancement, hopefully protecting the powerhead, while still allowing the operator to limp the boat/motor back to port.

Fuel injected motors that are equipped with a knock sensor allow the ignition module to protect against conditions that would create spark knock. The advanced timing which comes with performance applications can combine with poor fuel quality, extreme heat and excessive carbon build-up to create situations that would produce harmful spark knock. When the module detects such a spark knock, it can retard spark timing (or more precisely, limit the amount of advance) until the knock ceases.

YMIS SELF-DIAGNOSIS

◆ See Figures 104 thru 107

The control unit on the carbureted V6 powerheads and all EFI/HPDI systems includes a self-diagnosis system which constantly monitors circuits for shorted or open sensors. Since the system on EFI/HPDI motors is integrated into the fuel injection system, details can be found in the Fuel System section. The information on the self-diagnosis system contained in this section applies to the V6 carbureted motors.

4-34 IGNITION AND ELECTRICAL SYSTEMS

When a fault occurs, the control unit will set a trouble code which can be accessed using a special diagnostic flash harness available from Yamaha (#YB-06765). Interestingly enough, this is the same test harness as used on most fuel injected motors as well.

1. Install the flash harness inline between the oil level sensor and the control unit. In addition, connect the additional lead (usually Blue) to activate the diagnosis readout.
2. Start the engine and allow it to reach operating temperature.
3. With the engine running at idle, read the LED on the flash harness and interpret codes by counting the flashes. Only one code will be displayed at a time. The first digit of the 2-digit code will display first, with one, two, three or possibly even four flashes (the light will remain on and off for the same amount of time, about 0.3 seconds, when flashing a code digit). Count the number of times on and you've got the first digit.

■ These engines use a single digit code 1 to indicate normal operation. When present the single 0.3 second flash will be separated from the next single flash by a long 5 second pause.

4. There will be a longer pause between digits of the 2-digit code (the light will remain off for about 1.65 seconds), then the short on/off flashes will indicate the second digit of the 2-digit code. This second flash may consist of any number from one to five on these motors. A single code will continue to flash until the problem is corrected and then the next higher code (if present) will flash until corrected. This pattern will repeat until all faults are corrected.
5. Compare the codes with the accompanying code table to determine the defective circuit. Remember that a code does NOT necessarily mean a given component is at fault, it means that the ECM sees signal that is out of the normal, predetermined operating range. The problem may also lie with the wiring, another system or another component that would make a circuit read out of specification.

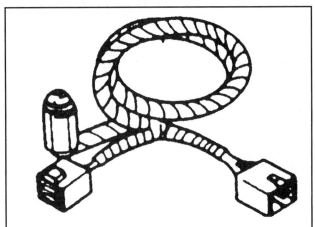

Fig. 104 The diagnostic flash harnesses consists of 3 connectors, an LED light and a processor

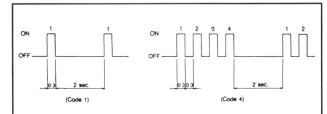

Fig. 106 Flash patterns can be read as follows to determine the appropriate fault code

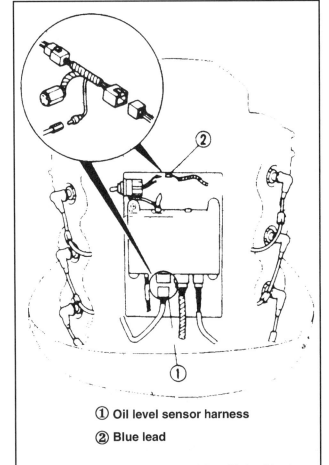

① Oil level sensor harness
② Blue lead

Fig. 105 The flash harness is connected inline with the oil level sensor harness. Another lead is connected to the lead (usually Blue) which activates diagnosis mode

Diagnosis code chart

Code	Symptoms
12	Incorrect charge coil input signal
13	Incorrect pulser coil input signal
14	No crank position sensor input signal
15	Incorrect engine cooling water temperature sensor input signal
33 ~ 44	Microcomputer processing information
(33)	Ignition timing is being slightly corrected (when starting a cold engine)
(41)	Overrevolution control (during ignition cutoff operation)
(42)	Overheat control/oil empty control
(43)	Buzzer sounding
(44)	Engine stop switch control operating

Fig. 107 Self-diagnostic trouble codes - 1997 and later 2.6L carbureted motors

IGNITION AND ELECTRICAL SYSTEMS   4-35

■ **Make sure all connections for that circuit and related components of the electrical system are clean and tight before troubleshooting the circuit. Bad wiring or connections can cause out of range signals and set trouble codes.**

6. Test the sensor or system components as described in this section. When troubleshooting, always start with the easiest checks/fixes and work toward the more complicated.

■ **Codes are only present during the actual fault with the engine running and are not stored in memory.**

7. Once all of the problems have been corrected, connect a source of cooling water to the flushing system, then start and run the engine to verify proper operation. If the problems are gone, the diagnostic indicator either will remain off or will flash a code 1 (depending upon the model).
8. Shut the ignition switch **OFF** and remove the diagnostic connector.

TROUBLESHOOTING

■ **For carbureted V6 engines, before proceeding with system troubleshooting, attempt to narrow the problem down using the YMIS Self-Diagnosis feature of the CDI unit. But remember, a code means there is a problem with a sensor circuit output, not necessarily the sensor itself.**

When working on these systems, keep in mind that most engine problems are NOT caused by the computer. If ignition timing changes are noticed at the same time as an engine variation, the computer is probably just reacting to a carburetion/fuel injection or basic ignition system problem, not a computer problem. Always check the fuel system fully before troubleshooting the computer.

■ **On EFI/HPDI engines the ignition and fuel systems are integrated, before proceeding, refer to the section on Electronic Fuel Injection for tips on troubleshooting the entire system, including the use of the EFI/HPDI Self-Diagnostic system.**

An intermittent problem can cause severe changes in ignition timing. For most systems, when the computer is bypassed, the ignition timing will be retarded and engine performance will be reduced. This will result in reduced fuel flow and any existing carburetion problem may not occur at this reduced fuel flow.

The first objective of troubleshooting the YMIS system is to determine if the problem is ignition timing related. A weak spark or no spark is usually caused by the basic ignition components (stator, pulser, ignition coil, CDI, or wiring), not the computer system.

The Yamaha high voltage spark has different characteristics than the spark produced by some other manufacturer's engines. When looking for a spark, especially when in direct sunlight, the Yamaha spark may be hardly visible. If using an aftermarket spark checker, it is possible for electrical radio interference from this checker to interfere with proper computer operation and cause an erratic spark. If you are unsure of your test results, repeat the test in the shade using a resistor spark plug.

If the computer system is suspected, experience has shown that the problem is usually not the computer unit itself. Most computer problems are caused by either a bad sensor, poor connections or grounds or basic engine problems.

Nothing can be more frustrating than an intermittent failure of a component. The resistance tests provided under certain components are procedures to test parts in a laboratory situation mostly at room temperature in ideal working conditions. Unfortunately, this does not provide 100% reliable test results. The sensor, switch, wire harness, and the connector, are all subjected to vibration and temperature extremes during actual operation of the powerhead. If the component could be tested in these operating conditions it might register readings which are borderline with the specifications. If the heat or the vibration becomes excessive, the part may temporarily fail. Perhaps even if tested as outlined in the various sections of this guide, the conditions may not be severe enough to induce the part to fail. The best advice is to attempt to duplicate the conditions under which the part failed before troubleshooting can be successful.

Other than the CDI output test, no test exists, either in resistance values or operational performance for the microcomputer. If all the sensor signals are processed correctly, all the sensor signals are processed correctly, all engine control circuits function, and the timing is correct then the microcomputer is alive and well.
If the following tests and conditions have been performed or met:
All tests performed on the sensors.
All test performed on the switches.
Operational checks on the system have been conducted and are satisfactory.
All connectors have been checked and rechecked.
The battery is ample in size and power.
Then it is time to consider the computer as a possible fault.

Additional Troubleshooting Tips for Carbureted Motors

50-70 Hp (849cc) and 75/80/90 Hp (1140cc) Models

When equipped with the YMIS these motors are unique in that there is only one pulser that normally controls cylinder #1 & #3. The ignition timing for cylinder #2 is entirely controlled by the computer. This normally means that when any problem occurs and the computer automatically goes into the bypass mode, cylinder #2 will not fire and timing will be fixed.
This system is also the only YMIS system uses a throttle sensor incorporated into the main control unit (activated by linkage which connects to the CDI unit itself). All electrical power is supplied by the charge coils and there are no connections to battery power.
Check for spark to #2. If there is no spark, check the yellow lead and crank position sensor. If there is spark, disconnect the linkage to the throttle sensor and check for timing changes while moving the throttle sensor.

105J-225 HP (2596cc)

This is a later version of the system originally used by Yamaha on the V6 carburetor line. The system uses a movable timer base, a crank position sensor, and only two pulsers to control all six cylinders. There is no throttle sensor or knock sensor, as there were on earlier models (and as there are on fuel injected models). Basic ignition timing is controlled by the movable timer base, but the computer will adjust timing for some specific conditions; Like engine start and warm up. The computer includes a diagnosis system. Troubleshooting should start with the Self-Diagnosis system, but don't get pigeonholed into looking for only electronic causes to poor operation.

Crankshaft Position Sensor (CPS)

DESCRIPTION & OPERATION

◆ See Figure 108

Yamaha carbureted motors equipped with CDI Micro ignitions, as well as EFI OX66 and HPDI motors are equipped with a Crankshaft Position Sensor (CPS) in order to give flywheel rotational (and thereby crankshaft/piston positioning) information to the CDI unit/ECM.
The CPS is mounted on the cylinder block and is positioned a critical (specified) distance from the outer rim of the flywheel. The inside face of the sensor has an electronic pickup device which is sensitive to changes in magnetic flux. The sensor generates a voltage which is sent to the CDI Unit or ECM, but the signal fluctuates in response to the flywheel teeth which are passing through the sensors magnetic field. Essentially the sensor electronically counts the number of flywheel teeth passing the pickup in a specified time and is therefore able to calculate engine rpm. The sensor is also able to detect crank phase angle by counting the number of teeth which pass the pickup from a specified point on the flywheel where the teeth differ magnetically.

OPERATIONAL CHECK

◆ See Figure 108

If the crankshaft position sensor is functioning properly the engine will remain in proper timing for varied rpm conditions. Once the powerhead is warmed to normal operating temperature, check the timing against the values provided in the Tune-Up Specifications chart found in the section for Maintenance and Tune-Up.

4-36 IGNITION AND ELECTRICAL SYSTEMS

Fig. 108 The crankshaft position sensor is located on top of the powerhead, right next to the flywheel

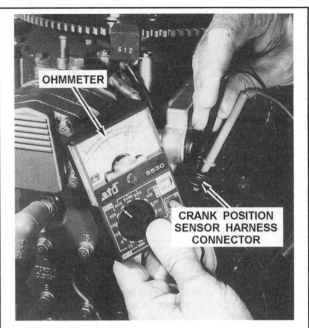

Fig. 109 Crankshaft position sensor resistance is checked at the sensor connector using an ohmmeter

If there is a problem with the sensor/circuit, a trouble-code will set on motors which incorporate a self-diagnostic system. However, for most carbureted motors (which do not have a self-diagnostic system) the symptoms of a defective crank position sensor are usually engine rpm surges and erratic timing shifts.

On all models, Yamaha provides a cranking (in some cases with load, meaning the circuit complete and in other cases without load meaning open circuit) and running sensor output voltage. For circuit complete tests you'll need to back-probe the sensor connector using the voltmeter or disconnect the harness and use jumper wires between the terminal halves for probing access while the circuit remains connected. For test values, please refer to the Crankshaft Position Sensor Specifications chart in this section. You'll notice that resistance specifications for the sensor coil windings are also provided for most models, although keep in mind that dynamic (cranking/running) output tests are almost always preferred to static (resistance) checks when it comes to accuracy.

■ It is really best to use jumper wires, as backprobing a sensor connector can lead to problems with corrosion or broken wire strands down the road.

To replace the sensor, simply remove the two securing screws and remove the sensor. Upon installation the new sensor must be properly adjusted (set with the appropriate gap between the sensor and flywheel), that is unless the sensor is designed for mounting in a fixed, non-adjustable position.

CHECKING RESISTANCE

◆ See Figure 109  MODERATE

The Crankshaft Position Sensor Specifications chart in this section gives resistance specifications for all sensors EXCEPT those used on HPDI motors and those found on the 150 hp EFI OX66 motor. What is strange about the later, is that it appears from other specifications to be the SAME sensor used on the 200 hp EFI OX66 motor (also produced on the same 2.6L powerhead). Unfortunately would could not verify this suspicion, but we'd venture a guess that if the sensor on the 150 checks within the specification given for the 200, it's probably good. Either way, a dynamic cranking/running output test is really a better method of sensor testing anyway.

Remember that resistance tests will yield slightly different results depending upon ambient/component temperature and the brand/type of meter being used. Furthermore, a component may test fine statically, but still exhibit intermittent failures under load. For this reason, though a resistance check may quickly condemn a component, don't take a pass on the resistance check as a guaranteed bill of health for the sensor.

1. Disconnect the leads from the sensor at the harness connector.
2. Using an ohmmeter measure the resistance across the two leads. On three wire sensors, refer to the Crankshaft Position Sensor Specifications chart in this section for the appropriate wire colors. For all models, refer to the chart to confirm the appropriate resistance specs.
3. On most models resistance should be 158-236 ohms at 68°F (20°C). However the sensor used on a few fuel injected motors will vary slightly, so check the chart. If the reading is well out of specification, the sensor is faulty and should be replaced. If the reading is borderline or the sensor is still suspect, perform a dynamic cranking/running voltage output check.
4. To replace the sensor, simply remove the two securing screws and remove the sensor. Install the new sensor and adjust, as applicable.

GAP ADJUSTMENT & SERVICE

◆ See Figures 108 and 110  MODERATE

Although all carbureted motors equipped with CDI Micro ignition, as well as all fuel injected (EFI OX66 and HPDI) motors utilize a CPS, not all of the Yamaha service literature mentions adjusting the sensor. This means that the either the gap is fixed on some models and not adjustable, or this is just an oversight in the Yamaha service literature.

On adjustable sensor, check the sensor gap by trying to insert a 0.02-0.06 in. (0.5-1.5mm) feeler gauge between the sensor and the flywheel. When measuring a gap with a feeler gauge, remember that the gap is equal to the size gauge that will fit through the gap with a slight drag. The next larger size gauge should not fit, while the next smaller gauge should fit without touching or dragging. If adjustment is necessary, loosen the 2 CPS mounting screws and gently slide the sensor against a 0.02-0.06 in. (0.5-1.5mm) feeler gauge inserted between the flywheel and sensor. Tighten the mounting screws and double-check the gap using the gauge set.

Thermo-sensor & Thermo-switch

Ok, let's try to explain this best we can. Most Yamaha outboards are equipped with either a thermo-switch (and on/off switch which is activated by temperature) or a thermo-sensor (a variable resistor whose values change with changes in temperatures). The breakdown goes as follows:

• The 20 hp and larger (except the 430cc 20/25 hp) motors utilize one thermo-switch per cylinder head for the over-heat warning system. When equipped with YMIS the CDI unit may also receive the switch signal to be used in limiting engine speed to protect the powerhead. In addition to the thermo-switches, the carbureted V6 engine also uses a thermo-sensor which provides data to the CDI unit for YMIS operation.

IGNITION AND ELECTRICAL SYSTEMS 4-37

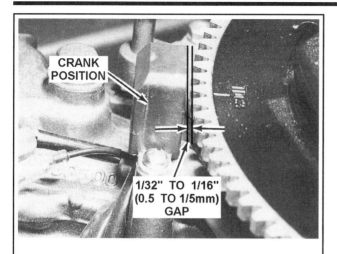

Fig. 110 Measure the gap between the CPS inner face and outer edge of the flywheel teeth

- All V4 and V6 fuel injected motors are equipped with both a thermo-sensor (known as a WTS) for the fuel and ignition system as well as a pair of thermo-switches (one at the top of each cylinder head) for use with the warning system.

This section deals with the thermo-sensors on CARBURETED motors only, as the WTS used on fuel injected motors is covered in the Fuel System section. Furthermore, since the thermo-switch is really more a component of the cooling and warning systems, they are covered (for all motors, carbureted and fuel injected) in the Lubrication and Cooling section.

DESCRIPTION & OPERATION

◆ See Figures 111, 112 and 113

On certain carbureted engines (identified earlier) a thermo-sensor is located somewhere on the powerhead for use with either the overheat warning system and/or the YMIS. This sensor is a thermistor, which is a variable resistor. A thermistor's resistance changes with temperature. There are generally 2 types of thermistors; a negative temperature coefficient thermistor and a positive temperature coefficient thermistor. The difference comes in how each type of thermistor responds to temperature changes. The resistance of a negative temperature coefficient sensor will change in the opposite direction of temperature changes (i.e. if the temperature goes UP, the resistance goes DOWN and vice versa). The resistance of a positive temperature coefficient sensor will change WITH the direction of temperature changes (meaning that as temperature goes UP, resistance goes UP and vice versa). Most temperature sensors used by Yamaha are of the negative coefficient design.

This information can be useful to the YMIS system for one of two reasons. For starters, it can alert the control module as to whether or not the engine has reached normal operating temperatures. A YMIS module may have the programming to limit rpm and spark advance until the motor reaches normal temperatures. Perhaps more importantly, the sensor can also be used to limit rpm and spark advance if an overheat signal is received, helping to slow and hopefully allow the powerhead to cool (preventing damage).

There is a distinct difference between a thermo-sensor (covered here) and a thermo-switch (covered in the Lubrication and Cooling section). A thermo-sensor was identified as a thermistor - an electronic device capable of detecting changes in temperature and transmitting these changes as voltage signals to the microcomputer. However, a thermo-switch is simply a bimetal type switch which has 2 settings ON and OFF and nothing in between. It cannot tell a control module at what temperature the engine is operating, it can only say whether or not it has reached a specified point which is considered overheating.

To locate the thermo-sensor on carbureted models so equipped, proceed as follows:

- On carbureted, V6 motors, the sensor is found in the starboard cylinder head bank, about half-way down the cylinder head (almost adjacent to the middle ignition coil).

OPERATIONAL THERMOSENSOR CHECK

◆ See Figures 111 thru 114

1. Mount the engine in a test tank, or on a boat in a body of water.
2. Connect a timing light and tachometer to the powerhead.
3. Start the engine and allow it to warm to operating temperature.
4. Aim the timing light at the timing pointer and ensure the initial timing is at specification.
5. Disconnect the thermo-sensor wire harness at the quick disconnect fitting and use a small jumper cable and connect the two leads of the harness (motor side of the harness not sensor side) together. The timing should now retard and the engine rpm decrease.
6. Remove the thermo-sensor from the motor. Reconnect the leads to the sensor and hold the sensing end of the sensor against an ice cube. The timing should now advance and the engine rpm increase.
7. To replace the sensor, simply unscrew or unbolt it from the powerhead.

CHECKING THERMO-SENSOR RESISTANCE

◆ See Figures 111, 112 and 115

■ Resistance tests can be performed while the sensor is installed in the powerhead, but you won't have the best control over the temperatures so for most accurate testing, remove the sensor from the block.

1. Disconnect the wire harness connector, at the quick disconnect fitting and remove the sensor from the powerhead.

Fig. 111 The thermo-sensor is usually threaded. . .

Fig. 112 . . .or bolted. . .into position on the powerhead

Fig. 113 In contrast, a Yamaha thermo-switch is usually press fit into a bore at the top of a cylinder head

4-38 IGNITION AND ELECTRICAL SYSTEMS

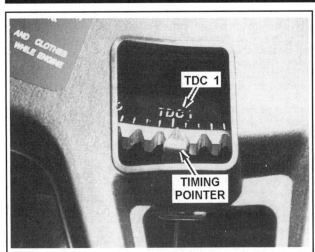

Fig. 114 Mark the timing locations on the flywheel prior to starting the thermo-sensor test

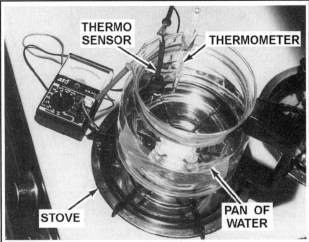

Fig. 115 Thermo-sensor resistance should decrease smoothly as the temperature rises and increase smoothly as it falls

2. The resistance of the sensor is monitored at different temperatures while heating the sensing end of the sensor in a body of water.

3. Place the container of water, at room temperature, on a stove. Secure the thermometer in the water in such a manner to prevent the bulb from contacting the sides or bottom of the pan (so you'll be assured to get the water temperature and not the temperature of the container).

4. Immerse the sensing ends of the sensor into the water up to the shoulder (suspend the sensor so it too is not touching the side or bottom of the container so it will become the same temperature as the water and thermometer).

5. Check resistance across the sensor terminals and compare the readings for the appropriate water temperature. The readings should be 128,000 ohms at 41°F (5°C), 54,000-69,000 ohms at 68°F (20°C) and 3,020-3,480 ohms at 212°F (100°C).

6. Resistance should rise and fall evenly, but in the opposite direction of the temperature. Remember, the thermo-sensor is a thermistor which characteristically varies resistance (increases or decreases voltage in a circuit) with a change in temperature. Slowly heat or chill the water and watch to make sure the sensor shows values within specification over the anticipated operating range.

7. If resistance is not within specification or does not fluctuate evenly, the sensor is faulty.

8. To replace the sensor, simply unscrew or unbolt it from the powerhead (unless of course, you already did to test the sensor).

CHARGING CIRCUIT

◆ See Figure 116

For many years, single-phase, full-wave charging systems were the dominant design. Most manufacturers used single-phase systems because they were simple and reliable. The drawback to these systems is low output. A typical single-phase lighting coil system is capable of only 10 to 15 amps. On many larger rigs, the electrical demand is more than 15 amps. New electronic and electrical devices are arriving on the market every day, so the demand for higher amperage output continues to rise.

In response to higher electrical demands, three-phase systems were introduced on the larger Yamaha outboards in 1990 and since they've found their way onto many of the smaller Yamaha motors. These systems produce 25 to 45 amps. At the heart of a three-phase charging system is an interconnected three coil winding. By using three coils instead of one (or two), output is more than doubled. The three-phase rectifier/regulator works in a similar way to the single-phase system. The difference is the addition of more diodes.

■ On many Yamahas the lighting coils are part of a 1-piece stator unit which is also an integral part of the ignition system. Even on motors that utilize separate charge and lighting coils, access to the coils themselves and testing procedures are similar enough that all are covered in the Ignition System section.

Yamaha Charging Systems

DESCRIPTION & OPERATION

The single-phase charging system found on inline motors provides basic battery maintenance. Single-phase, full wave systems like these are found on a variety of products. Many outboard engines, water vehicles, motorcycles, golf cars and snowmobiles use similar systems.

This charging system produces electricity by moving a magnet past a fixed coil. Alternating current is produced by this method. Since a battery cannot be charged by AC (alternating current), the AC current produced by the lighting coil is rectified or changed into DC (direct current) to charge the battery.

To control the charging rate an additional device called a regulator is used. When the battery voltage reaches approximately 14.6 volts the regulator sends the excess current to ground. This prevents the battery from overcharging and boiling away the electrolyte.

The charging system consists of the following components:
- A flywheel containing magnets
- The lighting coil or stator coil
- The battery, fuse assembly and wiring
- A regulator/rectifier

All of the V4 and V6 motors are equipped with a three-phase charging system. The charging system operates in essentially the same manner as the single-phase system. The differences in the system come in output and testing, as it uses 3 coils wired to each other in a Y configuration (and 3 stator coil wires for the charging system circuit all of which must be tested for shorts to ground and for proper output).

■ Some 3-phase systems may also have a battery isolator integrated into the charging system circuit for the use of multiple batteries.

Servicing charging systems is not difficult if you follow a few basic rules. Always start by verifying the problem. If the complaint is that the battery will not stay charged do not automatically assume that the charging system is at fault. Something as simple as an accessory that draws current with the key off will convince anyone they have a bad charging system. Another culprit is the battery. Remember to clean and service your battery regularly. Battery abuse is the number one charging system problem.

The regulator/rectifier is the brains of the charging system. This assembly controls current flow in the charging system. If battery voltage is below 14.6 volts the regulator sends the available current from the rectifier to the battery. If the battery is fully charged (is at or above 14.6 volts) the regulator diverts most of the current from the rectifier back to the lighting coil through ground.

IGNITION AND ELECTRICAL SYSTEMS

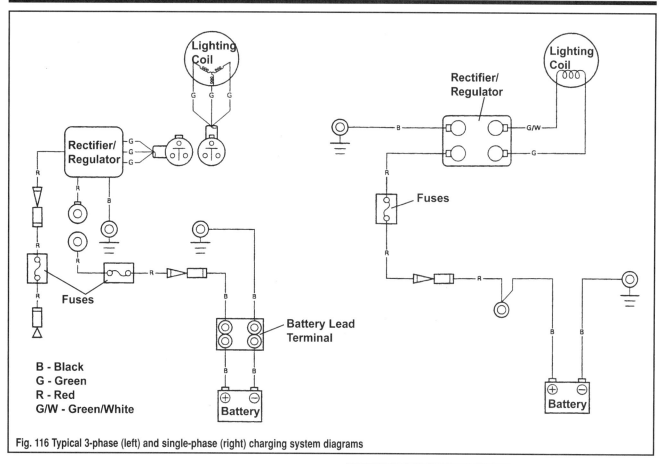

Fig. 116 Typical 3-phase (left) and single-phase (right) charging system diagrams

Do not expect the regulator/rectifier to send current to a fully charged battery. You may find that you must pull down the battery voltage below 12.5 volts to test charging system output. Running the power trim and tilt will reduce the battery voltage. Even a pair of 12 volt sealed beam lamps hooked to the battery will reduce the battery voltage quickly.

In the charging system the regulator/rectifier is both the hardest and easiest item to troubleshoot. This is normally accomplished by checking around it. Check the battery and charge or replace it as needed. Check the AC voltage output of the lighting coil. If AC voltage is low check the charge coil for proper resistance and insulation to ground and repair/replace, as necessary. If these check OK measure the resistance of the Black wire from the rectifier/regulator to ground and for proper voltage output on the Red lead coming from the rectifier/regulator going to the battery. If all the above check within specification replace the rectifier/regulator and verify the repair by performing a charge rate test. This same check around method is used on other components like the CDI unit or ECM.

TROUBLESHOOTING

◆ See Figure 117

There are several conditions you can find when troubleshooting a charging system:
- **Normal** - In a normal system the charge rate is correct. There could be a problem elsewhere, such as an electrical short, too many accessories or a bad battery. Whatever the problem, there is no need to work on the charging system. It is working correctly!
- **Overcharging** - The charging system continues to supply current to the battery even after the battery is fully charged. Usually this is caused by a bad regulator portion of the rectifier. However, in some cases it could also potentially be due to the wiring the regulator uses to determine battery voltage.
- **Undercharging** - The charging system is not producing current at all or is producing it at a reduced rate. In this circumstance the question is whether or not the lighting coils are producing what they should and, if they ARE, the fault is in the wiring or the regulator.

Diagnosis code chart

Code	Symptoms
12	Incorrect charge coil input signal
13	Incorrect pulser coil input signal
14	No crank position sensor input signal
15	Incorrect engine cooling water temperature sensor input signal
33 ~ 44	Microcomputer processing information
(33)	Ignition timing is being slightly corrected (when starting a cold engine)
(41)	Overrevolution control (during ignition cutoff operation)
(42)	Overheat control/oil empty control
(43)	Buzzer sounding
(44)	Engine stop switch control operating

Fig. 117 Charging system diagnosis

4-40 IGNITION AND ELECTRICAL SYSTEMS

- Battery drain during non-use - Isolate the battery draw to the motor or to the boat accessories by removing loads one at a time. Although it is rare, one of the diodes in the rectifier can short and slowly drain the battery.

Keep the above possibilities in mind when troubleshooting. By simplifying your troubleshooting approach you may get to the solution more quickly.

The first step in testing a charging system is to confirm there is a charging problem.

■ **The absolute number 1 charging system problem is a faulty battery!**

The real problem may be too many accessories or a draw with the key off. The charging system may really be OK, but if the draw on the system exceeds the charging output, the charging system will appear faulty.

There are two methods for checking three-phase charging systems, the DC amp check and the DC voltage check. The DC amp check method is usually preferred, but both methods are useful.

DC Amperage Check

1. Eliminate the battery as the problem by using a known good one. Now check for any amperage drain on the battery with an ammeter. Connect the ammeter between the positive battery cable and the positive post on the battery. Take your first reading with the key switch **OFF**. Your reading should be at or near zero (does your boat have a clock or radio with a memory?). Then take a reading with the key switch **ON**. This reading will vary according to how many accessories are connected to key-on power. Record the amperage reading.
2. Determine actual DC amperage output from the alternator. To determine alternator output, subtract the key-on/engine-off amperage draw from the maximum alternator output. Start this process by disconnecting the isolator lead from the accessory battery and protect it from grounding out. Next, connect a DC ammeter (0-40 amps) in series with the rectifier/regulator output lead at the starter solenoid. (Motor should be in the water for this test to provide adequate cool water supply).
3. Now, start and rev the motor to 3000 rpm. Observe the alternator output.

■ **When making the DC amperage output check, do not use a fully charged battery! A fully charged battery will not receive full system output. It may only show half the expected amperage output. Reduce the battery state of charge by operating the trim/tilt motor through several cycles. This reduces the charge on the battery and permits higher alternator output.**

DC Voltage Check

1. The DC voltage check begins just like the DC amperage check. Begin the DC voltage check by establishing the condition of the battery and by checking for any key-off amperage draw.
2. Verify that you have 12 volts through the fuse and Red wire. Reading voltage through the fuse establishes continuity and fuse condition.
3. Check the system carefully for good grounds at all the ground points for the Black wires. Remember that there are several splices within the harness that can corrode and create high resistance.
4. At this point move to the lighting coil wires and disengage the connector(s). Crank the engine and check the AC voltage output among the wires (across different combinations of wires when dealing with a 3-phase system). The voltage should be above battery voltage at idle and rise with engine rpm. Check each leg to ground. The AC reading should be nearly the same from each leg to ground. If not, shut off the engine, switch the meter to ohms and look for a coil leg that is open or shorted to ground.

■ **Refer to the Charging System Testing specifications charts in this section for details on lighting coil voltage output under various conditions.**

5. The rectifier/regulator is purposely left last. In reality, these checks aren't usually necessary. By the process of elimination you can decide if the rectifier is bad. If the stator, battery and fuse are all good, and if the Red and Black wires are OK, the only thing left is the rectifier/regulator assembly. Since you can't disassemble the unit to fix it, replace it.

Charging System Checks

Excessive charging

There is really only one cause for this type of failure, the regulator is not working (is not controlling voltage supplied to the battery), either because it is not getting a proper battery voltage signal or because it is broken. If there are no wiring problems, since there is no repair of regulator/rectifier, replace it.

Undercharging

If there is an undercharge condition after running the DC amperage check at the fuse assembly, then disconnect the stator coupling from the harness and perform AC voltage checks between the stator leads. Check between two stator leads at a time on 3-phase systems, but be sure to check across all possible combinations (3 separate checks). On 3-phase systems all three readings should be equal, within a volt or two.

■ **Refer to the Charging System Testing specifications charts in this section for details on lighting coil voltage output under various conditions.**

Stator/lighting coil shorts to ground can be checked by doing a voltage test between one stator lead and ground, engine running. There should be roughly half the normal stator voltage check reading.

If the readings are all within specification, the stator is working correctly. Proceed to the Red wire and Black wire checks.

If any or all readings are below normal, turn the engine **OFF** and check the stator windings using an ohmmeter. An isolated continuity check and a short to ground check should be done. If the stator is bad, replace it since it can't be repaired.

Battery

The battery is one of the most important (and vulnerable) components of the electrical system. In addition to providing electrical power to start the engine, it also provides power for operation of the running lights, radio, and electrical accessories.

Because of its job and the potential consequences of a failure (especially in an emergency), the best advice is to purchase a well-known brand, with an sufficient warranty period, from a reputable dealer.

The usual warranty covers a pro-rated replacement policy, which means the purchaser would be entitled to a consideration for the time left on the warranty period if the battery should prove defective before its time.

Do not consider a battery of less than the rating provided in the General Engine System Specifications chart from the Maintenance and Tune-Up section. For most Yamaha motors (except the smallest and largest, including all fuel injected motors) this means a battery of 70- amp/hour or 100-minute reserve capacity. But it never hurts to buy a battery of larger capacity. Especially as the boat and motor's electrical system ages, cables crack internally and resistance increases.

MARINE BATTERIES

◆ See Figure 118

Because marine batteries are required to perform under much more rigorous conditions than automotive batteries, they are constructed differently than those used in automobiles or trucks. Therefore, a marine battery should always be the No. 1 unit for the boat and other types of batteries used only in an emergency.

Marine batteries have a much heavier exterior case to withstand the violent pounding and shocks imposed on it as the boat moves through rough water and in extremely tight turns. The plates are thicker and each plate is securely anchored within the battery case to ensure extended life. The caps are spill proof to prevent acid from spilling into the bilges when the boat heels to one side in a tight turn, or is moving through rough water. Because of these features, the marine battery will recover from a low charge condition and give satisfactory service over a much longer period of time than any type intended for automotive use.

IGNITION AND ELECTRICAL SYSTEMS

Fig. 118 A fully charged battery, filled to the proper level with electrolyte, is the heart of the ignition and electrical systems. Engine cranking and efficient performance of electrical items depend on a full rated battery

✷✷ WARNING

Avoid the use an automotive type batteries with an outboard unit. It is not built for the pounding or deep cycling to which this use will subject it and it may be quickly damaged.

BATTERY CONSTRUCTION

◆ See Figure 119

A battery consists of a number of positive and negative plates immersed in a solution of diluted sulfuric acid. The plates contain dissimilar active materials and are kept apart by separators. The plates are grouped into elements. Plate straps on top of each element connect all of the positive plates and all of the negative plates into groups.

The battery is divided into cells holding a number of the elements apart from the others. The entire arrangement is contained within a hard plastic case. The top is a one-piece cover and contains the filler caps for each cell. The terminal posts protrude through the top where the battery connections for the boat are made. Each of the cells is connected to its neighbor in a positive-to-negative manner with a heavy strap called the cell connector.

BATTERY RATINGS

◆ See Figure 120

Three different methods are used to measure and indicate battery electrical capacity:
- Amp/hour rating
- Cold cranking performance
- Reserve capacity

The amp/hour rating of a battery refers to the battery's ability to provide a set amount of amps for a given amount of time under test conditions at a constant temperature. Therefore, if the battery is capable of supplying 4 amps of current for 20 consecutive hours, the battery is rated as an 80 amp/hour battery. The amp/hour rating is useful for some service operations, such as slow charging or battery testing.

Cold cranking performance is measured by cooling a fully charged battery to 0°F (-17°C) and then testing it for 30 seconds to determine the maximum current flow. In this manner the cold cranking amp rating is the number of amps available to be drawn from the battery before the voltage drops below 7.2 volts.

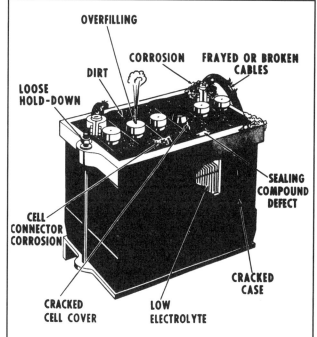

Fig. 119 A visual inspection of the battery should be made each time the boat is used. Such a quick check may reveal a potential problem in its early stages. A dead battery in a busy waterway or far from assistance could have serious consequences

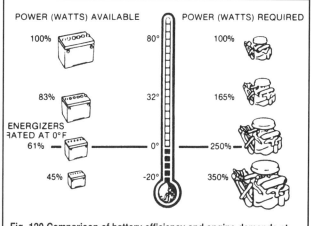

Fig. 120 Comparison of battery efficiency and engine demands at various temperatures

The illustration depicts the amount of power in watts available from a battery at different temperatures and the amount of power in watts required of the engine at the same temperature. It becomes quite obvious - the colder the climate, the more necessary for the battery to be fully charged.

Reserve capacity of a battery is considered the length of time, in minutes, at 80°F (27°C), a 25 amp current can be maintained before the voltage drops below 10.5 volts. This test is intended to provide an approximation of how long the engine, including electrical accessories, could operate satisfactorily if the stator assembly or lighting coil did not produce sufficient current. A typical rating is 100 minutes.

If possible, the new battery should have a power rating equal to or higher than the unit it is replacing.

BATTERY LOCATION

Every battery installed in a boat must be secured in a well protected, ventilated area. If the battery area lacks adequate ventilation, hydrogen gas, which is given off during charging, is very explosive. This is especially true if the gas is concentrated and confined.

Remember, the boat is going to take a pounding at some point and it is critical that the battery is secured so that wire terminals are not subjected to stresses which could break the cables or battery (and could provide for other dangerous conditions, including the potential for shorts, sparks and even an explosion of that hydrogen gas we just mentioned). There are many types of battery tray mounting assemblies available and you will normally find the perfect one for your application on the shelves of boat supply stores. Make sure the tie down bracket or strap is snug, but not over-tightened to the point of distorting or cracking the case.

BATTERY AND CHARGING SAFETY PRECAUTIONS

◆ See Figure 121

Always follow these safety precautions when charging or handling a battery:
• Wear eye protection when working around batteries. Batteries contain corrosive acid and produce explosive gas as a byproduct of their operation. Acid on the skin should be neutralized with a solution of baking soda and water made into a paste. In case acid contacts the eyes, flush with clear water and seek medical attention immediately.
• Avoid flame or sparks that could ignite the hydrogen gas produced by the battery and cause an explosion. Connection and disconnection of cables to battery terminals is one of the most common potential causes of sparks. Use extreme caution with jumper cables and make sure all switches are **OFF** before making any other connections.
• Always turn a battery charger **OFF**, before connecting or disconnecting the leads. When connecting the leads, connect the positive lead first, then the negative lead, to avoid sparks.
• When lifting a battery, use a battery carrier or lift at opposite corners of the base.
• Ensure there is good ventilation in a room where the battery is being charged.
• Do not attempt to charge or load-test a maintenance-free battery when the charge indicator dot is indicating insufficient electrolyte.
• Disconnect the negative battery cable if the battery is to remain in the boat during the charging process.
• Be sure the ignition switch is **OFF** before connecting or turning the charger **ON**. Sudden power surges can destroy electronic components.
• Use proper adapters to connect charger leads to batteries with non-conventional terminals.

BATTERY CHARGERS

Before using any battery charger, consult the manufacturer's instructions for its use. Battery chargers are electrical devices that change Alternating Current (AC) to a lower voltage of Direct Current (DC) that can be used to charge a marine battery. There are two types of battery chargers - manual and automatic.

A manual battery charger must be physically disconnected when the battery has come to a full charge. If not, the battery can be overcharged, and possibly fail. Excess charging current at the end of the charging cycle will heat the electrolyte, resulting in loss of water and active material, substantially reducing battery life.

■ As a rule, on manual chargers, when the ammeter on the charger registers half the rated amperage of the charger, the battery is fully charged. This can vary, and it is recommended to use a hydrometer to accurately measure state of charge.

Automatic battery chargers have an important advantage - they can be left connected (for instance, overnight) without the possibility of overcharging the battery. Automatic chargers are equipped with a sensing device to allow the battery charge to taper off to near zero as the battery becomes fully charged. When charging a low or completely discharged battery, the meter will read close to full rated output. If only partially discharged, the initial reading may be less than full rated output, as the charger responds to the condition of the battery. As the battery continues to charge, the sensing device monitors the state of charge and reduces the charging rate. As the rate of charge tapers to zero amps, the charger will continue to supply a few milliamps of current - just enough to maintain a charged condition.

Fig. 121 Explosive hydrogen gas is released from the batteries in a discharged state. This one exploded when something ignited the gas. Explosions can be caused by a spark from the battery terminals or jumper cables

✴✴ WARNING

Even with automatic storage chargers, don't assume the charger is working properly. When a battery is connected for any time longer than a day, check it frequently (at least daily for the first week and weekly thereafter). Feel the battery case for excessive heat, listen and look for gassing, check the electrolyte levels.

REPLACING BATTERY CABLES

Battery cables don't go bad very often, but like anything else, they can wear out. If the cables on your boat are cracked, frayed or broken, they should be replaced.

When working on any electrical component, it is always a good idea to disconnect the negative (-) battery cable. This will prevent potential damage to many sensitive electrical components.

Always replace the battery cables with one of the same length, or you will increase resistance and possibly cause hard starting. Smear the battery posts with a light film of dielectric grease, or a battery terminal protectant spray once you've installed the new cables. If you replace the cables one at a time, you won't mix them up.

■ Any time you disconnect the battery cables, it is recommended that you disconnect the negative (-) battery cable first. This will prevent you from accidentally grounding the positive (+) terminal when disconnecting it, thereby preventing damage to the electrical system.

Before you disconnect the cable(s), first turn the ignition to the **OFF** position. This will prevent a draw on the battery which could cause arcing. When the battery cable(s) are reconnected (negative cable last), be sure to check all electrical accessories are all working correctly.

IGNITION AND ELECTRICAL SYSTEMS   4-43

CRANKING CIRCUIT

Description And Operation

In the early days, all outboard engines were started by simply pulling on a rope wound around the flywheel. As time passed (and outboards became bigger and bigger) owners became more reluctant to use muscle power (or less capable, some of these motors are HUGE), so it was necessary to replace the rope starter with some form of power cranking system. Today, many small engines are still started by pulling on a rope, but most have a powered cranking motor installed.

The system utilized to replace the rope method was an electric cranking motor coupled with a mechanical gear mesh between the cranking motor and the powerhead flywheel, similar to the method used to crank an automobile engine.

As the name implies, the sole purpose of the cranking motor circuit is to control operation of the starter motor to turn or "crank" the powerhead until the engine is operating. The circuit includes a solenoid or magnetic switch to connect or disconnect the motor from the battery. The operator controls the switch with a key switch.

A neutral safety switch is normally installed into the circuit to permit operation of the cranking motor only if the shift control lever is in neutral. This switch is a safety device to prevent accidental engine start when the engine is in gear.

The cranking motor is a series wound electric motor which draws a heavy current from the battery. It is designed to be used only for short periods of time to crank the engine for starting. To prevent overheating the motor, cranking should not be continued for more than 30-seconds without allowing the motor to cool for at least three minutes. Actually, this time can be spent in making preliminary checks to determine why the engine fails to start (such as checking the safety lanyard, the fuel supply, the fuel line primer bulb, etc).

Power is transmitted from the cranking motor to the powerhead flywheel through a Bendix drive. This drive has a pinion gear mounted on screw threads. When the motor is operated, the pinion gear moves upward and meshes with the teeth on the flywheel ring gear.

When the powerhead starts, the pinion gear is driven faster than the shaft, and as a result, it screws out of mesh with the flywheel. A rubber cushion is built into the Bendix drive to absorb the shock when the pinion meshes with the flywheel ring gear. The parts of the drive must be properly assembled for efficient operation. If the drive is removed for cleaning or overhaul, take care to assemble the parts as shown in the accompanying illustrations in this section. If the screw shaft assembly is reversed, it will strike the splines and the rubber cushion will not absorb the shock.

The sound of the motor during cranking is a good indication of whether the cranking motor is operating properly or not. Naturally, temperature conditions will affect the speed at which the cranking motor is able to crank the engine. The speed when cranking a cold engine will be much slower than when the same starter and battery are used to crank a warm engine. An experienced operator will learn to recognize the favorable sounds of the powerhead cranking under various conditions.

Maintenance

 MODERATE

The cranking motor does not require periodic maintenance or lubrication. If the motor fails to perform properly, the checks outlined in the previous paragraph should be performed. The frequency of starts governs how often the motor should be removed and reconditioned. The manufacturer recommends removal and reconditioning (or replacement) every 1000 hours. Naturally, the motor will have to be replaced if corrective actions do not restore the motor to satisfactory operation.

Faulty Symptoms

If the cranking motor spins, but fails to crank the engine, the cause is usually a corroded or gummy Bendix drive. The drive should be removed, cleaned, and given an inspection.

If the cranking motor cranks the engine too slowly, the following are possible causes and the corrective actions that may be taken:
- Battery charge is low. Charge the battery to full capacity.
- High resistance connections at the battery, solenoid, or motor. Clean and tighten all connections. (A loose or bad ground/battery negative cable is often the culprit here).
- Undersize battery cables (creating too much resistance). Replace cables with sufficient size.
- Battery cables too long (creating too much resistance). Relocate the battery to shorten the run to the solenoid.

Starter Motor Circuit

◆ See Figures 122, 123 and 124

Before wasting too much time troubleshooting the cranking motor circuit, the following checks should be made. Many times, the problem will be corrected.
- Battery fully charged.
- Shift control lever in neutral (many models use a Neutral Safety Switch).
- Are there any blown fuses?
- All electrical connections clean and tight.
- Wiring in good condition, insulation not worn or frayed.

Two more areas may cause the powerhead to crank slowly even though the cranking motor circuit is in excellent condition: a tight or frozen powerhead and water in the lower unit. The following troubleshooting procedures are presented in a logical sequence, with the most common and easily corrected areas listed first in each problem area. The connection number refers to the numbered positions in the accompanying illustrations.

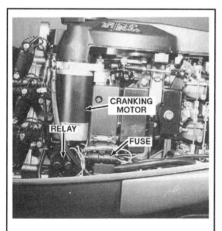

Fig. 122 Typical Yamaha electric starter circuit components

Fig. 123 Many late-model motors place relays and fuses in an electrical junction box

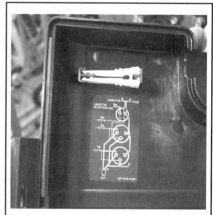

Fig. 124 The inside of the box is usually labeled to help with component identification

4-44 IGNITION AND ELECTRICAL SYSTEMS

TROUBLESHOOTING

◆ See Figure 125

The starter circuit may seem complicated at first, but it is actually pretty simple when it comes down to it. Power from the positive side of the battery is normally connected to one side of the Starter Relay/Solenoid (No. 7 in the accompanying diagram). A ground cable from the battery either connects to a common ground on the powerhead or to the Starter Motor itself. When the Starter Relay/Solenoid is activated (remember, a relay is essentially a remote controlled switch), it closes an internal switch connecting power from the positive battery cable to a cable (No. 7 to No. 1) which runs to the Starter Motor itself (No. 6).

The Starter Relay/Solenoid is activated when power is applied to it (No. 3) from a Starter Button or Ignition Switch (No. 5 and 4). In many cases the circuit contains a normally open Neutral Safety Switch which closes only when the switch plunger is depressed by the shift linkage (which should only occur when the linkage is in the neutral position).

Obviously since this circuit will vary slightly depending upon the motor and rigging you should also refer to the Wiring Diagram for the particular model on which you are working to verify components and wire colors.

✱✱ WARNING

Although we've given some resistance tests to illustrate how a component can be checked on or off the motor, DON'T perform resistance checks with the wires still connected. The meter could be damaged if you attempt to take resistance readings on a live circuit. If the component is still installed, use the voltage checks.

1. Cranking Motor Rotates Slowly

■ In this case you KNOW that the activation circuit is working. You don't know how well the power circuit is doing (i.e. is there too much resistance somewhere in the circuit) and you don't know if the problem is the starter motor itself or a mechanical problem with the powerhead.

    a. Battery charge is low. Charge the battery to full capacity.
    b. Electrical connections corroded or loose. Clean and tighten.
    c. Defective cranking motor. Perform an amp draw test. Lay an amp draw-gauge on the cable leading to the cranking motor. Turn the key on and attempt to crank the engine. If the gauge indicates an excessive amperage draw, the cranking motor must be replaced or rebuilt.

2. Cranking Motor Fails To Crank Powerhead

■ In this case, you don't know whether the problem is the activation portion of the circuit or the starter motor itself? Pick a spot in the circuit and trace the power back from the starter motor and solenoid toward the battery.

    a. Disconnect the cranking motor lead from the solenoid (No. 1) to prevent the powerhead from starting during the testing process.

■ This lead is to remain disconnected from the solenoid during tests No. 2-6.

    b. Use a voltmeter to check for approximate battery voltage at the output (starter) side of the starter relay/solenoid (No. 1) when the starter switch or button is turned to the start position. If there is sufficient voltage, the problem lies in the cable or starter (or ground circuit between the starter and the powerhead/battery). If there is no voltage, check the battery side of the solenoid (No. 7), just to verify that the solenoid is getting battery power to the starter circuit.

    c. Check the solenoid side of the ground circuit by disconnecting the Black ground wire from the No. 2 terminal. Connect a voltmeter between the No. 2 terminal and a common engine ground. Turn the key switch to the start position. Observe the voltmeter reading. If there is the slightest amount of reading, check the Black ground wire connection or check for an open circuit.

3. Test Cranking Motor Relay/Solenoid

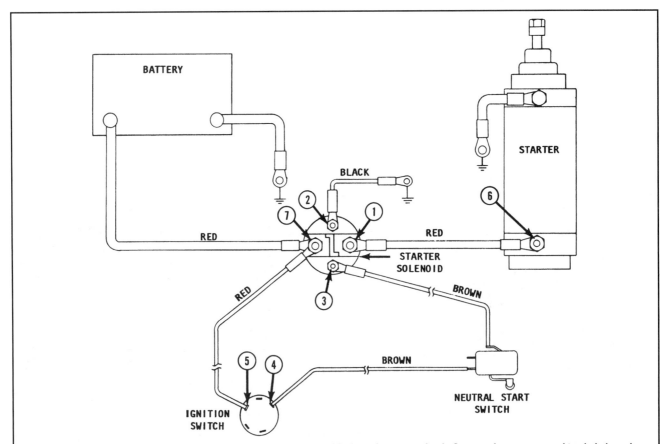

Fig. 125 Use this typical starting system diagram to help understand/troubleshoot the starter circuit. Step numbers correspond to circled numbers in the diagram

IGNITION AND ELECTRICAL SYSTEMS 4-45

■ The solenoid can easily be tested while it is on or off the motor. The basis for the test is to apply 12 volts across the activation circuit (No. 3 and No. 2 in the illustration) and then to check and see if there is continuity across the starter motor power supply circuit (No. 7 and No. 1). If there is continuity without power applied to the activation circuit the switch is stuck closed and faulty, or alternately, if there is no continuity with power applied the switch is stuck open and faulty.

    a. Check to see if the activation circuit (neutral safety switch and/or ignition switch/button) is supplying voltage to the starter solenoid. Connect a voltmeter between the engine common ground and the No. 3 terminal. Turn the ignition key switch to the start position. Observe the voltmeter reading. If the meter gives a significant voltage reading (something more than at least 0.3 volt), the solenoid is defective and must be replaced (this is true since you've already verified the ground and battery voltage to power the circuit). If however there is little or no voltage, trace the circuit back toward the ignition start button/switch (through the neutral start switch, if equipped).

  4. Test Neutral Start Switch

■ Like the relay/solenoid, the neutral start (also known as a neutral safety) switch can be checked either on or off the motor. When equipped, most Yamahas use a switch with a plunger which is normally open, but closes the switch contacts when the plunger is depressed. The most simple test you can perform on the switch is to check for resistance across the switch terminals when the plunger (or lever) is depressed vs. when it is released. If the switch closes (meter shows little/no resistance) when it is depressed and opens (meter shows no continuity, meaning infinite resistance) when released, the switch is operating properly.

    a. Connect a voltmeter between the common engine ground and the No. 4. Turn the ignition key switch to the start position. Observe the voltmeter. If there is any indication of a reading at No 4, but NOT at No. 3, the neutral start switch is open or the lead is open between the No. 3 and No. 4. Repair the switch or lead, as applicable. If there was no power at No. 4, it is time to check the switch/button and power supply to the switch.

  5. Test for power to the Ignition Switch/Button
    a. Connect a voltmeter between the common engine ground and No. 5.
    b. The voltmeter should indicate approximate battery voltage (about 12-volts). If the meter needle flickers (fails to hold steady), check the circuit between No. 5 and common engine ground. If meter fails to indicate voltage, replace the positive battery cable.
    c. If power is good to the button/switch, but not on the other (No 4, neutral switch) side, the button or switch is suspect.

  6. Test Large Red Power Supply Cables
    a. Connect the Red cable to the cranking motor solenoid.
    b. Connect the voltmeter between the engine common ground and No. 6.
    c. Turn the ignition key switch to the start position, or depress the start button.
    d. Observe the voltmeter. If there is no reading, check the Red cable for a poor connection or an open circuit. If there is any indication of a reading, and the cranking motor does not rotate, make sure the starter motor ground is good, otherwise the cranking motor must serviced or replaced.

Starter Motor Relay/Solenoid

DESCRIPTION & OPERATION

◆ See Figure 126

The starter cranking motor relay is actually a remote controlled switch located in the wiring between the battery and the powerhead. Although it can be tested, it cannot be repaired, therefore, if troubleshooting indicates the switch to be faulty, it must be replaced.

Before beginning any work on the relay, disconnect the positive (+)and negative (-) leads from the battery terminal for safety. Keep in mind that the positive lead, where it connects to the solenoid, should always be hot and it is too easy to accidentally ground the tool you are using to the powerhead. This not only will give you quite a shock (we've actually seen it all but weld a wrench in place), but the resultant sparks could create a dangerous condition.

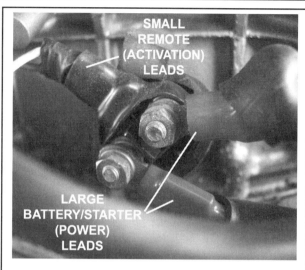

Fig. 126 Typical Yamaha starter relay (the large terminal leads are exposed as the boots are pulled back)

※※ WARNING

Disconnecting the battery leads is most important because the cranking motor relay lead will be disconnected and allowed to hang free. The other end of this lead is connected to the battery. If the leads are not disconnected from the battery and the free relay end should happen to come in contact with any metal part on the powerhead, sparks would fly and the end of the lead would be burned.

As detailed in the Starter Motor Circuit section, the starter relay/solenoid has the all important function of responding to a activation circuit in order to physically close the switch contacts that will apply battery power to the starter. The starter motor power circuit contains heavy gauge leads capable of large amp loads in order to power the motor. The activation circuit contains smaller gauge leads, since it only need carry sufficient amperage to activate the relay itself. The activation circuit applies battery power to the solenoid through a neutral safety switch (when equipped) and a starter button or switch.

TESTING

◆ See Figures 126, 127 and 128

※※ WARNING

The following tests must be conducted with the relay wiring disconnected (to prevent accidental starting and to prevent potential damage to the meter).

■ The wire colors provided in this procedure are true for most years and models, but if yours differs, check the Wiring Diagrams in this section. In some cases wires may not agree with either the procedure or the diagram as all or part of a harness may have been replaced previously on older outboards, or on motors rigged with an electrical starter after the factory.

The test is relatively simple. Since the relay is simply a remote controlled electric switch, first you check that the remote control (activation) circuit has continuity and the main switch (starter power) circuit does NOT. Then you apply 12 volts to that remote control circuit and make sure the main switch contacts close only when power is applied.

  1. Obtain an ohmmeter or DVOM set to read resistance. Check for continuity across the terminals for the small relay leads (The remote control/activation circuit). In most cases these are black and brown leads (black for ground and brown from the neutral or ignition switch). The meter

4-46 IGNITION AND ELECTRICAL SYSTEMS

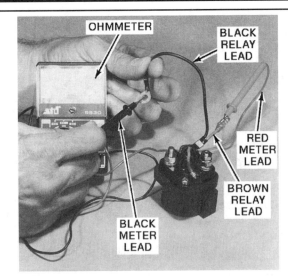

Fig. 127 Check for continuity across the small relay lead terminals (then check for continuity across the large relay lead terminals). There should only be continuity across the small terminals with no voltage applied

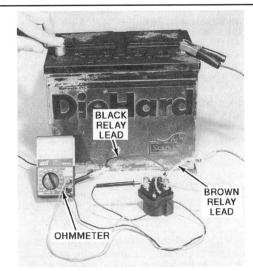

Fig. 128 Connect a 12-volt battery to the small relay lead terminals and recheck the large lead terminals for continuity. If the relay is good the contacts for the large terminals will close and there will be continuity as long as 12 volts are applied to the small terminals

should indicate continuity. If the meter registers no continuity, the relay is defective and must be replaced. No service or adjustment is possible.

2. Connect one test lead of the ohmmeter to each of the large relay terminals. There should be NO continuity.

3. Now, use a 12-volt battery connected to the remote control circuit of the relay in order to activate the relay while continuing to check for continuity across the large relay terminals. Connect the positive (+) lead from a fully charged 12-volt battery to the Brown lead and momentarily make contact with the ground lead from the battery to the Black lead. If a loud click sound is heard, and the ohmmeter indicates continuity, the solenoid is in serviceable condition. If, however, a click sound is not heard, and/or the ohmmeter does not indicate continuity, the solenoid is defective and must be replaced with a suitable marine type solenoid.

REMOVAL & INSTALLATION

◆ See Figure 126

■ The wire colors provided in this procedure are true for most years and models, but if yours differs, check the Wiring Diagrams in this section. In some cases wires may not agree with either the procedure or the diagram as all or part of a harness may have been replaced previously on older outboards, or on motors rigged with an electrical starter after the factory.

Most Yamaha starter relays are press-fit into a mounting bracket. But some are bolted into position - when they are bolted in position, use care not to over-tighten the fasteners and crack the case. The same goes for the terminal nuts, make sure they are snug but do not over-tighten them.

1. Push down the boots from the two large leads (they are usually Red) at the relay. Take time tag and/or note each of the leads and to which terminal they are connected, then remove each of the leads from the solenoid. Disconnect the smaller leads (usually Brown and Black).

2. Slide the relay from its mounting bracket. In some cases there will be a grounding bolt on the bracket behind the relay to release the Black relay grounding wire.

To Install:

3. Slide the relay back onto its mounting bracket. If applicable, secure the Black lead behind the relay by installing the grounding bolt on the bracket.

4. Connect the smaller relay leads (usually Brown and Black) as tagged during removal.

5. Connect the two large leads (usually Red) to the front of the cranking motor relay, as tagged during removal. One is from the terminal on the cranking motor, while the other is hot, coming from the positive battery terminal. This second lead also usually has a small Red lead attached to the relay terminal.

6. Install the elbow boots onto the relay leads.
7. If the work is complete, then reconnect the battery terminals. If further work is to be carried out on the cranking system, then leave the battery cables disconnected until the work is complete.

Starter Motor

DESCRIPTION

One basic type of cranking motor is used on all Yamaha powerheads covered here. However, the housing/mounting and exterior appearance of the motor itself will vary slightly from model-to-model. Therefore, the accompanying illustrations may differ slightly from the unit being serviced, but the procedures and maintenance instructions are valid.

As an example, the collar on the cranking motor for the 65J - 90 Hp (1140cc) models has been modified to provide an escape route for any water trapped around the pinion gear. However, that doesn't change the way the motor itself works.

Marine cranking motors are very similar in construction and operation to the units used in the automotive industry.

All Yamaha marine cranking motors use the inertia type drive assembly. This type assembly is mounted on an armature shaft with external spiral splines which mate with the internal splines of the drive assembly.

As with most starter motors, the housing is not designed to provide airflow which would be necessary for continuous or repeated operation. Therefore, never operate a cranking motor for more than 30 seconds without allowing it to cool for at least three minutes. Continuous operation without the cooling period can cause serious damage to or destroy the cranking motor.

REMOVAL & INSTALLATION

2-Cylinder Powerheads

◆ See Figures 129 and 130

1. Disconnect the battery cables for safety. It is a good idea to place a plastic bag over the cables or secure them away from the battery to prevent them from accidentally contacting the terminals again during the procedure.

2. Remove the cowling from the powerhead.

3. Some models also contain a hand-rewind starter or flywheel cover that might interfere with access to the starter bolts. If necessary, remove the hand-rewind starter or flywheel cover for access.

■ A few powerheads such as the 48 hp twin utilize a pinion cover that is bolted in place on top of the starter motor assembly. If equipped, unbolt and remove the cover for access.

IGNITION AND ELECTRICAL SYSTEMS 4-47

4. Disconnect the starter motor wiring. Unless the motor is grounded through the mounting there should be 2 large leads, one for power and the other for ground. But on some models, you cannot reach both leads while the starter is still installed. If necessary, you can disconnect the lead on the other end (at the starter relay or the powerhead) or you can wait until the starter is unbolted and repositioned to access the lead(s) at the starter.

5. Remove the mounting bolts and washers from the starter motor housing. Most starters on these size engines are secured by 2 bolts at the top of the motor (threaded downward). On some models one of the bolts may be partially obscured by the flywheel. However, by rotating the flywheel (CLOCKWISE when viewed from above, in the normal direction of rotation) until a cutout is directly over the bolt head, the bolt may be wiggled out.

6. Remove the cranking motor from the powerhead. If not done already, disconnect the large Red lead from the motor.

Fig. 129 The starters on most 2-cylinder powerheads are secured using 2 mounting bolts (typical)

Fig. 130 On some motors, the wiring may not be accessible until the motor is repositioned slightly

To Install:

7. Move the starter motor close to position on the powerhead. If access is not possible once the motor is bolted to powerhead, connect the Red lead to the cranking motor terminal before positioning it to the mounting bracket.

8. Hold the motor in position as you thread the 2 (or 3) retaining bolts and washers. Remember, if one of the bolts is partially obscured you may have to slowly rotate the flywheel (CLOCKWISE when viewed from above) until a cutout is directly over the bolt hole allowing the bolt to be wiggled into place and the threads started. Tighten the mounting bolts securely.

9. Connect the starter motor wiring (or alternately the wiring from the motor to the relay and powerhead, as applicable). Tighten the terminals securely, but be careful not to over-tighten and strip the bolts or crack the housing.

10. Mount the outboard unit in a test tank, on the boat in a body of water, or connect a flush attachment and hose to the lower unit.

✱✱ CAUTION

Water must circulate through the lower unit to the engine any time the engine is run to prevent damage to the water pump in the lower unit. Just five seconds without water will damage the water pump.

Never, again, never operate the engine at high speed with a flush device attached. The engine, operating at high speed with such a device attached, would runaway from lack of a load on the propeller, causing extensive damage.

11. Crank the powerhead with the starter motor and start the unit. Shut the powerhead down and restart it several times to check operation of the cranking motor.

3-Cylinder Inline Powerheads

◆ See Figures 131 thru 135

1. Disconnect the battery cables for safety. It is a good idea to place a plastic bag over the cables or secure them away from the battery to prevent them from accidentally contacting the terminals again during the procedure.

2. Remove the cowling from the powerhead.

3. Locate the starter motor, on many motors the flywheel cover should be removed for better access. When equipped, the flywheel cover is usually retained by one or more bolts and rubber mounting grommets/pins.

4. Disconnect the hot lead and ground lead (if equipped, since some of these motors ground the starter circuit through the starter mounting). Typically, the hot lead is connected toward the bottom of the starter and the ground lead toward the top, but this is not always the case, so watch the wire colors and tag them if necessary to ensure proper installation.

■ **On some motors, the starter relay is installed in a bracket at the base of the starter itself. On these models a grounding lead and mounting bolt is located behind the relay. If so, tag and disconnect the wiring from the relay as necessary to reposition it for access. Reach inside the rubber mounting ring and disconnect the Black lead with the eye connector from its grounding bolt securing the cranking motor bracket. The bracket to the rubber mounting ring will come away with the grounding bolt.**

5. Remove the bolts threaded through the starter and into the powerhead or mounting bracket. The mounting bolt patterns vary greatly from model-to-model. Most motors are secured by 2 or 3 mounting bolts as follows:

• On smaller (up to 50 hp) motors, there are usually 2 bolts threaded downward into the top of the starter motor. Though the bolts should be tightened to about one 22 ft. lbs. (30 Nm) upon installation for some models, most motors do not list a torque spec for these bolts, so do not over-tighten them.

• On the larger (50 hp and up) motors there are usually 3 bolts threaded sideways into the motor, 2 at the top and one at the bottom. However, the 2 bolts at the top are usually threaded at a 90 degree angle to each other, each through a different mounting boss on the starter motor housing. On these models the bolts should be tightened to 14 ft. lbs. (20 Nm) upon installation.

6. Once all the bolts are removed, carefully lift the motor free of the bracket.

7. Installation is essentially the reverse of the removal procedure. Tighten the bolts to specification (when the spec is available.) Visually inspect the condition of the cables and the terminals before installation.

4-48 IGNITION AND ELECTRICAL SYSTEMS

Replace any cables whose insulation has become cracked, brittle or otherwise damaged. Carefully clean all terminals and ground mounting points of any signs of corrosion. This is especially true for starter motors whose circuit grounds through the mount itself. Also, use care when reconnecting the cables. Make sure the cables are routed as noted during removal and are not pinched or damaged in anyway.

8. Once you are finished, mount the outboard unit in a test tank, on the boat in a body of water, or connect a flush attachment and hose to the lower unit.

✱✱ CAUTION

Water must circulate through the lower unit to the powerhead anytime the powerhead is operating to prevent damage to the water pump in the lower unit. Just five seconds without water will damage the water pump impeller.

Never, again, never operate the engine at high speed with a flush device attached. The engine, operating at high speed with such a device attached, would runaway from lack of a load on the propeller, causing extensive damage.

9. Crank the powerhead with the cranking motor and start the unit. Shut the powerhead down and restart it several times to check operation of the cranking motor.

V4 and V6 Powerheads

◆ See Figures 136 thru 141

1. Disconnect the battery cables for safety. It is a good idea to place a plastic bag over the cables or secure them away from the battery to prevent them from accidentally contacting the terminals again during the procedure.
2. Remove the cowling from the powerhead.
3. If equipped, remove the flywheel cover for better access.
4. For all EFI OX66/HPDI motors (except the 3.3L engine), remove the electrical junction box cover for access to starter wiring.
5. Tag and disconnect the ground (if equipped, since some of these motors ground through the housing, refer to the Wiring Diagrams for clarification) and hot (battery/relay power) leads for the starter. The attachment points vary from model-to-model, and on some motors it is easier to disconnect the lead(s) from the powerhead and/or relay (respectively) removing the starter motor with the leads attached.

Fig. 131 Many Yamaha starters are secured by 2 bolts threaded downward at the top of the housing - 40 hp shown

Fig. 132 On others, there are 2 or 3 bolts threaded sideways through a mounting boss - 90 hp shown

Fig. 133 Once the bolts are removed, lift off the starter

Fig. 134 On some older motors with the relay mounted under the starter...

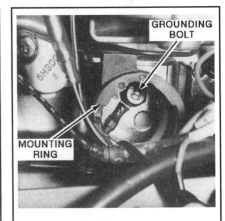

Fig. 135 ... you have to remove the relay for access to a ground bolt

IGNITION AND ELECTRICAL SYSTEMS 4-49

Fig. 136 Tag and disconnect the ground...

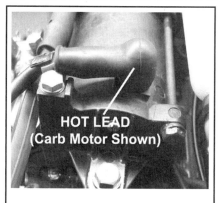
Fig. 137 ... and/or hot leads (as applicable)

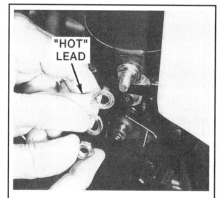
Fig. 138 Keep track of the terminal nuts and washers

Fig. 139 Don't be fooled, these bolts hold the starter together...

Fig. 140 ... mounting bolts are threaded sideways at the top...

Fig. 141 ... and bottom of the assembly

6. Remove the bolts threaded through the starter and into the powerhead or mounting bracket. The mounting bolt patterns and installation torque specifications (when provided) vary from model-to-model, however it appears that all motors use 3 mounting bolts:
• On HPDI motors, there are 2 bolts at the top of the starter motor and one at the bottom. All 3 bolts are threaded sideways into the mounting bracket. The bolts should be tightened to 21 ft. lbs. (29 Nm) upon installation.

■ On some motors the oil tank may interfere slightly with removal of one starter bolt, if necessary, reposition the tank for access.

• On carbureted and EFI OX66 motors, there are usually 3 bolts threaded sideways into the motor, 2 at the top and one at the bottom. However, the 2 bolts at the top are usually threaded at a 90 degree angle to each other, each through a different mounting boss on the starter motor housing. On these models the bolts should be tightened to 22 ft. lbs. (30 Nm) upon installation.

7. Once all the bolts are removed, carefully lift the motor free of the bracket.

8. Installation is essentially the reverse of the removal procedure. Tighten the bolts to specification (when the spec is available.) Visually inspect the condition of the cables and the terminals before installation. Replace any cables whose insulation has become cracked, brittle or otherwise damaged. Carefully clean all terminals and ground mounting points of any signs of corrosion. This is especially true for starter motors whose circuit grounds through the mount itself. Also, use care when reconnecting the cables. Make sure the cables are routed as noted during removal and are not pinched or damaged in anyway.

9. Once you are finished, mount the outboard unit in a test tank, on the boat in a body of water, or connect a flush attachment and hose to the lower unit.

✱✱ CAUTION

Water must circulate through the lower unit to the powerhead anytime the powerhead is operating to prevent damage to the water pump in the lower unit. Just five seconds without water will damage the water pump impeller.

Never, again, never operate the engine at high speed with a flush device attached. The engine, operating at high speed with such a device attached, would runaway from lack of a load on the propeller, causing extensive damage.

10. Crank the powerhead with the cranking motor and start the unit. Shut the powerhead down and restart it several times to check operation of the cranking motor.

OVERHAUL

◆ See Figures 142 thru 152    DIFFICULT

Although the wide availability of rebuilt starters on the marine parts market today dissuades most technicians and many DIYers from overhauling the starter, rebuild kits area normally available. If you decide to undertake this, keep close track of all starter components as you disassemble it. Digital cameras are relatively cheap these days, so if you've got access to one, work slowly taking pics as you go. They can be a great reference when it comes time for assembly. Be sure to keep track of all washers and shims so that they are placed in the same positions upon assembly.

■ Yamaha reuses essentially the same starter assembly on multiple powerheads. Besides small differences in the top or bottom housing, differences will occur with the thickness of shims used between armature washers from year-to-year or model-to-model. For this reason, it is critical that you keep track of all washers and shims for installation in the same positions.

4-50 IGNITION AND ELECTRICAL SYSTEMS

Before disassembling the starter motor, scribe matchmarks between the upper/lower (front/rear) covers and the starter housing (also sometimes known as the stator housing). This will ensure easy alignment during assembly.

Keep in mind that the magnets which make up a large part of the starter motor are quite strong. Once the through bolts are removed it can be difficult to separate the field frame/housing and the armature shaft. It is even more important to keep this in mind when assembling the starter as it is easy to badly pinch a finger during that part of the process. So take your time and watch the piggys.

Once disassembled, carefully clean all metal components with a mild solvent and either blow dry with compressed air or allow them to air dry. Thoroughly inspect all components as follows:

- Inspect the pinion gear for wear or damage to the teeth and replace, as necessary.
- Check the clutch for freedom of movement in one direction and stiff movement in the other (generally it is clockwise-free, counterclockwise-stiff).
- A dirty commutator may be cleaned using #600 grit sandpaper, but should be thoroughly cleaned afterwards using solvent and compressed air.
- Be sure to inspect the housing and end caps for signs of wear or damage such as cracks.

- If you have access to a set of precision V-blocks and a dial-gauge, check the commutator (armature shaft) run-out. Generally no more than 0.0020 in. (0.05mm) of run-out is considered serviceable.
- Carefully exam the mica undercut (depth of the grooves) on the end of the commutator. Build specifications vary greatly, however service limits are generally 0.01-0.03 in. (0.2-0.8mm). If the undercut depth is less than this spec, remove some metal from between the commutator segments using a hacksaw blade, then remove all metal and mica particles using compressed air (and safety goggles!).
- Inspect the brush lengths (height of each brush from top to bottom, if the point where the wire attaches is considered the side). Again, build specifications vary greatly, however service limits are generally 0.35-0.47 in. (9-12mm) on most larger motors. Some smaller Yamahas use proportionally smaller brushes (as small as 0.18-0.25 in./4.5-6.5mm), so if you are unsure as to whether or not the brushes are serviceable, check with a local parts supplier to see if you can measure the height of a replacement set (or just spring for the replacements, since you've come this far).
- Use an ohmmeter to inspect armature continuity. First check for continuity between each of the segments on the commutator, there should be continuity. Next, check between the commutator segments and the armature core (the thick main shaft of the armature), there should be NO

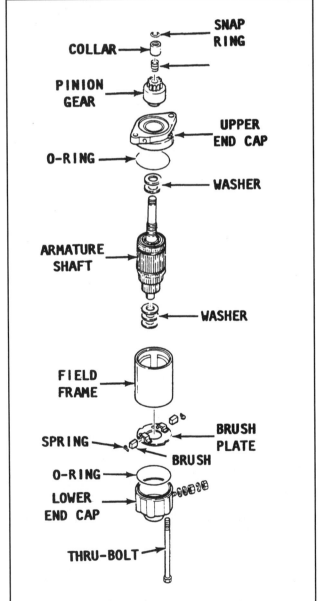

Fig. 142 Exploded view of a typical starter assembly found on the smallest Yamaha outboards

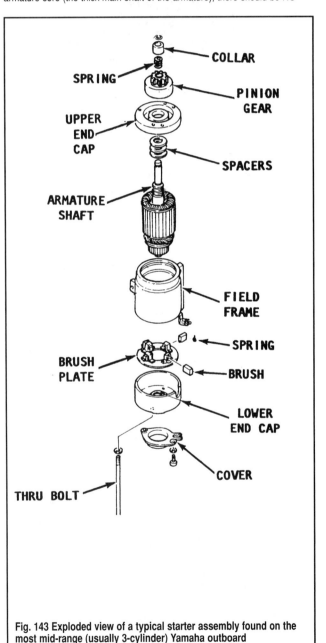

Fig. 143 Exploded view of a typical starter assembly found on the most mid-range (usually 3-cylinder) Yamaha outboard

IGNITION AND ELECTRICAL SYSTEMS 4-51

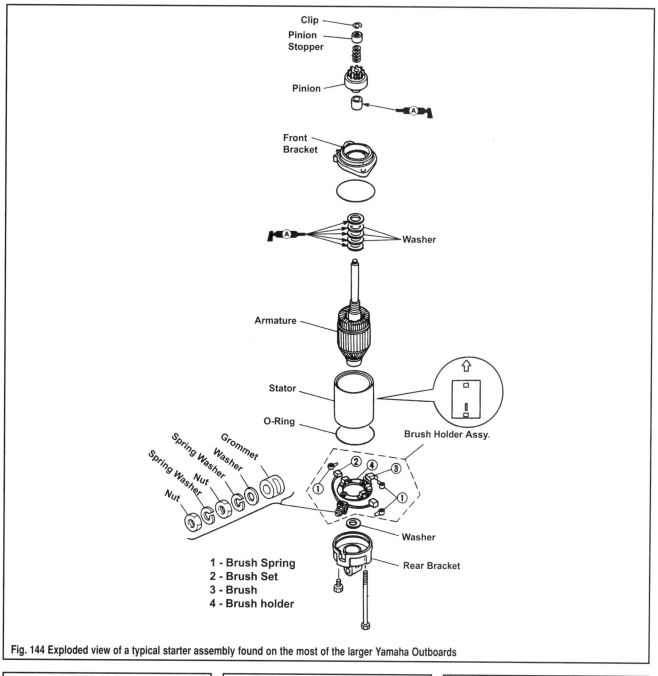

Fig. 144 Exploded view of a typical starter assembly found on the most of the larger Yamaha Outboards

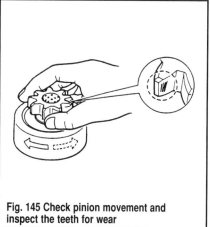

Fig. 145 Check pinion movement and inspect the teeth for wear

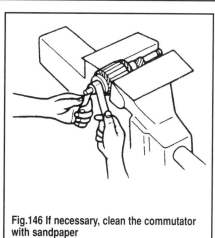

Fig. 146 If necessary, clean the commutator with sandpaper

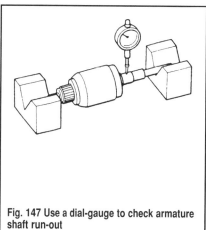

Fig. 147 Use a dial-gauge to check armature shaft run-out

4-52 IGNITION AND ELECTRICAL SYSTEMS

continuity. Finally check between the commutator segments and the armature shaft (the long, thin shaft at the other end of the armature from the commutator segments), again there should be NO continuity. Replace the armature if it fails this armature continuity test.

• Use an ohmmeter to inspect brush set continuity. Use the accompanying illustration of a typical brush set to help identify components.

Check that there is continuity between brush No. 1 and No. 2. Next make sure there is NO continuity between those brushes and brush No. 3. Also there should be no continuity between each brush holder (No. 4) and the brush assembly holder or brush plate (No. 5) as the holder is used to isolate the brushes from the plate.

Replace any components which do not appear to be serviceable.

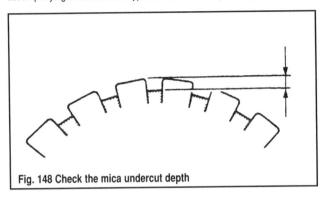

Fig. 148 Check the mica undercut depth

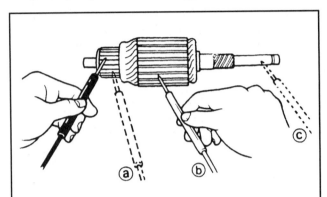

Fig. 150 Use a ohmmeter to check for continuity between the commutator segments (a), the segments and the armature core (b) and lastly, between the segments and armature shaft (c)

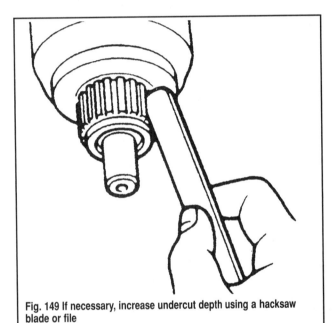

Fig. 149 If necessary, increase undercut depth using a hacksaw blade or file

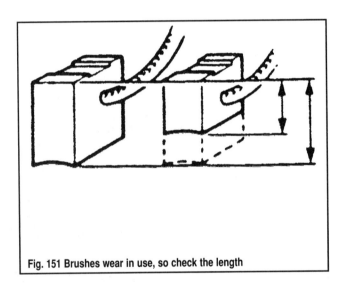

Fig. 151 Brushes wear in use, so check the length

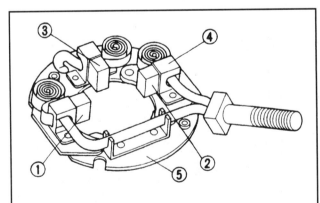

Fig. 152 Using an ohmmeter check brush continuity - typical brush plate assembly shown

IGNITION AND ELECTRICAL SYSTEMS

ELECTRICAL SWITCH/SOLENOID SERVICE

This short section provides testing procedures for other electrical parts installed on the powerhead. If a unit fails the testing, the faulty part must be replaced. In most cases, removal and installation is through attaching hardware.

Main Keyswitch

TESTING

◆ See Figures 153 and 154  MODERATE

The main keyswitch is normally located inside the control box or it is rigged to the dash of the boat. When installed in the control box, the box must be normally be opened to gain access to the main switch leads.

■ **The wire colors on this test are only applicable to Yamaha harnesses and keyswitches. Depending upon boat rigging, the craft on which you are working could deviate from these colors, at least on the switch side of the harness.**

Disconnect the five leads from the main switch at their quick disconnect fittings: the White, Black, Red, Pink or Yellow (depending upon the switch/model), and Brown leads. Obtain an ohmmeter or a DVOM set to read resistance. Check to be sure the switch is in the off position. Make contact with the Red meter lead to the White switch lead, and the Black meter lead to the Black switch lead. The meter should indicate continuity. Keep both meter leads in place. Rotate the switch to the on position, and then to the start position. If the meter indicates continuity in either or both switch positions, the switch is defective and must be replaced.

■ **Likewise, check for any continuity between the White or Black leads and each of the three additional wire leads. There should be no continuity with the Red, Yellow (or Pink) and/or Brown leads and the White or Black leads with the switch in ANY position.**

Move the Red meter lead to make contact with the Red switch lead, and the Black meter lead to make contact with the Yellow (or Pink) switch lead, with the switch OFF there should be no continuity. Now rotate the switch to the ON position. The meter should indicate continuity. Turn the switch back to the off position and then on to the start position. If the meter fails to indicate continuity in either the ON or START position, the switch is defective and must be replaced.

Keep the Red meter lead in contact with the Red switch lead, but move the Black meter lead to the Brown switch lead. There should be NO continuity in any switch position except START.

Keep the Black meter lead on the Brown switch lead and move the Red meter lead to the Yellow (or Pink) switch lead. Again, there should be no continuity in any position except START.

■ **If the meter fails to indicates continuity when specified OR shows continuity in other positions/lead combinations, the switch is defective and must be replaced.**

** CAUTION

A faulty main switch could cause the powerhead to start accidentally, while being serviced, and cause personal injury.

Once testing has been satisfactorily completed connect the five leads, color to color, at their quick connect fittings. Tuck the leads neatly inside the control box, away from moving parts, and replace the outer cover.

Kill Switch

TESTING

◆ See Figure 155  MODERATE

The Kill switch is normally located on the front panel of any powerhead not equipped with a control box. For powerheads equipped with a control box, the kill switch is usually mounted on the forward side of the box. The Kill switch must have the emergency tether in place before testing.

■ **Depending upon how the boat is rigged, the kill switch may be attached to the boat dash itself.**

Trace the Kill switch button harness containing two wires from the switch to their nearest quick disconnect fitting. The colors may vary for different models, but it shouldn't make a difference, because it is only a 2 wire switch and your tests will be with one meter lead attached to each wire. Generally the wires are White, Black or White/Black. For more details, refer to the Wiring Diagrams, in this section.

Disengage the two leads, then connect an ohmmeter or DVOM set to read resistance across the disconnected leads.

Verify the emergency tether is in place behind the Kill switch button. Check the meter, with the button released and the tether installed there should be NO continuity (continuity would shut down the powerhead if the switch was connected and the powerhead was operating).

Now, depress the kill button. The meter should indicate continuity.

Release the button (making sure the tether is still in place). The meter should now indicate NO continuity.

Remove the emergency tether, and the meter should again indicate continuity.

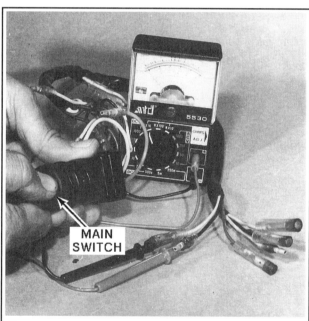

Fig. 153 Using an ohmmeter to test the main switch

Switch position	Lead color				
	White (W)	Black (B)	Red (R)	Pink or Yellow (P or Y)	Brown (Br)
OFF	O—	—O			
ON			O—	—O	
START			O—	—O—	—O

Fig. 154 Yamaha keyswitch testing, the diagram shows continuity between leads in various switch positions (NOTE, some switches use Yellow, instead of a Pink lead)

4-54 IGNITION AND ELECTRICAL SYSTEMS

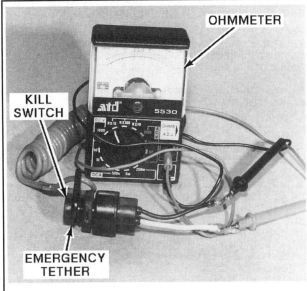

Fig. 155 Using an ohmmeter to test the kill switch with the emergency tether in place

First, obtain a piece of plastic from the cover of a container of margarine, whipped topping, or similar product.

Next, using the pattern shown on this page, cut out about four shapes, as shown. Stack the four cutouts together, secure them with a paper clip, or similar object, and then insert them behind the kill switch.

■ **If the material described is not available, obtain some other pliable material and cut the shape indicated. The thickness of the substitute tether device should be approximately 1/8 in. (3mm).**

Remember, use this device only in an emergency situation and purchase the proper tether from the local Yamaha dealer at the first opportunity. By substituting this home made tether, both the safety and the security features intended by the manufacturer have been lost.

Start Button

TESTING

◆ See Figures 157 and 158

Trace the start button harness containing two wires from the switch to their nearest quick disconnect fitting. The colors may vary for different models, but it shouldn't make a difference, because it is only a 2 wire switch and your tests will be with one meter lead attached to each wire. Generally the wires are Red and Brown. For more details, refer to the Wiring Diagrams, in this section.

Disengage the two leads and connect an ohmmeter across the disconnected leads. Depress the start button. The meter should register continuity. Release the button and the meter should now register NO continuity. Both tests must be successful. If the tests are not successful, the start button must be replaced. The start button is a one piece sealed unit and cannot be serviced.

All tests must be successful. If the switch fails any one test, the switch is defective and must be replaced. The switch is a one piece sealed unit and cannot be serviced.

Tether

TEMPORARY REPLACEMENT

◆ See Figure 156

If the boat owner loses the emergency tether and is unable to obtain one immediately from the local Yamaha dealer, an emergency substitute tether may be made using only a common knife and a couple pieces of plastic.

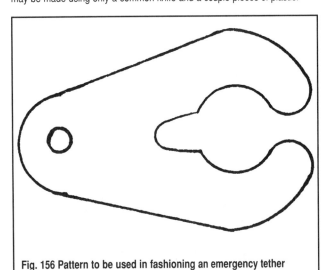

Fig. 156 Pattern to be used in fashioning an emergency tether

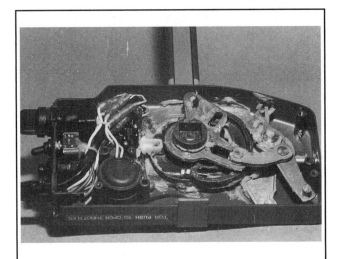

Fig. 157 The most difficult task involving testing of the remote control box components is stuffing the electrical leads back into the box in an orderly manner

IGNITION AND ELECTRICAL SYSTEMS 4-55

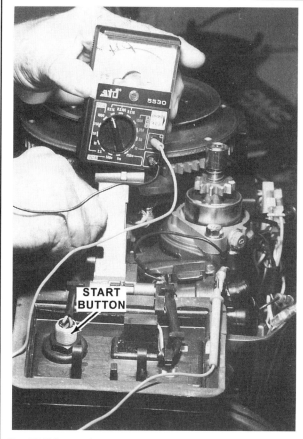

Fig. 158 Using an ohmmeter to test the start button

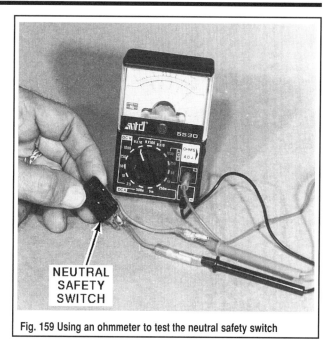

Fig. 159 Using an ohmmeter to test the neutral safety switch

Fig. 160 When powerhead mounted, the neutral safety switch is sometimes hidden, but is usually located on the axis of the shift lever just inside the lower cowling pan

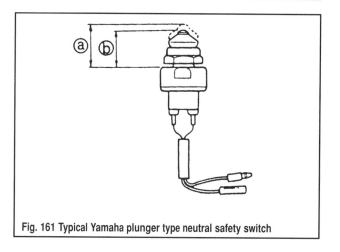

Fig. 161 Typical Yamaha plunger type neutral safety switch

Neutral Safety Switch

TESTING

◆ See Figures 159, 160 and 161

The neutral safety switch on some manual start models is replaced by the no-start-in-gear protection system located at the hand rewind starter. For other models the switch is either located on the powerhead or in the control box. In either case, it is positioned where it can be physically activated by movement of the shift linkage.

On all models so equipped, the switch is a normally open switch (should exhibit no continuity when the plunger or lever is released). Therefore, it is positioned so that the shift linkage will push inward on the plunger (or downward on the lever for some models) in any position EXCEPT neutral.

Switches of the plunger design normally have a plunger free-length (measured from the switch face) of 0.77-0.81 in. (19.5-20.5mm). Once the plunger is pushed inward the switch will show continuity across both contacts and plunger length should be reduced to about 0.73-0.77 in. (18.5-19.5mm)

To test the switch, trace the neutral safety switch leads from the switch to their nearest quick disconnect fitting. Though the colors may vary, both of these leads are usually Brown. Of course, it shouldn't make a difference, because it is only a 2 wire switch and your tests will be with one meter lead attached to each wire. For more details, refer to the Wiring Diagrams, in this section.

Disengage the two leads, then connect an ohmmeter or a DVOM set to read resistance across the two disconnected leads. When the shift lever is in the neutral position (the plunger or lever is free), the meter should indicate continuity.

When the lower unit is shifted into either forward or reverse gear (the plunger or lever is depressed), the meter should indicate no continuity.

4-56 IGNITION AND ELECTRICAL SYSTEMS

The switch must pass all three tests to verify the safety aspect of the switch is functioning properly. If the switch fails any one or more of the tests, the switch must be adjusted (if the physical positioning of the switch allows this) or replaced (more likely on Yamahas).

Remember this is a safety switch. A faulty switch may allow the powerhead to be started with the lower unit in gear - an extremely dangerous situation for the boat, crew, and passengers.

Warning Buzzer/Horn

TESTING

◆ See Figure 160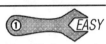

The buzzer or horn is a warning device to indicate low oil in the reservoir, an over rev. condition, or overheating of the powerhead (depending upon the model). The buzzer is usually located inside the control box (for remote units).

Remove the control box cover and identify the two leads from the buzzer, one is normally Yellow and the other is normally Pink. Of course, it shouldn't make a difference, because it is only a 2 wire component and your tests will be with one meter lead attached to each wire. For more details, refer to the Wiring Diagrams, in this section.

Disconnect the two buzzer leads at their quick disconnect fittings, and ease the buzzer out from between the four posts which anchor it in place.

Obtain a 12-volt battery. Momentarily connect the buzzer leads to the battery (generally the Yellow buzzer lead would connect to the negative battery terminal, while the Pink buzzer lead would connect to the positive terminal). As soon as the leads are connected, the buzzer should sound. If the buzzer is silent, or the sound emitted does not capture the helmsperson's attention immediately, the buzzer should be replaced. Service or adjustment is not possible. If the sound is satisfactory and immediate, install the buzzer between the four posts and connect the two leads matching color to color. Tuck the leads to prevent them from making contact with any moving parts inside the control box. Replace the cover.

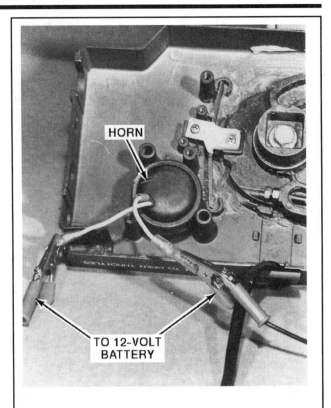

Fig. 160 The warning buzzer/horn is tested by applying 12 volts and listening for activation

IGNITION AND ELECTRICAL SYSTEMS 4-57

SPECIFICATIONS

IGNITION SYSTEM COMPONENT TESTING - SINGLE CYLINDER MODELS

HP	No. of Cyl	Engine Type	Model Year	Displace cu. in. (cc)	Pulser Coils Low Speed ① Wire leads (+)	Pulser Coils Low Speed ① Wire leads (-)	Pulser Coils Low Speed ① Resistance Ohms ②	Pulser Coils High Speed ① Wire leads (+)	Pulser Coils High Speed ① Wire leads (-)	Pulser Coils High Speed ① Resistance Ohms ②	Charge Coil Wire leads (+)	Charge Coil Wire leads (-)	Charge Coil Resistance Ohms ②	Ignition Coil / TCI Unit Primary Wire Leads (+)	Ignition Coil / TCI Unit Primary Wire Leads (-)	Ignition Coil / TCI Unit Primary Resistance Ohms ②	Ignition Coil / TCI Unit Secondary Resistance Ohms ②
2	1	IL 2-stroke	1997-02	2.6 (43)	-	-	-	-	-	-	Br	B	316-387	O	B	0.18-0.24	2700-3700
2	1	IL 2-stroke	1997-02	3.0 (50)	-	-	-	-	-	-	Br	B	316-387	O	B	0.18-0.24	2700-3700
3	1	IL 2-stroke	1997-02	4.3 (70)	W/R	B	29.7-36.3	W/G	B	279-341	Br	B	247.5-302.5	O	B	0.08-0.12	2080-3120
4	1	IL 2-stroke	1997-99	5.0 (83)	W/R	B	30-36	W/G	B	279-341	Br	B	248-303	B/W	B	0.14-0.22	3200-4800
4	1	CL 2-stroke	1997-02	6.3 (103)	W/R	B	30-36	W/G	B	279-341	Br	B	248-303	B/W	B	0.14-0.22	3200-4800
5	1	CL 2-stroke	1997-02	6.3 (103)	W/R	B	30-36	W/G	B	279-341	Br	B	248-303	B/W	B	0.14-0.22	3200-4800

NOTE: Unless stated otherwise, With Load tests are made with circuit connected as normal, while No Load tests are conducted with the component disconnected from the wiring harness

① Some Yamaha factory materials conflict with which wire colors and specification combinations belong to the low and high speed coils. But, in all instances, the wire colors and specifications remain as shown in this chart.

② Unless noted otherwise all resistance specifications are at an ambient temperature 68 degrees F (20 degrees C). Keep in mind that all resistance readings will vary with temperature and from meter-to-meter

③ Peak TCI unit cranking output at 300 rpm is 126 volts

Fig. 161 Ignition System Component Testing - Single Cylinder Models

IGNITION AND ELECTRICAL SYSTEMS

IGNITION SYSTEM COMPONENT TESTING - 2-STROKE INLINE ENGINES

HP	No. of Cyl	Engine Type	Model Year	Displace cu. in. (cc)	Pulser Coils Wire leads (+)	Pulser Coils Wire leads (-)	Pulser Coils Resistance Ohms ①	Pulser Coils Peak Voltage With Load	Pulser Coils Peak Voltage No Load	Pulser Coils V @ RPM	Charge Coil Wire leads (+)	Charge Coil Wire leads (-)	Charge Coil Resistance Ohms ①	Charge Coil Peak Voltage With Load	Charge Coil Peak Voltage No Load	Charge Coil V @ RPM	CDI Output Wire leads (+)	CDI Output Wire leads (-)	CDI Output Peak Voltage With Load	CDI Output Peak Voltage No Load	CDI Output V @ RPM	Ignition Coil Resistance Primary Ohms	Ignition Coil Resistance Secondary Ohms ①	Peak Voltage With Load	Peak Voltage No Load
6	2	CL 2-stroke	1997-00	10 (165)	W/R	B	92-112	-	-	-	Br	B	81-99	-	-	-	-	-	-	-	-	0.25-0.35	6800-10200	-	-
8	2	CL 2-stroke	1997-03	10 (165)	W/R	B	92-112	-	-	-	Br	B	81-99	-	-	-	O	B	-	-	-	0.25-0.35	6800-10200	-	-
9.9	2	IL 2-stroke	1997-03	15 (246)	W/R	B	352-528	-	-	-	Br	Blue	248-372	-	-	-	B/W	B	-	-	-	0.05-0.07	1680-2520	-	-
15	2	IL 2-stroke	1997-03	15 (246)	W/R	B	352-528	-	-	-	Br	Blue	248-372	-	-	-	B/W	B	-	-	-	0.05-0.07	1680-2520	-	-
20	2	IL 2-stroke	1997	24 (395)	②	B	311-381	5.5	5.5	15 @ 1500	Br	Blue	342-418	125	130	125 @ 1500	③	B	105	-	110 @ 1500	0.18-0.24	2720-3680	-	-
25	2	IL 2-stroke	1997-03	24 (395)	②	B	311-381	5.5	5.5	15 @ 1500	Br	Blue	342-418	125	130	125 @ 1500	③	B	105	-	110 @ 1500	0.18-0.24	2720-3680	5000	4000
C20	2	IL 2-stroke	1997-98	26 (430)	W/R	Gnd	-	5	-	-	Br	B	-	-	-	④	O	B	-	-	⑤	0.12-0.18	6800-10200	6000	22,000
C25	2	IL 2-stroke	1997-98	26 (430)	W/R	Gnd	-	5	-	-	Br	B	-	190	-	④	O	B	-	-	⑤	0.12-0.18	6800-10200	6000	22,000
25	2	IL 2-stroke	1997	30 (496)	②	B	311-381	4	4.5	15 @ 1500	Br	Blue	342-418	125	125	125 @ 1500	B/W	B	100	-	110 @ 1500	0.18-0.24	2700-3700	-	-
30	2	IL 2-stroke	1997	30 (496)	W/R	B	12.6-15.4	5.5	5.5	20 @ 1500	Br	B	121-147	125	90	180 @ 1500	O	B	115	45	175 @ 1500	0.08-0.10	2970-4030	6000	22,000
40	2	IL 2-stroke	1997	36 (592)	W/R	B	92-112	-	-	-	Br	B	81-99	-	-	-	O	B	-	-	-	0.12-0.18	4320-6480	8000	20,000
48	2	IL 2-stroke	1997-00	46 (760)	W/R	B	277-415	4	4	⑥	Br	Blue	164-246	175	210	⑦	B/W	B	135	0	⑨	0.46-0.62	5400-7200	10,000	30,000
25	3	IL 2-stroke	1997-02	30 (496)	②	B	277-415	4	4	⑥	Br	Blue	164-246	175	210	⑦	⑧	B	190	4.5	⑨	0.46-0.62	5400-7200	10,000	30,000
30	3	IL 2-stroke	1998-02	30 (496)	②	B	277-415	4	4	⑥	Br	Blue	164-246	175	210	⑦	⑧	B	135	0	⑨	0.46-0.62	5400-7200	10,000	30,000
30	3	IL 2-stroke	1997	30 (496)	②	B	277-415	4	4	⑥	Br	Blue	164-246	175	210	⑦	⑧	B	190	4.5	⑨	0.46-0.62	5400-7200	10,000	30,000
28J	3	IL 2-stroke	1998-02	43 (698)	②	B	168-252	3	4	⑥	Br	Blue	368-552	145	115	⑪	⑧	B	125	0	⑨	0.18-0.24	2720-3680	6000	30,000
35J	3	IL 2-stroke	1997-01	43 (698)	②	B	168-252	3	4	⑥	Br	Blue	368-552	145	115	⑪	⑧	B	125	0	⑨	0.18-0.24	2720-3680	6000	30,000
40	3	IL 2-stroke	1997-01	43 (698)	②	B	168-252	3	4	⑥	Br	Blue	368-552	145	115	⑪	⑧	B	125	0	⑨	0.18-0.24	2720-3680	6000	30,000
50	3	IL 2-stroke	1997-03	43 (698)	②	B	168-252	3	4	⑥	Br	Blue	368-552	145	115	⑪	⑧	B	125	0	⑨	0.18-0.24	2720-3680	6000	30,000
50	3	IL 2-stroke	1997-03	52 (849)	W/R	W/B	240-360	2.5	4.5	6.5 @ 1500	Br	Blue	136-204	120	110	160 @ 1500	B/W	B	105	0	145 @ 1500	0.18-0.24	3260-4880	12,000	20,000
60	3	IL 2-stroke	⑩	52 (849)	②	B	104-156	2.5	2.7	8 @ 1000	Br	Blue	132-198	140	170	230 @ 1000	B/W	B	100	0	200 @ 1000	0.20-0.24	3840-5760	12,000	20,000
70	3	IL 2-stroke	⑩	52 (849)	W/R	W/B	240-360	2.5	4.5	⑥	Br	Blue	136-204	150	120	⑫	B/W	B	⑬	0	⑨	0.18-0.24	3260-4880	-	-
70	3	IL 2-stroke	1997-03	52 (849)	W/R	W/B	240-360	2.5	4.5	⑥	Br	Blue	136-204	150	120	⑫	B/W	B	⑬	0	⑨	0.18-0.24	3260-4880	-	-
65J	3	IL 2-stroke	1997-01	70 (1140)	⑧	⑧	256-384	3	4	⑥	⑧	⑧	48-72	40	40	145 @ 1500	B/W	B	105	0	140 @ 1500	0.18-0.24	3260-4880	-	-
75	3	IL 2-stroke	⑫	70 (1140)	W/R	B	256-384	3	4	⑥	⑧	⑧	48-72	40	40	145 @ 1500	B/W	B	105	0	140 @ 1500	0.18-0.24	3260-4880	-	-
	3	IL 2-stroke	⑫	70 (1140)	W/R	B	241-362	5	7	⑥	⑧	⑧	64-96	60	55	170 @ 1500	Blue/R Blue/W	B	130	-	155 @ 1500	0.18-0.24	3200-4800	-	-
	3	IL 2-stroke	⑫	70 (1140)	W/R	B	264-396	2.5	0	7 @ 1500	⑧	⑧	840-1260	85	120	79 @ 1500	B/W	B	95	-	130 @ 1500	0.18-0.26	3840-5760	-	-
80	3	IL 2-stroke	1997-00	70 (1140)	W/R	W/B	240-360	5	7	⑥	⑧	⑧	64-96	60	55	170 @ 1500	B/W	B	130	-	155 @ 1500	0.18-0.24	3280-4920	-	-
85	3	IL 2-stroke	1997-00	70 (1140)	⑧	⑧	256-384	3	4	⑥	⑧	⑧	48-72	40	40	145 @ 1500	B/W	B	105	-	140 @ 1500	0.18-0.24	3260-4880	-	-
90	3	IL 2-stroke	1997-03	70 (1140)	W/R	B	242-362	5	7	⑥	⑧	⑧	64-96	60	55	170 @ 1500	B/W	B	130	-	155 @ 1500	0.18-0.24	3280-4920	-	-

NOTE: Unless stated otherwise, With Load tests are made with circuit connected as normal, while No Load tests are conducted with the component disconnected from the wiring harness

① Unless noted otherwise all resistance specifications are at an ambient temperature 68 degrees F (20 degrees C). Keep in mind that all resistance readings will vary with temperature and from meter-to-meter

② The pulser coil wire leads are normally W/R for the No. 1 cylinder and W/B for the No. 2 cylinder, and for triples, W/G for the No. 3 cylinder

③ The CDI out leads are normally either B/W for both cylinders or B/O for the No. 1 cylinder and B/W for the No. 2 cylinder

④ Specifications are a minimum 150 volts cranking and 165 volts with the engine running at 1500 rpm

⑤ Specification are a minimum 140 volts cranking and 155 volts with the engine running at 1500 rpm

⑥ Specifications are a minimum 11 volts @ 1500 rpm or 20 volts @ 3500 rpm

⑦ Factory specifications vary and conflict. The Yamaha tune-up guide states output should be a minimum of 205 volts @ 1500 rpm for all models, but the 1997 service manual gives 200 volts @ 1500 rpm and 130 volts @ 3500 rpm while 1998 and later service manuals state 205 volts @ 1500 rpm or 115 volts @ 3500 rpm. (Note the 1998 service manuals also reverse the resistance specs for load vs. no load)

⑧ Wire colors may vary, some models may have 3 B/W leads, while most have B/O for No. 1, B/W for No. 2 and B/Y for No. 3. Note that colors are for leads coming directly out of CDI unit as they may all connect to B/W leads before reaching the individual coils

IGNITION AND ELECTRICAL SYSTEMS

IGNITION SYSTEM COMPONENT TESTING - 2-STROKE INLINE ENGINES

HP	No. of Cyl	Engine Type	Model Year	Displace cu. in. (cc)	Pulser Coils					Charge Coil					CDI Output					Ignition Coil		
					Resistance Ohms ①	Wire leads (+)	Wire leads (−)	Peak Voltage With Load	Peak Voltage No Load	V @ RPM	Resistance Ohms ①	Wire leads (+)	Wire leads (−)	Peak Voltage With Load	Peak Voltage No Load	V @ RPM	Wire leads (+)	Wire leads (−)	Peak Voltage With Load	Peak Voltage No Load	V @ RPM	Resistance Primary Ohms ①	Resistance Secondary Ohms ①	Peak Voltage With Load	Peak Voltage No Load

① Specification is 180 volts @ 1500 rpm and 120 volts @ 3500 rpm for 1997 motors or 185 volts @ 1500 rpm or 15 volts @ 3500 rpm
② Specifications are a minimum 9 volts @ 1500 rpm or 15 volts @ 3500 rpm
③ Specifications are a minimum 160 volts @ 1500 rpm or 130 volts @ 3500 rpm
④ Specifications are a minimum 140 volts @ 1500 rpm or 110 volts @ 3500 rpm
⑤ Specifications vary by ignition system (which depends upon year and model). Use these specifications for models (like the E60 produced through 2000) which utilizes 3 pulser coils and no crankshaft position sensor
⑥ Specifications vary by ignition system (which depends upon year and model). Use these specifications for models equipped with a single pulser coil and a crankshaft position sensor (which is most models, except the E60)
⑦ Specifications are a minimum 6.5 volts @ 1500 rpm and 10 volts @ 3500 rpm
⑧ Specifications are a minimum 160 volts @ 1500 rpm and 120 volts @ 3500 rpm
⑨ Specifications are for readings taken on #1 and #3 coil wires only for 01-03 models, except the C60 and C70, readings should be 145 volts @ 1500 or 105 volts @ 3500 rpm
⑩ Wire leads are W/R (+) and W/Y (−) for the No. 1 pulser coil or W/B (+) and W/G (−) for the No. 2 pulser coil
⑪ Specifications are a minimum 8 volts @ 1500 rpm and 12 volts @ 3500 rpm
⑫ These models use both low and high speed coils, factory sources sometimes conflict as to which coil, the specs in the chart are normally for the low speed coil. Low speed coil Br and R wires (and a spec of 160 volts @ 3500 rpm). High speed coil: R and Blue wires, 428-642 ohms, 105 volts cranking, 125 volts unloaded (open circuit), 135 volts @ 1500 rpm and 160 volts @ 3500 rpm. Always verify the reading by checking the other set of wires against both sets of specs as well (noting that the voltage and resistance specs for a given coil reverse in some Yamaha publications, i.e. that the voltage readings in some publications correspond to the coil with the opposite resistance reading)
⑬ Specifications vary more by model than by year, use these specs for E75 models
⑭ Specifications vary more by model than by year, use these specs for C75 (which are the only 75 hp motors that utilize a single pulser coil and a crankshaft position sensor)
⑮ Specifications vary more by model than by year, use these specs for P75 models (which, like the E75 use dual pulser coils and no crank sensor)
⑯ Specifications are a minimum 14 volts @ 1500 rpm and 20 volts @ 3500 rpm
⑰ These models use both low and high speed coils, factory sources sometimes conflict as to which specs apply to which coil, the specs in the chart are normally for the low speed coil. Low speed coil: Br and R wires (and a spec of 150 volts @ 3500 rpm). High speed coil: R and Blue wires, 191-288 ohms, 100 volts loaded, 90 volts unloaded, 135 volts @ 1500 rpm and 135 volts @ 3500 rpm. Always verify the reading by checking the other set of wires against both sets of specs as well (noting that the voltage and resistance specs for a given coil reverse in some Yamaha publications, i.e. that the voltage readings in some publications correspond to the coil with the opposite resistance reading)
⑱ These models use both low and high speed coils, factory sources sometimes conflict as to which specs apply to which coil, the specs in the chart are normally for the low speed coil. Low speed coil: Br and Blue wires. High speed coil: R and Blue wires, 96-144 ohms, 45 volts loaded/unloaded, 130 volts @ 1500 rpm. Always verify the reading by checking the other set of wires against both sets of specs as well
⑲ Specifications are a minimum 14 volts @ 1500 rpm and 45 volts @ 3500 rpm

Fig 162 Ignition System Component Testing - Inline Engines

4-60 IGNITION AND ELECTRICAL SYSTEMS

IGNITION SYSTEM COMPONENT TESTING - V ENGINES - PART I

HP	No. of Cyl	Engine Type	Model Year	Displace cu. in. (cc)	Pulser Coils Wire leads (+)	Pulser Coils Wire leads (-)	Pulser Coils Resistance Ohms ①	Pulser Coils Peak Voltage With Load	Pulser Coils Peak Voltage No Load	Pulser Coils Peak Voltage V @ RPM	Pulser Coils Peak Voltage V @ RPM	Charge Coil Wire leads (+)	Charge Coil Wire leads (-)	Charge Coil Resistance Ohms ①	Charge Coil Low Speed Peak Voltage With Load	Charge Coil Low Speed Peak Voltage No Load	Charge Coil Low Speed Peak Voltage V @ RPM	Charge Coil Low Speed Peak Voltage V @ RPM	Charge Coil High Speed Peak Voltage With Load	Charge Coil High Speed Peak Voltage No Load	Charge Coil High Speed Peak Voltage V @ RPM	Charge Coil High Speed Peak Voltage V @ RPM
80J	4	90 LV 2-St	1997-01	106 (1730)	②	②	256-384	2.5	2.5	7 @ 1500	11 @ 3500	Br	R	592-888	160	170	165 @ 1500	170 @ 3500	45	45	165 @ 1500	170 @ 3500
100	4	90 LV 2-St	1997-02	106 (1730)	②	②	256-384	2.5	2.5	7 @ 1500	11 @ 3500	Br	R	592-888	160	170	165 @ 1500	170 @ 3500	45	45	165 @ 1500	170 @ 3500
115	4	90 LV 2-St	1997	106 (1730)	②	②	256-384	2.5	2.5	7 @ 1500	11 @ 3500	Br	R	592-888	160	170	165 @ 1500	170 @ 3500	45	45	165 @ 1500	170 @ 3500
115	4	90 LV 2-St	③	106 (1730)	②	②	288-432	2.5	3.0	8 @ 1500	12 @ 3500	Br	R	840-1260	105	95	160 @ 1500	160 @ 3500	30	30	160 @ 1500	160 @ 3500
130	4	90 LV 2-St	1997-03	106 (1730)	②	②	256-384	2.5	2.5	7 @ 1500	11 @ 3500	Br	R	592-888	160	170	165 @ 1500	170 @ 3500	45	45	165 @ 1500	170 @ 3500
140	4	90 LV 2-St	1997-02	106 (1730)	②	②	288-432	2.5	3.0	8 @ 1500	12 @ 3500	Br	R	840-1260	105	95	160 @ 1500	160 @ 3500	30	30	160 @ 1500	160 @ 3500
105J	6	90 LV 2-St	1997-00	158 (2596)	⑥	⑥	256-384	2.0	3.0	8 @ 1500	14 @ 3500	Br	R	592-888	160	140	165 @ 1500	165 @ 3500	55	40	165 @ 1500	165 @ 3500
150 Carb	6	90 LV 2-St	④	158 (2596)	⑥	⑥	256-384	2.0	3.0	8 @ 1500	14 @ 3500	Br	R	592-888	160	140	165 @ 1500	165 @ 3500	55	40	165 @ 1500	165 @ 3500
150 Carb	6	90 LV 2-St	⑤	158 (2596)	⑥	⑥	256-384	2.0	2.5	9.5 @ 1500	16 @ 3500	Br	R	428-642	90	80	150 @ 1500	150 @ 3500	30	30	150 @ 1500	150 @ 3500
150 EFI	6	90 LV 2-St	1999-03	158 (2596)	⑦	B	294-398	3.0	3.0	16 @ 1500	30 @ 3500	Br	R	224-336	110	85	165 @ 1500	165 @ 3500	110	85	165 @ 1500	165 @ 3500
175	6	90 LV 2-St	④	158 (2596)	⑥	⑥	256-384	2.0	3.0	8 @ 1500	14 @ 3500	Br	R	592-888	160	140	165 @ 1500	165 @ 3500	55	40	165 @ 1500	165 @ 3500
175	6	90 LV 2-St	⑤	158 (2596)	⑦	B	256-384	2.0	2.5	9.5 @ 1500	16 @ 3500	Br	R	428-642	90	80	150 @ 1500	150 @ 3500	30	30	150 @ 1500	150 @ 3500
200 Carb	6	90 LV 2-St	1997-99	158 (2596)	⑥	⑥	256-384	2.0	3.0	8 @ 1500	14 @ 3500	Br	R	592-888	160	140	165 @ 1500	165 @ 3500	55	40	165 @ 1500	165 @ 3500
200 EFI	6	90 LV 2-St	1999-03	158 (2596)	⑥	B	294-398	3.0	3.0	16 @ 1500	30 @ 3500	Br	R	224-336	110	85	150 @ 1500	150 @ 3500	110	85	150 @ 1500	150 @ 3500
225	6	90 LV 2-St	1997	158 (2596)	⑥	⑥	256-384	2.0	3.0	8 @ 1500	14 @ 3500	Br	R	592-888	160	140	165 @ 1500	165 @ 3500	55	40	165 @ 1500	165 @ 3500
150 HPDI	6	76 LV 2-St	2000-03	158 (2596)	⑧	B		5.0	5.0	20 @ 1500	35 @ 3500	-	-	-	-	-	-	-	-	-	-	-
175 HPDI	6	76 LV 2-St	2001-03	158 (2596)	⑧	B		5.0	5.0	20 @ 1500	35 @ 3500	-	-	-	-	-	-	-	-	-	-	-
200 HPDI	6	76 LV 2-St	2000-03	158 (2596)	⑧	B		5.0	5.0	20 @ 1500	35 @ 3500	-	-	-	-	-	-	-	-	-	-	-
200 EFI	6	76 LV 2-St	1998-03	191 (3130)	⑩	B	294-398	3.0	3.0	16 @ 1500	30 @ 3500	Br	R	224-336	115	90	150 @ 1500	150 @ 3500	115	90	150 @ 1500	150 @ 3500
225 EFI	6	76 LV 2-St	1997-03	191 (3130)	⑩	B	294-398	3.0	3.0	16 @ 1500	30 @ 3500	Br	R	224-336	115	90	150 @ 1500	150 @ 3500	115	90	150 @ 1500	150 @ 3500
250 EFI	6	76 LV 2-St	1997-01	191 (3130)	⑩	B	294-398	3.0	3.0	16 @ 1500	30 @ 3500	Br	R	224-336	115	90	150 @ 1500	150 @ 3500	115	90	150 @ 1500	150 @ 3500
250 EFI	6	76 LV 2-St	⑨	191 (3130)	⑧	B		3.0	4.0	18 @ 1500	33 @ 3500	Br	R		100	80	150 @ 1500	130 @ 3500	100	80	150 @ 1500	130 @ 3500
225 HPDI	6	76 LV 2-St	2003	204 (3342)	⑧	B	294-398	3.5	3.5	20 @ 1500	30 @ 3500	⑫	⑫	⑫	⑫	⑫	⑫	⑫	⑫	⑫	⑫	⑫
250 HPDI	6	76 LV 2-St	2003	204 (3342)	⑧	B	294-398	3.5	3.5	20 @ 1500	30 @ 3500	⑫	⑫	⑫	⑫	⑫	⑫	⑫	⑫	⑫	⑫	⑫

NOTE: Unless stated otherwise, With Load tests are made with circuit connected as normal, while No Load tests are conducted with the component disconnected from the wiring harness
PLEASE SEE IGNITION SYSTEM COMPONENT TESTING - V ENGINES - PART II for footnotes

Fig. 163 Ignition System Component Testing - V-Engines

IGNITION SYSTEM COMPONENT TESTING - V ENGINES - Part II

HP	No. of Cyl	Engine Type	Model Year	Displace cu. in. (cc)	CDI Output Wire leads (+)	CDI Output Wire leads (-)	Peak Voltage With Load	Peak Voltage No Load	V @ RPM	V @ RPM	Ignition Coil Wire leads (+)	Ignition Coil Wire leads (-)	Resistance Primary Ohms ①	Resistance Secondary Ohms ①	Peak Voltage With Load	Peak Voltage No Load
80J	4	90 LV 2-St	1997-01	106 (1730)	B/W	B	125	0-1	140 @ 1500	145 @ 3500	B/W	B	-	3040-4560	5000	11,000
100	4	90 LV 2-St	1997-02	106 (1730)	B/W	B	125	0-1	140 @ 1500	145 @ 3500	B/W	B	-	3040-4560	5000	11,000
115	4	90 LV 2-St	1997	106 (1730)	B/W	B	125	0-1	140 @ 1500	145 @ 3500	B/W	B	-	3040-4560	5000	11,000
	4	90 LV 2-St	③	106 (1730)	B/W	B	85	0-1	140 @ 1500	135 @ 3500	B/W	B	-	2000-3000	-	-
130	4	90 LV 2-St	1997-03	106 (1730)	B/W	B	125	0-1	140 @ 1500	145 @ 3500	B/W	B	-	3040-4560	5000	11,000
140	4	90 LV 2-St	1997-02	106 (1730)	B/W	B	85	0-1	140 @ 1500	135 @ 3500	B/W	B	-	2000-3000	-	-
105J	6	90 LV 2-St	1997-00	158 (2596)	B/W	B	130	-	145 @ 1500	145 @ 3500	B/W	Gnd	-	3280-4920	-	-
150 Carb	6	90 LV 2-St	④	158 (2596)	B/W	B	130	-	145 @ 1500	145 @ 3500	B/W	Gnd	-	3280-4920	-	-
150 Carb	6	90 LV 2-St	⑤	158 (2596)	B/W	B	65	-	140 @ 1500	135 @ 3500	B/W	Gnd	-	3280-4920	-	-
150 EFI	6	90 LV 2-St	1999-03	158 (2596)	⑨	B	100	80	150 @ 1500	130 @ 3500	-	-	-	-	-	-
175	6	90 LV 2-St	④	158 (2596)	B/W	B	130	-	145 @ 1500	145 @ 3500	B/W	Gnd	-	3280-4920	-	-
	6	90 LV 2-St	⑤	158 (2596)	B/W	B	65	-	140 @ 1500	135 @ 3500	B/W	Gnd	-	3280-4920	-	-
200 Carb	6	90 LV 2-St	1997-99	158 (2596)	B/W	B	130	-	145 @ 1500	145 @ 3500	B/W	Gnd	-	3280-4920	-	-
200 EFI	6	90 LV 2-St	1999-03	158 (2596)	⑨	B	100	80	150 @ 1500	130 @ 3500	-	-	-	-	-	-
225	6	90 LV 2-St	1997	158 (2596)	B/W	B	130	-	145 @ 1500	145 @ 3500	B/W	Gnd	-	3280-4920	-	-
150 HPDI	6	76 LV 2-St	2000-03	158 (2596)	⑨	R/Y	140	-	205 @ 1500	220 @ 3500	-	-	-	-	-	-
175 HPDI	6	76 LV 2-St	2001-03	158 (2596)	⑨	R/Y	140	-	205 @ 1500	220 @ 3500	-	-	-	-	-	-
200 HPDI	6	76 LV 2-St	2000-03	158 (2596)	⑨	R/Y	140	-	205 @ 1500	220 @ 3500	-	-	-	-	-	-
200 EFI	6	76 LV 2-St	1998-03	191 (3130)	⑪	B	100	85	150 @ 1500	130 @ 3500	B/W	Gnd	-	2700-3600	-	-
225 EFI	6	76 LV 2-St	1997-03	191 (3130)	⑪	B	100	85	150 @ 1500	130 @ 3500	B/W	Gnd	-	2700-3600	-	-
250 EFI	6	76 LV 2-St	1997-01	191 (3130)	⑪	B	100	85	150 @ 1500	130 @ 3500	B/W	Gnd	-	2700-3600	-	-
	6	76 LV 2-St	⑫	191 (3130)	⑨	B	84	10	140 @ 1500	110 @ 3500	B/W	B	0.18-0.24	2720-3680	-	-
225 HPDI	6	76 LV 2-St	2003	204 (3342)	⑬	⑬	160	-	260 @ 1500	260 @ 3500	R	B/W	1.78-2.53	8930-1208	-	-
250 HPDI	6	76 LV 2-St	2003	204 (3342)	⑬	⑬	160	-	260 @ 1500	260 @ 3500	R	B/W	1.78-2.53	8930-1208	-	-

NOTE: Unless stated otherwise, With Load tests are made with circuit connected as normal, while No Load tests are conducted with the component disconnected from the wiring harness

① Unless noted otherwise all resistance specifications are at an ambient temperature 68 degrees F (20 degrees C). Keep in mind that all resistance readings will vary with temperature and from meter-to-meter
② Wire leads are normally W/R (+) W/Y (-) for the No. 1 & 3 cyl and W/B (+) W/G (-) for the No. 2 & 4 cyl
③ Specifications vary more by model than by year. For oil injected models (except the 115BETO) use the specs for 1997. For pre-mix models and for the 115BETO use the specs listed in this row
④ Specifications vary more by model than by year. Specs listed in this row are for oil injected models
⑤ Specifications vary more by model than by year. Specs listed in this row are for pre-mix models, like the C150
⑥ Wire leads are normally W/R (+) W/G (-) and W/Y (+) W/Br (-)
⑦ Wire leads are normally W/R (+) W/G (-), W/B (+) W/Blue (-) and W/Y (+) W/Br (-)
⑧ There are separate wire leads for each cylinder, test between each of these and B: W/R, W/Y, W/G, W/B, W/Blue and W/Br
⑨ There are separate wire leads for each cylinder, test between each of these and B (EFI) or R/Y (HPDI): B/O, B/Y, B/Blue, B/Br, B/G, and B/W
⑩ Separate wire leads for each cylinder test between each of these and B: W/R, W/B, W/Y, W/G, W/Blue and W/Br
⑪ There are separate wire leads for each cylinder, test between each of these and a good ground: B/O, B, B/Y, B/G, B/Blue and B/Br
⑫ Specifications for 2002 and later motors vary more by model than year. For most models, refer to the 1997-01 row. For VX250 models use the specs in this row
⑬ Power is supplied by the stator coil on these models
⑭ The ignition is ECM controlled, power output is tested across the R and B/W wires
⑮ There are 3 separate pulser coil windings, the leads are W/R, W/B and W/G
⑯ The ignition is ECM controlled and output should be checked across 3 pairs of wires the B and each of the following: B/O, B/Y and B/W
⑰ The ignition coil primary resistance is checked between R/Y and each of the following: B/O, B/Y and B/W
⑱ The ignition coil secondary resistance is checked from secondary wire lead-to-secondary wire lead and should be 19,600-35,400 ohms

Fig 164 Ignition System Component Testing - V-Engines (Cont'd)

IGNITION AND ELECTRICAL SYSTEMS

CHARGING SYSTEM TESTING - 2-Stroke Motors

Model (Hp)	No. of Cyl	Engine Type	Year	Disp cu. in. (cc)	Lighting Coil Wire Colors	Lighting Coil Resistance Ohms ①	Voltage @ RPM Minimum	Voltage @ RPM Peak
6	2	CL 2-stroke	1997-00	10 (165)	G - G	0.36-0.44	12 @ 3000	13.5-16.5 @ 5500
8	2	CL 2-stroke	1997-03	10 (165)	G - G	0.36-0.44	12 @ 3000	13.5-16.5 @ 5500
9.9	2	IL 2-stroke	1997-03	15 (246)	G - G	0.16-0.24	11.5 @ 3000	14.0-17.5 @ 5500
15	2	IL 2-stroke	1997-03	15 (246)	G - G	0.16-0.24	11.5 @ 3000	14.0-17.5 @ 5500
20	2	IL 2-stroke	1997	24 (395)	G - G	0.30-0.36	4 @ cranking	13 @ 1500 (min)
25	2	IL 2-stroke	1997-03	24 (395)	G - G	0.30-0.36	4 @ cranking	13 @ 1500 (min)
20	2	IL 2-stroke	1997-98	26 (430)	G - G	0.38-0.46	5.5 cranking	9.5 @ 1500 (Min)
25	2	IL 2-stroke	1997-98	26 (430)	G - G	0.38-0.46	5.5 cranking	9.5 @ 1500 (Min)
25	2	IL 2-stroke	1997	30 (496)	G - G	0.31-0.37	11.5 @ 3000 ②	13.5-16.5 @ 5500 ②
30	2	IL 2-stroke	1997	30 (496)	G - G	0.31-0.37	11.5 @ 3000 ②	13.5-16.5 @ 5500 ②
40	2	IL 2-stroke	1997	36 (592)	G - G	0.23-0.29	4.1 cranking	17 @ 1500 (Min)
48	2	IL 2-stroke	1997-00	46 (760)	G - G	0.36-0.44	-	13.5-16.5 @ 1500
25	3	IL 2-stroke	1997	30 (496)	G - G	-	8.0 cranking	20 @ 1500 (Min) ③
	3	IL 2-stroke	1998-02	30 (496)	G - G	-	4.0 cranking	④
30	3	IL 2-stroke	1997	30 (496)	G - G	-	8.0 cranking	20 @ 1500 (Min) ③
	3	IL 2-stroke	1998-02	30 (496)	G - G	-	4.0 cranking	④
28J	3	IL 2-stroke	1997	43 (698)	G/W - G	0.56-0.84	12 @ 3000	13.5-16.5 @ 5500
	3	IL 2-stroke	1998-01	43 (698)	G/W - G	-	9.0 cranking	25 @ 1500-3500
35J	3	IL 2-stroke	1997	43 (698)	G/W - G	0.56-0.84	12 @ 3000	13.5-16.5 @ 5500
	3	IL 2-stroke	1998-01	43 (698)	G/W - G	-	9.0 cranking	25 @ 1500-3500
40	3	IL 2-stroke	1997	43 (698)	G/W - G	0.56-0.84	12 @ 3000	13.5-16.5 @ 5500
	3	IL 2-stroke	1998-03	43 (698)	G/W - G	-	9.0 cranking	25 @ 1500-3500
50	3	IL 2-stroke	1997	43 (698)	G/W - G	0.56-0.84	12 @ 3000	13.5-16.5 @ 5500
	3	IL 2-stroke	1998-03	43 (698)	G/W - G	-	9.0 cranking	25 @ 1500-3500
50	3	IL 2-stroke	1997-03	52 (849)	G/W - G	0.57-0.85	8.0 cranking	25 @ 1500-3500
60	3	IL 2-stroke	1997-03	52 (849)	G/W - G	0.57-0.85	8.0 cranking	25 @ 1500-3500
70	3	IL 2-stroke	1997-03	52 (849)	G/W - G	0.57-0.85	8.0 cranking	25 @ 1500-3500
65J	3	IL 2-stroke	1997-01	70 (1140)	G/W - G	0.36-0.54	8.0 cranking	25 @ 1500-3500 ⑤
75	3	IL 2-stroke	⑥	70 (1140)	G/W - G	0.40-0.60	10.0 cranking	25 @ 1500-3500
80	3	IL 2-stroke	1997-00	70 (1140)	G/W - G	0.40-0.60	10.0 cranking	25 @ 1500-3500
85	3	IL 2-stroke	1997-00	70 (1140)	G/W - G	0.36-0.54	8.0 cranking	25 @ 1500-3500 ⑤
90	3	IL 2-stroke	1997-03	70 (1140)	G/W - G	0.40-0.60	10.0 cranking	25 @ 1500-3500
80J	4	90 LV 2-stroke	1997-01	106 (1730)	G - G	0.40-0.60	7.0 cranking	35 @ 1500/85 @ 3500
100	4	90 LV 2-stroke	1997-02	106 (1730)	G - G	0.40-0.60	7.0 cranking	35 @ 1500/85 @ 3500
115	4	90 LV 2-stroke	1997	106 (1730)	G - G	0.40-0.60	7.0 cranking	35 @ 1500/85 @ 3500
	4	90 LV 2-stroke	⑦	106 (1730)	G/W - G ⑧	-	6.0 cranking	30 @ 1500/75 @ 3500
130	4	90 LV 2-stroke	1997-03	106 (1730)	G - G	0.40-0.60	7.0 cranking	35 @ 1500/85 @ 3500
140	4	90 LV 2-stroke	1997-02	106 (1730)	G/W - G ⑧	-	6.0 cranking	30 @ 1500/75 @ 3500
105J	6	90 LV 2-stroke	1997-00	158 (2596)	G - G	-	5.5 cranking	35 @ 1500/85 @ 3500
150 Carb	6	90 LV 2-stroke	⑨	158 (2596)	G - G	-	5.5 cranking	35 @ 1500/85 @ 3500
150 Carb	6	90 LV 2-stroke	⑩	158 (2596)	G - G	-	3.0 cranking	20 @ 1500/50 @ 3500
150 EFI	6	90 LV 2-stroke	1999-03	158 (2596)	G - G	-	-	14 @ 1500/14 @ 3500

Fig. 165 Charging System Testing

IGNITION AND ELECTRICAL SYSTEMS

CHARGING SYSTEM TESTING - 2-Stroke Motors

Model (Hp)	No. of Cyl	Engine Type	Year	Disp cu. in. (cc)	Lighting Coil Wire Colors	Lighting Coil Resistance Ohms ①	Voltage @ RPM Minimum	Voltage @ RPM Peak
175	6	90 LV 2-stroke	⑨	158 (2596)	G - G	-	5.5 cranking	35 @ 1500/85 @ 3500
	6	90 LV 2-stroke	⑩	158 (2596)	G - G	-	3.0 cranking	20 @ 1500/50 @ 3500
200 Carb	6	90 LV 2-stroke	⑨	158 (2596)	G - G	-	5.5 cranking	35 @ 1500/85 @ 3500
	6	90 LV 2-stroke	⑩	158 (2596)	G - G	-	3.0 cranking	20 @ 1500/50 @ 3500
200 EFI	6	90 LV 2-stroke	1999-03	158 (2596)	G - G	-	-	14 @ 1500/14 @ 3500
225	6	90 LV 2-stroke	1997	158 (2596)	G - G	-	5.5 cranking	35 @ 1500/85 @ 3500
150 HPDI	6	76 V 2-stroke	2000-03	158 (2596)	G - G	-	5.5 cranking ⑪	37 @ 1500/86 @ 3500
175 HPDI	6	76 V 2-stroke	2001-03	158 (2596)	G - G	-	5.5 cranking ⑪	37 @ 1500/86 @ 3500
200 HPDI	6	76 V 2-stroke	2000-03	158 (2596)	G - G	-	8.0 cranking ⑫	12 @ 1500-3500
200 EFI	6	76 V 2-stroke	1998-03	191 (3130)	G - G	-	6 cranking	14 @ 1500-3500
225 EFI	6	76 V 2-stroke	1997-03	191 (3130)	G - G	-	6 cranking	⑬
250 EFI	6	76 V 2-stroke	1997-03	191 (3130)	G - G	-	6 cranking	⑬
225 HPDI	6	76 V 2-stroke	2003	204 (3342)	G - G	-	5.5 cranking	40 @ 1500/90 @ 3500
250 HPDI	6	76 V 2-stroke	2003	204 (3342)	G - G	-	5.5 cranking	40 @ 1500/90 @ 3500

① Unless noted otherwise all resistance specifications are at an ambient temperature 68 degrees F (20 degrees C). Keep in mind that all resistance readings will vary with temperature and from meter-to-meter

② Specification is for lighting voltage, charging voltage is a minimum of 3 @ 3000 rpm and a max of 5-7 @ 5500 rpm

③ Readings should increase to a minimum of 40 volts @ 3500 rpm

④ Minimum output varies with test conditions. Specifications are 12 volts @ 1500 rpm or 25 volts @ 3500 rpm (loaded/circuit complete) and 18 volts @ 1500 rpm or 30 volts @ 3500 rpm (unloaded, circuit open)

⑤ Spec is for lighting coil output on models w/out voltage regulator, output for models w/ a regulator is 30 volts @ 1500 rpm or 75 volts @ 3500 rpm. Also charging system current is a min of 7 amps @ 3000 rpm / max of 9-11 amps @ 5500 rpm

⑥ Specifications vary more by model than year. Specs in chart apply to most models, except A and E models for which the specifications are the same as the 65J

⑦ Speciciations vary more by model than year. For oil injection models use the specs for 1997, for pre-mix models use the specs in this row

⑧ Though most models are G/W - G, the lighting coil wires on some of these pre-mix modes are G - G

⑨ Specifications vary more by model than by year, the specs in this row are for oil injected models

⑩ Specifications vary more by model than by year, the specs in this row are for pre-mix models

⑪ Spec is loaded (complete circuit), however specs in chart are from factory service manual and they conflict w/ those listed for in the factory tune-up guide. The specs in the tune-up guide are the same as those listed for the 200HPDI

⑫ Specification is loaded (complete circuit), spec for cranking without load (open circuit) is 7.5 volts

⑬ Specs conflict in different factory sources. The service manuals say 14 volts @ 1500-3500 rpm on all EXCEPT V/VX models for which they say 40 volts @ 1500 rpm / 84 volts @ 3500 rpm. However, the tune-up guide says 40 @ 1500 / 84 @ 3500 for all models

Fig. 166 Charging System Testing (Cont'd)

Crankshaft Position Sensor Specifications - 2-Stroke Motors

Model (Hp)	No. of Cyl	Engine Type	Year	Displace cu. in. (cc)	Wire Colors	Sensor Resistance	With Load	W/out Load	Sensor Output V @ rpm	V @ rpm
50	3	IL 2-stroke	1997-03	52 (849)	Blue/W - Blue/R	158-236	5	5	20 @ 1500	16 @ 3500
60	3	IL 2-stroke	1997-03	52 (849)	Blue/W - Blue/R	158-236	5	5	20 @ 1500	16 @ 3500
70	3	IL 2-stroke	1997-03	52 (849)	Blue/W - Blue/R	158-236	5	5	20 @ 1500	16 @ 3500
75	3	IL 2-stroke	1997-00	70 (1140)	Blue/W - Blue/R	158-236	5.5	5.5	25 @ 1500	20 @ 3500
80	3	IL 2-stroke	1997-00	70 (1140)	Blue/W - Blue/R	158-236	5.5	5.5	25 @ 1500	20 @ 3500
90	3	IL 2-stroke	1997-03	70 (1140)	Blue/W - Blue/R	158-236	5.5	5.5	25 @ 1500	20 @ 3500
105J	6	90 LV 2-st	1997-00	158 (2596)	G - G	158-236	2	3	5.5 @ 1500	6.0 @ 3500
150 Carb	6	90 LV 2-st	1997-03	158 (2596)	G - G	158-236	2	3	5.5 @ 1500	6.0 @ 3500
150 EFI	6	90 LV 2-st	1999-03	158 (2596)	G/W - G/Blue	-	0.5	6	3.0 @ 1500	4.0 @ 3500
175	6	90 LV 2-st	1997-00	158 (2596)	G - G	158-236	2	3	5.5 @ 1500	6.0 @ 3500
200 Carb	6	90 LV 2-st	1997-99	158 (2596)	G - G	158-236	2	3	5.5 @ 1500	6.0 @ 3500
200 EFI	6	90 LV 2-st	1999-03	158 (2596)	G/W - G/Blue	179-242	0.5	6	3.0 @ 1500	4.0 @ 3500
225	6	90 LV 2-st	1997	158 (2596)	G - G	158-236	2	3	5.5 @ 1500	6.0 @ 3500
150 HPDI	6	76 LV 2-st	2000-03	158 (2596)	G - Blue	-	4.0	4.5	13 @ 1500	20 @ 3500
175 HPDI	6	76 LV 2-st	2001-03	158 (2596)	G - Blue	-	4.0	4.5	13 @ 1500	20 @ 3500
200 HPDI	6	76 LV 2-st	2000-03	158 (2596)	G - Blue	-	4.0	4.5	13 @ 1500	20 @ 3500
200 EFI	6	76 LV 2-st	1998-03	191 (3130)	G/W - G/Blue	158-236	0.5	6	3.0 @ 1500	4.0 @ 3500
225 EFI	6	76 LV 2-st	1997-03	191 (3130)	G/W - G/Blue	158-236	0.5	6	3.0 @ 1500	4.0 @ 3500
250 EFI	6	76 LV 2-st	1997-03	191 (3130)	G/W - G/Blue	158-236	0.5	6	3.0 @ 1500	4.0 @ 3500
225 HPDI	6	76 LV 2-st	2003	204 (3342)	G/Blue - G/W	179-242	1.5	1.5	8.0 @ 1500	10 @ 3500
250 HPDI	6	76 LV 2-st	2003	204 (3342)	G/Blue - G/W	179-242	1.5	1.5	8.0 @ 1500	10 @ 3500

Fig. 167 Crankshaft Position Sensor Specifications

IGNITION AND ELECTRICAL SYSTEMS 4-65

WIRING DIAGRAMS

■ The following wiring diagrams represent the most common models of outboard engines covered in this manual. Models with accessory options may not be depicted here.

1-CYLINDER MOTORS..................................... 4-66
   2 HP (43 AND 50CC) MODELS 4-66
   3 HP ENGINES 4-66
   4/5 HP (83 AND 103CC) MODELS 4-67
2-CYLINDER MOTORS..................................... 4-67
   6/8 HP MODELS 4-67
   9.9/15 HP MODELS 4-68
      CHARGING SYSTEM 4-69
      IGNITION SYSTEM................................. 4-68
      STARTING SYSTEM................................ 4-69
   20/25 HP (395CC) MODELS 4-70
      CHARGING SYSTEM 4-71
      IGNITION SYSTEM................................. 4-70
      STARTING SYSTEM................................ 4-71
   20/25 HP (430CC) MODELS 4-72
   25/30 HP (496CC) MODELS 4-72
   40 HP MODELS...................................... 4-73
   48 HP MODELS...................................... 4-74
3-CYLINDER MOTORS..................................... 4-74
   25/30 HP (496CC) 4-74
   28J-50 HP (698CC) MODELS 4-76
      CHARGING SYSTEM 4-78
      ENRICHMENT CONTROL SYSTEM 4-79
      IGNITION SYSTEM................................. 4-76
      POWER TRIM/TILT SYSTEM 4-78
      STARTING SYSTEM................................ 4-77
   MOST 50-70 HP (849CC) MODELS (EXCEPT C60 AND
   MODELS W/ 3 PULSER COILS SUCH AS THE E60) 4-79
   E60 HP (849CC) MODELS W/3 PULSER COILS 4-80
      CHARGING SYSTEM 4-82
      IGNITION SYSTEM................................. 4-80
      STARTING SYSTEM................................ 4-81
   MOST 65J/75/80/90 HP (1140CC) MODELS (W/SINGLE
   PULSER COIL AND A CRANKSHAFT POSITION SENSOR) 4-83
   C75 HP (1140CC) MODELS (W/SINGLE PULSER COIL
   AND A CRANKSHAFT POSITION SENSOR)................. 4-84
   60J/E75/85 HP (1140CC) MODELS W/DUAL PULSER COILS 4-85
      CHARGING SYSTEM 4-86
      IGNITION SYSTEM................................. 4-85
      POWER TRIM/TILT SYSTEM 4-87
      STARTING SYSTEM................................ 4-86

V4 MOTORS... 4-88
   MOST 1997-98 80J-140 HP (1730CC) V4 MODELS W/OIL
   INJECTION AND POWER TRIM/TILT (EXCEPT 115BETO)...... 4-88
   1997-98 115BETO (1730CC) V4S (SOLD OUTSIDE THE
   U.S./CANADA), W/OIL INJECTION AND POWER TRIM/TILT..... 4-89
   1997-98 C115TR/115BET/140BET (1730CC) V4 PRE-MIX
   MODELS W/POWER TRIM/TILT 4-89
   1997-98 115BE (1730CC) V4 PRE-MIX MODELS W/MANUAL
   TILT (OUTSIDE THE U.S./CANADA) 4-90
   MOST 1999-03 80J-140 HP (1730CC) V4 MODELS W/OIL
   INJECTION AND POWER TRIM/TILT (EXCEPT 115BETO)...... 4-90
   1999-03 80J-140 HP (1730CC) V4 REMOTE CONTROL
   MODELS W/PRE-MIX (INCLUDING THE C115)............... 4-92
   1999-03 115BETO (1730CC) V4S (OUTSIDE
   THE U.S./CANADA), W/OIL INJECTION AND POWER TRIM/TILT . 4-91
   1999-03 115 HP (1730CC) V4 TILLER CONTROL MODELS
   W/PRE-MIX AND MANUAL TILT (INCLUDING THE E115AMH) ... 4-93
V6 MOTORS... 4-94
   MOST 105J-225 HP (2596CC) CARBURETED V6 OIL INJECTION
   MODELS (THESE MOTORS NORMALLY UTILIZE 2 PULSER
   COILS AND A CPS) 4-94
   MOST 105J-225 HP (2596CC) CARBURETED V6 PRE-MIX
   MODELS SUCH AS THE C150 (THESE MOTORS NORMALLY
   UTILIZE 3 PULSER COILS AND NO CPS).................. 4-95
   EFI OX66 (2596CC) V6 MODELS 4-95
      CHARGING SYSTEM 4-97
      FUEL INJECTION SYSTEM 4-96
      IGNITION SYSTEM................................. 4-95
      OIL FEED PUMP SYSTEM 4-98
      POWER TRIM/TILT SYSTEM 4-98
      STARTING SYSTEM................................ 4-97
   HPDI (2596CC) V6 MODELS 4-99
   200-250 HP (3130CC) V6 EFI MODELS
   (EXCEPT THE 250 HP VMAX).......................... 4-101
   250 HP (3130CC) V6 EFI VMAX MODELS................. 4-103
   225-250 HP (3342CC) V6 HPDI MODELS 4-106
RIGGING... 4-108
   TYPICAL YAMAHA REMOTE CONTROL BOX 4-108
   TYPICAL YAMAHA DIGITAL METER ASSEMBLY............. 4-109
   TYPICAL YAMAHA REMOTE HARNESS ASSEMBLY
   (MID-RANGE 3-CYLINDER SHOWN)...................... 4-110

4-66 IGNITION AND ELECTRICAL SYSTEMS

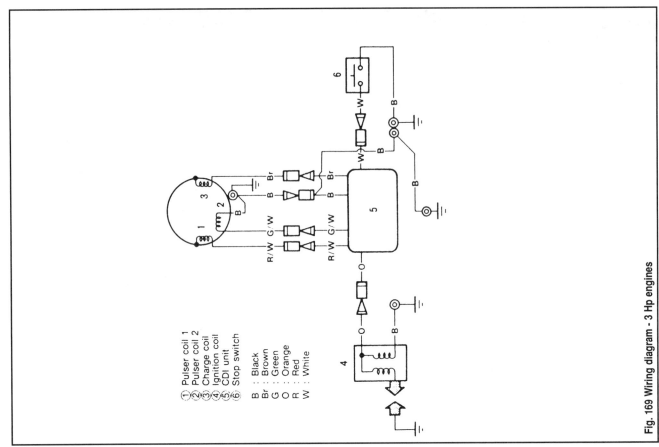

Fig. 169 Wiring diagram - 3 Hp engines

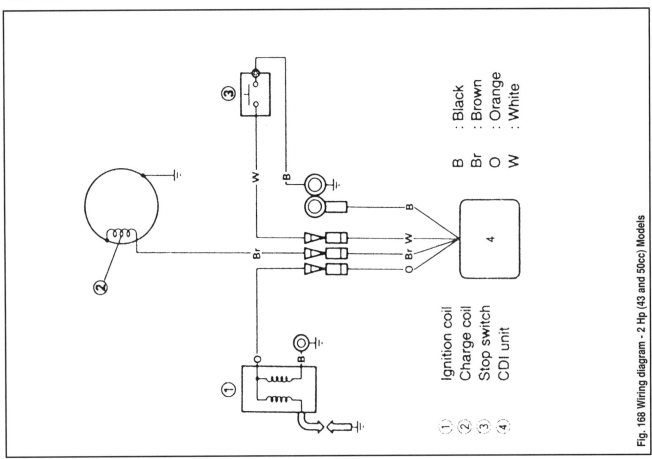

Fig. 168 Wiring diagram - 2 Hp (43 and 50cc) Models

IGNITION AND ELECTRICAL SYSTEMS

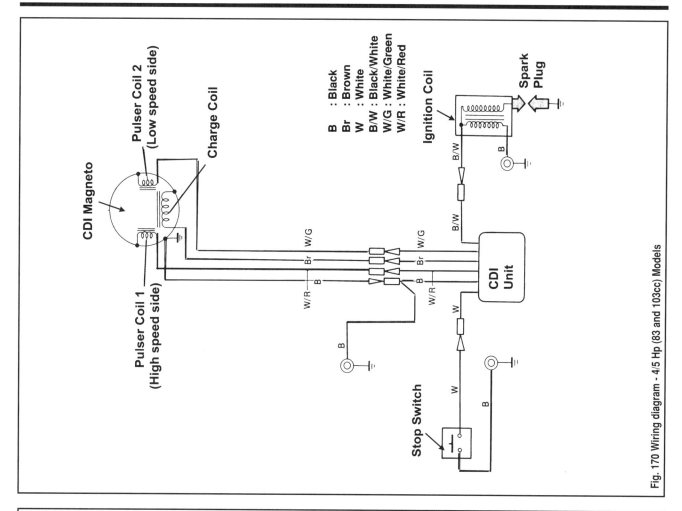

Fig. 170 Wiring diagram - 4/5 Hp (83 and 103cc) Models

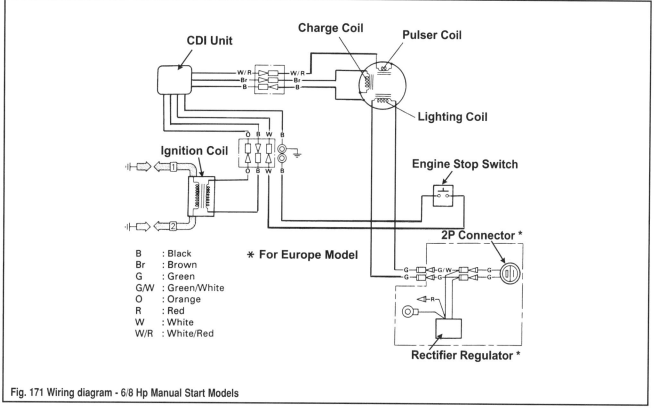

Fig. 171 Wiring diagram - 6/8 Hp Manual Start Models

4-68 IGNITION AND ELECTRICAL SYSTEMS

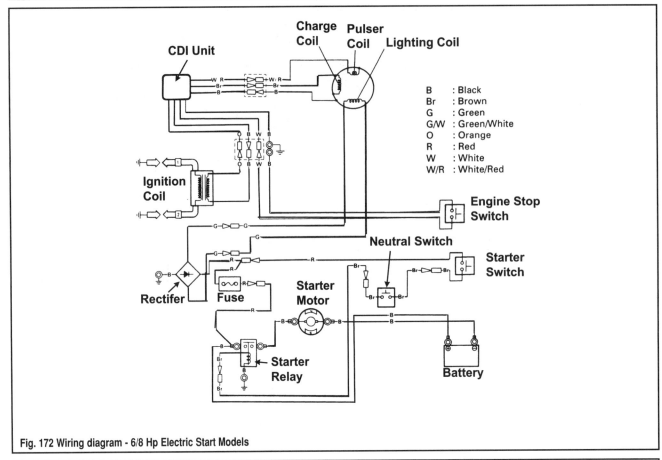

Fig. 172 Wiring diagram - 6/8 Hp Electric Start Models

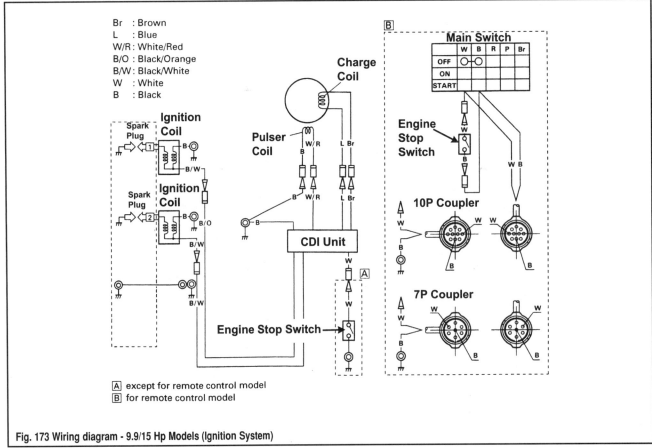

Fig. 173 Wiring diagram - 9.9/15 Hp Models (Ignition System)

IGNITION AND ELECTRICAL SYSTEMS 4-69

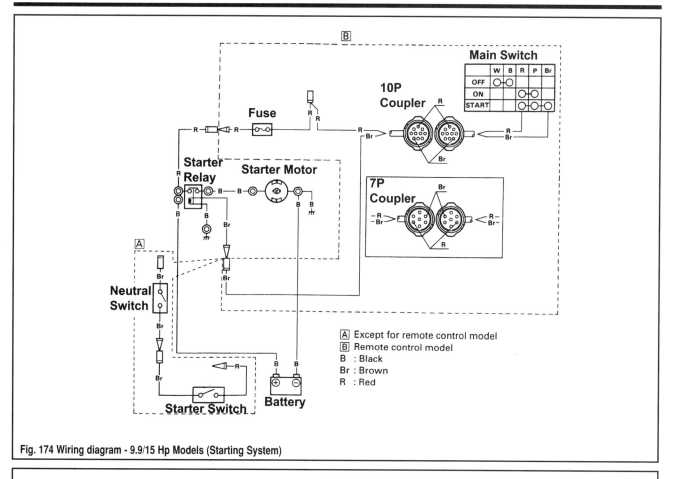

Fig. 174 Wiring diagram - 9.9/15 Hp Models (Starting System)

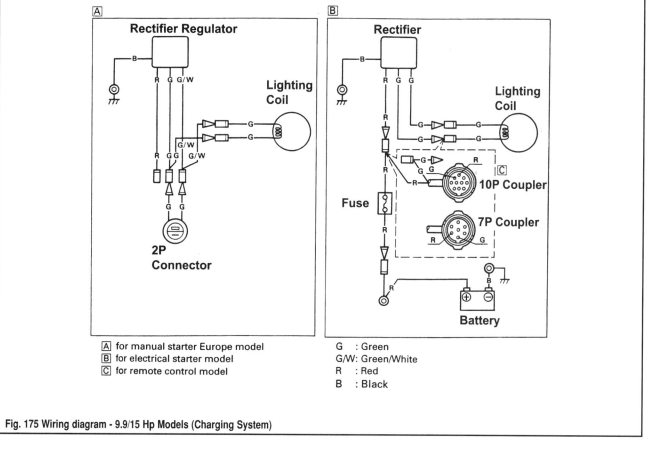

Fig. 175 Wiring diagram - 9.9/15 Hp Models (Charging System)

4-70 IGNITION AND ELECTRICAL SYSTEMS

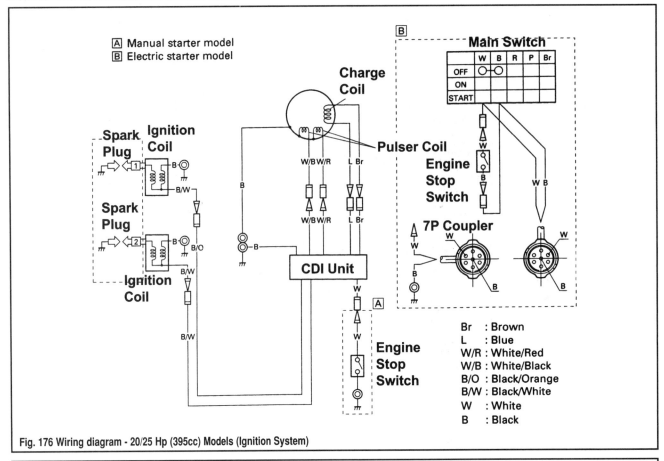

Fig. 176 Wiring diagram - 20/25 Hp (395cc) Models (Ignition System)

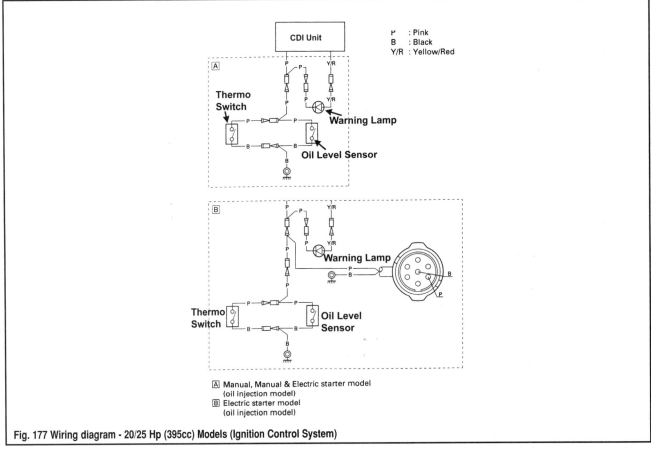

Fig. 177 Wiring diagram - 20/25 Hp (395cc) Models (Ignition Control System)

IGNITION AND ELECTRICAL SYSTEMS 4-71

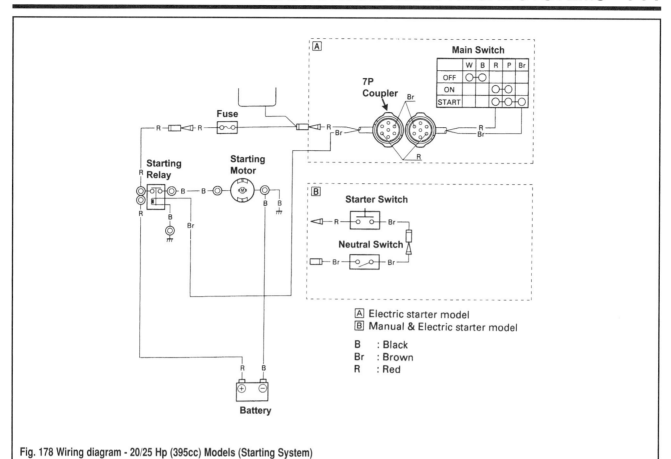

Fig. 178 Wiring diagram - 20/25 Hp (395cc) Models (Starting System)

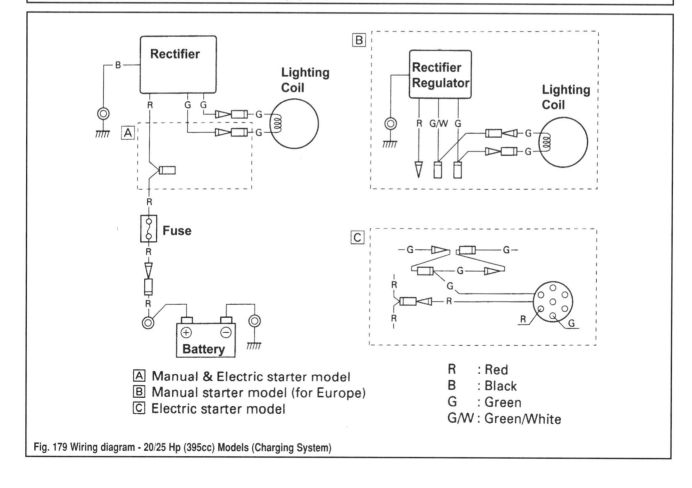

Fig. 179 Wiring diagram - 20/25 Hp (395cc) Models (Charging System)

4-72 IGNITION AND ELECTRICAL SYSTEMS

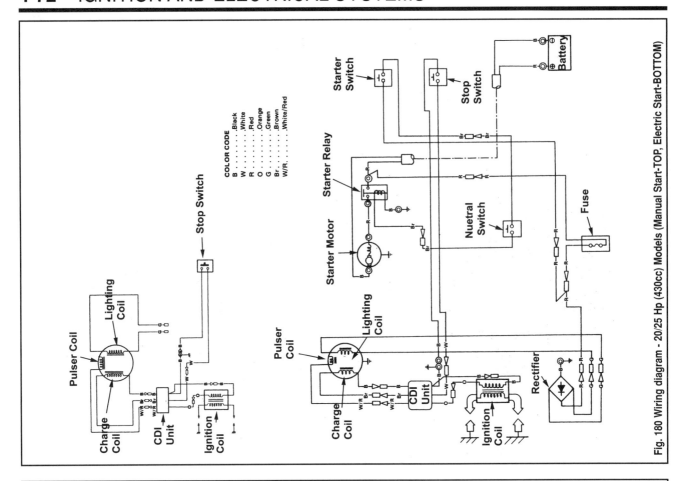

Fig. 180 Wiring diagram - 20/25 Hp (430cc) Models (Manual Start-TOP, Electric Start-BOTTOM)

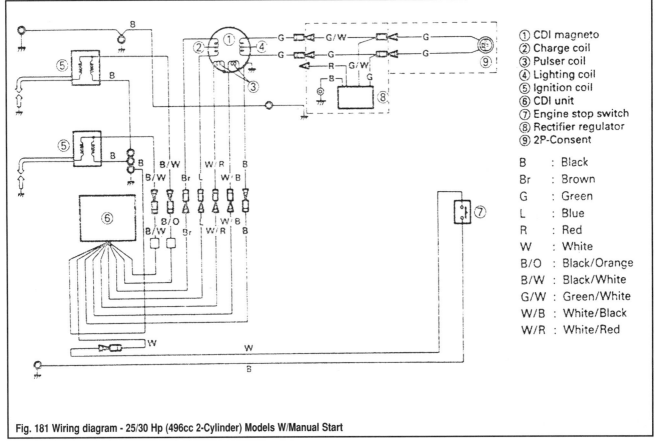

Fig. 181 Wiring diagram - 25/30 Hp (496cc 2-Cylinder) Models W/Manual Start

IGNITION AND ELECTRICAL SYSTEMS

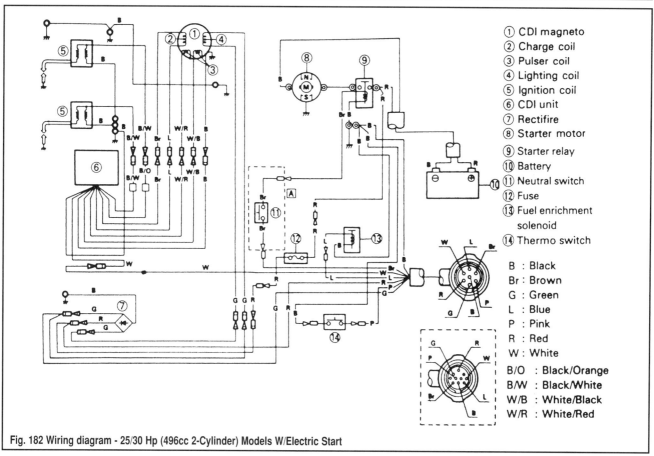

Fig. 182 Wiring diagram - 25/30 Hp (496cc 2-Cylinder) Models W/Electric Start

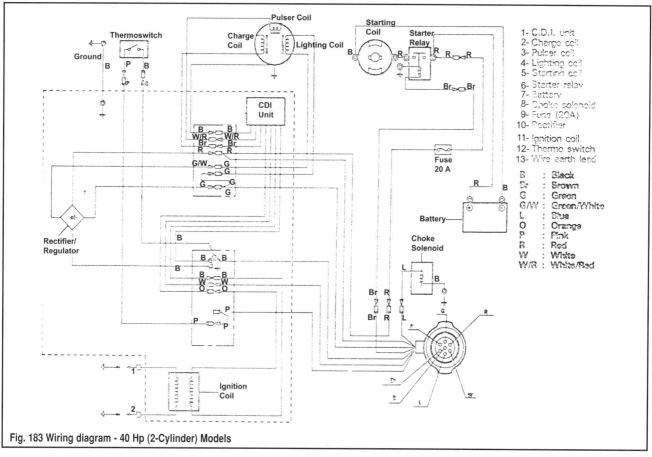

Fig. 183 Wiring diagram - 40 Hp (2-Cylinder) Models

4-74 IGNITION AND ELECTRICAL SYSTEMS

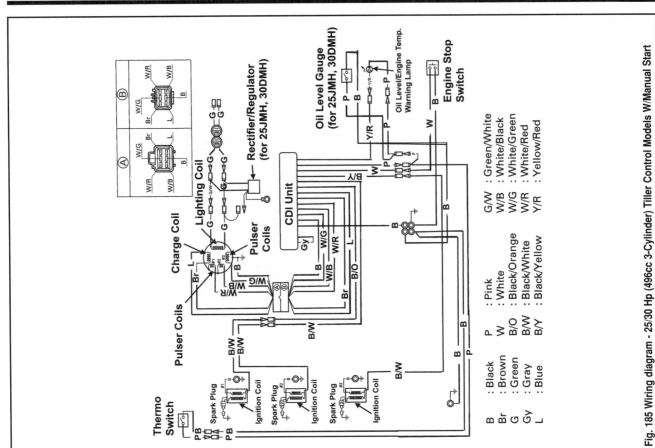

Fig. 185 Wiring diagram - 25/30 Hp (496cc 3-Cylinder) Tiller Control Models W/Manual Start

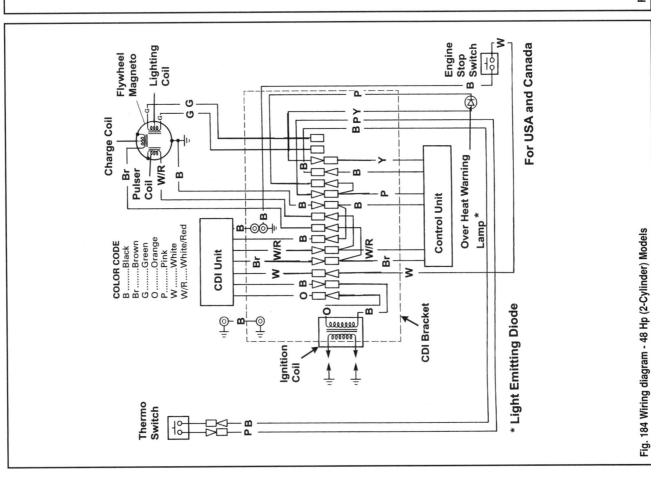

Fig. 184 Wiring diagram - 48 Hp (2-Cylinder) Models

IGNITION AND ELECTRICAL SYSTEMS 4-75

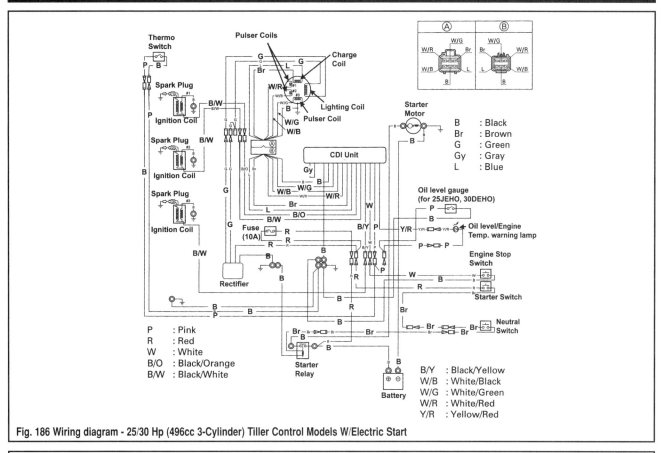

Fig. 186 Wiring diagram - 25/30 Hp (496cc 3-Cylinder) Tiller Control Models W/Electric Start

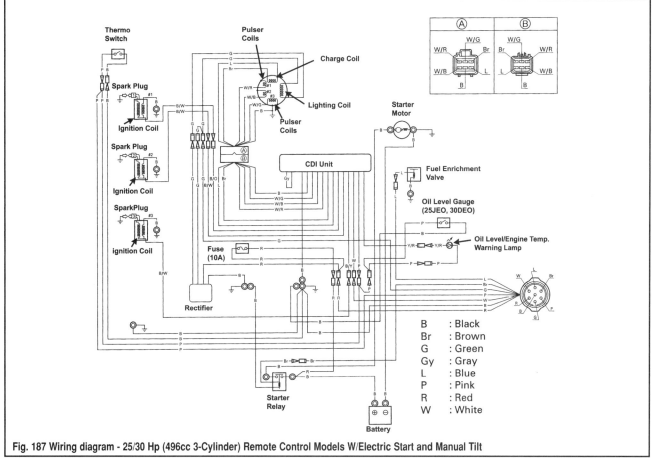

Fig. 187 Wiring diagram - 25/30 Hp (496cc 3-Cylinder) Remote Control Models W/Electric Start and Manual Tilt

4-76 IGNITION AND ELECTRICAL SYSTEMS

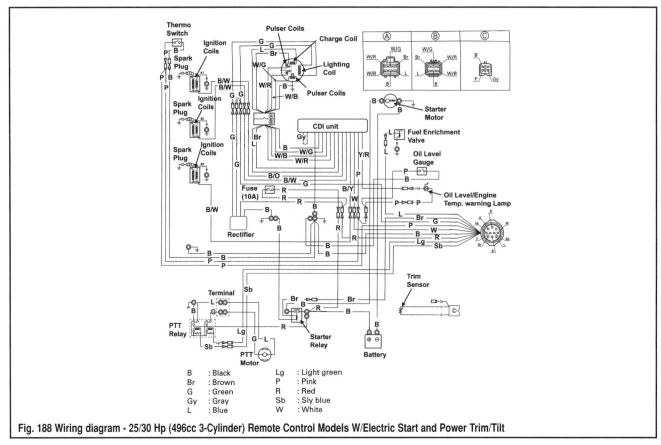

Fig. 188 Wiring diagram - 25/30 Hp (496cc 3-Cylinder) Remote Control Models W/Electric Start and Power Trim/Tilt

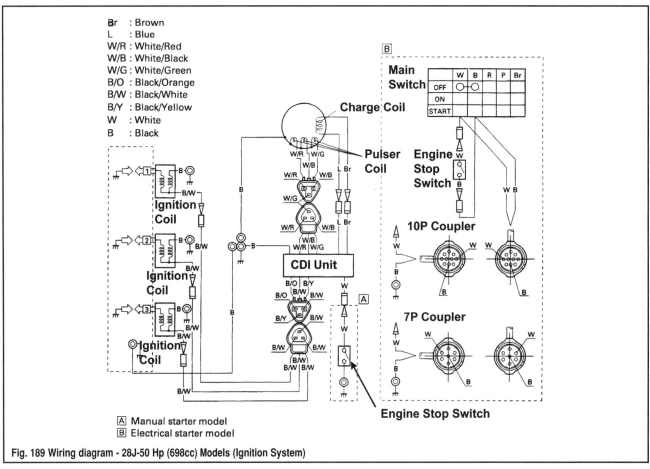

Fig. 189 Wiring diagram - 28J-50 Hp (698cc) Models (Ignition System)

IGNITION AND ELECTRICAL SYSTEMS

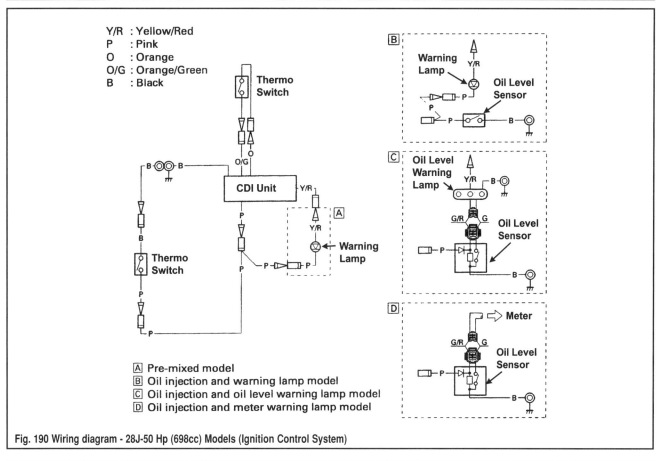

Fig. 190 Wiring diagram - 28J-50 Hp (698cc) Models (Ignition Control System)

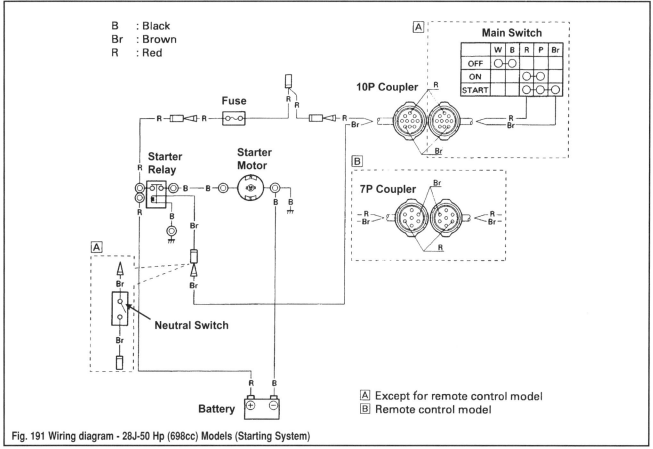

Fig. 191 Wiring diagram - 28J-50 Hp (698cc) Models (Starting System)

4-78 IGNITION AND ELECTRICAL SYSTEMS

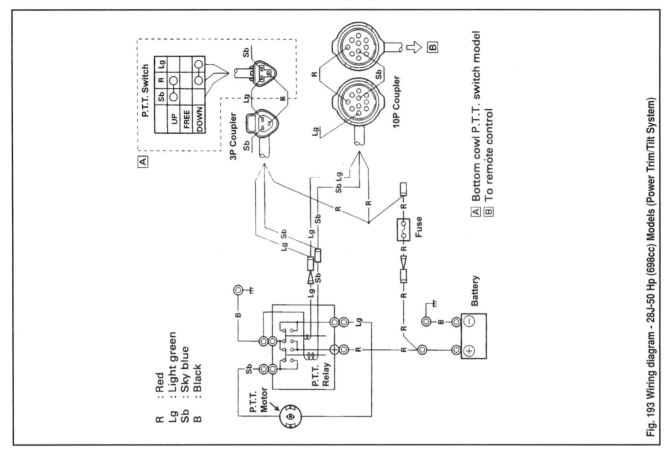

Fig. 193 Wiring diagram - 28J-50 Hp (698cc) Models (Power Trim/Tilt System)

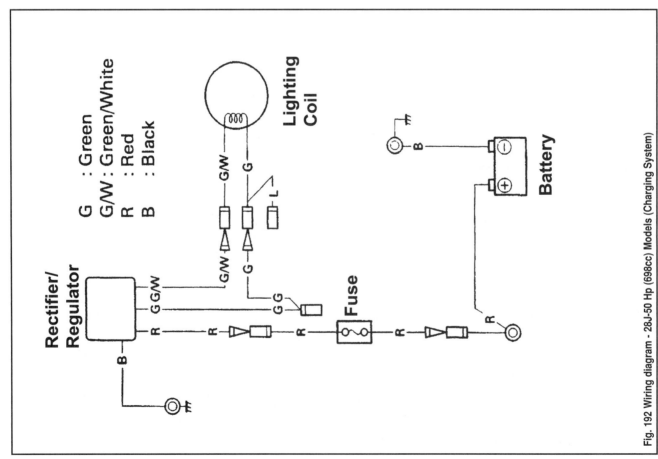

Fig. 192 Wiring diagram - 28J-50 Hp (698cc) Models (Charging System)

IGNITION AND ELECTRICAL SYSTEMS 4-79

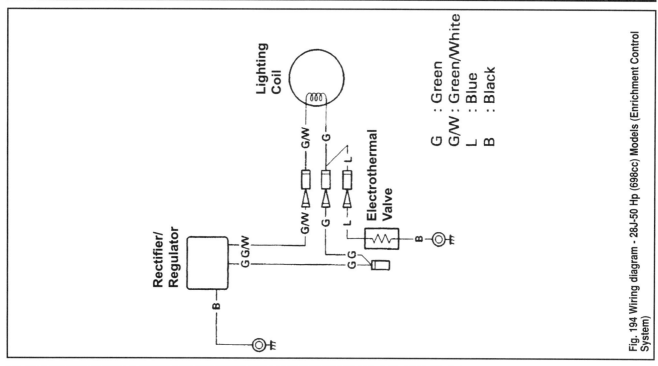

Fig. 194 Wiring diagram - 28J-50 Hp (698cc) Models (Enrichment Control System)

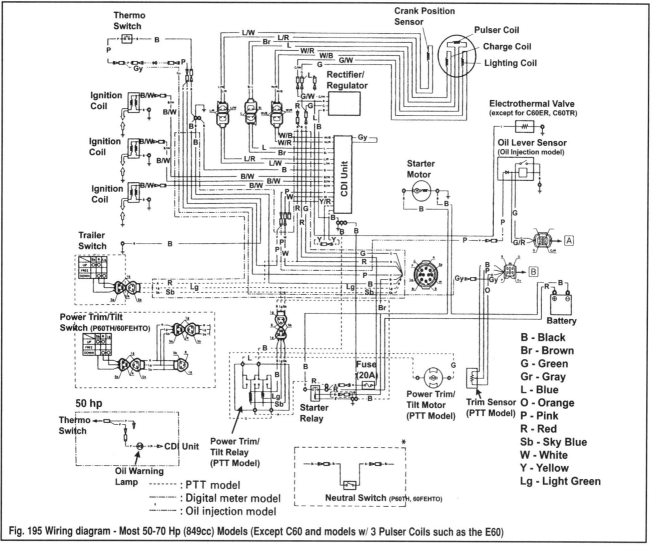

Fig. 195 Wiring diagram - Most 50-70 Hp (849cc) Models (Except C60 and models w/ 3 Pulser Coils such as the E60)

4-80 IGNITION AND ELECTRICAL SYSTEMS

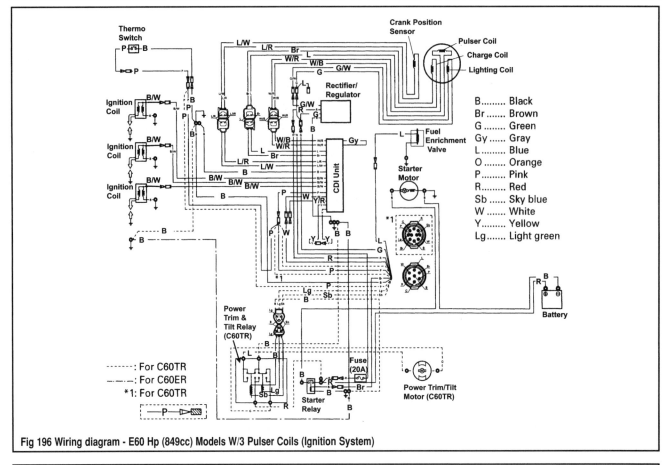

Fig 196 Wiring diagram - E60 Hp (849cc) Models W/3 Pulser Coils (Ignition System)

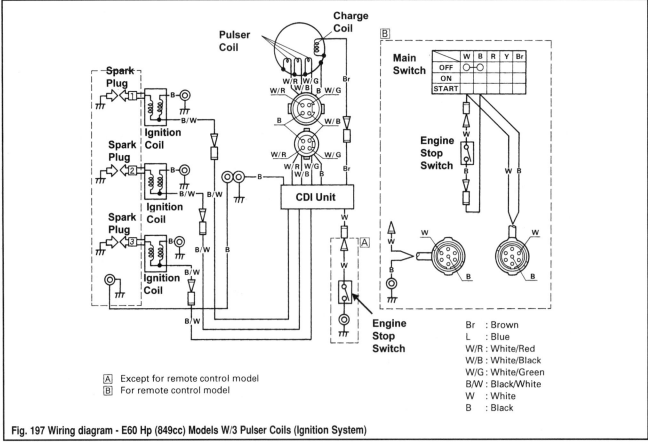

Fig. 197 Wiring diagram - E60 Hp (849cc) Models W/3 Pulser Coils (Ignition System)

IGNITION AND ELECTRICAL SYSTEMS 4-81

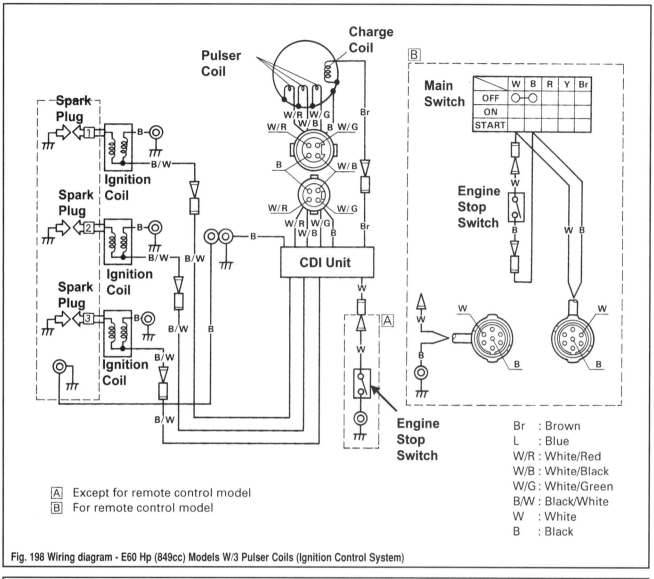

Fig. 198 Wiring diagram - E60 Hp (849cc) Models W/3 Pulser Coils (Ignition Control System)

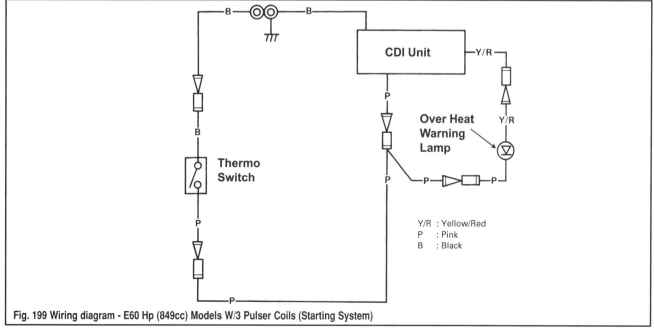

Fig. 199 Wiring diagram - E60 Hp (849cc) Models W/3 Pulser Coils (Starting System)

4-82 IGNITION AND ELECTRICAL SYSTEMS

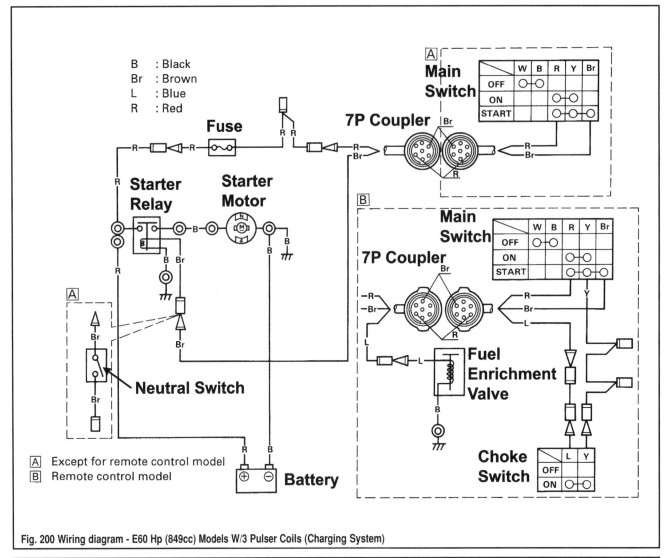

Fig. 200 Wiring diagram - E60 Hp (849cc) Models W/3 Pulser Coils (Charging System)

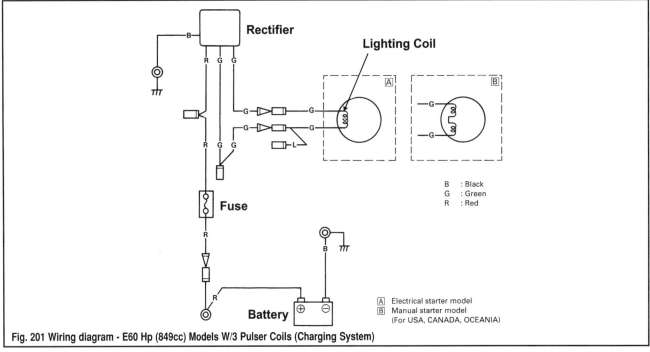

Fig. 201 Wiring diagram - E60 Hp (849cc) Models W/3 Pulser Coils (Charging System)

IGNITION AND ELECTRICAL SYSTEMS 4-83

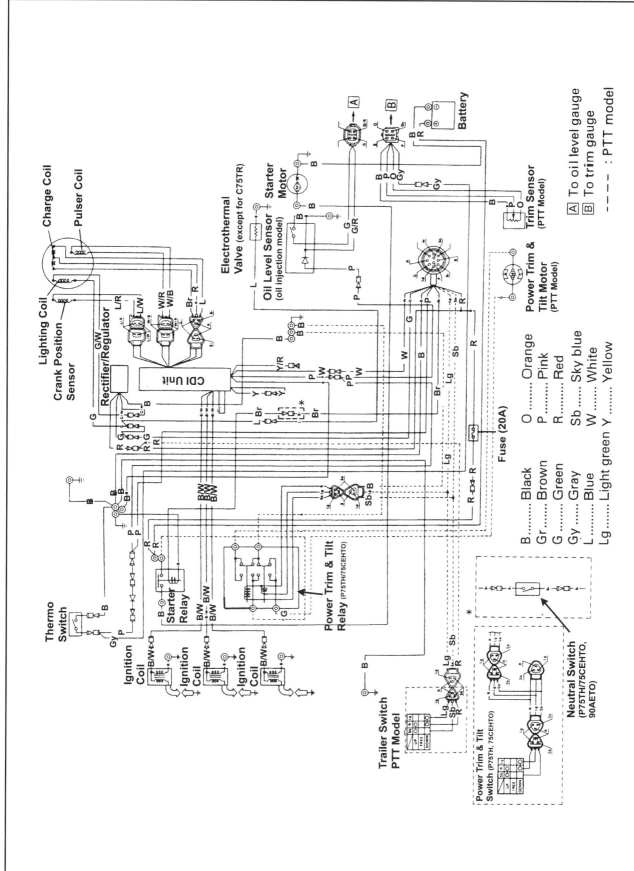

Fig. 202 Wiring diagram - Most 65J/75/80/90 Hp (1140cc) Models (W/Single Pulser Coil and a Crankshaft Position Sensor)

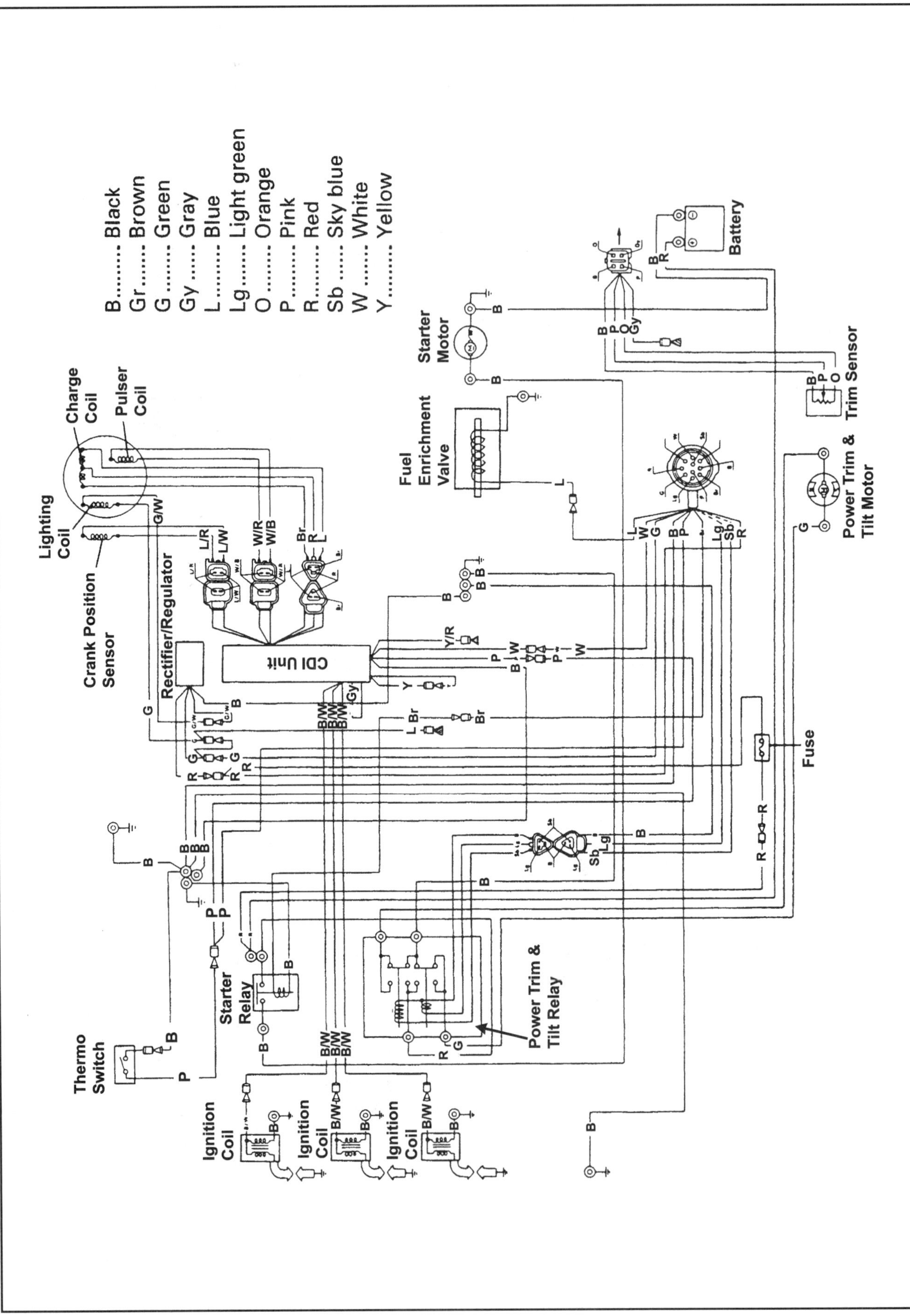

4-84 IGNITION AND ELECTRICAL SYSTEMS

Fig. 203 Wiring diagram - C75 Hp (1140cc) Models (W/Single Pulser Coil and a Crankshaft Position Sensor)

IGNITION AND ELECTRICAL SYSTEMS

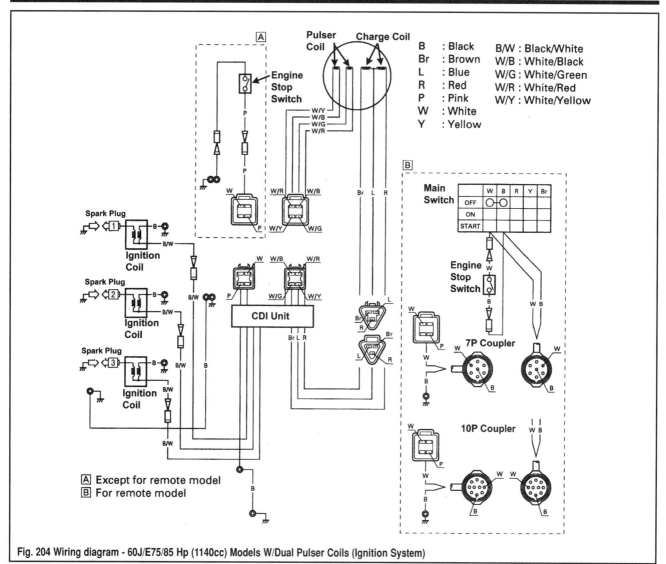

Fig. 204 Wiring diagram - 60J/E75/85 Hp (1140cc) Models W/Dual Pulser Coils (Ignition System)

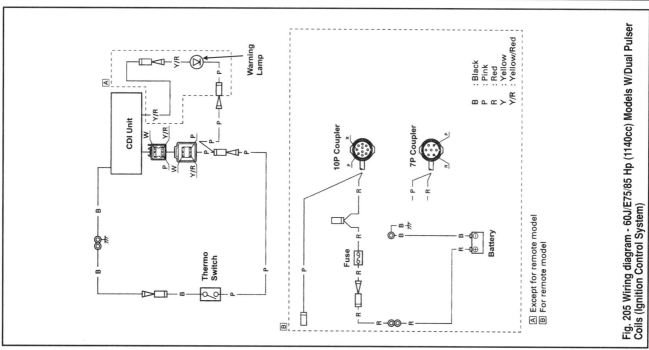

Fig. 205 Wiring diagram - 60J/E75/85 Hp (1140cc) Models W/Dual Pulser Coils (Ignition Control System)

4-86 IGNITION AND ELECTRICAL SYSTEMS

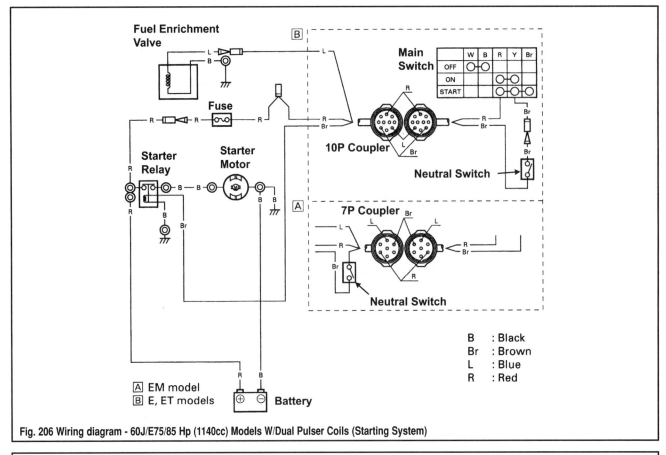

Fig. 206 Wiring diagram - 60J/E75/85 Hp (1140cc) Models W/Dual Pulser Coils (Starting System)

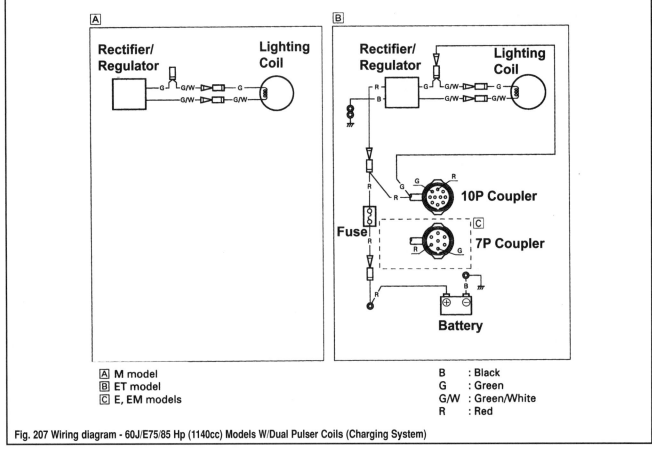

Fig. 207 Wiring diagram - 60J/E75/85 Hp (1140cc) Models W/Dual Pulser Coils (Charging System)

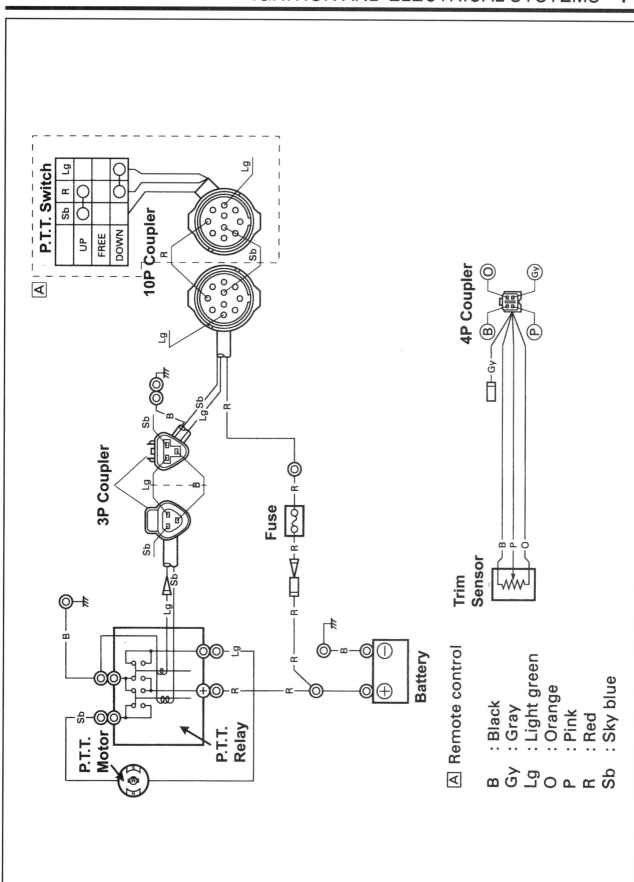

Fig. 208 Wiring diagram - 60J/E75/85 Hp (1140cc) Models W/Dual Pulser Coils (Power Trim/Tilt System)

4-88 IGNITION AND ELECTRICAL SYSTEMS

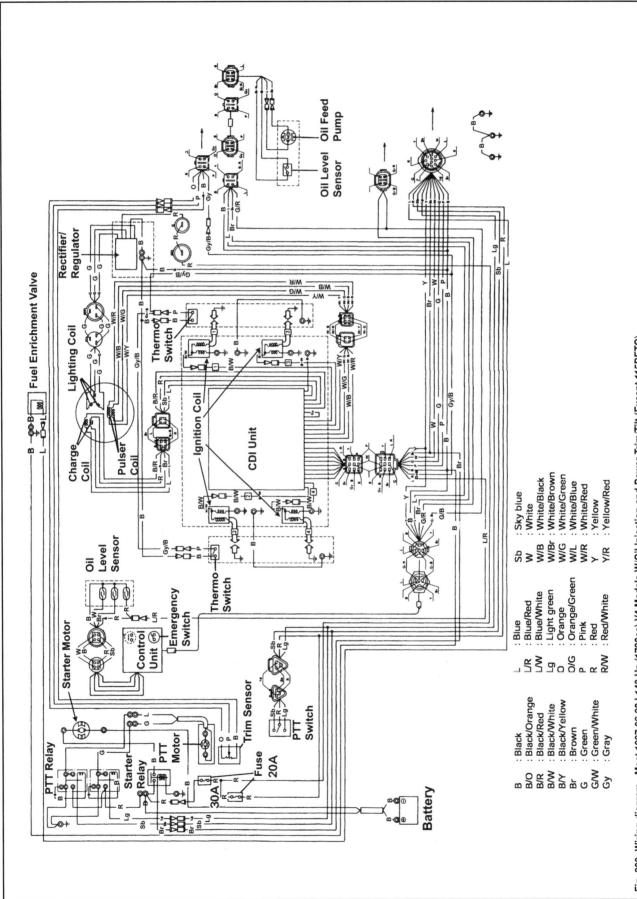

Fig. 209 Wiring diagram - Most 1997-98 80J-140 Hp (1730cc) V4 Models W/Oil Injection and Power Trim/Tilt (Except 115BETO)

IGNITION AND ELECTRICAL SYSTEMS 4-89

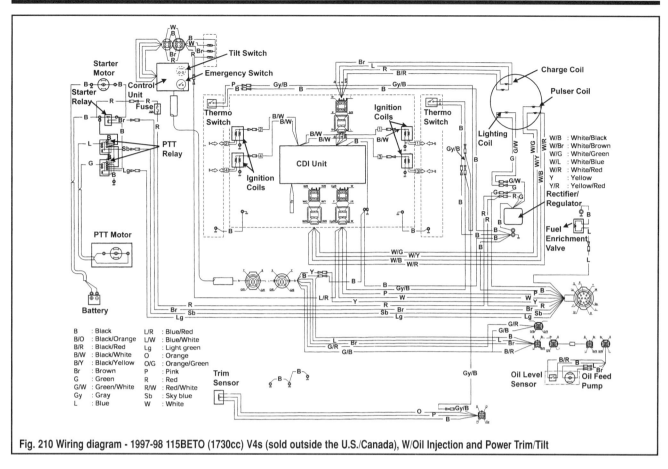

Fig. 210 Wiring diagram - 1997-98 115BETO (1730cc) V4s (sold outside the U.S./Canada), W/Oil Injection and Power Trim/Tilt

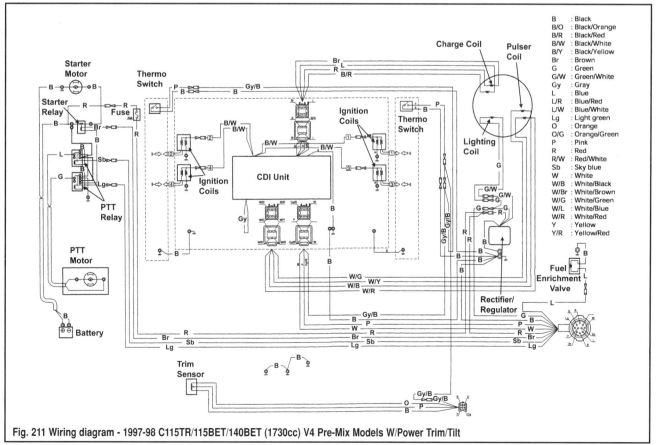

Fig. 211 Wiring diagram - 1997-98 C115TR/115BET/140BET (1730cc) V4 Pre-Mix Models W/Power Trim/Tilt

4-90 IGNITION AND ELECTRICAL SYSTEMS

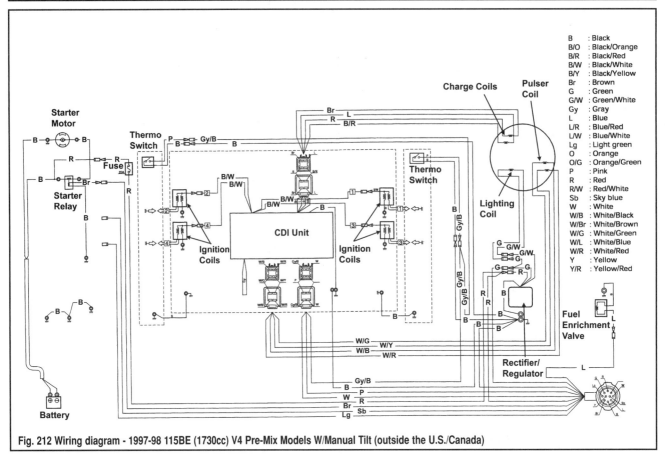

Fig. 212 Wiring diagram - 1997-98 115BE (1730cc) V4 Pre-Mix Models W/Manual Tilt (outside the U.S./Canada)

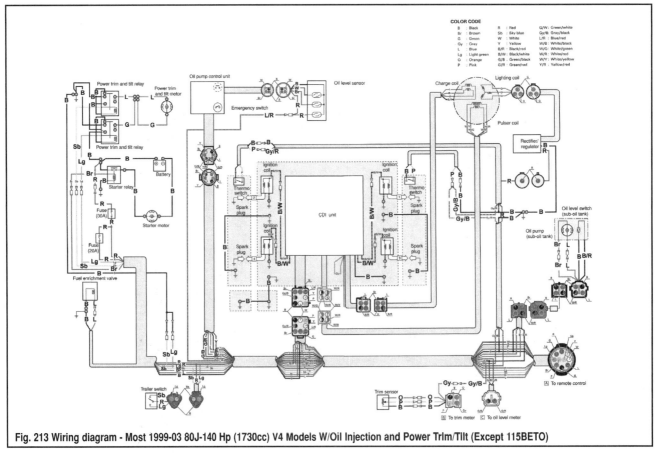

Fig. 213 Wiring diagram - Most 1999-03 80J-140 Hp (1730cc) V4 Models W/Oil Injection and Power Trim/Tilt (Except 115BETO)

IGNITION AND ELECTRICAL SYSTEMS

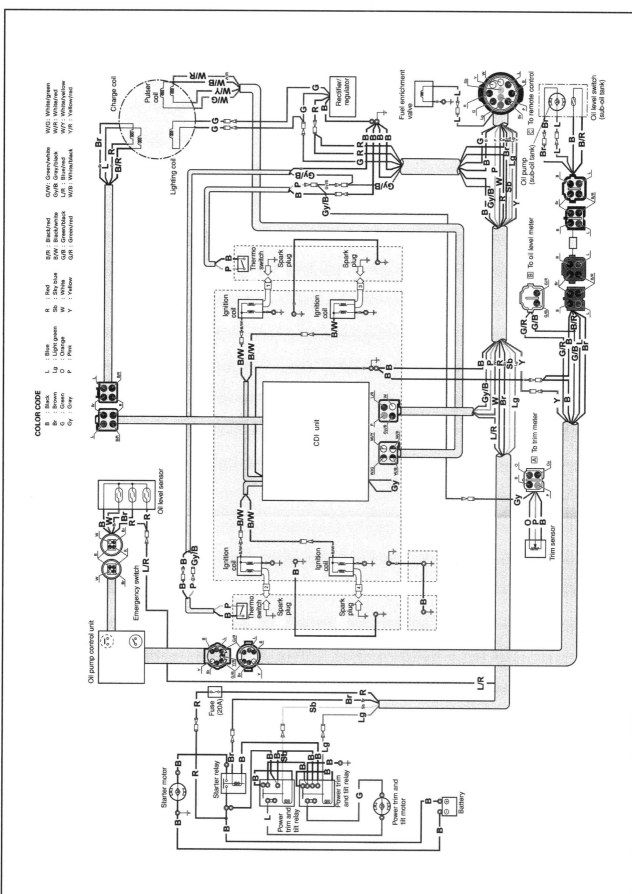

Fig. 214 Wiring diagram - 1999-03 115BETO (1730cc) V4s (outside the U.S./Canada), W/Oil Injection and Power Trim/Tilt

4-92 IGNITION AND ELECTRICAL SYSTEMS

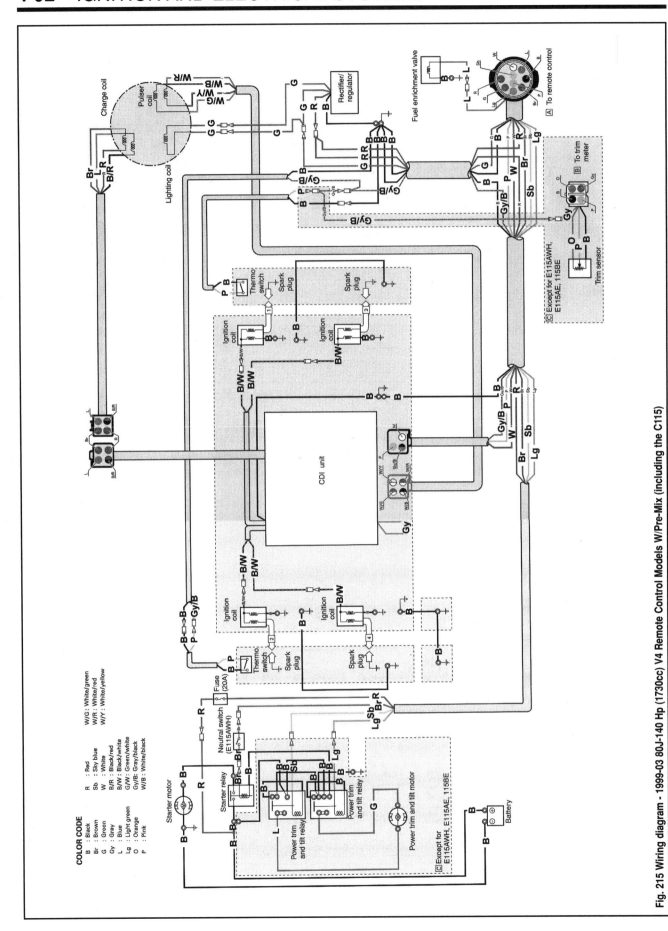

Fig. 215 Wiring diagram - 1999-03 80J-140 Hp (1730cc) V4 Remote Control Models W/Pre-Mix (including the C115)

IGNITION AND ELECTRICAL SYSTEMS

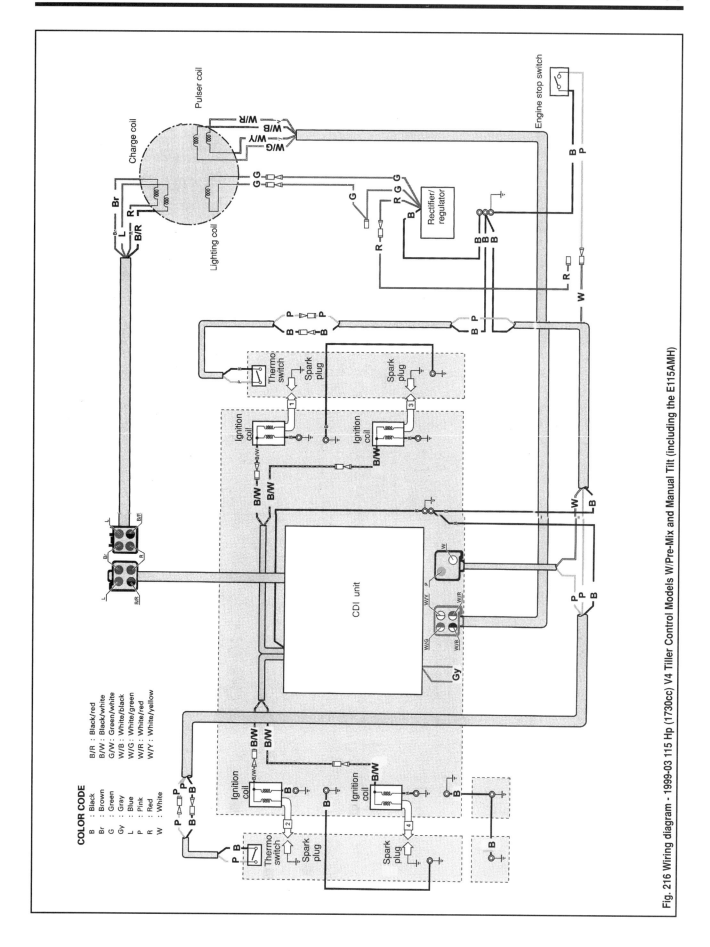

Fig. 216 Wiring diagram - 1999-03 115 Hp (1730cc) V4 Tiller Control Models W/Pre-Mix and Manual Tilt (including the E115AMH)

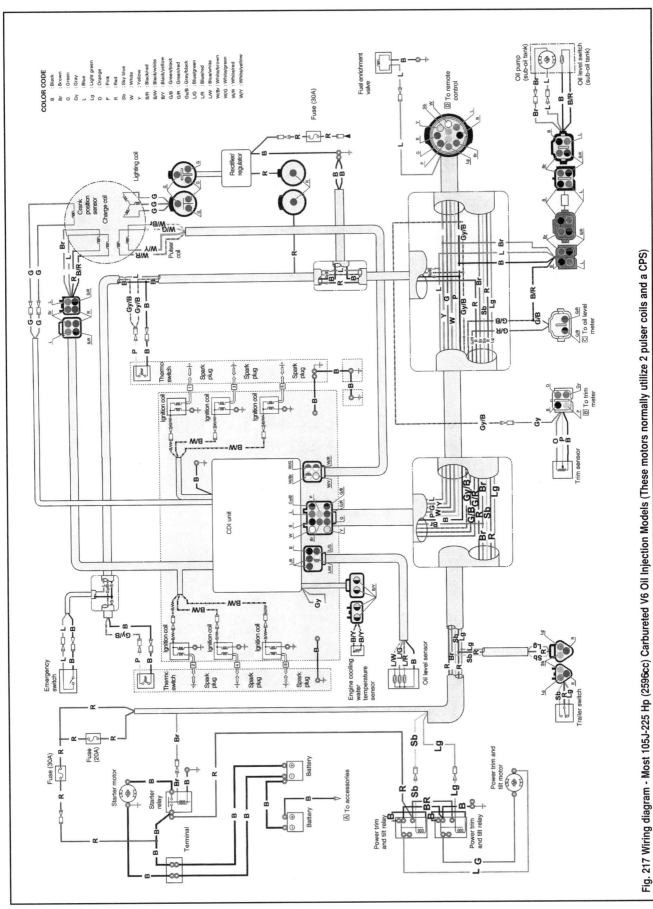

Fig. 217 Wiring diagram - Most 105J-225 Hp (2596cc) Carbureted V6 Oil Injection Models (These motors normally utilize 2 pulser coils and a CPS)

IGNITION AND ELECTRICAL SYSTEMS

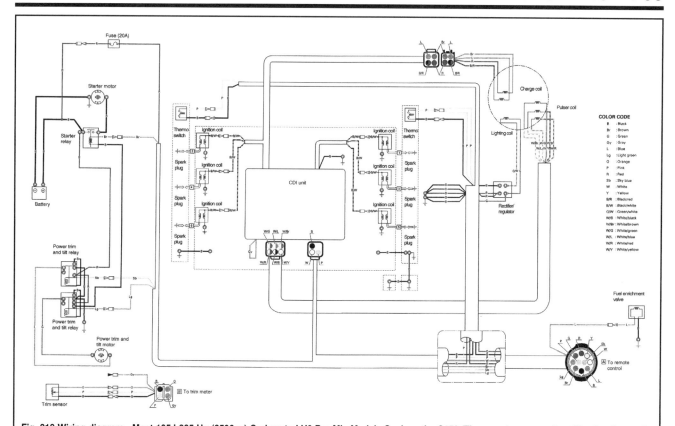

Fig. 218 Wiring diagram - Most 105J-225 Hp (2596cc) Carbureted V6 Pre-Mix Models Such as the C150 (These motors normally utilize 3 pulser coils and NO CPS)

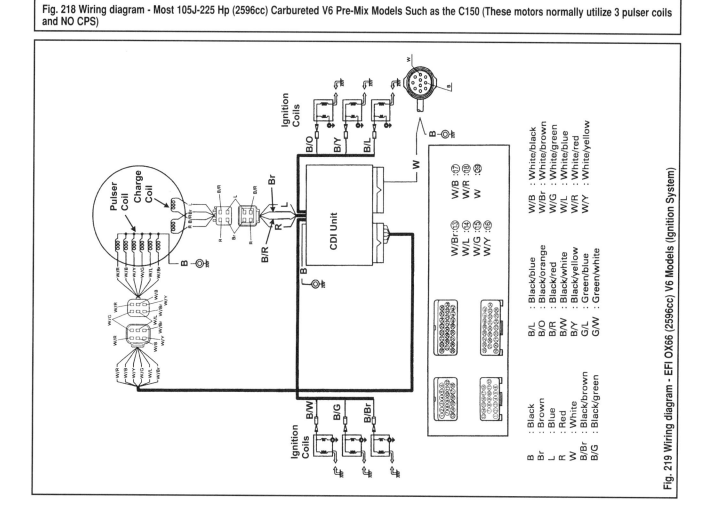

Fig. 219 Wiring diagram - EFI OX66 (2596cc) V6 Models (Ignition System)

4-96 IGNITION AND ELECTRICAL SYSTEMS

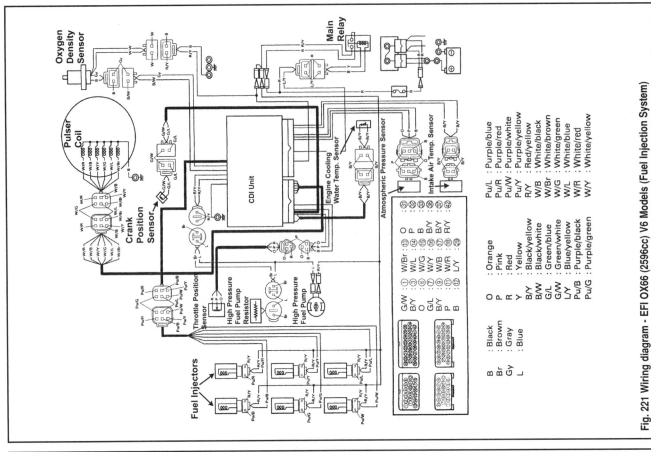

Fig. 221 Wiring diagram - EFI OX66 (2596cc) V6 Models (Fuel Injection System)

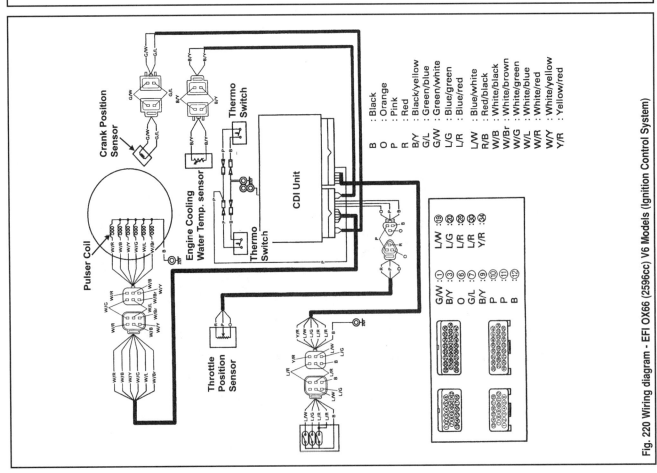

Fig. 220 Wiring diagram - EFI OX66 (2596cc) V6 Models (Ignition Control System)

IGNITION AND ELECTRICAL SYSTEMS 4-97

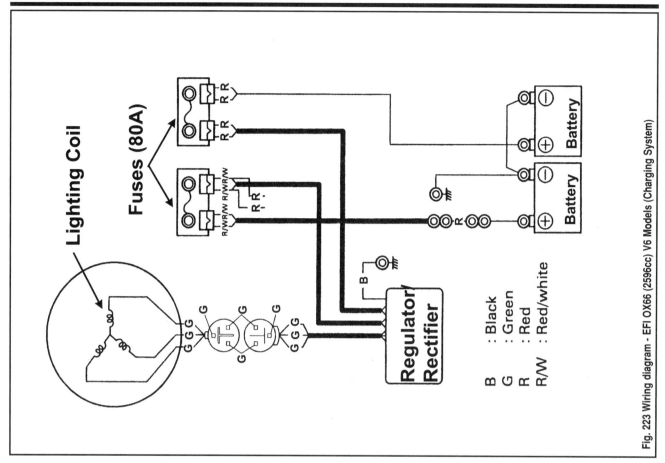

Fig. 223 Wiring diagram - EFI OX66 (2596cc) V6 Models (Charging System)

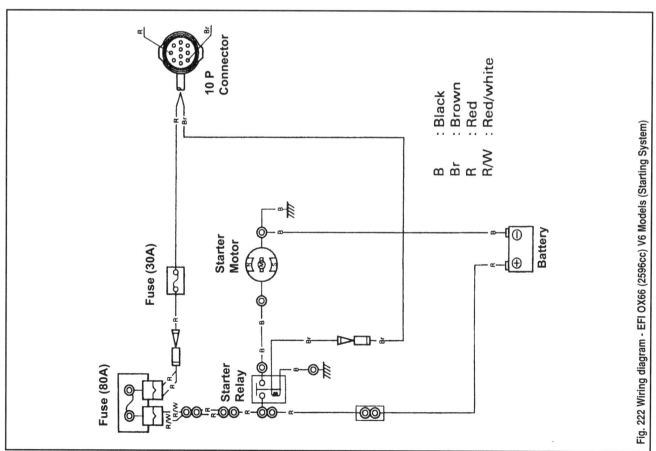

Fig. 222 Wiring diagram - EFI OX66 (2596cc) V6 Models (Starting System)

4-98 IGNITION AND ELECTRICAL SYSTEMS

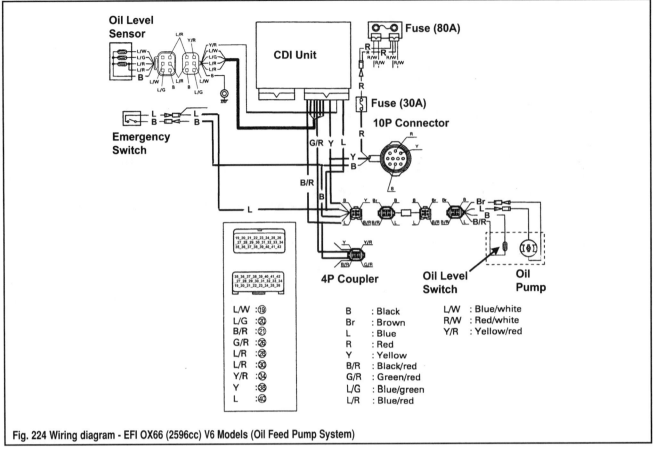

Fig. 224 Wiring diagram - EFI OX66 (2596cc) V6 Models (Oil Feed Pump System)

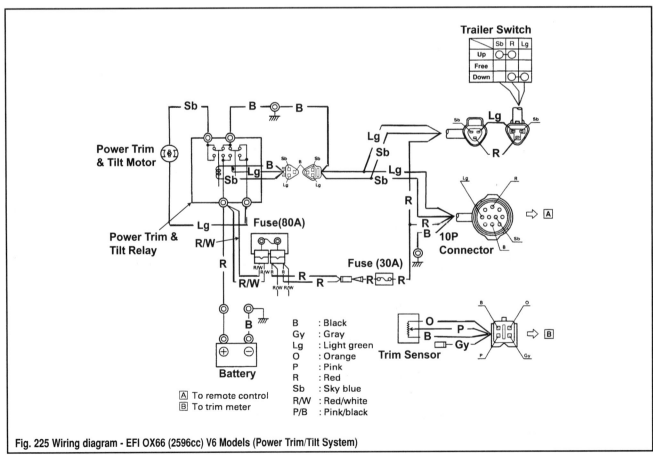

Fig. 225 Wiring diagram - EFI OX66 (2596cc) V6 Models (Power Trim/Tilt System)

IGNITION AND ELECTRICAL SYSTEMS 4-99

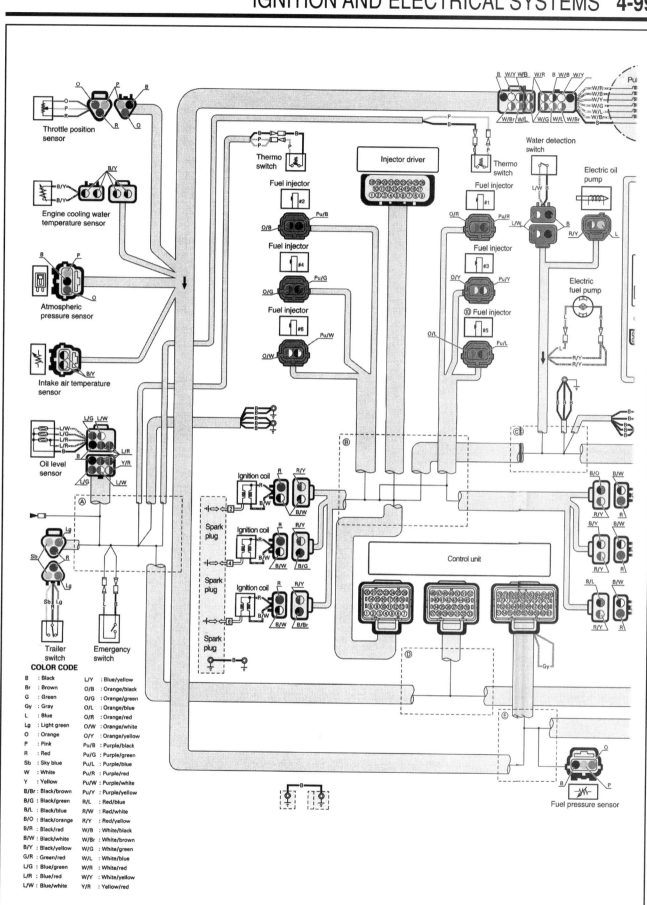

Fig. 226 Wiring diagram - HPDI (2596cc) V6 Models (Part 1)

4-100 IGNITION AND ELECTRICAL SYSTEMS

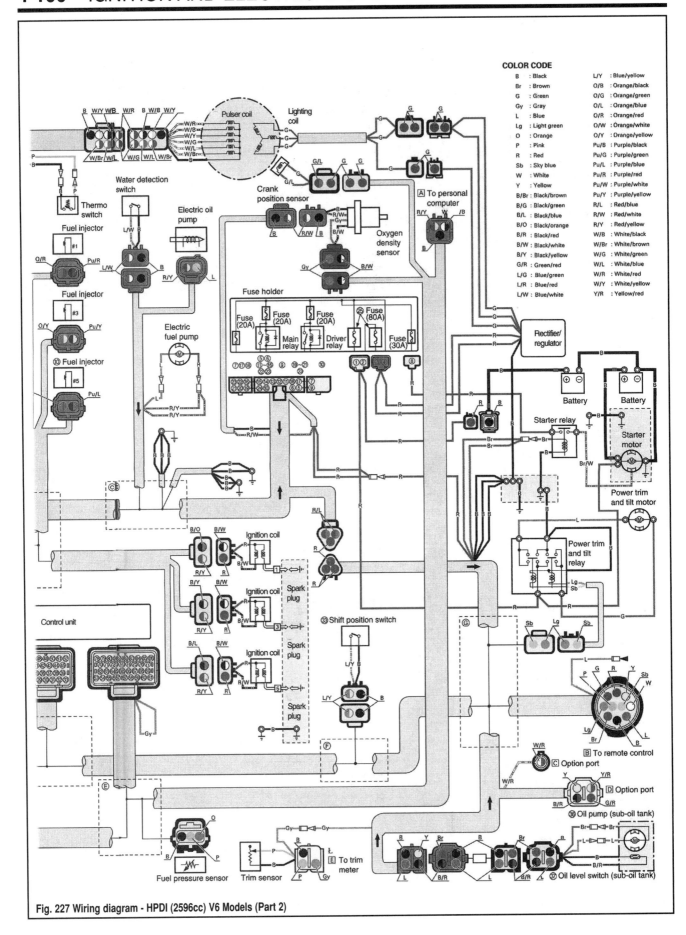

Fig. 227 Wiring diagram - HPDI (2596cc) V6 Models (Part 2)

IGNITION AND ELECTRICAL SYSTEMS

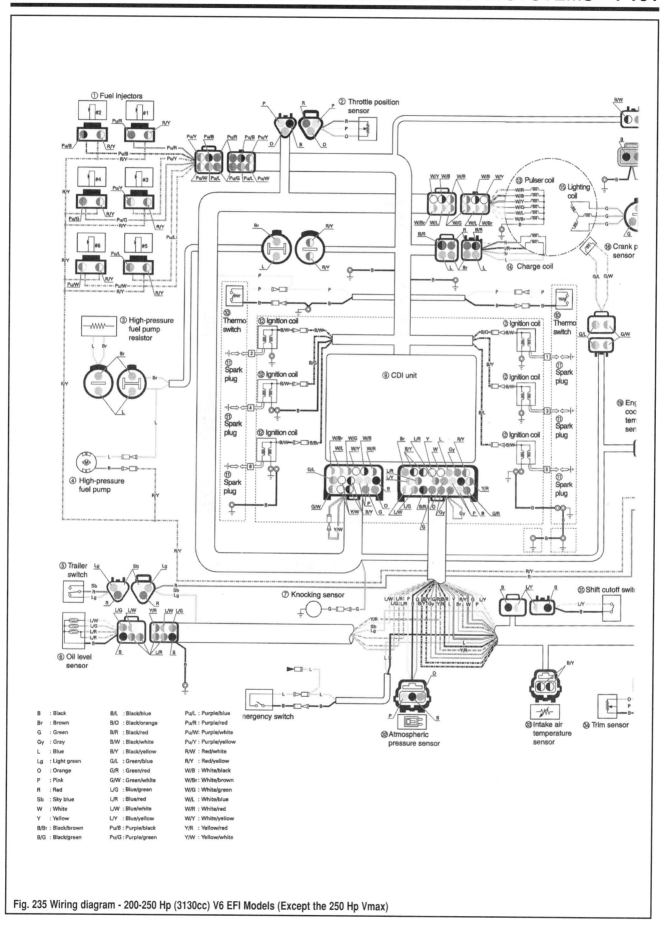

Fig. 235 Wiring diagram - 200-250 Hp (3130cc) V6 EFI Models (Except the 250 Hp Vmax)

4-102 IGNITION AND ELECTRICAL SYSTEMS

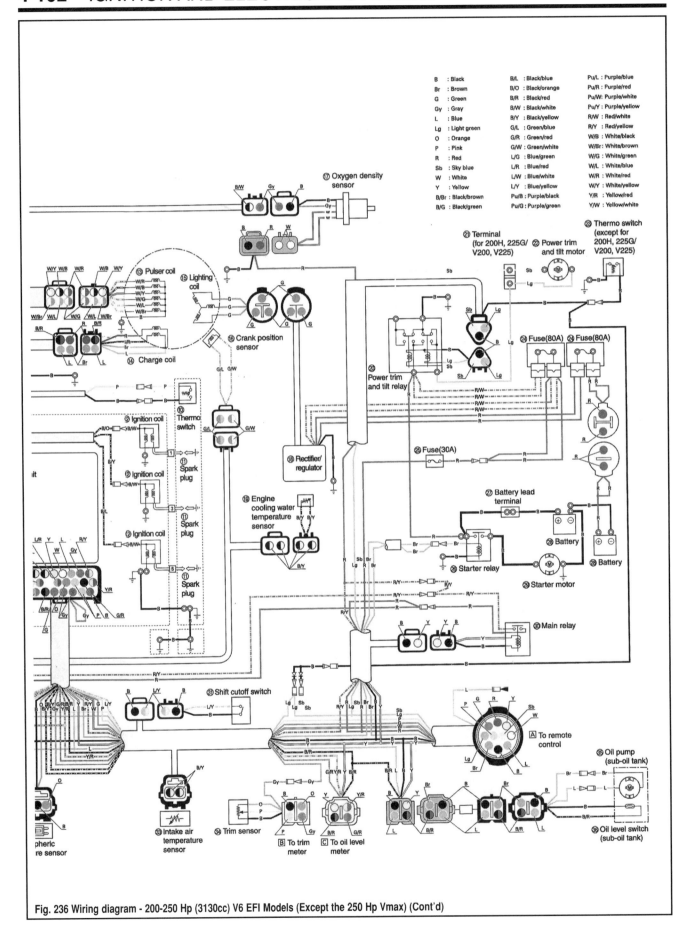

Fig. 236 Wiring diagram - 200-250 Hp (3130cc) V6 EFI Models (Except the 250 Hp Vmax) (Cont'd)

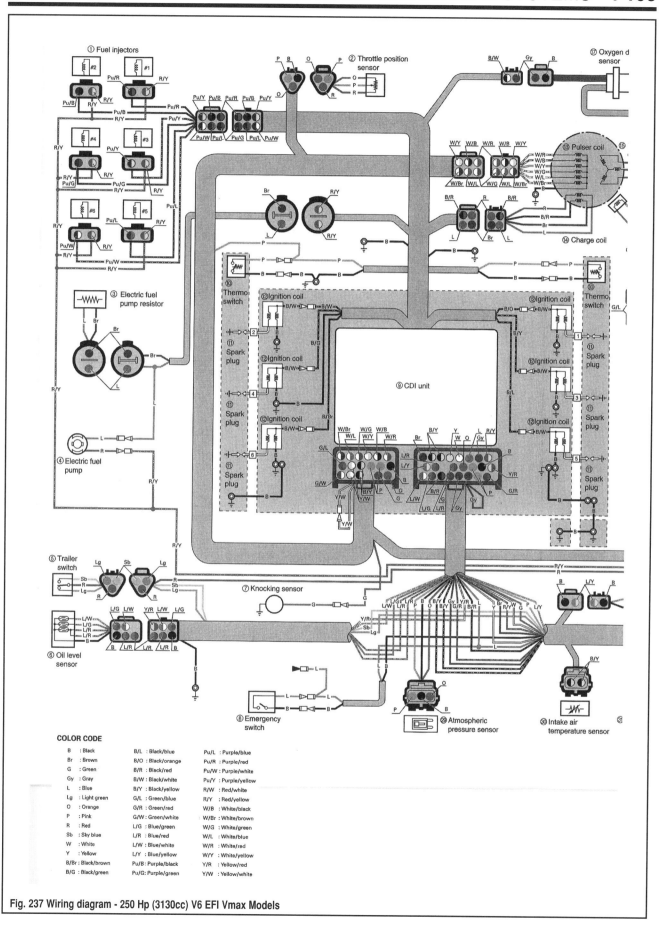

Fig. 237 Wiring diagram - 250 Hp (3130cc) V6 EFI Vmax Models

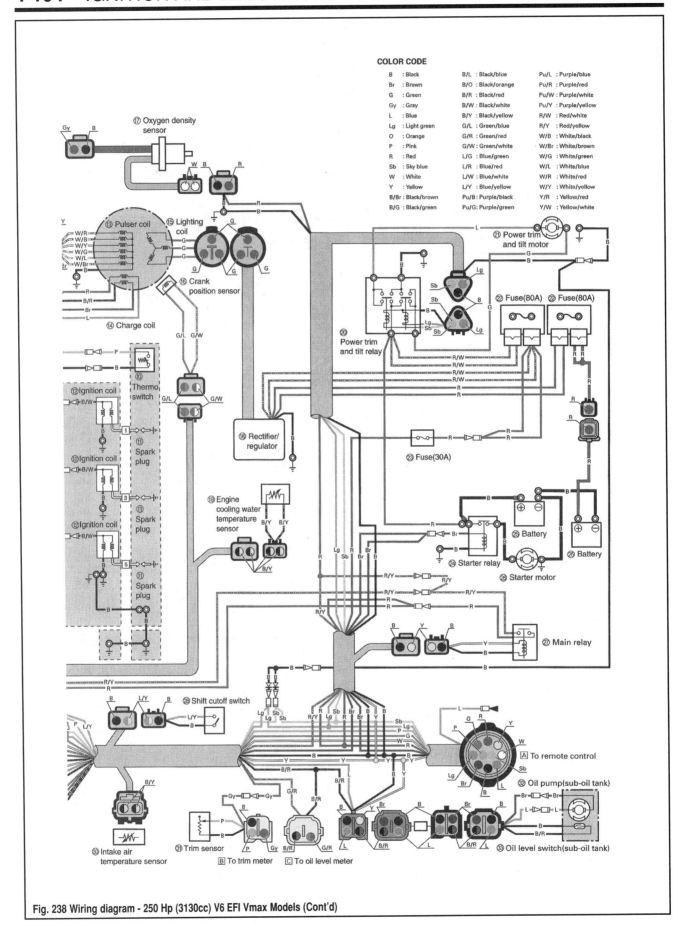

Fig. 238 Wiring diagram - 250 Hp (3130cc) V6 EFI Vmax Models (Cont'd)

IGNITION AND ELECTRICAL SYSTEMS

Wiring diagram

1. Electric fuel pump
2. Electric oil pump
3. Oil level sensor
4. Water detection switch
5. Throttle position sensor
6. Emergency switch
7. Atmospheric pressure sensor
8. Thermoswitch
9. Power trim and tilt switch
10. Fuel pressure sensor
11. Pulser coil
12. Stator coil
13. Crank position sensor
14. Fuel injector
15. Fuel pump relay
16. Injector driver
17. Ignition coil
18. Spark plug
19. ECM
20. Diode
21. Engine temperature sensor
22. Trim sensor
23. Oil pump (remote oil tank)
24. Oil level sensor (remote oil tank)
25. Shift cut switch
26. Shift position switch
27. Power trim and tilt relay
28. Power trim and tilt motor
29. Intake air temperature sensor
30. Starter relay
31. Starter motor
32. Starting battery
33. Accessory battery
34. Fuse holder
35. Fuse (20 A)
36. ECM main relay
37. Injector driver relay
38. Fuse (100 A)
39. Fuse (30 A)
40. Rectifier Regulator

A. To personal computer for diagnosis
B. To remote control box/switch panel
C. To oil level warning indicator
D. To trim meter
E. To diagnostic flash indicator (special service tool)

(*1) Isolator cable (optional)
(*2) Negative cable (commercially available)

Color code

B	Black
Br	Brown
G	Green
Gy	Gray
L	Blue
Lg	Light green
O	Orange
P	Pink
R	Red
Sb	Sky blue
W	White
Y	Yellow
B/Br	Black/brown
B/G	Black/green
B/L	Black/blue
B/O	Black/orange
B/R	Black/red
B/W	Black/white
B/Y	Black/yellow
G/L	Green/blue
G/R	Green/red
G/W	Green/white
L/G	Blue/green
L/R	Blue/red
L/W	Blue/white
L/Y	Blue/yellow
O/B	Orange/black
O/G	Orange/green
O/L	Orange/blue
O/R	Orange/red
O/W	Orange/white
O/Y	Orange/yellow
Pu/B	Purple/black
Pu/G	Purple/green
Pu/L	Purple/blue
Pu/R	Purple/red
Pu/W	Purple/white
Pu/Y	Purple/yellow
R/L	Red/blue
R/W	Red/white
R/Y	Red/yellow
W/B	White/black
W/Br	White/brown
W/G	White/green
W/L	White/blue
W/R	White/red
W/Y	White/yellow
Y/R	Yellow/red

Fig. 239 Keylist for the 225-250 Hp (3342cc) V6 HPDI Model Wiring Diagrams

4-106 IGNITION AND ELECTRICAL SYSTEMS

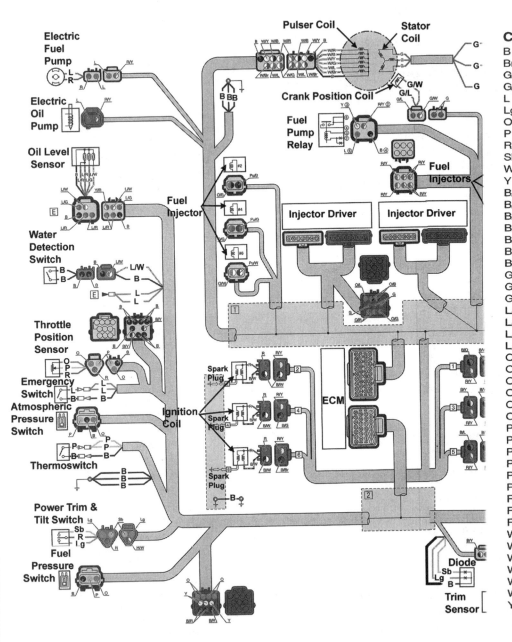

Fig. 240 Wiring diagram - 225-250 Hp (3342cc) V6 HPDI Models (Part 1)

IGNITION AND ELECTRICAL SYSTEMS

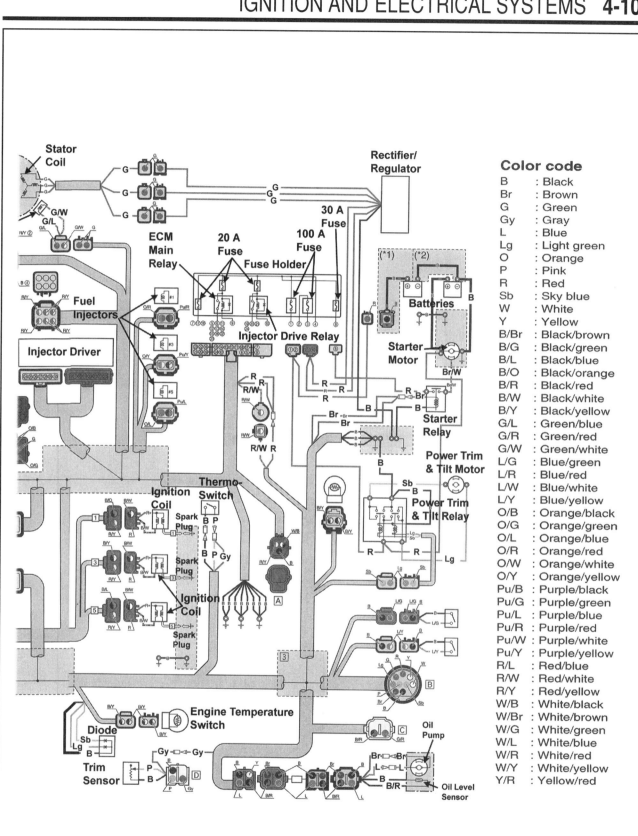

Fig. 241 Wiring diagram - 225-250 Hp (3342cc) V6 HPDI Models (Part 2)

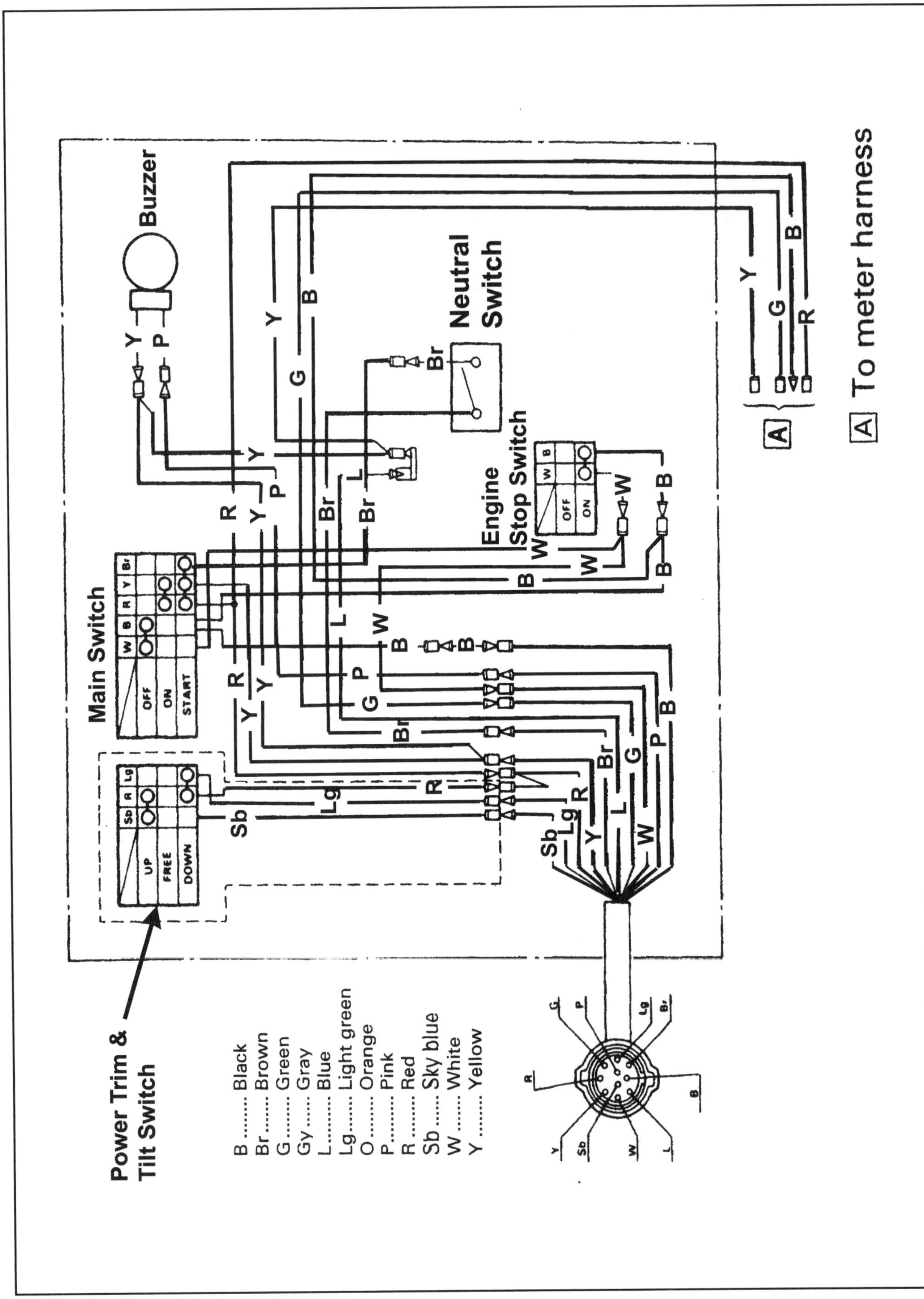

4-108 IGNITION AND ELECTRICAL SYSTEMS

Fig. 246 Wiring diagram - Typical Yamaha Remote Control Box

B ...... Black
Br ..... Brown
G ...... Green
Gy ..... Gray
L ...... Blue
Lg ..... Light green
O ...... Orange
P ...... Pink
R ...... Red
Sb ..... Sky blue
W ...... White
Y ...... Yellow

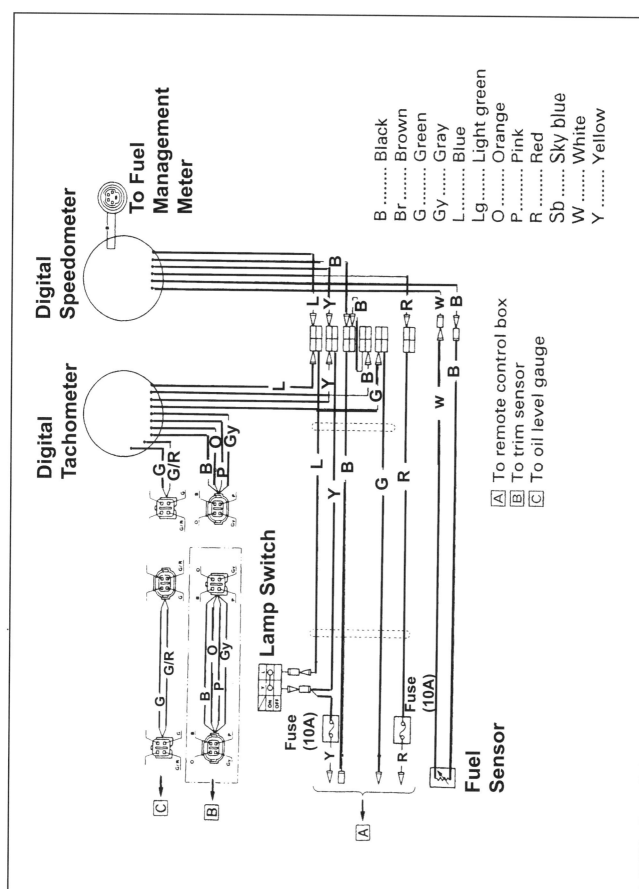

Fig. 247 Wiring diagram - Typical Yamaha Digital Meter Assembly

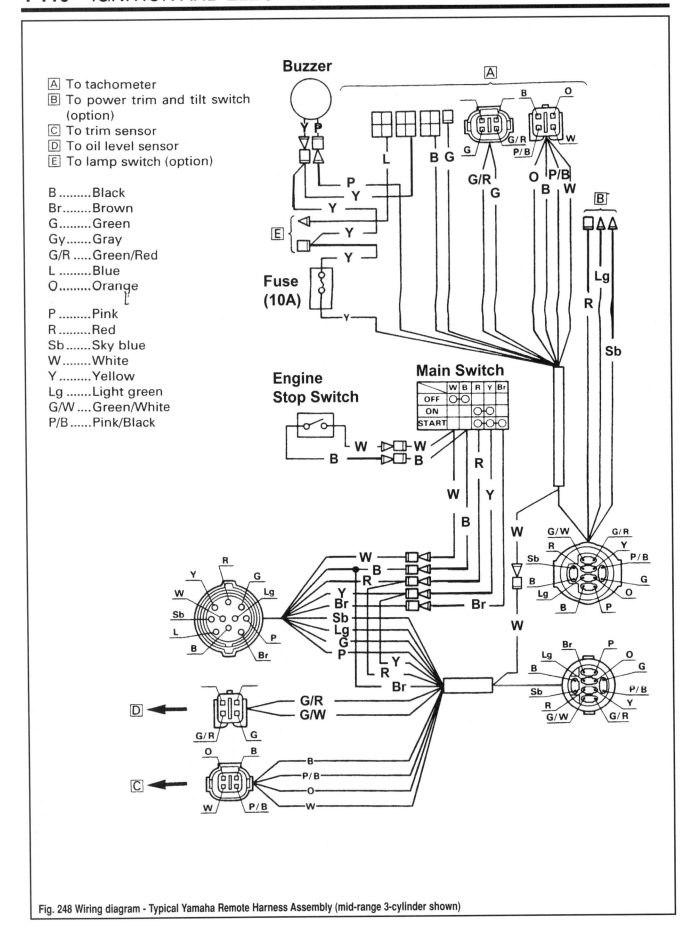

Fig. 248 Wiring diagram - Typical Yamaha Remote Harness Assembly (mid-range 3-cylinder shown)

COOLING SYSTEM **5-27**
   DESCRIPTION AND OPERATION 5-27
   THERMOSTAT 5-41
   THERMO-SWITCH 5-46
   TROUBLESHOOTING 5-28
   WATER PUMP 5-31
OIL INJECTION WARNING SYSTEM **5-20**
   GENERAL INFORMATION 5-20
   OPERATIONAL CHECK 5-23
   TROUBLESHOOTING 5-22
OIL PUMP ASSEMBLY 5-18
   BLEEDING THE OIL PUMP 5-19
   REMOVAL & INSTALLATION 5-18
PBS DIAGRAMS 5-10
POWERHEAD MOUNTED OIL TANK 5-16
   REMOVAL & INSTALLATION 5-16
PRECISION BLEND OIL INJECTION SYSTEMS **5-2**
   GENERAL INFORMATION AND TROUBLESHOOTING 5-2
   PBS DIAGRAMS 5-10
   POWERHEAD MOUNTED OIL TANK 5-16
   REMOTE OIL TANK 5-17
   OIL PUMP ASSEMBLY 5-18
REMOTE OIL TANK 5-17
   CLEANING & INSPECTING 5-18
   REMOVAL & INSTALLATION 5-17
   TESTING THE REMOTE TANK LIFT PUMP 5-17
SPECIFICATIONS **5-49**
   COOLING SYSTEM SPECIFICATIONS 5-49
THERMOSTAT 5-41
   EXPLODED VIEWS 5-43
   REMOVAL & INSTALLATION 5-41
THERMO-SWITCH 5-46
   CHECKING CONTINUITY 5-47
   DESCRIPTION & OPERATION 5-46
   OPERATIONAL CHECK 5-47
WATER PUMP 5-31
   EXPLODED VIEWS 5-35
   INSPECTION & OVERHAUL 5-33
   REMOVAL & INSTALLATION 5-31

5

LUBRICATION AND COOLING

PRECISION BLEND OIL INJECTION
SYSTEMS 5-2
OIL INJECTION WARNING SYSTEM 5-20
COOLING SYSTEM 5-27
SPECIFICATIONS 5-49

5-2 LUBRICATION AND COOLING

PRECISION BLEND OIL INJECTION SYSTEMS

Unlike 4-stroke engines, which contain a reservoir of oil that is recirculated during engine operation, mixing oil with the fuel lubricates 2-stroke engines. Internal engine components of 2-stroke motors are lubricated as the fuel/oil mixture passes through the crankcase and the cylinder.

Generally speaking, there are 2 methods of adding oil to a 2-stroke outboard. The first is to pre-mix oil with the gasoline whenever the fuel tank is filled. The pre-mix method is generally used on smaller (lower horsepower) motors and on many commercial outboards. It is easiest to perform this on portable fuel tanks that can be agitated to ensure proper mixture, but it can be successfully accomplished on larger built-in tanks, as long as care is taken to properly measure the amounts of fuel/oil being added. The better method, found on most larger displacement motors is to use an oil injection system.

General Information and Troubleshooting

◆ See Figures 1 thru 4

In the past, two-stroke outboard motors required the oil and gas to be mixed. This method supplied the engine with the oil necessary for lubrication. This system, while simple, has several drawbacks:

• The boat owner has to remember to mix oil with the gas. The owner also has to remember the correct mix ratio. This is a messy and sometimes confusing task for many customers.

• If the owner forgets to add oil the engine can be ruined. There is no simple way for the customer to verify or remember if oil has been added.

• Even when the oil is mixed at the correct ratio, this ratio is not perfect for all engine speeds. An engine may need as little as 200:1 (200 parts gas to 1 part oil) at idle and as much as 50:1 (50 parts gas to 1 part oil) at wide open throttle. Mix ratios must be set rich for safety. This means that at idle, the motor may be getting more oil than it needs. This can foul plugs and produces large clouds of smoke.

The various versions of the Precision Blend System (PBS) system used by Yamaha solve many of the problems associated with premixing. At the heart of the system is a mechanical oil pump driven by a brass gear on the crankshaft. As engine rpm changes, so does oil pump output. An arm on the side of the pump is connected to the carburetor or throttle body linkage. As the throttle valves open, the arm moves to increase the pump's stroke. The increased stroke produces a subsequent increase in oil output.

By having a pump that is sensitive to engine rpm and throttle valve opening, the engine receives the exact amount of oil at all times. This increases engine life and reduces oil consumption and smoke. On most systems the oil feed lines from the pump go directly into the intake manifold, however on EFI OX66 motors, the feed lines go to common manifold that delivers the oil supply to the fuel water separator. In all cases, the oil lines are routed in such a manner as to make any change in oil delivery is almost instantaneous.

Oil injection systems from some other manufacturers deliver oil to the carburetor float bowl. Oil mixed in the carburetor bowl will take several seconds to enter the engine. If engine rpm or load changes rapidly, the slower response of a bowl-mix system may leave the engine momentarily under-oiled or over-oiled.

The PBS system also reduces the mess associated with premixing the oil and gas. The fuel is in one tank and the oil in another. Another feature of the PBS is a low-oil level warning indicator. The oil level is microprocessor monitored. Typically, when the oil level is low, a warning indicator flashes and rpm is reduced to 2,000.

For safety's sake, a hold circuit is built into the PBS system. Once the rpm reduction has been activated, the system must be reset to bring back full rpm operation. Rpm will not automatically increase even if the oil tank is refilled. The reset for most late models is to return to idle, though on some older models the ignition key must be cycled from **OFF** to **ON**

Finally, a holding system protects the owner from an unexpected increase in rpm if he accidentally refills the PBS with the engine running.

SYSTEM COMPONENTS

■ The system components discussed in this section are inclusive of all possible PBS systems installed on Yamaha outboards, however not all components are installed on every outboard.

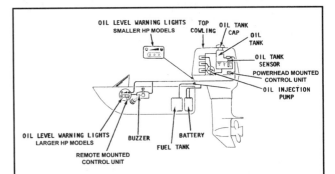

Fig. 1 Simplified functional diagram of a typical Yamaha oil injection system. The diagram depicts the relative location of the major components

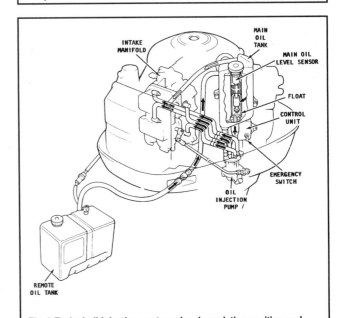

Fig. 2 Typical oil injection system showing relative position and function of major components - carbureted V4/V6 powerheads

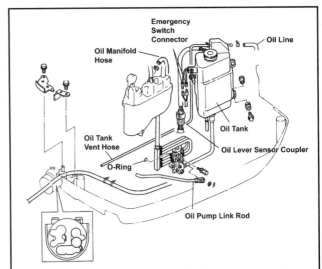

Fig. 3 Typical oil injection system for EFI OX66 powerheads. Note how the oil pump feeds a manifold that supplies oil to the fuel vapor separator assembly

LUBRICATION AND COOLING 5-3

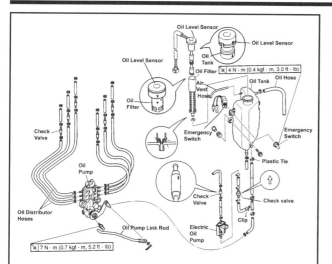

Fig. 4 The HPDI oil injection system uses an additional powerhead mounted electric oil pump which delivers oil to the fuel vapor separator assembly while individual oil lines also feed the cylinders directly

On a powerheads equipped with Yamaha Microcomputer Ignition System (YMIS), an extra lead from the oil injection control unit to the microcomputer is used. Other than this one difference, the oil injection systems installed on most carbureted powerheads are virtually identical.

Powerhead Oil Tank

◆ See Figures 5 and 6

Oil supply to the oil injection pump is gravity feed from a powerhead mounted oil reservoir tank. A breather for the tank is usually located next to the filler cap (or otherwise integrated into the cap). This breather must remain open at all times.

The main oil tank is mounted on the powerhead. The oil tank filler is easily accessible on top of the tank. On some models there is even a cutout in the cowling for access without removal of the engine top cover. A transparent plastic water/dust trap is normally located at the bottom of the main oil tank and is used to trap contaminates. On some late-models the trap is no longer a blind end hose, but is looped back upward to the tank itself. This both makes it easier to use the fitting as an oil tank drain (as the hose can be left attached to the fitting at the bottom of the tank) as well as gives the trap a larger capacity to hold water or contaminants.

Remote Oil Tank

◆ See Figure 7

Some models (typically the V4 and V6 powerheads) use a remote oil tank installed in a convenient location in the boat. The remote tank feeds the powerhead mounted oil tank with fresh oil when the level of the main tank drops below about 0.53 U.S. quarts (0.5 liters).

Oil Tank Sensors

◆ See Figures 8 and 9

An oil level sensor is mounted in the main oil tank. The sensor consists of a float sliding up and down in the sensor shaft between stops. The float rises and falls with the level of the oil. This sensor monitors the remaining oil level in the tank and, if the oil level falls to a dangerously low level or if the oil filter should become clogged, it sends signals to the oil level lights, the warning buzzer and on some powerheads, the control unit. The sensor signal sent to the control unit will cause the unit to reduce powerhead rpm.

On models with powerhead and remote oil tanks, the oil level in the main tank is replenished after it falls to about 0.53 U.S. quarts (0.5 liters). The oil tank sensor signals the feed pump in the remote tank to replenish the main tank. Oil will be pumped up from the remote tank to the main tank, until the level in the main tank reaches 0.9 U.S. quarts (0.85 liters). If there is no oil in the remote tank, the Yellow warning light will come on to inform the operator of the condition of the remote tank. Under this condition, the powerhead will operate strictly on the oil remaining in the main tank.

When the oil level in the main tank falls to about 0.3 U.S. quarts (0.3 liters), the oil level sensor will activate the Red warning light, cause the buzzer to sound and send a signal to the ignition control unit. The sensor signal sent to the control unit will cause the unit to reduce powerhead rpm.

There is no ignition cutout switch. Theoretically, it is possible to operate the powerhead when the main oil tank is dry, leading to overheating and seizure. However, the manufacturer has incorporated enough visual and audible danger signals to alert the operator well before any internal engine damage occurs.

■ If the warning system should operate and the buzzer sound, the powerhead must be shut down and the oil in the tank replenished. Filling the tank will cancel the engine speed restrictions.

Oil Injection Pump

◆ See Figures 10 thru 13

The oil injection pump is a positive displacement type unit that is normally driven by a worm gear and two short shafts from the lower end of the crankshaft. A gear pressed onto the lower end of the crankshaft drives the worm gear and a short shaft indexed with a second short shaft to the pump. This second short shaft to the pump drives the plunger cam. With each revolution of the plunger cam, the plunger moves up and down multiple times pumping oil to the cylinders.

A link (usually from the lower throttle valve) operates the oil injection pump lever. This lever affects the movement of the plunger cam by limiting the plunger stroke. In this manner, the amount of oil leaving the pump is regulated. As throttle movement is advanced and crankshaft rotation is increased, the amount of oil entering the intake manifold will increase. The mixing ratio depends upon the angle of the pump lever shaft. The pump lever shaft is directly connected to the throttle plate via a link rod. If the lever

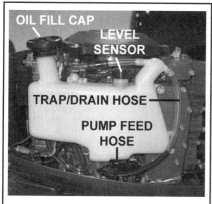

Fig. 5 A powerhead mounted oil tank is used to gravity feed the oil pump...

Fig. 6 ...most have a transparent water/dust trap and drain

Fig. 7 When equipped, a remote oil tank is installed on the boat

5-4 LUBRICATION AND COOLING

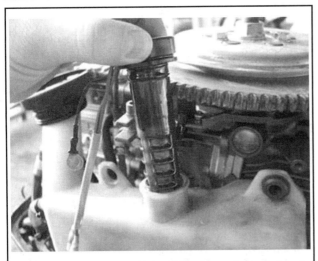

Fig. 8 An oil level sensor and integral filter is mounted to the top of the oil tank

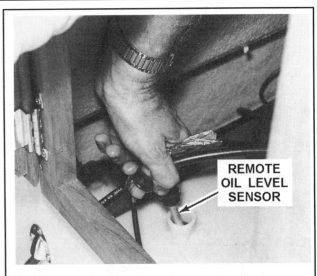

Fig. 9 Removing the oil level sensor from a remote oil tank

Fig. 10 Typical oil injection pump mounted on a smaller (40 hp) 3-cylinder powerhead

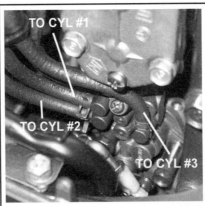

Fig. 11 Pump from a larger 3-cylinder (90 hp) powerhead

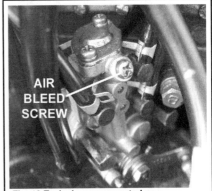

Fig. 12 Typical pump mounted on a carbureted V4 motors (V6 and EFI/HPDI similar)

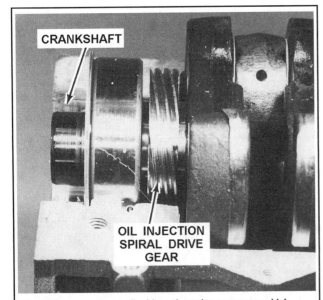

Fig. 13 The pump is usually driven through a worm gear which indexes with a spiral gear pressed onto the lower end of the driveshaft

angle is between 0° and 5°, the mixing ratio is usually about 200:1. If the lever angle is between 5° and 50°, the mixing ratio is usually somewhere between 200:1 and 50:1.

The oil leaves the pump and is delivered to the cylinders (or vapor separator assembly on EFI OX66 motors) through a series of 2-6 transparent hoses, depending upon the model.

Oil Feed (Lift) Pump

◆ See Figure 14

The oil feed pump is usually located at the rear of the remote oil tank. The unit is an electrically-operated gear type pump capable of delivering oil from the remote tank to the main tank.

The pump is activated to start and stop by signals from the oil sensor in the main tank. Under normal operation conditions, when the oil level in the main tank falls to a predetermined low level, the sensor sends a signal and the oil feed pump is activated. The pump will deliver oil to the main tank until the oil level in the main tank reaches a predetermined full quantity. At this time, another signal is sent to the oil feed pump and pumping from the remote tank ceases.

The oil feed pump will pump oil until only 1.6 U.S. quarts (1.5 liters) remain in the tank. However, a manual override of the main oil tank sensor will permit the pump to be activated and the remote tank drained of all oil. This override action is accomplished by activating the emergency switch on the control box.

LUBRICATION AND COOLING

Control Unit

◆ See Figure 15

Control of the electric feed pump for models with remote oil tanks is handled either by the CDI Unit/ECM or through a separate, usually powerhead mounted, control unit. Similarly, functions of the warning system may be CDI Unit/ECM controlled, or may occur in conjunction with a powerhead or remote mounted control unit. For model details, please refer to the Wiring Diagrams found in the Ignition and Electrical System section.

Whether or not the control unit is integrated into the CDI Unit/ECM, the control unit circuits receive input signals from the oil level sensors in both oil tanks and determines which light should illuminate on the display. The control unit also sends signals to the buzzer in the remote control box and to the CDI unit to reduce powerhead speed under specified conditions.

An emergency switch is normally mounted to the powerhead (often on/near the control unit itself or to a mounting bracket on/near the oil tank) which can be used to manually activate the oil lift pump transferring the contents from the remote tank to the powerhead mounted oil tank.

Tilt Switch

The tilt switch is a function of the control unit and is incorporated inside the unit. Should the powerhead be tilted during operation, the sensors in both tanks would register a false reading. This condition could cause unnecessary transfer of oil from the remote tank to the main tank, possibly filling and overflowing the tank. The tilt switch senses the angle of the powerhead at all times. Typically, if the unit is tilted beyond 35° to the vertical, the tilt switch cuts out the oil feed pump circuit until the powerhead returns to 20° or less.

Emergency Switch

◆ See Figure 15

An emergency switch is normally mounted to the powerhead (often on/near the control unit itself or to a mounting bracket on/near the oil tank) which can be used to manually activate the oil lift pump transferring the contents from the remote tank to the powerhead mounted oil tank.

A decal is often mounted on the control box or near the switch which describes the purpose of the switch. Because the decal may no longer be legible, the typical wording is given here as follows:

"Holding up the emergency switch pumps reserve oil into the main oil tank so you can continue cruising.

When the switch is held in the up position, the operator is able to manually override the signal from the main oil tank sensor and cause the oil feed pump to completely drain the remote tank.

The emergency switch should only be activated if the operator is unable to replenish the remote tank and is forced to use the tank's reserve capacity. If the emergency switch is released before all the oil is pumped to the main tank, the necessity of bleeding the oil feed pump may be avoided. If the remote tank is pumped completely dry, the feed pump must be purged of air

Buzzer

◆ See Figure 16

On most Yamaha remotes, the buzzer is installed in the remote control box and is used to warn the operator of a possible problem. The most common cause is low oil level in the remote tank. However, the buzzer can be potentially activated by signals from three different sensors - the over-rev circuit, the overheating thermo-switch, and the low oil sensor. If the buzzer sounds due to a signal from the over-rev or overheating circuits, the signal will be accompanied by a sudden (noticeable) drop in powerhead rpm and the Red warning light will come on.

If the buzzer sounds and the Red light comes on but powerhead rpm is not reduced, then the signal responsible for activating the buzzer and light is from the sensor in the remote oil tank. Adding oil to the tank will correct the problem.

⁂ WARNING

No matter what generates the signal to the warning buzzer, you're first step should always be to shut the powerhead down and determine what is wrong so you can remedy the situation before the powerhead is damaged.

Warning Light Display

The warning light display is located either on the front cowling of the powerhead or built into the tachometer. The warning lights and their significance are explained in the following paragraphs.

■ If the warning system fails (no lights come on) the oil level must be checked immediately. Operate the powerhead at reduced speed and with caution to get back to port until the unit can be properly serviced.

Remote Warning Lamp System

◆ See Figure 17

Yamaha has produced various tachometers with integrated warning light displays for installation to the boat dash on remote control motors. Modern Yamahas are usually equipped with an all digital display. The display lights vary slightly from one model to another, but the meaning of each should be relatively obvious from the symbols in the display.

The warning light display usually includes a temperature light (it looks like a thermometer immersed in liquid, not unlike most automotive temperature light/gauge displays). Immediately to the right of the temperature light there are usually 3 oil level related lights. These lights loosely correspond to the 3 lights of the 3 LED system on tiller models. One symbol (to the far right of the display) illuminates at start-up and usually remains on during engine operation to show the system has a proper oil level. Once the oil level drops to a certain point in the remote tank a second symbol (usually the oil can at the center) will illuminate or blink reminding you to check the level. Once the oil level in the powerhead mounted tank drops to a certain point the final LED will illuminate. This final led is usually a low oil symbol located to the left

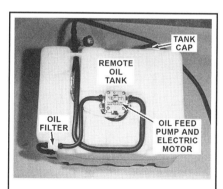

Fig. 14 Typical oil feed pump electric motor and oil filter

Fig. 15 Emergency switch from a typical carbureted V-motor (on the CDI unit bracket)

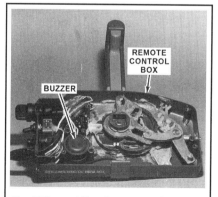

Fig. 16 On most Yamaha remotes, the buzzer and its harness are inside the control box

5-6 LUBRICATION AND COOLING

of the 3 lights. Illumination of the final light is normally accompanied by the warning buzzer, to indicate that oil level is dangerously low in the powerhead tank.

Single Lamp System

◆ See Figure 18

The single lamp system is used on the smallest of the tiller models. A low oil level in the powerhead mounted tank on these models will illuminate the LED, usually accompanied by a warning buzzer signal. The operator must immediately slow or stop the powerhead and remedy the situation to prevent damage to the powerhead.

Three Led System

◆ See Figure 19

Probably the most common display on tiller models equipped with an oil injection system uses three colored warning lights, a Green round light, a Yellow oil can or round light, and a Red round light. These lights warn the operator of a low oil or an overheating condition. Different combinations of lights with or without the buzzer can indicate a variety of conditions.

Green Light Comes On, Buzzer Does Not Sound - During normal operating conditions, if the powerhead has an adequate supply of oil, the Green light is always on. This Green light informs the operator there is more than about 0.5 U.S. quarts (0.5 liters) is in the main tank, and at least 1.6 U.S. quarts (1.5 liters) in the remote tank.

Green or Yellow Light Comes On, Buzzer Sounds - This condition should be accompanied with a sudden drop in engine rpm. The oil level in both tanks is acceptable. The problem lies in one or both banks of cylinders because an overheating condition is developing. The powerhead should be shut down immediately; the problem isolated; and corrective action taken. Continued operation of the powerhead with one of these lights on and the buzzer sounding could lead to serious and expensive internal damage and seizure.

Yellow Light Comes On, Buzzer Does Not Sound - If the Yellow light comes on, the operator is advised only a reserve quantity of oil remains in the remote tank. In order to pump this reserve oil, approximately 1.6 U.S. quarts (1.5 liters) to the main tank, the operator must hold the emergency switch in the up position. The Yellow light will remain on until the remote tank is replenished with oil.

Red Light Comes On, Buzzer Sounds - If this condition is accompanied with a sudden drop in powerhead rpm, the problem is an overheating condition. One or both banks of cylinders have experienced prolonged excessive temperatures and are in immediate danger of seizing. Shut the powerhead down immediately.

If this signal is not accompanied with a drop in powerhead rpm, then a dangerously low oil level condition exists, which requires the operator's immediate attention. There is almost no oil left. The main tank has less than 0.3 U.S. quarts (0.3 liters) and the remote tank has less than 1.6 U.S. quarts (1.5 liters).

■ **If the Red light is ON, the buzzer should be sounding. Check the operation of the buzzer as soon as possible if it does not sound.**

Green ad Red Lights Come On Buzzer Sounds - If the Green and Red lights both come on simultaneously, a problem has arisen in the transfer of oil from the remote tank to the main tank. There is adequate oil in the remote tank but either the oil feed pump has failed or there is a blockage in one of the oil lines. Again, this situation demands immediate action by the operator. Shut down the powerhead at once. In order to return the boat to its point of origin, the oil in the remote tank must be manually transferred to the main tank. Once the main tank has an adequate supply, the powerhead may be restarted.

■ **If the Green and Red lights are ON, the buzzer should be sounding. Check the operation of the buzzer as soon as possible if it does not sound.**

Hose Network

Oil from the main oil tank is gravity fed to the oil injection pump. Then, on most models oil from the pump is routed through individual oil hoses directly to the ports on the intake manifold. However, on EFI OX66 motors all the oil pump hoses which feed individual cylinders on other models are routed to a common manifold that feeds the fuel vapor separator. Some routing diagrams are provided under the service procedures in this section, however, it is always a good idea to note hose routing and tag all lines before removal.

PBS TROUBLESHOOTING

 DIFFICULT

This section is divided into the most common problem areas and complaints concerning the Precision Blend System. Troubleshooting the warning system itself is covered later in this section under Warning System.

Wire Colors

The Yamaha electrical systems are normally designed to use the ground side of any electrical component for control rather than the power side. This means the warning lights, oil transfer pump, and warning buzzer have power when the main switch is in the **ON** position. Also, if the Pink wire is shorted to ground, the buzzer and rpm reduction feature will operate.

The color codes of the wires for the PBS control and warning circuits are usually as follows (but double-check the model on which you are working using the Wiring Diagrams in the Ignition and Electrical System section to be sure):

- Red - Battery power +12 volts
- Yellow - Main switched power +12 volts
- Black - Ground
- Pink - Warning horn rpm reduction control

The color codes of the wires coming from the remote oil tank are usually as follows (follow the same advice on checking the Wiring Diagrams):

- Black - Ground lead for oil level sensor
- Black/Red - Oil level sensor signal
- Brown - Switched power +12 volts
- Blue - Control lead for transfer pump ground

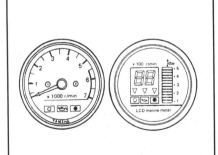

Fig. 17 The warning light display on remote models is usually part of the tachometer - older (left) and newer digital meter (right) faces

Fig. 18 A single LED type oil lever indicator is used on some tiller powerheads

Fig. 19 The oil warning display on this tiller model uses 3 colored lights

LUBRICATION AND COOLING

Main (Engine) Oil Tank Overflowing

General Troubleshooting Information

When the engine oil tank overflows, it could be caused by a mechanical problem, electrical problem, or both. Electrical problems are usually caused by the engine oil sender not turning the oil pump **OFF** when the oil tank is full.

On all most V4 engines, the White wire from the engine oil tank sends the signal to the oil control unit to shut off the oil transfer pump. Most V6 engines (including EFI/HPDI models) use a Blue wire with a White tracer. This wire connects to ground to activate the control unit to stop the transfer pump. If it is not grounded, check to see if the Black wire to the sending unit is grounded.

The only components left are the oil control unit and the wire harness. When the engine oil tank overfills, assuming you have not mounted the remote tank higher than the engine so that it just plain siphons, it is caused by the Blue wire to the transfer pump being grounded. Whenever the ignition key is in the **ON** position, the oil transfer pump has a positive 12 volts supplied to it on the Brown wire. If the Blue wire running to the pump is grounded, the oil transfer pump will run.

So why might the Blue wire be grounded? Well, the harness may have a screw through it, grounding the Blue wire or it is possible that the oil control unit (CDI Unit/ECM on some models) may have grounded the Blue wire for some reason.

Why would the oil control unit ground the Blue wire when the engine oil tank is full? Either the sender is telling it to ground the Blue wire or the control unit is grounded internally.

Here are two items to check before replacing the engine oil tank sending unit. Check if the filter is turned (on the sending unit) blocking the vent hole at the top. Also check if the White or Blue/White wire has continuity to the Black wire (in the sending unit) when the float is all the way in the up position. If the engine oil tank sending unit connects the White or Blue/White wire to the Black wire (in the sending unit) when the float is all the way up, there is nothing wrong with the sending unit. If all harnesses check out OK, odds are you have a bad oil control unit.

Overflowing Tanks with No Evidence of Defective Parts

◆ See Figure 20

Check the following items to prevent the needless replacement of parts when attempting to diagnose overflowing tank problems.

A very low battery voltage with main switch **ON** is usually caused when an operator leaves the main (key) switch in the **ON** position while the boat is in storage. After a time, the battery will discharge to the point where the voltage drops and the electronic circuitry on the engine cannot operate properly. When this happens, the transfer pump will start transferring oil to the engine oil tank, overflowing it. Operation returns to normal after the battery is charged.

An improper location of the filter screen on the engine oil level sender may cause an overflow condition. There is normally a tubular plastic screen installed around the oil level sender. This screen has a vent hole near the top of the tube. This hole lines up at the center of the flat area on the sender. It is to let air escape so oil may pass through the screen. The fitting at the bottom of the tube is offset and should line up with the oil tank outlet.

Sometimes when someone is inspecting an engine, they will twist or turn the top of the oil sender. The offset outlet of the screen will keep the filter screen from twisting, but the sender itself will twist easily, causing the filter screen tube to distort and bind the up and down action of the sensor float. If this does not occur, then the screen will twist so the vent hole is blocked. Either of these conditions will cause the sensor to malfunction and cause the engine oil tank to overflow.

Twisting the oil sender back to the correct position may not correct the problem because the screen inside the oil talk may be in the wrong position.

Remember the following when installing and aligning the screen:

1. The vent hole normally goes at the center of the flat area on the oil sender.
2. The offset outlet fitting aligns with the oil tank outlet.
3. There is an alignment detent hole on the screen that fits over a small bump on the sender just to the side of the flat area. This alignment bump is even with the arrow marking on the side of the sender cap.
4. Make sure the above mentioned arrow marking on the sender cap aligns with the arrow marking on the oil tank.
5. Make sure to check the screen for plugging and clean it, if necessary.

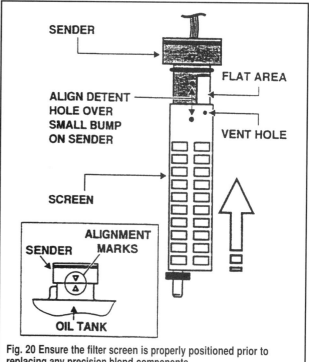

Fig. 20 Ensure the filter screen is properly positioned prior to replacing any precision blend components

Remote Oil Tank Location

When picking the mounting location for the remote oil tank, or when troubleshooting a condition where the engine oil tank is overflowing when the motor is not in use, note the relative height of the remote oil tank in relation to the engine oil tank. If the remote oil tank is located where the oil level is above the engine oil tank inlet, then (through normal gravity and siphoning) the remote oil will drain into the engine tank, causing it to overfill. This is especially possible with pontoon or house boats. If this occurs, the remote oil tank must be relocated below the level of the engine oil tank. This problem may also occur when a trailered boat is parked on a steep hill. Make sure you ask your customers about their boat storage location if they have this complaint.

Dual Engine Installations

A common problem on dual engine installations is one engine (e.g. port engine) begins to smoke heavily and stalls. This problem may occur on either engine in a dual installation, depending on the way the engines are used. If you come across this problem, make sure the harnesses and oil lines are matched up and connected to the proper engines.

During rigging or routine maintenance was the harness connecting the starboard engine to the sub-oil tank (boat mounted tank) connected instead to the main oil tank of the port engine? If so, when the oil level in the starboard main tank decreased, the control unit turned on the transfer pump to refill the tank of the starboard engine. Because of the improperly connected harness, the control unit sent the oil to the main tank forcing the oil to pass the in-line check valves and into the intake manifold; making the engine smoke and stall. Because the port tank never received any oil, the shut down of the transfer pump did not occur.

Oil Will Not Transfer

General Troubleshooting Information

◆ See Figure 21

Older model Yamaha motors used various methods of preventing oil transfer during certain circumstances. The earliest models used a Mercury switch in the powerhead mounted control unit to prevent transferring oil when the engine was tilted (top prevent overfilling). Some early models wired the trim sender to prevent transfer past a certain degree of tilt. However, on most 2.6L and larger late-model motors, oil will not be transferred unless the engine is running and this control comes from the CDI unit or ECM.

5-8 LUBRICATION AND COOLING

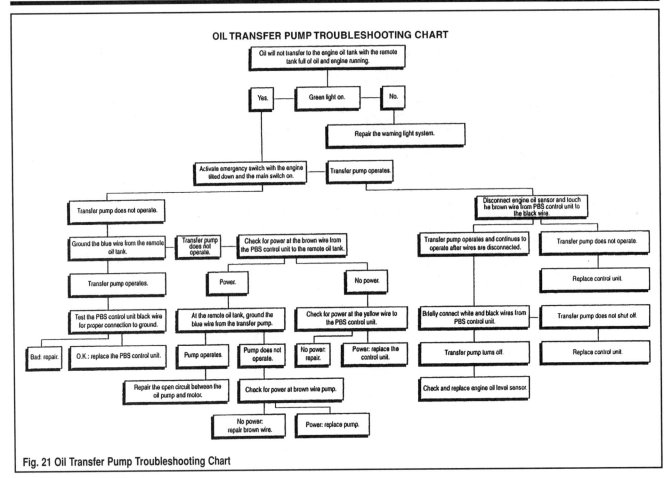

Fig. 21 Oil Transfer Pump Troubleshooting Chart

Engine Oil Sender Hold Circuit

◆ See Figure 22

Some V4 and V6 engines (usually just carbureted models) contain an rpm sensing hold circuit. The engine speed must be reduced below approximately 1300 rpm or the throttle must be returned to idle to reset the system, provided the oil tank has been refilled.

Although some models control this rpm hold inside the CDI computer unit, others use a hold circuit built into the engine oil level sensor. Hold circuit sensors can be identified by an "H" molded into the top of the sensor.

Sensors with the oil level hold circuits may be tested just like a diode. Ohmmeter polarity is important. If the hold circuit appears to be bad, reverse the ohmmeter leads and try again. The circuit should only function when polarity is correct.

■ **If you are using a digital voltmeter, select the diode setting if the meter has one.**

1. Position the float at the top of the sender.
2. Connect the positive lead of the ohmmeter (not necessarily the Red lead depending on your meter) to the warning system lead (Blue/Red or Red).
3. Connect the negative lead of the ohmmeter to the Black ground lead on the sensor.
4. Use the 10 scale on your meter. If you are using a digital voltmeter, select the diode setting if the meter has one.
5. At this time, the meter reading should be infinite or open. Move the float to the bottom of the sensor. You should now have some resistance reading between zero and infinity.
6. Move the float to the top of the sensor. The resistance reading should stay between 0 and infinity.
7. Momentarily disconnect and then reconnect the ohmmeter. The resistance reading should be infinite or open again.

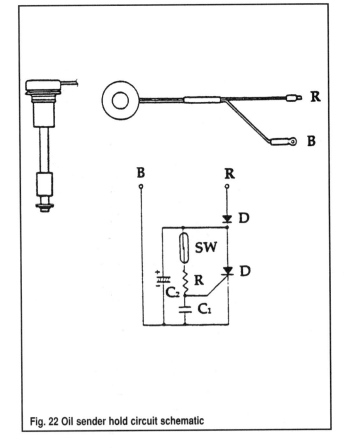

Fig. 22 Oil sender hold circuit schematic

LUBRICATION AND COOLING 5-9

Excessive Smoke At Idle

If the engine exhaust smoke is too thick at idle, check the following:
- Oil pump linkage, is it disconnected?
- Oil pump linkage for free movement
- Oil pump linkage adjustment
- Throttle valve synchronization and, if applicable, idle fuel adjustment
- Quality of oil being used
- Oil pump output check valves for proper sealing
- Engine recirculation check valves not sealing
- Leaking fuel pump diaphragms
- Leaking exhaust gaskets
- Excessive oil passing through vent hose to the air box. This may be caused by an oil sender not fully inserted, causing a high oil level in engine oil tank.
- Oil pump driveshaft seal leaking. Remove pump and check for out of position or damaged seal.

Oil Pump Check Valve(s)

◆ See Figure 23

Various check valves are used in most V4 and V6 systems. If one or more of the pump related check valves is suspected, it may be tested using a combination vacuum pump and gauge.

1. Note the positioning of the check valve so that it can be installed in the same direction. There is normally a directional arrow on the side of the valve itself. If not, scribe one to ensure proper installation.
2. Using a hand-held pressure pump, slowly apply about 11.4 psi (80kPa) of pressure to the valve, in the reverse of the normal directional flow. No air should leak from the opposite side of the valve.
3. Now reverse the connection and apply about 11.4 psi (80kPa) of pressure to the valve, in the normal directional flow. Air should not flow out the other side of the valve until this pressure or slightly higher is achieved.

■ Don't use unregulated or high-pressure air to test the check valve in either direction as the valve could become damaged. Also, if you don't have a hand pump but do have a decent set of lungs you SHOULD be able to check the valve by blowing through it. Air should only flow in one direction.

CHECKING OIL PUMP OUTPUT

◆ See Figure 24

Generally speaking, the output of the oil injection pump is tested by running the motor with one or more pump lines disconnected while you capture the oil discharge into a graded cylinder for precision measurement. The most important point we can make about this check is that, unless you are working on an HPDI motor, ATTACH A TANK OF PRE-MIX to prevent possible engine seizure. Probably the second most important point we can

make is that your measurements will become more accurate the longer you conduct the test. Yamaha recommends a minimum of 3 minutes (which is why they've given specs based on a 3 minute interval), but even Yamaha recommends performing a longer test. For ease of calculations, increase the test in 3 minute intervals so that the results can be divided by the number of additional 3 minute groups you've run the motor and compared to the specifications provided in the accompanying chart.

For all motors except the EFI OX66, test results are for the output to 1 cylinder during the given test period. You'll notice the EFI OX66 test results are quite a bit larger (actually exactly 6 times larger) than the results on a similar motor with the same pump. That is because on EFI OX66 motors the output is measured from the common oil manifold which is fed by the 6 oil pump output hoses which would feed individual cylinders on motors with other fuel systems.

Unfortunately, Yamaha did not appear to publish specifications for pump output on HPDI motors. They did say that the electric pump (which is used to feed the vapor separator on those powerheads) can be tested simply by disconnecting the line from the vapor separator with the engine running and observing output, but again, they did not say how MUCH output is expected.

One last point to keep in mind when testing pump output is that operating conditions are very specific. Since output may vary greatly with oil viscosity and temperature you must only use YAMAHA 2-stroke engine oil which is at an ambient temperature of about 50-86°F (10-30°C).

So, for all except HPDI motors, measure system output as follows:

1. For all except EFI OX66 motors, connect a portable fuel tank filled with a pre-mix solution of gasoline and 2-stroke engine oil in the proper pre-mix ratio for these motors. For more details, please refer to Engine Oil (2-Stroke) in the Maintenance and Tune-Up section.
2. Connect a suitable source of cooling water or launch the boat. In the case of the 250 hp EFI OX66 Vmax you will have to install the motor in a test tank OR launch the boat as engine operating speeds will be too great for unloaded flush fitting operation.

■ For all motors except the 250 hp EFI OX66 Vmax, the engine is run at 1500 rpm during the test. This can usually be accomplished on a flush fitting, unless the motor surges or speeds up due to a lack of engine load. If this is true, you'll have to mount it in a test tank or launch the craft to ensure proper test readings and to protect the powerhead.

3. If the pump was just removed/replaced or any of the oil lines disconnected, properly Bleed the Oil Pump of air, as detailed under Oil Pump, later in this section.
4. Disconnect the throttle link from the oil pump control lever, taking care not to disturb the length of the link so as not to upset adjustment when it is reinstalled.
5. Start and run the engine to normal operating temperature.
6. Shut the powerhead down momentarily so that you can disconnect the necessary oil line. On carbureted motors, disconnect an oil line from an intake manifold fitting or conversely disconnect the line at the pump (using a hose pincher to keep it from draining) and temporarily connect a test line to that pump fitting (with the other end of THAT line in the graded cylinder). On EFI OX66 motors, disconnect the common oil manifold supply line from the vapor separator and direct that line into the graded cylinder.
7. Manually open the oil pump control lever to the Wide Open Throttle (WOT) position in order to ensure full pump output.
8. Start and run the engine at a speed of 1500 rpm (except for 250 hp EFI OX66 Vmax motors) for at least 3 minutes (or for a multiple of 3 minutes) while you watch the oil entering the graded cylinder. On 250 hp EFI OX66 Vmax motors do the same thing, except run the motor at 5500 rpm.

■ Make sure the oil flow is steady and there are no air bubbles otherwise the system must be bled of air before proceeding and measuring the output.

9. Shut the powerhead down and reconnect the oil line. On carbureted motors, repeat the test for each cylinder.

■ When checking the output with a graded cylinder, make sure the wall surface of the cylinder is not coated with oil. If it is, the measurement may be incorrect.

10. Once the test is finished be sure to reconnect the oil pump link and double-check the adjustment.

Pumps which fail to perform as specified must be replaced in order to protect the powerhead.

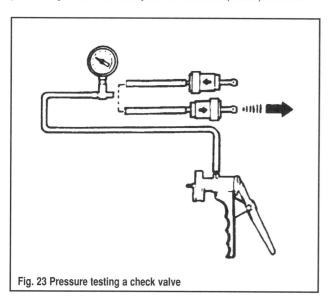

Fig. 23 Pressure testing a check valve

5-10 LUBRICATION AND COOLING

Oil Injection Pump Specifications - Precision Blend 2-Stroke Motors

Model (Hp)	No. of Cyl	Engine Type	Displacement cu. in. (cc)	Oil Pump ID Mark	Oil Pump Discharge Per 3 Minute Interval (Oz)	(CC)
20/25	2	IL 2-stroke	24 (395)	6L200	0.027-0.035	0.7-0.9
25/30	3	IL 2-stroke	30 (496)	n/a ①	0.014-0.034	0.4-1.0
28J-50	3	IL 2-stroke	43 (698)	63D00	0.037-0.071	1.1-2.1
50	3	IL 2-stroke	52 (849)	62F00	0.047-0.081	1.4-2.4
60-70	3	IL 2-stroke	52 (849)	6H302	0.057-0.091	1.7-2.7
65J-90	3	IL 2-stroke	70 (1140)	6H102	0.081-0.129	2.4-3.8
80J-140	4	90 LV 2-stroke	106 (1730)	6N600	0.084-0.132	2.5-3.9
80J-140	4	90 LV 2-stroke	106 (1730)	6N700	0.125-0.199	3.7-5.9
105J-225	6	Carb, 90 LV 2-stroke	158 (2596)	6R400	0.091-0.139	2.7-4.1
105J-225	6	Carb, 90 LV 2-stroke	158 (2596)	6R510	0.125-0.199	3.7-5.9
150	6	EFI OX66, 90 LV 2-stroke	158 (2596)	6R400	0.558-0.842	16.2-24.6
200	6	EFI OX66, 90 LV 2-stroke	158 (2596)	67H00	0.751-1.197	22.2-35.4
150	6	HPDI, 90 LV 2-stroke	158 (2596)	68H00	④	④
175	6	HPDI, 90 LV 2-stroke	158 (2596)	68L00	④	④
200	6	HPDI, 90 LV 2-stroke	158 (2596)	68F00	④	④
200-250	6	EFI OX66, 76 LV 2-stroke	191 (3130)	65L00 ②	1.075-1.603	31.8-47.4
250	6	EFI OX66, 76 LV 2-stroke	191 (3130)	69L00 ③	1.628-1.990	48.2-58.8
225-250	6	HPDI, 76 LV 2-stroke	204 (3342)	60V00	④	④

IL -Inline loop-charge   # LV - degree "V" loop-charge

* NOTE: Measurements are approximate using only YAMAHA oil at a temp of 50-86 degrees F (10-30 degrees C).

Also, Keep in mind that longer tests will be more accurate and only perform tests under specified conditions (with oil pump link disconnected and oil pump lever manually held wide open)

① Not available, but previous powerheads of this size used the 6J801
② This pump is used on all 3.1L models EXCEPT the 250 hp Vmax
③ This pump is only used on 250 hp Vmax
④ We could not locate any data from Yamaha on output for these pumps

Fig. 24 Oil Pump Output Specifications

PBS Diagrams

◆ See Figures 25 thru 35

Although some items like pump housing and hose routing will vary slightly, overall the Precision Blend System is pretty consistent from Yamaha motor to Yamaha motor. The major components are listed in this section with information for removal and installation. If oil delivery components are disconnected, the system should be bled of air to prevent potential powerhead damage, details can be found under Bleeding the Oil Pump.

The system diagrams included here show basic hose routing and component mounting. Further details on component servicing can be found in the procedures for Powerhead Mounted Oil Tank, Remote Oil Tank and Oil Pump, in this section.

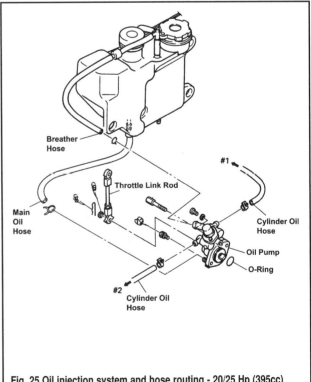

Fig. 25 Oil injection system and hose routing - 20/25 Hp (395cc) Models

LUBRICATION AND COOLING 5-11

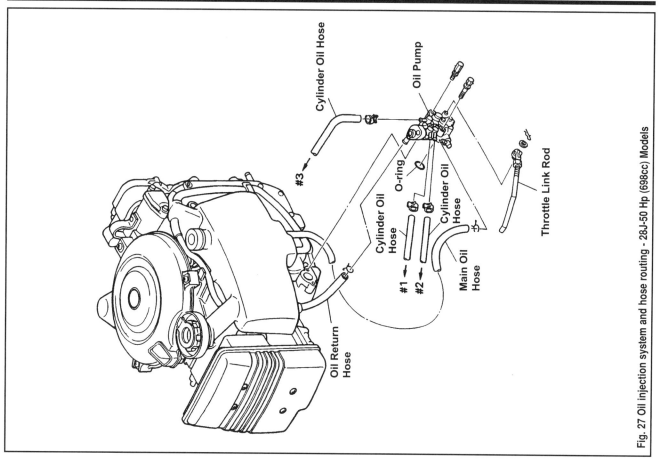

Fig. 27 Oil injection system and hose routing - 28J-50 Hp (698cc) Models

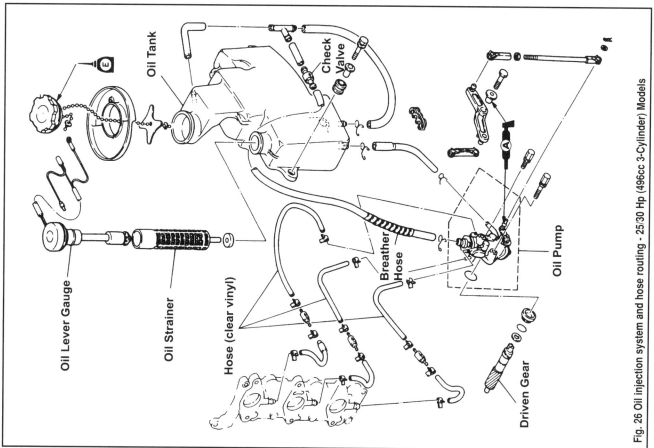

Fig. 26 Oil injection system and hose routing - 25/30 Hp (496cc 3-Cylinder) Models

5-12 LUBRICATION AND COOLING

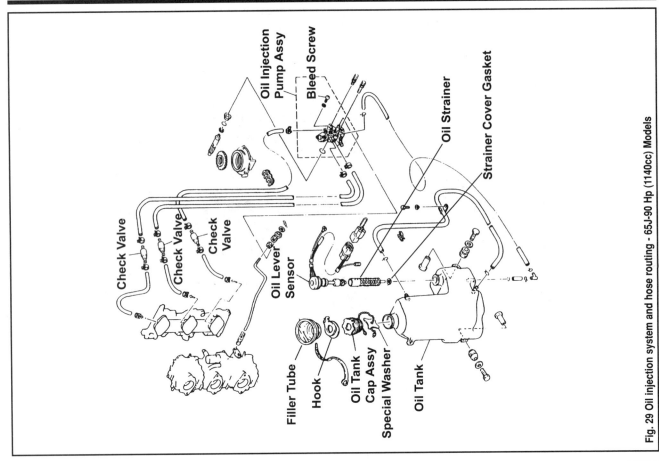

Fig. 29 Oil injection system and hose routing - 65J-90 Hp (1140cc) Models

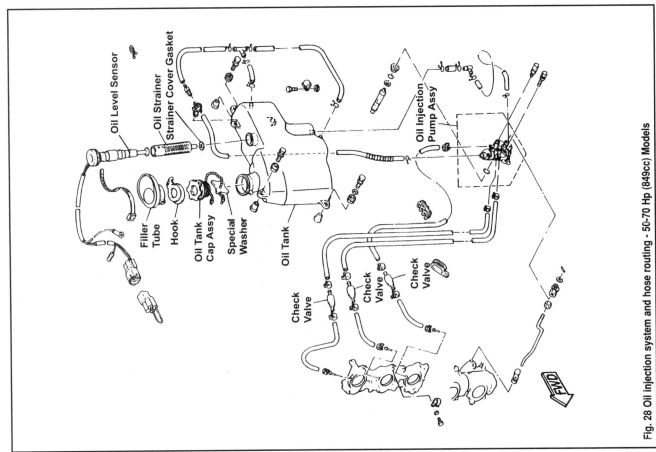

Fig. 28 Oil injection system and hose routing - 50-70 Hp (849cc) Models

LUBRICATION AND COOLING 5-13

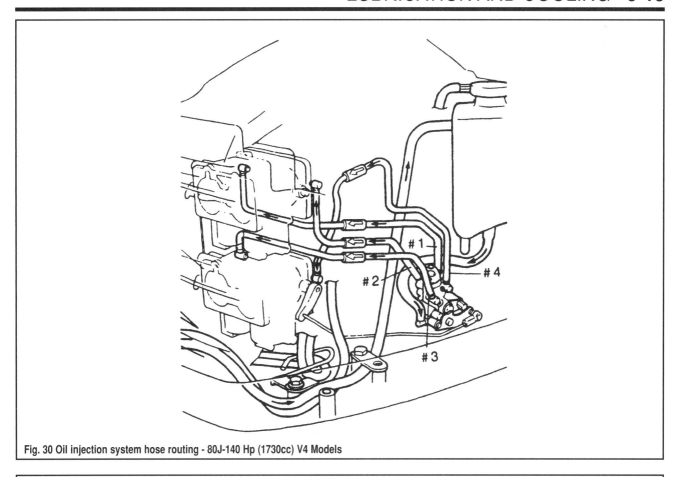

Fig. 30 Oil injection system hose routing - 80J-140 Hp (1730cc) V4 Models

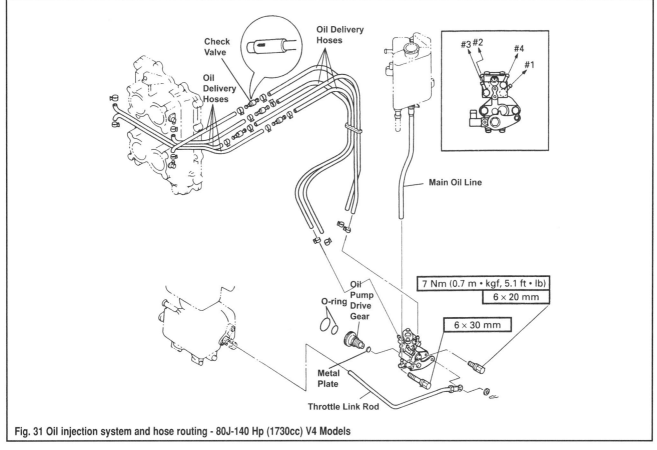

Fig. 31 Oil injection system and hose routing - 80J-140 Hp (1730cc) V4 Models

5-14 LUBRICATION AND COOLING

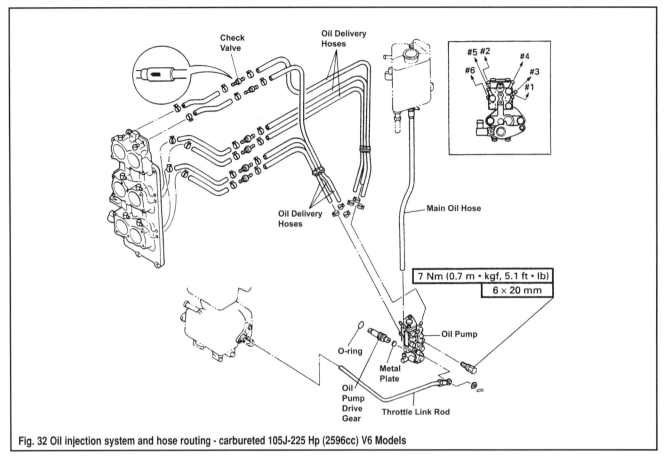

Fig. 32 Oil injection system and hose routing - carbureted 105J-225 Hp (2596cc) V6 Models

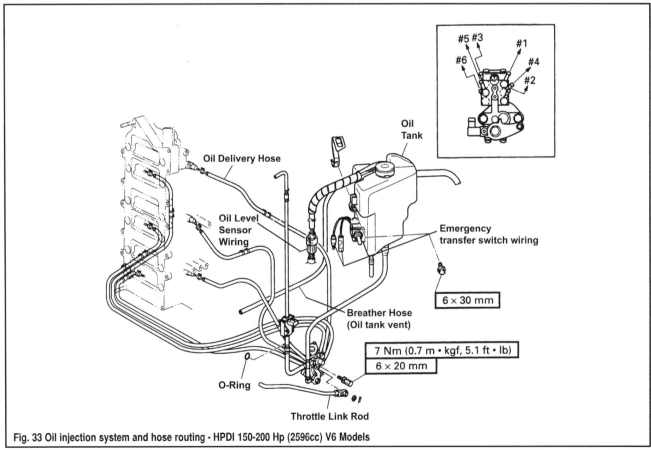

Fig. 33 Oil injection system and hose routing - HPDI 150-200 Hp (2596cc) V6 Models

LUBRICATION AND COOLING 5-15

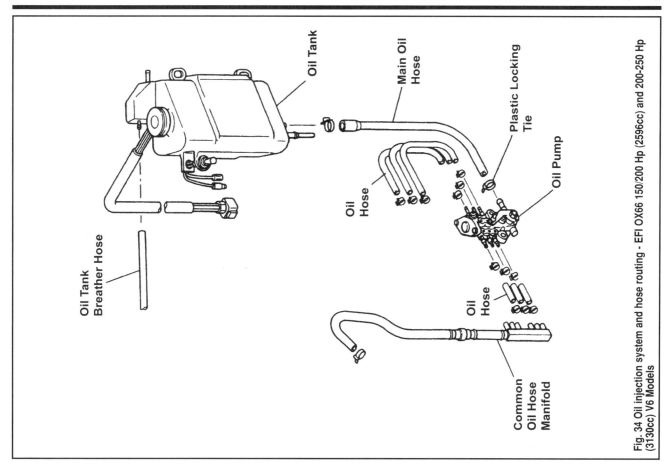

Fig. 34 Oil injection system and hose routing - EFI OX66 150/200 Hp (2596cc) and 200-250 Hp (3130cc) V6 Models

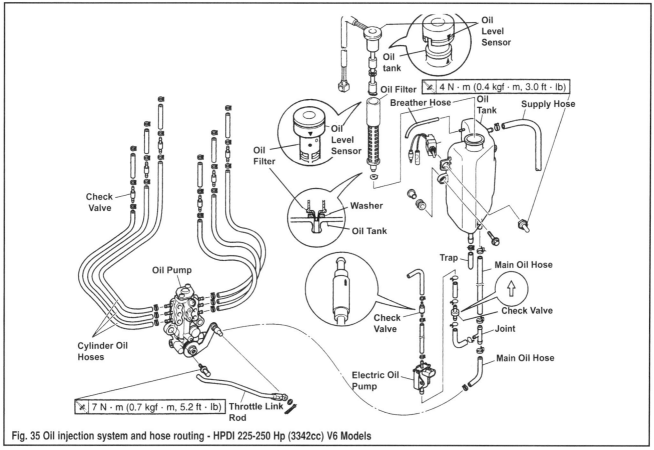

Fig. 35 Oil injection system and hose routing - HPDI 225-250 Hp (3342cc) V6 Models

5-16 LUBRICATION AND COOLING

Powerhead Mounted Oil Tank

REMOVAL & INSTALLATION

◆ See Figures 25 thru 35 and 36 thru 40

The oil tank is easily accessible and may be moved aside to facilitate removal of the flywheel or other powerhead mounted components, without draining the tank or disconnecting the oil lines.

However, complete removal may be necessary for additional access when servicing major powerhead components. In this case, the tank can be drained and/or lines can be temporarily pinched off or plugged to prevent oil loss.

1. Place some shop cloths in the bottom of the lower cowling to catch any oil which may drain during disconnection of oil supply lines.

■ Decide whether you want to pinch/plug the main oil line and/or drain the oil tank. For the first option it is best to use a dedicated hose pincher (available at most automotive and marine parts supply stores) as they are designed to minimize or prevent damage to a hose (as opposed to a pair of vise grips or pliers which will usually chew up a line).

2. If you've decided to drain the oil tank, either use the trap/drain hose (on some models a trap hose is routed back to the top of the tank so it can easily be disconnected and directed into a drain pan or suitable oil container). On models that use only a blind trap, it is usually easiest to drain the tank using the main oil feed hose, by disconnecting it from the pump (as directed in the next step).

■ If draining the tank, take note of the oil condition. The color should be clear. If the color is murky or milky, the oil is contaminated with water or other impurities. In this case the tank should be thoroughly cleaned and the old oil should be disposed of properly. Don't reuse contaminated oil under any condition.

3. Tag and disconnect the main oil supply line which leads from the main oil tank to the oil pump beneath the tank. Depending upon access you can disconnect it either at the tank or the pump. We like disconnecting it at the pump if the tank is still full of oil and you're using a hose pincher or using the main line to drain the tank. Squeeze the oil line to restrict the flow of oil while pulling it free of the fitting. Plug the line quickly with a suitable screw to prevent excessive loss of oil.

■ Oil line hose clamps are usually of the spring-tension type which are removed by compressing the wire clamp with a pair of pliers and gently pushing the clamp up along the oil line. However, some models may use wire ties which must be carefully cut away (so that you don't nick the oil line) and then replaced with new wire ties during installation.

4. Some models (usually 2-cylinder motors, smaller 3-cylinder motors and some V4/V6 motors) are equipped with an oil tank breather hose. When equipped, tag and disconnect the hose from the top of the oil tank. A few 3-cylinder models are equipped with an oil return hose, if so equipped, tag and disconnect it from the tank, but keep in mind that you'll likely have to pinch or plug this line depending upon location.

5. On V4 and V6 models, tag and disconnect the remote mounted oil tank supply line from the powerhead tank. Use a hose pincher and/or plug the line to minimize oil loss and prevent potential system contamination.

Fig. 36 A powerhead mounted oil tank is used on all oil-injected models

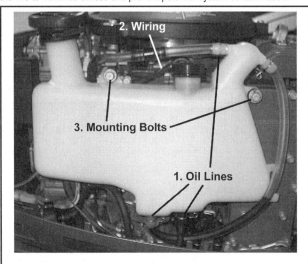

Fig. 37 Removal is a simple matter of disconnecting the 1. hoses, 2. wiring and 3. fasteners

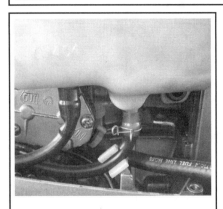

Fig. 38 Cut the wire ties or squeeze the spring clamps to disconnect the hoses...

Fig. 39 ...then plug lines to prevent drainage or contamination

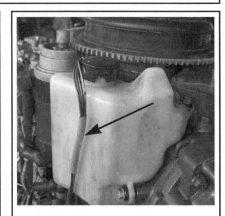

Fig. 40 Tag and disconnect the wiring, then remove the tank fasteners

LUBRICATION AND COOLING 5-17

6. Tag and disconnect the wiring for the oil level sensor.
7. For EFI/HPDI motors, tag and disconnect the wiring for the emergency transfer switch (which is normally mounted to the oil tanks on these models).
8. Loosen and remove the bolts (usually 2 or 3) securing the oil tank. Carefully lift the tank free of the powerhead. If the tank is still full of oil, be sure to set the tank upright in a safe place.

To Install:
9. Position the oil tank to the powerhead and secure the tank using the attaching bolts.
10. For EFI/HPDI motors, reconnect the emergency transfer switch wiring.
11. Reconnect the wiring for the oil level sensor.
12. If equipped, reconnect the return, breather hose and/or remote tank oil supply lines to the tank, as applicable and as tagged during removal.
13. Remove the plug or pincher (if used) and reconnect the main oil tank supply line.
14. Check around the tank to be sure no oil line is crimped or flattened. Make sure all hose/line connections are secured using the spring clamps or new wire ties, as applicable.
15. If not done already, refill the oil tank to the full line embossed on the tank. Use only Yamaha 2-stroke outboard oil if you are planning on checking pump output (otherwise any equivalent 2-stroke engine oil with a BIA certified rating TC-W3 is sufficient).

✱✱ WARNING

Any time the oil tank hose is disconnected, the oil injection pump must be purged (bled) of any trapped air. Failure to bleed the system could lead to powerhead seizure due to lack of adequate lubrication. For details, please refer to the Bleeding the Oil Pump procedure later in this section.

Remote Oil Tank

Because of the larger oil demand of larger motors, the oil injected versions of Yamaha V4 and V6 powerheads use a remote oil tank to feed the main powerhead mounted oil tank. These remote tanks use an electric lift pump to transfer oil to the powerhead mounted oil tank (where it can be gravity feed to the oil pump).

Although Yamaha provides a specific tank for use in boat rigging, there is no requirement that boat builders or dealers use only that tank. Additionally, the tank location, though Yamaha recommends it be close to the outboard, may vary greatly from rigging to rigging. The information provided is based on the typical installation of the Yamaha tank.

REMOVAL & INSTALLATION

◆ See Figures 41, 42 and 43

The most common oil tank mounting method is to position it in a holder or on the boat deck, with a securing strap across the top of the tank. Therefore the following removal procedures begin with this strap.

1. Loosen and remove the strap from the top of the remote oil tank. Most installations utilize a nut and long bolt to secure the top strap to the holder. Set aside the strap.
2. Disconnect the wire harness connector from the oil feed pump motor and the oil level sensor.
3. Place some shop cloths around the oil supply line hose joint to catch any oil which may drain from the hose. Snip the clamp with a pair of dykes (as the line is usually secured with a plastic wire tie). Squeeze the oil line to restrict the flow of oil while pulling it free of the fitting. Plug the line quickly with a suitable screw to prevent loss of oil.
4. Carefully lift out the remote tank.

To Install:
5. Prime the pump by removing the upper plate and filling the cavities/coating
the gear teeth with oil. Install the upper plate over the two small shafts. Check to be sure the O-ring is correctly seated in the groove of the pump body. Place the pump cover in position over the pump, with the screw holes aligned.
6. Secure the cover to the pump with the four Phillips head screws. Tighten the screws securely.
7. Position the tank into the boat.
8. Slide a new hose clamp onto the remote tank hose (unless you're using a wire tie, then it can obviously be positioned after the line is connected), and then connect the hose to the hose from the main tank. Tighten the hose clamp at the joint.
9. Reconnect the two halves of the wire harness connector together. Typically they will only fit one way.
10. Bring the strap across the tank and secure the strap to the holder (with the long bolt and nut, if applicable).
11. Test the remote tank lift pump, as detailed later in this section, in order to both purge the air from the pump and to ensure the system is working properly.

TESTING THE REMOTE TANK LIFT PUMP

◆ See Figures 44, 45 and 46

By connecting a 12-volt battery to the pump leads you can manually activate the pump in order to purge air/prime the pump and/or to check output. In most cases the pump wiring should contain a Brown wire female terminal in the harness connector for the ground portion of the circuit and a Blue wire female terminal connector for the power portion of the circuit. If the pump does not run or does not appear to pump properly, double-check the harness connection to make sure they are not reversed.

1. Make sure the remote tank contains at least 3/4 U.S. quart (0.7 liter).
2. Position a suitable container under the oil supply line (it should be disconnected from the outboard to prevent potentially overfilling the powerhead tank).
3. Obtain a 12-volt battery and two leads. Make contact with the lead from the negative battery terminal to the Brown female terminal in the harness connector. Make contact with the lead from the positive battery terminal to the Blue female terminal in the harness connector. The oil line should emit a strong, solid flow of oil.

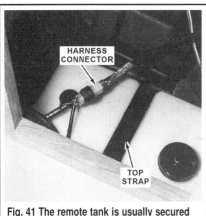

Fig. 41 The remote tank is usually secured by a strap

Fig. 42 Before installation, remove the cover to prime the pump...

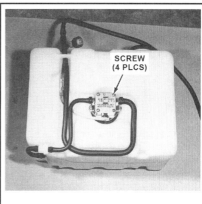

Fig. 43 ...then install the cover and secure using the screws

5-18 LUBRICATION AND COOLING

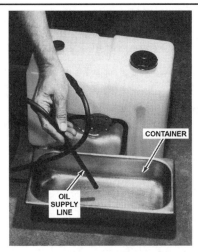

Fig. 44 Checking oil pump output

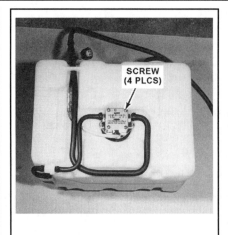

Fig. 45 The cover can be removed in order to inspect for clogs or wear

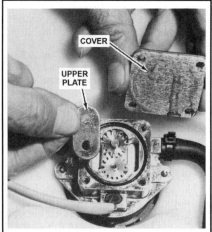

Fig. 46 Be sure to check the pump teeth as well as the cover O-ring

■ If your goal is the purge the line of air, continue to allow the flow of oil until there is a sold flow, then stop the pump and immediately pinch or plug the line to minimize the amount of air in it and reconnect the line to the outboard.

4. To check pump output, obtain a measuring cup or cylinder and measure the rate at which oil is delivered. The flow of oil should be about 0.2 U.S. quarts (0.2 liters) per minute. Perform the test over a three minute period and average the reading.

■ If the pump is allowed to empty the tank of oil and pump air, the feed pump must be purged of air.

If the flow of oil is less than specified check all oil lines for leaks, kinks, or restrictions. To check the pump itself, remove the four Phillips head screws securing the pump cover, then lift of the cover and upper plate to inspect for clogging, wear or damage.

■ If questionable test results were obtained and the pump was subsequently overhauled, go back and repeat the test outlined. If the flow of oil is still not within specifications and the lines have been checked for leaks, obstructions, and kinks, then the feed pump must be removed and replaced.

CLEANING & INSPECTING

If the cover is removed for inspection, check the two cavities of the pump body for any foreign material which might wedge between the teeth of the pump. Momentarily connect the Brown and Blue leads to the 12-volt battery. Check to ensure the teeth rotate smoothly. Inspect the condition of the O-ring in the groove of the pump body, replace if necessary.

Oil Pump Assembly

REMOVAL & INSTALLATION

◆ See Figures 25 thru 35 and 47 thru 52

The oil injection pump is normally located at one of the lowest points of the powerhead, inside the lower cowling. Some parts of the fuel system (carburetor, fuel pump, fuel filter housing) must sometimes be removed to gain access to the pump. Also, keep a rag handy any time an oil line is being disconnected. Preventing an oily mess is usually easier than cleaning one up.

1. Disconnect the negative battery cable and hang a reminder near the ignition that the oiling system is disconnected.

2. If necessary for access, remove or reposition the Powerhead Mounted Oil Tank. It is usually necessary to at least reposition the tank on most inline motors and a few of the V motors, when the tank is mounted almost directly over the pump.

3. Carefully disconnect the throttle valve-to-oil pump link rod from the oil pump. For most models this means carefully pulling out the tiny cotter pin from the shaft joint, then removing the washer and carefully prying the shaft from the oil pump lever.

4. For all except the EFI OX66 motor, wrap a small piece of masking tape around each of the oil supply lines and write the cylinder number on each piece of tape. On EFI OX66 motors the oil delivery lines all go to a common manifold, although placement on the manifold should not matter, to make sure there are no kinks or line problems it is probably a good idea to matchmark or tag their positions on the manifold.

■ Taking time now to identify the oil lines will ensure each will be connected back in its original location. Also, the lines may be disconnected at either end (cylinder/manifold or the pump). But, if the lines are being disconnected at the manifold take careful note of their routing for installation purposes. A digital camera is VERY handy for this.

5. The oil delivery lines on the pump are normally secured using spring-tension type clamps. To disconnect each oil line, gently squeeze the two loops of the wire clip together and at the same time gently pull up on the oil line to free the line from the fitting on the oil pump. This line will contain some oil, therefore, be prepared to plug the line with a bolt, or similar device.

6. If the powerhead mounted oil tank was not completely removed for access, tag and disconnect the main oil line. If the line is long enough, raise the end to a point higher than the oil level in the tank and secure it to prevent drainage, however, plug the line with a bolt to prevent contamination and minimize the danger of an oil spill.

7. Some models (like the 25/30 hp 3-cylinder motors) are equipped with an oil tank-to-oil pump breather hose. Tag and disconnect the hose from the pump (or tank, whichever is easier).

8. Remove the bolts (there are usually 2 rather odd-shaped bolts with long hex-heads protruding from the housing) securing the oil pump to the powerhead. Carefully pull the pump from the side of the powerhead.

9. On most models there is an O-ring sealing the pump to the powerhead. On a few motors, you can pull the driven gear from the bore in the powerhead, so make sure no components are unintentionally dislodged or lost. Inspect the condition of the pump drive/driven gear.

To Install:

10. If removed, push the driven gear shaft into the powerhead to index with the drive gear around the crankshaft.

11. If used, install a new O-ring on the oil injection pump. Check to be sure the shaft of the oil injection pump will index into the crankshaft driven gear. If the two are no longer aligned, rotate the slotted shaft on the pump to match the slot in the driven gear.

12. Install the pump and secure using the attaching bolts, then tighten the bolts securely.

LUBRICATION AND COOLING 5-19

Fig. 47 Disconnect the oil pump link...

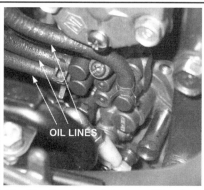

Fig. 48 ...then tag and disconnect the oil lines

Fig. 49 Loosen and remove the pump attaching bolts

Fig. 50 Bolt size and location will vary

Fig. 51 Remove the pump from the powerhead

Fig. 52 During installation, be sure to align the drive gear

13. If equipped, connect the breather hose to the pump or tank.
14. Pinch the line to prevent excessive oil loss, then remove the plug and reinstall the main oil line to the oil pump connector.
15. Install each of the oil supply lines onto their respective discharge fittings. If the lines were not identified as to which cylinder they originated from, refer to the PBS Diagrams to help identify the oil lines. If the lines were removed from the cylinders/manifold fittings, the same goes for properly routing the lines. If you're notes are not good enough (or weren't taken) refer to the PBS Diagrams to help determine proper hose routing.
16. Snap the oil injection link rod back onto the ball joint on the pump shaft. Slide the washer onto the shaft and install the tiny cotter pin.

■ If the length of the rod was accidentally altered, adjust the rod to the specified length. For details, refer to the Timing and Synchronization procedures found in the Maintenance and Tune-Up section.

BLEEDING THE OIL PUMP

◆ See Figures 53 thru 56    MODERATE

The lifeblood of a 2-stroke powerhead is the oil supply. The advantage of an oil-injection system is that the outboard can physically meter the oil supply to better meet engine operating conditions than can be accomplished with pre-mix. The disadvantage is that a failure of the system can destroy the powerhead.

After repairs are conducted on any mechanical/hydraulic part of the system, it is imperative that you bleed the system of air to make sure the oil supply will be uninterrupted and of the proper volume once the outboard is returned to service. Bleeding is actually a simple matter on these outboards.

In its most simple form, it is accomplished by running the motor on a portable tank of pre-mix while the system pump oil through the lines, ensuring all air bubbles are pushed out.

✱✱ WARNING

Yamaha warns NOT to run pre-mix, even during oil pump/system bleeding, on their HPDI motors.

The only things you'll need to perform this procedure on all but HPDI motors is a small portable fuel tank, associated lines and some pre-mix of the proper ratio (50:1 for MOST, but not all motors, so refer to the section on Maintenance and Tune-Up or your owner's manual to be sure).

n On a few models Yamaha claims that the system can be sufficiently bled by opening the bleed screw on the pump and allowing the system to gravity feed. However, that doesn't take into consideration that there are anywhere from 2-6 oil lines to the cylinders that may also be empty and should be bled as well. Since most of their motors, including most that ARE equipped with bleed screws, require some formal bleeding procedure other than JUST opening the bleed screw, we'd recommend you follow this procedure for ALL motors.

1. On all except HPDI motors, prepare about a ten minute supply of pre-mix using Yamaha 2-stroke engine oil in a portable fuel tank. If necessary you can usually jury-rig a setup to provide this mixture to the fuel pump during this purging procedure (using a small gas can, funnel and length of fuel hose) but a portable tank is handy to have around, so it's probably worth the investment.
2. Check the level of oil in the oil tank and replenish as necessary.
3. Connect a flush device to the lower unit.

5-20 LUBRICATION AND COOLING

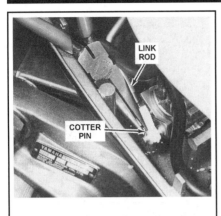

Fig. 53 Disconnect the oil pump link joint

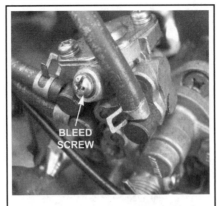

Fig. 54 If equipped, loosen the bleed screw

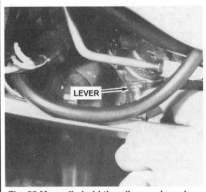

Fig. 55 Manually hold the oil pump lever in WOT

※※ WARNING

Never operate the powerhead over 1000 rpm with a flush device attached, because the engine may runaway due to no load on the propeller. A runaway engine could be severely damaged. Furthermore, Water must circulate through the lower unit to the powerhead anytime the powerhead is operating to prevent damage to the water pump in the lower unit. Just five seconds without water will damage the water pump impeller.

4. Disconnect the throttle valve-to-oil pump link rod from the pump lever. Once the engine is started you will manually hold the oil pump lever in the Wide Open Throttle (WOT) position.

5. If equipped, place a suitable cloth under the air bleed screw. Loosen or remove the air bleed screw and allow oil to flow from the opening until a bubble free flow of oil is obtained. On models equipped with a check valve ball on the main tank breather, a long thin object such as a nail or a toothpick can be used to depress the check valve ball to quicken the flow of oil through the bleed screw. On other models the bleed screw must remain loose or opened until after the engine is idling, so don't completely tighten the screw just to make sure you've removed as much air as possible.

6. Some models are equipped with clear plastic oil delivery lines. If so, you can watch for oil and air bubbles in the lines themselves. However, if not, disconnect one or more of the oil delivery lines from the manifold fitting and direct the line into a clean contain to catch escaping oil.

7. Start the engine and allow it to idle while you manually hold the oil pump lever in the WOT position to maximize oil flow. If you've got a bleed screw on the pump, watch and listen at the screw to confirm when there is no longer any air escaping past the screw head. Once there is only oil flowing from the screw, carefully tighten the screw in the bore.

■ Except for HPDI motors (which don't use pre-mix for this procedure) the fuel and oil pre-mix will be fed both through the system AND through the oil injection system directly into the cylinder. As the oil injection system purges air and comes up to full oil flow, the powerhead will be operating on a heavy mixture of oil (it may smoke or stumble a little). Actually, even HPDI motors will have an higher ratio of oil in the fuel than normal because of you are holding the oil pump lever open.

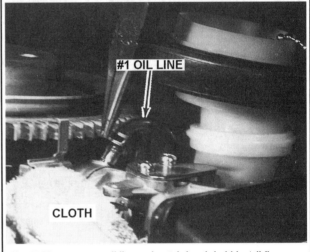

Fig. 56 Disconnect an oil line and watch for air bubbles/oil flow

8. Observe the flow of oil for five full minutes to verify the pump is functioning correctly. A steady slow pulsing flow with no air bubbles may be expected.

■ If there are any questions about pump condition or delivery rates, now is a good time to check the pump. For details, refer to Checking Oil Pump Output earlier in this section.

9. Once you are satisfied with oil flow, reconnect the line. Shut down the powerhead and remove the flushing device.

10. Reconnect the throttle valve link rod to the oil pump lever and secure using the washer and cotter pin.

■ If the length of the rod was accidentally altered, adjust the rod to the specified length. For details, refer to the Timing and Synchronization procedures found in the Maintenance and Tune-Up section.

OIL INJECTION WARNING SYSTEM

General Information

SINGLE LAMP SYSTEM

◆ See Figure 57

The single lamp warning system is the simplest system used by Yamaha. The oil tank is located under the engine cowling. Inside the oil tank is a sensor. This sensor has a single switch located near the bottom of the sensor shaft. The switch is closed when the sensor float is down near the bottom of the shaft.

When the switch is closed, the circuit is complete between ground (Black) and the warning system (Pink).

The system is powered by the CDI charge coil. No battery is needed. This also means that the system can only operate when the engine is running.

A Red warning light is wired in between the Capacitor Discharge Ignition (CDI) box and the oil tank sensor.

When the oil tank runs low, the float on the sensor rests down at the bottom. This closes the switch inside of it. Current from the CDI box now has a path to ground, and the warning light now glows to indicate a problem. The sensor switch closing also tells the CDI box to lower engine speed to 2,000 rpm.

LUBRICATION AND COOLING 5-21

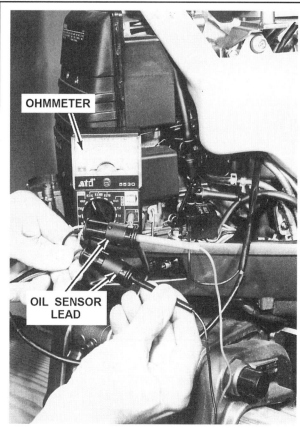

Fig. 57 Testing the oil level sensor with a multi-meter set to the 1000 ohm scale

Please note that the overheat sensor is also connected to the warning light. When the engine gets too hot, the overheat sensor closes its switch, completing the Pink lead circuit to ground. This completes the same circuit that the low oil circuit is on. The light comes on, the rpm is reduced. The same warning occurs for overheat or low oil. What would happen if the boat operator was at wide open throttle, the warning kicked in, and reduced the rpm to 2,000? If the operator stayed at the helm with the throttle open and sent someone to fill the oil tank what might happen? As the tank filled and the sensor switch went open, the engine rpm would return to wide open. This sudden acceleration could possibly injure someone.

To prevent a sudden return to a previous throttle setting, a holding circuit is normally built into the PBS. This circuit won't allow the engine to return to full throttle operation with the warning on until the ignition is cycled off and then back on. This resets the system.

THREE LED SYSTEM

This system uses three LEDs (light emitting diodes) to indicate the amount of oil left in the injector tank. This system is used on many of the three cylinder models. The oil tank is located underneath the cowling on the side of the powerhead.

The three colors are Green, Yellow and Red. Green indicates a full to near full tank Yellow indicates that about a third of the tank is left. Refilling should be done soon. Red indicates that the system is very low on oil and something must be done immediately. When the Red lamp comes on, the rpm reduction is switched on and the warning buzzer sounds.

The Yellow/Red wire from the CDI powers the system and no battery is required (like the single lamp system).

The three indicator system uses the following lights to signal the operator of the level in the oil tank:
- Green - Oil Tank Over 30% Full
- Yellow - Oil Tank 10-30% Full
- Red - Oil Tank Less Than 10%

The tank sensor has two switches built in. One switch is located near the top of the sensor shaft and the other switch is located near the bottom. One side of each switch is connected to ground through the Black wire. The other end of the upper switch is connected to a Green wire while the other end of the lower switch is connected to a Green/Red wire. When the oil tank level is near the full mark, the sensor's magnetic float closes the circuit on the upper switch. This completes the circuit between ground and the Green wire. The control unit receives this signal and turns on the Green LED lamp (full tank).

As oil is consumed, the oil level drops in the tank. The sensor float follows the oil level. When the float reaches a point between the two switches in the sensor neither switch is closed. This "no switch closed" signal triggers the control unit to shut off the Green and turn on the Yellow.

If the oil level continues to drop, the sensor float will drop down and close the lower switch. This completes the circuit between the Green/Red wire and ground. When the control unit gets this signal, the Yellow light is turned off, and the Red light is turned on. This also activates the warning system including rpm reduction. The Green/Red wire from the oil tank sensor is also connected to the Pink warning wire.

The whole system is operated by making and breaking the path to ground. As a switch is grounded or non-grounded, the control unit reads this and reacts. The control unit and lamps are located either on the engine pan or in the gauge. Either way, the electronics are the same. This system also uses the holding circuit discussed under single lamp systems to prevent unintended acceleration.

■ If a tank sensor has an "H" stamped on the top it has a holding circuit under the cap. Use the diode setting on your multimeter to test the Red lamp circuit. The 65J-90 hp motors may or may-not have an "H" stamped on the top of the oil level sensor. However, these units usually have a holding circuit built into them. The holding circuit is either in the oil level sensor or in the oil control unit itself.

REMOTE WARNING LAMP SYSTEMS

◆ See Figures 58 and 59

Yamaha's most advanced PBS system is used on the V4 and V6 engines. Larger motors consume more oil than would be conveniently stored in a powerhead mounted oil tank. An engine-mounted oil tank for V motors that would have sufficient capacity to satisfy customer demands would take up excessive space and add additional weight to the engine.

Therefore, a large remote tank is included in this system to supply oil to the smaller powerhead mounted tank (main tank) under the cowling. Oil is pumped from the remote tank (sub-tank or reservoir) to the main tank using an electric transfer pump.

Each tank has a sensor to monitor the oil level. A control unit takes the information from the sensors and decides when oil needs to be pumped from the remote tank to the main tank. The control unit also uses this sensor information to decide which light in the gauge to illuminate (Green, Yellow, or Red or LED backed icons in the remote display which represent oil level OK, oil level low, oil out). The control unit also kicks in the warning systems (rpm reduction, buzzer, etc.) when the Red light (oil out) is lit.

While very early Yamaha outboards used incandescent bulbs in the PBS gauge, the current PBS gauges use LCD indicators. An LCD arrow points at a colored block to indicate oil status.

Now, let's look at the two oil tank sensors and the control unit in more detail. The main tank sensor has three magnetic switches inside its shaft. If one of these switches closes, it completes the path to ground for that circuit.

The remote tank sensor has only one switch at the bottom of the sensor shaft. When the switch is closed, the circuit is complete to ground. This switch's color code is Black/Red.

Now, put everything together to see how it works. Start with both the main and remote tanks filled.

As the engine consumes oil, the float in the main tank drops. About halfway down the shaft, the magnetic float closes the switch on the Blue/Green wire. The control unit senses this connection and turns on the transfer pump. The transfer Pump already has I2-volts on one leg when the key is **ON**. The ground path is through the control unit. When the control unit turns the pump, it completes the path to ground.

The pump fills up the main tank. As the main tank sensor float moves up toward the top of the shaft, the Blue/Green switch opens and the Blue/White switch closes. The control unit senses this and turns the pump off. This cycle of turning the pump off and on continues as long as the oil supply in the remote tank is high enough to keep the remote tank switch closed.

During this time, the control unit turns on the Green light (oil level OK) at the gauge.

5-22 LUBRICATION AND COOLING

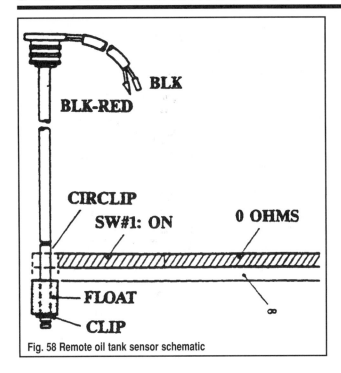
Fig. 58 Remote oil tank sensor schematic

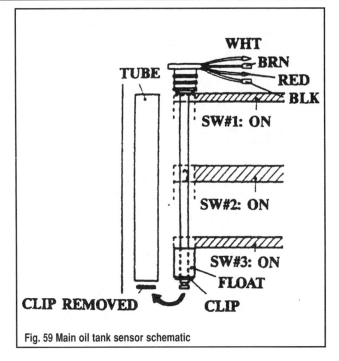
Fig. 59 Main oil tank sensor schematic

The remote tank level falls as oil is transferred to the main engine tank. When the remote tank level is down to less than a quarter tank, the sensor float drops to the bottom of the shaft. This opens the switch on the remote tank sensor.

When the remote tank switch opens, two things happen. First, the control unit turns on the Yellow low (oil low) light at the gauge. Next, the control unit blocks any call for automatic oil transfer from the remote to the main tank. The remaining oil in the remote tank can now only be transferred manually using the emergency override switch.

If the operator continues to run the engine without refilling the remote tank, the main engine tank will run out of oil. With no more oil coming from the remote tank, the main tank oil level drops to a point where the float closes the switch on the Blue/Red wire at the bottom of the sensor. This immediately turns on the Red light (oil out) at the gauge and activates the other warning systems.

Another feature of this system is the circuit to turn off the oil pump when the engine is tilted past a certain point (usually over 30 degrees). Without this, the angle of the tilted engine could let the pump stay on even after the main tank is filled.

When rigging is completed, fill the remote tank at least one-quarter full. Turn on the ignition switch and, within 5 seconds, the oil pump will run for 180 seconds. If you do not get the full 180 seconds of transfer, check to see if first switch has closed. If enough oil transferred to close the first switch, then pump shut off was normal.

During normal operation with the engine running, the second switch turns the transfer pump on. When the oil level reaches the top of the tank (first switch closed) or after 90 seconds of transfer, the pump shuts off. The transfer pump will not run if the remote tank is low (third switch closed) while the engine is running. The transfer pump will not run if the engine is **OFF**.

The oil transfer will not take place with the main tank sender disconnected. If the main tank sensor is disconnected with the key **ON** the buzzer will sound.

Troubleshooting

SINGLE LAMP SYSTEM

 MODERATE

This a very simple system to check. The following components are involved: Charge coil, Igniter (CDI) box and Warning lamp.

Oil tank sensor Troubleshooting strategy is problem dependent. What is wrong will determine you troubleshooting direction. Is the system staying on all the time? Not coming on at all? Gather information.

The oil tank sensor operation can be checked by running the motor and grounding, then ungrounding the Pink harness lead going to the sensor. When grounded, the warning system should engage. When ungrounded, the system should operate normally.

If the system works correctly when doing this check, investigate the sensor itself. Do not forget that the overheat sensor is also connected in this circuit. If there is an overheat, the operator will get the same warning. To isolate the low oil warning from the overheat, disconnect the overheat sensor Pink lead from the circuit. Now only the oil warning system is operating.

If the engine is running, it is probable that the charge coil is good. It is necessary for ignition.

The only component that can't be directly checked is the CDI box. Remember that all other components can be directly checked. If they are all good, then the CDI box is the only remaining source of the problem.

THREE LED SYSTEM

There are basically two parts to this system. The oil tank sensor and the lamp/control unit. The lamp assembly and the control unit come only as an assembly.

Begin troubleshooting by identifying the problem. Is the system totally dead? Is one light out? Are the lights dim?

Although most versions of this system currently in use no longer require a battery, be sure to check the battery to be sure it is not the cause or is not contributing to the problem. Also don't forget to check the grounds on the system. Total failure or dim lights (high resistance) can be a symptom of bad grounds.

Make sure the Yellow (or Yellow/Red) wire to the control unit has 12-volts and the Black wire is grounded.

A quick way to troubleshoot this system is to disconnect the sensor. With the power on and the tank sensor disconnected, the control unit is getting a "no switch closed." You should have a Yellow lamp. If this works ground the Green wire. The control unit is being told the upper sensor switch is closed. The Green lamp should come on. The last check is to ground the Green/Red wire. The control unit should read this as a closed lower sensor switch. The Red lamp should come on.

If these tests give the correct readings, the control unit and lamps are OK. A failure of one or all the lamps will require replacement of the brain/lamp assembly.

If the lamps are OK, the oil tank sensor needs to be resistance checked.

This system can be checked out quickly once you understand how it works. Other PBS Related Problems

LUBRICATION AND COOLING 5-23

Excessive smoke at idle is a common problem. The following items should be inspected. They are listed in the order of their probability.
- The oil pump linkage is disconnected
- The oil pump linkage is binding
- The oil pump linkage is out of adjustment
- The unit needs throttle valve synchronization and, on carbureted motors, idle fuel adjustment
- The oil is not Yamalube. Some oil brands smoke excessively when all adjustments are correct.
- The oil line check valves are leaking.

Some special notes regarding electrical problems bear mentioning. This is a "ground control" system. If you have a short to ground on a control (sensor) lead, the control unit will think a switch is closed. Watch for pinched wires. Check color codes and watch out for incorrect connections.

REMOTE (V4 & V6) WARNING LAMP SYSTEMS

As with any electrical problem, the best troubleshooting strategy begins with good preparation, knowledge of the system to be checked, and a Seloc manual. Don't forget to get details about the problem. Good information can vastly reduce troubleshooting time.

Note when the problem occurs and how long it occurs. Is only part of the system not working? Is the whole system dead? These questions can make a difference.

Remember that this system has only one component that is difficult to check and that is the control module. All other pieces have direct tests. Your goal is to check the components that can be checked. If they are good, the control module is the prime suspect.

Start off with the basics. Are all fuses in good shape? Is the battery fully charged? Is current getting to the precision blend system? Is there oil in both tanks?

A quick check for the tank sensors is to bypass them. Remember that as each sensor lead from the control unit is grounded, the unit activates or turns off a circuit.

Start with the main tank sensor. Disconnect the sensor leads from the main harness. We know that when the Blue/Green lead is grounded through its switch, the transfer pump turns on. The same thing can happen if we jump the harness side Blue/Green to the harness side Black. This is doing the same thing as closing the switch.

If the system is good, the transfer pump will now turn **ON** and stay on. If it doesn't, the wiring, pump, or the control module could be bad.

To turn the pump off, jump between the Blue/White and Black on the harness side. The pump should stop. Again, if it doesn't, check the wiring and the control module.

Next, activate the Red lamp and the warning system by jumping between the Blue/Red and Black wires. The Red lamp and other warning systems should activate. If not, check the wiring and the control module.

Remember that if the system wouldn't work before, and these bypass checks give proper operation, the main tank sensor is the problem. The beauty of these checks is that they were quick and didn't require taking the sensor out of the tank.

The remote tank sensor is the next item to check. It has only one switch on the Black/Red wire. Disconnect the sensor from the wiring harness. Jump between the Black/Red and Black wires on the harness side. The Green light should come on in automatic mode.

Disconnect the jumper. With the circuit open, the Yellow light should come on and the transfer pump should not work.

Like before, if the jumps allow the system to work when they wouldn't before, the sensor is at fault. If the jumps don't solve the problem, the failure is elsewhere in the system.

To check the transfer pump operation, try the emergency bypass switch first with the key on. If the pump runs, the problem is elsewhere. If the pump doesn't run, go directly to the pump. Connect the brown lead to the pump to the positive side of a battery and the negative lead to the Blue pump lead. If the pump works, the problem is elsewhere. If the pump still won't operate, the pump is bad.

If the pump, sensors, and wiring all check out, the only item left is the control module.

Don't forget that some systems are dependent upon trim angle and may use a signal from the trim sender to prevent oil transfer. Other motors use a control module which will not allow transfer unless the engine is running. A problem in either of these areas could also cause failures.

Operational Check

SINGLE LAMP SYSTEM

♦ See Figures 60 thru 63

You can test the entire system by slowly draining the tank and watching for the point where the light illuminates. If the system does not function properly check the level sensor, the light and/or the wiring, as necessary. After conducting this test, there will be a small amount of air in the oil lines so the system should be bled to protect the powerhead.

1. Tag and disconnect the spark plug leads from the spark plugs for safety.
2. Lift out the oil level gauge cap with the sensor attached. Wrap a shop towel around the sensor to prevent dripping oil on the deck. Using a vernier caliper or precision ruler, measure the distance from the cap lip to the bottom of the cap (this is the distance from the point at which the top of the cap contacts the oil tank when it is inserted, to the point where the float contacts the bottom of the cap when the tank is full. Typically this distance is ABOUT 1.92-2.03 in. (47.7-50.7mm). This number is necessary for the next step. Set the sensor aside.
3. The level sensor (actually an on/off switch) turns on when the top of the float drops 2.22-2.33 in. (56.3-59.3mm) below the bottom of the sensor cap. Add this dimension (we'd use the larger of both numbers) and adjust a vernier caliper to read the same dimension. The resulting dimension is usually in the neighborhood of about 4 1/4 in. (107mm) give or take a tenth of an inch or so. Make a mark on the oil tank which equals this distance below the top of the tank (where the sensor sits). This is the point on the tank at which the system MUST activate (it may activate slightly sooner) as the oil level drops.
4. Reinstall the sensor assembly. Turn the main switch to the **ON** position but do not start the powerhead.
5. Position a suitable container close by to receive the oil from the tank.
6. Disconnect the main oil line from the oil pump and use the line to slowly drain the oil from the main tank until the warning system operates. Usually on manual start models the light on the front cowling panel will come on, while on the electric start models the light on front cowling panel will come on AND the buzzer will sound. Either way, you can stop draining the oil as soon as the system operates (be sure to reconnect and secure the main oil line). It must operate by the time the oil level reaches the scribed mark (or VERY shortly thereafter) otherwise you must check the sensor, led and/or wiring. Make sure the float is not sticking.
7. To check the sensor, remove it from the tank again and connect an ohmmeter or DVOM set to read resistance. Move the float slowly by hand and make sure that the switch shows no continuity across the two wires UNTIL the float is about 2.22-2.33 in. (56.3-59.3mm) below the bottom of the sensor cap. Once the float reaches that point the switch contacts must close and the meter must show continuity or the sensor/wiring is bad.

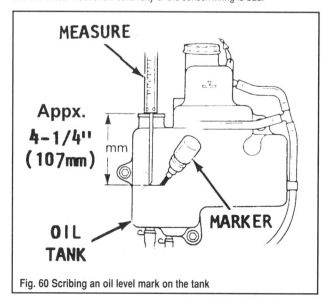

Fig. 60 Scribing an oil level mark on the tank

5-24 LUBRICATION AND COOLING

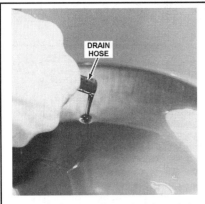
Fig. 61 Drain oil from the tank and watch for the warning light

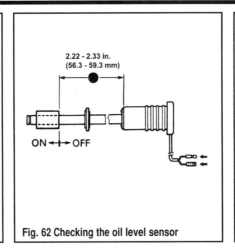
Fig. 62 Checking the oil level sensor

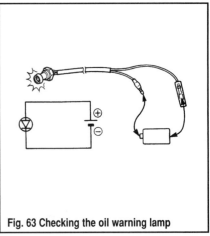
Fig. 63 Checking the oil warning lamp

8. To check the LED itself, disconnect the wiring and connect a 1.5 volt pen light battery. The LED is a diode, meaning it is sensitive to current polarity so you must hook up the battery in a specific direction (usually that is positive to the Yellow/Red lead and Negative to the Pink lead, but if the light doesn't illuminate, switch leads to make sure polarity is not the issue). If the light does not work, make sure the battery is charged and the wiring/connections are good, otherwise the LED is bad and must be replaced.

✲✲ WARNING

DO NOT use alkaline batteries/higher voltage batteries as they will burn out the diode.

9. Once you are finished testing/repairing the system, turn the main switch to the **OFF** position.
10. Reconnect the spark plug leads to the spark plugs, as tagged.
11. Top off the oil in the oil tank to the full mark.

✲✲ WARNING

Water must circulate through the lower unit to the powerhead anytime the powerhead is operating to prevent damage to the water pump in the lower unit. Just five seconds without water will damage the water pump impeller.

12. Start the engine and check the oil light on the front cowling panel. The light should remain off and the warning buzzer (if equipped) should not sound.

THREE LED SYSTEM 2

◆ See Figures 64, 65 and 66  MODERATE

The following procedures check the operation of the buzzer and the Green, Yellow, and Red lights to ensure they come on at the proper time. Correct operation of the buzzer and lights is essential to warn the operator of a low oil level in the tank. This simple procedure should be performed at the start of each season to verify the warning system is functioning correctly as a protection against expensive powerhead service work. Inadequate lubrication can almost destroy a powerhead.

After these tests are completed, the injection system should be bled.

1. Place a suitable container under the oil tank, then disconnect the main oil hose from the oil pump and direct the oil flow from the tank into a suitable clean container for temporary storage. Visually examine the oil for signs of moisture (a milky appearance) or other contamination.
2. Turn the ignition switch the **ON** position.
3. The Red light should come on and the buzzer should sound. Slowly begin adding the oil drained from the tank (if in good condition) or fresh 2-stroke engine oil to the tank. As the oil level begins to rise to a point about 1/4 full or so, the Red light should go off, the buzzer should stop sounding and the Yellow light should come on.
4. Continue slowly adding oil while watching the oil level and the lights. Once the level reaches about the 1/3 full point (or so) the Yellow light should go out and the Green light should come on. If this occurs, the system is working properly, refill the tank and properly bleed the system.
5. If there is a problem with the lights, perform the testing procedures outlined under Troubleshooting in this section and/or check the sensor as follows:

 a. Remove the sensor from the top of the tank.
 b. The sensor essentially contains two switch circuits which are activated/deactivated by float height. Use an ohmmeter or DVOM set to read resistance across the sensor leads for the first and second switch positions. Generally speaking the first switch position is responsible for illuminating the Green LED, the meter should show continuity once the float is a certain (small) distance from the bottom of the sensor cap. Similarly, the second (lower) switch position illuminates the Red LED, the meter should show continuity across a different combination of leads once the float is toward the bottom of its travel. Specifications vary slightly by model.
 c. For 28J-50 Hp (698cc) models the first switch contacts should be closed (meaning there is continuity) across the Green and Black wires from the TOP of float travel to a point about 1.29-1.41 in. (32.8-35.8mm) measured from the bottom of the sensor cap to the top of the float. Slightly below this point the switch contacts will open. Slide the float further down the sensor to a point about 2.24-2.35 in. (56.8-59.8mm) below the sensor cap. At this point the second switch contacts should close and there should be continuity across the Green/Red and Black leads. On some models there should also be contact between these leads and the Pink lead.
 d. For 50-70 Hp (849cc) models, there should be no continuity across Green and Black wires until the top of the float drops about 3.13-3.25 in. (79.5-82.5mm) below the bottom of the sensor cap. The second switch contacts should close showing continuity across the Green/Red and Black leads once the float drops to a point about 4.51-4.62 in. (114.5-117.3mm) below the cap.
 e. For 65J-90 Hp (1140cc) models, there should be no continuity across Green and Black wires until the top of the float drops about 0.23-0.35 in. (5.8-8.8mm) below the bottom of the sensor cap. The second switch contacts should close showing continuity across the Green/Red and Black leads once the float drops to a point about 1.67-1.78 in. (42.3-45.3mm) below the cap.

REMOTE (V4 & V6) WARNING LAMP SYSTEMS

 MODERATE

◆ See Figures 67 thru 73

The following procedures check the operation of the buzzer and the Green, Yellow, and Red lights to ensure they come on at the proper time. Correct operation of the buzzer and lights is essential to warn the operator of a low oil level in the tank. This simple procedure should be performed at the start of each season to verify the warning system is functioning correctly as a protection against expensive powerhead service work. Inadequate lubrication can almost destroy a powerhead.

After these tests are completed, the injection system should be purged of air (bled).

LUBRICATION AND COOLING 5-25

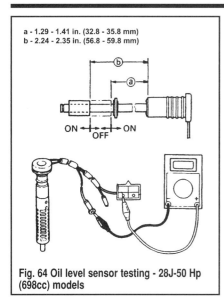

Fig. 64 Oil level sensor testing - 28J-50 Hp (698cc) models

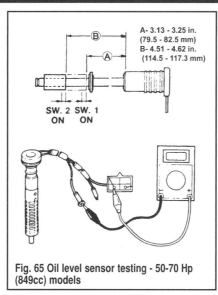

Fig. 65 Oil level sensor testing - 50-70 Hp (849cc) models

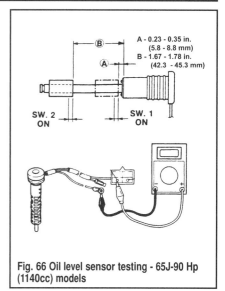

Fig. 66 Oil level sensor testing - 65J-90 Hp (1140cc) models

■ The following tests are performed with the powerhead not running, as both oil tanks are to be drained. Running the powerhead with inadequate lubrication will cause damage to internal moving parts and may cause seizure. The battery must be connected for the tests, but the spark plugs should be removed and the high tension leads grounded for safety and to make sure no attempts are made to start the motor.

1. Obtain a suitable container with a large enough capacity to hold all the oil in the powerhead and remote oil tanks. Place some shop cloths in the bottom of the lower cowling to catch any oil which may drain during the removal of the oil supply line. Remove the oil supply line leading from the main oil tank to the oil pump by compressing the wire clamp with a pair of needle-nose pliers and pushing the clamp up along the oil line. Squeeze the oil line to restrict the flow of oil while pulling it free of the fitting. Allow the tank to completely drain into the container. Install the line back onto the center fitting on the oil pump.

2. Remove the remote tank from the boat. Once the tank is removed from the boat, completely drain the oil from the tank. Install the empty tank into the boat.

3. Turn the main switch to the **ON** position. The Red light (or oil out icon) should illuminate and the buzzer should sound. The oil feed pump should not operate.

4. Begin to add oil to the remote tank. Usually at some point between one and two quarts, the transfer pump should operate as oil is moved from the remote tank to the powerhead mounted tank.

5. Continue to slowly ad oil to the remote tank. As the transfer continues, the Red light (oil out icon) will eventually go out as the level in the powerhead mounted tank reaches a certain point (generally more than 1/4 full but less than 1/3 full). At this point however, the Yellow light (or oil low icon) should come on.

6. Continue adding oil, as the powerhead oil tank passes the approximate 1/3 full mark the Yellow light (oil low icon) should eventually go out and the Green light (or oil ok icon) should come on.

7. If the lights work as indicated, shut the main switch **OFF**, the warning system is working as specified.

8. If there is a problem with the lights, perform the testing procedures outlined under Troubleshooting in this section and/or check the sensor as follows:

   a. Remove the sensor from the top of the tank.

   b. The powerhead mounted oil tank level sensor generally contains three switch circuits which are activated/deactivated by float height. The circuits are used both for the remote display LEDs (oil ok, low oil, oil out) and to activate the transfer pump from the remote tank (usually when the low oil signal is received). Use an ohmmeter or DVOM set to read resistance across the sensor leads for the each of the first, second and third switch positions. Specifications and circuit wire colors may vary slightly by model.

   c. For most 1999 and later 80J-140 Hp (1730cc) V4 models the first switch contacts should be closed (meaning there is continuity) across the White and Black wires when the TOP of the float drops to a point about 0.12-0.24 in. (3-6mm) measured from the bottom of the sensor cap. The second set of switch contacts should be closed across the Brown and Black wires, once the top of the float is 1.30-1.42 in. (33-36mm) below the bottom of the sensor cap. The third set of switch contacts should be closed across the Red and Black wires, once the top of the float is about 2.09-2.20 in. (53-56mm) below the sensor cap.

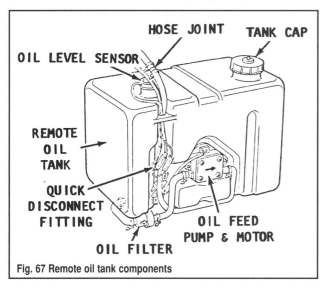

Fig. 67 Remote oil tank components

Fig. 68 The powerhead tank contains a 3-switch oil level sensor

5-26 LUBRICATION AND COOLING

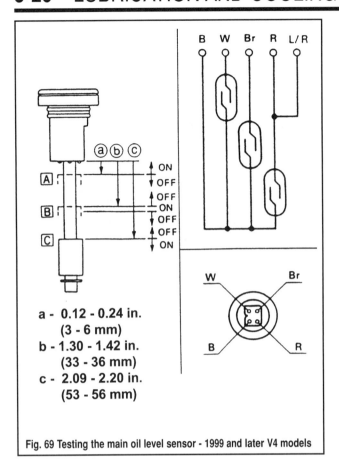

Fig. 69 Testing the main oil level sensor - 1999 and later V4 models

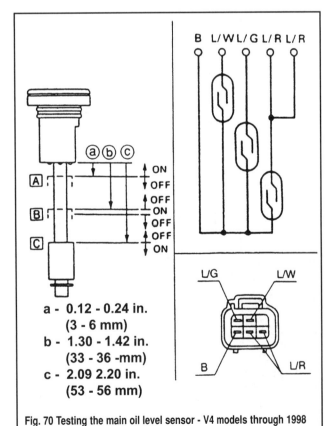

Fig. 70 Testing the main oil level sensor - V4 models through 1998 and most V6 models (measurements for 2.6L shown, see text for other dimensions)

■ The V4 motors THROUGH 1998 models use the same sensor (with regards to wire colors and float heights) as most carbureted V6 models. However, the Yamaha sources conflict. Some sources say the wire colors will match, while others say the first switch wire colors are Blue/Black and Black (instead of Blue/White and Black).

d. For 105J-225 Hp (2596cc) V6 models (carbureted, EFI OX66 or HPDI) and for 80J-140 Hp (1730cc) V4 models through 1998, the first switch contacts should be closed (meaning there is continuity) across the Blue/White and Black wires when the TOP of the float drops to a point about 0.12-0.24 in. (3-6mm) measured from the bottom of the sensor cap. The second set of switch contacts should be closed across the Blue/Green and Black wires, once the top of the float is 1.30-1.42 in. (33-36mm) below the bottom of the sensor cap. The third set of switch contacts should be closed across the Blue/Red and Black wires, once the top of the float is about 2.09-2.20 in. (53-56mm) below the sensor cap.

e. For EFI OX66 200-250 Hp (3130cc) V6 models (except the 250 hp Vmax), the first switch contacts should be closed (meaning there is continuity) across the Blue/White and Black wires when the TOP of the float drops to a point about 0.10-0.22 in. (2.5-5.5mm) measured from the bottom of the sensor cap. The second set of switch contacts should be closed across the Blue/Green and Black wires, once the top of the float is 1.28-1.40 in. (32.5-35.5mm) below the bottom of the sensor cap. The third set of switch contacts should be closed across the Blue/Red and Black wires, once the top of the float is about 2.99-3.11 in. (76-79mm) below the sensor cap.

f. For EFI OX66 250 hp (3130cc) Vmax V6 models, the first switch contacts should be closed (meaning there is continuity) across the Blue/White and Black wires when the TOP of the float drops to a point about 0.35-0.51 in. (9-13mm) measured from the bottom of the sensor cap. The second set of switch contacts should be closed across the Blue/Green and Black wires, once the top of the float is 1.54-1.69 in. (39-43mm) below the bottom of the sensor cap. The third set of switch contacts should be closed across the Blue/Red and Black wires, once the top of the float is about 3.25-3.41 in. (82.5-86.5mm) below the sensor cap.

g. For HPDI 225-250 Hp (3342cc) V6 models, the first switch contacts should be closed (meaning there is continuity) across the Blue/White and Black wires when the top third of the float drops to a point about 2.09-2.24 in. (53-57mm) measured from the middle of the sensor cap. The second set of switch contacts should be closed across the Blue/Green and Black wires, once the top third of the float is 3.27-3.43 in. (83-87mm) below the middle of the sensor cap. The third set of switch contacts should be closed across the Blue/Red and Black wires, once the top third of the float is about 4.98-5.14 in. (126.5-130.5mm) below the sensor cap.

9. The remote oil tanks on EFI OX66 and HPDI motors are also equipped with a level or oil out sensor. Similar to the oil level sensor for the powerhead mounted tanks, the sensor contains an internal switch (just one in this case) which is activated by the oil float. Basically when the float approaches the end of it's travel (about 5.91-6.02 in./150-153mm when measured from the underside of the sensor cap to the top or top third of the float), there should be continuity across the Black/Red and Black wires.

10. This system also uses an emergency switch to allow a manual transfer of oil (emptying the remote tank) to provide all reserves of 2-stroke oil to the powerhead mounted tank in order to get the boat back to port. During testing, be sure to check and make sure the switch will make the pump operational. This test should be performed before the powerhead mounted tank is topped off, if not you'll have to remove the level sensor from the tank for a moment to make sure the switch works. The switch itself is a simple on/off switch which should only have continuity when placed in the ON position. If necessary, use an ohmmeter or DVOM set to read resistance to check the switch.

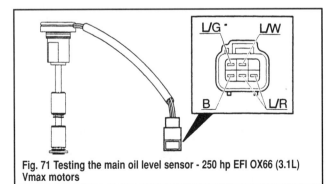

Fig. 71 Testing the main oil level sensor - 250 hp EFI OX66 (3.1L) Vmax motors

LUBRICATION AND COOLING 5-27

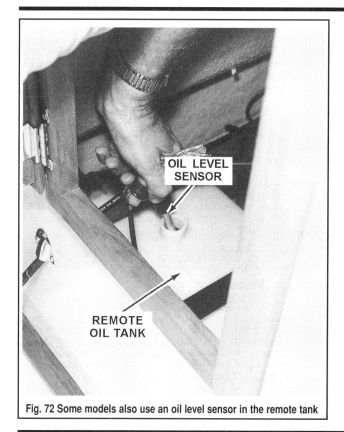

Fig. 72 Some models also use an oil level sensor in the remote tank

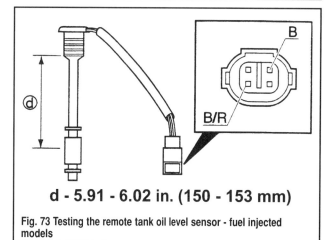

d - 5.91 - 6.02 in. (150 - 153 mm)

Fig. 73 Testing the remote tank oil level sensor - fuel injected models

✱✱ CAUTION

During these tests, the remote oil tank was completely drained allowing air to enter the feed pump. Therefore, the feed pump must be purged (bled) of any trapped air. Failure to bleed the system could lead to powerhead seizure due to lack of adequate lubrication. Therefore, bleed the system.

11. Refill the tank and properly bleed the system.

■ Be sure to check the oil filter at the remote tank periodically and replace, as necessary.

COOLING SYSTEM

◆ See Figure 74

All Yamaha outboard engines are equipped with a raw water cooling system, meaning that sea, lake or river water is drawn through a water intake in the gearcase lower unit and pumped through the powerhead by a water pump impeller. The exact mounting and location of the pump/impeller varies only slightly from model-to-model, but on all Yamahas it is mounted to the lower unit along the gearcase-to-intermediate section split line.

For many boaters, annual replacement of the water pump impeller is considered cheap insurance for a trouble-free boating season. This is probably a bit too conservative for most people, but after a number of trouble-free seasons, an impeller doesn't owe you anything and you should consider taking the time and a little bit of money necessary to replace it. Remember that should an impeller fail you'll be stranded. Not to mention that a worn impeller will simply supply less cooling water than required by specification, allowing the powerhead to run hot placing unnecessary stress on components and best or risking overheating the powerhead at worst. And, to make matters even worse, should an impeller disintegrate the broken bits will likely travel up through the system, potentially forming blocks/clogs in cooling passages.

■ While impeller replacement is no longer usually an annual task, most people we talk to still recommend the impeller be replaced every 2nd or 3rd season to ensure trouble-free operation.

All but the smallest motors (the 2 hp units) are equipped with a thermostat that restricts the amount of cooling water allowed into the powerhead until the powerhead reaches normal operating temperature. The purpose of the thermostat is to increase engine performance and reduce emissions by making sure the engine warms as quickly as possible to operating temperature and remains there during use under all conditions. Running a motor without a thermostat may prevent it from fully warming, not only increasing emissions and reducing fuel economy, but it will likely lead to carbon fouling, stumbling and poor performance in general. It can even damage the motor, especially if the motor is then run under load (such as full-throttle operation) without allowing it to thoroughly warm. A restricted thermostat can promote engine overheating. The good news is that should you be caught on the water with a restricted thermostat, you should be able to easily remove it and get back to shore, just make sure you replace it before the next outing.

■ Most V4 and V6 Yamaha motors are equipped with a separate thermostat for each cylinder bank. The most notable exception are the 4-stroke V6 motors.

The water intake grate and cooling passages throughout the powerhead and gearcase comprise the balance of the cooling system. Both components require the most simple, but most frequent maintenance to ensure proper cooling system operation. The water intake grate should be inspected before and after each outing to make sure it is not clogged or damaged. A damaged grate could allow debris into the motor that could clog passages or damage the water pump impeller (both conditions could lead to overheating the powerhead). Cooling passages have the tendency to become clogged gradually over time by debris and corrosion. The best way to prevent this is to flush the cooling system **after each use** regardless of where you boat (salt or freshwater). But obviously, this form of maintenance is even more important on vessels used in salt, brackish or polluted waters that will promote internal corrosion of the cooling passages.

■ Most V4 and V6 (along with a few of the inline motors) are equipped with a spring-loaded pressure relief valve which gives the cooling water an additional escape path to prevent system over-pressurization during warm-up or if there is a clog somewhere in the system.

Description and Operation

◆ See Figures 75 thru 78

The water pump uses an impeller driven by the driveshaft, sealing between an offset housing and lower plate to create a flexing of the impeller blades. The rubber impeller inside the pump maintains an equal volume of water flow at most operating speeds.

At low speeds the pump acts like a full displacement pump with the longer impeller blades following the contour of the pump housing. As pump speed

5-28 LUBRICATION AND COOLING

increases, and because of resistance to the flow of water, the impeller vanes bend back away from the pump housing and the pump acts like a centrifugal pump. If the impeller blades are short, they remain in contact throughout the full RPM range, supplying full pressure.

※※ WARNING

The outboard should never be run without water, not even for a moment. As the dry impeller tips come in contact with the pump housing or insert, the impeller will be damaged. In most cases, damage will occur to the impeller in seconds.

On most powerheads, if the powerhead overheats, a warning circuit is triggered by a temperature switch to signal the operator of an overheat condition. This should happen before major damage can occur. Reasons for overheating can be as simple as a plastic bag over the water inlet, or as serious as a leaking head gasket.

Whenever the powerhead is started and the cooling system begins pumping water through the powerhead, a water indicator stream will appear from a cooling system indicator in the engine cover. The water stream fitting commonly becomes blocked with debris (especially when lazy operators fail to flush the system after each use, yes we said LAZY, does this mean YOU?) and ceases flowing. This leads one to suspect a cooling system malfunction. Clean the opening in the fitting using a stiff piece of wire before testing or inspecting other cooling system components.

Whenever water is pumped through the powerhead it absorbs and removes excessive heat. This means that anytime a motor begins to overheat, there is not enough (or no) water flowing to the powerhead. This can happen for various reasons, including a damaged or worn impeller, clogged intake or water passages or a stuck closed/restricted thermostat. A sometimes overlooked cause of overheating is the inability of the linings of the cooling passage to conduct heat. Over time, large amounts of corrosion deposits will form, especially on engines that have not received sufficient maintenance. Corrosion deposits can insulate the powerhead passages from the raw water flowing through them.

Troubleshooting the Cooling System

◆ See Figures 75 thru 78

■ When troubleshooting the cooling system, especially for overheat conditions, the motor should be run in a test tank or on a launched vessel (to simulate normal running conditions.). Running the motor on a flushing device may provide both a higher volume of water than the system would deliver (and often times with water that is much colder than would be found in your local lake or bay).

A water-cooled powerhead has a lot of potential problems to consider when talking about overheating. The most overlooked tends to be the

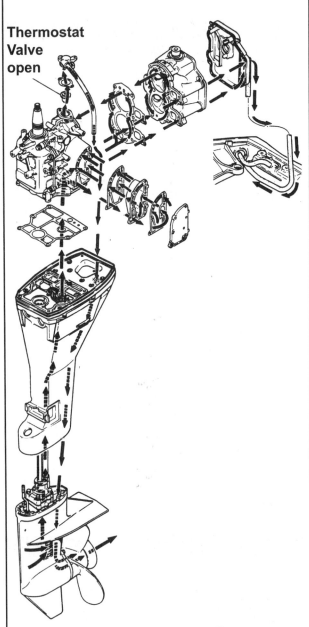

Fig. 74 Cut-away view of a typical outboard cooling system showing water flow - 9.9 hp 4-stroke shown, but the basic idea is the same for all Yamahas

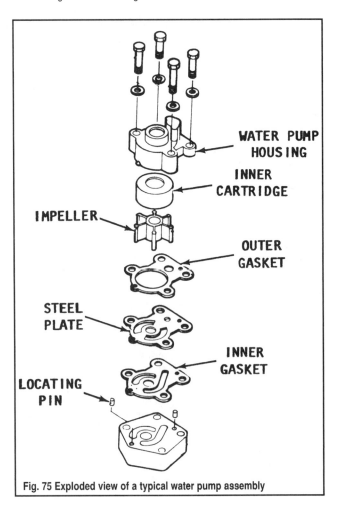

Fig. 75 Exploded view of a typical water pump assembly

LUBRICATION AND COOLING 5-29

Fig. 76 The water pump housing is mounted to the top of the gearcase lower unit, with the impeller just underneath the housing

Fig. 77 This cutaway of a Yamaha outboard shows the water pump housing, impeller, O-ring and the driveshaft which turns the impeller

Fig. 78 The water intake grate and cooling passages should be checked and cleaned with each use

simplest, clogged cooling passages or water intake grate. Although a visual inspection of the intake grate will go a long way, the cooling passage condition can really only be checked by operating the motor or disassembling it to observe the passages.

Damaged or worn cooling system components tend to cause most other problems. And since there are relatively few components, they are easy to discuss. The most obvious is a thermostat that is damaged or corroded, which will often cause the motor to run hot or cold (depending on the position in which the thermostat is stuck, closed or open). The other most obvious component is the water pump impeller is really the heart of the cooling system and it is easy to check, easy that is once it is accessed.

Periodic inspection and replacement of the water pump impeller is a mainstay for many mariners. There are those that wouldn't really consider launching their vessel at the beginning of a season without performing this check. If the water pump is removed for inspection, check the impeller and housing for wear, grooves or scoring that might prevent proper sealing. Check for grooves in the driveshaft where the seal rides. Any damage in these areas may cause air or exhaust gases to be drawn into the pump, putting bubbles into the water. In this case, air does not aid in cooling. When inspecting the pump, consider the following:

• Is the pump inlet clear and clean of foreign material or marine growth? Check that the inlet screen is totally open. How about the impeller?

• Try and separate the impeller hub from the rubber. If it shows signs of loosening or cracking away from the hub, replace the impeller.

• Has the impeller taken a set, and are the blade tips worn down or do they look burned? Are the side sealing rings on the impeller worn away? If so, replace the impeller.

Remember that the life of the powerhead depends on this pump, so don't reuse any parts that look damaged. Are any parts of the impeller missing! If so, they must be found. Broken pieces will migrate up the water tube into the water jacket passages and cause a restriction that could block a water passage. It can be expensive or time consuming to locate the broken pieces in the water passages, but they must be found, or major damage could occur.

■ **Impellers are made of highly durable materials these days and although some people still replace them annually, it is much more common to get 2 or 3 seasons out of an impeller before replacing it. And, even after 2-3 seasons, impellers are more often replaced out of preventive maintenance than for necessary repairs.**

The best insurance against breaking the impeller is to replace it at the beginning of each boating season (or after a couple of seasons, are you sensing a pattern here?), and to NEVER run it out of the water. If installing a metallic body pump housing, coat all screws with non-hardening sealing compound to retard galvanic corrosion. A water tube normally carries the water from the pump to the powerhead. Grommets seal the water tube to the water pump and exhaust housing at each end of the tube, and can deteriorate. Also, the water tube(s) should be checked for holes through the side of the tube, for restrictions, dents, or kinks.

Overheating at high RPM, but not under light load, may indicate a leaking head gasket. If a head gasket it leaking, water can go into the cylinder, or hot exhaust gases may go into the water jacket, creating exhaust bubbles and excessive heat. Remember that aluminum heads have a tendency to warp, and need to be checked (if not re-surfaced) each time they are removed. If necessary, they can be resurfaced by using emery paper and a surface block moving in a figure-eight motion. If this is done, you should also inspect the cylinders and pistons for damage. Other areas to consider are the exhaust cover gaskets and plate. Look for corrosion pin holes. This is rare, but if the outboard has been operated in salt water over the years, there may just be a problem.

If the outboard is mounted too high on the transom, air may be drawn into the water inlet or sufficient water may not be available at the water inlet. When underway the outboard anti-ventilation plate should be running at or near the bottom of the boat and parallel to the surface of the water. This will allow undisturbed water to come to the lower unit, and the water pick-up should be able to draw sufficient water for proper cooling.

Whenever the outboard has been run in polluted, brackish or saltwater, the cooling system should be flushed. Follow the instructions provided under Flushing the Cooling System in the Engine Maintenance section for more details. But in most cases, the outboard should be flushed for at least five minutes. This will wash the salt from the castings and reduce internal corrosion. If the outboard is small and there is no flushing tool that will fit, run the outboard in a tank, drum, or large, sturdy bucket.

5-30 LUBRICATION AND COOLING

There is no need to run in gear during the flushing operation. After the flushing job is done, rinse the external parts of the outboard off to remove the salt spray.

When service work is done on the water pump or lower unit, all the bolts that attach the lower unit to the exhaust housing, and bolts that hold the water pump housing (unless otherwise specified), should be coated with non-hardening gasket sealing compound to guard against corrosion. If this is not done, the bolts may become seized by galvanic corrosion and may become extremely difficult to remove the next time service work is performed.

Last but not least, check to be sure that the overheat warning system is working properly. On most motors, grounding the wire at the sending unit will cause the horn to sound and/or a light should turn on.

TESTING COOLING SYSTEM EFFICIENCY

◆ See Figure 79

If trouble is suspected, cooling system efficiency can be checked by running the motor in a test tank or on a launched boat (while an assistant navigates) and monitoring cylinder head temperatures. There are 2 common methods available to monitor cylinder head temperature, the use of a heat sensitive marker or an electronic pyrometer.

The Stevens Instrument company markets a product known as the Markal Thermomelt Stik®. This is a physical marker that can be purchased to check different heat ranges. The marker is designed to leave a chalky mark behind on a part of the motor that will remain solid and chalky until it is warmed to a specific temperature, at which point the mark will melt appearing liquid and glossy. When using a Thermomelt Stik or equivalent indicator, markers of 2 different heat specifications are necessary for this test. One, the low temperature range marker, should be rated just equal to or very slightly less than the range of the Thermostat Opening rating. The second, the high temperature range marker should be just equal to or slightly less than the Overheat Thermo-switch On rating.

■ To determine normal operating temperatures for you motor, please refer to the Cooling System Specifications chart in this section. Generally speaking, engines should run at or slightly above the Thermostat Opening Temperature or anywhere in the range below the Overheat Sensor Off temperature.

Generally speaking this means you'll want either a 125° (52°C) marker or a 158° (70°C) marker to determine if the motor is reaching normal operating temperature and a higher marker, at least a 163°F (73°C) marker to check for overheating.

Alternately, an electronic pyrometer may be used. Many DVOMs are available with thermo-sensor adapters that can be touched to the cylinder head in order to get a reading. Also, some instrument companies are now producing relatively inexpensive infra-red pyrometers (such as the Raytek® MiniTemp®) of a point-and-shoot design. These units are simply pointed toward the cylinder head while holding down the trigger and the electronic display will give cylinder head temperature. For ease of use and relative accuracy of information, it is hard to beat these infra-red pyrometers. Be sure to follow the tool manufacturer's instructions closely when using any pyrometer to ensure accurate readings.

■ The infra-red pyrometers are EXTREMELY useful when attempting to find a cooling system blockage. You can, in a matter of seconds, take temperature readings of all different parts/locations on the outboard. Once you've used a pyrometer, you won't go back to the markers.

To test the cooling system efficiency, obtain either a Thermomelt Stik (or equivalent temperature indicating marker) or a pyrometer and proceed as follows:

1. If available, install a shop tachometer to gauge engine speed during the test.
2. Make sure the proper propeller or test wheel is installed on the motor.
3. Place the motor in a test tank or on a launched craft.

■ In order to ensure proper readings, water temperature must be approximately 60-80°F (18-24°C).

4. Start and run the engine at idle until it warms, then run it and about 3000 rpm for **at least** five minutes.
5. Reduce engine speed to a low idle and as proceed as follows depending on the test equipment:
• If using Thermomelt Stiks, make 2 marks on the cylinder head (at the thermostat housing), one with the low-range marker and one with the high-range marker. Continue to operate the motor at idle. The low-range mark must turn liquid and glossy or the engine is being overcooled (if equipped, check the thermostat for a stuck open condition). The high-range mark must remain chalky, or the motor is overheating (if equipped, check the thermostat for a stuck closed condition and then check the cooling system passages and the water pump impeller).

■ For most models, temperature readings should be taken on the cylinder head or powerhead (at or near the thermostat housing). For details on thermostat opening temperature specifications please refer to the Cooling System Specifications chart in this section.

• If using a pyrometer, take temperature readings on the cylinder head and powerhead. The meter should indicate temperatures between the Thermostat Opening Temp and the Overheat Sensor Off specification (when available). On smaller motors which do not use a thermostat and/or an overheat sensor temperatures should usually be in approximately 125-155°F (53-67°C) range, otherwise it is likely that the engine is being over/under cooled. If equipped, check the thermostat first for either condition and then suspect the cooling water passages and/or the impeller.

6. Increase engine speed to Wide Open Throttle (WOT) rpm and continue to watch the markers or the reading on the pyrometer. The engine must not overheat at this speed either or the system components must be examined further.

■ When checking the engine at speed expect temperatures to vary slightly from the idle test. Some models will run slightly hotter and some slightly cooler due to the differences in volume of water delivered by the cooling system when compared with engine load. This is normal. In all cases (when equipped), temperatures should not exceed the Overheat Warning Sensor On temperature.

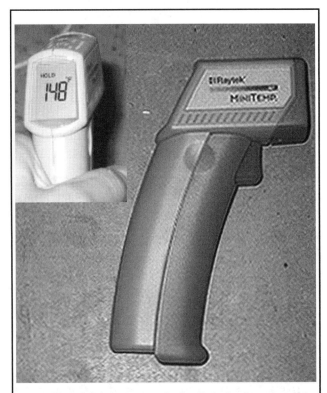

Fig. 79 By far the easiest way to check cylinder head temperature is with a hand-held pyrometer like the MiniTemp® from Raytek® pictured here

LUBRICATION AND COOLING 5-31

TESTING THE THERMOSTAT

◆ See Figure 80 and 81

All 3 hp and larger motors are equipped with a thermostat that restricts the amount of cooling water allowed into the powerhead until the powerhead reaches normal operating temperature. The purpose of the thermostat is to prevent the water from starting to cool the powerhead until the powerhead has warmed to normal operating temperature. In doing this the thermostat will increase engine performance and reduce emissions.

However, this means that the thermostat is vitally important to proper cooling system operation. A thermostat can fail by seizing in either the open or closed positions, or it can, due to wear or deterioration, open or close at the wrong time. All failures would potentially affect engine operation.

A thermostat that is stuck open or will not fully close, may prevent a powerhead from ever fully warming, this could lead to carbon fouling, stumbling, hesitation and all around poor performance. Although these symptoms could occur at any speed, they are more likely to affect most motors at idle when high water flow through the open orifice will allow for more cooling than the lower production of heat in the powerhead requires.

A thermostat that is stuck closed will usually reveal itself right away as the engine will not only come up to temperature quickly, but the temperature warning circuit should be triggered shortly thereafter. However, a thermostat that is stuck partially closed may be harder to notice. Cooling water may reach the powerhead and keep it within a normal operating temperature range at various engine rpm, but allow heat to build up at other rpm. Generally speaking, engines suffering from this type of thermostat failure will show symptoms at part or full throttle, but problems can occur at idle as well. Symptoms, besides overheating, may include hesitation, stumbling, increased noise and smoke from the motor and, general, poor performance.

Testing a thermostat is a relatively easy proposition. Simply remove the thermostat from the powerhead and suspend it in a container of water, then heat the water watching for the thermostat element to move (open) and noting at what temperature it accomplishes this. If you suspect a faulty or inoperable thermostat and cannot seem to verify proper opening/closing temperatures, it may be a good idea (especially since you've already gone through the trouble of removing the thermostat) to simply replace it (it's a relatively low cost part, that performs an important function). Doing so should remove it from suspicion for at least a couple of seasons.

1. Locate and remove the thermostat from the powerhead, as detailed in this section.
2. Suspend the thermostat and a thermometer in a container of water. For most accurate test results, it is best to hang the thermostat and a thermometer using lengths of string so that they are not touching the bottoms or sides of the container (this ensures that both components remain at the same temperature as the water and not the container).
3. Slowly heat the water while observing the thermostat for movement. The moment you observe movement, check the thermometer and note the temperature. If the water begins to boil (reaches about 212°F/100°C at normal atmospheric pressure) and NO movement has occurred, discontinue the test and throw the piece of junk thermostat away (if you are SURE there was no movement).
4. Remove the source of heat and allow the water to cool (you can speed this up a little by adding some cool water to the container, but if you're using a glass container, don't add too much or you'll risk breaking the container). Observe the thermostat again for movement as the water cools. When movement occurs, check the thermometer and record the temperature.
5. Compare the opening temperature with the specifications provided in the Cooling System Specifications chart in this section.
6. Specifications for closing temperatures are not specifically provided by the manufacturer, but typically a thermostat must close at a temperature near, but below the temperature for the opening specification. A slight modulation (repeated opening and closing) of the thermostat will normally occur at borderline temperatures during engine operation keeping the powerhead in the proper operating range.
7. Replace the thermostat if it does not operate as described, or if you are unsure of the test results and would like to eliminate the thermostat as a possible problem. Refer to the removal and installation procedure for Thermostat in this section for more details.

Water Pump

On all Yamaha motors, the water pump is attached to the inside of the gearcase at the gearcase-to-intermediate housing split-line. The pump itself usually consists of a composite material impeller attached to a portion of the driveshaft that runs through a pump housing and/or cover. Usually there is a replaceable insert installed into the cover which is replaced along with the impeller as a matched set whenever the pump requires service.

REMOVAL & INSTALLATION

◆ See Figures 75, 76, 77 and 82 thru 87

■ For more specific art showing the water pump and related components, please refer to the Lower Unit section.

Since the water pump is mounted at the top of the lower unit (at the case split line) the entire lower unit must be removed from the outboard for access. This is not as difficult as it may sound, but it varies by model so for details, please refer to the Lower Unit section.

■ In order to remove the water pump impeller from 2 hp models, the driveshaft must first be removed from the lower unit. On these models the impeller and pump housing should only be removed from the lower end of the driveshaft. Therefore, the circlip securing the pinion gear to the driveshaft must first be removed. This can only be accomplished by removing the lower unit bearing carrier cap and removing the propeller shaft. Again, for details, please refer to the Gearcase Overhaul procedures found in the Lower Unit section.

1. Remove the lower unit from the intermediate housing for access to the water pump housing.

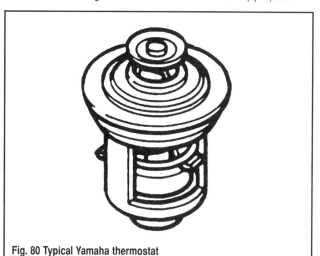

Fig. 80 Typical Yamaha thermostat

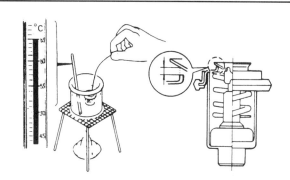

Fig. 81 Test a thermostat by heating it and observing at what temperatures it opens/closes

LUBRICATION AND COOLING

■ If the water tube came out with the lower unit, remove the tube from the pump housing grommet (so the grommet can be removed and replaced).

2. Loosen and remove the bolts securing the water pump housing to the lower unit, then slide or work the cover up off the driveshaft.

■ Depending on whether or not the impeller pulls off the driveshaft with the housing, either remove it from the housing using a pair of needle-nose pliers or gently pry it upward from the shaft using a prybar.

3. If the impeller did not come off with the cover, work the impeller off the driveshaft.
4. Remove the impeller drive pin or drive key from the driveshaft.

■ Although people have been known to reuse some of the serviceable water pump components, it just doesn't make sense. If you've come this far to fix a water pump problem (or for other service requirements) it is a good idea to replace the impeller, insert, grommet(s) and all O-rings or gaskets to ensure proper water pump operation.

5. Disassemble and service the water pump housing by removing the insert (cartridge), grommet(s), gasket(s) and/or O-ring(s), as applicable depending upon the model being serviced as follows:
• 2 Hp (43 and 50cc) Models - Remember that the driveshaft must be removed on this model to service the water pump (from the bottom end of the shaft). For more details, please refer to the Gearcase Assembly Overhaul procedures in the Lower Unit section. In addition, remember the following additional pointers. Remove the water tube grommet and driveshaft O-ring from the top of the pump housing, then remove the cartridge outer plate and 2 driveshaft oil seals from inside the pump housing. Remove the insert and, if necessary, the oil seal protector from the top of the gearcase. During installation, be sure to install the 2 seals inside the pump housing with their lips facing downward and positioned deep enough that there is still about 4.0-4.5mm of clearance beneath the seal lips. Remember to apply marine grease to the O-ring, grommet, oil seals and impeller. Apply a Gasket Maker sealant to the water pump housing-to-cartridge outer plate mating surface. And lastly, be sure to position the housing using the 2 dowel pins.
• 3 Hp Models - Above the impeller there is a large O-ring under the cartridge in the water pump housing. Below the impeller there is the outer plate cartridge and a lower plate (through which the shift rod is also installed). Also, on these models there is a large replaceable water tube grommet on top of the pump housing. The shift rod assembly, utilizes multiple washers and seals mounted through the lower plate. Although the lower plate is installed using a gasket which must be coated with sealant or Gasket Maker, Yamaha does not specify Yamabond for this motor (although it is likely that Yamabond will work just fine). During installation apply a light coat of marine grade grease to the O-ring(s), cartridge insert, impeller and the impeller mating surface of the outer cartridge plate. During installation, align the projection on the top of the cartridge with the hole in the water pump housing.
• 4/5 Hp (83 and 103cc) and 6/8 Hp 2-Stroke Models - On these models there is a water tube grommet mounted to the top of the pump, an insert cartridge in the pump housing and a lower cartridge installed with dowel pins (and the water pump housing bolts) to a lower plate. The lower plate is sealed to the gearcase with a gasket, but removal is not necessary and it should not be disturbed by water pump removal. During installation apply a light coating of marine grease to the water tube grommet, insert cartridge, the impeller and the impeller mating surface of the lower cartridge. Also, although the water pump housing is installed to the lower cartridge using a gasket, no mention is made of sealant. If the old gasket was dry sealed, use the same method during installation.
• 9.9/15 Hp 2-Stroke Models - On these models there is a water tube grommet mounted to the top of the pump, a housing O-ring and a lower cartridge (also called a cartridge outer plate) installed with dowel pins and 2 retaining bolts to a lower plate on the gearcase. Although most Yamaha illustrations do not show an insert in the water pump for the 2-stroke

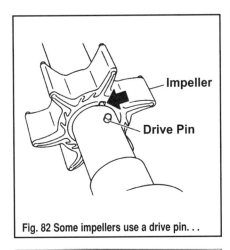

Fig. 82 Some impellers use a drive pin...

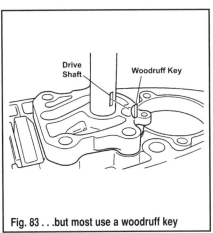

Fig. 83 ...but most use a woodruff key

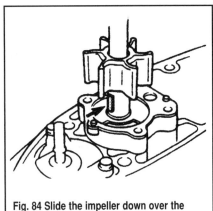

Fig. 84 Slide the impeller down over the pin/key

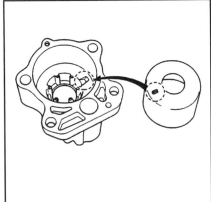

Fig. 85 Install the insert, indexing any tabs as shown

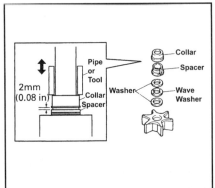

Fig. 86 On 115 hp and larger motors, install the spacer and collar as shown

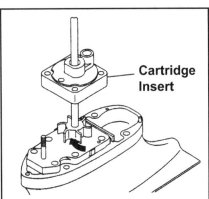

Fig. 87 Slide the pump housing down over the impeller while turning the driveshaft clockwise

LUBRICATION AND COOLING 5-33

versions, our experience leads us to believe that a cartridge insert is also normally mounted in the water pump housing. During installation apply a light coating of marine grease to the water tube grommet, housing O-ring, insert cartridge (if used) and the impeller.

• 20/25 Hp (395cc), 20/25 Hp (430cc) and 25/30 Hp (496cc 2-Cylinder) Models - Although water pump housing designs may vary slightly on these models, most are equipped with a water tube seal rubber grommet and cover on top of the pump. The water pump housing is mounted to a bearing carrier on top of the gearcase. Between the housing and carrier are the replaceable components of an O-ring seal, cartridge insert, impeller and lower cartridge plate. During installation the grommet, O-ring, cartridge and impeller should all be coated using a suitable marine grade grease.

• 40 Hp and 48 Hp (2-Cylinder) Models - Designs for these models may vary slightly. These models are usually, but not always equipped with a replaceable water tube grommet on top of the water pump housing. The housing itself contains a 1 or 2 piece insert and the impeller. The housing and impeller are sealed to an outer plate cartridge using a gasket. The outer plate is then secured to a lower water pump housing using a second gasket (held in position during assembly by at least one dowel pin). The lower water pump housing itself is ALSO sealed to the lower gearcase using a gasket. You do not need to remove and replace the gaskets on the lower housing or outer plate cartridge, as long as they are undisturbed, however once removed, the gaskets must be replaced. During installation be sure to coat the impeller and water tube mounting points with marine grade grease.

• 25-90 Hp 3-Cylinder Models (except 28J-50 Hp 698cc Models) - Although the size and shape of the water pump housing used on 3-cylinder, 2-stroke motors will vary slightly from model-to-model, the basic design and service procedures are the same. These models are usually, but not always equipped with a replaceable water tube grommet on top of the water pump housing. The housing itself contains an insert (which is 2-piece on a few models) and the impeller. The housing and impeller are sealed to an outer plate cartridge using a gasket. The outer plate is then secured to either a lower water pump housing or oil seal housing using a second gasket (held in position during assembly by at least one dowel pin). The lower water pump housing/oil seal housing itself is ALSO sealed to the lower gearcase using a gasket. You do not need to remove and replace the gaskets on the lower housing or outer plate cartridge, as long as they are undisturbed, however once removed, the gaskets must be replaced. During installation be sure to coat the insert, impeller, impeller mating surface of the outer plate and water tube mounting points with marine grade grease.

• 28J-50 Hp (698cc) Models - These models appear to use the same water pump assembly as high-thrust 25 hp 4-stroke motors. On top of the pump is a plastic collar and rubber spacer (also called a seal cover and seal rubber grommet on some models). Under the housing there is an insert cartridge, O-ring, impeller and lower cartridge plate. The lower plate contains 2 dowel pins and is sealed to the gearcase using a gasket (which should be replaced if the plate is disturbed during this procedure). The lower plate gasket is normally installed dry (but if it was sealed from the factory, we'd recommend using gasket making sealant on it). During installation be sure to coat the water tube attaching point, O-ring and impeller with marine grade grease.

• 115-250 Hp Models - The water pumps used on all 115 hp and larger models (including all V4 and V6 Models) are of the same basic design. There is a grommet and spacer mounted to the top of the pump housing. Inside the housing there is a small O-ring, an impeller insert and then a large O-ring. The impeller is held in position on the driveshaft by a combination of components listed (from the top down) as follows: A collar, spacer, washer, wave washer and second washer. When installed properly the collar is a tight fit on the spacer but leaving a 0.08 in. (2mm) gap as shown in the accompanying illustration. Underneath the impeller on these models is a lower impeller plate that is sealed to the gearcase using a gasket and at least one (sometimes 2) dowel pin. During installation, apply a light coating of marine grade grease to the impeller, insert, and O-rings, then slide the impeller carefully down over the driveshaft and into position over the impeller woodruff drive key. Next install the washer, wave washer, washer assembly followed by the spacer and finally the collar. Use a smooth length of pipe (slipped over the driveshaft, but small enough to evenly contact the diameter of the collar) to carefully tap the collar down over the spacer until there is only a 0.08 in. (2.0mm) gap between the collar and spacer. As usual, when replacing the insert, index the tab or tabs on the top of the insert with the hole(s) in the water pump housing and take care not to damage the small O-ring which is should already be installed in the housing at that point.

6. Thoroughly inspect the impeller, cover, insert and impeller plate for signs of damage or wear and replace, as necessary. For more details, please refer to Inspection & Overhaul, in this section.

■ Regardless of condition, it is always a good idea to replace the impeller (as well as the insert and the O-rings) to ensure proper cooling system operation.

To install:

7. Replace all O-rings and gaskets according to the specific model instructions earlier in this procedure. Coat all appropriate components with marine grade grease.

8. If the lower housing and/or lower cartridge plate was removed, install it now using a new gasket or O-ring, as applicable. Most Yamaha motors use a gasket, installed dry, however if the old gasket was coated with sealant, take this as a sign that you can/should use sealant this time.

9. Apply a dab of marine grade grease to the impeller drive pin or woodruff key (this will help hold it in position) and insert it into the driveshaft.

10. Carefully slide the impeller down the driveshaft and over the pin or woodruff key.

11. If not done already, assembly the water pump housing. Install any O-rings or grommets and, if removed, the cartridge insert. Most Yamaha motors use one or more raised tabs on the top of the insert which must be indexed with a corresponding hole in the water pump housing to ensure proper positioning.

12. On 115 hp and larger motors, be sure to install the washers, spacer and collar over top of the impeller as detailed earlier in this procedure.

13. If a gasket is used to seal the housing (as opposed to an O-ring), place the gasket on the lower cartridge insert at this time.

■ On some motors (except the 115 hp and larger models which use the washers, spacer and collar assembly) it may be possible to pre-install the impeller into the housing, but then you've got the issue of locating it properly over the drive pin/woodruff key as the housing is slid into position. Still, if you're having a problem positioning the housing down over the impeller this might be a suitable alternative.

14. Carefully slide the water pump housing down over the driveshaft until it is JUST above the impeller. Then, while slowly rotating the driveshaft CLOCKWISE when viewed from above, slowly slide the housing down over top of the impeller. The driveshaft must be turned so the blades are inserted into the housing facing the normal direction of rotation.

✱✱ WARNING

Once the impeller is installed into the water pump NEVER turn the driveshaft counterclockwise, as the impeller vanes will likely become bent back in the wrong direction and could be damaged either by this motion or by the sudden movement bending them back to the proper direction when the outboard is cranked or started.

15. Once the impeller is started correcting into the housing, continue to turn the driveshaft slowly CLOCKWISE and seat the pump to the gearcase.

16. Apply a light coating of Loctite® 572 or suitable threadlocking material to the threads of the water pump housing retaining bolts. The threadlock will not only keep the bolts from loosening in service, but will also help prevent the possibility of galvanic corrosion locking them in place when it comes time to service the pump again.

17. Install the water retaining bolts and tighten them securely to fasten the pump to the gearcase.

18. Install the Lower Unit, as detailed in the Lower Unit section. Be careful to properly align the shift linkage, water tube and/or driveshaft as the gearcase is raised into position.

INSPECTION & OVERHAUL

◆ See Figures 88 and 89

■ Replace the impeller, gaskets and any O-rings/seal whenever the water pump is removed for inspection or service. There is no reason to use questionable parts. Keep in mind that damage to the powerhead caused by an overheating condition (and the subsequent trouble that can occur from becoming stranded on the water) quickly overtakes the expense of a water pump service kit.

Let's face it, you've gone through the trouble of removing the water pump for a reason. Either, you've already had cooling system problems and you're looking to fix it, or you are looking to perform some preventive maintenance.

5-34 LUBRICATION AND COOLING

Although the truth is that you can just remove and inspect the impeller, replacing only the impeller (or even reusing the impeller if it looks to be in good shape). But WHY would you? The cost of a water pump rebuild kit is very little when compared with the even the time involved to get this far. If you've misjudged a component, or an O-ring (which by the way, never reseal quite the same way the second time), then you'll be taking this lower unit off again in the very near future to replace these parts. And, at best this will be because the warning system activated or you noticed a weak coolant indicator streams or, at worst, it will be cause you're dealing with the results of an overheated powerhead.

In short, if there is one way to protect you and your engine, it is to replace the impeller and insert (if used) along with all O-ring seals, grommets and gaskets, anytime the pump housing is removed. If not, take time to thoroughly clean and inspect the old impeller, housing and related components before assembly and installation. New components should be checked against the old. Seek explanations for differences with your parts supplier (but keep in mind that some rebuild kits may contain upgrades or modifications.)

1. Remove and disassemble the water pump assembly as detailed in this section.

2. If used, carefully remove all traces of gasket or sealant material from components. If some material is stubborn, use a suitable solvent in the next steps to help clean material. Avoid scraping whenever possible, especially on plastic components whose gasket surfaces are easily scored and damaged.

3. Clean all metallic components using a mild solvent (such as Simple Green® cut with water or mineral spirits), then dry using compressed air (or allow them to air dry).

4. Clean plastic components using isopropyl alcohol.

■ Skip the next step, we really mean it, don't INSPECT the impeller, REPLACE IT. Ok, we've been there before, if you absolutely don't want to replace the impeller. Let's say it's only been used one season or so and you're here for another reason, but you're just being thorough and checking the pump, then perform the next step.

5. Check the impeller for missing, brittle or burned blades. Inspect the impeller side surfaces and blade tips for cracks, tears, excessive wear or a glazed (or melted) appearance. Replace the impeller if these defects are found. Next, squeeze the vanes toward the hub and release them. The vanes should spring back to the extended position. Replace the impeller if the vanes are set in a curled position and do not spring back when released.

6. The water pump impeller should move smoothly up or downward on the driveshaft. If not, it could become wedged up against the housing or down against the impeller plate, causing undue wear to the top or bottom of the blades and hub. Check the impeller on the driveshaft and, if necessary, clean the driveshaft contact surface (inside the impeller hub) using emery cloth.

7. Inspect the water pump body or the lining inside the water pump body, as applicable, for burned, worn or damaged surfaces. Replace the impeller lining, if equipped, or the water pump body if any defects are noted.

8. Visually check the water pump housing (and insert, if equipped), along with the impeller wear plate for signs of overheating including warpage (especially on the plate) or melted plastic. Some wear is expected on the impeller plate and, if equipped, on the housing insert, but deep grooves (with edges) are signs of excessive wear requiring component replacement.

■ A groove is considered deep or edged if it catches a fingernail.

9. Check the water tube grommets and seals for a burned appearance or for cracked or brittle surfaces. Replace the grommets and seals if any of these defects are noted.

After 60 seconds at 1500 rpm.

After 90 seconds at 1500 rpm.

After 30 seconds at 2000 rpm.

After 45 seconds at 2000 rpm.

After 60 seconds at 2000 rpm.

Fig. 88 The impeller can be damaged by as little as 30-90 seconds of engine operation without a suitable source of cooling water

Fig. 89 A typical water pump/impeller replacement kit - it's worth the $

LUBRICATION AND COOLING 5-35

WATER PUMP EXPLODED VIEWS

◆ See Figures 90 thru 102

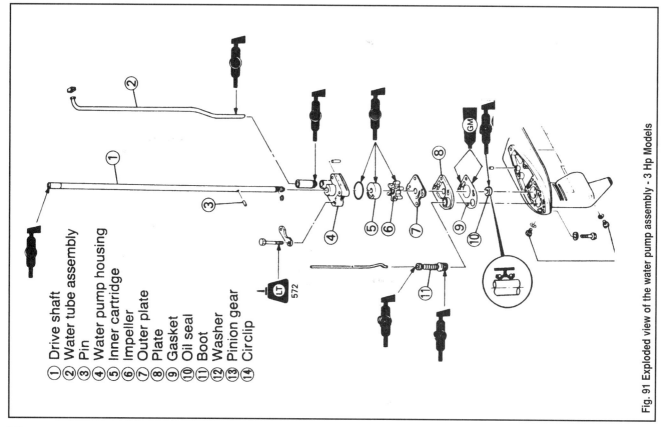

Fig. 91 Exploded view of the water pump assembly - 3 Hp Models

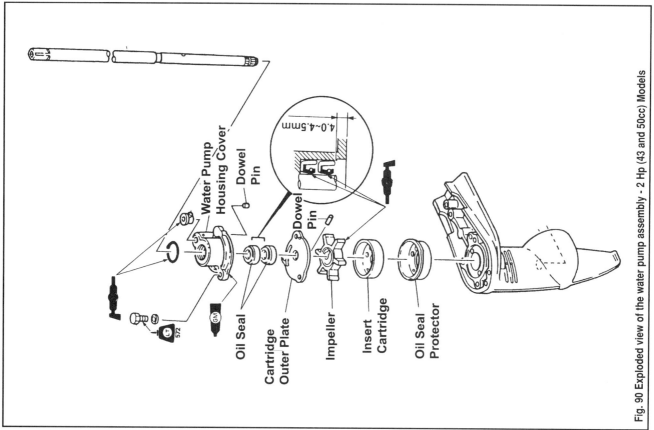

Fig. 90 Exploded view of the water pump assembly - 2 Hp (43 and 50cc) Models

5-36 LUBRICATION AND COOLING

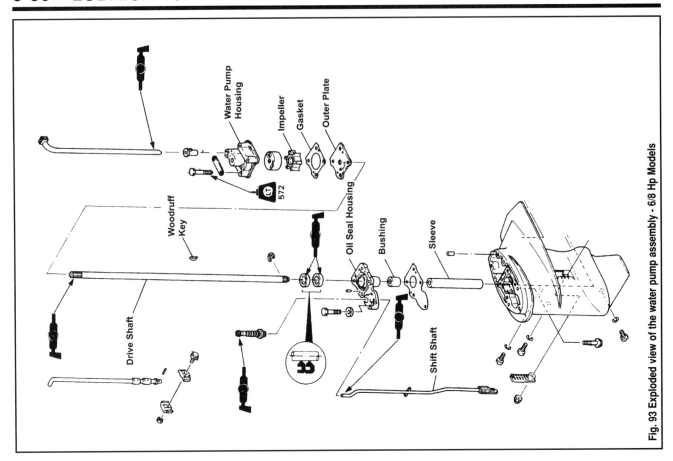

Fig. 93 Exploded view of the water pump assembly - 6/8 Hp Models

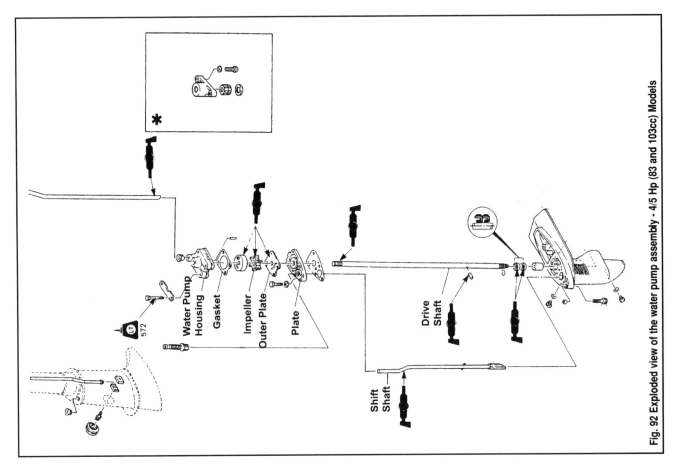

Fig. 92 Exploded view of the water pump assembly - 4/5 Hp (83 and 103cc) Models

LUBRICATION AND COOLING 5-37

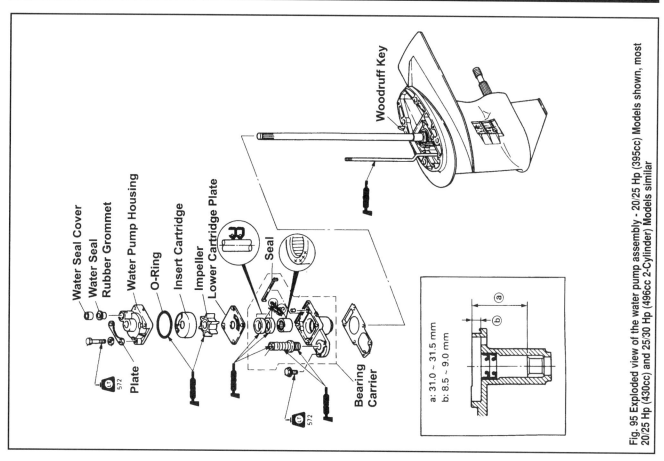

Fig. 95 Exploded view of the water pump assembly - 20/25 Hp (395cc) Models shown, most 20/25 Hp (430cc) and 25/30 Hp (496cc 2-Cylinder) Models similar

a: 31.0 ~ 31.5 mm
b: 8.5 ~ 9.0 mm

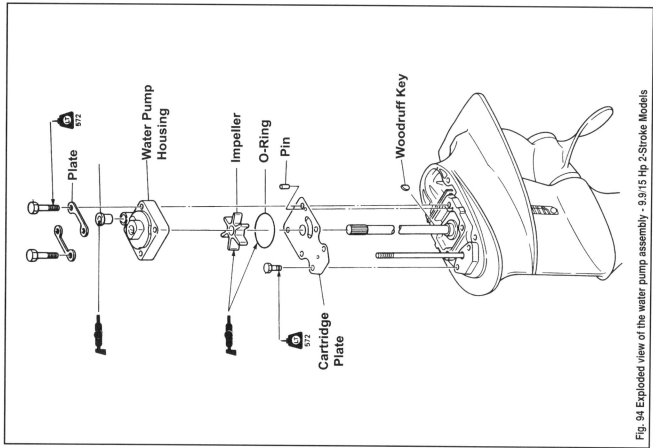

Fig. 94 Exploded view of the water pump assembly - 9.9/15 Hp 2-Stroke Models

5-38 LUBRICATION AND COOLING

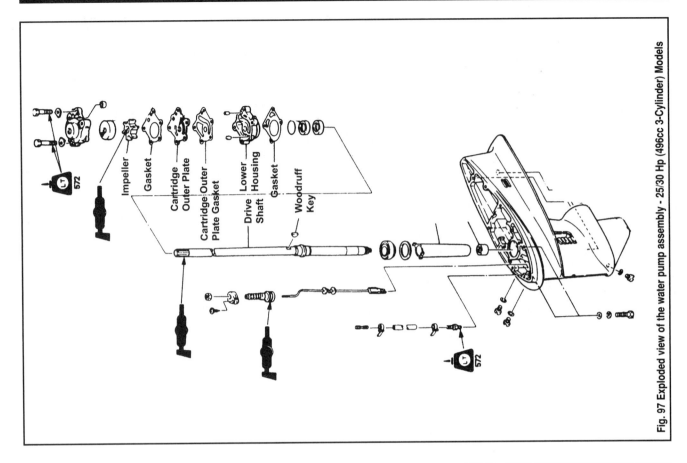

Fig. 97 Exploded view of the water pump assembly - 25/30 Hp (496cc 3-Cylinder) Models

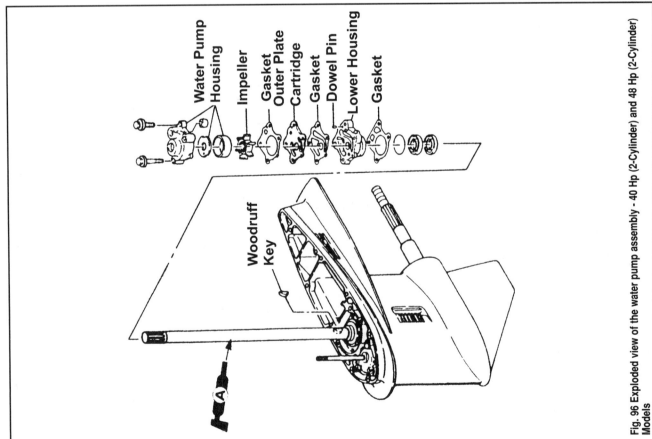

Fig. 96 Exploded view of the water pump assembly - 40 Hp (2-Cylinder) and 48 Hp (2-Cylinder) Models

LUBRICATION AND COOLING 5-39

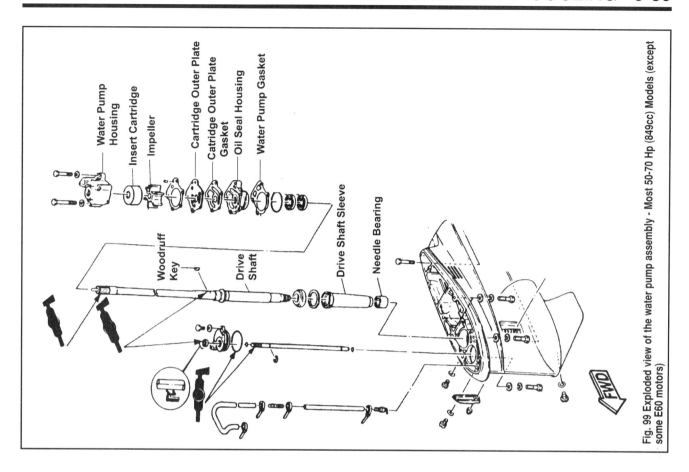

Fig. 99 Exploded view of the water pump assembly - Most 50-70 Hp (849cc) Models (except some E60 motors)

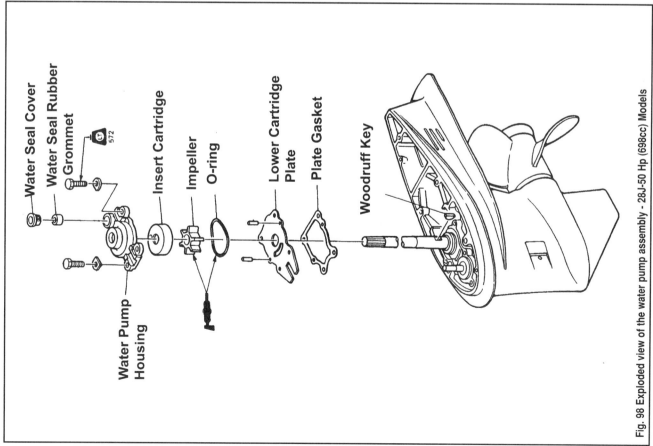

Fig. 98 Exploded view of the water pump assembly - 28J-50 Hp (698cc) Models

5-40 LUBRICATION AND COOLING

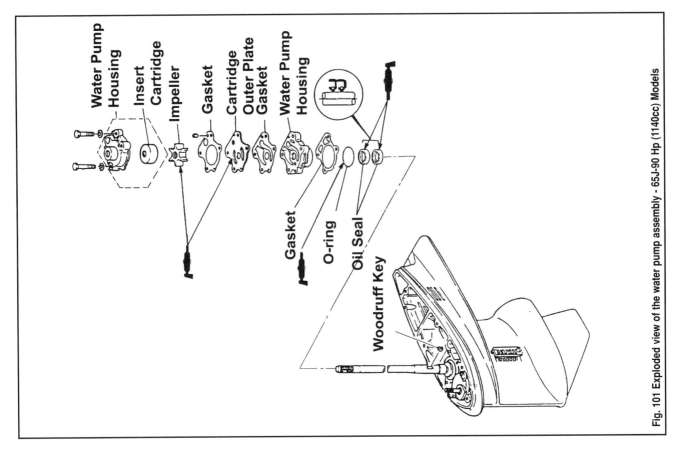

Fig. 101 Exploded view of the water pump assembly - 65J-90 Hp (1140cc) Models

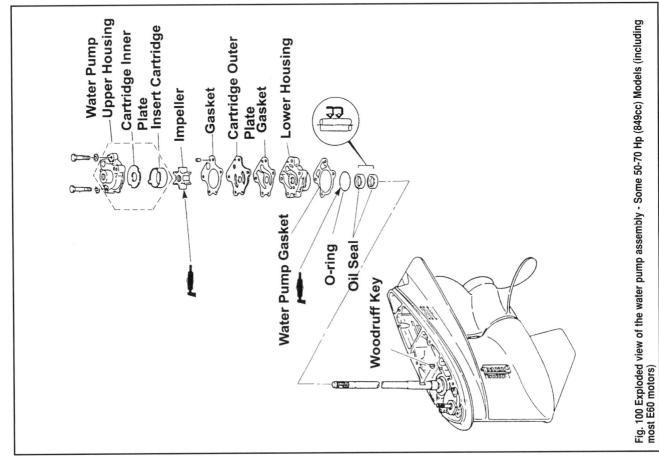

Fig. 100 Exploded view of the water pump assembly - Some 50-70 Hp (849cc) Models (including most E60 motors)

LUBRICATION AND COOLING 5-41

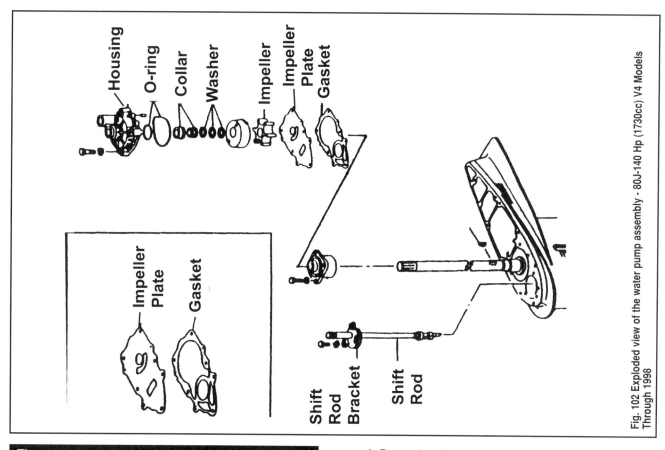

Fig. 102 Exploded view of the water pump assembly - 80J-140 Hp (1730cc) V4 Models Through 1998

Thermostat

All 3 hp and larger motors are equipped with a thermostat that restricts the amount of cooling water allowed into the powerhead until the powerhead reaches normal operating temperature. The purpose of the thermostat is to prevent cooling water from reaching the powerhead until the powerhead has warmed sufficiently. In doing this the thermostat will increase engine performance and reduce emissions.

On all models so equipped, the thermostat components are mounted somewhere on the powerhead, in a cooling passage, under an access cover that is sealed using a gasket or an O-ring. On many smaller Yamaha models they are on the top or side of the powerhead itself (block), however probably on MOST Yamaha models they are mounted to the top or top side of the cylinder head cover (or cylinder head on some models).

REMOVAL & INSTALLATION

◆ See Figures 103 thru 107   MODERATE

■ For more specific art showing the thermostat and cover mounting, please refer to the Powerhead section.

On all models, the thermostat assembly is mounted under a cover on the powerhead. The size, shape and location of this cover, including the number of components and seals found underneath varies slightly by model. However, even with this said the service procedures are virtually identical for all Yamahas.

■ On some motors, the cover can be installed facing different directions. For these models, matchmark the cover to the mating surface or otherwise make a note of cover orientation to ensure installation facing the proper direction. This is especially important for covers to which hoses attach (that have been removed).

1. Disconnect the negative battery cable (if equipped) and/or remove the spark plug wire(s) from the plug(s) and ground them on the powerhead for safety.

2. Remove the engine top cover for access.
3. Locate the thermostat housing on the powerhead and remove any interfering components (flywheel or side cover), then remove the cover and thermostat as follows:
• 3 Hp Models - The thermostat is located under a small cover on top of the powerhead, just behind the cylinder head split line. The cover is usually secured by 2 bolts and a gasket.
• 4/5 Hp (83 and 103cc) Models - The thermostat is located under the exhaust outer cover on the side of the powerhead. Because the outer and inner covers are secured by the same fasteners and are each sealed with a gasket it is a good idea to remove both and replace both gaskets if the thermostat is removed. When loosening the cover fasteners follow the reverse of the torque sequence (there is usually one molded into the cover)

Fig. 103 On Yamahas the thermostats are mounted at the top of the cylinder head (pictured), exhaust cover or powerhead - cutaway of an 8 hp motor shown

5-42 LUBRICATION AND COOLING

or use multiple passes of a crossing pattern. During installation be sure to coat the bolts threads with a threadlocking compound and then to tighten the bolts using at least 2 passes of a torque sequence to 6.5 ft. lbs. (9 Nm).

■ **On 4/5 hp motors, removing the exhaust cover gives you a change to inspect the crankcase/combustion chamber for excessive deposits, so it's not a bad thing.**

- 6/8 Hp Models - The thermostat is located under the top of the cylinder head cover (which must be removed for access, which isn't a bad thing, since it gives you the opportunity to visually inspect the pistons/combustion chambers and lets you re-torque the head upon installation). Be sure to apply a light coating of engine oil on the threads of the cylinder head bolts and tighten them using multiple passes of a clockwise spiraling sequence that starts on the outside of the cylinder head and works inward (unless a different sequence is molded on the cover). Be sure to tighten the bolts to 5.8 ft. lbs. (8 Nm).
- 9.9/15 Hp Models - The thermostat is located under a small, irregular-shaped cover at the top of the cylinder head. It is secured by 4 bolts and a gasket. There is normally a plain washer mounted between the thermostat and the cover. On some models there may be a cylinder head anode located under the cover as well, if so, it should be inspected/replaced anytime the thermostat is serviced.
- 20/25 Hp (395cc), 20/25 Hp (430cc) and 25/30 Hp (496cc 2-Cylinder) Models - On all models the thermostat is located at the top of the cylinder head. On most it is located under a small, almost rectangular cover directly on top of the cylinder head. The cover is usually secured by 2 bolts and a gasket.
- 40 Hp and 48 Hp (2-Cylinder) Models - The thermostat is normally located under a small, irregular-shaped cover at the top of the cylinder head. For most models it is secured by 4 bolts and a gasket. However, use some caution when removing the last cover bolt, on some models the cover also conceals a spring loaded pressure relief valve assembly (spring, valve and grommet) which tends to push the cover outward. Take care not to loose any components.
- 25/30 Hp (496cc 3-Cylinder) Models - The thermostat is located on top of the powerhead, just behind the cylinder head and exhaust cover mating surfaces. It is under a small, almost rectangular cover which is secured by 2 bolts and a gasket.
- 28J-50 Hp (698cc) Models - The thermostat is located at the top side of the powerhead, directly above the exhaust cover mating surface. It is under a small, almost rectangular cover which is secured by 2 bolts and a gasket.
- 50-70 Hp (849cc) Models - The thermostat is located at the top side of the powerhead, under a small irregular-shaped cover which is bolted to the inner exhaust cover (however the outer exhaust cover does not overlap and so the thermostat cover can be removed without disturbing either). The cover is secured by 2 bolts and a gasket.
- 65J-90 Hp (1140cc) Models - The thermostat is normally located under a small, irregular-shaped cover at the top of the cylinder head. For most models it is secured by 4 bolts and a gasket. However, use some caution when removing the last cover bolt as the cover also conceals a spring loaded pressure relief valve assembly (spring, valve and grommet) which tends to push the cover outward. Take care not to loose any components.
- 80J-140 Hp (1730cc) V4 Models - The thermostat is located under a small, irregular-shaped cover at the top of the cylinder head. For most models it is secured by 2 bolts and a gasket. On these models, there is usually a separate thermostat for each cylinder head.
- 105J-225 Hp (2596cc) V6 Models - The thermostat is located under a small, irregular-shaped cover at the top of the cylinder head. For most models it is secured by 4 bolts and a gasket. On these models, there is usually a separate thermostat for each cylinder head. There may also be a pressure relieve valve assembly under the cover, so remove the last bolt while holding the cover in position, then remove the cover slowly.
- 200-250 Hp (3130cc) and 225-250 Hp (3342cc) V6 Models - There is one thermostat for each cylinder bank. Each is located on top of the powerhead, about halfway between the cylinder head mating surface and the center of the flywheel. There is usually JUST enough access to remove the bolts and the cover without removing the flywheel, however, if difficulty is encountered the flywheel can be removed for additional access. Each cover is secured by 2 bolts and gasket.

4. Loosen and remove the bolts (as noted in the previous step) securing the cover, then carefully pull the cover from the powerhead. If necessary, tap around the outside of the cover using a rubber or plastic mallet to help loosen the seal.

5. Check if the seal or gasket was removed with the cover. When a composite gasket and/or sealant was used, make sure all traces of gasket and sealant material are removed from the cover and the powerhead mounting surface. For installation purposes, take note of the direction in which the thermostat is facing before removal.

■ **Most Yamaha motors do NOT use sealant on the thermostat housing gasket. Also, the thermostats on most Yamaha motors are installed with the flange side facing the cover.**

6. Visually inspect the thermostat for obvious damage including corrosion, cracks/breaks or severe discoloration from overheating. Make sure any springs have not lost tension. If necessary, refer to the Testing the Thermostat in this section for details concerning using heat to test thermostat function.

Fig. 104 Most (but not all) covers are secured using 2 or 4 bolts - V configuration motor shown. . .

Fig. 105 . . .though shape and location will vary slightly from model-to-model - 3-cylinder, 2-stroke shown

LUBRICATION AND COOLING 5-43

To install:

7. Install each of the thermostat components in the reverse of the removal procedure. Replace any gaskets, seals and/or O-rings (on a few models you may have to reinstall the pressure valve assembly in the correct order, normally grommet, valve and spring). Pay close attention to the direction each component is installed.

■ You've got to love Yamaha. When a powerhead torque value is important or a torque sequence is necessary, it is normally molded on the powerhead component itself. Check the thermostat housing cover for numbers near the bolt holes and, if present, follow that order during installation and tightening. Also, check the cover itself for an embossed torque specification and, if you've got a torque wrench available, follow it as well.

8. Install the thermostat housing cover using a new gasket or seal. Tighten the bolts until snug. On motors with 4 housing bolts be sure to use multiple passes of a criss-cross pattern to tighten the bolts (this will help make sure the cover is not cracked or warped during installation).

9. Connect the negative battery cable and/or spark plug lead(s), then verify proper cooling system operation.

THERMOSTAT MOUNTING EXPLODED VIEWS

◆ See Figures 108 thru 121

Fig. 106 When necessary, a torque sequence is normally molded onto the cover

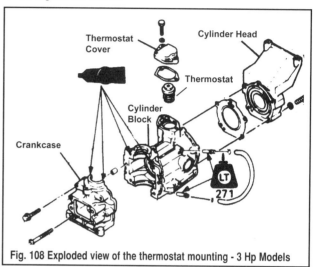

Fig. 108 Exploded view of the thermostat mounting - 3 Hp Models

Fig. 107 The corrosion found in this housing shows you why a thermostat could fail

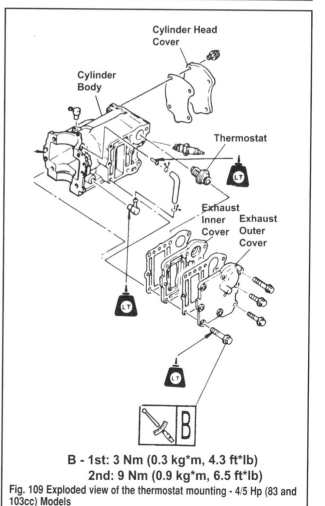

B - 1st: 3 Nm (0.3 kg*m, 4.3 ft*lb)
2nd: 9 Nm (0.9 kg*m, 6.5 ft*lb)

Fig. 109 Exploded view of the thermostat mounting - 4/5 Hp (83 and 103cc) Models

5-44 LUBRICATION AND COOLING

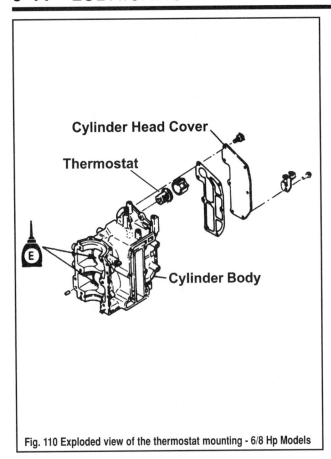

Fig. 110 Exploded view of the thermostat mounting - 6/8 Hp Models

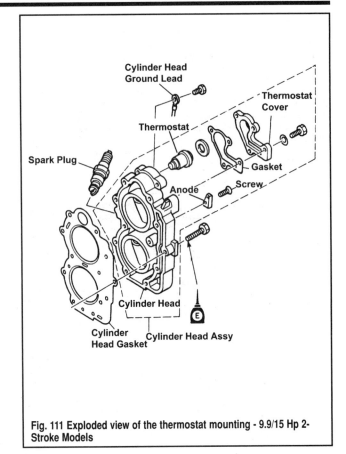

Fig. 111 Exploded view of the thermostat mounting - 9.9/15 Hp 2-Stroke Models

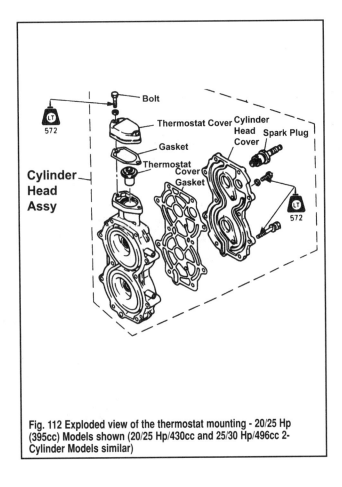

Fig. 112 Exploded view of the thermostat mounting - 20/25 Hp (395cc) Models shown (20/25 Hp/430cc and 25/30 Hp/496cc 2-Cylinder Models similar)

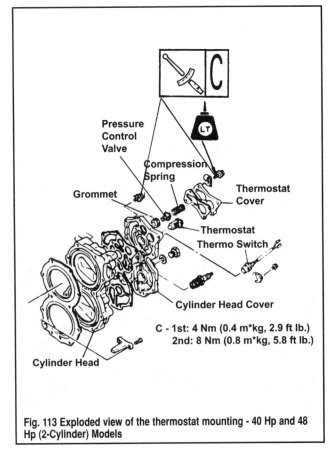

Fig. 113 Exploded view of the thermostat mounting - 40 Hp and 48 Hp (2-Cylinder) Models

LUBRICATION AND COOLING 5-45

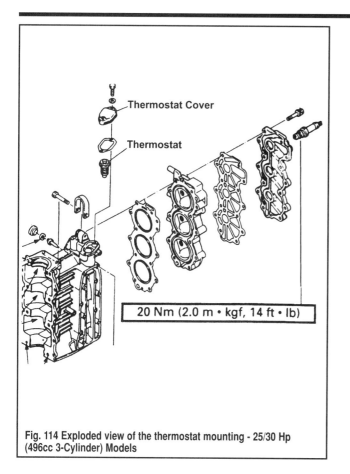

Fig. 114 Exploded view of the thermostat mounting - 25/30 Hp (496cc 3-Cylinder) Models

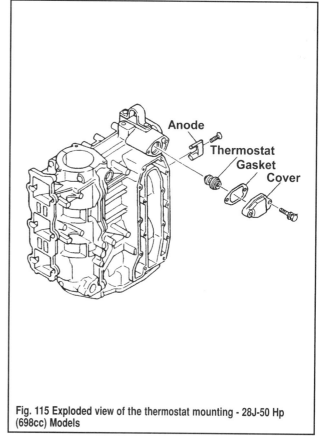

Fig. 115 Exploded view of the thermostat mounting - 28J-50 Hp (698cc) Models

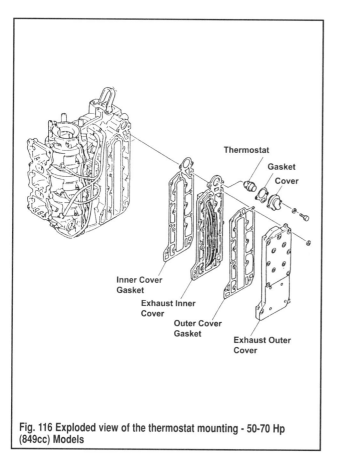

Fig. 116 Exploded view of the thermostat mounting - 50-70 Hp (849cc) Models

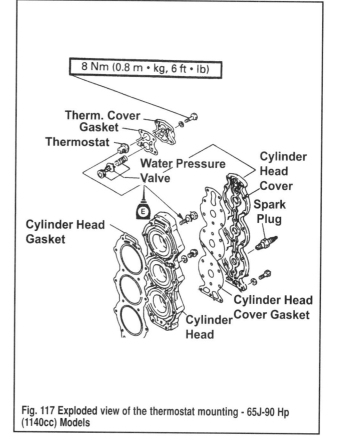

Fig. 117 Exploded view of the thermostat mounting - 65J-90 Hp (1140cc) Models

5-46 LUBRICATION AND COOLING

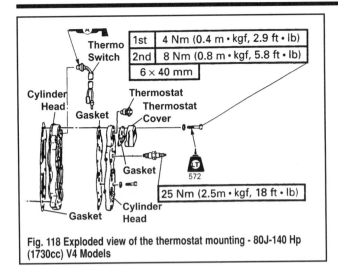

Fig. 118 Exploded view of the thermostat mounting - 80J-140 Hp (1730cc) V4 Models

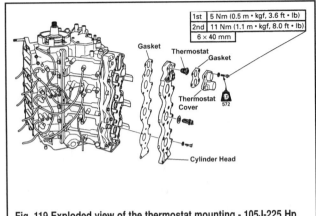

Fig. 119 Exploded view of the thermostat mounting - 105J-225 Hp (2596cc) V6 Models (HPDI shown, but carb and EFI OX66 very similar)

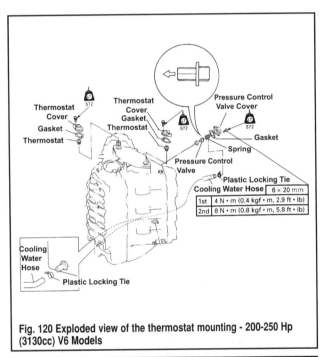

Fig. 120 Exploded view of the thermostat mounting - 200-250 Hp (3130cc) V6 Models

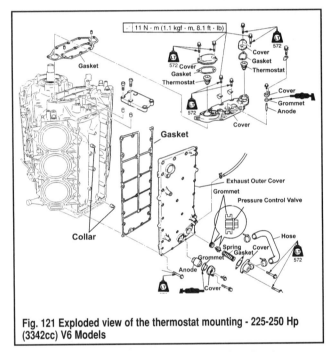

Fig. 121 Exploded view of the thermostat mounting - 225-250 Hp (3342cc) V6 Models

Thermo-switch

DESCRIPTION & OPERATION

◆ See Figures 122 thru 124

Most Yamaha outboards are equipped with either a thermo-switch (and on/off switch which is activated by temperature) and/or a thermo-sensor (a variable resistor whose values change with changes in temperatures). The breakdown goes as follows:

- The 20 hp and larger (except the 430cc 20/25 hp) 2-stroke motors utilize a thermo-switch for the overheat warning system. When equipped with YMIS the CDI unit may also receive the switch signal to be used in limiting engine speed to protect the powerhead.
- All V4 and V6 fuel injected motors are equipped with both a thermo-sensor (known as a WTS) for the fuel and ignition system as well as a pair of thermo-switches (usually there is one at the top of each cylinder head) for use with the warning system.

This section deals with the thermo-switches only (on/off temperature activated switches). For information on the variable resistance temperature sensors such as the WTS used on fuel injected motors, please refer to the Fuel System section. For more information on the variable resistance temperature sensors used on some CARBURETED motors, please refer to the Ignition and Electrical System section.

A thermo-switch is a bimetallic type on/off switch. Typically a thermo-switch will consist of two different type metals attached together in a small strip. Because each metal has a different coefficient of expansion - each will expand a different amount within a given temperature range, and the strip will bend rather than extend. Depending upon whether the switch is designed to be normally closed or normally open in a given temperature range will determine how the metal switch and contacts are arranged.

Yamaha motors use a normally open switch assembly that provides no continuity across the switch contacts until a given temperature is reached. The internals of the switch are designed to bend in such a manner that the switch contacts will only close at or above a certain temperature. In this way, only once the switch physically reaches a given temperature will the switch contacts close, activating the warning circuit.

The thermo-switch therefore only sends an ON or OFF signal, not a variable voltage as in the thermo-sensor signal. On V configuration powerheads covered in this manual, two thermo-switches are installed - one on each of the cylinder banks. A thermo-switch is required on each cylinder bank, because it is possible for the temperature of one bank to vary considerably from the other bank. The cause for such a difference may be due to blockage of a cooling passage or blockage of an oil hose leading to one cylinder. Either one of these conditions will contribute to increased friction, raising the temperature in or around a particular cylinder. The failure of a mechanical part within a cylinder, such as a broken ring, will cause added friction and increased temperature in the area.

LUBRICATION AND COOLING 5-47

If the powerhead temperature exceeds a specified level, the thermo-switch will close a warning circuit activating a buzzer. Generally, the buzzer will continue to sound until the temperature drops to an acceptable level.

The thermo-switches used by Yamaha have two calibrations. Each has an overheat warning ON temperature setting (the temperature at which the switch contacts will close), but they also have an overheat warning OFF temperature (which is usually significantly lower than the ON temperature). The switches are designed not to close until a certain point has been reached, but once reached, they remain closed until the powerhead (switch) has cooled well below the activation point. For details, please refer to the Cooling System Specifications charts in this section.

■ **On some models a separate thermo-sensor is used to signal the CDI unit or control module to limit engine RPM in order to protect the powerhead.**

OPERATIONAL THERMOSWITCH CHECK

♦ See Figures 122 thru 124

1. Identify the leads (usually Pink and Black, but check the Wiring Diagrams in the Ignition and Electrical section for the motor on which you are working to be sure) from one of the two thermo-switches.
2. Disconnect the two leads at the quick disconnect fittings.
3. Start and run the motor (because you'll be running at above idle, it is really best to use a test tank or launch the craft to prevent the possibility of engine over-speed). Increase engine speed to just above 2500 rpm.
4. Touch the two leads from the switch together. The warning buzzer should sound continuously and, on most models, the powerhead rpm should drop to 2000 rpm.
5. Reconnect the wiring to the switch leads, back the throttle down to idle rpm and the buzzer should be silent.
6. On V configuration motors, repeat this test for the second thermo-switch installed in the other cylinder bank.
7. If the thermo-switch does not operate as specified, either the switch or its circuit are faulty.

CHECKING THERMOSWITCH CONTINUITY

♦ See Figures 122 thru 125

Locate and remove the thermo-switch from the powerhead. The thermo-switch resembles a condenser, usually with Pink and Black leads (but check the Wiring Diagrams from the Ignition and Electrical System section for the motor on which you are working). The thermo-switch can be easily removed from the powerhead, by pulling it out of its recess. The thermo-switch is a very fragile sensor and should never be dropped or handled in a rough manner.

Fig. 122 On most Yamaha motors there is a thermo-switch mounted to the cylinder head/head cover

Fig. 123 To remove a thermo-switch, grasp and gently pull it from the mounting bore

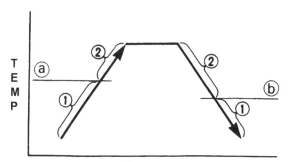

Fig. 124 Graphical display of Yamaha thermo-switch operation showing switch position (on/off, i.e. continuity/discontinuity) in relation to temperature and time

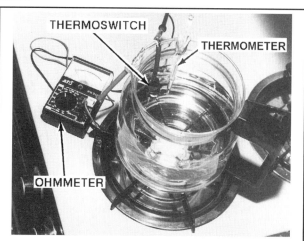

Fig. 125 Heating a container of water to check the performance of the thermo-switch

5-48 LUBRICATION AND COOLING

Remove the thermo-switch from the block and disconnect the two leads at their quick disconnect fittings. Since the switch should have no continuity until it is heated past a point which could damage the powerhead, there are no resistance tests to be performed on this switch while it is installed.

■ Ok, that last statement was too much of a generalization. If the warning system is activated and remains activated on a cold motor, you can check the thermo-switch for continuity across the contacts. If there is continuity even with a cold motor, then the switch is bad, get rid of it!

Testing is a simple matter of watching for continuity across the switch contacts (using an ohmmeter or DVOM set to read resistance across the switch wiring) as the switch is slowly heated to activation temperature. The best method for this is to suspend the switch in water which is slowly heated to temperature while you watch the meter. Obtain a thermometer, an ohmmeter, and a suitable pan in which to boil water.

Place the container of water, at room temperature, on a stove. Secure the thermometer in the water in such a manner to prevent the thermometer bulb from contacting the bottom or sides of the pan (this ensures the thermometer will read the temperature of the water and NOT the pan, which could be considerably higher).

Immerse the sensing end of the thermo-switch into the water up to the shoulder and suspend it, also NOT touching the pan for the same reason as the thermometer.

Make contact with the Red meter lead to the Pink thermo-switch lead, and the Black meter lead to the Black thermo-switch lead. The meter should indicate no continuity, as long as the water temperature is below the Overheat Thermo-switch On temperature listed in the Cooling System Specifications charts in this section.

Slowly heat the water, while watching the thermometer and meter. When the temperature rises above the activation point listed in the chart, the meter should indicate continuity.

■ A slight variation is allowable above or below the stated temperatures in the charts. But keep in mind that too great a variation could either allow damage to the powerhead by operating it under conditions that are too hot or could allow activation of the warning system when the powerhead is not in danger.

Remove the pan from heat and allow the sensor to slowly cool (you can add a little cool water or, if using a metal pan, immerse the pan itself in a second pan of cooler water to help speed the cooling process). Watch the meter and the thermometer as the water cools. The switch contacts must open (showing no continuity) at or near the switch Off temperature listed in the charts.

If the meter reading is acceptable, the thermo-switch is functioning correctly. If the meter readings are not acceptable, the thermo-switch can be easily replaced with a new unit.

LUBRICATION AND COOLING

SPECIFICATIONS

COOLING SYSTEM SPECIFICATIONS - 2-Stroke Motors

Model (Hp)	No. of Cyl	Engine Type	Year	Displace cu. in. (cc)	Thermostat Open Temp. Degrees F(C)	Overheat Thermoswitch Degrees F(C) On	Overheat Thermoswitch Degrees F(C) Off
2	1	IL 2-stroke	1997-02	2.6 (43)	-	-	-
2	1	IL 2-stroke	1997-02	3.0 (50)	-	-	-
3	1	IL 2-stroke	1997-02	4.3 (70)	118-126 (48-52)	-	-
4	1	CL 2-stroke	1997-99	5.0 (83)	118-126 (48-52)	-	-
4	1	CL 2-stroke	1997-02	6.3 (103)	118-126 (48-52)	-	-
5	1	CL 2-stroke	1997-02	6.3 (103)	118-126 (48-52)	-	-
6	2	CL 2-stroke	1997-00	10 (165)	118-126 (48-52)	-	-
8	2	CL 2-stroke	1997-03	10 (165)	118-126 (48-52)	-	-
9.9	2	IL 2-stroke	1997-03	15 (246)	118-126 (48-52)	-	-
15	2	IL 2-stroke	1997-03	15 (246)	118-126 (48-52)	-	-
20	2	IL 2-stroke	1997	24 (395)	118-126 (48-52)	199 (93)	181 (83)
25	2	IL 2-stroke	1997-03	24 (395)	118-126 (48-52)	199 (93)	181 (83)
20	2	IL 2-stroke	1997-98	26 (430)	136-144 (58-62)	-	-
25	2	IL 2-stroke	1997-98	26 (430)	136-144 (58-62)	-	-
25	2	IL 2-stroke	1997	30 (496)	118-126 (48-52)	230-247 (106-114)	191-221 (88-102)
30	2	IL 2-stroke	1997	30 (496)	118-126 (48-52)	230-247 (106-114)	191-221 (88-102)
40	2	IL 2-stroke	1997	36 (592)	136-144 (58-62)	205-219 (96-104)	163-189 (73-87)
48	2	IL 2-stroke	1997-00	46 (760)	136-144 (58-62)	171-181 (77-83)	127-153 (53-67)
25	3	IL 2-stroke	1997-02	30 (496)	118-126 (48-52)	194-205 (90-96)	169-194 (76-90)
30	3	IL 2-stroke	1997-02	30 (496)	118-126 (48-52)	194-205 (90-96)	169-194 (76-90)
28J	3	IL 2-stroke	1997-01	43 (698)	118-126 (48-52)	194-205 (90-96)	169-194 (76-90)
35J	3	IL 2-stroke	1997-01	43 (698)	118-126 (48-52)	194-205 (90-96)	169-194 (76-90)
40	3	IL 2-stroke	1997-03	43 (698)	118-126 (48-52)	194-205 (90-96)	169-194 (76-90)
50	3	IL 2-stroke	1997-03	43 (698)	118-126 (48-52)	194-205 (90-96)	169-194 (76-90)
50	3	IL 2-stroke	1997-03	52 (849)	118-126 (48-52)	183-194 (84-90)	140-165 (60-74)
60	3	IL 2-stroke	1997-03	52 (849)	118-126 (48-52)	183-194 (84-90)	140-165 (60-74)
70	3	IL 2-stroke	1997-03	52 (849)	118-126 (48-52)	183-194 (84-90)	140-165 (60-74)
65J	3	IL 2-stroke	1997-01	70 (1140)	118-126 (48-52)	183-194 (84-90)	140-165 (60-74)
75	3	IL 2-stroke	1997-00	70 (1140)	118-126 (48-52)	183-194 (84-90)	140-165 (60-74)
80	3	IL 2-stroke	1997-00	70 (1140)	118-126 (48-52)	183-194 (84-90)	140-165 (60-74)
85	3	IL 2-stroke	1997-00	70 (1140)	118-126 (48-52)	183-194 (84-90)	140-165 (60-74)
90	3	IL 2-stroke	1997-03	70 (1140)	118-126 (48-52)	183-194 (84-90)	140-165 (60-74)
80J	4	90 LV 2-st	1997-01	106 (1730)	118-126 (48-52)	183-194 (84-90)	140-165 (60-74)
100	4	90 LV 2-st	1997-02	106 (1730)	118-126 (48-52)	183-194 (84-90)	140-165 (60-74)
115	4	90 LV 2-st	1997-03	106 (1730)	118-126 (48-52)	183-194 (84-90)	140-165 (60-74)
130	4	90 LV 2-st	1997-03	106 (1730)	118-126 (48-52)	183-194 (84-90)	140-165 (60-74)
140	4	90 LV 2-st	1997-02	106 (1730)	118-126 (48-52)	183-194 (84-90)	140-165 (60-74)
105J	6	90 LV 2-st	1997-00	158 (2596)	118-126 (48-52)	183-194 (84-90)	140-165 (60-74)
150 Carb	6	90 LV 2-st	1997-03	158 (2596)	118-126 (48-52)	183-194 (84-90)	140-165 (60-74)
150 EFI	6	90 LV 2-st	1999-03	158 (2596)	118-126 (48-52)	183-194 (84-90)	140-165 (60-74)
175	6	90 LV 2-st	1997-00	158 (2596)	118-126 (48-52)	183-194 (84-90)	140-165 (60-74)

COOLING SYSTEM SPECIFICATIONS - 2-Stroke Motors

Model (Hp)	No. of Cyl	Engine Type	Year	Displace cu. in. (cc)	Thermostat Open Temp. Degrees F(C)	Overheat Thermoswitch Degrees F(C) On	Overheat Thermoswitch Degrees F(C) Off
200 Carb	6	90 LV 2-st	1997-99	158 (2596)	118-126(48-52)	183-194(84-90)	140-165(60-74)
200 EFI	6	90 LV 2-st	1999-03	158 (2596)	118-126(48-52)	183-194(84-90)	140-165(60-74)
225	6	90 LV 2-st	1997	158 (2596)	118-126(48-52)	183-194(84-90)	140-165(60-74)
150 HPDI	6	76 LV 2-st	2000-03	158 (2596)	118-126(48-52)	183-194(84-90)	140-165(60-74)
175 HPDI	6	76 LV 2-st	2001-03	158 (2596)	118-126(48-52)	183-194(84-90)	140-165(60-74)
200 HPDI	6	76 LV 2-st	2000-03	158 (2596)	118-126(48-52)	183-194(84-90)	140-165(60-74)
200 EFI	6	76 LV 2-st	1998-03	191 (3130)	118-126(48-52)	183-194(84-90)	140-165(60-74)
225 EFI	6	76 LV 2-st	1997-03	191 (3130)	118-126(48-52)	183-194(84-90)	140-165(60-74)
250 EFI	6	76 LV 2-st	1997-03	191 (3130)	118-126(48-52)	183-194(84-90)	140-165(60-74)
225 HPDI	6	76 LV 2-st	2003	204 (3342)	118-126(48-52)	183-194(84-90)	140-165(60-74)
250 HPDI	6	76 LV 2-st	2003	204 (3342)	118-126(48-52)	183-194(84-90)	140-165(60-74)

BLEED SYSTEM	6-69
INSPECTION	6-69
BREAK-IN PROCEDURES	6-79
CONNECTING RODS	6-73
INSPECTION	6-73
CRANKSHAFT	6-71
INSPECTION	6-71
MEASURING BEARING CLEARANCE USING PLASTIGAGE	6-71
CYLINDER BLOCK	6-75
BLOCK & CYLINDER HEAD WARPAGE	6-77
INSPECTION	6-75
HONING CYLINDER WALLS	6-77
EXHAUST COVER	6-69
INSPECTION	6-69
GENERAL INFORMATION	6-2
ASSEMBLY SEALANTS AND LUBRICANTS	6-3
CLEANLINESS	6-3
IMPORTANT POINTS FOR OVERHAUL	6-2
POWERHEAD COMPONENTS	6-3
REED VALVE SERVICE	6-3
PISTONS	6-74
INSPECTION	6-74
OVERSIZE PISTONS & RINGS	6-75
PISTON RING SIDE CLEARANCE	6-75
RING END-GAP CLEARANCE	6-74
POWERHEAD BREAK-IN	**6-79**
BREAK-IN PROCEDURES	6-79
POWERHEAD MECHANICAL	**6-2**
GENERAL INFORMATION	6-2
2 HP (43 AND 50CC) AND 3 HP MODELS	6-4
4/5 HP (83 AND 103CC) MODELS	6-10
6-48 HP (2-CYLINDER) POWERHEADS	6-15
25-90 HP (3-CYLINDER) POWERHEADS	6-25
V4 AND V6 POWERHEADS	6-36
POWERHEAD REFINISHING	**6-67**
BLEED SYSTEM	6-69
CONNECTING RODS	6-73
CRANKSHAFT	6-71
CYLINDER BLOCK	6-75
EXHAUST COVER	6-69
GENERAL INFORMATION	6-67
PISTONS	6-74
REED BLOCK	6-69
REED BLOCK	6-69
INSPECTION	6-69
SPECIFICATIONS	**6-80**
1-CYLINDER MOTORS	6-80
2-CYLINDER MOTORS	6-83
3-CYLINDER MOTORS	6-89
V4 MOTORS	6-93
V6 MOTORS	6-94
2 HP (43 AND 50CC) AND 3 HP MODELS	6-4
CLEANING & INSPECTION	6-9
DISASSEMBLY & ASSEMBLY	6-5
REMOVAL & INSTALLATION	6-4
4/5 HP (83 AND 103CC) MODELS	6-10
CLEANING & INSPECTION	6-15
DISASSEMBLY & ASSEMBLY	6-11
REMOVAL & INSTALLATION	6-10
6-48 HP (2-CYLINDER) POWERHEADS	6-15
REMOVAL & INSTALLATION	6-15
DISASSEMBLY & ASSEMBLY	6-17
CLEANING & INSPECTION	6-25
25-90 HP (3-CYLINDER) POWERHEADS	6-25
CLEANING & INSPECTION	6-36
DISASSEMBLY & ASSEMBLY	6-26
REMOVAL & INSTALLATION	6-25
V4 AND V6 POWERHEADS	6-36
ASSEMBLY	6-60
CLEANING & INSPECTION	6-67
DISASSEMBLY	6-43
REMOVAL & INSTALLATION	6-36

6

POWERHEAD

POWERHEAD MECHANICAL	**6-2**
POWERHEAD REFINISHING	**6-67**
POWERHEAD BREAK-IN	**6-79**
SPECIFICATIONS	**6-80**

6-2 POWERHEAD

POWERHEAD MECHANICAL

General Information

You can compare the major components of an outboard with the engine and drivetrain of your car or truck. In doing so, the powerhead is the equivalent of the engine and the gearcase is the equivalent of your drivetrain (the transmission/transaxle). The powerhead is the assembly that produces the power necessary to move the vehicle, while the gearcase is the assembly that transmits that power via gears, shafts and a propeller (instead of tires).

Speaking in this manner, the powerhead is the "engine" or "motor" portion of your outboard. It is an assembly of long-life components that are protected through proper maintenance. Lubrication, the use of high-quality oils and proper fuel/oil ratios are the most important ways to preserve powerhead condition. Similarly, proper tune-ups that help maintain proper air/fuel mixture ratios and prevent pinging, knocking or other potentially damaging operating conditions are the next best way to preserve your motor. But, even given the best of conditions, components in a motor begin wearing the first time the motor is started and will continue to do so over the life of the powerhead.

Eventually, all powerheads will require some repair. The particular broken or worn component, plus the age and overall condition of the motor may help dictate whether a small repair or major overhaul is warranted. As much as you can generalize about mechanical work, the complexity of the job will vary with 2 major factors:
- The age of the motor (the older OR less well maintained the motor is) the more difficult the repair
- The larger and more complex the motor, the more difficult the repair.

Again, these are generalizations and, working carefully, a skilled do-it-yourself boater can disassemble and repair a 250 hp, fuel injected V6 powerhead, as well as a seasoned professional. But both DIYers and professionals must know their limits. These days, many professionals will leave portions of machine work (from cylinder block and piston disassembly, cleaning and inspection to honing and assembly up to a machinist). This is not because they are not capable of the task, but because that's what a machinist does day in and day out. A machinist is naturally going to be more experienced with the procedures.

If a complete powerhead overhaul is necessary on your outboard, we recommend that you find a local machine shop that has both an excellent reputation and that specializes in marine work. This is just as important and handy a resource to the professional as a DIYer. If possible, consult with the machine shop before disassembly to make sure you follow procedures or mark components, as they would desire. Some machine shops would prefer to perform the disassembly themselves. In these cases, you can usually remove the powerhead from the gearcase and deliver the entire unit to the shop for disassembly, inspection, machining and assembly.

If you decide to perform the entire overhaul yourself, proceed slowly, taking care to follow instructions closely. Consider using a digital camera (if available) to help document component mounting during the removal and disassembly procedures. This can be especially helpful if the overhaul or rebuild is going to take place over an extended amount of time. If this is your first overhaul, don't even THINK about trying to get it done in one weekend, YOU WON'T. It is better to proceed slowly, asking for help when necessary from your trusted parts counterman or a tech with experience on these motors.

Keep in mind that anytime pistons, rings and bearings have been replaced, the powerhead must be broken-in again, as if it were a brand-new motor. Once a major overhaul is completed, refer to the section on Powerhead Break-In for details on how to ensure the rings set properly without damage or scoring to the new cylinder wall or the piston surfaces. Careful break-in of a properly overhauled motor will ensure many years of service for the trusty powerhead.

IMPORTANT POINTS FOR OVERHAUL

Repair Procedures

Service and repair procedures will vary slightly between individual models, but the basic instructions are quite similar. Special tools may be called out in certain instances. These tools may be purchased from the local marine dealer.

✱✱ CAUTION

When removing the powerhead, always secure the gearcase in a suitable holding fixture to prevent injury or damage if the powerhead releases from the gearcase suddenly (which often occurs on outboards that have been in service for some time or that are extensively corroded).

On most of the single cylinder motors, and a FEW of the twins, the powerhead is actually light enough to be lifted by hand. BUT, it is really a better idea to use an engine hoist when lifting the powerhead assembly. This not only makes sure that the powerhead is secured at all times, but helps prevent injuries, which could occur if the powerhead releases suddenly. Also, keep in mind the components such as the driveshaft could be damaged if the powerhead is removed or installed at an angle other than perfectly perpendicular to the gearcase. It is a lot easier to align the powerhead when it is supported, than when you are holding the powerhead and trying to raise it from or lower it into position.

✱✱ WARNING

If powerhead removal is difficult, first check to make sure there are no missed fasteners between the gearcase (midsection/exhaust housing/intermediate housing, as applicable) and the powerhead itself. If none are found, apply suitable penetrating oil, like WD-40® or our new favorite, PB Blaster® to the mating surfaces. Give a few minutes (or a few hours) for the oil to work, then carefully pry the powerhead free while lifting it from the gearcase. Be careful not to damage the mating surfaces by using any sharp-tipped prybars or other by prying on thin/weak gearcase or powerhead bosses/surfaces.

When working on a powerhead, use either a very sturdy workbench or an engine stand to hold the assembly. Keep in mind that larger HP motors utilize powerheads that can weigh several hundred pounds.

Although removal and installation is relatively straightforward on most models, the overhaul procedures can be quite involved. Whatever portion you decide to tackle, always, always, ALWAYS take good notes and tag as many parts/hoses/connections as you can during the removal process. As they are removed, arrange components along the worksurface in the same orientation to each other as they are when installed.

Torque Sequence and Values

◆ See Figures 1 and 2

You've just GOT to love Yamaha. We've worked on a lot of engines from a lot of manufacturers, but Yamaha has to be one of the best for designing serviceable motors. Not only are most of the components laid out logically for service, but Yamaha has gone as far as giving the technician some of the necessary information RIGHT ON the motor itself. Most major powerhead components (usually including Cylinder head covers, cylinder heads, exhaust covers, intake manifolds, thermostat housings, etc) have a torque sequence molded into the cover itself. Take a close look at some of these components and you see small numbers next to each of the retaining bolts. When present, this is a torque sequence that should be followed using multiple passes.

A torque sequence means that when installing that component you tighten bolt No. 1 a couple of turns, then No. 2, then 3, and so on. Once the bolts start to snug, you can proceed one of two ways. If you don't have a torque value, you tighten the bolts securely, a 1/4 turn or so at a time still following that sequence until you feel they are securely fastened. However, if you have do have a torque value, you tighten the bolts just to the point when they start to snug, then set a torque wrench to the proper value and continue to tighten each bolt in 1-2 passes of that sequence until the proper amount of torque is placed on each of the bolts.

To take a moment and praise Yamaha even a step further, they've been kind enough to also mold the torque value on these same components. Now, this is where my praise ends, because you'll have to keep a calculator handy. Someone didn't tell the Yamaha engineers that, at least in the US, our torque wrenches say Foot Pounds (ft. lbs.) and usually the common metric conversion of Newton Meters (Nm). So they must have gone with the measurement they work with the most, Kilogram Meter (KGM). Luckily, it's a

POWERHEAD SYSTEMS 6-3

Fig. 1 Most Yamaha torque sequences are molded onto the component - and many of them follow a spiraling pattern such as this cylinder head cover

Fig. 2 Yamaha was also kind enough to mold the torque value into most powerhead components

simple enough conversion (simple enough with a calculator). To convert a KGM value to ft. lbs. simply multiply the KGM value by 7.233058 (of course 7.2 is probably accurate enough for most torque wrenches). To convert a KGM spec to Nm, simply multiply by 9.80665 (again, 9.8 should be sufficient). For example, the exhaust cover pictured in the accompanying illustration shows a torque value spec of 0.8 KGM, which is 5.7864464 ft. lbs. (5-6 ft. lbs. should be sufficient for a typical torque wrench). The same 0.8 KGM converts to 7.84532 Nm (or about 7-8 Nm).

POWERHEAD COMPONENTS

Service procedures for the carburetors, fuel pumps, starter, and other powerhead components are given in their respective Sections of this technical guide.

REED VALVE SERVICE

◆ See Figures 3, 4 and 5

The reeds on Yamaha two-stroke powerheads are contained in an externally mounted reed valve block, found under the carburetor(s) or throttle bodies. Therefore, the powerhead need not be completely disassembled in order to replace a worn or broken reed. Follow the appropriate portions of the overhaul procedure for access to the reed valves if a repair is suspected or necessary.

CLEANLINESS

◆ See Figure 6

Make a determined effort to keep parts and the work area as clean as possible. Parts must be cleaned and thoroughly inspected before they are assembled, installed, or adjusted. Use proper lubricants, or their equivalent, whenever they are recommended.

ASSEMBLY SEALANTS AND LUBRICANTS

Over the years Yamaha seems to vary or change their recommendations for fastener and component installation, not only from model-to-model but from year-to-year on even the same model. Some items, such as a head gasket, NEVER receive sealant, but others, such as the powerhead-to-intermediate housing gasket is sometimes installed using a gasket making sealant applied to both sides and other times installed dry. Similarly, some bolts are installed dry, some are to be coated with engine oil and others with

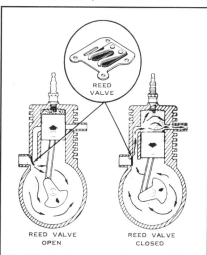

Fig. 3 Reed valves are used to control the flow of air and fuel into the crankcase and eventually into the cylinder

Fig. 4 The reeds are located in an externally mounted housing behind the intake manifold - two-cylinder powerhead shown

Fig. 5 The powerhead need not be disassembled to replace a worn/broken reed - V4 and V6 powerheads shown

6-4 POWERHEAD

Fig. 6 The exterior and interior of the powerhead must be kept clean, well lubricated and properly tuned for the owner to receive maximum enjoyment

Loctite. Some seals should be lubricated with clean engine oil and others with marine grade grease. And lastly, internal engine components are usually coated with engine oil, but many bearing surfaces may also be coated with a molybdenum disulfide engine assembly lube.

We honestly don't believe that coating a seal with grease instead of oil or vice versa is going to cause any problem. Nor do we believe that Loctite on bolt threads instead of oil is ever an issue. When it comes to engine oil vs. engine assembly lube, the assembly lube may have some advantages and can almost always be substituted for oil on bearing surfaces. Pay attention to the instructions and accompanying illustrations. When an illustration calls for engine oil, there is a small symbol of an oil can. When the application calls for grease the illustration will show a small grease gun, however the same grease gun (sometimes with a letter M in the middle) is meant to mean an engine assembly lube when directed towards an internal engine component other than a seal.

2 Hp (43 and 50cc) and 3 Hp Models

REMOVAL & INSTALLATION

◆ See Figures 7 and 8

These powerheads are small enough that you CAN remove them from the intermediate housing without removing most fuel and electrical components. However, Yamaha specifically directs their technicians to remove those components on the 3 hp models and does not on the 2 hp models. It is not clear whether or not there is a good reason for this difference or if it is just because different engineers worked on the literature for the 3 hp motor than worked on the literature for the 2 hp motor.

We advise that you consider why the powerhead is being removed. If you are planning on disassembling the powerhead for inspection or overhaul, then you'll have to remove the fuel and electrical components anyway. If this is the case, removing them before powerhead removal is probably a good idea, if only to protect them from potential damage when lifting and moving the powerhead itself. However, the choice remains yours whether or not you follow all of the initial steps for stripping the powerhead of these components.

1. Remove the cowling/apron from the powerhead for access.
2. If desired, remove the hand-rewind starter assembly and flywheel from the powerhead.
3. If desired, remove the electrical components from the powerhead.
4. If desired, remove the fuel components from the powerhead.
5. If the fuel components were NOT removed, disconnect the throttle cable.
6. From the underside of the powerhead-to-intermediate housing mounting flange, loosen and remove the bolts (usually 6) securing the powerhead.

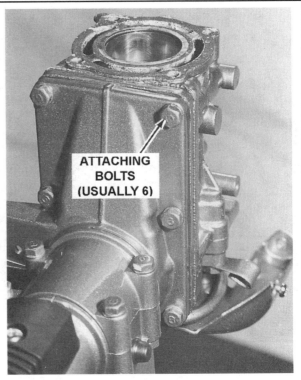

Fig. 7 The powerhead is normally secured using 6 bolts

7. Tap the side of the powerhead with a soft head mallet to shock the gasket seal, and then lift the powerhead from the intermediate housing.

■ On 3 hp motors there is a LONG shaft exhaust manifold installed to the bottom of the powerhead that protrudes through the intermediate housing, be sure to lift the powerhead straight upward until the manifold clears the intermediate housing.

8. Remove all traces of gasket and sealant from the mating surfaces.
9. There is usually a locating dowel pin in the mating surface, remove and take care not to drop or lose the dowel pin(s).

To Install:
10. Make sure any dowel pins used by the powerhead are in position on the intermediate housing.
11. For 2 hp motors, apply a light coating of a Gasket Maker compound to the mating surface on the intermediate housing.
12. Place the new gasket in position over the dowel pin(s) and, on 2 hp motors, the gasket making material.

■ Make sure the gasket is positioned properly and not obscuring any water passage.

13. Apply a light coating of marine grade grease to the driveshaft-to-crankshaft splines
14. Carefully install the powerhead onto the intermediate housing, using the dowel pin(s) to align the two surfaces. If necessary, slowly rotate the flywheel/crankshaft (clockwise) to align the crankshaft and driveshaft splines. Alternately the propeller shaft may be rotated (if the gearcase is in gear), but again, only in the normal direction of rotation to prevent potential damage to the water pump impeller.

■ If there is only 1 dowel pin used on the gasket mating surface, make certain that the gasket is not cocked during installation.

15. Once the powerhead is properly seated (the crankshaft is properly splined to the driveshaft) it is time to install the retaining bolts. Apply a light coating of Loctite® 572 to the threads of the bolts that secure the powerhead to the intermediate housing. Tighten the bolts in two stages starting with the center two bolts then the two on both ends. Tighten the bolts alternately and evenly until secure. On 3 hp motors Yamaha provides a torque specification of 5.8 ft. lbs. (8 Nm), but no spec is listed for 2 hp motors.

POWERHEAD SYSTEMS 6-5

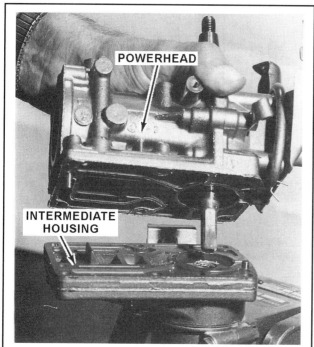

Fig. 8 Carefully lift the powerhead (this one is already stripped) from the housing

gasket.

5. On 3 hp models, there is a dual oil seal assembly or an oil seal and bearing assembly pressed into the driveshaft housing at top of the exhaust manifold. The seal(s) really should be removed and replaced as long as you are here, but at the very least inspect the seal(s) at this time and replace, if there are any signs of wear or damage. If so, remove the seal(s), taking care not to score or damage the sealing surfaces. If equipped with a driveshaft bearing in this portion of the manifold, rotate the bearing feeling for rough spots and replace, as necessary.

6. Using a criss-cross pattern, loosen and remove the bolts securing the intake manifold/reed valve assembly to the powerhead. On 2 hp motors there is a horseshoe shaped bracket over top of the reed valve mounting plate. On 3 hp motors there is a single intake manifold assembly, to the outside of which is bolted the carburetor and to the inside of which mounts the reed assembly.

7. Carefully lift off the intake manifold/reed valve assembly and remove all traces of gasket from the mating surfaces. Note and remember the direction of the reeds as an aid during installation. On some models it is possible to install this housing backwards by mistake.

8. After the reed housing has been removed, inspect the reed valve (check the lift height and, if applicable, warpage). Warpage is the amount that the valve tip lifts when the valve is gently held in the closed position, while lift is the total height of the tip off the valve base when fully released. If the valve is distorted or cracked, a proper seal cannot be obtained. Therefore, the reed valve must be replaced.

9. The reed stop is retained by two screws. If the reed valve must be removed, simply remove the reed stop and lift the valve out of the housing.

10. Remove the bolts holding the two halves of the crankcase together. For 2 hp motors there are only 2 bolts, loosen them gradually, alternating between each bolt after a turn or two. On 3 hp motors, gradually loosen the 6 crankcase bolts using a spiraling pattern that starts at the outer bolts and moves towards the center (reverse of the torque sequence).

■ On 2 hp motors, note the position of the wire clip that is positioned on one of the crankcase bolts as an aid during installation.

11. Carefully separate the two crankcase halves. It may be necessary to shock one half with a soft headed mallet in order to jar the gasket sealing qualities loose. Slowly pull the two halves apart, keeping track of the crankshaft which may remain with the cylinder half, or may remain with the opposite half, pulling the piston from the cylinder as the assembly is separated.

12. Slide the lower (2 hp) and/or upper (all motors) crankcase oil seals and the washer(s) free of the crankshaft.

✱✱ CAUTION

The piston pin lockrings are made of spring steel and may slip out of the pliers or pop out of the groove with considerable force. Therefore, wear eye protection glasses while removing the piston pin lockrings in the next step.

13. Remove the G-lockring from both ends of the piston pin using a pair of needle nose pliers (note that these lockrings are often distorted slightly during removal and it is usually best to replace them). Slide out the piston pin, and then the piston may be separated from the connecting rod. Push out the caged roller bearings from the small end of the connecting rod.

■ On 2 hp motors, the rod and crank pin are normally pressed into the counterweights and can only be separated using a hydraulic press. On 3 hp motors, the bearings are usually seated using a driver, but can normally be removed without a bearing separator, however the separator tool may be used if they are stuck.

Slowly rotate both bearing races. If rough spots are felt, the bearings will have to be pressed free of the crankshaft.

14. If necessary, obtain special bearing separator tool, Yamaha P/N YB-06219 and an arbor press. Press each bearing from the crankshaft. Be sure the crankshaft is supported, because once the bearing breaks loose, the crankshaft is free to fall.

15. Gently spread the top piston ring enough to pry it out and up over the top of the piston. No special tool is required to remove the piston rings. Remove the lower ring in a similar manner. These rings are extremely brittle and have to be handled with care if they are intended for further service.

16. If removed, install the fuel components to the powerhead.
17. If removed, install the electrical components to the powerhead.
18. If removed, install the flywheel, followed by the hand-rewind starter assembly.
19. Reconnect the throttle cable. Refer to the Timing and Synchronization adjustments in the Engine and Maintenance section for adjustment procedures.

■ Be sure to run the motor without the apron or cowling installed in order to check for potential fuel or oil leaks. Remedy any leaks before proceeding.

20. Install the apron or cowling to the powerhead.

■ If the powerhead was rebuilt or replaced with a remanufactured unit, don't forget to follow the proper Break-In procedures.

DISASSEMBLY & ASSEMBLY

◆ See Figures 9 thru 30

■ Refer to the Engine Specifications chart in this section for overhaul and inspection dimensions/limits.

■ The accompanying photos are from the tear-down of a 2 hp motor. Although the 3 hp models are similar, some components differ visually from the counterpart on the 2 hp models.

1. If not done during powerhead removal, strip the powerhead of the hand-rewind starter, flywheel and the fuel and electrical components. For details, please refer to the appropriate sections of this guide.
2. Using the reverse of the torque sequence, gradually loosen and remove the 4 (2 hp) or 6 (3 hp) bolts securing the cylinder head to the block.
3. Remove the cylinder head and the gasket. The head may have to be tapped lightly with a soft mallet to shock the gasket seal free of the powerhead. Remove the gasket material from the mating surfaces of the head and the block.
4. For 3 hp motors, gradually loosen and remove the exhaust manifold retaining bolts using a spiraling pattern that starts at the outside bolts and works inward (the reverse of the torque sequence). Separate the exhaust manifold from the bottom of the powerhead, then remove and discard the

6-6 POWERHEAD

Fig. 9 Loosen and remove the cylinder head bolts...

Fig. 10 ...then remove the cylinder head and gasket

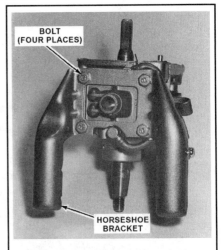

Fig. 11 Loosen and remove the reed valve assembly retaining bolts...

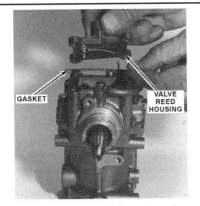

Fig. 12 ...then remove the housing and gasket from the crankcase

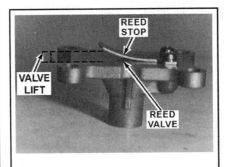

Fig. 13 Inspect the reed valve lift height and warpage

Fig. 14 The reed valve is bolted to the assembly

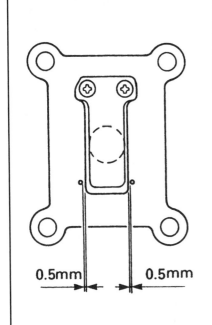

Fig. 15 If removed, make sure the valve is centered upon installation

Fig. 16 Loosen and remove the crankcase bolts...

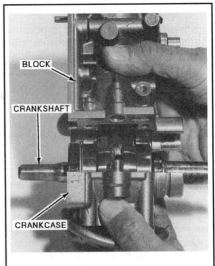

Fig. 17 ...then carefully separate the crankcase halves

POWERHEAD SYSTEMS 6-7

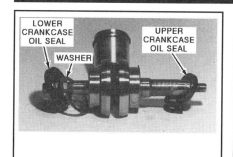

Fig. 18 Remove the crankshaft seals

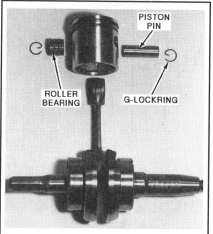

Fig. 19 Remove the lockring and wrist pin to separate the piston

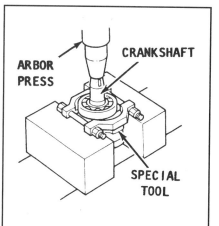

Fig.20 On some models you'll need a bearing separator to remove the crankshaft bearings

Fig. 21 Carefully remove the rings from the piston

To Assemble:

16. Install a new set of piston rings onto the piston. No special tool is necessary for installation however, take care to spread the ring only enough to clear the top of the piston. The rings are extremely brittle and will snap if spread beyond their limit. Align the ring gap over the locating pin. When equipped with mark on the lower ring should face upward and the sloped surface of the tapered upper (keystone) ring should face upward.

17. If the main bearings were removed, support the crankshaft and press the bearings onto the shaft one at a time. Take note of the bearing size embossed on one side of the bearing. The side with the marking must face away from the crankshaft throw.

✸✸ CAUTION

The piston pin lockrings are made of spring steel and may slip out of the pliers or pop out of the groove with considerable force. Therefore, wear eye protection glasses while installing the piston pin lockrings in the next step.

18. Apply a thin coating of marine grade grease to the inside surface of the upper end of the piston rod. Slide the caged roller bearings into the small end of the rod. Move the piston over the rod end with The word **UP** on the piston crown facing toward the tapered (flywheel) end of the crankshaft. Shift the piston to align the holes in the piston with the rod end opening. Slide the piston pin through the piston and connecting rod. Center the pin in the piston. Install a G-ring at each end of the piston pin.

■ The word UP is embossed on the piston crown. This word must face the threaded (flywheel) end of the crankshaft when the piston and rod assembly is installed into the cylinder.

19. If applicable, install the large washer onto the squared end of the crankshaft. Apply a light coating of marine grade grease to the crankshaft seals before installation. If applicable, slide the lower crankshaft oil seal onto the squared end of the crankshaft with the lip facing toward the connecting rod. For all models, slide the upper crankshaft oil seal onto the threaded end of the crankshaft with the lip of the seal facing toward the connecting rod.

20. Coat the piston and rings with engine oil. Check to be sure the ring gaps are centered over the locating pins. Lower the piston into the cylinder bore and seat the crankshaft assembly onto the crankcase. The rings will slide into the cylinder, provided each ring end-gap is centered properly over the locating pin. Rotate the two main bearings until the two indexing pins (if equipped) are recessed into the square notches at the mating surface of the crankcase.

21. Apply a bead of Gasket Maker to the crankcase mating surfaces. Check to be sure the dowel bushings are in place on either side of the crankcase. Press the two halves together. Wipe away any excess gasket making compound. Hold the two halves together and at the same time rotate the crankshaft. If any binding is felt, separate the two halves before the sealing agent has a chance to set. Verify the crankshaft is properly seated and the locating pins are in their recesses. Bring the two halves together as described in the first portion of this step.

22. Coat the threads of the crankcase bolts as follows, depending upon the model: For 2 hp motors coat the threads of the crankcase bolt using Loctite®572 or equivalent threadlocking compound. For 3 hp motors, coat the threads with clean engine oil.

23. Install the crankcase bolts. For 2 hp motors, be sure to install the bolt with the clip on the side of the crankcase which has the hose (and orient the clip as noted during removal).

24. Tighten the crankcase bolts alternately and evenly (2 hp motors) or using the proper torque sequence (in a spiraling pattern starting at the center and working outward for 3 hp motors) to the specified torque. Tighten the bolts first to 3.6 ft. lbs. (5 Nm) and then to 8 ft. lbs. (11 Nm).

■ A torque wrench is essential to correctly assemble the powerhead. Never attempt to assemble the crankcase halves without a torque wrench. Attaching bolts must be tightened to the required torque value in three progressive stages, following the specified tightening sequence.

25. Check to be sure each piston ring has spring tension. This is accomplished by carefully pressing on each ring with a screwdriver extended through the transfer port. Take care not to burr the piston rings while checking for spring tension. If spring tension cannot be felt (the ring fails to return to its original position), the ring was probably broken during the piston and crankshaft installation process. Should this occur, new rings must be installed.

■ If the reed valve stop or reed valve was replaced, a new valve and stop should be positioned with the same orientation as the one which was removed. On some motors this means that the cut on the valve is in the lower corner when the notch in the housing is facing up. Tighten the screws alternately to avoid warping the valve.

6-8 POWERHEAD

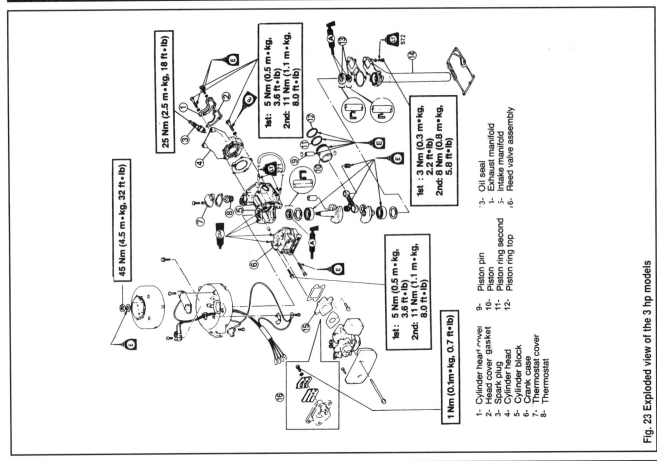

Fig. 23 Exploded view of the 3 hp models

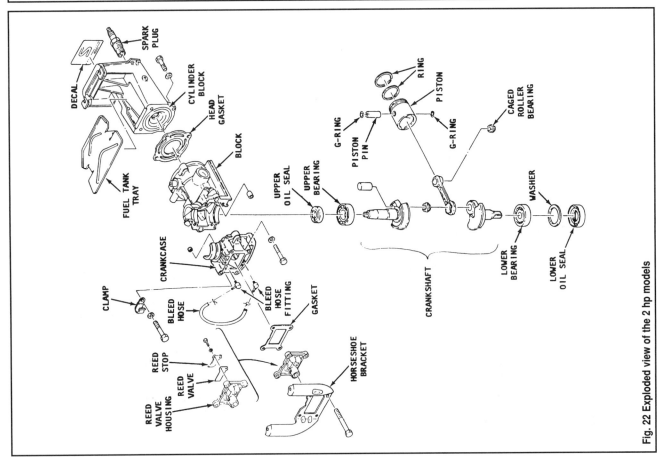

Fig. 22 Exploded view of the 2 hp models

POWERHEAD SYSTEMS 6-9

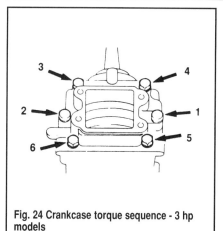

Fig. 24 Crankcase torque sequence - 3 hp models

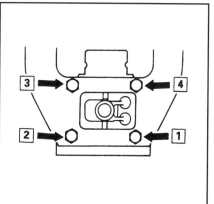

Fig. 25 Reed valve assembly torque sequence - 2 hp models

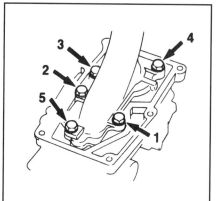

Fig. 26 Exhaust manifold head torque sequence - 3 hp models

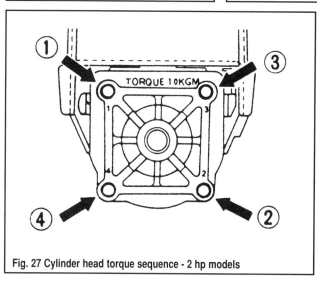

Fig. 27 Cylinder head torque sequence - 2 hp models

26. Assemble the reed valve assembly to the intake manifold. On 2 hp motors, it should be oriented with the notch in the housing facing upward toward the threaded end of the crankshaft.

27. Install the intake manifold (horseshoe bracket on 2 hp motors)/reed valve assembly over to the powerhead using a new gasket. Install the retaining bolts and tighten alternately and evenly. On 2 hp motors, be sure to follow the clockwise torque sequence starting at the lower right corner of the housing. On 2 hp motors, tighten the bolts in 2 stages to the same torque as the crankcase bolts. Tighten the bolts on 3 hp motors securely, but no specific torque is provided.

28. For 3 hp motors, if equipped and removed, apply a light coating of 2-stroke engine oil to the driveshaft bearing and use a driver to carefully install it to the exhaust manifold. If removed, apply a suitable coating of marine grade grease to the lips of the replacement seal(s), then use a suitable driver (smooth edged socket or length of pipe of a suitable diameter) to carefully tap the seal(s) into position.

29. For 3 hp motors, install the exhaust manifold to the bottom of the powerhead using a new gasket. Coat the threads of the exhaust manifold bolts using Loctite®572 or equivalent threadlocking compound. Then, install and tighten the bolts using 2 passes of the torque sequence starting at the inner bolts and working toward the outer bolts. Tighten the bolts first to 2.2 ft. lbs. (3 Nm) and then to 5.8 ft. lbs. (8 Nm).

30. Install a new head gasket without any sealing substance. Position the head in place over the gasket.

31. Coat the threads of the cylinder head bolts as follows, depending upon the model: For 2 hp motors coat the threads of the bolts using Loctite®572 or equivalent threadlocking compound. For 3 hp motors, coat the threads with clean engine oil.

32. Install and tighten the bolts using 2 passes of the crossing torque sequence to the specified torque. First tighten the bolts to 3.6 ft. lbs. (5 Nm) and then to 8 ft. lbs. (11 Nm).

33. Install the Powerhead, as detailed in this section.

34. Once the powerhead is fully assembled, if cylinder block components were replaced, especially the pistons and/or rings, be sure to operate the engine as directed for new component break-in. For details, please refer to Powerhead Break-In, in this section.

CLEANING & INSPECTION

◆ See Figures 22 and 23

Cleaning and inspecting the components is virtually the same for any two-stroke outboard and varies mostly by specifications (which are listed in the Engine Specifications charts) or by component type. A section detailing the proper procedures, sorted mostly by component, can be found under Powerhead Refinishing.

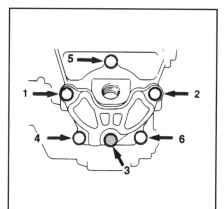

Fig. 28 Cylinder head torque sequence - 3 hp models

Fig. 29 Piston rings are installed with gaps over locating pins

Fig. 30 If equipped, be sure the piston UP mark faces the flywheel

6-10 POWERHEAD

4/5 Hp (83 and 103cc) Models

REMOVAL & INSTALLATION

◆ See Figures 31 and 32

These powerheads are small enough that you CAN remove them from the intermediate housing without removing most fuel and electrical components. Whether or not to strip the powerhead before removal is really your call on these motors. However, we advise that you consider why the powerhead is being removed. If you are planning on disassembling the powerhead for inspection or overhaul, then you'll have to remove the fuel and electrical components anyway. If so, removing them before powerhead removal is probably a good idea, if only to protect them from potential damage when lifting and moving the powerhead itself. However, the choice remains yours whether or not you follow all of the initial steps for stripping the powerhead of these components.

1. Remove the engine top cover from the powerhead for access.
2. If desired, remove the hand-rewind starter assembly and flywheel from the powerhead. If the hand-rewind starter is NOT being removed, remove the clip and disconnect the link for the start-in-gear protection wire.
3. If desired, remove the electrical components from the powerhead. If the electrical components are NOT being removed, disconnect the black (ground) lead from the from the powerhead terminal, then disconnect the white lead and remove the engine stop switch.

■ The ignition bracket on these models usually has one or more bolts threaded upward from underneath the cowling. If so, they must be removed, regardless of whether or not the ignition components are being stripped from the powerhead.

4. If desired, remove the fuel components from the powerhead. If the fuel components were NOT removed, disconnect the throttle linkage.
5. From the underside of the cowling, remove the seven bolts, four on one side and three on the other side, threaded upward and securing the powerhead to the intermediate housing.

■ Two dowel pins are used to mate the powerhead perfectly to the intermediate housing. These dowel pins may remain with the powerhead or they may stay with the intermediate housing. Take care when removing the powerhead, to prevent the dowel pins from falling down into the intermediate housing.

6. Grasp the powerhead with both hands, pull upward, and make an attempt to rock the powerhead to break the gasket seal. Lift the powerhead straight upward until the oil seal housing clears the intermediate housing.
7. Remove all traces of gasket and sealant from the mating surfaces.

To Install:

8. Check to be sure the two dowel pins on the intermediate housing are in place forward and aft of the driveshaft area. Coat both sides of the powerhead gasket with Permatex®, or equivalent gasket making sealant material. Position the gasket in place with the dowel pins passing up through the correct holes in the gasket. The dowel pins and gasket making material SHOULD keep the gasket from cocking during assembly, but keep an eye on it anyway.
9. Apply a coating of marine grade grease to the lower end of the driveshaft. Install the powerhead onto the intermediate housing using the dowel pins to align the two surfaces. The propeller (if in gear) or crankshaft (if not) may have to be rotated slightly to permit the crankshaft to index with the driveshaft and allow the powerhead to seat properly.
10. Apply a threadlocking compound to the threads of the seven bolts securing the powerhead to the intermediate housing. Tighten the bolts in two stages starting with the center bolts and working toward the outer bolts. Tighten them first to 2.2 ft. lbs. (3 Nm) and then to 5.8 ft. lbs. (8 Nm).
11. If removed, install the fuel components to the powerhead. Reconnect the throttle linkage.
12. If removed, install the electrical components to the powerhead. Install the ignition coil bracket to the powerhead and secure. Don't forget the bolts that thread upward from underneath the cowling. Reconnect the engine stop switch wiring.
13. If removed, install the flywheel, followed by the hand-rewind starter assembly. Reconnect the start-in-gear protection linkage.
14. Refer to the Timing and Synchronization adjustments in the Engine and Maintenance section for adjustment procedures.

■ Be sure to run the motor without the top cover installed in order to check for potential fuel or oil leaks. Remedy any leaks before proceeding.

15. Install the engine top cowling.

■ If the powerhead was rebuilt or replaced with a remanufactured unit, don't forget to follow the proper Break-In procedures.

Fig. 31 Remove the powerhead mounting bolts...

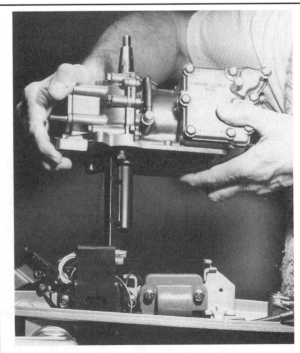

Fig. 32 ... then carefully separate from the housing

POWERHEAD SYSTEMS

DISASSEMBLY & ASSEMBLY

◆ See Figures 33 thru 51

■ Refer to the Engine Specifications chart in this section for overhaul and inspection dimensions/limits.

1. Remove the two bolts and the two nuts, and then lift the reed valve housing clear of the two studs. NOTE and remember the direction of the reeds as an aid to installation. Remove and discard the gasket.

2. Remove the bolts securing the magneto housing to the top of the powerhead assembly. Remove and discard the O-ring.

3. Remove the bolt and washer securing the oil seal housing over the lower end of the crankshaft. Remove and discard the O-ring.

■ The upper oil seal is pressed into the bottom of the magneto housing, while a dual lower seal assembly is pressed into the lower oil seal housing. Both housing are also equipped with an O-ring. There is really no reason to be here and NOT replace these. So carefully remove and discard the seals and O-rings, noting the direction that the seal lips face for installation purposes.

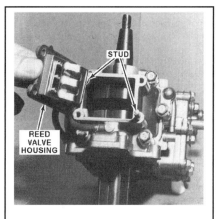

Fig. 33 Remove the reed valve housing

Fig. 34 Remove the magneto and lower (shown) seal housings...

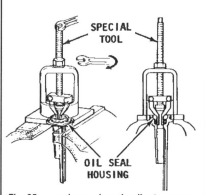

Fig. 35 ...and use a jawed puller to remove the oil seals

Fig. 36 Loosen and remove the cylinder head cover bolts...

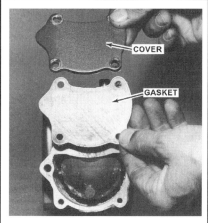

Fig. 37 ...then remove the cover and gasket

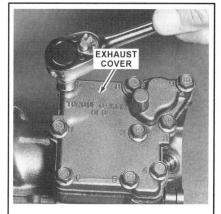

Fig. 38 Loosen and remove the exhaust cover bolts...

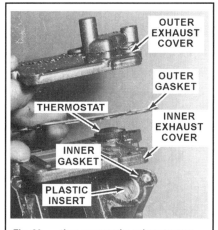

Fig. 39 ...then remove the exhaust cover and thermostat components

Fig. 40 Loosen the crankcase bolts...

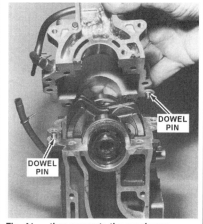

Fig. 41 ...then separate the crankcase halves

POWERHEAD

4. Obtain a bearing/oil seal puller tool with bridge and internal jaws. To remove the upper seal, assemble the jawed puller so the internal jaws grab the seal from underneath the housing and pull out from the same (powerhead) side. To remove the oil seals from the lower housing, secure the oil seal housing in a vise equipped with soft jaws. Insert the expanding jaws of the tool under the edge of the oil seal. Tighten the top nut against the collar of the tool housing, and then rotate the center shaft to raise the jaws and pull the seal free.

5. Remove the bolts (usually 4) securing the cylinder head cover to the block. Loosen the bolts using a criss-cross pattern.

6. Lift off the cylinder head cover. Remove and discard the cover gasket.

7. Remove the bolts (usually 9) securing the exhaust cover to the powerhead. Loosen the bolts using the reverse of the torque sequence either molded onto the cover or illustrated later in this section.

■ Keep close track of the exhaust cover bolts as they are removed. The cover on these motors is usually equipped with 3 different lengths of bolts and they must be installed back in their original locations to ensure proper sealing.

8. Lift off the outer exhaust cover, the outer gasket, the thermostat, the inner exhaust cover, the inner gasket and finally the plastic thermostat insert.

9. Remove the bolts (usually 6) securing both halves of the crankcase together. Loosen the bolts using the reverse of the torque sequence either molded onto the cover or illustrated later in this section.

10. Insert two small pry-tools between the projections, provided for this purpose, on both sides of the crankcase. Pry on both sides at the same time to move the two crankcase halves apart. Take care not to lose the two dowel pins used to align the two halves perfectly.

11. Tap the crankshaft lightly with a soft head mallet to unseat it from the crankcase.

12. Lift out the crankshaft assembly from the cylinder block. The connecting rod is an integral part of the crankshaft assembly and cannot be separated without using an arbor press to disassemble the crankshaft.

13. Slide the washer free of the crankshaft lower portion of the crankshaft.

■ The piston pin G-lockrings are made of spring steel and may slip out of the pliers or pop out of the groove with considerable force. Therefore, WEAR eye protection glasses while removing the piston pin lockrings in the next step.

14. Remove the G-lockrings from both ends of the piston pin using a pair of needle-nose pliers (note that these lockrings are often distorted slightly during removal and it is usually best to replace them). Slide out the piston pin, and then the piston may be separated from the connecting rod. Push the caged roller bearings free of the connecting rod end.

■ The rod and crank pin are pressed into the counterweights and can only be separated using a hydraulic press.

Fig. 42 Carefully unseat the crankshaft and pull upward. . .

Fig. 43 . .withdrawing the piston from the bottom of the block

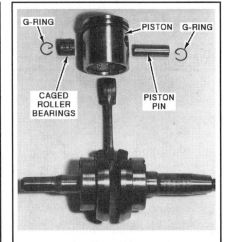

Fig. 44 Removing the piston from the connecting rod

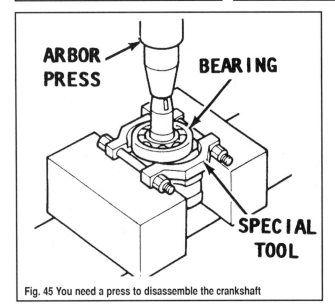

Fig. 45 You need a press to disassemble the crankshaft

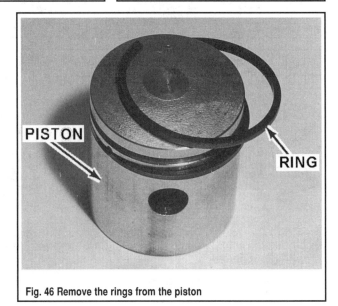

Fig. 46 Remove the rings from the piston

POWERHEAD SYSTEMS 6-13

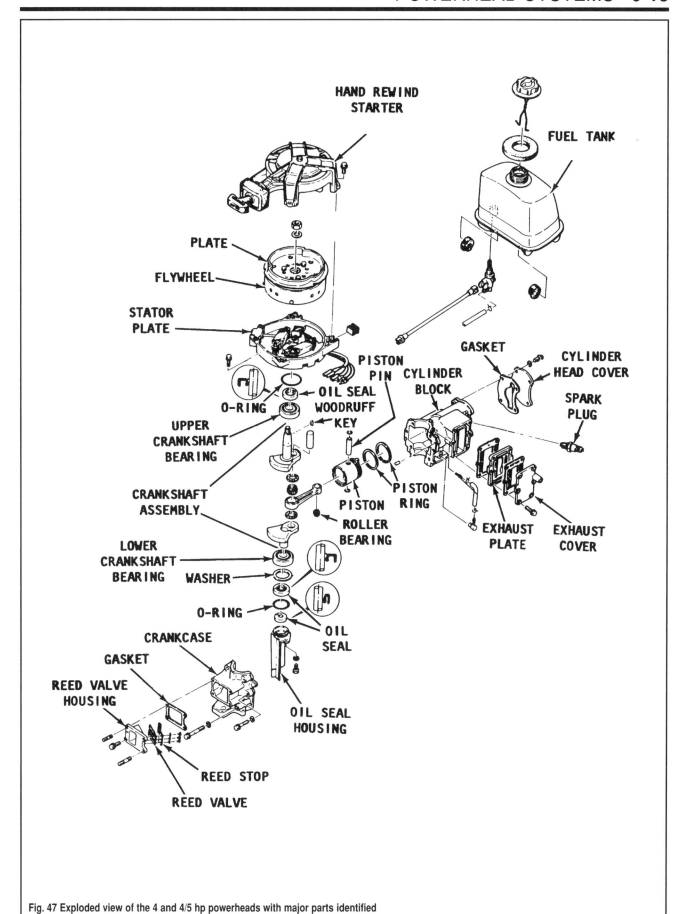

Fig. 47 Exploded view of the 4 and 4/5 hp powerheads with major parts identified

6-14 POWERHEAD

15. Slowly rotate both bearing races. If rough spots are felt, the bearings will have to be pressed free of the crankshaft.

16. Obtain special bearing separator tool (#YB-06219) and an arbor press. Press each bearing from the crankshaft. Be sure the crankshaft is supported, because once the bearing breaks loose, the crankshaft is free to fall.

17. Gently spread the top piston ring enough to pry it out and up over the top of the piston. No special tool is required to remove the piston rings. Remove the lower ring in a similar manner. These rings are extremely brittle and have to be handled with care if they are intended for further service (but once again, you're here right, take the time/expense to install a new set of rings).

To Assemble:

■ Piston rings on these models are normally installed with any markings facing upward. Also, the pistons for these models normally contain a locating pin for one or more of the piston rings. When equipped, be certain to install the piston with the gap straddling the ring.

Fig. 48 After installation, use a screwdriver to check the rings through the exhaust port

18. Install a new set of piston rings onto the piston. No special tool is necessary for installation. However, take care to spread the ring only enough to clear the top of the piston. The rings are extremely brittle and will snap if spread beyond their limit. Align the ring gap over the locating pin.

19. If the main bearings were removed, support the crankshaft and press the bearings onto the shaft one at a time. Take note of the bearing size embossed on one side of the bearing. The side with the marking must face away from the crankshaft throw.

■ The piston pin lockrings are made of spring steel and may slip out of the pliers or pop out of the groove with considerable force. Therefore, wear eye protection glasses while installing the piston pin lockrings in the next step.

20. Apply a thin coating of marine grade grease to the inside surface of the upper end of the piston rod. Slide the caged roller bearings into the small end of the rod. Move the piston over the rod end with The word **UP** on the piston crown facing toward the tapered end of the crankshaft. Shift the piston to align the holes in the piston with the rod end opening. Slide the piston pin through the piston and connecting rod. Center the pin in the piston. Install a G-ring at each end of the piston pin.

21. Install the large washer onto the squared end of the crankshaft.

22. Coat the piston and rings with engine oil. Check to be sure the ring gaps are centered over the locating pins. Lower the piston into the cylinder bore and seat the crankshaft assembly onto the crankcase. The rings will slide into the cylinder, provided each ring end-gap is centered properly over the locating pin. Rotate the two main bearings until the two indexing pins are recessed into the square notches at the mating surface of the crankcase.

23. Check to be sure each piston ring has spring tension. This is accomplished by carefully pressing on each ring with a screwdriver extended through the transfer port. Press gently, taking care not to burr the piston rings while checking for spring tension. If spring tension cannot be felt (the ring fails to return to its original position), the ring was probably broken during the piston and crankshaft installation process. Should this occur, the piston must be removed and new rings must be installed.

24. Lay down a bead of Gasket Maker sealant to the crankcase mating surfaces. Check to be sure the dowel bushings on both halves of the crankcase are in place on either side of the crankcase.

25. Press the two halves together. Wipe away any excess sealant. Hold the two halves together and at the same time rotate the crankshaft. If any binding is felt, separate the two halves before the sealing agent has a chance to set. Verify the crankshaft is properly seated and the locating pins are in their recesses. Bring the two halves together as described in the first portion of this step.

26. If applicable, install the drain hose onto the crankcase and position the metal bracket of the in-gear-protection system against bolt holes No. 1 and No. 5.

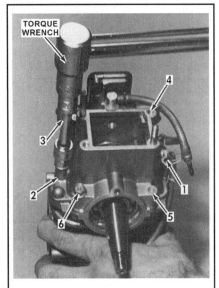

Fig. 49 Crankcase fastener torque sequence

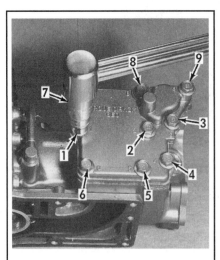

Fig. 50 Exhaust cover fastener torque sequence

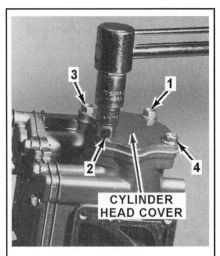

Fig. 51 Cylinder head cover fastener torque sequence

POWERHEAD SYSTEMS

■ **A torque wrench is essential to correctly assemble the powerhead. Never attempt to assemble a powerhead without a torque wrench. Attaching bolts must be tightened to the required torque value in two progressive stages, following the specified tightening sequence. Tighten all bolts to about 1/2 the torque value, then repeat the sequence tightening to the full torque value.**

27. Install the crankcase attaching bolts and tighten using multiple passes of the proper torque sequence first to 4.3 ft. lbs. (6 Nm) and then to 8.7 ft. lbs. (12 Nm).
28. If applicable, slide the plastic thermostat insert into the block. Then install the following components in the order given: first, the inner gasket, then the inner plate, the thermostat, the outer gasket, and finally the outer cover.

■ **Three different length bolts are used to install the exhaust cover. You can identify them using the torque sequence. Bolt No. 9 is 1.38 in. (35mm), bolt No. 3 is 0.63 in. (16mm), and the other seven are 0.98 in. (25mm). Be sure the proper bolts are installed in the proper locations.**

29. Apply a light coating of Loctite® or equivalent threadlocking compound to the threads of the exhaust cover bolts. Install and tighten the bolts using multiple passes of the torque sequence to 2.2 ft. lbs. (3 Nm) and then to 6.5 ft. lbs. (9 Nm).
30. Position a new cylinder head cover gasket in place, and then install the cover. Secure the cover with the four attaching bolts. Tighten the bolts using multiple passes of the torque sequence to 2.2 ft. lbs. (3 Nm) and then to 6.5 ft. lbs. (9 Nm).
31. If the oil seals inside the lower oil seal housing or magneto housing were removed, coat the lips of a new seals with Yamaha all-purpose grease, or equivalent water resistant grease. Press each new seal into the housing using the appropriate mandrel. On the magneto housing the seal lips face downward toward the powerhead, while on the lower oil seal housing both sets of seal lips should face upward toward the powerhead.
32. Install a new O-ring into the outer groove of the oil seal housing and the magneto housing. Coat the outer surfaces of the O-rings with the same grease as for the seal.
33. Install the lower oil seal housing and tap it lightly with a soft head mallet to be sure it is fully seated. Secure the housing in place with the attaching bolt.
34. Install the magneto housing to the top of the powerhead and secure using the retaining bolts.
35. Position the reed valve housing onto the crankcase over the two studs. The reed valve opening must face the starboard side of the block. Install and tighten the two bolts and two nuts to specification.
36. Install the Powerhead, as detailed in this section.
37. Once the powerhead is fully assembled, if cylinder block components were replaced, especially the pistons and/or rings, be sure to operate the engine as directed for new component break-in. For details, please refer to Powerhead Break-In, in this section.

CLEANING & INSPECTION

Cleaning and inspecting the components is virtually the same for any two-stroke outboard and varies mostly by specifications (which are listed in the Engine Specifications charts) or by component type. A section detailing the proper procedures, sorted mostly by component, can be found under Powerhead Refinishing.

6-48 Hp (2-Cylinder) Powerheads

Procedural steps are given to remove and disassemble virtually all items of the powerhead. However, as the work moves along, if certain items (i.e. bearings, bushings, seals, etc.) are found to be fit for further service, simply skip the disassembly steps involved. Proceed with the required tasks to disassemble the necessary components.

REMOVAL & INSTALLATION

◆ See Figures 52 thru 56

These powerheads are small enough that you CAN remove them from the intermediate housing without removing most fuel and electrical components. Whether or not to strip the powerhead before removal is really your call on these motors. However, we advise that you consider why the powerhead is being removed. If you are planning on disassembling the powerhead for inspection or overhaul, then you'll have to remove the fuel and electrical components anyway. If so, removing them before powerhead removal is probably a good idea, if only to protect them from potential damage when lifting and moving the powerhead itself. However, the choice remains yours whether or not you follow all of the initial steps for stripping the powerhead of these components.

1. Remove the engine top cover from the powerhead for access.
2. If desired, remove the hand-rewind starter assembly and flywheel from the powerhead. On most models, if the hand-rewind starter is NOT being removed you'll have to disconnect the cable for the neutral safety starter lockout.

■ **On models with 2 push-pull throttle cables, be sure to tag the cables for easy identification during installation.**

3. Disconnect the shift and/or throttle linkage, as applicable. The linkage varies motor and tiller vs. remote control models, however there is usually a cable or linkage connection which you can undo without disturbing the cable/linkage adjustment. Do your best to leave these settings untouched, as it will give you a good starting point for the adjustments which must be made after installation.

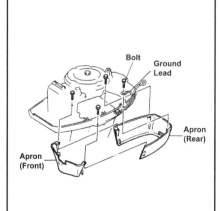

Fig. 52 On some motors, aprons must be removed to access the bolts

Fig. 53 Once accessed, loosen the powerhead mounting bolts...

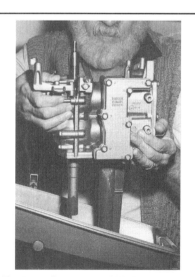

Fig. 54 ... then grasp and lift the powerhead from the intermediate housing

6-16 POWERHEAD

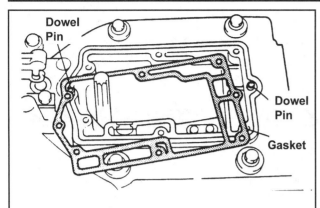

Fig. 55 Although shapes/locations vary, gaskets are held in place by 2 dowel pins

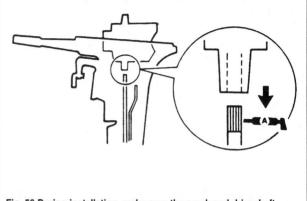

Fig. 56 During installation, make sure the crank and driveshaft splines mesh

■ On some models, like the 9.9/15 hp motors, it may be easier to unbolt and reposition the control pulley bracket assembly, disconnecting the linkage, but leaving the cables attached. Also on the 9.9/15 hp motors, there is a coupler locknut on the shift rod linkage which goes through the lower cowling, be sure to loosen the coupler to disconnect the linkage.

4. If desired, remove the electrical components from the powerhead. If the electrical components are NOT being removed, disconnect the engine stop switch and warning lamp wiring (tiller models) or the remote and battery wiring (remote models), as applicable.
5. If desired, remove the fuel components from the powerhead. If the fuel components were NOT removed, disconnect the fuel line.
6. On 6/8 hp motors, remove the throttle wire stay bracket from the lower front of the powerhead to provide the necessary clearance for powerhead removal.
7. Disconnect the cooling system indicator hose.
8. Tilt the lower unit to the full up position and lock it in place. Remove the bolts (usually 6, but some larger powerheads may have as many as 8 bolts so check carefully) securing the powerhead to the intermediate housing. The powerhead may have to be rotated to gain access to the front bolts. After the bolts have been removed, lower the unit to the full down position.

✳✳ WARNING

If the unit is several years old, or if it has been operated in salt water, or has not had proper maintenance, or shelter, or any number of other factors, then separating the powerhead from the intermediate housing may not be a simple task. An air hammer may be required on the bolts to shake the corrosion loose, heat may have to be applied to the casting to expand it slightly, or other devices employed in order to remove the powerhead. One very serious condition would be the driveshaft frozen with the lower end of the crankshaft. In this case a circular plug type hole must be drilled and a torch used to either free or cut the driveshaft.

9. Lift the powerhead straight up until the exhaust manifold (and oil seal housing, if applicable) is free of the intermediate housing. A piece of wood may be inserted between the powerhead and the intermediate housing as a means of using leverage to force them apart. Let's assume the powerhead will come free on the first attempt.

■ Take care not to lose the two dowel pins. The pins may come away with the powerhead or they may stay in the intermediate housing. Be especially careful not to drop them into the lower unit.

10. Remove all traces of gasket and, if used, sealant from the mating surfaces.

■ Not all models require sealant on the surfaces of the powerhead base gasket. Take note at this time whether or not sealant was originally used to help determine if it should be used during installation.

To Install:
11. Check to be sure the two dowel pins on the intermediate housing are in place.

■ Whether or not the powerhead base gasket is coated with sealant during installation seems to vary greatly by model. For the most part it appears the smallest (6/8 hp) and largest (40, 48 hp) motors tend to sealant, while most of the mid-range motors do not. However, your best indicator is whether or not sealant was present during removal. Also, keep in mind that there is normally no harm in the use of sealant on these motors, even if not present previously. In all cases, make sure the gasket mating surfaces are clean and free of gasket material and any damage, nicks or deep scratches.

12. If sealant was used on the old gasket, coat both sides of a NEW powerhead gasket using Permatex®, or equivalent gasket making sealant material.
13. Position the gasket in place with the dowel pins passing up through the correct holes in the gasket. The dowel pins and gasket making material SHOULD keep the gasket from cocking during assembly, but keep an eye on it anyway.
14. Apply a coating of marine grade grease to the lower end of the driveshaft. Install the powerhead onto the intermediate housing using the dowel pins to align the two surfaces. The propeller (if in gear) or crankshaft (if not) may have to be rotated slightly to permit the crankshaft to index with the driveshaft and allow the powerhead to seat properly.

■ You shouldn't have to rotate the crankshaft or driveshaft very far to mesh the splines, however, if you do turn it more than a hair, to prevent potential water pump impeller damage, always rotate the crankshaft or driveshaft in the normal direction of rotation. The crankshaft/driveshaft normally rotates CLOCKWISE when viewed from above.

15. Apply a light coating of Loctite®572, or an equivalent threadlocking compound to the threads of the powerhead mounting bolts, then carefully thread the bolt from underneath the cowling. Tighten the bolts securely starting with the center bolts and working toward the outer bolts.
16. Connect the cooling system indicator hose.
17. On 6/8 hp motors, install the throttle wire stay bracket to the lower front of the powerhead.
18. If removed, install the fuel components to the powerhead. Reconnect the fuel line.
19. If removed, install the electrical components to the powerhead. Reconnect the engine stop switch and warning lamp wiring (tiller models) or the remote and battery wiring (remote models), as applicable.
20. Reconnect the shift and/or throttle linkage, as applicable.

■ When installing the various link rods, check to be sure the adjustable end is fully snapped onto the ball joints on the levers.

21. If removed, install the flywheel, followed by the hand-rewind starter assembly. Reconnect the start-in-gear protection linkage.
22. Refer to the Timing and Synchronization adjustments in the Engine and Maintenance section for adjustment procedures.

POWERHEAD SYSTEMS 6-17

■ Be sure to run the motor without the top cover installed in order to check for potential fuel or oil leaks. Remedy any leaks before proceeding.

23. Install the engine top cowling.

■ If the powerhead was rebuilt or replaced with a remanufactured unit, don't forget to follow the proper Break-In procedures.

DISASSEMBLY & ASSEMBLY

♦ See Figures 57 thru 84

Remember, when loosening the retainers on manifolds, covers and other major components, always try to follow the reverse of the torque sequence indicated in the assembly procedure (later in this section) or molded on the component itself. You'll notice that the torque sequence on many Yamaha components is a clockwise spiraling pattern that starts and the inner bolts and works towards the outer bolts. When in doubt, use this pattern and you'll be fine.

■ Most of the photos included in this section are from the teardown of a 6/8 hp motor, so expect the shapes of some components to vary slightly on other motors. However, regardless of minor differences in component shape, most of the disassembly and overhaul procedures are similar across Yamaha 2-cylinder motors. Differences, when significant, are mentioned in the steps of the procedure.

■ Because of the high temperatures and pressures developed, the sealing surfaces of the cylinder head and the block are the most prone to water leaks. No sealing agent is recommended because it is almost impossible to apply an even coat of sealer. An even coat would be essential to ensure an air/water tight seal.

Some head gaskets are supplied with a tacky coating on both surfaces applied at the time of manufacture. This tacky substance will provide an even coating all around. Therefore, no further sealing agent is required.

However, if a slight water leak should be noticed following completed assembly work and powerhead start up, do not attempt to stop the leak by tightening the head bolts beyond the recommended torque value. Such action will only aggravate the problem and most likely distort the head.

Furthermore, tightening the bolts, which are case hardened aluminum, may force the bolt beyond its elastic limit and cause the bolt to fracture. bad news, very bad news indeed. A fractured bolt must usually be drilled out and the hole re-tapped to accommodate an oversize bolt, etc. Avoid such a situation.

Probable causes and remedies of a new head gasket leaking are:
• Sealing surfaces not thoroughly cleaned of old gasket material. Disassemble and remove all traces of old gasket.
• Damage to the machined surface of the head or the block. The remedy for this damage is the same as for the next case.
• Permanently distorted head or block. Spray a light even coat of any type metallic spray paint on both sides of a new head gasket. Use only metallic paint - any color will do. Regular spray paint does not have the particle content required to provide the extra sealing properties this procedure requires.

Assemble the block and head with the gasket while the paint is till tacky. Install the head bolts and tighten in the recommended sequence and to the proper torque value and no more!

Allow the paint to set for at least 24 hours before starting the powerhead.

Consider this procedure as a temporary band aid type solution until a new head may be purchased or other permanent measures can be performed.

Under normal circumstances, if procedures have been followed to the letter, the head gasket will not leak.

■ Refer to the Engine Specifications chart in this section for overhaul and inspection dimensions/limits.

Fig. 57 If equipped, remove the lower external oil seal housing...

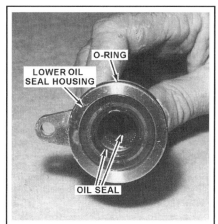

Fig. 58 ...then inspect the condition of the seal(s)

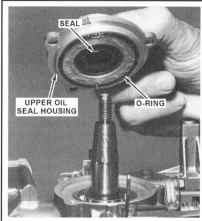

Fig. 59 Some models have an external upper oil seal housing

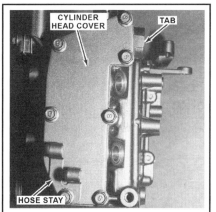

Fig. 60 Remove the cylinder head cover bolts...

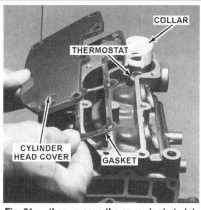

Fig. 61 ...then remove the cover (note t-stat under cover on 6/8 hp motors)

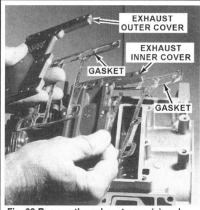

Fig. 62 Remove the exhaust cover(s) and gasket(s)

6-18 POWERHEAD

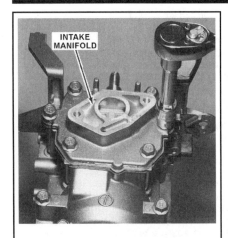

Fig. 63 Remove the intake manifold bolts...

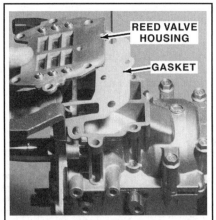

Fig. 64 ... then remove the manifold and reed valve assembly

Fig. 65 Loosen the crankcase bolts in the reverse of the torque sequence

Fig. 66 Remove the crankshaft and piston assembly

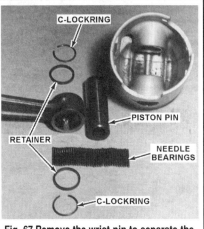

Fig. 67 Remove the wrist pin to separate the piston from the con rod

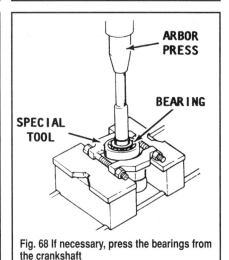

Fig. 68 If necessary, press the bearings from the crankshaft

24. Remove the powerhead from the intermediate housing and strip the powerhead removing the hand-rewind starter, flywheel and all fuel/electrical components. For details, please refer to the appropriate sections of this guide.

■ Most of these motors utilize upper and lower seal housings which cannot be accessed until the crankcase is split open. However, there are a few exceptions, such as the 6/8 hp motors.

25. On 6/8 hp and 9.9/15 hp motors (or other models, if equipped), remove the bolt securing the lower oil seal housing. Tap the housing lightly with a soft head mallet to jar it loose. Inspect the condition of the oil seals in the lower oil seal housing. Make a determination if they are fit for further service. If they are not, remove the oil seals from the lower oil seal housing, noting the direction the lips face (usually downward, away from the crankshaft) for installation purposes. Remove and discard the housing O-ring.

26. Most motors have a cylindrical exhaust manifold bolted to the powerhead (usually using about 3 or 4 bolts). Remove the bolts and remove the manifold. It may be necessary to tap the manifold lightly with a soft head mallet to jar it loose. Remove and discard the gasket.

■ On some motors, the upper oil seal housing is integral with the stator plate.

27. On 6/8 hp motors (or other models, if equipped), unbolt and remove the upper oil seal housing from the top of the powerhead and crankshaft. Inspect the condition of the oil seal in the housing. Make a determination if it is fit for further service. If not, remove the oil seal from the housing, noting the direction the lips face (usually downward, toward the crankshaft) for installation purposes. Remove and discard the housing O-ring.

■ We say to make a determination if seals are fit for further service, because there are times a powerhead which has only been in service a short time is disassembled. However, if you're rebuilding/repairing a powerhead that has been in service for any length of time, it is usually best to just go ahead and replace these seals. You've taken the trouble to come this far, why start scrimping now?

28. On 9.9 hp and larger models, remove the attaching bolts, and then the thermostat cover and thermostat components. If the cover is stuck fast, tap it lightly with a soft head mallet to jar it free. Note the direction the thermostat faces as an aid during installation. Remove and discard the cover gasket.

29. Remove the cylinder head cover bolts (using the reverse of the torque sequence), then remove the cover and discard the gasket. If the cover is stuck to the cylinder head, insert a small pry-tool between the tabs provided for this purpose and pry the two surfaces apart. Never pry at gasket sealing surfaces. Such action would very likely damage the sealing surface of an aluminum powerhead.

30. On 6/8 hp models, remove thermostat assembly from beneath the cylinder head cover.

■ The exhaust cover should always be removed during a powerhead overhaul. Many times water in the powerhead is caused by a leaking exhaust cover gasket or plate.

31. Loosen the exhaust cover bolts using the reverse of the torque sequence. Remove the exhaust cover and gasket. For most models you'll also have to remove an exhaust inner cover and another gasket. Discard all gasket(s). If the cover is stuck to the powerhead, insert a small pry-tool between the tabs provided for this purpose and pry the two surfaces apart. Never pry at gasket sealing surfaces. Such action would very likely damage the sealing surface of an aluminum powerhead.

POWERHEAD SYSTEMS 6-19

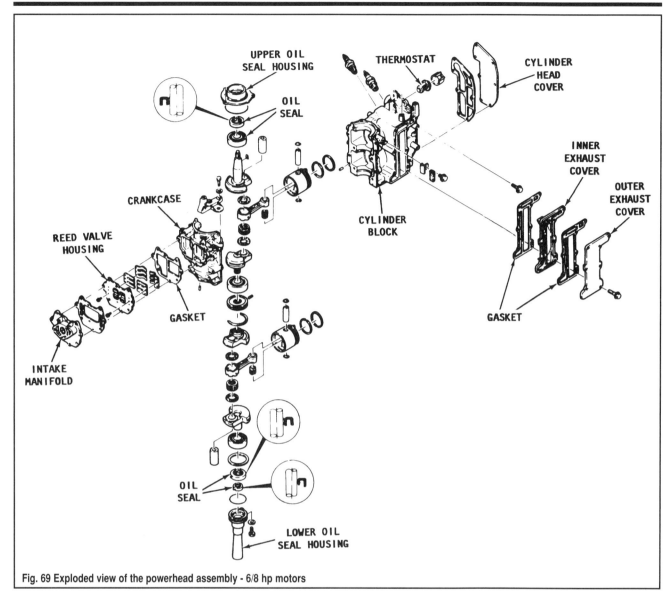

Fig. 69 Exploded view of the powerhead assembly - 6/8 hp motors

■ The 9.9/15 hp models do not have an intake manifold. The carburetor is mounted directly onto the reed valve assembly. Also, although the 6/8 hp motors DO utilize an intake manifold, the bolts are also used to retain the crankcase and should be removed in sequence with the remaining bolts (unless JUST the reed valves are being serviced). If you're separating the gear case halves, wait and remove the reed valve assembly on these motors after the crankcase fasteners have all been loosened using the reverse of the torque sequence.

32. On all but 6/8 hp and 9.9/15 hp models, remove the bolts and then the intake manifold. When removing the bolts - notice their lengths and locations as an aid during assembling. Remove and discard the gasket.

■ Take great care in the next step when handling reed valve assemblies. Once the assembly is removed, keep it away from sunlight, moisture, dust, and dirt. Sunlight can deteriorate valve seat rubber seals. Moisture can easily rust stoppers overnight. Dust and dirt - especially sand or other gritty material can break reed petals if caught between stoppers and reed petals.

33. Remove the reed valve assembly. Discard the gasket under the reed valve assembly.
34. Loosen and remove the crankcase bolts using multiple passes in the reverse of the torque sequence. On many motors these bolts are of various sizes. If so for the motor on which you are working, pay special attention to the locations and sizes of bolts during disassembly.

35. To remove the crankcase, insert a small pry-tool between the tabs provided for this purpose and pry the two surfaces apart. Never pry at gasket sealing surfaces. Such action would very likely damage the sealing surface of an aluminum powerhead.
36. Separate the two halves of the crankcase. Take care not to lose the two dowel pins. These pins may remain in either half when the crankcase is separated.

■ This procedure removes the crankshaft along with the pistons and connecting rods, which is necessary on MOST (but not all) of the 2-cylinder motors since the connecting rods are pressed onto or integral with the crankshaft. However, on the few larger models which utilize connecting rod bolted to the crankshaft, if the pistons are not to be removed for any reason, matchmark and remove the connecting rod caps at this point in order to leave the pistons behind. If the pistons are left behind, temporarily reinstall the cylinder head cover to make sure they do not fall from the cylinder block during service.

37. Tap the tapered end of the crankshaft, with a soft head mallet, to jar it free from the block. Slowly lift the crankshaft assembly straight up and out of the block (along with the pistons on most models).
38. Models that did not have an external upper and/or lower oil seal housing will have one installed to the crankshaft. If so, remove the housing(s) with seal(s) from the crankshaft at this point. On a few models the seal is mounted in a bearing housing, some of which are pressed onto the crankshaft and must be carefully separated with an arbor press.

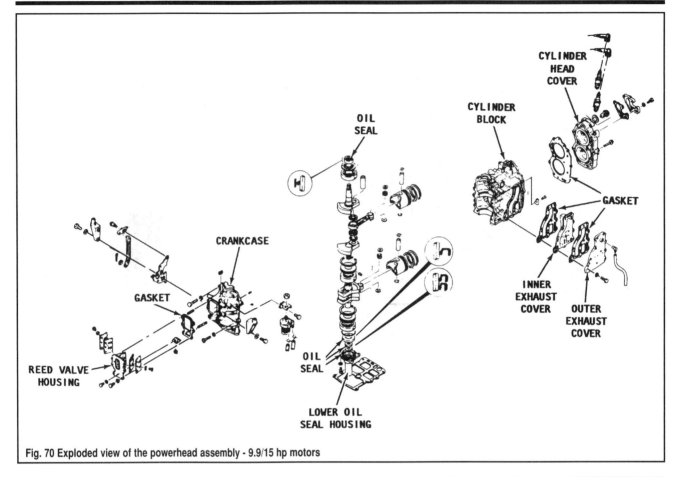

Fig. 70 Exploded view of the powerhead assembly - 9.9/15 hp motors

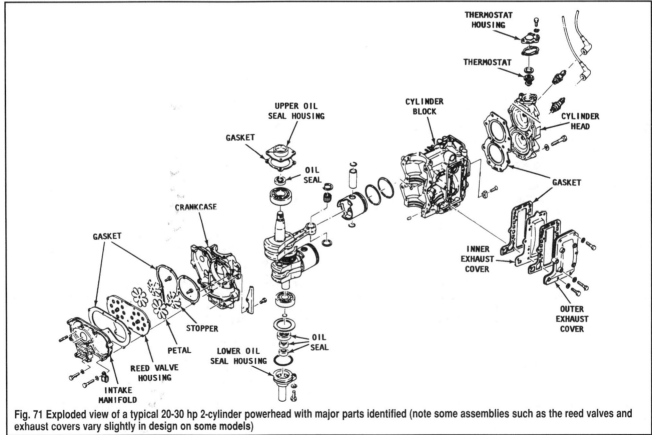

Fig. 71 Exploded view of a typical 20-30 hp 2-cylinder powerhead with major parts identified (note some assemblies such as the reed valves and exhaust covers vary slightly in design on some models)

POWERHEAD SYSTEMS 6-21

A		160 Nm (16 m·kg, 116 ft·lb)
B	1st:	15 Nm (1.5 m·kg, 11 ft·lb)
	2nd:	30 Nm (3.0 m·kg, 22 ft·lb)
C	1st:	4 Nm (0.4 m·kg, 2.9 ft·lb)
	2nd:	8 Nm (0.8 m·kg, 5.8 ft·lb)
D	1st:	4 Nm (0.4 m·kg, 2.9 ft·lb)
	2nd:	8 Nm (0.8 m·kg, 5.8 ft·lb)
E	1st:	17 Nm (1.7 m·kg, 12 ft·lb)
	2nd:	32 Nm (3.2 m·kg, 23 ft·lb)
F		8 Nm (0.8 m·kg, 5.8 ft·lb)
G	1st:	6 Nm (0.6 m·kg, 4.3 ft·lb)
	2nd:	12 Nm (1.2 m·kg, 8.7 ft·lb)
H	1st:	20 Nm (2.0 m·kg, 14 ft·lb)
	2nd:	40 Nm (4.0 m·kg, 29 ft·lb)

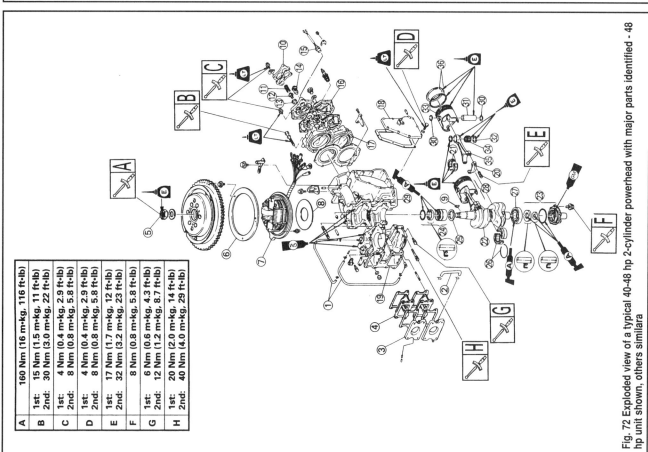

Fig. 72 Exploded view of a typical 40-48 hp 2-cylinder powerhead with major parts identified - 48 hp unit shown, others similar

Extent of removal:
① Thermostat removal
② Crankshaft removal
③ Piston removal
④ Power unit disassembly

Extent of removal	Order	Part name	Q'ty	Remarks
	1	Drainless hose	2	
	2	Intake manifold hose	1	
	3	Intake manifold	2	
	4	Reed valve	2	
	5	Nut	1	
	6	Magneto base retainer	1	Refer to the "DISASSEMBLY POINTS".
	7	Magneto base	1	
	8	Magneto washer	1	
	9	Woodruff key	1	
	10	Thermostat cover	1	Refer to the "DISASSEMBLY POINTS".
	11	Compression spring	1	
	12	Pressure control valve	1	
	13	Grommet	1	Refer to the "DISASSEMBLY POINTS".
	14	Thermostat [for 55B (C55*1)]	1	
	15	Thermo switch [except for 55BM (–)/55BEM (–)]	1	
	16	Cylinder head cover	1	Refer to the "DISASSEMBLY POINTS".
	17	Cylinder head	1	
	18	Exhaust cover	1	
	19	Crankcase	1	
	20	Connecting rod cap	2	
	21	Big end bearing	2	Refer to the "DISASSEMBLY POINTS".
	22	Crankshaft	1	
	23	Oil seal housing	1	
	24	Needle bearing housing	1	
	25	Crank shim	1	
	26	Crankshaft main bearing	1	Refer to the "DISASSEMBLY POINTS".
	27	Crankshaft ball bearing	1	
	28	Piston and connecting rod	2	
	29	Cylinder body	1	
	30	Piston pin clip	4	
	31	Piston pin	2	Refer to the "DISASSEMBLY POINTS".
	32	Washer	4	
	33	Piston	2	
	34	Small end needle bearing	56	
	35	Connecting rod	2	
	36	Piston ring	2 set	Refer to the "DISASSEMBLY POINTS".

Fig. 73 Keylist for the 48 hp powerhead exploded view

6-22 POWERHEAD

Fig. 74 Install the piston ring with the gap over the locating pin

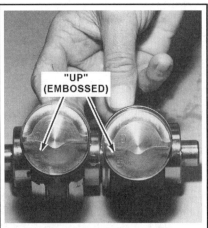

Fig. 75 Make sure the pistons face the proper end of the crankshaft...

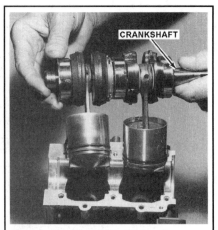

Fig. 76 ..then install the piston and crankshaft assembly

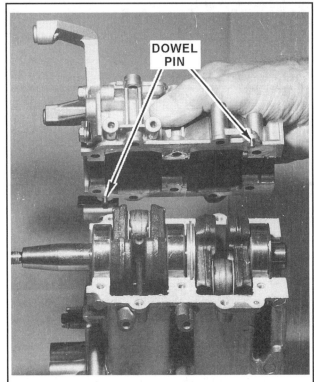

Fig. 77 Make sure the dowel pins are in position and install the crankcase halves

39. Scribe a mark on the inside of the piston skirt or to the piston dome and connecting rod (if being removed) to identify the top and bottom of the piston and rod before removal from the connecting rod as described in the next step.

✱✱ WARNING

The piston pin lockrings are made of spring steel and may slip out of the pliers or pop out of the groove with considerable force. Therefore, wear eye protection glasses while removing the piston pin lockrings in the next step.

40. Remove the lockring from both ends of the piston pin using a pair of needle nose pliers. Discard the lockrings. These rings often stretch during removal and should not be used a second time. Some models utilize free (uncaged) needle bearings and two retainers that will fall away from the connecting rod small end, while others are equipped with a one-piece caged needle bearing assembly.

■ New needle bearings should be installed in the connecting rods, even through they may appear to be in serviceable condition. New bearings will ensure lasting service after the overhaul work is completed. If it is necessary to install the used bearings, keep them separate and identified to ensure they will be installed onto the same crank pin throw and with the same connecting rod from which they were removed.

41. For models whose connecting rods are bolted to the crankshaft, alternately and evenly loosen the connecting rod cap bolts, then remove the caps and separate the rods from the crankshaft. Keep track of the caged bearing halves. Once the assemblies are removed, immediately reassemble the bearings and caps onto their respective rods to ensure proper orientation. Also, remember that the caps and rods should be match-marked to each other and marked to show which side faces the flywheel.

42. Slowly rotate the crankshaft bearing races. If rough spots are felt, the bearings will have to be pressed free of the crankshaft.

43. Obtain special bearing separator tool (#YB-6219) and an arbor press. Press each bearing from the crankshaft. Be sure the crankshaft is supported, because once the bearing breaks loose, the crankshaft is free to fall.

■ Good shop practice dictates to replace the rings during a powerhead overhaul. And we're sure you're growing tired of listening to us say this, but you've come this far, so unless the rings were just replaced a month ago, we'd consider replacing them now. However we'll reluctantly add that if the rings are to be used again, expand them only enough to clear the piston and the grooves because used rings are brittle and break very easily.

44. Gently spread the top piston ring enough to pry it out and up over the top of the piston. No special tool is required to remove the piston rings. Remove the lower ring in a similar manner. These rings are extremely brittle and have to be handled with care if they are intended for further service.

To Assemble:

■ Remember, whenever possible tighten all bolts in using the proper torque sequence and to the proper specification. When possible we've included it in the procedure, but on many Yamahas both the sequence and value is molded into the casting on critical components. You'll notice that most Yamaha torque sequences are clockwise spirals starting somewhere near the center of the component and working outward.

45. Install a NEW set of piston rings onto the piston. No special tool is necessary for installation. However, take care to spread the ring only enough to clear the top of the piston. The rings are extremely brittle and will snap if spread beyond their limit. Align the ring gap over the locating pin.

46. If the main bearings were removed, support the crankshaft and press the bearings onto the shaft one at a time. Press only on the inner race. Pressing on the cage, the ball bearings, or the outer race may destroy the bearing. Take note of the bearing size embossed on one side of the bearing. The side with the marking must face away from the crankshaft throw.

POWERHEAD SYSTEMS 6-23

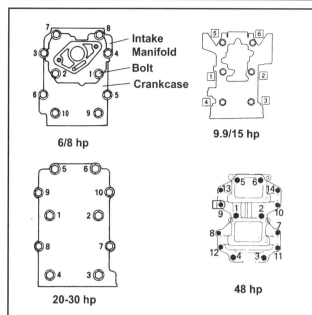

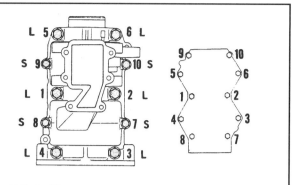

Fig. 78 Typical crankcase torque sequences (note the spiraling patterns) - Remember they are normally embossed on the crankcase)

Fig. 79 Typical Yamaha crankcase torque sequences (they are normally molded onto the cover) - 9.9/15 and 20-30 hp motors shown (note the similar clockwise spiraling pattern starting at the center)

■ The piston pin lockrings are made of spring steel and may slip out of the pliers or pop out of the groove with considerable force. Therefore, wear eye protection glasses while installing the piston pin lockrings in the next step.

47. Select the set of needle bearings removed from the No. 1 piston or obtain a new set of bearings.
48. Coat the inner circumference of the small end of the No. 1 connecting rod with marine grade grease. For models equipped with loose needle bearings, position the needles one by one around the circumference. Dab some lubricant on the sides of the rod and stick the retainers in place. For models equipped with a caged needle assembly, simply insert the pin through the caged assembly.
49. Position the end of the rod with the needle bearings and retainers in place up into the piston. The word **UP** on the piston crown must face toward the tapered (upper) end of the crankshaft. Slide the piston pin through the piston and connecting rod. Center the pin in the piston. Install the lockring at each end of the piston pin.
50. Repeat the piston assembly steps to install the No. 2 needle bearing set and the piston onto the No. 2 connecting rod.

51. If upper or lower seal housings were removed from the crankshaft assembly, apply a light coating of marine grade grease to the lips of the NEW seal(s) and install the housing(s) to the crankshaft at this time.

■ Again, this procedure assumes that the pistons and connecting rods are connected to the crankshaft. On models where the con rods bolt to the shaft you've got a choice. You can install the pistons and crankshaft as an assembly by bolting the connecting rods to the shaft before proceeding, OR you can install the pistons to the bores first, and bolt the connecting rods to the crankshaft only AFTER the crankshaft has been placed in the block. This second method is more popular with most engines on which the connecting rods bolt in place and we recommend you probably want to go with that, that way you break up the piston installation and crankshaft into 2 steps.

52. Coat the upper sides of each piston with 2-stroke engine oil. Hold the crankshaft at right angles to the cylinder bores and slowly lower one piston at a time into the appropriate cylinder. The upper edge of the each cylinder bore has a slight taper to squeeze in the rings around the locating pin and allow the piston to center the bore.

■ If difficulty is experienced in fitting the piston into the cylinder, do not force the piston. Such action might result in a broken piston ring. Raise the crankshaft and make sure the ring end-gap is aligned with the locating pin.

53. Push the crankshaft assembly down until it seats in the block. On some models the crankshaft bearings will have locating tangs which must be properly fit into bores or grooves in the block. Rotate the upper, center, and lower bearings until their locating pins fit into the recesses in the block.

Fig. 80 Check through the exhaust port for broken rings

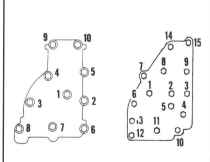

Fig. 81 Typical Yamaha exhaust cover torque sequences (again, they are normally molded onto the cover) - note the spiraling pattern

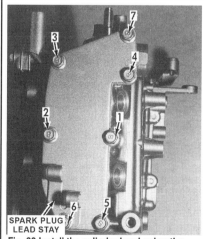

Fig. 82 Install the cylinder head using the correct (molded) torque sequence - 6/8 hp motors shown (see that spiral pattern again?)

6-24 POWERHEAD

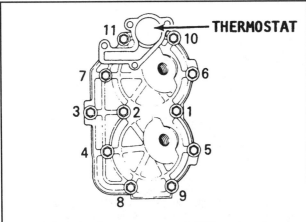

Fig. 83 Cylinder head torque sequence - 9.9/15 hp motors shown

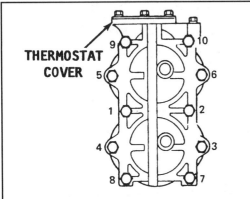

Fig. 84 Cylinder head torque sequence - typical 20 hp or larger motor shown

54. For models on which the connecting rods bolt into place, if the crankshaft was installed after the pistons, apply a light coating of engine oil or engine assembly lube to the connecting rod bearings, then pull the connecting rods up into contact with the crankshaft. Install the connecting rod cap with the appropriate bearing half (making sure all matchmarks are aligned) then tighten the bolts alternately and evenly first to 12 ft. lbs. (17 Nm) and then to 23 ft. lbs. (32 Nm).

55. Apply a thin bead of Permatex® or an equivalent gasket making sealant around both surfaces of the crankcase and block. Check to be sure the two dowel pins are in place and install the crankcase to the block using the dowels to align them.

56. Make sure the crankcase retaining bolt threads are clean and free of dirt or debris. Apply a light coating of Loctite®242 or equivalent threadlock to the bolt threads.

■ On 6/8 hp models the reed valve housing and intake manifold share fasteners with the crankcase, therefore before installing and tightening the crankcase bolts you should install the following parts onto the intake port in the order given: the gasket, the reed valve housing, another gasket, and finally the intake manifold.

57. Install and tighten the crankcase attaching bolts in the proper sequence shown in the accompanying illustration or molded into the crankcase casting. Tighten the bolts in at least two passes to the specified torque molded on the crankcase or, if not, to the value listed here:
- 6/8 hp motors - first to 4.3 ft. lbs. (6 Nm) and then to 8 ft. lbs. (11 Nm).
- 9.9/15 hp motors - first to 11 ft. lbs. (15 Nm) and then to 22 ft. lbs. (30 Nm).
- 20-35 hp motors - M6 bolts first to 3.6 ft. lbs. (5 Nm) and then to 8.0 ft. lbs. (11 Nm). M8 bolts first to 11 ft. lbs. (15 Nm) and then to 20 ft. lbs. (28 Nm).
- 40-48 hp motors - Small bolts (usually M6) first to 4.3 ft. lbs. (6 Nm) and then to 8.7 ft. lbs. (12 Nm). Large bolts (usually M8) first to 14 ft. lbs. (20 Nm) and then to 29 ft. lbs. (40 Nm).

58. Rotate the crankshaft by hand to be sure the crankshaft does not bind.

✳✳ WARNING

If binding is felt, it will be necessary to remove the crankcase and reseat the crankshaft and also to check the positioning of any bearing or seal housing locating pins. If binding is still a problem after the crankcase has been installed a second time, the cause might very well be a broken piston ring.

59. Check to be sure each piston ring has spring tension. This is accomplished by carefully pressing on each ring with a screwdriver extended through the exhaust ports, as shown in the accompanying illustration. If spring tension cannot be felt (the spring fails to return to its original position), the ring was probably broken during the piston and crankshaft installation process. Take care not to burr the piston rings while checking for spring tension.

60. Install the exhaust cover(s) and gasket(s).

■ The manufacturer recommends no sealing agent be applied to any of the gasket sealing surfaces.

61. Install and tighten the exhaust cover attaching bolts in the sequence shown to the specified torque value. For all except the 9.9/15 hp motors, tighten the bolts first to 2.9 ft. lbs. (4 Nm) and then to 5.8 ft. lbs. (8 Nm). For 9.9/15 hp motors, tighten the bolts first to 4.3 ft. lbs. (6 Nm) and then to 8.7 ft. lbs. (12 Nm).

62. On 9.9/15 hp models, install the reed valve assembly to the crankcase and secure using the retaining bolts.

63. For 20 hp and larger, install the intake manifolds and reed valve assemblies using a new gasket (or new gaskets, as applicable).

64. On 6/8 hp models, insert the thermostat components into the cylinder head.

■ The manufacturer recommends no sealing agent whatsoever be used on either side of the gasket. However, the manufacturer recommends that the head bolt threads be coated with clean engine oil 6-15 hp motors or with Loctite or an equivalent threadlocking compound on 20 hp and larger motors.

65. Install the cylinder head using a new gasket. Treat the bolt threads with oil or threadlock (as applicable), then install and tighten the bolts using multiple passes of the torque sequence to the specifications as follows:
- 6/8 hp motors - first to 2.9 ft. lbs. (4 Nm) and then to 5.8 ft. lbs. (8 Nm).
- 9.9/15 hp motors - first to 5.8 ft. lbs. (8 Nm) and then to 12 ft. lbs. (17 Nm).
- 20-35 hp motors - first to 11 ft. lbs. (15 Nm) and then to 20 ft. lbs. (28 Nm).
- 40-48 hp motors - first to 11 ft. lbs. (15 Nm) and then to 22 ft. lbs. (30 Nm).

For all except the 6/8 hp motors (on which you've already installed this), install the thermostat assembly using a new gasket. For details, please refer to the Lubrication and Cooling section.

66. On models equipped with external upper and/or lower seal housings, service the seals (if removed) before installing the housing. Install new oil seals in the housing using a suitable driver. For lower housings, the seals are installed with the lips facing downward (into the housing, away from the crankshaft). For upper housing, seals are installed with the lips facing outward (downward, toward the crankshaft when the housing is installed). Pack the seal lip using a suitable marine grade grease.

■ When two seals are installed, the lips of both seals face in the same direction. Yamaha engineers have concluded it is better to have both seals face the same direction rather than have them back to back as directed by some other outboard manufacturers. In this position, with both lips facing toward the water after oil seal installation, the engineers feel the seals will be more effective in keeping water out of the oil seal housing. Any lubricant lost will be negligible.

67. If the upper and/or lower seal housings are equipped, install and grease a NEW O-ring, then install the seal housing to the powerhead. Secure using the retainer(s).

68. Most motors have a cylindrical exhaust manifold which bolts to the bottom of the powerhead. If equipped, install the manifold using a new gasket and tighten the attaching bolts securely.

POWERHEAD SYSTEMS 6-25

CLEANING & INSPECTION

Cleaning and inspecting the components is virtually the same for any two-stroke outboard and varies mostly by specifications (which are listed in the Engine Specifications charts) or by component type. A section detailing the proper procedures, sorted mostly by component, can be found under Powerhead Refinishing.

25-90 Hp (3-Cylinder) Powerheads

REMOVAL & INSTALLATION

◆ See Figures 85 thru 90

Starting with the 3-cylinder powerheads the Yamaha motors are becoming large enough that stripping the powerhead before removal (for both the weight savings and prevention of damage to the powerhead mounted components of the fuel and electrical systems) is advisable. Although a lifting device is not absolutely necessary, one is still advisable to make for easy removal and to help when positioning the powerhead over the mating surface and crankshaft splines.

Consider why the powerhead is being removed. If you are planning on disassembling the powerhead for inspection or overhaul, then you'll have to remove the fuel and electrical components anyway. If this is the case, removing them before powerhead removal is a good idea, if only to protect them from potential damage when lifting and moving the powerhead itself. As usual, we've pretty much only included the steps in this procedure which are necessary in order to remove the assembled powerhead. Additional steps should be taken if you desire to strip the powerhead of all fuel and electrical components before proceeding. If so, refer to components in the Fuel System and Ignition and Electrical System sections for details.

1. Remove the engine top cover for access to the powerhead.
2. If equipped, remove the hand-rewind starter assembly.
3. Although not absolutely necessary (except to strip the powerhead of components) remove the flywheel for increased access.
4. If not completely stripping the powerhead of electrical components, remove/disconnect the following:
   a. On electric start models, tag and disconnect the battery cables from the powerhead (at the starter and starter solenoid).
   b. For remote control models, disconnect the main engine harness connector at the front of the cowling.
   c. For manual start models, disconnect the engine stop switch wiring.
   d. For tiller control models, disconnect the engine overheat warning lamp wiring.
   e. If equipped with power trim/tilt, disconnect the motor lead and/or switch lead coupler, as applicable. Disconnect the trailer tilt switch connector.
   f. On oil injection models, tag and disconnect the wiring for the oil level sensor and/or oil level warning lamp, as applicable.
   g. Either gently pull the thermo-switch from the cylinder head cover and set it aside or disconnect the switch from the leads, as desired.
5. Tag and disconnect the throttle and shift cables. For most models this involves pulling the retaining clip and carefully separating the cable end from the linkage ball stud.
6. If not completely stripping the powerhead of fuel and oil system components, remove/disconnect the following:
   a. If equipped, disconnect the choke lever rod.
   b. Disconnect and plug the main fuel supply line.
7. Disconnect the cooling system indicator hose from the fitting on either end.
8. Loosen the fasteners and remove the apron from around the lower cowling (for access to the powerhead mounting fasteners). Apron fasteners vary slightly according to model, but generally are as follows:
   - The 25/30 and 28J-50 hp model, usually have two screws securing the apron around the lower cowling. The screws are threaded horizontally through the split apron on one end.
   - The 50-70 hp and 65J-90 hp models, usually have four attaching bolts threaded downward from the inside the cowling.

■ On some models components like the shift and/or throttle control linkage brackets are bolted to both the powerhead AND the intermediate housing. If the powerhead seems stuck but you THINK you've removed all the fasteners, double-check that the fasteners from another component aren't holding it in place.

9. Remove the bolts securing the powerhead to the intermediate housing as follows:
   - The 25/30 hp models have six bolts, three on each side.
   - The 28J-50 hp models have eight bolts, four on each side, but pay attention because 2 are usually of a different size.
   - The 50-70 hp models have eight bolts, three on each side, plus another two bolts at the forward end of the underside of the lower cowling. Like the 28J-50 hp models, pay attention to bolt positioning because 2 are usually shorter than the others.
   - The 65J-90 hp models have eleven bolts, five on each side, plus one at the aft end of the underside of the lower cowling. REALLY pay attention to bolt positioning on these models as 8 are of one length, but 2, found at opposite corners, are of different lengths (from each other as well).

■ The powerhead may be difficult to dislodge from the intermediate housing because of a tight sealing gasket and the fact that some assemblies may have used joining compound (either from the factory or when serviced in the field). Prying up on the powerhead by using a long piece of wood and leverage on the edge of the lower cowling is an acceptable method by the manufacturer. If this method is employed, CARE must be exercised not to damage either the powerhead or the cowling. Once the powerhead has broken free of the intermediate housing, proceed to the next step.

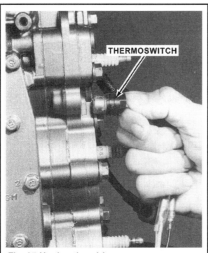

Fig. 85 Unplug the wiring or remove electrical parts. . .

Fig. 86 . . .then tag and disconnect fuel and water hoses

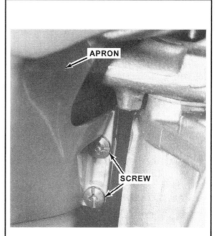

Fig. 87 On smaller motors, the apron is secured using 2 external bolts. . .

6-26 POWERHEAD

Fig. 88 ... but on most models apron fasteners are found inside the cowling

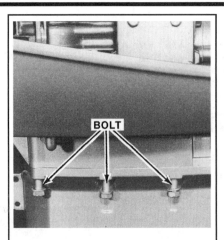

Fig. 89 Loosen and remove the bolts from either side of the powerhead...

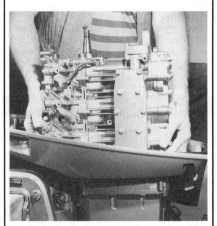

Fig. 90 ...then carefully remove the powerhead

✻✻ WARNING

If the unit is several years old, or if it has been operated in salt water, or has not had proper maintenance, or shelter, or any number of other factors, then separating the powerhead from the intermediate housing may not be a simple task. An air hammer may be required on the bolts to shake the corrosion loose, heat may have to be applied to the casting to expand it slightly, or other devices employed in order to remove the powerhead. One very serious condition would be the driveshaft frozen with the lower end of the crankshaft. In this case a circular plug type hole must be drilled and a torch used to cut the driveshaft.

10. Let's assume the powerhead will come free on the first attempt. Take care not to lose the dowel pins (there are USUALLY 2, but some models may only be equipped with 1 pin). The pin(s) may come away with the powerhead or may stay in the intermediate housing. Be especially careful not to drop a dowel pin into the lower unit.

11. Raise the powerhead free of the intermediate housing. An alternate method, and easier on your back muscles, is to use a hook through the eye provided and a lifting device, such as a chain hoist.

12. Place the powerhead on a suitable work surface. Again, take care not to loose or damage the alignment dowel pin(s).

13. Carefully remove all traces of gasket from the powerhead mating surfaces.

To Install:

■ No mention is made in Yamaha literature regarding the use of sealant on the gasket mating surfaces for these powerhead. However, your BEST indicator is what you found during disassembly. If sealant was used previously, it is probably for a reason (either from the factory, or because the powerhead was disassembled previously, possibly even to correct a leak at this seam), so we'd probably apply a coating to both sides of the replacement gasket. However if no sealant was used previously, and there have been no indications of leaks or other problems with this surface, we'd install the gasket dry.

14. Check to be sure the dowel pin(s) is(are) in place, then position the gasket on the intermediate housing mating surface.

15. Apply a light coating of marine grade grease to the driveshaft splines.

16. Carefully lower the powerhead onto the intermediate housing making sure the gasket remains in position on the dowel pin(s) while making sure to index the crankshaft and driveshaft splines. If the splines do not align, shift the lower unit into forward gear, have an assistant rotate the propeller just a whisker clockwise until the splines do index. Check to be sure the powerhead is fully down on the intermediate housing - the mating surfaces hard against each other.

17. Although not called for by the manufacturer on most models apply a light coating of Loctite® or an equivalent threadlocking compound to the threads of the powerhead retaining bolts. This will not only keep them from loosening in service, but will potentially protect against corrosion that could lock them in place should removal ever be required again.

18. Install the bolts in the positions noted during removal and tighten them securely. No torque specification is provided for the smallest (25/30 hp) 3-cylinder motors, however for the rest of the models these bolts should be tightened to 15 ft. lbs. (21 Nm).

19. Install the apron to the lower cowling and secure using the retaining screws. Be sure to tighten the fasteners securely, but do not overtighten and crack the apron.

20. Reconnect the cooling system indicator hose to the fitting.

21. If the powerhead was stripped of fuel and oil components, install them at this time. Either way, reconnect the fuel supply line and, if equipped, reconnect the choke link rod.

22. If the powerhead was stripped of electrical components, install them at this time. Reconnect all wiring leads as tagged during removal.

23. If removed, install the flywheel.

24. If equipped, install the hand-rewind starter assembly.

25. Reconnect the throttle and shift cables.

26. Refer to the Timing and Synchronization adjustments in the Engine and Maintenance section for cable/linkage adjustment procedures.

■ Be sure to run the motor without the top cowling installed in order to check for potential fuel or oil leaks. Remedy any leaks before proceeding.

27. Install the top cowling to the powerhead.

■ If the powerhead was rebuilt or replaced with a remanufactured unit, don't forget to follow the proper Break-In procedures.

DISASSEMBLY & ASSEMBLY

◆ See Figures 91 thru 121

■ Refer to the Engine Specifications chart in this section for overhaul and inspection dimensions/limits.

■ The accompanying photos are from the tear-down of a typical 3-cylinder motor. Although most models are very similar in design and service, some components will differ visually.

Remember, when loosening the retainers on manifolds, covers and other major components, always try to follow the reverse of the torque sequence indicated in the assembly procedure (later in this section) or molded on the component itself. Similarly, when tightening bolts, on these same components, use the proper torque sequence and tighten the bolts to the proper specification. When possible we've included it in the procedure, but on most Yamahas both the sequence and value is molded into the casting on critical components. You'll notice that most Yamaha torque sequences are clockwise spirals starting somewhere near the center of the component.

Detailed procedures are given to assemble and install virtually all parts of the powerhead. Therefore, if certain parts were not removed or disassembled because the part was found to be fit for further service, simply skip the particular step involved and continue with the required tasks to return the powerhead to operating condition.

■ **Because of the high temperatures and pressures developed, the sealing surfaces of the cylinder head and the block are the most prone to water leaks. No sealing agent is recommended because it is almost impossible to apply an even coat of sealer. An even coat would be essential to ensure an air/water tight seal.**

Some head gaskets are supplied with a tacky coating on both surfaces applied at the time of manufacture. This tacky substance will provide an even coating all around. Therefore, no further sealing agent is required.

However, if a slight water leak should be noticed following completed assembly work and powerhead start up, do not attempt to stop the leak by tightening the head bolts beyond the recommended torque value. Such action will only aggravate the problem and most likely distort the head.

Furthermore, tightening the bolts, which are case hardened aluminum, may force the bolt beyond its elastic limit and cause the bolt to fracture. Bad news, very bad news indeed. A fractured bolt must usually be drilled out and the hole re-tapped to accommodate an oversize bolt, etc. Avoid such a situation.

Probable causes and remedies of a new head gasket leaking are:
• Sealing surfaces not thoroughly cleaned of old gasket material. Disassemble and remove all traces of old gasket.
• Damage to the machined surface of the head or the block. The remedy for this damage is the same as for the next case.
• Permanently distorted head or block. Spray a light even coat of any type metallic spray paint on both sides of a new head gasket. Use only metallic paint - any color will do. Regular spray paint does not have the particle content required to provide the extra sealing properties this procedure requires.

Assemble the block and head with the gasket while the paint is till tacky. Install the head bolts and tighten in the recommended sequence and to the proper torque value and no more!

Allow the paint to set for at least 24 hours before starting the powerhead.

Consider this procedure as a temporary band aid type solution until a new head may be purchased or other permanent measures can be performed.

Under normal circumstances, if procedures have been followed to the letter, the head gasket will not leak.

1. If not done during powerhead removal, strip the powerhead of the hand-rewind starter and/or flywheel cover, the flywheel and the fuel and electrical components. For details, please refer to the appropriate sections of this guide.

■ **Take care in the next step when handling reed valve assemblies. Once the assembly is removed, keep it away from sunlight, moisture, dust, and dirt. Sunlight can deteriorate valve seat rubber seals. Moisture can easily rust stoppers overnight. Dust and dirt - especially sand or other gritty material can break reed petals if caught between stoppers and reed petals.**

Prepare a storage area in which to place the reed valves to keep them isolated from elements (dirt, moisture, direct sunlight, etc) while further work is being performed on the powerhead.

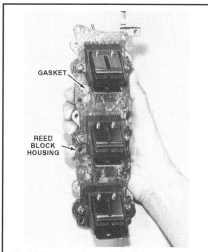

Fig. 91 Remove the reed block housing from the crankcase (intake)

Fig. 92 Remove the outer exhaust cover and gasket. . .

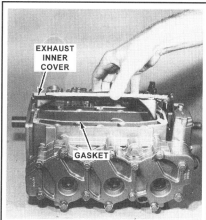

Fig. 93 . . .followed by the inner exhaust cover and gasket

Fig. 94 Remove the cylinder head cover and gasket. . .

Fig. 95 . . . then remove the cylinder head and gasket

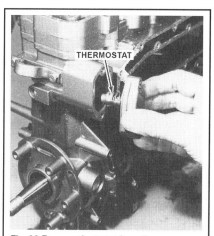

Fig. 96 Remove the t-stat from the powerhead (shown) or cylinder head cover

2. Remove the bolts, and then remove the reed valve housing manifold. Remove and discard the gasket. Set the housing aside, intact as a unit. Further work on the reed valves will be performed later in the Powerhead Refinishing section of this chapter.

■ The exhaust cover should always be removed during a powerhead overhaul. Many times, water in the powerhead is caused by a leaking exhaust cover gasket or plate.

3. Loosen the exhaust cover bolts using multiple passes in the reverse of the torque sequence (this essentially involves a counterclockwise spiraling pattern that starts at the upper right bolt and works back toward the center bolts). Remove the bolts securing the outer cover, and then remove the cover and the gasket.

■ If the inner or outer cover is stuck to the powerhead, insert a small prytool between the tabs that are normally provided for this purpose, and pry the two surfaces apart. Never pry at a gasket sealing surface. Such action would very likely damage the sealing surface of an aluminum powerhead. If no tabs are present, carefully tap around the edges of the cover using a plastic or rubber mallet.

4. Remove the inner cover and gasket. Discard the outer and inner cover gaskets.

■ On all but the 65J-90 hp motors, the cylinder head cover bolts are also used to retain the cylinder head itself. However, on 28J-50 hp motors keep in mind that there is also a separate set of 4 cylinder head cover bolts that run down the center of the cover itself. These must be removed before the cover can be separated from the head itself, but the cover and head can be removed as an assembly once the outer 14 bolts are removed on these models.

5. Remove the cylinder head cover bolts using the reverse of the torque sequence (which essentially comes down to a counterclockwise spiraling pattern that usually starts at the upper right bolt and works towards the center bolts, but there are some exceptions so check the illustrations and any sequences molded on to the cover). If you are separating the cover from the head on 28J-50 hp motors (recommended for overhaul purposes) loosen the 4 center cover bolts starting with the two center bolts and then moving to the top and bottom bolts. Once all of the bolts have been completely loosened, carefully remove the cover. Remove and discard the gasket.

■ If the cover is stuck to the cylinder head, insert a small pry-tool between the tabs normally provided for this purpose and pry the two surfaces apart. Again, never pry at the gasket sealing surfaces, because the sealing surface of an aluminum powerhead could be damaged.

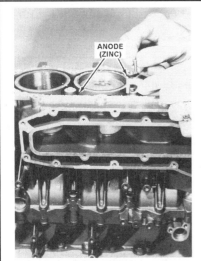

Fig. 97 Some models have anodes in the water jacket...

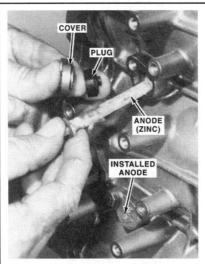

Fig. 98 ...while others mount them under covers in the side of the powerhead

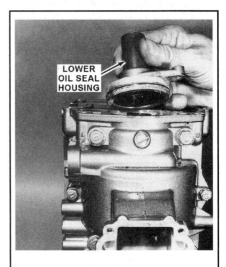

Fig. 99 Remove the lower oil seal housing...

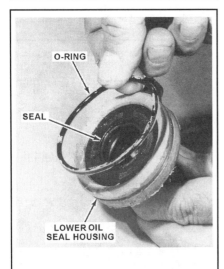

Fig. 100 ...then remove and discard the old O-ring

Fig. 101 Loosen the crankcase retaining bolts and separate the case from the cylinder block

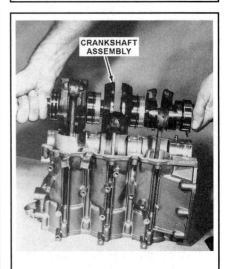

Fig. 102 On most models you remove the crankshaft and pistons as an assembly

POWERHEAD SYSTEMS 6-29

6. The cylinder head on 65J-90 hp models is bolted to the cylinder block using a separate set of bolts from the cylinder head cover. Carefully loosen the head bolts using multiple passes in the reverse of the torque sequence. This results in 2 counterclockwise spirals, first a small spiral on only the inner 4 bolts (starting at the upper right bolt) and then a second, larger spiral that covers all the outer bolts around the perimeter of the head, again starting at the top right bolt and working toward the center bolts.

7. Remove the cylinder head. If the head is stuck to the powerhead block, insert a small pry-tool between the tabs normally provided for this purpose and carefully pry the two surfaces apart. Again, never pry at the gasket sealing surfaces, because the sealing surface of an aluminum powerhead could be damaged.

■ **If the cylinder head is stubborn and cannot be broken free of the block, tap the head lightly with a soft head mallet to jar it free.**

8. If not done already, remove the attaching bolts, and then remove the thermostat cover (it may already be done, because on some models the thermostat cover is bolted to the cylinder head cover and uses the same retaining bolts as the cover). If the cover is stuck to the cylinder head or powerhead, tap it lightly with a soft head mallet to jar it free. Lift out the thermostat and at the same time, note the direction the thermostat faces, as an aid during installation. Remove and discard the gasket.

■ **On 65J-90 hp models there is also a water pressure valve installed beneath the thermostat cover. Keep track of all components when servicing the thermostat, for more details, please refer to the section on Lubrication and Cooling.**

9. Some models, (usually versions of the 28J-50 hp and 65J-90 hp motors) are equipped with a couple of anodes under the cylinder head cover, in the water jacket. If so, loosen the Phillips head screws, and remove the wedge shaped anodes from their mounts in the water jacket.

10. Other models (usually versions of the 25/30 hp and 50-70 hp motors) are equipped with 2 long anodes inserted into the starboard side of the cylinder block. If applicable, remove the bolts, covers and rubber plugs to gain access to these anodes.

11. Remove the bolts securing the lower oil seal housing. Some models have only one bolt, other models have more than one. Tap the housing lightly with a soft head mallet to jar it loose.

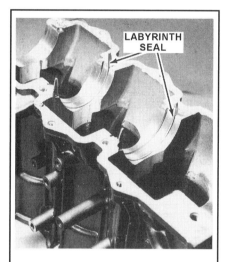

Fig. 103 Some models use labyrinth seal circlips positioned in the crankcase/block

Fig. 104 Mark the pistons before removing them from the connecting rods

Fig. 105 The connecting rods on 65J-90 hp models attach to the crankshaft with caps

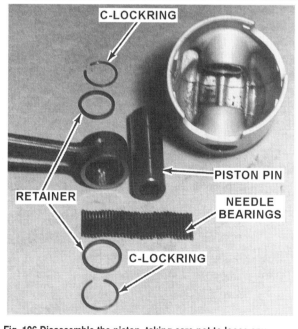

Fig. 106 Disassemble the piston, taking care not to loose any needle bearings

Fig. 107 Crankshaft bearings should be removed only if damaged

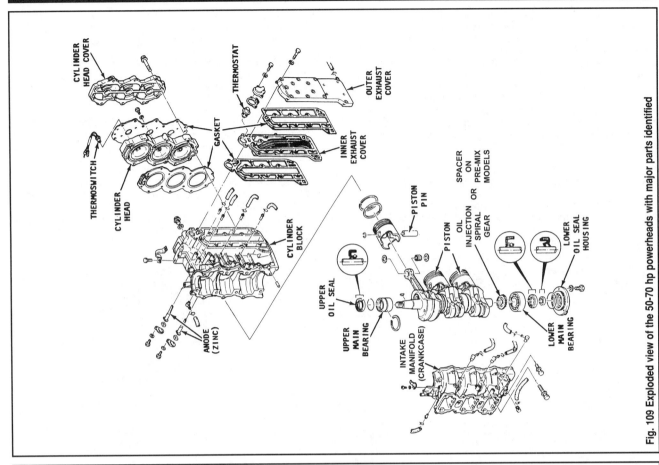

Fig. 109 Exploded view of the 50-70 hp powerheads with major parts identified

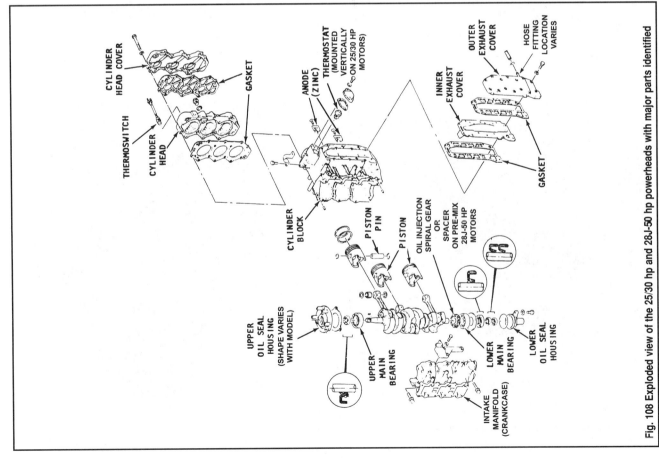

Fig. 108 Exploded view of the 25/30 hp and 28J-50 hp powerheads with major parts identified

POWERHEAD SYSTEMS 6-31

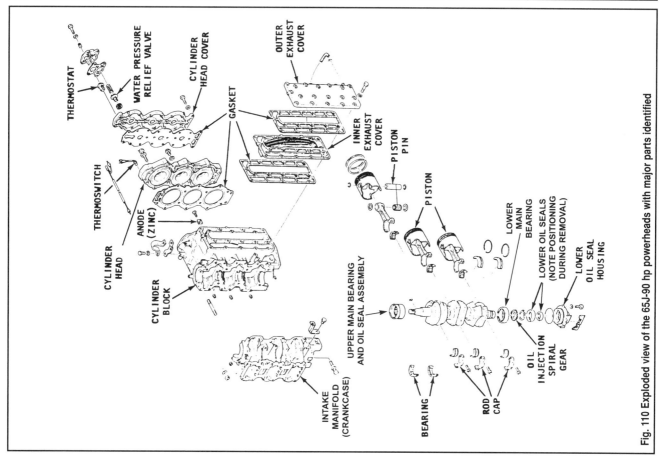

Fig. 110 Exploded view of the 65J-90 hp powerheads with major parts identified

12. Inspect the condition of the oil seals in the lower oil seal housing. Make a determination if they are fit for further service. If they are not, first note their positioning (specifically in which direction the seal lips are facing, since there is some variance from model-to-model). Then, use a suitable slide hammer an expanding jaw attachment to remove the oil seals from the lower oil seal housing.

■ Normally the 25/30 hp and 28J-50 hp units have one large seal and two identical small seals behind the large seal. Meanwhile, the 50-70 hp and 65J-90 hp units have one large seal and one small seal behind the large seal.

13. Remove and discard the outer O-ring from the lower seal housing.

■ Although most early-model 3-cylinder Yamaha powerheads utilized an external upper crankshaft oil seal, most late-model powerheads utilize a seal or seal and bearing housing installed inside the crankcase. The external type should be removed now, but the internal type cannot be serviced until the crankcase is split.

14. All of these motors utilize crankcase bolts of various sizes. So pay special attention to the locations and sizes of bolts during disassembly. Loosen and remove the crankcase bolts using multiple passes in the reverse of the torque sequence (which is essentially 2 counterclockwise spirals, one for each set of bolts, starting first with the lower left bolt of the inner/longer crankcase bolts working toward the center and then proceeding with the lower left bolt of the outer, flange bolts, again working toward the center).

■ The 25/30 hp, 28J-50 hp, and 50-70 hp models all utilize 14 bolts of two different sizes, while the 65J-90 hp models have 20 bolts of two different sizes.

15. To remove the crankcase, insert a small prytool between the tabs provided for this purpose and pry the two surfaces apart. Never pry at gasket sealing surfaces. Such action would very likely damage the sealing surface of an aluminum powerhead.

16. Separate the two halves of the crankcase. Take care not to lose the dowel pins (there are normally 2). These pins may remain in either half when the crankcase is separated.

■ For most motors, this procedure removes the crankshaft along with the pistons and connecting rods. This is necessary on all but the 65J-90 hp motors, since the connecting rods are pressed onto or integral with the crankshaft. However, the largest 3-cylinder models (65J-90 hp) utilize connecting rods which are bolted to the crankshaft using rod caps and bearings, if the pistons are not to be removed for any reason, matchmark and remove the connecting rod caps at this point in order to leave the pistons behind. If the pistons are left behind, temporarily reinstall the cylinder head cover to make sure they do not fall from the cylinder block during service.

17. On all models except the 65J-90 hp, remove the crankshaft, bearings and pistons as follows:

   a. Tap the tapered end of the crankshaft, with a soft head mallet, to jar it free from the block. Slowly lift the crankshaft assembly straight up and out of the block along with the pistons.

   b. If equipped (in late-model powerheads they are mostly found on 28J-50 hp motors), slide the labyrinth seal circlips from the two grooves in the crankcase.

   c. Scribe a mark on the inside of the piston skirt to identify the top, center, and bottom piston before removal from the connecting rod.

18. On the 65J-90 hp model, proceed as follows:

   a. Obtain a marker and matchmark both halves of each connecting rod to ensure the mating halves will be brought together with the cap aligned and facing the original direction - during assembly and also to ensure each rod will be installed in its original location. Numbers such as 1 and 1, 2 and 2, 3 and 3 would be excellent. Include marks on the piston to ensure they are numbered and you can tell which side is facing up (although Yamaha usually marks the latter on the piston itself).

■ Each rod cap must be kept with its connecting rod to ensure they remain as matched sets and the cap must be installed in its original direction - not 180° out. If new parts are being used, the connecting rod and rod cap must be installed in the same direction from which they were separated when removed from the package.

6-32 POWERHEAD

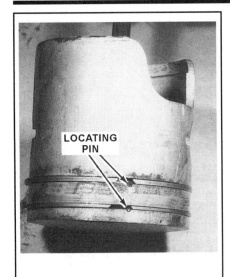

Fig. 111 Carefully install the rings with their gaps over the locating pins

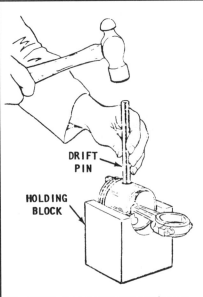

Fig. 112 On some motors you may have to carefully tap the piston pin into place

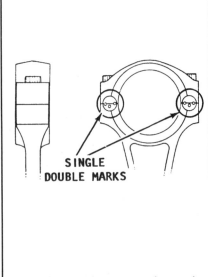

Fig. 113 On 65J-90 hp motors, make sure the cap and connecting rod marks align properly

b. Remove the connecting rod bolts and rod caps. Lift out both sets of caged needle bearings from around the crankshaft journal. Keep the bearings with the connecting rod cap to ensure they will be installed in their original locations.

c. Tap the crankshaft lightly with a soft head mallet to jar it free from the block. Lift the crankshaft out of the block.

■ Most 3-cylinder powerheads utilize an upper seal or seal and bearing housing. On most motors this housing contains an internal oil seal and an external O-ring. The housing may also be secured using a large C-clip. If present, pull the clip straight outward from the side of the housing before attempting to remove the housing itself.

19. If equipped, remove the circlip from the upper oil seal/bearing housing.

20. Remove the upper oil seal housing from the top of the crankshaft. Remove and discard the O-ring around the housing.

21. Inspect the condition of the oil seal in the upper oil seal housing. Make a determination if the seal is fit for further service as the seal would be destroyed during removal. Therefore, remove the seal only if it is damaged and has lost its sealing qualities.

22. If seal replacement is necessary, confirm the seal positioning, on these motors it should be faced with the lips downward, toward the crankshaft, but you never now. Next use a slide hammer puller with expanding jaw attachment to free the seal from the housing.

23. On 65J-90 hp motors, pull each piston and connecting rod assembly from the bottom - not through the top of the block. A ridge might have formed on the top of the cylinder bore. This ridge may have to be removed with a ridge reamer.

■ On 25/30 hp, 28J-50 hp, and 50-70 hp models, the connecting rods are not normally removed from the crankshaft unless it is determined there is a problem in this area. Therefore, the following tasks will proceed with the crankshaft and connecting rods remaining as an assembly.

The connecting rod axial play and side clearance will be determined during cleaning and inspection and Powerhead Refinishing. If excessive clearances are found, then the occasion will arise when the connecting rod journal will be pressed from the crankshaft throw in order to replace the connecting rod.

✱✱ CAUTION

The piston pin C-lockrings are made of spring steel and may slip out of the pliers or pop out of the groove with considerable force. Therefore, wear eye protection glasses while removing the piston pin lockrings in the next step.

24. Remove the C-lockring from both ends of the piston pin using a pair of needle nose pliers. Discard the C-lockrings. These rings often stretch during removal and should not be used a second time. The two bearing retainers will fall away from the connecting rod small end.

25. Although there are a few exceptions, most late-model Yamaha's do NOT use interference fit piston pins. On these models, the pin should pull free of the piston with little or no effort. However, if the pin is stuck or if the motor on which you are working is equipped with an interference fit design pin, proceed as follows:

a. Press the pin free of the piston using an arbor press. Be sure to catch all loose needle bearings as they fall free of the piston pin bore.

b. If an arbor press or holding block is not available, heat the piston with a small torch. Keep the torch moving to prevent overheating in any one area. Heating the piston will cause the metal to expand ever so slightly, but ease the task of driving the pin free. Assume a sitting position in a chair, on a box, whatever.

c. Next lay a couple of towels over your legs. Hold your legs tightly together to form a cradle for the piston above your knees. Set the piston between your legs.

d. Now, drive the piston pin free using a drift pin with a shoulder. The drift pin and the shoulder will ride on the edge of the piston pin. Use sharp hard blows with a hammer. Your legs will absorb the shock without damaging the piston. If this method is used on a regular basis during the busy season, your legs will develop black and blue areas, but no problem, the marks will disappear in a few days.

■ New needle bearings should be installed in the connecting rods, even through they may appear to be in serviceable condition. New bearings will ensure lasting service after the overhaul work is completed. If it is necessary to install the used bearings, keep them separate and identified to ENSURE they will be installed onto the same crank pin throw and with the same connecting rod from which they were removed.

26. On all models except 65J-90 hp, slowly rotate both bearing races on the crankshaft. If rough spots are felt, the bearings will have to be removed (and in most cases, pressed) free of the crankshaft. If necessary, proceed as follows:

a. Obtain special bearing separator tool (#YB6219) and an arbor press. Press each bearing from the crankshaft. Be sure the crankshaft is supported, because once the bearing breaks loose, the crankshaft is free to fall.

■ Good shop practice dictates replacing the rings during a powerhead overhaul. However, if the rings are to be used again, expand them only enough to clear the piston and the grooves, because used rings are brittle and break very easily.

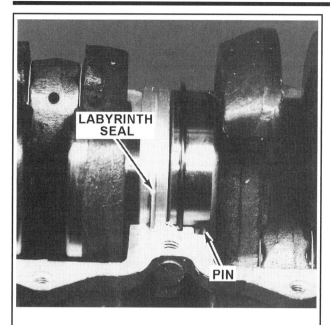

Fig. 114 On models so equipped, install the labyrinth seal circlips

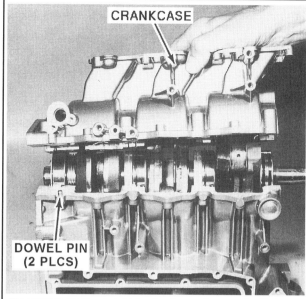

Fig. 115 Use the dowel pin(s) to align the crankcase and cylinder block during installation

27. Gently spread the top piston ring enough to pry it out and up over the top of the piston. No special tool is required to remove the piston rings. Remove the middle ring in a similar manner. These rings are extremely brittle and have to be handled with care if they are intended for further service.

To Assemble:

■ Remember to align all matchmarks made during removal. Components such as the pistons, connecting rods, rod end caps (if equipped) must be positioned with the same sides facing the flywheel end of the crankshaft during assembly.

28. Install a new set of piston rings onto the piston, with the embossed marks (if used) facing up. No special tool is necessary for installation. However, take care to spread the ring only enough to clear the top of the piston. The rings are extremely brittle and will snap if spread beyond their limit. Align the ring gap over the locating pin.

■ Take note of the bearing size embossed on one side of the bearing. The side with the marking must normally face away from the crankshaft throw (unless otherwise instructed with the replacement bearing).

29. On all models except 65J-90 hp, if the main bearings were removed, support the crankshaft and press the bearings onto the shaft one at a time. Press only on the inner race. Pressing on the cage, the ball bearings, or the outer race may destroy the bearing. Hold off on upper bearing installation at this time if the bearing housing also contains the upper oil seal.

✷✷ WARNING

The piston pin lockrings are made of spring steel and may slip out of the pliers or pop out of the groove with considerable force. Therefore, wear eye protection glasses while installing the piston pin lockrings in the next step.

30. Select the set of needle bearings removed from the No. 1 piston or obtain a new set of bearings.
31. Coat the inner circumference of the small end of the No. 1 connecting rod with Yamaha marine grade grease, or equivalent multipurpose water resistant lubricant. Position the needle bearings one by one around the circumference. Dab some lubricant on the sides of the rod and stick the retainers in place.

■ When installing the piston pin, stop pressing if you feel any binding start to occur, which might indicate that a needle has slipped out of position and is being bound by the pin.

32. Position the end of the rod with the needle bearings and retainers in place up into the piston. The word **UP** on the piston crown must face toward the tapered (upper) end of the crankshaft. Insert the piston pin through the piston and connecting rod.

■ Before pressing the piston pin into place, hold the piston and rod assembly near the cylinder block and check to be sure both will be facing in the proper direction when they are installed.

33. If the piston/pin are of an interference fit design:
 a. Carefully press the pin into position in the piston and connecting rod using an arbor press. Center the pin in the piston. Install the C-lockring at each end of the piston pin.
 b. If an arbor press is not available, the piston may be heated and the piston pin chilled, and then the pin may be driven in using a holding block.
 c. The pin may also be installed with the piston on your lap. To make the task easier, with the holding block or in your lap, heat the piston, either in a container of boiling water for about 10 minutes, or for about a minute using a small bottle torch. At the same time, place the piston pin in a cold area, refrigerator, some ice, or cold water. Heating the piston will expand the metal in the piston ever so slightly and chilling the pin will shrink the metal slightly. This exceedingly small amount of change in the metal will ease the task of driving the pin into place. If a bottle torch is used to heat the piston, keep the torch moving to prevent excessive heating of any one area.
 d. If not already done, pack the piston pin needle bearing cage with needle bearing grease. Load the bearing cage with needles and insert it into the end of the rod.
 e. Slide the rod into the piston boss and check a second time to be sure the piston is being installed correctly.
 f. Now, assume a sitting position and lay a couple towels over your lap. Hold your legs tightly together to form a cradle for the piston above your knees. Set the piston between your legs. Drive the piston pin through the piston using a drift pin with a shoulder. The drift pin will fit into the hole through the piston pin and the shoulder will ride on the end of the piston pin. Use sharp hard blows with a hammer. Your legs will absorb the shock without damaging the piston. If this method is used on a regular basis during the busy season, your legs will develop black and blue areas, but no problem, the marks will disappear in a few days.
 g. Continue to drive the piston pin through the piston until the groove in the piston pin for the lockring is visible at both ends. Install a C-lockring onto each end of the piston pin.
34. Repeat the previous step(s), as necessary, to install the No. 2 needle bearing set and the piston onto the No. 2 connecting rod, and then the No. 3 needle bearing set onto the No. 3 connecting rod.
35. If equipped, slide the labyrinth seal circlips into the two grooves of the crankcase.

6-34 POWERHEAD

36. Coat the cylinder bores with a good grade of engine oil before installing the pistons (65J-90 hp) or piston and crankshaft assembly (all other models).
37. Before installing the piston into the cylinder, make the following test.
38. For 65J-90 hp motors, install the pistons, crankshaft and connecting rods as follows:

   a. Run your finger along the top rim of the cylinder. If the surface of the block and the surface of the bore have a sharp edge (meaning no sign of a ridge), the piston may be installed from the top. However, if the slightest groove or ridge is felt on the rim, the piston must be installed from the lower end of the cylinder. Attempting to install the piston from the top would not be wise as the piston ring would likely bottom on the ridge. If force is used the ring may very well break.

   b. For top installation, use a ring compressor to compress the aligned rings and then use the end of a wooden mallet handle to gently tap the piston down into the cylinder bore. The word **UP** embossed on the piston crown (and any marks made during removal for alignment purposes) must face toward the flywheel end of the block.

   c. Installation of the piston from the lower end of the cylinder bore is an easy matter and can be accomplished without the use of special tools. Verify the ring end-gaps are properly aligned and the word up, embossed on the piston crown, faces toward the flywheel end of the block.

   d. Three hands are better than two for this task. Simply compress each ring, one by one and at the same time push the piston up into the cylinder bore.

   e. Repeat either procedure for the other pistons.

   f. After all three pistons have been installed, slide each piston up and down in the cylinder bore several times. Check for binding. Listen for scratching noises. Scratching (or any other spooky noise) may indicate a ring was broken during installation.

■ **If the old roller bearings are to be installed for further service, each must be installed in the same location from which it was removed. Normally a tang on the bearing will slide into the groove in the crankshaft.**

   g. Place the lower half roller bearings in their original positions on the connecting rod caps. The locating pin hole for the first and third bearing outer race/cage should face upward, toward the tapered end of the crankshaft, when installed. Apply a light coating of marine grade grease to each of the bearing assemblies.

   h. For most models, which contain an internal upper oil seal assembly. If the seal was removed, use a suitable driver to install a replacement to the seal or bearing housing. A light coating of clean engine oil or grease will help ease installation. Be sure to face the seal lips in the same direction (usually downward, facing the crankshaft) as noted during removal.

   i. Pack the lip of the upper oil seal with Yamaha marine grade grease, or an equivalent water resistant grease. Install the upper bearing, with the seal installed to the crankshaft.

   j. Position the crankshaft into place in the block.

   k. Place the upper half of the connecting rod roller bearings in their original positions, as recorded during disassembling, on the crankshaft journals. Place the connecting rod caps over the caged roller bearings in their original locations.

■ **On most models the manufacturer recommends against using the rod cap bolts a second time. However that may not be true for all years/models, so if in doubt, check with your local marine parts supplier. Also, the manufacturer recommends a specific method of tightening the rod cap bolts, be followed as outlined in the next step.**

   l. Work on just one rod at a time, thread in the connecting rod cap bolts a few turns each. Now, tighten the cap bolts alternately and evenly, first to 8.7 ft. lbs. (12 Nm) and then finally to 25 ft. lbs. (35 Nm). Check to be sure the single mark on each rod/end cap is centrally located between the two matching embossed marks on the opposite rod/end cap, as shown in the accompanying illustration.

   m. If the marks are aligned, as described and shown, loosen each bolt one half turn, and then tighten the bolt again to the first and then final specified torque values.

   n. If the marks are not aligned, as described and shown, remove the cap bolts, and then the upper caged roller bearing half. Reinstall the upper caged roller bearing half and double-check the alignment. If the second attempt fails to align the marks as described and shown, the crankshaft must be removed (or at least repositioned) and the lower caged roller bearing half checked for correct alignment.

39. On all models except 65J-90 hp, install the crankshaft and piston assembly as follows:

■ **Another set of hands can be really helpful here, not absolutely necessary or anything like that, just really helpful.**

   a. Coat the upper sides of each piston with clean engine oil. Hold the crankshaft at right angles to the cylinder bores and slowly lower one piston at a time into the appropriate cylinder. The upper edge of each cylinder bore has a slight taper to squeeze in the rings around the locating pin and allow the piston to center in the bore.

■ **If difficulty is experienced in fitting the piston into the cylinder, do not force the piston. Such action might result in a broken piston ring. Raise the crankshaft and make sure the ring end-gap is aligned with the locating pin.**

   b. Push the crankshaft assembly down until it seats in the block. If equipped, fit the labyrinth seal circlips into the grooves in the block. Rotate the four bearings until their locating pins fit into the recesses in the block.

40. Some models have a circlip retaining the upper seal/bearing housing, if equipped be sure to install the clip.
41. Check to be sure the dowel pins (there are usually 2) are in place, then apply a thin bead of Yamabond No. 4 Permatex® around both surfaces of the crankcase and block. Carefully install the crankcase to the block, aligning the 2 components using the dowel pin(s).

■ **Yamaha has made various recommendations over the years regarding bolt threads. On some models, they make no mention of lubrication or threadlock. On others the recommend engine oil or Loctite. If there are traces of oil or a threadlocking compound on the old crankcase bolt threads, then you've got your answer. However, there is usually no reason that you could not use Loctite® or an equivalent threadlock to both keep the bolts from loosening in service and to help prevent corrosion of the bolt threads in service.**

42. Install and finger-tighten the crankcase attaching bolts using the proper clockwise, spiraling torque sequence as shown in the accompanying illustrations. The numbering sequence is usually embossed around the matting surface of the crankcase on both sides of the block. Tighten the bolts to the specified torque molded on the crankcase or, if not, to the value listed here:
   - 25/30 hp and 28J-50 hp motors - small bolts first to 3.6 ft. lbs. (5 Nm) and then to 8.0 ft. lbs. (11 Nm). Larger bolts first to 11 ft. lbs. (15 Nm) and then to 20 ft. lbs. (28 Nm).
   - 50-70 hp and 65J-90 hp motors - Small bolts first to 7.2 ft. lbs. (10 Nm) and then to 14 ft. lbs. (20 Nm). Large bolts first to 14 ft. lbs. (20 Nm) and then to 29 ft. lbs. (40 Nm).

43. Rotate the crankshaft by hand to be sure the crankshaft does not bind.

■ **If binding is felt, it will be necessary to remove the crankcase and reseat the crankshaft and also to check the positioning of the labyrinth seal circlips (if used) and/or the bearings, bearing locating pins, as equipped. If binding is still a problem after the crankcase has been installed a second time, the cause might very well be a broken piston ring.**

44. Check to be sure each piston ring has spring tension. This is accomplished by carefully pressing on each ring with a screwdriver extended through the exhaust ports. If you cannot feel spring tension on the ring (the ring fails to return to its original position), the ring was probably broken during the piston and crankshaft installation process.

✴✴ WARNING

Take care not to burr the piston rings while checking for spring tension.

45. If the oil seals in the lower oil seal housing were removed during disassembling, install new oil seals in the housing using a handle and the appropriate driver. Seal installation on these models will vary, so always install the new seals with the lips facing the same directions as noted during removal. Secure the lower oil seal housing in a vice equipped with soft jaws.

POWERHEAD SYSTEMS 6-35

■ In general most of the lower housing oil seals on these motors are installed with their lips facing downward. However, the smaller seal on some models (mounted further down in the lower seal housing) is sometimes installed with the lips facing upward toward the crankshaft and toward the lips of the larger seal installed higher up in the same housing.

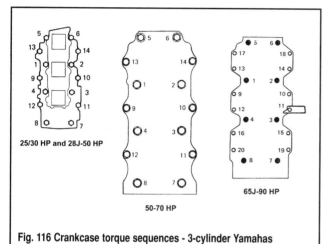

Fig. 116 Crankcase torque sequences - 3-cylinder Yamahas

46. Pack each seal lip with Yamaha marine grease, or equivalent water resistant lubricant, as soon as the seal is installed.

47. Apply engine oil around the groove in the oil housing and around the new O-ring. Install the O-ring around the housing. Install the housing over the lower end of the crankshaft. Tap lightly around the circumference to seat the housing properly into the crankcase. Align the hole(s) and install and tighten the securing bolt(s).

48. If equipped with wedge-shaped anodes which mount in the water jackets (this is usually the case on 28J-50 hp and 65J-90 hp models), insert the anodes into the water jacket of the block and secure the anodes using the Phillips head screws. These anodes will only be visible and accessible the next time the powerhead is overhauled. Therefore, if the anodes show any sign of deterioration in this location, a new anode should be installed.

49. If equipped with long anodes mounted under covers on the side of the powerhead (this is usually the case on 25/30 hp and 50-70 hp models), insert the two long anodes into the side of the block. Cover each anode with a rubber plug (apply a light coating of marine grade grease to the plug). Install the covers. Tighten the cover attaching bolts securely.

50. Position a new head gasket in place on the powerhead.

■ The manufacturer recommends no sealing agent be used on either side of the head gasket.

51. Position the cylinder head onto the powerhead.

52. For 65J-90 hp motors, apply a light coating of engine oil to the threads of the cylinder head retaining bolts, then install the bolts and tighten them using a single clockwise spiraling pattern that starts at the flange bolts

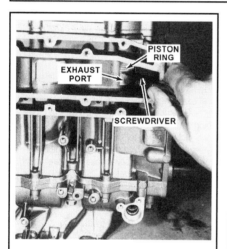

Fig. 117 Check each of the piston rings through the exhaust cover opening to make sure no ring has broken during piston installation

Fig. 118 If removed, install new oil seals to the upper or lower seal housings

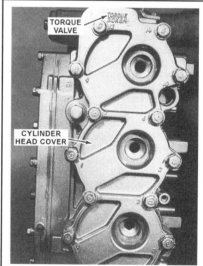

Fig. 119 On Yamaha motors, most torque sequences and specs are molded onto the components

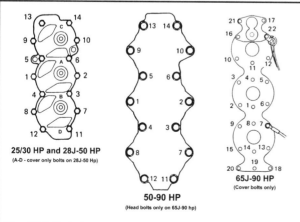

Fig. 120 Cylinder head and cover torque sequences - 3-cylinder Yamahas

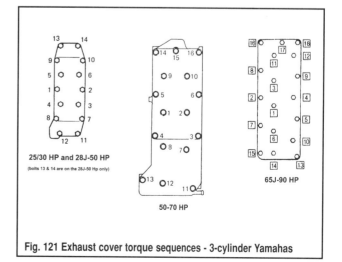

Fig. 121 Exhaust cover torque sequences - 3-cylinder Yamahas

toward the center (vertically speaking) of the cylinder head and works to the outer bolts (at the top and bottom of the head). The pattern continues and finished with the 4 bolts at the center of the head. Tighten the bolts using multiple passes of this pattern first to 11 ft. lbs. (15 Nm) and then either to 17 ft. lbs. (23 Nm) for the smaller bolts or to 22 ft. lbs. (30 Nm) for the larger bolts.

■ **On some models the thermostat is installed into the top of the cylinder head cover and utilizes some of the same bolts. On these models follow the next step at this time. On models where the thermostat assembly is installed in the powerhead itself you can follow this step now, or skip it and come back as soon as you've finished with the cylinder head cover. It's up to you.**

53. Insert the thermostat into the bore with the spring end going in first. Some models, such as the 65J-90 hp motors also have a pressure relief valve installed next to the thermostat. Install the thermostat housing using a new gasket. If installed on the powerhead itself, install and tighten the bolts at this time. If it is installed with the cylinder head, prepare the head itself for installation and install/tighten all bolts in the next step.

■ **Just like crankcase cover bolts, Yamaha has made various recommendations regarding installing head bolts dry, coated with oil or coated with Loctite. Generally speaking, they suggest Loctite® 572 on 25/30 hp and 28J-50 hp motors, engine oil on 50-70 hp motors and dry on 65J-90 hp motors, however, if in doubt keep in mind that the use of a high quality threadlocking compound is rarely an error.**

54. Install a new cylinder head cover gasket, and then position the cylinder head cover over the gasket (holding the thermostat assembly in place on models where the assembly installs into the cylinder head cover).

55. Install the head bolts and tighten them using the proper torque sequence (usually a rough clockwise spiral the starts at the inner bolts and works outwards). Remember that the 28J-50 hp motors use both the 14 head/cover bolts along the perimeter of the head and cover, plus 4 cover only bolts down the center of the cover. The cover only bolts are tightened in a separate sequence starting at the center and then working to the top and bottom bolts. Tighten the bolts to specification, as follows:
 • 25/30 hp and 28J-50 hp motors - tighten all head and/or cover bolts first to 11 ft. lbs. (15 Nm) and then to 20 ft. lbs. (28 Nm).
 • 50-70 hp motors - first to 11 ft. lbs. (15 Nm) and then to 23 ft. lbs. (32 Nm).
 • 65J-90 hp motors (remember the head is already bolted to the cylinder block on these models, so these bolts are JUST for the cover) - to 6 ft. lbs. (8 Nm).

■ **On most models, the bolt torque value and the tightening sequence is embossed on the cylinder head cover.**

56. The same differences in recommendations for coating the threads of exhaust bolts exist for the threads of the exhaust cover bolts. Generally speaking, Yamaha suggests Loctite® 572 on 25/30 hp and 28J-50 hp motors, engine oil on 50-70 hp motors and dry on 65J-90 hp motors, however, if in doubt keep in mind that the use of a high quality threadlocking compound is rarely an error.

57. Install the first gasket, the inner exhaust cover, the second gasket, and then the outer exhaust cover. Install the exhaust cover retaining bolts and tighten them using the proper torque sequence (usually a rough clockwise spiral the starts at the inner bolts and works outwards). Tighten the bolts to specification, as follows:
 • 25/30 hp, 28J-50 hp and 50-70 hp motors - first to 2.9 ft. lbs. (4 Nm) and then to 5.8 ft. lbs. (8 Nm).
 • 65J-90 hp motors - first to 7 ft. lbs. (9 Nm) and then to 13 ft. lbs. (18 Nm).

58. Install the reed valve manifold housing. The manufacturer recommends no sealant be applied to the gasket at this location. Install and tighten the attaching bolts to the specified torque value using multiple passes of a clockwise spiraling torque sequence (are you surprised by this?) that starts at the center bolts and works outward. Tighten the bolts to specification as follows:
 • 25/30 hp, 28J-50 hp and 50-70 hp motors - first to 2.9 ft. lbs. (4 Nm) and then to 5.8 ft. lbs. (8 Nm).
 • 65J-90 hp motors - first to 4 ft. lbs. (6 Nm) and then to 9 ft. lbs. (12 Nm).

59. If desired before powerhead installation install the fuel and electrical components, the flywheel and the hand-rewind starter and/or flywheel cover. However, we recommend waiting until the powerhead is installed on the intermediate housing to install these components.

60. Install the powerhead assembly. Be sure to follow the proper break-in procedures if bearings, rings or other wear parts were replaced.

CLEANING & INSPECTION

Cleaning and inspecting the components is virtually the same for any two-stroke outboard and varies mostly by specifications (which are listed in the Engine Specifications charts) or by component type. A section detailing the proper procedures, sorted mostly by component, can be found under Powerhead Refinishing.

V4 and V6 Powerheads

REMOVAL & INSTALLATION

◆ See Figures 122 thru 139

These powerheads are large and heavy, don't even think about attempting to remove them from the intermediate housing without using a suitable lifting device, such as an engine hoist. In addition, although the powerheads may be removed with most of the fuel and electrical components installed, it is a good idea to consider stripping the powerhead before removal (for the additional maneuverability gained from the weight savings and to help prevent potential damage to the powerhead mounted components of the fuel and electrical systems).

Consider why the powerhead is being removed. If you are planning on disassembling the powerhead for inspection or overhaul, then you'll have to remove the fuel and electrical components anyway. If this is the case, removing them before powerhead removal is a good idea, if only to protect them from damage when lifting and moving the powerhead itself. As usual, we've tried to pretty much only included the steps in this procedure which are necessary in order to remove the assembled powerhead. Additional steps should be taken if you desire to strip the powerhead of all fuel and electrical components before proceeding. If so, refer to components in the Fuel System and Ignition and Electrical System sections for details.

1. On fuel injected motors properly relieve the fuel system pressure, as detailed in the Fuel System section.
2. Disconnect the negative battery cable from the battery for safety.
3. Remove the engine top cover for access to the powerhead.
4. Remove the flywheel from the powerhead for increased access to components and lift brackets on some models (and to prevent potential damage to the flywheel). Flywheel removal is necessary to keep it from interfering with the lift chains on some powerheads, but removal is advisable on all motors.
5. Tag and disconnect the throttle and shift cables. For most models this involves pulling the retaining clip and carefully separating the cable end from the linkage ball stud.
6. For carbureted motors, disconnect the choke rod from the lower front (front, starboard side of the motor/cowling) of the carburetor assembly.
7. For 2.6L EFI motors, remove the CDI unit cover for access.
8. If not completely stripping the powerhead of electrical components, remove/disconnect the following:
 a. On electric start models (yes, believe it or not, there are actually a couple of manual start V motors), tag and disconnect the battery cables from the powerhead (usually at the starter/starter solenoid, but one or both of the cables may instead be connected to a junction block on some models).
 b. For carbureted and 2.6L EFI motors, loosen the bolt securing the bracket (carb only) and ground lead (all) to the rear of the power head (right below or next to the lower port spark plug). Reposition the ground lead out of the way.
 c. The 80J-140 Hp (1730cc) V4 and 105J-225 Hp (2596cc) V6 models, have a wire loom cable guide on the port side of the motor, in the base cowling about halfway up the side of the motor, under the fuel filter, loosen the bolt and remove the guide.
 d. On EFI/HPDI motors, remove the electrical junction box cover from the rear, starboard side of the motor. On 3.1L EFI motors you'll also want to remove the adjacent fuse cover.

POWERHEAD SYSTEMS 6-37

e. For carbureted remote control models, disconnect the remote control coupler located along the cowling at the front, starboard side of the motor.

■ If you have difficulty locating some of these connections, refer to the accompanying illustrations and/or to the Wiring Diagrams in the Ignition and Electrical System section for additional information.

f. For all models equipped with Power Trim/Tilt, tag and disconnect the trim sensor coupler and/or connector, located along the base of the cowling, on the mid-to-front starboard side of the motor.

g. If equipped with PTT, tag and disconnect the motor leads (usually one Blue and one Greed lead) from the relay assembly. On some 3.1L motors there are three leads (one Sky Blue, one Light Green and one Black).

h. For 2.6L HPDI and 3.1L EFI motors, locate the shift position or shit cutout switch coupler toward the front of the starboard side of the motor, along the base cowling. Tag and disconnect the switch coupler.

i. For 3.1L EFI motors, there is a ground wire at the base of the powerhead, just below the lower, port spark plug. Locate the lead, then remove the bolt and disconnect it from the powerhead.

9. Tag and disconnect the main fuel supply line at the fuel filter. Plug the openings to prevent fuel loss and system contamination.

10. For oil injection models, tag and disconnect the main oil supply line at the tank. Plug the openings to prevent oil loss and system contamination.

■ Some pilot or cooling water hoses are retained without clamps, while others use a wire tie and/or a spring-type clamp.

11. Tag and disconnect the cooling water and/or pilot water (cooling system indicator) hose, as follows:

• For carburetor motors both hoses are connected to adjacent fittings at the rear center of the powerhead.

• For fuel injected 2.6L motors, a pilot water outlet-to-regulator/rectifier hose connects to a fitting on the lower, starboard side, rear of the motor. On EFI models, there is also a cooling water hose on a fitting at the lower center, rear of the powerhead.

• On 3.1L motors, a pilot water hose is connected to a fitting at the lower, rear of the cowling (similar to the 2.6L pilot hose).

• On 3.3L motors, a cooling water hose is located under the junction box cover, near the PTT relay assembly.

12. On 2.6L HPDI motors, remove the clip and shift rod lever bushing from the shift assembly (near the shift position switch on the starboard side of the motor, just below the starter).

13. Disconnect the shift rod assembly, depending upon the model, as follows:

• For carbureted motors, on the mid-starboard side of the motor, locate the shift rod assembly, remove the two 6mm and two 8mm bolts from the assembly.

• For fuel injected 2.6L motors, toward the front of the port side, remove the 2 bolts and disconnect the shift rod assembly.

• For 3.1L motors, at the mid-starboard side of the motor, remove the retaining clip and bushing and disconnect the shift lever.

• For 3.3L motors, on the starboard side of the motor, remove the retaining clip, then disconnect the shift lever.

14. For all, except HPDI models, remove the lower apron by first loosening the 2 bolts securing the front of the apron to the rear, then removing the front piece. Next, loosen the 4 bolts (2 threaded upward from below at the front and 2 threaded downward from above at the rear of the powerhead) securing the rear apron to the cowling, then remove the rear apron.

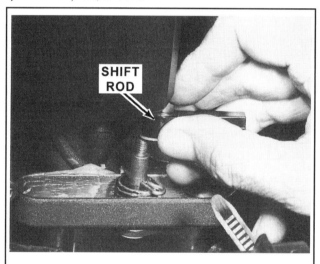

Fig. 122 Tag and disconnect the throttle and shift cables. . .

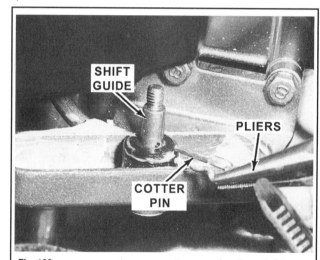

Fig. 123 . . .on some motors you must remove the clip and shift bushing

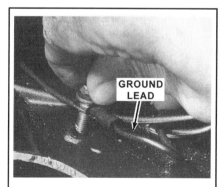

Fig. 124 Disconnect all noted wiring, be careful not to miss ground leads

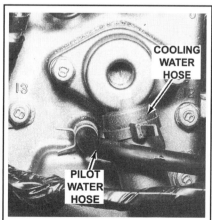

Fig. 125 Tag and disconnect the cooling and/or pilot water hoses

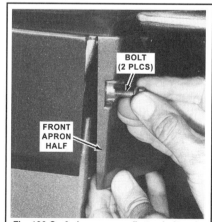

Fig. 126 On 2 piece aprons, first unbolt and remove the front piece. . .

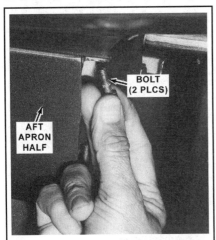

Fig. 127 . . .then unbolt and remove the aft piece

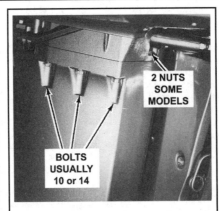

Fig. 128 Loosen and remove the powerhead fasteners. . .

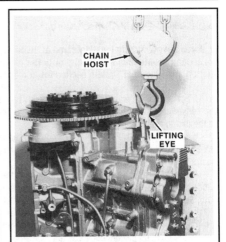

Fig. 129 . . .then use an engine hoist to remove the powerhead

15. For 2.6L HPDI motors, remove the apron and, if necessary, the exhaust expansion chamber, as follows:

   a. Loosen the screw threaded horizontally into the starboard side of the apron and outboard.

   b. Loosen the 2 bolts threaded downward into the apron (at the rear of the powerhead) and the 2 bolts threaded upward from the apron to the cowling at the front of the powerhead. Carefully open and remove the apron, while keeping track of the collars used on these 4 bolts.

   c. If necessary, remove the hose joint from the apron and the rubber seal from the joint or the exhaust expansion chamber hose.

   d. Remove the 9 bolts securing the exhaust expansion chamber to the underside of the powerhead/cowling, then remove the expansion chamber and hose. Keep track of the 4 rubber seals positioned in the top of the housing and the square rubber seal positioned between the expansion chamber housing and the intermediate housing.

   e. Loosen the clip, then disconnect the cooling water hose.

16. For 3.3L motors, remove the four 6mm bolts from the sides of the apron, then loosen the screw and clamp at the front of the apron split line. Remove the apron for access to the powerhead retaining bolts.

■ Most models use powerhead retaining bolts of various lengths, be sure to keep track of the mounting position for each bolt as it is being removed.

17. Remove the bolts securing the powerhead to the intermediate housing as follows:

• The 80J-140 Hp (1730cc) V4 and 105J-225 Hp (2596cc) V6 models utilize 2 nuts on the front underside of the powerhead, along with 4 short and 6 long bolts. The short bolts are at the center rear and at the front corners of the intermediate housing-to-powerhead mounting flange, while the 6 long bolts are located, 3 to a side working their way back to the rear of the powerhead.

• The 200-250 Hp (3130cc) V6 models utilize 6 short and 8 long bolts. The short bolts are at the center rear and at the two front corners of the intermediate housing-to-powerhead mounting flange, while the 8 long bolts are located, 4 to a side working their way back to the rear of the powerhead.

• The 225-250 Hp (3342cc) V6 models utilize 6 short and 8 long bolts. Pay close attention to the bolt locations on these models (as they may vary), but normally they are the same as for the 3.1L motors.

18. Attach a suitable lifting device (hook and high-strength) chain to the lift bracket(s). When multiple lift brackets are provided, adjust the chains to make sure the powerhead is lifted STRAIGHT upward and off the driveshaft splines.

■ The powerhead may be difficult to dislodge from the intermediate housing because of a tight sealing gasket and the fact that some assemblies may have used joining compound (either from the factory or when serviced in the field). Prying up on the powerhead by using a long piece of wood and leverage on the edge of the lower cowling is an acceptable method by the manufacturer. If this method is employed, CARE must be exercised not to damage either the powerhead or the cowling. Once the powerhead has broken free of the intermediate housing, proceed to the next step.

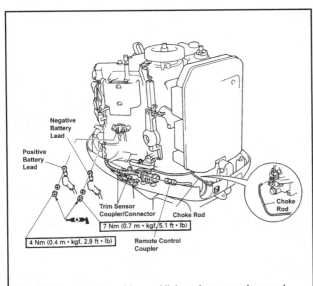

Fig. 130 Disconnecting wiring and linkage for powered removal - Carbureted motors

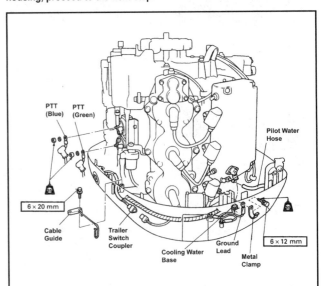

Fig. 131 Disconnecting wiring for powered removal - Carbureted motors

POWERHEAD SYSTEMS 6-39

✱✱ WARNING

If the unit is several years old, or if it has been operated in salt water, or has not had proper maintenance, or shelter, or any number of other factors, then separating the powerhead from the intermediate housing may not be a simple task. An air hammer may be required on the bolts to shake the corrosion loose, heat may have to be applied to the casting to expand it slightly, or other devices employed in order to remove the powerhead. One very serious condition would be the driveshaft frozen with the lower end of the crankshaft. In this case a circular plug type hole must be drilled and a torch used to cut the driveshaft.

19. Let's assume the powerhead will come free on the first attempt. Take care not to lose the dowel pins (there are USUALLY 2). The pin(s) may come away with the powerhead or may stay in the intermediate housing. Be especially careful not to drop a dowel pin into the lower unit.
20. Raise the powerhead free of the intermediate housing.
21. Place the powerhead on a suitable work surface or on an engine stand. Again, take care not to loose or damage the alignment dowel pin(s).
22. Carefully remove all traces of gasket from the powerhead mating surfaces. On 3.3L motors it may be necessary to remove the shift rod assembly in order to remove the gasket.

To Install:

23. Check to be sure the dowel pins are in place, then position the gasket (and the shift rod assembly on 3.3L motors) on the intermediate housing mating surface.
24. Apply a light coating of marine grade grease to the driveshaft splines.
25. Carefully lower the powerhead onto the intermediate housing making sure the gasket remains in position on the dowel pins while making sure to index the crankshaft and driveshaft splines. If the splines do not align, shift the lower unit into forward gear, have an assistant rotate the propeller just a whisker clockwise until the splines do index. Check to be sure the powerhead is fully down on the intermediate housing - the mating surfaces hard against each other.
26. For all except the 3.3L engine, apply a light coat of Loctite® 572 or an equivalent threadlocking compound to the threads of the powerhead and apron fasteners. On the 3.3L engine, only apply the threadlocking compound to the powerhead fasteners and use Loctite® 572 for the long bolts, but use Loctite®648 on the threads of the short bolts.
27. Install the powerhead mounting bolts in the positions noted during removal, then tighten them to 15 ft. lbs. (21 Nm) for all models except the 3.3L motor, on which they should be tightened to 23.6 ft. lbs. (32 Nm).
28. Install the apron (and the exhaust expansion chamber on HPDI motors, if removed) to the lower cowling and secure using the retaining screws. Be sure to tighten the fasteners securely, but do not over-tighten and crack the apron.
29. Reconnect the shift rod assembly.
30. On 2.6L HPDI motors, apply a light coating of grease to the shift rod lever bushing, then install the bushing and secure using the retaining clip.
31. Reconnect the cooling water and/or pilot water (cooling system indicator) hose.
32. For oil injection models, reconnect the main oil supply line to the tank.
33. Reconnect the main fuel supply line to the fuel filter.
34. If the powerhead was not completely stripped of electrical components, reconnect the following:
 a. For 3.1L EFI motors, there is a ground wire at the base of the powerhead, just below the lower, port spark plug. Reconnect the lead and secure using the mounting bolt.
 b. For 2.6L HPDI and 3.1L EFI motors, locate the shift position or shit cutout switch coupler toward the front of the starboard side of the motor,

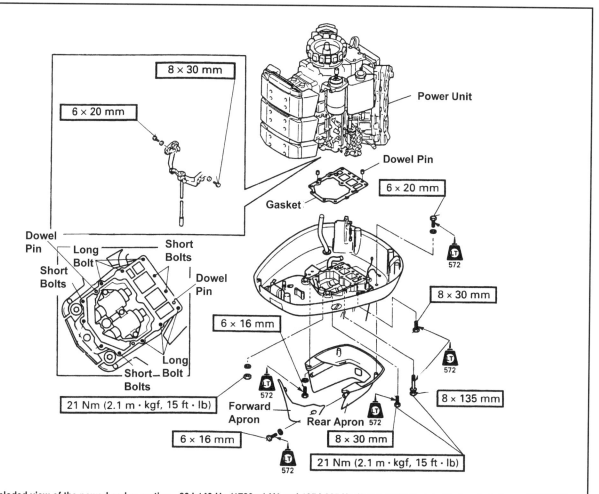

Fig. 132 Exploded view of the powerhead mounting - 80J-140 Hp (1730cc) V4 and 105J-225 Hp (2596cc) V6 Models (Carbureted shown, EFI/HPDI Similar)

along the base cowling. Reconnect both ends of the switch coupler as tagged during removal.

   c. If equipped with PTT, reconnect the motor leads (usually one Blue and one Greed lead) to the relay assembly, as tagged during removal. On some 3.1L motors there are three leads (one Sky Blue, one Light Green and one Black).

   d. For all models equipped with Power Trim/Tilt, reconnect the trim sensor coupler and/or connector, located along the base of the cowling, on the mid-to-front starboard side of the motor.

   e. For carbureted remote control models, reconnect the remote control coupler located along the cowling at the front, starboard side of the motor.

   f. On EFI/HPDI motors, install the electrical junction box cover to the rear, starboard side of the motor. On 3.1L EFI motors also install the adjacent fuse cover.

   g. The 80J-140 Hp (1730cc) V4 and 105J-225 Hp (2596cc) V6 models, have a wire loom cable guide on the port side of the motor, in the base cowling about halfway up the side of the motor, under the fuel filter, install the guide and secure using the retaining bolt.

   h. For carbureted and 2.6L EFI motors, install the ground lead and, if applicable, bracket, using the retaining bolt to the rear of the power head (right below or next to the lower port spark plug).

   i. On electric start models (yes, believe it or not, there are actually a couple of manual start V motors), reconnect the battery cables to the powerhead (usually at the starter/starter solenoid, but one or both of the cables may instead be connected to a junction block on some models) as noted during removal.

35. For 2.6L EFI motors, install the CDI unit cover.
36. Install the flywheel and cover.
37. For carbureted motors, reconnect the choke rod to the lower front (front, starboard side of the motor/cowling) of the carburetor assembly.
38. Reconnect the negative battery cable, then properly pressurize the fuel system and check for leaks.
39. Reconnect the throttle and shift cables.
40. Refer to the Timing and Synchronization adjustments in the Engine and Maintenance section for cable/linkage adjustment procedures.

■ Be sure to run the motor without the top cowling installed in order to double-check for potential fuel or oil leaks. Remedy any leaks before proceeding.

41. Install the top cowling to the powerhead.

■ If the powerhead was rebuilt or replaced with a remanufactured unit, don't forget to follow the proper Break-In procedures.

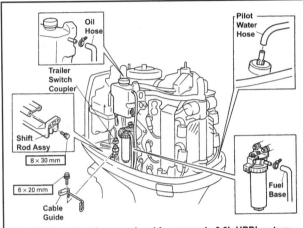

Fig. 133 Preparing the powerhead for removal - 2.6L HPDI motors

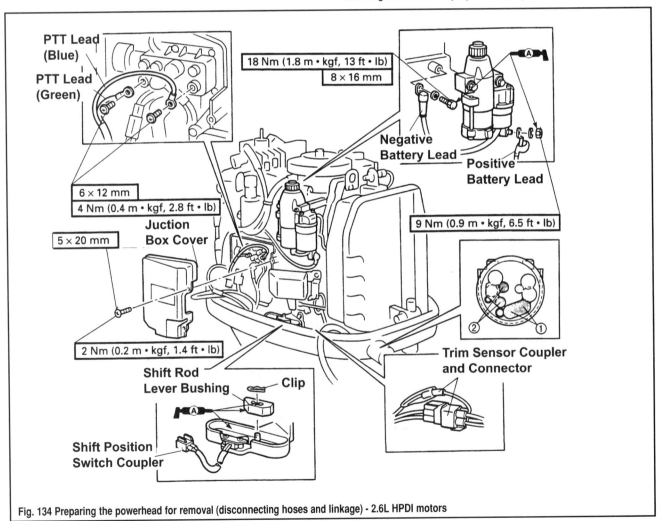

Fig. 134 Preparing the powerhead for removal (disconnecting hoses and linkage) - 2.6L HPDI motors

POWERHEAD SYSTEMS 6-41

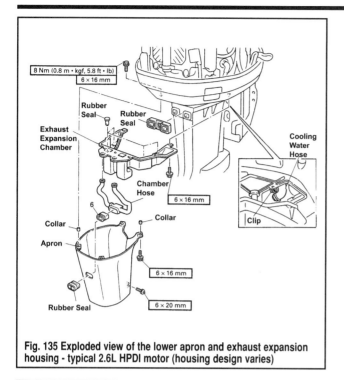

Fig. 135 Exploded view of the lower apron and exhaust expansion housing - typical 2.6L HPDI motor (housing design varies)

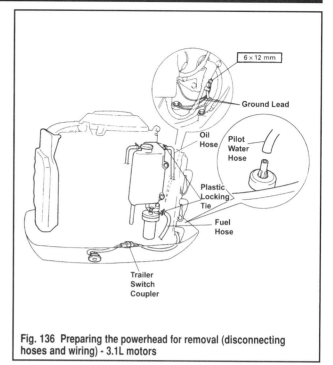

Fig. 136 Preparing the powerhead for removal (disconnecting hoses and wiring) - 3.1L motors

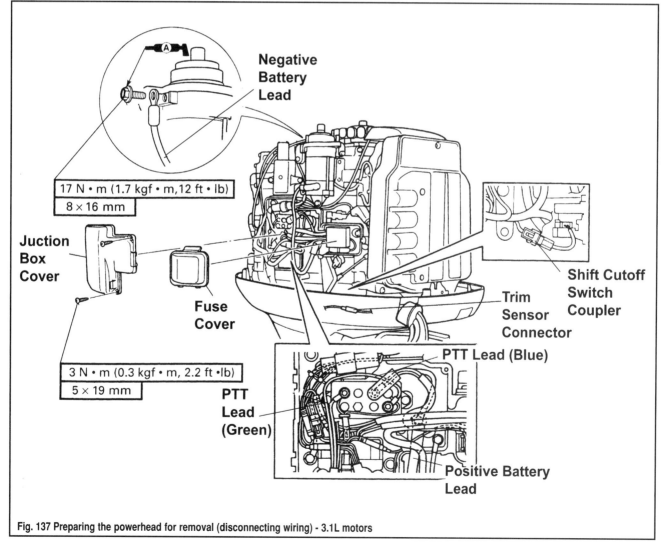

Fig. 137 Preparing the powerhead for removal (disconnecting wiring) - 3.1L motors

6-42 POWERHEAD

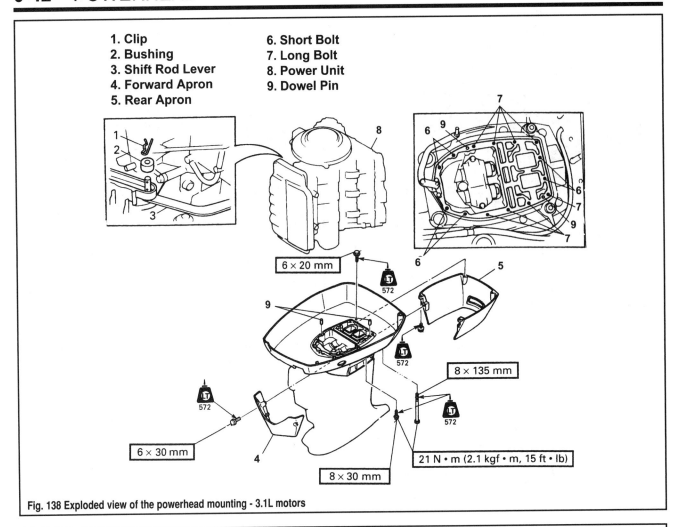

Fig. 138 Exploded view of the powerhead mounting - 3.1L motors

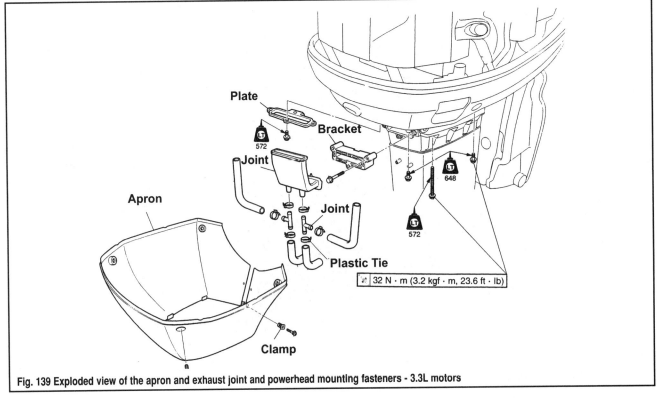

Fig. 139 Exploded view of the apron and exhaust joint and powerhead mounting fasteners - 3.3L motors

POWERHEAD SYSTEMS 6-43

DISASSEMBLY

Getting Started/Important Points for this Overhaul

■ Most of the accompanying photos are from tear-downs of a typical Yamaha V4 and V6 motors. Although most models are very similar in design and service, some components will differ visually. Illustrations have been provided showing most of the differences, so when in doubt, refer to the illustrations for the model on which you are working.

The following procedures cover removal and installation of virtually all components of the powerhead. However, if the determination can be made certain seals, bearings, etc. are fit for further service, leave a sleeping dog lie, simply skip the steps involved for such items and proceed with the necessary work. In certain instances (such with some seals and/or bearings) the component would be destroyed or at least rendered unfit for service during the removal process.

■ Refer to the Engine Specifications chart in this section for overhaul and inspection dimensions/limits.

Remember, when loosening the retainers on manifolds, covers and other major components, always try to follow the reverse of the torque sequence indicated in the assembly procedure (later in this section) or molded on the component itself. Similarly, when tightening bolts, on these same components, use the proper torque sequence and tighten the bolts to the proper specification. When possible we've included it in the procedure, but on most Yamahas both the sequence and value is molded into the casting on critical components. You'll notice that most Yamaha torque sequences are clockwise spirals starting somewhere near the center of the component.

■ Because of the high temperatures and pressures developed, the sealing surfaces of the cylinder head and the block are the most prone to water leaks. No sealing agent is recommended because it is almost impossible to apply an even coat of sealer. An even coat would be essential to ensure an air/water tight seal.

Lower Oil Seal Housing

◆ See Figures 140 thru 144

1. Remove the powerhead from the intermediate housing, as detailed earlier in this section.

■ If a suitable engine stand is not available you'll have to set the powerhead down on a work surface. However, most (though not all) motors utilize an externally mounted lower oil seal housing. In order to make sure the powerhead sits securely on a workbench, start disassembly by removing the external lower oil seal housing leaving a fairly flat surface on the bottom of the powerhead. A couple of 1/2 inch pieces of wood wedged under the powerhead at strategic locations will stabilize the block.

2. On 3.1L and 3.3L engines, remove the balancer from the bottom of the crankshaft. Use the Yamaha Flywheel Holder (#YB-16139), or equivalent, inserted into the damper and, on some models, a crankcase bolt hole) to keep the damper from turning while you loosen and remove the nut. Then use a universal puller to free the damper from the crankshaft. On 3.1L motors you should fabricate a 1.38 in. (35mm) diameter round metal plate to insert between the puller tool and the crankshaft in order to prevent possible shaft damage.

3. Remove the bolts (usually 4) securing the lower oil seal housing to the bottom of the block, then remove the housing.

4. Use a couple of small pry-tools to carefully lever the housing away

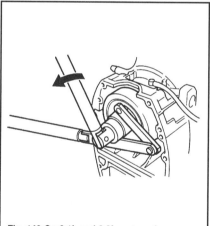

Fig. 140 On 3.1L and 3.3L motors, loosen the crankshaft damper nut...

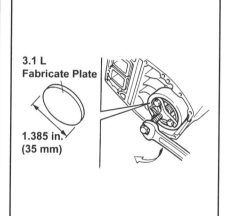

Fig. 141 ...then use a puller to remove the damper

Fig. 142 Unbolt the lower seal housing...

Fig. 143 ...then remove the housing...

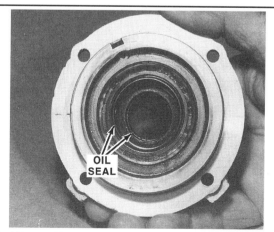

Fig. 144 ...and inspect the seals

6-44 POWERHEAD

from the block or tap the housing lightly with a soft head mallet to jar it loose. Remove and discard the outer O-ring.

■ **Removal of the seals destroys their sealing qualities. Therefore, they cannot be installed a second time. Be absolutely sure a new seal is available before removing the old seal in the next step.**

5. Inspect the condition of the seal or seals in the lower oil seal housing (and damper/nut on 3.1L and 3.3L motors). If the seals appear to be damaged and replacement is required, proceed with removal. Start by verifying seal positioning. Seal installation varies slightly by model, as follows:
- On 1.7L and 2.6L motors there are normally 2 seals both of which are installed from the crankshaft side of the housing. The seal lips are normally faced downward away from the crankshaft.
- On 3.1L and 3.3L motors, there are usually 2 seals. One seal is in the crankshaft damper or damper nut, and is positioned with the facing upward toward the crankshaft. The other is mounted in the top of the oil seal housing, and its lips face downward, away from the crankshaft. There may be a large circlip which must be removed for access to the seal in the housing.

6. Position the powerhead upright to continue with disassembly.

Intake Manifold and Reed Valve Assemblies

◆ See Figures 145 thru 152

1. If not done during powerhead removal, strip the powerhead of the flywheel, along with the remaining fuel and electrical components. For details, please refer to the appropriate sections of this guide.

2. For carbureted motors, if necessary, remove the timing pointer, fuel enrichment valve (V6) and/or the hose clamp (V4) from the intake manifold. Also on carbureted motors, tag and disconnect the intake manifold air vent hose.

3. Tag each and every one of the oil supply and, for carbureted motors, the bleed/recirculation hoses (most of these lines are of different lengths), then disconnect the lines from the manifold assembly (and crankcase/cylinder block).

■ **Take care in the next step when handling reed valve assemblies. Once the assembly is removed, keep it away from sunlight, moisture, dust, and dirt. Sunlight can deteriorate valve seat rubber seals. Moisture can easily rust stoppers overnight. Dust and dirt - especially sand or other gritty material can break reed petals if caught between stoppers and reed petals. Make special arrangements to store the reed valves to keep them isolated from elements while further work is being performed on the powerhead.**

4. Remove the bolts securing the intake manifold/reed valve assembly to the block. There are 12 bolt on V4 models, 16 bolts on carbureted V6 models and 2 bolts on EFI/HPDI motors. For all carbureted models, be sure to loosen the manifold bolts using multiple passes following the reverse of the torque sequence.

5. Carefully separate the intake manifold from the crankcase. If necessary on carbureted models you may obtain a long wide pry-tool to separate the manifold and reed valve assembly from the block and from each other. Insert the tool between the tabs normally provided for this purpose, and pry the two surfaces apart. Never pry at a gasket sealing surface. Such action would very likely damage the sealing surface of an aluminum powerhead.

6. Remove and discard the gasket. Set the reed valves aside. Further work on the reed valves may be performed during cleaning and inspection.

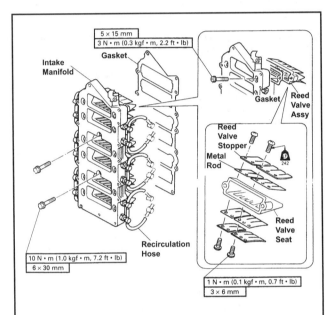

Fig. 148 Typical intake manifold and reed valve assembly from a fuel injected motor

Fig. 145 Tag and disconnect any oil lines at the intake manifold

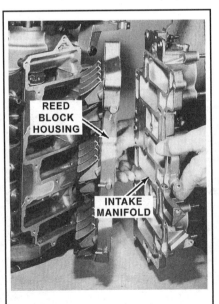

Fig. 146 Remove the intake manifold/reed valve assembly

Fig. 147 While removed, protect the reed valves from damage

POWERHEAD SYSTEMS 6-45

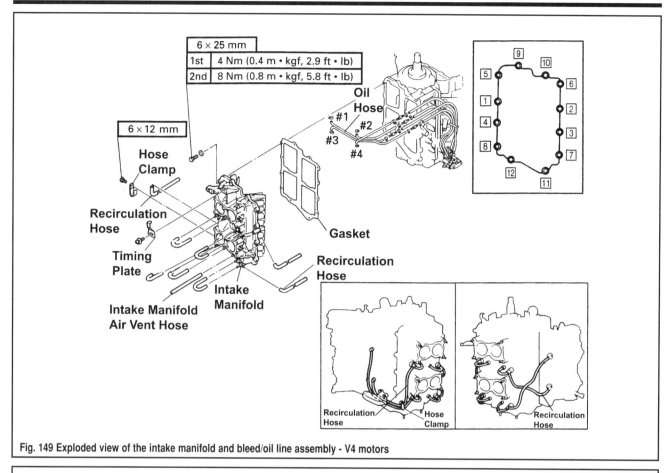

Fig. 149 Exploded view of the intake manifold and bleed/oil line assembly - V4 motors

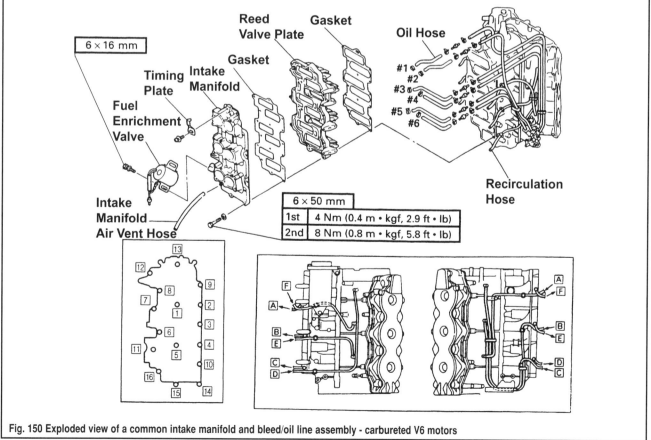

Fig. 150 Exploded view of a common intake manifold and bleed/oil line assembly - carbureted V6 motors

6-46 POWERHEAD

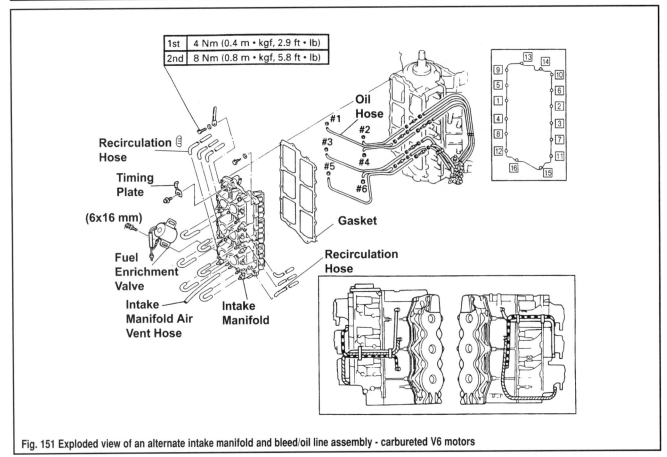

Fig. 151 Exploded view of an alternate intake manifold and bleed/oil line assembly - carbureted V6 motors

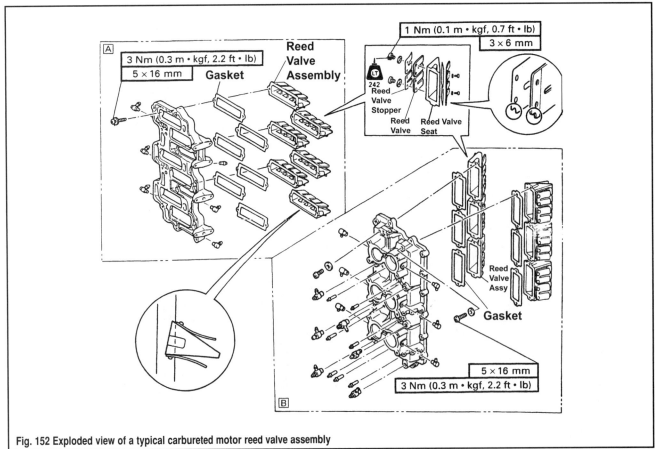

Fig. 152 Exploded view of a typical carbureted motor reed valve assembly

POWERHEAD SYSTEMS

Exhaust Cover

◆ See Figures 153 thru 158

■ The exhaust cover should always be removed during a powerhead overhaul. Many times, water in the powerhead is caused by a leaking exhaust cover gasket or plate. If the inner or outer cover (some models use only a single cover) is stuck to the powerhead, insert a small prytool between the tabs normally provided for this purpose, and pry the two surfaces apart. Never pry at a gasket sealing surface. Such action would very likely damage the sealing surface of an aluminum powerhead.

1. Using multiple passes in the reverse of the molded or illustrated torque pattern, remove the bolts securing the exhaust cover.
2. Remove the cover and the gasket. If applicable, remove the inner cover and gasket. Discard the outer and inner cover gaskets.
3. On the 1.7L and 2.6L powerheads, remove the two bolts securing the bypass/pressure control valve cover to the exhaust cover. Lift the cover up, and then remove the spring and water pressure valve. Observe how the short end of the insert faces outward, as an assist during assembling.
4. On 3.3L motors, loosen the bolt securing the anode assembly to the lower center of the exhaust cover. Remove the anode cover, then loosen the bolt and remove the anode and grommet from the cover.

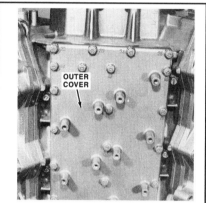

Fig. 153 Loosen the exhaust cover retaining bolts...

Fig. 154 ...then remove the cover(s) and gasket(s) assembly

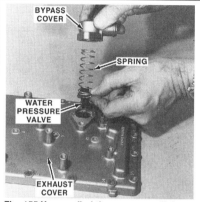

Fig. 155 You can find the pressure relief valve on some covers

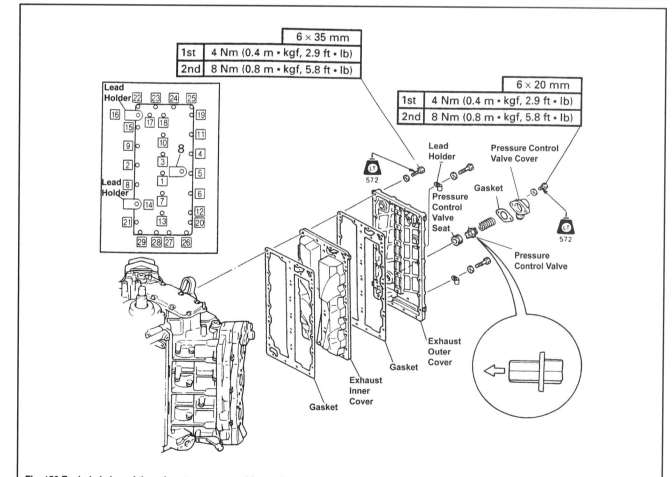

Fig. 156 Exploded view of the exhaust cover assembly - carbureted 2.6L shown (1.7L and fuel injected 2.6L similar, but watch bolt torques and patterns, on fuel injected models, be sure to re-torque bolts 1 and 2 at the end of the sequence

6-48 POWERHEAD

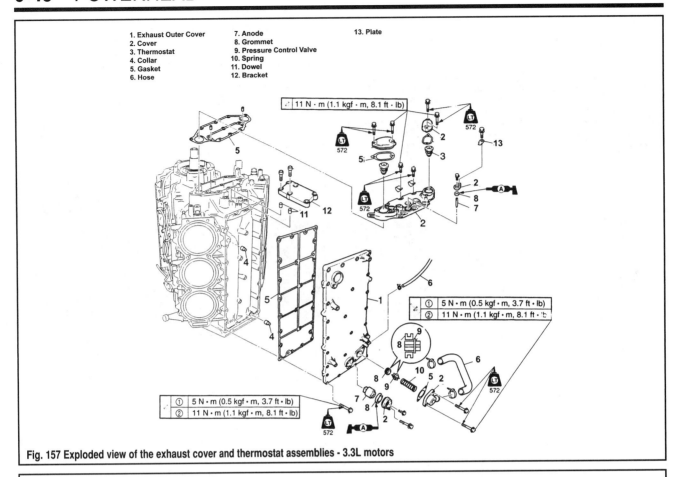

Fig. 157 Exploded view of the exhaust cover and thermostat assemblies - 3.3L motors

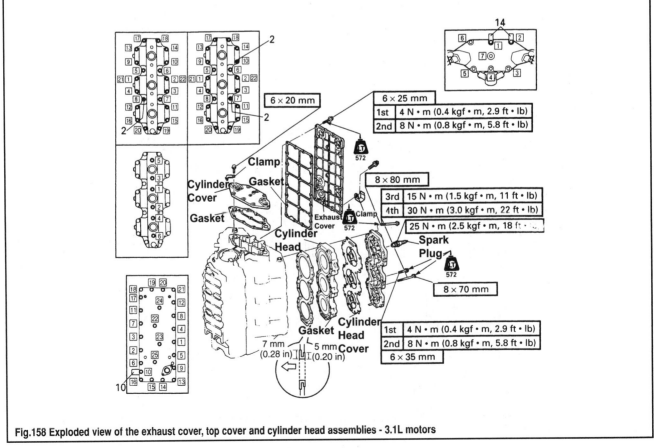

Fig.158 Exploded view of the exhaust cover, top cover and cylinder head assemblies - 3.1L motors

POWERHEAD SYSTEMS 6-49

Cylinder Heads

◆ See Figures 159 thru 165

1. If not done already, remove the spark plugs.

■ On 1.7L and 2.6L motors a thermostat is installed at the top of the cylinder head cover and is removed along with the cover (as the thermostat cover is also secured by the head cover bolts). On larger models the thermostat is mounted either to the cylinder head itself or the block under a separate cover. During overhaul it must at least be removed and tested, but should probably just be replaced.

2. Remove the cylinder head cover bolts using multiple passes in the reverse of the torque sequence. For models with the thermostat installed in the cover, once the bolts are loose enough to spin with your finger, you can remove the upper 4 which also hold the thermostat cover to the head cover, then remove the thermostat assembly and gasket. Once all of the head bolts are removed, carefully remove the head cover. Remove and discard the gasket. If the cover is stuck to the cylinder head, insert a small pry-tool between the tabs normally provided for this purpose and pry the two surfaces apart. Again, never pry at the gasket sealing surfaces, because the sealing surface of an aluminum powerhead could be damaged.

■ Most motors use separate cylinder head bolts, however on 3.1L motors the cylinder head cover bolts are threaded through the head and into the block, so no additional bolts are used.

3. On all except 3.1L motors, loosen and remove the cylinder head bolts using multiple passes in the reverse of the torque sequence.
4. Insert a long wide pry-tool between the tabs normally provided for this purpose on the cylinder head and block. Carefully pry the two surfaces apart using the tabs provided for this purpose. Never pry at a gasket sealing surface. Such action would very likely damage the sealing surface of an aluminum powerhead.

■ If the cylinder head is stubborn and cannot be broken free of the block, tap the head lightly with a soft head mallet to jar it free.

5. Some models are equipped with small wedge-shaped anodes mounted in the water jackets around the cylinders. If so, remove the Phillips head screws, and then remove the anodes from in the water jacket.

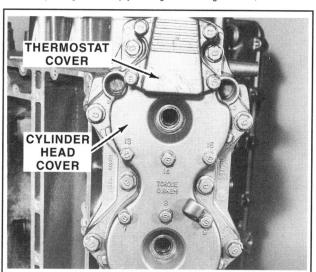

Fig. 159 Loosen the cylinder head bolts (on some motors the bolts also secure the thermostat cover)

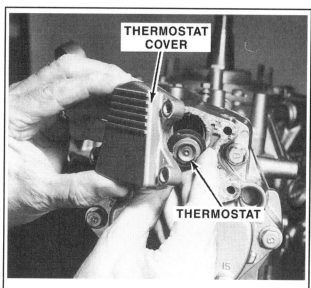

Fig. 160 Remove the thermostat assembly

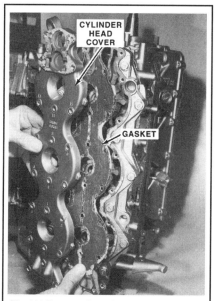

Fig. 161 Remove the cylinder head cover and gasket. . .

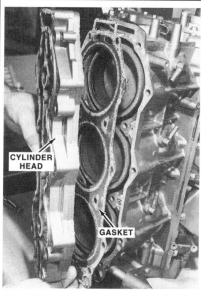

Fig. 162 . . .then remove the cylinder head and gasket

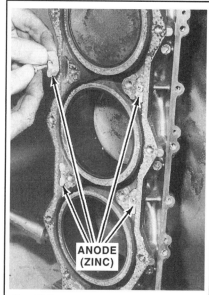

Fig. 163 Some models are equipped with anodes in the water jackets

6-50 POWERHEAD

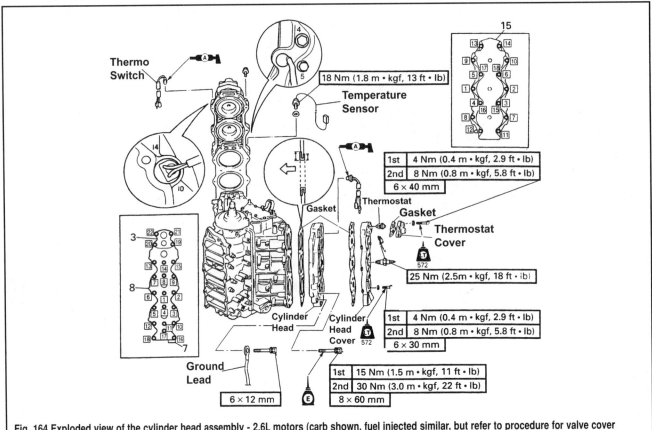

Fig. 164 Exploded view of the cylinder head assembly - 2.6L motors (carb shown, fuel injected similar, but refer to procedure for valve cover torque values on HPDI motors)

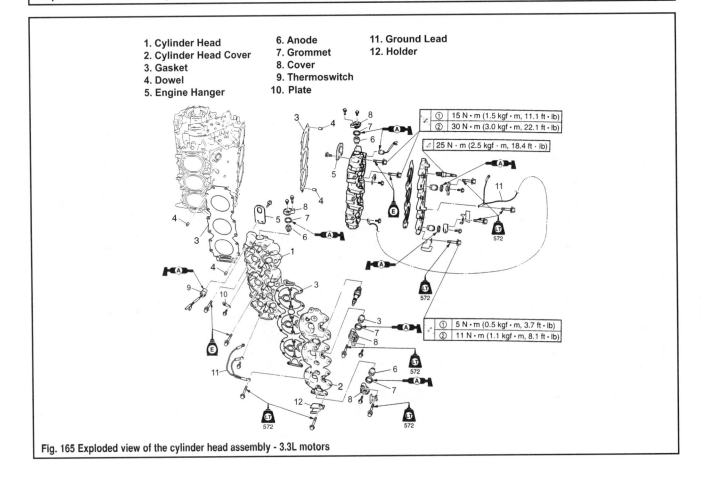

Fig. 165 Exploded view of the cylinder head assembly - 3.3L motors

POWERHEAD SYSTEMS

Crankcase and Upper Oil Seal Housing

◆ See Figures 166 thru 174

1. On fuel injected models, tag and disconnect the various bleed/recirculation hoses located on the crankcase. Pay very close attention to hose routing for installation purposes. Start by comparing the current routing to the accompanying illustrations then if possible, take one or more digital pictures for reference.
2. If applicable, remove the oil pump from the cylinder block.
3. On models with an external upper oil seal housing loosen and remove the bolts at this time, but do not attempt to remove the housing yet.

✱✱ WARNING

For 1.7L and 2.6L motors, do not attempt to remove the upper oil seal housing, at this time. The housing must be removed from the crankshaft after the crankshaft has been lifted from the block. To attempt removal at this time would only damage the housing.

4. If equipped, loosen and remove the bolts securing the engine top cover, then remove the cover plate and gasket.
5. Using multiple passes in the reverse of the torque sequence loosen and remove the crankcase retaining bolts. Most powerheads utilize bolts of at least 2 different lengths, so keep track of the bolts and their positions. Also, generally, the reverse of the torque sequence would be a counterclockwise spiraling pattern that starts with the outer flange bolts and works toward the center of the flange, then continues with the top and bottom inner bearing bolts and works towards the center of these longer bearing bolts.
6. After all bolts have been removed, insert a small pry-tool between the tabs normally provided for this purpose and CAREFULLY pry the two surfaces apart. Never pry at gasket sealing surfaces. Such action would very likely damage the sealing surface of an aluminum powerhead. If no tabs are provided, tap along the perimeter using a rubber mallet and, if necessary, carefully use a thin plastic putty knife to try and cut the sealant between the crankcase and cylinder block.
7. Separate the two halves of the crankcase. Take care not to lose the two dowel pins. These pins may remain in either half when the crankcase is separated.

■ It appears that 1.7L and 2.6L motors utilize a large partially external upper oil seal housing, while the 3.1L and 3.3L motors utilize an upper oil seal housing/bearing assembly which is installed over the crankshaft, just inside the cylinder block.

8. For 1.7L and 2.6L motors, tap the edge of the oil seal housing with a soft head mallet to jar it free of the crankshaft. Remove the upper oil seal housing from the top of the crankshaft.
9. For 3.1L and 3.3L motors, carefully slide the upper roller bearing/seal housing from the top of the crankshaft (NOT the upper ball bearing which is pressed onto the shaft on these models).
10. For all motors, take a moment to service the oil upper oil seal housing. The 1.7L and 2.6L motors use 2 external O-rings, while the 3.1L and 3.3L motors use a single external O-ring. In all cases, remove and discard the old O-ring(s) around the housing.
11. Inspect the condition of the oil seal(s) in the upper oil seal housing. Make a determination if the seal is fit for further service. The seal will be destroyed during removal. Therefore, remove the seal only if it is damaged and has lost its sealing qualities. If necessary, remove the seal using either a slide hammer and expanding jaw attachment or other suitable seal puller. Before removal, note seal lip positioning for installation purposes. The upper seals and housings vary slightly from model-to-model as follows:
• On 1.7L and 2.6L motors there is normally 1 seal which is installed from the top of the upper seal housing. The seal lips are faced downward toward the crankshaft.
• On 3.1L and 3.3L motors, there is one seal mounted in the top roller bearing, with its lips facing downward.

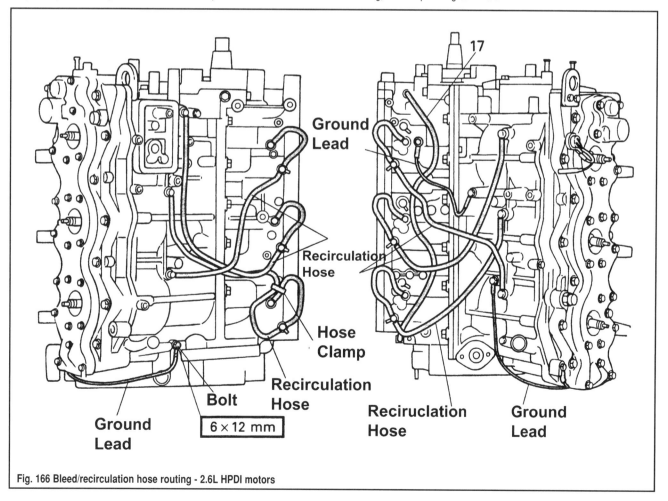

Fig. 166 Bleed/recirculation hose routing - 2.6L HPDI motors

6-52 POWERHEAD

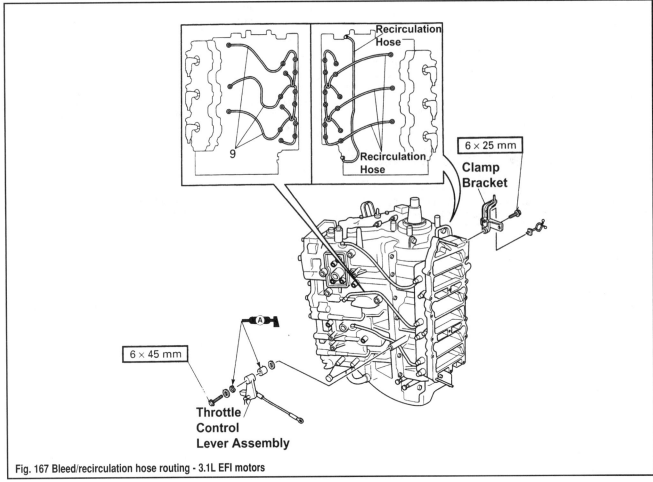

Fig. 167 Bleed/recirculation hose routing - 3.1L EFI motors

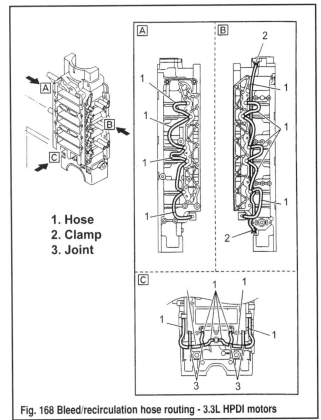

1. Hose
2. Clamp
3. Joint

Fig. 168 Bleed/recirculation hose routing - 3.3L HPDI motors

■ Remove the upper roller or ball bearing that is pressed onto the crankshaft only if it is in question or is no longer fit for service.

12. On 1.7L and 2.6L motors, a needle bearing (or 2 bearings on HPDI models) is (are) pressed into the upper oil seal housing. To remove the bearing(s), first obtain Yamaha's special bearing installer tool (#YB6205). Using the shaft of the installer, remove the bearing(s) from the housing using a hydraulic press. Press the bearing(s) downwards free of the housing.

Crankshaft and Bearing Removal

◆ See Figures 175 thru 188

■ Most parts inside the cylinder block are matched sets and should not be moved from one cylinder to another. This includes the bearings and pistons as well as each of the connecting rod/end cap combination. For this reason, prior to disassembly it is essential to matchmark each of the connecting rods to their end caps (this ensures that not only will each cap be installed on the proper rod, but facing the correct direction). Also, be sure to label each piston and connecting rod as to what cylinder it belongs and identify the factory marks and/or place an arrow or some indicator facing upward to ensure they are installed in the same direction. Yamaha often marks pistons with UP on the dome and connecting rods with the letter Y or word YAMAHA and/or alignment dots at the con rod-to-cap mating surface, all of which should face the flywheel.

1. Take time to mark the cylinder number on both halves of the connecting rod caps. This mark should be made with a marker, whiteout, paint, or any substance which will adhere to a metal surface. Under no circumstances should the mark be a series of notches, or gouges, or even scribed. All these can cause stress risers, and under heavy engine load can cause parts to crack and fail.

POWERHEAD SYSTEMS 6-53

Fig. 169 If equipped remove the upper seal housing bolts

Fig. 170 Remove the engine top cover

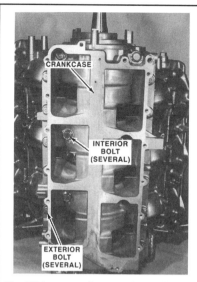

Fig. 171 Loosen and remove the crankcase retaining bolts. . .

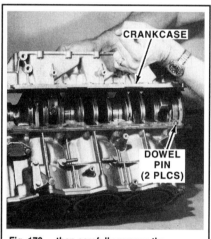

Fig. 172 . .then carefully remove the crankcase from the block

Fig. 173 Remove the upper seal housing from the crankshaft

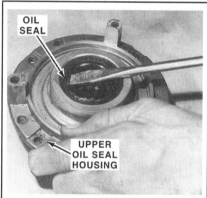

Fig. 174 If necessary, remove the seal from the upper housing

■ In most Yamaha motors of this size, these connecting rods were cast as one piece. Holes were then drilled where they would normally separate, and then the two halves were physically broken apart. Therefore, each mating surface is unique, one of a kind. These rods can never be resized and can never be made to match another set.

■ It has been our experience that most Yamaha connecting rod bolts require the use of a thin-walled, 12-point socket (usually 8mm in size) - a rare animal - not in every one's tool kit. Snap-on Tool Company (one of our FAVORITE TOOL COMPANIES!) markets such a socket. Without modification, this socket can be used successfully to remove the rod bolts. If this particular socket is not available for 8mm size bolt heads, a 5/16 inch 12 point socket will USUALLY suffice. However, the external diameter of most sockets, with the exception of the first one listed, will be too great to fit between the bolt and the cap. As a last resort, socket may be ground down to fit, but not without its disadvantages: First, grinding away material from the external wall will weaken the socket and the walls may break away. Secondly, upon installation, the rod bolts are tightened to the specified torque value. If the grinding is not symmetrical, any high spots on the external socket wall may flake off when used and fall down into the cylinder bore.

    2. Alternately and evenly loosen the connecting rod cap bolts on each of the rods. Lift off the cap and caged roller bearing beneath the cap.

■ Cleanliness is the password when handling bearings. Take care to prevent any dirt, lint or other contaminants from getting onto the bearings or in the cages. If the bearings are to be used again, store them either in a numbered container to ensure they will be installed with the same rod and cap from which they were removed OR as soon as the crankshaft is removed, return the bearings and cap to the appropriate connection rod (we prefer this second method). Never mix roller bearings from one rod to another. Never mix used roller bearings with new bearings. If just one bearing is unfit for further service, the entire set must be replaced.

    New bearings really should be installed in the connecting rods, even if the old bearings appear to be in serviceable condition. New bearings will ensure lasting service after the overhaul work is completed. If it is necessary to install the used bearings, keep them separate and identified to ensure they will be installed onto the same crank pin throw and with the same connecting rod from which they were removed.

    3. Continue removing the rod caps and bearing halves until all have been removed from the crankshaft.

    4. Tap the crankshaft lightly with a soft head mallet to jar it free from the block. Lift the crankshaft out of the block. Immediately return the connecting rods and bearings back to the their respective connecting rods. Cover the ends of the rod caps with shop rags to prevent potential damage to the cylinder walls.

6-54 POWERHEAD

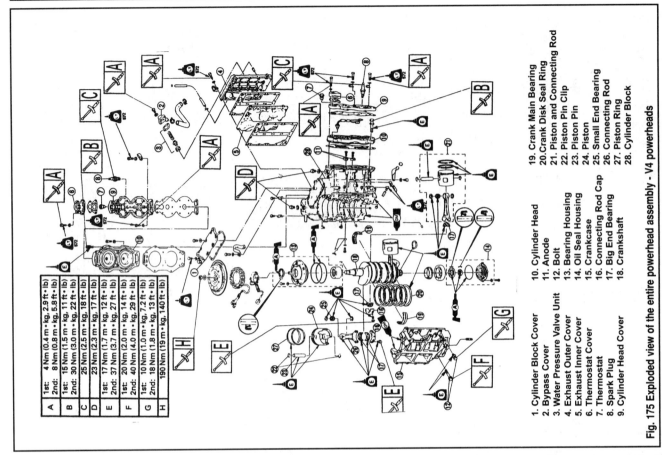

Fig. 175 Exploded view of the entire powerhead assembly - V4 powerheads

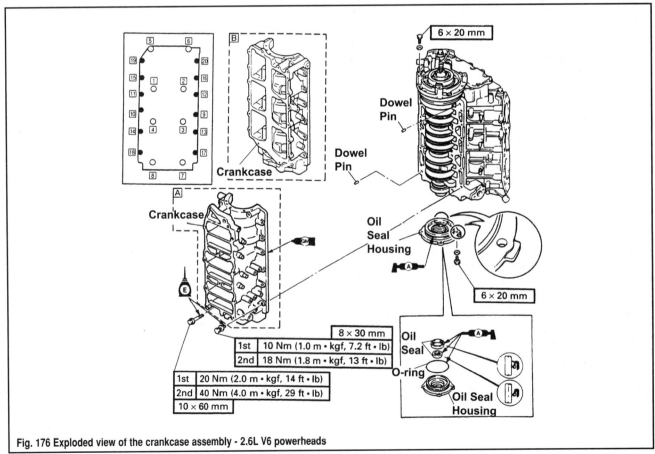

Fig. 176 Exploded view of the crankcase assembly - 2.6L V6 powerheads

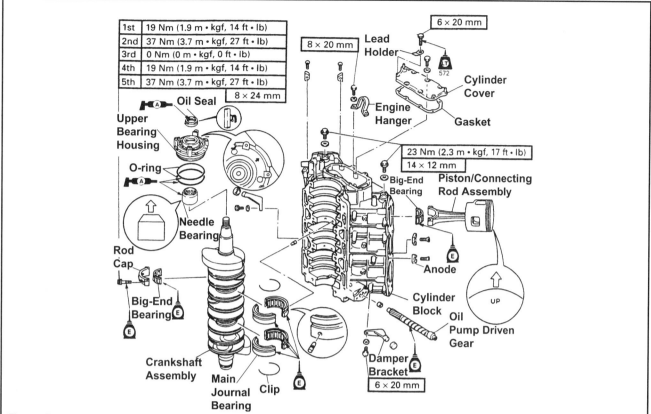

Fig. 177 Exploded view of the cylinder block and crankshaft - V4 powerheads (carbureted shown, EFI/HPDI very similar, though some models use dual bearings in the upper housing)

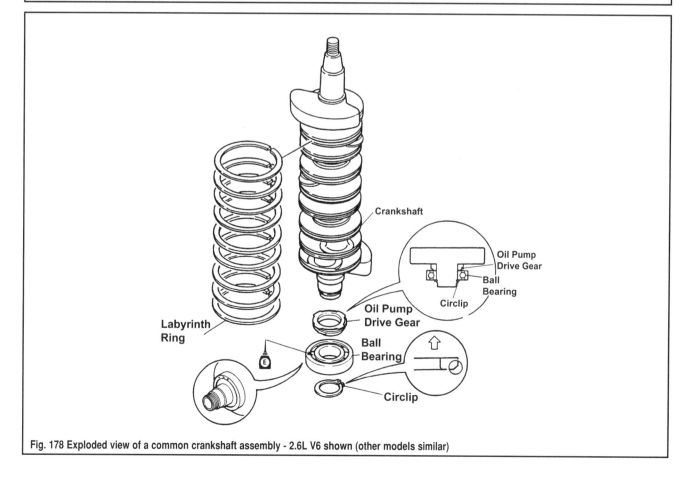

Fig. 178 Exploded view of a common crankshaft assembly - 2.6L V6 shown (other models similar)

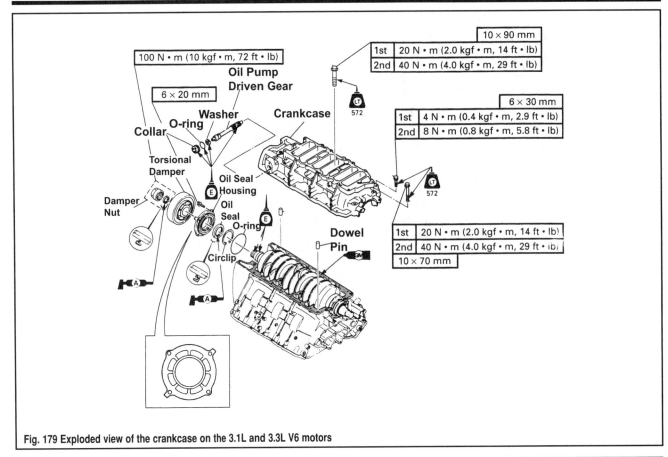

Fig. 179 Exploded view of the crankcase on the 3.1L and 3.3L V6 motors

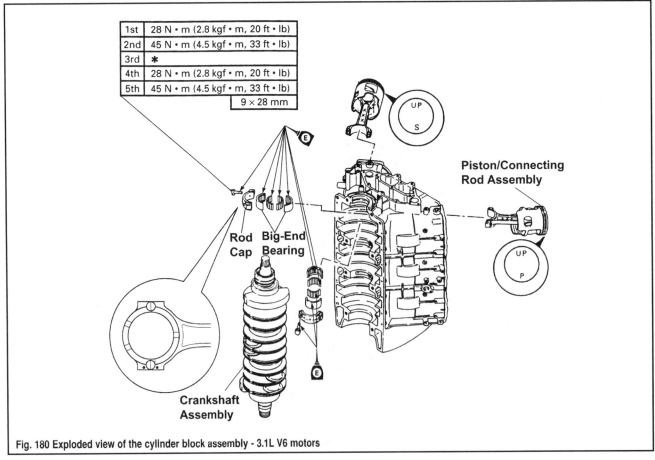

Fig. 180 Exploded view of the cylinder block assembly - 3.1L V6 motors

POWERHEAD SYSTEMS 6-57

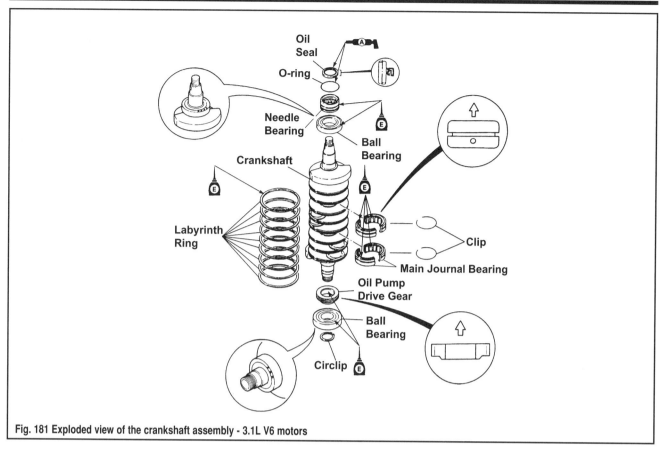

Fig. 181 Exploded view of the crankshaft assembly - 3.1L V6 motors

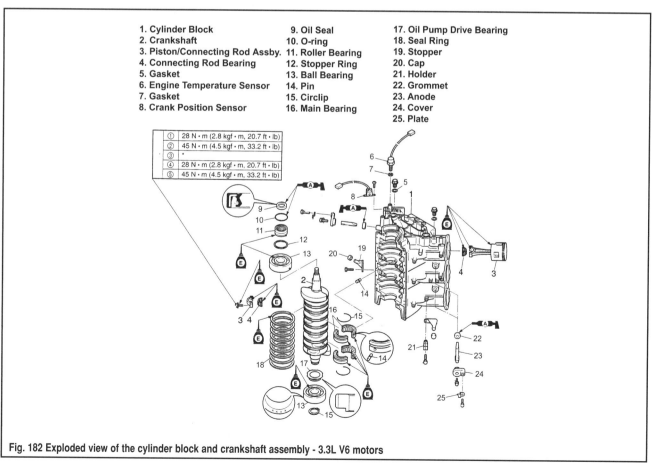

Fig. 182 Exploded view of the cylinder block and crankshaft assembly - 3.3L V6 motors

6-58 POWERHEAD

Fig. 183 Check for or make ID marks on the con rods/caps

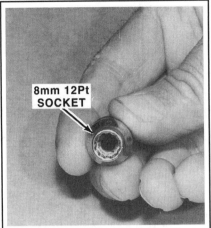

Fig. 184 Most con rod bolts require a 12-point socket

Fig. 185 Alternately loosen the bolts for each con rod cap

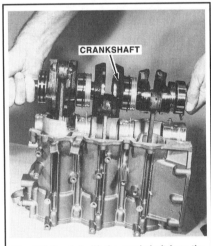

Fig. 186 Carefully lift the crankshaft from the block

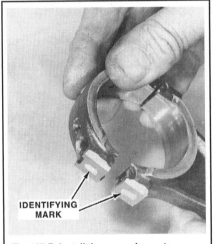

Fig. 187 Reinstall the con rod caps to prevent potential mix-up

Fig. 188 Remove the clips and main bearings from the crank

5. Most motors use a circlip to secure each of the main bearings. Remove the large circlip from around each set of main bearings. Remove the two halves of the bearing shell and the two halves of the roller bearing. If the main bearings are to be reused, keep each set together, so they may be installed back into their original locations.

■ New main bearings should be installed in the crankcase/cylinder block, even though they may appear to be in serviceable condition. New bearings will ensure lasting service after the overhaul work is completed. If it is necessary to install the used bearings, keep them separate and identified to ensure they will be installed in the same location from which they were removed.

Pistons and Rings

◆ See Figures 189 thru 196

1. Pull each piston and connecting rod assembly from the bottom - not through the top of the block. A ridge might have formed on the top of the cylinder bore. If so, this ridge may have to be removed with a ridge reamer before installation (otherwise you'd risk breaking a piston ring).

2. Identify the factory marks and/or make an identifying mark on the outside edge of each rod I beam and a matching mark on the inside of each piston skirt. The identification mark must match the cylinder from which the piston and rod were removed.

■ This mark should be made with a marker, whiteout, paint or any substance which will adhere to a metal surface. Under no circumstances should the mark be a series of notches, or gouges, or even scribed. All these can cause stress risers, and under heavy engine load can cause parts to crack and fail.

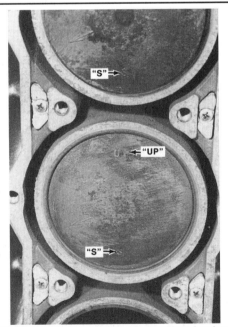

Fig. 189 Before removal, note and factory markings for piston orientation. . .

POWERHEAD SYSTEMS 6-59

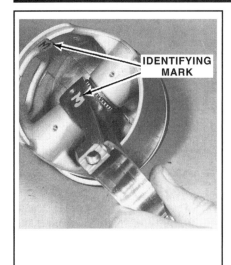

Fig. 190 . . .and add additional marks as necessary to prevent piston/rod mix-up

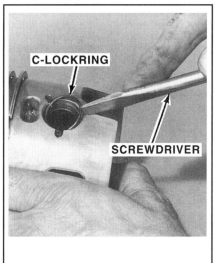

Fig. 191 Carefully remove the piston pin lockrings. . .

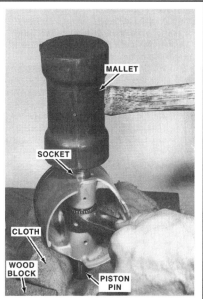

Fig. 192 . . .then use a driver to carefully tap out the pin

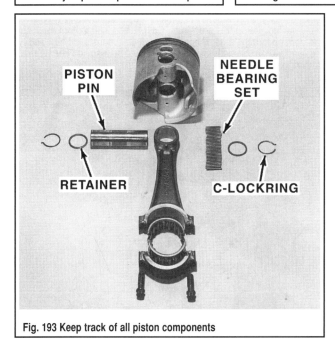

Fig. 193 Keep track of all piston components

Fig. 194 You can remove the piston rings by hand

3. Remove the C-lockring from both ends of the piston pin using a pair of needle nose pliers or an awl. Discard the C-lockrings. These rings normally stretch during removal and cannot be used a second time.

✱✱ WARNING

The piston pin C-lockrings are made of spring steel and may slip out of the pliers or pop out of the groove with considerable force. Therefore, wear eye protection glasses while removing the piston pin lockrings in the next step.

4. Slide the piston pin free of the piston. Although most Yamaha piston pins are not of an interference fit, tolerances are usually close enough to require the use of a driver to carefully tap (using a mallet) or push (using an arbor press) the pin free of the piston bore. Position the piston on a padded surface and use a long socket or suitable driver to push on the end of the pin. If equipped with loose needle bearings (as opposed to caged bearings), be sure to catch or recover all of the needles, as well as the two small spacers, as they fall free of the piston pin bore.

■ Good shop practice dictates to replace the rings during a powerhead overhaul. However, if the rings are to be used again, expand them only enough to clear the piston and the grooves, because used rings are brittle and break very easily.

5. Gently spread the top piston ring enough to pry it out and up over the top of the piston. No special tool is required to remove the piston rings. Remove the remaining ring(s) in a similar manner. These rings are extremely brittle and have to be handled with care if they are intended for further service.

Crankshaft and Bearing Disassembly

◆ See Figures 197, 198 and 199

1. On oil injected powerheads, inspect the bronze oil injection worm gear at the lower end of the crankshaft. If the gear needs to be replaced, the lower main bearing must be pulled from the crankshaft first. The bronze worm gear may then be pulled.

6-60 POWERHEAD

2. Slowly rotate the bearing race on the bottom (and on some models on the top) of the crankshaft. If rough spots are felt, the bearing(s) will have to be pressed free of the crankshaft. Most lower bearings are secured using a snapring. Use a pair of snapring pliers to remove the snapring from around the lower end of the crankshaft.

3. If the bearing or bearing and oil injection worm gear must be replaced, use a bearing separator and either universal puller or an arbor press to force the bearing off the crankshaft. When using an arbor press, be sure the crankshaft is supported, because once the bearing breaks loose, the crankshaft is free to fall. Relocate the bearing separator and press the bronze oil injection gear from the crankshaft.

4. Gently spread each of the crankshaft sealing (labyrinth) rings enough to ease it out and up over the top of the groove and into a journal space. Then, once again separate the two ends enough to clear the journal, and lift the ring free. Take care not to scratch the highly polished journal surface as each ring is removed.

ASSEMBLY

Crankshaft and Bearing Assembly

◆ See Figures 175 thru 182 and 200

■ Remember to align all matchmarks made during removal. Components such as the pistons, connecting rods, rod end caps (if equipped) must be positioned with the same sides facing the flywheel end of the crankshaft during assembly.

1. Install the crankshaft sealing (labyrinth) rings by spreading each ring enough to clear the journal adjacent to its groove. Once the ring is around the journal, spread the ends gently again and slide it up and over the upper edges and into its groove.

■ Perform the next few steps only if the ball bearing(s) was(were) removed during disassembly. If the ball bearing was not disturbed, apply a coat of clean engine oil to the inner bearing surfaces. The upper bearings for 1.7L and 2.6L motors are installed in an external seal housing later in this procedure. On 3.1L and 3.3L motors there are two upper bearings, the ball bearing which can be installed now and the needle bearing/seal housing which should be installed later in this procedure.

2. Install the bronze oil injection gear over the lower end of the crankshaft. Using an arbor press, slowly and evenly install the gear down until it seats against the shoulder on the shaft.

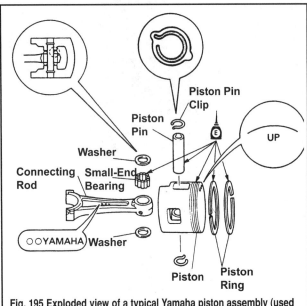

Fig. 195 Exploded view of a typical Yamaha piston assembly (used on most V4 and V6 powerheads)

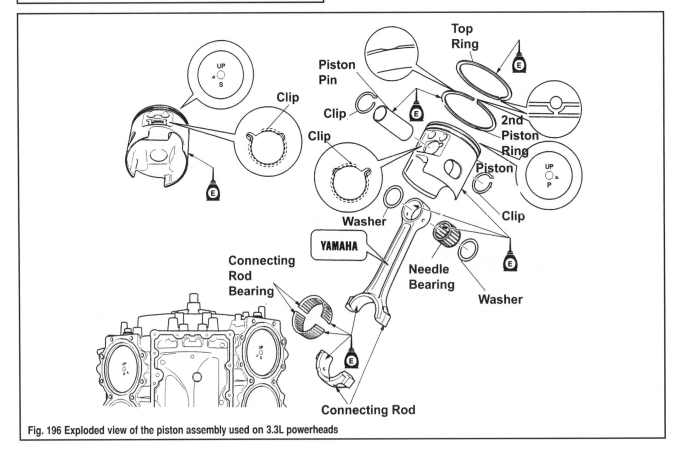

Fig. 196 Exploded view of the piston assembly used on 3.3L powerheads

POWERHEAD SYSTEMS 6-61

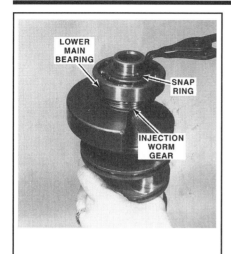

Fig. 197 If necessary, remove the oil pump gear and/or lower bearing

Fig. 198 Use a bearing separator and puller or press to remove pressed on crankshaft bearings

Fig. 199 Remove the labyrinth seal rings from the crankshaft

3. Position the ball bearing over the shaft with the embossed marks (if present) on the bearing surface facing upward towards the press shaft. Now, use a suitable mandrel and press against the inner bearing race. Continue to press the bearing into place until the bearing is seated against the oil injection gear (lower bearing) or crankshaft throw (upper bearing), as applicable.

✹✹ WARNING

Take care to ensure the mandrel is pressing on the inner race and not on the outer race or the ball bearings. Such action would destroy the bearing.

4. For lower bearings, use a pair of snapring pliers to install the ring into the crankshaft groove.

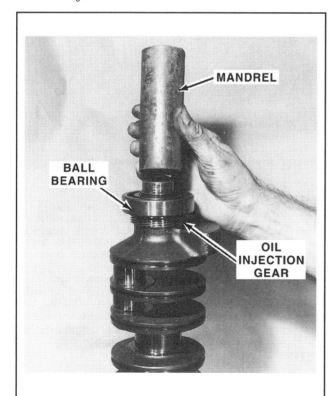

Fig. 200 Use a suitable driver to install pressed bearings

Pistons and Rings

◆ See Figures 195 and 196 and 201 thru 205

■ Be sure to coat each of the piston components with clean engine oil during assembly.

1. Install a new set of piston rings onto the piston, with the embossed marks (if present) facing up. No special tool is necessary for installation. However, take care to spread the ring only enough to clear the top of the piston. The rings are extremely brittle and will snap if spread beyond their limit. Align the ring gap over the locating pin.

✹✹ CAUTION

The piston pin lockrings are made of spring steel and may slip out of the pliers or pop out of the groove with considerable force. Therefore, wear eye protection glasses while installing the piston pin lockrings in the next step.

2. Gather all the components needed to assemble the piston for installation into No. 1 cylinder bore: the piston, two C-lockrings, two retainers, the piston pin, needle bearing set, and the connecting rod.
3. Select the set of needle bearings removed from the No. 1 piston or obtain a new set of bearings.

■ Most Yamaha models use loose needle bearings, though a few are equipped with caged bearing sets.

4. Install the needle bearings to the connecting rod small ends. For loose bearing models, coat the inner circumference of the small end of the each connecting rod using a marine grade grease. Position the needle bearings one by one around the circumference.
5. Position the end of the rod with the needle bearings and retainers in place up into the piston with the word **Yamaha**, letter **Y** or other appropriate marking facing the same direction as the word **UP** on the piston crown. The word **UP** must face toward the flywheel (upper) end of the crankshaft.
6. Dab some lubricant on the sides of the rod and stick the retainers in place.
7. Insert the piston pin into the piston and slowly press the pin into position through the piston and connecting rod. Make sure one of the needles (on loose needle models) or retainers do not cock and bind the pin as it is installed. Once inserted, center the pin in the piston.
8. Install the C-lockring at each end of the piston pin.
9. Repeat these last 5 steps to assemble the remaining connecting rods and pistons.
10. Coat the cylinder bores with a good grade of engine oil. Before installing the piston into the cylinder, make the following test:

6-62 POWERHEAD

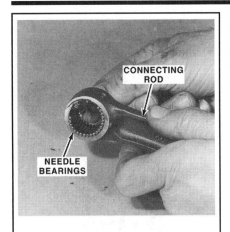

Fig. 201 Insert the needle bearings to the rod small end (use grease to help hold them)

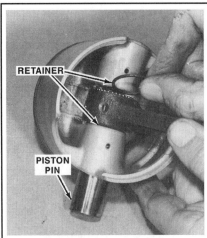

Fig. 202 Insert each retainer as the pin is slowly slide into position

Fig. 203 If there is a ridge, use a reamer on the crankcase bore before piston installation

Fig. 204 When present, place the ring gap(s) over the locating pin(s)

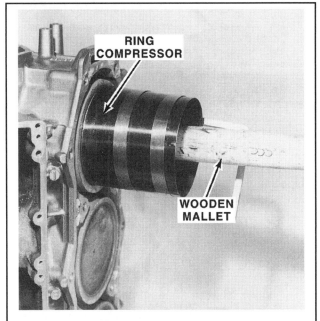

Fig. 205 Install each piston using a ring compressor

11. Run your finger along the top rim of the cylinder. If the surface of the block and the surface of the bore have a sharp edge, the piston may be installed from the top. If the slightest groove or ridge is felt on the rim, the ridge must be removed using a ridge reamer. Attempting to install the piston from the top without removing the ridge, will not be successful because the piston ring will bottom on the ridge. If force is used the ring may very well break.

12. If not done already, align each ring gap over its locating pin.

■ The word UP embossed on the piston crown must face toward the flywheel end of the block. When present, the embossed letter P must face port side, and the embossed letter S must face starboard.

13. Obtain a ring compressor (applying a light coat of oil to the inside of the ring compressor, similar to how you coated the cylinder bore). Compress the aligned rings and then use the end of a wooden mallet handle to gently tap the piston down into the cylinder bore. Repeat for the remaining pistons.

14. After all pistons have been installed, slide each piston up and down in the cylinder bore several times. Check for binding. Listen for scratching noises. Scratching, or any other spooky noise, may indicate a ring was broken during installation.

Crankshaft, Bearings and Upper Oil Seal Housing

◆ See Figures 175 thru 182 and 206, 207 and 208

1. On 1.7L and 2.6L motors, a needle bearing (or 2 bearings on HPDI models) is (are) pressed into the upper oil seal housing. If removed, install NEW bearings to the housing using. Obtain a special driver (#YB6205) or an equivalent sized driver which pushes only on the portion of the bearing cage that contacts the housing. Support the upper oil seal housing in an arbor press with the O-ring end facing upward. Lower the needle bearing into the housing from the O-ring end. Although recent factory information doesn't mention it, the older Yamahas used to install the bearing with the stamped mark on the bearing facing downward - away from the driver. Seat the bearing squarely into the bore, until the shoulder of the driver seats against the shoulder of the housing. The bearing will then be correctly seated in the housing. Invert the housing and leave it in place on the arbor press for the next step.

■ On some models which use an external bearing/seal housing the bearing has a tapered side, which should be installed with the taper faced away from the crankshaft (toward the housing).

POWERHEAD SYSTEMS 6-63

Fig. 206 Apply a light coat of grease and install the seal(s) to the oil housing

Fig. 207 Grease the O-rings and position the upper oil seal housing over the crank

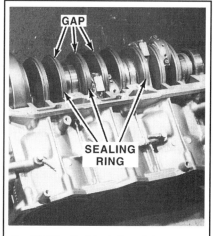

Fig. 208 Align the labyrinth seal ring gaps with their locating pins

2. Pack the lips of a new upper seal with Yamalube Grease, or equivalent marine grade water resistant lubricant. Install the new seal into the upper housing (1.7L and 2.6L motors) or to the upper bearing/seal assembly (3.1L and 3.3L motors) with the seal lips facing downward (toward the crankshaft when the housing and seal are installed).

3. Install a NEW O-ring (2 on 1.7L and 2.6L motors) around the upper seal housing assembly. Apply a light coat of marine grade grease to the O-ring, then position the assembly aside for positioning right before the crankshaft is installed to the cylinder block.

■ If the old roller bearings are to be installed for further service, each must be installed in the same location from which it was removed. On some V4 crankcases there are center bearing locating holes between the No. 2 and No. 3 cylinders. On some V6 crankcases, these holes are located between the No. 2 and No. 3 cylinders, also between the No. 4 and No. 5 cylinders. When present, these locating holes must match tangs on the main bearing shells during installation, or the crankcase will not seat properly.

4. Remove each of the bearing caps from their respective connecting rods, making sure each of the bearing halves are in their proper positions, then place each of the pistons toward the top of the cylinder bore in a position which will allow the crankshaft to be positioned into the block.

5. Oil and install the main bearing halves and clips around the main journals of the crankshaft, hold them together while lowering the crankshaft down onto the connecting rod bearings.

6. Slide the upper bearing/seal housing over the tapered end of the crankshaft and carefully lower the crankshaft into the cylinder block. On 1.7L and 2.6L motors, rotate the installed housing until the arrow embossed on the top points directly towards the centerline of the cylinder block.

■ In days gone by the manufacturer did not recommend reusing rod cap bolts. However no mention of this is made in current Yamaha service publications. We'll leave the decision to you, but keep how important a job the rod caps perform.

7. One at a time, carefully draw connecting rods up into contact with the crankshaft, then install the cap and bearing half while aligning the matchmarks. Install the connecting rod cap bolts, then tighten them alternately and evenly as follows:
• First tighten the bolts to 14 ft. lbs. (19 Nm) for 1.7L and 2.6L motors, or to 20 ft. lbs. (28 Nm) for 3.1L and 3.3L motors.
• Second, tighten the bolts to 27 ft. lbs. (37 Nm) for 1.7L and 2.6L motors, or to 33 ft. lbs. (45 Nm) for 3.1L and 3.3L motors.
• Third, loosen both bolts to remove all torque from them. Double-check to make sure that all matchmarks are aligned and the rod cap is perfectly centered and aligned with the rod.
• Fourth, tighten the bolts to 14 ft. lbs. (19 Nm) for 1.7L and 2.6L motors, or to 20 ft. lbs. (28 Nm) for 3.1L and 3.3L motors.
• Fifth (and finally), tighten the bolts to 27 ft. lbs. (37 Nm) for 1.7L and 2.6L motors, or to 33 ft. lbs. (45 Nm) for 3.1L and 3.3L motors.

■ If the cap and rod do not align properly, remove the cap bolts, and then the upper caged roller bearing half. Install the upper caged roller bearing half again. Check the alignment. If the second attempt fails to align the marks as described, the crankshaft must be removed and the lower caged roller bearing half checked for correct alignment.

8. Arrange the end-gaps of all the crankshaft sealing rings to face upward. If used, install the main bearing locating pins.

Lower Oil Seal Housing and Crankcase

◆ See Figures 166, 167, 168 and 209 thru 212

1. If the oil seal(s) in the lower oil seal housing were removed during disassembly, install new oil seals in the housing using a handle and suitable driver. Be sure to position the seals as noted during removal. Seal installation generally varies with model as follows:
• On 1.7L and 2.6L motors there are 2 seals both of which are installed from the crankshaft side of the housing. The seal lips are normally faced downward away from the crankshaft.
• On 3.1L and 3.3L motors, there are usually 2 seals. One seal is in the crankshaft damper or damper nut, and is positioned with the facing upward toward the crankshaft. The other is mounted in the top of the oil seal housing, and its lips face downward, away from the crankshaft. If equipped, install the large circlip into the housing, after the seal is installed. Install the damper/nut seal at this time, but put it aside for use after the crankcase is installed.

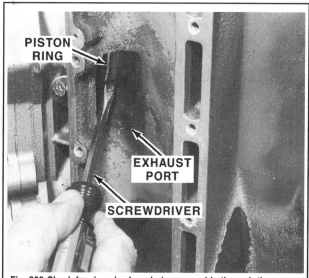

Fig. 209 Check for rings broken during assembly through the exhaust ports

6-64 POWERHEAD

2. Install a NEW O-ring to the lower seal housing, then apply a light coating of marine grade grease to the O-ring and to the lower housing seal (and the damper/nut seal on 3.1L and 3.3L motors). Position the seal housing to the cylinder block. On some models (like the 3.3L motors) there is a stamped arrow and/or a stamped **F** on the housing which should be pointed outward, toward the crankcase once the case is installed.

3. Apply a thin bead of Gasket Maker sealant around both surfaces of the crankcase and block. Check to be sure the two dowel pins are in place, then carefully lower the crankcase onto the block.

4. Apply a light coat of engine oil to the threads of the crankcase retaining bolts for all motors except the 3.1L powerheads, on which the threads should be instead coated with Loctite®572, or an equivalent threadlocking compound.

5. Install and finger-tighten the long and short crankcase attaching bolts. Then, tighten the bolts using multiple passes of the embossed or illustrated torque sequence as follows:
 • For 1.7L and 2.6L motors, tighten the short bolts first to 7.2 ft. lbs. (10 Nm), then to 13 ft. lbs. (18 Nm) and long bolts first to 14 ft. lbs. (20 Nm), then to 29 ft. lbs. (40 Nm).
 • For 3.1L and 3.3L motors use 2 separate torque sequences. The tighten the longer center bolts using the sequence, then tighten the shorter flange bolts. Tighten long bolts first to 15 ft. lbs. (20 Nm), then to 29 ft. lbs. (40 Nm) and the short bolts first to 3 ft. lbs. (4 Nm), then to 6 ft. lbs. (8 Nm).

6. Check to be sure each piston ring has spring tension. This is accomplished by carefully pressing on each ring with a screwdriver extended through the exhaust ports, as shown in the accompanying illustration. If spring tension cannot be felt (the spring fails to return to its original position), the ring was probably broken during the piston and crankshaft installation process. Take care not to burr the piston rings while checking for spring tension.

✱✱ WARNING

Once the crankcase bolts are installed and properly tightened, slowly and carefully rotate the crankshaft. If binding is felt, it will be necessary to remove the crankcase and reseat the crankshaft and also to check the positioning of the crankshaft sealing rings and the bearing locating pins. If binding is still a problem after the crankcase has been installed a second time, the cause might very well be a broken piston ring.

7. Check the upper and lower oil seal housings to make sure they are installed properly. On models with external upper seal housing there is normally an embossed arrow which must be positioned facing the center line of the cylinder block (on the cylinder head side). On some models such as the 3.3L a mark is embossed on the LOWER oil seal housing which should be pointed the opposite direction of the upper housing mark (toward the crankcase).

8. Install the upper and/or lower oil seal housing retaining bolts and tighten securely, alternating in a star pattern from side to side.

9. If removed, install the engine top cover using a new gasket. Tighten the bolts securely using the molded or illustrated sequence (or if not present, using a clockwise spiraling pattern that starts at the center and works outward).

10. If applicable, install the oil pump to the cylinder block.

11. On fuel injected models, reconnect the various bleed/recirculation hoses located on the crankcase. Pay very close attention to hose routing for installation purposes. Use the accompanying illustrations or digital pics you took before starting for a reference.

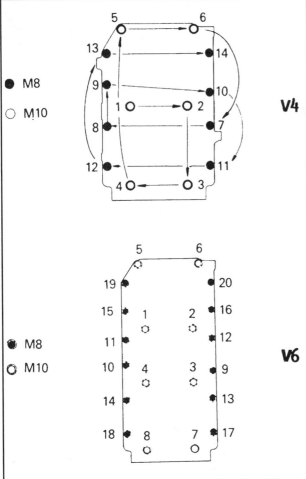

Fig. 210 Crankcase torque sequences - 80J-140 Hp (1730cc) V4 and 105J-225 Hp (2596cc) V6 Motors

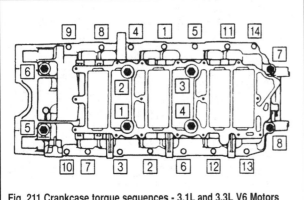

Fig. 211 Crankcase torque sequences - 3.1L and 3.3L V6 Motors

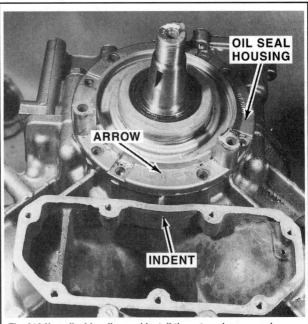

Fig. 212 If applicable, align and install the external upper seal housing

POWERHEAD SYSTEMS 6-65

Cylinder Heads

◆ See Figures 164, 165 and 213 thru 216

1. If used, insert the wedge shaped anodes into the water jacket of the block. Secure the anodes with Phillips head screws. These anodes will only be visible and accessible the next time the powerhead is overhauled. Therefore, if the anodes show ANY sign of deterioration in this location, a NEW anode should be installed.
2. Position a new head gasket in place on the powerhead.

■ Because of the high temperatures and pressures developed, the sealing surfaces of the cylinder head and the block are the most prone to water leaks. No sealing agent is recommended because it is almost impossible to apply an even coat of sealer. An even coat would be essential to ensure an air/water tight seal.

3. Install the cylinder head onto the powerhead.
4. For all models, except 3.1L motors (which use the cylinder cover bolts to also secure the head), apply a light coat of engine oil to the threads of the cylinder head retaining bolts for then install and tighten them using multiple passes of the illustrated or embossed sequence. Tighten the bolts first to 11 ft. lbs. (15 Nm) and then to 22 ft. lbs. (30 Nm).

5. Position a new cylinder head cover gasket in place on the cylinder head, then install the cover (on some models this includes the thermostat assembly, whose cover should also be installed at the same time, using a new gasket). Apply a light coating of Loctite®572, or equivalent to the threads of the cylinder head cover retaining bolts. Install and tighten the attaching bolts following multiple passes of the embossed or illustrated sequence. Tighten the bolts as follows, depending upon the model:
• For 1.7L and 2.6L motors (except HPDI models), tighten the bolts first to 2.9 ft. lbs. (4 Nm) and then to 5.8 ft. lbs. (8 Nm)
• For 2.6L and 3.3L HPDI motors, tighten the short bolts first to 3.6 ft. lbs. (5 Nm), then to 8 ft. lbs. (11 Nm).
• For 3.1L motors, tighten the short bolts first to 2.9 ft. lbs. (4 Nm), then to 5.8 ft. lbs. (8 Nm) and the long bolts (be careful, there are 2 different length long bolts) first to 11 ft. lbs. (15 Nm), then to 22 ft. lbs. (30 Nm).

6. On models where the thermostats are installed to the cylinder head or block (as opposed to the cylinder head cover), install them now, using a new gasket for each cover. For details, please refer to the Lubrication and Cooling section.
7. Install the spark plugs.

Exhaust Cover

◆ See Figures 156, 157 and158 and 217 thru 219

1. On 3.3L motors, apply a light coating of marine grade grease to the grommet for the exhaust cover anode. Install the anode and grommet to the anode cover plate and tighten the short retaining bolt. Install the anode and cover assembly to the exhaust cover and secure using the longer retaining bolt.
2. On the 1.7L and 2.6L powerheads, install the bypass/pressure control valve assembly to the exhaust cover. Install the valve with the longer end inserted into the exhaust cover, then place the spring over the short end of the valve. Install a new gasket and install the cover. Tighten the bolts securely.
3. Apply a light coating of Loctite® 572 to the threads of the exhaust cover retaining bolts. Install the exhaust cover using a new gasket (or inner and outer covers, each with a new gasket, as applicable). Install the exhaust cover retaining bolts and tighten them using multiple passes of the molded or illustrated torque sequence to specification, as follows:
• For 1.7L, 2.6L and 3.1L motors, tighten the bolts first to 2.9 ft. lbs. (4 Nm) and then to 5.8 ft. lbs. (8 Nm).
• For 3.3L motors, tighten the bolts first to 3.7 ft. lbs. (5 Nm) and then to 8.1 ft. lbs. (11 Nm).

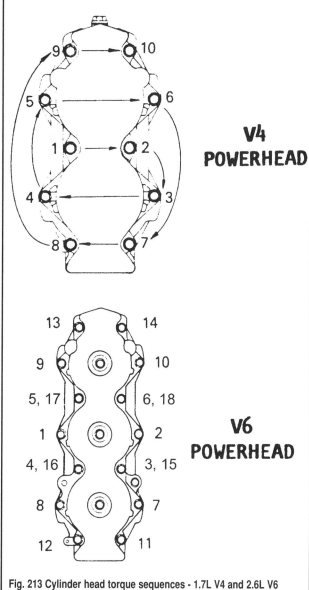

Fig. 213 Cylinder head torque sequences - 1.7L V4 and 2.6L V6 Motors

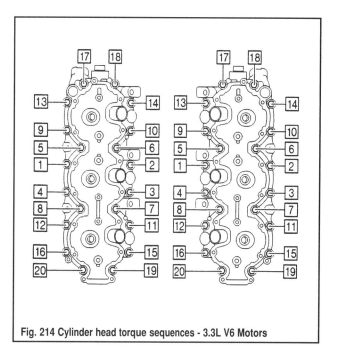

Fig. 214 Cylinder head torque sequences - 3.3L V6 Motors

6-66 POWERHEAD

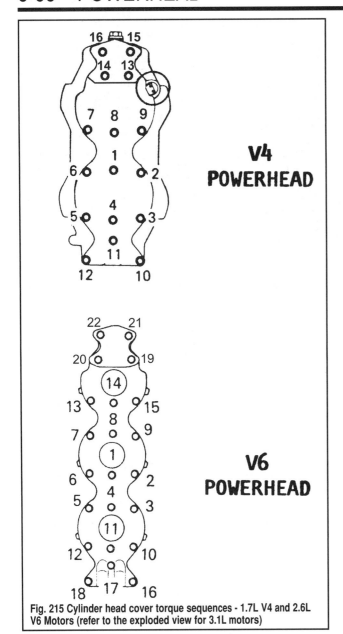

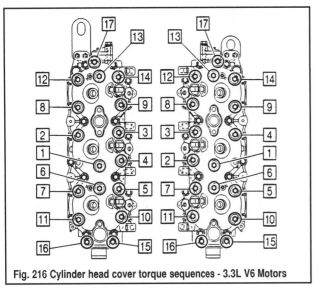

Fig. 215 Cylinder head cover torque sequences - 1.7L V4 and 2.6L V6 Motors (refer to the exploded view for 3.1L motors)

Fig. 216 Cylinder head cover torque sequences - 3.3L V6 Motors

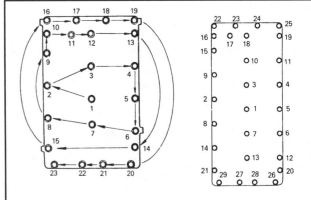

Fig. 217 Exhaust cover torque sequences - 1.7L V4 (top) and 2.6L V6 Motors except HPDI (bottom)

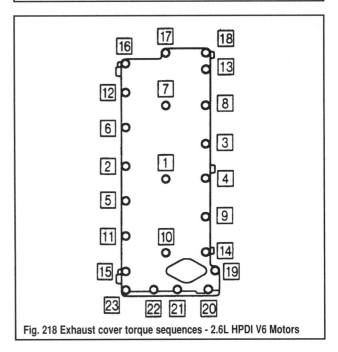

Fig. 218 Exhaust cover torque sequences - 2.6L HPDI V6 Motors

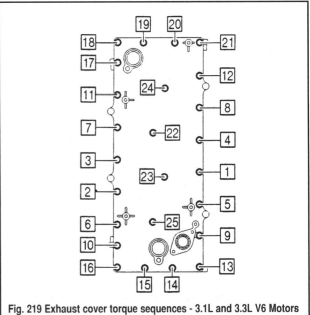

Fig. 219 Exhaust cover torque sequences - 3.1L and 3.3L V6 Motors

POWERHEAD SYSTEMS

Intake Manifold and Reed Valves

◆ See Figures 146, 147 and 148 and 220

1. Install the intake manifold/reed valve assembly to the crankcase.
2. On fuel injected models install and tighten the 2 intake manifold retaining bolts to 7.2 ft. lbs. (10 Nm) for all except 3.3L motors on which the bolts should be tightened to 9.6 ft. lbs. (13 Nm).
3. On carbureted models, install and tighten the 14 (V4) or 16 (V6) intake manifold retaining bolts. Tighten the bolts first to 2.9 ft. lbs. (4 Nm) and then to 5.8 ft. lbs. (8 Nm), using multiple passes of the molded or illustrated torque sequence.
4. Reconnect the oil supply and, for carbureted models, the bleed/recirculation hoses to the manifold, crankcase and cylinder block assembly (most of these lines are of different lengths), as tagged and/or noted during removal.
5. For carbureted motors, reconnect the intake manifold air vent hose, then, as necessary, install the timing pointer, fuel enrichment valve (V6) and/or the hose clamp (V4) to the intake manifold.
6. On 3.1L and 3.3L engines, install the balancer to the bottom of the crankshaft. Use the Yamaha Flywheel Holder (#YB-16139), or equivalent, inserted into the damper and, on some models, a crankcase bolt hole) to keep the damper from turning while you tighten the retaining nut to 73.8 ft. lbs. (100 Nm).
7. Either install the remaining fuel and electrical components which were stripped for overhaul or wait until the powerhead assembly is positioned on and bolted to the intermediate housing. The choice is yours. At the very least, you probably want to hold off on flywheel installation if it might interfere with a powerhead lift bracket.
8. Install the powerhead assembly. Be sure to follow the proper break-in procedures if bearings, rings or other wear parts were replaced.

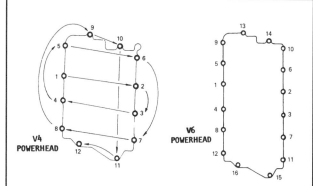

Fig. 220 The most common carbureted engine intake manifold torque sequences (refer to the intake manifold exploded views for more details)

CLEANING & INSPECTION

◆ See Figures 175 thru 182

Cleaning and inspecting the components is virtually the same for any two-stroke outboard and varies mostly by specifications (which are listed in the Engine Specifications charts) or by component type. A section detailing the proper procedures, sorted mostly by component, can be found under Powerhead Refinishing.

POWERHEAD REFINISHING

General Information

The success of the overhaul work is largely dependent on how well the cleaning and inspecting tasks are completed. If some parts are not thoroughly cleaned, or if an unsatisfactory unit is allowed to be returned to service through negligent inspection, the time and expense involved in the work will not be justified with peak powerhead performance and long operating life. Therefore, the procedures in the following sections should be followed closely and the work performed with patience and attention to detail.

■ **When measuring components, remember that specifications may vary with temperature. Unless otherwise noted, specifications are for components at an ambient room temperature of 68°F (20°C).**

CLEANING

◆ See Figures 221 thru 226

All powerhead components must be clean and free of gasket material, oil and carbon deposits before they are inspected. Take your time when cleaning components. Before using solvent to clean something, make sure the chemical is compatible with the material of which the component is constructed. Also, check each component for matchmarks or ID marks before cleaning (as some solvents may remove even permanent marker). If necessary, re-ID the component as soon it has been cleaned and dried (before moving onto the next component). Whenever you match-marked 2 components in order to ensure exact alignment during installation, if possible, fasten them together during cleaning (in case the marks come off). In this way, they can be re-marked after the cleaning process is through.

When it comes to inspection, this section will provide information on how to check various components. But, keep in mind that not all components will be found on all motors. If in doubt whether a component is used (or should be checked) refer to the Disassembly, Assembly or Overhaul procedures (as applicable). Also, be sure to check the Specifications Chart for your motor. If a component is not listed in the specification chart, it is either not used, or, does not have a specific tolerance for which it must be inspected. Whether or not a tolerance is provided, all components must be clean and free of obvious defects (deep cracks, scoring, excessive carbon deposits that cannot be removed, warpage, etc). If in doubt whether or not a component is serviceable, seek someone with more Yamaha engine rebuilding experience than yourself.

Generally speaking, the larger the hp model, the more involved will be the cleaning and inspection process. This is not because of size, as much as it has to do with the larger motors being designed to produce more hp and sell for much more money, so they are usually designed with more components/systems.

When inspection involves precision measurements, not only must the components be clean, but they must be measured roughly at room temperature, using the appropriate measurement equipment to ensure accurate results.

It is very important to keep in mind that all wear components (pistons, rings, shafts, springs, valves, bearings, etc) **must** be reinstalled in their original locations whenever they are being reused. Wear patterns form on all contact surfaces during use. Mismatching wear patterns will accelerate wear, while matching wear patterns helps ensure a durable and reliable repair.

✱✱ WARNING

Avoid removing excessive amounts of metal when removing carbon deposits.

1. Use a blunt-tip scraper or dulled chisel to loosen carbon deposits from various components of the combustion chamber and ports/valves. Work slowly and carefully to prevent damaging or excessively scoring the surfaces. Then use a Scotchbrite® pad and mild solvent to remove most/all of the remaining deposits. Remove deposits from the following components, as applicable:
   • Remove carbon from the combustion chambers in the cylinder head.
   • Remove all carbon deposits from the areas around exhaust ports.
   • Remove carbon deposits or corrosion from the exhaust cover(s).
   • Remove carbon deposits from the top of the piston(s). When working on the piston domes, use a light touch to prevent scratching, or worse, gouging the piston.
   • Remove carbon from the ring grooves either using a ring groove cleaner, or, better yet, using a broken piece of the piston ring with an angle ground on the end. When using a ring to clean the piston grooves, use the ring actually removed from the groove (if it is being replaced) or one from the same groove on another piston.

Fig. 221 Although a blunt chisel is preferred, a wire brush can be used WITH CARE to clean most carbon deposits

Fig. 222 Pistons CAN be cleaned while still installed, but again use care, and leave NO metallic deposits behind

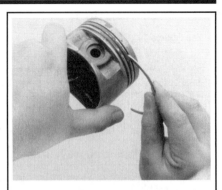

Fig. 223 The preferred method for cleaning piston grooves is to use the filed end of a broken ring...

Fig. 224 ... but a ring groove cleaning tool can be used, with care to prevent damaging the piston

Fig. 225 The cylinder walls must show an obvious cross-hatching (tiny grooves, criss-crossed in a pattern around the bore)...

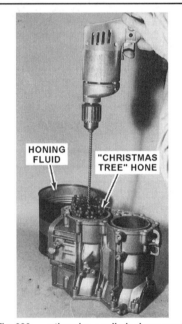

Fig. 226 ... otherwise a cylinder hone must be used to break the smooth glazed surface into a cross-hatch pattern

■ Deposits on pistons can be removed while they are still installed in the bores. This is handy when the cylinder head is removed for service without completely disassembling the crankcase. If this is to be accomplished, position the piston to be cleaned at TDC and cover the other piston bore(s) using rags and plastic. Thoroughly clean all debris using solvent and compressed air (WHILE WEARING SAFETY GLASSES) before moving to the next piston.

✶✶ WARNING

Wire brushes are not recommended for cleaning piston domes since particles of steel could become lodged into the piston surface. If this occurs, they could glow hot when the piston is returned to service, causing pre-ignition or detonation that could damage the piston and combustion chamber.

✶✶ WARNING

When cleaning piston ring grooves, use the same caution as with the pistons. Do NOT remove excessive amount of material or the piston will be damaged beyond use. Some ring groove cleaning tools are heavy duty and will easily remove too much material, so use them with care. Believe it or not some manufacturer recommend the filed broken ring method.

2. Inspect all water passages for corrosion deposits, debris or blockage. Remove debris and clean corrosion as needed and accessible. Use low-pressure compressed air to blow out all water passages.

3. Clean and degrease all regularly oiled surfaces (including the crankshaft, pistons and connecting rods) using solvent or degreaser. Use low-pressure compressed air to remove all build-up from shaft and rod oil holes.

4. Remove all traces of gasket or sealant using a spray gasket remover or equivalent gasket removing solvent. Whenever possible, avoid the use of gasket scrapers to help avoid the possibility of scoring and damaging the gasket mating surfaces.

■ Honing the cylinder walls can wait until after measurements are taken to determine if boring for oversized pistons/rings will be necessary. If honing is pushed off until after inspection, be sure to follow the remaining steps in order to finish cleaning the block, then repeat the appropriate steps again, after honing.

5. Check the cylinder walls for glazing (a smooth, glassy appearance) and, if found, hone the cylinder walls using a medium grit cylinder hone. Use a slow rpm while raising and lowering the hone through the cylinder in order to cross-hatch the cylinder walls for maximum oil retention.

POWERHEAD SYSTEMS 6-69

⚠️ WARNING

Use the cylinder hone slowly, carefully and as little as possible to avoid the possibility of removing too much material from the cylinder walls. If the bores are over-honed, it could cause the pistons to be below specification for the new cylinder measurement, or worse, cause the cylinder bores to be overspec for what pistons are available.

6. Wash the entire cylinder block, crankcase and head using warm, soapy water to remove all traces of contaminants. Use low-pressure compressed air to blow dry all passageways.

7. Apply a light coating of clean engine oil to all machined surfaces that are not about to be measured right away. When you return to the task of measuring components that have been oiled, use a solvent covered rag to wipe away the oil before measurements are taken.

8. Cover all components using a plastic sheet to keep dust, dirt or debris from contaminating the cleaned and especially the oiled surfaces.

Reed Block

INSPECTION

◆ See Figures 227, 228 and 229

Disassemble the reed block housing by first removing the screws securing the reed stoppers and reed petals to the housing. After the screws are removed, lift the stoppers and petals from the housing.

Clean the gasket surfaces of the housing. Check the surfaces for deep grooves, cracks or any distortion which could cause leakage.

Replace the reed block housing if it is damaged. The reed petals should fit flush against their seats, and not be preloaded against their seats or bend away from their seats.

If the reed petal is distorted beyond its Warpage (bending) limit, rarely, if ever, can the petal be successfully straightened. Therefore, it must be replaced.

The valve stopper clearance (Valve Lift or Valve Height) is the distance between the bottom edge of the stopper and the top of the reed petal.

The valve stopper can sometimes be successfully bent to achieve the required clearance, if not, it must be replaced.

■ A few early-model Yamahas have a published reed valve Thickness specification which can be measured using a micrometer.

Do not remove the reed valves unless they, or the stoppers, are to be replaced. If servicing a 9.9 hp unit or larger, the reeds must be replaced in sets.

Apply Loctite® to the threads of the reed retaining screws. Tighten each screw to specification gradually, starting in the center and working outwards across the reed block.

Exhaust Cover

INSPECTION

◆ See Figures 230 and 231

The exhaust covers are one of the most neglected items on any outboard powerhead. Seldom are they checked and serviced. Many times a powerhead may be overhauled and returned to service without the exhaust covers ever having been removed.

One reason the exhaust covers are not removed is because the attaching bolts usually become corroded in place. This means they are very difficult to remove, but the work should be done. Heat applied to the bolt head and around the exhaust cover will help in removal. However, some bolts may still be broken. If the bolt is broken it must be drilled out and the hole tapped with new threads.

The exhaust covers are installed over the exhaust ports to allow the exhaust to leave the powerhead and be transferred to the exhaust housing. If the cover was the only item over the exhaust ports, they would become so hot from the exhaust gases they might cause a fire or a person could be severely burned if they came in contact with the cover.

Therefore, depending upon the inner plumbing of the powerhead, MOST are equipped with an inner plate to help dissipate the exhaust heat. Two gaskets are normally installed - one on either side of the inner plate. Water is channeled to circulate between the exhaust cover and the inner plate. This circulating water cools the exhaust cover and prevents it from becoming a hazard.

A thorough cleaning of the inner plate behind the exhaust covers should be performed during a major powerhead overhaul. If the integrity of the exhaust cover assembly is in doubt, replace the inner plate.

Bleed System

INSPECTION

◆ See Figures 232 thru 235

■ For V4 and V6 bleed/recirculation hose schematics, please refer to the exploded views and illustrations accompanying the V4/V6 powerhead overhaul procedures.

The bleed system consists of one or more hoses depending on the model being serviced. These hoses transfer unburned fuel/oil mixture from one cylinder to another. The fuel is pulsed by crankcase vacuum. This system prevents accumulation of unburned fuel in the lower cylinder and transfers the fuel and oil mixture to the crankshaft upper main bearing for lubrication.

The accompanying illustrations show the common routing of bleed hoses for most Yamaha powerheads equipped with this system. On V4 and V6

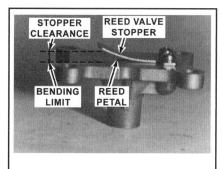

Fig. 227 A single cylinder motor reed valve housing with a single reed petal

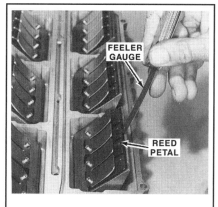

Fig. 228 Using a feeler gauge to measure valve warpage (reed petal clearance)

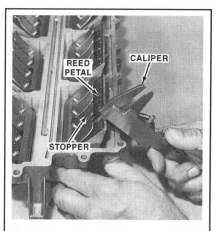

Fig. 229 Measuring the Valve stopper Height (the lift or clearance between the stopper and the reed petal)

6-70 POWERHEAD

Fig. 230 The exhaust cover must be removed...

Fig. 231 ... and inspected for warpage or clogged passages in order to ensure proper powerhead operation

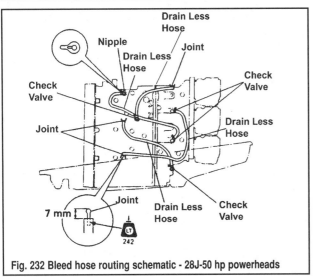
Fig. 232 Bleed hose routing schematic - 28J-50 hp powerheads

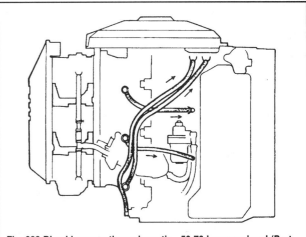
Fig. 233 Bleed hose routing schematic - 50-70 hp powerhead (Port Side)

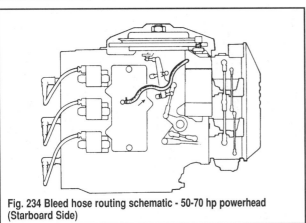
Fig. 234 Bleed hose routing schematic - 50-70 hp powerhead (Starboard Side)

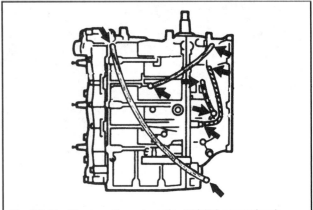
Fig. 235 Bleed hose routing schematic - 65J-90 hp powerheads

POWERHEAD SYSTEMS 6-71

models there are too many illustrations to include here, so please refer to the Powerhead, Overhaul procedures.

Check the condition of each rubber bleed hose. Replace the hose if it shows signs of deterioration or leakage. Check the operation of the check valves. The air/fuel mixture should be able to pass through the valve in only one direction.

Defective check valves cannot be serviced. If defective, they must be replaced.

Crankshaft

INSPECTION

◆ See Figures 236 thru 238

Clean the crankshaft with solvent and wipe the journals dry with a lint free cloth. Inspect the main journals and connecting rod journals for cracks, scratches, grooves, or scores. Inspect the crankshaft oil seal surface for nicks, sharp edges or burrs which might damage the oil seal during installation or might cause premature seal wear. Always handle the crankshaft carefully to avoid damaging the highly finished journal surfaces. Blow out all oil passages with compressed air. The oil passageway leads from the rod to the main bearing journal. Take care not to blow dirt into the main bearing journal bore.

If equipped, inspect the bronze oil injection worm gear at the lower end of the crankshaft. If the gear needs to be replaced, the lower main bearing must be pulled from the crankshaft first (a procedure which normally makes the bearing unfit for further service). The bronze worm gear may then be pulled.

Inspect the internal splines at one end and threads at the other end for signs of abnormal wear. Check the crankshaft for run-out by supporting it on two V-blocks at the main bearing surfaces.

Install a dial indicator gauge above the main bearing journals. Rotate the crankshaft and measure the run-out (or the out-of-round) and the taper at both ends and in the two center journals (center journal on all two-cylinder models).

If V-blocks or a dial indicator are not available, a micrometer may be used to measure the diameter of the journal. Make a second measurement at right angles to the first. Check the difference between the first and second measurement for out-of-round condition. If the journals are tapered, ridged, or out-of-round by more than the specification allows, the journals should be reground, or the crankshaft replaced.

Any out-of-round or taper shortens bearing life. Good shop practice dictates new main bearings be installed with a new or reground crankshaft.

■ Some models utilize pressed together crankshaft assemblies. On these models, normally the connecting rods would only be pressed from the crankshaft if either the crankshaft and/or the connecting rods were to be replaced. Therefore, the connecting rod axial play is checked at the piston end to determine the amount of wear at the crankshaft end of the connecting rod.

Specifications on some models are provided for different crankshaft widths. These measurements are made to determine if excessive wear has caused the need to replace the crankshaft and/or connecting rods. When specifications are provided they vary greatly by year and model. Also, the measurements may be around a single crankpin, the distance between to crankpins or the distance around all crankpins. Be sure to use a precision instrument when making these measurements. Refer to the Engine Specifications chart to see which measurements apply.

■ If the measurement is not within specification, check the Connecting Rod Big End Side Clearance and Small-End Free-play measurements of the connecting rod(s) involved. This measurement will give an indication if the counterweight is walking on the crank pin or if the clearance is due to worn parts. Normally the connecting rod would wear before the edge of the counterweight because the rod is manufactured from a much softer material.

Inspect the crankshaft oil seal surfaces to be sure they are not grooved, pitted, or scratched. Replace the crankshaft if it is severely damaged or worn. Check all crankshaft bearing surfaces for rust, water marks, chatter marks, uneven wear, or overheating. Clean the crankshaft surfaces with 320-grit wet/dry sandpaper dampened with solvent. Never spin-dry a crankshaft ball bearing with compressed air.

Clean the crankshaft and, if equipped, crankshaft ball bearing with solvent. Dry the parts, but not the ball bearing, with compressed air. Check the crankshaft surfaces a second time. Replace the crankshaft if the surfaces cannot be cleaned properly for satisfactory service. If the crankshaft is to be installed for service, lubricate the surfaces with light oil. Do not lubricate the crankshaft ball bearing at this time.

On models so equipped, after the crankshaft has been cleaned, grasp the outer race of the crankshaft ball bearing installed on the lower end of the crankshaft, and attempt to work the race back and forth. There should not be excessive play. A very slight amount of side play is acceptable because there is only about 0.001 inch (0.025mm) clearance in the bearing.

Lubricate the ball bearing with light oil. Check the action of the bearing by rotating the outer bearing race. The bearing should have a smooth action and no rust stains. If the ball bearing sounds or feels rough or catches, the bearing should be removed and discarded.

For models equipped with bearing inserts, use a strip of Plastigage® or some equivalent compressible gauging material across the center of the crankshaft journal (lengthwise from flywheel-to-driveshaft end of the journal) and bolt the crankcase/bearing retainers in position, then remove again to exam the gauging material. For more details, please refer to Measuring Bearing Clearance Using Plastigage® later in this section.

MEASURING BEARING CLEARANCE USING PLASTIGAGE®

◆ See Figures 239, 240 and 241

Many of the larger Yamaha motors use automotive style bearing inserts for the crankshaft and connecting rods. Besides a visual inspection for obvious defects or flaws, the most important check that can be made to

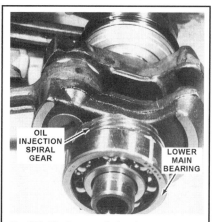

Fig. 236 The lower main bearing on some engines secures the bronze oil injection gear

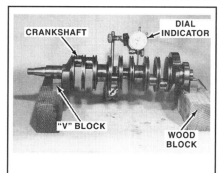

Fig. 237 Crankshaft set up with V-blocks and a dial indicator to measure run-out

Fig. 238 You can also use a micrometer to check run-out and taper on the main bearing journals (and connecting rod bearing journals depending upon what specs are available)

6-72 POWERHEAD

bearing inserts is to measure their installed oil clearance (the distance between the bearing and crankshaft journal) to determine if they are still fit for service. Even when selecting replacement bearings using color or number codes provided on the motor components, this check must be made to ensure the assembled motor will be within design tolerances.

To check bearing oil clearances you need a way of measuring a tiny gap (thousands of an inch or hundredths of a millimeter) between otherwise inaccessible components. The basic method is to apply a strip of a gauging material such as Plastigage® to the crankshaft journal and then to bolt the crankcase/main bearing retainer and/or connecting rod cap together. When the components are bolted together the gauging material is compressed (squished) between the bearing and journal. The key is that the material is designed to expand at a uniform rate so that you can then remove the bearing and measure the width of the squished gauging material. The width of the material will correspond to a specific clearance measurement (a width-to-clearance conversion scale is provided with the gauging material).

Using Plastigage® is a simple, inexpensive, and fairly reliable method to measure main bearing and connecting rod bearing clearance. However, Plastigage® is soluble in oil, so clean the crankshaft journal thoroughly of all traces of oil, and then turn the crankshaft until the oil hole is down away from the cap to prevent getting any oil on the Plastigage®. To measure bearing clearances, proceed as follows:

9. Obtain a package of Plastigage® from the local auto or marine parts outlet.
10. Cut a one inch (2.5cm) section of the bead for each journal/bearing combination that is being measured.
11. Place each one inch section on the crankshaft journal surface at the main bearings and on each connecting rod bearing surface.

■ DO NOT rotate the crankshaft until the measurements have been completed. Such action will smear the Plastigage® bead and the measurement will be useless.

12. Install the connecting rod caps along with the bearing inserts and tighten to specification (for details, refer to the Powerhead, Overhaul procedures found earlier in this section). **Take your time** to be sure each is installed back into the same position from which it was removed according to the marks made during disassembling.

✱✱ WARNING

For motors whose connecting rod caps are angle-torqued and should not be reused, you can either install the old bolts (JUST for the test) or the new bolts, but in either case, tighten the bolts to specification BUT refrain from the angle-torquing step to prevent the possibility of breaking the old bolts or stretching the new bolts beyond use.

13. Install the main bearings and tighten the crankcase/bearing retainers to specification (again, for details, please refer to the Powerhead, Overhaul procedures for the particular model on which you are working).
14. Remove the crankcase/bearing retainer bolts, the crankcase, the connecting rod bolts, and the connecting rod caps to gain access to the Plastigage®.
15. Measure the flattened Plastigage® at its widest point using the scale printed on the Plastigage® envelope. The number within the graduation which most closely corresponds to the width of the Plastigage® indicates the bearing clearance in thousands of an inch (or hundredths of a millimeter, depending upon the scale). A metric scale in mm is normally given on the back of the package. The widest point of the Plastigage® is the minimum clearance and the narrowest point is the maximum clearance. The difference between the readings indicates the taper of the journals. Compare the Plastigage® measurements with the specifications found in the Engine Specifications chart in this section.

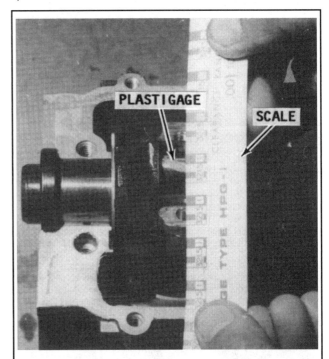

Fig. 240 . . .then remove the crankcase and compare it to the scale

Fig. 239 To check bearing clearances, apply a strip of gauge material to each journal and bolt the bearings/retainers in place. . .

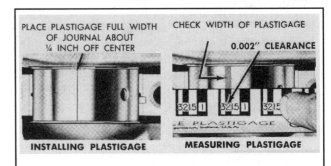

Fig. 241 Once the crankcase and/or rod caps are properly tightened and subsequently removed again compare the now flattened gauging material to the scale that came with the package

POWERHEAD SYSTEMS 6-73

16. If the clearance is not within specifications, the bearings or the crankshaft, or the connecting rods **must** be replaced as required. Obviously, you'd start with new bearing inserts to see if one can be obtained to return that particular journal to specified clearances.

17. Once you are finished checking and selecting bearings, clean all traces of the Plastigage® from the bearing surfaces, and then follow necessary steps again (oiling the bearing surfaces and skipping the Plastigage®) to make a final installation of the connecting rod caps and the crankcase. Check to be **sure** the crankshaft rotates freely. If any binding is discovered, the thrust washer tabs may be misaligned or one or more of the connecting rod caps may be installed backwards.

Connecting Rods

INSPECTION

◆ See Figures 242 thru 247

■ Specifications available for the connecting rods vary by model, so check the Engine Specifications charts to determine which steps are applicable to the motor on which you are working.

Each connecting rod should be cleaned and, once cleaned, should be thoroughly inspected to make sure it is not damaged, bent or excessively worn. Obvious defects, such as visible warpage or cracks are signs of unserviceability. On many motors, the connecting rod small-end (piston end) free-play should be measured using a dial gauge, while connecting rod big-end (crankshaft end) must be checked for Side Clearance using a feeler gauge. On some models measurements may be provided for small end diameter (which will correspond to a measurement slightly larger than piston pin diameter). To thoroughly inspect the connecting rods, proceed as follows:

1. Clean the inside diameter of the piston pin end of the connecting rod with crocus cloth. Clean the connecting rod only enough to remove marks. Do not continue, once the marks have disappeared.

2. Assemble the piston end of the connecting rod with the loose needle bearings, piston pin, retainers, and C-lockrings. Insert the piston pin and check for vertical play. The piston pin should have no noticeable vertical play.

3. If the pin is loose or there is vertical play check for and replace the worn part/s.

4. Inspect the piston pin and matching rod end for signs of heat discoloration. Overheating is identified as a bluish bearing surface color and is caused by inadequate lubrication or by operating the powerhead at excessive high rpm.

5. On all bolt together connecting rods, check the bearing surface of the rod and rod cap for signs of chatter marks. This condition is identified by a rough bearing surface resembling a tiny washboard. The condition is caused by a combination of low-speed low-load operation in cold water. The condition is aggravated by inadequate lubrication and improper fuel.

Under these conditions, the crankshaft journal is hammered by the connecting rod. As ignition occurs in the cylinder, the piston pushes the connecting rod with tremendous force. This force is transferred to the connecting rod journal. Since there is little or no load on the crankshaft, it bounces away from the connecting rod. The crankshaft then remains immobile for a split second, until the piston travel causes the connecting rod to catch up to the waiting crankshaft journal, then hammers it. In some instances, the connecting rod crank pin bore becomes highly polished.

While the powerhead is running, a whir and/or chirp sound may be heard when the powerhead is accelerated rapidly - say from idle speed to about 1500 rpm, then quickly returned to idle. If chatter marks are discovered, the crankshaft and the connecting rods should be replaced.

6. Inspect the bearing surface of the rod and rod cap for signs of uneven wear and possible overheating. Overheating is identified as a bluish bearing surface color and is caused by inadequate lubrication or by operating the powerhead at excessive high rpm.

7. To check Small-End Free-play, set up the crankshaft in the V-blocks. Set up a dial indicator to touch the flat surface of the piston end of the rod. Now, hold the crankshaft steady in the V-blocks and at the same time, rock the piston end of the rod along the same axis as the crankshaft. If the dial indicator needle moves through more than specified, the play is considered excessive. The rod must be pressed from the crankshaft, the journal

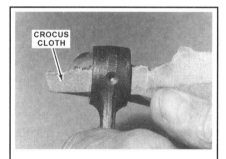

Fig. 242 Cleaning the piston end of a connecting rod with crocus cloth

Fig. 243 Checking the piston end of a connecting rod for vertical free-play using a piston pin

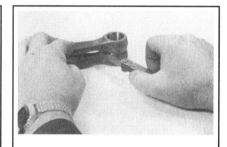

Fig. 244 Testing two rods for warpage at the piston end using a feeler gauge

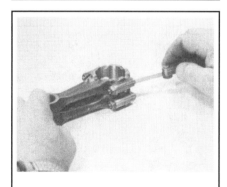

Fig. 245 Testing two rods for warpage at the crankshaft end using a feeler gauge

Fig. 246 Visually inspect the connecting rod for signs of damage

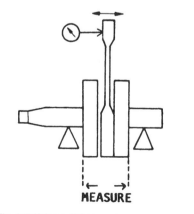

Fig. 247 Using a dial gauge to measure connecting rod Small End Free-play

6-74 POWERHEAD

checked and a determination made as to whether the crankshaft and/or the rod must be replaced. A new rod may be purchased, but on some models it must be pressed onto the crankshaft throw with a hydraulic press.

8. Use a feeler gauge to check Big-End Side Clearance on the crankshaft/connecting rod assembly by inserting a feeler gauge between the connecting rod and the counterweight of the crankshaft.

If the Free-play or Side-Clearance measurement is not within specification, measure the crank width (if specifications are provided). Such a measurement may give an indication if the counterweight is walking on the crank pin or if the clearance is due to worn parts. The connecting rod would wear before the edge of the counterweight because the rod is manufactured from a much softer material.

Keep in mind that even if only one connecting rod shows excessive Free-play or Side-Clearance, good shop practice dictates the entire set of rods be replaced. Rods are sold in sets.

9. Some models utilize replaceable automotive style connecting rod cap with plain bearing inserts. On these models there is normally an Oil Clearance provided in the engine specification chart. Be sure to measure this clearance using Plastigage® or an equivalent gauging material as detailed earlier in this section under Crankshaft, Measuring Bearing Clearances using Plastigage®.

Pistons

INSPECTION

◆ See Figures 248 thru 253

Inspect each piston for evidence of scoring, cracks, metal damage, cracked piston pin boss, or worn pin boss. Be especially critical during inspection if the outboard unit has been submerged. If the piston pin is bent, the piston and pin must be replaced as a set for two reasons. First, a bent pin will damage the boss when it is removed. Secondly, a piston pin is not sold as a separate item.

Check the piston ring grooves for wear, burns distortion, or loose locating pins. During an overhaul, the rings should be replaced to ensure lasting repair and proper powerhead performance after the work is completed. Clean the piston dome, ring grooves, and the piston skirt. Clean carbon deposits from the ring grooves using the recessed end of a broken piston ring.

Never use a rectangular ring to clean the groove for a tapered ring, or use a tapered ring to clean the groove for a rectangular ring.

Unless you are VERY cautious, avoid the use an automotive type ring groove cleaner, because such a tool may loosen the piston ring locating pins.

Clean carbon deposits from the top of the piston using a soft wire brush, carbon removal solution or by sand blasting. If a wire brush is used, Take care not to burr or round machined edges. Clean the piston skirt with crocus cloth.

Install the piston pin through the first boss only. Check for vertical free-play. There should be no vertical free-play. The presence of play is an indication the piston boss is worn. The piston is manufactured from a softer material than the piston pin. Therefore, the piston boss will wear more quickly than the pin.

Excessive piston skirt wear cannot be visually detected. Therefore, good shop practice dictates the piston skirt diameter be measured with a micrometer.

Piston skirt diameter should be measured at right angles to the piston pin axis at a point above the bottom edge of the piston. Refer to the Engine Specifications chart for Measuring Point specifications.

RING END-GAP CLEARANCE

◆ See Figures 254 and 255

Before the piston rings are installed onto the piston, the ring end-gap clearance for each ring must be determined. The purpose of the piston rings is to prevent the blow-by of gases in the combustion chamber. This cannot be achieved unless the correct oil film thickness is left on the cylinder wall.

This thin coating of oil acts as a seal between the cylinder wall and the

Fig. 248 This piston seized at high rpm (the connecting rod ripped the lower part of the piston from the dome)

Fig. 249 This pitting is a result of foreign matter in the cylinder

Fig. 250 Cleaning the ring groove with a broken ring

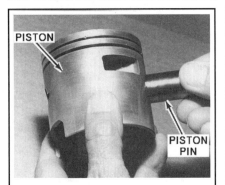

Fig. 251 There should be no play on the piston pin

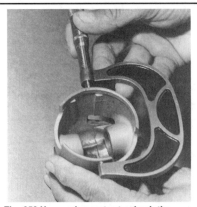

Fig. 252 Use a micrometer to check the piston skirt diameter

Fig. 253 The rings on this piston became stuck due to lack of lubrication

face of the piston ring. An excessive end-gap will allow blow-by and the cylinder will lose compression. An inadequate end-gap will scrape too much oil from the cylinder wall and limit lubrication. Lack of adequate lubrication will cause excessive heat and wear.

Ideally the ring end-gap measurement should be taken after the cylinder bore has been measured for wear and taper and after any corrective work, such as boring or honing, has been completed.

If the ring end-gap is measured with a taper to the cylinder wall, the diameter at the lower limit of ring travel will be smaller than the diameter at the top of the cylinder.

If the ring is fitted to the upper part of a cylinder with a taper, the ring end-gap will not be great enough at the lower limit of ring travel. Such a condition could result in a broken ring and/or damage to the cylinder wall and/or damage to the piston and/or damage to the cylinder head.

If the cylinder is to be only honed, not bored, or if only cleaned, not honed, the ring end-gap should be measured at the lower limit of ring travel.

Some manufacturers give the precise depth the ring should be inserted into the cylinder - usually just above the ports - and assumes the cylinder walls are parallel with no taper.

Rings may be inserted from the top or bottom (depending upon the measurement point). In most cases the piston can be inserted and used to square the ring up in the bore. Once the ring is in the proper position, measure the ring end-gap with a feeler gauge.

If the end-gap is greater than the amount listed in the Engine Specification chart, replace the entire ring set.

If the end-gap is less than the amount listed, carefully file the ends of the ring - just a little at a time - until the correct end-gap is obtained.

■ Inspect the piston ring locating pins to be sure they are tight. Most Yamaha pistons have at least one groove that has a locating pin. When used, there is only one locating pin in each ring groove. If the locating pins are loose, the piston must be replaced.

Fig. 254 Using a feeler gauge to check ring end-gap

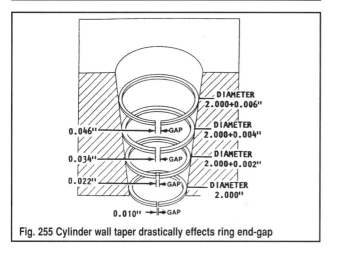
Fig. 255 Cylinder wall taper drastically effects ring end-gap

PISTON RING SIDE CLEARANCE

◆ See Figure 256

After the rings are installed on the piston, the clearance in the grooves needs to be checked with a feeler gauge. Check the clearance between the ring and the upper land and compare your measurement with the specifications. Ring wear forms a step at the inner portion of the upper land. If the piston grooves have worn to the extent to cause high steps on the upper land, the step will interfere with the operation of new rings and the ring clearance will be too much. Therefore, if steps exist in any of the upper lands, the piston should be replaced.

On a plain ring this clearance may be measured either above or below the ring. On a Keystone ring - one with a taper at the top - the clearance must be measured below the ring.

OVERSIZE PISTONS & RINGS

Scored cylinder blocks can be saved for further service by boring and installing oversize pistons and piston rings. However, in most cases, if the scoring is over 0.0075 inch (0.13mm) deep, the block cannot be effectively re-bored for continued use.

Oversize pistons and rings are not available for all powerheads. Check with the parts department at your local Yamaha dealer for the model being serviced.

If oversize pistons are not available, the local marine shop may have the facilities to knurl the piston, making it larger.

A honed cylinder block is just surface finished. It is not parallel and therefore if an oversized or knurled piston is installed in this bore, the piston will soon seize-at the lower narrow end of the bore.

Cylinder Block

INSPECTION

◆ See Figures 257 thru 260

Inspect the cylinder block and cylinder bores for cracks or other damage. Remove carbon with a fine wire brush on a shaft attached to an electric drill or use a carbon remover solution.

✴✴ WARNING

If the cylinder block is to be submerged in a carbon removal solution, components like the crankcase bleed system must be removed from the block to prevent damage to hoses and check valves.

Use an inside micrometer or telescopic gauge and micrometer to check the cylinders for wear. Check the bore for out-of-round and/or oversize bore. If the bore is tapered, out-of-round or worn more than the wear limit specified by the manufacturer, the cylinders should be re-bored - provided oversize pistons and rings are available.

Check with the Yamaha dealer prior to boring. If oversize pistons and matching rings are not available, the block must be replaced.

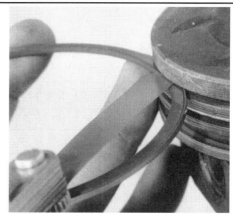
Fig. 256 Using a feeler gauge to check piston ring side clearance

6-76 POWERHEAD

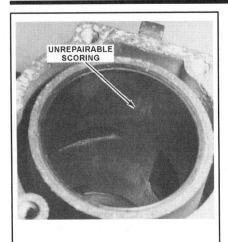

Fig. 257 This cylinder was scored beyond repair when a piston ring broke and worked its way into the combustion chamber

Fig. 258 Check the cylinder bore for taper using an inside micrometer. Measure near the near the top, in the middle and a near the bottom

Fig. 259 Using a drill mounted wire brush to clean carbon deposits from the cylinder head

■ Oversize piston weight is approximately the same as a standard size piston. Therefore, it is not necessary to re-bore all cylinders in a block just because one cylinder requires boring.

Cylinder sleeves are an integral part of the die cast cylinder block and cannot be replaced. In other words, the cylinder cannot be sleeved.

Four inside cylinder bore measurements must be taken for each cylinder to determine an out-of-round condition, the maximum taper, and the maximum bore diameter.

In the accompanying illustration, measurements D1 and D2 are diameters measured at about 0.8 inch (20mm) from the top of the cylinder at right angles to each other. Measurements D3 and D4 are diameters measured at about 2.4 inch (60mm) from the top of the cylinder at right angles to each other.

Fig. 260 Top view and cross section of a typical cylinder to indicate where measurements should be made for wear, taper and out of round

Out-of-Round

◆ See Figure 260

Measure the cylinder diameter at D1 and D2. The manufacturer requires the difference between the two measurements should be less than 0.002 inch (0.050mm) for most models. Refer to the Engine Specifications chart.

Maximum Taper

◆ See Figure 260

Measure the cylinder diameter at D1, D2, D3, and D4. Take the largest of the D1 or D2 measurements and subtract the smallest measurement at D3 or D4. The result is the cylinder taper - it should be less than 0.003 inch (0.08mm) for most models. Refer to the Engine Specifications chart.

Bore Wear Limit

◆ See Figure 260

The maximum cylinder diameter D1, D2, D3, and D4 must not exceed the bore wear limits (when indicated in the Engine Specifications chart) before the bore is re-bored for the first time. These limits are only imposed on original parts because the sealing ability of the rings would be lost, resulting in power loss, increased powerhead noise, unnecessary vibration, piston slap, and excessive oil consumption.

The limits above the standard bore are usually 0.003 to 0.005 inch (0.08mm to 0.127mm) on all powerheads except V4 and V6, or 0.007 inch (0.1mm) on V4 and V6 powerheads. For details, refer to the Engine Specifications chart.

Piston Clearance

◆ See Figure 260

Piston clearance is the difference between a maximum piston diameter and a minimum cylinder bore diameter. If this clearance is excessive, the powerhead will develop the same symptoms as for excessive cylinder bore wear - loss of ring sealing ability, loss of power, increased powerhead noise, unnecessary vibration, and excessive oil consumption.

Maximum piston diameter was described earlier in this section. Minimum cylinder bore diameter is usually determined by measurement D3 or D4 also described earlier in this section.

If the piston clearance exceeds the limits in the specifications chart, either the piston or the cylinder block must be replaced.

Calculate the piston clearance by subtracting the maximum piston skirt diameter from the maximum cylinder bore measurement and compare the results for the model being serviced.

POWERHEAD SYSTEMS 6-77

HONING CYLINDER WALLS

◆ See Figures 261 and 262

Hone the cylinder walls lightly to seat the new piston rings, as outlined in this section. If the cylinders have been scored, but are not out-of-round or the bore is rough, clean the surface of the cylinder with a cylinder hone as described in the following procedures.

■ If overheating has occurred, check and resurface the spark plug end of the cylinder block, if necessary. This can be accomplished with 240-grit sandpaper and a small flat block of wood.

To ensure satisfactory powerhead performance and long life following the overhaul work, the honing work should be performed with patience, skill, and in the following sequence:

1. Follow the hone manufacturer's recommendations for use of the hone and for cleaning and lubricating during the honing operation. A Christmas tree hone may also be used.
Pump a continuous flow of honing oil into the work area. If pumping is not practical, use an oil can. Apply the oil generously and frequently on both the stones and work surface.
2. Begin the stroking at the smallest diameter. Maintain a firm stone pressure against the cylinder wall to assure fast stock removal and accurate results.
3. Expand the stones as necessary to compensate for stock removal and stone wear. The best cross hatch pattern is obtained using a stroke rate of 30 complete cycles per minute. Again, use the honing oil generously.
4. Hone the cylinder walls only enough to deglaze the walls.
5. After the honing operation has been completed, clean the cylinder bores with hot water and detergent. Scrub the walls with a stiff bristle brush and rinse thoroughly with hot water. The cylinders must be thoroughly cleaned to prevent any abrasive material from remaining in the cylinder bore. Such material will cause rapid wear of new piston rings, the cylinder bore, and the bearings.

6. After cleaning, swab the bores several times with engine oil and a clean cloth, and then wipe them dry with a clean cloth. Never use kerosene or gasoline to clean the cylinders.
7. Clean the remainder of the cylinder block to remove any excess material spread during the honing operation.

■ If oversize pistons are not available, the local marine shop may have the facilities to knurl the piston, making it larger. If installing oversize or knurled pistons, it should be remembered that a honed cylinder block is just surface finished. It is not parallel and therefore if an oversized or knurled piston is installed in this bore, the piston will soon seize at the lower narrow end of the bore. When installing oversize or knurled pistons, the block must be bored oversize.

BLOCK & CYLINDER HEAD WARPAGE

◆ See Figures 263 thru 267

First, check to be sure all old gasket material has been removed from the contact surfaces of the block and the cylinder head. Clean both surfaces down to shiny metal, to ensure a true measurement.

Next, place a straightedge across the gasket surface. Check under the straightedge with a suitably sized feeler gauge (see the Engine Specifications chart). The warpage limit is normally 0.004 inch (0.1mm) on Yamaha motors. Move the straightedge to at least eight different locations. If the feeler gauge can pass under the straightedge - anywhere contact with the other is made - the surface will have to be resurfaced.

The block or the cylinder head may be resurfaced by placing the warped surface on 400-600 grit wet sandpaper, with the sandpaper resting on a flat machined surface. If a machined surface is not available a large piece of glass or mirror may be used. Do not attempt to use a workbench or similar surface for this task. A workbench is never perfectly flat and the block or cylinder head will pickup the imperfections of the surface and the warpage will be made worse.

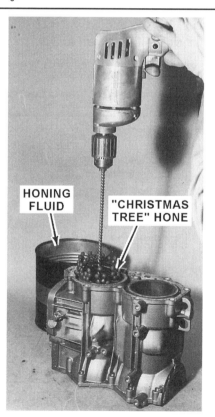

Fig. 261 Refinishing the cylinder wall using a ball hone. Always keep the home moving and constantly flood the cylinder with honing oil

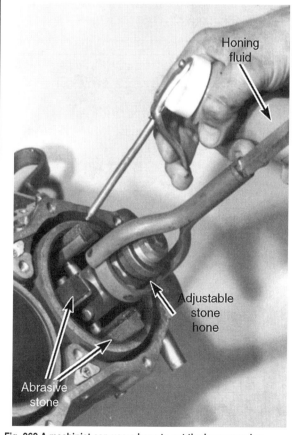

Fig. 262 A machinist can use a hone to cut the bore oversize

Sand - work - the warped surface on the wet sandpaper using large figure 8 motions. Rotate the block, or head, through 180° (turn it end for end) and spend an equal amount of time in each position to avoid removing too much material from one side.

If a suitable flat surface is not available, the next best method is to wrap 400-600 grit wet sandpaper around a large file. Draw the file as evenly as possible in one sweep across the surface. Do not file in one place. Draw the file in many, many directions to get as even a finish as possible.

As the work moves along, check the progress with the straightedge and feeler gauge. Once the 0.004 inch (0.1mm) feeler gauge will no longer slide under the straightedge, consider the work completed.

If the warpage cannot be reduced using one of the described methods, the block or cylinder head should be replaced (or resurfaced by a machine shop, if possible).

Assembling a Warped Cylinder Head or Block

If the warpage cannot be reduced and it is not possible to obtain new items - and the warped part must be assembled for further use - there is a strong possibility of a water leak at the head gasket. In an effort to prevent a water leak, follow the instructions outlined in the following paragraphs.

■ **Because of the high temperatures and pressures developed, the sealing surfaces of the cylinder head and the block are the most prone to water leaks. No sealing agent is recommended because it is almost impossible to apply an even coat of sealer. An even coat would be essential to ensure an air/water tight seal.**

Never, never, use automotive type head gasket sealer. The chemicals in the sealer will cause electrolytic action and eat the aluminum faster than you can get to the bank for money to buy a new cylinder block.

Some head gaskets are supplied with a tacky coating on both surfaces applied at the time of manufacture. This tacky substance will provide an even coating all around. Therefore, no further sealing agent is required.

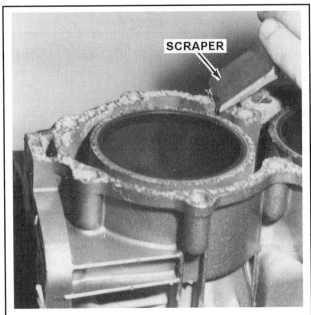

Fig. 263 All traces of gasket material must be removed to ensure accuracy

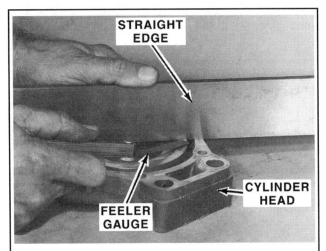

Fig. 265 Use a straightedge and feeler gauge to check for cylinder head warpage

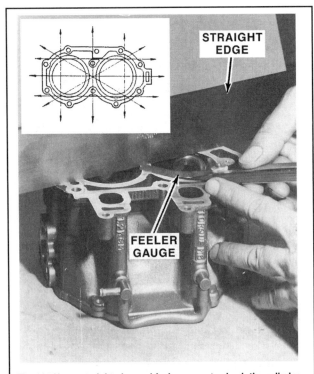

Fig. 264 Use a straightedge and feeler gauge to check the cylinder head-to-powerhead mating surfaces

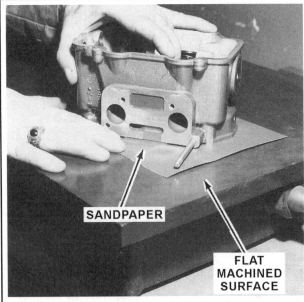

Fig. 266 Use only a flat machined surface and sandpaper to correct warpage on a cylinder head

However, if a slight water leak should be noticed following completed assembly work and powerhead start up, do not attempt to stop the leak by tightening the head bolts beyond the recommended torque value. Such action will only aggravate the problem and most likely distort the head.

Furthermore, tightening the bolts, which are case hardened aluminum, may force the bolt beyond its elastic limit and cause the bolt to fracture. bad news, very bad news indeed. A fractured bolt must usually be drilled out and the hole re-tapped to accommodate an oversize bolt, etc. Avoid such a situation.

Probable causes and remedies of a new head gasket leaking are:

a. Sealing surfaces not thoroughly cleaned of old gasket material. Disassemble and remove all traces of old gasket.

b. Damage to the machined surface of the head or the block. The remedy for this damage is the same as for the next case.

c. Permanently distorted head or block. Spray a light even coat of any type metallic spray paint on both sides of a new head gasket. Use only metallic paint - any color will do. Regular spray paint does not have the particle content required to provide the extra sealing properties this procedure requires.

Assemble the block and head with the gasket while the paint is till tacky. Install the head bolts and tighten in the recommended sequence and to the proper torque value and no more!

Allow the paint to set for at least 24 hours before starting the powerhead.

Consider this procedure as a temporary band aid type solution until a new head may be purchased or other permanent measures can be performed.

Under normal circumstances, if procedures have been followed to the letter, the head gasket will not leak.

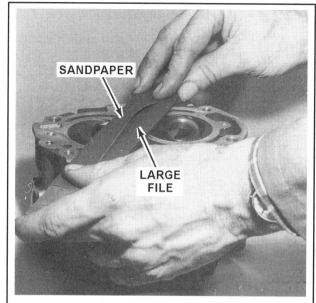

Fig. 267 When a flat machined surface is not available, a large flat file wrapped with sandpaper may be used

POWERHEAD BREAK-IN

Break-in Procedures

Anytime a new or rebuilt powerhead is installed (this includes a powerhead whose wear components such as pistons and rings or main bearings the motor must undergo proper break-in.

By following break-in procedures largely consisting of specific engine operating limitations during the first 10 hours of operation, you will help you will help ensure a long and trouble-free life. Failure to follow these recommendations may allow components to seat improperly, causing accelerated wear and premature powerhead failure.

On all motors, special attention is required to the engine oil during initial break-in. Pay close attention to the special fuel/oil mixture requirements. Especially during break-in, pay close attention to all pre and post operation checks. This goes double when checking for fuel, oil or water leaks. At each start-up and frequently during operation, check for presence of the cooling indicator stream.

At the completion of break-in, double-check the tightness of all exposed engine fasteners.

Even though Yamaha technically ends the break-in period of a motor after the first 10 hours of operation, it is a good idea to continue top pay closer attention to the motor until the completion of the entire first 20 hours of engine operation (and completion of a 20 hour service). During break-in, one of the most important things you can do is to **vary** the engine speed. This allows parts to wear in under conditions throughout the powerband, not just at idle or mid-throttle.

■ **During break-in, check your hourmeter or a watch frequently and be sure to change the engine speed at least every 15 minutes (that means between every 2-3 tenths on the hourmeter).**

Be sure to **always** allow the engine to reach operating temperature before setting the throttle anywhere above idle. This means you should always start and run the motor for at least 5 minutes before advancing the throttle.

※※ **WARNING**

NEVER run the engine out of the water, unless a flush fitting is used to provide a source of cooling. Remember that the water pump can be destroyed in less than a minute just from a lack of water. The powerhead will suffer damage in very little time as well, but even if it is not overheated out of water, reduced cooling from a damaged water pump impeller could destroy it later. Don't risk it.

■ **Although a flush fitting can be used to run an engine, it CANNOT be used to break-in a motor, as you cannot run an engine much above idle on a flush fitting without risking damage from a runaway powerhead (the motor running overspeed).**

During engine break-in all motors (except HPDI models) must use a higher ratio of oil in the fuel, this is true whether or not they are equipped with the Precision Blend oil injection system. Yamaha motors are normally designed to operate with either a 100:1 or a 50:1 fuel:oil ratio (for details on your motor please refer to the owners manual or to the General Engine System Specifications chart found in the Maintenance and Tune-Up section). However, during break-in all motors require a 25:1 ratio. Pre-mix models achieve this ratio directly by mixing a greater amount of oil in the fuel tank. Precision Blend oil injection and EFI OX66 models do this by running a 50:1 ratio of pre-mix in the fuel tank to supplement the 50:1 ratio of oil that is normally produced by the oil injection system (effectively mixing double the amount of oil, changing the ratio thereby to 25:1).

■ **Don't run the higher fuel:oil ratio any longer than absolutely necessary to clear the fuel tank after the 10 hours. Doing so could lead to excessive carbon fouling of the spark plugs and excessive carbon deposits in the combustion chambers.**

On Precision Blend oil injection models top off the oil tank and place a piece of tape or make a mark on the side of the tank housing to note the oil level. Watch this level and make sure it is dropping slowly throughout the break-in period. This will ensure you the system is working. Also, it is a good habit to get into as it will help you gauge system and engine operation long after engine break-in.

※※ **WARNING**

Do not operate the engine at full throttle except for very short periods, until after 10 hours of operation.

To properly break-in a powerhead, proceed as follows:

1. Each time you start the motor, IMMEDIATELY check to be sure the water pump is operating. If the water pump is operating, a fine stream will be discharged from the exhaust relief hole at the rear of the driveshaft housing. Also, each time you start the powerhead, allow it to idle until it reaches normal operating temperature (at least 5 minutes) before increasing speed above idle.

■ **If the cooling water indicator stream is missing, shut the powerhead down and figure out why before proceeding.**

POWERHEAD

2. The first time you start the motor, allow it to idle for **10** minutes. Make sure the motor runs at the slowest possible speed (a fast idle in neutral is best).

3. For the remainder of the first hour operate the motor in gear at various speeds, but DO NOT exceed 1/2 throttle. The one exception is if you have an easily planning boat, accelerate at full throttle (or close to it) JUST long enough to get the boat up onto plane, then immediately reduce speed below 3000 rpm. Vary engine speeds below 1/2 throttle, but try to keep the boat on plane (as that lessons the amount of load on the motor).

4. For the second hour of operation accelerate at full throttle onto plane, then reduce speed to 3/4 throttle (approximately 4000 rpm). Continue to vary the engine speed and operate the motor under different load conditions. At least a couple of times during the hour, run the engine AT FULL THROTTLE for ONE MINUTE, then allow at least 10 minutes of operation at 3/4 throttle or less to allow the motor to cool.

■ Build good habits now. It is NEVER good for the motor to operate at full throttle for any length of time then suddenly drop off to idle. You should always decelerate the motor gradually, allowing the pistons to cool as the speed is reduced.

5. For the remainder of the first 10 hours (for hours 3-10) continue the ritual of warming the motor and varying engine speed. Full throttle operation should NOT exceed 5 minutes allowing sufficient time to slowly cool the pistons and operate under other conditions before the next full throttle run.

6. After the 10th hour of operation, run the engine as normal and on a normal fuel:oil ratio. Before removing the pre-mix from the fuel tank on Precision Blend models verify that the system has been working by making a visual check of oil level.

■ While the engine is operating during the initial period, check the fuel, exhaust, and water systems for leaks. Tilt the motor up out of the water at the end of each day to closely inspect the drive unit and look for signs of damage or leakage (of course, if you're trailering, this should be done to protect the skeg before pulling it out on the trailer anyway, right?).

SPECIFICATIONS

ENGINE SPECIFICATIONS - 2 HP (43cc) ENGINES

Component	U.S. (in.) ①	Metric (mm) ①
Cylinder Bore		
Standard Bore Diameter	1.535-1.536	39.00-39.02
Wear Limit	1.54	39.1
Out-of-round		
Service Limit	0.002	0.05
Taper		
Service Limit	0.003	0.08
Cylinder Head		
Gasket Surface Warpage Limit	0.004	0.1
Crankshaft		
Crank Width	1.098-1.100	27.90-27.95
Runout		
Limit	0.001	0.02
Connecting Rod		
Big End Side Clearance	0.012-0.024	0.30-0.60
Small End Freeplay Limit	0.08	2.0
Piston		
Standard Size		
1984-94	1.5271-1.5350	38.80-39.00
1995+	1.5341-1.5349	39.967-38.986
Measuring Point (above bottom of skirt)		
1984-94	0.197	5.0
1995+	0.39	10.0
Clearance		
1984-94	0.0012-0.0016	0.03-0.04
1995+	0.0012-0.0014	0.030-0.035
Service Limit (1995 and later)	0.0033	0.085
Oversize		
1st	1.545	39.25
2nd	1.555	39.50
Offset (1984-94 only)	0	0
Piston Pin (specs are for 1995 and later models only)		
Pin Boss Bore	0.3939-0.3943	10.004-10.015
Pin Diameter	0.3935-0.3937	9.996-10.000
Piston Rings		
Width (measured from outer-to-inner edge of ring)	0.0671-0.0749	1.7-1.9
Thickness (measured vertically top-to-bottom of ring)	0.0779-0.0786	1.97-1.99
End Gap		
Standard Installed Gap	0.004-0.012	0.10-0.30
Service Limit (1995 and later)	0.020	0.50
Ring Side Clearance	0.001-0.003	0.03-0.07
Reed Valve		
Thickness (1984-94 only)	0.0051-0.0067	0.13-0.17
Lift (height)	0.23-0.25	4.0-8.0
Warpage (limit)	0.012	0.03

① Unless otherwise noted

Fig. 268 Engine Specifications - 2 hp (43cc) 1-Cylinder Motors

POWERHEAD SYSTEMS 6-81

ENGINE SPECIFICATIONS - 2 HP (50cc) ENGINES

Component	U.S. (in.) [1]	Metric (mm) [1]
Cylinder Bore		
Standard Bore Diameter	1.6535-1.6543	42.00-42.02
Wear Limit	1.66	42.10
Out-of-round Service Limit	0.002	0.05
Taper Service Limit	0.003	0.08
Cylinder Head		
Gasket Surface Warpage Limit	0.004	0.1
Crankshaft		
Crank Width	1.098-1.100	27.90-27.95
Runout Limit	0.001	0.02
Connecting Rod		
Big End Side Clearance	0.012-0.024	0.30-0.60
Small End Freeplay Limit	0.08	2.0
Piston		
Standard Size	1.6524-1.6535	41.97-42.00
Measuring Point (above bottom of skirt)	0.39	10.0
Clearance	0.0012-0.0014	0.030-0.035
Service Limit	0.0033	0.085
Oversize		
1st	1.663	42.25
2nd	1.673	42.50
Piston Pin		
Pin Boss Bore	0.3939-0.3943	10.004-10.015
Pin Diameter	0.3935-0.3937	9.996-10.000
Piston Rings		
Width (measured from outer-to-inner edge of ring)	0.07	1.8
Thickness (measured vertically top-to-bottom of ring)	0.08	2.0
End Gap Standard Installed Gap	0.004-0.012	0.10-0.30
Service Limit	0.020	0.50
Ring Side Clearance		
Top (keystone ring)	0.001-0.002	0.03-0.06
Bottom (barrel ring)	0.001-0.003	0.03-0.07
Reed Valve		
Lift (height)	0.23-0.25	4.0-8.0
Warpage (limit)	0.012	0.03

[1] Unless otherwise noted

Fig. 269 Engine Specifications - 2 hp (50cc) 1-Cylinder Motors

ENGINE SPECIFICATIONS - 3 HP (70cc) ENGINES

Component	U.S. (in.) [1]	Metric (mm) [1]
Cylinder Bore		
Standard Bore Diameter	1.811-1.812	46.00-46.02
Out-of-round Service Limit	0.002	0.05
Taper Service Limit	0.003	0.08
Cylinder Head		
Gasket Surface Warpage Limit	0.004	0.1
Crankshaft		
Crank Width	1.41-1.42	35.9-36.0
Runout Limit	0.001	0.03
Connecting Rod		
Big End Side Clearance	0.012-0.024	0.30-0.60
Small End Freeplay Limit	0.08	2.0
Piston		
Standard Size	1.8096-1.8106	45.965-45.990
Measuring Point (above bottom of skirt)	0.39	10.0
Clearance	0.0012-0.0014	0.030-0.035
Service Limit	0.004	0.1
Oversize (check availability)		
1st	1.821	46.25
2nd	1.831	46.50
Piston Rings		
Width (measured from outer-to-inner edge of ring)	0.08	2.0
Thickness (measured vertically top-to-bottom of ring)	0.06	1.5
End Gap Standard Installed Gap	0.004-0.012	0.1-0.3
Ring Side Clearance		
Top Ring	0.0008-0.0024	0.02-0.06
Bottom Ring	0.0012-0.0028	0.03-0.07
Reed Valve		
Thickness	0.008	0.2
Lift (height)	0.152-0.168	3.8-4.2
Warpage (limit)		
Models through 1991	0.060	1.5
1992 and later models	0.008	0.2

[1] Unless otherwise noted

Fig. 270 Engine Specifications - 3 hp (70cc) 1-Cylinder Motors

ENGINE SPECIFICATIONS - 4 HP (83cc) ENGINES

Component	U.S. (in.) ①	Metric (mm) ①
Compression		
Standard Compression	78-96 psi	556-680 kPa
Cylinder Bore		
Standard Bore Diameter	1.9685-19697	50.00-50.03
Wear Limit (1993 and later models)	1.9720	50.1
Out-of-round Service Limit	0.002	0.05
Taper Service Limit	0.003	0.08
Crankshaft		
Crank Width	1.571-1.573	39.90-39.95
Runout Limit	0.0012	0.03
Connecting Rod		
Big End Side Clearance	0.008-0.028	0.2-0.7
Small End Diameter (1993 and later models)	0.5906-0.5910	15.000-15.011
Small End Freeplay Limit	0.08	2.0
Piston		
Standard Size	1.9673-1.9685	49.97-50.00
Measuring Point (above bottom of skirt)	0.39	10.0
Clearance	0.0012-0.0014	0.030-0.035
Service Limit	0.0039	0.1
Offset (thru 1992)	0.020	0.5
Offset Direction (1993 and later models)	Intake Side	Intake Side
Piston Pin (1993 and later models) Pin Diameter	0.4723-0.4724	11.996-12.000
Pin Boss Inner Diameter in Piston	0.4726-0.4730	12.004-12.015
Piston Rings		
Width (measured from outer-to-inner edge of ring) Top Ring	0.079	2.0
Bottom Ring	0.079	2.0
Thickness (measured vertically top-to-bottom of ring)		
End Gap Installed Gap	0.006-0.014	0.15-0.35
Limit	0.022	0.55
Ring Side Clearance Top Ring	0.0008-0.0024	0.02-0.06
Bottom Ring	0.0012-0.0028	0.03-0.07
Reed Valve		
Thickness Plastic Reed	0.016	0.4
Stainless Reed	0.008	0.2
Lift (height)	0.272-0.288	6.8-7.2
Warpage (limit)	0.008	0.2

Fig. 271 Engine Specifications - 4 hp (83cc) 1-Cylinder Motors

ENGINE SPECIFICATIONS - 4/5 HP (103cc) ENGINES

Component	U.S. (in.) ①	Metric (mm) ①
Compression		
Standard Compression	113-138 psi	795-971 kPa
Cylinder Bore		
Standard Bore Diameter	2.1260-2.1268	54.00-50.02
Wear Limit (1993 and later models)	2.130	54.1
Out-of-round Service Limit	0.002	0.05
Taper Service Limit	0.003	0.08
Crankshaft		
Crank Width	1.571-1.573	39.90-39.95
Runout Limit	0.0012	0.03
Connecting Rod		
Big End Side Clearance	0.008-0.028	0.2-0.7
Small End Diameter (1993 and later models)	0.5906-0.5910	15.000-15.011
Small End Freeplay Limit	0.08	2.0
Piston		
Standard Size	2.1248-2.1268	53.97-54.00
Measuring Point (above bottom of skirt)	0.39	10.0
Clearance	0.0012-0.0014	0.030-0.035
Service Limit	0.0039	0.1
Offset (thru 1992)	0.020	0.5
Offset Direction (1993 and later models)	Intake Side	Intake Side
Piston Pin (1993 and later models) Pin Diameter	0.4723-0.4724	11.996-12.000
Pin Boss Inner Diameter in Piston	0.4726-0.4730	12.004-12.015
Piston Rings		
Width (measured from outer-to-inner edge of ring) Top Ring	0.079	2.0
Bottom Ring	0.087	2.2
Thickness (measured vertically top-to-bottom of ring)	0.079	2.0
End Gap Installed Gap	0.006-0.014	0.15-0.35
Limit	0.022	0.55
Ring Side Clearance Top Ring	0.0008-0.0024	0.02-0.06
Bottom Ring	0.0012-0.0028	0.03-0.07
Reed Valve		
Thickness Plastic Reed	0.016	0.4
Stainless Reed	0.008	0.2
Lift (height)	0.272-0.288	6.8-7.2
Warpage (limit)	0.008	0.2

Fig. 272 Engine Specifications - 4/5 hp (103cc) 1-Cylinder Motors

ENGINE SPECIFICATIONS - 6/8 HP (165cc) ENGINES

Component	U.S. (in.) ①	Metric (mm) ①
Compression		
Standard Compression	153 psi	1079 kPa
Cylinder Bore		
Standard Bore Diameter	1.9685-1.9697	50.00-50.03
Wear Limit (1993 and later models)	1.972	50.1
Out-of-round		
Service Limit	0.002	0.05
Taper		
Service Limit	0.003	0.08
Crankshaft		
Crank Width		
Width around 1 crankpin	1.571-1.573	39.90-39.95
Total width (both crankpins)-1993 and later	4.004-4.016	101.7-102.0
Runout		
Limit	0.0012	0.03
Connecting Rod		
Big End Side Clearance	0.008-0.028	0.2-0.7
Small End Freeplay Limit	0.08	2.0
Piston		
Standard Size	1.9667-1.9677	49.955-49.980
Measuring Point (above bottom of skirt)	0.39	10.0
Clearance	0.0016-0.0018	0.040-0.045
Service Limit (1993 and later)	0.0037	0.095
Offset	0.020	0.5
Offset Direction	Exhaust Side	Exhaust Side
Piston Pin (1993 and later models)		
Pin Diameter	0.4723-0.4724	11.996-12.000
Pin Boss Inner Diameter in Piston	0.4726-0.4730	12.004-12.015
Piston Rings		
Both (Top and Bottom) Rings		
Width (measured from outer-to-inner edge of ring)		
Models through 1992	0.075-0.083	1.99-2.01
1993 and later models	0.079	2.0
Thickness (measured vertically top-to-bottom of ring)		
Models through 1992	0.0778-0.0786	1.97-1.99
1993 and later models	0.079	2.0
End Gap		
Installed Gap	0.006-0.014	0.15-0.35
Limit (1993 and later)	0.022	0.55
Ring Side Clearance		
Top Ring	0.0008-0.0024	0.02-0.06
Bottom Ring	0.0012-0.0028	0.03-0.07
Reed Valve		
Thickness (models through 1992)	0.060	0.15
Lift (height)	0.169-0.185	4.3-4.7
Warpage (limit)		
Models through 1992	0.024	0.6
1993 and later models	0.008	0.2

Fig. 273 Engine Specifications - 6/8 hp (165cc) 2-Cylinder Motors

ENGINE SPECIFICATIONS - 9.9/15 HP (246cc) ENGINES

Component	U.S. (in.) [1]	Metric (mm) [1]
Cylinder Bore		
Standard Bore Diameter	2.205-2.206	56.00-56.02
Wear Limit (1993 and later models)	2.21	56.1
Out-of-round Service Limit	0.002	0.05
Taper Service Limit	0.003	0.08
Cylinder Head		
Gasket Surface Warpage Limit	0.004	0.1
Crankshaft		
Crank Width		
Width around 1 crankpin	1.846-1.848	46.90-46.95
Total width around both crankpins (1993-95)	4.713-4.424	119.7-120.0
Width between crankpins (1996 and later)	1.020-1.028	25.90-26.10
Runout Limit	0.0012	0.03
Connecting Rod		
Big End Side Clearance		
Models through 1992 (no spec for 1993-95)	0.008-0.028	0.2-0.7
1996 and later models	0.012-0.031	0.3-0.8
Small End Freeplay Limit	0.08	2.0
Small End Diameter (1993 and later models)	0.7087-0.7091	18.000-18.011
Piston		
Standard Size		
Models through 1993	2.2023-2.2043	55.940-55.990
1994 and later models	2.2024-2.2041	55.940-55.985
Measuring Point (above bottom of skirt)	0.39	10.0
Clearance	0.0014-0.0016	0.035-0.040
Service Limit (1993 and later)	0.0035	0.09
Offset (1993-95 only)	0.04	1.0
Offset Direction (1993-95 only)	Intake Side	Intake Side
Piston Pin (1993 and later models)		
Pin Diameter	0.5510-0.5512	13.996-14.000
Pin Boss Inner Diameter in Piston	0.5513-0.5518	14.004-14.015
Oversize (check availability)		
1st	2.215	56.25
2nd	2.224	56.50
Piston Rings		
Both (Top and Bottom) Rings		
Width (measured from outer-to-inner edge of ring)		
Models through 1992	0.0960-0.1004	1.5-3.5
1993 and later models	0.10	2.5
Thickness (measured vertically top-to-bottom of ring)		
Models through 1992	0.0778-0.0786	1.97-1.99
1993 and later models	0.079	2.0

Fig. 274 Engine Specifications - 9.9/15 hp (246cc) 2-Cylinder Motors

ENGINE SPECIFICATIONS - 9.9/15 HP (246cc) ENGINES

Component	U.S. (in.) [1]	Metric (mm) [1]
End Gap		
Installed Gap	0.006-0.014	0.15-0.35
Limit (1993 and later)	0.022	0.55
Ring Side Clearance		
Top Ring		
Models through 1992	0.0016-0.0310	0.04-0.08
1993 and later models	0.0008-0.0024	0.02-0.06
Bottom Ring	0.0016-0.0310	0.04-0.08
Reed Valve		
Thickness (models through 1992)	0.008	0.20
Lift (height)		
9.9 hp		
Models through 1992	0.05	1.3
1993-03 (Europe only)	0.0508-0.0516	1.2-1.4
1993-95 (Except Europe)	0.0626-0.0634	1.5-1.7
1996-03 (Except Europe)	0.0236-0.0315	0.6-0.8
15 hp		
Models through 1992	0.16	4.0
1993-95	0.153-0.161	3.9-4.1
1996 and later models	0.232-0.240	5.9-6.1
Warpage (limit)	0.008	0.2

[1] Unless otherwise noted

Fig. 275 Engine Specifications - 9.9/15 hp (246cc) 2-Cylinder Motors (Cont'd)

POWERHEAD SYSTEMS

ENGINE SPECIFICATIONS - 20/25 HP (395cc) ENGINES

Component	U.S. (in.)①	Metric (mm)①
Cylinder Bore		
Standard Bore Diameter	2.638-2.639	67.00-67.02
Wear Limit (1996 and later models)	2.642	67.1
Out-of-round Service Limit	0.002	0.05
Taper Service Limit	0.003	0.08
Cylinder Head		
Gasket Surface Warpage Limit	0.004	0.1
Crankshaft		
Crank Width		
Width around 1 crankpin		
Models through 1991	1.97	50
1992 and later models	1.965-1.967	49.90-49.95
Total width around both crankpins (1988-95)	5.46-5.47	138.7-139.0
Width between crankpins (1996 and later)	1.531-1.539	38.90-39.10
Runout Limit	0.0012	0.03
Connecting Rod		
Big End Side Clearance (1994 and later models)	0.008-0.028	0.2-0.7
Small End Freeplay Limit (1994 and later models)	0.08	2.0
Small End Diameter (1996 and later models)	0.8671-0.8675	22.024-22.035
Piston		
Standard Size		
Models through 1992	2.637-2.638	66.98-67.00
1993 and later models	2.636-2.637	66.955-66.980
Measuring Point (above bottom of skirt)	0.39	10.0
Clearance	0.0016-0.0018	0.040-0.045
Service Limit	0.004	0.1
Piston Pin (1996 and later models)		
Pin Diameter	0.7085-0.7087	17.995-18.000
Pin Boss Inner Diameter in Piston	0.7088-0.7093	18.004-18.015
Oversize (check availability)		
1st	2.648	67.25
2nd	2.657	67.50
Piston Rings		
Both (Top and Bottom) Rings		
Width (measured from outer-to-inner edge of ring)	0.10	2.6
Thickness (measured vertically top-to-bottom of ring)	0.06	1.5
End Gap		
Installed Gap	0.016-0.024	0.40-0.60
Limit (1996 and later)	0.031	0.80
Ring Side Clearance		
Top Ring	0.0008-0.0024	0.02-0.06
Bottom Ring	0.0012-0.0027	0.03-0.07
Reed Valve		
Thickness (models through 1995)	0.008	0.20
Lift (height)	0.228-0.244	5.8-6.2
Warpage (limit)		
Models through 1991	0.059	1.5
1992 and later models	0.008	0.2

① Unless otherwise noted

Fig. 276 Engine Specifications - 20/25 hp (395cc) 2-Cylinder Motors

ENGINE SPECIFICATIONS - 20/25 HP (430cc) ENGINES

Component	U.S. (in.)①	Metric (mm)①
Cylinder Bore		
Standard Bore Diameter	2.638-2.639	67.00-67.02
Wear Limit (1996 and later models)	2.642	67.1
Out-of-round Service Limit	0.002	0.05
Taper Service Limit	0.003	0.08
Cylinder Head		
Gasket Surface Warpage Limit	0.004	0.1
Crankshaft		
Crank Width		
Width around 1 crankpin	2.122-2.124	53.90-53.95
Total width around both crankpins (through 1995)	5.62-5.63	142.7-143.0
Width between crankpins (1996 and later)	1.373-1.382	34.88-35.10
Runout Limit	0.0012	0.03
Connecting Rod		
Big End Side Clearance		
Models through 1995	0.075-0.082	1.90-2.10
1996 and later models	0.008-0.028	0.2-0.7
Small End Freeplay Limit	0.08	2.0
Small End Diameter (1996 and later models)	0.8663-0.8668	22.004-22.017
Piston		
Standard Size - 20 hp Motors		
USA, Canada, Oceania & NV	2.6299-2.6378	66.80-67.00
Except USA, Canada, Oceania & NV	2.6360-2.6370	66.955-66.980
Standard Size - 25 hp Motors		
Models through 1995	2.637-2.638	66.80-67.00
1996 and later models	2.6299-2.6378	66.80-67.00
Measuring Point (above bottom of skirt)	0.39	10.0
Clearance		
Models through 1995	0.0024-0.0025	0.060-0.065
1996 and later models	0.0016-0.0018	0.040-0.045
Service Limit (1996 and later models)	0.0037	0.095
Piston Pin (1996 and later models)		
Pin Diameter	0.7085-0.7087	17.995-18.000
Pin Boss Inner Diameter in Piston	0.7088-0.7093	18.004-18.015
Oversize (check availability)		
1st	2.648	67.25
2nd	2.657	67.50
Piston Rings		
Both (Top and Bottom) Rings		
Width (measured from outer-to-inner edge of ring)		
20 hp motors		
USA, Canada, Oceania and NV	0.10	2.6
Except USA, Canada, Oceania and NV	0.12	3.0
25 hp motors	0.10	2.6

① Unless otherwise noted

Fig. 217 Engine Specifications - 20/25 hp (430cc) 2-Cylinder Motors

6-86 POWERHEAD

ENGINE SPECIFICATIONS - 25/30 HP (496cc) ENGINES

Component	U.S. (in.) ①	Metric (mm) ①
Cylinder Bore		
Standard Bore Diameter	2.8346-2.8354	72.00-72.02
Wear Limit (1993 and later models)	2.839	72.1
Out-of-round Service Limit	0.002	0.05
Taper Service Limit	0.003	0.08
Cylinder Head		
Gasket Surface Warpage Limit	0.004	0.1
Crankshaft		
Crank Width		
Width around 1 crankpin	2.240-2.242	56.90-56.95
Total width around both crankpins (1989 and later)	6.051-6.063	153.7-154.0
Width between crankpins (1993 and later)	1.571-1.579	39.90-40.10
Runout		
Limit		
Except 1989-92	0.0012	0.03
1989-92	0.0020	0.05
Connecting Rod		
Big End Side Clearance - 25 Hp Models		
Models through 1987	0.015-0.019	0.38-0.48
1996 and later	0.008-0.028	0.20-0.70
Big End Side Clearance - 30 Hp Models		
Models through 1986	0.000-0.000	0.0-0.0
1989-92	n/a	n/a
1993 and later models	0.008-0.028	0.2-0.7
Small End Freeplay Limit (1993 and later models)	0.08	2.0
Piston		
Standard Size	2.832-2.833	71.94-71.96
Measuring Point (above bottom of skirt)	0.39	10.0
Clearance	0.0024-0.0026	0.060-0.065
Service Limit (1989 and later models)	0.0039	0.1
Piston Pin (1993 and later models)		
Pin Diameter	0.7833-0.7835	19.895-19.900
Pin Boss Inner Diameter in Piston	0.7836-0.7841	19.904-19.915
Oversize (check availability)		
1st	2.84	72.25
2nd	2.85	72.50
Piston Rings		
Both (Top and Bottom) Rings		
Width (measured from outer-to-inner edge of ring)		
Top Ring	0.12	3.0
Bottom Ring		
Models through 1992	0.13	3.2
1993 and later models	0.12	3.0
Thickness (measured vertically top-to-bottom of ring)		
Except 1989-92 Top Ring	0.080	2.0
1989-92 Top Ring	0.059	1.5

① Unless otherwise noted

Fig. 219 Engine Specifications - 25/30 hp (496cc) 2-Cylinder Motors

ENGINE SPECIFICATIONS - 20/25 HP (430cc) ENGINES

Component	U.S. (in.) ①	Metric (mm) ①
Thickness (measured vertically top-to-bottom of ring)		
20 hp motors		
USA, Canada, Oceania and NV	0.06	1.5
Except USA, Canada, Oceania and NV	0.08	2.0
25 hp motors	0.06	1.5
End Gap		
Installed Gap (Limit)		
20 hp motors		
USA, Canada, Oceania and NV	0.016-0.024 (0.031)	0.40-0.60 (0.80)
Except USA, Canada, Oceania and NV	0.012-0.020 (0.028)	0.30-0.50 (0.70)
25 hp motors	0.016-0.024 (0.031)	0.40-0.60 (0.80)
Ring Side Clearance		
Top Ring		
Models through 1995	0.0012-0.0020	0.03-0.05
1996 and later models	0.0010-0.0024	0.02-0.06
Bottom Ring	0.0012-0.0028	0.03-0.07
Reed Valve		
Thickness (models through 1995)	0.006	0.15
Lift (height)	0.197-0.217	5.0-5.5
Warpage (limit)	0.0079	0.20

① Unless otherwise noted

Fig. 218 Engine Specifications - 20/25 hp (430cc) 2-Cylinder Motors

Fig. 219 Engine Specifications - 25/30 hp (496cc) 2-Cylinder Motors (Cont'd)

POWERHEAD SYSTEMS

ENGINE SPECIFICATIONS - 25/30 HP (496cc) ENGINES

Component		U.S. (in.)①	Metric (mm)①
End Gap			
Installed Gap			
Models through 1987		0.008-0.016	0.20-0.40
1989 and later models		0.008-0.014	0.20-0.35
Limit (1993-95 models only)		0.022	0.55
Ring Side Clearance (no spec 1993 and later)			
Top Ring		0.0008-0.0024	0.02-0.06
Bottom Ring	25 hp Models	0.0016-0.0032	0.04-0.08
	30 hp Models	0.0012-0.0028	0.03-0.07
Reed Valve			
Thickness (models through 1995 only)			
25 hp models		0.008	0.20
30 hp models			
Lift (height)			
25 hp models			
Models through 1986		0.006	0.15
1989 and later models		0.008	0.20
30 hp models			
Models through 1987		0.188-0.212	4.7-5.3
1996 and later models		0.067-0.091	1.7-2.3
Warpage (limit)			
25 hp models			
Models through 1986		0.21	5.4
1989-92		0.19-0.21	4.7-5.3
1993 and later models		0.197-0.217	5.0-5.5
30 hp models			
Models through 1986		0.008	0.2
1989-92		0.012	0.3
1993 and later models		0.024	0.6
		0.008	0.2

① Unless otherwise noted

Fig. 220 Engine Specifications - 25/30 hp (496cc) 2-Cylinder Motors (Cont'd)

ENGINE SPECIFICATIONS - 40 HP (592cc) ENGINES

Component		U.S. (in.)①	Metric (mm)①
Cylinder Bore			
Standard Bore Diameter		2.953-2.954	75.00-75.02
Wear Limit (1996 and later models)		2.96	75.1
Out-of-round			
Service Limit		0.002	0.05
Taper			
Service Limit		0.003	0.08
Cylinder Head			
Gasket Surface Warpage Limit		0.004	0.1
Crankshaft			
Crank Width			
Width around 1 crankpin (1989-95 only)		2.372-2.382	60.25-60.50
Total width around both crankpins			
1989-95		6.050-6.063	153.68-154.00
1996 and later		6.150-6.170	156.30-166.70
Runout			
Limit			
1989-95		0.001	0.02
1996 and later		0.002	0.05
Connecting Rod (1996 and later only)			
Big End Side Clearance		0.005-0.010	0.12-0.26
Small End Freeplay Limit		0.08	2.0
Piston			
Standard Size		2.9506-2.9516	74.945-74.970
Measuring Point (above bottom of skirt)		0.39	10.0
Clearance			
Service Limit		0.0020-0.0022	0.050-0.055
		0.0041	0.105
Piston Pin (1996 and later models)			
Pin Diameter		0.7833-0.7835	19.895-19.900
Pin Boss Inner Diameter in Piston		0.7833-0.7835	19.895-19.900
Offset (models through 1995)		Exaust Side 0.059	Exaust Side 1.5
Oversize (check availability)			
1st		2.84	72.25
2nd		2.85	72.50
Piston Rings			
Both (Top and Bottom) Rings			
Width (measured from outer-to-inner edge of ring)		0.102	2.6
Thickness (measured vertically top-to-bottom of ring)			
Top Ring		0.059	1.5
Bottom Ring		0.079	2.0
End Gap			
Installed Gap			
Both (Top and Bottom) Rings		0.012-0.020	0.30-0.50
Limit (1996 and later)		0.028	0.70
Ring Side Clearance			
Top Ring		0.0008-0.0024	0.02-0.06
Bottom Ring		0.0012-0.0028	0.03-0.07
Reed Valve			
Thickness (models through 1995 only)		0.008	0.20
Lift (height)		0.19-0.21	4.8-5.2
Warpage (limit)			
1989-95		0.028	0.7
1996 and later models		0.008	0.2

① Unless otherwise noted

Fig. 221 Engine Specifications - 40 hp (592cc) 2-Cylinder Motors

6-88 POWERHEAD

ENGINE SPECIFICATIONS - 48/55 HP (760cc) ENGINES

Component		U.S. (in.) ①	Metric (mm) ①
Cylinder Bore			
Standard Bore Diameter		3.228-3.229	82.00-82.02
Wear Limit (1995 and later models)		3.23	82.1
Out-of-round Service Limit		0.002	0.05
Taper Service Limit		0.003	0.08
Cylinder Head			
Gasket Surface Warpage Limit		0.004	0.1
Crankshaft			
Crank Width			
Width around 1 crankpin (1995 and later only)		2.44-2.46	61.9-62.5
Total width around both crankpins		6.45-6.46	163.08-164.2
Runout Limit	1989-94	0.001	0.02
	1995 and later	0.002	0.05
Connecting Rod (1995 and later only)			
Big End Side Clearance		0.005-0.010	0.12-0.26
Small End Inner Diameter		0.9803-0.9808	24.900-24.912
Small End Freeplay Limit		0.08	2.0
Piston			
Standard Size		3.2258-3.2266	81.935-81.955
Measuring Point (above bottom of skirt)		0.39	10.0
Clearance	1989-94	0.0024-0.0026	0.060-0.065
	1995 and later	0.0026-0.0028	0.065-0.070
Service Limit		0.0047	0.12
Piston Pin (1995 and later models)			
Pin Diameter		0.7833-0.7835	19.895-19.900
Pin Boss Inner Diameter in Piston		0.7836-0.7841	19.904-19.915
Offset		Exaust Side 0.059	Exaust Side 1.5
Oversize (check availability)			
1st		3.238	82.25
2nd		3.248	82.50
Piston Rings			
Both (Top and Bottom) Rings			
Width (measured from outer-to-inner edge of ring)			
	1989-94	0.098	2.5
	1995 and later	0.130	3.2
Thickness (measured vertically top-to-bottom of ring)		0.079	2.0
End Gap			
Installed Gap (Both Top and Bottom Rings)		0.016-0.024	0.40-0.60
Limit (1995 and later)		0.031	0.80

Fig. 222 Engine Specifications - 48 hp (760cc) 2-Cylinder Motors

ENGINE SPECIFICATIONS - 48/55 HP (760cc) ENGINES

Component		U.S. (in.) ①	Metric (mm) ①
Ring Side Clearance			
1989-94 Top Ring		0.0012-0.0026	0.020-0.065
Bottom Ring		0.0012-0.0028	0.030-0.070
1995 and later		0.0026-0.0028	0.065-0.070
Reed Valve			
Thickness		0.010	0.25
Lift (height)			
48 Hp Models	1989-94	0.112-0.128	2.8-3.2
55 Hp Models	1995	0.332-0.348	8.4-8.8
		0.382-0.398	9.7-10.1
Warpage (limit)			
	1989-94	0.020	0.5
	1995 and later models	0.008	0.2

① Unless otherwise noted

Fig. 223 Engine Specifications - 48 hp (760cc) 2-Cylinder Motors (Cont'd)

POWERHEAD SYSTEMS 6-89

ENGINE SPECIFICATIONS - 25/30 HP (496cc) 3 CYL ENGINES

Component		U.S. (in.)①	Metric (mm)①
Ring Side Clearance			
	1987-91	0.0028-0.0043	0.07-0.11
	1992 and later Top Ring	0.0020-0.0035	0.05-0.09
	1992 and later Bottom Ring	0.002-0.004	0.05-0.09
Reed Valve			
Thickness (1987-92 only)		0.008	0.20
Lift (height)		0.094-0.106	2.50-2.80
Warpage (limit)			
	1987-91	0.035	0.9
	1992 and later models	0.010	0.2

① Unless otherwise noted

Fig. 225 Engine Specifications - 25/30 hp (496cc) 3-Cylinder Motors (Cont'd)

ENGINE SPECIFICATIONS - 25/30 HP (496cc) 3 CYL ENGINES

Component		U.S. (in.)①	Metric (mm)①
Cylinder Bore			
Standard Bore Diameter		2.3425-2.3433	59.50-59.52
Wear Limit (1998 and later models)		2.35	59.6
Out-of-round Service Limit		0.002	0.05
Taper Service Limit		0.003	0.08
Cylinder Head			
Gasket Surface Warpage Limit		0.004	0.1
Crankshaft			
Crank Widths			
Width around 1 crankpin (1992 and later only)		1.965-1.976	49.90-49.95
Width around end and center crankpins (1993 and later)		5.06-5.08	128.6-129.0
Total width around all 3 crankpins (1992 and later only)		8.17-8.19	207.5-208.1
Runout Limit	1987-97 only	0.002	0.05
Connecting Rod (1992 and later only)			
Big End Side Clearance		0.008-0.028	0.20-0.70
Small End Inner Diameter (1998 and later only)		0.7876-0.7878	20.004-20.007
Small End Freeplay Limit		0.08	2.0
Piston			
Standard Size		2.3407-2.3417	59.455-59.480
Measuring Point (above bottom of skirt)		0.39	10.0
Clearance			
	1987-91	0.0014-0.0016	0.035-0.040
	1992 and later	0.0016-0.0018	0.040-0.045
Piston Pin (1998 and later models)			
Pin Diameter		0.6825-0.6287	15.965-15.970
Pin Boss Inner Diameter in Piston		0.6289-0.6293	15.974-15.985
Oversize (check availability)			
	1st	2.352	59.75
	2nd	2.362	60.00
Piston Rings			
Both (Top and Bottom) Rings			
Width (measured from outer-to-inner edge of ring)			
	1987-91 (top ring / bottom ring)	0.083 / 0.098	2.1 / 2.4
	1992 and later	0.09	2.4
Thickness (measured vertically top-to-bottom of ring)			
	1987-91 (top ring / bottom ring)	0.059 / 0.079	1.5 / 2.0
	1992 only	0.08	2.0
	1993 and later	0.07	1.9
End Gap			
Installed Gap (Both Top and Bottom Rings)		0.006-0.012	0.15-0.30

Fig. 224 Engine Specifications - 25/30 hp (496cc) 3-Cylinder Motors

6-90 POWERHEAD

ENGINE SPECIFICATIONS - 25J/30J/40/50 HP (698cc) 3 CYL ENGINES

Component	U.S. (in.)①	Metric (mm)①
Cylinder Bore		
Standard Bore Diameter	2.638-2.639	67.00-67.02
Wear Limit (1994 and later models)	2.642	67.10
Out-of-round Service Limit	0.002	0.05
Taper Service Limit	0.003	0.08
Cylinder Head		
Gasket Surface Warpage Limit	0.004	0.1
Crankshaft		
Crank Widths		
Width around 1 crankpin	2.122-2.124	53.90-53.95
Width around end & center crankpins (1993-94 only)	5.539-5.551	140.70-141.00
Width between end & center crankpins (1995 and later)	1.294-1.303	32.88-33.10
Total width around all 3 crankpins (through 1994 Only)	8.955-8.978	227.46-228.05
Runout Limit 1984-94	0.002	0.05
1995-03	0.001	0.03
Connecting Rod (1989 and later only)		
Big End Side Clearance	0.008-0.028	0.20-0.70
Small End Inner Diameter (1995 and later only)	0.8663-0.8665	22.005-22.008
Small End Freeplay Limit	0.08	2.0
Piston		
Standard Size		
1984-87	2.6376-2.6378	66.994-67.000
1988	2.6360-2.6364	66.955-66.965
1989-1992	2.6352-2.6362	66.935-66.960
1993-03	2.6354-2.6378	66.940-67.000
Measuring Point (above bottom of skirt)	0.39	10.0
Clearance	0.0024-0.0026	0.060-0.065
Piston Pin (1995 and later models)		
Pin Diameter	0.7085-0.7087	17.995-18.000
Pin Boss Inner Diameter in Piston	0.7090-0.7093	18.008-18.015
Oversize (check availability)		
1st	2.648	67.25
2nd	2.657	67.50
Piston Rings		
Both (Top and Bottom) Rings		
Width (measured from outer-to-inner edge of ring)	0.10	2.6
Thickness (measured vertically top-to-bottom of ring)		
1984-88 (top ring / bottom rings)	0.06 / 0.08	1.5 / 2.0
1989-03	0.08	2.0
End Gap		
Installed Gap (Both Top and Bottom Rings)	0.016-0.024	0.4-0.6
Limit (1994 and later)	0.031	0.80

Fig. 226 Engine Specifications - 25J/30J/40/50 hp (698cc) 3-Cylinder Motors

ENGINE SPECIFICATIONS - 25J/30J/40/50 HP (698cc) 3 CYL ENGINES

Component		U.S. (in.)①	Metric (mm)①
Ring Side Clearance			
Top Ring	1984-88	0.001-0.002	0.03-0.05
	1989-03 (Includes Bottom Ring 1989)	0.002-0.003	0.04-0.08
Bottom Ring/Rings (Except 1989)		0.001-0.003	0.03-0.07
Reed Valve			
Thickness (1984-92 only)		0.008	0.20
Lift (height)			
	1984-88	0.06-0.07	1.55-1.85
	1989-92	0.23-0.24	5.84-6.10
	1993-03	0.23-0.25	5.84-6.35
Warpage (limit)			
	1987-91	0.035	0.9
	1992 and later models	0.010	0.2

① Unless otherwise noted

Fig. 227 Engine Specifications - 25J/30J/40/50 hp (698cc) 3-Cylinder Motors (Cont'd)

ENGINE SPECIFICATIONS - 50/60/70 HP (849cc) 3 CYL ENGINES

Component		U.S. (in.) ①	Metric (mm) ①
Compression			
Standard Compression		124 psi	853 kPa
Cylinder Bore			
Standard Bore Diameter		2.834-2.835	72.00-72.02
Out-of-round	Service Limit	0.002	0.05
Taper	Service Limit	0.003	0.08
Cylinder Head			
Gasket Surface Warpage Limit		0.004	0.1
Crankshaft			
Crank Widths			
Width around 1 crankpin		2.280-2.281	57.90-57.95
Width around end & center crankpins (1993 and later)		5.972-5.984	151.70-152.00
Total width around all 3 crankpins		9.67-9.69	245.5-246.1
Runout	Limit		
	1984-93	0.002	0.05
	1994-03	0.001	0.03
Connecting Rod (1993 and later only)			
Big End Side Clearance		0.008-0.028	0.20-0.70
Small End Freeplay Limit		0.08	2.0
Piston			
Standard Size	1984-88	2.8329-2.8332	71.955-71.965
	1989-03	2.8325-2.8335	71.945-71.970
Measuring Point (above bottom of skirt)		0.40	10.0
Clearance		0.0020-0.0022	0.050-0.055
Oversize (check availability)			
1st		2.844	72.25
2nd		2.854	72.50
Piston Rings			
Both (Top and Bottom) Rings			
Width (measured from outer-to-inner edge of ring)			
Top Ring (& 2nd Ring 1990 and later)		0.120	3.0
Bottom Rings (1984-89 only)		0.098	2.5
Thickness (measured vertically top-to-bottom of ring)			
1984-89 (top and both bottom rings)		0.06	1.5
1990-03 (top and 2nd ring only)		0.08	2.0
End Gap (Top, and Bottom Rings -89, 2nd only 1990+)			
Installed Gap		0.012-0.020	0.3-0.5
Ring Side Clearance			
Top Ring	1984-89	0.0008-0.0024	0.02-0.06
	1990-03	0.0012-0.0028	0.03-0.07
Bottom Ring(s) - Only measure 2nd ring 1993-03		0.0012-0.0028	0.03-0.07

Fig. 228 Engine Specifications - 50/60/70 hp (849cc) 3-Cylinder Motors

ENGINE SPECIFICATIONS - 50/60/70 HP (849cc) 3 CYL ENGINES

Component		U.S. (in.) ①	Metric (mm) ①
Reed Valve			
Thickness (1984-92 only)		0.008	0.20
Lift (height)	1984-88	0.39	9.9
	1989-91	0.332-0.348	8.4-8.8
	1992-03		
Warpage (limit)	50/60 Hp Motors	0.11-0.13	2.8-3.2
	70 Hp Motors	0.38-0.40	9.7-10.1
	1989-91	0.020	0.5
	1984-88 and 1992-03	0.010	0.2

① Unless otherwise noted

Fig. 229 Engine Specifications - 50/60/70 hp (849cc) 3-Cylinder Motors (Cont'd)

ENGINE SPECIFICATIONS - 65J/75/80/85/90 HP (1140cc) 3 CYL ENGINES

Component		U.S. (in.)①	Metric (mm)①
Compression			
Standard Compression		134 psi	922 kPa
Cylinder Bore			
Standard Bore Diameter		3.228-3.229	82.00-82.02
Wear Limit (1996 and later)		3.23	82.1
Out-of-round		0.002	0.05
Taper	Service Limit	0.003	0.08
Cylinder Head			
Gasket Surface Warpage Limit		0.004	0.1
Crankshaft			
Crank Widths			
Total Width around all 3 Crankpins (1997 and later) ②		11.19-11.21	284.2-284.8
Runout Limit	1984-92	0.0008	0.02
	1993	0.0012	0.03
	1994-03	0.0020	0.05
Connecting Rod			
Big End Side Clearance	1984-88	0.008-0.013	0.20-0.33
	1989-92	no spec	no spec
	1993-03	0.005-0.010	0.12-0.26
Small End Inner Diameter (1997+ 65J/75/85 only) ③		0.9803-0.9808	24.900-24.912
Small End Freeplay Limit (1994 and later only)		0.08	2.0
Piston			
Standard Size		3.23	82.00
		3.2266-3.2269	81.955-81.965
		3.2258-3.2268	81.935-81.960
Measuring Point (above bottom of skirt)		0.39	10.0
Clearance			
All, except 1997 or later 65J/85 and 75 A or E models		0.0024-0.0026	0.060-0.065
1997 or later 65J/85 and 75 A or E models ③			
US or World Spec Motors		0.0026-0.0028	0.065-0.070
European Spec Motors		0.0024-0.0026	0.060-0.065
Limit (1997 and later motors only)			
All, except 1997 or later 65J/85 and 75 A or E models		0.0045	0.115
1997 or later 65J/85 and 75 A or E models ③			
US or World Spec Motors		0.0047	0.120
European Spec Motors		0.0045	0.115
Piston Pin (1997 or later 65J/85 and 75 A or E only) ③			
Pin Diameter		0.7833-0.7835	19.895-19.900
Pin Boss Inner Diameter in Piston		0.7836-0.7841	19.904-19.915
Offset (except 1997 or later 65J/85 or 75 A or E models)			
Measurement		0.039	1.0
Direction		Exhaust Side	Exaust Side

Fig. 230 Engine Specifications - 65J/75/80/85/90 hp (1140cc) 3-Cylinder Motors

ENGINE SPECIFICATIONS - 65J/75/80/85/90 HP (1140cc) 3 CYL ENGINES

Component	U.S. (in.)①	Metric (mm)①
Oversize (check availability)		
1st (usually, except US)	3.238	82.25
2nd	3.248	82.50
Piston Rings (Top and 2nd - unless otherwise noted)		
Width (measured from outer-to-inner edge of ring)		
1984-89	0.01	2.5
1990-93	0.11	2.8
1994-03	0.13	3.2
	0.08	2.0
Thickness (measured vertically top-to-bottom of ring)		
End Gap		
Installed Gap (Usually just Top and 2nd Rings)	0.016-0.024	0.4-0.6
Limit (1997 and later 65J/85 and 75 A or E models) ③	0.031	0.80
Side Clearance (Just Top and 2nd Rings after 1992)		
All, except 1997 or later 65J/85 and 75 A or E models		
1984-89 (Top Ring)	0.0012-0.0026	0.030-0.065
1984-89 (2nd and Bottom Rings)	0.0012-0.0028	0.030-0.070
1990-93 (Top Ring)	0.0016-0.0031	0.040-0.080
1990-93 (2nd and Bottom Rings)	0.0008-0.0024	0.020-0.060
1994-03	0.0012-0.0024	0.030-0.060
1997 or later 65J/85 and 75 A or E models ③		
US or World Spec Motors (Top / Second) ④	0.002-0.003 / 0.001-0.002	0.05-0.08 / 0.03-0.06
European Spec Motors (Top & Second)	0.001-0.002	0.03-0.06
Reed Valve		
Thickness (1984-92 only)	0.008	0.20
Lift (height)		
1984-88	0.39	9.9
1989-1991	0.332-0.348	8.4-8.8
1992-03	0.38-0.40	9.7-10.1
Warpage (limit)		
1989-91	0.020	0.5
1984-88, and 1992-03	0.010	0.2

① Unless otherwise noted
② Specification applies to all 75 hp motors except A and E models (75AM, 75AEM, 75AE, 75AET)
③ Specification only applies to 1997 and later 65J/85 and A or E model 75 hp motors (75AM, 75AEM, 75AE, 75AET)
④ Specifications are for all 75 hp A or E models, except the 75AM for which the ring clearance is same as Europe spec motors

Fig. 231 Engine Specifications - 65J/75/80/85/90 hp (1140cc) 3-Cylinder Motors (Cont'd)

POWERHEAD SYSTEMS 6-93

ENGINE SPECIFICATIONS - 80J/100/115/130/140 HP (1730cc) 90 DEG V4 ENGINES

Component		U.S. (in.)①	Metric (mm)①
Compression			
Standard Compression		no spec avail.	no spec avail.
80J / 115 hp		136 psi	941 kPa
130 / 140 hp		149 psi	1030 kPa
Cylinder Bore			
Standard Bore Diameter		3.543-3.544	90.00-90.02
Wear Limit (1993 and later models)		3.550	90.10
Out-of-round			
Service Limit		0.002	0.05
Taper			
Service Limit		0.003	0.08
Cylinder Head			
Gasket Surface Warpage Limit		0.004	0.1
Crankshaft			
Runout			
Limit	1984-92	0.001	0.02
	1993	0.001	0.03
	1994-03	0.002	0.05
Connecting Rod (1988 and later only)			
Big End Side Clearance	1988-92	0.0079-0.0126	0.20-0.32
	1993-03	0.0050-0.0100	0.12-0.26
Small End Freeplay Limit		0.08	2.0
Piston			
Standard Size	1984-95	3.5400-3.5410	89.920-89.940
	1996-03	3.5402-3.5407	89.920-89.935
Measuring Point (above bottom of skirt)		0.394	10.0
Clearance	1984-87	0.0033-0.0035	0.085-0.090
	1987-03	0.0031-0.0033	0.080-0.085
Oversize (check availability)			
1st		3.553	90.25
2nd		3.563	90.50
Piston Rings			
Both (Top and 2nd) Rings			
Width (measured from outer-to-inner edge of ring)		0.110	2.8
Thickness (measured vertically top-to-bottom of ring)		0.079	2.0
End Gap			
Installed Gap		0.012-0.016	0.30-0.40
Limit (1993 and later)		0.024	0.60

Fig. 232 Engine Specifications - 80J/100/115/130/140 hp (1730cc) 90° V4 Motors

ENGINE SPECIFICATIONS - 80J/100/115/130/140 HP (1730cc) 90 DEG V4 ENGINES

Component		U.S. (in.)①	Metric (mm)①
Ring Side Clearance			
Top Ring	1984-87	0.0012-0.0026	0.030-0.065
	1988-03	0.0010-0.0020	0.020-0.060
2nd Ring	1984-87	0.0016-0.0030	0.040-0.075
	1988-03	0.0010-0.0020	0.020-0.060
Reed Valve			
Thickness (1984-87)			
Lift (height)		0.01	0.25
Warpage (limit)	1984-87	0.25-0.26	6.3-6.7
	1988-03	0.25-0.27	6.2-6.8
	1987-92	0.035	0.9
	1993 and later models	0.010	0.2

① Unless otherwise noted

Fig. 233 Engine Specifications - 80J/100/115/130/140 hp (1730cc) 90° V4 Motors (Cont'd)

ENGINE SPECIFICATIONS - 105J/140J/150/175/200/220/225 HP (2596cc) V6 ENGINES

Component		U.S. (in.) ①	Metric (mm) ①
Compression			
Standard Compression			
Carburetted	150 hp	108 psi	745 kPa
	175 hp	129 psi	892 kPa
	200 hp	129 psi	892 kPa
	225 hp	121 psi	834 kPa
EFI (OX66)		No spec avail.	No spec avail.
HPDI (all)		94 psi	650 kPa
Cylinder Bore			
Standard Bore Diameter		3.543-3.544	90.00-90.02
Limit (1993 and later only)		3.547	90.10
Out-of-round	Service Limit	0.002	0.05
Taper	Service Limit	0.003	0.08
Cylinder Head			
Gasket Surface Warpage Limit		0.004	0.1
Crankshaft			
Runout 1984-92		0.001	0.02
1993-03		0.002	0.05
Connecting Rod (1990 and later only)			
Big End Side Clearance			
1990-92	Except ProV200	0.0079-0.0126	0.20-0.32
	ProV200	0.0039-0.0075	0.10-0.19
1993-03		0.005-0.010	0.12-0.26
Small End Freeplay Limit		0.08	2.0
Piston			
Standard Size			
1984-89		3.54	90
1990-93		3.5402-3.5409	89.920-89.940
1994-95	Except V6Excel (220/225 hp)	3.5402-3.5409	89.920-89.940
	V6 Excel (220/225 hp)	3.5398-3.5403	89.910-89.930
1996-03 (carb and cyl #'s 2-6 of EFI motors)		3.5392-3.5400	89.895-89.915
1999 and later HPDI motors and cyl # 1 for EFI motors		3.5372-3.5381	89.845-89.869
Measuring Point (above bottom of skirt)		0.40	10.0
Clearance			
1984-89			
Except V6 Excel (220/225 hp)		0.0033-0.0035	0.085-0.090
V6 Excel (220/225 hp)		0.0023-0.0026	0.060-0.065
1990-93		0.0031-0.0033	0.080-0.085
1994-95			
Except P150/P175 - Build/Limit		0.0031-0.0033 / 0.0053	0.080-0.085 / 0.135
P150/P175 - Build/Limit		0.0035-0.0037 / 0.0057	0.090-0.095 / 0.145
1996-03 (carb and cyl #'s 2-6 of EFI motors)		0.0039-0.0042	0.100-0.106
1999 and later HPDI motors and cyl # 1 for EFI motors		0.0059-0.0061	0.150-0.156

Fig. 234 Engine Specifications - 105J/140J/150/175/200/220/225 hp (2596cc) 90° V6 Motors

ENGINE SPECIFICATIONS - 105J/140J/150/175/200/220/225 HP (2596cc) V6 ENGINES

Component		U.S. (in.) ① Starboard / Port — Exhaust Side	Metric (mm) Starboard / Port — Exhaust Side
Offset (models through 1989 only)		0 / 0.059	0 / 1.5
Oversize (check availability)			
1984-95	1st	3.553	90.25
	2nd	3.563	90.50
1996-03 (carb and cyl #'s 2-6 of EFI motors)	1st	3.549	90.15
	2nd	3.559	90.40
1999 and later HPDI motors and cyl # 1 for EFI motors	1st	3.548	90.11
	2nd	3.557	90.36
Piston Rings			
Both (Top and 2nd) Rings			
Width (measured from outer-to-inner edge of ring)		0.110	2.8
Thickness (measured vertically top-to-bottom of ring)		0.079	2.0
End Gap			
Installed Gap			
1984-89 (except V6Excel 200/225 hp)		0.012-0.020	0.30-0.50
1990-89 (& earlier V6 Excel 200/225hp)		0.012-0.016	0.30-0.40
Limit (1993 and later)		0.024	0.60
Ring Side Clearance			
Top Ring	1984-89	0.0012-0.0026	0.030-0.065
	1990-03	0.0010-0.0020	0.020-0.060
2nd Ring	1984-89	0.0016-0.0030	0.040-0.075
	1990-03	0.0010-0.0020	0.020-0.060
Reed Valve			
Thickness (1984-89 only)		0.010	0.25
Lift (height)			
1984-88			6.3-6.7
Except V6Excel 200/225 hp		0.250-0.260	6.2-6.8
V6 Excel 200/225hp		0.240-0.260	6.2-6.8
1990-92		0.244-0.268	6.2-6.8
1993		0.215-0.230	5.45-5.85
1994-95		0.216-0.232	5.5-5.9
Except P175 and all 200-225 hp motors		0.244-0.268	6.2-6.8
P175 and all 200-223 hp motors			
1996-03	Carbureted Motors	0.25-0.27	6.20-6.80
	Fuel Injected Motors	0.34-0.36 ②	8.65-9.35 ②
Warpage (limit)			
1984-92		0.035	0.9
1994-95	Except P200	0.010	0.2
	P200	0.012	0.3
1993 and 1996-03		0.010	0.2

① Unless otherwise noted
② Specification is for all motors except 150-175 HPDI, for which the spec is 0.31-0.33 in. (7.8-8.4mm)

Fig. 235 Engine Specifications - 105J/140J/150/175/200/220/225 hp (2596cc) 90° V6 Motors (Cont'd)

POWERHEAD SYSTEMS

ENGINE SPECIFICATIONS - 200/225/250 HP (3130cc) V6 ENGINES

Component		U.S. (in.) ①	Metric (mm) ①
Compression			
Standard Compression			
200 / 225 hp		No spec avail. 107 psi	No spec avail. 750 kPa
250 hp			
Cylinder Bore			
Standard Bore Diameter		3.543-3.544	90.00-90.02
Limit (1994 and later only)		3.547	90.10
Out-of-round Service Limit		0.002	0.05
Taper Service Limit		0.003	0.08
Cylinder Head			
Gasket Surface Warpage Limit		0.004	0.1
Crankshaft			
Runout			
1990-97		0.001	0.02
1998-03		0.002	0.05
Connecting Rod			
Big End Side Clearance		0.005-0.010	0.12-0.26
Big End Radial Play (1990-91 Only)		0.0008-0.0016	0.02-0.04
Small End Diameter (1996-98 Only)		1.2205-1.2209	31.000-31.012
Small End Freeplay Limit		0.08	2.0
Piston			
Standard Size			
1990-93		3.5380-3.5390	89.87-89.89
1994	225 HP	3.5376-3.5384	89.855-89.875
	250 HP	3.5380-3.5388	89.865-89.885
1995		3.5374-3.5382	89.85-89.89
1996-97			
	Starboard Side	3.5374-3.5382	89.85-89.87
	Port Side	3.5370-3.5378	89.84-89.86
1998-03		3.5370-3.5378	89.84-89.86
Measuring Point (above bottom of skirt)		0.40	10.0
Clearance			
1990-92		0.0053-0.0055	0.135-0.140
1993			
	Starboard Side	0.0047-0.0049	0.120-0.125
	Port Side	0.0053-0.0055	0.135-0.140
1994-95		0.0057-0.0059	0.145-0.151
1996-97	225 HP	0.0053-0.0055	0.135-0.140
	250 HP		
1998-03			
	Starboard Side	0.0057-0.0059	0.145-0.151
	Port Side	0.0061-0.0063	0.155-0.161
Limit			
1995-98		0.0079	0.201
1999-03		0.0063	0.161

Fig. 236 Engine Specifications - 200/225/250 hp (3130cc) 76° V6 Motors

ENGINE SPECIFICATIONS - 200/225/250 HP (3130cc) V6 ENGINES

Component		U.S. (in.) ①	Metric (mm) ①
Oversize (check availability)			
1990-91	1st	3.5488-3.5494	90.140-90.155
	2nd	3.5587-3.5593	90.390-90.405
1992	1st	3.557-3.558	90.370-90.390
1993	1st	3.557-3.558	90.370-90.390
1994-03			
1st (except V/VX models)		3.549	90.15
1st (V/VX models)		3.5468-3.5476	90.090-90.110
2nd (except V/VX models)		3.559	90.40
2nd (V/VX models)		3.5566-3.5574	90.340-90.360
Piston Rings			
Both (Top and 2nd) Rings			
Width (measured from outer-to-inner edge of ring)			
1990-91		0.130-0.138	3.30-3.50
1992-93		0.106-0.114	2.70-2.90
1994-03		0.110	2.8
Thickness (measured vertically top-to-bottom of ring)			
1990-91		0.077-0.078	1.95-1.97
1992-93		0.0776-0.0783	1.97-1.99
1994-03		0.079	2.0
End Gap			
Installed Gap			
1990-91		0.012-0.017	0.30-0.42
1992-03		0.012-0.016	0.30-0.40
Limit (1996 and later)		0.024	0.60
Ring Side Clearance			
Both Top and 2nd Rings			
1990-91		0.0016-0.0031	0.040-0.080
1992-03		0.0010-0.0020	0.020-0.060
Reed Valve			
Lift (height)			
1990-96		0.343-0.366	8.7-9.3
1997-03			
Except 225 S & L Models		0.34-0.36	8.7-9.3
225 S & L Models		0.30-0.32	7.6-8.2
Warpage (limit)			
1990-03		0.010	0.2

① Unless otherwise noted

Fig. 267 Engine Specifications - 200/225/250 hp (3130cc) 76° V6 Motors (Cont'd)

ENGINE SPECIFICATIONS - 225/250 HP (3342cc) V6 ENGINES

Component	U.S. (in.) ①	Metric (mm) ①
Cylinder Bore		
Standard Bore Diameter	3.6614-3.6622	93.000-93.020
Limit	3.6654	93.100
Out-of-round		
Service Limit	0.002	0.05
Taper		
Service Limit	0.003	0.08
Cylinder Compression - Minimum	81 psi	560 kPa
Cylinder Head		
Gasket Surface Warpage Limit	0.004	0.1
Crankshaft		
Crankshaft Journal Diameter	2.3219-2.3225	58.975-58.991
Crankpin Diameter	1.5939-1.5945	40.485-40.500
Runout - Max	0.0008	0.02
Connecting Rod		
Big End Side Clearance	0.0047-0.0102	0.12-0.26
Small End Diameter	1.2205-1.2209	31.000-31.012
Small End Freeplay Limit	0.08	2.0
Piston		
Standard Size	3.6551-3.6559	92.840-92.860
Measuring Point (above bottom of skirt)	0.40	10.0
Clearance	0.0061-0.0063	0.155-0.161
Piston Pin Boss (bore inner diameter)	1.0238-1.0242	26.004-26.015
Piston Pin (outer diameter)	1.0234-1.0236	25.995-26.000
Piston Rings		
Both (Top and 2nd) Rings		
Width (measured from outer-to-inner edge of ring)	0.106-0.114	2.7-2.9
Thickness (measured vertically top-to-bottom of ring)	0.0807	2.05
End Gap		
Top Ring	0.0118-0.0157	0.30-0.40
Bottom Ring	0.0118-0.0177	0.30-0.45
Ring Side Clearance (Both Top and 2nd Rings)	0.0008-0.0024	0.02-0.06
Reed Valve		
Lift (height)	0.31	8
Warpage (limit)	0.008	0.2

① Unless otherwise noted

Fig. 268 Engine Specifications - 225/250 hp (3342cc) 76° V6 Motors

EXHAUST GASES... 7-3
GEARCASE (LOWER UNIT)... 7-4
   REMOVAL & INSTALLATION... 7-4
      2 AND 3 HP (1-CYLINDER) MODELS... 7-4
      4-40 HP 1-, 2- AND 3-CYLINDER MODELS (EXCEPT 40/50 HP/698CC)... 7-6
      48 HP (2-CYLINDER), 40/50 HP (3-CYLINDER) & ALL V4/V6 MODELS... 7-8
GEARCASE OVERHAUL ... **7-10**
   OVERHAUL TIPS... 7-10
   2 AND 3 HP (1-CYLINDER) MODELS... 7-11
   4-40 HP 1-, 2- AND 3-CYLINDER MODELS (EXCEPT 40/50 HP/698CC)... 7-19
   48 HP 2-CYLINDER, 40/50 HP AND LARGER 3-CYLINDER & ALL V4/V6 MODELS... 7-43
GENERAL INFORMATION... 7-2
JET DRIVE ... **7-78**
   DESCRIPTION & OPERATION... 7-78
   JET DRIVE ADJUSTMENTS... 7-88
   JET DRIVE ASSEMBLY... 7-79
   MODEL IDENTIFICATION & SERIAL NUMBERS... 7-78
JET DRIVE ASSEMBLY... 7-79
   ASSEMBLY... 7-85
   CLEANING & INSPECTION... 7-80
   REMOVAL & DISASSEMBLY... 7-79
JET DRIVE ADJUSTMENTS... 7-88
   GATE LINKAGE ADJUSTMENTS... 7-88
   TRIM ADJUSTMENT... 7-92
LOWER UNIT ... **7-2**
   EXHAUST GASES... 7-3
   GEARCASE (LOWER UNIT)... 7-4
   GENERAL INFORMATION... 7-2
   SHIFTING PRINCIPLES... 7-2
   TROUBLESHOOTING... 7-4
OVERHAUL TIPS... 7-10
SHIFTING PRINCIPLES... 7-2
   COUNTER-ROTATING UNIT... 7-3
   STANDARD ROTATING UNIT... 7-2
TROUBLESHOOTING... 7-4
2 AND 3 HP (1-CYLINDER) MODELS... 7-11
   ASSEMBLY... 7-15
   CLEANING & INSPECTING... 7-13
   DISASSEMBLY... 7-11
4-40 HP 1-, 2- AND 3-CYLINDER MODELS (EXCEPT 40/50 HP/698CC)... 7-19
   ASSEMBLY... 7-29
   CLEANING & INSPECTING... 7-24
   DISASSEMBLY... 7-19
   GEARCASE SHIMMING... 7-36
48 HP 2-CYLINDER, 40/50 HP AND LARGER 3-CYLINDER & ALL V4/V6 MODELS... 7-43
   ASSEMBLY... 7-55
   CLEANING & INSPECTING... 7-52
   DISASSEMBLY... 7-43
   GEARCASE EXPLODED VIEWS... 7-52
   GEARCASE SHIMMING... 7-66

7

LOWER UNIT

LOWER UNIT... 7-2
GEARCASE OVERHAUL... 7-10
JET DRIVE... 7-78

LOWER UNIT

Generally speaking, the lower unit is considered to be the part of the outboard below the exhaust housing. The unit contains the propeller shaft, the driven and pinion gears, the driveshaft from the powerhead and the water pump. Torque is transferred from the powerhead's crankshaft to the gearcase by a driveshaft. A pinion gear at the bottom of the driveshaft meshes with a drive gear in the gearcase to change the vertical power flow into a horizontal flow through the propeller shaft. The power head driveshaft rotates clockwise continuously when the engine is running, but propeller rotation is controlled by the gear train shifting mechanism.

The lower units normally found on all but a few small motors are equipped with shifting capabilities. The forward and reverse gears together with the clutch, shift assembly and related linkage are all housed within the lower unit.

On Yamaha outboards with a reverse gear, a sliding clutch engages the appropriate gear in the gearcase when the shift mechanism is placed in forward or reverse. This creates a direct coupling that transfers the power flow from the pinion to the propeller shaft. A similar clutch mechanism is used on a handful of small hp motors that contain both a forward and neutral shift position, but no reverse gear.

Two types of lower units are available. The first type is a conventional propeller driven lower unit which includes both the standard model and the counter-rotating model installed on some outboards.

Counter-rotating models are designated with an "L" in front of the horsepower model designation. On these models, the propeller rotates in the opposite direction than on regular models and is used when dual engines are mounted in the boat to equalize the propeller's directional churning force on the water. This allows the operator to maintain a true course instead of being pulled off toward one side while underway.

The second type of lower unit available is a jet drive propulsion system. Water is drawn in from the forward edge of the lower unit and forced out under pressure to propel the boat forward.

The lower unit is easily removed without removing the entire outboard from the boat.

Shifting Principles

STANDARD ROTATING UNIT

Non-Reverse Type

◆ See Figure 1

There are 2 types of non-reversing type lower units which are normally only used on a handful of the small hp (2 and 3 hp models). The first is a direct drive unit - the pinion gear on the lower end of the driveshaft is in constant mesh with the forward gear. The second is a Forward-Neutral shifting unit, in which shift assembly allows the drive gear on the propshaft to move in and out of contact with the pinion gear on the lower end of the driveshaft.

In both cases, reverse action of the propeller is accomplished by the operator swinging the engine with the tiller handle 180° and holding it in this position while the boat moves sternward. When the operator is ready to move forward again, they simply swing the tiller handle back to the normal forward position.

Reverse Type

◆ See Figures 2, 3 and 4

The standard lower unit is equipped with a clutch dog permitting operation in neutral, forward and reverse. Although some units use a step shifter mechanism, most larger units use a rotating shifter cam. On these larger units unit when the outboard is shifted from neutral to reverse, the shift rod is rotated. This action moves the plunger and clutch dog toward the reverse gear. When the unit is shifted from reverse to neutral and then to forward gear, the rotation of the shift rod is reversed and the clutch dog is moved toward the forward gear.

The shift mechanism may vary slightly on some models but typically consists of a cam on the shift rod, a shifter (plunger), a shift slide and a compression spring. Historically on Yamahas the shift slide is a hollow tube, closed at one end. The closed end of this tube is necked and the head or knob of the tube fits into the shifter. The shifter has a vertical hole all the way through it. The shift rod passes through this hole in the shifter. The cam in

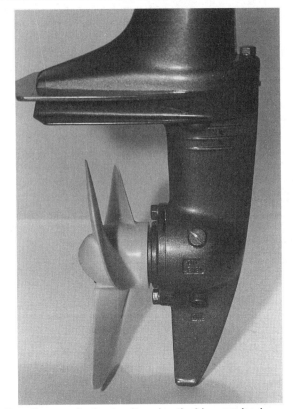

Fig. 1 Non reversing lower unit used on the 2 hp powerhead

the shift rod engages the shifter at the side notch, so any rotating movement in the shift rod will translate to a back and forth movement in the shifter and the shift slide.

In the neutral position, the cam is centered in the shifter.

When the lower unit is shifted into forward gear, the shift cam rotates in a counterclockwise direction and pulls on the shifter and shift slide. Because the shifter and shift slide are connected, they are moved out of the propeller shaft - moving the clutch dog toward the forward gear. The shift into forward gear is complete.

When the lower unit is shifted into reverse gear, the unit first moves into the neutral position. The shift cam rotates in a clockwise direction. The cam pushes the shifter and the shift slide in toward the propeller shaft. The clutch dog moves toward the reverse gear and the shift movement to reverse gear is complete.

The pinion gear on the lower end of the driveshaft is in constant mesh with both the forward and reverse gear. These three gears constantly rotate anytime the powerhead is operating.

A sliding clutch dog is mounted on the propeller shaft. A shifting motion at the control box will translate into a back and forth motion at the clutch dog via a series of shift mechanism components. When the clutch dog is moved forward, it engages with the forward gear. Because the clutch is secured to the propeller shaft with a pin, the shaft rotates at the same speed as the clutch. The propeller is thereby rotated to move the boat forward.

When the clutch dog is moved aft, the clutch engages only the reverse gear. The propeller shaft and the propeller are thus moved in the opposite direction to move the boat sternward.

When the clutch dog is in the neutral position, neither the forward nor reverse gear is engaged with the clutch and the propeller shaft does not rotate.

From this explanation, an understanding of wear characteristics can be appreciated. The pinion gear and the clutch dog receive the most wear, followed by the forward gear, with the reverse gear receiving the least wear. All three gears, the forward, reverse and pinion, are spiral bevel type gears.

A mixture of ball bearings, tapered roller bearings and caged or loose needle bearings is used in each unit. The type bearing used is normally indicated in the procedures.

LOWER UNIT

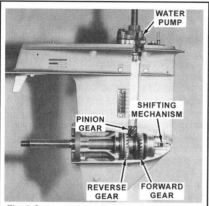

Fig. 2 Cutaway view of a lower unit with major parts, including the shift mechanism and water pump, identified. Note how the forward, reverse and pinion gears are all bevel cut

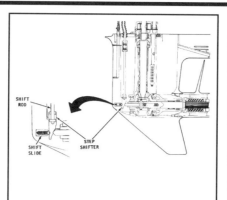

Fig. 3 Cross section of a typical lower unit showing a step shifter mechanism

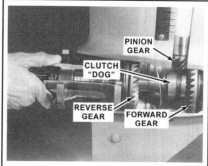

Fig. 4 Cut away view of a standard rotation lower unit with major parts identified

COUNTER-ROTATING UNIT

◆ See Figures 5 thru 9

■ Counter-rotational shifting is accomplished without modification to the shift cable at the shift box. The normal setup is essential for correct shifting. The only real special equipment the counter-rotating unit requires is the installation of a left-hand propeller.

The same shifting mechanisms used on standard units may also be found on their counter-rotating counterparts. There are a few minor differences between the standard and counter-rotating units. The minor differences are two-fold, one that with the counter-rotating shift mechanism being a mirror image of the standard shift mechanism. Mirror image shifting mechanisms produce counter-rotation of the propeller shaft.

The second minor difference being that the forward and reverse gears are reversed in position when compared to standard units (well, there's no actual difference in the gears, but the gears that perform the functions are reversed): what would be the forward gear on a standard unit becomes the reverse gear on a counter-rotating unit and what would normally be the reverse gear on a standard unit, becomes the forward gear on a counter-rotating unit. The pinion gear remains the same and driveshaft rotation remains the same as on a standard lower unit.

However for some early-model units (mostly through 1998 or so) a larger physical difference exists in the propeller shaft. Some early-model counter-rotating units utilize a two piece propeller shaft.

On a standard lower unit, the cam on the shift shaft is normally located on the starboard side of the shifter. Therefore, when the rod is rotated counterclockwise, the clutch dog is pulled forward and the forward gear is engaged.

On a typical counter-rotating lower unit, the cam on the shift rod is normally located on the port side of the shifter. Therefore, when the rod is rotated counterclockwise, the clutch dog is pushed back and the gear in the aft end of the housing (which normally is the reverse gear) is engaged. In this manner, the rotation of the propeller shaft is reversed. The same logic applies to the selection of reverse gear.

Exhaust Gases

◆ See Figure 10

At low powerhead speed, the exhaust gases from the powerhead escape from an idle hole in the intermediate housing on many motors. As powerhead rpm increases to normal cruising rpm or high-speed rpm, these gases are forced down through the intermediate housing and lower unit, then out with cycled water through the propeller.

Fig. 5 Cutaway view of a counter-rotating lower unit with the clutch dog in the neutral position

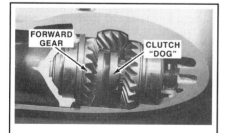

Fig. 6 Cutaway view of a counter-rotating lower unit with the clutch dog in the forward position

Fig. 7 Cutaway view of a counter-rotating lower unit with the clutch dog in the reverse position

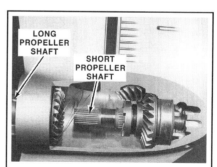

Fig. 8 On older Yamaha motors the counter-rotating lower unit used a two piece propeller shaft. The longer section attached to the propeller, the shorter section was internal

7-4 LOWER UNIT

Water for cooling is pumped to the powerhead by the water pump and is expelled with the exhaust gases. The water pump impeller is installed on the driveshaft. Therefore, the water output of the pump is directly proportional to powerhead rpm.

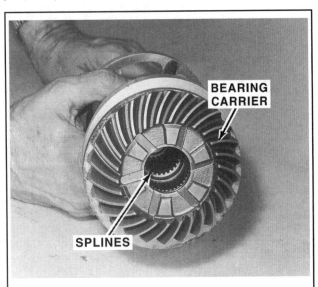

Fig. 9 Bearing carrier of an older counter-rotating lower unit with the internal splines for the short section of the propeller shaft clearly visible

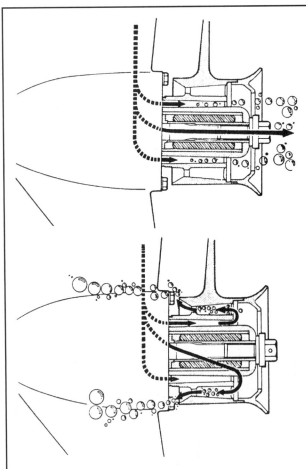

Fig. 10 Cross section of the lower unit showing route of the exhaust gases with the unit in forward (top) and reverse (bottom) gears

Troubleshooting

◆ See Figures 11 and 12

Troubleshooting (other than disassembly and inspection) must be done before the unit is removed from the powerhead to permit isolating the problem to one area. Always attempt to proceed with troubleshooting in an orderly manner. The shot-in the-dark approach will only result in wasted time, incorrect diagnosis, frustration and unnecessary replacement of parts.

The following procedures are presented in a logical sequence with the most prevalent, easiest and less costly items to be checked listed first.

1. Check the propeller and the rubber hub. See if the hub is shredded or loosen and slipping. If the propeller has been subjected to many strikes against underwater objects, it could slip on its hub. If the hub appears to be damaged, replace it with a new hub. Replacement of the hub must be done by a propeller rebuilding shop equipped with the proper tools and experience for such work.

2. Verify the ignition switch is **OFF**, to prevent possible personal injury, should the engine start. Shift the unit into reverse gear and at the same time have an assistant turn the propeller shaft to ensure the clutch is fully engaged.

3. If the shift handle is hard to move, the trouble may be in the lower unit shift rod or in the cable itself, requiring an adjustment or a repair, or in the shift box.

4. Disconnect the remote control cable at the engine and then remove the remote control shift cable. Operate the shift lever. If shifting is still hard, the problem is in the shift cable or control box.

5. If the shifting feels normal with the remote control cable disconnected, the problem must be in the lower unit. To verify the problem is in the lower unit, have an assistant turn the propeller and at the same time move the shift cable back and forth. Determine if the clutch engages properly.

Gearcase (Lower Unit)

REMOVAL & INSTALLATION

Gearcase removal or installation is a relatively straightforward procedure on most models. Generally speaking on all but the smallest of the Yamaha outboards it involves disconnecting the shift linkage and unbolting the gearcase from the intermediate housing, then carefully lowering the gearcase (along with the shift rod and/or driveshaft and water tube) straight down and off the motor. Because the water pump housing is mounted to the top of the gearcase, between the case and intermediate housing, removal is necessary for impeller inspection and/or replacement.

Because the gearcase is a sealed unit when it comes to gear oil, the fluid only needs to be drained if the case itself is being opened for further service (some form of seal or internal parts replacement). However, if this is one of the reasons the case is being removed, it is usually easier to go ahead and drain the gearcase oil before the assembly is removed.

2 and 3 Hp (1-Cylinder) Models

◆ See Figures 13 thru 18

On these models the gearcase is bolted to the intermediate housing with a handful (usually 2-3) bolts, mostly threaded upward from the underside of the anti-cavitation plate. If the model contains shift linkage, the linkage is accessed by removing a cover or covers from the side of the intermediate housing.

1. If the gearcase is to be overhauled or resealed, drain the gearcase fluid as detailed in this Maintenance and Tune-Up section. After the lubricant has drained, temporarily reinstall both the drain and oil level screws.

■ As the lubricant drains, catch some with your fingers from time to time and rub it between your thumb and finger to determine if any metal particles are present. If any significant amount of metal is detected in the lubricant, the unit must be completely disassembled, inspected and the damaged parts replaced. Check the color of the lubricant as it drains. A whitish or creamy color indicates the presence of water in the lubricant. Check the drain pan for signs of water separation from the lubricant. The presence of any water in the gear lubricant is bad news. The unit must be completely disassembled, inspected and the cause of the problem determined and corrected.

LOWER UNIT 7-5

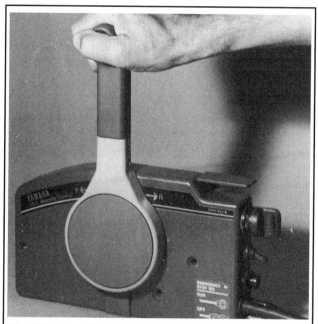

Fig. 11 Test the shift mechanism to see that the shifter fully engages...

Fig. 12 ... if it does not check cable adjustment, or try shifting the mechanism by hand

2. Remove the propeller for better access to the gearcase retaining bolt(s) threaded upward from underneath the anti-cavitation plate. For details, please refer to the Maintenance and Tune-Up section.

3. On models equipped with a **Forward-Neutral** or **Forward-Neutral-Reverse** shifter assembly, remove the round side cover(s) from the intermediate housing for access to the shifter bolt. Loosen and remove the shifter linkage bolt.

4. Loosen and remove the bolts securing the gearcase to the intermediate housing. On the smallest of motors (2 hp models) there are only 2 bolts, one threaded upward from under the anti-cavitation plate just behind the propeller and one threaded downward from a lip on the intermediate housing at the front of the outboard. Most other models use at 2- 3 bolts, all threaded upward from under the anti-cavitation plate. You may have to remove the bolt securing the anode as well since (even though usually not), it might thread all the way through to the intermediate housing. Of course, there is no harm in removing and inspecting the anode at this time anyway.

5. Separate the lower unit from the intermediate housing by pulling carefully STRAIGHT downward. When the two units are separated watch for and save any dowel pins (there are usually 2) used to help secure the gearcase. Normally the water tube will come out of the grommet and remain with the intermediate housing, while the driveshaft will remain with the lower unit.

■ On 2 hp motors, the anti-cavitation plate itself is not integral with the housing or gearcase, but is instead held in position between the 2 when the gearcase is bolted into position. Once the gearcase is removed the plate should come free of the intermediate housing and should be inspected for damage.

To Install:

6. Apply very a light coating of Yamaha Marine Grease or an equivalent water resistant lubricant to the splines on the driveshaft. On 2 hp motors, apply just a dab of, to the indexing pin on the mating surface of the lower unit, then install the anti-cavitation plate onto the lower surface of the intermediate housing.

7. Begin to bring the intermediate housing and the lower gear housing together. As the two units come closer, rotate the propeller shaft slightly to index the upper end of the lower driveshaft tube with the upper rectangular driveshaft. At the same time, feed the water tube into the water tube seal (and, if equipped, align the Neutral shifter shaft with the upper half of the shaft located in the intermediate housing).

8. Push the lower gear housing and the intermediate housing together. The dowel pin(s) will index with a matching hole as the two pieces are compressed together.

9. Apply Loctite® 572 or an equivalent threadlocking compound to the threads of the bolts used to secure the lower unit to the intermediate housing. A threadlocking compound can also be used on the anode retaining bolt. Install and tighten the bolts securely to 5.8 ft. lbs. (8 Nm).

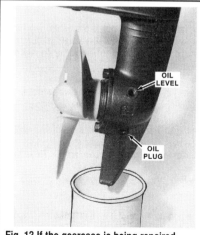

Fig. 13 If the gearcase is being repaired, start by draining the gearcase oil

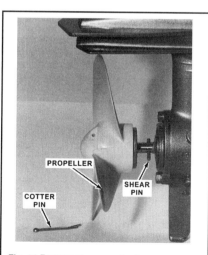

Fig. 14 Remove the propeller for access

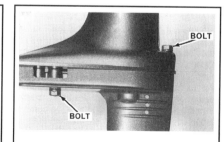

Fig. 15 Remove the gearcase bolts (2 hp motors shown)

7-6 LOWER UNIT

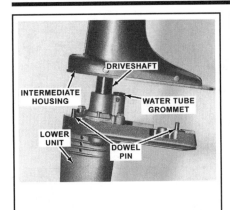

Fig. 16 Carefully lower the gearcase straight downward off the housing

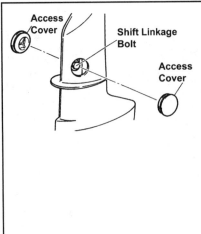

Fig. 17 Typical shift linkage access

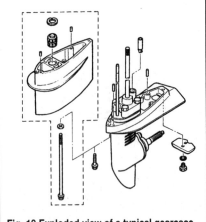

Fig. 18 Exploded view of a typical gearcase mounting

10. For models with a shifter assembly, install and tighten the shifter linkage bolt, then install the side cover(s) to the intermediate housing.

11. Apply Yamaha All Purpose Grease, or equivalent anti-seize compound to the propeller shaft.

■ The compound will prevent the propeller from freezing to the shaft and permit the propeller to be removed, without difficulty, the next time removal is required.

12. Install the propeller, as detailed in the Maintenance and Tune-Up section.

13. If drained for service, properly refill the gearcase with lubricant, as detailed in the Maintenance and Tune-Up section.

4-40 Hp 1-, 2- and 3-Cylinder Models (Except 40/50 Hp/698cc)

◆ See Figures 19 thru 26  MODERATE

On these models the gearcase is bolted to the intermediate housing with 4-5 bolts threaded upward from the underside of the anti-cavitation plate. Generally, if equipped with a trim tab, the tab must be removed in order to access one of the bolts. The shift linkage on these models is normally connected by a set of nuts (one connecting nut and one locknut) which attach to the threaded ends of the upper and lower shift rods. Shift linkage is found at the front of the gearcase and is normally accessible somewhere on the front of the outboard, above the anti-cavitation plate (though occasionally under an intermediate housing front cover or under the engine cowling).

1. If the gearcase is to be overhauled or resealed, drain the gearcase fluid as detailed in this Maintenance and Tune-Up section. After the lubricant has drained, temporarily reinstall both the drain and oil level screws.

■ As the lubricant drains, catch some with your fingers from time to time and rub it between your thumb and finger to determine if any metal particles are present. If any significant amount of metal is detected in the lubricant, the unit must be completely disassembled, inspected and the damaged parts replaced. Check the color of the lubricant as it drains. A whitish or creamy color indicates the presence of water in the lubricant. Check the drain pan for signs of water separation from the lubricant. The presence of any water in the gear lubricant is bad news. The unit must be completely disassembled, inspected and the cause of the problem determined and corrected.

2. Remove the propeller for better access to the gearcase retaining bolt(s) threaded upward from underneath the anti-cavitation plate. For details, please refer to the Maintenance and Tune-Up section.

3. Locate and disconnect the shifter linkage at the shift rod union (at the front of the intermediate housing, just above the gearcase, possibly under a small cover on some models). Most models utilize threaded shift rods with a locknut and an adjusting nut (it's easy to tell the difference, the locknut is the small one). On these models mark the position of the adjuster nut on the threads, loosen the locknut, and then unthread the adjuster nut so the 2 rods are disconnected. Some motors utilize a different method of joining the shift linkage. For instance the 6/8 hp motors use a bolt and nut which secure 2 halves of a shift rod connector that bolt around the rods. On other motors, such as the 20/25 hp (395cc) twins there is a single locknut that secures the looped (not threaded) upper shift linkage to the threaded top of the gearcase shift rod.

■ On some of the larger models so equipped the trim tab obscures access to one of the gearcase retaining bolts.

4. For most 25/30 hp 3-cylinders, matchmark the position of the trim tab to the gearcase, as an aid to installing it back in its original location, then loosen the retaining bolt and remove the trim tab from the underside of the anti-cavitation plate.

5. Some models, such as the 25/30 hp 3-cylinders, utilize a speedometer pilot water hose at the top front of the gearcase. If equipped, carefully cut the wire tie and gently pull the two ends of the tube apart, at their connector.

6. Remove the 4 bolts threaded upward (2 on each side) from the gearcase to the intermediate housing. On models with a trim tab that was removed for access, remove the 5th bolt from the center of the gearcase, under the trim tab mounting area.

7. Carefully separate the lower unit from the intermediate housing by pulling STRAIGHT downward from the intermediate housing. Watch for and save the dowel pins (usually 2) when the two units are separated. The water tube will normally come out of the grommet and remain with the intermediate housing. The driveshaft will remain with the lower unit. The lower portion of the shift linkage should remain with the lower housing.

To Install:

8. Apply a dab of Yamaha Marine Grease, or equivalent water resistant lubricant to the driveshaft splines (the sides of the driveshaft's upper end, NOT the top surface itself).

■ An excessive amount of lubricant on top of the driveshaft will be trapped in the clearance between the top of the driveshaft and bottom of the crankshaft. This trapped lubricant may not allow the driveshaft to fully engage with the crankshaft.

9. Apply some of the same lubricant to the end of the water tube in the intermediate housing, and to the dowel pin(s) on the mating surface of the lower unit.

10. Set the gearcase in gear so that you can rotate the propeller shaft slightly to assist with crankshaft-to-driveshaft spline alignment. Yamaha recommends that both the shifter and the gearcase be placed in **Reverse** for ease of installation.

11. Begin to bring the intermediate housing and lower gear housing together.

■ The next step takes time and patience. If it's your first time, success will probably not be achieved on the first attempt. Two items must mate at the same time before the lower unit can be seated against the intermediate housing.

LOWER UNIT

- The top of the driveshaft on the lower unit indexes with the lower end of the crankshaft.
- The water tube in the intermediate housing slides into the grommet on the water pump housing.

12. As the two units come closer, rotate the propeller shaft slightly to index the splines on the upper end of the lower driveshaft with the crankshaft splines. Be sure to turn the propeller shaft only in the correct direction of rotation for the gear in which the gearcase is set. (This means that the propeller shaft is turned clockwise if the shifter is in forward or counterclockwise if the shifter is in reverse. This is done to make sure the water pump impeller is not damaged). At the same time, feed the water tube into the water tube grommet and make sure the upper and lower shift rods are positioned so they can be reconnected.

13. Push the lower unit housing and the intermediate housing together. The dowel pin(s) should index as the surfaces mate.

■ If all items appear to mate properly, but the lower unit seems locked in position about 4 inches (10cm) away from the intermediate housing and it is not possible, with ease, to bring the two housings closer, the driveshaft has missed the cylindrical lower oil seal housing leading to the crankshaft. Move the lower unit out of the way. Shine a flashlight up into the intermediate housing and find the oil seal housing. Now, on the next attempt, find the edges of the oil seal housing with the driveshaft before trying to mate anything else - the water tube and the top of the shift rod. If the driveshaft can be made to enter the oil seal housing, the driveshaft can then be easily indexed with the crankshaft.

14. Install a finger-tighten 2 of the gearcase retainers just to hold the assembly in place while you turn your attention to the shifter assembly.

■ The 6/8 hp motor uses a 2 piece shift linkage connector which is bolted together. The bolt should be tightened to 7.2 ft. lbs. (10 Nm) on these models.

15. Reconnect the shift linkage and secure using the nut, adjuster nut and locknut or connector and retaining bolt as applicable. On some models there are specifications for adjuster nut installation as follows:
- On 20/25 hp (395cc) twins, place the loop of the shifter linkage over the exposed top end of the gearcase shifter cam lever, then install and tighten the retaining nut. Make sure there is about 0.14-0.18 in. (3.5-4.5mm) between the bottom of the loop and the top of the gearcase.
- On 4/5 hp and most 9.9 hp or larger motors, no specification is given for the distance the adjuster nut is threaded onto each rod. Make sure both the gearcase and shifter are in Reverse and then install the adjuster nut until you reach the alignment mark made during removal. If there is no mark, try to make sure there is about an equal amount of thread from each rod in the adjuster nut. Tighten the locknut to hold the adjuster in position once you are happy with adjustment.

16. Apply a light coating of Loctite® 572 or an equivalent threadlocking compound to the threads of the gearcase retaining bolts. Remember, one bolt passes through the area covered by the trim tab. Install and finger-tighten the bolts as you apply the compound, then withdraw the 2 bolts you'd previously installed to hold the gearcase in place so that you can apply the compound to those threads as well. Once all retainers are installed either tighten them securely. A torque specification is not available for most 4-15 hp motors, however the retaining bolts on all larger models should be tightened to 29 ft. lbs. (39 Nm).

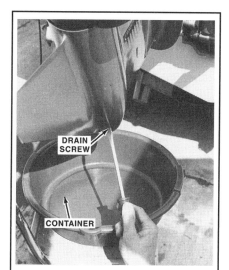

Fig. 19 If necessary, drain the gearcase before removal

Fig. 20 Metal shards in the oil indicates a need for overhaul

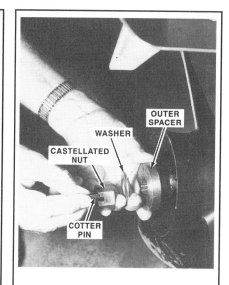

Fig. 21 Remove the propeller for access...

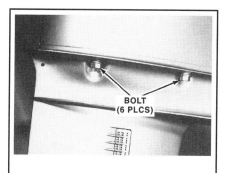

Fig. 22 ...then remove the gearcase retaining bolts

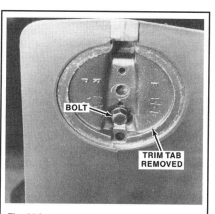

Fig. 23 Larger models have a bolt under the trim tab

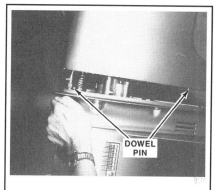

Fig. 24 Carefully lower the gearcase from the intermediate housing

7-8 LOWER UNIT

17. Apply a light coating of Loctite® 572 or an equivalent threadlocking compound to the threads trim tab retaining bolt, then install the trim tab, aligning the matchmarks made earlier and tighten the bolt securely.

18. On models equipped with a speedometer pickup tube, slide the two ends of the pilot tube together at the push-on fitting. Secure the tube using a new wire tie.

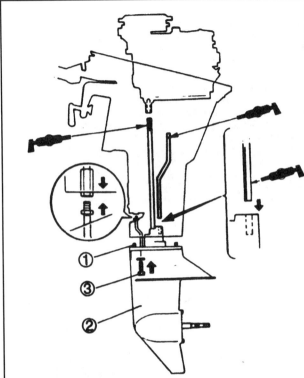

Fig. 25 Typical gearcase mounting, showing most common shift linkage

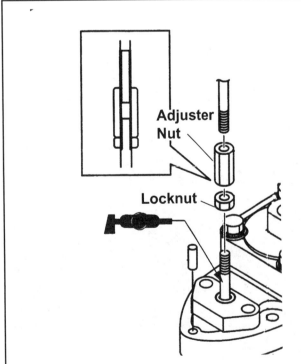

Fig. 26 Threaded shift linkage like this may have the lock nut on the top or the bottom, depending upon the model

19. Operate the shift lever through all gears. The shifting should be smooth and the propeller should rotate in the proper direction when the flywheel is rotated by hand in a clockwise direction. Naturally the propeller should not rotate when the unit is in neutral.

20. Apply Yamaha All Purpose Grease, or equivalent anti-seize compound to the propeller shaft.

■ The compound will prevent the propeller from freezing to the shaft and permit the propeller to be removed, without difficulty, the next time removal is required.

21. Install the propeller, as detailed in the Maintenance and Tune-Up section.

22. If drained for service, properly refill the gearcase with lubricant, as detailed in the Maintenance and Tune-Up section.

48 Hp (2-Cylinder), 40/50 Hp (3-Cylinder) & All V4/V6 Models

◆ See Figures 19, 20, 21 and 27 thru 30

On these models the gearcase is bolted to the intermediate housing with anywhere from 5 to 8 bolts and/or nuts threaded upward from the underside of the anti-cavitation plate. Generally, if equipped with a trim tab, the tab must be removed in order to access one of the bolts. The shift linkage on these models works by rotation and not by lifting or pressing downward as on the smaller Yamaha gearcases. As a result the top of the gearcase shift linkage is actually splined for connection with the linkage found in the intermediate housing and can therefore simply be pulled apart when the gearcase is unbolted. Many of these models are also equipped with a speedometer pickup/water tube at the front of the gearcase which must be disconnected before gearcase removal.

1. If the gearcase is to be overhauled or resealed, drain the gearcase fluid as detailed in this Maintenance and Tune-Up section. After the lubricant has drained, temporarily reinstall both the drain and oil level screws.

■ As the lubricant drains, catch some with your fingers from time to time and rub it between your thumb and finger to determine if any metal particles are present. If any significant amount of metal is detected in the lubricant, the unit must be completely disassembled, inspected and the damaged parts replaced. Check the color of the lubricant as it drains. A whitish or creamy color indicates the presence of water in the lubricant. Check the drain pan for signs of water separation from the lubricant. The presence of any water in the gear lubricant is bad news. The unit must be completely disassembled, inspected and the cause of the problem determined and corrected.

2. Most models utilize a speedometer pilot water hose at the top front of the gearcase. If equipped, carefully cut the wire tie and gently pull the hose off the fitting.

3. Remove the propeller for better access to the gearcase retaining bolt(s) threaded upward from underneath the anti-cavitation plate. For details, please refer to the Maintenance and Tune-Up section.

■ The trim tab obscures access to one of the gearcase retaining bolts.

4. Matchmark the position of the trim tab to the gearcase, as an aid to installing it back in its original location.

■ Generally speaking, on smaller motors (most in the 40-60 hp range, but not including the 48 hp motor) the trim tab is retained by a bolt threaded upward from underneath the anti-cavitation plate. However, on most larger motors the tab is installed by a bolt threaded downward through the intermediate housing flange (it can be found by removing a small rubber cap/grommet).

5. For larger outboards, remove the plastic cap above the trim tab for access to the trim tab retaining bolts.

6. For all motors, use the correct size socket (and a long extension when threaded from above the tab) to remove the bolt and the trim tab.

7. Remove the bolts (or nuts) threaded upward (there are usually 2 or 3 on each side) securing the gearcase to the intermediate housing. Also, remove the bolt from the center of the gearcase which was exposed by removing the trim tab.

LOWER UNIT 7-9

Fig. 27 The trim tab hides one of the gearcase bolts

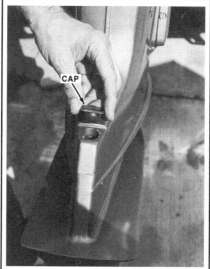
Fig. 28 On most models, access the tab bolt from the top

Fig. 29 Remove the bolts and carefully lower the gearcase

■ On models with studs and nuts instead of bolts retaining the gearcase there is normally a intermediate housing extension piece which mounts between the housing and the gearcase. This extension is normally free to come off once the gearcase is removed, so take care that it does not drop and become damaged.

8. Carefully separate the lower unit from the intermediate housing by pulling STRAIGHT downward from the intermediate housing. Watch for and save the dowel pins (usually 2) when the two units are separated. The water tube will normally come out of the grommet and remain with the intermediate housing. The driveshaft will remain with the lower unit. The lower portion of the shift linkage should remain with the lower housing.

■ On V4 and V6 models there is a normally a rubber exhaust seal mounted to the top of the gearcase or bottom of the intermediate housing. On the larger V6 models there may also be a seal plate between the seal and water pump housing. Keep track of these components for installation purposes. Also, take a good look at the rubber seal before deciding that it is fit for further service.

To Install:

9. If removed on V4 and V6 models install the plate (if equipped) and the rubber exhaust seal. Place a light coating of marine grade grease on the rubber seal before installation.
10. Apply a dab of lubricant such as a Molybdenum lubricant or Yamaha Marine Grease, or equivalent water resistant lubricant to the driveshaft splines (the sides of the driveshaft's upper end, NOT the top surface itself).

■ An excessive amount of lubricant on top of the driveshaft will be trapped in the clearance between the top of the driveshaft and bottom of the crankshaft. This trapped lubricant may not allow the driveshaft to fully engage with the crankshaft.

11. Apply some of the same lubricant to the end of the water tube in the intermediate housing, and to the dowel pin(s) on the mating surface of the lower unit.
12. Set the gearcase in gear so that you can rotate the propeller shaft slightly to assist with crankshaft-to-driveshaft spline alignment. Be sure to place both the gearcase and the shifter in the same gear for ease of installation.
13. Begin to bring the intermediate housing and lower gear housing together.

■ The next step takes time and patience. If it's your first time, success will probably not be achieved on the first attempt. Three items must mate at the same time before the lower unit can be seated against the intermediate housing.

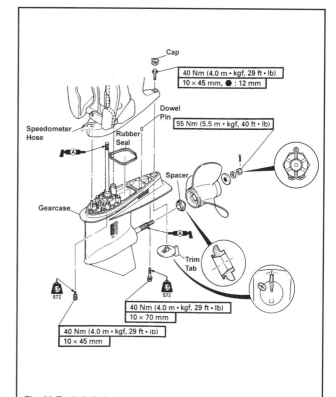
Fig. 30 Exploded view of a typical gearcase mounting - V4 and V6 (except 3.1L and larger) shown

• The top of the driveshaft on the lower unit indexes with the lower end of the crankshaft.
• The water tube in the intermediate housing slides into the grommet on the water pump housing.
• The top splines of the lower shift rod in the lower unit slide into the internal splines of the upper shift rod in the intermediate housing.

14. As the two units come closer, rotate the propeller shaft back and forth EVER-SO-SLIGHTLY to index the upper end of the lower driveshaft tube with the crankshaft. At the same time, feed the water tube into the water tube grommet and feed the lower shift rod into the upper shift rod.
15. Push the lower unit housing and the intermediate housing together. The dowel pin(s) should index ensuring the gearcase is properly aligned.

■ **If all items appear to mate properly, but the lower unit seems locked in position about 4 inches (10cm) away from the intermediate housing and it is not possible, with ease, to bring the two housings closer, the driveshaft has missed the cylindrical lower oil seal housing leading to the crankshaft. Move the lower unit out of the way. Shine a flashlight up into the intermediate housing and find the oil seal housing. Now, on the next attempt, find the edges of the oil seal housing with the driveshaft before trying to mate anything else - the water tube and the top of the shift rod. If the driveshaft can be made to enter the oil seal housing, the driveshaft can then be easily indexed with the crankshaft.**

16. Apply a light coating of Loctite® 572, or an equivalent threadlocking compound to the threads of the fasteners used to secure the lower unit to the intermediate housing. Remember, one bolt passes through the area covered by the trim tab. Install and tighten the bolts to 26-29 ft. lbs. (37-40 Nm) for all except the largest V6 motors (the 3.1L or larger models) on which the retainers should be tightened to 35 ft. lbs. (48 Nm).

■ **Although Yamaha recommends threadlocking compound on most of their gearcase retainers, they occasionally do not specifically call for it. We prefer to ALWAYS use it in these applications because it offers to benefits. Besides helping to prevent the fasteners from loosening in service threadlocking compound also insulates the fastener threads from the threads of the housing or stud, helping to prevent corrosion that could seize the fastener in place. In this way it also acts as something of an anti-seize that also holds the bolt in place, a win, win situation if you ask us.**

17. Apply a light coating of Loctite® 572 or an equivalent threadlocking compound to the threads trim tab retaining bolt, then install the trim tab, aligning the matchmarks made earlier and tighten the bolt securely. On larger models where the bolt is threaded from above the tab, install the rubber grommet into the intermediate housing over the bolt.

18. If equipped reconnect the speedometer pilot hose and secure using a new wire tie.

19. Operate the shift lever through all gears. The shifting should be smooth and the propeller should rotate in the proper direction when the flywheel is rotated by hand in a clockwise direction. Naturally the propeller should not rotate when the unit is in neutral.

20. Apply Yamaha All Purpose Grease, or equivalent anti-seize compound to the propeller shaft.

■ **The compound will prevent the propeller from freezing to the shaft and permit the propeller to be removed, without difficulty, the next time removal is required.**

21. Install the propeller, as detailed in the Maintenance and Tune-Up section.

22. If drained for service, properly refill the gearcase with lubricant, as detailed in the Maintenance and Tune-Up section.

GEARCASE OVERHAUL

Overhaul Tips

All Yamaha gearcases use the same basic inner design in that pinion gear is splined to the bottom of a vertical driveshaft and secured by a retaining nut. The driveshaft uses the pinion gear to transfer the rotational power from the crankshaft to the propeller shaft. Both the driveshaft and propeller shaft ride in replaceable bushing (on the smaller gearcases) or bearings and a bath of gear oil. Gear oil is kept inside the case itself (and water is kept out) by propeller shaft and driveshaft housing seals at either end (bottom and top) of the gearcase.

The size and type of the shafts, bearings, gears and seals will vary slightly from model-to-model. Also, all gearcases except the 2 hp contain some form of Neutral or Neutral and Reverse shifting capabilities. Generally speaking one of 2 major designs are employed, a vertical push-pull shifter retaining by nuts on threaded rods (or a bolted coupler assembly) is used by most smaller gearcases, while a horizontally ratcheting shifter which connected through splines to the shift lever in the intermediate housing is used on most larger outboards.

Like the gearcase removal procedures we've divided the overhaul procedures into 3 major categories. First a procedure for 2 and 3 hp (1 cylinder) motors covers the smallest Yamaha gearcases (most of which do NOT contain a Reverse gear). Second, a procedure for 4-40 hp (1-, 2- and 3-cylinder) models (except 40/50 hp/698cc motors) covers most small and intermediate sized gearcases that do contain Reverse gears AND push-pull shifter assemblies. Lastly we've provided a procedure which applies to 48 hp (2-cylinder) models, 40/50 hp and larger (3-cylinder) motors as well as all V4 and V6 models which includes the larger and more complicated gearcases that contain Reverse gears and splined, ratcheting shifter assemblies.

We've included diagrams and photographs from as many gearcases on which we could realistically get our hands, however keep in mind that subtle differences will occur from model-to-model or year-to-year. To ensure a successful overhaul a few general notes should be followed throughout the overhaul procedure, as follows:

- As with all overhaul procedures, make sure the work area is clean and free of excessive dirt or moisture.
- Take your time when removing and disassembling components taking a moment with each to either matchmark its positioning or note its orientation in relation to other components. This goes ESPECIALLY for the placement of any shims which might be used. Absent specific instructions for checking and re-shimming a gearcase (which occurs when bearings, gears and/or shafts are replaced on some gearcases) all shims must be returned to their original positions during assembly.
- Seal placement and orientation will vary on some models. Before removing any seal be sure to first note the direction in which the seal lips are facing and be sure to install the replacement seal (and you ALWAYS replace seals when they are removed) facing the same direction.
- Unless otherwise directed seal lips are packed with a light coating of Yamaha Marine Grease or an equivalent water resistant grease. When used, O-rings are also coated with marine grease during installation.
- Needle bearings which are pressed in place are generally destroyed by the removal process and should be replaced when separated from a gear, shaft or bore. Ball Bearings can usually be freed without damage, but use common sense. Once the bearing is free, clean it thoroughly and then rotate the inner/outer cage feeling for rough spots. Replace any bearing that is not in top shape.
- The correct size driver must be used for bearing or seal installation. Generally speaking a driver must be smooth (especially for seals, to prevent it from cutting the sealing surface) and should be sized to spread as much of the force over the seal or bearing surface as possible. When a bearing is installed ON a shaft, the driver MUST contact the inner race (never push on only the outer race in this case or the bearing will be damaged). When a bearing is installed IN a bore, the driver MUST contact the outer race (in this circumstance, do NOT push only on the inner race or the bearing will be damaged).
- Driveshaft bearings or bushings are often pressed straight into position from above and/or pulled into position from below. Although Yamaha often has a special tool available to help draw a bushing or bearing into position in the lower portion of the gearcase, all you REALLY need to accomplish this is a sufficiently long threaded bolt with a nut and a washer that is the size of the driver you'd want to contact that bushing or bearing. You'll also need a nut and a plate to place on top of the gearcase. Place the threaded rod through one nut and plate and then down into the gearcase through the bushing/bearing (so the plate is resting against the top of the gearcase, the upper nut is resting against the plate and the threads of the rod are just protruding down into the pinion mounting area of the gearcase.). Next install the washer and lower nut on the bottom threads of the rod to support the bearings or busing. Slowly draw the bearing or bushing into position in the gearcase by holding the threaded rod from turning while at the same time turning the upper nut to draw the entire bolt, bushing/bearing, washer and nut assembly upward until the bushing or bearing seats. Then loosen the nut on the bottom of the threaded rod in the gearcase and remove the threaded rod, plate, nut and washer.

In most cases the water pump, driveshaft oil seal or propeller shaft oil seal may be replaced without complete gearcase disassembly. However, there are always exceptions, such as the 2 hp powerhead, where Yamaha requires that the driveshaft be removed in order to service the water pump (pulling the pump housing off the bottom of the shaft instead of the top).

Although Yamaha calls for various special tools for gearcase overhaul MOST of them come down to the appropriate sized driver (see the bullet about drivers earlier in this section), OR a universal puller or a slide hammer generally with inside jaws. Whenever possible, we prefer the use of a puller over a slide hammer, just because the force is applied more linearly, however there are simply times when the impact of a slide hammer makes the job easier, so use common sense when determining which tool you'd prefer to use for bearing, race or potentially even shaft removal.

Lastly, we've tried to provide the best procedures possible for gearcase overhaul, and in the most logical order. However, in most cases procedures for some of the earlier procedures can be performed alone, without additional procedures. For instance water pump or propeller shaft bearing carrier/seal housing procedures can normally be performed completely independently of each other (again, there are exceptions such as the 2 hp motor). However, since the driveshaft oil seal housing is found UNDER the water pump, you'd obviously have to remove the pump first for access.

2 and 3 Hp (1-Cylinder) Models

■ In order to remove the water pump impeller from 2 hp models, the driveshaft must first be removed from the lower unit. On these models the impeller and pump housing should only be removed from the lower end of the driveshaft. Therefore, the circlip securing the pinion gear to the driveshaft must first be removed. This can only be accomplished by removing the lower unit bearing carrier cap and removing the propeller shaft.

DISASSEMBLY

◆ See Figures 31 thru 39

■ For gearcase exploded views, please refer to Cleaning & Inspection in this section.

1. Remove the Propeller from the gearcase, as detailed under Maintenance and Tune-Up.
2. Drain the gearcase oil and remove the gearcase from the outboard as detailed earlier in this section under Gearcase (Lower Unit).
3. For all except 2 hp motors, if necessary to access the driveshaft or upper oil seal assembly, remove the Water Pump, as detailed in the Lubrication and Cooling system section.
4. Remove the two bolts securing the bearing/seal carrier cap.
5. Either rotate the cap 90° so the flanges protrude past the side of the gearcase and then gently tap the bearing carrier cap using a rubber mallet to break it free from the lower unit OR, some models contain a small slit at the lower portion of the cap-to-gearcase sealing surface where you can use a small pry-tool to carefully pry the cap free.

■ For all except the 2 hp motors, the propeller shaft MAY come free with the bearing/seal cap. If it does, no big deal, just carefully separate it from the cap for attention later.

6. Remove the cap from the lower unit (on some models the entire propeller shaft assembly will come out with the cap, this is fine). Remove and discard the bearing/seal carrier O-ring. For the record, the O-ring CAN be reused on some models, however it is a cheap part and it is not worth the risk of a water leak ruining the gearcase.

■ Perform the next step only if the seal has been damaged and is no longer fit for service. Removal of the seal destroys its sealing qualities and it cannot be installed a second time. Therefore, be absolutely sure a new seal is available before removing the old seal in the next step.

7. If necessary, disassemble the bearing/seal carrier. On most models the seal(s) can be removed using either a small pry-tool (or hooked seal removal tool) or by using a slide hammer with expanding jaw attachment. If equipped with a bushing, it is usually removed with a driver, however bearings must be removed either using a puller or the same slide-hammer with expanding jaw attachment which can be used to remove the seals. Components and access varies as follows:
 • For 2 hp motors, a single seal is used on the cap which is removed or installed from the propeller side of the cap. Before removal note the seal lip positioning, it is normally installed with the lips facing the propeller.
 • For 3 hp motors, dual seals are installed or removed from the propeller side of the cap. Note the seal lips normally both face toward the propeller. Yamaha recommends using a small slide hammer to grasp and remove both seals at once, but they can normally be carefully pried free.
8. For all except 2 hp motors, if the propeller shaft did not come out with the bearing/seal cap, remove it from the gearcase at this time. Carefully disassemble the propeller shaft and shifter assembly. On most models this involves removing the shift plunger from the end of the shaft, then using a small flat-bladed tool (such as a small screwdriver) inserted from the side of the shaft to carefully compress the spring while you push the dog clutch out of the side of the shaft's opening.

■ Take careful note as to the dog clutch positioning on some models there is an orientation mark on the clutch which must be installed in the same position.

9. For 2 hp motors, remove the two bolts securing the water pump housing cover to the lower unit. Raise the water pump cover a bit to clear the indexing pin. Leave the water pump cover in this position at this time.
10. At the bottom end of the driveshaft (inside the gearcase) carefully pry the circlip free of the groove in the shaft using a small pry-tool. This clip holds the pinion gear onto the driveshaft. The clip may not come free on the first try, but have patience and it will come free. With one hand, remove driveshaft up and out of the lower unit housing and at the same time, with the other hand, catch the pinion gear, washer and any shim material from behind the gear. The shim material is critical to obtaining the correct backlash during installation. Using the old shim material will save considerable time over the alternative of starting with no shim material or by guessing on the shim amount.

Fig. 31 Remove the bolts securing the prop shaft bearing/seal cap

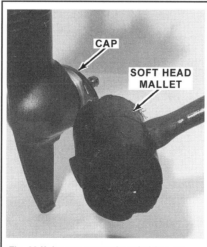

Fig. 32 If the cap cannot be pried free, try spinning it slightly a tapping on the cap ears

Fig. 33 Remove and discard the O-ring from the prop shaft bearing seal/cap

7-12 LOWER UNIT

■ Remember removal of an oil seal destroys its sealing qualities and it cannot be installed a second time. Therefore, remove the oil seal(s) only if it is unfit for further service and be absolutely sure a new seal is available before removing the old seal.

11. On 2 hp models removal of the pinion gear will free the forward gear and propeller shaft for removal at this time. Remove the assembly and retain the forward gear shim.

12. Inspect the condition of the upper oil seals for the gearcase. If replacement is required, carefully pry them free or use a slide hammer and internal jaw adapter to remove the seal. For all models except the 2 hp, remove them oil seal or seals from the top of the gearcase housing (we say except the 2 hp, because the dual seal assembly used on that motor is found inside the water pump housing, for more details, please refer to Water Pump, in the Lubrication and Cooling section). As usual, note the seal orientation. The 3 hp motors use dual lipped seals so orientation is not as big a deal, but they are not completely symmetrical and have an outer casing lip that faces downward toward the gearcase.

13. Remove the Forward gear from the gearcase for all except 2 hp motors (since on those models the gear already came out with the propeller shaft).

14. On most of these gearcases the driveshaft rides in 2 or 3 bushing which may be removed for replacement purposes only (since the action of removing them will make them unfit for further service). Lower bushings can normally be driven out of the gearcase using a long rod/driver inserted from the top at a slight angle. Tap a lower bushing down and into the gearcase where it can be fished out and removed. Upper bushings may be pulled upward and out using a small internal jawed puller and slide-hammer or using a threaded rod assembled with a suitably sized washer, plate and a couple of nuts (as described earlier under Overhaul Tips, it's essentially the same setup that can be used to install lower bushings). A few models may use 3 bushings a lower, upper and middle bushing, however most other models only use a lower and upper.

15. If necessary for replacement remove the Forward Propeller Shaft Bearing assembly from the nose of the gearcase.

Forward Propeller Shaft Bearing

◆ See Figures 40, 41 and 42

■ Removal of the forward propeller shaft bearing in the next step will almost always destroy the bearing. Therefore, remove the bearing only if it is no longer fit for service and be absolutely sure a replacement bearing is available prior to removal.

Rotate the forward propeller shaft bearing. If any roughness is felt, the bearing must be removed and a new one installed.

1. Removal of the bearing is accomplished using a slide hammer with jaw expander attachment (#YB-6096), however if one if not available there is an alternative that we've found usually works.

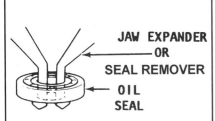

Fig. 34 If seal replacement is necessary, use a seal remover or an expanding jaw puller (shown)

Fig. 35 Unbolts and remove the water pump from the gearcase...

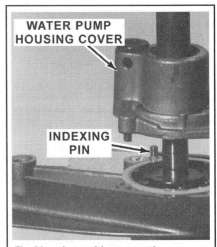

Fig. 36 ... but on 2 hp motors, the pump comes off WITH the driveshaft

Fig. 37 Carefully free the circlip and then remove the pinion gear with any washers or shims

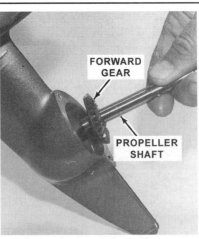

Fig. 38 This will free the forward gear from the gearcase (and for 2 hp motors, the propeller shaft)

Fig. 39 If necessary, remove the oil seal from the top of the gearcase either for replacement or for access to driveshaft bushings

LOWER UNIT

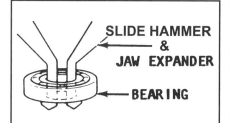

Fig. 40 The best method for bearing removal is a slide hammer with expanding jaw puller

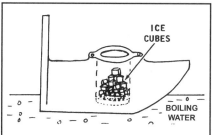

Fig. 41 An alternate method is to heat the case while chilling the bearing...

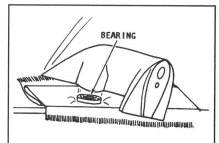

Fig. 42 ... then invert and jar the case to hopefully free the bearing

2. On 3 hp motors, if used, the forward gear shim pack is located in front of the forward bearing (between the bearing and gearcase, NOT between the bearing and gear). Retain the shim pack for reuse. On this powerhead there is only 1 shim pack available and although adjustment is first made at the pinion gear and pinion shim pack adjustment can ALSO be made by including or excluding the forward gear bearing shim. Unfortunately when components are changed, measurements may also vary, however we recommend installing the replacement under the same circumstances as the original which was removed (that is, installing it with or without the shim pack depending upon whether or not you found one during removal).

■ The following 3 steps are to be performed only if the forward propeller shaft bearing must be removed and a suitable bearing puller is not available. The procedure will change the temperature between the bearing retainer and the housing substantially, hopefully about 80°F (50°C). This change will contract one metal - the bearing retainer and expand the other metal - the housing, giving perhaps as much as 0.003 inch (.08mm) clearance to allow the bearing to fall free. Read the complete steps before commencing the work because three things are necessary: a freezer, refrigerator, or ice chest; some ice cubes or crushed ice; and a container large enough to immerse about 1-1/2 in. (3.8cm) of the forward part of the lower gear housing in boiling water.

3. After all parts, including all seals, grommets, etc., have been removed from the lower gear housing, place the gear housing in a freezer, preferably overnight. If a freezer is not available try an electric refrigerator or ice chest. The next morning, obtain a container of suitable size to hold about 1-1/2 in. (3.8cm) of the forward part of the lower gear housing. Fill the container with water and bring to a rapid boil. While the water is coming to a boil, place a folded towel on a flat surface for padding.

4. After the water is boiling, remove the lower gear housing from its cold storage area. Fill the propeller shaft cavity with ice cubes or crushed ice. Hold the lower gear housing by the trim tab end and the lower end of the housing. Now, immerse the lower unit in the boiling water for about 20 or 30 seconds.

5. Quickly remove the housing from the boiling water; dump the ice; and with the open end of the housing facing downward, slam the housing onto the padded surface. Presto, the bearing should fall out.

■ If the bearing fails to come free, try the complete procedure a second time. Failure on the second attempt will require the use of a bearing puller.

CLEANING & INSPECTING

◆ See Figures 43 thru 46

✱✱ WARNING

If an old impeller is installed be sure the impeller is installed in the same manner from which it was removed - the blades will rotate in the same direction. Never turn the impeller over thinking it will extend its life. On the contrary, the blades would crack and break after just a short time of operation.

Clean all water pump parts with solvent and then dry them with compressed air. Inspect the water pump cover and base for cracks and distortion. If possible, always install a new water pump impeller while the lower unit is disassembled. A new impeller will ensure extended satisfactory service and give peace of mind to the owner. If the old impeller must be returned to service, never install it in reverse to the original direction of rotation. Installation in reverse will cause premature impeller failure.

Inspect the ends of the impeller blades for cracks, tears and wear. Check for a glazed or melted appearance, caused from operating without sufficient water. If any question exists, as previously stated, install a new impeller if at all possible.

Inspect the bearing surface of the propeller shaft. Check the shaft surface for pitting, scoring, grooving, imbedded particles, uneven wear and discoloration.

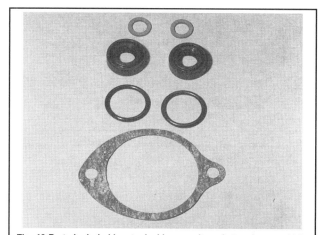

Fig. 43 Parts included in a typical lower unit gasket replacement kit

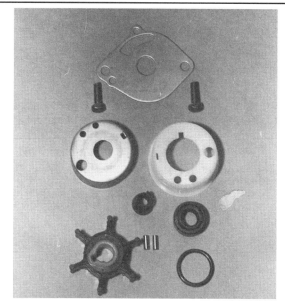

Fig. 44 Parts included in a typical water pump repair kit (2 hp powerhead shown)

7-14 LOWER UNIT

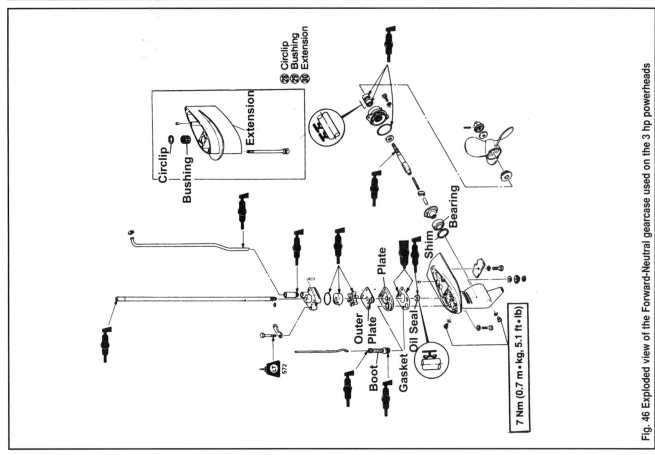

Fig. 46 Exploded view of the Forward-Neutral gearcase used on the 3 hp powerheads

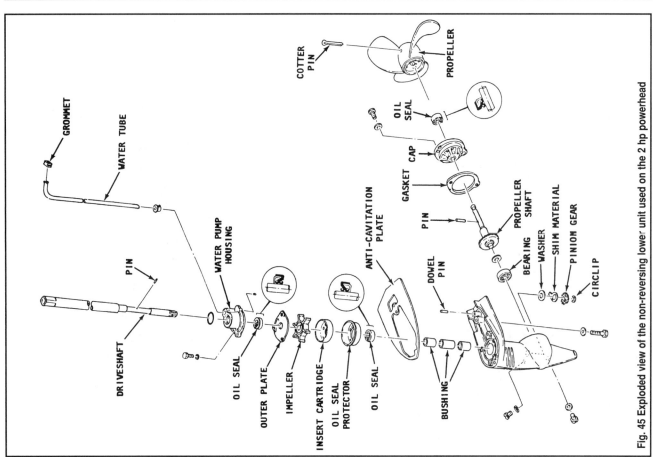

Fig. 45 Exploded view of the non-reversing lower unit used on the 2 hp powerhead

Check the straightness of the propeller shaft with a set of V-blocks. Rotate the propeller on the blocks

Good shop practice dictates installation of new O-rings and oil seals regardless of their appearance.

Clean the pinion gear and the propeller shaft with solvent. Dry the cleaned parts with compressed air.

Check the pinion gear and the drive gear for damage or abnormal wear.

ASSEMBLY

◆ See Figures 43 thru 46

The following procedures are intended to guide your step-by-step through reinstallation and setup of the major gearcase components. If one or more of the components was not removed you can skip that section and proceed to the next following component.

Forward Propeller Shaft Bearing

◆ See Figures 47 and 48

This first section applies only if the forward propeller shaft bearing was removed from the housing.

On 3 hp motors you've got a quandary. As noted earlier, for some reason Yamaha decided to place the forward gear shim pack IN FRONT of the forward bearing (between the bearing and gearcase, NOT between the bearing and gear). On this powerhead there is only 1 shim pack available and although adjustment is first made at the pinion gear and pinion shim pack adjustment can ALSO be made by including or excluding the forward gear bearing shim. Unfortunately when components are changed, measurements may also vary, however we recommend installing the replacement under the same circumstances as the original which was removed (that is, installing it with or without the shim pack depending upon whether or not you found one during removal).

■ If the gear lash is later checked and it is determined that varying the pinion shims cannot bring it within tolerance, you may have to remove the bearing you're about to install and either ad or remove the shim pack. Unfortunately you might destroy the replacement bearing in the process and the replacement for the replacement might have a slightly different tolerance which could again change the need to use the shim pack. Are you as confused as we are why Yamaha did it this way on the 3 hp motor? Suffice it to say, most people we've talked to were fortunate NOT to have to change the use of this shim pack and, when adjustment was necessary, they succeeded by changing the pinion gear shims. We hope you have similar luck. One more suggestion, if you MUST remove the bearing for gear lash adjustment, you MIGHT have success with the ice and boiling water method we outlined in the removal procedure and this should NOT damage the bearing.

1. On 3 hp motors, if a shim pack was removed with the original forward propeller shaft bearing, install it (or a replacement shim pack, since they should only come in one size) in the housing.

2. Place the propeller shaft forward bearing squarely into the housing with the side embossed with the bearing size facing outward (back toward the gear and propeller shaft). Drive the bearing into the housing until it is fully seated using a suitable driver that contacts the appropriate portion of the bearing cage (the outer cage on this bearing should be in contact with the gearcase, so the driver should apply force to that portion of the bearing cage).

3. If the bearing is an unusually tight fit you may use the boiling water trick to help ease installation as follows:

a. Place a new forward propeller shaft bearing in a freezer, refrigerator, or ice chest, preferably overnight. The next morning, boil water in a container of sufficient size to allow about 1-1/2 in. (3mm) of the forward part of the lower gear housing to be immersed. While the water is coming to a boil, place a folded towel on a flat surface for padding. Immerse the forward part of the gear housing in the boiling water for about a minute.

b. Quickly remove the lower gear housing from the boiling water and place it on the padded surface with the open end facing upward. At the same time, have an assistant bring the bearing from the cold storage area. Continue working rapidly. Carefully place the bearing squarely into the housing as far as possible, again with the embossed side facing outward. Push the bearing into place. Obtain a blunt punch or piece of tubing to bear

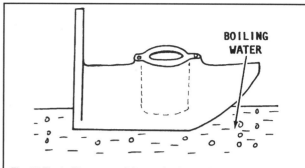

Fig. 47 The boiling water trick can also be used to ease bearing installation

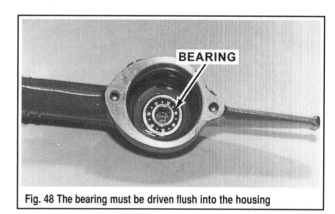

Fig. 48 The bearing must be driven flush into the housing

on the complete circumference of the retainer. Carefully tap the bearing retainer all the way into its forward position - until it bottoms-out. Tap evenly around the outer perimeter of the retainer, shifting from one side to the other to ensure the bearing is going squarely into place. The bearing must be properly installed to receive the forward end of the propeller shaft.

Driveshaft Bushings

◆ See Figure 49

■ The following steps are to be performed if any of the driveshaft bushings were removed.

Lower bushings are installed into a gearcase by drawing the bushing upward into position from underneath. This is normally accomplished using a long threaded bushing installer (#YB-06029). However, it is possible to fabricate this tool using a long threaded bolt, a suitably sized washer, plate and a couple of nuts (as described earlier under Overhaul Tips).

1. Apply a light coating of gearcase or engine oil to the outside of the bushing to help ease installation.

2. To install a lower bushing:

a. Place a nut and a plate toward the top of the long threaded bushing installer, then position the installer into the gearcase from the top until the threads protrude into the lower cavity of the lower unit, as shown.

b. Now, if not already down, lower the plate so it sits on the top of the gearcase and thread the nut down until it is snug against the plate.

c. Slide the first bushing over the threads at the lower end of the driveshaft and then install the bushing retainer (or suitably sized washer and nut) onto the threads.

d. Generally these tools work by holding the rod from turning with one wrench, while you turn the nut against the plate with another wrench. This will slowly draw the entire threaded rod along with the bushing retainer (or nut and washer and the bottom) along with the bushing upward into the gearcase. Continue turning the nut until the bushing is drawn up into the lower unit and seats properly.

3. To install an upper bushing, start the bushing into the opening on the top of the gearcase by hand, then use a suitable driver to gently push or tap the bushing squarely into position. Make sure the bushing either fully seats against the stop in the gearcase.

7-16 LOWER UNIT

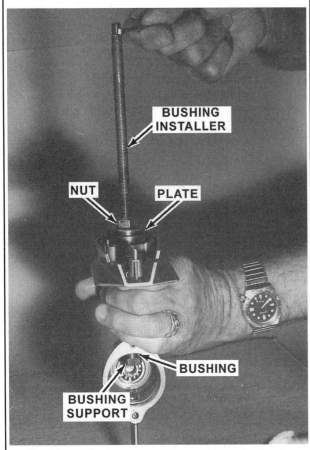

Fig. 49 Use a bushing installer or threaded rod with nuts and washers to lower gearcase bushings

Upper Oil Seal

◆ See Figure 50

■ This procedure does not apply to the 2 hp model, since the oil seals are installed in the top of the Water Pump housing, please refer to the section on Lubrication and Cooling for seal installation on those models.

Once removed, oil seals MUST be replaced, like many bearings, the very action of removing the seal will make it unfit for further use.

1. Obtain a suitable driver or seal installer. If one is not available a socket or smooth piece of pipe of the right size will suffice. The correct size driver, socket etc is one that will place the force on the outside of the seal (the portion of the seal that contacts the gearcase).

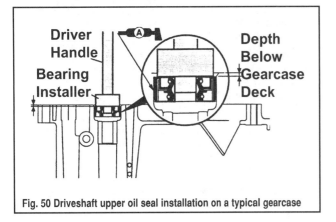

Fig. 50 Driveshaft upper oil seal installation on a typical gearcase

2. Apply a light coating of marine grade grease to the outer diameter of the oil seal casing.

■ When two seals are used in this application it is usually easier and safer to install one seal at a time. Seat the first one with the driver, then install and seat the second seal against the first one.

3. Place the oil seal over the end of the installer with the lips facing as noted during removal. The 3 hp motors use dual lipped seals so orientation is not as big a deal, but they are not completely symmetrical and have an outer casing lip that faces downward toward the gearcase.
4. Lower the seal installer and the seal squarely into the seal recess. Tap the end of the handle with a hammer until the seal is fully seated or the seal is installed to a depth of 0.06-0.08 in. (1.5-2.0mm) below the surface of the gearcase deck.
5. After installation, pack the seal with Yamaha all-purpose marine grease or equivalent water resistant lubricant.

Propeller Shaft and Forward Gear

◆ See Figure 51

■ The propeller shaft and forward gear are integral on 2 hp motors, so they must be installed together. On all other models the forward gear is installed separately from the propeller shaft, which inserts through the gear and bearing and engages using a clutch dog/shifter assembly. The propeller shaft on these other models is installed later, AFTER the driveshaft and pinion gear.

1. On shifting models, if the propeller shaft was disassembled, take a moment to prepare it for installation. Install and compress the spring, then install the clutch dog. On some models there is a marking on the clutch which must be positioned so it faces upward when the propeller shaft is installed. Once the clutch dog is in position, install the shift plunger in the end of the shaft.
2. For 2 hp motors, place the shim material saved during disassembly onto the propeller shaft ahead (on the gearcase and forward bearing side) of the forward gear. The shim material should give the same amount of backlash between the pinion and forward gear as before disassembling (unless other components were replaced).
3. Install either the propeller shaft/forward gear (2 hp motors) or just the forward gear (3 hp motors) into the lower unit housing. Push the propeller shaft and/or gear into the forward propeller shaft bearing as far as possible.

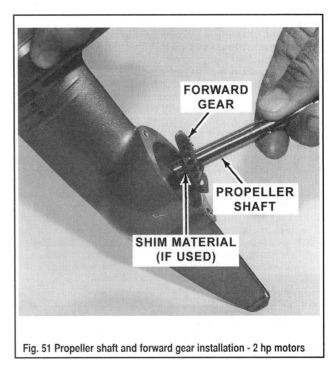

Fig. 51 Propeller shaft and forward gear installation - 2 hp motors

LOWER UNIT 7-17

Driveshaft (and Pinion Gear)

◆ See Figures 52, 53 and 54  DIFFICULT

■ On 2 hp motors the Water Pump assembly is installed from the BOTTOM of the driveshaft, therefore you must prepare it for installation at this time so that it can be positioned on the driveshaft and installed to the top of the gearcase as the driveshaft is installed. During installation, be sure to install the 2 seals inside the pump housing with their lips facing downward and positioned deep enough that there is still about 4.0-4.5mm of clearance beneath the seal lips. Remember to apply marine grease to the O-ring, grommet, oil seals and impeller. Apply a Gasket Maker sealant to the water pump housing-to-cartridge outer plate mating surface. And lastly, be sure to position the housing using the 2 dowel pins.

1. For 2 hp motors, prepare the water pump for installation (oil seals, impeller, etc), then slide the water pump housing cover over the lower end of the driveshaft. Apply a coating of gasket making sealant to both sides of the water pump plate. Next, install the plate with the hole in the plate indexed over the pin on the water pump housing. Dab some Yamaha all-purpose marine grease, or equivalent water resistant lubricant, onto the dowel pin and insert it into the hole in the driveshaft. Slide the water pump impeller onto the driveshaft with the notch in the impeller indexed over the dowel pin. Slide the metal cup over the impeller and then the plastic cup over the metal cup.

2. Lower the driveshaft (or driveshaft and water pump assembly on 2 hp motors) into the lower unit until the splines on the lower end of the driveshaft protrude into the lower unit cavity. On 2 hp motors, do not mate the cover with the lower unit surface at this time. Leave some space, as shown.

3. Slide the thrust washer and any shim material saved during disassembly onto the lower end of the driveshaft. Slide the pinion gear up onto the end of the driveshaft. The splines of the pinion gear will index with the splines of the driveshaft and the gear teeth will mesh with the teeth of the forward gear. Rotating the pinion gear slightly will permit the splines to index and the gears to mesh. Now, comes the hard part. Snap the circlip into the groove on the end of the driveshaft to secure the pinion gear in place. If the first attempt is not successful, try again. Take a break, have a cup of coffee, tea, (a beer, but JUST ONE) whatever, then give it another go. With patience, the task can be accomplished.

4. For 2 hp motors, apply some Loctite, or equivalent, to the two water pump cover retaining bolts, install and tighten the bolts alternately and evenly until well snugged.

Propeller Shaft Bearing and/or Seal Carrier

◆ See Figures 55, 56 and 57  DIFFICULT

At this point the gearcase is all but ready for final assembly, except that for most motors (not including the 2 hp) you still need to actually install the propeller shaft assembly and you need to prepare the bearing/seal carrier for installation.

■ Replacement oil seals should be coated and packed with a suitable marine grade grease. Replacement bearings or bushings should be coated lightly with clean engine or gear oil.

1. Start by installing any oil seal or bearing that was removed from the bearing/seal cap. Use a suitable driver to install a replacement bearing, bushing and/or seal(s), depending upon the model as follows:
- For 2 hp motors, a single seal is used in the cap which is installed from the propeller side of the cap. The seal lips are normally faced toward the propeller.
- For 3 hp motors, dual seals are installed from the propeller side of the

Fig. 52 On 2 hp motors, install the water pump housing with the driveshaft

Fig. 53 Install the pinion gear with any shims, then secure with the circlip

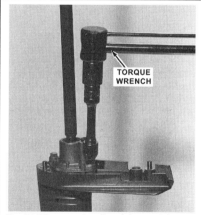

Fig. 54 After securing the pinion on 2 hp motors, install the water pump bolts

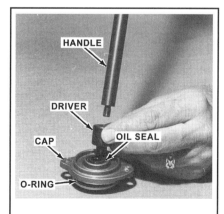

Fig. 55 Use a driver to install replacement bushings, bearings and/or seals

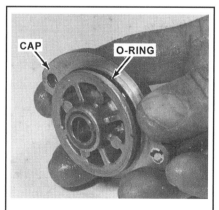

Fig. 56 Grease and install a new O-ring...

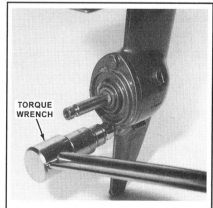

Fig. 57 ...then install the cap and tighten the retaining bolts

7-18 LOWER UNIT

cap. The seal lips normally both face toward the propeller. Once installed the outer of the 2 seals must be at a depth of 0.04-0.06 in. (1.0-1.5mm) below the edge of the bushing/seal cap.

   2. Apply a light coating of marine grade grease to a NEW O-ring, then install the O-ring to the groove in the bearing/bushing and seal cap.

   3. On models with shifting capability you can either pre-position the propeller shaft in the gearcase THEN install the cap OR, we usually find it is easier to insert the shaft through the cap, then install the shaft and cap as an assembly. Remember that these models will normally utilize one or more thrust washers on the shaft. Assemble the cap and shaft.

   4. Install the bearing carrier cap (and propeller shaft on shifting models) to the lower unit, then carefully seat the cap making sure it is flush with the gearcase evenly all the way around.

   5. Apply Loctite, or equivalent, to the threads of the bearing carrier attaching bolts. Install and tighten the bolts securely alternating back and forth until they are tight.

   6. If bearings, bushings, gears or shafts were replaced, proceed to Gear Backlash.

Gear Backlash

◆ See Figures 58 and 59

Backlash between two gears is the amount of movement one gear will make before turning its mating gear when either gear is rotated back and forth.

■ **Yamaha recommends you check some gearcases right side up and others inverted. On these motors, they recommend that the 3 hp gearcase be inverted during lash measurement.**

To perform this check you'll need a dial gauge on a magnetic or threaded base, as well as a gear lash indicator. The lash indicator is essentially a hose clamp with a straight lever attached at a 90 degree angle to the clamp body. The clamp is attached to the driveshaft in a position that any movement in the shaft will result in movement at the lever. The dial indicator is then installed to the gearcase itself in a position contacting the lever to read the amount of movement. The only thing that stops you from fabricating the indicator yourself if the fact that the amount of lever movement (when compared to shaft rotation) increases dramatically as you move outward away from the clamp further down the lever. And Yamaha simply directs you to use the "mark" provided on the indicator when measuring lash. For both motors you can use the Yamaha Indicator (#YB-06265).

■ **Although Yamaha recommends taking this measurement at the driveshaft, conversely it is just as possible to hold the driveshaft from turning and measure the lash with the indicator and dial gauge on the prop shaft.**

   1. Install the indicator on the driveshaft. If the hose clamp for the indicator is too large to fit the propeller shaft snugly, insert some kind of packing into the clamp to allow it to be tightened. The clamp must be secure to permit no movement of the indicator on the shaft.

   2. Install the gauge to the gearcase. Position the dial gauge against the mark on the indicator tool.

   3. For 3 hp motors, place the shifter in the **forward** position.

   4. Grab a hold of the propeller shaft in order to keep it from turning and push in on it gently to make sure the gear is preloaded, meanwhile, pull outward gently on the driveshaft while twisting it in one direction (either clockwise or counterclockwise, it doesn't matter at this time) and zero the dial gauge.

   5. Now, continue to hold/push the propeller shaft to prevent the shaft from rotating. With the other hand gently pull/rock the driveshaft back and forth. In this way, you'll move the driveshaft back and forth between gear contact points with the forward gear (which can't move because you're holding the propeller shaft). Note the maximum deflection of the dial gauge needle, this is the amount of gear lash present in the gearcase for the current shims in use. The 2 hp motor should have gear lash of 0.011-0.04 in. (0.27-0.99mm). The 3 hp gearcase should have a gear lash of 0.006-0.048 (0.15-1.22mm).

   6. If gear lash is within specifications, you're done here. If lash is OUT of specification you'll have to adjust it depending upon the gearcase, as follows:

• For 2 hp motors there are shims available for the forward gear in 0.30, 0.40 and 0.50mm thicknesses. If gear lash is LESS than 0.27mm use the following formula, (0.63mm minus measurement) multiplied by 0.47 = replacement shim thickness. If gear lash is MORE than 0.99mm use the following formula, (measurement minus 0.63) multiplied by 0.47 = replacement shim thickness. Remove the cap, pinion gear and propeller shaft again for access to the forward gear shim, then use the closest replacement shim as determined by the earlier formula. Reassemble and recheck lash.

• For 3 hp motors if the measurement is out of range you should first try to change the pinion gear shim. There are shims available in the following thicknesses, 0.079 in. (2.0mm), 0.083 in. (2.1mm), 0.087 in. (2.2mm) and 0.091 in. (2.3mm). Add a thicker or thinner shim from what is currently in use, based on whether lash is too low or too high, then reassemble and recheck. There is only 1 forward gear shim size available and if lash cannot be brought within spec by adjusting the pinion gear shim pack, then you must remove the forward bearing from the gearcase and either take away the existing shim or use one (if one is not installed), then reassemble the gearcase and recheck. Unfortunately, you may damage the bearing during removal, so use extreme care (and try our alternate ice and boiling water method of removal to hedge your bet).

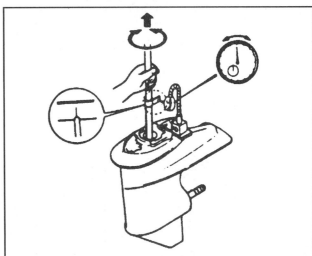

Fig. 58 Yamaha recommends reading lash at the driveshaft. . .

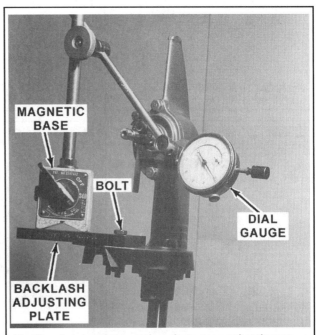

Fig. 59 . . . but sometimes people prefer to measure it at the prop shaft

LOWER UNIT 7-19

■ Remember, adding or removing shim material will affect the forward gear backlash as follows: Adding shim material decreases backlash. Removing shim material increases backlash.

7. If the correct backlash cannot be obtained by adding or removing shim material, it may be necessary to replace either the pinion gear, the forward gear, or the forward gear bearing.

8. For shifting models, before installing the gearcase, check the action of the shifter to make sure it engages and disengages smoothly without hitches. Slowly rotate the driveshaft CLOCKWISE when viewed from above (this is necessary on 2 hp motors to prevent potential damage to the water pump impeller).

9. If possible, pressure test the gearcase. Yamaha only provides specs for the 3 hp motor, but the theory behind the test should work for all models. With the oil fill/drain screw installed and tightened, attach a threaded pressure tester to the oil level hole, then use a hand pump to apply 14.22 psi (100 kPa) of pressure to the gearcase. Watch that the pressure remains steady for at least 10 seconds. If pressure falls before that point one or more of the seals require further attention. You can immerse the gearcase in water and look for bubbles to determine where the leak is occurring. Be sure of gearcase integrity before installation.

10. If not done already, install the Water Pump to the top of the gearcase as detailed in the Lubrication and Cooling section.

11. Install the Gearcase to the intermediate housing, as detailed in this section.

12. Install the Propeller and properly refill the Gearcase Lubricant, both as detailed in the Maintenance and Tune-Up section.

4-40 Hp 1-, 2- and 3-Cylinder Models (Except 40/50 Hp/698cc)

DISASSEMBLY

◆ See Figures 60, 61 and 62

■ For gearcase exploded views, please refer to Cleaning & Inspection in this section.

1. Remove the Propeller from the gearcase, as detailed under Maintenance and Tune-Up.

2. Drain the gearcase oil and remove the gearcase from the outboard as detailed earlier in this section under Gearcase (Lower Unit).

3. If necessary to access the driveshaft or upper oil seal assembly, remove the Water Pump assembly from the top of the gearcase, as detailed in the Lubrication and Cooling section.

■ On all motors the impeller and water pump housing mounts to a water pump plate on top of the gearcase and driveshaft oil seal housing.

4. Remove the water pump plate from the top of the gearcase. For most models the water pump housing bolts were holding the plate in position, but for some gearcases the plate may be secured with bolts of its own. If so loosen and remove the bolts, then carefully lift the plate from the top of the gearcase.

5. If the water pump plate was sealed to the oil seal housing using a gasket, remove and discard the gasket at this time. Generally speaking only the gearcases for the following motors used gaskets under the water pump plate: 9.9/15 hp and 25/30 hp 3-cylinder motors.

■ The driveshaft oil seal housing can be removed from MOST (but not all) models without removing the driveshaft itself from the gearcase. This is because on most motors the upper portion of the driveshaft is a smooth shaft of constant outer diameter, so the oil seal housing can be slid upward and off the top of the driveshaft. Unfortunately the driveshafts used on some gearcases (such as the ones typically found on 9.9/15 hp and 25/30 hp 3-cylinder motors) are of a larger outer-diameter than the housing, requiring that the housing be removed from the bottom of the shaft. Therefore, on some models it will be necessary to remove the propeller shaft and pinion gear, but on most, if there is no problem with the propeller shaft bearing or seal, you CAN skip the Oil Seal Housing procedure.

Propeller Shaft Bearing/Seal Carrier Cap

◆ See Figures 63 thru 71

The rear of the propeller shaft rides in a bearing/seal carrier cap which mounts in the lower rear of the gearcase. As the name suggest, the purpose of the carrier is to house oil seals as well as one or more propeller shaft bearings.

Seal or bearing replacement is a relatively straightforward task that can occur with the gearcase still installed on the outboard (if it is first drained of oil or tilted fully upward). Seal/bearing carrier removal is required for most gearcase services (other than water pump and driveshaft oil seal on most models).

The bearing/seal carrier is secured to the gearcase by one of two methods. On most of the models covered in this procedure the carrier is bolted to the rear of the gearcase using 2 flange bolts. However on some of the larger motors, such as the 25/30 hp 3-cylinder models the carrier is instead secured by a large ring nut and tabbed washer. On these models a special tool is required to remove the ring nut. The tool looks like a very deep socket (since it fits over the propeller shaft) that contains 4 tabs which contact and transmit rotational force against the ring nut. Since the nut is

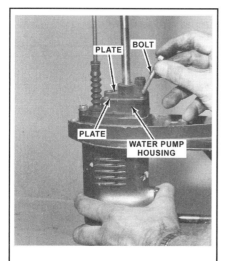

Fig. 60 Start gearcase overhaul by unbolting...

Fig. 61 ...and removing the water pump assembly...

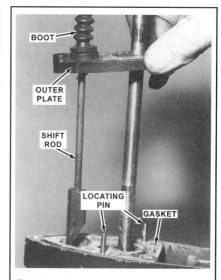

Fig. 62 ...then remove the water pump plate

7-20 LOWER UNIT

tightened to a specification of 65 ft. lbs. (90 Nm), we cannot recommend the use of another spanner to safely loosen the nut.

1. For all except 25/30 hp 3-cylinder, loosen and remove the 2 flange bolts which are securing the bearing/seal carrier. Either rotate the cap 90° so the flanges protrude past the side of the gearcase and then gently tap the bearing carrier with a rubber mallet to break it free from the lower unit OR, some models contain a small slit at the lower portion of the cap-to-gearcase sealing surface where you can use a small pry-tool to carefully pry the cap free.

2. For 25/30 hp 3-cylinders, straighten the tabs on the lock-washer which are bent over the ring nut. There is normally the word **OFF** embossed on the ring nut along with an arrow indicating the correct direction to loosen the nut. Using a suitable ring nut spanner wrench (#YB-06075) or equivalent, rotate the ring in the direction indicated (normally counterclockwise) until the nut is free. Remove the tabbed lock-washer. Attempt to remove the propeller shaft. If the shaft and the bearing carrier will not come out easily, they will have to be pulled.

■ The propeller shaft MAY come free with the bearing/seal cap. If it does, no big deal, just carefully separate it from the cap for attention later.

3. Remove the bearing seal/carrier from the lower unit (on some models the entire propeller shaft assembly will come out with the cap, this is fine). Remove and discard the bearing/seal carrier O-ring(s). For the record, the O-ring(s) CAN be reused on some models, however it is a cheap part and it is not worth the risk of a water leak ruining the gearcase.

Fig. 63 On most gearcases, you can gently pry or tap...

Fig. 64 ...then pull the bearing carrier free (often with the prop shaft)

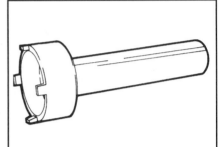

Fig. 65 A special tool is needed on carriers retained by a ring nut

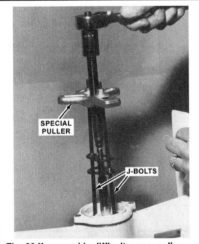

Fig. 66 If removal is difficult, use a puller with J-bolts

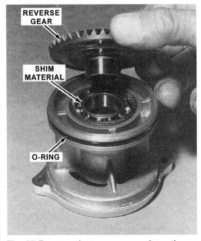

Fig. 67 Remove the reverse gear from the carrier, saving any shim material

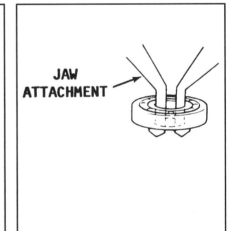

Fig. 68 Use internal jaw pullers to remove most bearings or seals

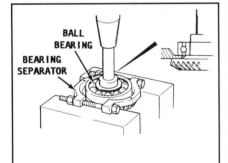

Fig. 69 On some models you need a bearing separator to remove the bearing from the reverse gear

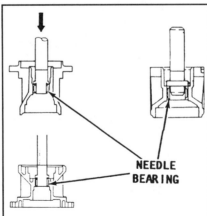

Fig. 70 Use a driver to remove the needle bearing...

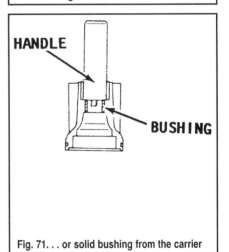

Fig. 71 ...or solid bushing from the carrier

LOWER UNIT 7-21

■ On models whose bearing carrier uses the ring nut for retention, take care not to lose the small key fitted into the side of the bearing carrier.

4. On stubborn carriers, you can use a universal type puller (#YB-06234) and a set of J-bolts which hook to the carrier in order to free the assembly from the gearcase. Hook the ends of the J-bolts into the bearing carrier ribs across from each other. Insert the threaded ends through the puller and then install the washers and nuts to take up the slack. Rotate the center threaded shaft clockwise, pressing against the propeller shaft to separate the bearing carrier from the lower unit. Remove the tools and withdraw the bearing carrier along with the reverse gear assembly.

■ If bearing service is not required, it may not be necessary to remove the reverse gear assembly at all, as long as the oil seals are accessible from the opposite end of the housing. It depends on the extent of the rebuild or reseal you are performing.

5. If possible, remove the reverse gear from the bearing carrier. Watch for and save any shim material from the back side of the reverse gear. The shim material is critical in obtaining the correct backlash during assembling. Using the old shim material will save considerable time, especially starting with no shim material. If the gear won't come free, proceed with the next step.

6. On most models the reverse ball bearing assembly is pressed into the bearing carrier and on many the ball bearing is pressed onto the back of the reverse gear. If this is the case, the gear and bearing assembly must be separated from the carrier (or from each other) either using a slide hammer with jaw expander attachment or using a bridged puller and a bearing separator. Use a slide hammer with jaw expander attachment and pull the bearing or the bearing and reverse gear from the bearing carrier. When using the tool check to be sure the jaws are hooked onto the inner race. If the bearing and gear come out as an assembly you'll need to use a bearing separator and a shop press to separate the two.

7. Inspect the condition of the two seals in the bearing carrier. If the seals appear to be damaged and replacement is required, note the direction the seal lips are facing and then remove them from the housing. In almost all cases these gearcases are equipped with dual oil seals whose lips are both faced in the same direction, towards the propeller. Seal removal can be accomplished by various means. On most models you can use the same slide hammer and jaw attachments which were used on the gear or bearing assembly to remove the seals. On some models you can drive the seals out from the opposite side of the carrier. And, on a couple of models a hooked seal remover or small prytool may be used to free the seal. Whatever method you use, just be sure not to score or damage the inside diameter of the carrier (the sealing surface).

8. On all models except 4/5 hp, and 6/8 hp models, a caged needle bearing set is pressed into the bearing carrier. To remove the bearing set use a suitable driver to push the bearing set free of the bearing carrier. In most, but not all cases, the bearing should be driven in the direction of the oil seals. Inspect both sides of the bearing to determine the proper direction.

9. On the 6/8 hp models, a solid bushing is installed in the carrier, between the bearing set and the oil seals. Using a suitable driver, carefully push the bushing free of the bearing carrier. The direction the bushing is to be driven out does not matter, unless the bushing has a shoulder. In this case, drive the bushing from the flush side.

Driveshaft

◆ See Figures 72 thru 75

The driveshaft is used to transmit the clockwise (when viewed from the top) rotational motion of the crankshaft through the pinion gear to the propeller shaft (either through the Forward or Reverse gears). The driveshaft is splined at both ends, the top splines mating with the lower end of the crankshaft and the bottom end mating with the pinion gear. The pinion gear itself is held in place either by a circlip (4/5 hp and 6/8 hp motors) or by a retaining nut (8/9.9 hp and larger motors).

Exact positioning of the pinion gear on the shaft (so that it meshes properly with the Forward and Reverse gears) is often adjusted through the use of different thickness shim packs. Although, adjustments may also take place at the gears themselves with additional shim pack (or instead of a shim pack used on the pinion gear). Also, keep in mind that the exact placement of the shim pack in relation to other potential gear components such as a plate washer and/or thrust bearing will vary, so pay CLOSE attention to the order of removal of components mounted above the pinion gear. In other words, on some models there will only be a gear, on others a gear and a shim pack. On still others there will be a shim pack above or below a thrust bearing. And, on a few there may even be a shim, thrust washer and plate washer (usually in that order from bottom-to-top).

Before the driveshaft can be removed, the pinion gear at the lower end of the shaft must be removed.

■ Driveshaft removal is necessary for access gearcase mounted components like driveshaft bearings or bushings and/or the Forward gear/bearing. However, on some models (where the oil seal housing must be removed from the bottom of the driveshaft), shaft removal may also be necessary.

1. On 4/5 and 6/8 hp models, pry the circlip free from the end of the driveshaft with a thin pry-tool. This clip holds the pinion gear onto the lower end of the driveshaft. The clip may not come free on the first try, but have patience and it will come free. With one hand, lift the driveshaft up and out of the lower unit housing and at the same time, with the other hand, catch the pinion gear and SAVE any shim material from behind the gear. The shim material is critical to obtaining the correct backlash during installation. Using the old shim material will save considerable time, especially starting with no shim material.

■ In most cases, when working with tools, a nut is rotated to remove or install it to a particular bolt, shaft, etc. In the next two steps, the reverse is usually required because there is very little-to-no room to move a wrench inside the lower unit cavity. Normally, the nut on the lower end of the driveshaft is held steady and the shaft is rotated until the nut is free.

2. Obtain a pinion nut wrench (#YB-06078). This wrench is used to prevent the nut from rotating while the driveshaft is rotated. It's a LONG handled wrench of the same size as the pinion nut. You may be able to use a wrench or a breaker bar and socket already in your box. Also, you'll need the correct driveshaft holder (it's essentially a nut with internal splines that fits on top of the driveshaft so that you can hold or turn it using a wrench or large socket). On 15 hp and smaller motors (with a nut retained pinion gear), obtain a driveshaft tool (#YB-06228). On 20 hp and larger models, obtain another driveshaft tool (#YB-06079).

3. Now, hold the pinion nut steady with the tool and at the same time install the proper driveshaft tool, for the model being serviced, on top of the driveshaft. With both tools in place, one holding the nut and the other on the driveshaft, rotate the driveshaft counterclockwise to break the nut free.

■ Save any shim material from behind the pinion gear. The shim material is critical to obtaining the correct backlash during installation. Using the old shim material will save considerable time, especially starting with no shim material.

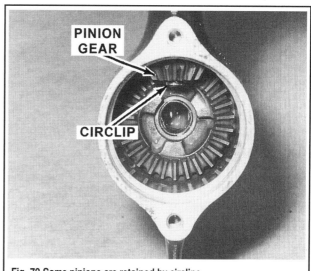

Fig. 72 Some pinions are retained by circlips...

7-22 LOWER UNIT

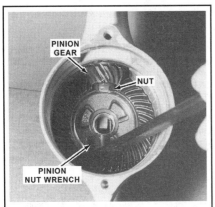

Fig. 73 For nut retained pinions, hold the nut from turning...

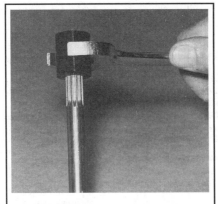

Fig. 74 ...while rotating the driveshaft counterclockwise...

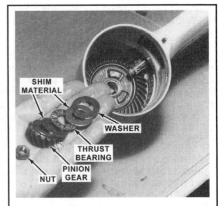

Fig. 75 ...then remove the nut and gear with any shims or washers

4. Remove the pinion nut and gently pull up on the driveshaft and at the same time rotate the driveshaft. The pinion gear, followed by any shim material and washer or thrust bearing will come free from the lower end of the driveshaft. As we noted earlier, pay CLOSE attention to the order in which these components come free from the driveshaft (not easy to do unless you hold them in position of pull them down one at a time).

5. Pull the driveshaft up out of the lower unit housing. Keep the parts from the lower end of the driveshaft in order, as an aid during assembling.

6. On some larger models there is a tapered roller bearing in the center of the driveshaft. If it is unfit for further service, press the bearing free using the proper size mandrel. Take care not to bend or distort the driveshaft because of its length.

Driveshaft Oil Seal Housing

◆ See Figures 76 and 77

At the very top of the gearcase is a small housing that contains the oil seals. On some models that housing may also contain the driveshaft's upper bearing or bushing assembly. The size a shape of the housing varies GREATLY from model-to-model and may be tiny as a small round housing just a little bit larger than the seal or as large as a plate which stretches from the driveshaft to the shift shaft (and also containing a bushing or boot for the shifter assembly). Servicing the oil seal and bearing or bushing is essentially accomplished in the same manner as one the propeller bearing/seal carrier.

On motors where the oil seal housing may be slid over the top of the driveshaft, you do not NEED to remove the pinion gear and driveshaft if you are ONLY going to service the seals. However, it is a little more difficult to protect the seals during installation, so if you've got the time, it's never a bad idea to pull the driveshaft. Your choice. If you decide NOT to pull the driveshaft, take great care when freeing the housing to pull straight upward and, more importantly, when installing to keep the housing squared to the driveshaft so as not to damage the seals (or bearings/bushing on some models).

■ On the 4/5 hp motorss, the oil seals and driveshaft bushing are all installed directly in the top of the gearcase, not a separate removable housing. However, they are serviced in the same manner.

1. Most oil seal housings are secured by the water pump housing and/or plate bolts. However, a few are also secured by one or two additional bolts. This is usually true housings that also contain a passage for the shifter shaft and the additional bolt(s) is(are) normally found at the front of the oil seal housing, near the shifter. Remove any bolt or bolts that are still threaded through the housing assembly.

2. Carefully lift upward to remove the seal housing from the top of the gearcase. Remove and discard the housing gasket.

■ All of these models use dual oil seals which are positioned with their lips facing upwards (away from the gearcase). Double-check this to make sure there is no variance on your gearcase before removal. Also, remember that removal of the seals from the housing will destroy their sealing qualities and they cannot be installed a second time. Therefore, remove a seal only if it is unfit for further service and be absolutely sure a new seal is available before removing the old seal.

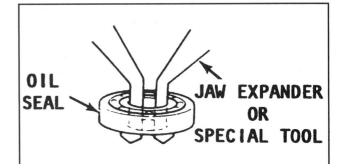

Fig. 76 Use a seal puller, pry-tool or expanding jaw puller to remove the oil seals

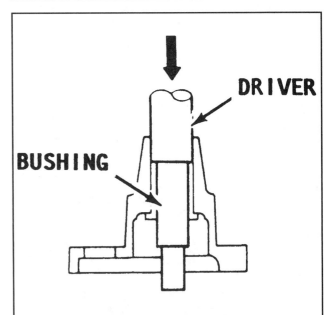

Fig. 77 If applicable, use a suitable driver to remove the bushing (or needle bearing) from the housing

3. Seal removal can be accomplished by various means. On most models you can use a hooked seal remover or small pry-tool in order to free the each seal. However, if access is difficult you can always use a slide hammer and expanding jaw attachment to remove the seals. Sometimes you can drive them out from the other side of the housing, but use care not to damage the housing or the bushing (if used and if it is not being replaced). Whatever method you use, just be sure not to score or damage the inside diameter of the carrier (the sealing surface).

LOWER UNIT 7-23

4. On 6-15 hp models there is also a solid bushing in the housing. If the bushing must be replaced use the appropriate sized driver to carefully push it from the oil seal housing.

■ **The direction the bushing is to be driven out does not matter, UNLESS the bushing has a shoulder. In this case, drive the bushing from the flush side.**

5. On some 20 hp and larger 2-cylinder motors there is a needle bearing in the housing. Only remove this bearing if it is going to be replaced. To remove the bearing use a suitable driver (#YB-06229) and needle bearing attachment (#YB-06082) to push the bearing from the top out through the bottom of the housing.

Driveshaft Bushings/Bearings

◆ See Figures 78, 79 and 80

The driveshaft for all models rides in multiple bushings or bearings. For many models the upper bushing or bearing is installed in the oil seal housing. However, for some models the upper bushing (4/5 hp motors) or upper bearing (25/30 hp 3-cylinder) is mounted directly in the gearcase.

All of these gearcases use a lower driveshaft bushing (4/5 and 6/8 hp motors) or a lower shaft bearing (8/9.9 hp and larger motors) which is pressed directly into the gearcase. In addition, most gearcases have some sort of driveshaft sleeve mounted between the upper and lower bearing assemblies. The sleeves are easily removed (by hand) once the upper bearing or bushing is out of the way (this includes models where the bearing or bushing is installed in the oil seal housing).

Upper bearings or bushings are removed from the top of the gearcase using expanding jaw adapters attached to either a slide-hammer or a bridged puller, while the lower bearing or bushing is normally removed by pushing it into the propeller shaft cavity from above using a suitable driver.

If the driveshaft bushing, upper or lower, on some models, or the needle bearing on other models, is unfit for service, perform the following steps, depending on the model being serviced.

■ **Although the ball bearings used on the upper end of some driveshafts/gearcases can usually be removed and reinstalled without damage, all needle bearings and bushings should be left undisturbed unless they are damaged and in need of replacement. The act of removing most bearings (at least needle bearings, but including ball bearings that are mishandled) and bushings normally makes them unfit for further service.**

1. On 4/5 hp motors, if not done already, remove the oil seals from the top of the gearcase.

■ **A few models, such as the 25/30 hp, 3-cylinder, place the pinion gear shim directly beneath the upper bearing. Watch for and save any shim material from behind the bearing. The shim material is critical to obtaining the correct backlash during installation. Using the old shim material will save considerable time, especially starting with no shim material.**

2. On models where the upper driveshaft bushing or bearing is mounted directly to the gearcase, use a slide hammer with expanding jaw attachment (or a bridged puller) to remove the bearing/bushing from the top of the gearcase. Check the bottom of the bearing and the top of the mounting in the gearcase for shim material and, if found, retain for installation purposes.

3. Grasp and remove the driveshaft sleeve from the top of the gearcase. Although some are completely round with no identifying marks or variance in shape, many are equipped with a tab, flat or marking which must be faced forward during installation. Check the sleeve as it is removed and, if applicable, note the direction of installation for use later.

4. If removal is necessary, use a suitable long-handled driver to push the lower bushing or needle bearing downward (into the propeller shaft cavity).

Forward Gear

◆ See Figures 81 thru 85

All models contain a Forward gear that rotates in a dedicated forward bearing. The 4/5 hp motor normally uses a one-piece bearing and race assembly that is pressed into the front of the gearcase and a gear which can be removed or installed through the bearing by hand. All 6 hp and larger motors are normally equipped with a tapered roller bearing that is pressed onto the Forward gear. The assembly can be removed or installed by hand in a race which is pressed into the nose cone of the gearcase. Shimming material is placed between the bearing race and the mounting surface in the nose cone on these models.

1. Reach into the gearcase and remove the Forward gear (4/5 hp motors) or the Forward gear and bearing assembly (6 hp and larger motors).

2. On 4/5 hp motors, if the forward ball bearing is no longer fit for service, use a slide hammer and jaw attachment (#YB-6096) to pull the ball bearing set from the lower unit.

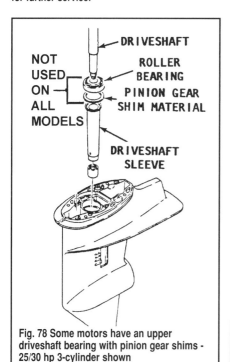

Fig. 78 Some motors have an upper driveshaft bearing with pinion gear shims - 25/30 hp 3-cylinder shown

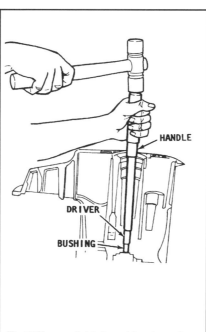

Fig. 79 Use a suitably long driver to push the lower bushing...

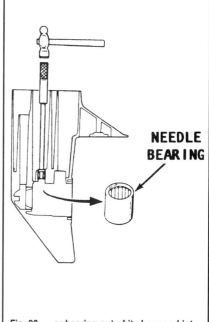

Fig. 80 ...or bearing out of its bore and into the gearcase cavity

7-24 LOWER UNIT

■ If a two piece bearing is to be replaced, the bearing and the race must be replaced as a matched set. Remove the bearing only if it is unfit for further service.

3. On 6 hp and larger motors, if the Forward bearing is not longer fit for service, proceed as follows:

 a. Position a bearing separator between the forward gear and the tapered roller bearing. Using a hydraulic press, separate the gear from the bearing.

 b. Obtain a slide hammer with a jaw attachment. Use the tool to pull the bearing race from the lower unit housing. Be sure to hold the slide hammer at right angles to the driveshaft while working the slide hammer. Watch for and save any shim material found in front of the forward gear bearing race (deeper into the gearcase nose cone). The shim material is critical to obtaining the correct backlash during installation. Using the old shim material will save considerable time, especially starting with no shim material.

Clutch Dog

◆ See Figures 86, 87 and 88

The clutch dog and shifting assembly may be removed from the propeller shaft for cleaning, inspection and replacement, as necessary. The clutch dog itself is splined to the shaft to allow for movement back and forth to engage/disengage the Forward and Reverse gears. Movement along the shaft is limited by a cross-pin. A shift plunger and spring are used to maintain contact with the linkage found in the gearcase nose cone, in front of the Forward gear bearing.

If propeller shaft shifter component disassembly is necessary, proceed as follows:

1. Insert an awl under the end loop of the cross pin ring and carefully pry the ring free of the clutch dog.

2. Use a long pointed punch to press out the cross pin, then remove the plunger and spring from the end of the propeller shaft.

3. Slide the clutch dog from the shaft. Observe how the clutch dog was installed. It must be installed in the same direction. Most clutch dogs are embossed with a marking (such as **F** to denote which edge faces front, toward the Forward gear).

CLEANING & INSPECTING

◆ See Figures 89 thru 95

Good shop practice requires installation of new O-rings and oil seals regardless of their appearance.

Clean all water pump parts with solvent and then dry them with compressed air. Inspect the water pump housing and oil seal housing for cracks and distortion, possibly caused from overheating. Inspect the inner and outer plates and water pump cartridge for grooves and/or rough surfaces. If possible, always install a new water pump impeller while the lower unit is disassembled. A new impeller will ensure extended satisfactory service and give peace of mind to the owner. If the old impeller must be returned to service, never install it in reverse to the original direction of rotation. Installation in reverse will cause premature impeller failure.

If installation of a new impeller is not possible, check the seal surfaces. All must be in good condition to ensure proper pump operation. Check the upper, lower and ends of the impeller vanes for grooves, cracking and wear. Check to be sure the indexing notch of the impeller hub is intact and will not allow the impeller to slip.

Fig. 81 The forward gear will just pull free of the gearcase...

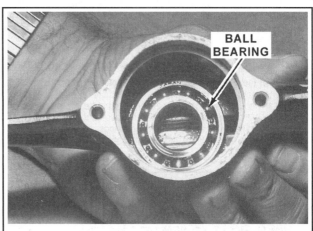

Fig. 82 ...leaving behind a pressed ball bearing on 4/5 hp motors..

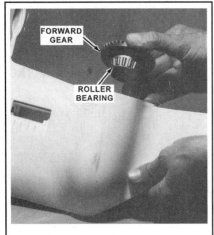

Fig. 83 ...but on 6 hp or larger motors a tapered roller is pressed to the gear...

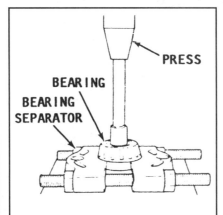

Fig. 84 ...on these motors, use a bearing separator to remove the bearing from the gear

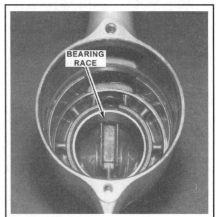

Fig. 85 A slide hammer and puller jaws are needed to remove the race and shim material

LOWER UNIT 7-25

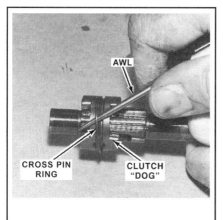

Fig. 86 To remove the clutch dog, carefully remove the cross pin ring...

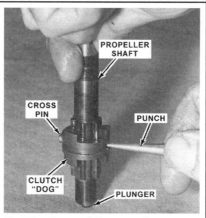

Fig. 87 ...then push the cross pin out, freeing the dog

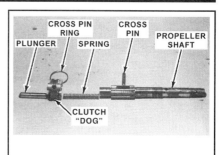

Fig. 88 Exploded view of shift parts on the propeller shaft

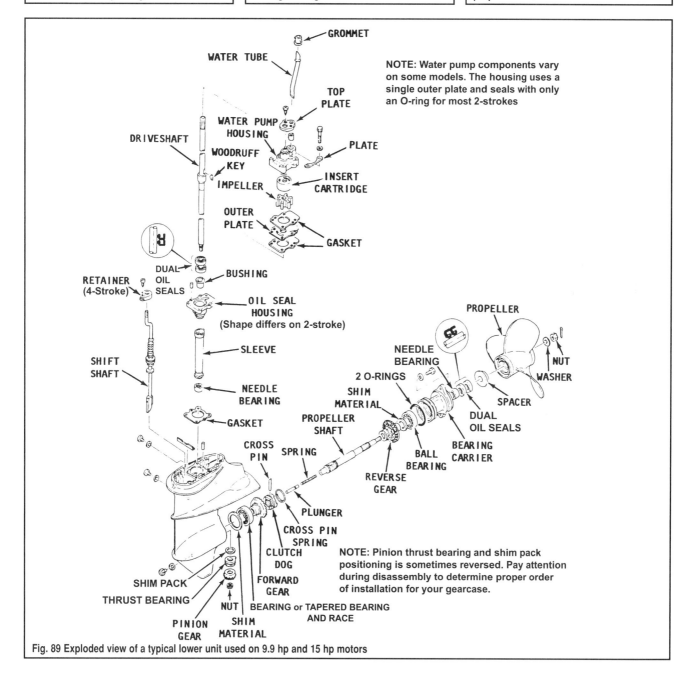

Fig. 89 Exploded view of a typical lower unit used on 9.9 hp and 15 hp motors

7-26 LOWER UNIT

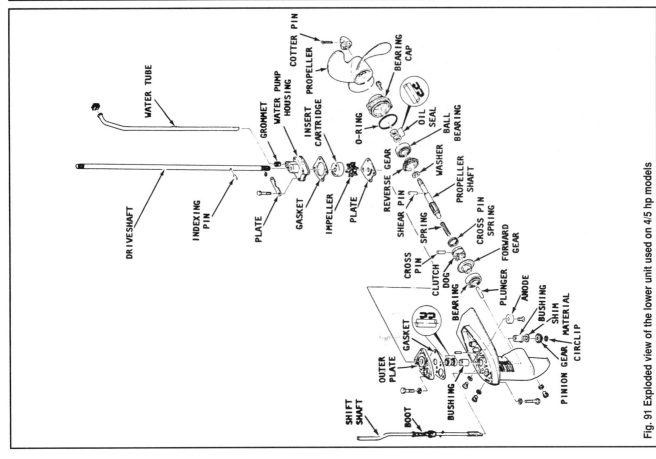

Fig. 91 Exploded view of the lower unit used on 4/5 hp models

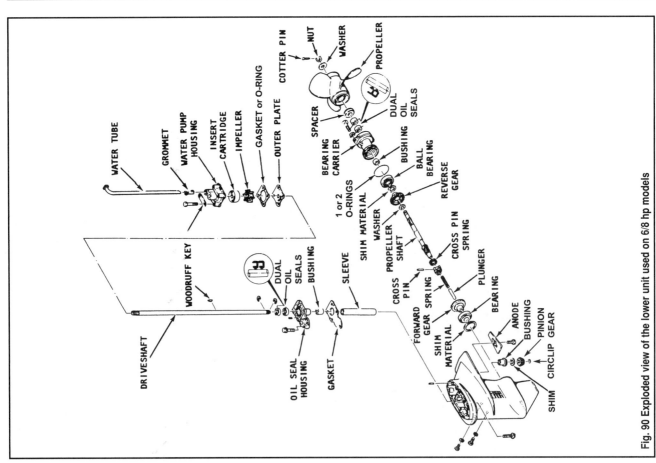

Fig. 90 Exploded view of the lower unit used on 6/8 hp models

LOWER UNIT 7-27

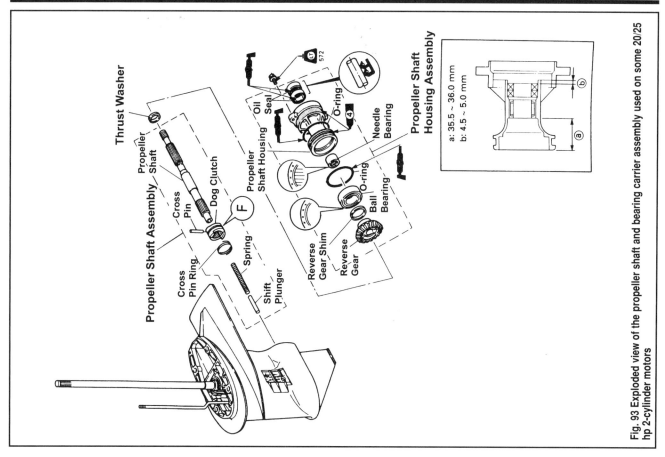

Fig. 93 Exploded view of the propeller shaft and bearing carrier assembly used on some 20/25 hp 2-cylinder motors

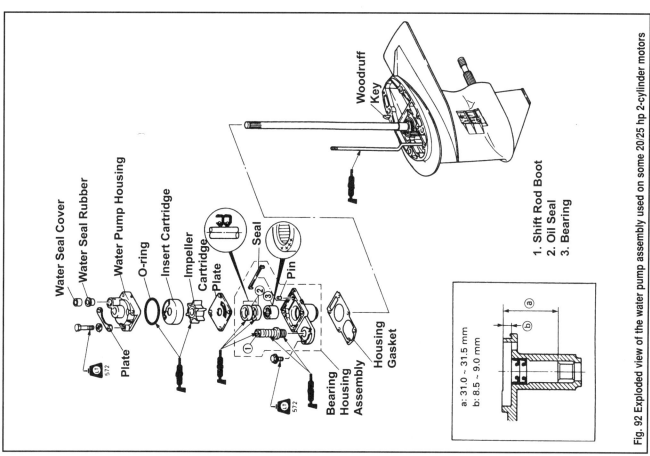

Fig. 92 Exploded view of the water pump assembly used on some 20/25 hp 2-cylinder motors

7-28 LOWER UNIT

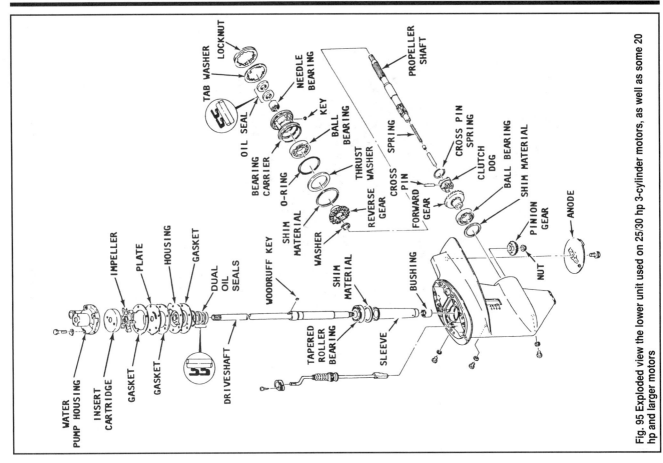

Fig. 95 Exploded view the lower unit used on 25/30 hp 3-cylinder motors, as well as some 20 hp and larger motors

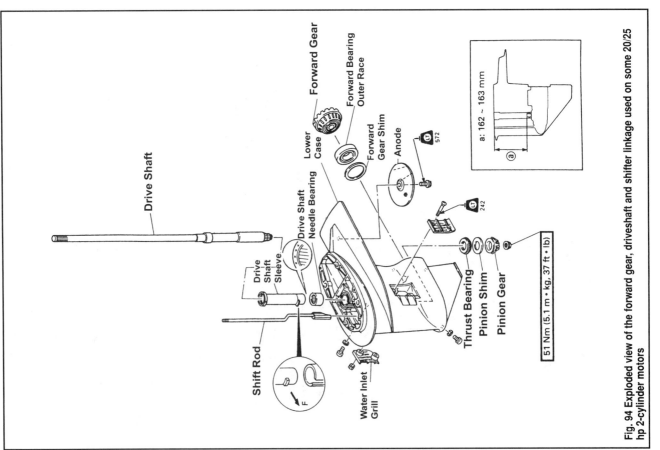

Fig. 94 Exploded view of the forward gear, driveshaft and shifter linkage used on some 20/25 hp 2-cylinder motors

LOWER UNIT 7-29

Clean around the Woodruff key or impeller pin. Clean all bearings with solvent, dry them with compressed air and inspect them carefully. Be sure there is no water in the air line. Hold the bearing to keep it from turning and direct the air stream through the bearing. After the bearings are clean and dry, lubricate them with oil. Do not lubricate tapered bearing cups until after they have been inspected.

✲✲ WARNING

Never spin a bearing with compressed air. Such action is highly dangerous and may cause the bearing to score from lack of lubrication.

Inspect all ball bearings for roughness, scratches and bearing race side wear. Hold the outer race and work the inner bearing race in-and-out, to check for side wear.

Determine the condition of tapered bearing rollers and inner bearing race, by inspecting the bearing cup for pitting, scoring, grooves, uneven wear, imbedded particles and discoloration caused from overheating. Always replace tapered roller bearings and their race as a set.

Clean the forward gear with solvent and then dry it with compressed air. Inspect the gear teeth for wear. Under normal conditions the gear will show signs of wear but it will be smooth and even.

Clean the bearing carrier or cap with solvent and then dry it with compressed air. Check the gear teeth of the reverse gear for wear. The wear should be smooth and even.

Check the clutch dogs to be sure they are not rounded-off, or chipped. Such damage is usually the result of poor operator habits and is caused by shifting too slowly or shifting while the engine is operating at high rpm. Such damage might also be caused by improper shift rod adjustments.

Rotate the reverse gear and check for catches and roughness. Check the bearing for side wear of the bearing races.

Inspect the roller bearing surface of the propeller shaft. Check the shaft surface for pitting, scoring, grooving, embedded particles, uneven wear and discoloration caused from overheating.

Clean the driveshaft with solvent and then dry it with compressed air. Inspect the bearing for roughness, scratches, or side wear. If the bearing shows signs of such damage, it should be replaced. If the bearing is satisfactory for further service coat it with oil.

Inspect the driveshaft splines for excessive wear. Check the oil seal surfaces above and below the water pump drive pin or Woodruff key area for grooves. Replace the shaft if grooves are discovered.

Inspect the driveshaft bearing surface above the pinion gear splines for pitting, grooves, scoring, uneven wear, embedded metal particles and discoloration caused by overheating.

Inspect the propeller shaft oil seal surface to be sure it is not pitted, grooved, or scratched. Inspect the roller bearing contact surface on the propeller shaft for pitting, grooves, scoring, uneven wear, embedded metal particles and discoloration caused from overheating.

Inspect the propeller shaft splines for wear and corrosion damage. Check the propeller shaft for straightness.

As equipped, inspect the following parts for wear, corrosion, or other signs of damage:
- Shift shaft boot
- Shift shaft retainer
- Shift cam
- All bearing bores for loose fitting bearings.
- Gear housing for impact damage.
- Threads for cross-threading and corrosion damage.
- If used, check the circlip to be sure it is not bent or stretched. If the clip is deformed, it must be replaced.
- Check the pinion nut corners for wear or damage. This nut is a special locknut. Therefore, do not attempt to replace it with a standard nut. Obtain the correct nut from an authorized Yamaha dealer.

ASSEMBLY

◆ See Figures 89 thru 95

The following procedures are intended to guide your step-by-step through reinstallation and setup of the major gearcase components. If one or more of the components was not removed you can skip that section and proceed to the next following component.

Gearcase shimming procedures vary slightly from model-to-model and they depend upon the components that are being replaced. However, generally speaking:
- NO SHIMMING is required if the original case and original inner parts (gears, bearings and shafts) are installed.
- NUMERIC CALCULATION RESHIMMING should be performed on 20 hp and larger gearcases when the original inner parts (gears, bearings and shafts) are installed in a NEW case. Many original and replacement cases are marked with various gear measurements. Typically an embossed F (forward gear), R (reverse gear) and/or P (pinion) gear measurements. When these marks are present the differences between the marks on the new case vs. the marks on the original case can be used to determine new shim sizes BEFORE assembly (potentially saving you from having to disassemble/reassemble the gearcase a second time).
- SPECIAL TOOL MEASUREMENT RESHIMMING is performed prior to assembly on certain models and should be used when a shaft, bearing/bushing or gear is replaced. In this case, special tools are used to measure pre-assembled or partially assembled components and determine if a change in shim sizes is necessary. A special tool procedure is available for the gearcase found on the 9.9/15 hp motors. A complete pinion gear shimming tool set is used on gearcases whose driveshaft is equipped with a pressed on tapered roller bearing. In addition, one or more shimming gauges are available for all 20 hp and larger motors.
- BACKLASH MEASUREMENT RESHIMMING should be performed and adjusted when one or more of the original inner components (gears, bearings and shafts) are installed. Backlash measurement is normally performed on an almost fully assembled gearcase using a dial gauge to read the amount of play in the driveshaft. This procedure can check the placement of all 3 gears through this measurement. Though adjustment is much more hit and miss than the other methods, since you need to then re-disassemble the gearcase, swap shims, re-assemble and re-check, then repeat until you get it right. Fortunately, formulas are available to use the measured backlash in helping to pick shim sizes and small shim changes are all that is normally necessary. It is rare that you would have to disassemble the gearcase more than one additional time.

■ **If the case, gear(s), bearing(s) and/or shaft(s) have been replaced, refer to the Gearcase Shimming procedures later in this section BEFORE starting reassembly to determine if there are any changes to shims that can be made before your start assembly so that you can potentially avoid having to disassemble the gearcase a second time to change shims after checking backlash.**

Throughout installation you will need to make a couple of choices. If you have simply replaced O-rings and seals (not bearings, shafts or gears) then there is probably little chance that the gear lash will require adjustment. If so you can lubricate all components and apply sealant or threadlocking compounds as called for in the procedures. However, if you have replaced components such as the bearings, shafts or gears you may wish to check gear contact patterns using a Machine Dye or suitable powder, and all components should then be temporarily installed dry. Also, when components such as these are replaced you'll need to check and possibly adjust gear lash, though reusing shims which were removed during disassembly can often give you a good starting point and MAY prevent you from having to partially disassemble the gearcase again (but only if you've chosen to lubricate components during installation). As we said, the choice is yours and can only be made based upon the circumstances of your overhaul or reseal and the condition of the unit before tear-down.

Unless otherwise noted, upon final installation be sure to lubricate all bearings with gear oil, shafts with oil or a marine grade grease and seals with oil, packing their lips with a suitable marine grade grease.

Clutch Dog

◆ See Figures 96 and 97

1. Most clutch dogs are embossed with an **F** on the face or side (at one end) of the dog itself. During installation on the propeller shaft, the marking on the clutch dog must face the forward gear (toward the front of the gearcase).

2. Slide the spring down into the propeller shaft. Insert a narrow screwdriver into the slot in the shaft and compress the spring until approximately 1/2 in. (12mm) is obtained between the top of the slot and the screwdriver.

7-30 LOWER UNIT

3. Hold the spring compressed and at the same time, slide the clutch dog over the splines of the propeller shaft with the hole in the dog aligned with the slot in the shaft. Again, the embossed letter **F** on the dog must face toward the forward gear.

4. Insert the cross pin into the clutch dog and through the space held open by the screwdriver. Center the pin and then remove the screwdriver allowing the spring to pop back into place.

5. Fit the cross pin ring into the groove around the clutch dog to retain the cross pin in place.

6. Insert the flat end of the plunger into the propeller shaft, with the rounded end protruding to permit the plunger to slide along the cam of the shift rod.

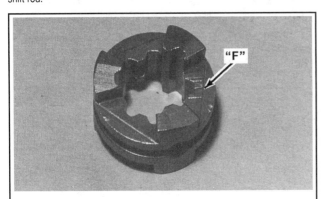

Fig. 96 Most clutch dogs contain an embossed marking which must be faced forward when properly installed. . .

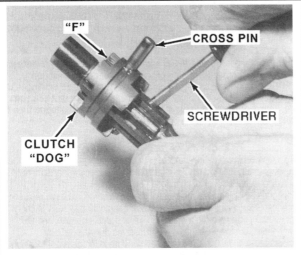

Fig. 97 . . . holding the spring with a screwdriver, slide the clutch dog into place and secure using the cross pin

Forward Gear Bearing

◆ See Figures 98 thru 101

As noted earlier, all models contain a Forward gear that rotates in a dedicated forward bearing. The 4/5 hp motor normally uses a one-piece bearing and race assembly that is pressed into the front of the gearcase and a gear which can be installed through the bearing by hand. All 6 hp and larger motors are normally equipped with a tapered roller bearing that is pressed onto the Forward gear. The assembly can be installed by hand in a race which is pressed into the nose cone of the gearcase. Shimming material is placed between the bearing race and the mounting surface in the nose cone on these models.

The next 2 steps apply only if the forward gear bearing was removed.

1. On 4/5 hp models, position the ball bearing squarely into the housing with the embossed side facing outward (back towards the propeller). Using a suitable driver (which contacts the outside of the bearing), carefully drive the bearing into the housing until it is fully seated.

2. On 6 hp and larger motors, if the Forward bearing and/or race was removed, proceed as follows:

a. Insert the shim material saved during disassembly into the lower unit bearing race cavity. Depending upon what components were changed, the shim material should give the same amount of backlash between the pinion gear and the forward gear as before disassembling (or at least put you in the right ballpark).

b. Insert the bearing race squarely into the gearcase nose cone, then use a suitable driver to gently tap the race into the housing until it is fully seated.

■ Remember the bearing and race are a set here. Although you CAN replace a damaged race and keep the original bearing, there is a good chance that a damaged race will also have a damaged bearing and both should be replaced. Of course, if the bearing is replaced the race MUST also be replaced.

Fig. 98 On 4/5 hp motors, a ball bearing assembly is driven directly into the gearcase

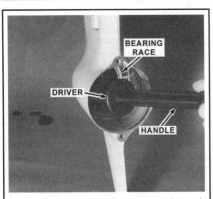

Fig. 99 On 6 hp and larger motors, a forward gear bearing race is driven into the gearcase. . .

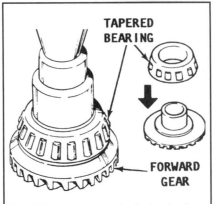

Fig. 100 . . . and a tapered roller bearing is pressed onto the gear itself. . .

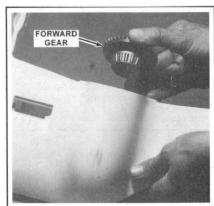

Fig. 101 . . .the gear and tapered roller bearing is inserted by hand

LOWER UNIT 7-31

c. Position the forward gear tapered bearing over the forward gear. Use a suitable mandrel and press the bearing flush against the shoulder of the gear. Always press on the inner race, never on the cage or the rollers.

3. If you wish to visually check gear contact patterns, obtain a suitable substance which can be used to indicate a wear pattern on the forward and pinion gears as they mesh. Machine dye may be used and if this material is not available, Desenex® Foot Powder (obtainable at the local Drug Store/Pharmacy), or equivalent, may be substituted. Desenex® is a white powder available in an aerosol container. Before assembling either gear, apply a light film of the dye, Desenex®, or equivalent, to the driven side of the gear. After the gears are assembled and rotated several times, they will be disassembled and the wear pattern can be examined. The substance will be removed from the gears prior to final assembly.

4. If this is the final assembly, be sure to lubricate the bearing with fresh gear oil.

5. Position the forward gear assembly into the forward bearing (4/5 hp) or race (6 hp and larger).

Driveshaft Bushings/Bearings

◆ See Figures 102 thru 105

As we stated earlier, the driveshaft for all models rides in multiple bushings or bearings. For many models the upper bushing or bearing is installed in the oil seal housing. However, for some models the upper bushing (4/5 hp motors) or upper bearing (25/30 hp 3-cylinder) is mounted directly in the gearcase.

Also, all of these gearcases use a lower driveshaft bushing (4/5 and 6/8 hp motors) or bearing (8/9.9 hp and larger motors) which is pressed directly into the gearcase. In addition, most gearcases have some sort of driveshaft sleeve mounted between the upper and lower bearing assemblies. The sleeves are easily removed (by hand) once the upper bearing or bushing is out of the way (this includes models where the bearing or bushing is installed in the oil seal housing).

The lower bearing or bushing is usually (but not always) installed from above using a driver and a long driver handle to position it to the proper depth. In some cases however, the lower bearing/bushing must be installed from below, using a threaded rod and a driver to pull it upward into the bore from the bottom (from inside the gearcase propeller shaft cavity).

■ Yamaha often has a special tool available to help draw a bushing or bearing upward and into position in the lower portion of the gearcase, all you REALLY need to accomplish this is a sufficiently long threaded bolt with a nut and a washer that is the size of the driver you'd want to contact that bushing or bearing. You'll also need a nut and a plate to place on top of the gearcase. Place the threaded rod through one nut and plate and then down into the gearcase through the bushing/bearing (so the plate is resting against the top of the gearcase, the upper nut is resting against the plate and the threads of the rod are just protruding down into the pinion mounting area of the gearcase.). Next install the washer and lower nut over the bottom threads of the rod to support the bearings or busing. Slowly draw the bearing or bushing into position in the gearcase by holding the threaded rod from turning while at the same time turning the upper nut to draw the entire bolt, bushing/bearing, washer and nut assembly upward until the bushing or bearing seats. Then loosen the nut on the bottom of the threaded rod in the gearcase and remove the threaded rod, plate, nut and washer.

Fig. 102 Some lower bushings/bearings are drawn into the gearcase from below using a threaded rod and driver (or nuts and washers)

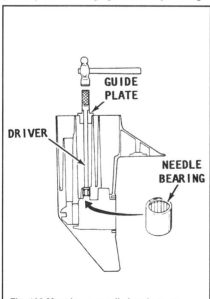

Fig. 103 Most lower needle bearings are installed into the gearcase from above using a driver...

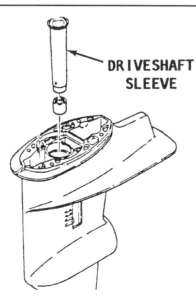

Fig. 104 On models with a shaft mounted upper bearing a race is installed in the gearcase

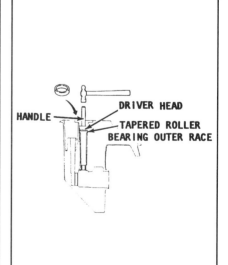

Fig. 105 On models with a shaft mounted upper bearing a race is installed in the gearcase

7-32 LOWER UNIT

Upper bearings or bushings are always installed from the top of the gearcase simply using a suitable driver to seat them.

■ **When bearings are installed from above using a driver, the only potential issue is that if you are not using the specific Yamaha driver for that gearcase would be the depth to which the bushing/bearing is installed. On some models there are shoulders against which the bushing or bearing seats. If not, we will provide a dimension to measure installed depth, ensuring a proper installation.**

If the driveshaft bushing, upper or lower, on some models, or the needle bearing on other models, was removed because they were unfit for service, perform the following steps, depending on the model being serviced.

• The 4/5 hp models have two bushings for the driveshaft. One is located at the top of the lower unit housing and the other just above the pinion gear at the lower end. Use a threaded rod with nuts and washers/plates to draw the lower bushing into the gearcase from the propeller shaft cavity. Then use a driver to install the upper bushing from the top of the case. Lastly, on these motors the dual oil seals are installed directly into the top of the gearcase (both with their lips facing upward, toward the water pump). Install the 2 seals at the same time using a suitable driver. The seals should be installed to a depth of 0.04-0.06 in. (1.0-1.5mm).

• The 6/8 hp motors have two bushing for the driveshaft. One (the upper) is located in the Oil Seal Housing, and details on its installation can be found under that procedure. The other (the lower) is positioned just above the pinion gear. The lower bushing is installed from underneath by pulling it upward using a threaded rod. The bushing driver is slightly larger than the bushing and will stop at the correct depth in the gearcase as it contacts a slightly tapered flange. Once the lower bushing is installed, slide the driveshaft sleeve into position from above.

• The driveshafts on 9.9/15 hp motors use a lower needle bearing assembly that is driven down into the gearcase from above using a suitable driver and long driver handle. The bearing installation depth is determined by the use of a special tool, or by measuring the distance the top edge of the bearing sits below the gearcase deck. The proper installation depth should be 6.80-6.82 in. (172.7-173.2mm). A driveshaft sleeve is then installed over top of the bearing. Standard versions of these models should also have a replaceable upper bushing that is mounted in the oil seal housing.

• For most 20 hp and larger 2-cylinder motors the driveshaft rotates in 2 needle bearing assemblies. The upper, along with the dual-oil seals, is mounted in the Oil Seal Housing, and installation is covered under that component. The lower bearing is normally driven into position from the top of the gearcase, to a specified depth. Position the bearing with any stamped markings facing upward, then use a suitable driver and long driver handle to carefully tap the bearing into the gearcase. When properly installed the top edge of the bearing (with the markings) will be at a depth of 6.38-6.42 in. (162-163mm) below the surface of the gearcase deck. Once the lower needle bearing is installed, slide the driveshaft sleeve into place with the small tab positioned at the lower front of the tube.

• All 25/30 hp 3-cylinder motors, as well as some 20 hp or larger motors are equipped with a lower needle bearing assembly, as well as an upper tapered roller bearing that is pressed onto the driveshaft. As a matter of fact, these gearcases are serviced in the exact same manner. Be sure to drive the lower needle bearing to a depth of 7.74-7.78 in. (196.5-197.5mm), then install the driveshaft sleeve, shim pack and tapered roller bearing race to the gearcase.

Driveshaft Oil Seal Housing

◆ See Figures 106 and 107

With the exception of the 4/5 hp motor, on which the driveshaft oil seals are installed directly into a bore in the gearcase itself, all motors utilize some form of oil seal housing which mounts to the top of the gearcase. On some motors this housing may also include an upper driveshaft bushing or bearing assembly.

■ **Even through the oil seals are installed directly in the top of the gearcase on 4/5 hp motors, not in a separate removable housing, they are still serviced in the same manner.**

1. On 20 hp or larger motors that contain a needle bearing in the oil seal housing, use a suitable driver to install the bearing from the top of the housing. Make sure the driver contacts the outer bearing cage to make sure the bearing is not damaged and install the bearing to a depth of 1.22-1.24 in. (31.0-31.5mm) measured from the top of the housing to the top of the bearing cage.

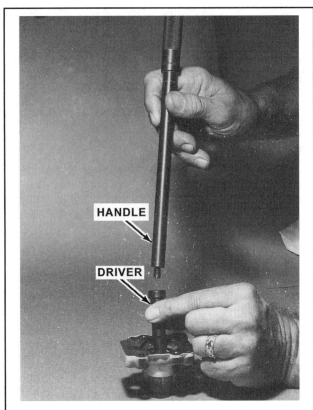

Fig. 106 Bushings or bearings are installed to the oil seal housing using a driver...

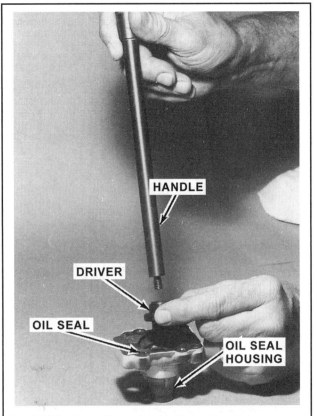

Fig. 107 Oil seals are then installed in the housing also using a driver

LOWER UNIT 7-33

2. On 6-15 hp motors equipped with a bushing in the oil seal housing, use a driver to carefully install the bushing to the housing. Most, but not all, bushings are installed from the oil seal (top) side of the housing. However on some models, such as the 6/8 hp motors the bushing is installed from the opposite (lower) side. Also, most housings do not give an installation depth for the bushing, which probably indicates that the bushing is installed against a shoulder or stop in the housing. Again, on 6/8 hp motors a specification IS provided and the bushing should be installed to a depth of 0.20 in. (5mm) from the BOTTOM of the oil seal housing. On all other motors either carefully seat the bushing against the stop, or drive the bushing down into the lower portion of the housing, making sure it does not extend beyond the lower edge of the bore.

■ **Remember to apply a light coating of gear oil to the oil seal lips and cases before installation. Then after installation, pack the seal lips with a small amount of marine grade grease.**

3. Oil seal installation varies only slightly by model. For all gearcases covered in this procedure a dual oil seal assembly is installed with BOTH sets of lips facing upward, toward the water pump, away from the gearcase. All oil seals are installed from the top of the housing. If no dimension is provided, drive the seals into the housing until they are at least flush with or slightly below the top deck of the housing mating surface. For the following models, drive the seals slowly downward until the TOP edge of the top most seal is the specified distance below the housing deck:

- On 4/5 hp motors (if not done already during driveshaft bushing service), install the 2 seals at the same time using a suitable driver. The seals should be installed to a depth of 0.04-0.06 in. (1.0-1.5mm).
- On 6/8 hp motors, install the seals to a depth of 0.12-0.14 in (3.0-3.5mm).
- On 9.9/15 hp motors, install the seals to a depth of 0.28-0.31 in. (7.0-8.0mm).
- On most 20 hp and larger 2-cylinder motors (models equipped with upper needle bearings that are mounted in the oil seal housing), install the seals to a depth of 0.33-0.35 in. (8.5-9.0).

■ **If you are going to measure gear backlash you may wish to leave the oil seal housing off the motor until after lash is measured (and potentially the gearcase is partially re-disassembled for adjustment). However, on models where the upper driveshaft bearing or bushing is installed in the oil seal housing, it is a good idea to at least temporarily install the housing to make sure there is no additional lateral play on the driveshaft. It really shouldn't affect your lash readings, but it does not hurt to be sure.**

4. If you are ready to install the oil seal housing position it to the top of the gearcase. Some models use dowels and/or a gasket, and if applicable, be sure they are already in position.

Driveshaft (and Pinion Gear)

◆ See Figures 108, 109 and 110

The driveshaft can be installed to the gearcase with or without the upper seal housing in place. In either case you need to take great care when sliding the splines through the seals to make sure the seals are not damaged.

For all except 4/5 hp and 6/8 hp motors the driveshaft pinion is secured by a retaining nut. In order to install and tighten this nut you'll need to obtain a pinion nut wrench (#YB-06078). This wrench is used to prevent the nut from rotating while the driveshaft is rotated. It's a LONG handled wrench of the same size as the pinion nut. You may be able to use a wrench or a breaker bar and socket already in your box. Also, you'll need the correct driveshaft holder (it's essentially a nut with internal splines that fits on top of the driveshaft so that you can turn the driveshaft using a torque wrench and large socket in order to tighten the pinion nut). On 15 hp and smaller motors (with a nut retained pinion gear), obtain a driveshaft tool (#YB-06228). On 20 hp and larger models, obtain a different driveshaft tool (#YB-06079).

1. Lower the driveshaft down through the sleeve in the upper end of the lower unit. If the upper seal housing is already on the gearcase, take extra care not to damage the seals with the splines on the lower end of the shaft.

2. If you are looking to visually check gear mesh patterns, coat the pinion gear with a fine spray of Desenex®, as instructed in the opening paragraphs for Assembly and again under Forward Gear Bearing. Handle the gear carefully to prevent disturbing the powder.

■ **When installing the pinion gear onto the driveshaft, it may be necessary to rotate the driveshaft slightly to allow the splines on the driveshaft to index with the internal splines of the pinion gear. The teeth of the pinion gear will index with the teeth of the forward gear.**

3. Hold the driveshaft in place with one hand and at the same time assemble the parts onto the lower end of the driveshaft with the other hand in the same order as they were removed. Hopefully they were kept in order on the work surface, but since the order varies slightly by model, here's how to assemble the pinion gear and shim pack and/or thrust washer/bearing:

- On 4/5 hp and 6/8 hp motors, install the shim washer followed by the pinion gear and the circlip.

■ **On 4/5 hp and 6/8 hp models, slipping the circlip into the groove on the end of the driveshaft is not an easy task, but have patience and success will be the reward.**

- On 9.9/15 hp motors, install the thrust bearing, followed by the shim, pinion gear and retaining nut. Tighten the nut to 19 ft. lbs. (26 Nm).

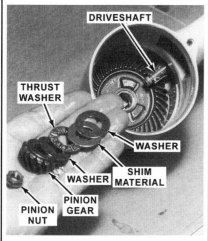

Fig. 108 Assemble the pinion gear and nuts/washers/shims as noted during removal...

Fig. 109 ...then use a wrench to hold the pinion nut as you tighten it by turning the driveshaft

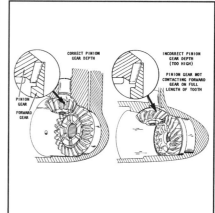

Fig. 110 Visually check the pinion gear depth before proceeding with assembly

7-34 LOWER UNIT

- On most 20 hp and larger 2-cylinder motors (whose gearcase does NOT use a tapered roller upper bearing pressed to the driveshaft), install the thrust bearing, followed by the shim, pinion gear and nut. Tighten the nut to 27 ft. lbs. (51 Nm).
- On 25 hp and larger motors (all motors whose gearcase uses a tapered roller bearing that is pressed to the driveshaft) the pinion gear shim was already installed underneath the tapered bearing race. For these motors, install the pinion gear, followed by the retaining nut (there are no shims, washers or other bearings). Tighten the retaining nut to 36 ft. lbs. (50 Nm).

■ **Where used, the same amount of shim material which was removed during disassembly should be reused. However, if the gearcase itself or a shaft, bearing or gear was replaced, the Gearcase Shimming procedure(s) must be followed to ensure proper lash adjustment.**

How well the pinion gear can mesh with the forward or reverse gear is partially a factor of how far the pinion gear protrudes into the gearcase. Before proceeding with gearcase assembly, take a moment to visually inspect the pinion gear depth.

4. Push downward slightly on the top of the driveshaft (pre-loading it inward toward the gearcase). At the same time, shine a flashlight into the gearcase and look at the pinion gear-to-forward gear contact areas. Look at the pinion gear tooth engagement with the forward gear teeth to be sure contact is made the full length of the tooth. If the pinion gear depth is not correct, the amount of shim material behind the pinion gear must be adjusted. Refer to the Gearcase Shimming procedures for more information Pinion Gear selection.

■ **On 8/9.9 and 15 hp motors there is a pinion gear shimming tool which can be inserted into the gearcase and used, along with a feeler gauge, to check the exact pinion gear depth setting. Obviously if this is necessary, this must be done BEFORE the propeller shaft and bearing seal/carrier cap is installed.**

Propeller Shaft Bearing/Seal Carrier Cap

◆ See Figures 111 thru 119

With the exception of potential Gearcase Shimming procedures which may need to be performed if certain parts were replaced, the Bearing/Seal Carrier (Cap) and propeller shaft installation is the last major piece of the gearcase overhaul.

If the bearing/seal carrier cap was disassembled to replace a bearing, bushing and/or oil seal it must be properly prepared for installation. This means you'll need to use suitably sized drivers to install replacement seals, bearing or bushing, depending upon what was removed. As usual, remember the driver must contact the part of the seal, bearing or bushing that is in contact with the housing. On most installations that means the outside of the seal, bearing or bushing, HOWEVER on some models the reverse gear bearing is installed ONTO the gear and not INTO the housing. When a gear bearing is pressed onto the outer diameter of a shaft on the gear, the driver needs to contact the INNER race instead.

■ **Remember to oil the contact surfaces of bearings, bushings or seals before installation. Be sure to pack the seal lips with a suitable marine grade grease once they are installed.**

Keep in mind that seals, bushing or bearing installation varies slightly from model-to-model. Although all seals used on these gearcases are of the dual variety and install with BOTH of their seal lips facing the rear of the outboard (the propeller), they do not all install from the same side of the Bearing/Seal Carrier Cap. Also, although some parts will install against a shoulder inside the carrier, others must be driven to a specified depth. If disassembled, prepare the bearing/seal carrier cap and the propeller shaft components for installation, as follows:

- For 4/5 hp motors, start by installing the dual oil seals. Drive the 2 seals in at the same time, from the bearing side of the cap. Once the seals are installed position the roller bearing with the embossed side facing the driver (facing the reverse gear or gearcase side of the cap) and carefully tap the bearing into position until it stops.
- For 6/8 hp motors, start by installing the solid bushing to the carrier from the oil seal (propeller side). Gently tap or press the bushing into place, then install each oil seal individually (again from the propeller side), checking the installation depth on each seal as you go. The first (inner) seal installed should go to a depth of 0.39-0.41 in. (10.0-10.5mm) measured from the propeller THRUST BEARING surface (NOT the absolute top of the carrier housing) to the top of the seal. The second (outer) seal installed should go to a depth of 0.12-0.14 in. (3.0-3.5mm) measured the same way. Lastly, turn the carrier over and press the reverse gear ball bearing into position, making sure the embossed markings on the bearing face outward (toward the driver).
- For 9.9/15 hp motors, start by installing the needle bearing to the carrier. Install the needle bearing from the oil seal (propeller) side. Either way, position the bearing so the embossed side faces toward the propeller side and gently tap or press it into place (until it is flush with the mounting surface for the reverse gear ball bearing). Next install the oil seals until the top seal is at the proper depth when measured from the propeller THRUST BEARING surface (NOT the absolute top of the carrier housing) to the top of the seal. The seals are properly installed when the second (outer) installed is at a depth of 0.12-0.14 in. (3.0-3.5mm). Lastly, turn the carrier over and press the reverse gear ball bearing into position, making sure the embossed markings on the bearing face the gearcase (outward toward the driver).
- On most 20 hp and larger 2-cylinder motors (whose gearcase does NOT use a tapered roller upper bearing pressed to the driveshaft), start by installing the needle bearing to the carrier from the gearcase (reverse gear) side of the carrier. Make sure any embossed markings on the bearing are facing outward toward the gearcase side (driver). Press until the top of the bearing reaches a depth of 1.40-1.42 in. (35.5-36.0mm) from the gearcase end of the housing. Next, invert the carrier and install the oil seals until the top (outer) of the 2 seals is 0.18-0.20 in. (4.5-5.0mm) from the top of the seal bore. The reverse gear bearing on these motors is pressed directly onto a bearing, which is in turn pressed into the housing. You can press the bearing on the gear and then the assembly into the housing, but it is usually easier to first press the bearing into the housing and then the gear to the bearing/housing assembly. Make sure that you face the embossed side of the bearing toward the gearcase (driver) side of the housing and, before installing the gear, position the reverse gear shim material.
- On 25/30 hp 3-cylinder models (and 20 hp or larger 2-cylinder motors whose gearcase uses a tapered roller upper bearing pressed to the driveshaft), start by installing the needle bearing to the carrier from the propeller (oil seal) side of the carrier. Make sure any embossed markings on the bearing are facing outward toward the propeller side (driver). Press until the top of the bearing reaches a depth of 0.94 in. (24mm) from the propeller end of the housing. Next, invert the carrier and install the oil seals until the top (outer) of the 2 seals is 0.270-0.272 in (6.85-6.90mm) from the top of the seal bore. The reverse gear bearing on these motors is pressed directly onto a bearing, which is in turn pressed into the housing. Yamaha recommends assembling the gear, shim and ball bearings, then installing the whole assembly to the bearing carrier. Make sure that you face the embossed side of the bearing toward the reverse gear (gearcase) side of the housing.

5. The bearing carrier itself is sealed to the gearcase bore using one or two O-rings (depending upon the model). If this is the final assembly (and not a temporary assembly before Gearcase Shimming or to check gear contact patterns), for each O-ring used, install a NEW O-ring to ensure proper seal, then apply a light coating of marine grade grease.

6. If this assembly is to check gear contact patterns, coat the teeth of the reverse gear with a fine spray of Desenex®, as mentioned earlier. Handle the gear carefully to prevent disturbing the powder.

7. If applicable, place the washer over the reverse gear, then slide the threaded end of the assembled propeller shaft into the bearing carrier.

8. Install the propeller shaft and bearing carrier into the gearcase making sure the shaft inserts through the forward gear and into contact with the shifter. If equipped, align the keyway in the lower unit housing with the keyway in the bearing carrier. Insert the key into both grooves and then push the bearing carrier into place in the lower unit housing.

9. If this is the final installation (again, not temporary for shimming or checking contact patterns), apply a light coating of Loctite® 572 or an equivalent threadlocking compound to the bearing/seal carrier cap retaining screws.

■ **Some bearing carriers on larger models are retained by a ring nut and lockwasher. The threads of the ring nut and gearcase on these models should be clean and dry, there is no need to use a threadlocking compound.**

10. Install and tighten the bearing/seal carrier cap retaining bolts or ring nut and tighten to specification as follows:
- For all 4-20 hp motors and any larger motors that do NOT use a ring nut, tighten the bolts to 5.8 ft. lbs. (8 Nm).

LOWER UNIT 7-35

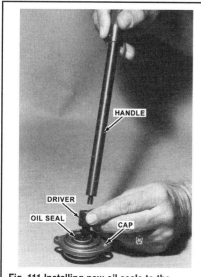

Fig. 111 Installing new oil seals to the bearing/seal carrier cap on 4/5 hp motors

Fig. 112 Install the ball-bearing and O-ring to the bearing/seal carrier cap - 4/5 hp motors

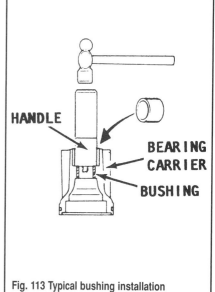

Fig. 113 Typical bushing installation

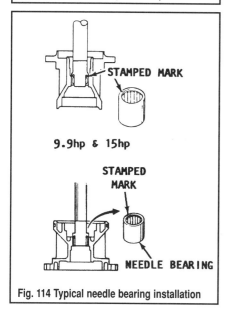

Fig. 114 Typical needle bearing installation

Fig. 115 Oil seals on all of these gearcases are installed with the lips facing the propeller

Fig. 116 Be sure the shim is installed between the reverse gear and ball bearing

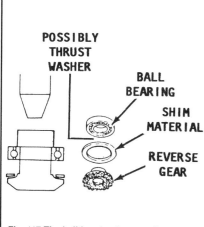

Fig. 117 The ball bearing is normally pressed onto the gear and then the assembly is pressed into the bearing carrier

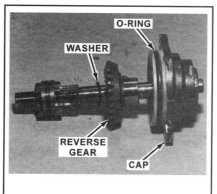

Fig. 118 Insert the propeller shaft into the bearing carrier...

Fig. 119 ... then install the assembly to the gearcase

7-36 LOWER UNIT

- For motors equipped with a single ring nut, install the ring nut with the embossed marks facing outward, away from the bearing carrier, then tighten the nut to 65 ft. lbs. (90 Nm) using the special tool (#YB-6075) or an equivalent ring nut wrench. If this is the final installation, bend the ears of the claw washer over the ring nut to lock it in position.

11. If components other than seals were replaced (such as the case itself, gears, bearings and/or shafts) proceed to the appropriate section of the Gearcase Shimming procedures to check and/or adjust gear lash.

12. If this is final assembly, install the Water Pump, as detailed in the Lubrication and Cooling section.

13. Refill the gearcase, install the propeller and install the lower unit assembly (in whatever order you'd prefer).

GEARCASE SHIMMING

■ **Since Yamaha shims (and gearcase markings) are all in mm we have kept most of the procedures in metric measurements only.**

In order for the gearcase to function properly the pinion gear teeth must fully engage with the teeth of the forward or reverse gear (depending on shifter position). Because build tolerances for the gearcase, gears, shafts and bearings/bushings will vary slightly the manufacturer has positioned shims strategically at various points in the gearcase to give you the ability to make up for the differences in these tolerances. In this way, a gear that is not fully engaging can be repositioned slightly by the use of a thicker shim. Similarly, a gear that does not fully disengage can be repositioned slightly further away by using a thinner shim.

Gearcase shimming procedures vary slightly from model-to-model and they depend upon the components that are being replaced. However, generally speaking:

- NO SHIMMING is required if the original case and original inner parts (gears, bearings and shafts) are installed.
- NUMERIC CALCULATION RESHIMMING should be performed on 20 hp and larger gearcases when the original inner parts (gears, bearings and shafts) are installed in a NEW case. Many original and replacement cases are marked with various gear measurements. Typically an embossed F (forward gear), R (reverse gear) and/or P (pinion) gear measurements. When these marks are present the differences between the marks on the new case vs. the marks on the original case can be used to determine new shim sizes BEFORE assembly (potentially saving you from having to disassemble/reassemble the gearcase a second time).
- SPECIAL TOOL MEASUREMENT RESHIMMING is performed prior to assembly on certain models and should be used when a shaft, bearing/bushing or gear is replaced. In this case, special tools are used to measure pre-assembled or partially assembled components and determine if a change in shim sizes is necessary. A special tool procedure is available for the gearcase found on the 9.9/15 hp motors. A complete pinion gear shimming tool set is used on gearcases whose driveshaft is equipped with a pressed on tapered roller bearing. In addition, one or more shimming gauges are available for all 20 hp and larger motors.
- BACKLASH MEASUREMENT RESHIMMING should be performed and adjusted when one or more of the original inner components (gears, bearings and shafts) are installed. Backlash measurement is normally performed on an almost fully assembled gearcase using a dial gauge to read the amount of play in the driveshaft. This procedure can check the placement of all 3 gears through this measurement. Though adjustment is much more hit and miss than the other methods, since you need to then re-disassemble the gearcase, swap shims, re-assemble and re-check, then repeat until you get it right. Fortunately, formulas are available to use the measured backlash in helping to pick shim sizes and small shim changes are all that is normally necessary. It is rare that you would have to disassemble the gearcase more than one additional time.

Follow the appropriate Shimming or Backlash measurement procedure, depending upon your needs as outlined above.

Numeric Calculation Re-shimming

◆ See Figure 120

The gearcase for all 20 hp and larger models SHOULD be marked with measurements representing the case's deviation from standard. Typically an embossed F (forward gear), R (reverse gear) and/or P (pinion) and a + or - with a number. The codes can be found on the underside of the anti-cavitation plate, usually underneath the trim tab (so if you look under there and can't find them, remove the trim tab and check again).

The stamped number should be interpreted as hundredths of a mm. This means that if there is a marking such as F +5, it means that the case requires a shim that it 5/100mm (0.05mm) larger than standard. The + or - accompanying the number tells you whether you should add (+) or take away (-) the number from the standard measurement. If there is NO number after the F, R or P, assume the value is 0 and the case meets standard tolerance. If the number is there, but illegible, you'll have to work from 0 and make adjustments for that shim by checking lash.

Now, when moving all of the internal components to a new case, compare the numbers on the 2 cases and adjust the shims accordingly. If the numbers are all the same (lets say they both are F +5, R-1 and P+8), you've got it easy, use the original shims and make no changes. But, if the new case is, lets say an F+3, then subtract the new value from the old value to determine how much of a change you need to make to the forward gear shim. In this example, 5-3=2. This means the NEW gearcase needs a shim that is 0.02mm SMALLER than the forward gear shim which was originally used. Measure the old shim and obtain a new shim which is about 0.02mm smaller for reassembly.

Once you've selected the proper shims, assemble the gearcase as detailed earlier in this section and, before the water pump is installed, check the gearcase Backlash as detailed later in this section. As long as the backlash is within spec, no further adjustment is necessary. However, if backlash is out of specification you'll need to further adjust one or more of the shims to correct the problem.

Special Tool Measurement Re-shimming

On some models one ore more special tools are available to assist in determining shim sizes before installation is complete. A pinion height gauge is available for the gearcase found on 9.9/15 hp motors. A complete pinion gear shimming tool set is used on gearcases whose driveshaft is equipped with a pressed on tapered roller bearing. In addition, one or more shimming gauges are available for all 20 hp and larger motors.

9.9/15 Hp Motors

◆ See Figures 121 thru 125

On the US and Canadian versions of these gearcases, a pinion height gauge is available (#YB-34232) for standard thrust models or (#YB-6299) for high-thrust models, in order to help determine pinion shim size. Then, forward and reverse gear shim selection is made by checking backlash and adjusting the shims according to backlash readings.

On world production models (those made for sale/use outside the US and Canada), pinion gear shim selection is determined using a formula (and a sliding caliper to measure the thickness of the current shim and bearing). Following pinion gear shim selection on world production models, additional measurement are taken using a flat shimming plate (#YB-90890), which is 10mm thick is used to take measurements (again using the sliding caliper) to help determine forward and reverse shim thickness. Of course backlash readings should still be taken to verify shim selection on the world production models.

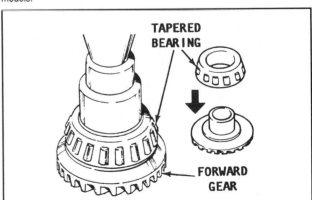

Fig. 120 The gearcase on 20 hp and larger models should be marked with measurements representing the cases deviation from standard

LOWER UNIT 7-37

1. On US and Canadian models, make sure the gearcase is assembled to the point where the driveshaft and pinion are in place. Insert the appropriate pinion height gauge, then use a feeler gauge between the pinion gear and the gauge itself to check clearance. Clearance should be 1.15-1.25mm on all standard thrust models, or 0.45-0.55mm on high-thrust models. If the measurement is out of specification, remove the pinion gear and measure the shim material that was installed. Once you've got the old shim thickness, and the clearance measurement, proceed as follows to select the ideal new shim thickness which would put you toward the center of the range:

- For standard thrust models, if the clearance was LESS than 1.15mm, subtract the measurement from 1.20mm. The result is the amount you need to DECREASE the pinion gear shim. For example, if the measurement was 1.10mm, subtract it from 1.20mm to get 0.10mm. The new shim should be about 0.10mm SMALLER than the shim used during the measurement.
- For standard thrust models, if the clearance was MORE than 1.25mm, subtract 1.20mm from the measurement. The result is the amount you need to INCREASE the pinion gear shim. For example, if the measurement was 1.35mm, subtract 1.20mm from it to get 0.15mm. The new shim should be about 0.15mm LARGER than the shim used during the measurement.
- For high-thrust models, if the clearance was LESS than 0.45mm, subtract the measurement from 0.50mm. The result is the amount you need to DECREASE the pinion gear shim. For example, if the measurement was 0.30mm, subtract it from 0.50mm to get 0.20mm. The new shim should be about 0.20mm SMALLER than the shim used during the measurement.
- For high-thrust models, if the clearance was MORE than 0.55mm, subtract 0.50mm from the measurement. The result is the amount you need to INCREASE the pinion gear shim. For example, if the measurement was 0.65mm, subtract 0.50mm from it to get 0.15mm. The new shim should be about 0.15mm LARGER than the shim used during the measurement.

2. For World production models (except USA and Canada), proceed as follows:

   a. Pinion shim selection is made by measuring the combined thickness of the thrust bearing and plate washer, then subtracting that measurement from a standard spec of 6.05mm for standard models or 5.99mm for high-thrust models. In other words, the shim material, plate washer and thrust bearing together must be a combined total thickness equal to the standard spec.

   b. Now measure the thickness of the forward gear bearing (at the very outer race of the bearing cage). Take the measurement at 2-3 spots around the bearing cage, then average the measurements. Take the measurement and subtract it from the required bearing and shim thickness spec of

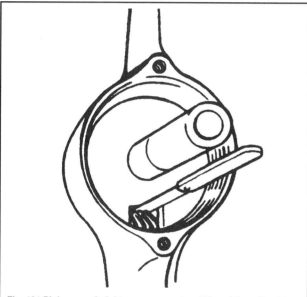

Fig. 121 Pinion gear height measurement on US and Canadian high-thrust models

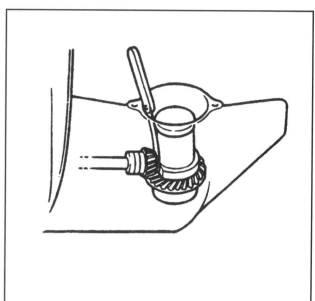

Fig. 122 Pinion gear height measurement on US and Canadian standard thrust models

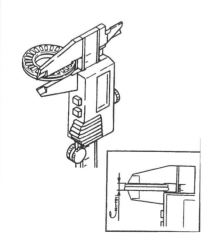

Fig.123 On world production models measure the thrust bearing and plate washer thickness

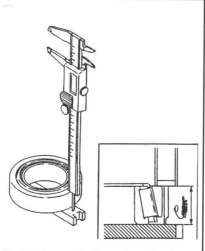

Fig. 124 Measure the forward gear bearing thickness to determine forward gear shim size

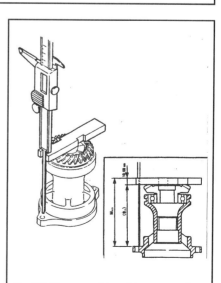

Fig. 125 Also on world models measure the assembled bearing carrier-to-gear thickness

7-38 LOWER UNIT

16.60mm for standard thrust models and 16.50mm for high-thrust models. The result is the ideal forward gear shim size, select one that is as close as possible.

c. To determine the reverse gear shim size, first place the shimming plate (or other 10mm thick piece of flat stock) across the back of the assembled propeller shaft bearing carrier and gear assembly. Measure the distance from the top of the plate to the back (gearcase NOT propeller) side of the carrier mounting flange (through which the carrier retaining bolts are threaded). Subtract this measurement from the required carrier mount-to-gear total thickness of 80.57mm for standard thrust models or 81.00mm for standard thrust models. The result is the ideal reverse gear shim size that would place the reverse gear at a suitable distance in front of the mounting flange when the assembly is installed in the gearcase.

3. Once you've selected the proper shims, assemble the gearcase as detailed earlier in this section and, before the water pump is installed, check the gearcase Backlash as detailed later in this section. As long as the backlash is within spec, no further adjustment is necessary. However, if backlash is out of specification you'll need to further adjust one or more of the shims to correct the problem.

Shim Selection on 20 Hp and Larger Motors Without A Tapered Roller Driveshaft Bearing

◆ See Figures 126, 127 and 128

On some 20 hp and larger 2-cylinder motors the driveshaft does NOT have a tapered roller bearing pressed into position on the shaft. Instead the gearcase uses bearings (usually needle) which are pressed into the upper seal housing. On these models shim selection is performed in different manners for each gear. Pinion gear selection is performed by a simple thickness measurement on the thrust bearing and washer (and a calculation that includes the standard spec and stamped gearcase deviation). The forward gear shim selection is performed using various shim gauge tools. Reverse gear shim selection is performed by measuring backlash after assembly and adjusting the shims accordingly. The only real difference between the US/Canadian and World Production models for these gearcases is that they two groups use a different shim gauge for the forward gear.

1. First determine the pinion gear shim thickness, as follows:
a. Check the stamped code on the underside of the anti-ventilation plate (it is stamped in the trim tab mounting area, so if you haven't already, matchmark and remove the tab). The code F followed by a + or - and a number is the gearcases forward bearing mounting deviation from the standard spec, in hundredths of a mm. This means that if there is a marking such as F +5, it means the case requires a shim that is 5/100mm (0.05mm) larger than standard. The + or - accompanying the number tells you whether you should add (+) or take away (-) the number from the standard measurement. If there is NO number after the F, assume the value is 0 and the case meets standard tolerance. If the number is there, but illegible, you'll have to work from 0 and make adjustments for that shim by checking lash.
b. Use a sliding caliper to measure the combined thickness of the pinion gear thrust bearing.
c. Now, take the standard specification of 6.5mm and to it add or subtract the deviation. This gives you the TOTAL thickness that the thrust bearing, washer AND shim must be for proper bearing/gear placement. To determine the necessary shim thickness, subtract the measured thickness of the thrust bearing and washer from this total calculated thickness.

2. Next, use the appropriate gauge to measure the forward gear bearing and determine the necessary shim thickness (also using the case deviation from standard markings) as follows:
a. For US and Canadian models, start by determining what the specified clearance between the gauge and bearing/shim assembly should be using the standard spec and the stamped gearcase deviation. On these models take the STANDARD spec of 1.5mm and SUBRACT from it the deviation (if the deviation was POSITIVE number) OR ADD to it the deviation (if the deviation was a NEGATIVE number, because subtracting a negative is really adding the positive value of the same number, right!). This new number is the size of the feeler gauge that should pass between the shimming gauge and proper bearing/shim assembly. Place the forward gear bearing and shim together on a flat worksurface (shim on top), then position the Yamaha shimming gauge YB-34468-5 over top of them. Using a feeler gauge, measure the gap between the tab protruding from the underside of the gauge and the shim on the bearing. The gap must equal the measurement you calculated earlier or the shim must be adjusted to get it as close as possible.
b. On World Production models, measure the thickness of the forward gear bearing (at the very outer race of the bearing cage). Take the measurement at 2-3 spots around the bearing cage, then average the measurements. Next, calculate the deviation from spec for the gearcase by taking the standard spec of 17.50mm and adding or subtracting the stamped deviation. In this case a positive deviation would be ADDED to 17.50, while a negative deviation would be subtracted from it. For example, if the stamping was +5 then the spec for this gearcase would be 17.50mm plus 0.05mm or 17.55mm. Now, take the resulting spec for this gearcase and subtract the bearing thickness measurement. The result is the ideal forward gear shim size which, when combined with the bearing thickness will equal the spec for this particular gearcase, select a shim that is as close as possible.

3. Once you've selected the proper pinion and forward gear shims, assemble the gearcase as detailed earlier in this section and, before the water pump is installed, check the gearcase Backlash as detailed later in this section. The reverse gear shim thickness should be changed/adjusted based upon these measurements. A good starting point for reassembly is to use either the old shim from the previous assembly OR use the smallest of the new shims that are available and work you way upward. As long as the backlash is within spec, no further adjustment is necessary. However, if backlash is out of specification you'll need to further adjust one or more of the shims to correct the problem.

Shim Selection on US and Canadian 20 Hp and Larger Motors W/Tapered Roller Driveshaft Bearing

◆ See Figures 129 thru 133

On US and Canadian models that use a tapered roller bearing pressed onto the driveshaft (this includes most of the 20 hp and larger motors), a special pinion gauge block, adaptor plate and base is used to measure clearance. This is basically a method of pre-assembling the driveshaft, bearing, pinion gear and selected shims outside the gearcase in order to precisely check the assembled distance between the bearing and the pinion. The components of the gauge set used vary slightly from model-to-model and are detailed in the procedure.

On these models both the forward and reverse gear bearing shim thicknesses are determined using a shimming gauge, the bearing and a calculation that includes the stamped gearcase deviation from standard

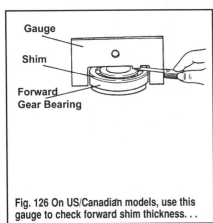

Fig. 126 On US/Canadian models, use this gauge to check forward shim thickness...

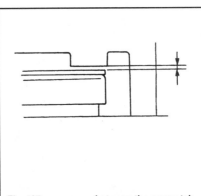

Fig. 127 ...measure between the gauge tab and the shim

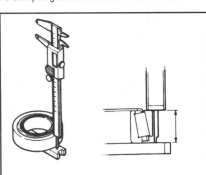

Fig. 128 On World Production models measure the forward bearing on a flat surface

LOWER UNIT 7-39

1. First determine the pinion gear shim thickness, as follows:

   a. Check the stamped code on the underside of the anti-ventilation plate (it is stamped in the trim tab mounting area, so if you haven't already, matchmark and remove the tab). The code **P** followed by a **+** or **-** and a number is the gearcase's pinion bearing mounting deviation from the standard spec, in hundredths of a mm. This means that if there is a marking such as **P** +5, it means that the case requires a shim that it 5/100mm (0.05mm) larger than standard. The **+** or **-** accompanying the number tells you whether you should add (**+**) or take away (**-**) the number from the standard measurement. If there is NO number after the P, assume the value is 0 and the case meets standard tolerance. If the number is there, but illegible, you'll have to work from 0 and make adjustments for that shim by checking lash.

   b. For all motors, assemble the driveshaft, bearing, shim and pinion gear in the gauge base (#YB-34432-11) along with the adapter plate (#YB-34432-10) and the appropriate gauge block. For all motors, use gauge block (#YB-34432-16).

   c. Once assembled, use a feeler gauge to check the gap between the top of the pinion gear and the bottom of the gauge block.

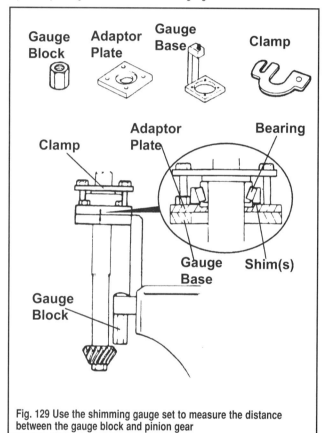

Fig. 129 Use the shimming gauge set to measure the distance between the gauge block and pinion gear

   d. Determine the necessary shim size using the following equation. Take the measured gap and add the standard spec of 0.2mm. Now take the sum of those two numbers and SUBTRACT from it the stamped value. Meaning if the value was P+5, you would subtract 0.05mm from the sum. However, if the value was P-5 (was a negative deviation from norm) you would SUBRACT a NEGATIVE (which means to add the positive value), so you would actually ADD 0.05 to the sum. The result is the thickness of the necessary pinion shim.

2. Next use the appropriate gauge to measure the forward gear bearing and the assembled reverse gear/bearing assembly and determine the necessary shim thicknesses (also using the case deviation from standard markings) as follows:

   a. Place the forward gear bearing a flat work surface (without the shim), then position the Yamaha shimming gauge (#YB-06344) over top of it. Using a feeler gauge, measure the gap between the tab protruding from the underside of the gauge and the bearing. This measurement will be used along with the standard spec and gearcase deviation from spec marking to determine the size of the shim which should be used.

   b. To determine the size of the necessary forward gear shim ADD the standard spec to the measurement AND the gearcase deviation. The standard spec is 0.1mm. For example, if you were working on a motor with a gearcase marking of P+5 AND measured a clearance of 0.15mm you would add 0.15mm (measured clearance), plus 0.1mm (standard spec) and 0.05 (gearcase deviation) to come up with a shim size of 0.30mm.

   c. Place the assembled propeller shaft bearing carrier and reverse gear/bearing assembly with the propeller side facing downward on a suitable work surface. Place the appropriate shimming gauge over the top of the reverse gear so that the gauge legs contact the housing. Use the same gauge as for the forward gear bearing (#YB-06344), the gauge is short and will contact the housing right behind the gear.

   d. Hold the shimming gauge firmly against the housing while using a feeler gauge to measure the gap between the protrusion on the underside of the shimming gauge and the gear itself. For all motors the desired gap should be equal to the standard spec of 1.0mm MINUS the gearcase deviation stamp. Now remember, if the spec is a R-5 you would actually ADD 0.05mm to the standard spec because subtracting a negative is the same as adding the positive, remember? Adjust the shim size(s) until the specified feeler gauge is a proper fit.

3. Once you've selected the proper shims, assemble the gearcase as detailed earlier in this section and, before the water pump is installed, check the gearcase Backlash as detailed later in this section. As long as the backlash is within spec, no further adjustment is necessary. However, if backlash is out of specification you'll need to further adjust one or more of the shims to correct the problem.

Shim Selection on World Production 20 Hp and Larger Motors W/Tapered Roller Driveshaft Bearing

◆ See Figure 124 and 134 thru 136

On World Production models that use a tapered roller bearing pressed onto the driveshaft (this includes most of the 20 hp and larger motors), a different special pinion gauge set (consisting only of a driveshaft holder and height gauge) than used by US and Canadian models. But the concept is the same in that using the gauge set is a method of pre-assembling the driveshaft, bearing, pinion gear and selected shims outside the gearcase in

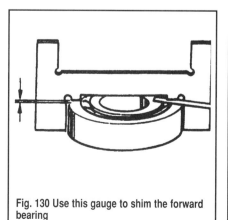

Fig. 130 Use this gauge to shim the forward bearing

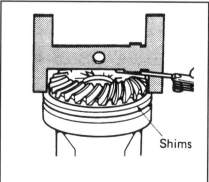

Fig. 131 The same gauge is used to measure the reverse gear shim

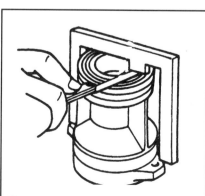

Fig. 132 However a different gauge is used for the reverse gear on the F25

7-40 LOWER UNIT

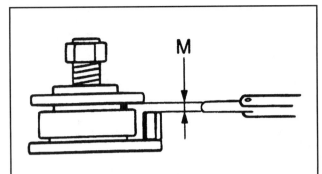

Fig. 133 ...measure clearance between the gauge pin and press plate

order to precisely check the assembled distance between the bearing and the pinion.

On these models both the forward and reverse gear bearing shim thicknesses are determined calculated based on measurements taken of the bearing or assembled gear/bearing housing assembly and the stamped gearcase deviation from standard.

1. To select the pinion gear bearing shim, proceed as follows:

    a. Check the stamped code on the underside of the anti-ventilation plate (it is stamped in the trim tab mounting area, so if you haven't already, matchmark and remove the tab). The code **P** followed by a **+** or **-** and a number is the gearcase's pinion bearing mounting deviation from the standard spec, in hundredths of a mm. This means that if there is a marking such as **P** +5, it means that the case requires a shim that it 5/100mm (0.05mm) larger than standard. The **+** or **-** accompanying the number tells you whether you should add (**+**) or take away (**-**) the number from the standard measurement. If there is NO number after the P, assume the value is 0 and the case meets standard tolerance. If the number is there, but illegible, you'll have to work from 0 and make adjustments for that shim by checking lash.

    b. Assemble the driveshaft, bearing and pinion gear (without any shim) in the Yamaha driveshaft holder (#90890-06517) along with the pinion height gauge (#90890-06702).

    c. Once assembled, use a sliding caliper to check the gap between the top of the pinion gear and the bottom of the height gauge. Rotate the driveshaft in the bearing and repeat this measurement at 2-3 spots around the pinion, then average the results.

    d. Now, determine the thickness of the shim which is necessary by using the following formula and standard spec. It varies slightly from model-to-model but in all cases you are subtracting a series of numbers, including the stamped gearcase deviation spec, the standard for the given gearcase

and the measurement taken in the previous step. Remember that if the stamped deviation spec was a negative number that when you subtract a negative you are actually adding the absolute value (adding the positive of that number). Calculate the necessary pinion gear shim thickness which is equal to the measurement, MINUS 29.3mm, MINUS the stamped gearcase deviation.

■ **On these gearcases, Yamaha recommends that you use the smallest or nearest small shim available during calculations and adjustments.**

2. Next, determine the thickness of the forward gear bearing shim by measuring the bearing itself (at the very outer race of the bearing cage). Take the measurement at 2-3 spots around the bearing cage, then average the measurements. Use the measurement in the appropriate equation for your gearcase (along with the standard spec for that model and the stamped gearcase deviation, although in this case you are adding the deviation so a negative deviation does reduce the amount of shimming material needed).

Calculate the necessary pinion gear shim thickness which is equal to the standard spec for this gearcase of 16.6mm, MINUS the measurement, PLUS the stamped gearcase deviation.

3. Finally use a measurement and a calculation to determine the thickness of the reverse gear bearing shim. You are going to need the Yamaha shimming plate (#90890-06701) OR a piece of perfectly flat stock that is 10mm thick.

4. Place the plate flat over the assembled reverse gear/bearing assembly and measure the distance from the TOP of the plate down to the top of the carrier. Use this measurement in the following equation of measurement, MINUS the standard spec of 26mm, MINUS the stamped gearcase deviation. We probably don't have to remind you YET again that if the stamped deviation is already a negative number it means you would ADD the positive value of the same number, right?

5. Once you've selected the proper shims, assemble the gearcase as detailed earlier in this section and, before the water pump is installed, check the gearcase Backlash as detailed later in this section. As long as the backlash is within spec, no further adjustment is necessary. However, if backlash is out of specification you'll need to further adjust one or more of the shims to correct the problem.

Backlash

◆ See Figures 136, 137 and 138

Backlash (also known as lash or play) is the acceptable clearance between two meshing gears, in order to take into account possible errors in machining, deformation due to load, expansion due to heat generated in the lower unit and center-to-center distance tolerances. A no backlash condition

Fig. 134 Use the pinion height gauge to for the pinion shimming measurement

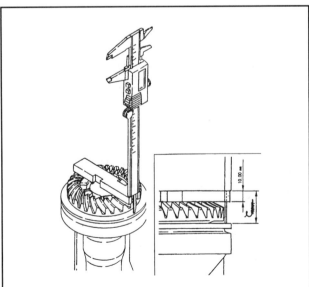

Fig. 135 Measure from the top of the 10mm shimming plate to the carrier

LOWER UNIT 7-41

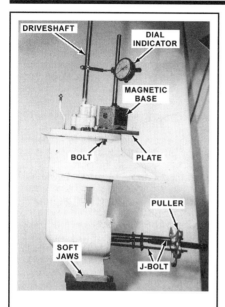

Fig. 136 Yamaha gearcase set-up for Forward gear lash measurement

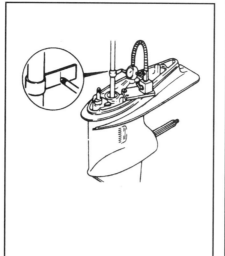

Fig. 137 Position the dial gauge so it contacts the mark on the lash indicator...

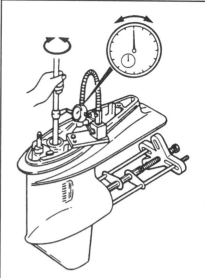

Fig. 138 ... turn the driveshaft one way, zero the gauge, then rotate to the opposite stop

is unacceptable, as such a condition would mean the gears are locked together or are too tight against each other which would cause phenomenal wear and generate excessive heat from the resulting friction.

Excessive backlash which cannot be corrected with shim material adjustment indicates worn gears. Such worn gears must be replaced. Excessive backlash is usually accompanied by a loud whine when the lower unit is operating in neutral gear.

The backlash measurements are taken before the water pump is installed. If the amount of backlash needs to be adjusted, the lower unit must be partially disassembled to change the amount of shim material behind one or more of the gears. Of course, it usually does not make sense to change shims on BOTH the pinion AND either the forward or reverse gears at the same time, since the changes will have an aggregate affect on each other's readings and you're just making it harder on yourself.

As a general rule, if the lower unit was merely disassembled, cleaned and then assembled with only a new water pump impeller, new gaskets, seals and O-rings, there is no reason to believe the backlash would have changed. Therefore, it is safe to say this procedure may be skipped.

However, if any one or more of the following components were replaced, the gear backlash should be checked for possible shim adjustment (even if shimming calculations were performed):
- New lower unit housing - check forward and reverse gear shim material.
- New forward gear bearing - check forward gear backlash.
- New pinion gear - check pinion gear depth and forward/reverse gear backlash.
- New forward gear - check forward gear backlash.
- New reverse gear - check reverse gear backlash.
- New bearing carrier - check reverse gear backlash.
- New thrust washer on 20 hp and larger models - check reverse gear backlash.

■ **The lower unit backlash is measured with the unit inverted (upside down) for all 4-15 hp models or in the upright position for most 20 hp and larger models (except motors that use a driveshaft with a tapered roller bearing).**

To perform this check you'll need a dial gauge on a magnetic or threaded base, as well as a gear lash indicator. The lash indicator is essentially a hose clamp with a straight lever attached at a 90 degree angle to the clamp body. The clamp is attached to the driveshaft in a position that any movement in the shaft will result in movement at the lever. The dial indicator is then installed to the gearcase itself in a position contacting the lever to read the amount of movement. The only thing that stops you from fabricating the indicator yourself if the fact that the amount of lever movement (when compared to shaft rotation) increases dramatically as you move outward

away from the clamp further down the lever. And Yamaha simply directs you to use the "mark" provided on the indicator when measuring lash. For all motors you can use the Yamaha Indicator (#YB-06265).

1. Position the gearcase in a suitable work stand either upright or inverted, depending upon the model. The lower unit backlash is measured with the unit inverted (upside down) for all 4-15 hp models or in the upright position for most 20 hp and larger models (except motors that use a driveshaft with a tapered roller bearing).

■ **Some people prefer to set up the backlash indicator and dial gauge while the gearcase is still upright (just because it is easier than working on it while it is upside down). If so, perform the next couple of steps first, then invert the gearcase on 4-15 hp motors.**

2. Make sure the gearcase is in **neutral** by slowly spinning the driveshaft and checking for a lack of motion at the propeller shaft.

■ **We're not really sure why Yamaha chooses to preload gears by pushing on the prop shaft when checking Forward gear backlash and pulling on the prop shaft when checking Reverser gear backlash. Some other manufacturers (and on some other Yamahas) the preferred method is simply to shift the gearcase into the gear whose lash is being checked. This is always an alternative if you can't seem to get the prop shaft and gear loaded sufficiently to check lash.**

3. Since you normally start by checking Forward gear backlash, install a bearing carrier puller and J bolts onto the ribs of the carrier. With the bearing carrier attaching bolts or the bearing carrier locknut secured, tighten the puller just enough to push on the propeller shaft preloading it.

■ **The bearing carrier puller is used to both keep the propeller shaft from turning and to preload the forward gear, thus freeing up one of your hands. However, you can simply push inward and hold the shaft (or have an assistant hold the shaft) while taking readings.**

4. Obtain and install backlash indicator gauge (#YB6265) onto the driveshaft. If the hose clamp for the indicator is too large to fit the propeller shaft snugly, insert some kind of packing into the clamp to allow it to be tightened. The clamp must be secure to permit no movement of the indicator on the shaft.

■ **Yamaha recommends using the backlash adjusting plate (#YB-07003) to mount the dial indicator to the case. Although it is not absolutely necessary, it does provide a level, fixed position to which the dial gauge can be mounted. If you do NOT use this plate, make sure the gauge is secure and will not move at all during lash measurement.**

7-42 LOWER UNIT

5. Install the gauge to the gearcase. Position the dial gauge against the mark on the indicator tool.

6. If you are not using a puller to hold the propeller shaft from turning, grab a hold of the propeller shaft in order to keep it from turning, then twist the driveshaft in one direction (either clockwise or counterclockwise, it doesn't matter at this time) and zero the dial gauge.

■ **See the specifications listed below before attempting to read gear backlash. On some models other special conditions are necessary, such as pulling or pushing on the driveshaft as it is rotated or installing the propeller without the spacer to pre-load the reverse gear when checking reverse gear lash.**

7. Now, continue to hold the propeller shaft to prevent the shaft from rotating. With the other hand gently rock the driveshaft back and forth. In this way, you'll move the driveshaft back and forth between gear contact points with the forward or reverse gear (which can't move because you're holding the propeller shaft). Note the maximum deflection of the dial gauge needle, this is the amount of gear lash present in the gearcase for the current shims in use and compare them to the specifications, as follows:

- On 4/5 hp motors, with the gearcase inverted and while pulling downward slightly on the driveshaft, backlash should be about 0.28-0.71mm (0.0110-0.0280 in.) for either the forward or reverse gears. However, when checking the reverse gear, be sure to remove the puller and manually pull outward on the propeller shaft itself to load the gear.

■ **When making gear lash adjustments. Remember, too much lash means there is too much space between the gear in question and the pinion, therefore ADDING shim material will DECREASE backlash. Too little lash means there is not a large enough gap, therefore REMOVING shim material will INCREASE backlash.**

- On 6/8 hp motors, with the gearcase inverted and while pulling downward slightly on the driveshaft, forward or reverse gear backlash should be about 0.25-0.75mm (0.0098-0.0295 in.). The puller which is used to preload the forward gear should be tightened by hand just until the forward gear restricts movement sufficiently to measure lash. However, when checking the reverse gear, be sure to remove the puller and manually pull outward on the propeller shaft itself to load the gear. If lash is LESS than 0.25mm, determine the amount of shim to remove by subtracting the measurement from 0.50mm and dividing the result by 1.6mm. If the amount of lash is MORE than 0.75mm, determine the amount of shim to add by subtracting 0.50mm from the measurement and dividing the result by 1.6mm.

- On 9.9/15 hp motors, with the gearcase inverted and while pulling downward slightly on the driveshaft, forward gear backlash should be about 0.19-0.86mm (0.0070-0.0340 in.). The center bolt on the puller which is used to preload the forward gear should be tightened to 3.6 ft. lbs. (5 Nm). If forward gear lash is LESS than 0.19mm, determine the amount of shim to remove by subtracting the measurement from 0.53mm and dividing the result by 2.10mm. If the amount of lash is MORE than 0.86mm, determine the amount of shim to add by subtracting 0.53mm from the measurement and dividing the result by 2.10mm. To check reverse gear backlash, remove the puller and install the propeller without the front side spacer, then tighten the nut to 3.6 ft. lbs. (5 Nm) to preload the reverse gear, then take the readings in the same manner, while pulling downward on the driveshaft. Reverse gear backlash should be about 0.95-1.65mm (0.0370-0.0640 in.). If lash is LESS than 0.95mm, determine the amount of shim to remove by subtracting the measurement from 1.30mm and dividing the result by 2.52mm. If the amount of lash is MORE than 1.65mm, determine the amount of shim to add by subtracting 1.30mm from the measurement and dividing the result by 2.52mm.

- On 20 hp and larger motors (NOT equipped with a tapered roller bearing pressed onto the driveshaft), with the gearcase inverted and while pulling downward slightly on the driveshaft, forward gear backlash should be about 0.32-0.53mm (0.0130-0.0210 in.). The center bolt on the puller which is used to preload the forward gear should be tightened to 3.6 ft. lbs. (5 Nm) so the gear restricts movement sufficiently to measure lash. If forward gear lash is LESS than 0.32mm, determine the amount of shim to remove by subtracting the measurement from 0.43mm and then multiplying the result by 0.47mm. If the amount of lash is MORE than 0.53mm, determine the amount of shim to add by subtracting 0.43mm from the measurement and multiplying the result by 0.47mm. To check reverse gear backlash, turn the gearcase upright, then remove the puller and install the propeller without the spacer, tighten the nut to 3.6 ft. lbs. (5 Nm) to preload the reverse gear. Take the readings in the same manner as forward gear lash except that the gearcase remains upright, so you will be pulling upward on the driveshaft (which is still outward from the gearcase). Reverse gear backlash should be about 0.85-1.17mm (0.0340-0.0460 in.). If lash is LESS than 0.85mm, determine the amount of shim to remove by subtracting the measurement from 1.01mm and multiplying the result by 0.47mm. If the amount of lash is MORE than 1.17mm, determine the amount of shim to add by subtracting 1.01mm from the measurement and multiplying the result by 0.47mm.

- On 25/30 hp 3-cylinder motors, and 20 hp and larger motors that ARE equipped with a tapered roller bearing pressed onto the driveshaft, all measurements are taken with the gearcase upright. The center bolt on the puller which is used to preload the forward gear should be tightened to 3.6 ft. lbs. (5 Nm) so the gear restricts movement sufficiently to measure lash. The forward gear backlash should be about 0.20-0.50mm (0.0080-0.0200 in.). If lash is LESS than 0.20mm, determine the amount of shim to remove by subtracting the measurement from 0.35mm and then dividing the result by 2.0mm. If the amount of lash is MORE than 0.50mm, determine the amount of shim to add by subtracting 0.35mm from the measurement and dividing the result by 2.0mm. To check reverse gear backlash, remove the puller and install the propeller with the front side facing backwards, tighten the nut to 3.6 ft. lbs. (5 Nm) to preload the reverse gear. Take the readings in the same manner as forward gear. Reverse gear backlash should be about 0.70-1.00mm (0.0280-0.0390 in.). If lash is LESS than 0.70mm, determine the amount of shim to remove by subtracting the measurement from 0.85mm and then dividing the result by 2.0mm. If the amount of lash is MORE than 1.00mm, determine the amount of shim to add by subtracting 0.85mm from the measurement and then dividing the result by 2.0mm.

■ **If the lash is too low on BOTH the forward and reverse gears, the pinion height may be too low. If the lash is too high on BOTH gears, the pinion may be too high in the gearcase. Check the installation of the bearings/bushings and the shim materials or spacers used.**

8. If adjustment is necessary, partially disassemble the gearcase as necessary to access the forward or reverse gear shims, then ad or take away shim material, as necessary to bring backlash with spec. However, if the backlash specification cannot be reached by adding and removing shim material, the gears may have to be replaced.

9. If possible, pressure test the gearcase. With the oil fill/drain screw installed and tightened, attach a threaded pressure tester to the oil level hole, then use a hand pump to apply 14.22 psi (100 kPa) of pressure to the gearcase. Watch that the pressure remains steady for at least 10 seconds. If pressure falls before that point one or more of the seals require further attention. You can immerse the gearcase in water and look for bubbles to determine where the leak is occurring. Be sure of gearcase integrity before installation.

✱✱ WARNING

Don't over-pressurize the gearcase as this could CAUSE leaks.

10. Install the Water Pump to the top of the gearcase as detailed in the Lubrication and Cooling section.
11. Install the Gearcase to the intermediate housing, as detailed in this section.
12. Install the Propeller and properly refill the Gearcase Lubricant, both as detailed in the Maintenance and Tune-Up section.

Gear Mesh Pattern

◆ See Figure 139

As noted earlier, you can use a machinist dye or other powder to check the gear mesh pattern. This is useful when determining if there is too much wear on a gear set to return them to service (especially if a dial gauge set and lash indicator is not available to check gear backlash).

The basic method used to check a gear mesh pattern is to coat the gears with a dye or powder, then install the gears, carefully rotate the shafts/gears and carefully disassemble the case again to examine the gears.

All models should be held with the lower unit in the upright (normal) position.

Grasp the driveshaft and pull upward. At the same time, rotate the propeller shaft counterclockwise through about six or eight complete

LOWER UNIT

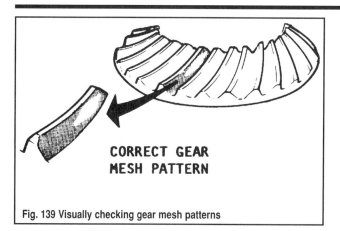

Fig. 139 Visually checking gear mesh patterns

revolutions. This action will establish a wear pattern on the gears with the dye/Desenex® powder.

1. Disassemble the unit and compare the pattern made on the gear teeth with the accompanying illustrations. The pattern should almost be oval on the drive side and be positioned about halfway up the gear teeth.

If the pattern appears to be satisfactory, clean the dye or powder from the gear teeth and assemble the unit one final time.

If the pattern does not appear to be satisfactory, add or remove shim material, as required. Adding or removing shim material will move the gear pattern towards or away from the center of the teeth.

After the gear mesh pattern is determined to be satisfactory, assemble the bearing carrier one final time.

48 Hp 2-Cylinder, 40/50 Hp and Larger 3-Cylinder & All V4/V6 Models

The overhaul procedure the gearcases used on the largest of Yamaha motors (48 hp 2-cylinder, 40/50 hp and larger 3-cylinder, V4 and V6 units is essentially the same. Differences between the units have been called out in the procedures as necessary.

Counter-rotating models are, for the most part, the same as the normal rotation models of the same HP. There are 2 exceptions. First off, many counter-rotation models utilize a 2-piece propeller shaft assembly. And, secondly, whereas the Reverse gear on normal rotation models is mounted in the bearing carrier (at the rear of the gearcase) and the Forward gear is mounted in the nose cone (at the front of the gearcase), the positions of these two gears are SWITCHED on counter-rotation models. Throughout this procedure, we will call the gears by their names as if they were mounted in a normal rotation model, so unless we specify otherwise, when working on a counter-rotation model, remember that what we are calling the Reverse gear is really the Forward gear and vice versa.

DISASSEMBLY

◆ See Figures 140, 141 and 142

■ For gearcase exploded views, please refer to Cleaning & Inspection in this section.

1. Remove the Propeller from the gearcase, as detailed under Maintenance and Tune-Up.
2. Drain the gearcase oil and remove the gearcase from the outboard as detailed earlier in this section under Gearcase (Lower Unit).
3. If necessary to access the driveshaft or upper oil seal assembly, remove the Water Pump assembly from the top of the gearcase, as detailed in the Lubrication and Cooling section.

■ On all motors the impeller and water pump housing mounts to a water pump plate on top of the gearcase and driveshaft oil seal housing.

4. Remove the water pump plate from the top of the gearcase (on most models the bolts which secured the water pump housing also secured the plate). Remove and discard the plate gasket.

■ The driveshaft oil seal housing can be removed from these models without removing the driveshaft itself from the gearcase. This is because the upper portion of the driveshaft is a smooth shaft of constant outer diameter, so the oil seal housing can be slid upward and off the top of the driveshaft. Therefore, if there is no problem with the propeller shaft bearing or seal, you CAN skip the Oil Seal Housing procedure.

Shift Rod

◆ See Figures 143 and 144

The shifter assembly on these models rotates back and forth (as opposed to the smaller Yamaha gearcases on which it lifts up and down). On some models a straight 1-piece shifter shaft is splined at both ends, the bottom of which splines to a removable shift cam inside the housing which can be removed independently of the propeller shaft. However, most models (including most V4 and V6 models) utilize a shaft that is only splined on the top end. For these models the lower end of the shaft contains a pivot pin that is inserted through or around a shift joint that attaches to the propeller shaft. On models that utilize a pivot pin, the shifter assembly must be removed from the gearcase before the propeller shaft can be safely withdrawn.

Either way, the shifter should be removed for inspection and O-ring/seal replacement.

1. Check to ensure the lower unit is in neutral gear. On most models the shift rod cannot (or should not) be removed if the unit is in forward or reverse. If the shifter is in gear, rotate it clockwise or counterclockwise, as

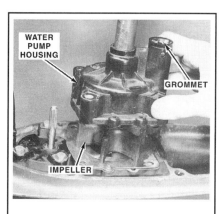

Fig. 140 Start gearcase overhaul by removing the water pump

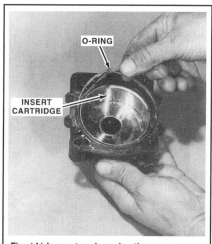

Fig. 141 Inspect and service the pump components

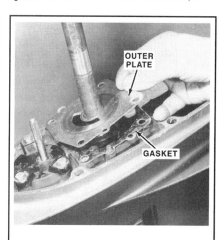

Fig. 142 Remove the water pump plate and gasket for access

7-44 LOWER UNIT

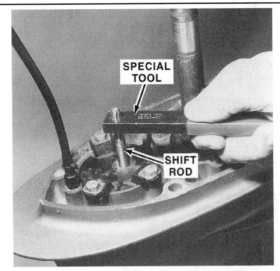

Fig. 143 Make sure the gearcase is in Neutral, rotate the shifter if necessary. . .

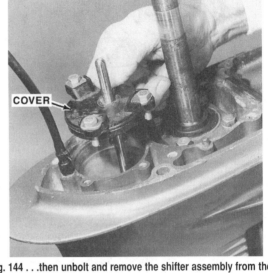

Fig. 144 . . .then unbolt and remove the shifter assembly from the gearcase

necessary to disengage the gear. A special tool with splines that are matched to the top of the shift shaft is available to assist in rotation the shaft. If an alternate tool is used, take care not to damage the shaft or splines.

2. Loosen and remove the shift rod housing retaining bolts (there are normally either 2 or 3 bolts securing the housing to the gearcase). Carefully lift straight up to disengage the lower end of the shift rod from the shift cam or shift rod joint (as applicable) on most motors.

3. Remove the shift rod and housing, then remove and discard the O-ring(s).

■ As with most oil seals, do NOT remove the oil seal from the shift rod housing unless you are planning to replace it.

4. Separate the shift rod from the shift rod housing so that you can remove the oil seal. Note the positioning of the seal lips (they are normally faced downward toward the gearcase), then carefully remove and discard the old seal.

Driveshaft Oil Seal Housing

◆ See Figures 145 thru 148

At the very top of the gearcase is a small housing that contains the oil seals. On some models the housing also contains the driveshaft's upper needle bearing. The size a shape of the housing varies slightly from model-to-model, but there are generally 3 designs used on these gearcases.

The smallest of these motors (typically the 40/50 hp 3-cylinder motors) are equipped with a small round housing just a little bit larger than the seal. The housing is not secured by any bolts, but sits in the gearcase, just above a bearing that is pressed onto the center of the driveshaft. Dual oil seals are mounted into the bottom of this housing, usually with the lips of both seals facing upward toward the water pump. The housing itself is sealed to the crankcase using an O-ring.

■ We say usually in the previous paragraph because some Yamaha factory sources conflict. In some service manuals the seals are pictured at one point with the both lips facing upwards toward the water pump, but in another illustration with both lips facing downward toward the gearcase. For this reason it is CRITICAL on these motors to pay attention to and note seal lip direction BEFORE removing the seals from the housing! However, on almost all other Yamaha gearcases, the seal lips do face UPWARD, so in a pinch, use that as a guide.

Mid-range motors (including the 48 hp 2-cylinder and 3-cylinder motors [except the 40/50 model]) utilize an irregularly-shaped housing which matches the shape of the water pump plate. This housing is normally secured to the gearcase by the water pump assembly bolts. Dual oil seals are also mounted into the bottom of this housing, both with the lips facing upward toward the water pump. This housing is sealed to the top of the gearcase using a gasket and, in most (but not all) cases, is also sealed to the driveshaft bore using an O-ring.

Fig. 145 If applicable, remove the upper housing retaining bolts

Fig. 146 Carefully lift the housing out of the case and off the driveshaft

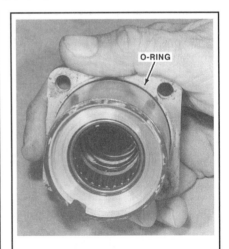

Fig. 147 Remove and discard the housing O-ring (and gasket, if equipped)

LOWER UNIT 7-45

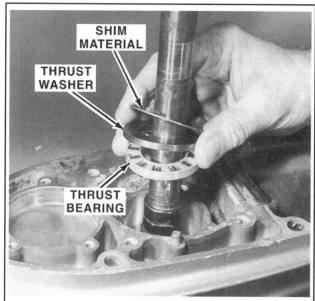

Fig. 148 On models with a bearing in the housing remove the shim, washer and thrust bearing

The largest of these motors (including all V4 and V6 motors) utilize a small oil seal and bearing housing that is square on top and round on the bottom. For these motors dual oil seals are mounted into the top of the housing with both their sets of lips facing upward toward the water pump, while a needle bearing is pressed into the housing from below. The housing is secured into the top of the gearcase normally using 4 bolts and is sealed to the gearcase using an O-ring.

With the exception of how the housing is secured and the side from which seals are installed, all housings are basically serviced in the same general manner. To remove and service the housing, proceed as follows:

On these motors the oil seal housing may be slid over the top of the driveshaft, therefore you do not NEED to remove the pinion gear and driveshaft if you are ONLY going to service the seals. However, it is a little more difficult to protect the seals during installation, so if you've got the time, it's never a bad idea to pull the driveshaft. It's your choice. If you decide NOT to pull the driveshaft, take great care when freeing the housing to pull straight upward and, more importantly, when installing to keep the housing squared to the driveshaft so as not to damage the seals (and bearings on some models).

1. For all V4/V6 motors, remove the four bolts securing the oil seal housing.
2. Carefully lift upward to remove the seal housing from the top of the gearcase. If removal is difficult, use 2 small pry-tools to gently pry up on both sides of the oil seal housing. Lift the housing up and free of the driveshaft. If the housing is stubborn and cannot be removed you can always disassemble the lower portion of the gearcase and remove the driveshaft (this can be especially helpful on the smallest of motors covered here as the housing sits on top of a bearing and shoulder on the driveshaft and MUST come free of the gearcase if the driveshaft is removed).

■ The small round oil housing used on the smallest of motors in this section sometimes contains an oil seal cover (it looks something like a small washer on top of the housing). If it comes loose on models so equipped, be sure to retain it for installation.

3. The housing on most motors is sealed with an O-ring. If so, remove and discard the old O-ring. On the mid-range motors listed earlier, the housing is also sealed with a gasket, remove and discard the old housing gasket.

■ On MOST of these models use dual oil seals which are positioned with their lips facing upwards (away from the gearcase). Double-check this to make sure there is no variance on your gearcase before removal. Also, remember that removal of the seals from the housing will destroy their sealing qualities and they cannot be installed a second time. Therefore, remove a seal only if it is unfit for further service and be absolutely sure a new seal is available before removing the old seal. The same goes for the needle bearing used on the largest of these motors.

4. Seal removal can be accomplished by various means. On most models you can use a hooked seal remover or small pry-tool in order to free the each seal. However, if access is difficult you can always use a slide hammer and expanding jaw attachment to remove the seals. Sometimes you can drive them out from the other side of the housing, but use care not to damage the housing or the bearing (on the larger models if it is not being replaced). Whatever method you use, just be sure not to score or damage the inside diameter of the carrier (the sealing surface).

5. On all V4/V6 models, if the driveshaft needle bearing must be replaced, position the housing with the seal installation surface facing upwards, then use a suitable driver to carefully tap or push the needle bearing out the bottom of the housing. Also on these models there is normally shim material, a plate washer and thrust bearing positioned on top of the driveshaft (right below where the housing mounts to the gearcase). If possible, remove the shims, washer and thrust bearing by sliding them up/off the shaft. Keep the shim material for reuse or reference during installation.

Propeller Shaft Bearing Carrier

♦ See Figures 149 thru 161

The rear of the propeller shaft rides in a bearing and seal carrier which mounts in the lower rear of the gearcase. As the name suggest, the purpose of the carrier is to house oil seals as well as propeller shaft bearings (usually 2, a large diameter ball bearing at the front of the housing and a smaller diameter needle bearing towards the center or rear of the housing).

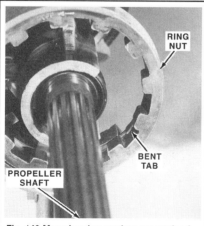

Fig. 149 Many bearing carriers are retained by a ring nut...

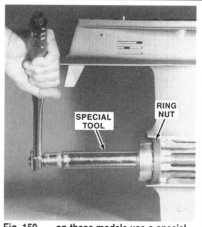

Fig. 150 ...on these models use a special lockring nut wrench...

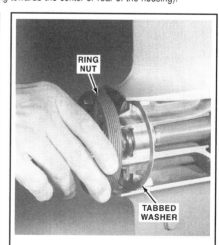

Fig. 151 ...to loosen and remove the nut

7-46 LOWER UNIT

Seal or bearing replacement is a relatively straightforward task that can usually occur with the gearcase still installed on the outboard (if it is first drained of oil or tilted fully upward) and as long as the carrier can be carefully slide back, off the propeller shaft. Bearing carrier removal is required for most gearcase services (other than water pump and driveshaft oil seal on most models).

The bearing/seal carrier is secured to the gearcase by one of two methods. On about half of the models covered in this procedure (some of the smallest and largest engines) the carrier is bolted to the rear of the gearcase using 2 flange bolts. However on most of the mid-range engines (and a few of the smaller and larger motors) the carrier is instead secured by a large ring nut and tabbed washer. On these models a special tool is required to remove the ring nut. The tool looks like a very deep socket (since it fits over the propeller shaft) that contains multiple tabs which contact and transmit rotational force against the ring nut. Since the nut is tightened to a specification of somewhere between 76-105 ft. lbs. (105-145 Nm) depending upon the model in question, we cannot recommend the use of another spanner to safely loosen the nut. For 48 hp 2-cylinder motors, use lockring nut wrench (#YB-06048-1), for all other models use (#YB-34447).

■ If you're working on a counter-rotation model, remember the actual gear positions are opposite of a normal rotation model. On counter-rotation models, the Forward gear is in the bearing carrier and the Reverse gear is in the gearcase nose cone.

1. For models equipped with a bolt retained bearing carrier, proceed as follows:
   a. For V4/V6 motors that utilize a bolted bearing carrier, loosen and remove the 2 bolts securing the retaining ring over the carrier. Remove the retaining ring for access to the carrier flange mounting bolts.
   b. Loosen and remove the 2 flange bolts which secure the bearing carrier.

2. For models equipped with a lockring retained bearing carrier, proceed as follows:
   a. Straighten the tabs on the lockwasher which are bent over the ring nut.
   b. There is normally the word **OFF** embossed on the ring nut along with an arrow indicating the correct direction to loosen the nut. Using a suitable ring nut spanner wrench, rotate the ring in the direction indicated (normally counterclockwise) until the nut is free.
   c. Remove the tabbed lockwasher.

■ Lockring mounted bearing carriers normally utilize a tiny key to help keep the carrier from turning inside the gearcase. The key, is usually installed into the underside of the carrier, be sure to locate and retain it as the carrier is removed.

3. Attempt to remove the bearing carrier. If the carrier will not come out easily, it will have to be pulled. On most models Yamaha recommends or illustrates the use of a universal type puller, however on some V6 models they show the use of a slide hammer, the choice is really yours as both should work on all of these gearcases. But proceed slowly and carefully to make sure you do not damage anything.

■ The propeller shaft (or rear half of the shaft on some counter-rotating models) MAY come free with the bearing carrier. If it does, no big deal, just carefully separate it from the carrier for attention later.

4. Watch for and save any shim material or spacer/thrust washer from the back side of the reverse gear (forward gear on counter-rotation units).

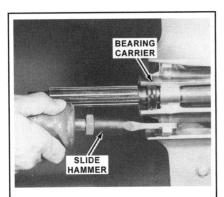

Fig. 152 If necessary use a puller or slide hammer to free the carrier

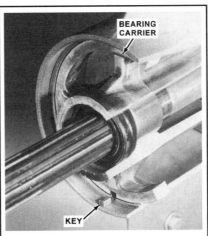

Fig. 153 Locknut style carriers use a key in the bottom of the housing

Fig. 154 Watch for possible shim material behind the gearcase mounted gear

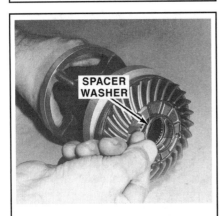

Fig. 155 The carrier mounted gear may have a spacer washer

Fig. 156 Remove and discard the O-ring(s) from the carrier

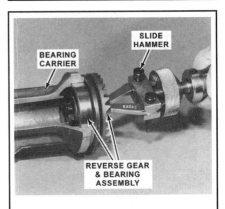

Fig. 157 Remove the gear (or gear/bearing assembly) from the carrier

LOWER UNIT

Fig. 158 If the gear/bearing come out together, use a bearing separator to free them

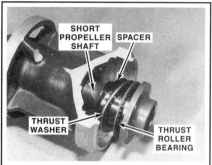

Fig. 159 Counter-rotating models often use a rear/short propeller shaft mounted in the carrier

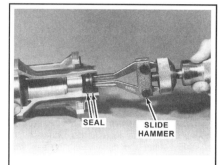

Fig. 160 You can remove JUST the oil seals for replacement...

The shim material is critical in obtaining the correct backlash during assembling. Using the old shim material will save considerable time, especially starting with no shim material.

5. Remove and discard the O-ring(s) around the bearing carrier. There may be as few as 1 or as many as 3 O-rings depending upon the model.

■ The reverse gear (forward on counter-rotation units) and ball bearing assembly is pressed into the bearing carrier. In addition, the ball bearing is pressed onto the back of the reverse gear. Usually all of these components must be separated using a slide hammer and/or bearing separator.

6. The reverse gear (forward on counter-rotation units) and ball bearing assembly is pressed into the bearing carrier. In addition the ball bearing itself was pressed onto the gear. Removal usually involves first pulling the gear from the bearing and housing, and then pulling the bearing from the housing. However, the bearing has been known to come out with the gear, depending upon the method used for removal. It's your choice as you CAN use either a slide hammer with internal expanding jaw attachment or you can use a bridged puller with our without a bearing separator. If the gear and bearing come out as an assembly, use a bearing separator to free the bearing from the gear. If they do NOT come out as an assembly, use the puller or slide-hammer to then extract the bearing from the carrier.

■ On the largest of V6 counter-rotation models the propeller shaft is part of the carrier, retained by a tapered roller bearing, race, tabbed washer and ring nut. On these models, once the forward gear is removed, turn the carrier around and install it back to the gearcase backwards (with the forward gear mounting area pointing outward). Use a ring nut wrench (like #YB-06578) to then loosen the ring nut. The propeller shaft, shims, bearing and race can then be removed using a shop press.

7. If the oils seals AND needle bearing are all being replaced, the easiest way to remove them is to drive them out the propeller side of the housing all together by tapping with a suitable driver from the gearcase side of the housing. However, DO NOT remove the needle bearing unless it is being replaced, as removal normally makes it unfit for further service. In the later case, if only the oil seals must be replaced, use a hooked seal removal tool, small pry bar or at worst, an internal expanding jawed puller attachment and a puller or slide-hammer to remove the seals.

Propeller Shaft and Clutch Dog

◆ See Figures 162 thru 168

The job of the propeller shaft is straight forward enough, change the direction of the clockwise, vertical rotating driveshaft to clockwise (or counterclockwise) horizontal rotation in order to drive the propeller. At all times the pinion gear on the end of the driveshaft is in contact with and rotating 2 gears (a forward and reverse)... but because of their positioning the gears are rotating in different directions. A clutch dog assembly which is splined to the propeller shaft is positioned in the middle of the 2 gears. A shift mechanism is attached to the clutch dog with a cross pin. This mechanism extends forward from the propeller shaft, through the gear and bearing assembly in the gearcase nose cone. The shift rod mechanism works by rotating forward or backward, transmitting this force through the shifter in the propeller shaft to the clutch dog, moving it toward one or the other gear.

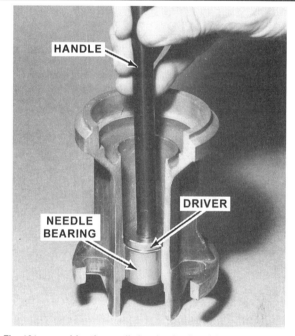

Fig. 161 ... or drive the needle bearing (and seals) out together

When the clutch dog comes in contact with one of the gears it locks into place and begins to spin with that gear, transmitting the force of the gear to the propeller shaft through the clutch dog-to-shaft splines.

Propeller shaft disassembly involves removal of the clutch dog and internal shift components, which vary slightly from model-to-model. Most smaller gearcases (generally found on 60 hp and smaller motors) contain a simple spring, plunger and slider mechanism. However there are some exceptions, which uses a shift mechanism more similar to the larger motors.

Conversely, motors larger than 60 hp (again, there are a few exceptions) utilize a slightly more complicated shift mechanism that includes a shift rod joint, joint slider, spring (sometimes with spring nuts and/or washers on either end) and anywhere from 2-6 check balls. At minimum 2 check balls are installed at the front end of the joint slider, though many models also 2 balls of the same size at the rear of the joint slider. Still other models use 2 larger check balls positioned in the propeller shaft behind the joint slider and on either side of the spring.

Work slowly and carefully, making sure no check balls, springs, spring nuts, etc are lost when disassembling the shift mechanism. For more details, refer to the exploded views found in the Cleaning and Inspection section.

1. If the shaft did not come out with the bearing carrier, pull the propeller shaft, clutch and shifter assembly straight back and free of the lower unit.

■ If you're working on a counter-rotation model, remember the actual gear positions are opposite of a normal rotation model. On counter-rotation models, the Forward gear is in the bearing carrier and the Reverse gear is in the gearcase nose cone.

7-48 LOWER UNIT

2. Insert an awl under the end loop of the cross pin ring and carefully pry the ring free of the clutch dog.

3. Use a long pointed punch to press out the cross pin.

4. Slide the clutch dog from the shaft.

■ Observe how the clutch dog was installed. On some models it must be installed in the same direction from which it was removed. On these models the clutch dogs are embossed with a marking (such as F to denote which edge faces front, toward the Forward gear). On other models you might notice that both shoulders of the dog are an equal width and the dog is not stamped. On those models the dog may be reinstalled either way, usually with the least worn side facing the forward gear. Remember this fact, as an aid during assembling.

5. Carefully disassemble the balance of the shifting components, by withdrawing them from the propeller shaft. On some models (those equipped with a separate shim cam that splines onto the bottom of the shift rod), this might only be a shift plunger, slider and spring, but on others (those with shift pins on the bottom of the shift rods) there will be other components, including a shifter (pin receiver), separate slider, 2-6 shifter check balls, spring and possible a few spring nuts and/or washers. For models equipped with 2 or more shifter check balls, slowly withdraw the shift slider, recovering the balls as they JUST pull free of the propeller shaft. Remember there may be as many as 4 mounted 2 on each end of the shifter, plus there may be 2 balls with the spring (one on either side of the spring).

■ As the work proceeds in the disassembly of the shift slide, the check balls used in various shifters are often of different sizes (but grouped in pairs of similar sizes). Take care to store these balls separately to avoid confusion during assembly.

Driveshaft and Bearings

◆ See Figures 169 thru 175

The driveshaft is used to transmit the clockwise (when viewed from the top) rotational motion of the crankshaft through the pinion gear to the propeller shaft (either through the Forward or Reverse gears). The driveshaft is splined at both ends, the top splines mating with the lower end of the crankshaft and the bottom end mating with the pinion gear.

Exact positioning of the pinion gear on the shaft (so that it meshes properly with the Forward and Reverse gears) is adjusted through the use of different thickness shim packs. There are essentially 2 configurations for shims on these models. For many of the larger gearcases which contain a needle bearing for the driveshaft in the driveshaft oil housing, there is normally a shim, followed by a thrust washer and thrust bearing all mounted toward the top of the driveshaft, just underneath the housing. They may have been removed earlier during Driveshaft Oil Seal Housing service.

Most other gearcases found on these models use a tapered roller bearing that is pressed directly on the driveshaft. For these models a bearing race and shim pack may be found in the gearcase, just above the driveshaft sleeve.

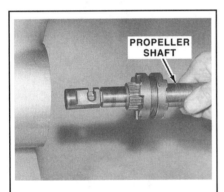

Fig. 162 If not already done, pull the prop shaft from the gearcase

Fig. 163 To remove the dog clutch, first remove the cross pin ring...

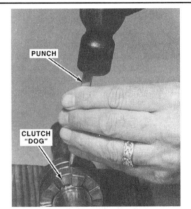

Fig. 164 ...then carefully push the pin from the dog and shift slide

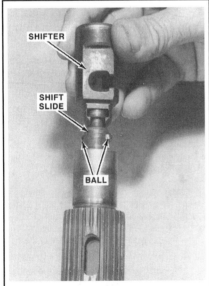

Fig. 165 Remove the shift plunger or shifter (as equipped)

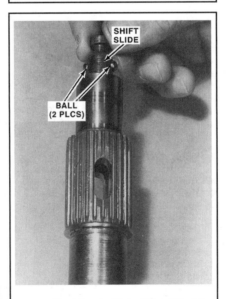

Fig. 166 If equipped, withdraw the shift slide just enough to remove the check balls...

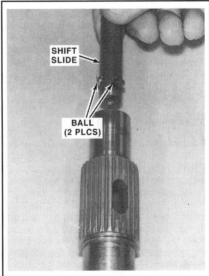

Fig. 167 ...on some models you'll find a second set of check balls at the back of the slide

LOWER UNIT 7-49

Regardless of the upper bearing configuration, all models use a needle bearing assembly at the lower end of the driveshaft. Many early models and some late-models used a bearing with loose needles, however the majority of Yamahas these days use only a one-piece needle bearing assembly.

■ Driveshaft removal is necessary for access gearcase mounted components like driveshaft bearings or bushings and/or the Forward (Reverse on counter-rotating units) gear/bearing.

Before the driveshaft can be freed from the gearcase, the pinion gear at the lower end of the shaft must first be removed. On all models a driveshaft holder tool (essentially a large nut with slides over the driveshaft's splines in order to provide a means of holding or turning the driveshaft) must be used. This tool varies slightly by model as follows:
- For all 40/50 motors use Yamaha #YB-06079.
- For 50/60/70 hp (849cc) motors use Yamaha #YB-06049 on US and Canadian models or #90890-06158 for World Production models.
- For 48 hp and 75/80/85/90 hp motors, use Yamaha #YB-06151.
- For all V4 and V6 models, use Yamaha #YB-06201.

■ In most cases, when working with tools, a nut is rotated to remove or install it to a particular bolt, shaft, etc. In the next step, the reverse is required because there is really is too little room to move a wrench inside the lower unit cavity. The nut on the lower end of the driveshaft is held steady and the shaft is rotated until the nut is free.

1. Place a large, long handled wrench or breaker bar with socket (usually 22mm on these motors) over the pinion nut. Slide the driveshaft holder tool over the splines on the top of the driveshaft and place a large wrench or socket over the tool. Now, hold the pinion nut steady with the tool, rotate the driveshaft counterclockwise to break the nut free.

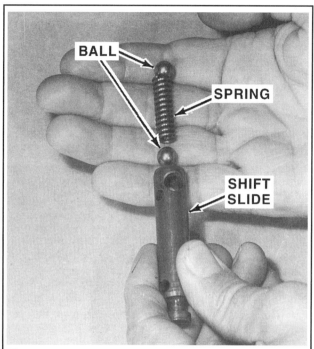

Fig. 168 Remove the shift spring (with large check balls on some models)

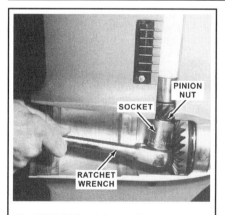

Fig. 169 Hold the pinion nut from turning. . .

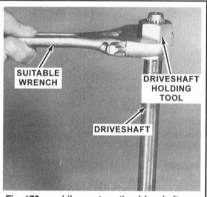

Fig. 170 . . .while you turn the driveshaft using the holder tool

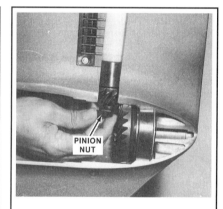

Fig. 171 Remove the pinion nut and. . .

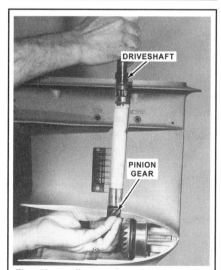

Fig. 172 . . .pull up gently on the driveshaft to free the pinion gear

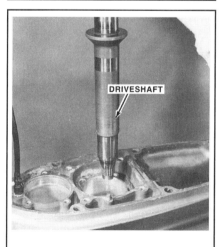

Fig. 173 Carefully remove the driveshaft from the housing

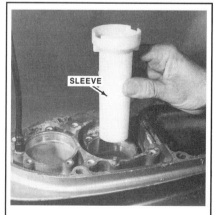

Fig. 174 Once any shims and thrust bearings or races are removed, withdraw the driveshaft sleeve. . .

7-50 LOWER UNIT

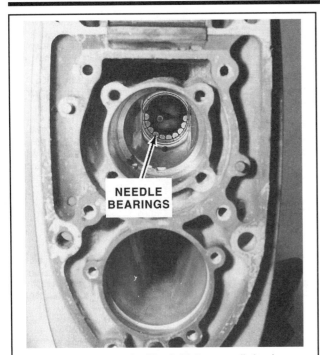

Fig. 175 ...for access to the driveshaft's lower needle bearing

2. Remove the pinion nut.
3. Gently pull up on the driveshaft and at the same time rotate the driveshaft. The pinion gear will come free from the lower end of the driveshaft.
4. Pull the driveshaft up out of the lower unit housing.
5. If the oil seal housing and the components beneath it were not removed the housing at this point will still be on the driveshaft and can now be easily removed. Go back and perform the steps which were omitted.
6. On models with a tapered roller bearing pressed onto the driveshaft, use a suitable internal expanding jawed puller attachment on either a bridged puller or a slide-hammer to remove the bearing race from the gearcase. Keep the shim pack found under the race for installation purposes.
7. Lift out the driveshaft sleeve for access to the lower needle bearing.

■ For all 2 and 3-cylinder motors, you'll need to install the replacement needle bearing at the same depth below the top of the gearcase as the current bearing. Because specifications may vary slightly on gearcases or in case you don't have all the necessary special bearing depth tools for Yamaha, take a measurement of the installed height of the needle bearing set (from top of the gearcase to the top of the bearing) before removal for reference during installation.

8. The needle bearing set (with loose needle on some models) is pressed into the driveshaft housing. Use a suitable driver and long driver handle carefully push the needle bearing assembly down and into the gearcase propeller shaft cavity.

Nose Gear/Bearing

All gearcases use one gear mounted to the propeller shaft bearing carrier assembly and one gear which turns the opposite direction mounted just forward of the driveshaft in the gearcase nose cone.
On standard rotation models the Reverse gear is found at the propeller shaft bearing carrier, while the Forward gear is mounted in the nose cone.
On counter-rotation models the Forward gear is found in the propeller shaft bearing carrier, while the Reverse gear is mounted in the nose cone.

■ Gear and bearing designs used in the nose cone vary slightly between standard and counter-rotation models

Forward Gear (Standard Rotation Units)

◆ See Figures 176 thru 179

On most standard rotation units a tapered roller bearing is pressed onto the back of the Forward gear (found in the nose cone). The assembly is inserted into a race that is pressed into the gearcase (on top of the shim material necessary for proper Forward gear placement). If the tapered roller bearing is pressed off the gear, it should be replaced. And, anytime the bearing is replaced, the race must also be replaced.

■ Most of the larger standard-rotation gearcases ALSO use one or two needle bearings (or a bushing) pressed inside the Forward gear. Like most needle bearings, they must be replaced if removed.

1. Lift out the Forward gear and tapered roller bearing assembly from the gearcase mounted race. Inspect the gear and bearing assembly to determine if further disassembly is necessary for component replacement.
2. If necessary, use an expanding jaw attachment with a suitable slide hammer or puller to remove the bearing race from the lower unit housing. Be sure to hold the tool at right angles when pulling the race.
3. Watch for and save any shim material found behind the forward gear bearing race. The shim material is critical to obtaining the correct backlash during installation. Using the old shim material will save considerable time, especially starting with no shim material.

■ Remove the bearing only if it is unfit for further service.

4. If necessary to replace the bearing, position a bearing separator between the Forward gear and the tapered roller bearing. Using a hydraulic press, separate the gear from the bearing.
5. On models equipped with one or more needle bearings pressed into the gear, if replacement is necessary, use an expanding jaw attachment on a puller or slide hammer to remove the bearing(s) from the gear.
6. On models equipped with a bushing pressed into the gear, if replacement is necessary, use a suitable driver to push it out from the rear of the gear.

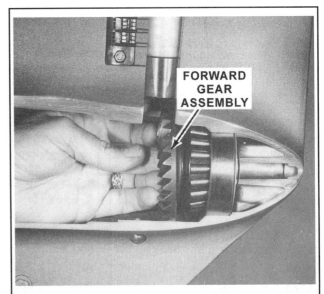

Fig. 176 Remove the forward gear/bearing assembly from the nose cone

LOWER UNIT 7-51

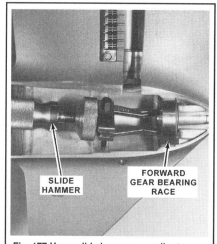

Fig. 177 Use a slide hammer or puller to remove the forward gear bearing race. . .

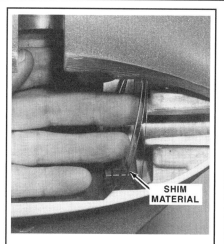

Fig. 178 . . .then remove and retain the shim material

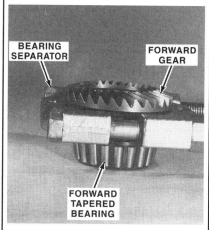

Fig. 179 Use a bearing separator to free the gear and bearing

Reverse Gear (Counter-Rotation Units)

◆ See Figures 180 and 181

■ If you're working on a counter-rotation model, remember the actual gear positions are opposite of a normal rotation model. On counter-rotation models, the Forward gear is in the bearing carrier and the Reverse gear is in the gearcase nose cone.

On most counter-rotation units instead of the tapered roller bearing (found on other models) the Reverse gear in the nose cone uses a needle bearing and retainer or roller bearing that is pressed into the gearcase itself. Reverse gear shim material is found mounted underneath the bearing and retainer assembly (similar to the material found under the tapered bearing race on standard rotation models).

In addition, the Reverse gear is normally equipped with a thrust bearing with a large washer and/or spacer on the shaft of the gear (between the gear and bearing assembly). On models that use a roller bearing pressed into the gearcase (as opposed to a needle bearing and retainer) a roller bearing inner race is pressed onto the back of the Reverse.

These gearcases ALSO use one or two needle bearings pressed inside the Forward gear. Like most needle bearings, they must be replaced if removed.

1. lift out the reverse gear, then lift out the thrust bearing. Take care not to confuse this thrust bearing with the one previously removed. If these thrust bearing are to be reused, they must be installed in their original locations, because each develops a unique wear pattern. If a bearing is installed in a different location, the wear on the individual tiny roller bearings would be greatly accelerated and lead to premature failure of the bearing.
2. Lift out the large thrust washer and/or spacer, as equipped.
3. Obtain a slide hammer (or puller) with expanding jaw attachment to pull the needle bearing and retainer or roller bearing (as applicable) from the lower unit housing. Be sure to hold the tool right angles to the driveshaft while working.
4. Watch for and save any shim material found behind the reverse gear bearing. The shim material is critical to obtaining the correct backlash during installation. Using the old shim material will save considerable time, especially starting with no shim material.
5. If necessary due to bearing replacement on roller bearing models, position a bearing separator between the Reverse gear and the inner bearing race (on the back of the gear). Using a hydraulic press, carefully separate the gear from the race.
6. If necessary due to bearing replacement on needle bearing models, use a suitable driver and a hydraulic press (or hammer) to push (or tap) the needle bearing out of the retainer.
7. On models equipped with one or more needle bearings pressed into the gear, if replacement is necessary, use an expanding jaw attachment on a puller or slide hammer to remove the bearing(s) from the gear.

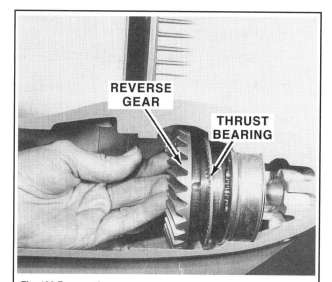

Fig. 180 Remove the reverse gear and thrust bearing from the nose cone. . .

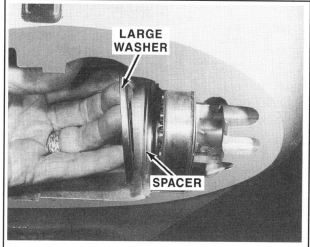

Fig. 181 . . .then remove the thrust washer and spacer

7-52 LOWER UNIT

CLEANING & INSPECTING

Gearcase Exploded Views

◆ See Figures 182, 183 and 184

Because of the shear number of gearcases available on these models it is impossible to include exploded view of each and every minor variation. The vast majority of gearcases used on these models follow the same basic design with small variances in the type or placement of shims, washers and bearings. Small differences also may be found on some shifter mechanisms. We've included a couple of the most representative exploded views for these gearcases. But, during disassembly, keep comparing what you're pulling out of the gearcase with what is pictured here and note any differences for assembly purposes.

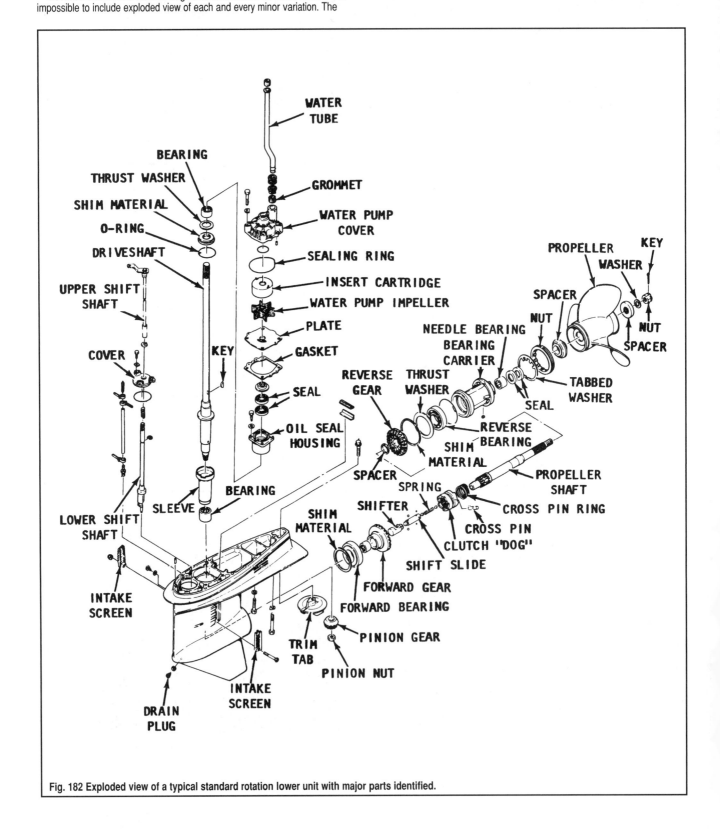

Fig. 182 Exploded view of a typical standard rotation lower unit with major parts identified.

LOWER UNIT 7-53

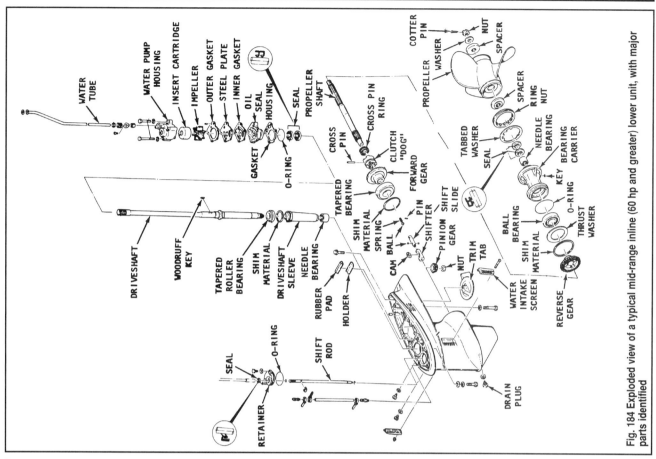

Fig. 184 Exploded view of a typical mid-range inline (60 hp and greater) lower unit, with major parts identified

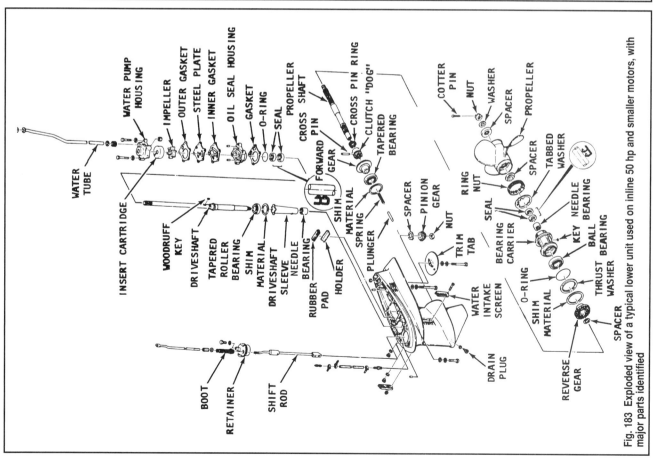

Fig. 183 Exploded view of a typical lower unit used on inline 50 hp and smaller motors, with major parts identified

7-54 LOWER UNIT

Checking Gearcase Components

◆ See Figures 185 thru 193

■ If you're working on a counter-rotation model, remember the actual gear positions are opposite of a normal rotation model. On counter-rotation models, the Forward gear is in the bearing carrier and the Reverse gear is in the gearcase nose cone.

Good shop practice requires installation of new O-rings and oil seals regardless of their appearance.

Clean all water pump parts with solvent and then dry them with compressed air. Inspect the water pump housing and oil seal housing for cracks and distortion, possibly caused from overheating. Inspect the plate and water pump cartridge for grooves and/or rough surfaces. If possible, always install a new water pump impeller while the lower unit is disassembled. A new impeller will ensure extended satisfactory service and give peace of mind to the owner. If the old impeller must be returned to service, never install it in reverse to the original direction of rotation. Installation in reverse will cause premature impeller failure.

If installation of a new impeller is not possible, check the seal surfaces. All must be in good condition to ensure proper pump operation. Check the upper, lower and ends of the impeller vanes for grooves, cracking and wear. Check to be sure the indexing notch of the impeller hub is intact and will not allow the impeller to slip.

Clean around the Woodruff key. Clean all bearings with solvent, dry them with compressed air and inspect them carefully. Be sure there is no water in the air line. Hold the bearing to keep it from turning and direct the air stream through the bearing. After the bearings are clean and dry, lubricate them with oil. Do not lubricate tapered bearing cups until after they have been inspected.

✱✱ WARNING

Never spin a bearing with compressed air. Such action is highly dangerous and may cause the bearing to score from lack of lubrication.

Inspect all ball bearings for roughness, scratches and bearing race side wear. Hold the outer race and work the inner bearing race in-and-out, to check for side wear.

Determine the condition of tapered bearing rollers and inner bearing race, by inspecting the bearing cup for pitting, scoring, grooves, uneven wear, imbedded particles and discoloration caused from overheating. Always replace tapered roller bearings and their race as a set.

Clean the forward gear with solvent and then dry it with compressed air. Inspect the gear teeth for wear. Under normal conditions the gear will show signs of wear but it will be smooth and even.

Clean the bearing carrier with solvent and then dry it with compressed air. Check the gear teeth of the reverse gear for wear. The wear should be smooth and even.

Check the clutch dogs to be sure they are not rounded-off, or chipped. Such damage is usually the result of poor operator habits and is caused by shifting too slowly or shifting while the engine is operating at high rpm. Such damage might also be caused by improper shift rod adjustments.

Rotate the reverse gear and check for catches and roughness. Check the bearing for side wear of the bearing races.

Inspect the bearing surfaces of the propeller shaft. Check the shaft surface for pitting, scoring, grooving, embedded particles, uneven wear and discoloration caused from overheating.

Clean the driveshaft with solvent and then dry it with compressed air. Inspect the driveshaft splines for excessive wear. Check the oil seal surfaces above and below the water pump drive pin or Woodruff key area for grooves. Replace the shaft if grooves are discovered.

Inspect the driveshaft bearing surface above the pinion gear splines for pitting, grooves, scoring, uneven wear, embedded metal particles and discoloration caused by overheating.

Inspect the propeller shaft oil seal surface to be sure it is not pitted, grooved, or scratched. Inspect the bearing contact surface on the propeller shaft for pitting, grooves, scoring, uneven wear, embedded metal particles and discoloration caused from overheating.

Inspect the propeller shaft splines for wear and corrosion damage. Check the propeller shaft for and driveshaft for run-out.

Place each shaft on V blocks and measure the run-out with a dial indicator gauge at a point midway between the V blocks. Typically the maximum acceptable run-out for a propeller shaft is around 0.0008 in. (0.02mm), while the maximum acceptable run-out for the driveshaft is about 0.02 in. (0.5mm). However, keep in mind that Yamaha does not publish specs for most of these gearcases, so you may have to use a judgment call.

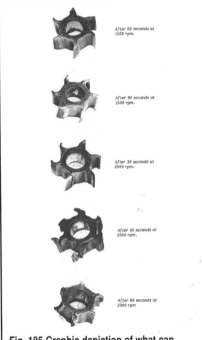

Fig. 185 Graphic depiction of what can happen to the water pump impeller if water is not circulated through the lower unit to the engine when the engine is running

Fig. 186 Grooves on the cross pin or marked wear patterns on the outer roller bearing races are evidence of premature failure of these or associated parts

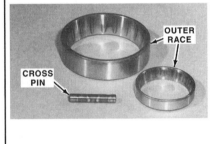

Fig. 187 A new needle bearing (left) along side a used needle bearing after the bearing was removed

LOWER UNIT 7-55

As equipped, inspect the following parts for wear, corrosion, or other signs of damage:
- Shift shaft assembly
- Shift cam
- All bearing bores for loose fitting bearings.
- Gear housing for impact damage.
- Threads for cross-threading and corrosion damage.
- Check the pinion nut corners for wear or damage. This nut is a normally a special locknut. Therefore, do not attempt to replace it with a standard nut. Obtain the correct nut from an authorized Yamaha dealer.

If the lower unit case is to be repainted. Mask off the threads engaging the cover nut. If this nut is installed against painted threads, a false torque value will be obtained upon tightening the nut and it is possible the nut could back off with continued use.

ASSEMBLY

◆ See Figures 182 thru 184

The following procedures are intended to guide your step-by-step through reinstallation and setup of the major gearcase components. If one or more of the components was not removed you can skip that section and proceed to the next following component.

Gearcase shimming procedures vary slightly from model-to-model and they depend upon the components that are being replaced. However, generally speaking:
- NO SHIMMING is required if the original case and original inner parts (gears, bearings and shafts) are installed.
- NUMERIC CALCULATION RESHIMMING should be performed on all gearcases when the original inner parts (gears, bearings and shafts) are installed in a NEW case. Most original and replacement cases are marked with various gear measurements. Typically an embossed F (forward gear), R (reverse gear) and/or P (pinion) gear measurements. When these marks are present the differences between the marks on the new case vs. the marks on the original case can be used to determine new shim sizes BEFORE assembly (potentially saving you from having to disassemble/reassemble the gearcase a second time).
- SPECIAL TOOL MEASUREMENT RESHIMMING is performed prior to assembly on all models and should be used when a specific shaft, bearing/bushing or gear is replaced. In this case, special tools are used to measure pre-assembled or partially assembled components and determine if a change in shim sizes is necessary. If the special shimming tools are not available, you can often reuse shim sizes from the previous assembly to at least get you in the ball-park and then use the Backlash Measurement Reshimming procedures to dial it in.
- BACKLASH MEASUREMENT RESHIMMING should be performed and adjusted when one or more of the original inner components (gears, bearings and shafts) are installed. Backlash measurement is normally performed on an almost fully assembled gearcase using a dial gauge to read the amount of play in the driveshaft. This procedure can check the placement of all 3 gears through this measurement. Though adjustment is much more hit and miss than the other methods, since you need to then re-disassemble the gearcase, swap shims, re-assemble and re-check, then repeat until you get it right. Fortunately, formulas are available to use the measured backlash in helping to pick shim sizes and small shim changes are all that is normally necessary. It is rare that you would have to disassemble the gearcase more than one additional time.

■ If the case, gear(s), bearing(s) and/or shaft(s) have been replaced, refer to the Gearcase Shimming procedures later in this section BEFORE starting reassembly to determine if there are any changes to shims that can be made before your start assembly so that you can potentially avoid having to disassemble the gearcase a second time to change shims after checking backlash.

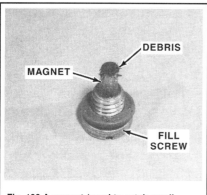

Fig. 188 A magnet (used to catch small metallic parts in the lubricant) is an integral part of the fill screw

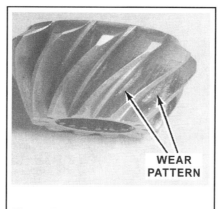

Fig. 189 Excessive pinion wear, probably caused by lack of lubrication

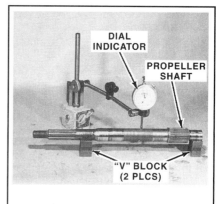

Fig. 190 Using a dial indicator to measure the prop shaft run-out

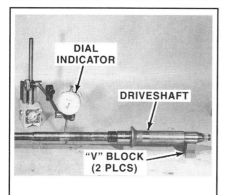

Fig. 191 Using a dial indicator and V-blocks to measure driveshaft run-out

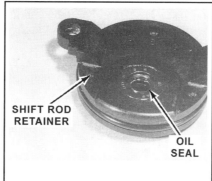

Fig. 192 Oil seals should not be removed unless they are unfit for further service

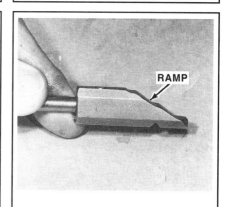

Fig. 193 This shifter ramp should be carefully inspected, as grooves, chips or other damage will affect shift operation

7-56 LOWER UNIT

Throughout installation you will need to make a couple of choices. If you have simply replaced O-rings and seals (not bearings, shafts or gears) then there is probably little chance that the gear lash will require adjustment. If so you can lubricate all components and apply sealant or threadlocking compounds as called for in the procedures. However, if you have replaced components such as the bearings, shafts or gears you may wish to check gear contact patterns using a Machine Dye or suitable powder, and all components should then be temporarily installed dry. Also, when components such as these are replaced you'll need to check and possibly adjust gear lash, though reusing shims which were removed during disassembly can often give you a good starting point and MAY prevent you from having to partially disassemble the gearcase again (but only if you've chosen to lubricate components during installation). As we said, the choice is yours and can only be made based upon the circumstances of your overhaul or reseal and the condition of the unit before tear-down.

Unless otherwise noted, upon final installation be sure to lubricate all bearings with gear oil, shafts with oil or a marine grade grease and seals with oil, packing their lips with a suitable marine grade grease.

Assembly generally means preparing the gearcase (by installing any bearings or races which were pressed or pulled out of the case itself) and the various sub-assemblies such as the propeller shaft, bearing carrier, oil seal housing etc. Then assembly means carefully bringing these components back together in the proper order inside the gearcase. The order in which the sub-assemblies are readied for installation doesn't matter, as long as they are not needed in the gearcase at a certain point. Typically speaking, first you would install bearings or races into the gearcase itself, then prepare all of the sub-assemblies. The nose cone gear must be installed into the gearcase BEFORE the driveshaft (since the pinion gear which holds the driveshaft in position will also hold the nose cone gear in position. On dual-propeller models this nose-cone gear is actually mounted to the inner propeller shaft, so the inner propeller shaft must be installed BEFORE the driveshaft. Also, the water pump assembly, generally the last component installed, is left OFF the gearcase until after backlash has been check and adjusted.

Nose Gear/Bearing

All gearcases use one gear mounted to the propeller shaft bearing carrier assembly and one gear which turns the opposite direction mounted just forward of the driveshaft in the gearcase nose cone.

■ **Gear and bearing designs used in the nose cone vary slightly between standard and counter-rotation models**

On standard rotation models the Reverse gear is found at the propeller shaft bearing carrier, while the Forward gear is mounted in the nose cone.
On counter-rotation models the Forward gear is found in the propeller shaft bearing carrier, while the Reverse gear is mounted in the nose cone.

■ **On some smaller gearcases covered in this section (though including some powerheads all the way up to some 90 hp motors) the shifter cam is a small, tab-like component that splines onto the end of the shift shaft and sits otherwise freely in the nose cone. On some of these models installation of the shift shaft to the cam can be very difficult once the nose cone gear is retained by the driveshaft and pinion nut. On these models, you may wish to assemble and install the shift-shaft at this time or sometime before the driveshaft is installed.**

Reverse Gear (Counter-Rotation Units)

◆ See Figures 194 and 195

■ **If you're working on a counter-rotation model, remember the actual gear positions are opposite of a normal rotation model. On counter-rotation models, the Forward gear is in the bearing carrier and the Reverse gear is in the gearcase nose cone.**

On most counter-rotation units instead of the tapered roller bearing (found on other models) the Reverse gear located in the nose cone uses a needle bearing and retainer or roller bearing that is pressed into the gearcase itself. Reverse gear shim material is found mounted underneath the bearing and retainer assembly (similar to the material found under the tapered bearing race on standard rotation models).

In addition, the Reverse gear is normally equipped with a thrust bearing with a large washer and/or spacer on the shaft of the gear (between the gear and bearing assembly). On models that use a roller bearing pressed into the gearcase (as opposed to a needle bearing and retainer) a roller bearing inner race is pressed onto the back of the Reverse.

These gearcases ALSO use one or two needle bearings pressed inside the Forward gear. Like most needle bearings, they must be replaced if removed.

1. On models equipped with one or more needle bearings pressed into the gear, install the new bearing(s) using a suitable driver. The bearings are normally installed from the rear of the gear (not the toothed side) and normally with any embossed markings on the bearing facing back towards the driver. Install the bearings to the specified depth in the gear depending upon the model, as follows:

• On all V4 motors, install the bearing to a depth of 0.098-0.138 in. (2.5-3.5mm) below the rear edge of the gear.
• On 2.6L V6 motors, install the first bearing to a depth of 0.827-0.843 in. (21.0-21.4mm) and the second bearing 0.177-0.193 in. (4.5-4.9mm) below the rear edge of the gear.
• On 3.1L V6 motors, install the first bearing to a depth of 0.82-0.83 in. (20.7-21.2mm) and the second bearing 0.17-0.18 in. (4.3-4.7mm) below the rear edge of the gear.

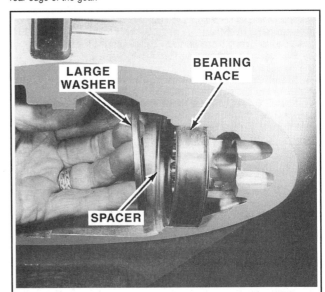

Fig. 194 If you're ready to install the reverse gear position the large washer or spacer. . .

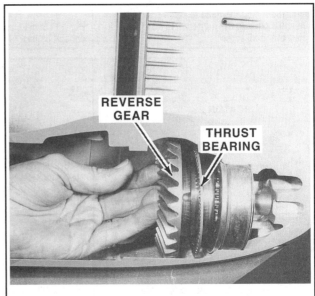

Fig. 195 . . .followed by the thrust bearing and the gear assembly

LOWER UNIT 7-57

- On 3.3L V6 motors, install the first bearing to a depth of 0.817-0.836 in. (20.75-21.25mm) and the second bearing 0.167-0.187 in. (4.25-4.75mm) below the rear edge of the gear.

2. On roller bearing models (normally only 2.6L V6 engines), if the bearing was replaced, use a hydraulic press to carefully push the replacement bearing inner race onto the back of the gear. Be sure to position the bearing with any markings embossed on the race facing back towards the driver.

3. On needle bearing models (normally all except 2.6 V6 engines), if the gearcase retainer mounted bearing is being replaced, use a hydraulic press to install the bearing into the retainer. Position the bearing so any embossed markings are facing away from the retainer (toward the gear). Install the bearing to the specified depth in the retainer depending upon the model, as follows:

- On all V4 motors, install the bearing to a depth of 0.030-0.049 in. (0.75-1.25mm) below the edge of the retainer.
- On 3.1L V6 motors, seat the bearing to the housing. No depth specification is given, but it should normally be flush or slightly below the retainer's surface.
- On 3.3L V6 motors, install the bearing so that it is flush, give or take 0.010 in. (0.25mm) meaning a depth of between -0.010 and +0.010 in. (-0.25 and +0.25mm) from the edge of the retainer.

4. Position the roller bearing or needle bearing and retainer to the gearcase, along with the shims that were removed during disassembly OR with the proper calculated shims if the gearcase has been replaced (for details please refer to Gearcase Shimming, later in this section).

■ Roller bearings should be positioned into the nose cone with any embossed markings facing back toward the gear.

5. Use a suitable long handled driver to gently tap the bearing or bearing and retaining assembly into position until they are seated in the gearcase.

■ If the driveshaft is ready for installation (the bearings were not removed or are already replaced), proceed with the next step in anticipation of driveshaft installation. However, if you still need to install the lower driveshaft bearing, you'll probably want to hold off on positioning the gear and wait until you ARE ready to install the driveshaft.

6. Position the reverse gear in the nose cone bearing/race along with the thrust washer and plane washer/spacer (the order of insertion to the case it washer/spacer, thrust washer and finally gear).

■ If you wish to visually check gear contact patterns, obtain a suitable substance which can be used to indicate a wear pattern on the forward and pinion gears as they mesh. Machine dye may be used and if this material is not available, Desenex® Foot Powder (obtainable at the local Drug Store/Pharmacy), or equivalent may be substituted. Desenex® is a white powder available in an aerosol container. Before assembling the gears, apply a light film of the dye, Desenex®, or equivalent, to the driven side of the each. After the gears are assembled and rotated several times, they will be disassembled and the wear pattern can be examined. The substance will be removed from the gears prior to final assembly.

Forward Gear (Standard Rotation Units)

◆ See Figures 196, 197 and 198

On most standard rotation units a tapered roller bearing is pressed onto the back of the Forward gear (found in the nose cone). The assembly is inserted into a race that is pressed into the gearcase (on top of the shim material necessary for proper Forward gear placement). If the tapered roller bearing is pressed off the gear, it should be replaced. And, anytime the bearing is replaced, the race must also be replaced.

■ Most of the larger standard-rotation gearcases ALSO use one or two needle bearings (or possibly even a bushing) pressed inside the Forward gear. Like most needle bearings, they must be replaced if removed.

1. On models equipped with one or more needle bearings (or a bushing) pressed into the gear, install the new bearing(s) or bushing using a suitable driver. The bearings are normally installed from the rear of the gear (not the toothed side) and normally with any embossed markings on the bearing facing back towards the driver. Install the bearing(s) to the specified depth in the gear depending upon the model, as follows:

- On all V4 motors, install the bearing(s) so the bearing (or top/outer bearing if there are two) is at a depth of 0.098-0.138 in. (2.5-3.5mm) below the rear edge of the gear.
- On 2.6L V6 motors, install the first bearing to a depth of 0.827-0.843 in. (21.0-21.4mm) and the second bearing 0.177-0.193 in. (4.5-4.9mm) below the rear edge of the gear.
- On 3.1L V6 motors install the bushing to a depth of 0.09-0.10 in. (2.3-2.7mm) below the rear edge of the gear on all except 250 hp Vmax models, on which the bushing should be installed 0.44-0.46 in. (11.25-11.75mm) below the rear edge of the gear.

2. If the tapered roller bearing was removed from the outside of the Forward gear use a hydraulic press to carefully push the replacement bearing into position on the back of the gear.

3. Position the forward gear bearing race to the gearcase, along with the shims that were removed during disassembly OR the proper calculated shims if the gearcase has been replaced (for details please refer to Gearcase Shimming, later in this section).

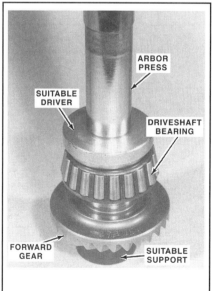

Fig. 196 Use a press to install the new bearing onto the forward gear

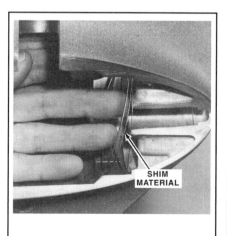

Fig. 197 Position the shim material...

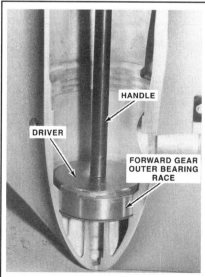

Fig. 198 ...then use a driver to seat the bearing race

7-58 LOWER UNIT

4. Use a suitable long handled driver to gently tap the bearing race into position until it is seated in the gearcase.

■ If the driveshaft is ready for installation (the bearings were not removed or are already replaced), proceed with the next step in anticipation of driveshaft installation. However, if you still need to install the lower driveshaft bearing, you'll probably want to hold off on positioning the gear and wait until you ARE ready to install the driveshaft.

5. Position the forward gear and bearing assembly into the bearing race in the nose cone.

■ If you wish to visually check gear contact patterns, obtain a suitable substance which can be used to indicate a wear pattern on the forward and pinion gears as they mesh. Machine dye may be used and if this material is not available, Desenex® Foot Powder (obtainable at the local Drug Store/Pharmacy), or equivalent may be substituted. Desenex® is a white powder available in an aerosol container. Before assembling the gears, apply a light film of the dye, Desenex®, or equivalent, to the driven side of the each. After the gears are assembled and rotated several times, they will be disassembled and the wear pattern can be examined. The substance will be removed from the gears prior to final assembly.

Driveshaft Bearings

◆ See Figures 199, 200 and 201

As noted earlier during disassembly, the driveshaft is used to transmit the clockwise (when viewed from the top) rotational motion of the crankshaft through the pinion gear to the propeller shaft (either through the Forward or Reverse gears). The driveshaft is splined at both ends, the top splines mating with the lower end of the crankshaft and the bottom end mating with the pinion gear.

Exact positioning of the pinion gear on the shaft (so that it meshes properly with the Forward and Reverse gears) is adjusted through the use of different thickness shim packs. There are essentially 2 configurations for shims on these models. For many of the larger gearcases which contain a needle bearing for the driveshaft in the driveshaft oil housing, there is normally a shim, followed by a thrust washer and thrust bearing all mounted toward the top of the driveshaft, just underneath the housing. They may have been removed earlier during Driveshaft Oil Seal Housing service.

Most other gearcases found on these models use a tapered roller bearing that is pressed directly on the driveshaft. For these models a bearing race and shim pack may be found in the gearcase, just above the driveshaft sleeve.

Regardless of the upper bearing configuration, all models use a needle bearing assembly at the lower end of the driveshaft. Many early models and some late-models used a bearing with loose needles, however the majority of Yamahas these days use only a one-piece needle bearing assembly (however there are still a couple of large gearcases that use loose needle bearings).

Additionally the lower needle bearing is installed in one of two ways, depending upon the model. For all smaller gearcases (used on 2 and 3-cylinder motors) the bearing is driven in from the top of the gearcase to a specified depth. However, for larger gearcases (used on all V4/V6 models) the bearing must be pulled up into position. either used a special tool or a long threaded bolt with appropriately sized washers and nuts.

On the larger gearcases, although Yamaha often has a special tool available to help draw the bearing into position in the lower portion of the gearcase, all you REALLY need to accomplish this is a sufficiently long threaded bolt with a nut and a washer that is the size of the driver you'd want to contact that bushing or bearing. You'll also need a nut and a plate to place on top of the gearcase. Place the threaded rod through one nut and plate and then down into the gearcase through the bushing/bearing (so the plate is resting against the top of the gearcase, the upper nut is resting against the plate and the threads of the rod are just protruding down into the pinion mounting area of the gearcase.). Next install the washer and lower nut over the bottom threads of the rod to support the bearings or busing. Slowly draw the bearing or bushing into position in the gearcase by holding the threaded rod from turning while at the same time turning the upper nut to draw the entire bolt, bushing/bearing, washer and nut assembly upward until the bushing or bearing seats. Then loosen the nut on the bottom of the threaded rod in the gearcase and remove the threaded rod, plate, nut and washer.

1. For models whose gearcase contains loose needle bearings, apply a light coating of marine grade grease to the bearing housing. The loose needles are normally installed after the housing is pulled into position, HOWEVER, if you have something which can be used as a bushing (a piece of plastic pipe or something that is just a little smaller in outer diameter than the driveshaft, but still big enough to allow the bearing installation rod to go through the center) the needles can be positioned now.

2. For large gearcases (as noted earlier) where the needle bearing (or bearing housing) is installed from below, assemble an installation tool and use the tool to carefully draw the bearing up into the gearcase until it is seated. On some models there may be a part number embossed on one end of the bearing, make sure the embossed surface is positioned facing downward, toward the gear. Once the bearing (or bearing housing) is pulled into position, loosen the retainer on the bottom and remove the tool. If not done earlier on loose needle bearing models, apply a light coating of grease to the inside of the bearing carrier and install the needles. Keep an eye on the needles as the driveshaft is installed later to make sure none are dislodged.

■ On smaller gearcases, always compare the specification to the depth measured before disassembly to make sure there are no gearcase design variances from published spec.

Fig. 199 Install the needle bearing assembly into the lower gearcase

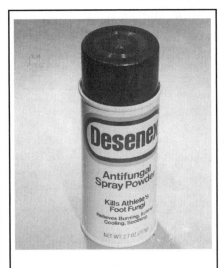

Fig. 200 If you want to check gear mesh, apply a light coating of powder to the gears.

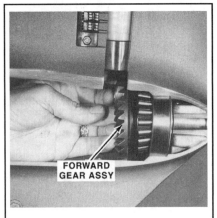

Fig. 201 . . .then position the nose cone gear so you're ready to install the driveshaft

LOWER UNIT 7-59

3. For smaller gearcases (as noted earlier) use a suitable driver and long driver handle to carefully tap the bearing in from the top. Since the bearing must be installed to a specific depth, you should probably insert the driver and handle into the bore, then measure and mark on the driver a spot JUST above the spec, so you'll know to measure and double-check the installed bearing height right before you finish tapping the bearing into position. Install the bearing to the suitable depth as follows:

- On 48 hp 2-cylinder motors, Yamaha does NOT publish a bearing depth spec for US or Canadian models, so use the spec measured during bearing removal. That SHOULD place the bottom of the bearing ABOUT flush with the lower end of driveshaft bore (where it begins to taper outward). On world production models Yamaha does publish a bearing depth spec, HOWEVER, it is not clear whether or not it includes the thickness of the depth plate. For this reason the depth of 7.89-7.93 in. (200.5-201.5mm) should be compared with the measurement taken during removal.
- On 40/50 hp (698cc) 3-cylinder motors, position the bearing with the embossed numbers facing upward, toward the driver, then carefully tap it squarely into the housing to a depth of 7.19-7.20 in. (182.5-183.0mm) below the gearcase deck.
- On 50-90 hp (849cc and 1140cc) motors, the bearing is normally positioned with the embossed numbers facing upward and set to a depth (from the top of the deck) of 7.81-7.83 in. (198.3-198.8mm) for 50 hp models or to 7.39-7.43 in. (187.6-188.6mm) for 60-90 hp models.

4. Slide the driveshaft sleeve into position from the upper end of the lower unit. Some sleeves contain a notch or tab at the front which should be faced toward the front of the gearcase.

5. On models with a tapered roller bearing pressed onto the driveshaft, position the shim pack (either which was removed during disassembly or determined by Numerical Calculation or Special Tool Measurement shimming, as detailed under Gearcase Shimming) followed by the bearing race into the top of the gearcase. Using a suitable driver, carefully seat the race.

6. If the tapered bearing was removed from the driveshaft, on models so equipped, use a suitable driver to press the replacement bearing onto the shaft. A "suitable" driver in this case will have an outer-diameter JUST larger than the driveshaft so that it ONLY contacts the inner bearing race (and does NOT contact the rollers or outer race). The driver must also obviously fit over the driveshaft.

7. As mentioned earlier, if you want to visually check gear mesh, obtain a suitable substance which can be used to indicate a wear pattern on the forward and pinion gears as they mesh. Machine dye may be used and if this material is not available, Desenex® Foot Powder (obtainable at the local Drug Store/Pharmacy), or equivalent may be substituted. Desenex® is a white powder available in an aerosol container. Before assembling either gear, apply a light film of the dye, Desenex®, or equivalent, to the driven side of the gear. After the gears are assembled and rotated several times, they will be disassembled and the wear pattern can be examined. The substance will be removed from the gears prior to final assembly.

■ **If this is final assembly, already with the needed shims, be sure to oil all gears and bearings.**

8. Position the forward gear (reverse gear on counter-rotating models or forward gear along with the inner prop shaft on dual-propeller models) and tapered bearing assembly into the gearcase. On dual-propeller models you may have to skip ahead and reassemble the clutch dog and shifter mechanism (if this inner propeller shaft was disassembled, for more details, refer to the Exploded Views found earlier in this section, under Cleaning and Inspection).

■ **On counter-rotating lower units, remember to insert the spacer or washer and thrust bearing to rest up against the reverse gear bearing race.**

You're now ready to install the driveshaft and pinion gear.

Pinion Gear and Driveshaft

◆ See Figures 202 thru 208

1. Lower the driveshaft down through the sleeve in the upper end of the lower unit. On models with loose needle bearings it is usually a good idea to apply a fresh coat of grease to the accessible portion of the needles to help hold them in place as you slide the driveshaft down into position. Also on models with loose needles, work slowly making sure no needle becomes dislodged and/or damaged.

■ **If you're planning on visually checking gear mesh, coat the pinion gear with a fine spray of Desenex®. Handle the gear carefully to prevent disturbing the powder.**

2. Insert the pinion gear into the gearcase and raise the pinion gear to allow the driveshaft to pass fully through the gear. It may be necessary to rotate the driveshaft slightly to allow the splines on the driveshaft to index with the internal splines of the pinion gear. The teeth of the pinion gear will index with the teeth of the forward gear (reverse gear on counter-rotating units).

3. Once the pinion gear is in place, start the threads of the pinion gear nut. Tighten the nut as much as possible by rotating the driveshaft with one hand and holding the nut with the other hand.

4. Tighten the pinion gear nut to specification either using a wrench or pinion nut holder to keep the nut steady with a torque wrench on the driveshaft holder OR by turning the driveshaft with a wrench and holding the pinion nut with the torque wrench and socket. However, the later option is not possible on dual-propeller models, as the inner propeller shaft will normally prevent anything thicker than a thin wrench from accessing the pinion nut. Tighten the nuts to spec as follows:

- On all 40-50 hp 2 motors, tighten the pinion nut to 54 ft. lbs. (75 Nm).

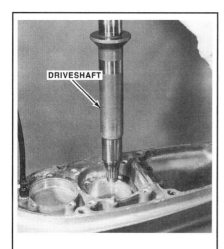

Fig. 202 Carefully lower the driveshaft into the gearcase...

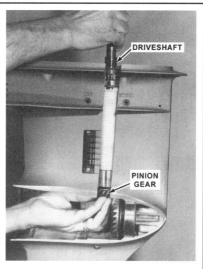

Fig. 203 ...turning slightly to mesh with the pinion gear splines

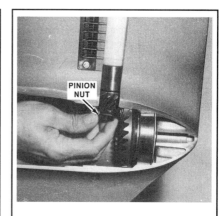

Fig. 204 Thread the pinion gear nut on by hand, then tighten...

7-60 LOWER UNIT

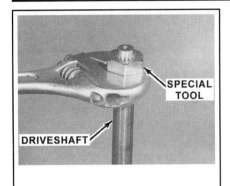

Fig. 205 . . . using the driveshaft holder. . .

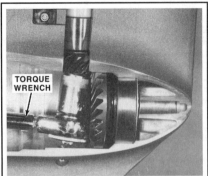

Fig. 206 . . . and a pinion nut wrench or socket

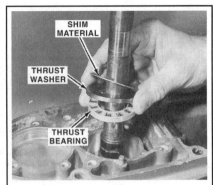

Fig. 207 On non-tapered bearing models, position the thrust bearing assembly

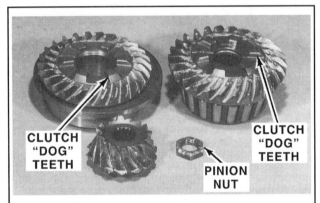

Fig. 208 Pinion nut torque is CRITICAL. This damage may have been caused by failure to tighten the pinion gear nut. The nut came loose, allowing the pinion gear to lower, drastically changing the backlash and gear mesh pattern. Notice the clutch dog engaging teeth were not affected, indicating the gear damage was not caused by a speed shift.

- On V4 and 2.6L V6 motors, tighten the pinion nut to 69 ft lbs. (95 Nm).
- On 3.1L V6 motors, tighten the pinion nut to 105 ft. lbs. (145 Nm).
- On 3.3L V6 motors, tighten the pinion nut to 103 ft. lbs. (142 Nm).

✱✱ WARNING

The pinion nut torque is CRITICAL for the well-being of the gearcase. Should the nut come loose in service, the gear would fall striking the forward and reverse gears (wreaking havoc on the gears and teeth, possibly damaging other parts of the case and components as well).

5. For models where the upper driveshaft bearing is located in the oil seal housing (as opposed to tapered roller bearing models where the bearing was pressed onto the shaft), position the thrust bearing, thrust washer and shim material (either that which was removed or that which was calculated or measured during Gearcase Shimming, as detailed later in this section) over the driveshaft and against the driveshaft sleeve.

Driveshaft Oil Seal Housing

◆ See Figures 209, 210 and 211

As noted during disassembly there are basically 3 types of oil seal housings used on these motors. However, all 3 types have a few things in common. For starters, dual oil seals (both of whose seal lips are NORMALLY positioned facing upwards, towards the water pump) are used to seal the driveshaft. Additionally, an O-ring is used to help seal the housing to the gearcase. The rest of the design features vary, some use a gasket in addition to the O-ring, others do not. Some contain a needle bearing pressed into the housing, others do not.

In all cases, seals, the gasket and/or O-ring should be replaced to ensure your gearcase components are protected. Additionally, care must be given to make sure the new seals are not damaged by the driveshaft splines when the housing is slid into position. Otherwise, it is a relatively straight forward reseal and installation.

1. On all V4/V6 models, if the driveshaft needle bearing was removed, use a suitable installer to drive the replacement bearing in from the underside of the housing. Install the bearing to the proper depth below the bottom edge of the oil seal housing. The bearing should be installed to a depth of 0.226-0.246 in. (5.75-6.25mm) for all 4-cylinder and 2.6L V6 motors or to a depth of 0.167-0.187 in. (4.25-4.75mm) for 3.1L and 3.3L motors.

■ Remember to pack the seal lips with grease upon installation.

2. If the oil seals were removed, place the housing on the work surface with the seal bore facing upward (this means with the housing upright on the largest motors such as the V4/V6 models, but the housing inverted on smaller motors, where the gearcase uses a driveshaft with a pressed on tapered roller bearing). Install the replacement oil seals using a suitable driver, making sure their lips are facing upward (toward the water pump, unless a variance was noted during disassembly) and the seals are installed to the proper depth as follows depending upon the type of oil seal housing used by this gearcase:

- On the smallest of these motors (typically the 40/50 hp 3-cylinder motors) which are equipped with a small round housing just a little bit larger than the seal (not retained by bolts) and a driveshaft with a tapered roller bearing pressed onto the shaft, the oil seals are installed from UNDERNEATH the housing. As stated earlier during disassembly some Yamaha sources conflict as to which direction the seal lips should be facing, but we think it should normally be upward toward the water pump. On these models, drive the seals into position until the outer of the two (bottom in this case, since they are installed from underneath the housing) is 0.00-0.02 in. (0.0-0.5mm) below the edge of the housing.
- On mid-range motors (including the 48 hp 2-cylinder, 3-cylinder motors [except the 40/50 model]) which utilize an irregularly-shaped housing that matches the shape of the water pump plate, both seals are installed from underneath the housing, with their lips facing upward. On MOST of these gearcases Yamaha does NOT give a dimension for seal installation depth, meaning that the seals should be installed under they are seated or flush.
- On the largest of these motors (including all V4 and V6 motors) a small oil seal and bearing housing mounts to the top of the gearcase. On these the housings are designed so the seals are installed from the TOP of the housing (since the bearing would be in the way from the bottom). As usual, both seal lips should be positioned so they are facing upward. On all models they should be installed so the top most seal is 0.010-0.030 in. (0.25-0.75mm) below the top edge of the housing.

3. Install a new O-ring around the oil seal housing and coat the O-ring with marine grade grease.

■ If you are going to check Gear Backlash on mid-range models where the seal housing follows the shape of the water pump, it may be easier to leave the housing off until after checking lash.

4. If equipped, position a new oil seal housing gasket.

LOWER UNIT 7-61

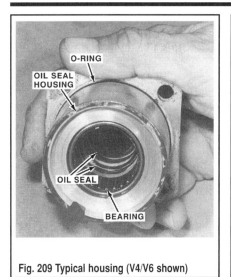

Fig. 209 Typical housing (V4/V6 shown)

Fig. 210 When installing the housing, take care not to damage the seals

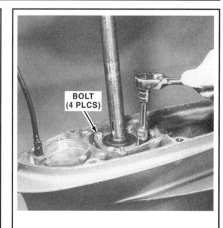

Fig. 211 On the largest motors, the housing has its own retaining bolts

5. CAREFULLY lower the oil seal housing down over the driveshaft making sure not to damage the seals on the driveshaft splines. Keep the housing squared to the driveshaft and gearcase, then tap it lightly to seat it properly in the lower unit.

6. On all but the smallest of gearcases noted earlier, check to be sure the housing bolt holes align. On housings that use their own retaining bolts (as opposed to housings which are retained by water pump retaining bolts), install and tighten the bolts securely.

Propeller Shaft and Clutch Dog

◆ See Figures 212 thru 219

The job of the propeller shaft is straight forward enough, change the direction of the clockwise, vertical rotating driveshaft to clockwise (or counterclockwise) horizontal rotation in order to drive the propeller. At all times the pinion gear on the end of the driveshaft is in contact with and rotating 2 gears (a forward and reverse). . . but because of their positioning the gears are rotating in different directions. A clutch dog assembly which is splined to the propeller shaft is positioned in the middle of the 2 gears. A shift mechanism is attached to the clutch dog with a cross pin. This mechanism extends forward from the propeller shaft, through the gear and bearing assembly in the gearcase nose cone. The shift rod mechanism works by rotating forward or backward, transmitting this force through the shifter in the propeller shaft to the clutch dog, moving it toward one or the other gear. When the clutch dog comes in contact with one of the gears it locks into place and begins to spin with that gear, transmitting the force of the gear to the propeller shaft through the clutch dog-to-shaft splines.

Propeller shaft assembly involves installation of the clutch dog and internal shift components, which vary slightly from model-to-model. Most smaller gearcases (generally found on 60 hp and smaller motors) contain a simple spring, plunger and slider mechanism. However there are some exceptions, which use a shift mechanism more similar to the larger motors.

Conversely, motors larger than 60 hp (again, there are a few exceptions) utilize a slightly more complicated shift mechanism that includes a shift rod joint, joint slider, spring (sometimes with spring nuts and/or washers on either end) and anywhere from 2-6 check balls. At minimum 2 check balls are installed at the front end of the joint slider, though many models also 2 balls

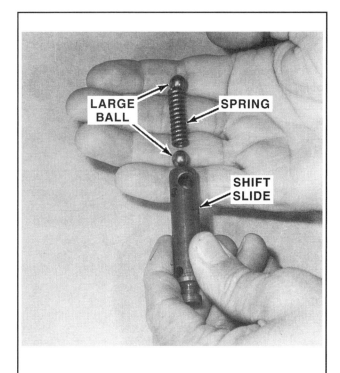

Fig. 212 Install the compression spring and any balls, washers or spacers

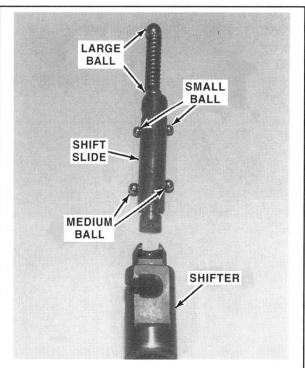

Fig. 213 If equipped with side check balls, use grease to hold them in place. . .

7-62 LOWER UNIT

of the same size at the rear of the joint slider. Still other models use 2 larger check balls positioned in the propeller shaft behind the joint slider and on either side of the spring.

Work slowly and carefully, making sure no check balls, springs, spring nuts, etc are lost when assembling the shift mechanism. For more details, refer to the exploded views found in the Cleaning and Inspection section.

1. Install the compression spring into the shift slide or end of the prop shaft (as applicable). If the compression spring is surrounded by large check balls, washers or spacers, be sure to position them properly as it is inserted.
2. For models equipped with 2 or more side mounted check balls, apply just a dab of grease to the two smallest balls to help keep them in place and then insert them into the holes nearest the open end of the shift slide.
3. Insert the shift slide into the propeller shaft with the large hole in the slide aligned with the slot in the propeller shaft.
4. For models which use check balls on the shift slide, continue to insert the shift slide into the propeller shaft until both small balls enter the shaft - then stop. If the shifter uses 2 more check balls at the opposite end of the shaft apply a dab of grease to those balls (they may be the same size as the small balls or they may be medium sized, depending upon the model) to help keep them in place and then insert them into the holes nearest the necked end of shift slide. Continue to insert the shift slide into the propeller shaft until both medium balls are about to enter the shaft - then stop.
5. For models with a separate shifter joint, hook the shifter onto the knob on the end of the shift slide. Push in the shift slide until seated. Again, on models with check balls work slowly feeling for the balls to click slightly into place in the grooves inside the shaft. Stop the action when a click is heard, indicating the two small balls have moved into the neutral groove in the propeller shaft. There is only one groove, so wiggle the slide back and forth until the ball is definitely seated in the groove.
6. Slide the clutch dog onto the propeller shaft, with the hole in the dog aligned with the two holes (hopefully already aligned), of the shaft and shift slide. Check the clutch dog for markings, often there is a stamped mark (usually a **F** on the clutch dog which should be positioned facing forward, toward the shifter).
7. Insert the cross pin into the hole and push it through until the pin ends are flush with the clutch dog groove surface.
8. Wrap the cross pin retaining ring around the groove to retain the cross pin. The ring may be wrapped in either direction. Check to be sure the ring turns do not overlap one another.
9. On some motors it is easier to guide the assembled propeller shaft into the lower unit at this time, on others it is easier to wait and install it as an assembly with the propeller shaft bearing carrier (a lot depends on the gearcase and whether the shaft came out with the carrier). On dual-propeller models you should be reading this step THEN going back to perform driveshaft installation.

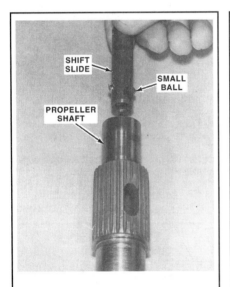

Fig. 214 . . .while you carefully insert the assembly to the prop shaft. . .

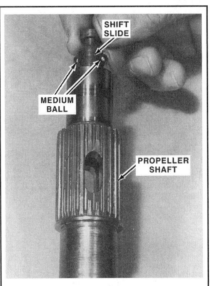

Fig. 215 . . .making sure the balls click lightly into place

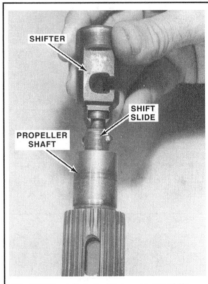

Fig. 216 If equipped with a separate shift joint, attach it to the slide

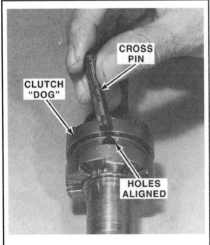

Fig. 217 Install the clutch dog and secure with the cross pin. . .

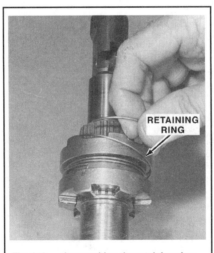

Fig. 218 . . .then position the retaining ring to secure the pin

Fig. 219 If you're ready, install the assembled prop shaft

LOWER UNIT 7-63

Shift Rod

◆ See Figure 220

The shifter assembly on these models rotates back and forth (as opposed to the smaller Yamaha gearcases on which it lifts up and down). On some models a straight 1-piece shifter shaft is splined at both ends, the bottom of which splines to a removable shift cam inside the housing which can be removed independently of the propeller shaft. However, on most models (including most V4 and V6 models) utilize a shaft that is only splined on the top end. For these models the lower end of the shaft contains a pivot pin that is inserted through or around a shift joint that attaches to the propeller shaft.

The best time for shifter assembly installation will vary with the design of the lower shift mechanism. For models where the shift rod slides into a splined cam installation occurs either right before the propeller shaft is installed (if the cam sits freely in the housing, so you can insert your hand or a tool to position the cam while the shift rod splines are mated to it) or right AFTER the propeller shaft is installed (on models where the cam rides in a small housing on the end of the prop shaft shifter assembly. On models that utilize a pivot pin, the shifter assembly must be installed AFTER the propeller shaft is in place (because the rod/pin must be lowered into or around the shift slider, again depending upon the design).

■ **For most of the gearcases covered here that means shift rod installation CAN occur at this point, but there are exceptions where the shift rod should already be installed or cannot be installed until later, after the bearing carrier. If necessary come back to this procedure after the propeller shaft and bearing carrier are in position.**

1. If removed, install a new oil seal with the lips facing the same direction as noted during removal (or as shown in the exploded views under Cleaning & Inspection, earlier in this section).
2. Install new housing O-ring(s) and apply a light coating of marine grade grease.
3. Lower the shift rod down into the lower unit and through the hole in the shifter. For splined units, make sure the splines on the bottom of the rod engage with the splines on the shift cam. On pin-type shifters, the cam on the shift rod must normally face to starboard (for standard lower units) or to port (for counter-rotating lower units) and the cam or pin on the end of the rod will insert directly into the hole in the shift slider.
4. Once the shift rod is properly engaged on the lower end, install and tighten the bolts that secure the housing.

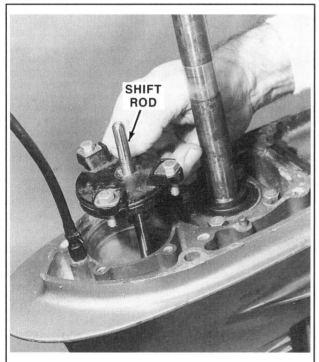

Fig. 220 The timing for shifter installation varies with shifter design

Propeller Shaft Bearing Carrier

◆ See Figures 221 thru 229

■ **If you're working on a counter-rotation model, remember the actual gear positions are opposite of a normal rotation model. On counter-rotation models, the Forward gear is in the bearing carrier and the Reverse gear is in the gearcase nose cone.**

With the exception of potential Gearcase Shimming procedures which may need to be performed if certain parts were replaced, the bearing carrier and propeller shaft installation is the last major piece of the gearcase overhaul.

If the bearing carrier cap was disassembled to replace a bearing, bushing and/or oil seal it must be properly prepared for installation. This means you'll need to use suitably sized drivers to install replacement seals, bearing or bushing, depending upon what was removed. As usual, remember the driver must contact the part of the seal, bearing or bushing that is in contact with the housing. On most installations that means the outside of the seal, bearing or bushing, HOWEVER on some installations a gear bearing is installed ONTO the gear and not INTO the housing. When a gear bearing is pressed onto the outer diameter of a shaft on the gear, the driver needs to contact the INNER race instead.

■ **Remember to oil the contact surfaces of bearings, bushings or seals before installation. Be sure to pack the seal lips with a suitable marine grade grease once they are installed.**

Keep in mind that seals, bushing or bearing installation varies slightly from model-to-model. Although all seals used on these gearcases are of the dual variety they are all installed with BOTH of their seal lips facing the rear of the outboard (the propeller).

All models (even including the counter-rotation and dual-propeller units) are equipped with a needle bearing and dual seal assembly that installs from the propeller side of the carrier. However, that's where the similarities end, as the counter and dual-prop models utilize slightly different components on the other end of the housing. If disassembled, prepare the bearing/seal carrier cap and the propeller shaft components for installation, as follows:

■ **Unless otherwise noted, install all bearings with any embossed markings facing back toward the propeller.**

- For 48 hp 2-cylinder motors, start by installing the needle bearing through the propeller side of the housing until seated. On world production models the top of the bearing should be roughly 1.06 in. (27mm) below the surface of the housing. Install the 2 seals over top of the needle bearing. Position the thrust washer on the reverse gear (with the chamfered side facing toward the gear), then use a suitable driver to carefully tap the ball bearing into position on the gear. Place the bearing carrier over the assembled gear and use a shop press to install the gear and bearing assembly to the carrier.
- For 40/50 hp (698cc) 3-cylinder motors, start by installing the needle bearing through the propeller side of the housing until seated. On US and Canadian models, if you have access to the Yamaha bearing attachment (#YB-06111) you would drive the bearing into position until the flange on the bearing tool is 0.12-0.14 in. (3.0-3.5mm) from the contact surface of the carrier. On world production models the top of the bearing should be roughly 0.91-0.93 in. (23.0-23.5mm) below the surface of the housing (but this dimension includes the thickness of the Yamaha bearing depth plate). Install the 2 seals over top of the needle bearing so that the top of the 2 seals is 0.16-0.18 in. (4.0-4.5mm) below the end of the carrier. Position the gear shim on the reverse gear, then use a suitable driver to carefully tap the ball bearing into position on the gear. Place the bearing carrier over the assembled gear and use a shop press to install the gear and bearing assembly to the carrier.
- For 50-90 hp (849cc and 1140cc) 3-cylinder motors, start by installing the needle bearing through the propeller side of the housing until seated. On US and Canadian models, if you have access to the Yamaha bearing attachment YB-06152 you would drive the bearing into position until the flange on the bearing tool is just in contact with the surface of the carrier. For world production models, use (#90890-06614) for 50 hp motors or (#90890-06611) for 60-90 hp motors along with the Yamaha bearing depth plate and

7-64 LOWER UNIT

install so the top of the driver is about 1.06 in. (20.7mm) below the surface of the housing for 50 hp models or 1.00 in. (25.5mm) below the surface of the housing on 60-90 hp models. Install the 2 seals over top of the needle bearing so that the top of the outer seal is approximately 0.20 in. (5.0mm) below the end of the housing. Position the thrust washer and/or spacer on the reverse gear (note that if the washer is beveled, position it with the chamfered side facing toward the gear), then use a suitable driver to carefully tap the ball bearing into position on the gear. Place the bearing carrier over the assembled gear and use a shop press to install the gear and bearing assembly to the carrier.

- For standard rotation models of the V4 and 2.6L V6 motors, start by installing the needle bearing through the propeller side of the housing until the top of the bearing is about 0.974-0.994 in. (24.75-25.25mm) below the surface of the housing. Install the 2 seals over top of the needle bearing so that the top of the outer seal is approximately 0.187-0.207 in. (4.75-5.25mm) below the end of the housing. Position the thrust washer on the reverse gear, then use a suitable driver to carefully tap the ball bearing into position on the gear. Place the bearing carrier over the assembled gear and use a shop press to install the gear and bearing assembly to the carrier.

- For counter-rotation models of the V4 motors, start by installing the first of the 2 carrier needle bearings through the gearcase end of the housing until the top of the bearing is about 1.348-1.368 in. (34.25-34.75mm) below the surface of the housing (gearcase end). Then, invert the housing so the propeller end is facing upward and install the other needle bearing so the top of the bearing is about 0.974-0.994 in. (24.75-25.25mm) below the propeller end of the housing. Install the 2 seals over top of the needle bearing so that the top of the outer seal is approximately 0.187-0.207 in. (4.75-5.25mm) below the end of the housing. Position the thrust washer on the forward gear, then use a suitable driver to carefully tap the ball bearing into position on the gear. Place the bearing carrier over the assembled gear and use a shop press to install the gear and bearing assembly to the carrier. Finally, prepare and install the forward gear assembly. Insert the needle bearing into the back of the gear and press until it is seated with the top of the bearing about 0.40-0.42 in. (10.25-10.75mm) below the rear surface of the gear. Install the rear propeller shaft, tapered roller bearing and outer race to the gear end of the carrier housing using a driver, then use a shop press to install the forward gear and thrust washer to the assembly.

- For counter-rotation models of the 2.6L V6 motors, start by installing the needle bearing through the propeller side of the housing until the top of the bearing is about 0.974-0.994 in. (24.75-25.25mm) below the surface of the housing. Install the 2 seals over top of the needle bearing so that the top of the outer seal is approximately 0.187-0.207 in. (4.75-5.25mm) below the end of the housing. Next, prepare and install the forward gear assembly. Using a suitable driver, install the thrust washer, tapered bearing outer race and bearing to the back of the forward gear. Position the gear onto the propeller shaft, then install the Clutch Dog assembly (as detailed earlier in this section). From the propeller end of the shaft slide the thrust bearing, then washer (with the flat side of the washer towards the thrust bearing and the chamfered side facing toward the propeller) and finally the propeller shaft shim (if used and of proper size) over the shaft. Insert the propeller shaft to the bearing carrier.

- For dual-propeller models of the 2.6L V6 motors, start by preparing the outer driveshaft assembly by installing the needle bearing in the propeller side end of the shaft. Seat the bearing to a depth of 1.368-1.388 in. (34.75-35.25mm) inside the end of the shaft. Next install the dual oil seals inside the end of the shaft. UNLIKE most other Yamaha seals, these two seals are installed BACK-TO-BACK, meaning the lips on the first (inner) seal are installed facing inward toward the shaft and the lips of the second (outer) seal are installed facing outward. Next, turn your attention to the bearing carrier itself by installing the rear needle bearing on the propeller end of the housing until the top of the bearing is about 0.974-0.994 in. (24.75-25.25mm) below the surface of the housing. Install the 2 seals over top of the needle bearing so that the top of the outer seal is approximately 0.187-0.207 in. (4.75-5.25mm) below the end of the housing. (In case you're wondering, YES, both of these seal lips are positioned the usual Yamaha way, facing toward the propeller). Invert the bearing carrier and install the front needle bearing, pressing until the top of the bearing is about 0.030-0.049 in. (0.75-1.25mm) below the gearcase edge of the carrier. Finally, prepare and install the outer propeller shaft and bearing assembly. To prevent damage to the oil seals, wrap the entire threaded and splined portion of the shaft with tape. Slide the thrust washer, flat washer and, if used, propeller shaft shim over the propeller end of the shaft and up into position on the shaft shoulder, then carefully insert the shaft into the carrier. Using a suitable driver, install the tapered bearing outer race and bearing, then install the thrust washer and rear gear using a driver to press them into position in the assembly.

- For standard rotation models of the 3.1L and 3.3L V6 motors, start by installing the needle bearing through the propeller side of the housing until the top of the bearing is about 0.986-1.006 in. (25.05-25.55mm) below the surface of the housing. Install the 2 seals over top of the needle bearing so that the top of the outer seal is approximately 0.187-0.207 in. (4.75-5.25mm) below the end of the housing. Use a suitable driver to carefully tap the ball bearing into position in the bearing carrier, then position the reverse gear shim on the gear and use a driver or press to install the gear to the bearing and carrier assembly.

- For counter-rotation models of the 3.1L V6 motors, start by installing the first of the 2 carrier needle bearings through the gearcase end of the housing until the top of the bearing is about 1.762-1.781 in. (44.75-45.25mm) below the surface of the housing (gearcase end). Then, invert the housing so the propeller end is facing upward and install the other needle bearing so the top of the bearing is about 0.986-1.006 in. (25.05-25.55mm) below the propeller end of the housing. Install the 2 seals over top of the needle bearing so that the top of the outer seal is approximately 0.187-0.207 in. (4.75-5.25mm) below the end of the housing. Next prepare the forward gear for installation by inserting the new needle bearing and bushing, then driving it into position until the top of the bearing is about 0.46-0.48 in. (11.8-12.2mm) below the rear of the gear. Install the rear propeller shaft (along with the thrust bearing, flat washer and, if/as equipped, propeller shaft shim), tapered roller bearing and bearing outer race to the carrier and seat using a suitable driver. Install the claw washer and ring nut, using YB-06048 or an equivalent ring nut wrench to tighten it to 80 ft. lbs. (110 Nm). In order to hold the bearing carrier steady for this you can temporarily install it backwards into the bottom of the gearcase. Lastly, use a shop press to install the forward gear onto the back of the carrier and rear propeller shaft assembly.

- For counter-rotation models of the 3.3L V6 motors, start by installing the needle bearing into the propeller end of the carrier so the top of the bearing is about 0.986-1.006 in. (25.05-25.55mm) below the propeller end of the housing. Install the 2 seals over top of the needle bearing so that the top of the outer seal is approximately 0.187-0.207 in. (4.75-5.25mm) below the end of the housing. Next, prepare and install the forward gear assembly. From the propeller end of the shaft slide the thrust bearing, then washer and finally the propeller shaft shim (if used and of proper size) over the shaft. Insert the propeller shaft to the bearing carrier. Using a shop press, install the tapered roller bearing and outer race to the carrier and shaft assembly. In order to hold the carrier steady for ring nut installation, temporarily install the bearing carrier in the reverse of the normal position (with the gearcase side facing outward) into the bottom of the gearcase. Install the claw washer and ring nut, then tighten the nut to 80 ft. lbs. (108 Nm) using (#YB-06578) or an equivalent ring nut wrench. Remove the carrier from the gearcase, then use a shop press and driver to install the forward gear shim, the gear and the clutch dog (making sure the embossed V on the clutch dog is facing the gear) onto the front of the propeller shaft. Finally, install the Clutch Dog assembly and shifter joint (as detailed earlier in this section).

1. If it is time for final assembly (not if you are just assembling to check shimming and backlash) install a new O-ring into each groove provided in the bearing carrier (these carriers may use 1-3 O-rings, depending upon the model. Apply a light coating of marine grade grease to the O-ring(s).

2. If this assembly is to check gear contact patterns, coat the teeth of the reverse gear with a fine spray of Desenex®, as mentioned earlier. Handle the gear carefully to prevent disturbing the powder.

3. Install the bearing carrier or carrier and propeller shaft assembly (depending upon model and method of installation) into the lower unit. If applicable, align the keyway in the lower unit housing with the keyway in the bearing carrier. Insert the key into both grooves and then push the bearing carrier into place in the lower unit housing.

■ **On some assemblies, such as the dual-propeller models, the bearing carrier is equipped with a stamping that should be faced upward. Of course, since many carriers use a keyway, it is easy to figure out correct installation and orientation.**

4. If this is the final installation (again, not temporary for shimming or checking contact patterns), apply a light coating of Loctite® 572 or an equivalent threadlocking compound to the bearing carrier retaining screws or a light coating of marine grade grease to the threads of the lockring (as equipped).

■ **Many of the bearing carriers on the larger models are retained by a ring nut and lock-washer. On these models install a tabbed washer against the bearing carrier and then install the ring nut with the embossed marks facing outward, away from the bearing carrier.**

LOWER UNIT 7-65

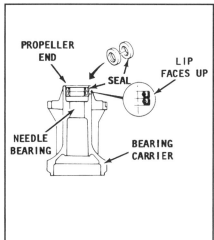

Fig. 221 All carries use a needle bearing and 2 seals at the propeller end

Fig. 222 Most standard rotation units use a ball bearing pressed onto the gear...

Fig. 223 ...then the gear and bearing is pressed into the carrier

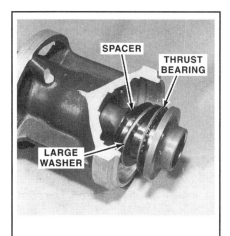

Fig. 224 Most counter-rotating units use a shim, washer and thrust bearing on the prop shaft...

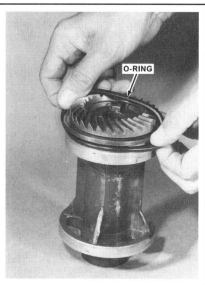

Fig. 225 ...and many use a spacer after the gear

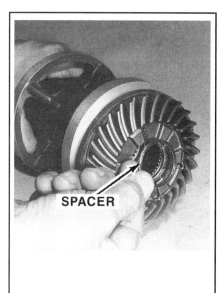

Fig. 226 These carriers may be equipped with 1-3 O-rings

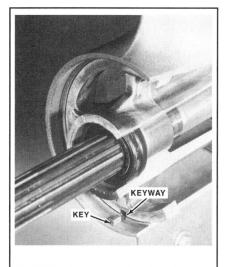

Fig. 227 Lockring retained carriers also utilize a key to secure the carrier

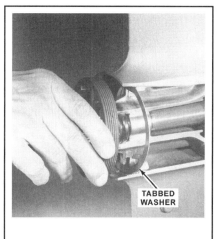

Fig. 228 On lockring retained carriers install the tabbed washer and ring nut...

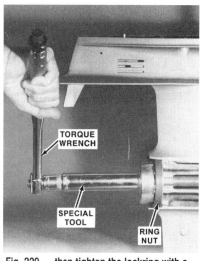

Fig. 229 ...then tighten the lockring with a special ring nut tool

7-66 LOWER UNIT

5. Install the bearing carrier retaining bolts or ring nut and tighten to specification as follows:
- For 48 hp 2-cylinder motors, tighten the ring nut to 94 ft. lbs. (130 Nm).
- For 40/50 hp (698cc) 3-cylinder motors tighten the retaining bolts to 11 ft. lbs. (16 Nm).
- For 50-90 hp (849cc and 1140cc) 3-cylinder motors tighten the ring nut to either 94 ft. lbs. (130 Nm) for 50 hp motors, or to 105 ft. lbs. (145 Nm) for 60-90 hp motors.
- For V4 and 2.6L V6 motors, tighten the ring nut to 105 ft. lbs. (145 Nm).
- For 3.1L V6 motors tighten the carrier retaining bolts to 17 ft. lbs. (24 Nm) for all except Vmax models, on which the carrier bolts should be tightened to 21 ft. lbs. (29 Nm). Then for all models, install the ring cap and tighten cap bolts securely.
- For 3.3L V6 motors tighten the carrier retaining bolts to 22 ft. lbs. (30 Nm), then, install the ring cap and tighten cap bolts securely.

6. If components other than seals were replaced (such as the case itself, gears, bearings and/or shafts) proceed to the appropriate section of the Gearcase Shimming procedures to check and/or adjust gear lash.

7. If this is final assembly, install the Water Pump, as detailed in the Lubrication and Cooling section.

8. Refill the gearcase, install the propeller and install the lower unit assembly (in whatever order you'd prefer).

GEARCASE SHIMMING

◆ See Figure 230

■ Since Yamaha shims (and gearcase markings) are all in mm we have kept most of the procedures in metric measurements only.

In order for the gearcase to function properly the pinion gear teeth must fully engage with the teeth of the forward or reverse gear (depending on shifter position). Because build tolerances for the gearcase, gears, shafts and bearings/bushings will vary slightly the manufacturer has positioned shims strategically at various points in the gearcase to give you the ability to make up for the differences in these tolerances. In this way, a gear that is not fully engaging can be repositioned slightly by the use of a thicker shim. Similarly, a gear that does not fully disengage can be repositioned slightly further away by using a thinner shim.

Gearcase shimming procedures vary slightly from model-to-model and they depend upon the components that are being replaced. However, generally speaking:
- NO SHIMMING is required if the original case and original inner parts (gears, bearings and shafts) are installed after cleaning, inspection and/or seal replacement.

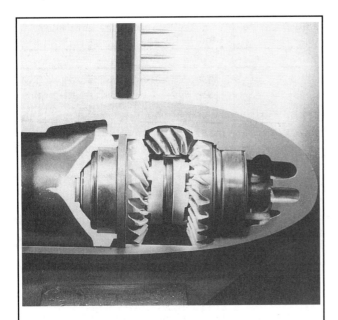

Fig. 230 Gearcase shimming is all about getting the forward and reverse gears to mesh properly with the pinion gear!

- NUMERIC CALCULATION RESHIMMING should be performed on all gearcases when the original inner parts (gears, bearings and shafts) are installed in a NEW case. Most original and replacement cases are marked with various gear measurements. Typically an embossed F (forward gear), R (reverse gear) and/or P (pinion) gear measurements. When these marks are present the differences between the marks on the new case vs. the marks on the original case can be used to determine new shim sizes BEFORE assembly (potentially saving you from having to disassemble/reassemble the gearcase a second time).
- SPECIAL TOOL MEASUREMENT RESHIMMING is performed prior to assembly on all models and should be used when a specific shaft, bearing/bushing or gear is replaced. In this case, special tools are used to measure pre-assembled or partially assembled components and determine if a change in shim sizes is necessary. If the special shimming tools are not available, you can often reuse shim sizes from the previous assembly to at least get you in the ball-park and then use the Backlash Measurement Reshimming procedures to dial it in.
- BACKLASH MEASUREMENT RESHIMMING should be performed and adjusted when one or more of the original inner components (gears, bearings and shafts) are installed. Backlash measurement is normally performed on an almost fully assembled gearcase using a dial gauge to read the amount of play in the driveshaft. This procedure can check the placement of all 3 gears through this measurement. Though adjustment is much more hit and miss than the other methods, since you need to then re-disassemble the gearcase, swap shims, re-assemble and re-check, then repeat until you get it right. Fortunately, small shim changes are all that is normally necessary and it is rare that you would have to disassemble the gearcase more than one additional time.

Follow the appropriate Shimming or Backlash measurement procedure, depending upon your needs as outlined above.

Numeric Calculation Reshimming

◆ See Figure 120

The gearcase for all of these models SHOULD be marked with measurements representing the case's deviation from standard. Typically an embossed F (forward gear), R (reverse gear) and/or P (pinion) and a + or - with a number. The codes can be found on the underside of the anti-cavitation plate, usually underneath the trim tab (so if you look under there and can't find them, remove the trim tab and check again).

The stamped number should be interpreted as hundredths of a mm. This means that if there is a marking such as F +5, it means that the case requires a shim that it 5/100mm (0.05mm) larger than standard. The + or - accompanying the number tells you whether you should add (+) or take away (-) the number from the standard measurement. If there is NO number after the F, R or P, assume the value is 0 and the case meets standard tolerance. If the number is there, but illegible, you'll have to work from 0 and make adjustments for that shim by checking lash.

Now, when moving all of the internal components to a new case, compare the numbers on the 2 cases and adjust the shims accordingly. If the numbers are all the same (lets say they both are F +5, R-1 and P+8), you've got it easy, use the original shims and make no changes. But, if the new case is, lets say an F+3, then subtract the new value from the old value to determine how much of a change you need to make to the forward gear shim. In this example, 5-3=2. This means the NEW gearcase needs a shim that is 0.02mm SMALLER than the forward gear shim which was originally used. Measure the old shim and obtain a new shim which is about 0.02mm smaller for reassembly.

Once you've selected the proper shims, assemble the gearcase as detailed earlier in this section and, before the water pump is installed, check the gearcase Backlash as detailed later in this section. As long as the backlash is within spec, no further adjustment is necessary. However, if backlash is out of specification you'll need to further adjust one or more of the shims to correct the problem.

Special Tool Measurement Re-shimming

On all models one or more special tools (pinion height gauges, bearing press plates/gauges and/or shimming gauges) are available to assist in determining shim sizes before installation is complete. All shim changes must still be verified through backlash measurement, but when available, the use of shimming gauges may help select the right shim the first time out, saving you un-necessary disassembly/reassembly of the gearcase during overall.

LOWER UNIT 7-67

Tool sets will vary first and foremost by market. The tools used for USA and Canadian models are slightly different than those that are available for World Production models. Also, tools will vary slightly from model-to-model.

The one constant about re-shimming procedures is that you will pre-assemble some portion of the shaft and/or bearings or gears and then measure a dimension of the assembly using the special tool. The measurement is then applied to a standard formula (which often includes the gearcase deviation from standard measurements) and the result is the desired shim size (or amount of change to the shim which was used during the measurements).

Pinion Gear Shim Selection

◆ See Figures 231 thru 237

For all models, the quickest way to achieve proper gear mesh and pinion gear height is the use of Yamaha's pinion height gauge set. Different tools, formulas and specifications are used for US/Canadian models than are used for World Production models, however the actual measurement procedure is the similar for all. This is basically a method of pre-assembling the driveshaft, bearing (tapered roller on some or thrust bearing on others depending on how the gearcase is constructed), pinion gear and selected shims outside the gearcase in order to precisely check the assembled distance between the bearing and the pinion. The components of the gauge set used vary slightly from model-to-model and are detailed in the procedure.

■ Unless otherwise noted, all US and Canadian models use components of Yamaha's gauge set (#YB-34432). Parts of this set will be represented as -## (such as -9 or -10) for short-hand, but the full number of each part would be #YB-34432-## (meaning #YB-34432-9 or #YB-34432-10).

■ The pinion gear shim selection procedure varies slightly from model-to-model. The variances are essentially 2 different ways around a square (2 equivalent processes that approach the problem from the opposite directions). However, because of the limited specifications provided by Yamaha you must adhere to the method for the particular model in question.

Determine the pinion gear shim thickness, as follows:
1. Check the stamped code on the underside of the anti-ventilation plate (it is stamped in the trim tab mounting area, so if you haven't already, matchmark and remove the tab). The code **P** followed by a **+** or **-** and a number is the gearcase's pinion bearing mounting deviation from the standard spec, in hundredths of a mm. This means that if there is a marking such as **P** +5, it means that the case requires a shim that it 5/100mm (0.05mm) larger than standard. The **+** or **-** accompanying the number tells you whether you should add (**+**) or take away (**-**) the number from the standard measurement. If there is NO number after the P, assume the value is 0 and the case meets standard tolerance. If the number is there, but illegible, you'll have to work from 0 and make adjustments for that shim by checking lash.

■ All measurements should be taken a no less than 3 points around the bearing and the measurements should be averaged.

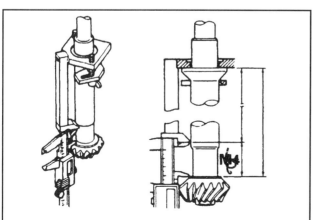

Fig. 231 On most worldwide units you measure from the pinion to the height gauge. . .

2. For US and Canadian Models, assemble the appropriate portions of the gauge set (#YB-34432) along with the driveshaft, bearing, shim, pinion and pinion gear nut (tighten the nut to the spec listed in the Pinion Gear and Driveshaft procedure), then take the measurement using a feeler gauge between the top of the gear and the bottom of the gauge block and calculate the shim changes depending upon the model as follows:

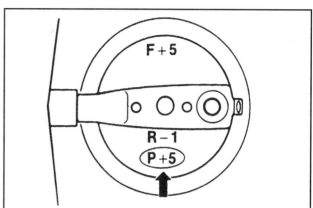

Fig. 232 All pinion shimming formulas use the stamped gearcase P deviation

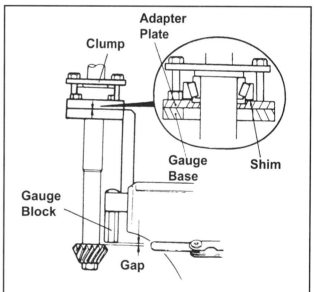

Fig. 233 Checking the assembled driveshaft on US & Canadian models that utilizes a bearing pressed onto the driveshaft

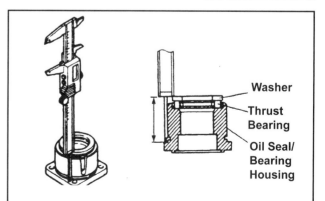

Fig. 234 . . .but on models where the bearing mounts in a housing, you also need to check the distance from the mounting flange to the thrust washer

7-68 LOWER UNIT

- For 48 hp 2-cylinder motors, assemble the gauge block -8, adapter plate -10, gauge base -11-A and clamp -17-A. If the original shim is not available start measurement with the largest shim (0.50mm). The measured gap must be 0.20mm MINUS the **P** stamped deviation (remember that if the standard deviation is a negative number, that subtracting a negative is the same as adding the positive or absolute value of that number). If not add or subtract shim material to achieve the specified gap.
- For 40/50 hp (698cc) motors, assemble the gauge block -9, adapter plate -10, gauge base -11 and clamp -17. If the original shim is not available start measurement with the largest shim (0.50mm). The measured gap must be 0.30mm PLUS the **P** stamped deviation. If not add or subtract shim material to achieve the specified gap.
- For 50-90 hp (849cc and 1140cc) motors, assemble the gauge block -9, adapter plate -10, gauge base -11-A and clamp -17-A. If the original shim is not available start measurement with the largest shim (0.50mm). The measured gap must be 0.20mm PLUS the **P** stamped gearcase deviation. If not add or subtract shim material to achieve the specified gap.
- For V4 motors, assemble the gauge block -6 (for all except the B115/115F which use block -97) and gauge base -11 along with the driveshaft, upper oil seal/bearing housing, thrust washer, thrust bearing and shim. If the original shim is not available start measurement with the largest shim (0.50mm). The measured gap must be 1.00mm PLUS the **P** stamped gearcase deviation. If not add or subtract shim material to achieve the specified gap.
- For 2.6L V6 motors, assemble the gauge block -7 (for all except the dual-propeller models which use block -97) and gauge base -11 along with the driveshaft, upper oil seal/bearing housing, thrust washer, thrust bearing and shim. If the original shim is not available start measurement with the largest shim (0.50mm). The measured gap must be 1.00mm PLUS the **P** stamped gearcase deviation. If not add or subtract shim material to achieve the specified gap.
- For 3.1L V6 Motors, assemble the driveshaft (along with the driveshaft oil seal/bearing housing, thrust bearing and shims, pinion gear and nut) in Yamaha pinion height gauge (#YB-06441) and tighten the pinion nut to spec, then use a feeler gauge to check the gap between the pinion gear and the gauge. If the original shim is not available start measurement with the largest shim (0.50mm). The measured gap must be 1.00mm PLUS the **P** stamped gearcase deviation. If not add or subtract shim material to achieve the specified gap.
- For 3.3L V6 motors, assemble the driveshaft (along with the driveshaft oil seal/bearing housing, thrust bearing and shims, pinion gear and nut) in Yamaha pinion height gauge (#YB-06441) and tighten the pinion nut to spec, then use a feeler gauge to check the gap between the pinion gear and the gauge. If the original shim is not available start measurement with the largest shim (0.50mm). The measured gap must be 0.55mm, PLUS the **P** stamped gearcase deviation. If not add or subtract shim material to achieve the specified gap.

3. For World Production Models, assemble the appropriate portions of the listed gauge set along with the driveshaft, bearing, shim, pinion and pinion nut (tighten the nut to the spec listed in the Pinion Gear and Driveshaft procedure), then take the measurement with a feeler gauge and calculate the shim changes depending upon the model as follows:

- For 48 hp 2-cylinder motors, assemble the driveshaft components along with height gauge (#90890-06702), then use a sliding caliper to measure the distance from the top of the pinion gear to the underside of the height gauge. To select the proper shim, take the measurement, then SUBRTRACT 40.2mm and SUBTRACT the **P** stamped gearcase deviation (remember that if the standard deviation is a negative number, that subtracting a negative is the same as adding the positive or absolute value of that number). If the resulting value is zero, no shim change is necessary otherwise add or subtract the amount shim equal to the value.
- For 40/50 hp (698cc) motors, assemble the driveshaft components along with height gauge (#90890-06702), then use a sliding caliper to measure the distance from the top of the pinion gear to the underside of the height gauge. To select the proper shim, take the measurement and SUBRTRACT 11.3mm then ADD the **P** stamped gearcase deviation. If the resulting value is zero, no shim change is necessary otherwise add or subtract the amount shim equal to the value.
- For 50-90 hp (849cc and 1140cc) motors, assemble the driveshaft components along with height gauge 90890-06702, then use a sliding caliper to measure the distance from the top of the pinion gear to the underside of the height gauge. To select the proper shim, take the measurement, then SUBRTRACT 40.2mm (for 50 hp motors) or 31.5mm (for 60-90 hp motors), and SUBTRACT the **P** stamped gearcase deviation (remember that if the standard deviation is a negative number, that subtracting a negative is the same as adding the positive or absolute value of that number). If the resulting value is zero, no shim change is necessary otherwise add or subtract the amount shim equal to the value.
- For V4 motors, you're going to take 2 measurements. For the first, invert the driveshaft upper oil seal/bearing housing, then place the thrust bearing and washer in that order on top of the inverted housing (in their normal positions relative to the housing). Now, use a sliding caliper to measure the distance from the top of the thrust washer to the flange on the oil seal/bearing housing (the portion of the flange that would contact the gearcase when right-side-up and installed). Now for the second measurement, assemble the driveshaft components along with height gauge (#90890-06702). After the wing nuts contact the fixing plate tighten them an additional 1/4 turn. Next, use a sliding caliper to measure the distance from the top of the pinion gear to the underside of the height gauge. To select the proper shim, take the specification 62.5mm, then ADD the **P** stamped gearcase deviation. Now from that sum, SUBTRACT both measurements to determine proper shim thickness.
- For 2.6L V6 motors, you're going to take 2 measurements. For the first, invert the driveshaft upper oil seal/bearing housing, then place the thrust bearing and washer in that order on top of the inverted housing (in their normal positions relative to the housing). Now, use a sliding caliper to measure the distance from the top of the thrust washer to the flange on the oil seal/bearing housing (the portion of the flange that would contact the gearcase when right-side-up and installed). Now for the second measurement, assemble the driveshaft components along with height gauge (#90890-06702). After the wing nuts contact the fixing plate tighten them an additional 1/4 turn. Next, use a sliding caliper to measure the distance from the top of the pinion gear to the underside of the height gauge. To select the proper shim, take the specification 80.0mm, then ADD the **P** stamped gearcase deviation. Now from that sum, SUBTRACT both measurements to determine proper shim thickness.
- For 3.1L V6 motors, you're going to take 2 measurements. For the first, invert the driveshaft upper oil seal/bearing housing, then place the thrust bearing and washer in that order on top of the inverted housing (in their normal positions relative to the housing). Now, use a sliding caliper to

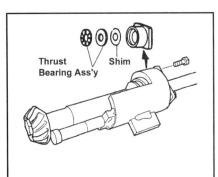

Fig. 235 For US & Canadian models that utilize a bearing housing, a different pinion height gauge is available...

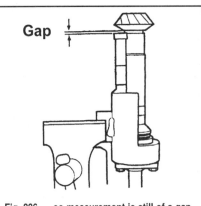

Fig. 236 ...so measurement is still of a gap using a feeler gauge

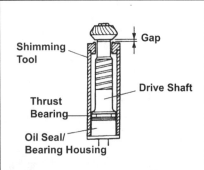

Fig. 237 The largest gearcases utilize a different height gauge, but it works the same way as the gauges on other US and Canadian models

LOWER UNIT 7-69

measure the distance from the top of the thrust washer to the flange on the oil seal/bearing housing (the portion of the flange that would contact the gearcase when right-side-up and installed). Now for the second measurement, assemble the driveshaft components along with height gauge (#90890-06702). After the wing nuts contact the fixing plate tighten them an additional 1/4 turn. Next, use a sliding caliper to measure the distance from the top of the pinion gear to the underside of the height gauge. To select the proper shim, take the specification 82.0mm, then ADD the **P** stamped gearcase deviation. Now from that sum, SUBTRACT both measurements to determine proper shim thickness.

4. Once you've selected the proper shims, assemble the gearcase as detailed earlier in this section and, before the water pump is installed, check the gearcase Backlash as detailed later in this section. As long as the backlash is within spec, no further adjustment is necessary. However, if backlash is out of specification you'll need to further adjust one or more of the shims to correct the problem.

Nose Gear Shim Selection

◆ See Figures 238 thru 243

For all models, the quickest way to achieve proper nose gear (Forward gear on standard units or Reverse on counter-rotating units) backlash is the use of Yamaha's gear height gauge set. Different tools, formulas and specifications are used for US/Canadian models than are used for World Production models, however the actual measurement procedure is basically the same in that you are checking the actual manufactured height of the bearing.

On US and Canadian models this means placing the bearing and shim in a special height gauge (#YB-34446) along with various components such as a press plate, compression spring and gauge pin, then checking the gap between the plate and pin.

On World Production models this means placing the bearing on a flat work surface and measuring the height of the bearing from one end of the bearing cage to the other edge of a roller (not to the other end of the cage, unless otherwise noted).

■ Unless otherwise noted, all US and Canadian models use components of Yamahas gauge set (#YB-34446). Parts of this set will be represented as -## (such as -1 or -7) for short-hand, but the full number of each part would be YB-34446-## (meaning #YB-34446-1 or #YB-34446-7).

■ The nose gear shim selection procedure varies slightly from model-to-model. The variances are essentially 2 different ways around a square (2 equivalent processes that approach the problem from the opposite directions). However, because of the limited specifications provided by Yamaha you must adhere to the method for the particular model in question.

Determine the nose gear shim thickness, as follows:

1. Check the stamped code on the underside of the anti-ventilation plate (it is stamped in the trim tab mounting area, so if you haven't already, matchmark and remove the tab). The code **F** followed by a + or - and a number is the gearcase's forward bearing mounting deviation from the standard spec, in hundredths of a mm. This means that if there is a marking such as F +5, it means that the case requires a shim that it 5/100mm (0.05mm) larger than standard. The + or - accompanying the number tells you whether you should add (+) or take away (-) the number from the standard measurement. If there is NO number after the **F**, assume the value is 0 and the case meets standard tolerance. If the number is there, but illegible, you'll have to work from 0 and make adjustments for that shim by checking lash.

■ All measurements should be taken a no less than 3 points around the bearing and the measurements should be averaged.

2. For US and Canadian Models, assemble the appropriate portions of the gauge set (#YB-34446), along with the bearing and gear. Install the bearing and shim into the gauge set (tightening the nut 4 FULL turns after it contacts the spring, unless otherwise noted). Measure the gap between the press plate and the gauge pin using a feeler gauge, then calculate the shim changes depending upon the model as follows:

• For 48 hp 2-cylinder motors, assemble the bearing and shim along with height gauge -1, compression spring -3, press plate -5, and the gauge pin -7. If the original shim is not available, start using the largest 0.50mm shim. The gap should be 1.00mm, PLUS the stamped **F** deviation. If not add or subtract shim material to achieve the specified gap.

• For 40/50 hp (698cc) motors, assemble the bearing and shim along with height gauge -1, compression spring -3, press plate -4, and the gauge pin -7. If the original shim is not available, start using the largest 0.50mm shim. The gap should be 0.06mm, PLUS the stamped **F** deviation. If not add or subtract shim material to achieve the specified gap.

• For 50-90 hp (849cc and 1140cc) motors, assemble the bearing and shim along with height gauge/base plate -1, compression spring -3, press plate -5, and the gauge pin -7. If the original shim is not available, start using the largest 0.50mm shim. The gap should be 1.69mm (for all except E60/E60H models which are 1.50mm), PLUS the stamped **F** deviation. If not add or subtract shim material to achieve the specified gap.

• For standard-rotation V4 motors, assemble the bearing and shim along with height gauge/base plate -1, compression spring -3, press plate -5, and the gauge pin -7. If the original shim is not available, start using the largest 0.50mm shim. The gap should be 1.80mm, PLUS the stamped **F** deviation. If the measurement is out of spec add or subtract shim material to achieve the specified gap.

• For counter-rotation V4 motors, make a stack out of the reverse gear washer, thrust bearing, roller bearing and shim, then use a micrometer or sliding caliper to measure the thickness of the stack. The measured value must be equal to 25.30mm, PLUS the stamped **F** deviation (yes, they don't know during manufacture and measurement that the gearcase will be used on a counter-rotating unit so the F stamp is still for the nose gear). If the measurement is out of spec add or subtract shim material to achieve the specified gap.

• For standard rotation and dual-propeller 2.6L V6 motors, assemble the bearing and shim along with height gauge/base plate -1, compression spring -3, press plate -4, and the gauge pin -8 (except for dual-prop models which use gauge pin -7). If the original shim is not available, start using the largest 0.50mm shim. The gap should be 1.60mm (for standard rotation models) or

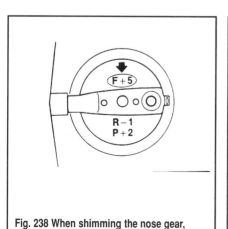

Fig. 238 When shimming the nose gear, always start with the stamped gearcase F deviation

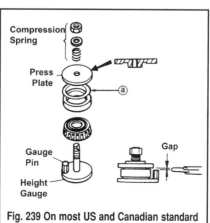

Fig. 239 On most US and Canadian standard rotation models, you'll need a height gauge set...

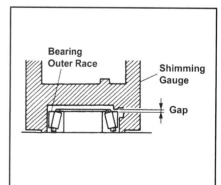

Fig. 240 ...except on the largest gearcases which use this shimming tool

1.00mm (for dual-prop models), PLUS the stamped **F** deviation. If the measurement is out of spec add or subtract shim material to achieve the specified gap.

- For counter-rotation 2.6L V6 motors, make a stack out of the reverse gear washer, thrust bearing, roller bearing and shim, then use a micrometer or sliding caliper to measure the thickness of the stack. The measured value must be equal to 29.10mm, PLUS the stamped **F** deviation (yes, they don't know during manufacture and measurement that the gearcase will be used on a counter-rotating unit so the F stamp is still for the nose gear). If the measurement is out of spec add or subtract shim material to achieve the specified gap.

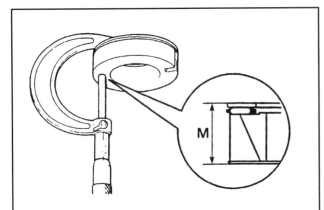

Fig. 241 For most counter-rotation units you make and measure a stack of the bearing and shim (US & Canada)...

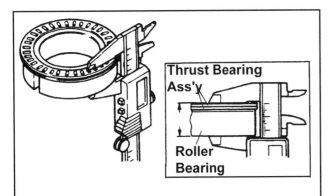

Fig. 242 ...or the bearing and thrust bearing assembly (worldwide production)

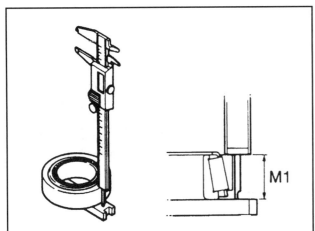

Fig. 243 On standard rotation worldwide units you normally measure bearing height from a particular side of the bearing case, to the opposite end (thin end) of the roller

- For standard rotation 3.1L V6 motors, place the forward gear bearing and shim down on a perfectly flat work surface with the tapered portion of the bearing (wider portion of the roller) downward. Place the Yamaha shimming gauge (#YB-06439) over top of the bearing and use a feeler gauge to check the gap between the edge of the bearing and the tab on the underside of the gauge. The gap must be equal to 0.60mm, MINUS the **F** stamped gearcase deviation (remember that if the deviation is a negative number that subtracting a negative is the same as adding the absolute value). If not add or subtract shims as necessary to achieve the specified gap.

- For counter-rotating 3.1L and 3.3L V6 motors, make a stack out of the reverse gear washer, thrust bearing, roller bearing and shim, then use a micrometer or sliding caliper to measure the thickness of the stack. The measured value must be equal to 30.60mm, PLUS the stamped **F** deviation (yes, they don't know during manufacture and measurement that the gearcase will be used on a counter-rotating unit so the F stamp is still for the nose gear). If the measurement is out of spec add or subtract shim material to achieve the specified gap.

- For standard rotation 3.3L V6 motors, place the forward gear bearing and shim down on a perfectly flat work surface with the tapered portion of the bearing (wider portion of the roller) downward. Place the Yamaha shimming gauge (#YB-06439) over top of the bearing and use a feeler gauge to check the gap between the edge of the bearing and the tab on the underside of the gauge. The gap must be equal to 0.50mm, MINUS the **F** stamped gearcase deviation (remember that if the deviation is a negative number that subtracting a negative is the same as adding the absolute value). If not add or subtract shims as necessary to achieve the specified gap.

3. For World Production Models, place the bearing down on a completely flat work surface, then measure the distance from the top of the roller (not the bearing cage) to the portion of the bearing that is in contact with the work surface. Calculate the shim changes depending upon the model as follows:

- For 48 hp 2-cylinder motors, the forward gear shim should be equal to the 22.7mm, PLUS the **F** stamped gearcase deviation, MINUS the bearing measurement.

- For 40/50 hp (698cc) motors, the forward gear shim should be equal to the 22.75mm, PLUS the **F** stamped gearcase deviation, MINUS the bearing measurement.

- For 50-90 hp (849cc and 1140cc) motors, the forward gear shim should be equal to the 22.70mm (for 50 hp models) or 24.50mm (for 60-90 hp models), PLUS the **F** stamped gearcase deviation, MINUS the bearing measurement.

- For standard rotation V4 motors, the forward gear shim should be equal to the 24.60mm, PLUS the **F** stamped gearcase deviation, MINUS the bearing measurement.

- For counter-rotating V4 motors, make a stack out of the reverse gear ball/roller bearing, thrust bearing and washer, then measure the thickness of the stack using a micrometer or sliding caliper. The proper reverse gear shim thickness is then calculated by taking the specification of 25.3mm, PLUS the **F** stamped gearcase deviation (the F stamping on the gearcase indicates the nose cone bearing/gear mounting deviation since MOST gearcases are used on standard rotation units), MINUS the stack measurement.

- For standard rotation 2.6L V6 motors, the forward gear shim should be equal to the 28.60mm, PLUS the **F** stamped gearcase deviation, MINUS the bearing measurement.

- For counter-rotating 2.6L V6 motors, make a stack out of the reverse gear roller bearing, thrust bearing and washer, then measure the thickness of the stack using a micrometer or sliding caliper. The proper reverse gear shim thickness is then calculated by taking the specification of 29.1mm, PLUS the **F** stamped gearcase deviation (the F stamping on the gearcase indicates the nose cone bearing/gear mounting deviation since MOST gearcases are used on standard rotation units), MINUS the stack measurement.

- For standard rotation 3.1L V6 motors, the forward gear shim should be equal to the 29.5mm, PLUS the **F** stamped gearcase deviation, MINUS the bearing measurement.

- For counter-rotating 3.1L V6 motors, make a stack out of the reverse gear roller bearing, thrust bearing and washer, then measure the thickness of the stack using a micrometer or sliding caliper. The proper reverse gear shim thickness is then calculated by taking the specification of 30.6mm, PLUS the **F** stamped gearcase deviation (the F stamping on the gearcase indicates the nose cone bearing/gear mounting deviation since MOST gearcases are used on standard rotation units), MINUS the stack measurement.

LOWER UNIT 7-71

4. Once you've selected the proper shims, assemble the gearcase as detailed earlier in this section and, before the water pump is installed, check the gearcase Backlash as detailed later in this section. As long as the backlash is within spec, no further adjustment is necessary. However, if backlash is out of specification you'll need to further adjust one or more of the shims to correct the problem.

Bearing Carrier Gear Shim Selection

◆ See Figures 244 thru 249

For all models, the quickest way to achieve proper bearing carrier gear (Reverse gear on standard units or Forward on counter-rotating units) backlash is the use of Yamaha's gear height gauge. Different tools, formulas and specifications are used for US/Canadian models than are used for World Production models, however the actual measurement procedure is basically the same in that you are checking the actual assembled height of the gear on the carrier.

On US and Canadian models this means placing the gauge set over the gear in contact with the carrier and using a feeler gauge between the tab on the underside of the gauge and the gear teeth.

On World Production models this usually means placing a flat-stock shimming plate over top of the gear and then measuring the distance from the TOP of the shimming plate to the top edge (thrust plate) of the carrier. Although you could substitute another piece of flat-stock, it would have to be the exact same thickness as the shimming plate.

■ The bearing carrier gear shim selection procedure varies slightly from model-to-model. The variances are essentially 2 different ways around a square (2 equivalent processes that approach the problem from the opposite directions). However, because of the limited specifications provided by Yamaha you must adhere to the method for the particular model in question.

Determine the bearing carrier gear shim thickness, as follows:
1. Check the stamped code on the underside of the anti-ventilation plate (it is stamped in the trim tab mounting area, so if you haven't already, matchmark and remove the tab). The code R (or F for counter-rotation units) followed by a + or - and a number is the gearcase's forward bearing

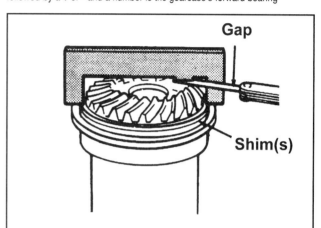

Fig. 246 On many US and Canadian models check the gap between a shim gauge and the gear assembled in the carrier...

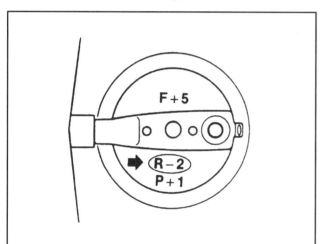

Fig. 244 When shimming the prop shaft bearing carrier, always start with the stamped gearcase R deviation

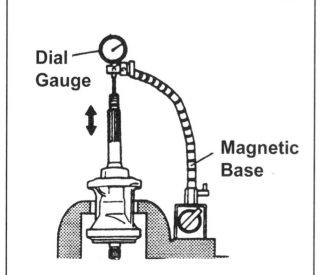

Fig. 245 On all counter-rotating units, you'll also need to check propeller shaft free-play to determine shaft-to-carrier shimming

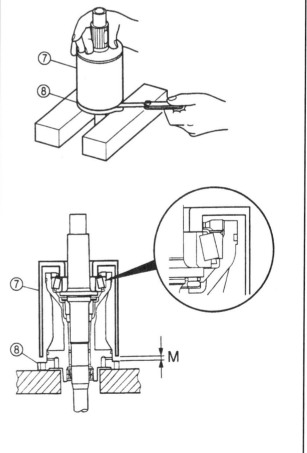

Fig. 247 On the largest counter-rotating units, the gap is checked between the gauge and carrier flange

mounting deviation from the standard spec, in hundredths of a mm. This means that if there is a marking such as R +5, it means that the case requires a shim that it 5/100mm (0.05mm) larger than standard. The **+** or **-** accompanying the number tells you whether you should add (**+**) or take away (**-**) the number from the standard measurement. If there is NO number after the **R**, assume the value is 0 and the case meets standard tolerance. If the number is there, but illegible, you'll have to work from 0 and make adjustments for that shim by checking lash.

■ **All measurements should be taken a no less than 3 points around the bearing and the measurements should be averaged.**

2. For US and Canadian Models, place the shimming gauge squarely over top of the gear with the legs in contact with the bearing carrier and press down firmly on the gauge as you insert a feeler gauge between the gear and the tab on the shimming gauge. Measure and calculate the shim changes depending upon the model as follows:

• For 48 hp 2-cylinder motors, use gauge (#YB-34468-3). The gap between the gear and the tab on the gauge should be 1.10mm MINUS the stamped **R** deviation (remember that if the standard deviation is a negative number, that subtracting a negative is the same as adding the positive or absolute value of that number). If not add or subtract shim material to achieve the specified gap.

• For 40/50 hp (698cc) motors, no shim measurement tool is available for these models. All shim changes should be made using Backlash measurements. If the original shim is not available, start adjustment using the largest of the shims (0.50mm).

• For 50-90 hp (849cc and 1140cc) motors, use gauge (#YB-34468-5) for 50-70 hp models or (#YB-34468-3) for 75-90 hp models. The gap between the gear and the tab on the gauge should be 1.10mm (for 50-70 hp models) or 1.50mm (for 75-90 hp models), MINUS the stamped **R** deviation (remember that if the standard deviation is a negative number, that subtracting a negative is the same as adding the positive or absolute value of that number). If not add or subtract shim material to achieve the specified gap.

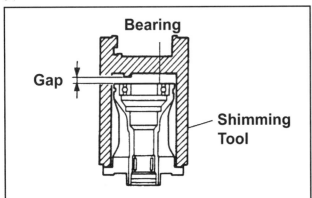

Fig. 248 . . .though on the largest gearcases, you'll check between the gauge and bearing (no gear)

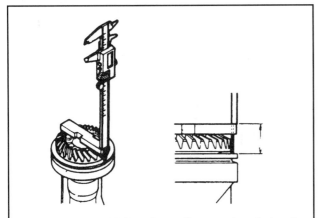

Fig. 249 For most worldwide units, you'll measure from the top of a shimming plate to the thrust plate on the carrier

• For V4 motors, use gauge (#YB-34468-2). The gap between the gear and the tab on the gauge should be 1.80mm (for the reverse gear on standard rotation units) or 2.70mm (for the forward gear on counter-rotation units), MINUS the stamped **R** deviation (remember that if the standard deviation is a negative number, that subtracting a negative is the same as adding the positive or absolute value of that number). If not add or subtract shim material to achieve the specified gap.

On counter-rotation units there is an additional shim found in the propeller shaft bearing carrier (between the carrier and the thrust bearing and on usually a thrust washer). To determine the proper shim size to use here, assemble the propeller shaft with tapered roller bearing, thrust bearing, washer (if used), old shim and the bearing carrier, then place the assembly gearcase side down in a vise. Attach a dial gauge to the very end of the propeller shaft and check shaft free-play by pulling and pushing on the shaft. Free-play must be 0.25-0.35mm (0.010-0.014 in.) or it must be adjusted using the propeller shaft shim.

• For 2.6L V6 motors, the gauge and spec varies slightly from model-to-model. For MOST models use gauge (#YB-34468-1). The exceptions are the 150L/PX150 models which use (#YB-34468-6) and all counter-rotation units which use (#YB-34468-2). The gap between the gear and the tab on the gauge should be a standard spec (which varies by model), MINUS the stamped **R** deviation (remember that if the standard deviation is a negative number, that subtracting a negative is the same as adding the positive or absolute value of that number). For most standard rotation units the standard spec is 1.80mm, however the 150L/PX150 models are 1.30mm and the Z150Q/VZ150 models are 0.90mm. Counter-rotation units (remember this is the forward gear on these units) are also 1.30mm. Dual-propeller models are 1.60mm. If the gap does not equal this spec minus the deviation, add or subtract shim material to achieve the specified gap.

Furthermore, on counter-rotation units there is an additional shim found in the propeller shaft bearing carrier (between the carrier and the thrust bearing and thrust washer). To determine the proper shim size to use here, assemble the propeller shaft with tapered roller bearing, thrust bearing, washer, old shim and the bearing carrier, then place the assembly gearcase side down in a vise. Attach a dial gauge to the very end of the propeller shaft and check shaft free-play by pulling and pushing on the shaft. Free-play must be 0.25-0.35mm (0.010-0.014 in.) or it must be adjusted using the propeller shaft shim.

• For 3.1L V6 motors, use the Yamaha shimming gauge (#YB-06439) for standard rotation units or (#YB-06440) for counter-rotating units. In all cases, the gauge fits over the partially assembled propeller shaft carrier (the bearing is installed but NOT the gear) to contact the mounting flange of the carrier and provide a gap at a tab over the bearing (at the other end from the mounting flange) which can then be measured with a feeler gauge. The gap must be equal to 0.50mm (for standard rotation units) or 0.60mm (for counter-rotating units), MINUS the stamped **R** deviation (remember that if the standard deviation is a negative number, that subtracting a negative is the same as adding the positive or absolute value of that number). If the gap does not equal this spec minus the deviation, add or subtract shim material to achieve the specified gap.

Furthermore, on counter-rotation units there is an additional shim found in the propeller shaft bearing carrier (between the carrier and the thrust bearing/thrust washer). To determine the proper shim size to use here, assemble the propeller shaft with tapered roller bearing, thrust bearing, washer, old shim and the bearing carrier (all retained by the claw washer and ring nut and tightened to spec), then place the assembly gearcase side down in a vise. Attach a dial gauge to the very end of the propeller shaft and check shaft free-play by pulling and pushing on the shaft. Free-play must be 0.25-0.35mm (0. 0.010-0.014 in.) or it must be adjusted using the propeller shaft shim.

• For standard rotation 3.3L V6 motors, use the Yamaha shimming gauge (#YB-06439). The gauge fits over the partially assembled propeller shaft carrier (the bearing is installed but NOT the gear) to contact the mounting flange of the carrier and provide a gap at a tab over the bearing (at the other end from the mounting flange) which can then be measured with a feeler gauge. The gap must be equal to 0.50mm, MINUS the stamped **R** deviation (remember that if the standard deviation is a negative number, that subtracting a negative is the same as adding the positive or absolute value of that number). If the gap does not equal this spec minus the deviation, add or subtract shim material to achieve the specified gap.

• For counter-rotation 3.3L V6 motors, you'll need to assemble the rear propeller shaft and bearing carrier assembly (without the gear) before measurements can be taken. For details, refer to the Assembly procedures found earlier in this section, but once the propeller shaft is installed to the bearing carrier along with the old shim material, thrust washer, thrust

LOWER UNIT 7-73

bearing, tapered roller bearing and outer race, the claw washer and ring nut should be installed and tightened to spec. First, use Yamaha shimming gauge (#YB-06440-A) placed over the rear (gearcase side) of the housing and measure the gap between the gauge and the bearing carrier mounting flange. This gap should be equal to 2.50mm, PLUS the stamped **R** deviation (remember that even though this is the forward bearing for this gearcase, most cases are manufactured and measured assuming the reverse bearing is installed in the carrier, and so are stamped assuming this as well). Add or remove forward gear shims to achieve the proper gap.

Furthermore, on these counter-rotation units there is an additional shim found in the propeller shaft bearing carrier (between the carrier and the thrust bearing/thrust washer). To determine the proper shim size to use here, place the assembled carrier with the gearcase side down in a vise. Attach a dial gauge to the very end of the propeller shaft and check shaft free-play by pulling and pushing on the shaft. Free-play must be 0.25-0.35mm (0.010-0.014 in.) or it must be adjusted using the propeller shaft shim.

3. For World Production Models, use place the flat-stock shimming plate over the gear (assembled into the bearing carrier already) and measure the distance from the TOP of the shimming plate to the thrust plate on top of the carrier. Calculate the shim changes depending upon the model as follows:
- For 48 hp 2-cylinder motors, shim changes are equal to the measurement MINUS 25.4mm, MINUS the **R** stamped gearcase deviation (remember that if the standard deviation is a negative number, that subtracting a negative is the same as adding the positive or absolute value of that number).
- For 40/50 hp (698cc) motors, no shim measurement tool is available for these models. All shim changes should be made using Backlash measurements. If the original shim is not available, start adjustment using the largest of the shims (0.50mm).
- For 50-90 hp (849cc and 1140cc) motors, shim changes are equal to the measurement, MINUS 25.4mm (for 50 hp models) or 28.0mm (for 60-70 hp models) or 26.0mm (for 75-90 hp models), MINUS the **R** stamped gearcase deviation (remember that if the standard deviation is a negative number, that subtracting a negative is the same as adding the positive or absolute value of that number).
- For V4 motors, shim changes are equal to the measurement, MINUS 27.4mm (for standard rotation units) or 28.1mm (for counter-rotation units), MINUS the **R** stamped gearcase deviation (remember that if the standard deviation is a negative number, that subtracting a negative is the same as adding the positive or absolute value of that number).

On counter-rotation units there is an additional shim found in the propeller shaft bearing carrier (between the carrier and the thrust bearing/thrust washer). To determine the proper shim size to use here, assemble the propeller shaft with tapered roller bearing, thrust bearing, washer (if used), old shim and the bearing carrier, then place the assembly gearcase side down in a vise. Attach a dial gauge to the very end of the propeller shaft and check shaft free-play by pulling and pushing on the shaft. Free-play must be 0.25-0.35mm (0.010-0.014 in.) or it must be adjusted using the propeller shaft shim.

- For 2.6L V6 motors, shim changes are equal to the measurement, MINUS the standard spec (which varies slightly by model), MINUS the **R** stamped gearcase deviation (remember that if the standard deviation is a negative number, that subtracting a negative is the same as adding the positive or absolute value of that number). For most standard rotation units the standard spec is 29.0mm, however the 150L/PX150 models and Z150Q/VZ150 models are 29.9mm. Counter-rotation units (remember this is the forward gear on these units) are also 29.5mm.

On counter-rotation units there is an additional shim found in the propeller shaft bearing carrier (between the carrier and the thrust bearing/thrust washer). To determine the proper shim size to use here, assemble the propeller shaft with tapered roller bearing, thrust bearing, washer, old shim and the bearing carrier, then place the assembly gearcase side down in a vise. Attach a dial gauge to the very end of the propeller shaft and check shaft free-play by pulling and pushing on the shaft. Free-play must be 0.25-0.35mm (0.010-0.014 in.) or it must be adjusted using the propeller shaft shim.

- For 3.1L V6 motors, you'll need to take a measurement (or two on counter-rotating models), then use that measurement, along with the standard specification for the model, the **R** stamped gearcase deviation AND a second stamped deviation (this time it is found on the side of the propeller shaft bearing carrier and identified as **A**. All of these numbers are placed into the appropriate formula to determine necessary shim sizes. For all motors, place the bearing on the Yamaha shimming plate (#90890-06701). Now with the bearing in this position for standard rotation units measure the thickness of the bearing along the outer edge of the race and use the measurement in the following formula, 21.0mm, PLUS the **R** stamped gearcase deviation, MINUS **A** stamped bearing carrier deviation (remember if the stamped deviation is a negative that subtracting a negative is the same as adding the positive or absolute value), MINUS the measurement. The result is the necessary shim size.

For counter-rotating units, make sure the bearing is sitting on the shimming plate with the wider tapered end toward the bottom and measure the thickness of the bearing long the inner edge of the race. Next on counter-rotating units, invert the bearing (wider edge of the rollers on top) and place the shimming plate on TOP of the bearing, then measure the distance from the TOP of the shimming plate to the top of the OUTER bearing race. Finally determine the necessary bearing thickness with the following formula, 6.50mm, PLUS the **R** stamped gearcase deviation, MINUS **A** stamped bearing carrier deviation (remember if the stamped deviation is a negative that subtracting a negative is the same as adding the positive or absolute value), MINUS the first measurement (total inner race bearing thickness), PLUS the second measurement (the distance from the top of the outer race to the top of the shimming plate).

Furthermore, on counter-rotation units there is an additional shim found in the propeller shaft bearing carrier (between the carrier and the thrust bearing/thrust washer). To determine the proper shim size to use here, assemble the propeller shaft with tapered roller bearing, thrust bearing, washer, old shim and the bearing carrier (all retained by the claw washer and ring nut and tightened to spec), then place the assembly gearcase side down in a vise. Attach a dial gauge to the very end of the propeller shaft and check shaft free-play by pulling and pushing on the shaft. Free-play must be 0.25-0.35mm (0.010-0.014 in.) or it must be adjusted using the propeller shaft shim.

4. Once you've selected the proper shims, assemble the gearcase as detailed earlier in this section and, before the water pump is installed, check the gearcase Backlash as detailed later in this section. As long as the backlash is within spec, no further adjustment is necessary. However, if backlash is out of specification you'll need to further adjust one or more of the shims to correct the problem.

Backlash

◆ See Figures 250 and 251

■ If you're working on a counter-rotating unit, remember the gear positions are reversed so you don't get too confused. The nose-cone gear, referred to as the Forward gear throughout this procedure, is actually the Reverse gear on counter-rotating gearcases. Similarly, the propeller shaft bearing carrier gear, referred to as the Reverse gear throughout this procedure, is actually the Forward gear.

Backlash (also known as lash or play) is the acceptable clearance between two meshing gears, in order to take into account possible errors in machining, deformation due to load, expansion due to heat generated in the lower unit and center-to-center distance tolerances. A no backlash condition is unacceptable, as such a condition would mean the gears are locked together or are too tight against each other which would cause phenomenal wear and generate excessive heat from the resulting friction.

Excessive backlash which cannot be corrected with shim material adjustment indicates worn gears. Such worn gears must be replaced. Excessive backlash is usually accompanied by a loud whine when the lower unit is operating in neutral gear.

The backlash measurements are taken before the water pump is installed. If the amount of backlash needs to be adjusted, the lower unit must be partially disassembled to change the amount of shim material behind one or more of the gears. Of course, it usually does not make sense to change shims on BOTH the pinion AND either the forward or reverse gears at the same time, since the changes will normally have an aggregate affect on each other's readings and you're just making it harder on yourself.

As a general rule, if the lower unit was merely disassembled, cleaned and then assembled with only a new water pump impeller, new gaskets, seals and O-rings, there is no reason to believe the backlash would have changed. Therefore, it is safe to say this procedure may be skipped.

However, if any one or more of the following components were replaced, the gear backlash should be checked for possible shim adjustment (even if shimming calculations were performed):

7-74 LOWER UNIT

- New lower unit housing - check forward and reverse gear backlash.
- New forward gear bearing - check forward gear backlash.
- New pinion gear - check forward/reverse gear backlash.
- New forward gear - check forward gear backlash.
- New reverse gear - check reverse gear backlash.
- New bearing carrier - check reverse gear backlash.
- New thrust washers - check the affected gear's backlash.

■ The lower unit backlash is generally measured with the unit right side up for all inline motors, except all V4/V6 motors, where the backlash is measured with the gearcase inverted (upside down).

To perform this check you'll need a dial gauge on a magnetic or threaded base, as well as a gear lash indicator. The lash indicator is essentially a hose clamp with a straight lever attached at a 90 degree angle to the clamp body. The clamp is attached to the driveshaft in a position that any movement in the shaft will result in movement at the lever. The dial indicator is then installed to the gearcase itself in a position contacting the lever to read the amount of movement. The only thing that stops you from fabricating the indicator yourself if the fact that the amount of lever movement (when compared to shaft rotation) increases dramatically as you move outward away from the clamp further down the lever. And Yamaha simply directs you to use the "mark" provided on the indicator when measuring lash. For all motors you can use the Yamaha Indicator (#YB-06265).

In order to perform this check it is necessary to properly pre-load the Forward or Reverse gears. The method of achieving this is similar on all models. For the nose cone gear (Forward on most units) this is done by attaching a universal puller to the bearing carrier and tightening it lightly against the propeller shaft. For the propeller shaft bearing carrier gear this is normally achieved by installing the propeller in some particular fashion (reversed, without a specific spacer or with a spacer repositioned or facing some specific direction). Specifics will be given in the bullet lists for each model.

1. Position the gearcase in a suitable workstand either upright or inverted, depending upon the model. The lower unit backlash is measured with the unit inverted (upside down) for all V4/V6 models or in the upright position for all inline motors.

■ Some people prefer to set up the backlash indicator and dial gauge while the gearcase is still upright (just because it is easier than working on it while it is upside down). If so, perform the next couple of steps first, then invert the gearcase.

2. Make sure the gearcase is in **neutral** by slowly spinning the driveshaft and checking for a lack of motion at the propeller shaft. Although most gearcases are checked with the shifter in Neutral, there are some exceptions which are listed later in this procedure.

■ We're not really sure why Yamaha chooses to preload gears by pushing on the prop shaft when checking Forward gear backlash and pulling on the prop shaft when checking Reverser gear backlash. Some other manufacturers (and on some other Yamahas) the preferred method is simply to shift the gearcase into the gear whose lash is being checked. This is always an alternative if you can't seem to get the prop shaft and gear loaded sufficiently to check lash.

3. Since you normally start by checking Forward gear backlash, install a bearing carrier puller and J bolts onto the ribs of the carrier. With the bearing carrier attaching bolts or the bearing carrier locknut secured, tighten the puller just enough to push on the propeller shaft preloading it (if a specific torque is required on a given model, it will be listed in the bullet list later in this procedure with the backlash specifications).

■ The bearing carrier puller is used to both keep the propeller shaft from turning and to preload the forward gear, thus freeing up one of your hands. However, you can simply push inward and hold the shaft (or have an assistant hold the shaft) while taking readings.

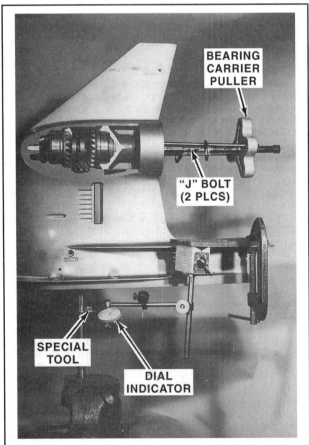

Fig. 250 Use a puller to preload the gear when checking forward gear lash

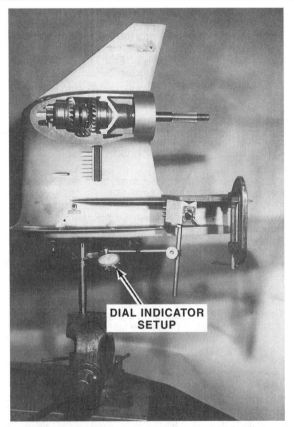

Fig. 251 When checking reverse gear lash most models either install the propeller minus a particular spacer and/or shift the gearcase into reverse to preload the gear

4. Obtain and install backlash indicator gauge (#YB-06265) onto the driveshaft. If the hose clamp for the indicator is too large to fit the propeller shaft snugly, insert some kind of packing into the clamp to allow it to be tightened. The clamp must be secure to permit no movement of the indicator on the shaft.

5. Install the dial gauge to the gearcase and position the pointer against the mark on the indicator tool.

6. If you are not using a puller to hold the propeller shaft from turning, grab a hold of the propeller shaft in order to keep it from turning, then twist the driveshaft in one direction (either clockwise or counterclockwise, it doesn't matter at this time) and zero the dial gauge.

■ **See the specifications listed below before attempting to read gear backlash. On some models other special conditions are necessary, such as pulling or pushing on the driveshaft as it is rotated or installing the propeller without the spacer to pre-load the reverse gear when checking reverse gear lash.**

7. Now, continue to hold the propeller shaft to prevent the shaft from rotating. With the other hand gently rock the driveshaft back and forth. In this way, you'll move the driveshaft back and forth between gear contact points with the forward or reverse gear (which can't move because you're holding the propeller shaft). Note the maximum deflection of the dial gauge needle, this is the amount of gear lash present in the gearcase for the current shims in use and compare them to the specifications, as follows:

- For 48 hp 2-cylinder motors, check backlash with the gearcase upright and while pulling upward lightly on the driveshaft. The center bolt on the puller which is used to preload the forward gear should be tightened to 3.6 ft. lbs. (5 Nm) so the gear restricts movement sufficiently to measure lash. The forward gear backlash should be about 0.09-0.27mm (0.004-0.011 in.). If lash is LESS than 0.09mm, determine the amount of shim to remove by subtracting the measurement from 0.18mm and then multiplying the result by 0.55mm. If the amount of lash is MORE than 0.27mm, determine the amount of shim to add by subtracting 0.18mm from the measurement and then multiplying the result by 0.55mm. To check reverse gear backlash, remove the puller and install the propeller facing backwards and tighten the nut to 3.6 ft. lbs. (5 Nm) to preload the reverse gear. Take the readings in the same manner as forward gear. Reverse gear backlash should be about 0.90-1.26mm (0.035-0.050 in.). If lash is LESS than 0.90mm, determine the amount of shim to remove by subtracting the measurement from 1.08mm and then multiplying the result by 0.55mm. If the amount of lash is MORE than 1.26mm, determine the amount of shim to add by subtracting 1.08mm from the measurement and then multiplying the result by 0.55mm.

■ **When making gear lash adjustments. Remember, too much lash means there is too much space between the gear in question and the pinion, therefore ADDING shim material will DECREASE backlash. Too little lash means there is not a large enough gap, therefore REMOVING shim material will INCREASE backlash.**

- For 40/50 hp (698cc) motors, check backlash with the gearcase upright and while pulling upward lightly on the driveshaft. The center bolt on the puller which is used to preload the forward gear should be tightened to 3.6 ft. lbs. (5 Nm) so the gear restricts movement sufficiently to measure lash. The forward gear backlash should be about 0.18-0.45mm (0.007-0.018 in.). If lash is LESS than 0.18mm, determine the amount of shim to remove by subtracting the measurement from 0.31mm and then multiplying the result by 0.56mm. If the amount of lash is MORE than 0.45mm, determine the amount of shim to add by subtracting 0.31mm from the measurement and then multiplying the result by 0.56mm. To check reverse gear backlash, remove the puller and install the propeller facing backwards and tighten the nut to 3.6 ft. lbs. (5 Nm) to preload the reverse gear. Take the readings in the same manner as forward gear. Reverse gear backlash should be about 0.71-0.98mm (0.028-0.039 in.). If lash is LESS than 0.71mm, determine the amount of shim to remove by subtracting the measurement from 0.85mm and then multiplying the result by 0.56mm. If the amount of lash is MORE than 0.98mm, determine the amount of shim to add by subtracting 0.85mm from the measurement and then multiplying the result by 0.56mm.

- For 50 hp (849cc) motors, check backlash with the gearcase upright and while pulling upward lightly on the driveshaft. The center bolt on the puller which is used to preload the forward gear should be tightened to 3.6 ft. lbs. (5 Nm). The forward gear backlash should be about 0.08-0.25mm (0.003-0.009 in.). If lash is LESS than 0.08mm, determine the amount of shim to remove by subtracting the measurement from 0.17mm and then multiplying the result by 0.60mm. If the amount of lash is MORE than 0.25mm, determine the amount of shim to add by subtracting 0.17mm from the measurement and then multiplying the result by 0.60mm. To check reverse gear backlash, remove the puller and install the propeller facing backwards, then tighten the propeller nut to 3.6 ft. lbs. (5 Nm). Take the readings in the same manner as forward gear. Reverse gear backlash should be about 0.84-1.17mm (0.033-0.046 in.). If lash is LESS than 0.84mm, determine the amount of shim to remove by subtracting the measurement from 1.01mm and then multiplying the result by 0.60mm. If the amount of lash is MORE than 1.17mm, determine the amount of shim to add by subtracting 1.01mm from the measurement and then multiplying the result by 0.60mm.

- For 60/70 hp (849cc) motors, check backlash with the gearcase upright and while pulling upward lightly on the driveshaft. The center bolt on the puller which is used to preload the forward gear should be tightened to 3.6 ft. lbs. (5 Nm). The forward gear backlash should be about 0.009-0.28mm (0.004-0.011 in.). If lash is LESS than 0.09mm, determine the amount of shim to remove by subtracting the measurement from 0.19mm and then multiplying the result by 0.53mm. If the amount of lash is MORE than 0.28mm, determine the amount of shim to add by subtracting 0.19mm from the measurement and then multiplying the result by 0.53mm. To check reverse gear backlash, remove the puller and install the propeller facing backwards, then tighten the propeller nut to 3.6 ft. lbs. (5 Nm). Take the readings in the same manner as forward gear. Reverse gear backlash should be about 0.75-1.13mm (0.030-0.044 in.). If lash is LESS than 0.75mm, determine the amount of shim to remove by subtracting the measurement from 0.94mm and then multiplying the result by 0.53mm. If the amount of lash is MORE than 1.13mm, determine the amount of shim to add by subtracting 0.94mm from the measurement and then multiplying the result by 0.53mm.

- For 75-90 hp (1140cc) motors, check backlash with the gearcase upright and while pulling upward lightly on the driveshaft. The center bolt on the puller which is used to preload the forward gear should be tightened to 3.6 ft. lbs. (5 Nm). The forward gear backlash should be about 0.08-0.25mm (0.003-0.009 in.). If lash is LESS than 0.08mm, determine the amount of shim to remove by subtracting the measurement from 0.17mm and then multiplying the result by 0.60mm. If the amount of lash is MORE than 0.25mm, determine the amount of shim to add by subtracting 0.17mm from the measurement and then multiplying the result by 0.60mm. To check reverse gear backlash, remove the puller and install the propeller facing backwards, then tighten the propeller nut to 3.6 ft. lbs. (5 Nm). Take the readings in the same manner as forward gear. Reverse gear backlash should be about 0.67-1.00mm (0.026-0.039 in.). If lash is LESS than 0.67mm, determine the amount of shim to remove by subtracting the measurement from 0.84mm and then multiplying the result by 0.60mm. If the amount of lash is MORE than 1.00mm, determine the amount of shim to add by subtracting 0.84mm from the measurement and then multiplying the result by 0.60mm.

- For standard-rotation V4 motors, check backlash with the gearcase inverted (upside down) and with the backlash indicator attached to the 22.4mm (0.88 in.) diameter portion of the driveshaft. The center bolt on the puller which is used to preload the forward gear should be tightened to 7.2 ft. lbs. (10 Nm). The forward gear backlash should be about 0.10-0.31mm (0.008-0.012 in.). If lash is LESS than 0.20mm, determine the amount of shim to remove by subtracting the measurement from 0.26mm and then multiplying the result by 0.58mm. If the amount of lash is MORE than 0.31mm, determine the amount of shim to add by subtracting 0.26mm from the measurement and then multiplying the result by 0.58mm.

Reverse gear backlash is checked by removing the puller and then installing the propeller, but without the spacer that normally installs between the propeller and the gearcase. Tighten the propeller retaining nut to 7.2 ft. lbs. (10 Nm) in order to preload the reverse gear. Otherwise check reverse gear backlash in the same way as the forward gear, it should be about 0.50-0.73mm (0.020-0.029 in.). If lash is LESS than 0.50mm, determine the amount of shim to remove by subtracting the measurement from 0.62mm and then multiplying the result by 0.58mm. If the amount of lash is MORE than 0.73mm, determine the amount of shim to add by subtracting 0.62mm from the measurement and then multiplying the result by 0.58mm.

- For counter-rotation V4 motors, check backlash with the gearcase inverted (upside down) and with the backlash indicator attached to the 22.4mm (0.88 in.) diameter portion of the driveshaft. The center bolt on the puller which is used to preload the forward gear should be tightened to 7.2 ft. lbs. (10 Nm). The forward gear backlash should be about 0.15-0.30mm (0.006-0.012 in.). If lash is LESS than 0.15mm, determine the amount of

7-76 LOWER UNIT

shim to remove by subtracting the measurement from 0.23mm and then multiplying the result by 0.58mm. If the amount of lash is MORE than 0.30mm, determine the amount of shim to add by subtracting 0.23mm from the measurement and then multiplying the result by 0.58mm.

Reverse gear backlash is checked by removing the puller and then installing the propeller, but without the spacer that normally installs between the propeller and the gearcase. Tighten the propeller retaining nut to 7.2 ft. lbs. (10 Nm), then shift the gearcase into Reverse and SLOWLY turn the driveshaft clockwise until the clutch dog fully engages Reverse gear ensuring the gear is fully preloaded. Now turn the shift rod back into Neutral and turn the driveshaft about 30° counterclockwise. Finally turn the shift rod BACK into Reverse and slowly check backlash from one stop to the other, it should be about 0.50-0.70mm (0.020-0.028 in.). If lash is LESS than 0.50mm, determine the amount of shim to remove by subtracting the measurement from 0.60mm and then multiplying the result by 0.58mm. If the amount of lash is MORE than 0.70mm, determine the amount of shim to add by subtracting 0.60mm from the measurement and then multiplying the result by 0.58mm.

- For standard-rotation V4 motors, check backlash with the gearcase inverted (upside down) and with the backlash indicator attached to the 22.4mm (0.88 in.) diameter portion of the driveshaft. The center bolt on the puller which is used to preload the forward gear should be tightened to 3.6 ft. lbs. (5 Nm) for models through 1998 or to 7.2 ft. lbs. (10 Nm) for 1999 and later models. In addition, on models through 1998 Yamaha recommended shifting the gearcase into Forward gear. The forward gear backlash should be about 0.32-0.50mm (0.013-0.020 in.). If lash is LESS than 0.32mm, determine the amount of shim to remove by subtracting the measurement from 0.41mm and then multiplying the result by 0.63mm (for models through 1998) or by 1.25 (for 1999 and later models). If the amount of lash is MORE than 0.50mm, determine the amount of shim to add by subtracting 0.41mm from the measurement and then multiplying the result by 0.63mm (for models through 1998) or by 1.25 (for 1999 and later models).

Reverse gear backlash is checked by removing the puller and then installing the propeller, but without the spacer that normally installs between the propeller and the gearcase. Tighten the propeller retaining nut to the same spec as the center bolt on the puller. On models through 1998, shift the gearcase into Reverse, but for 1999 and later models leave the shifter in Neutral. Otherwise check reverse gear backlash in the same way as the forward gear, it should be about 0.80-1.17mm (0.031-0.046 in.). If lash is LESS than 0.80mm, determine the amount of shim to remove by subtracting the measurement from 0.99mm and then multiplying the result by 0.63mm (for models through 1998) or by 1.25 (for 1999 and later models). If the amount of lash is MORE than 1.17mm, determine the amount of shim to add by subtracting 0.99mm from the measurement and then multiplying the result by 0.63mm (for models through 1998) or by 1.25 (for 1999 and later models).

- For counter-rotation V4 motors, check backlash with the gearcase inverted (upside down) and with the backlash indicator attached to the 22.4mm (0.88 in.) diameter portion of the driveshaft. The center bolt on the puller which is used to preload the forward gear should be tightened to 3.6 ft. lbs. (5 Nm) for models through 1998 or to 7.2 ft. lbs. (10 Nm) for 1999 and later models. The forward gear backlash should be about 0.32-0.45mm (0.013-0.018 in.). If lash is LESS than 0.32mm, determine the amount of shim to remove by subtracting the measurement from 0.39mm and then multiplying the result by 0.63mm (for models through 1998) or by 1.25 (for 1999 and later models). If the amount of lash is MORE than 0.45mm, determine the amount of shim to add by subtracting 0.39mm from the measurement and then multiplying the result by 0.63mm (for models through 1998) or by 1.25 (for 1999 and later models).

Reverse gear backlash is checked by removing the puller and then installing the propeller, but without the spacer that normally installs between the propeller and the gearcase. Tighten the propeller retaining nut to the same spec as the center bolt on the puller. On models through 1998, shift the gearcase into Reverse, but for 1999 and later models leave the shifter in Neutral. Otherwise check reverse gear backlash in the same way as the forward gear, it should be about 0.80-1.12mm (0.031-0.044 in.). If lash is LESS than 0.80mm, determine the amount of shim to remove by subtracting the measurement from 0.96mm and then multiplying the result by 0.68mm (for models through 1998) or by 1.25 (for 1999 and later models). If the amount of lash is MORE than 1.12mm, determine the amount of shim to add by subtracting 0.96mm from the measurement and then multiplying the result by 0.68mm (for models through 1998) or by 1.25 (for 1999 and later models).

- For standard rotation 2.6L V6 motors (except 150L/PX150 and Z150Q/VZ150 models), check backlash with the gearcase inverted (upside down) and with the backlash indicator attached to the 22.4mm (0.88 in.) diameter portion of the driveshaft. The center bolt on the puller which is used to preload the forward gear should be tightened 7.2 ft. lbs. (10 Nm). The forward gear backlash should be about 0.25-0.46mm (0.010-0.018 in.). If lash is LESS than 0.25mm, determine the amount of shim to remove by subtracting the measurement from 0.36mm and then multiplying the result by 0.54mm. If the amount of lash is MORE than 0.46mm, determine the amount of shim to add by subtracting 0.36mm from the measurement and then multiplying the result by 0.54mm.

Reverse gear backlash is checked by removing the puller and then installing the propeller, but without the spacer that normally installs between the propeller and the gearcase. Tighten the propeller retaining nut to the same spec as the center bolt on the puller. Check reverse gear backlash in the same way as the forward gear, it should be about 0.74-1.29mm (0.029-0.051 in.). If lash is LESS than 0.74mm, determine the amount of shim to remove by subtracting the measurement from 1.02mm and then multiplying the result by 0.54mm. If the amount of lash is MORE than 1.29mm, determine the amount of shim to add by subtracting 1.02mm from the measurement and then multiplying the result by 0.54mm.

- For 150L/PX150 and Z150Q/VZ150 models of the standard rotation 2.6L V6 motors, check backlash with the gearcase inverted (upside down) and with the backlash indicator attached to the 22.4mm (0.88 in.) diameter portion of the driveshaft. The center bolt on the puller which is used to preload the forward gear should be tightened 7.2 ft. lbs. (10 Nm). The forward gear backlash should be about 0.71-1.01mm (0.028-0.040 in.). If lash is LESS than 0.71mm, determine the amount of shim to remove by subtracting the measurement from 0.86mm and then multiplying the result by 0.54mm. If the amount of lash is MORE than 1.01mm, determine the amount of shim to add by subtracting 0.86mm from the measurement and then multiplying the result by 0.54mm.

Reverse gear backlash is checked by removing the puller and then installing the propeller, but without the spacer that normally installs between the propeller and the gearcase. Tighten the propeller retaining nut to the same spec as the center bolt on the puller. Check reverse gear backlash in the same way as the forward gear, it should be about 0.79-1.38mm (0.031-0.054 in.). If lash is LESS than 0.79mm, determine the amount of shim to remove by subtracting the measurement from 1.09mm and then multiplying the result by 0.54mm. If the amount of lash is MORE than 1.38mm, determine the amount of shim to add by subtracting 1.09mm from the measurement and then multiplying the result by 0.54mm.

- For counter-rotation 2.6L V6 motors, check backlash with the gearcase inverted (upside down) and with the backlash indicator attached to the 22.4mm (0.88 in.) diameter portion of the driveshaft. The center bolt on the puller which is used to preload the forward gear should be tightened to 7.2 ft. lbs. (10 Nm). The forward gear backlash should be about 0.21-0.43mm (0.008-0.017 in.). If lash is LESS than 0.21mm, determine the amount of shim to remove by subtracting the measurement from 0.32mm and then multiplying the result by 0.54mm. If the amount of lash is MORE than 0.43mm, determine the amount of shim to add by subtracting 0.32mm from the measurement and then multiplying the result by 0.54mm.

Reverse gear backlash is checked by removing the puller and then installing the propeller, but without the spacer that normally installs between the propeller and the gearcase. Tighten the propeller retaining nut to 3.6 ft. lbs. (5 Nm), then shift the gearcase into Reverse and SLOWLY turn the driveshaft clockwise until the clutch dog fully engages Reverse gear ensuring the gear is fully preloaded. Now turn the shift rod back into Neutral and turn the driveshaft about 30° counterclockwise. Finally turn the shift rod BACK into Reverse and slowly check backlash from one stop to the other, it should be about 0.97-1.29mm (0.038-0.051 in.). If lash is LESS than 0.97mm, determine the amount of shim to remove by subtracting the measurement from 1.13mm and then multiplying the result by 0.54mm. If the amount of lash is MORE than 1.29mm, determine the amount of shim to add by subtracting 1.13mm from the measurement and then multiplying the result by 0.54mm.

- For dual-propeller 2.6L V6 motors, check backlash with the gearcase inverted (upside down) and with the backlash indicator attached to the 22.4mm (0.88 in.) diameter portion of the driveshaft. Attach a puller so that it pushes on the against the inner propeller shaft, then tighten the center bolt on the puller which to 7.2 ft. lbs. (10 Nm) in order to preload the forward gear in the lower unit. The forward gear backlash should be about 0.19-0.59mm (0.007-0.023 in.). If lash is LESS than 0.19mm, determine the amount of shim to remove by subtracting the measurement from 0.39mm and then multiplying the result by 0.50mm. If the amount of lash is MORE than 0.59mm, determine the amount of shim to add by subtracting 0.39mm from the measurement and then multiplying the result by 0.50mm.

LOWER UNIT 7-77

Reverse gear backlash is checked by removing the puller and then installing it again, but this time using an adapter that slides OVER the inner propeller shaft and applies force from the center screw to the OUTER propeller shaft only. The correct diameter length of pipe and a flat-washer should be sufficient to fabricate this adapter, then tighten the center screw of the puller to 7.2 ft. lbs. (10 Nm) in order to properly preload the Reverse gear in the lower unit. Otherwise, check reverse gear backlash in the same way as the forward gear, it should be about 0.39-0.70mm (0.015-0.027 in.). If lash is LESS than 0.39mm, determine the amount of shim to remove by subtracting the measurement from 0.55mm and then multiplying the result by 0.50mm. If the amount of lash is MORE than 0.70mm, determine the amount of shim to add by subtracting 0.55mm from the measurement and then multiplying the result by 0.50mm.

• For standard rotation 3.1L V6 motors (except the 250 hp Vmax), check backlash with the gearcase inverted (upside down) and with the backlash indicator attached to the 22.4mm (0.88 in.) diameter portion of the driveshaft. The center bolt on the puller which is used to preload the forward gear should be tightened 7.2 ft. lbs. (10 Nm). The forward gear backlash should be about 0.19-0.40mm (0.007-0.016 in.). If lash is LESS than 0.19mm, determine the amount of shim to remove by subtracting the measurement from 0.30mm and then multiplying the result by 0.78mm. If the amount of lash is MORE than 0.40mm, determine the amount of shim to add by subtracting 0.30mm from the measurement and then multiplying the result by 0.78mm.

Reverse gear backlash is checked by removing the puller and then installing the propeller, but without the spacer that normally installs between the propeller and the gearcase. Tighten the propeller retaining nut to the same spec as the center bolt on the puller. Check reverse gear backlash in the same way as the forward gear, it should be about 0.64-0.93mm (0.025-0.037 in.). If lash is LESS than 0.64mm, determine the amount of shim to remove by subtracting the measurement from 0.79mm and then multiplying the result by 0.78mm. If the amount of lash is MORE than 0.93mm, determine the amount of shim to add by subtracting 0.79mm from the measurement and then multiplying the result by 0.78mm.

• For standard rotation 250 hp Vmax 3.1L V6 motors, check backlash with the gearcase inverted (upside down) and with the backlash indicator attached to the 22.4mm (0.88 in.) diameter portion of the driveshaft. The center bolt on the puller which is used to preload the forward gear should be tightened 7.2 ft. lbs. (10 Nm). The forward gear backlash should be about 0.19-0.38mm (0.007-0.015 in.). If lash is LESS than 0.19mm, determine the amount of shim to remove by subtracting the measurement from 0.29mm and then multiplying the result by 0.81mm. If the amount of lash is MORE than 0.38mm, determine the amount of shim to add by subtracting 0.29mm from the measurement and then multiplying the result by 0.81mm.

Reverse gear backlash is checked by removing the puller and then installing the propeller, but without the spacer that normally installs between the propeller and the gearcase. Tighten the propeller retaining nut to the same spec as the center bolt on the puller. Check reverse gear backlash in the same way as the forward gear, it should be about 0.62-0.90mm (0.024-0.035 in.). If lash is LESS than 0.62mm, determine the amount of shim to remove by subtracting the measurement from 0.76mm and then multiplying the result by 0.81mm. If the amount of lash is MORE than 0.90mm, determine the amount of shim to add by subtracting 0.76mm from the measurement and then multiplying the result by 0.81mm.

• For counter-rotation 3.1L V6 motors, check backlash with the gearcase inverted (upside down) and with the backlash indicator attached to the 22.4mm (0.88 in.) diameter portion of the driveshaft. The center bolt on the puller which is used to preload the forward gear should be tightened to 7.2 ft. lbs. (10 Nm). The forward gear backlash should be about 0.32-0.52mm (0.013-0.020 in.). If lash is LESS than 0.32mm, determine the amount of shim to remove by subtracting the measurement from 0.42mm and then multiplying the result by 0.78mm. If the amount of lash is MORE than 0.52mm, determine the amount of shim to add by subtracting 0.42mm from the measurement and then multiplying the result by 0.78mm.

Reverse gear backlash is checked by removing the puller and then shifting the gearcase into Reverse and SLOWLY turning the driveshaft clockwise until the clutch dog fully engages Reverse gear ensuring the gear is fully preloaded. Now turn the shift rod back into Neutral and turn the driveshaft about 30° counterclockwise. Finally turn the shift rod BACK into Reverse and slowly check backlash from one stop to the other, it should be about 0.64-0.93mm (0.025-0.037 in.). If lash is LESS than 0.64mm, determine the amount of shim to remove by subtracting the measurement from 0.79mm and then multiplying the result by 0.78mm. If the amount of lash is MORE than 0.93mm, determine the amount of shim to add by subtracting 0.79mm from the measurement and then multiplying the result by 0.78mm.

• For standard rotation 3.3L V6 motors, check backlash with the gearcase inverted (upside down) and with the backlash indicator attached to the 22.4mm (0.88 in.) diameter portion of the driveshaft. The center bolt on the puller which is used to preload the forward gear should be tightened until it keeps the propeller shaft from turning, thus preloading the forward gear. The forward gear backlash should be about 0.13-0.42mm (0.005-0.017 in.). If lash is LESS than 0.13mm, determine the amount of shim to remove by subtracting the measurement from 0.28mm and then multiplying the result by 0.78mm. If the amount of lash is MORE than 0.42mm, determine the amount of shim to add by subtracting 0.28mm from the measurement and then multiplying the result by 0.78mm.

Reverse gear backlash is checked by removing the puller and then installing the propeller, but with the spacer that normally installs between the propeller and the gearcase moved to sit between the propeller and small washer normally used with the propeller nut (this installation leaves out the large thrust washer normally found between the small nut washer and the propeller, did you get all that?). Tighten the propeller while rotating the driveshaft until it holds the shaft from turning more than between gears. Check reverse gear backlash in the same way as the forward gear, it should be about 0.64-0.93mm (0.025-0.037 in.). If lash is LESS than 0.64mm, determine the amount of shim to remove by subtracting the measurement from 0.79mm and then multiplying the result by 0.78mm. If lash is MORE than 0.93mm, determine the amount of shim to add by subtracting 0.79mm from the measurement and then multiplying the result by 0.78mm.

• For counter-rotating 3.3L V6 motors, check backlash with the gearcase inverted (upside down) and with the backlash indicator attached to the 22.4mm (0.88 in.) diameter portion of the driveshaft. The center bolt on the puller which is used to preload the forward gear should be tightened until it keeps the propeller shaft from turning, thus preloading the forward gear. The forward gear backlash should be about 0.32-0.64mm (0.013-0.025 in.). If lash is LESS than 0.32mm, determine the amount of shim to remove by subtracting the measurement from 0.48mm and then multiplying the result by 0.78mm. If the amount of lash is MORE than 0.64mm, determine the amount of shim to add by subtracting 0.48mm from the measurement and then multiplying the result by 0.78mm.

Reverse gear backlash is checked by removing the puller and then shifting the gearcase into Reverse and SLOWLY turning the driveshaft clockwise until the clutch dog fully engages Reverse gear ensuring the gear is fully preloaded. Now turn the shift rod back into Neutral and turn the driveshaft about 30° counterclockwise. Finally turn the shift rod BACK into Reverse and slowly check backlash from one stop to the other, it should be about 0.57-0.96mm (0.022-0.038 in.). If lash is LESS than 0.57mm, determine the amount of shim to remove by subtracting the measurement from 0.77mm and then multiplying the result by 0.78mm. If lash is MORE than 0.96mm, determine the amount of shim to add by subtracting 0.77mm from the measurement and then multiplying the result by 0.78mm.

■ **If the lash is too low on BOTH the forward and reverse gears, the pinion height may be too low. If the lash is too high on BOTH gears, the pinion may be too high in the gearcase. Check the installation of the bearings/bushings and the shim materials or spacers used.**

8. If adjustment is necessary, partially disassemble the gearcase as necessary to access the forward or reverse gear shims, then ad or take away shim material, as necessary to bring backlash with spec. However, if the backlash specification cannot be reached by adding and removing shim material, the gears may have to be replaced.

9. If possible, pressure test the gearcase. With the oil fill/drain screw installed and tightened, attach a threaded pressure tester to the oil level hole, then use a hand pump to apply 14.22 psi (100 kPa) of pressure to the gearcase. Watch that the pressure remains steady for at least 10 seconds. If pressure falls before that point one or more of the seals require further attention. You can immerse the gearcase in water and look for bubbles to determine where the leak is occurring. Be sure of gearcase integrity before installation.

7-78 LOWER UNIT

※※ WARNING
Don't over-pressurize the gearcase as this could CAUSE leaks.

10. Install the Water Pump to the top of the gearcase as detailed in the Lubrication and Cooling section.
11. Install the Gearcase to the intermediate housing, as detailed in this section.
12. Install the Propeller and properly refill the Gearcase Lubricant, both as detailed in the Maintenance and Tune-Up section.

Gear Mesh Pattern

◆ See Figure 252

As noted earlier, you can use a machinist dye or other powder to check the gear mesh pattern. This is useful when determining if there is too much wear on a gear set to return them to service (especially if a dial gauge set and lash indicator is not available to check gear backlash).

The basic method used to check a gear mesh pattern is to coat the gears with a dye or powder, then install the gears, carefully rotate the shafts/gears and carefully disassemble the case again to examine the gears.

All models should be held with the lower unit in the upright (normal) position.

Grasp the driveshaft and pull upward. At the same time, rotate the propeller shaft counterclockwise through about six or eight complete revolutions. This action will establish a wear pattern on the gears with the dye/Desenex® powder.

Disassemble the unit and compare the pattern made on the gear teeth with the accompanying illustrations. The pattern should almost be oval on the drive side and be positioned about halfway up the gear teeth.

If the pattern appears to be satisfactory, clean the dye or powder from the gear teeth and assemble the unit one final time.

If the pattern does not appear to be satisfactory, add or remove shim material, as required. Adding or removing shim material will move the gear pattern towards or away from the center of the teeth.

After the gear mesh pattern is determined to be satisfactory, assemble the bearing carrier one final time.

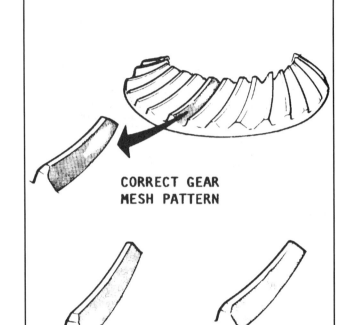

Fig. 252 Visually checking gear mesh patterns

JET DRIVE

Description & Operation

◆ See Figure 253

The jet drive unit is designed to permit boating in areas that would be impossible for a boat equipped with a conventional propeller drive system. The housing of the jet drive barely extends below the hull of the boat allowing passage in ankle deep water, white water rapids and over sand bars or in shoal water which would foul a propeller drive.

The jet drive provides reliable propulsion with a minimum of moving parts. Simply stated, water is drawn into the unit through an intake grille by an impeller driven by a driveshaft off the crankshaft of the powerhead. The water is immediately expelled under pressure through an outlet nozzle directed away from the stern of the boat.

As the speed of the boat increases and reaches planing speed, the jet drive discharges water freely into the air and only the intake grille makes contact with the water.

The jet drive is provided with a gate arrangement and linkage to permit the boat to be operated in reverse. When the gate is moved downward over the exhaust nozzle, the pressure stream is reversed by the gate (it bounces off or is turned around the thrust the opposite direction by the gate) and the boat moves sternward.

Conventional controls are used for powerhead speed, movement of the boat, shifting and power trim and tilt.

Model Identification & Serial Numbers

◆ See Figure 254

A model letter identification is stamped on the rear, port side of the jet drive housing. A serial number for the unit is stamped on the starboard side of the jet drive housing, as indicated in the accompanying illustration.

Fig. 253 Jet drive mounted to the intermediate housing of a 40 hp powerhead (a 28J model)

LOWER UNIT 7-79

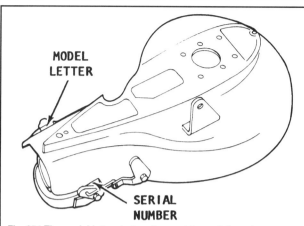

Fig. 254 The model letter designation and the serial numbers are embossed on the jet drive housing

Since their introduction by Yamaha in 1984, jet drive units have been installed on various 3-cylinder and V4 and V6 units, however starting in 2002 all jet drives were moved from 2-stroke platforms to 4-stroke platforms, specifically inline 3 and 4-cylinder powerheads.

Typically four different size jet drives are used with Yamaha outboards in these model years: Model Z and Model Y, Model AA4 and AA6. These letters are embossed on the port side of the jet drive housing.

Model Z is normally used with the 28J and 35J (Model 40 hp and 50 hp motors respectively with jet drives). Model Y is usually used with the 65J (90 hp motors). Model AA4 is typically used on the 80J (115 hp V4), while the Model AA6 is typical used only on the largest of jet drive motors, the 105J and 140J (150 hp and 200 hp V6 units, respectively).

For the most part, jet drive units are identical in design, function and operation. Differences lie in size and securing hardware, so just be sure to take the drive model number and the powerhead model number with you when ordering parts.

Jet Drive Assembly

REMOVAL & DISASSEMBLY

♦ See Figures 255 thru 268

1. Unbolt the retainer securing the shift cable to the shift cable support bracket. There are normally 2 bolts.

2. Remove the locknut, bolt and washer securing the shift cable to the shift arm. Try not to disturb the length of the cable.

3. Remove the six bolts securing the intake grille to the jet casing.

4. Ease the intake grille from the jet drive housing.

5. Pry the tab or tabs of the tabbed washer away from the nut to allow the nut to be removed.

6. Loosen and then remove the nut.

7. Remove the tabbed washer and spacers. Make a careful count of the spacers behind the washer. If the unit is relatively new, there could be as many as eight spacers stacked together. If less than eight spacers are removed from behind the washer, the others will be found behind the jet impeller, which is removed in the following step. A total of eight spacers will normally be found.

8. Remove the jet impeller from the shaft. If the impeller is frozen to the shaft, obtain a block of wood and a hammer. Carefully tap the impeller in a clockwise direction to release the shear key.

9. Slide the nylon sleeve and shear key free of the driveshaft and any spacers found behind the impeller. Make a note of the number of spacers at both locations - behind the impeller and on top of the impeller, under the nut and tabbed washer

10. Typically 1 external bolt and 4 internal bolts are used to secure the jet drive to the intermediate housing. The external bolt is located at the aft end of the anti-cavitation plate.

11. The four internal bolts are located inside the jet drive housing, as indicated in the accompanying illustration. Remove the five attaching bolts.

■ DO NOT mistake the 4 bearing housing bolts for the 4 internal jet drive assembly retaining bolts. The jet drive bolts are found outboard of the bearing housing.

12. Lower the jet drive from the intermediate housing. Remove the locating pin from the forward starboard side (or center forward, depending on the model being serviced) of the upper jet housing.
■ There area a number of locating pins (as many as 6 on some models) to be removed in the following steps. Make careful note of the size and location of each when they are removed, as an assist during assembling.

13. Remove the locating pin from the aft end of the housing. This pin and the one removed in the previous step should be of identical size.

14. Remove the four bolts and washers from the water pump housing.

15. Pull the water pump housing, the inner cartridge and the water pump impeller, up and free of the driveshaft. Remove the Woodruff key from its recess in the driveshaft. Next, remove the outer gasket, the steel plate and the inner gasket.

16. Remove the two small locating pins and lift the aluminum spacer up and free of the driveshaft.

Remove the driveshaft and bearing assembly from the housing.

■ Model Z installed has two #10-24x5/8 in. screws and lock-washers securing the driveshaft assembly to the housing. Model Y has four 1/4-20x7/8 in. bolts and lock-washers securing the driveshaft assembly to the housing.

17. Remove the large thick adaptor plate from the intermediate housing. This plate is secured with seven bolts and lock-washers. Lower the adaptor plate from the intermediate housing and remove the two small locating pins, one on the forward port side and another from the last aft hole in the adaptor plate. Both pins should be of identical size.

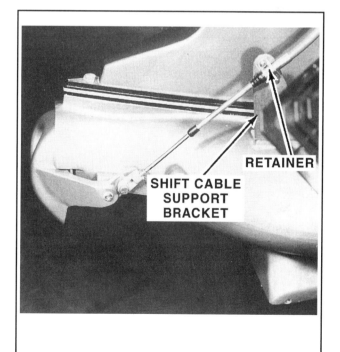

Fig. 255 Unbolt the shift cable from the support bracket...

7-80 LOWER UNIT

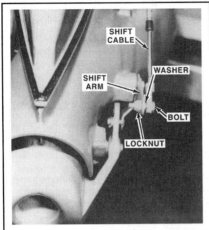

Fig. 256 . . .then disconnect the cable from the shift arm

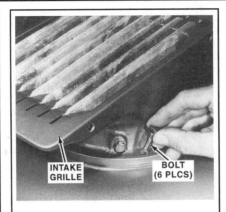

Fig. 257 Unbolt and remove the water intake grille for access to the impeller

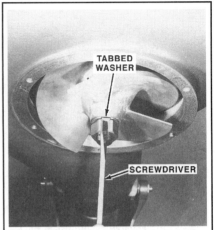

Fig. 258 To remove the impeller, bend back the locktab. . .

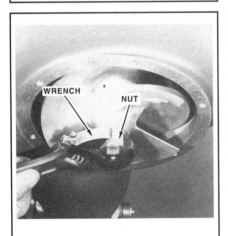

Fig. 259 . .then loosen and remove the nut

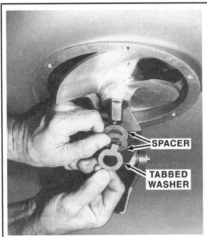

Fig. 260 Note the location of the tabbed washer and shims. . .

Fig. 261 . . .then lower the impeller from the drive

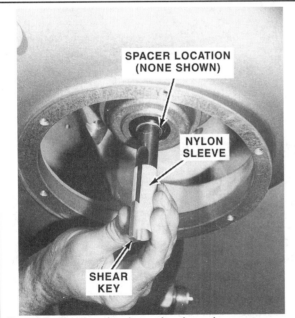

Fig. 262 Remove the nylon sleeve, shear key and any spacers which were above the impeller

CLEANING & INSPECTION

◆ See Figures 269 and 270

Wash all parts, except the driveshaft assembly, in solvent and blow them dry with compressed air. Rotate the bearing assembly on the driveshaft to inspect the bearings for rough spots, binding and signs of corrosion or damage.

Saturate a shop towel with solvent and wipe both extensions of the driveshaft.

Bearing Assembly

◆ See Figures 270 and 271

Lightly wipe the exterior of the bearing assembly with the same shop towel. Do not allow solvent to enter the three lubricant passages of the bearing assembly. The best way to clean these passages is not with solvent - because any solvent remaining in the assembly after installation will continue to dissolve good useful lubricant and leave bearings and seals dry. This condition will cause bearings to fail through friction and seals to dry up and shrink - losing their sealing qualities.

The only way to clean and lubricate the bearing assembly is after installation to the jet drive - via the exterior lubrication fitting.

If the old lubricant emerging from the hose coupling is a dark, dirty, gray color, the seals have already broken down and water is attacking the bearings. If such is the case, it is recommended the entire driveshaft bearing assembly be taken to the dealer for service of the bearings and seals.

LOWER UNIT 7-81

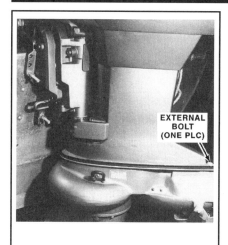

Fig. 263 To unbolt the drive remove any external bolts (usually 1)...

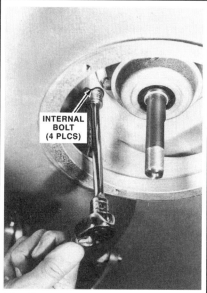

Fig. 264 ...and any internal bolts (usually 4 around the outside of the impeller housing)

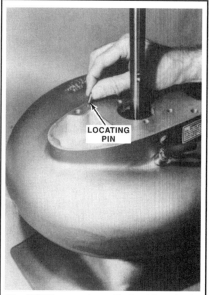

Fig. 265 When you lower the jet drive, check for the dowel (locating) pins...

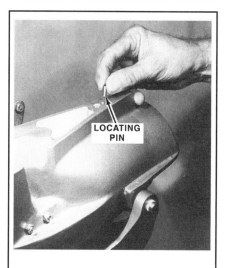

Fig. 266 ...there is usually one at each end

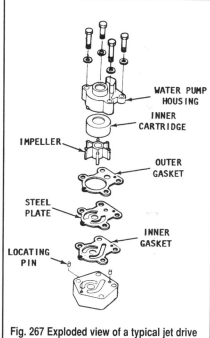

Fig. 267 Exploded view of a typical jet drive water pump assembly

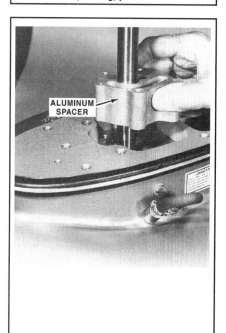

Fig. 268 If necessary, remove the pump shaped aluminum spacer

Disassembling the Bearing Carrier

◆ See Figures 272 and 273

A slightly involved procedure must be followed to dismantle the bearing carrier including "torching" off the bearing housing. Naturally, excessive heat will destroy the seals and possibly damage the bearings. Therefore, this procedure should be avoided unless you are planning (or at least open to the possibility) of a complete overhaul to the Jet Drive. Otherwise, use great care or leave this part of the service work to someone who is more familiar with Jet Drives.

However, if you have the experience and the shop tooling required to perform the torch heating and bearing press tasks, use the following procedures and illustrations to perform this work. Refer to the exploded view provided under the Cleaning & Inspecting heading in this section to help identify internal components, but remember that assemblies may vary slightly. A photographic exploded view is also provided in this section.

The following procedures pickup the work after the water pump assembly has been removed from

1. Remove the bolts (usually 4) securing the bearing carrier assembly in the jet pump housing. Lift out and discard the four small O-rings from the bolt holes on top of the bearing carrier flange.

※※ **CAUTION**

The retaining ring is under considerable force when installed. Use caution and wear eye protection when removing or installing this retaining ring.

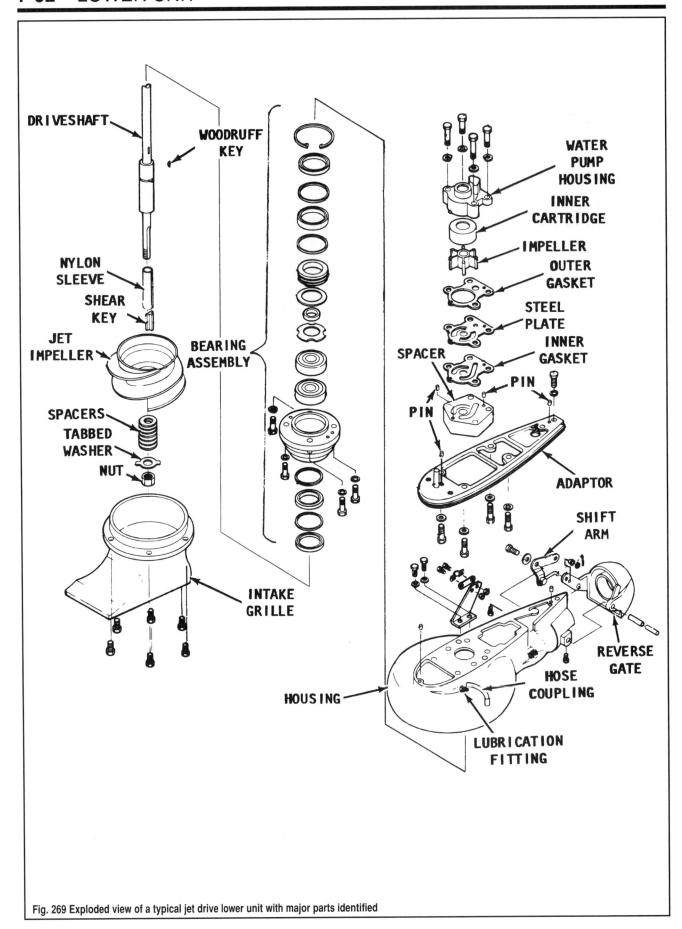

Fig. 269 Exploded view of a typical jet drive lower unit with major parts identified

LOWER UNIT 7-83

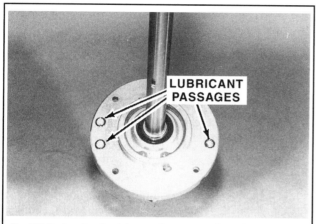

Fig. 270 During cleaning take care to prevent solvent from entering the lubrication passages

Fig. 271 Cleaning and lubricating the bearing assembly is best accomplished by completely replacing the old lubricant

2. Reach inside the top of the bearing carrier with a pair of snap ring pliers and remove the Truarc retaining ring.

3. Apply heat evenly all around the bearing carrier using a hand held propane torch or equivalent item. Apply only enough heat to the bearing carrier so it is very warm to the touch. Now, invert the assembly in a bearing press and support the carrier between two plates as shown in the accompanying illustration. Press on the impeller end of the driveshaft and drive the shaft and bearings out the bottom of the carrier. Have an assistant standing by to catch the shaft as it falls free of the bearing carrier, otherwise the splined end of the shaft could be damaged if it strikes a concrete floor.

■ Use gloves or hand protection when handling the components such as the shaft, as should still be a bit hot.

4. Remove the outer seal in the bearing carrier by placing the blunt end of a punch down through the bearing carrier onto the seal metal case. Tap the seal down and out the bottom of the bearing carrier. Next, grasp the tab end of the spiral retaining ring below the inner seal. Unwind the spiral retaining ring from the bearing carrier.

Again, place the blunt end of a punch down through the bearing carrier onto the seal metal case. Tap the seal down and out the bottom of the

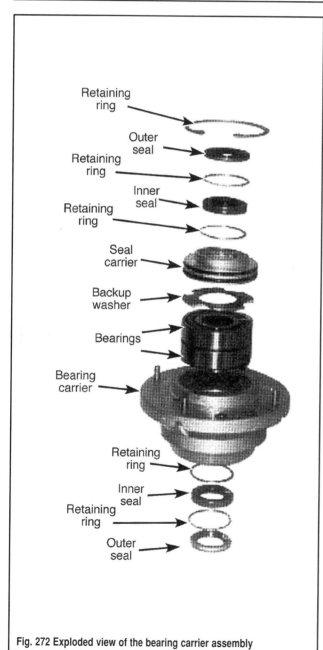

Fig. 272 Exploded view of the bearing carrier assembly

Fig. 273 After heating, use a shop press to separate the driveshaft from the carrier

7-84 Lower Unit

bearing carrier. Discard both seals and remove the remaining spiral retaining ring inside the bearing carrier. Thoroughly clean the bearing carrier in cleaning solvent and blow dry with compressed air.

Disassembling the Driveshaft

◆ See Figures 274 and 275

This procedure picks up where Disassembling the Bearing Carrier leaves off, at least, after the part where the driveshaft is removed from the carrier.

1. Slide the seal carrier and backup (thrust) washer off the end of the driveshaft.
2. Remove the two O-rings from the outer grooves and discard the O-rings.
3. Carefully, and closely, examine the seals inside the seal carrier. If there is any doubt as to the sealing action of the seals, replace the seal carrier with a new unit. Individual serviceable parts for the seal carrier are not available.
4. Inspect the driveshaft bearings carefully for roughness, dragging, or excessive clearance. If the bearings are defective or are removed from the driveshaft for any reason, they must be replaced with new bearings.

✱✱ WARNING

The driveshaft bearings need not be removed, unless they are unfit for further service. Once they are removed, they cannot be installed and used a second time.

5. To remove the bearings, first remove the Truarc snapring using safety glasses and a suitable pair of snapring pliers, then support the shaft in a shop press (with the impeller end up) using a bearing separator and press the bearings off the shaft. Discard both driveshaft bearings.
6. If necessary, remove and discard the retaining ring in the shaft itself.

Driveshaft and Associated Parts

Inspect the threads and splines on the driveshaft for wear, rounded edges, corrosion and damage.

Carefully check the driveshaft to verify the shaft is straight and true without any sign of damage.

Inspect the jet drive housing for nicks, dents, corrosion, or other signs of damage. Nicks may be removed with No. 120 and No. 180 emery cloth.

Reverse Gate

Inspect the gate and its pivot points. Check the swinging action to be sure it moves freely the entire distance of travel without binding.

Inspect the slats of the water intake grille for straightness. Straighten any bent slats, if possible. Use the utmost care when prying on any slat, as they tend to break if excessive force is applied. Replace the intake grille if a slat is lost, broken, or bent and cannot be repaired. The slats are spaced evenly and the distance between them is critical, to prevent large objects from passing through and becoming lodged between the jet impeller and the inside wall of the housing.

Jet Impeller

◆ See Figures 276 and 277

The jet impeller is a precisely machined and dynamically balanced aluminum spiral. Observe the drilled recesses at exact locations to achieve this delicate balancing.

Excessive vibration of the jet drive may be attributed to an out-of-balance condition caused by the jet impeller being struck excessively by rocks, gravel or cavitation burn.

The term cavitation burn is a common expression used throughout the world among people working with pumps, impeller blades and forceful water movement.

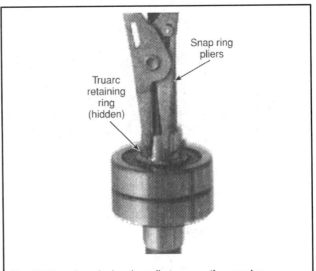

Fig. 274 To replace the bearings, first remove the snapring...

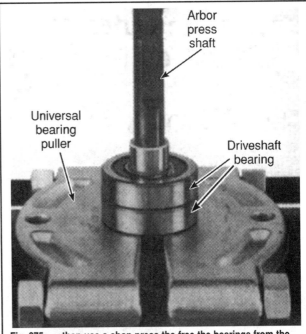

Fig. 275 ...then use a shop press the free the bearings from the shaft

Burns on the jet impeller blades are caused by cavitation air bubbles exploding with considerable force against the impeller blades. The edges of the blades may develop small dime size areas resembling a porous sponge, as the aluminum is actually eaten by the condition just described.

Excessive rounding of the jet impeller edges will reduce efficiency and performance. Therefore, the impeller should be inspected at regular intervals.

If rounding is detected, the impeller should be placed on a work bench and the edges restored to as sharp a condition as possible, using a file. Draw the file in only one direction. A back-and-forth motion will not produce a smooth edge. Take care not to nick the smooth surface of the jet impeller. Excessive nicking or pitting will create water turbulence and slow the flow of water through the pump.

Inspect the shear key. A slightly distorted key may be reused although some difficulty may be encountered in assembling the jet drive. A cracked shear key should be discarded and replaced with a new key.

LOWER UNIT

Fig. 276 The slats of the grille must be carefully inspected and any bent slats straightened for maximum jet drive performance

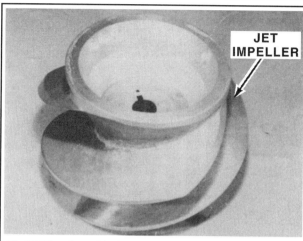

Fig. 277 The edges of the jet impeller should be kept as sharp as possible for maximum jet drive efficiency

Water Pump

Clean all water pump parts with solvent and then blow them dry with compressed air. Inspect the water pump housing for cracks and distortion, possibly caused from overheating. Inspect the steel plate, the thick aluminum spacer and the water pump cartridge for grooves and/or rough spots. If possible always install a new water pump impeller while the jet drive is disassembled. A new water pump impeller will ensure extended satisfactory service and give peace of mind to the owner. If the old water pump impeller must be returned to service, never install it in reverse of the original direction of rotation. Installation in reverse will cause premature impeller failure.

If installation of a new water pump impeller is not possible, check the sealing surfaces and be satisfied they are in good condition. Check the upper, lower and ends of the impeller vanes for grooves, cracking and wear. Check to be sure the indexing notch of the impeller hub is intact and will not allow the impeller to slip.

ASSEMBLY

◆ See Figure 278

1. Identify the two small locating pins used to index the large thick adaptor plate to the intermediate housing. Insert one pin into the last hole aft on the topside of the plate. Insert the other pin into the hole forward toward the port side.

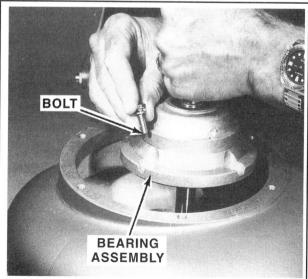

Fig. 278 If removed, install the bearing assembly into the jet drive housing

2. Lift the plate into place against the intermediate housing with the locating pins indexing with the holes in the intermediate housing. Secure the plate with the five (or seven) bolts.

■ On the five bolt model, one of the five bolts is shorter than the other four. Install the short bolt in the most aft location.

3. Tighten the long bolts to a torque value of 22 ft. lbs. (30 Nm). Tighten the short bolt to a torque value of 11 ft. lbs. (15 Nm).
4. Place the driveshaft bearing assembly into the jet drive housing. Rotate the bearing assembly until all bolt holes align. There is only one correct position.

Model Z has two securing screws and lock-washer. Tighten these screws just good and snug by hand.

Model Y has four securing bolts and lock-washer. Tighten these four bolts to a torque value of 5 ft. lbs. (7 Nm).

■ If installing a new jet impeller, place all eight spacers at the lower or nut end of the impeller, then check and, if necessary, adjust the impeller clearance by moving one or more of the shims above the impeller.

Shimming Jet Impeller

◆ See Figures 279 and 280

1. The clearance between the outer edge of the jet drive impeller and the water intake housing cone wall should be maintained at approximately 1/32 in. (0.79mm). This distance can be visually checked by shining a flashlight up through the intake grille and estimating the distance between the impeller and the casing cone, as indicated in the accompanying illustrations. But, it is not humanly possible to accurately measure this clearance by eye. Close observation between outings is fine to maintain a general idea of impeller condition, but, at least annually, the clearance must be measured using a set of long feeler gauges.

■ After continued use, the clearance will increase. The spacers previously removed are used to position the impeller along the driveshaft with a desired clearance of 1/32 in. (0.79mm) between the jet impeller and the housing wall.

2. A total of either eight (Model Z, Model AA4 and Model AA6) or nine (Model Y) spacers are normally used depending on the model being serviced. When new, all spacers are normally located at the tapered (or nut) end of the impeller. As the clearance increases, the spacers are transferred from the tapered (nut) end and placed at the wide (intermediate housing) end of the jet impeller.

7-86 LOWER UNIT

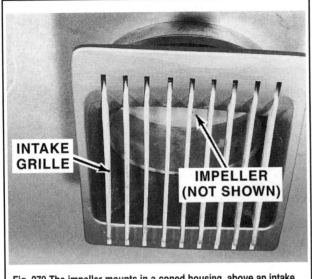

Fig. 279 The impeller mounts in a coned housing, above an intake grate

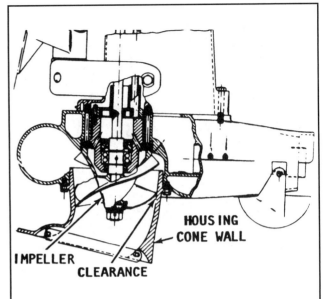

Fig. 280 The depth in which the impeller sits in the tapered housing determines impeller clearance, so adjustment is performed using shims

■ This procedure is most easily accomplished while the jet drive is removed from the intermediate housing, but it can be done with the drive attached to the outboard.

3. Secure the driveshaft with the attaching hardware. Installation of the shear key and nylon sleeve is not vital to this procedure. Place the unit on a convenient work bench. Shine a flashlight through the intake grille into the housing cone and eyeball the clearance between the jet impeller and the cone wall, as indicated in the accompanying line drawing. Move spacers one-at-a-time from the tapered end to the wide end to obtain a satisfactory clearance. Dismantle the driveshaft and note the exact count of spacers at both ends of the bearing assembly. This count will be recalled later during assembly to properly install the jet impeller.

Water Pump Assembly

◆ See Figures 281 and 282

1. Place the aluminum spacer over the driveshaft with the two holes for the indexing pins facing upward. Fit the two locating pins into the holes of the spacer.

Fig. 281 If removed, install the water-pump shaped spacer to the jet drive

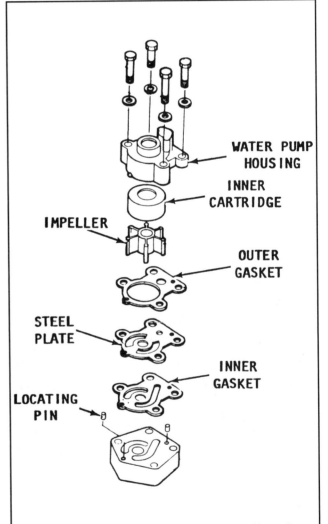

Fig. 282 Exploded view of a typical water pump assembly used on these jet drives

LOWER UNIT

■ The manufacturer recommends no sealant be used on either side of the water pump gaskets.

2. Slide the inner water pump gasket (the gasket with two curved openings) over the driveshaft. Position the gasket over the two locating pins. Slide the steel plate down over the driveshaft with the tangs on the plate facing downward and with the holes in the plate indexed over the two locating pins.

3. Check to be sure the tangs on the plate fit into the two curved openings of the gasket beneath the plate. Now, slide the outer gasket (the gasket with the large center hole) over the driveshaft. Position the gasket over the two locating pins.

4. Fit the Woodruff key into the driveshaft. Just a dab of grease on the key will help to hold the key in place. Slide the water pump impeller over the driveshaft with the rubber membrane on the top side and the keyway in the impeller indexed over the Woodruff key. Take care not to damage the membrane. Coat the impeller blades with Yamalube Grease or equivalent water resistant lubricant.

5. Install the insert cartridge, the inner plate and finally the water pump housing over the driveshaft. Rotate the insert cartridge counterclockwise over the impeller to tuck in the impeller vanes. Seat all parts over the two locating pins.

■ On some models, two different length bolts are used at this location. The Model Y uses special D shaped washers. The Model Z uses plain washers.

6. Tighten the four bolts to 11 ft. lbs. (15 Nm).

Jet Drive Installation

 DIFFICULT

◆ See Figures 283 thru 294

1. Install one of the small locating pins into the aft end of the jet drive housing.

2. Install the other small locating pin into the forward starboard side (or center forward end, depending on the model being serviced).

3. Raise the jet drive unit up and align it with the intermediate housing, with the small pins indexed into matching holes in the adaptor plate. Install the internal bolts (normally 4).

4. Install the one external bolt at the aft end of the anti-cavitation plate. Tighten all bolts to 11 ft. lbs. (15 Nm).

5. Place the required number of spacers up against the bearing housing. Slide the nylon sleeve over the driveshaft and insert the shear key into the slot of the nylon sleeve with the key resting against the flattened portion of the driveshaft.

6. Slide the jet impeller up onto the driveshaft, with the groove in the impeller collar indexing over the shear key.

7. Place the remaining spacers over the driveshaft. The number of spacers will be nine less the number used in previous step, for the Model Y and eight minus the number used during jet drive installation.

8. Tighten the nut to 17 ft. lbs. (23 Nm). If neither of the two tabs on the tabbed washer aligns with the sides of the nut, remove the nut and washer. Invert the tabbed washer. Turning the washer over will change the tabs by approximately 15°. Install and tighten the nut to the required torque value. The tabbed washer is designed to align with the nut in one of the two positions described.

9. Bend the tabs up against the nut to prevent the nut from backing off and becoming loose.

10. Install the intake grille onto the jet drive housing with the slots facing aft. Install and tighten the six securing bolts. Tighten 1/4 in. bolts to 5 ft. lbs. (7 Nm). Tighten 5/16 in. bolts to 11 ft. lbs. (15 Nm).

11. Slide the bolt through the end of the shift cable, washer and into the shift arm. Install the locknut onto the bolt and tighten the bolt securely.

12. Install the shift cable against the shift cable support bracket and secure it in place with the two bolts.

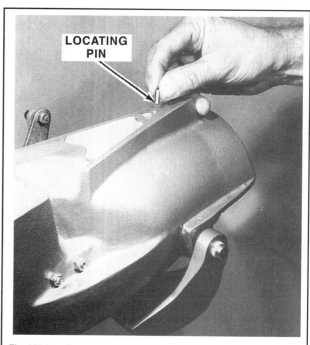

Fig. 283 Install the locating pins...

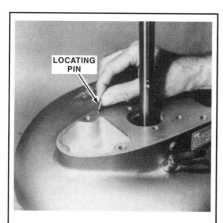

Fig. 284 ...at either end of the housing as noted during disassembly

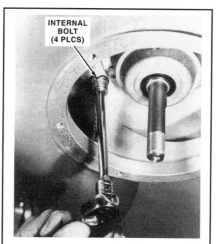

Fig. 285 Install and tighten the 4 internal drive mounting bolts...

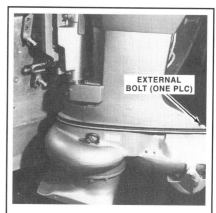

Fig. 286 ...along with any external bolts (there is usually just 1)

7-88 LOWER UNIT

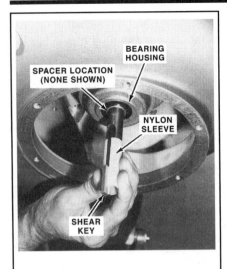

Fig. 2873 Install any shims, the nylon sleeve and shear key...

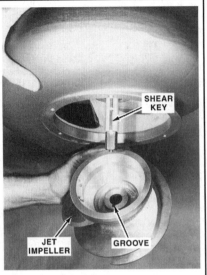

Fig. 288 ...then slide the impeller into position...

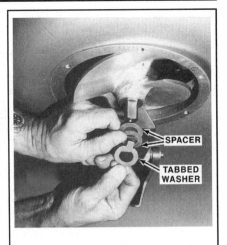

Fig. 289 ...followed by the shims and tabbed washer

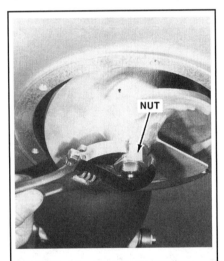

Fig. 290 Install and tighten the impeller retaining nut...

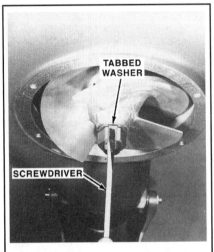

Fig. 291 ...then bend the washer tab into place to secure the nut

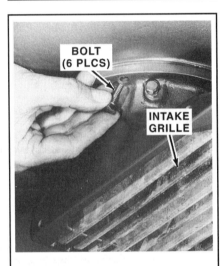

Fig. 292 Install the intake grille

Jet Drive Adjustments

GATE LINKAGE ADJUSTMENTS

There are 2 potential reverse gate linkage types used on Yamaha Jet Drives. The first, predominantly found on early-model jet drives, uses a shift arm that rides on a pivot. The second, normally found on late-model jet drives, uses a roller that rides within a shift cam. Use the accompanying illustrations to determine the type of gate linkage, then follow the appropriate set of adjustment procedures.

Shift Arm/Pivot Linkage (Early-Models)

 MODERATE

Cable Alignment and Free-play

◆ See Figures 295 and 296

1. Move the shift lever downward into the forward position. The leaf spring should snap over on top of the lever to lock it in position.
2. Remove the locknut, washer and bolt from the threaded end of the shift cable. Push the reverse gate firmly against the rubber pad on the underside of the jet drive housing.
3. Check to be sure the link between the reverse gate and the shift arm is hooked into the LOWER hole on the gate.
4. Hold the shift arm up until the link rod and shift arm axis form an imaginary straight line, as indicated in the accompanying illustration. Adjust the length of the shift cable by rotating the threaded end, until the cable can be installed back onto the shift arm without disturbing the imaginary line. Pass the nut through the cable end, washer and shift arm. Install and tighten the locknut.

Neutral Stop Adjustment

◆ See Figures 297, 298 and 299

In the forward position, the reverse gate is neatly tucked underneath and clear of the exhaust jet stream.
In the reverse position, the gate swings up and blocks the jet stream deflecting the water in a forward direction under the jet housing to move the boat sternward.
In the neutral position, the gate assumes a happy medium - a balance between forward and reverse when the powerhead is operating at IDLE speed. Actually, the gate is deflecting some water to prevent the boat from moving forward, but not enough volume to move the boat sternward.

LOWER UNIT 7-89

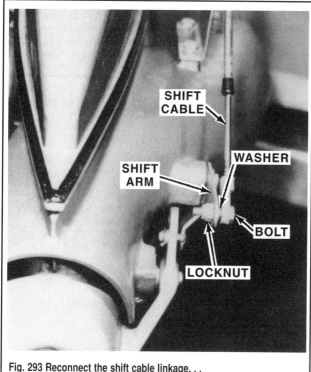

Fig. 293 Reconnect the shift cable linkage...

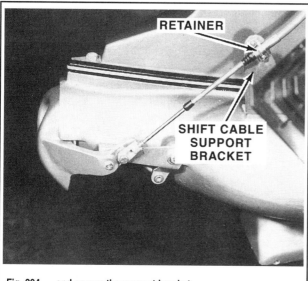

Fig. 294 ...and secure the support bracket

✳✳ WARNING

The gate must be properly adjusted for safety of boat and passengers. Improper adjustment could cause the gate to swing up to the reverse position while the boat is moving forward causing serious injury to boat or passengers.

1. Loosen, but do not remove the locknut on the neutral stop lever. Check to be sure the lever will slide up and down along the slot in the shift lever bracket.

■ The following procedure must be performed with the boat and jet drive in a body of water. Only with the boat in the water can a proper jet stream be applied against the gate for adjustment purposes.

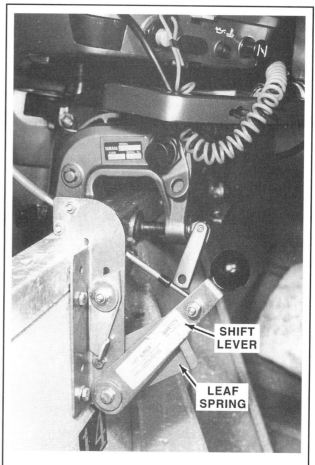

Fig. 295 When the shift lever moves down, the leaf spring should snap over and lock it in position for safety

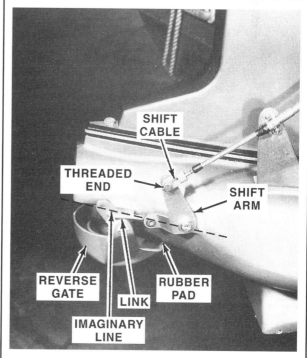

Fig. 296 Adjust the threaded end of the shift cable so that the reverse gate sits properly against the rubber pad as shown

7-90 LOWER UNIT

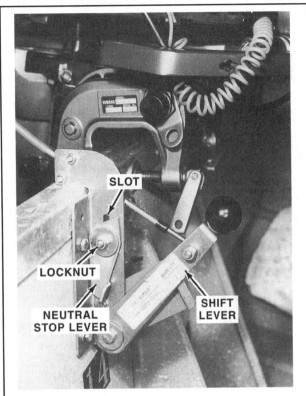

Fig. 297 Loosen the locknut for the neutral stop lever...

✶✶ CAUTION

Water must circulate through the lower unit to the powerhead anytime the powerhead is operating to prevent damage to the water pump in the lower unit. Just five seconds without water will damage the water pump impeller.

    2. Start the powerhead and allow it to operate only at IDLE speed. With the neutral stop lever in the down position, move the shift lever until the jet stream forces on the gate are balanced. Balanced means the water discharged is divided in both directions and the boat moves neither forward nor sternward. The gate is then in the neutral position with the powerhead at idle speed.

    3. Move the neutral stop lever up against the shift lever until the stop lever barely makes contact with the shift lever. Tighten the locknut to maintain this new adjusted position. Shut down the powerhead.

■ The reverse gate may not swing to the full up position in reverse gear after the previous steps have been performed. Do not be concerned. This condition is acceptable, because water pressure in reverse will close the gate fully under normal operation.

Cam and Roller Linkage (Late-Models)

Gate Adjustment

◆ See Figures 300, 301 and 302

    To adjust the reverse gate on the jet drive housing, proceed as follows:
    1. Set the shifter handle to the **neutral** detent.
    2. Hold the reverse gate up and check to make sure there is about a 15/32 in. (12mm) gap between the top of the gate and the water flow passage.

Fig. 298 ...and set the shift lever until the gate is in a balanced position...

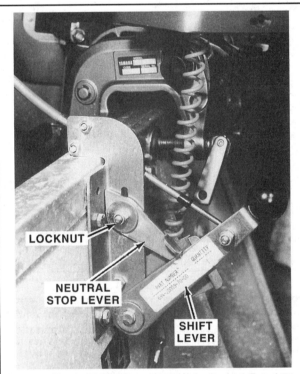

Fig. 299 ...then position the stop lever just barely against the shift lever and tighten the locknut

LOWER UNIT

3. If adjustment is necessary, have an assistant continue to hold the reverse gate up and check the clearance, while you loosen the cam screw and rotate the eccentric nut until the clearance is correct.
4. Tighten the cam screw, then recheck the adjustment.
5. Move the shifter handle to the **forward** detent.
6. With the gate now in the **forward** detent try to lift up on the reverse gate by hand (moving it back toward **neutral**). The gate should NOT move. If necessary, adjust the eccentric nut again, so the **neutral** gap is correct, but the gate will not move upward in the **forward** position.

Remote Cable Adjustment

◆ See Figures 300 thru 304

Proper cable adjustment is necessary in order to keep water pressure (caused by boat movement) from moving the reverse gate upward blocking normal thrust.

✳✳ CAUTION

Should be reverse gate shift suddenly and unexpectedly over the thrust nozzle, the boat will stop violently, shifting and possibly ejecting passengers and cargo. Proper adjustment is critical.

This procedure starts with the cable disconnected, for replacement, re-rigging or installation of a jet drive assembly.

1. Pull upward on the shift cam (1) until the roller is in the far end of the **forward** range (2).
2. Shift the remote control to the **forward** position. Temporarily, push the insert the cable guide onto the shift cam stud, then pull firmly on the cable casing to remove all free-play. Next, adjust the cable trunnion until it aligns with the anchor bracket, then remove the cable from the cam stud.
3. Connect the cable trunnion to the anchor bracket, then turn 90° to lock the trunnion in position.
4. Push the cable guide back onto the cam stud, then install the washer and finger-tighten the locknut.
5. Shift the remote control to the **neutral** position (3). The cam roller must snap into the **neutral** detent when you pull upward on the reverse gate with moderate pressure. If the gate does not shift properly into **neutral**, lengthen the cable slightly and recheck the adjustment.

■ Remember, it is CRITICAL that the reverse gate remains locked into the FORWARD position when the remote is shifted into FORWARD. You should NOT be able to move the gate out of this position by hand.

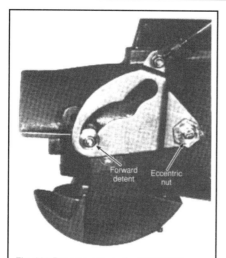

Fig. 300 Reverse gate in the FORWARD position with the cam roller at the lower end of the cutout in the bracket

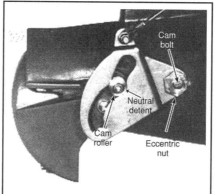

Fig. 301 Reverse gate in the NEUTRAL position, with the cam roller indexed into the neutral detent

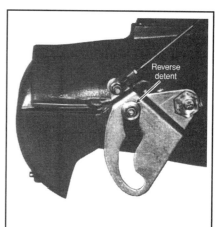

Fig. 302 Reverse gate in the REVERSE position, with the cam roller at the upper end of the cutout in the bracket

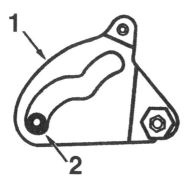

Fig. 303 When adjusting the remote cable/shift rod first move the cam (1) to the FORWARD position (2) . . .

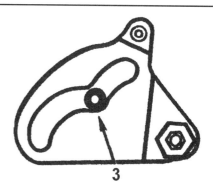

Fig. 304 . . . next, check that the cam snaps into the NEUTRAL position (3)

7-92 LOWER UNIT

6. Tighten the cam locknut securely, then loosen it 1/8-1/4 of a turn in order to allow free movement of the cam.
7. Secure the cable loosely to the steering cable using a wire tie, but be sure NOT to restrict shift cable movement.

■ With the engine running in NEUTRAL and the warm-up lever raised, the engine will run rough due to exhaust restriction from the reverse gate. This is normal.

Tiller Shift Rod Adjustment

◆ See Figures 300 thru 304

Proper shift rod adjustment is necessary in order to keep water pressure (from boat movement) from moving the reverse gate upward blocking normal thrust.

✳✳ CAUTION

Should be reverse gate shift suddenly and unexpectedly over the thrust nozzle, the boat will stop violently, shifting and possibly ejecting passengers and cargo. Proper adjustment is critical.

This procedure starts with the shift rod (and connector pin) disconnected, for replacement, re-rigging or installation of a jet drive assembly.
1. Pull upward on the shift cam (1) until the roller is in the far end of the **forward** range (2).
2. Move the shift handle to the **forward** position, then thread the connector onto the shift rod until it is aligned with the hole in the shift cam.
3. Install the connector pin and then secure using the cotter pin and washer.
4. Move the shifter to the **neutral** position (3). The cam roller must snap into the **neutral** detent when you pull upward on the reverse gate with moderate pressure. If the gate does not shift properly into **neutral**, remove the connector pin again and rotate the connector in order to lengthen the shift rod VERY slightly, then recheck the adjustment.

■ Remember, it is CRITICAL that the reverse gate remains locked into the FORWARD position when the handle is shifted into FORWARD. You should NOT be able to move the gate out of this position by hand.

5. Tighten the adjuster locknut securely.

■ With the engine running in NEUTRAL and the twist grip in the START position or higher, the engine will run rough due to exhaust restriction from the reverse gate. This is normal.

TRIM ADJUSTMENT

◆ See Figure 305

During operation, if the boat tends to pull to port or starboard, the flow fins may be adjusted to correct the condition. These fins are located at the top and bottom of the exhaust tube.
If the boat tends to pull to starboard, bend the trailing edge of each fin approximately 1/16 in. (1.5mm) toward the starboard side of the jet drive. Naturally, it follows logically that if the boat tends to pull to port, you should bend the fins toward the port side.

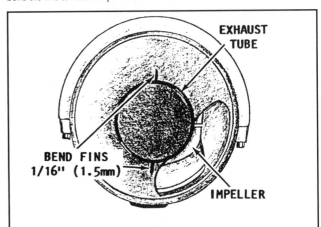

Fig. 305 Fins at the top and bottom of the exhaust tube allow for minor port/starboard trim thrust adjustment

TRIM & TILT SYSTEMS	8-2
INTRODUCTION	8-2
HYDRO TILT LOCK SYSTEM	8-2
DESCRIPTION & OPERATION	8-2
LOWERING OUTBOARD UNIT	8-2
RAISING OUTBOARD UNIT	8-2
UNDERWATER STRIKE	8-3
SERVICING THE HYDRO TILT SYSTEM	8-3
SINGLE TILT RAM POWER TRIM/TILT SYSTEMS	8-3
DESCRIPTION & OPERATION	8-3
MANUAL OPERATION	8-4
SHOCK ABSORBER ACTION	8-4
TILT DOWN OPERATION	8-4
TILT UP OPERATION	8-3
MAINTENANCE	8-4
CHECKING HYDRAULIC FLUID LEVEL	8-4
PURGING AIR FROM THE SYSTEM	8-5
PTT ASSEMBLY	8-9
COMPONENT REPLACEMENT	8-11
REMOVAL & INSTALLATION	8-9
TILT MOTOR	8-8
TESTING	8-8
TILT RELAY	8-6
TESTING	8-6
TILT SWITCH	8-7
TESTING	8-7
TRIM SENSOR	8-9
TESTING	8-9
TROUBLESHOOTING	8-6
LARGE MOTOR POWER TRIM/TILT SYSTEM	8-13
DESCRIPTION & OPERATION	8-13
MANUAL OPERATION	8-14
SHOCK ABSORBER ACTION	8-14
TILT DOWN	8-14
TILT UP	8-14
TRIM DOWN	8-14
TRIM UP	8-13
MAINTENANCE	8-15
CHECKING HYDRAULIC FLUID LEVEL	8-15
PURGING AIR FROM THE SYSTEM	8-15
POWER TRIM/TILT SWITCH	8-19
TESTING	8-19
PTT ASSEMBLY (DUAL TRIM RAM)	8-21
ASSEMBLY	8-25
CLEANING & INSPECTION	8-23
DISASSEMBLY	8-22
REMOVAL & INSTALLATION	8-21
TRIM SENSOR	8-20
ADJUSTMENT	8-20
TESTING	8-20
TRIM/TILT MOTOR	8-20
TESTING	8-20
TRIM/TILT RELAY	8-18
TESTING	8-18
TROUBLESHOOTING	8-15
HYDRAULIC TESTING	8-16
PRELIMINARY INSPECTION	8-15
SYMPTOM DIAGNOSIS	8-16

8

TRIM & TILT

TRIM & TILT SYSTEMS	8-2
HYDRO TILT LOCK SYSTEM	8-2
SINGLE TILT RAM POWER TRIM/TILT SYSTEMS	8-3
LARGE MOTOR POWER TRIM/TILT SYSTEM	8-13

8-2 TRIM & TILT

TRIM & TILT SYSTEMS

Introduction

All outboard installations are equipped with some means of raising or lowering (pivoting), the complete unit for efficient operation under various load, boat design, water conditions, and for trailering to and from the water.

The correct trim angle ensures maximum performance and fuel economy as well as a more comfortable ride for the crew and passengers.

The most simple form of tilt is a mechanical tilt adjustment consisting of a series of holes in the transom mounting bracket through which an adjustment pin passes to secure the outboard unit at the desired angle.

Such a mechanical arrangement works quite well for the smaller units, but with larger (and heavier) outboard units some form of assist or power system is required. A simple hydraulic tilt assist system, known as the Yamaha Hydro Tilt system is used on larger motors that are NOT equipped with a power trim-tilt assembly. The main component of the Hydro Tilt system is a gas filled shock absorber that is used to both provide mechanical lift assistance as well as act to protect the outboard and transom from shock should the motor strike an underwater object.

For most 40 hp and larger motors (as well as a few smaller models such as some 25 or 30 hp motors) some form of a Power Trim/Tilt (PTT) system is used. The Yamaha power systems are hydraulically operated and electrically controlled from the helmsperson's position. There are basically 2 forms of the PTT system. The first uses a single trim/tilt rod to control all functions and provide shock absorption, while the second, found primarily on larger motors, uses a tilt rod for tilting and shock absorption, while using two dedicated trim cylinders for fine trim adjustment.

■ **The single tilt rod PTT system is normally found on smaller motors, including most medium sized motors (such as the 28J-50 hp, 698cc models). The large motor PTT system (with single tilt rod and dual trim rods) is normally found on most 60 hp and larger motors.**

All trim and tilt systems are installed between the two large clamp brackets. On power systems, the trim/tilt relay is usually mounted in the upper cowling pan where it is fairly well protected from moisture.

All power trim/tilt systems contain a manual release valve to permit movement of the outboard unit in the remote event the trim/tilt system develops a malfunction, either hydraulic or electrical (or in case the battery or power source fails), preventing use of power.

This section covers three different types of trim/tilt units which may be installed on Yamaha outboards. Each system is described in a separate section. Troubleshooting, filling the system with hydraulic fluid and purging (bleeding) procedures are included, as applicable.

HYDRO TILT LOCK SYSTEM

Description & Operation

◆ See Figure 1

The Hydro Tilt Lock system is a mechanical assist lift and lock system for tilting medium-to-large sized outboards that are NOT equipped with a PTT system. It consists of a single shock absorber and tilt lever assembly.

■ **The Hydro Tilt Lock system is normally found on medium-to-large sized commercial outboards or motors designed for high-reliability and low price (lacking extra features such as electric start) for reasons of simplicity and low expense of production cost.**

The shock absorber contains a high pressure gas chamber located in the upper portion of the cylinder bore above the piston assembly. The piston contains a down relief valve and an absorber relief valve. Below the piston assembly, the lower cylinder bore contains an oil chamber. This lower chamber is connected to the upper chamber above the piston by a hydraulic line with a manual check valve. This check valve is located about half way down the hydraulic line.

This manual check valve is activated by the tilt lever, when the lever is rotated from the lock (down) position to the tilt (up) position. The check valve cam rotates and pushes the manual check valve push rod against the check valve. This action opens the check valve and allows hydraulic fluid to flow from the lower chamber through the hydraulic line, past the open manual check valve and into the upper gas chamber.

RAISING OUTBOARD UNIT

When the outboard unit is tilted up, the volume below the piston decreases, and at the same time, the volume above the piston increases until the piston has reached the bottom of its stroke. In this position all fluid is contained above the piston.

The tilt lever is then rotated to the lock (down) position to engage with the clamp bracket. When the tilt lever is in the lock position, the manual valve push rod rests on a flat spot of the manual valve cam and releases pressure on the check valve. Releasing pressure on the check valve closes off the hydraulic line and the flow of hydraulic fluid. The outboard unit is now in the trailering position.

LOWERING OUTBOARD UNIT

To lower the outboard unit from the full up and locked position, the tilt lever is again rotated from the lock (down) position to the tilt (up) position. The manual check valve cam rotates and pushes the manual check valve push rod against the check valve and opens the valve.

■ **When the manual check is open, the valve will allow hydraulic fluid to flow in only. One direction - from the lower chamber to the upper chamber.**

As the outboard unit is tilted down, the piston moves up and compresses the fluid in the upper chamber. The fluid pressure overcomes the down relief valve spring and opens the relief valve. The valve in the open position permits hydraulic fluid to flow through the piston from the upper chamber to the lower chamber.

During normal cruising, the tilt lever is set in the lock (down) position. The manual check valve is closed to prevent the outboard unit from being tilted up by water pressure against the propeller when the unit is in reverse gear. When the unit is in forward gear, the outboard is held in position by the tilt pin through the swivel bracket.

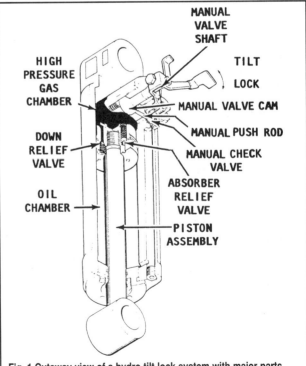

Fig. 1 Cutaway view of a hydro tilt lock system with major parts identified

TRIM & TILT

UNDERWATER STRIKE

In the event the outboard lower unit should strike an underwater object while the boat is underway, the piston would be forced down. The hydraulic fluid below the piston would be under pressure with no escape because the manual check valve is closed. The valve is closed because the tilt lever is in the lock (down) position.

To prevent rupture of the hydraulic line, a safety relief valve is incorporated in the piston. This relief valve permits fluid to pass through the piston from the lower chamber to the upper chamber through the absorber relief valve. After the outboard has passed the obstacle, the fluid returns to the lower chamber through the piston and the down relief valve, because the piston is pushed up.

Servicing the Hydro Tilt System

◆ See Figure 2

Service procedures for the Hydro Tilt Lock system are normally confined to periodic inspection for signs of extreme corrosion (especially on the shock absorber rod) and signs of oil leaks. Although small amounts of corrosion MAY be polished away in some cases, extreme corrosion and/or oil leaks will normally require that the entire unit be replaced.

Although it is possible to remove the unit from the steering and clamp bracket assembly on a few motors, for MOST motors it will require disassembly of the clamp bracket itself, a potentially time consuming task.

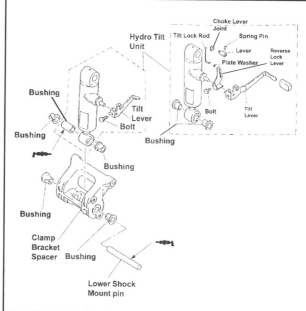

Fig. 2 Exploded view of a typical hydro tilt lock system

SINGLE TILT RAM POWER TRIM/TILT SYSTEMS

Description & Operation

◆ See Figures 3, 4 and 5

■ **The single tilt rod PTT system is normally found on smaller motors, including most medium sized motors (such as the 28J-50 hp, 698cc models).**

The PTT system found on most medium sized Yamaha outboards incorporates a single hydraulic cylinder and piston. Although on some older units this system was used only for tilt, on most modern Yamahas the system allows for some trim adjustment of the motor once lowered into operating position by the system. The system consists of an electric motor mounted on top of a gear driven hydraulic pump, a small fluid reservoir (which is an integral part of the pump) and a single hydraulic piston and cylinder used to move the outboard unit up or down, as required.

■ **The positioning of the tilt cylinder and the pump are reversed on some models. That is to say the pump may be found on either side (port or starboard) or the hydraulic ram.**

Unlike other power trim and tilt units, all hydraulic circuits are routed inside the unit.

■ **Three safety relief valves are incorporated into the hydraulic passageways as protection against excessive pressurization. Each of these valves has a different pressure release factor. The valves are not interchangeable. The up relief valve and the down relief valve are normally located, one on each side of the pump. The third, main relief valve, is normally found above the main valve assembly.**

Each valve is secured in place with an Allen head screw accessible from the exterior of the pump. The distance the Allen head screws are sunk into the pump housing is critical. Therefore, do not remove and examine the valves without good cause. If a valve is accidentally removed, refer to Troubleshooting, in this section.

TILT UP OPERATION

When the up portion of the tilt switch on the remote control handle is depressed, the electric motor rotates (usually in a clockwise direction). The drive gear, on the end of the motor shaft, indexed with the driven gear act as an oil pump. This action is very similar to the action in an automobile oil circulation pump.

The hydraulic fluid is forced through a series of valves into the lower chamber of the cylinder. The fluid fills the lower chamber and forces the piston upward and the outboard unit rises. As the piston continues to extend, oil in the upper chamber is routed back through the suction side of the pump until the tilt piston reaches the top of its stroke.

Fig. 3 Typical single tilt ram PTT unit installed on a 40 hp motor

8-4 TRIM & TILT

TILT DOWN OPERATION

When the down portion of the tilt switch on the remote control handle is depressed, the electric motor rotates in the opposite direction (therefore normally counterclockwise). The drive gear on the end of the motor shaft indexed with the driven gear are now rotating in the opposite direction. This action forces the fluid into the upper cylinder chamber under pressure causing the piston to retract, lowering the outboard. The fluid under pressure beneath the piston is routed back through a series of valves to the pump until the tilt piston reaches the bottom of its stroke and the outboard unit is in the full down position.

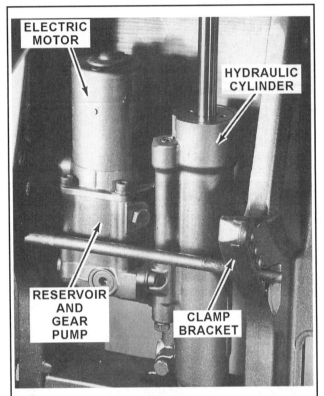

Fig. 4 PTT unit with major parts identified (note the position of the motor and cylinder is reversed on many late-model outboards)

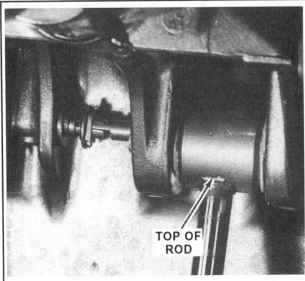

Fig. 5 Mounting for the upper piston end of the power tilt unit

MANUAL OPERATION

◆ See Figures 6 and 7

※※ WARNING

If outboard unit is in the up position when the manual release valve is opened, the outboard will drop to the full down position rapidly. Therefore, ensure all persons stand clear.

The outboard unit may be raised or lowered manually should the battery fail to provide sufficient current to operate the electric motor or should electric/hydraulic components of the PTT system fail for any other reason. A manual relief valve is provided to permit manual operation.

This manual relief valve is located on the lower end of the gear pump beneath the electric motor either facing aft (toward the motor/gearcase) or facing sideways (so that it is accessed through one of the clamp brackets). On most older Yamaha motors the valve faced aft and contained an Allen (hex key) head. On these models the valve usually contains left-hand threads, meaning it is OPENED or LOOSENED by turning CLOCKWISE. However, on MOST late-model Yamaha outboards (where the valve faces to one side, and is accessed through a clamp bracket), the valve is equipped with a slotted head and contains normal right-hand threads, meaning it is OPENED or LOOSENED by turning COUNTERCLOCKWISE.

Either way, opening the valve releases pressure in both the upper and lower cylinder chambers. With a complete loss of pressure, the piston may be moved up or down in the cylinder without resistance.

After the outboard unit has been moved to the desired position, the valve should be rotated in the opposite direction (counterclockwise for Allen head valves, clockwise for slotted valves) to close the valve and lock the outboard against movement.

SHOCK ABSORBER ACTION

The lower end of the tilt piston is capped with a free piston. This free piston normally moves up and down with the tilt piston. In the event the outboard lower unit should strike an underwater object while the boat is underway, the tilt piston would be suddenly and forcibly moved upward.

The free piston also moves upward but at a much slower rate than the tilt piston. The action of the tilt piston separating from the free piston causes two actions. First, the hydraulic fluid in the upper chamber above the piston is compressed and pressure builds in this area. Second, a vacuum is formed in the area between the tilt piston and the free piston.

This vacuum in the area between the two pistons sucks fluid from the upper chamber. The fluid fills the area slowly and the shock of the lower unit striking the object is absorbed. After the object has been passed, the weight of the outboard unit tends to retract the piston. The fluid between the tilt piston and the free piston is compressed and forced through check valves to the reservoir until the free piston reaches its original neutral position.

Maintenance

CHECKING HYDRAULIC FLUID LEVEL

◆ See Figures 8 and 9

■ If one of the hydraulic valves has been removed, or the system opened for any other reason, the system must be purged of any trapped air. To purge (bleed) the system, please refer to Purging Air From The System in this section.

The fluid in the power trim/tilt reservoir should be checked periodically to ensure it is full and is not contaminated. To check the fluid, tilt the motor upward to the full tilt position, then manually engage the tilt support for safety and to prevent damage.

In order to keep contaminants from entering the system when the plug is removed, clean the area around the fill plug (found on the reservoir, just below the motor on these models). Loosen and remove the plug using a suitable socket or wrench and make a visual inspection of the fluid. It should seem clear and not milky. The level is proper if, with the motor at full tilt, the level is even with the bottom of the filler plug hole.

TRIM & TILT 8-5

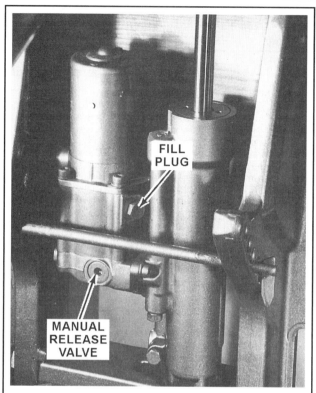

Fig. 6 Fill plug and manual relief valve on an early-model PTT system

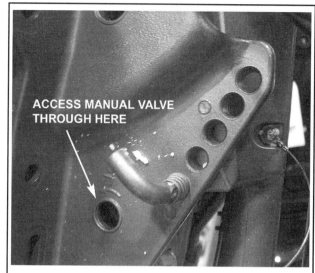

Fig. 7 On late-model PTT systems, the manual plug is accessed through the clamp bracket

■ Yamaha or Yamalube power trim and tilt fluid is a highly refined hydraulic fluid. This product has a high detergent content and additives to keep seals pliable. A high grade automatic transmission fluid, Dexron® II may also be used if the Yamalube fluid is not available.

PURGING AIR FROM THE SYSTEM

◆ See Figures 6 thru 9

Air must be purged (bled) from the system if any of the following conditions exist:
  a. Erratic motion is felt during operation of the system.
  b. Fluid has been contaminated with moisture or other foreign material.
  c. A component in the system has been replaced.
  d. The system has been opened for any reason.
  e. The outboard, when been left in the full up position, lowers slowly over a period of time.

■ When the reservoir is full to a normal level it should be even with the bottom of the plug threads when the outboard is in the fully upward, tilted position, meaning that when the outboard is lowered the level may be above the fill plug. One way to start refilling an empty or nearly empty system is to add fluid while the outboard is down, then open the manual release valve and manually lift the engine, drawing fluid into the system as the piston inside the tilt cylinder moves upward (as the ram is extended).

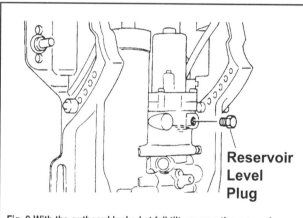

Fig. 8 With the outboard locked at full tilt, remove the reservoir plug. . .

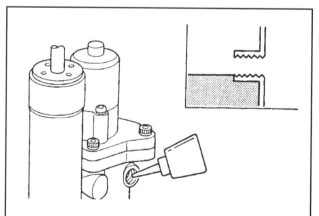

Fig. 9 . . .and make sure the fluid level is even with the bottom of the plug bore

1. Begin by making sure there is fluid inside the reservoir, even with the outboard in the normal running position (not tilted). Clean the area around the fill plug to prevent contaminants entering the system when the plug is loosened or removed. Slowly start to loosen the fill plug, if fluid starts to run (or spray) out, stop loosening the plug and retighten it until the outboard is tilted. However, if no fluid escapes, go ahead and remove the plug, then add fluid until it is level with the bottom of the screw threads. Lightly reinstall the reservoir plug.

2. Open the manual release valve (either by rotating the Allen head of the valve clockwise, or the slotted head counterclockwise, as applicable), rotating it slowly in the proper direction until it stops.

3. With the manual release valve open, slowly pull the outboard upward by hand, extending the tilt piston. As the tilt piston is extended, the piston in the cylinder will draw fluid into the cylinder, thus lowering the level in the chamber.

4. Carefully release the outboard, allowing it to lower by its own weight, then once the outboard is lowered fully, retighten the manual release valve.

8-6 TRIM & TILT

5. Allow the PTT system fluid to settle for about 5 minutes.
6. Push the PTT switch (either on the remote or the side of the motor) allowing the system to move the motor to the fully upward position.
7. Push the tilt stop lever(s) to support the outboard.
8. Again, allow the PTT system fluid to settle for about 5 minutes.
9. Remove the reservoir cap and check the fluid level. Add fluid, as necessary until it is even with the bottom of the fill plug, then install and tighten the reservoir cap.
10. Repeat as necessary until the fluid remains at the correct level.

Troubleshooting

◆ See Figure 10

If an operational problem develops and the determination is made the trouble is in the hydraulic system, only a few simple checks may be made. Valves, check valves, and relief valves, are all an integral part of the tilt unit.

Before assuming a serious fault in the system exists, make the checks listed below. Conduct an operational performance after each check has been completed for possible correction of the problem.

1. Check the quantity and quality of the hydraulic fluid in the reservoir. With the outboard unit in the full down position, the level of fluid should reach to the bottom of the fill plug threads. Add fluid, as required. If the fluid is murky drain all fluid from the system through the two drain plug openings. Fill the system with Yamalube Power Trim/Tilt Fluid or a good grade of automatic transmission fluid (Dexron or Type F). Purge the system of air.
2. Check the manual release valve to verify it is just snug (either counterclockwise on Allen head valves which are left-hand thread, or clockwise for slotted head valves).
3. Check the battery for a full charge.
4. Verify the quick disconnect plugs on the up/down relay are tight and making good contact. The relay is mounted in the bottom cowling pan, starboard side, next to the cranking motor relay.
5. Verify the wires are matched properly, color to color.
6. Test the up/down relay, the electric motor, and tilt button.

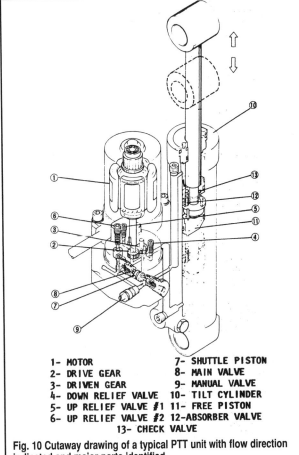

1- MOTOR
2- DRIVE GEAR
3- DRIVEN GEAR
4- DOWN RELIEF VALVE
5- UP RELIEF VALVE #1
6- UP RELIEF VALVE #2
7- SHUTTLE PISTON
8- MAIN VALVE
9- MANUAL VALVE
10- TILT CYLINDER
11- FREE PISTON
12- ABSORBER VALVE
13- CHECK VALVE

Fig. 10 Cutaway drawing of a typical PTT unit with flow direction indicated and major parts identified

Tilt Relay

TESTING

◆ See Figures 11, 12 and 13

The PTT motor runs either clockwise or counterclockwise depending upon which direction you want the motor to move and also depending upon how the power is applied to the motor's terminals. If you apply positive current to one terminal and negative to the other, the motor will run in one direction. If you switch and apply positive and negative to the opposite terminals, the motor will run in the other direction. Control of this circuit is achieved through one or more PTT relays. The job of the relay, like all relays, is to close certain switch circuits depending upon the signals received from the tilt switch on either the engine or the remote.

There are generally 3 major designs of PTT relays that you might find on these motors. The first and most common relay used on motors in this size range is a rectangular piece that looks like it has 8 terminals and a 3 wire harness that comes off one end (though the harness may use a single connector or individual wire bullets/terminals). Four of the visible terminals (identified by position) and the 3 wire harness are used during testing to with an ohmmeter (and a 12-volt battery) to determine if the relay is operating properly. For testing purposes, the exposed terminal at one end of the relay (where the wiring harness attaches) will be referred to as Terminal 1. This terminal is also identified because it is diagonally adjacent to the relay Negative terminal (the black ground wire was connected to it before removal). At the far diagonal end of the relay from Terminal 1 is the position for Terminal 2. Again, Terminal 2 is easily identified because it is diagonally adjacent to the relay Positive terminal (the terminal to which the red power wire was attached before removal). For more details, please see the accompanying illustration.

A few models, such as a handful of the motors including the 25/30 hp (496cc) 3-cylinder, utilize a rectangular relay box that does not have exposed terminals, but which contains a 6 wire harness (usually one 2-wire connector and 4 individual ring terminals or bullets). On these relays wire colors are used to identify proper test connections, but testing is otherwise the same, using an ohmmeter and a 12-volt battery.

Lastly, some models may be equipped with 2 individual round-bodied relays (like the type once more commonly found on the large motor PTT system). Both wire color and terminal positions are used to test these relays in much the same fashion as the rectangular bodied units.

■ On all relays, testing normally occurs with the relay completely removed from the powerhead, OR at least with all wiring disconnected (even if the relay is still physically bolted in position).

1. If not done already, tag and disconnect the wiring from the PTT relay(s).
2. Obtain a 12-volt battery with jumper leads and an ohmmeter for continuity testing.
3. First check terminal-to-terminal continuity as detailed by relay type:
• For rectangular, exposed terminal type relays, use the ohmmeter to confirm there is continuity between the following pairs of wires: Terminal 1 and the Negative terminal, Terminal 2 and the Negative terminal, The Sky Blue wire and the Black wire and finally, the Light Green wire and the Black wire. Next use the ohmmeter to confirm there is NO continuity between Terminal 1 and the Positive terminal, as well as between Terminal 2 and the Positive terminal.
• For rectangular, non-exposed terminal type relays (such as those used on the 25/30 hp [496cc] 3-cylinder motors), perform the SAME tests as listed for the exposed terminal relays, EXCEPT do it by wire color. Use the ohmmeter to confirm there is continuity between the following pairs of wires: Sky Blue and Black, Light Green and Black, Blue and Black, and finally, Green and Black. Next use the ohmmeter to confirm there is NO continuity between Blue and Red, as well as between Green and Red.
• For models with 2 round-bodied relays, all testing occurs with a battery connected, so skip this and go on to the next step.
4. If the initial relay continuity tests pan out, you still need to check how the relay functions once power is applied to it. Remember, this is how a relay works, it throws an internal switch, changing a condition of non-continuity to continuity once power is applied to another portion of the relay. Connect a 12-volt battery to the proper terminals of the relay and use the ohmmeter to

TRIM & TILT

Fig. 11 Using an ohmmeter to test the power trim/tilt relay

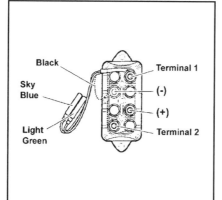

Fig. 12 Terminal identification on rectangular, exposed terminal PTT relays

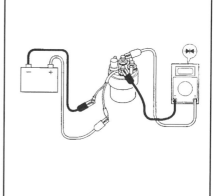

Fig. 13 Testing the round-bodied style PTT relays

check for continuity changes. Again, terminal or wire connections vary slightly by relay type as follows:

• For rectangular, exposed terminal type relays, if there are 3 wires coming off the harness make the battery connections on the wires only (if not, make the Negative battery connection directly to the Negative terminal on the relay). First connect the positive battery lead to the Light Green wire and the Negative battery lead to the Black wire, then use the ohmmeter to check between Terminal 1 and the Positive relay terminal, there should now be continuity when there was none before. Move the Positive battery lead from the Light Green wire to the Sky Blue wire, then check for continuity between relay Terminal 2 and the Positive relay terminal. Again, there should now be continuity when there was none before.

• For rectangular, non-exposed terminal type relays (such as those used on the 25/30 hp [496cc] 3-cylinder motors), connect the positive battery lead to the Light Green wire and the Negative battery lead to the Black wire, then use the ohmmeter to check between the Green and Red wires, there should now be continuity when there was none before. Move the Positive battery lead from the Light Green wire to the Sky Blue wire, then check for continuity between the Blue and Red Wires. Again, there should now be continuity when there was none before.

• For models with 2 round-bodied relays, connect an ohmmeter across the relay terminals as shown in the accompanying illustration. Next, connect the negative lead from the battery to the Black relay lead and the positive lead from the battery to the Sky Blue or Light Green lead on each relay. The ohmmeter should ONLY show continuity with the battery connected as noted.

5. If the relay should fail to show continuity when noted or should continuity on the switched circuits not open after power is removed, the relay must be replaced.

Tilt Switch

TESTING

◆ See Figures 14 and 15

There is normally one power tilt switch located at the top of the remote control handle. The harness from the switch is routed down the handle to the base of the control box, and then into the box through a hole. The control box cover must be removed to gain access to the quick disconnect fittings and permit testing of the switch.

An auxiliary tilt switch is also normally found mounted on the exterior surface of the starboard side lower cowling pan. This switch is convenient for performing tests on the tilt unit, when the need to observe the unit in operation is required.

In either case, the tilt switch is a simple, 3 wire, 3 position switch and testing is relatively straightforward. There should be no continuity between any of the leads when the switch is released, however there should be continuity between 2 of the leads when the switch is pressed in the UP position, and there should be continuity between a different pair of leads when the switch is pressed in the DOWN position.

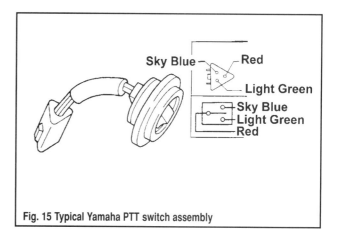

Fig. 14 Using an ohmmeter to test the power trim/tilt switch

Fig. 15 Typical Yamaha PTT switch assembly

8-8 TRIM & TILT

1. Disconnect the three leads for the switch: the Red, Light Green, and Sky Blue, leads either at their quick disconnect fittings or using the single switch connector (depending on the type of switch).

2. Connect the Red ohmmeter lead to the Red switch lead, and keep this connection for the following two resistance tests.

3. Connect the Black meter lead to the Sky Blue switch lead, then depress and hold the upper portion of the toggle switch (the UP button). The meter should indicate continuity. Release the switch to the neutral position. The meter should indicate no continuity. Depress the lower portion of the switch. The meter should still indicate no continuity.

4. With the Red meter lead still connected to the Red switch lead, move the Black meter lead to make contact with the Light Green switch lead. Depress the lower portion of the toggle switch. The meter should indicate continuity. Release the switch to the neutral position. The meter should indicate no continuity. Depress the upper portion of the switch. The meter should still indicate no continuity.

5. If the switch fails one or both resistance tests, replace the switch.

Tilt Motor

TESTING

◆ See Figure 16

To run the motor without a fully hydraulic load, set the manual release valve to the manual tilt position. This means make sure it is rotated approximately three full turns from the power tilt position, as evidenced by the embossed words and directional arrow on the housing (remember Allen head valves are normally left-hand thread and turn clockwise to loosen, while slotted head valves are normally regular, right-hand thread and turn counterclockwise to loosen).

1. Disconnect the 2 leads from the PTT motor at their quick disconnect fittings (the leads are normally Blue and Green on these Yamaha motors). Momentarily make contact with the two disconnected leads to the posts of a fully charged battery. Make the contact only as long as necessary to hear the electric motor rotating.

2. Reverse the leads on the battery posts and again listen for the sound of the motor rotating (when the leads are reversed the motor operates in the opposite direction). The motor should rotate with the leads making contact with the battery in either direction.

3. If the motor fails to operate in one or both directions, remove the electric motor for service or replacement as follows:

   a. First, place a suitable container under the unit to catch the hydraulic fluid as it drains.

   b. Next, disconnect the two lower hydraulic lines from the bottom of the housing (if equipped) or remove one or more of the valves, then remove the fill plug from the reservoir. Permit the fluid to drain into the container.

   c. After the fluid has drained, disconnect the electrical leads at the harness plug, and then remove the electric motor through the attaching hardware.

■ This motor is very similar in construction and operation to a cranking motor. The arrangement of the brushes differs in order to allow the trim/tilt motor to operate in opposite directions, but otherwise, it is almost identical to the cranking motor.

   d. Check armature commutator continuity, brush condition etc, in the same manner as the Yamaha electric starter motors. Repair or replace motor components, as necessary.

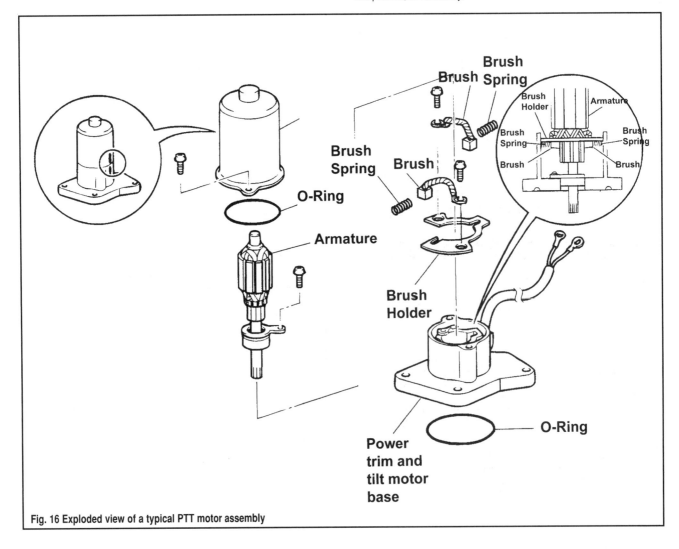

Fig. 16 Exploded view of a typical PTT motor assembly

TRIM & TILT

Trim Sensor

TESTING

◆ See Figure 17

Most models are equipped with a trim sensor attached to the tilt bracket. The job of the sensor is to provide feedback to the operator concerning the degree of engine tilt. The sensor consists of one or two potentiometer circuits. A potentiometer is basically a device whose electrical resistance varies with mechanical movement. In this case, the sensor contains a small arm which moves with the up or down movement of the outboard.

Sensor testing is relatively straightforward and is conducted strictly with an ohmmeter. Start by disconnecting the sensor wiring and identifying the appropriate terminals for testing (this varies slightly by model). Then connect and ohmmeter and move the sensor arm through the full range of movement (either by hand with the sensor removed from the outboard or by tilting the outboard itself). Watch the meter for acceptable resistance within the specified range (again, this varies slightly by model) and make sure there are not sudden opens or shorts where continuity drops or spikes. All changes to resistance should be relatively linear.

■ Remember that resistance specifications will vary slightly by meter and even more by temperature. All specs given are taken with a high quality Digital Ohmmeter, at an ambient/component temperature of approximately 68°F (20°C).

When testing the trim sensor, look for values within the following ranges, based on model:
- For 25/30 hp (496cc) 3-cylinder, motors, check resistance across the Pink and Black wires, it should vary from about 582-873 ohms through the range of motion.
- For 28J-50 hp (698cc) 3-cylinder motors, first check resistance across the Pink and Black wires, it should vary from about 360-540 ohms through the range of motion. Then check resistance across the Orange and Black wires, it should vary from about 800-1200 ohms through the range of motion.

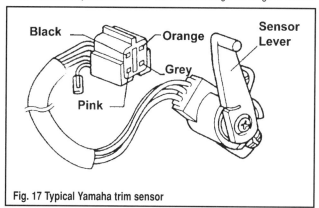

Fig. 17 Typical Yamaha trim sensor

PTT Assembly

REMOVAL & INSTALLATION

◆ See Figures 18 and 19

1. Tilt the outboard to the fully upward position then secure in place using the tilt lever stop (trailering lock).
2. Tag and disconnect the wiring for the PTT motor. Also, note the mounting point and disconnect the ground lead(s) from the unit. There are often 2 ground leads mounted to the bottom of the assembly and, on some models the bolts retaining the ground leads also secure an anode to the bottom of the assembly.

■ It will be necessary to cut one or more wire ties in order to reposition and free the necessary wiring. On some models, such as the 25/30 hp (496cc) 3-cylinder motor there may be as many as 5 wire ties.

3. If equipped with a tilt rod assembly (such as most 40/50 hp models), note the location (through which holes it passes in the bracket), then remove the rod.

■ Removal of some pins will require the removal of the bushings.

4. Remove the pivot components securing the PTT assembly at the bottom and top of the unit. The components vary slightly by model as follows:
   - For 25/30 hp (496cc) 3-cylinder motors, start at the bottom, removing the bolt and washer from each side of the clamp bracket, then carefully remove the shaft (pin). Then move to the top to locate a single circlip on the starboard side of the top attaching point for the tilt ram, remove the clip, then withdraw the shaft pin to free the PTT unit (obviously you'll need to support the unit as you pull this pin).
   - For 28J-50 hp (698cc) 3-cylinder motors, start at the bottom, removing the nut and plane washer from the port side of the clamp bracket, then carefully withdraw the stud bolt along with the large nut from the starboard side of the clamp bracket. Next, move to the top of the tilt ram and remove the 1 or 2 circlips securing the shaft pin. Support the unit as you carefully pull the pin free of the tilt ram.

5. If necessary remove the bushing(s) from the PTT unit and/or clamp bracket for installation purposes. Grease the replacement bushings to help ease installation.
6. Before installing the unit apply a light coating of marine grade grease to the inner diameter of each bushing.
7. Installation is essentially the reverse of the removal process. In most cases it really doesn't matter whether you remove or install the upper or lower shaft pin first, just make sure you properly secure the pins using the circlips or bolts, as applicable. For models that utilize bolts or nuts to secure the lower shaft mounting pin, apply a light coating of Loctite®242 or equivalent threadlock to the bolts before installation.

8-10 TRIM & TILT

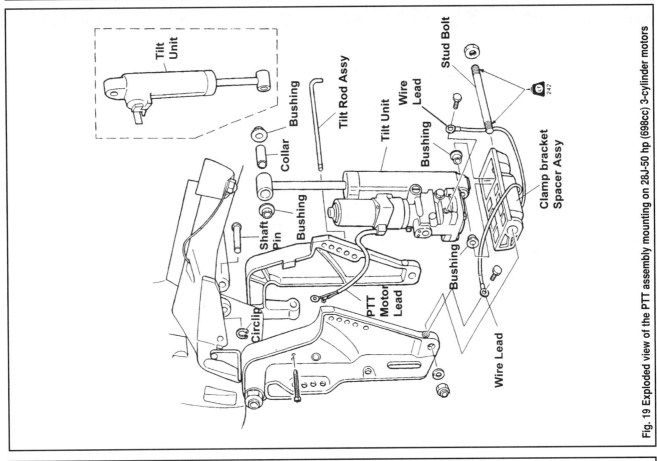

Fig. 19 Exploded view of the PTT assembly mounting on 28J-50 hp (698cc) 3-cylinder motors

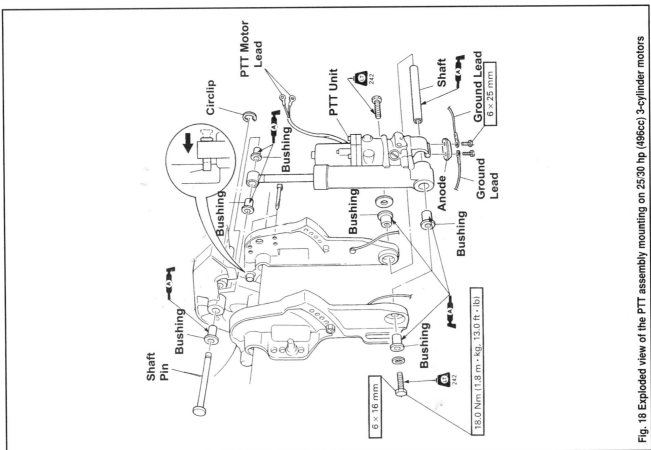

Fig. 18 Exploded view of the PTT assembly mounting on 25/30 hp (496cc) 3-cylinder motors

TRIM & TILT

COMPONENT REPLACEMENT

◆ See Figures 20 thru 24

 MODERATE

Service procedures for the power tilt system are confined to basic maintenance and some minor rebuilding. Rebuilding the unit involves removal of the end cap, removing the piston and replacing the O-rings.

A spanner wrench is required to remove the end cap. Even with the tool, removal of the end cap is not a simple task. The elements, especially if the unit has been used in a salt water atmosphere, will have their corrosive affect on the threads. The attempt with the special tool to break the end cap loose may very likely elongate the two holes provided for the tool. Once the holes are damaged, all hope of removing the end are lost. The only solution in such a case is to replace the unit.

If disassembly is required, work slowly keeping close track of the positioning for all components. Refer to the accompanying exploded views for help in identifying components and positions, but keep in mind that pump, cylinder and valve assemblies will vary greatly.

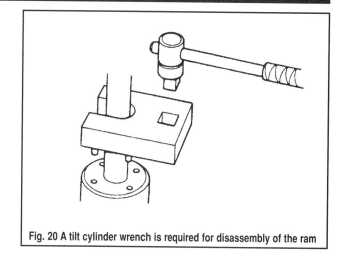

Fig. 20 A tilt cylinder wrench is required for disassembly of the ram

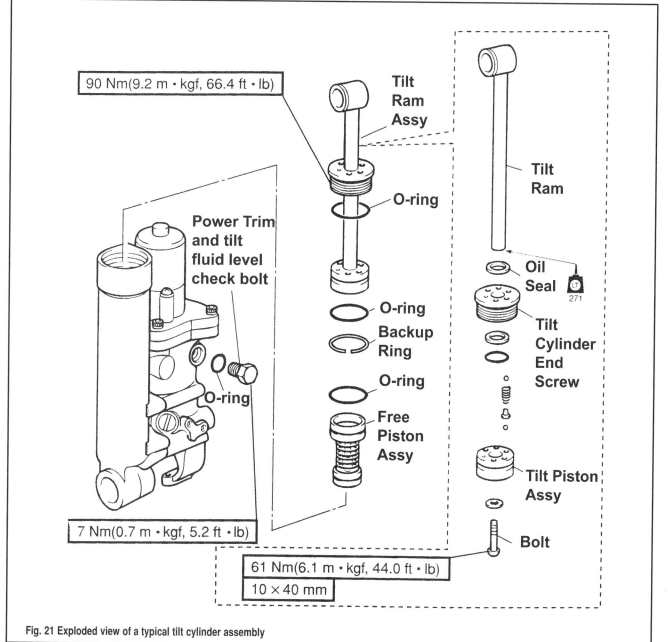

Fig. 21 Exploded view of a typical tilt cylinder assembly

8-12 TRIM & TILT

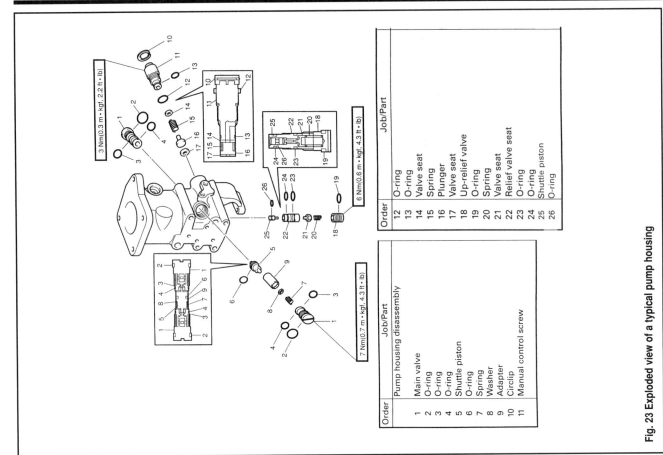

Fig. 23 Exploded view of a typical pump housing

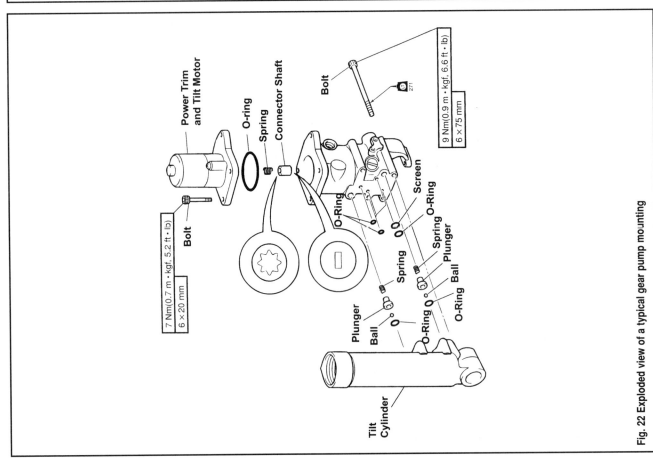

Fig. 22 Exploded view of a typical gear pump mounting

TRIM & TILT

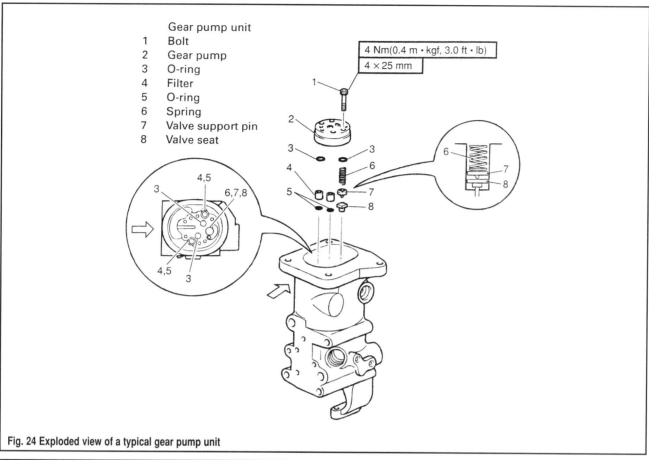

Fig. 24 Exploded view of a typical gear pump unit

Gear pump unit
1. Bolt
2. Gear pump
3. O-ring
4. Filter
5. O-ring
6. Spring
7. Valve support pin
8. Valve seat

LARGE MOTOR POWER TRIM/TILT SYSTEM

Description & Operation

◆ See Figures 25 and 26

■ The large motor PTT system (with single tilt rod and dual trim rods) is normally found on most 60 hp and larger motors.

The multi-cylinder power trim/tilt system consists of a housing with an electric motor, gear driven hydraulic pump, hydraulic reservoir, two trim cylinders, and one tilt cylinder. Like the tilt cylinder on the single ram systems, the large tilt cylinder performs a double function as tilt ram and also as a shock absorber, should the lower unit strike an underwater object while the boat is underway.

The necessary valves, check valves, relief valves, and hydraulic passageways are incorporated internally and externally for efficient operation. A manual release valve is provided, on the side of the housing (usually to the port side of late-model motors), to permit the outboard unit to be raised or lowered should the battery fail to provide the necessary current to the electric motor or if a malfunction should occur in the hydraulic system.

The gear driven pump operates in much the same manner as an oil circulation pump installed on motor vehicles. However, the gears may revolve in either direction, depending on the desired direction of cylinder movement, up or down. One side of the pump is considered the suction side, and the other the pressure side, when the gears rotate in a given direction. These sides are reversed, the suction side becomes the pressure side and the pressure side becomes the suction side when gear movement is changed to the opposite direction.

Reversing of pump direction is achieved simply by swapping polarity on the pump wiring. When power is applied to one side, the pump will operate in a given direction. However, if you provide power to the other motor wire, the pump will operate in the opposite direction.

Depending on the model, one or two relays for the electric motor are located at the bottom cowling pan, where they are fairly well protected from moisture.

■ Yamaha engineers have been constantly working to improve the operational performance of the trim and tilt system installed on their outboard units. Therefore, many of the units will appear to be quite similar but internal hydraulic passages and check valves have been changed as well as the external routing of hydraulic lines.

The basic principles described apply to all multi-cylinder power trim/tilt units, but the specific location and number of components, may vary, and the routing of hydraulic lines may not be exactly as viewed on the unit being serviced. As such, it is critical that you note the placement of all internal check valves or seals if the unit is being disassembled for overhaul.

TRIM UP

◆ See Figure 27

When the up portion of the trim/tilt switch on the remote control handle (or on the cowling) is depressed, the up circuit, through the relay, is closed and the electric motor rotates (usually in a clockwise direction).

The pump sucks in fluid from the reservoir through a check valve and forces the fluid out the pressure side of the pump.

From the pump, the pressurized fluid passes through a series of valves to the lower chamber of both trim cylinders and the pistons are extended. The outboard unit rises. The fluid in the upper chamber of the pistons is routed back to the reservoir as the piston is extended. When the desired position for trim is obtained and the switch is released, the outboard is held stationary.

On most motors, a trim sender unit is installed on the port side clamp bracket. This unit sends a signal to an indicator on the control panel to advise the helmsperson of the relative position of the outboard unit.

If the switch is not released when the trim cylinders are fully extended, the tilt cylinder will continue to move the outboard unit upward until the fully tilted position is reached and there is no more physical travel left in the system.

8-14 TRIM & TILT

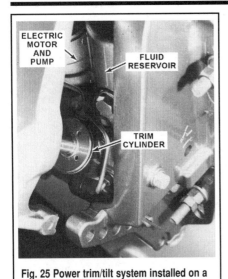

Fig. 25 Power trim/tilt system installed on a typical Yamaha (90 hp model shown)

Fig. 26 Cutaway view of a typical Yamaha PTT system

Fig. 27 Close-up of the auxiliary PTT switch found on the cowling of most outboards

TRIM DOWN

When the down portion of the trim/tilt switch on the remote control handle, or the auxiliary switch, is depressed, the down circuit, through the relay, is closed and the electric motor rotates in the opposite direction (therefore normally counterclockwise). The pressure side of the pump now becomes the suction side and the original suction side becomes the pressure side.

Fluid is forced through a series of check valves to the upper chamber of each trim cylinder and the pistons begin to retract, moving the outboard unit downward. Fluid from the lower chamber of each trim cylinder is routed back through the pump and a relief valve to the reservoir.

TILT UP

The first phase of the tilt up movement is the same as for the trim up function.

When the pistons of the trim cylinders are fully extended, fluid is forced through the system to the lower chamber of the tilt cylinder. Pressure increases in the lower chamber and the piston is extended, raising the outboard unit.

As fluid pressure in the upper chamber of the tilt cylinder increases, the fluid is routed through check valves back to the pump and the reservoir.

When the tilt piston is fully extended, fluid pressure in the lower chamber of the trim cylinders increases. This increase in pressure opens an up relief valve and the fluid is routed to the reservoir.

■ **When the tilt piston becomes fully extended, the outboard is in the full up position. The sound of the electric motor and the pump will have a noticeable change. The switch on the remote control handle should be released immediately.**

If the switch is not released, the motor will continue to rotate, the pump will continue to pump, but the up relief valve will open and the pressurized fluid will be routed to the reservoir.

If the boat is underway when the tilt cylinder is extended and powerhead rpm is increased beyond a very slow speed, the forward thrust of the propeller will increase the pressure on the tilt piston. This increase in pressure will cause the up relief valve to open and the outboard unit will begin a downward movement.

TILT DOWN

When the down portion of the trim/tilt switch on the remote control handle, or the auxiliary trim/tilt switch, is depressed, the down circuit is closed through the relay and the electric motor rotates in the opposite direction it would when the up switch is pressed (usually counterclockwise), as in the case of Trim Down.

The hydraulic pump sucks fluid from the reservoir. Fluid, under pressure is then routed to the upper chamber of the tilt cylinder and the piston begins to retract and the outboard unit moves downward.

Fluid in the lower chamber of the tilt cylinder is routed through the lower chamber of each trim cylinder and then back to the pump. When the outboard unit makes physical contact with the ends of the trim cylinders, the trim cylinders also retract until the outboard is in the full down position.

MANUAL OPERATION

◆ See Figure 7

✱✱ WARNING

If outboard unit is in the up position when the manual release valve is opened, the outboard will drop to the full down position rapidly. Therefore, ensure all persons stand clear.

The outboard unit may be raised or lowered manually should the battery fail to provide sufficient current to operate the electric motor or should electric/hydraulic components of the PTT system fail for any other reason. A manual relief valve is provided to permit manual operation.

This manual relief valve is located on the lower end of the gear pump beneath the electric motor either facing aft (toward the motor/gearcase) or facing sideways (so that it is accessed through one of the clamp brackets). On some older Yamaha motors the valve faced aft and often contained left-hand threads, meaning it is OPENED or LOOSENED by turning CLOCKWISE. However, on MOST late-model Yamaha outboards (where the valve faces to one side, and is accessed through a clamp bracket), the valve is equipped normal right-hand threads, meaning it is OPENED or LOOSENED by turning COUNTERCLOCKWISE.

Either way, opening the valve releases pressure in both the upper and lower cylinder chambers. With a complete loss of pressure, the piston may be moved up or down in the cylinder without resistance.

After the outboard unit has been moved to the desired position, the valve should be rotated in the opposite direction to close the valve and lock the outboard against movement.

SHOCK ABSORBER ACTION

The lower end of the tilt piston is capped with a free piston. This free piston normally moves up and down with the tilt piston - goes along for the ride.

In the event the outboard lower unit should strike an underwater object while the boat is underway, the tilt piston would be suddenly and forcibly extended, moved upward.

The free piston also moves upward but at a much slower rate than the tilt piston. The action of the tilt piston separating from the free piston causes two actions. First, the hydraulic fluid in the upper chamber above the piston is compressed and pressure builds in this area. Second, a vacuum is formed in the area between the tilt piston and the free piston.

This vacuum in the area between the two pistons sucks fluid from the upper chamber. The fluid fills the area slowly and the shock of the lower unit striking the object is absorbed. After the object has been passed the weight

TRIM & TILT 8-15

of the outboard unit tends to retract the piston. The fluid between the tilt piston and the free piston is compressed and forced through check valves to the reservoir until the free piston reaches its original neutral position.

Maintenance

CHECKING HYDRAULIC FLUID LEVEL

◆ See Figures 28 and 29

■ If one of the hydraulic valves has been removed, or the system opened for any other reason, the system must be purged of any trapped air. To purge (bleed) the system, please refer to Purging Air From The System in this section.

The fluid in the power trim/tilt reservoir should be checked periodically to ensure it is full and is not contaminated. To check the fluid, tilt the motor upward to the full tilt position, then manually engage the tilt support for safety and to prevent damage.

In order to keep contaminants from entering the system when the plug is removed, clean the area around the fill plug (found on the reservoir, just below the motor on these models). Loosen and remove the plug using a suitable socket or wrench and make a visual inspection of the fluid. It should seem clear and not milky. The level is proper if, with the motor at full tilt, the level is even with the bottom of the filler plug hole.

■ Yamaha or Yamalube power trim and tilt fluid is a highly refined hydraulic fluid. This product has a high detergent content and additives to keep seals pliable. A high grade automatic transmission fluid, Dexron® II may also be used if the Yamalube fluid is not available.

PURGING AIR FROM THE SYSTEM

◆ See Figure 30

Air must be purged (bled) from the system if any of the following conditions exist:
 a. Erratic motion is felt during operation of the system.
 b. Fluid has been contaminated with moisture or other foreign material.
 c. A component in the system has been replaced.
 d. The system has been opened for any reason.
 e. The outboard, when been left in the full up position, lowers slowly over a period of time.

■ When the reservoir is full to a normal level it should be even with the bottom of the plug threads when the outboard is in the fully upward, tilted position, meaning that when the outboard is lowered the level may be above the fill plug. One way to start refilling an empty or nearly empty system is to add fluid while the outboard is down, then open the manual release valve and manually lift the engine, drawing fluid into the system as the piston inside the tilt cylinder moves upward (as the ram is extended). Another method is to use is to refill the reservoir while the PTT unit is still completely removed from the outboard, and then use a battery with jumpers to cycle the motor and rams through their full motion of travel a couple of times, topping off the reservoir, as necessary.

1. Begin by making sure there is fluid inside the reservoir, even with the outboard in the normal running position (not tilted). Clean the area around the fill plug to prevent contaminants entering the system when the plug is loosened or removed. Slowly start to loosen the fill plug, if fluid starts to run (or spray) out, stop loosening the plug and retighten it until the outboard is tilted. However, if no fluid escapes, go ahead and remove the plug, then add fluid until it is level with the bottom of the screw threads. Lightly reinstall the reservoir plug.

2. Open the manual release valve (either by rotating the Allen head of the valve clockwise, or the slotted head counterclockwise, as applicable), rotating it slowly in the proper direction until it stops.

3. With the manual release valve open, slowly pull the outboard upward by hand, extending the tilt piston. As the tilt piston is extended, the piston in the cylinder will draw fluid into the cylinder, thus lowering the level in the chamber.

4. Carefully release the outboard, allowing it to lower by its own weight, then once the outboard is lowered fully, retighten the manual release valve.

5. Allow the PTT system fluid to settle for about 5 minutes.

6. Push the PTT switch (either on the remote or the side of the motor) allowing the system to move the motor to the fully upward position.

7. Push the tilt stop lever(s) to support the outboard.

8. Again, allow the PTT system fluid to settle for about 5 minutes.

9. Remove the reservoir cap and check the fluid level. Add fluid, as necessary until it is even with the bottom of the fill plug, then install and tighten the reservoir cap.

10. Repeat as necessary until the fluid remains at the correct level.

Troubleshooting

■ When moving through the listed troubleshooting procedures, always stop and check the system after each task. The problem may have been corrected, intentionally, or not.

PRELIMINARY INSPECTION

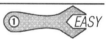

1. Check to be sure the manual release head is tightened snugly in the proper position.

2. Verify the hydraulic reservoir is filled with fluid. The level of fluid should reach to the lower edge of the fill opening. Replenish as required with Yamaha or Yamalube power trim and tilt fluid is a highly refined hydraulic fluid. This product has a high detergent content and additives to keep seals pliable. A high grade automatic transmission fluid, Dexron® II may also be used if the Yamalube fluid is not available.

※※ WARNING

The PTT system is pressurized! Unless instructed otherwise, do not remove fill screw unless outboard unit is raised to full up position. Tighten fill screw securely before lowering outboard.

3. Inspect the hydraulic system, lines and fittings, for leaks. If an external leak is discovered, correct the condition.

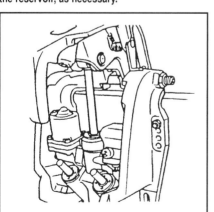

Fig. 28 To check fluid level, raise and lock the outboard to full tilt. . .

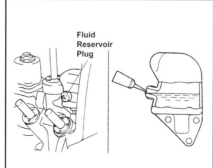

Fig. 29 . . .then remove the reservoir plug and make sure the fluid level is even with the bottom of the bore

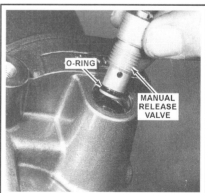

Fig. 30 A few manual relief valves used on early-model PTT system have left-hand threads

8-16 TRIM & TILT

4. Purge air from the system, if there is any indication the hydraulic fluid contains air.

5. To check for air in the system, first check to be sure the manual release valve is snug in the power tilt position. Next, activate the up circuit and raise the outboard slightly with the trim cylinders. Now, exert a heavy, steady, downward force on the lower unit.

6. If the trim pistons retract into the trim cylinders more than about 1/8 in. (3.2mm) there is air in the system.

7. Purge (bleed) air from the system.

8. If the system fails to hold the outboard unit in the full tilted (trailering) position, service the tilt cylinder.

9. If the system fails to hold the outboard unit in the desired trim position, service the trim cylinders.

10. If the hydraulic pump whines during operation, there is probably air in the system. Purge air from the system.

11. If the electric motor makes strange sounds or seems to be laboring, the electric motor may require service.

■ **If a problem is encountered with the PTT system, it is important to determine, if possible, whether the malfunction is in the hydraulic system or in an electrical circuit.**

SYMPTOM DIAGNOSIS

Symptoms of a Hydraulic Problem

The most common problem in the hydraulic system is failure of an O-ring to hold pressure.

- **Outboard Behaves Abnormally** - Ensure the battery is adequately charged; the fluid level in the reservoir reaches the lower edge of the fill opening, with the outboard fully raised; the hydraulic fluid is not contaminated (murky color); the system does not contain air; the unit does not physically bind somewhere due to an accident.
- **Outboard Fails to Trim Up or Down** - Ensure the unit has adequate fluid in the reservoir; manual release valve is tightened counterclockwise snugly to the power tilt position, approximately three full turns from the manual tilt position. Once cause may be hydraulic pump failure; O-rings in trim cylinders failing to hold pressure; trim cylinders damaged due to accident.
- **Outboard Trims/Tilts Up but Fails to Trim/Tilt Down** - Manual release valve is leaking; O-rings in trim cylinder or in the tilt cylinder failing to hold pressure; main valve has sticky or damaged shuttle piston; sticky or contaminated check valves; down relief valve has weak spring; damaged check ball or seat; or contamination is holding a valve open.
- **Outboard Trims/Tilts Down but Fails to Trim/Tilt Up** - Manual release valve is leaking; O-rings in trim cylinders or tilt cylinder failing to hold pressure; shuttle piston in main valve assembly is sticking; contaminated check valves; up relief valve has damaged seat or contamination holds the valve open.
- **Outboard Shudders When Shifted From One Gear to Another** - Hydraulic system contaminated with air or foreign matter; internal cylinder leaks; O-rings failing to hold pressure.
- **Outboard Fails to Hold Set Trim or Tilt Position** - O-rings in trim and/or tilt cylinder failing to hold pressure; check valves in tilt piston contaminated requiring cleaning; external leak - fitting or part; manual release valve damaged; shuttle piston in main valve assembly sticking or contaminated check valve; up relief valve damaged or contaminated causing a slow leak.
- **Outboard Tilts Up When Unit In Reverse Gear** - Tilt piston has leaky absorber valve or metering valve; main valve assembly has sticky or damaged shuttle piston or leaky check valves; manual release valve is leaking; O-rings in tilt piston failing to hold pressure.
- **Outboard Makes Excessive Noise** - Hydraulic fluid level is low; fluid is contaminated with air.
- **Outboard Begins To Trail Out When Throttle Backed Off at High Speed** - Manual release valve not tightened snugly counterclockwise to power tilt position; O-rings in tilt cylinder failing to hold pressure.
- **Outboard Fails to Hold Trim Position When Unit Operating In Reverse** - Manual release valve not tightened snugly counterclockwise to power tilt position; O-rings in trim cylinders failing to hold pressure.
- **Outboard Moves With Jerky Motion** - System contaminated with air; internal leaks in cylinders.
- **Outboard Fails to Reach Full Down Position** - System contaminated with air or internal leaks in cylinders.

Symptoms of an Electrical Problem

If any of the following problems are encountered, troubleshoot the electrical system.

- Outboard trims up and down, but the electrical motor grinds.
- Outboard will not trim up or down.
- Outboard trims up, but will not trim down.
- Outboard trims down, but will not trim up.

HYDRAULIC TESTING

The health of the PTT unit's hydraulic system (check valves, pistons, O-rings, seals and motor) can be checked using a high-pressure hydraulic gauge (extremely high pressure on some systems, so pre-read the procedure) and a 12-volt battery with jumpers to run the motor. Although Yamaha does not specify as to whether or not the test MUST occur with the PTT unit removed from the outboard, all of their illustrations show it performed this way. In addition, the mounting point for gauges on most PTT units is the manual valve bore, which often cannot be accessed with the PTT unit installed. However, if access is possible, you'll still want to at least remove the outboard load from the unit as it is tested. To do this tilt and lock the outboard in the full tilt position, then disconnect the tilt ram from the outboard itself.

Inline 3-Cylinder Motors

◆ See Figure 31

■ **Yamaha recommends using 2 sets of hoses and 2 pressure gauges so that both of the trim and tilt pressures can be measured at the same time. When connecting the gauges, pour hydraulic fluid into each pipe to minimize the amount of air that could enter the system.**

When performing this test, attach a 12-volt battery to the PTT motor leads using jumpers. Attach the positive jumper to the Green motor lead and negative to the Blue motor lead in order to run the PTT motor downward. Reverse the connections (Green to negative and Blue to positive) in order to run the motor upward.

When operating the pump motor, make sure the PTT has reached full UP or full DOWN positions before noting system pressures. However, don't run the motor excessively in either position. The motor should NOT be operated for more than 3 seconds in the full DOWN position or more than 14 seconds in the full UP position.

1. If not already in this position, use a screwdriver to close the manual release valve (this usually means by rotating it counterclockwise on these motors).

2. Place a small rag or suitable container under the fittings to catch any spilled hydraulic fluid.

3. Disconnect the hydraulic line from the bottom of the reservoir and upper chamber of the tilt cylinder. Install the Yamaha pressure gauge (#YB-6181), or equivalent, to the reservoir and the reservoir and tilt cylinder fittings. This gauge will measure tilt cylinder operating pressures.

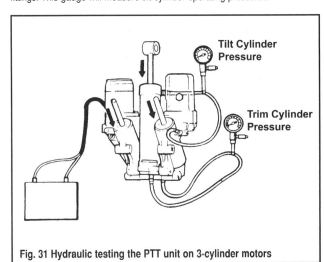

Fig. 31 Hydraulic testing the PTT unit on 3-cylinder motors

TRIM & TILT 8-17

4. Bleed the system of any air. For details, refer to Purging Air From The System earlier in this section. Check the hydraulic fluid level and top off as necessary.

5. Operate the PTT motor fully upward once, then operate it fully downward, observing hydraulic pressures, they will vary slightly based on the stamped identifier for the PTT system:
- For the 6H308, tilt operating pressure should be 583-782 psi (4100-5500 kPa).
- For the 6H1-15, tilt operating pressure should be 313-540 psi (2200-3800 kPa).
- For the 62F-02, tilt operating pressure should be 299-526 psi (2100-3700 kPa).

6. Now reverse the battery leads to the PTT motor to operate it fully upward, observing hydraulic pressures. As the tilt cylinder moves upward the gauge should read 0-71 psi (0-500 kPa). But once the fully upward position is reached, gauge readings should drop to 0 psi (0 kPa).

7. Remove the pressure gauge, reconnect the hydraulic line to the reservoir and tilt cylinder upper chamber and repeat the test on the lower chambers of the tilt/trim cylinder. This is where Yamaha recommends using 2 gauges to read both pressures simultaneously, and to prevent having to re-bleed the system again and again, but if you've only got one gauge, read one side, then the other.

8. Bleed the system of any air. For details, refer to Purging Air From The System earlier in this section. Check the hydraulic fluid level and top off as necessary.

9. Operate the PTT motor fully upward once, then operate it fully downward, observing hydraulic pressures, regardless of the stamped identifier, fully downward operating pressure should be 85-157 psi (600-1100 kPa).

10. Now reverse the battery leads to the PTT motor to operate it fully upward, observing hydraulic pressures. As the trim cylinders move upward the gauge should read 0-71 psi (0-500 kPa). But once the fully upward position is reached, gauge readings will vary slightly based on the stamped identifier for the PTT system:
- For the 6H308, tilt operating pressure should be 500-640 psi (3500-4500 kPa).
- For the 6H1-15, tilt operating pressure should be 1280-1565 psi (9000-11,000 kPa).
- For the 62F-02, tilt operating pressure should be 825-953 psi (5690-6570 kPa).

11. Once the tests are finished lower the PTT unit, then remove the pressure gauge and reconnect the hydraulic lines. Be sure to tighten the fittings securely. Bleed and refill the reservoir, as necessary.

V4 and V6 Motors (Except 3.3L Models)

◆ See Figures 32 and 33

For these motors you'll need an extremely high pressure Hydraulic Pressure Gauge (#90890-06776), the Up Relief Valve (#90890-06773) and the Down Relief Valve (#90890-06774) or their equivalents. The up or down relief valve is installed into the manual release valve bore (depending upon the portion of the system being testing), then the gauge is attached to the valve.

When performing this test, attach a 12-volt battery to the PTT motor leads using jumpers. Attach the positive jumper to the Green motor lead and negative to the Blue motor lead in order to run the PTT motor downward. Reverse the connections (Green to negative and Blue to positive) in order to run the motor upward.

1. Run the motor to place the tilt and trim rams at the full UP position.
2. Install the Hydraulic Pressure Gauge to the Up Relief Valve, then tighten the gauge to 6.5 ft. lbs. (9 Nm).

■ In the next step, minimize the amount of fluid lost and the amount of air that might enter the system by having the gauge and relief valve attachment ready to install AS SOON AS the manual release valve is removed from its bore.

3. Remove the circlip securing the manual release valve in its bore, then loosen and remove the valve. Quickly install the Up Relief Valve and Pressure Gauge assembly, then tighten them to 2.9 ft. lbs. (4 Nm).
4. Connect a 12-volt battery to the PTT motor using jumpers in order to run the trim and tilt rams to the fully downward position, then reverse the motor leads and run the rams to the fully extended (upward) position and observe hydraulic pressure, it should be 1421-1711 psi (9800-11,800 kPa / 9.8-11.8 mPa).
5. Lower the PTT motor partially from the full up position, then carefully release pressure and remove the gauge assembly. Temporarily reinstall the manual release valve to minimize the amount of fluid loss. Meanwhile remove the Gauge from the Up Relief Valve attachment and connect it to the Down Relief Valve attachment, again, tighten the gauge to 6.5 ft. lbs. (9 Nm).
6. Remove the manual release valve from the bore again and quickly install the Down Relief Valve and Pressure Gauge assembly, then tighten them to 2.9 ft. lbs. (4 Nm).
7. Run the PTT motor to the full up position, then check and top off the fluid reservoir, as necessary. If too much fluid was lost, bleed the system of air by following the steps for Purging Air From The System, located earlier in this section.
8. Connect a 12-volt battery to the PTT motor using jumpers in order to run the trim and tilt rams to the fully downward position and observe hydraulic pressure, it should be 856-1276 psi (5900-8,800 kPa / 5.9-8.8 mPa).
9. Run the PTT assembly back to the fully extended position, then carefully release pressure and remove the gauge assembly. Quickly install the manual release valve and secure using the circlip.
10. Check the fluid level and bleed the system, as necessary.

3.3L V6 Motors

◆ See Figures 34 and 35

For these motors you'll need an extremely high pressure Hydraulic Pressure Gauge (#YB-06580). The gauge will install in place of one of the external hydraulic lines either at the motor connection (when testing downward pressure) or at the fitting on the base of the assembly (when testing upward pressure).

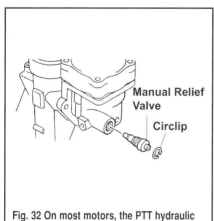

Fig. 32 On most motors, the PTT hydraulic pressure is tested...

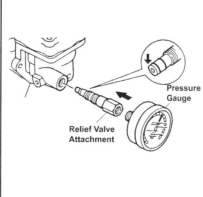

Fig. 33 ...through the manual release valve bore

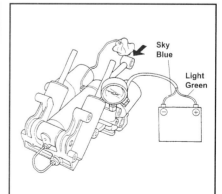

Fig. 34 Testing the PTT unit on a 3.3L motor for downward pressure

When performing this test, attach a 12-volt battery to the PTT motor leads using jumpers. Attach the positive jumper to the Green motor lead and negative to the Blue motor lead in order to run the PTT motor downward. Reverse the connections (Green to negative and Blue to positive) in order to run the motor upward.

1. Run the motor to place the tilt and trim rams at the full UP position.
2. Place a small rag or suitable container under the fittings to catch any spilled hydraulic fluid.
3. Loosen both of the hydraulic fittings for the line that runs from the bottom of the motor to the top of the tilt cylinder. Have the pressure gauge ready to install, then disconnect the line from the motor and install the pressure gauge in its place. Position the disconnected line end in a plastic bag or a small rag to catch any fluid which might leak out.
4. Connect a 12-volt battery to the PTT motor using jumpers in order to run the trim and tilt rams to the fully downward position and observe hydraulic pressure, it should be 682-972 psi (4700-6700 kPa / 4.7-6.7 mPa).
5. Reverse the battery connections and run the PTT cylinders to the full up (extended) positions.
6. Remove the pressure gauge and reconnect the hydraulic line to the motor, then carefully retighten both hydraulic fittings to 11 ft. lbs. (15 Nm).
7. Reconnect the battery jumpers to run the PTT cylinders to the full down (retracted) positions.
8. Loosen both of the hydraulic fittings for the line that runs from the bottom of the tilt cylinder to the bottom of the PTT assembly. Have the pressure gauge ready to install, then disconnect the line from the PTT assembly (not the bottom of the tilt cylinder) and install the pressure gauge in its place. Position the disconnected line end in a plastic bag or a small rag to catch any fluid which might leak out.
9. Connect a 12-volt battery to the PTT motor using jumpers in order to run the trim and tilt rams to the fully upward position and observe hydraulic pressure, it should be 1639-1929 psi (11,300-13,300 kPa / 11.3-13.3 mPa).
10. After noting the gauge reading, reverse the battery connection and run the PTT rams back to the fully downward (retracted position).
11. Remove the pressure gauge and reconnect the hydraulic line to the PTT assembly, then carefully retighten both hydraulic fittings to 11 ft. lbs. (15 Nm).
12. Run the PTT rams to the full upward position again, then remove the reservoir plug to check and possibly, top off the system, as necessary. If necessary, bleed the PTT system of air.

Trim/Tilt Relay

TESTING

◆ See Figures 36 thru 40

The PTT motor runs either clockwise or counterclockwise depending upon which direction you want the motor to move and also depending upon how the power is applied to the motor's terminals. If you apply positive current to one terminal and negative to the other, the motor will run in one direction. If you switch and apply positive and negative to the opposite terminals, the motor will run in the other direction. Control of this circuit is achieved through one or more PTT relays. The job of the relay, like all relays, is to close certain switch circuits depending upon the signals received from the tilt switch on either the engine or the remote.

There are generally 2 major designs of PTT relays that you might find on these motors. The first and most common relay used on late-model Yamahas is a rectangular piece that looks like it has 8 terminals and either a 2 or 3 wire harness that comes off one end (though the harness may use a single connector or individual wire bullets/terminals). Four of the visible terminals (identified by position) and the 2 or 3 wire harness are used during testing to with an ohmmeter (and a 12-volt battery) to determine if the relay is operating properly. For testing purposes, the terminal identification is slightly different on the 2 and 3 wire harness relays.

The 3 wire harness rectangular relay is normally found on 3-cylinder motors, and all EFI OX66 V6 motors (including 2.6L and 3.1L models). On these relays the exposed terminal at one end of the relay (where the wiring harness attaches, but on the opposite side of the relay on the same end) will be referred to as Terminal 1. This terminal is also identified because it is diagonally adjacent to the relay Negative terminal (the black ground wire was connected to it before removal). At the far diagonal end of the relay from Terminal 1 is the position for Terminal 2. Again, Terminal 2 is easily identified

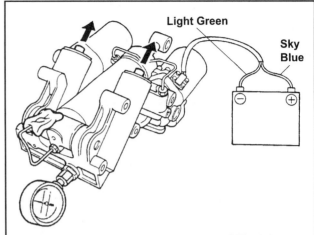

Fig. 35 Checking upward hydraulic pressure on a PTT unit for a 3.3L motor

because it is diagonally adjacent to the relay Positive terminal (the terminal to which the red power wire was attached before removal). For more details, please see the accompanying illustration.

The 2 wire harness rectangular relay is normally found on HPDI V6 motors (including 3.1L and 3.3L models). On these relays the exposed terminal on the wire harness end of the relay is also on the same side as the wiring harness itself and is referred to in the testing procedure as Terminal 1. This terminal is also easily identified since it is RIGHT next to the Positive relay terminal (the terminal to which the red power wire was attached before removal). At the far end of the relay (not diagonally on this type of relay), on the same side of the relay as Terminal 1 is Terminal 2. The other way to identify Terminal 2 is that it is diagonally adjacent to the relay Negative terminal (the terminal to which the black ground wire was connected before removal). For more details, please see the accompanying illustration.

A few motors use the old-Yamaha style round bodied relay, actually, they use 2 identical round-bodied relays to perform this function. Generally these can be found on all V4 motors and on carbureted versions of the V6 motors. Both wire color and terminal positions are used to test these relays in much the same fashion as the rectangular bodied units. To identify the terminals, look down on top of the relay with the 4 terminals positioned at 12, 3, 6 and 9 o'clock. Turn the relay so the wiring harness is coming out of the relay between the terminals at 9 and 12 o'clock. In this position the 12 o'clock terminal is normally covered and not exposed, the 6 o'clock terminal is considered Terminal 1, the 3 o'clock terminal is considered Terminal 2 and the 9 o'clock terminal is considered Terminal 3.

■ On all relays, testing normally occurs with the relay completely removed from the powerhead, OR at least with all wiring disconnected (even if the relay is still physically bolted in position).

1. If not done already, tag and disconnect the wiring from the PTT relay(s).
2. Obtain a 12-volt battery with jumper leads and an ohmmeter for continuity testing.
3. First check terminal-to-terminal continuity as detailed by relay type:

• For rectangular, 2 and 3-wire connector relays, use the ohmmeter to confirm there is continuity between the following pairs of wires: Terminal 1 and the Negative terminal, Terminal 2 and the Negative terminal, The Sky Blue wire and the Black wire and finally, the Light Green wire and the Black wire. Next use the ohmmeter to confirm there is NO continuity between Terminal 1 and the Positive terminal, as well as between Terminal 2 and the Positive terminal.

• For models with 2 round-bodied relays (like the carbureted V4 and V6 motors), use an ohmmeter to confirm there is continuity between the following pairs of wires: The colored lead (Sky Blue or Light Green depending upon which relay you are testing) and the Black lead, the colored lead and Terminal 1. Also check for continuity between Terminal 1 and Terminal 3. Next use the ohmmeter to confirm there is NO continuity between the colored lead and Terminal 2, as well as between Terminal 2 and Terminal 3.

4. If the initial relay continuity tests pan out, you still need to check how the relay functions once power is applied to it. Remember, this is how a relay

TRIM & TILT 8-19

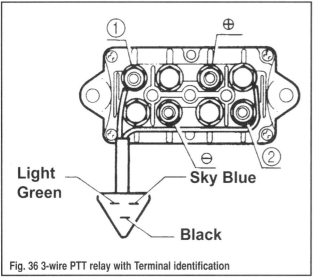

Fig. 36 3-wire PTT relay with Terminal identification

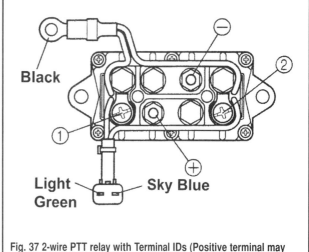

Fig. 37 2-wire PTT relay with Terminal IDs (Positive terminal may also be one to the right)

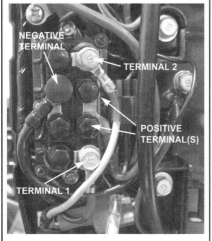

Fig. 38 The Red and Black wires help identify Positive and Negative terminals

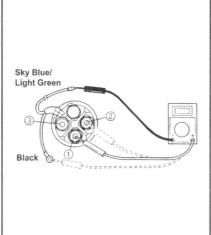

Fig. 39 Typical round-bodied PTT relay with Terminal IDs

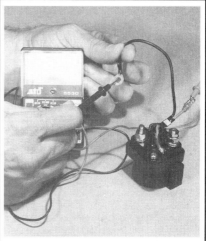

Fig. 40 Using an ohmmeter to test a round-bodied PTT relay

works, it throws an internal switch, changing a condition of non-continuity to continuity once power is applied to another portion of the relay. Connect a 12-volt battery to the proper terminals of the relay and use the ohmmeter to check for continuity changes. Again, terminal or wire connections vary slightly by relay type as follows:

• For rectangular, 2 and 3-wire connector relays, if there are 3 wires coming off the harness make the battery connections on the wires only (if not, make the Negative battery connection directly to the Negative terminal on the relay or the black wire coming off the Negative terminal). First connect the positive battery lead to the Light Green wire and the Negative battery lead to the Black wire. Then use the ohmmeter to check between Terminal 1 and the Positive relay terminal, there should now be continuity when there was none before. Move the Positive battery lead from the Light Green wire to the Sky Blue wire, then check for continuity between relay Terminal 2 and the Positive relay terminal. Again, there should now be continuity when there was none before.

• For models with 2 round-bodied relays (like the carbureted V4 and V6 motors), connect the positive battery connection to the colored wire (Sky Blue or Light Green depending upon which relay you are testing) and the negative battery connection to the Black wire. Then use an ohmmeter to check for continuity between Terminal 2 and Terminal 3, there should be continuity as long as 12-volts are applied to the wires as noted. Also, double-check that there is NO longer continuity between Terminal 1 and Terminal 3 (at least, not as long as 12-volts are applied to the noted wires).

5. If the relay should fail to show continuity when noted or should continuity on the switched circuits not open after power is removed, the relay(s) must be replaced.

Power Trim/Tilt Switch

TESTING

◆ See Figure 41

There is normally one power tilt switch located at the top of the remote control handle. The harness from the switch is routed down the handle to the base of the control box, and then into the box through a hole. The control box cover must be removed to gain access to the quick disconnect fittings and permit testing of the switch.

An auxiliary tilt switch is also normally found mounted on the exterior surface of the starboard side lower cowling pan. This switch is convenient for performing tests on the tilt unit, when the need to observe the unit in operation is required.

In either case, the tilt switch is a simple, 3 wire, 3 position switch and testing is relatively straightforward. There should be no continuity between any of the leads when the switch is released, however there should be continuity between 2 of the leads when the switch is pressed in the UP position, and there should be continuity between a different pair of leads when the switch is pressed in the DOWN position.

1. Disconnect the three leads for the switch: the Red, Light Green, and Sky Blue, leads either at their quick disconnect fittings or using the single switch connector (depending on the type of switch).

8-20 TRIM & TILT

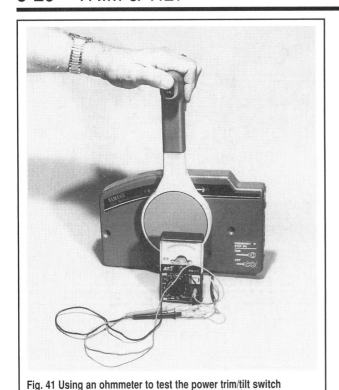

Fig. 41 Using an ohmmeter to test the power trim/tilt switch

2. Connect the Red ohmmeter lead to the Red switch lead, and keep this connection for the following two resistance tests.

3. Connect the Black meter lead to the Sky Blue switch lead, then depress and hold the upper portion of the toggle switch (the UP button). The meter should indicate continuity. Release the switch to the neutral position. The meter should indicate no continuity. Depress the lower portion of the switch. The meter should still indicate no continuity.

4. With the Red meter lead still connected to the Red switch lead, move the Black meter lead to make contact with the Light Green switch lead. Depress the lower portion of the toggle switch. The meter should indicate continuity. Release the switch to the neutral position. The meter should indicate no continuity. Depress the upper portion of the switch. The meter should still indicate no continuity.

5. If the switch fails one or both resistance tests, replace the switch.

Trim/Tilt Motor

TESTING

◆ See Figure 42

To run the motor without a fully hydraulic load, set the manual release valve to the manual tilt position. This means make sure it is rotated approximately three full turns from the power tilt position, as evidenced by the embossed words and directional arrow on the housing.

1. Disconnect the 2 leads from the PTT motor at their quick disconnect fittings (the leads are normally Blue and Green on these Yamaha motors). Momentarily make contact with the two disconnected leads to the posts of a fully charged battery. Make the contact only as long as necessary to hear the electric motor rotating.

2. Reverse the leads on the battery posts and again listen for the sound of the motor rotating (when the leads are reversed the motor operates in the opposite direction). The motor should rotate with the leads making contact with the battery in either direction.

3. If the motor fails to operate in one or both directions, remove the electric motor for service or replacement as follows:

    a. First, place a suitable container under the unit to catch the hydraulic fluid as it drains.

    b. Next, disconnect the two lower hydraulic lines from the bottom of the housing (if equipped) or remove one or more of the valves, then remove the fill plug from the reservoir. Permit the fluid to drain into the container.

    c. After the fluid has drained, disconnect the electrical leads at the harness plug, and then remove the electric motor through the attaching hardware.

■ This motor is very similar in construction and operation to a cranking motor. The arrangement of the brushes differs in order to allow the trim/tilt motor to operate in opposite directions, but otherwise, it is almost identical to the cranking motor.

    d. Check armature commutator continuity, brush condition etc, in the same manner as the Yamaha electric starter motors. Repair or replace motor components, as necessary.

Trim Sensor

ADJUSTMENT

◆ See Figure 43

The trim sensor is equipped with a lever that moves in relation to the position of the outboard. Since the sensor is a potentiometer (an electrical resistor whose value varies with the physical position of its lever) the exact positioning of the sensor on the outboard will can be set so the trim/tilt gauge displays properly. Adjustment will vary slightly from gauge-to-gauge, however all adjustment involves loosening the sensor mounting screws and rotating the housing slightly so the gauge reads properly.

Most trim tilt gauges are designed to only display the trim range of motor movement. This is the portion of movement which is controlled by the duel trim rods, not the tilt ram. In order to check adjustment compare the position of the motor at the top and bottom ends of the TRIM motion (not TILT) with the display on the gauge. If adjustment is necessary, loosen the screws and reposition the sensor slightly until the gauge properly reflects motor trim position.

■ If during adjustment the gauge shows a sudden spike or jump in position, test the sensor for potential resistance problems.

If the boat is equipped with Yamaha's digital meter, the following procedure can normally be used to ensure proper adjustment:

1. Raise the outboard to the full tilt position and support it in this position.
2. Loosen the 2 screws securing the trim sensor so it can move slightly.
3. Fully tilt the outboard to the full down position.
4. Set the main switch to the ON position to power the Yamaha digital meter.
5. With the outboard in the full down position, use a screwdriver and adjust the trim sensor so only one segment on the trim indicator on the digital meter goes on. Tighten the sensor screws while holding the outboard in this position.
6. Raise the outboard to the full tilt position, watching the display on the digital meter as the motor moves through the trim range. Make sure the trim sensor does not move from the original position. If necessary, tweak the adjustment to ensure proper display.

TESTING

◆ See Figure 43

Most models are equipped with a trim sensor attached to the tilt bracket. The job of the sensor is to provide feedback to the operator concerning the degree of engine tilt. The sensor consists of one or two potentiometer circuits. A potentiometer is basically a device whose electrical resistance varies with mechanical movement. In this case, the sensor contains a small arm which moves with the up or down movement of the outboard.

Sensor testing is relatively straightforward and is conducted strictly with an ohmmeter. Start by disconnecting the sensor wiring and identifying the appropriate terminals for testing (this varies slightly by model). Then connect and ohmmeter and move the sensor arm through the full range of movement (either by hand with the sensor removed from the outboard or by tilting the outboard itself). Watch the meter for acceptable resistance within the specified range (again, this varies slightly by model) and make sure there are not sudden opens or shorts where continuity drops or spikes. All changes to resistance should be relatively linear.

TRIM & TILT 8-21

■ Remember that resistance specifications will vary slightly by meter and even more by temperature. All specs given are taken with a high quality Digital Ohmmeter, at an ambient/component temperature of approximately 68°F (20°C).

When testing the trim sensor, look for values within the following ranges, based on model:

- For 3-cylinder motors, check resistance across the Pink and Black wires, it should vary from about 360-540 ohms through the range of motion, then check resistance across the Orange and Black wires, it should vary between 800-1200 ohms.
- For all V4 motors as well as most 2.6L V6 motors (except for 2001 and later HPDI models), check resistance across the Pink and Black wires, it should vary from about 582-873 ohms through the range of motion. Next check resistance across the Orange and Black wires, it should vary between 800-1200 ohms.
- For 2001 and later 2.6L HPDI motors, check resistance across the Pink and Black wires, it should vary from about 10-309 ohms through the range of motion.
- For 3.1L V6 motors (except the 250 hp Vmax), first check resistance across the Pink and Black wires, it should vary through the range of motion from about 494-742 ohms for all except the 220H, 225G/V200, V225 for which the reading should vary from about 582-873 ohms. Then for all motors (except the 250 hp Vmax still), check resistance across the Orange and Black wires, it should vary from about 800-1200 ohms through the range of motion.
- For 3.1L 250 hp Vmax motors, check resistance across the Pink and Black wires, it should vary from about 10-309 ohms through the range of motion.
- For all 3.3L V6 motors, check resistance across the Pink and Black wires, it should vary from about 9-11 ohms at one end of the sensor arm's travel to about 247-387 ohms at the other end of travel.

PTT Assembly (Dual Trim Ram)

REMOVAL & INSTALLATION

◆ See Figures 44 and 45

1. Tilt the outboard to the fully upward position then secure in place using the tilt lever stop (trailering lock).

2. Tag and disconnect the wiring for the PTT motor. Also, note the mounting point and disconnect the ground lead(s) from the unit. There are often 1 or 2 ground leads mounted to the bottom of the assembly and, on some models the bolts retaining the ground leads also help secure an anode to the bottom of the assembly.

■ It will be necessary to cut one or more wire ties in order to reposition and free the necessary wiring.

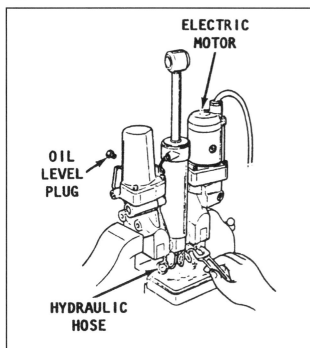

Fig. 42 Draining the hydraulic system prior to removing the electric motor for testing and/or servicing

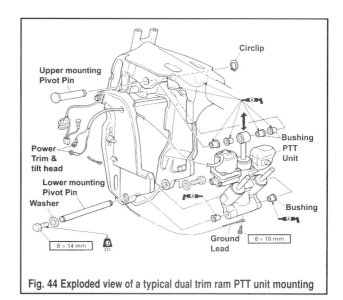

Fig. 44 Exploded view of a typical dual trim ram PTT unit mounting

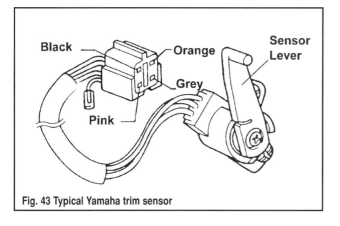

Fig. 43 Typical Yamaha trim sensor

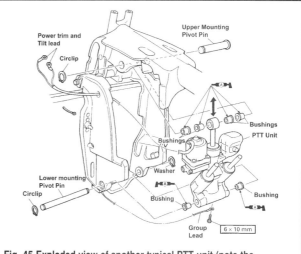

Fig. 45 Exploded view of another typical PTT unit (note the differences in pin mounting)

3. If equipped, remove the anode from the bottom of the PTT assembly.

■ **Removal of some pins will require the removal of the bushings.**

4. The PTT unit is secured by two pivot pin assemblies, one mounted on top and one of the bottom. The types of pins used vary greatly from model-to-model. Some are secured at one or both ends by a circlip, while others by a bolt or nut at either end. If there is only a bolt, circlip, nut etc at ONE end, it must be removed before the shaft can be withdrawn from the opposite end. If there are bolts, nuts, clips etc on both ends, then usually the pin can be withdrawn from either side. However, keep in mind that some models may actually use a captive nut on one end and obviously for those, the shaft must be withdrawn from the side of the captive nut. Start at one end (top or bottom of the PTT assembly usually doesn't matter) and remove the pivot shaft, then support the assembly as you remove the shaft from the other end.

5. If necessary remove the bushing(s) from the PTT unit and/or clamp bracket for installation purposes. Grease the replacement bushings to help ease installation.

6. Before installing the unit apply a light coating of marine grade grease to the inner diameter of each bushing.

7. Installation is essentially the reverse of the removal process. In most cases it really doesn't matter whether you remove or install the upper or lower shaft pin first, just make sure you properly secure the pins using the circlips or bolts, as applicable. For models that utilize bolts or nuts to secure the lower shaft mounting pin, apply a light coating of Loctite®241 or equivalent threadlock to the bolts before installation.

DISASSEMBLY

◆ See Figures 46 thru 51

The following procedures provide detailed instructions to service most accessible parts of the power trim/tilt system. Many of the instructions require the end cap of the tilt cylinder and/or the end caps of the trim cylinders to be removed. Potentially, this is not an easy task as salt or environmental exposure and corrosion often take their toll on motors over the years.

✱✱ WARNING

If the holes provided in the end cap for the special tool are damaged, the trim/tilt unit will probably have to be replaced.

If disassembly is required, work slowly keeping close track of the positioning for all components. Refer to the exploded views under Cleaning & Inspection for help in identifying components and positions, but keep in mind that pump, cylinder and valve assemblies will vary greatly.

■ **Some of the accompanying illustrations were made with the trim/tilt unit on the work bench for photographic clarity. However, the work described, with the exception of the tilt cylinder removal, may be performed without removing the unit from the clamp bracket.**

Do your BEST to keep parts identified as they are removed. Many parts may appear similar, but components should be installed into the same location from which they are removed.

1. If attempting to service the unit while installed on the outboard, power or manually, raise the outboard unit to the full up position and lock it in place.

2. Obtain a suitable container to receive the hydraulic fluid from the trim/tilt system.

3. If equipped, remove the two lines from the bottom of the trim/tilt housing in order to drain the hydraulic fluid. Remove the fill plug from the reservoir and allow the hydraulic fluid to drain into the container.

■ **If there are no lines on the bottom of the PTT unit, fluid will drain through one or more of the valves when they are removed a little later in this procedure.**

4. After the fluid has drained, replace the reservoir plug and loosely install the fluid line or valve body to assist in preventing contamination from entering the reservoir.

5. Obtain Yamaha special tool (#YB-06175) or an equivalent spanner to unthread the tilt/trim rod caps. Observe the pattern of pins and the words etched on each side of the special tool. The side with the three pins and the word TRIM is used to remove/install the end cap on the trim cylinders. The side with the four pins and the word tilt is used for the end cap on the tilt cylinder.

6. Using the trim side of the special tool, index the pins into the recess holes provided in the end cap. Index a socket wrench adaptor into the square hole in the tool and remove the end cap from the trim cylinder to be serviced. This is not an easy task, but under certain favorable conditions it can be accomplished.

7. Remove the end cap and, if equipped, the spring. Most late-models do not have the spring, but when present, the spring is partially attached to the end cap and will come free with the cap.

8. Withdraw the trim piston straight up and out of the cylinder. Repeat the last few steps in order to remove the other trim piston.

■ **If the tilt piston must be removed from the cylinder for servicing, if not done already, the trim/tilt unit must be removed from the clamp bracket assembly. Clamp the unit in a vise equipped with soft jaws or pad the jaws with a couple pieces of wood. The vise will provide stability and access to almost all parts.**

9. Using the tilt side of the same tool as for the trim cylinder end cap, index the pins into the recess holes provided in the end cap. Remove the end cap in the same manner with a socket adaptor and wrench as was used for the end cap of the trim cylinders.

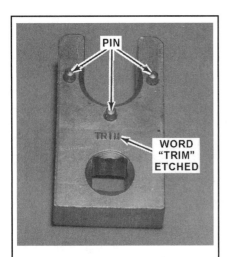

Fig. 46 A special spanner is used to remove the tilt or trim ram caps

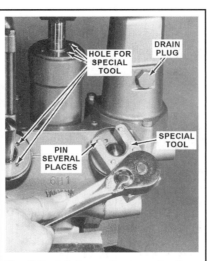

Fig. 47 Carefully loosen the caps using the spanner...

Fig. 48 ...then unthread and remove the cap (with the spring IF equipped)

TRIM & TILT 8-23

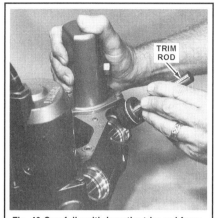

Fig. 49 Carefully withdraw the trim rod from the bore, keeping track of all components

Fig. 50 Removal of the tilt end cap and piston are similar

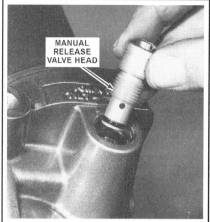

Fig. 51 Carefully unthread the valve bodies

10. Withdraw the tilt piston straight up and out of the cylinder.

11. Remove the circlip and then remove the manual release valve. The inner parts of the manual release valve, the ball, release rod, seat, spring, and pin, are all normally secured by the valve seat screw. Removal of this screw is extremely difficult. Without good cause, an attempt to remove the seat screw should not be made. Individual replacement parts are not available for the items behind the screw. Therefore, the only gain in removing them would be cleaning. In most cases replacement of the O-ring in the lower groove of the manual release head will solve a problem in this area.

■ In the majority of cases, service of the hydraulic items removed thus far will solve any rare problems encountered with the trim/tilt system.

12. The reservoir can be removed through the attaching hardware and cleaned if the system was considered contaminated with foreign material, which is highly unlikely, because the system is a closed system. The only route for entry of foreign material would be through the fill opening.

13. If necessary, remove the remaining valve bodies, seals and valve components. Take it slow, noting the positioning of each piece for installation purposes.

CLEANING & INSPECTION

◆ See Figures 52 thru 56

Keep the work area as clean as possible to prevent contamination through foreign material entering the system on the parts to be installed.

Clean all parts thoroughly with solvent and blow them dry with compressed air.

Carefully inspect the trim pistons and the tilt piston for any sign of damage.

Purchase, if available, new O-rings and discard the old items, but only after the replacement O-ring is verified as correct for the intended installation.

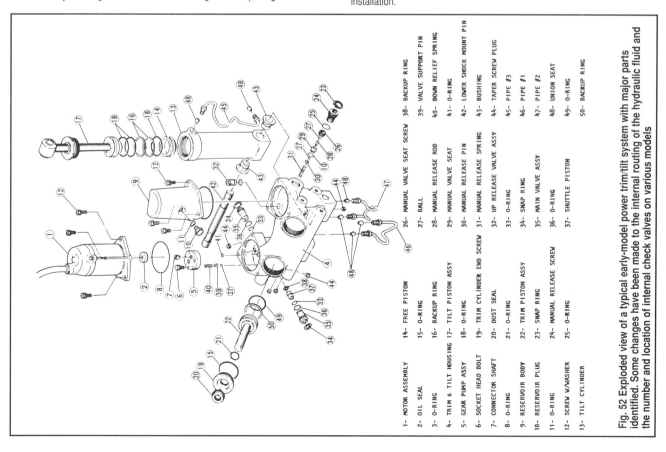

Fig. 52 Exploded view of a typical early-model power trim/tilt system with major parts identified. Some changes have been made to the internal routing of the hydraulic fluid and the number and location of internal check valves on various models

8-24 TRIM & TILT

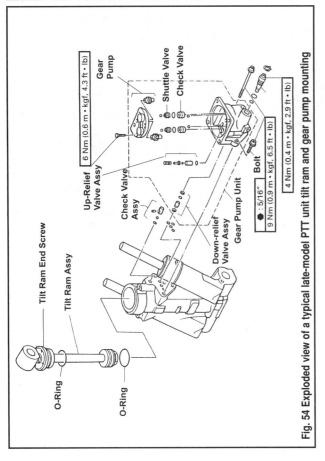

Fig. 54 Exploded view of a typical late-model PTT unit tilt ram and gear pump mounting

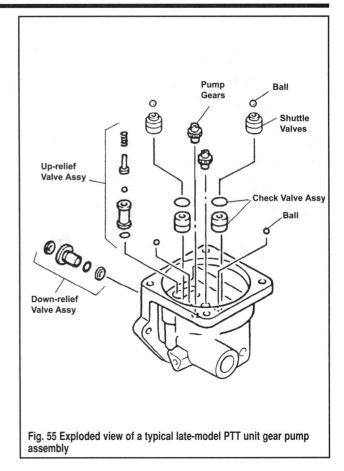

Fig. 55 Exploded view of a typical late-model PTT unit gear pump assembly

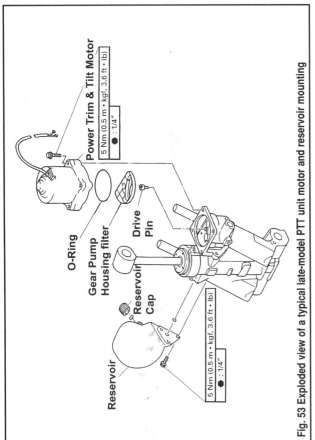

Fig. 53 Exploded view of a typical late-model PTT unit motor and reservoir mounting

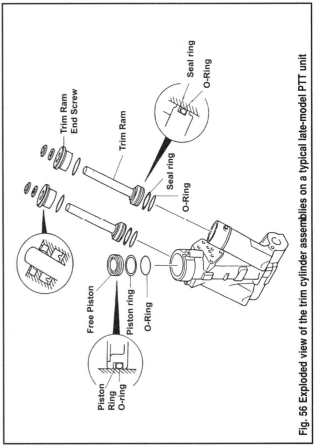

Fig. 56 Exploded view of the trim cylinder assemblies on a typical late-model PTT unit

TRIM & TILT

ASSEMBLY

◆ See Figures 53 thru 63

Good shop practice dictates new O-rings be installed anytime the unit is disassembled and the rings are exposed.

1. Coat all new O-rings with Yamalube Power Trim and Tilt Fluid or a good grade of automatic transmission fluid, and then install the O-rings into the grooves in the valve bodies, manual release heads, trim/tilt rods, etc.

2. If removed, install all valve bodies and related components in the exact order and to the exact positions from which they were removed.

3. Insert and thread the head into the manual release valve opening. Remember some units are standard right hand threads and others have left hand threads. Tighten the head just snug because it will be rotated for power tilt and manual tilt operation.

4. Secure the head in the trim/tilt housing with a circlip or Tru-arc snapring, depending on the trim/tilt unit being serviced.

5. Check to be sure the back-up rings are in place in the groove of the tilt piston and the free piston, or install the rings if they were removed. Apply a coating of Yamalube Power Trim and Tilt Fluid to new O-rings, and then install the O-rings into the grooves of the tilt piston, free piston, and the tilt piston end cap.

■ Normally, with the piston rod facing up the O-ring of the free piston must be installed under the back-up ring. The O-ring of the tilt piston must be installed on top of the back-up ring.

6. Insert the tilt piston into the piston and thread the end cap into the housing.

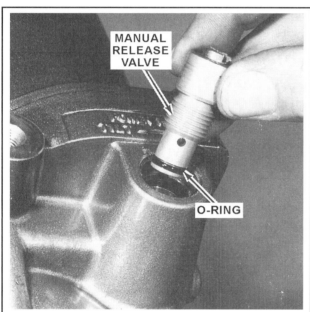

Fig. 57 Install the valve bodies in their original positions, using new O-rings

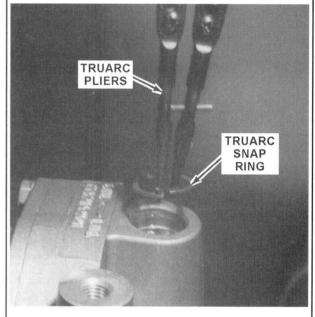

Fig. 58 Where used, carefully install the snaprings

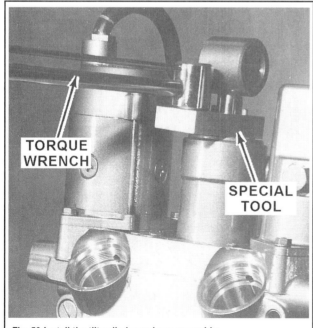

Fig. 59 Install the tilt cylinder and cap assembly

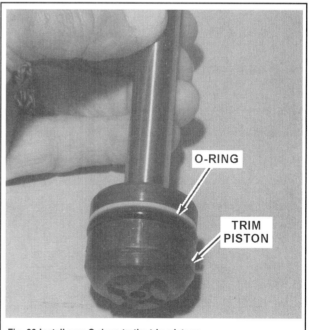

Fig. 60 Install new O-rings to the trim pistons...

8-26 TRIM & TILT

7. Clamp the trim/tilt housing in a vise equipped with soft jaws or pad the jaws with a couple pieces of wood. Using the tilt side of Yamaha special tool (#YB-06175) with the pins indexed into the recess holes of the end cap, tighten the end cap securely.

8. Check to be sure the back-up ring is properly installed in the trim piston groove, or install the ring if it was removed. Coat a new O-ring with Yamalube Power Trim and Tilt Fluid, and then install the O-ring into the groove of the trim piston. With the trim piston rod facing up, the O-ring must normally be installed UNDER the back-up ring.

9. Insert the trim piston straight down into the cylinder. Push the piston as far down as possible. Repeat the past few steps for the other trim piston.

10. Coat a new O-ring with Yamalube Power Trim and Tilt Fluid, and then install the O-ring into the groove of the trim end cap. On early models so equipped, check to be sure the spring is properly seated. If the two seals in the end cap were removed, coat new seals with hydraulic fluid, and then insert them into the individual grooves of the end cap.

11. Slide the spring down over the trim piston rod, and then thread the end cap into the trim/tilt housing.

12. Using the trim side of Yamaha special tool (#YB-06175) with the pins indexed into the recess holes of the end cap, tighten the end cap securely. Again, repeat the necessary steps for the other trim cylinder.

13. If the trim/tilt unit was not removed from the clamp brackets, fill the system with Yamalube Power Trim/Tilt Fluid or a good grade of automatic transmission fluid. Purge the system of air following the procedure found earlier in this section.

14. If the trim/tilt unit was removed from the clamp bracket, the system is filled and purged with the unit on the bench. Fill the system with Yamalube Power Trim/Tilt Fluid or a good grade of automatic transmission fluid and purge the system of air as follows:

    a. First pull the tilt cylinder to its fully extended position. Check the fluid level and add fluid as required to bring the level up to the bottom of the fill opening.

    b. Next, rotate the manual release valve to the power tilt position, approximately three full turns from the manual position.

    c. Now, obtain a 12 volt battery and connect the Blue lead from the trim/tilt unit to the positive lead from the battery. Now, momentarily make contact with the Green lead from the unit to the negative post of the battery. The trim cylinders should move to the fully extended position. Check the fluid level and add fluid, as required.

    d. Next, connect the Green lead from the unit to the positive post on the battery. Momentarily make contact with the Blue lead from the trim/tilt unit to the negative post of the battery. The trim cylinders and the tilt cylinder should retract to the full down position.

    e. Repeat the extension and retraction of the cylinders two or three times until the fluid level remains stable when checked with the tilt ram in the fully extended position.

15. To check for air in the system, extend the tilt cylinder, and then apply a downward heavy force manually to the end of the piston. The piston should feel solid. If the piston retracts more than about 1/8 in. (3.2mm), the system still contains air. Perform the purging sequence until the tilt cylinder is solid.

16. If removed, install the PTT assembly to the outboard.

17. Perform an operational check of the system.

Fig. 62 Carefully thread the end cap (and spring, if equipped), by hand...

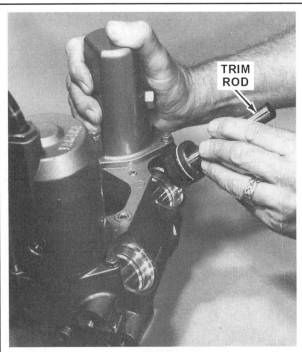

Fig. 61 ...then carefully insert them into the bores

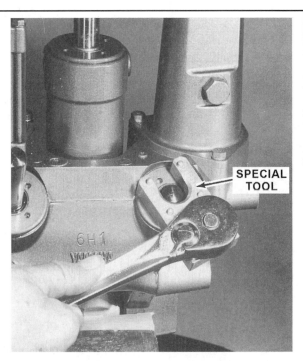

Fig. 63 ...then use the spanner to tighten the cap securely

CHOKE SWITCH	9-2
TESTING	9-2
CONTROL BOX ASSEMBLY	9-4
REMOVAL & DISASSEMBLY	9-4
ACCELERATION LEVER	9-6
CLEANING & INSPECTION	9-6
ASSEMBLY & INSTALLATION	9-6
CONTROL BOX	9-12
FREE ACCELERATION LEVER	9-9
TRIM/TILT BUTTON AND HARNESS	9-7
KILL SWITCH	9-4
TESTING	9-4
NEUTRAL SAFETY SWITCH	9-3
TESTING	9-3
REMOTE CONTROL BOX	9-2
DESCRIPTION & OPERATION	9-2
START BUTTON	9-3
TESTING	9-3
WARNING BUZZER/HORN	9-2
TESTING	9-2
REMOTE CONTROLS	**9-2**
CHOKE SWITCH	9-2
CONTROL BOX ASSEMBLY	9-4
KILL SWITCH	9-4
NEUTRAL SAFETY SWITCH	9-3
REMOTE CONTROL BOX	9-2
START BUTTON	9-3
WARNING BUZZER/HORN	9-2

9

REMOTE CONTROLS

REMOTE CONTROLS 9-2

REMOTE CONTROLS

Remote Control Box

DESCRIPTION & OPERATION

◆ See Figures 1 and 2

■ The type of remote control mated to the outboard is determined by the boat manufacturer or the rigging dealer. Because there are such a range of remote control units that could be mated to Yamaha outboards, there is no way to cover them all in detail here. This section deals with a typical Yamaha remote control unit. Illustrations are of the most common example of this unit. Although the procedures apply to most Yamaha remotes and many aftermarket units, care and observation must be used when working on any unit as differences may exist in assemblies. When a component or mounting differs from the components covered here, use common sense, or contact the manufacturer of the unit for more information.

The remote control unit allows the helmsperson to control throttle operation and shift movements from a location other than where the outboard unit is mounted.

In most cases, the remote control box is mounted approximately halfway forward (midship) on the starboard side of the boat or on the starboard side of the center console (depending upon the style/type of boat).

The Yamaha control unit normally houses a key switch, a choke switch, a kill switch, a neutral safety switch, a warning horn, and the necessary wiring and cable hardware to connect the control box to the outboard unit.

A safety feature is incorporated in the unit. The control arm can be shifted out of the **NEUTRAL** position if and only if the neutral lever is squeezed into the control arm. This feature prevents the arm from being accidentally moved from Neutral into either forward or reverse gear. Unintentional movement of the shift lever could be dangerous, resulting in personal injury to the operator, passengers, or the boat.

Starting from the upright **NEUTRAL** position, when the control arm is moved forward to about 30° from the vertical position, the unit shifts into forward gear. At this point the throttle plate is fully closed. As the control arm is moved past the 30° position, forward and downward, the throttle will open, until the wide open position, approximately 90° from the vertical position, is reached.

To shift into reverse gear, the control arm is first returned to the full upright position (Neutral) momentarily. From the upright position, the control arm is moved aft about 30°, and the unit shifts in reverse gear. At this point, the throttle plate is fully closed. If the arm is moved further aft and downward, the throttle will be opened until the wide open position is reached at about 60°.

The remote control unit is equipped with a free acceleration lever. This lever can be moved up to open the throttle and down to close the throttle. This lever is utilized only during powerhead startup and when the control arm is in the full upright (Neutral) position. When the free acceleration lever is not in the full down, idling position, the control lever cannot be moved from the **NEUTRAL** position.

Choke Switch

TESTING

◆ See Figure 3

■ On MOST late model control boxes, the choke switch is integrated into the main switch for easier operation. On these, the choke circuit operates when the main starter switch is depressed as it is rotated to start.

Some early-model control boxes utilize a separate choke switch. On these models the choke switch is a spring loaded toggle type switch located in the forward side of the control box. The control box must be opened to gain access to the switch leads. To check the switch, proceed as follows:

1. Disconnect the leads (normally Blue and Yellow, but check the Wiring Diagrams to be sure) from the choke switch at the nearest quick disconnect fitting.
2. Select the 1000 ohm scale on the meter. Make contact with the meter leads, one to each of the disconnected leads.
3. In the normal **OFF** position the meter should indicate no continuity. Move the switch to the **ON** position. The meter should indicate continuity.
4. If the switch fails either of these two tests, it must be replaced. The switch cannot be serviced or adjusted.

Warning Buzzer/Horn

TESTING

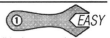

The buzzer is a warning device to indicate low oil in the reservoir, an over rev. condition, or overheating of the powerhead. The buzzer is normally located inside the control box on remote control powerheads.

1. Remove the control box cover and identify the two leads from the buzzer, one is normally Yellow and the other is normally Pink, but check the Wiring Diagrams to be sure.
2. Disconnect these two leads at their quick disconnect fittings, and ease the buzzer out from between the four posts which anchor it in place.
3. Obtain a 12-volt battery. Connect the Yellow buzzer lead to the negative battery terminal. Momentarily make contact with the Pink buzzer lead to the positive battery terminal.
4. As soon as the Pink buzzer lead makes contact with the positive battery terminal, the buzzer should sound. If the buzzer is silent, or the sound emitted does not capture the helmsperson's attention immediately, the buzzer should be replaced. Service or adjustment is not possible.
5. If the sound is satisfactory and immediate, install the buzzer between the four posts and connect the two leads matching color to color. Tuck the leads to prevent them from making contact with any moving parts inside the control box. Replace the cover.

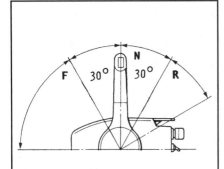

Fig. 1 Shift lever positioning for the remote control unit

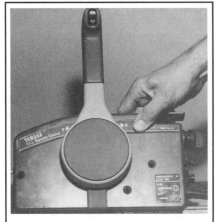

Fig. 2 A typical Yamaha remote control box

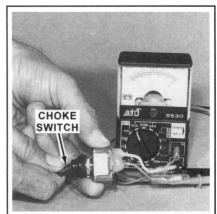

Fig. 3 Using an ohmmeter to test the choke switch

REMOTE CONTROLS

Neutral Safety Switch

TESTING

◆ See Figures 4 and 5

> ※※ **CAUTION**
>
> Remember this is a safety switch. A faulty switch may allow the powerhead to be started with the lower unit in gear - an extremely dangerous situation for the boat and persons aboard. And just imagine what might happen if someone was in the water nearby and a kid turned the switch?

1. Trace the neutral safety switch leads from the switch to their nearest quick disconnect fitting. Both of these leads are usually Brown, but may vary for different models. Refer to the wiring diagram in the Ignition and Electrical System section for the proper color identification.
2. Disconnect the two leads and connect an ohmmeter across the two disconnected leads. When the shift lever is in the **NEUTRAL** position, the meter should register continuity.
3. When the lower unit is shifted to either forward or reverse, the meter should register no continuity.
4. The switch must pass all three tests to indicate the safety switch is functioning properly. If the switch fails any one of the tests, the switch must be replaced.

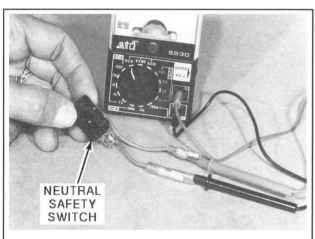

Fig. 4 Using an ohmmeter to test the neutral safety switch

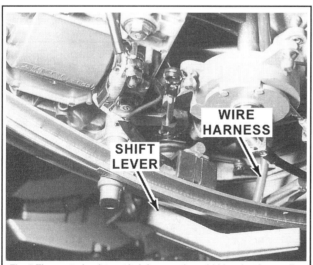

Fig. 5 The neutral safety switch is sometimes hidden, but is usually located on the axis of the shift lever just inside the lower cowling pan

Start Button

TESTING

◆ See Figures 6 and 7

1. Trace the start button harness containing two wires from the switch to their nearest quick disconnect fitting. The colors may vary for different models. Refer to the Wiring Diagrams in the Ignition and Electrical System section to confirm proper color identification.
2. Disconnect the two leads and connect an ohmmeter across the disconnected leads. Depress the start button. The meter should register continuity. Release the button and the meter should now register no continuity.
3. Both tests must be successful. If the tests are not successful, the start button must be replaced. The start button is a one piece sealed unit and cannot be serviced.

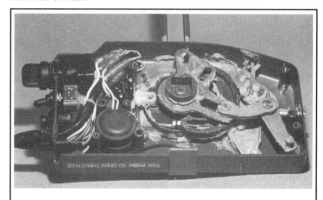

Fig. 6 The hardest part of control box service is stuffing the electrical leads back into the box so they are not damaged

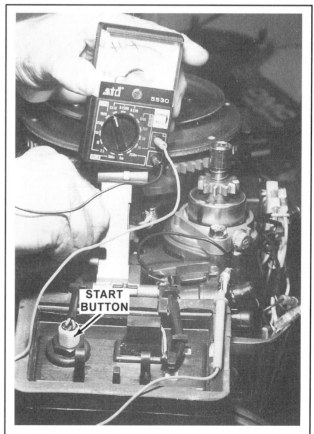

Fig. 7 Using an ohmmeter to test the start button

9-4 REMOTE CONTROLS

Kill Switch

TESTING

1. Trace the kill switch button harness containing two wires from the switch to their nearest quick disconnect fitting. The colors may vary for different models, but the hot wire is usually White/Black. Refer to the Wiring Diagrams in the Ignition and Electrical System section to confirm proper color identification.

Disconnect the two leads and connect an ohmmeter across the disconnected leads. Verify the emergency tether is in place behind the kill switch button. Select the 1000 ohm scale on the meter.

2. Depress the kill button. The meter should register continuity. Release the button. The meter should now register no continuity.

3. Both tests must be successful. If the switch fails either test, the switch is defective and must be replaced. The switch is a one piece sealed unit and cannot be serviced.

Control Box Assembly

REMOVAL & DISASSEMBLY

◆ See Figures 8 thru 18

1. Removal of the control box is accomplished by first disconnecting the throttle and shift cables at one end, and then disconnecting the main wiring harness at the quick disconnect fitting. Next, the mounting screws (usually 3) securing the control box to the boat are removed.

■ Disconnect the cables from the control box usually involves following the next 2 steps. So if you're going to remove the cable from the box end (as opposed to the powerhead end), then perform those steps before fully removing the box from the boat.

2. Move the control arm to the full upright (Neutral) position. Remove the Phillips head screws (usually 5) securing the upper and lower parts of the back plate.

3. Pry off the circlip retaining the throttle cable end to the throttle arm and the circlip retaining the shift cable end to the shift arm. Take care not to lose these two small circlips. Lift off the cable ends from the arms and be careful not to alter their length at the turnbuckles. Remove both cable ends from the box.

■ One of the most difficult tasks during assembling of the control unit is attempting to return all the wires back into their original positions.

Some wires are tucked neatly under switches, others are routed into neat bundles secured with plastic retainers, some are looped and double back, but believe-it-or-not, with patience and some good words, they all fit into one side of the box away from moving parts.

■ Therefore, it would be most advantageous to take a digital or Polaroid picture of the unit as an aid during assembling.

4. Remove the retaining nuts on the main key switch, the electric choke switch, and the kill switch (as equipped) on the side of the control box. Ease these three switches out of their grommets and holes. Disconnect the following leads (wire colors may vary slightly, refer to the Wiring Diagrams for more details) at their quick disconnect fittings:
• The wire colors Pink and Yellow leads to the neutral safety switch (if mounted in the housing).

Fig. 8 Loosen the screws and remove the cover...

Fig. 9 ...to disconnect the cables or access the internal components

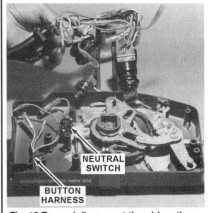

Fig. 10 Tag and disconnect the wiring, then carefully remove the harness

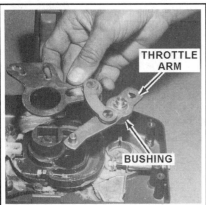

Fig. 11 Lift out the hinged throttle arm assembly (and bushing)

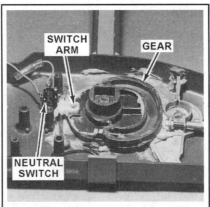

Fig. 12 If applicable, unscrew and remove the neutral switch

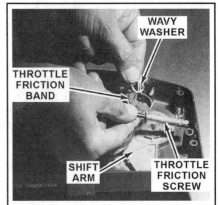

Fig. 13 Remove the throttle friction band assembly

REMOTE CONTROLS

- For PTT models, the Yellow, Green, and Black leads encased in a small harness leading to the up/down button on the upper control arm.

5. The entire harness, switches, and horn may now be lifted free of the control box, as shown in the accompanying illustration. The horn is normally not secured with hardware, but is simply retained between four bosses.

6. Lift out the throttle arm, consisting of three pieces hinged together. A plastic bushing will probably remain on the underneath side of the throttle arm. This bushing was indexed with the throttle friction bands.

7. If mounted in the housing, remove the Phillips head screws (usually 2) securing the neutral switch, and then remove the switch. Remove the Phillips head screw and switch arm from the gear.

8. Lift the two halves of the throttle friction band up and out of the control box. The throttle friction screw, with the circlip attached, will come away with the band.

9. Lift out the small wavy washer from the center of the shift arm.

10. Remove the Phillips screw securing the spring retainer, and then remove the spring retainer. Grasp one end of the leaf spring pack with a pair of needle nose pliers. Pull up on the pack and at the same time allow the pack to straighten to release tension on the springs. Lift out the detent roller from under the gear.

11. Loosen the bolt in the center of the gear 1/2 turn only. After the center bolt has been loosened, lightly tap on the bolt to free it from the gear. Now, remove the bolt.

12. Turn the control box over. Three things are now to be performed, almost simultaneously. Support the gear - now underneath - with one hand and at the same time unsnap the bottom harness cover from the front plate, as the control arm is lifted from the box. Remove the attaching screws, and then the **NEUTRAL** position plate.

13. Support the gear and at the same time turn the control box over, and then lift the gear free of the box.

14. Lift off the shift arm, a large flat washer, and two bushings from the front plate. This step concludes the disassembly of the inside of the front plate.

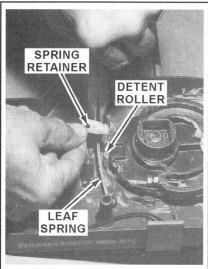

Fig. 14 Remove the spring retainer, release spring tension and free the detent roller

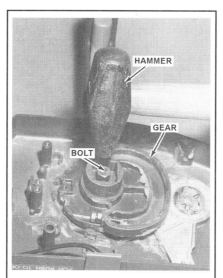

Fig. 15 After loosening it slightly, tap the gear center bolt to free the gear

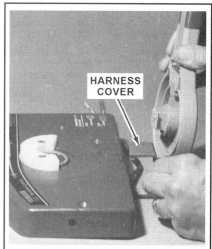

Fig. 16 Remove the control arm and neutral plate from the box, while supporting the gear underneath

Fig. 17 Turn the box over and remove the gear

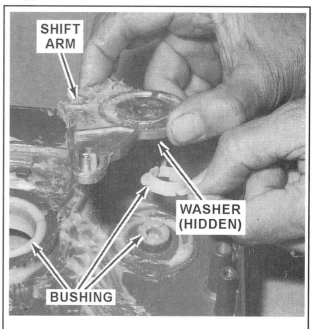

Fig. 18 Remove the shift arm and bushings

REMOTE CONTROLS

Acceleration Lever

◆ See Figures 19 thru 25

1. Place the upper back plate on the work bench with the lever facing down. Remove the Phillips screw and lift off the detent roller retainer.
2. Pry out the detent roller from the free acceleration disc.
3. Turn the back plate over and remove the Phillips screws (usually 2) securing the free acceleration lever to the disc. Lift off the lever and the wavy washer under the lever.
4. Turn the back plate over again, and then lift off the free acceleration disc and another wavy washer.
5. Remove the **NEUTRAL** position lever retainer and slide the lever from the arm. Remove the small spring between the top of the lever and the top of the arm.
6. Remove the small Phillips screws (usually 2) securing the handle to the control arm.
7. Pry up on the front disc and separate the disc from the control lever arm. Unthread the harness, to the up/down button, from the cavity of the arm.
8. Grasp the handle in one hand and the shaft of the control arm in the other. Gently slide them apart about 2 inches (5cm) or until the trim/tilt button with the harness attached can be removed from the handle.

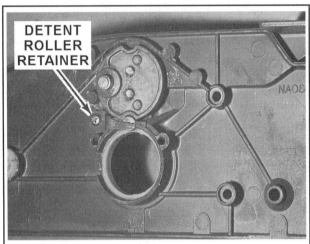

Fig. 19 Remove the detent roller retainer...

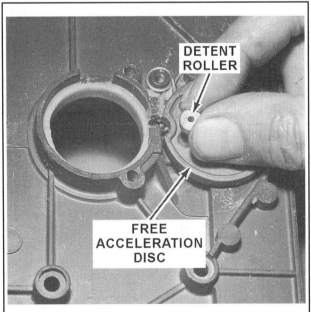

Fig. 20 ...then carefully pry the roller free of the acceleration disc

CLEANING & INSPECTION

◆ See Figures 26 and 27

Clean all metal parts with solvent, and then blow them dry with compressed air.

Never allow nylon bushings, plastic washers, nylon retainers, wiring harness retainers, and the like, to remain submerged in solvent more than just a few moments. The solvent will cause these type parts to expand slightly. They are already considered a tight fit and even the slightest amount of expansion would make them very difficult to install. If force is used, the part is most likely to be distorted.

Inspect the control housing plastic case for cracks or other damage that would allow moisture to enter and cause problems with the mechanism.

Carefully check the teeth on the gear and shift arm for signs of wear. Inspect all ball bearings for nicks or grooves which would cause them to bind and fail to move freely.

Closely inspect the condition of all wires and their protective insulation. Look for exposed wires caused by the insulation rubbing on a moving part, cuts and nicks in the insulation and severe kinking which could cause internal breakage of the wires.

Inspect the bosses on both ends of the leaf spring. If either end shows signs of failure, the front plate must be replaced.

Inspect the edges of the cut-out in the **NEUTRAL** position plate. Replace this plate if the corners of the cut-out show any sign of rounding. If rounded, a slight pressure on the **NEUTRAL** position lever could throw the lower unit into gear. Check and double check all components of the **NEUTRAL** position system from the spring at the top of the lever down to the extension at the bottom of the lever which indexes into the cut-out of the **NEUTRAL** position plate.

ASSEMBLY & INSTALLATION

The following procedures provide complete detailed instructions to assemble all parts of the remote control unit. If certain areas were not disturbed during disassembling, simply bypass the steps involved and proceed with the work.

Again, remember this is based on a common Yamaha remote control unit, but not all remotes (Yamaha or not) will have the exact same construction and internal components.

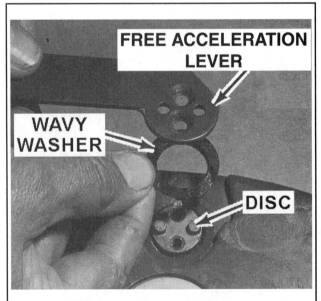

Fig. 21 Remove the screws and the free acceleration lever (with washer) from the disc

REMOTE CONTROLS 9-7

Trim/Tilt Button and Harness

◆ See Figures 28 thru 31

1. Feed the harness wires for the power tilt/trim button down from the top in between the handle and the arm. Push the face of the button into the handle and observe the button from the operator's position. The word **UP** should be up and the **DN** should be down. Slide the control lever arm up into the handle.

2. Continue to feed the harness wires along the length of the arm - in and out of the cavity of the arm. Check to be sure the harness lies flat with no kinks. A kink in the wire will damage the harness over a period of time. Snap the front disc over the cavity.

3. Secure the arm to the handle with the Phillips head screws.

4. Hold the return spring in place in the recess at the top of the **NEUTRAL** position lever and at the same time slide the lever along the length of the arm up into the handle. Install the lever retainer across the lever and secure it to the arm with the small Phillips head screws.

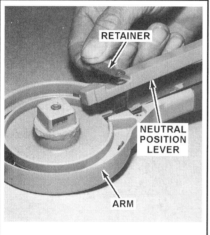

Fig. 22 Remove the retainer and slide the neutral position lever from the arm

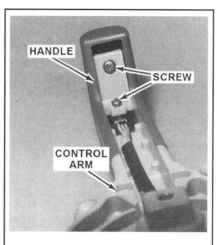

Fig. 23 Remove the screws securing the handle to the control arm

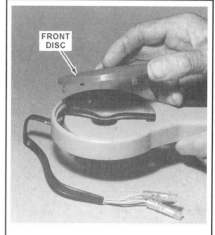

Fig. 24 Carefully pry the front disc free (for access to the PTT switch wiring)...

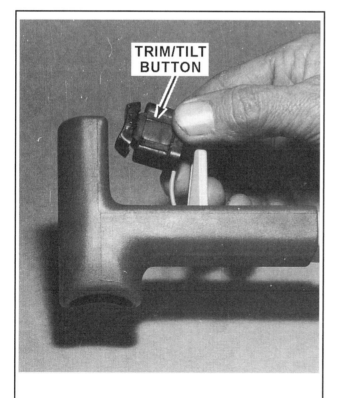

Fig. 25 ...then carefully free the switch from the handle

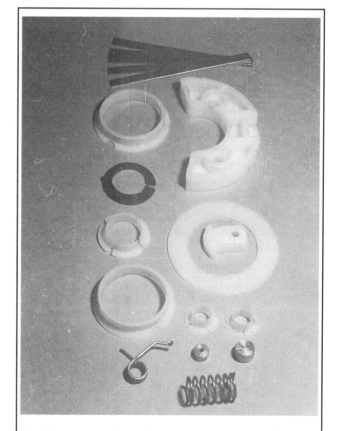

Fig. 26 Just some of the small metal and nylon parts from the interior of a remote control box

9-8 REMOTE CONTROLS

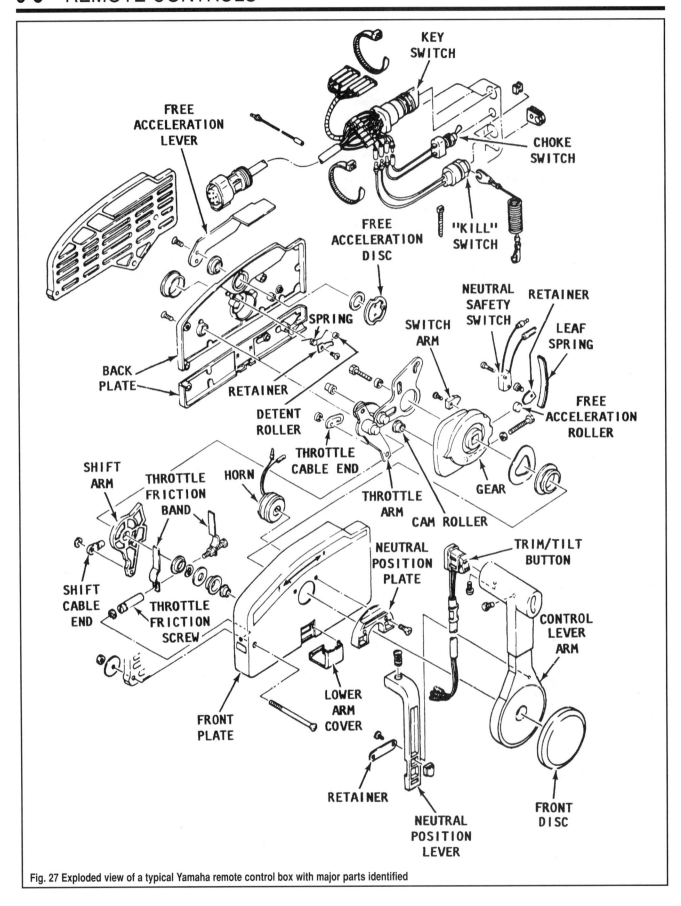

Fig. 27 Exploded view of a typical Yamaha remote control box with major parts identified

REMOTE CONTROLS 9-9

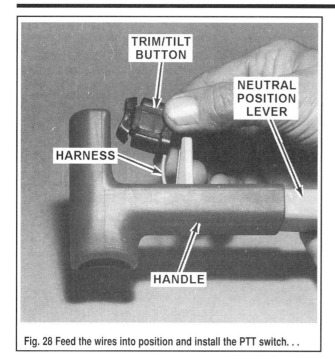

Fig. 28 Feed the wires into position and install the PTT switch...

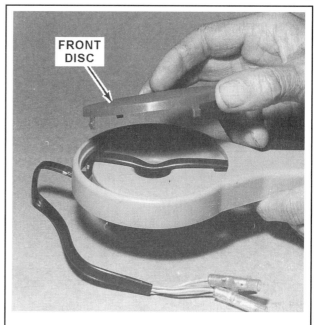

Fig. 29 ...then snap the front disc into place over the wires

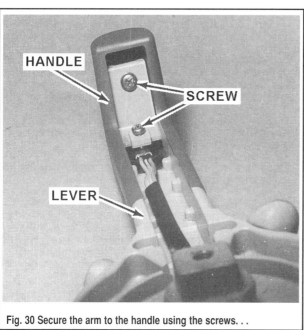

Fig. 30 Secure the arm to the handle using the screws...

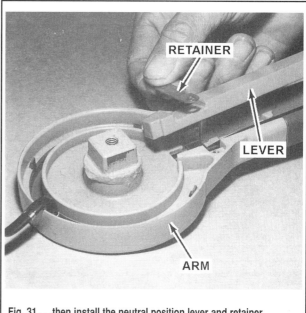

Fig. 31 ...then install the neutral position lever and retainer

Free Acceleration Lever

◆ See Figures 32 thru 46

1. Apply just a dab of Yamalube or an equivalent marine grade grease to the wavy washer and the free acceleration disc. Place the wavy washer and the free acceleration disc over the smaller hole in the inside of the top back plate. The single large post on the disc must face away from the notch cut in the plate for the free acceleration arm.

2. Hold the washer and disc (the Yamalube will help), in place - from falling away - and at the same time, turn the plate over on the work bench.

3. Place another wavy washer and the free acceleration arm over the installed disc.

■ The arm can only be installed one way because of the groove cut into the face of the upper back plate to accommodate the arm. If the free acceleration disc has been installed correctly in the previous step, the two beveled holes in the arm will align with the two threaded holes in the disc. Also, the two round holes in the arm will index over the two posts on the disc.

4. Install and tighten the two Phillips head screws. Turn the plate over again with the inside facing upward.

5. Install the upturned portion of the spring between the installed disc and the center boss, as shown. Hook the coiled part of the spring over the threaded post and at the same time direct the free end of the spring to anchor between the threaded and plain posts.

9-10 REMOTE CONTROLS

6. The detent roller must be positioned over the upturned end of the spring and be seated into the curved notch of the free acceleration disc. A small screwdriver inserted along the spring under the disc may aid in pushing the end into place to allow the roller to drop down over the spring end and against the disc.

7. Install the retainer over the detent roller. Secure the roller in place with the Phillips head screw threaded into the post with the spring around it.

8. Place the small bushing onto the shift arm boss on the inside back plate. Next, place the large bushing onto the same boss. The small and large bushing must be installed face-to-face. After both bushings are in position, slide the large flat washer onto the boss. Now, install the shift arm over the flat washer with the post for the shift cable facing upward - pointed toward the lower part of the case.

9. Lower the shift gear into the case with the post indexing into the notch in the shift gear. Support the shift gear inside the cover and turn the front plate over.

10. Place the **NEUTRAL** position plate onto the front plate and secure it in place with the two Phillips head screws. Position the control arm over the front plate with the **NEUTRAL** position lever indexed into the cutout of the **NEUTRAL** position plate.

✷✷ WARNING

If the NEUTRAL position lever is not seated properly, a slight pressure on the handle could shift the lower unit into gear - forward or reverse. Such action could cause serious injury to crew, passengers, or the boat.

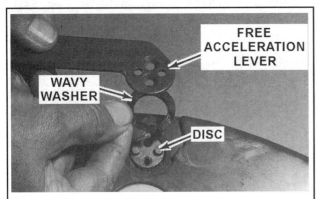

Fig. 32 Install the free acceleration lever to the disc using a new wavy washer

11. Guide the wire harness into the groove below the arm and snap on the lower arm cover to secure the harness in place.

12. Support the control arm and turn the front plate over. Install the washer and bolt into the square recess of the gear. Tighten the bolt to a torque value of 5.8 ft. lbs. (8 Nm).

13. Slide the free acceleration roller into the center left notch in the shift gear. Grasp the pack of leaf springs with a pair of needle nose pliers. Insert one end of the pack into the lower boss. Bend the pack around the roller. Guide the other end into the upper boss. Install the small metal leaf spring retainer onto the threaded post to the left of the leaf spring. Secure the retainer in place with a Phillips head screw.

14. Install the throttle friction band over the shift arm. Move the two ends of the band into the boss on the front plate. Center the opening of the band squarely over the shift arm. Insert a wavy washer into the center of the band on the shift arm.

15. Install the neutral safety switch with the tab on the switch facing toward the shift gear. Secure the switch in place with the two Phillips head screws. Position the white plastic switch arm into the slot provided on the shift gear. Install and tighten the Phillips head screw to secure the arm to the gear.

16. Install the throttle arm. Try not to disturb the wavy washer in the center of the throttle friction band when the bushing on the underneath side of the throttle arm indexes with the band.

■ If you think some of the other assembling procedures were tricky on this unit, stand by! This next one will put you to the test.

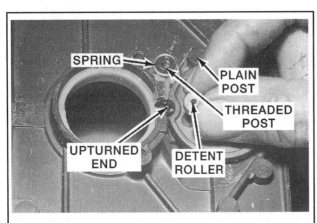

Fig. 33 Install the spring and detent roller. . .

17. Connect the neutral safety switch leads, usually Pink and Yellow (male) at the quick disconnect fittings - color-to-color. For models with PTT, connect the button harness leads Yellow (female), Green and Black at the quick disconnect fittings - color to color. Position the horn between the four tall posts next to the neutral safety switch. Install the main key switch, the electric choke switch, and the kill switch into their respective openings on the side of the remote control box. Arrange the mess of wires neatly.

■ Remember, some wires are tucked neatly under switches, others are routed into neat bundles secured with plastic retainers, and some are looped then doubled back. Believe-it-or-not, with patience and some good words, they will all fit into one side of the box. The wires must be clear of moving parts.

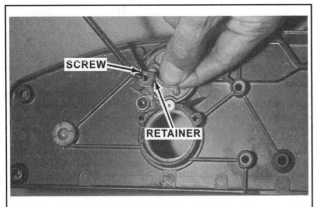

Fig. 34 . . .then secure using the detent roller retainer

18. Secure the electrical switches to the side of the box with the retaining nuts. Slide the rubber boot over the choke switch. If possible, secure a tie-wrap around the bundle of wires to prevent loose wires from rubbing against and interfering with moving parts.

19. Thread the cable joint 0.3 in. (8mm) onto the shift cable. Hold the joint to prevent it from rotating and at the same time hook the cable end onto the shift arm post. Secure the cable joint with the restraining circlip.

20. Install the throttle cable in the same manner as the shift cable. Bring both halves of the back plate together with the front plate. Secure it all together with the five Phillips head screws.

REMOTE CONTROLS 9-11

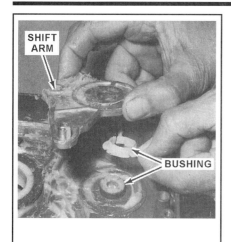

Fig. 35 Install the shift arm and bushings...

Fig. 36 ...then position the shift gear and support...

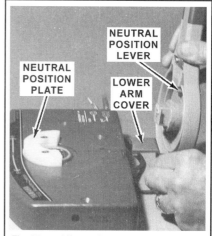

Fig. 37 ...while you install the neutral plate and the control arm

Fig. 38 Now support the control arm as you install and tighten the retaining bolt through the square recess in the gear

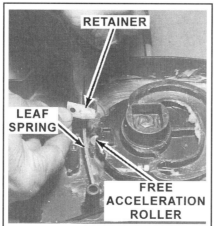

Fig. 39 Install the free acceleration roller, then position the leaf springs and secure the retainer

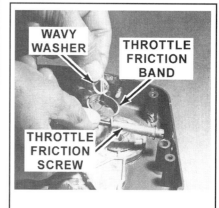

Fig. 40 Install the throttle friction assembly using a new wavy washer

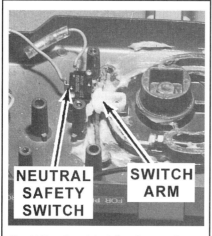

Fig. 41 If equipped, install and secure the neutral switch assembly

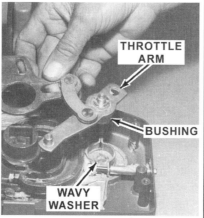

Fig. 42 Carefully install the throttle arm assembly and bushing to the throttle friction band (and wavy washer)

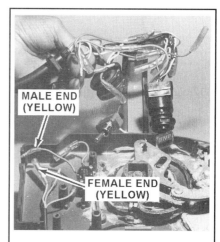

Fig. 43 Install the wiring harness as tagged during removal...

REMOTE CONTROLS

21. Check operation of the remote control lever to and between each of the three positions - Neutral, forward, and reverse. The lever should move smoothly and with a definite action. The shift cable and the throttle cable should move in and out of the sheathing without any indication of binding. The movement can be checked at the powerhead end of the cable.

Control Box

22. If the cable ends were removed from the powerhead:

a. Guide the cables along a selected path and secure them in place with the retainers. Do not bend any cable into a diameter smaller than 16 in. (40cm).

b. Connect the shift cable to the shift lever on the powerhead. Insert the outer throttle cable wire into the throttle cable bracket. Connect the cable joint onto the throttle control attachment.

c. Do not confuse the cables. If in doubt, operate the control lever on the control box. The cable moving first is the shift cable. Tilt the outboard unit from full up to the full down position, and from hard over starboard to hard over port, to verify the cables move smoothly without binding, bending, or buckling.

23. Install the control box to the boat and secure using the retainers.
24. Align the mark on the wire harness plug with the mark on the coupler plug, and then connect the plugs.
25. Install the safety tether to the back of the kill switch button.
26. Remember, the powerhead will not start without the tether in place.

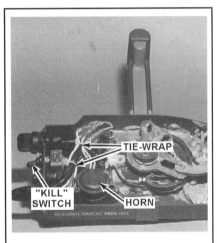

Fig. 44 . . .then carefully position and secure the wiring as noted during removal

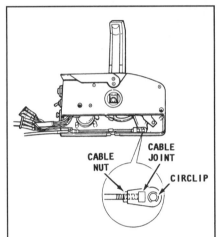

Fig. 45 Reconnect the cables to the control unit. . .

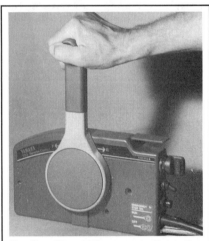

Fig. 46 . . .then confirm movement/operation of the throttle/shifter lever

10 HAND REWIND STARTER

HAND REWIND STARTER 10-2

HAND REWIND STARTER	10-2
2 Hp (43 and 50cc) and 3 Hp Models	10-2
4-50 Hp Models (Except 9.9/15 hp and 48 Hp Motors)	10-6
9.9/15 Hp Models	10-11
48 Hp and Larger Models	10-13
2 Hp (43 and 50cc) and 3 Hp Models	10-2
ASSEMBLY & INSTALLATION	10-3
CLEANING & INSPECTION	10-3
REMOVAL & DISASSEMBLY	10-2
4-50 Hp Models (Except 9.9/15 hp and 48 Hp Motors)	10-6
ADJUSTMENT	10-10
ASSEMBLING & INSTALLATION	10-9
CLEANING & INSPECTION	10-7
REMOVAL & DISASSEMBLY	10-6
9.9/15 Hp Models	10-11
ASSEMBLY & INSTALLATION	10-11
CLEANING & INSPECTION	10-11
REMOVAL & DISASSEMBLY	10-11
48 Hp and Larger Models	10-13
ASSEMBLY & INSTALLATION	10-18
CLEANING & INSPECTION	10-17
REMOVAL & DISASSEMBLY	10-13

HAND REWIND STARTER

2 Hp (43 and 50cc) and 3 Hp Models

REMOVAL & DISASSEMBLY

♦ See Figures 1 thru 11

■ For exploded views, please refer to Cleaning & Inspection, later in this section.

1. Remove the engine cover/cowling assembly. On models with a 2-piece cowling there may be as many as 10 screws of varying shapes and sizes threaded into the cover halves. Be sure to keep careful track of the screws and their mounting points.
2. Remove the mounting bolts (usually 3) securing the legs at the corners of the hand rewind starter to the powerhead. On 3 hp motors the rope guide is a separate piece from the starter housing, but does not need to be unbolted from the starter housing in order to remove the assembly.

■ A no start-in-gear protection device is not incorporated on these single-cylinder powerheads.

3. Lift the hand rewind starter free of the powerhead.

■ If more play is needed on the rope you can remove the handle at this time, just keep tension on the rope or knot it further back to keep it from fully retracting until the rope is inserted into the notch in the next step.

4. Rotate the starter sheave (drum) to align the notch in the sheave with the starter handle. With the sheave in this position, pull the rope out a little and hook it into the notch. Carefully turn the sheave clockwise until the spring has lost all its tension. Control the rotation with the rope secured in the notch to prevent the sheave from free wheeling.
5. Invert the hand starter and remove the sheave (drum) retainer bolt with the proper size socket.
6. Lift the drive plate free of the drive pawl. The drive pawl spring will remain attached to the plate. Observe how the drive pawl spring was indexed over the peg on the drive pawl.
7. Using a small screwdriver, hold the back of the return spring to allow the drive pawl to be removed.

■ There are usually two recesses which appear to be identical, but actually are mirror images of each other. The recess must be marked to ensure proper installation.

8. Carefully note, and mark the recess holding the return spring before removing the spring. After marking the recess, lift out the spring.

✳✳ WARNING

The rewind spring is a potential hazard. The spring is under tremendous tension when it is wound - a real tiger in a cage! If the spring should accidentally be released, severe personal injury could result from being struck by the spring with force. Therefore, the following steps must be performed with care to prevent personal injury to self and others in the area.

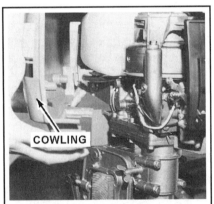
Fig. 1 Remove the engine cover/cowling assembly...

Fig. 2 ...then unbolt the hand rewind starter..

Fig. 3 ...and lift it from the powerhead

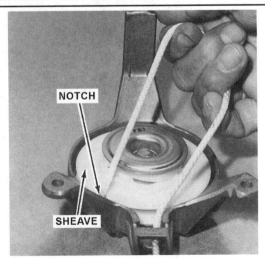
Fig. 4 Pull the rope out and insert it into the notch, then turn the drum clockwise to relieve spring tension

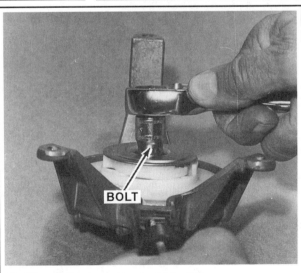
Fig. 5 Loosen the sheave (drum) retaining bolt...

HAND REWIND STARTER 10-3

Do not attempt to remove the spring unless it is unfit for service and a new spring is to be installed.

9. Carefully lift the sheave (drum) free of the starter housing, hopefully leaving the rewind spring still wound tightly inside the housing.

■ On some assemblies, the spring will come away with the drum. If so, you can remove the spring by holding the drum down near the floor with the spring facing downward and inserting a small screwdriver through the hole provided in the sheave drum to free the spring. Be sure to wear gloves and safety glasses while performing this step.

10. If the spring remained in the housing and must be removed, obtain two pieces of wood, a short 2-4 in. (5-10cm) will work fine. Place the two pieces of wood approximately 8 in. (20 cm) apart on the floor. Stand the housing on its legs between the two pieces of wood with the spring side facing down. Stand behind the wood, keeping away from the openings as the spring unwinds with considerable force and the housing will jump off the floor. Tap the housing a moderate blow with a soft mallet. The spring will fall and unwind almost instantly and with much force.

■ The accompanying illustration shows the spring being released from the same type, but different model rewind unit. Therefore, the principle is exactly the same. The procedure outlined in this step may be followed with safety.

11. If the rope is to be replaced, first push the rope back through the opening in the sheave, and then untie the knot and pull the rope free.

CLEANING & INSPECTION

◆ See Figures 12 and 13

Wash all parts except the rope and the handle in solvent, and then blow them dry with compressed air.

Remove any trace of corrosion and wipe all metal parts with an oil dampened cloth.

Inspect the rope. Replace the rope if it appears to be weak or frayed. If the rope is frayed, check the holes through which the rope passes for rough edges or burrs. Remove the rough edges or burrs with a file and polish the surface until it is smooth. Inspect the starter spring end hooks. Replace the spring if it is weak, corroded or cracked. Inspect the inside surface of the sheave rewind recess for grooves or roughness. Grooves may cause erratic rewinding of the starter rope.

Coat the entire length of the used rewind spring (a new spring will be coated with lubricant from the package), with low-temperature lubricant.

ASSEMBLY & INSTALLATION

◆ See Figures 12 thru 25

■ If the pull rope requires replacement use a 51.2 in. (13cm) length for 2 hp motors or 65 in. (16.5cm) length for 3 hp motors.

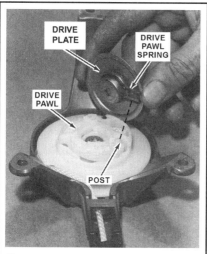

Fig. 6 . . .then remove the drive plate with pawl spring

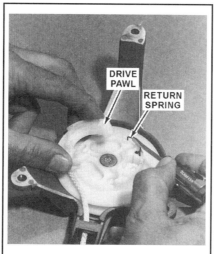

Fig. 7 Pull back on the return spring in order to remove the drive pawl

Fig. 8 Note the spring mounting location before removal

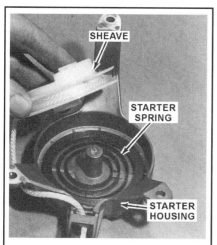

Fig. 9 Lift the sheave (drum) from the housing to expose the rewind spring

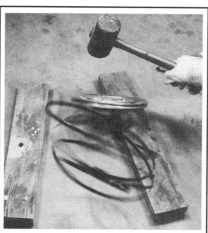

Fig. 10 Take CARE, as the spring may unwind violently when released from housing

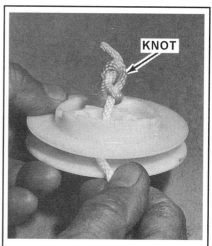

Fig. 11 To remove the rope, push it back sufficiently to untie or cut the knot

10-4 HAND REWIND STARTER

The authors, the manufacturers, and almost anyone else who has handled the spring from this type rewind starter strongly recommend a pair of safety goggles or a face shield be worn while the spring is being installed. As the work progresses a tiger is being forced into a cage - in some cases up to 14 ft. (4.3 m) of spring steel wound into about 4 in. (10.2cm) circumference. If the spring is accidentally released, it will lash out with tremendous ferocity and very likely could cause personal injury to the installer or other persons nearby.

■ **The rewind starter may be assembled with a new rewind spring or a used one. Procedures for assembling are not the same because the new spring will arrive held in a steel hoop already wound, lubricated, and ready for installation. The used spring must be manually wound into its recess.**

The situation may arise when it is only necessary to replace a broken spring. The following procedure outlines the tasks required to replace the spring. A new spring is already properly wound and will arrive in a special hoop. This hoop is designed to be used as an aid to installing the new spring.

1. To install a new spring, hook the outer end of the spring onto the insert in the starter housing, then place the spring into the housing. Seat the spring and then carefully remove the steel hoop.

2. A used spring naturally will not be wound. Therefore, to install an unwound used spring, proceed as follows:

■ **Wear a good pair of sturdy leather work gloves while winding and installing the spring. The spring will develop tension and the edges of the spring steel are extremely sharp. The gloves will help prevent cuts to the hands and fingers.**

a. Loop the spring loosely into a coil, as shown, to enable it to be handled and fed into its recess safely.

b. Insert the hook on the end of the spring into the notch of the recess. Feed the spring around the inner edge of the recess and at the same time rotate the housing counterclockwise. The spring will be slippery with lubrication. Work slowly and with definite movements to prevent losing control of the spring. Proceed with great care. Guide the spring into place.

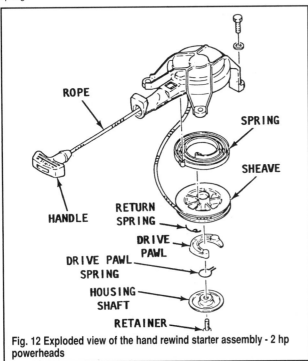

Fig. 12 Exploded view of the hand rewind starter assembly - 2 hp powerheads

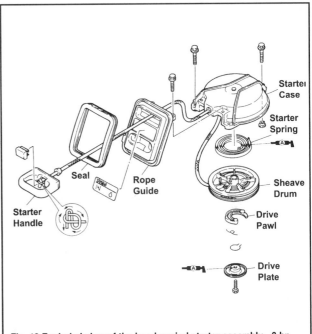

Fig. 13 Exploded view of the hand rewind starter assembly - 3 hp powerheads

Fig. 14 Use a heavy glove when installing a used spring

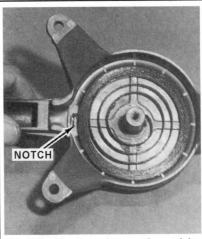

Fig. 15 Position the spring over the notch in the housing

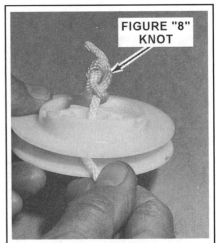

Fig. 16 Install the rope through the drum and tie a figure 8 knot

HAND REWIND STARTER 10-5

3. Thread one end of the rope through the sheave and tie a figure 8 knot, as shown. Thread the other end of the rope through the starter handle and again tie a figure 8 knot in the end.

4. Wind the rope 3-1/2 turns (for 2 hp motors) or 1-1/2 turns (for 3 hp motors) in a counterclockwise direction around the sheave. Position the rope at the notch on the outer edge. Lower the sheave into the starter housing.

5. Insert a small screwdriver or an awl into the access window of the sheave and push the inner end of the spring into the recess under the sheave. This is not an easy task and may not be accomplished on the first try. If too much trouble is encountered, take a break, have a cup of coffee, cup of tea, whatever, and then try again. With patience, much patience, it can be done.

6. Now, insert the return spring into the recess of the sheave which was previously marked during removal. If the recess was not marked during removal, rotate the sheave until both recesses are on the right and insert the spring into the upper one.

7. Hold back the return spring with a small screwdriver and at the same time install the drive pawl.

8. Place the drive plate, with the drive pawl spring attached, over the drive pawl. Check to be sure the spring is centered over the post of the drive pawl.

9. Apply a light coating of Loctite® 271 or equivalent threadlocking compound to the drive plate retaining screw, then install and tighten the screw securely.

10. Pick up the slack in the rope, and then hold the rope firmly in the notch of the sheave. While the rope is being firmly held in the notch, wind the sheave counterclockwise as necessary, normally 3 turns (for 2 hp motors) or 6 turns (for 3 hp motors), using the starter rope. You can also wind it slowly until it can be wound no further, then ease the sheave just a little at a time until the notch in the sheave aligns with the starter handle. Slowly release the sheave and allow the rope to feed around the sheave as it unwinds, pulling in the slack rope.

■ **Check the action of the rewind starter before proceeding with the installation. If all the rope is not taken in around the sheave as the spring unwinds, repeat this step, and then check again.**

11. Position the rewind starter on top of the powerhead with the legs in place on bracket arms.

12. Secure the starter legs with the attaching bolts (usually there are 3) and tighten them securely.

13. Slowly confirm the operation of the starter assembly.

14. Install the engine cover/cowling.

Fig. 17 Wind the rope 1-1/2 or 3-1/2 turns depending upon the model

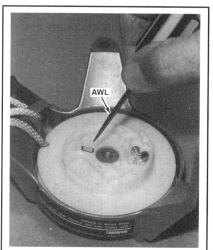

Fig. 18 Use an awl to connect the spring to the drum

Fig. 19 Insert the return spring to the marked hole. . .

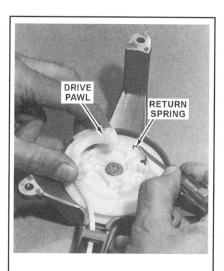

Fig. 20 . . .then install the drive pawl

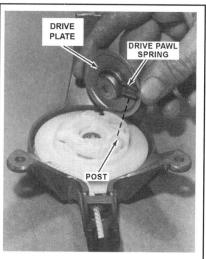

Fig. 21 Position the drive plate (with drive pawl spring). . .

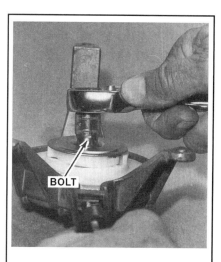

Fig. 22 . . .then install and tighten the retaining bolt

10-6 HAND REWIND STARTER

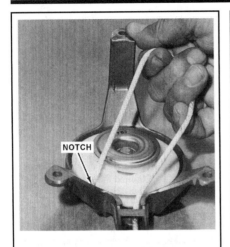

Fig. 23 Use the rope to wind up the spring

Fig. 24 Install the starter assembly to the powerhead...

Fig. 25 ...and tighten the bolts (usually 3) securely

4-50 Hp Models (Except 9.9/15 hp and 48 Hp Motors)

REMOVAL & DISASSEMBLY

◆ See Figures 30 thru 39

① EASY

■ For exploded views, please refer to Cleaning & Inspection, later in this section.

1. Remove the engine top cover.
2. If equipped, disconnect start-in-gear protection cable from the starter assembly. For most models this involves, unscrewing the plastic nut and then pulling the shift interlock cable free of the starter housing. Remove the plunger and spring from the cable end, because they are easily lost.
3. Remove the bolts (usually 3) securing the starter legs to the powerhead, and then remove the starter.
4. Place the starter assembly upside down on a suitable work surface.
5. Carefully pry the circlip from the pawl post using a narrow pry-tool.

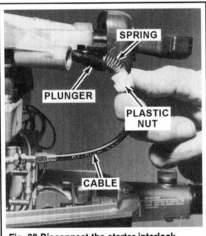

Fig. 30 Disconnect the starter interlock cable from the starter...

Fig. 31 ...then unbolt and remove the hand rewind starter assembly

Fig. 32 For disassembly, invert the starter on a workbench...

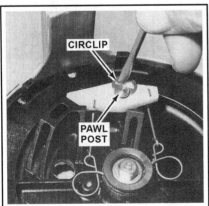

Fig. 33 ...then carefully pry the circlip off the pawl

Fig. 34 Remove the pawl and spring as an assembly

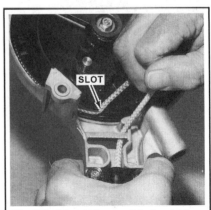

Fig. 35 Pull the starter rope into the slot to hold against spring tension...

HAND REWIND STARTER 10-7

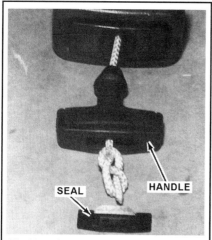

Fig. 36 . . .then remove the rope from the starter handle

Fig. 37 Remove the sheave drum center bolt and washer. . .

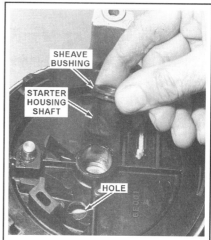

Fig. 38 . . .then remove the starter housing shaft

6. Lift the pawl, with the spring attached free of the sheave (drum).

7. Rotate the sheave to align the slot in the sheave with the starter handle, as shown. Lift out a portion of rope, feed the rope into the slot and with a controlled motion allow the sheave to rotate in a clockwise direction until the tension on the rewind spring is completely released. Do not allow the sheave to spin without control.

8. Pry the seal or cover from the handle and push out the knot in the end of the rope. Untie the knot and pull the handle free of the rope.

9. Remove the bolt and washer from the center of the sheave.

10. Remove the sheave bushing and starter housing shaft from the sheave.

■ If the only work to be performed on the hand rewind starter is to replace the rope, it is best not to disturb the sheave and spring beneath the sheave. However, if either the sheave or the starter rewind spring is to be replaced the rope may be left in place until the sheave is removed from the starter housing.

11. Hold the sheave against the starter housing to prevent the spring from disengaging from the sheave and carefully rotate the sheave to allow the rope hole to align with the starter handle. Pull the knotted end out of the sheave until all of the rope is free.

■ Wear a good pair of heavy gloves and safety glasses while performing the following tasks.

Do not attempt to remove the spiral spring unless it is unfit for service and a new spring is to be installed.

✱✱ WARNING

The spiral rewind spring is a potential hazard. The spring is under tremendous tension when it is wound - a real tiger in a cage! If the spring should accidentally be released, severe personal injury could result from being struck by the spring with force. Therefore, the following steps must be performed with care to prevent personal injury to self and others in the area.

12. Insert a screwdriver into the hole in the sheave, push down on the section of spring visible through the hole. At the same time gently lift up on the sheave and hold the spring down to confine it in the housing and prevent it from escaping uncontrolled. If the rope has not been removed from the sheave, remove it at this time.

■ The accompanying illustration shows the spring being released from the same type but different model rewind spring. Therefore, the principle is exactly the same. The procedure outlined in the next step may be followed with safety.

13. Obtain two pieces of wood, a short 2-4 in. (5-10cm) will work fine. Place the two pieces of wood approximately 8 in. (20 cm) apart on the floor. Center the housing on top of the wood with the spring side facing down.

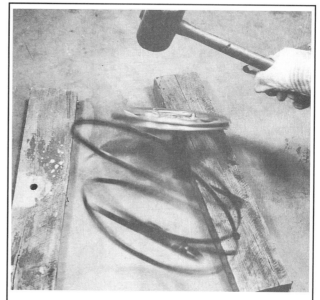

Fig. 39 This is one safe way of releasing the spiral spring

Check to be sure the wood is not touching the spring.

14. Stand behind the wood, keeping away from the openings as the spring unwinds with considerable force. Tap the sheave with a soft mallet. The spring retainer plate will drop down releasing the spring. The spring will fall and unwind almost instantly and with force.

CLEANING & INSPECTION

◆ See Figures 40 thru 44

Wash all parts except the rope and the handle in solvent, and then blow them dry with compressed air.

Remove any trace of corrosion and wipe all metal parts with an oil dampened cloth.

Inspect the rope. Replace the rope if it appears to be weak or frayed. If the rope is frayed, check the holes through which the rope passes for rough edges or burrs. Remove the rough edges or burrs with a file and polish the surface until it is smooth. Inspect the starter spring end hooks. Replace the spring if it is weak, corroded or cracked. Inspect the inside surface of the sheave rewind recess for grooves or roughness. Grooves may cause erratic rewinding of the starter rope.

Coat the entire length of the used rewind spring (a new spring will be coated with lubricant from the package), with low-temperature lubricant.

10-8 HAND REWIND STARTER

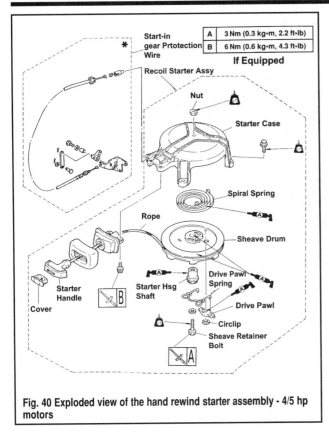

Fig. 40 Exploded view of the hand rewind starter assembly - 4/5 hp motors

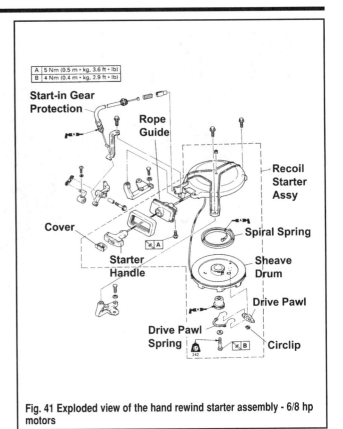

Fig. 41 Exploded view of the hand rewind starter assembly - 6/8 hp motors

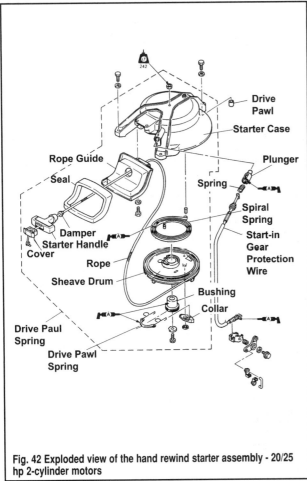

Fig. 42 Exploded view of the hand rewind starter assembly - 20/25 hp 2-cylinder motors

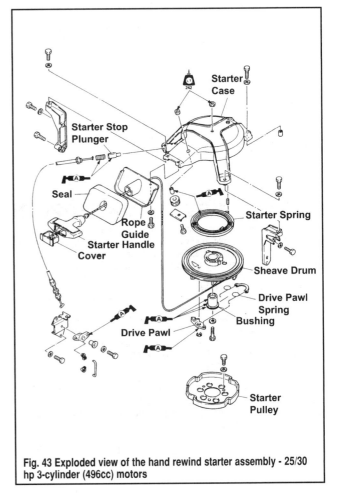

Fig. 43 Exploded view of the hand rewind starter assembly - 25/30 hp 3-cylinder (496cc) motors

HAND REWIND STARTER 10-9

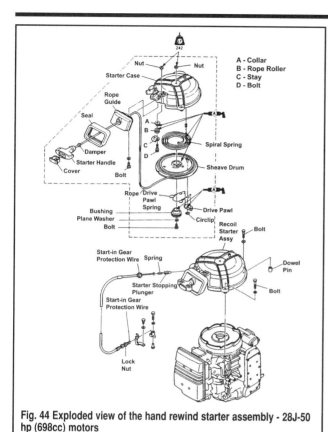

Fig. 44 Exploded view of the hand rewind starter assembly - 28J-50 hp (698cc) motors

ASSEMBLING & INSTALLATION

◆ See Figures 40 thru 57

If the pull rope requires replacement use the proper length, depending upon the model being serviced:
• For 4/5 hp and 6/8 hp motors, the rope should be 72.8 in. (18.5cm).
• For 20/25 hp, 2-cylinder motors, the rope should be 76.8 in. (19.5cm).
• For 25/30 hp (496cc) 3-cylinder motors, the rope should be 75.8 in. (19.25cm).
• For 28J-50 hp (698cc) 3-cylinder motors, the rope should be 82.5 in. (20.95cm).

The authors, the manufacturers, and almost anyone else who has handled the spring from this type rewind starter strongly recommend a pair of safety goggles or a face shield be worn while the spring is being installed. As the work progresses a tiger is being forced into a cage - in some cases up to 14 ft. (4.3 m) of spring steel wound into about 4 in. (10.2cm) circumference. If the spring is accidentally released, it will lash out with tremendous ferocity and very likely could cause personal injury to the installer or other persons nearby.

■ **The rewind starter may be assembled with a new rewind spring or a used one. Procedures for assembling are not the same because the new spring will arrive held in a steel hoop already wound, lubricated, and ready for installation. The used spring must be manually wound into its recess.**

The situation may arise when it is only necessary to replace a broken spring. The following procedure outlines the tasks required to replace the spring. A new spring is already properly wound and will arrive in a special hoop. This hoop is designed to be used as an aid to installing the new spring.

1. Apply a light coating of Yamaha All-Purpose Grease, or equivalent anti-seize lubricant to the inside surface of the starter housing. Wind the old spring loosely in one hand in a clockwise direction, as shown.
2. To install a new spring, hook the outer end of the spring onto the insert in the starter housing, then place the spring into the housing. Seat the spring and then carefully remove the steel hoop.
3. A used spring naturally will not be wound. Therefore, to install an unwound used spring, proceed as follows:

■ **Wear a good pair of sturdy leather work gloves while winding and installing the spring. The spring will develop tension and the edges of the spring steel are extremely sharp. The gloves will help prevent cuts to the hands and fingers.**

   a. Loop the spring loosely into a coil, as shown, to enable it to be handled and fed into its recess safely.

4. Hook the outer end of the spring onto the starter housing post. Rotate the sheave clockwise and at the same feed the spring into the housing in a counterclockwise direction. Continue working the spring into the housing until the entire length has been confined.
5. Insert one end of the rope through the hole in the starter sheave. Tie a figure 8 knot in the end of the rope leaving about one inch (2.5cm) beyond the knot. Tuck the end of the rope beyond the knot into the groove next to the knot.
6. Wind the rope in a clockwise direction around the sheave, the number of turns varies slightly by model, but it should ending at the slot in the sheave. For 4-8 hp motors, wind the rope 2 1/2 turns, for 25/30 hp (496cc) 3-cylinder motors, wind the rope 2 1/4 turns, for all other motors (20/25 hp 2-cylinder motors as well as 28J-50 hp/696cc 3-cylinder motors) wind the rope 1 9/10 turns around the sheave. Lower the sheave into the starter housing. At the same time, use a small screwdriver through the hole to guide the inner loop of the spring onto the post on the underneath side of the sheave.
7. On units with spring and sheave undisturbed, align the hole in the edge of the sheave with the starter handle. Thread the rope through the hole and up through the top side. Tie a figure 8 knot in the end which was just brought through, leaving about one inch (25cm). Tuck the short free end into the groove next to the hole.
8. Without rotating the sheave, feed the rope between the sheave and the edge of the starter housing in a clockwise direction. Push the rope into place with a narrow screwdriver. Continue feeding and tucking the rope for 1-1/2 turns, ending with the rope at the slot of the sheave.

Fig. 45 Use heavy leather gloves when working with the spring

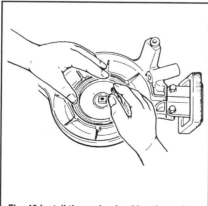

Fig. 46 Install the spring hooking the end onto the starter housing post

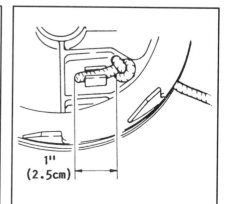

Fig. 47 When installing a rope, leave about 1 inch at the outside of the knot

10-10 HAND REWIND STARTER

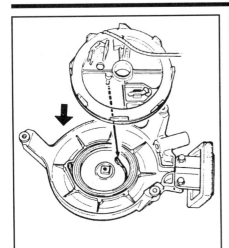

Fig. 48 Carefully install the drum, positioning spring on the post

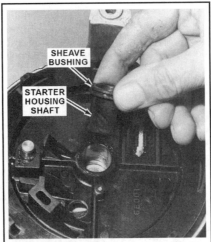

Fig. 49 Install the starter housing shaft (with bushing if applicable). . .

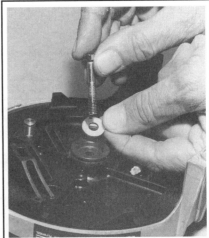

Fig. 50 . . .then secure using the bolt and washer

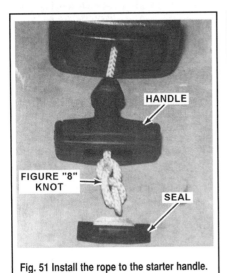

Fig. 51 Install the rope to the starter handle.

Fig. 52 . . .then use the rope to carefully wind the spring

Fig. 53 Install the pawl and spring. . .

Slide the sheave bushing into the starter housing shaft. Insert the shaft and bushing into the center of the sheave.

9. Coat the threads of the center bolt with Loctite® or an equivalent threadlocking compound. Install the washer and bolt, then tighten the bolt securely.

10. Thread the rope through the starter handle housing and through the handle. Tie a figure 8 knot in the rope as close to the end as practical. Pull the knot back into the handle recess, and then install the seal in the handle to hide the rope knot.

11. Lift up a portion of rope, and then hook it into the slot of the sheave. Hold the handle tightly and at the same time rotate the sheave counterclockwise until the spring beneath is wound tight. The number of turns will vary slightly by model. Rotate the sheave about 2 turns for 4/5 hp and 25/30 hp (496cc) 3-cylinder motors, about 3 turns for 6/8 hp motors, or about 2 1/2 turns for the remaining motors (20/25 hp 2-cylinder motors as well as 28J-50 hp/696cc 3-cylinder motors). Slowly release the tension on the sheave and allow it to rewind clockwise while the rope is taken up as it feeds around the sheave.

12. With the beveled end of the pawl facing to the left, hook each end of the pawl spring into the two small holes in the pawl, from the underneath side of the pawl, and with the pattern of the spring, as shown. The short ends of the spring will then be on the upper surface of the pawl. Move the spring up against the center of the sheave shaft, and then slide the center of the pawl onto the pawl post, as indicated in the accompanying illustration.

13. Snap the circlip into place over the pawl post to secure the pawl in place.

14. Check the action of the rewind starter before further installation work proceeds. Pull out the starter rope with the handle, and then allow the spring to slowly rewind the rope. The starter should rewind smoothly and take up all the rope to lightly seat the handle against the starter housing.

15. Position the rewind starter in place on the powerhead. Apply Loctite® to the threads of the attaching bolts. Secure the starter legs to the powerhead with the bolts, and tighten them securely.

16. Slip the starter stop cable end through the spring and then into the recess of the plunger. Hold these parts together and slide them into the starter housing. Tighten the plastic nut snugly.

■ If the cable adjustment at the other end was undisturbed, the no-start-in-gear protection system should perform satisfactorily. When the unit is not in neutral, the plunger should push out to lock the sheave and prevent it from rotating. This means an attempt to pull on the rope with the lower unit in any gear except neutral should fail.

ADJUSTMENT

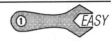

If the no-start-in-gear protection system fails to function properly, first remove the rewind hand starter from the powerhead. Make an adjustment on the length of the cable at the two locknuts at either side of the bracket to bring the plunger flush with the inner surface of the starter housing. Install the starter on the powerhead and again check the no-start-in-gear system. If the system still fails to function correctly, replace the cable.

HAND REWIND STARTER 10-11

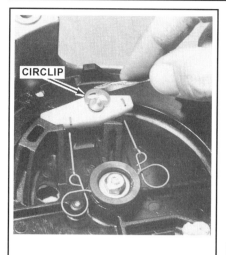

Fig. 54 . . .then secure using the C-clip

Fig. 55 Check the finished assembly to make sure the drum rotates smoothly. . .

Fig. 56 . . .then position and secure it to the powerhead

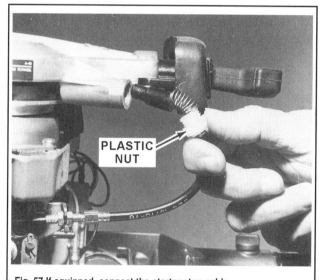

Fig. 57 If equipped, connect the starter stop cable

9.9/15 Hp Models

REMOVAL & DISASSEMBLY

◆ See Figures 58 and 59

1. Remove the engine top cover.
2. If equipped, disconnect start-in-gear protection cable at one end. It is usually easier to disconnect it from the shift linkage end and leave it attached to the starter assembly. However, on some models it may be easier to disconnect it from the starter housing.
3. Remove the 3 mounting bolts securing the corners of the hand rewind starter to the powerhead.
4. Lift the hand rewind starter free of the powerhead. Keep track of the 3 bolt/starter mounting collars.
5. If necessary to remove the start-in-gear protection cable from models so equipped, remove the cotter pin, plane washer and reel stopper from the top of the housing. Next free the stopper arm from underneath the housing and finally remove the spring back on top of the housing.
6. Invert the starter assembly on a suitable work surface
7. Turn the sheave drum until the cutout in the outer side of the drum faces the starter handle, then pull out the starter rope from the side of the housing (between the drum and handle). Hook the rope into the cutout in the side of the drum. Using the rope caught in the cutout, slowly wind the sheave drum clockwise until the spring tension is released.

8. Loosen and remove the drive plate retaining screw, then lift the drive plate and drive spring from the assembly.

■ When removing the drum, be sure to turn it upside down in order to keep the starter spring from popping up at you. If the spring is not being replaced, do not leave the drum out of the housing any longer than necessary OR use something (like a large heavy block of wood or strong wire ties) to help lock the spring in place.

9. Carefully pull the drum assembly from the starter housing (there is a bushing that may come out with the drum or may be left in the housing). Which steps you follow from this point will depend upon which components require service or replacement. The spring, the rope or other starter components can each be individually replaced.
10. Remove the circlip from the spring side of the drum, then remove the drive pawl, return spring and spring from the pawl assembly.
11. Wearing heavy leather gloves to protect your hands, slowly and carefully remove the return spring from the start housing by unwinding it, one turn of the winding each time.
12. If rope replacement is necessary, carefully push the knots out of the drum and starter handle, then untie or cut the rope to free it at each end. For some models, a damper is installed on the rope at the handle end, be sure to keep track of the damper positioning for installation purposes.

CLEANING & INSPECTION

◆ See Figures 58 and 59

Remove any trace of corrosion and wipe all metal parts with an oil dampened cloth.

Inspect the rope. Replace the rope if it appears to be weak or frayed. If the rope is frayed, check the holes through which the rope passes for rough edges or burrs. Remove the rough edges or burrs with a file and polish the surface until it is smooth. Inspect the starter spring end hooks. Replace the spring if it is weak, corroded or cracked. Inspect the inside surface of the sheave rewind recess for grooves or roughness. Grooves may cause erratic rewinding of the starter rope.

Coat the entire length of the used rewind spring (a new spring will be coated with lubricant from the package), with low-temperature lubricant.

ASSEMBLY & INSTALLATION

◆ See Figures 58 and 59

■ If the pull rope requires replacement be sure to use one which measures the same length as the original. Yamaha specifies a rope which is 70.9 in. (18cm) in length. When cutting a rope to the specified length, be sure to burn the ends so that it will not continue to unwind. Also, apply a light coating of marine grade grease to the new rope.

10-12 HAND REWIND STARTER

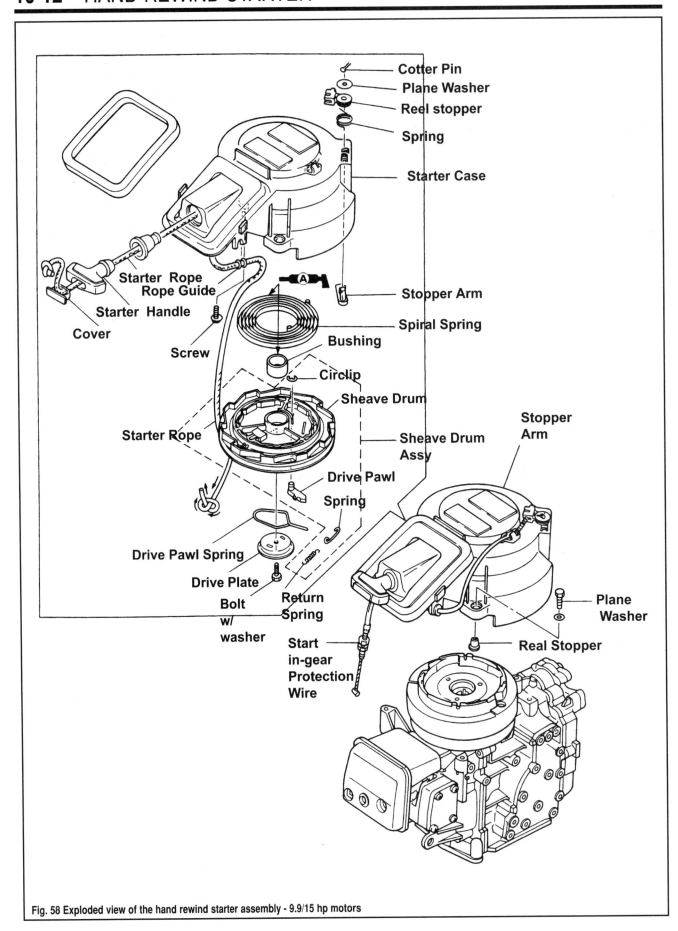

Fig. 58 Exploded view of the hand rewind starter assembly - 9.9/15 hp motors

HAND REWIND STARTER 10-13

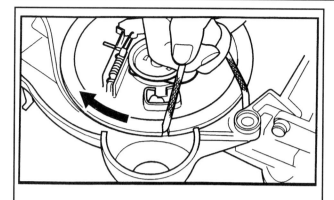

Fig. 59 Using the starter rope to unwind the starter spring

The authors, the manufacturers, and almost anyone else who has handled the spring from this type rewind starter strongly recommend a pair of safety goggles or a face shield be worn while the spring is being installed. As the work progresses a tiger is being forced into a cage - in some cases up to 14 ft. (4.3 m) of spring steel wound into about 4 in. (10.2cm) circumference. If the spring is accidentally released, it will lash out with tremendous ferocity and very likely could cause personal injury to the installer or other persons nearby.

■ **Replacement spiral springs are normally supplied pre-wound and secured by wire loop(s). If using a replacement spring, position it on the drum sheave before removing the wire loop(s)**

1. If removed, install the spring, return spring and drive pawl to the sheave drum and secure using the circlip.
2. If using a replacement spring, position it to the starter housing, then carefully remove the wire loop(s).
3. If using an unwound spring, carefully wind it into the starter housing, starting at the outside and working inward, one turn at a time.
4. If removed, pass a new rope through the rope guide, then insert the ends into the drum sheave and starter handle. Tie figure 8 knots in either end of the rope to secure them to the sheave and handle.

■ **When tying the figure 8 knots, be sure to leave about 0.20-0.39 in. (5-10mm) at the end of the starter rope beyond the knot in order to ensure it will not loosen.**

5. Position the starter rope in the cutout on the drum, then carefully install the drum assembly to the starter housing and over the spring. While installing the drum be sure to position the inner end of the spring over the retainer post on the drum.
6. If removed, install the drive spring to the drive plate.
7. Insert the tip of the drive spring through the coil portion of the return spring, then install the drive plate to the sheave drum.
8. Install the drive plate retaining screw and tighten securely.
9. Using the starter rope in the drum cutout, slowly wind the starter spring by turning the sheave 5 turns COUNTERCLOCKWISE. Once pre-wound, remove the rope from the cutout and allow the spring to slowly wind it back into the housing.
10. If removed, install the start-in-gear protection device components, insert the stopper arm through the bottom, then install the spring, reel stopper and plane washer. Secure the assembly with a cotter pin.
11. Make sure the start-in-gear protection device is released, then check drum rotation and spring return using the starter rope and handle.
12. Install the starter assembly to the top of the powerhead. Secure the starter assembly using the 3 retaining bolts.
13. Reconnect the start-in-gear protection wire to the starter housing or the shift linkage (as applicable).
14. Slowly confirm the operation of the starter assembly.
15. Install the engine cover.

48 Hp and Larger Models

REMOVAL & DISASSEMBLY

◆ See Figures 60 thru 66

1. Remove the engine top cover.
2. Disconnect start-in-gear protection cable from the starter housing. Start by removing the bolt and washer securing the cable guide to the housing, then disconnect the cable end. On most models the end is secured by a cotter pin and washer.
3. Remove the mounting bolts (there are normally 3 on these models) and washers threaded downward to the powerhead or mounting bracket securing the corners of the hand rewind starter.

■ **Keep track of the bolt locations since on most models there are 2 different length bolts used**

4. Remove the 2 bolts and washers threaded horizontally through the rope guide bracket to the powerhead.

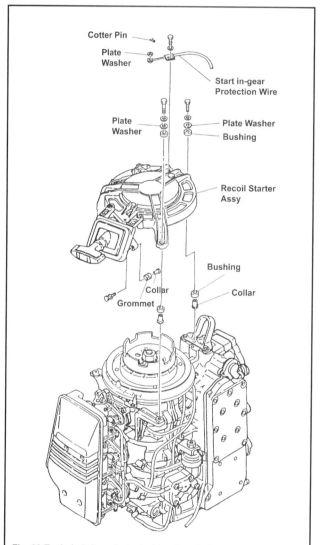

Fig. 60 Exploded view of a typical hand rewind starter mounting (849cc 3-cylinder motor shown)

10-14 HAND REWIND STARTER

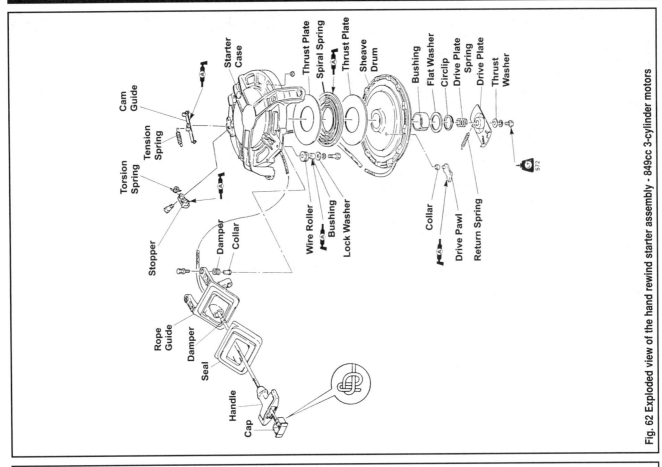

Fig. 62 Exploded view of the hand rewind starter assembly - 849cc 3-cylinder motors

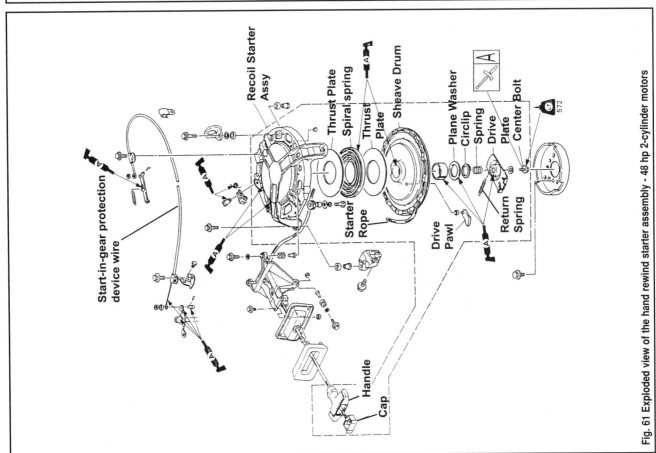

Fig. 61 Exploded view of the hand rewind starter assembly - 48 hp 2-cylinder motors

HAND REWIND STARTER 10-15

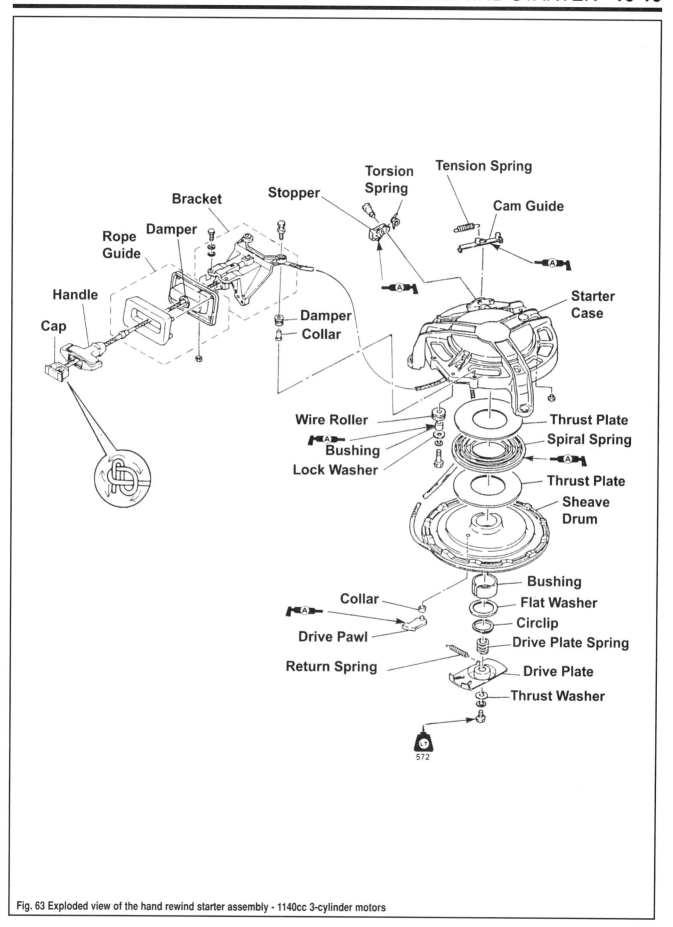

Fig. 63 Exploded view of the hand rewind starter assembly - 1140cc 3-cylinder motors

10-16 HAND REWIND STARTER

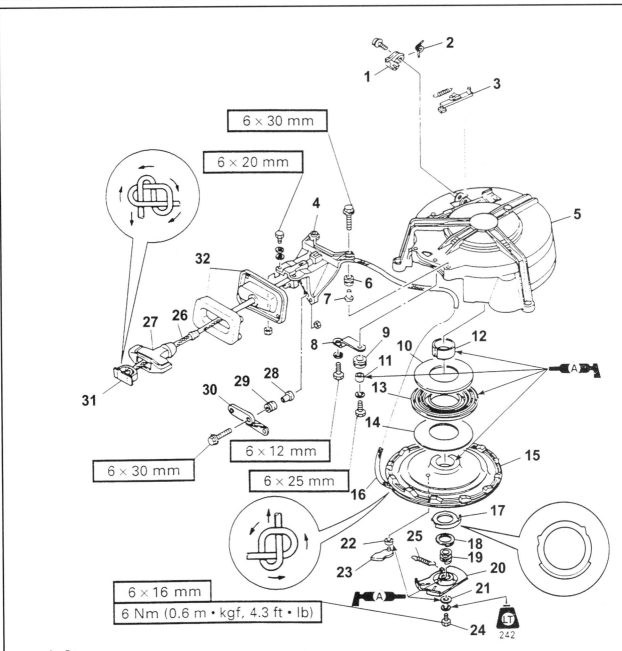

1- Stopper
2- Torsion Spring
3- Guide Cam
4- Starter Rope Guide Bracket
5- Recoil Starter Case
6- Grommet
7- Collar
8- Roller Holder
9- Roller
10 - Thrust Plate
11- Collar
12- Bushing
13- Spiral Spring
14- Thrust Plate
15- Sheave Drum
16- Starter Rope
17- Washer
18- Circlip
19- Drive Plate Spring
20- Drive Plate
21- Thrust Washer
22- Bushing
23- Drive Pawl
24- Bolt
25- Return Spring
26- Damper
27- Starter Handle
28- Collar
29- Grommet
30- Stay
31- Cap
32- Rope Guide

Fig. 64 Exploded view of the hand rewind starter assembly - V4 motors

HAND REWIND STARTER 10-17

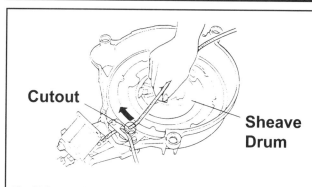

Fig. 65 A cutout in the drum is used to wind the spring up or down

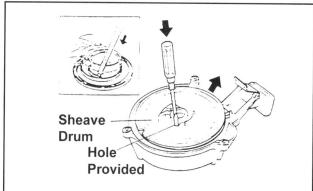

Fig. 66 A hole is provided in the drum to help push the spring off the drum

■ Most of the starter mounting bolts (threaded vertically and horizontally) utilize a collar or grommet to isolate the assembly. Keep track of all washers, collars or grommets to ensure proper installation. Also, give a quick check to the condition of all mounting and isolating components to make sure none show any signs of damage or excessive wear.

  5. Lift the hand rewind starter free of the powerhead.
  6. Invert the starter assembly on a suitable work surface
  7. Turn the sheave drum until the cutout in the outer side of the drum faces the starter handle, then pull out the starter rope from the side of the housing (between the drum and handle). Hook the rope into the cutout in the side of the drum. Using the rope caught in the cutout, slowly wind the sheave drum clockwise until the spring tension is released.
  8. Loosen and remove the drive plate retaining screw, along with the washer and in some cases a separate thrust washer.

  9. Unhook the return spring, then remove the drive plate, return spring and drive plate spring, then remove the circlip and washer. On some models there is a bushing which is accessible from this side of the drum, if so remove the bushing.
  10. Remove the drive pawl and pawl bushing or collar.

■ When removing the drum, be sure to turn it upside down in order to keep the starter spring from popping up at you. If the spring is not being replaced, do not leave the drum out of the housing any longer than necessary OR use something (like a large heavy block of wood or strong wire ties) to help lock the spring in place.

  11. Insert a slotted screwdriver through the hole provided in the sheave drum in order to push down on the spiral spring, then carefully pull the drum assembly from the starter housing (while still applying pressure with the screwdriver to free the spring from the drum and hopefully leave it behind in the housing. Which steps you follow from this point will depend upon which components require service or replacement. The spring, the rope or other starter components can each be individually replaced.
  12. Wearing heavy leather gloves to protect your hands, slowly and carefully remove the thrust plate, then slowly remove the return spring from the start housing by unwinding it, one turn of the winding each time. Under the spring there is usually another thrust plate, if necessary remove it from the housing as well. On models where the bushing was not accessed on the drive pawl side of the drum there is normally a bushing mounting on this side, if necessary remove it for inspection or replacement.
  13. If rope replacement is necessary, carefully push the knots out of the drum and starter handle, then untie or cut the rope to free it at each end. A damper is installed on the rope at the handle end of some models, if used, be sure to keep track of the damper positioning for installation purposes.
  14. Remove any other components from the housing necessary for replacement, such as the start in gear protection assembly or the rope guide assembly.

CLEANING & INSPECTION

◆ See Figures 60 thru 64

Remove any trace of corrosion and wipe all metal parts with an oil dampened cloth.
Inspect the rope. Replace the rope if it appears to be weak or frayed. If the rope is frayed, check the holes through which the rope passes for rough edges or burrs. Remove the rough edges or burrs with a file and polish the surface until it is smooth. Inspect the starter spring end hooks. Replace the spring if it is weak, corroded or cracked. Inspect the inside surface of the sheave rewind recess for grooves or roughness. Grooves may cause erratic rewinding of the starter rope.
Coat the entire length of the used rewind spring (a new spring will be coated with lubricant from the package), with low-temperature lubricant.

HAND REWIND STARTER

ASSEMBLY & INSTALLATION

◆ See Figures 60 thru 65 and 67, 68

■ If the pull rope requires replacement be sure to use one which is the same length as the rope which was removed. Yamaha provides a specification (for all except the V4 models) of 90.6 in. (23cm) in length. Unfortunately we could not locate a spec for the V4, so on those, you must measure the rope (or pieces) to be certain you've got the correct replacement length. When cutting a rope to the length, be sure to burn the ends so that it will not continue to unwind. Also, apply a light coating of marine grade grease to the new rope.

The authors, the manufacturers, and almost anyone else who has handled the spring from this type rewind starter strongly recommend a pair of safety goggles or a face shield be worn while the spring is being installed. As the work progresses a tiger is being forced into a cage - in some cases up to 14 ft. (4.3 m) of spring steel wound into about 4 in. (10.2cm) circumference. If the spring is accidentally released, it will lash out with tremendous ferocity and very likely could cause personal injury to the installer or other persons nearby.

■ Replacement spiral springs are normally supplied pre-wound and secured by wire loop(s). If using a replacement spring, position it on the drum sheave before removing the wire loop(s)

1. If removed, install any housing components such as the start in gear protection assembly or the rope guide assembly.
2. For models with a bushing that mounts on the housing side of the sheave, install the bushing at this time.
3. Position the upper thrust plate in the housing.
4. If using a replacement spring, position it to the starter housing, then carefully remove the wire loop(s).
5. If using an unwound spring, carefully wind it into the starter housing, starting at the outside and working inward, one turn at a time.
6. Install the lower thrust plate over the spring.
7. If removed, pass a new rope through the rope guide (and damper, if used), then insert the ends into the drum sheave and starter handle. Tie figure 8 knots in either end of the rope to secure them to the sheave and handle.

■ When tying the figure 8 knots, be sure to leave about 0.79-1.18 in. (20-30mm) at the end of the starter rope beyond the knot in order to ensure it will not loosen.

8. Pre-wind the rope around the sheave either 1-1/2 turns for all 3-cylinder motors, or 2 1/2 turns for 2-cylinder and V4 motors, then position the starter rope in the cutout on the drum. Holding the rope in this position, carefully install the drum assembly to the starter housing and over the spring. While installing the drum be sure to position the inner end of the spring over the retainer post on the drum.
9. For models where the bushing mounts on the underside of the sheave, install it now.
10. Install the washer and circlip, then install the drive plate spring, drive plate and return spring.
11. Apply a light coating of Loctite® 572 (for all except V4 models), Loctite® 242 (for V4 models) or an equivalent threadlocking compound to the threads of the drive plate retaining bolt. Install the bolt along with the washer and separate thrust washer (if applicable), then tighten the bolt securely.
12. Using the starter rope in the drum cutout, slowly wind the starter spring by turning the sheave 6 turns (for all except V4 motors) or 3 turns (for V4 motors) COUNTERCLOCKWISE. Once pre-wound, remove the rope from the cutout and allow the spring to slowly wind it back into the housing.
13. Make sure the start in gear protection device is released, then check drum rotation and spring return using the starter rope and handle.
14. Install the starter assembly to the top of the powerhead making sure all grommets or collars are in position as noted during removal. Secure the starter assembly using the 3 vertical and 2 horizontal retaining bolts. Tighten the bolts securely.
15. Reconnect the start-in-gear protection to the starter housing.
16. Slowly confirm the operation of the starter assembly.
17. Install the engine cover.

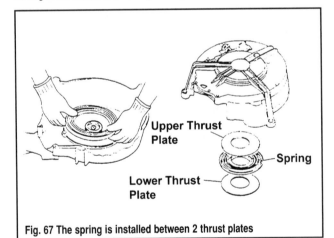

Fig. 67 The spring is installed between 2 thrust plates

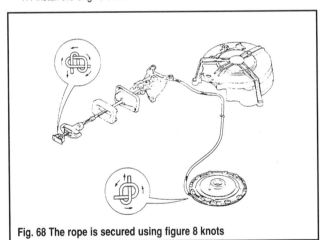

Fig. 68 The rope is secured using figure 8 knots

MASTER INDEX

ANCHORS	1-10
ANODES (ZINCS)	2-25
INSPECTION	2-26
SERVICING	2-26
BASIC ELECTRICAL THEORY	4-2
BASIC OPERATING PRINCIPLES	1-13
BATTERIES	2-27
MAINTENANCE	2-27
STORAGE	2-29
TESTING	2-28
BLEED SYSTEM	6-69
BOAT MAINTENANCE	**2-27**
BATTERIES	2-27
FIBERGLASS HULL	2-29
INTERIOR	2-30
BOATING EQUIPMENT (NOT REQUIRED BUT RECOMMENDED)	**1-10**
ANCHORS	1-10
BAILING DEVICES	1-10
COMPASS	1-11
FIRST AID KIT	1-10
OAR/PADDLE (SECOND MEANS OF PROPULSION)	1-10
TOOLS AND SPARE PARTS	1-12
VHF-FM RADIO	1-10
BOATING SAFETY	**1-4**
COURTESY MARINE EXAMINATIONS	1-10
REGULATIONS FOR YOUR BOAT	1-4
REQUIRED SAFETY EQUIPMENT	1-6
BOLTS, NUTS AND OTHER THREADED RETAINERS	1-27
BREAK-IN PROCEDURES	6-79
CARBURETED FUEL SYSTEM	**3-11**
DESCRIPTION AND OPERATION	3-12
TROUBLESHOOTING THE CARBURETED FUEL SYSTEM	3-14
CARBURETOR IDENTIFICATION	3-17
CARBURETOR SERVICE	3-17
CARBURETOR IDENTIFICATION	3-17
2 HP MODELS	3-17
3 HP MODELS	3-22
4/5 HP (83CC AND 103CC) MODELS	3-24
6/8 HP AND 9.9/15 HP MODELS	3-28
20-48 HP (2-CYLINDER) MODELS	3-35
3-CYLINDER POWERHEADS	3-40
V4 & V6 POWERHEADS	3-49
CDI/TCI SYSTEMS	4-8
CDI/TCI UNIT AND IGNITION COILS	4-30
CDI UNIT TEST CHARTS	4-18
CHARGING CIRCUIT	**4-38**
BATTERY	4-40
YAMAHA CHARGING SYSTEMS	4-38
CHOKE SWITCH	9-2
CLEARING A SUBMERGED MOTOR	**2-87**
COMBUSTION	3-6
COMPRESSION TESTING	2-32
COMPRESSION CHECK	2-32
LOW COMPRESSION	2-33
CONNECTING RODS	6-73
CONTROL BOX ASSEMBLY	9-4
ASSEMBLY & INSTALLATION	9-6
CLEANING & INSPECTION	9-6
REMOVAL & DISASSEMBLY	9-4
CONTROLLED COMBUSTION SYSTEM	3-58
COOLING SYSTEM	5-27
DESCRIPTION AND OPERATION	5-27
FLUSHING JET DRIVES	2-13
FLUSHING THE COOLING SYSTEM	2-11
THERMOSTAT	5-41
THERMO-SWITCH	5-46
TROUBLESHOOTING	5-28
WATER PUMP	5-31
CRANKING CIRCUIT	**4-43**
DESCRIPTION AND OPERATION	4-43
FAULTY SYMPTOMS	4-43
MAINTENANCE	4-43
STARTER MOTOR	4-46
STARTER MOTOR CIRCUIT	4-43
STARTER MOTOR RELAY/SOLENOID	4-45
CRANKSHAFT	6-71
CRANKSHAFT POSITION SENSOR (CPS)	4-35
CRANKSHAFT POSITION SENSOR (CPS)/PULSER COIL	3-82
CYLINDER BLOCK	6-75
ELECTRICAL COMPONENTS	4-2
ELECTRICAL SYSTEM CHECKS	2-38
CHECKING THE BATTERY	2-38
CHECKING THE INTERNAL WIRING HARNESS	2-39
CHECKING THE STARTER MOTOR	2-39
ELECTRICAL SYSTEM PRECAUTIONS	4-7
ELECTRICAL SWITCH/SOLENOID SERVICE	**4-53**
KILL SWITCH	4-53
MAIN KEYSWITCH	4-53
NEUTRAL SAFETY SWITCH	4-55
START BUTTON	4-54
TETHER	4-54
WARNING BUZZER/HORN	4-56
ELECTRICAL TESTING	4-6
ELECTRONIC FUEL INJECTION SYSTEMS	**3-66**
CONTROLLED COMBUSTION SYSTEM	3-58
CRANKSHAFT POSITION SENSOR (CPS)/PULSER COIL	3-82
ENGINE CONTROL MODULE (ECM/CDI)	3-79
FUEL INJECTION AND ELECTRONIC COMPONENT LOCATIONS	3-89
FUEL INJECTION BASICS	3-55
FUEL PRESSURE REGULATOR	3-79
FUEL RAIL AND INJECTORS (EFI)	3-74
HIGH-PRESSURE (MECHANICAL) FUEL PUMP, FUEL RAILS AND INJECTORS (HPDI)	3-72
KNOCK SENSOR (3.1L EFI OX66 MOTORS)	3-86
HPDI DRIVE BELT AND SPROCKETS	3-68
HPDI INJECTOR DRIVER	3-70
NEUTRAL OR SHIFT POSITION SWITCHES	3-89
OXYGEN SENSOR (EFI OX66 AND SOME HPDI)	3-87
PRESSURE SENSORS (APS AND FPS)	3-85
SELF DIAGNOSTIC SYSTEM	3-60
SENSOR AND CIRCUIT RESISTANCE/OUTPUT TESTS	3-61
TEMPERATURE SENSORS (AIR AND WATER/ENGINE)	3-83
THROTTLE BODY AND INTAKE ASSEMBLY	3-62
THROTTLE POSITION SENSOR (TPS)	3-82
TROUBLESHOOTING ELECTRONIC FUEL INJECTION	3-59
YAMAHA EFI OX66 INJECTION	3-56
YAMAHA HIGH PRESSURE DIRECT INJECTION (HPDI)	3-57
VAPOR SEPARATOR TANK AND HIGH-PRESSURE FUEL PUMP	3-64
WATER DETECTION SENSOR (HPDI)	3-88
ELECTRONIC IGNITION SYSTEMS	2-38
ELECTRONIC TOOLS	1-23
ENGINE CONTROL MODULE (ECM/CDI)	3-79
ENGINE COVERS	2-10
ENGINE IDENTIFICATION	2-2
ENGINE MODEL & SERIAL NUMBERS	2-3
ENGINE MAINTENANCE	**2-10**
ANODES (ZINCS)	2-25
COOLING SYSTEM	2-11
ENGINE COVERS	2-10
ENGINE OIL	2-13
FUEL FILTER	2-15

MASTER INDEX

JET DRIVE IMPELLER	2-24
PROPELLER	2-18
ENGINE OIL	2-13
CHECKING	2-15
FILLING	2-14
RECOMMENDATIONS	2-13
EXHAUST COVER	6-69
EXHAUST GASES	7-3
FASTENERS, MEASUREMENTS AND CONVERSIONS	**1-27**
BOLTS, NUTS AND OTHER THREADED RETAINERS	1-27
STANDARD AND METRIC MEASUREMENTS	1-27
TORQUE	1-27
FIBERGLASS HULL	2-29
FIRST AID KIT	1-10
FLYWHEEL AND STATOR PLATE	4-23
FUEL	3-2
ALCOHOL-BLENDED FUELS	3-3
CHECKING FOR STALE/CONTAMINATED FUEL	3-3
HIGH ALTITUDE OPERATION	3-3
OCTANE RATING	3-3
THE BOTTOM LINE WITH FUELS	3-3
VAPOR PRESSURE	3-3
FUEL AND COMBUSTION BASICS	**3-2**
COMBUSTION	3-6
FUEL SYSTEM SERVICE CAUTIONS	3-2
FUEL	3-2
FUEL SYSTEM PRESSURIZATION	3-4
FUEL SYSTEM SERVICE CAUTIONS	3-2
FUEL FILTER	2-15
FUEL INJECTION AND ELECTRONIC COMPONENT LOCATIONS	3-89
2.6L EFI OX66	3-89
2.6L HPDI	3-92
3.1L EFI OX66	3-94
3.3L HPDI	3-97
FUEL LINES AND FITTINGS	3-8
SERVICE	3-10
TESTING	3-8
FUEL PRESSURE REGULATOR	3-79
REMOVAL & INSTALLATION	3-79
TESTING	3-79
FUEL PUMP (LOW-PRESSURE) SERVICE	**3-99**
FUEL PUMPS	3-99
DESCRIPTION & OPERATION	3-99
GENERAL INFORMATION	3-100
OVERHUAL	3-103
REMOVAL & INSTALLATION	3-102
TESTING	3-100
FUEL RAIL AND INJECTORS (EFI)	3-74
REMOVAL & INSTALLATION	3-78
TESTING	3-74
FUEL SYSTEM CHECKS	2-39
FUEL SYSTEM PRESSURIZATION	3-4
PRESSURIZING THE FUEL SYSTEM (CHECKING FOR LEAKS)	3-5
RELIEVING FUEL SYSTEM PRESSURE (EFI & HPDI MOTORS)	3-5
FUEL SYSTEM SERVICE CAUTIONS	3-2
FUEL TANK	3-6
FUEL TANK AND LINES	**3-6**
FUEL LINES AND FITTINGS	3-8
FUEL TANK	3-6
GEARCASE (LOWER UNIT)	7-4
REMOVAL & INSTALLATION	7-4
2 AND 3 HP (1-CYLINDER) MODELS	7-4
4-40 HP 1-, 2- AND 3-CYLINDER MODELS (EXCEPT 40/50 HP/698CC)	7-6
48 HP (2-CYLINDER), 40/50 HP (3-CYLINDER) & ALL V4/V6 MODELS	7-8
GEARCASE (LOWER UNIT) OIL	2-7
CHECKING LEVEL & CONDITION	2-7
DRAINING AND FILLING	2-8
OIL RECOMMENDATIONS	2-7
GEARCASE OVERHAUL	**7-10**
OVERHAUL TIPS	7-10
2 AND 3 HP (1-CYLINDER) MODELS	7-11
4-40 HP 1-,2- AND 3-CYLINDER MODELS (EXCEPT 40/50 HP/698CC)	7-19
48 HP 2-CYLINDER, 40/50 HP AND LARGER 3-CYLINDER & ALL V4/V6 MODELS	7-43
GENERAL INFORMATION	**2-2**
BEFORE/AFTER EACH USE	2-4
ENGINE IDENTIFICATION	2-2
MAINTENANCE COVERAGE IN THIS MANUAL	2-2
MAINTENANCE EQUALS SAFETY	2-2
OUTBOARDS ON SAIL BOATS	2-2
HAND REWIND STARTER	**10-2**
2 Hp (43 and 50cc) and 3 Hp Models	10-2
4-50 Hp Models (Except 9.9/15 hp and 48 Hp Motors)	10-6
9.9/15 Hp Models	10-11
48 Hp and Larger Models	10-13
HAND TOOLS	1-19
HIGH-PRESSURE (MECHANICAL) FUEL PUMP FUEL RAILS AND INJECTORS (HPDI)	3-72
HOW TO USE THIS MANUAL	**1-2**
AVOIDING THE MOST COMMON MISTAKES	1-3
AVOIDING TROUBLE	1-2
CAN YOU DO IT?	1-2
DIRECTIONS AND LOCATIONS	1-2
MAINTENANCE OR REPAIR?	1-2
PROFESSIONAL HELP	1-3
PURCHASING PARTS	1-3
WHERE TO BEGIN	1-2
HPDI DRIVE BELT AND SPROCKETS	3-68
REMOVAL & INSTALLATION	3-68
HPDI INJECTOR DRIVER	3-70
REMOVAL & INSTALLATION	3-71
TESTING	3-70
HYDRO TILT LOCK SYSTEM	**8-2**
DESCRIPTION & OPERATION	8-2
SERVICING THE HYDRO TILT SYSTEM	8-3
IDLE SPEED ADJUSTMENT	2-43
IGNITION AND CHARGING SYSTEM TROUBLESHOOTING	4-12
IGNITION SYSTEMS	**4-8**
CDI UNIT TEST CHARTS	4-18
CDI/TCI SYSTEMS	4-8
CDI/TCI UNIT AND IGNITION COILS	4-30
CRANKSHAFT POSITION SENSOR (CPS)	4-35
FLYWHEEL AND STATOR PLATE	4-23
IGNITION AND CHARGING SYSTEM TROUBLESHOOTING	4-12
THERMO-SENSOR & THERMO-SWITCH	4-36
YAMAHA MICROCOMPUTER IGNITION SYSTEMS (YMIS)	4-33
INTERIOR	2-30
JET DRIVE	**7-78**
DESCRIPTION & OPERATION	7-78
JET DRIVE ADJUSTMENTS	7-88
JET DRIVE ASSEMBLY	7-79
MODEL IDENTIFICATION & SERIAL NUMBERS	7-78
JET DRIVE ADJUSTMENTS	7-88
GATE LINKAGE ADJUSTMENTS	7-88
TRIM ADJUSTMENT	7-92
JET DRIVE ASSEMBLY	7-79
ASSEMBLY	7-85
CLEANING & INSPECTION	7-80
REMOVAL & DISASSEMBLY	7-79

MASTER INDEX 10-21

JET DRIVE BEARING	2-9
JET DRIVE IMPELLER	2-24
KILL SWITCH	9-4
KILL SWITCH	4-53
KNOCK SENSOR (3.1L EFI OX66 MOTORS)	3-86
LARGE MOTOR POWER TRIM/TILT SYSTEM	**8-13**
DESCRIPTION & OPERATION	8-13
MAINTENANCE	8-15
POWER TRIM/TILT SWITCH	8-19
PTT ASSEMBLY (DUAL TRIM RAM)	8-21
TRIM SENSOR	8-20
TRIM/TILT MOTOR	8-20
TRIM/TILT RELAY	8-18
TROUBLESHOOTING	8-15
LOWER UNIT	**7-2**
EXHAUST GASES	7-3
GEARCASE (LOWER UNIT)	7-4
GENERAL INFORMATION	7-2
SHIFTING PRINCIPLES	7-2
TROUBLESHOOTING	7-4
LUBRICANTS	2-6
LUBRICATION	**2-5**
GEARCASE (LOWER UNIT) OIL	2-7
JET DRIVE BEARING	2-9
LUBRICANTS	2-6
LUBRICATING THE MOTOR	2-6
LUBRICATION INSIDE THE BOAT	2-7
POWER TRIM/TILT RESERVOIR	2-9
MAIN KEYSWITCH	4-53
MEASURING TOOLS	1-25
NEUTRAL OR SHIFT POSITION SWITCHES	3-89
NEUTRAL SAFETY SWITCH	4-55
NEUTRAL SAFETY SWITCH	9-3
OIL INJECTION WARNING SYSTEM	**5-20**
GENERAL INFORMATION	5-20
OPERATIONAL CHECK	5-23
TROUBLESHOOTING	5-22
OIL PUMP ASSEMBLY	5-18
OXYGEN SENSOR (EFI OX66 AND SOME HPDI)	3-87
PBS DIAGRAMS	5-10
PISTONS	6-74
POWER TRIM/TILT RESERVOIR	2-9
FLUID LEVEL/CONDITION	2-9
RECOMMENDED LUBRICANT	2-9
POWERHEAD BREAK-IN	**6-79**
BREAK-IN PROCEDURES	6-79
POWERHEAD MECHANICAL	**6-2**
GENERAL INFORMATION	6-2
2 HP (43 AND 50CC) AND 3 HP MODELS	6-4
4/5 HP (83 AND 103CC) MODELS	6-10
6-48 HP (2-CYLINDER) POWERHEADS	6-15
25-90 HP (3-CYLINDER) POWERHEADS	6-25
V4 AND V6 POWERHEADS	6-36
POWERHEAD MOUNTED OIL TANK	5-16
REMOVAL & INSTALLATION	5-16
POWERHEAD REFINISHING	**6-67**
BLEED SYSTEM	6-69
CONNECTING RODS	6-73
CRANKSHAFT	6-71
CYLINDER BLOCK	6-75
EXHAUST COVER	6-69
GENERAL INFORMATION	6-67
PISTONS	6-74
REED BLOCK	6-69
PRECISON BLEND OIL INJECTION SYSTEMS	**5-2**
GENERAL INFORMATION AND TROUBLESHOOTING	5-2
PBS DIAGRAMS	5-10
POWERHEAD MOUNTED OIL TANK	5-16
REMOTE OIL TANK	5-17
OIL PUMP ASSEMBLY	5-18
PRESSURE SENSORS (APS AND FPS)	3-85
PREPPING THE MOTOR	2-44
PROPELLER	2-18
GENERAL INFORMATION	2-18
INSPECTION	2-21
REMOVAL & INSTALLATION	2-21
RE-COMMISSIONING	2-86
REED BLOCK	6-69
INSPECTION	6-69
REED VALVES	**3-104**
REED VALVE BLOCK	3-104
REGULATIONS FOR YOUR BOAT	1-4
REMOTE CONTROL BOX	9-2
REMOTE CONTROLS	**9-2**
CHOKE SWITCH	9-2
CONTROL BOX ASSEMBLY	9-4
KILL SWITCH	9-4
NEUTRAL SAFETY SWITCH	9-3
REMOTE CONTROL BOX	9-2
START BUTTON	9-3
WARNING BUZZER/HORN	9-2
REMOTE OIL TANK	5-17
REQUIRED SAFETY EQUIPMENT	1-6
SAFETY IN SERVICE	**1-13**
SAFETY TOOLS	1-17
SELF DIAGNOSTIC SYSTEM	3-60
READING	3-60
TROUBLE CODES	3-62
SENSOR AND CIRCUIT RESISTANCE/OUTPUT	3-61
SHIFTING PRINCIPLES	7-2
COUNTER-ROTATING UNIT	7-3
STANDARD ROTATING UNIT	7-2
SHOP EQUIPMENT	**1-17**
SINGLE TILT RAM POWER TRIM/TILT SYSTEMS	**8-3**
DESCRIPTION & OPERATION	8-3
MAINTENANCE	8-4
PTT ASSEMBLY	8-9
TILT MOTOR	8-8
TILT RELAY	8-6
TILT SWITCH	8-7
TRIM SENSOR	8-9
TROUBLESHOOTING	8-6
SPARK PLUG WIRES	2-38
SPARK PLUGS	2-33
SPECIAL TOOLS	1-23
SPECIFICATIONS	**1-28**
CAPACITIES	2-96
CARBURETOR SET-UP	3-105
CHARGING SYSTEM	4-62
CONVERSION FACTORS	1-28
COOLING SYSTEM	5-49
ENGINE	2-91
GENERAL ENGINE - 1-CYLINDER MOTORS	6-80
GENERAL ENGINE - 2-CYLINDER MOTORS	6-83
GENERAL ENGINE -3-CYLINDER MOTORS	6-89
GENERAL ENGINE - V4 MOTORS	6-93
GENERAL ENGINE - V6 MOTORS	6-94
IGNITION SYSTEM	4-57
LUBRICATION	2-95
MAINTENANCE INTERVALS	2-90
METRIC BOLTS - TYPICAL TORQUE VALUES	1-28
TUNE-UP	2-88

MASTER INDEX

U.S. STANDARD BOLTS - TYPICAL TORQUE VALUES	1-28
START BUTTON	4-54
START BUTTON	9-3
STARTER MOTOR	4-46
STARTER MOTOR CIRCUIT	4-43
STARTER MOTOR RELAY/SOLENOID	4-45
STORAGE	**2-84**
RE-COMMISSIONING	2-86
WINTERIZATION	2-84
SYNCHRONIZATION	2-44
TEMPERATURE SENSORS (AIR AND WATER/ENGINE)	3-83
TEST EQUIPMENT	4-4
TETHER	4-54
THERMOSTAT	5-41
EXPLODED VIEWS	5-43
REMOVAL & INSTALLATION	5-41
THERMO-SENSOR & THERMO-SWITCH	4-36
THERMO-SWITCH	5-46
THROTTLE BODY AND INTAKE ASSEMBLY	3-62
THROTTLE POSITION SENSOR (TPS)	3-82
TIMING	2-44
TIMING AND SYNCHRONIZATION	**2-44**
GENERAL INFORMATION	2-44
PREPPING THE MOTOR	2-44
SYNCHRONIZATION	2-44
TIMING	2-44
2 HP MODEL	2-45
3 HP MODEL	2-45
4/5 HP (83 AND 103CC) MODELS	2-46
6/8 HP MODELS	2-47
9.9/15 HP MODELS	2-49
20/25 HP (395CC) MODELS	2-51
20/25 HP (430CC) MODELS	2-53
25/30 HP (496CC 2-CYLINDER) MODELS	2-54
40 HP (2-CYLINDER) MODEL	2-54
48 HP (2-CYLINDER) MODELS	2-57
25/30 HP (496CC 3-CYLINDER) MODELS	2-61
28J-50 HP (698CC) MODELS	2-66
50-70 HP (849CC) AND 65J-90 HP (1140CC) MODELS	2-68
E60 (849CC) AND E75, E75A, 85A, E60J (1140CC) MOTORS	2-70
V4 AND V6 CARBURETED MOTORS	2-74
V6 EFI (OX66) AND HPDI MOTORS	2-79
TOOLS	**1-19**
ELECTRONIC TOOLS	1-23
HAND TOOLS	1-19
MEASURING TOOLS	1-25
OTHER COMMON TOOLS	1-22
SPECIAL TOOLS	1-23
TOOLS AND SPARE PARTS	1-12
TRIM & TILT SYSTEMS	**8-2**
HYDRO TILT LOCK SYSTEM	8-2
INTRODUCTION	8-2
LARGE MOTOR POWER TRIM/TILT SYSTEM	8-13
SINGLE TILT RAM POWER TRIM/TILT SYSTEMS	8-3
TROUBLESHOOTING	**1-13**
BASIC OPERATING PRINCIPLES	1-13
CARBURETED FUEL SYSTEM	3-14
ELECTRICAL SYSTEMS	4-6
ELECTRONIC FUEL INJECTION	3-59
LOWER UNIT	7-4
TUNE-UP	**2-30**
COMPRESSION TESTING	2-32
ELECTRICAL SYSTEM CHECKS	2-38
ELECTRONIC IGNITION SYSTEMS	2-38
FUEL SYSTEM CHECKS	2-39
IDLE SPEED ADJUSTMENT	2-43
INTRODUCTION TO TUNE-UPS	2-30
SPARK PLUG WIRES	2-38
SPARK PLUGS	2-33
TUNE-UP SEQUENCE	2-31
TUNE-UP SEQUENCE	2-31
UNDERSTANDING AND TROUBLESHOOTING ELECTRICAL SYSTEMS	**4-2**
BASIC ELECTRICAL THEORY	4-2
ELECTRICAL COMPONENTS	4-2
ELECTRICAL SYSTEM PRECAUTIONS	4-7
ELECTRICAL TESTING	4-6
TEST EQUIPMENT	4-4
TROUBLESHOOTING ELECTRICAL SYSTEMS	4-6
WIRE AND CONNECTOR REPAIR	4-7
VAPOR SEPARATOR TANK AND HIGH-PRESSURE FUEL PUMP	3-64
WARNING BUZZER/HORN	4-56
WARNING BUZZER/HORN	9-2
WATER DETECTION SENSOR (HPDI)	3-88
WATER PUMP	5-31
EXPLODED VIEWS	5-35
INSPECTION & OVERHAUL	5-33
REMOVAL & INSTALLATION	5-31
WINTERIZATION	2-84
PREPPING FOR STORAGE	2-84
WIRE AND CONNECTOR REPAIR	4-7
WIRING DIAGRAMS	**4-65**
INDEX	4-65
1-CYLINDER MOTORS	4-66
2-CYLINDER MOTORS	4-67
3-CYLINDER MOTORS	4-74
V4 MOTORS	4-88
V6 MOTORS	4-94
YAMAHA CHARGING SYSTEMS	4-38
YAMAHA EFI OX66 INJECTION	3-56
YAMAHA HIGH PRESSURE DIRECT INJECTION (HPDI)	3-57
YAMAHA MICROCOMPUTER IGNITION SYSTEMS (YMIS)	4-33
YMIS SELF-DIAGNOSIS	4-33
TROUBLESHOOTING	4-35

ENGINE FINDER

The following listing contains all engines covered in this manual

Engine	Years
2 hp, 1 cyl (43cc), 2-stroke	1997-02
2 hp, 1 cyl (50cc), 2-stroke	1997-02
3 hp, 1 cyl (70cc), 2-stroke	1997-02
4 hp, 1 cyl (83cc), 2-stroke	1997-99
4 hp, 1 cyl (103cc), 2-stroke	1997-02
5 hp, 1 cyl (103cc), 2-stroke	1997-02
6 hp, 2 cyl (165cc), 2-stroke	1997-00
8 hp, 2 cyl (165cc), 2-stroke	1997-03
9.9 hp, 2 cyl (246cc), 2-stroke	1997-03
15 hp, 2 cyl (246cc), 2-stroke	1997-03
20 hp, 2 cyl (395cc), 2-stroke	1997
20 hp, 2 cyl (430cc), 2-stroke	1997-98
25 hp, 2 cyl (395cc), 2-stroke	1997-03
25 hp, 2 cyl (430cc), 2-stroke	1997-98
25 hp, 2 cyl (496cc), 2-stroke	1997
25 hp, 3 cyl (496cc), 2-stroke	1997-02
28 Jet, 3 cyl (698cc), 2-stroke	1997-01
30 hp, 2 cyl (496cc), 2-stroke	1997-02
35 Jet, 3 cyl (698cc), 2-stroke	1997-01
40 hp, 2 cyl (592cc), 2-stroke	1997
40 hp, 3 cyl (698cc), 2-stroke	1997-03
48 hp, 2 cyl (760cc), 2-stroke	1997-00
50 hp, 3 cyl (698cc), 2-stroke	1997-03
50 hp, 3 cyl (849cc), 2-stroke	1997-03
60 hp, 3 cyl (849cc), 2-stroke	1997-03
65 Jet, 3 cyl (1140cc), 2-stroke	1997-01
70 hp, 3 cyl (849cc), 2-stroke	1997-03
75 hp, 3 cyl, (1140cc), 2-stroke	1997-00
80 hp, 3 cyl, (1140cc), 2-stroke	1997-00
80 Jet, V4 (1730cc), 2-stroke	1997-01
85 hp, 3 cyl, (1140cc), 2-stroke	1997-00
90 hp, 3 cyl, (1140cc), 2-stroke	1997-03
100 hp, V4 (1730cc), 2-stroke	1997-02
105 Jet, V6 (2596cc), 2-stroke	1997-00
115 hp, V4 (1730cc), 2-stroke	1997-03
130 hp, V4 (1730cc), 2-stroke	1997-03
140 hp, V4 (1730cc), 2-stroke	1997-02
150 hp, V6 (2596cc), 2-stroke	1997-99
150 hp, 76 deg. V6 (2596cc), 2-stroke	2000-03
150 hp, 90 deg. V6 (2596cc), 2-stroke	2000-03
175 hp, V6 (2596cc), 2-stroke	1997-03
200 hp, V6 (2596cc), 2-stroke	1997-99
200 hp, 76 deg. V6 (2596cc), 2-stroke	2000-03
200 hp, 90 deg. V6 (2596cc), 2-stroke	2000-03
200 hp, 76 deg. V6 (3130cc), 2-stroke	1998-03
225 hp, 76 deg. V6 (3130cc), 2-stroke	1997-03
225 hp, 76 deg. V6 (3342cc), 2-stroke	2003
225 hp, 90 deg. V6 (2596cc), 2-stroke	1997
250 hp, V6 (3130cc), 2-stroke	1997-03
250 hp, V6 (3342cc), 2-stroke	2003